Alan Turing

HIS WORK AND IMPACT

Alan Turing

HIS WORK AND IMPACT

Edited by

S. BARRY COOPER
University of Leeds, UK
and
JAN VAN LEEUWEN
Utrecht University, The Netherlands

AMSTERDAM • BOSTON • HEIDELBERG • LONDON
NEW YORK • OXFORD • PARIS • SAN DIEGO
SAN FRANCISCO • SINGAPORE • SYDNEY • TOKYO

Elsevier
225 Wyman Street, Waltham, MA 02451, USA
The Boulevard, Langford Lane, Kidlington, Oxford OX5 1GB, UK
Radarweg 29, PO Box 211, 1000 AE Amsterdam, The Netherlands

Library of Congress Cataloging-in-Publication Data
Cooper, S. B. (S. Barry)
The selected works of A.M. Turing: his work and impact / S. Barry Cooper and J. van Leeuwen.
p. cm.
Includes bibliographical references.
ISBN 978-0-12-386980-7
1. Turing, Alan Mathison, 1912–1954. 2. Mathematicians–Great Britain–Biography. 3. Computer science–Mathematics. 4. Enigma cipher system. 5. Logic, Symbolic and mathematical.
I. Leeuwen, J. van (Jan) II. Title.
QA29.T8C65 2012
510.92–dc23
[B]

2012006868

British Library Cataloguing in Publication Data
A catalogue record for this book is available from the British Library

For information on all Elsevier publications
visit our web site at *elsevierdirect.com*

Printed and bound in USA
12 13 14 15 10 9 8 7 6 5 4 3 2 1

ISBN: 978-0-12-386980-7

List of Contributors

Alastair Abbott
Department of Computer Science, The University of Auckland, Auckland, New Zealand

Samson Abramsky
Department of Computer Science, Oxford University, Oxford, UK

Henk Barendregt
Institute for Computing and Information Sciences, Faculty of Science, Radboud University, Nijmegen, The Netherlands

Craig Bauer
Department of Physical Sciences, York College of Pennsylvania, York, PA, USA

Anthony Beavers
Philosophy & Cognitive Science, The University of Evansville, Evansville, IN, USA

Verónica Becher
Departamento de Computación, Facultad de Ciencias Exactas y Naturales, Universidad de Buenos Aires and CONICET, Ciudad Universitaria, Buenos Aires, Argentina

Nelson H. F. Beebe
Department of Mathematics, University of Utah, Salt Lake City, UT, USA

Henri Berestycki
University of Chicago, and L'École des hautes études en sciences sociales (EHESS), Paris, France

Meurig Beynon
Department of Computer Science, University of Warwick, Coventry, UK

Mark Bishop
Department of Computing, Goldsmiths, University of London, New Cross, London, UK

Lenore Blum
School of Computer Science, Carnegie Mellon University, Pittsburgh, PA, USA

Rodney Brooks
Emeritus Professor of Robotics, Computer Science and Artificial Intelligence Laboratory, MIT, Cambridge, USA

Cristian Calude
Department of Computer Science, The University of Auckland, Auckland, New Zealand

Gregory Chaitin
Professor of Mathematics, Federal University of Rio de Janeiro (UFRJ), Rio de Janeiro, Argentina

Jack Copeland
School of Social and Political Sciences, University of Canterbury, Christchurch, New Zealand

Martin Davis
Department of Mathematics, University of California, Berkeley, CA, USA

Daniel Dennett
Center for Cognitive Studies, Tufts University, Medford, MA, USA

Artur Ekert
Mathematical Institute, University of Oxford, Oxford, UK and Centre for Quantum Technologies, National University of Singapore, Singapore

Solomon Feferman
Department of Mathematics, Stanford University, Stanford, CA, USA

Luciano Floridi
School of Humanities, University of Hertfordshire, Hertfordshire, and Department of Computer Science, University of Oxford, Oxford, UK

Juliet Floyd
Philosophy Department, Boston University, Boston, MA, USA

Lance Fortnow
School of Computer Science, Georgia Institute of Technology, Atlanta, USA

Einar Fredriksson
IOS Press, Amsterdam, The Netherlands

Nicholas Gessler
Information Science & Information Studies, Duke University, Durham, NC, USA

Rainer Glaschick
Paderborn, Germany

David Harel
Department of Computer Science and Applied Mathematics, The Weizmann Institute of Science, Rehovot, Israel

John Harper
Retired, Honorary Fellow of the British Computer Society

Dennis Hejhal
School of Mathematics, University of Minnesota, Minneapolis, MN, USA

Andrew Hodges
Wadham College, Oxford University, Oxford, UK

Douglas Hofstadter
College of Arts and Sciences, Center for Research on Concepts and Cognition, Indiana University, Bloomington, IN, USA

Toby Howard
School of Computer Science, The University of Manchester, Manchester, UK

Cliff Jones
School of Computing Science, Newcastle University, Newcastle, UK

Richard Jozsa
DAMTP, Centre for Mathematical Sciences, University of Cambridge, Cambridge, UK

Jan van Leeuwen
Department of Information and Computing Sciences, Utrecht University, Utrecht, The Netherlands

David Levy
Intelligent Toys Ltd, London, UK

Philip Maini
Centre for Mathematical Biology, Mathematical Institute, Oxford University, Oxford, UK

Giulio Manzonetto
LIPN - Institut Galilée, Université Paris-Nord, Villetaneuse, France

Hans Meinhardt
Max-Planck-Institut für Entwicklungsbiologie, Tübingen, Germany

Peter Millican
Faculty of Philosophy, Oxford University, Oxford, UK

James D. Murray
Professor Emeritus, Mathematical Institute, University of Oxford, Oxford, UK

Andrew Odlyzko
School of Mathematics, University of Minnesota, Minneapolis, MN, USA

Christos Papadimitriou
EECS Department, University of California, Berkeley, CA, USA

Rinus Plasmeijer
Institute for Computing and Information Sciences, Faculty of Science, Radboud University, Nijmegen, The Netherlands

Antonino Raffone
Department of Psychology, Sapienza University of Rome, Rome, Italy

Michael Rathjen
School of Mathematics, University of Leeds, Leeds, UK

Bernard Richards
Emeritus Professor of Medical Informatics, University of Manchester, Manchester, UK

Anthony Edgar "Tony" Sale, FBCS (30 January 1931–28 August 2011)
Founder member Bletchley Park Trust, leader Colossus rebuild project

Peter T. Saunders
Department of Mathematics, King's College, Strand, London, UK

Klaus Schmeh
Gelsenkirchen, Germany

Huma Shah
School of Systems Engineering, University of Reading, Reading, UK

Wilfried Sieg
Department of Philosophy, Carnegie Mellon University, Pittsburgh, PA, USA

Aaron Sloman
School of Computer Science, University of Birmingham, Birmingham, UK

Alan Slomson
School of Mathematics, University of Leeds, Leeds, UK

Paul Smolensky
Department of Cognitive Science, The Johns Hopkins University, Baltimore, MD, USA

Robert I. Soare
Department of Mathematics, The University of Chicago, Chicago, IL, USA

Ludwig Staiger
Institut für Informatik, Martin-Luther-Universität, Halle, Germany

Michael Stay
Google Inc., CA, USA, and Department of Computer Science, The University of Auckland, Auckland, New Zealand

Karl Svozil
Institut für Theoretische Physik, Technische Universität Wien, Vienna, Austria

Jonathan Swinton
Physiomics plc, Oxford Science Park, Oxford, UK

Christof Teuscher
ECE Department, Portland State University, Portland, OR, USA

K. Vela Velupillai
Department of Economics, University of Trento, Trento, Italy

Tom Vickers
Retired, Manager computing service that developed and used the Pilot ACE, at National Physical Laboratory (NPL), Teddington, London, UK

Paul Vitányi
CWI, Amsterdam, The Netherlands

Kevin Warwick
School of Systems Engineering, University of Reading, Reading, UK

Frode Weierud
Le Pre Vert, Prevessin-Moens, France

Philip Welch
School of Mathematics, University of Bristol, Bristol, UK

Jiří Wiedermann
Institute of Computer Science, Academy of Sciences of the Czech Republic, Prague, Czech Republic

Stephen Wolfram
Wolfram Research, Champaign, IL, USA

Mike Yates
Emeritus Professor of Mathematics, University of Manchester, Manchester, UK

Sandy L. Zabell
Department of Mathematics, Northwestern University, Evanston, IL, USA

Introduction

This is an unusual book. Its unusualness and complexity is appropriate for such an unusually inventive scientist, who was *personally* unique, as are so many creative thinkers.

Of the writings of Alan Turing selected here – which comprise most of those to be found in the Collected Works – a number have attracted a special interest, recognition and impact. And this is reflected in the number of commentaries accompanying his "computable numbers" paper, or the late great pieces on morphogenesis and the Turing test. But the collective power and energy of Turing is in the theoretical coherence of this collection of diverse writings. They are diverse in content, in style, in discipline, conveying different facets of a basic quest for understanding of 'how the world computes'.

You will find here no anonymous papers by committees of researchers. Even the occasional unpublished writing by joint authors on closer inspection, turns out to be written by one man. The rewards of a visceral engagement with these original writings are on various levels. A researcher should always have first-hand experience of any writings referred to. But with Turing the sense of the man behind the formal words is ever present. The organic involvement with the technical material, the sense of its emergence – an important concept in relation to Turing – from some more basic level of thought, is ever with us.

And just as the work and the person are unusually at one, there is a personal organic involvement with the writings from many of those paying tribute to Turing's thinking in this volume. We have tried to tap a wide spectrum of responses to Turing, people touched in many different ways by this strangely appealing man.

You will find here much to fascinate or surprise, both from Alan Turing and his commentators. The book intends to show the great value and impact Alan Turing's work continues to have.

There is a living heterogeneity to the content, formatted by a major academic publisher, with the editors aiming at something with at least a hint of the newspaper's immediacy and reporting of events in progress.

In this context, we hope our readers will excuse some rough edges. If you go to the "Afterword" first, you will see a candid description of the history of the "Collected Works of A.M. Turing", from which this book grew. That invaluable four-volume work took over 40 years to complete. The present single volume, containing most of the Turing works and much else, had to be completed in less than three years, much of it under pressure from an anxious publisher, and with doubly anxious editors watching the pages of the calendar turn towards 2012 – and past. We are very grateful to the publisher for initiating this major contribution to the Turing centenary celebrations, and to the contributors and editorial support team at Elsevier for their enthusiasm for the project and their patience with, and understanding of, the difficulties and delays.

With a few more years, we might have done much better, though the result might have been less interesting, and certainly less timely! We took a decision early on to *not* try to subsume the Collected Works. The Collected Works continues to have its own unique place in the Turing scholar's library, its value as an artefact matching the facsimile reproductions of Turing's papers. And the editorial work is by thinkers much closer to Turing and his contemporaries than us, and more often than not no longer available to update their work.

So we have not tried to reproduce the style of an archive, rather aiming at a book to be read, to be dipped into for pure pleasure, to be enjoyed and browsed in office, on train or bus, or accompanying the proposer to some distant scientific meeting or place of relaxation. The rekeying of the historical items presented special challenges, but we hope the benefits in terms of readability and sense of contemporaneity made them worth taking on.

One omission from the Collected Works, spread as it was over four volumes and a decade of publishing, was a seriously comprehensive bibliography. This has been commented on by a number of people, and was something we were anxious to rectify. Turing's biographer, Andrew Hodges, has provided a characteristically careful and insightful summary of the literature at his "The Alan Turing Bibliography" webpage: http://www.turing.org.uk/sources/biblio.html.

We did think of asking Andrew for permission to reprint this. And then came an unexpected discovery, which was the gargantuan work – "Bibliography of Publications of Alan Mathison Turing" – by Professor Nelson Beebe of the University of Utah. This is a bibliography whose scope and attention to detail, and current updated status, is beyond anything we could have provided. For details, see the Bibliography page at the end of this book.

Sadly, we never met Alan Turing, though we have talked to those who did, some of them represented in this book. What we hope the reader will share with us is the excitement of an ongoing exploration of 'how the world computes', and of a distinct sense of Turing's visionary presence accompanying us as we carry forward, in many different ways, his uncompleted work. In the much-quoted words of the great man himself, from his 1950 *Mind* paper on *Computing Machinery and Intelligence*:

We can only see a short distance ahead, but we can see plenty there that needs to be done.

S. Barry Cooper
Jan van Leeuwen

Spring 2013

Contents

Part I

How Do We Compute? What Can We Prove?

Alan Mathison Turing by Max Newman

(Bibliographic Memoirs of the Fellows of the Royal Society, vol. 1 (Nov. 1955), pp. 253–263)

Andrew Hodges Contributes

A Comment on Newman's Biographical Memoir

Newman had to comply with official secrecy and said virtually nothing regarding Turing's work from 1939 to 1945. Although the words 'Foreign Office' would have conveyed 'codes and ciphers' to all but the most naive readers, nothing went beyond this to convey scale or significance or scientific content. Indeed Newman's account went further than *suppressio veri* and led into a *suggestio falsi.* The expression 'mild routine' probably reinforced the prevalent impression of Bletchley Park as the resort of leisured time-wasters. Turing's work had been far from routine, involving real-time day and night work on the U-boat messages, and hair-raising missions to France, the United States, and Germany. It also required great intellectual originality. Newman could probably have given a clue to its content by making a reference to I. J. Good's 1950 book *Probability and the weighing of evidence*. But there was no such hint, and the 1955 reader could never have guessed that Newman had headed the section that used the most advanced electronic technology and Turing's statistical theory to break Hitler's messages.

A more surprising feature of Newman's account is the claim that 'the designers' of 'the new automatic computing machines' had worked in ignorance of Turing's universal machine. This is an odd expression since Turing himself was one such designer, as Newman's reference to 'the first plan of the ACE' makes clear, and obviously he knew of his own theory. Moreover, this plan was a very early one submitted to the NPL in March 1946. Newman can therefore only have meant that von Neumann's report of June 1945 was written in ignorance of Turing's work. The origin of the digital computer is a major point of interest in the history of science, and it seems strange that Newman lent his authority to such an oblique and vague comment on it, with an implicit assertion about von Neumann that is at variance with other evidence. Newman's statement is also misleading in its implication that Turing only turned his attention to computers in the summer of 1945 after learning of von Neumann's design. As it happens, Newman had actually written to von Neumann on 8 February 1946 with a sharply worded statement about British developments, asserting their early start and intellectual independence.[1] Already he was applying to the Royal Society for a large grant to fund what became the Manchester computer. 'By about 18 months ago', he wrote, 'I had decided to try my hand at starting up a machine unit... This was before I knew anything of the American work... I am of course in close touch with Turing...' The date of '18 months ago' is that of August 1944. In the light of what was revealed over 20 years later, it seems obvious that the success of the electronic Colossus after D-Day prompted discussion between Turing and Newman of how the logic of the universal machine could be implemented in a practical form. All this pre-1945 history was obliterated by Newman's account in 1955. It is of course very possible that the overpowering nature of official secrecy deterred Newman from giving even the faintest hint of his own and Turing's wartime experience at Bletchley Park. Unfortunately this omission contributed to a distortion of the historical record.

---

[1] Letter in the von Neumann archive, Library of Congress, Washington D.C. Quoted by A. Hodges *Alan Turing: the enigma,* p. 341.

By courtesy of the National Portrait Gallery, London

ALAN MATHISON TURING

1912–1954

The sudden death of Alan Turing on 7 June 1954 deprived mathematics and science of a great original mind at the height of its power. After some years of scientific indecision, since the end of the war, Turing had found, in his chemical theory of growth and form, a theme that gave the fullest scope for his rare combination of abilities, as a mathematical analyst with a flair for machine computing, and a natural philosopher full of bold original ideas. The preliminary report of 1952, and the account that will appear posthumously, describe only his first rough sketch of this theory, and the unfulfilled design must remain a painful reminder of the loss that his early death has caused to science.

Alan Mathison Turing was born in London on 23 June 1912, the son of Julius Mathison Turing, of the Indian Civil Service, and of Ethel Sara Turing (*née* Stoney). The name 'Turing' is of Scottish, perhaps ultimately of Norman origin, the final *g* being an addition made by Sir William Turing, of Aberdeenshire, in the reign James VI and I. The Stoneys, an English-Irish family of Yorkshire origin, produced some distinguished physicists and engineers in the nineteenth century, three of whom became Fellows of the Society; and Edith A. Stoney was one of the early women equal-to-wranglers at Cambridge (bracketed with 17th Wrangler, 1893).

Alan Turing's interest in science began early and never wavered. Both at his preparatory schools and later at Sherborne, which he entered in 1926, the contrast between his absorbed interest in science and mathematics, and his indifference to Latin and 'English subjects' perplexed and distressed his teachers, bent on giving him a well-balanced education. Many of the characteristics that were strongly marked in his later life can already be clearly seen in remembered incidents of this time: his particular delight in problems, large or small, that enabled him to combine theory with the kind of experiments he could carry out with his own hands, with the help of whatever apparatus was at hand; his strong preference for working everything out from first principles instead of borrowing from others—a habit which gave freshness and independence to his work, but also undoubtedly slowed him down, and later on made him a difficult author to read. At school homemade experiments in his study did not fit well into the routine of the house: a letter from his housemaster mentions 'Heaven knows what witches' brew blazing on a naked wooden window sill'. But before he left school his abilities, and his obvious seriousness of purpose, had won him respect and affection, and even tolerance for his own peculiar methods.

In 1931 he entered King's College, Cambridge, as a mathematical scholar. A second class in Part I of the Tripos showed him still determined not to spend time on subjects that did not interest him. In Part II he was a Wrangler, with 'b^*', and he won a Smith's Prize in 1936. He was elected a Fellow of King's in 1935, for a dissertation on the Central Limit Theorem of probability (which he discovered anew, in ignorance of recent previous work).

It was in 1935 that he first began to work in mathematical logic, and almost immediately started on the investigation that was to lead to his best known results, on computable numbers and the

---

Royal Society Memoir (Max Newman)
Reproduced from the Bibliographic Memoirs of the Fellows of the Royal Society, vol. 1 (November 1955) pp. 253–263 by kind permission of the Royal Society and of Edward and William Newman.

'Turing machine'. The paper attracted attention as soon as it appeared and the resulting correspondence led to his spending the next two years (1936–8) in Princeton, working with Professor Alonzo Church, the second of them as Proctor Fellow.

In 1938 Turing returned to Cambridge; in 1939 the war broke out. For the next six years he was fully occupied with his duties for the Foreign Office. These years were happy enough, perhaps the happiest of his life, with full scope for his inventiveness, a mild routine to shape the day, and a congenial set of fellow-workers. But the loss to his scientific work of the years between the ages of 27 and 33 was a cruel one. Three remarkable papers written just before the war, on three diverse mathematical subjects, show the quality of the work that might have been produced if he had settled down to work on some big problem at that critical time. For his work for the Foreign Office he was awarded the O.B.E.

At the end of the war many circumstances combined to turn his attention to the new automatic computing machines. They were in principle realizations of the 'universal machine' which he had described in the 1937 paper for the purpose of a logical argument, though their designers did not yet know of Turing's work. Besides this theoretical link, there was a strong attraction in the many-sided nature of the work, ranging from electric circuit design to the entirely new field or organizing mathematical problems for a machine. He decided to decline an offer of a Cambridge University Lectureship, and join the group that was being formed at the National Physical Laboratory for the design, construction and use of a large automatic computing machine. In the three years (1945–8) that this association lasted he made the first plan of the ACE, the N.P.L's automatic computer, and did a great deal of pioneering work in the design of sub-routines.

In 1948 he was appointed to a Readership in the University of Manchester, where work was beginning on the construction of a computing machine by F. C. Williams and T. Kilburn. The expectation was that Turing would lead the mathematical side of the work, and for a few years he continued to work, first on the design of the sub-routines out of which the larger programmes for such a machine are built, and then, as this kind work became standardized, on more general problems of numerical analysis. From 1950 onward he turned back for a while to mathematics and finally to his biological theory. But he remained in close contact with the Computing Machine Laboratory, whose members found him ready to tackle the mathematical problems that arose in their work, and what is more, to find the answers, by that combination of powerful mathematical analysis and intuitive short cuts that showed him at heart more of an applied than a pure mathematician.

He was elected to the Fellowship of the Society in 1951.

For recreation he turned mostly to those 'home-made' projects and experiments, self-contained both in theory and practice, that have already been mentioned: they remained a ruling passion up to the last hours of his life. The rule of the game was that everything was to be done with the materials at hand, and worked out from data to be found in the house, or in his own head. This sort of self-sufficiency stood him in good stead in starting on his theory of 'morphogenesis', where the preliminary reading would have drowned out a more orthodox approach. In everyday life it led to a certain fondness for the gimcrack, for example the famous Bletchley bicycle, the chain of which would stay on if the rider counted his pedal-strokes and executed a certain manoeuvre after every seventeen strokes.

After the war, feeling in need of violent exercise, he took to long distance running, and found that he was very successful at it. He won the 3 miles and 10 miles championships of his club (the Walton Athletic Club), both in record time, and was placed fifth in the A.A.A. Marathon race in 1947. He thought it quite natural to put this accomplishment to practical use from time to time, for example by running some nine miles from Teddington to a technical conference at the Post Office Research Station in North London, when the public transport proved tedious.

In conversation he had a gift for comical but brilliantly apt analogies, which found its full scope in the discussions on 'brains *v.* machines' of the late 1940's. He delighted in confounding those who, as he thought, too easily assumed that the two things are separated by an impassable gulf, by challenging them to produce an examination paper that could be passed by a man, but not by a machine. The unexpected element in human behaviour he proposed, half seriously, to imitate by a random element, or roulette-wheel, in the machine. This, he said, would enable proud owners to say 'My machine' (instead of 'My little boy') 'said such a funny thing this morning'.

Those who knew Turing will remember most vividly the enthusiasm and excitement with which he would pursue any idea that caught his interest, from a conversational hare to a difficult scientific problem. Nor was it only the pleasure of the chase that inspired him. He would take the greatest pains over services, large or small, to his friends. His colleagues in the computing machine laboratory found him still as ready as ever with his help for their problems when his own interests were fully engaged with his bio-chemical theory; and, as another instance, he gave an immense amount of thought and care to the selection of the presents which he gave to his friends and their children at Christmas.

His death, at a time when he was fully absorbed in his scientific work, was a great and sad loss to his friends, as well as to the wider world of science.

Scientific work

The varied titles of Turing's published work disguise its unity of purpose. The central problem with which he started, and to which he constantly returned, is the extent and the limitations of mechanistic explanations of nature. All his work, except for three papers in pure mathematics (1935*b*, 1938*a* and *b*) grew naturally out of the technical problems encountered in these inquiries. His way of tackling the problem was not by philosophical discussion of general principles, but by mathematical proof of certain limited results: in the first instance the impossibility of the too sanguine programme for the complete mechanization of mathematics, and in his final work, the possibility of, at any rate, a partial explanation of the phenomena of organic growth by the 'blind' operation of chemical laws.

1. *Mathematical logic*

The Hilbert decision-programme of the 1920's and 30's had for its objective the discovery of a general process, applicable to any mathematical theorem expressed in fully symbolical form, for deciding the truth or falsehood of the theorem. A first blow was dealt at the prospects of finding this new philosopher's stone by Gödel's incompleteness theorem (1931), which made it clear that truth or falsehood of *A* could not be equated to provability of *A* or not-*A* in any finitely based logic, chosen once for all; but there still remained in principle the possibility of finding a mechanical process for deciding whether *A*, or not-*A*, or neither, was formally provable in a given system. Many were convinced that no such process was possible, but Turing set out to demonstrate the impossibility rigorously. The first step was evidently to give a definition of 'decision process' sufficiently exact to form the basis of a mathematical proof of impossibility. To the question 'What is a "mechanical" process?' Turing returned the characteristic answer 'Something that can be done by a machine, and he embarked on the highly congenial task of analyzing the general notion of a computing machine. It is difficult to-day to realize how bold an innovation it was to introduce talk about paper tapes and patterns punched in them, into discussions of the foundations of mathematics. It is worth while quoting from his paper (1937*a*) the paragraph in which the computing machine is first introduced, both for the sake of its content and to give the flavour of Turing's writings.

'1. *Computing machines*

'We have said that the computable numbers are those whose decimals are calculable by finite means. This requires rather more explicit definition. No real attempt will be made to justify the definitions given until we reach §9. For the present I shall only say that the justification lies in the fact that the human memory is necessarily limited.

'We may compare a man in the process of computing a real number to a machine which is only capable of a finite number of conditions $q_1, q_2, \ldots, q_R$ which will be called "*m*-configurations". The machine is supplied with a "tape" (the analogue of paper) running through it, and divided into sections (called "squares") each capable of bearing a "symbol". At any moment there is just one square, say the *r*-th, bearing the symbol $\mathfrak{S}(r)$ which is "in the machine". We may call this square the "scanned square". The symbol on the scanned square may be called the "scanned symbol". The "scanned symbol" is the only one of which the machine is, so to speak, "directly aware". However, by altering its *m*-configuration the machine can effectively remember some of the symbols which it has "seen" (scanned) previously. The possible behaviour of the machine at any moment is determined by the *m*-configuration q_n and the scanned symbol $\mathfrak{S}(r)$. This pair $q_n, \mathfrak{S}(r)$ will be called the "configuration": thus the configuration determines the possible behaviour of the machine. In some of the configurations in which the scanned square is blank (*i.e.* bears no symbol) the machine writes down a new symbol on the scanned square: in other configurations it erases the scanned symbol. The machine may also change the square which is being scanned, but only by shifting it one place to right or left. In addition to any of these operations the *m*-configuration may be changed. Some of the symbols written down will form the sequence of figures which is the decimal of the real number which is being computed. The others are just rough notes to "assist the memory". It will only be these rough notes which will be liable to erasure.

'It is my contention that these operations include all those which are used in the computation of a number. The defence of this contention will be easier when the theory of the machines is familiar to the reader.'

In succeeding paragraphs he gave arguments for believing that a machine of this kind could be made to do any piece of work which could be done by a human computer obeying explicit instructions given to him before the work starts. A machine of the kind he had described could be made for computing the successive digits of π, another for computing the successive prime numbers, and so forth. Such a machine is completely specified by a table, which states how it moves from each of the finite sets of possible 'configurations' to another. In the computations mentioned above, of π and of the successive primes, the machine may be supposed to be designed for its special purpose. It is supplied with a blank tape and started off. But we may also imagine a machine supplied with a tape already bearing a pattern which will influence its subsequent behaviour, and this pattern might be the table, suitably encoded, of a particular computing machine, *X*. It could be arranged that this tape would cause the machine, *M*, into which it was inserted to behave like machine *X*. Turing proved the fundamental result that there is a 'universal' machine, *U* (of which he gave the table), which can be made to do the work of any assigned special-purpose machine, that is to say to carry out any piece of computing, if a tape bearing suitable 'instructions' is inserted into it. The machine *U* is so constructed that, presented with a tape bearing any arbitrary pattern it will move through a determinate, in general endless, succession of configurations; and it may or may not print at least one digit, 0 or 1. If it does, the pattern is 'circle-free'. It is therefore a problem, for which a decision process might be sought, to determine from inspection of a tape, whether or not it is circle-free. By means of a Cantor diagonal argument, Turing showed that no instruction-tape will cause the machine *U* to solve this problem, i.e. no pattern *P* is such that *U*, when presented with *P* followed by an arbitrary pattern Υ, will print 0 if Υ is 'circle-free', and 1 if it is not. If Turing's thesis is accepted, that the existence of a *method* for solving such a problem means the existence of a machine (or an instruction-tape for the universal machine *U*) that will solve it, it follows that the discovery of a process for discriminating between circle-free and other tapes is an insoluble problem, in an absolute and inescapable sense. From this basic insoluble problem it was not difficult to infer that the Hilbert programme of finding a decision method for the axiomatic system, Z, of elementary number-theory, is also impossible.

In the application he had principally in mind, namely, the breaking down of the Hilbert programme, Turing was unluckily anticipated by a few months by Church, who proved the same result by means of his 'λ-calculus'. An offprint arrived in Cambridge just as Turing was ready to send off his manuscript. But it was soon realised that Turing's 'machine' had a significance going far beyond this particular application. It was shown by Turing (1937*b*) and others that the definitions of 'general recursive' (by Gödel in 1931 and Kleene in 1935), 'λ-definable' (by Church in 1936) and 'computable' (Turing, 1937*a*) have exactly the same scope, a fact which greatly strengthened the belief that they describe a fundamentally important body of functions. Turing's treatment has the merit of making a particularly convincing case for the acceptance of these and no other processes, as genuinely constructive; and it turned out to be well adapted for use in finding other insoluble problems, e.g., in the theory of groups and semi-groups.

Turing's other major contribution to this part of mathematical logic, the paper (1939) on systems logic based on ordinals, has received less attention than (1937*a*), perhaps owing to its difficulty. The method of Gödel for constructing an undecidable sentence in any finitely based logic, L, i.e. a sentence expressible, but neither provable nor disprovable, in L, has led to the consideration of infinite families of 'logics', L_α, one for each ordinal α, where $L_{\alpha+1}$ is formed from L_α by the adjunction as an axiom of a sentence undecidable for L_α, if such exist, and L_α for limit ordinals α has as 'provable formulae', the union of the sets $P\beta(\beta < \alpha)$, where $P\beta$ is the set of provable formulae in $\mathrm{L}\beta$. The process must terminate for some $\gamma < \omega_1$, since the total set of formulae (which does not change) is countable. This procedure opens up the possibility of finding a logic that is complete, without violating Gödel's principle, since L_α may not be finitely based if α is a limit ordinal. Rosser investigated this possibility in 1937, using the 'classical' non-constructive theory of ordinals. Turing took up the proposal, but with the proviso that, although some non-constructive steps must be made if a complete logic is to be attained, a strict watch should be kept on them. He first introduced a new theory of constructive ordinals, or rather of formulae (of Church's λ-calculus) representing ordinals; and he showed that the problem of deciding whether a formula represents an ordinal (in a plausible sense) is insoluble, in the sense of his earlier paper. A formula $\mathbf{L}$ of the λ-calculus is a *logic* if it gives a means of establishing the truth of number-theoretic theorems; formally, if $\mathbf{L}(\mathbf{A})$ conv. $\mathbf{2}$ implies that $\mathbf{A}(\mathbf{n})$ conv. $\mathbf{2}$ for each $\mathbf{n}$ representing a natural number. The *extent* of $\mathbf{L}$ is the set of $\mathbf{A}$'s such that $\mathbf{L}(\mathbf{A})$ conv. $\mathbf{2}$, i.e., roughly speaking, the set of $\mathbf{A}$'s for which $\mathbf{L}$ proves $\mathbf{A}(\mathbf{n})$ is true for all $\mathbf{n}$. An *ordinal logic* is now defined to be a formula $\mathbf{\Lambda}$, such that $\mathbf{\Lambda}(\mathbf{\Omega})$ is a logic whenever $\mathbf{\Omega}$ represents an ordinal; and $\mathbf{\Lambda}$ is *complete* if every number-theoretic theorem that is true, is probable in $\mathbf{\Lambda}(\mathbf{\Omega})$ for some $\mathbf{\Omega}$, i.e. if given $\mathbf{A}$ such that $\mathbf{A}(\mathbf{n})$ conv. $\mathbf{2}$ for each $\mathbf{n}$ representing a natural number, $\mathbf{\Lambda}(\mathbf{\Omega}, \mathbf{A})$ conv. $\mathbf{2}$ for some $\mathbf{\Omega}$ (depending on $\mathbf{A}$). It is next shown, by an example, that formulae $\mathbf{\Omega}_1$, $\mathbf{\Omega}_2$ may represent the same ordinal, but yet make $\mathbf{\Lambda}(\mathbf{\Omega}_1)$ and $\mathbf{\Lambda}(\mathbf{\Omega}_2)$ different logics, in the sense that they have different extents. An ordinal logic for which this cannot happen is *invariant*. It is only in invariant logics that the 'depth' of a theorem can be measured by the size of the ordinal required for its proof. The main theorems of the paper state (1) that complete ordinal logics and invariant ordinal logics exist, (2) that no complete and invariant ordinal logic exists.

This paper is full of interesting suggestions and ideas. In §4 Turing considers, as a system with the minimal departure from constructiveness, one in which number-theoretic problems of some class are arbitrarily assumed to be soluble: as he puts it, 'Let us suppose that we are supplied with some unspecified means of solving number-theoretic problems; a kind of oracle, as it were.' The availability of the oracle is the 'infinite' ingredient necessary to escape the Gödel principle. It also obviously resembles the stages in the construction of a proof by a mathematician where he 'has an idea', as distinct from making mechanical use of a method. The discussion of this and related matters in §11 ('The purpose of ordinal logics') throws much light on Turing's views on the place of intuition in mathematical proof. In the final rather difficult §12 the idea adumbrated by Hilbert in 1922 of recursive definitions of order-types other than ω received its first detailed exposition.

Besides these two pioneering works, and the papers (1937*b*, *c*), arising directly out of them, Turing published four papers of predominantly logical interest. (A) The paper (1942*a*), with M. H. A. Newman, on a formal question in Church's theory of types. (B) A 'practical form of type-theory' (1948*b*) is intended to give Russell's theory of types a form that could be used in ordinary mathematics. Since the more flexible Zermelo-von Neumann set-theory has been generally preferred to type-theory by mathematicians, this paper has received little attention. It contains a number of interesting ideas, in particular a definition of 'equivalence' between logical systems (p. 89). (C) The use of dots as brackets (1942*b*), an elaborate discussion of punctuation in symbolic logic. Finally, (D) contains the proof (1950*a*) of the insolubility of the word-problem for semi-groups with cancellation. A finitely generated semi-group *without* cancellation is determined by choosing a finite set of pairs of words, $(A_i, B_i)(i = 1_1 \ldots k)$ of some alphabet, and declaring two words to be 'equivalent' if they can be proved so by the use of the equations $PA_iQ = PB_iQ$, where P and Q can be arbitrary words (possibly empty). The word-problem for such a semi-group is to find a process which will decide whether or not two given words are equivalent. The insolubility of this problem can be brought into relation with the fundamental insoluble machine-tape problem. The table of a computing machine states, for each configuration, what is the configuration that follows it. Since a configuration can be denoted by a 'word', in letters representing the internal configurations and tape-symbols, this table gives a set of pairs words which, when suitably modified, determine a semi-group with insoluble word problem. So much was proved by E. L. Post in 1947. The question becomes much more difficult if the semi-group is required to satisfy the cancellation laws, '$AC = BC$ implies $A = B$' and '$CA = CB$ implies $A = B$' since now a condition is imposed on account of its mathematical interest, and not because it arises naturally from the machine interpretation. This was the step taken by Turing in (1950*a*). (For a helpful discussion and analysis of this difficult paper see the long review by W. W. Boone, *J. Symbolic Logic*, **17** (1952) 74.)

2. *Three mathematical papers*

Shortly before the war Turing made his only contributions to mathematics proper.

The paper 1938*a* contains an interesting theorem on the approximation of Lie groups by finite groups: if a (connected) Lie group, L, can for arbitrary $\varepsilon > 0$ be ε-approximated by a finite group whose multiplication law is an ε-approximation to that of L, in the sense that the two products of any two elements are within ε of each other, then L must be both compact and abelian. The theory of representations of topological groups is used to apply Jordan's theorem on the abelian invariant subgroups of finite groups of linear transformations.

Paper (1938*b*) lies in the domain of classical group theory. Results of R. Baer on the extensions of a group are re-proved by a more unified and simpler method.

Paper (1943)—submitted in 1939, but delayed four years by war-time difficulties—shows that Turing's interest in practical computing goes back at least to this time. A method is given for the calculation of the Riemann zeta-function, suitable for values of t in a range not well covered by the previous work of Siegel and Titchmarsh. The paper is a straightforward but highly skilled piece of classical analysis. (The post-war paper (1953*a*) describes an attempt to apply a modified form of this process, which failed owing to machine trouble.)

3. *Computing machines*

Apart from the practical 'programmer's handbook', only two published papers (1948*a* and 1950*b*) resulted from Turing's work on machines. When binary fractions of fixed length are used (as they must be on a computing machine) for calculations involving a very large number of multiplications and divisions, the necessary rounding-off of exact products introduces cumulative errors, which gradually consume the trustworthy digits as the computation proceeds. The paper (1948*a*) investigates questions of the following type: how many figures of the answer are trustworthy if k figures are retained in solving n linear equations in n unknowns? The answer depends on the method of solution, and a number of different ones are considered. In particular it is shown that the ordinary method of successive elimination of the variables does not lead to the very large errors that had been predicted, save in exceptional cases which can be specified.

The other paper (1950*b*) arising out of his interest in computing machines is of a very different nature. This paper, on computing machines and intelligence, contains Turing's views on some questions about which he had thought a great deal. Here he elaborates his notion of an 'examination' to test machines against men, and he examines systematically a series of arguments commonly put forward against the view that machines might be said to think. Since the paper is easily accessible and highly readable, it would be pointless to summarize it. The conversational style allows the natural clarity of Turing's thought to prevail, and the paper is a masterpiece of clear and vivid exposition.

The proposals (1953*b*) for making a computing machine play chess are amusing, and did in fact produce a defence lasting 30 moves when the method was tried against a weak player; but it is possible that Turing underestimated the gap that separates combinatory from position play.

4. *Chemical theory of morphogenesis*

For the following account of Turing's final work I am indebted to Dr N. E. Hoskin, who with Dr B. Richards is preparing an edition of the material for publication.

The work falls into two parts. In the first part, published (1952) in his lifetime, he set out to show that the phenomena of morphogenesis (growth and form of living things) could be explained by consideration of a system of chemical substances whose concentrations varied only by means of chemical reactions, and by diffusion through the containing medium. If these substances are considered as form-producers (or 'morphogens' as Turing called them) they may be adequate to determine the formation and growth of an organism, if they result in localized accumulations of form-producing substances. According to Turing the laws of physical chemistry are sufficient to account for many of the facts of morphogenesis (a view similar to that expressed by D'Arcy Thompson in *Growth and form*).

Turing arrived at differential equations of the form

$$\frac{\partial X_i}{\partial t} = f_i(X_1 \ldots, X_n) + \mu \nabla^2 X_i, \quad (i = 1, \ldots, n)$$

for n different morphogens in continuous tissue; where f_i is the reaction function giving the rate of growth of X_i, and $\nabla^2 X_i$ is the rate of diffusion of X_i. He also considered the corresponding equations for a set of discrete ceils. The function f_i involves the concentrations, and in his 1952 paper Turing considered the X_i's as variations from a homogeneous equilibrium. If, then, there are only small departures from equilibrium, it is permissible to linearize the f_i's, and so linearize the differential equations. In this way he was able to arrive at the conditions governing the onset of instability. Assuming initially a state of homogeneous equilibrium disturbed by random disturbances at $t = 0$, he discussed the various forms instability could take, on a continuous ring of tissue. Of the forms discussed the most important was that which eventually reached a pattern of stationary waves. The botanical situation corresponding to this would be an accumulation of the relevant morphogen in several evenly distributed regions around the ring, and would result in the main growth taking place at these points. (The examples cited are the tentacles *Hydra* and whorled leaves.) He also tested the theory by obtaining numerical solutions of the equations, using the electronic computer at Manchester. In the numerical example, in which two morphogens were supposed to be present in a ring of twenty cells, he found that a three or four lobed pattern would result. In other examples he found two-dimensional patterns, suggestive of dappling; and a system on a sphere gave results indicative of gastrulation. He also suggested that stationary waves in two dimensions could account for the phenomena of phyllotaxis.

In his later work (as yet unpublished) he considered quadratic terms in the reaction functions in order to take account of larger departures from the state of homogeneous equilibrium. He was attempting to solve the equations in two dimensions on the computer at the time of his death. The work is in existence, but unfortunately is in a form that makes it extremely difficult to discover the results he obtained. However, B. Richards, using the same equations, investigated the problem in the case where the organism forms a spherical shell and also obtained numerical results on the computer. These were compared with the structure of *Radiolaria*, which have spikes on a basic spherical shell, and the agreement was strikingly good. The rest of this part of Turing's work is incomplete, and little else can be obtained from it. However, from Richards's results it seems that consideration of quadratic terms is sufficient to determine practical solutions, whereas linear terms are really only sufficient to discuss the onset of instability.

The second part of the work is a mathematical discussion of the geometry of phyllotaxis (i.e. of mature botanical structures). Turing discussed many ways of classifying phyllotaxis patterns and suggested various parameters by which a phyllotactic lattice may be described. In particular, he showed that if a phyllotactic system is Fibonacci in character, it will change, if at all, to a system which has also Fibonacci character. This is in accordance with observation. However, most of this section was intended merely as a description preparatory to his morphogenetic theory, to account for the facts of phyllotaxis; and it is clear that Turing did not intend it to stand alone.

The wide range of Turing's work and interests have made the writer of this notice more than ordinarily dependent on the help of others. Among many who have given valuable information I wish to thank particularly Mr R. Gandy, Mr J. H. Wilkinson, Dr B. Richards and Dr N. E. Hoskin; and Mrs Turing, Alan Turing's mother, for constant help with biographical material.

BIBLIOGRAPHY

1935*a*. On the Gaussian Error Function (King's College Fellowship Dissertation)
1935*b*. Equivalence of left and right almost periodicity. J. Lond. Math. Soc. **10**, 284.
1937*a*. On computable numbers, with an application to the Entscheidungsproblem. Proc. Lond. Math. Soc. (2), **42**, 230.
1937*b*. Computability and λ-definability. J. Symbolic Logic, **2**, 153.
1937*c*. The *p*-function in λ-K-conversion. J. Symbolic Logic, **2**, 164.
1937*d*. Correction to 1937*a*. Proc. Lond. Math. Soc. (2), **43**, 544.
1938*a*. Finite approximations to Lie groups. Ann. Math. Princeton, **39**, 105.
1938*b*. The extensions of a group. Comp. Math. **5**, 357.
1939. Systems of logic based on ordinals. Proc. Lond. Math. Soc. (2), **45**, 161.
1942*a*. (With M. H. A. Newman) A formal theorem in Church's theory of types. J. Symbolic Logic, **7**, 28.
1942*b*. The use of dots as brackets in Church's system, J. Symbolic Logic, **7**, 146.
1943*. A method for the calculation of the zeta-function. Proc. Lond. Math. Soc. (2), **48**, 180.
1948*a*. Rounding-off errors in matrix processes. Quart. J. Mech. App. Math. **1**, 287.
1948*b*. Practical forms of type-theory. J. Symbolic Logic, **13**, 80.
1950*a*. The word problem in semi-groups with cancellation. Ann. Math. **52**, 491.
1950*b*. Computing machinery and intelligence. Mind **59**, 433.
1950*c*. Programmers' Handbook for the Manchester electronic computer
1952. The chemical basis of morphogenesis. Phil. Trans. B **237**, 37.
1953*a*. Some calculations of the Riemann zeta-function. Proc. Lond. Math. Soc. (3), **3**, 99.
1953*b*. Digital computers applied to games: chess, pp. 288–295 of Faster than thought, ed. B. V. Bowden. Pitman, London.
1954. Solvable and unsolvable problems. Sci. News **31**, 7.
[A second paper on morphogenesis is being prepared for publication by N. E. Hoskin and B. Richards, based on work left by Turing.]

* Received four years earlier (7 March 1939).

On Computable Numbers, with an Application to the Entscheidungsproblem

(Proc. Lond. Math. Soc., ser. 2 vol. **42** (1936–37), pp. 230–265)

– A Correction

(ibid. vol. **43** (1937), pp. 544–546)

Christos Papadimitriou on —

ALAN AND I

During my sad college years, I often dreamed of Alan. At the time I did not know it was Alan Turing that I was dreaming of, but it was. I was studying a subject that did not excite me (electrical engineering), in the inflexible educational system of an oppressive society (Greece of the colonels). I had no access to a proper scientific library. My life as a fledgling scientist was one of frustration, blind longing, and episodes of false epiphany. A few subjects (systems theory, communication theory), even though they were taught at school in the most mundane way, enabled one to imagine a courageous intellectual universe in which questions of the most fundamental nature are confronted rigorously and head on, and I was aching to enter that universe.

I don't remember when, in which outdated textbook, handed to me by whom, I got my first glimpse of the Turing machine. I did not get it all at once, but I knew immediately that this abstract device is an important exemplar of the higher sphere I had been dreaming. I looked up in the dictionary the verb 'to ture' (I really did). I sought more information, every book I opened those days I opened it on the index page where 'Turing machine' should be.

Eventually I did put it all together, how a British mathematician named Alan Turing answered through his machine the world's most fundamental question, 'what can be computed?' and did so with amazing rigor, elegance, imagination and economy. But those days I was thinking of the Turing machine as a singular breakthrough, the end of a story, something of the past. Two years later, in 1973 – after a year in the Greek army that was even bleaker than my tertiary studies – I was fortunate to find myself at Princeton (Turing's Princeton, by the way, where I lived for a year in room 2B of the Graduate College rumored to be Alan's room 35 years before). At Princeton I was introduced to the Theory of Computation, the rich and vibrant scientific tradition essentially built on Turing's formalism. I remember how grateful I felt that my prayers had been answered, so to speak, and I was finally entering the realm of my dreams, far more elegant and exciting than I could ever imagine. And 5 years later, whilst teaching at Harvard with my friend Harry Lewis, we wrote a textbook on the subject.

But even though my life now revolved around his intellectual heritage, the truth is, I did not know Alan. I understood next to nothing about the man's life, personality, and breadth of achievement. In 1983, 2 years after our textbook was published, I read one of the books that have influenced me most: Turing's powerful and definitive biography by Andrew Hodges. Alan became my hero, a giant and relentless intellect, a fascinating and complex personality, a man of immense accomplishment, impact, and tragedy. When, more than a decade later, the second edition of our textbook was published, Harry and I decided that we must have Alan Turing's image on the cover. Since the mid-1980s, every time I teach about Turing machines and undecidability, I stop to tell the class about Alan, about his ingenious work and about his tragic end. Once in a while I teach a course at Berkeley on 'Reading the Classics', and in it we spend a month on Turing, because I believe that every graduate student should be exposed directly to the exacting self-conscious greatness of Alan's opus.

There is a scene in Gibson's *Neuromancer*, at the very end of Part Three, where the hero returns to his hotel room to find it full of cops. *'Turing',* they tell him. *'You are under arrest'.* They mean 'Turing police', a fictional force bent on rooting out AI from the planet, but it does not matter, for me this line, read literally, contained the germ of an idea: What if Alan Turing were alive and turned up some place, unbidden? This strange fantasy lived inside me for a few years.

Turing is not my only intellectual hero. In poetry, my hero is Constantine Cavafy, a Greek from Alexandria who died when Alan Turing was 21. He wrote some of the greatest poems of the twentieth century, a stunning opus sharply divided between subtle historical metaphor and rather unsubtle eroticism. (I do not know why both my heroes happen to be homosexuals.) In 1997 I was in Greece, and I went to see a film titled *Cavafy*. I liked it so-so, but I remember coming out of the theatre impressed by the director's gesture: to honor one's hero by creating a work of art bearing his name. And then, right there in the theatre lobby, I had a vision: there was a blue paperback hovering above, and the title in front read *Turing (A Novel).*

Writing a novel had never occurred to me before. I had never written short stories or poetry. I had of course noticed over the years that writing was not my weakest point, and neither was it for me the hated chore that comes after research, as it was for many of my colleagues. That night I thought about a plot.

This was 1997, when it was slowly becoming apparent to many computer scientists that the true object of our science is not the computer, but the Internet (by which I mean both the network of networks and the World Wide Web). In a sense, the Internet is the ultimate legacy of Alan Turing. The reason it spread like wildfire ever since a physicist named Tim Burners-Lee invented 'click' in 1989 is because there were millions of computers on the desks of people at that time, and these computers were all universal and so, in addition to everything else, they could be easily made to click. But universality was a minority opinion among computer dreamers during the 1930s and 1940s. By making his universal machine so compelling, Turing influenced deeply von Neumann and the way computers turned out to be. Universality and software would have probably taken root at some point in computers no matter what, but we can only speculate about the setbacks and delays this would have required without Turing.

But why did Turing envision universality? The reason is, he did not set out just to answer the question 'What can (and what cannot) be computed?' per se, as I inaccurately mentioned above. If he had only wanted to establish the existence of unsolvable problems, Turing could just use diagonalisation or growth, and the universal Turing machine might never come to light. Fortunately, he wanted to do something more ambitious and specific – and central to the scientific agenda of the time – he wanted to show that the *Entscheidungsproblem* (the decision problem for sentences not of arithmetic, but even of first-order logic) cannot be solved by computers. In other words, his goal was to sharpen Gödel's negative result, to extinguish the last glimmer of hope left by the Incompleteness Theorem. And to accomplish this he needed the universal Turing machine.

The night after I saw that movie, all this was in my mind. If, through this long chain of logic, the Internet is Alan's ultimate creation, why would not the Internet return the favor, and bring its creator back to life? The thought did not let me sleep. The Internet does confer a lame kind of immortality (just search for 'Alan Turing' on the web). Imagining a more explicit form, an Internet spirit residing somewhere and everywhere, a kind of impromptu, hacked-up SETI, was only a modest step forward. And if I were to bring my hero back to life, why wouldn't I load him with gifts of gratitude, especially focusing on things he missed in life? I could give him, for example, a happy love life – after all obstacles are overcome, of course – make him a gifted teacher, give him a faithful pupil who would be a Greek man my age, yes, an archeologist perhaps, pining for a lost love, for an American woman, a software wiz maybe, why not? When I was a child, I happened to visit the island of Corfu with my father during the same summer Alan Turing was there. As the night advanced, the plot thickened. In the morning I recounted the story to my wife Martha, as I always do when I want to rid myself of an idea – because she can be a pitiless critic – but she went extra soft on it, she actually liked it. Two days later I was flying to California, and during that flight I drafted the first chapter – taking place, as it happens, in an airplane. For the next two and a half years I wrote every day, usually the first hours of my day, until the book was finished.

This book was a watershed. Whilst writing it, I understood things about myself that surprised me utterly. One of them was, I would be writing fiction again. Once more, Alan Turing had changed my life.

Alan inspires my papers and my stories, he fires my talks and my courses, inhabits my memories and my dreams. And because he's so intimate, impossible to examine anew and from a distance in order to discern something fresh, I could only speak here of this intimacy.

ON COMPUTABLE NUMBERS, WITH AN APPLICATION TO THE ENTSCHEIDUNGSPROBLEM

By A. M. TURING

[Received 28 May, 1936.—Read 12 November, 1936.]

The "computable" numbers may be described briefly as the real numbers whose expressions as a decimal are calculable by finite means. Although the subject of this paper is ostensibly the computable *numbers*, it is almost equally easy to define and investigate computable functions of an integral variable or a real or computable variable, computable predicates, and so forth. The fundamental problems involved are, however, the same in each case, and I have chosen the computable numbers for explicit treatment as involving the least cumbrous technique. I hope shortly to give an account of the relations of the computable numbers, functions, and so forth to one another. This will include a development of the theory of functions of a real variable expressed in terms of computable numbers. According to my definition, a number is computable if its decimal can be written down by a machine.

In §§9, 10 I give some arguments with the intention of showing that the computable numbers include all numbers which could naturally be regarded as computable. In particular, I show that certain large classes of numbers are computable. They include, for instance, the real parts of all algebraic numbers, the real parts of the zeros of the Bessel functions, the numbers π, e, etc. The computable numbers do not, however, include all definable numbers, and an example is given of a definable number which is not computable.

Although the class of computable numbers is so great, and in many ways similar to the class of real numbers, it is nevertheless enumerable. In §8 I examine certain arguments which would seem to prove the contrary. By the correct application of one of these arguments, conclusions are reached which are superficially similar to those of Gödel[1]. These results have valuable applications. In particular, it is shown (§11) that the Hilbertian Entseheidungsproblem can have no solution.

In a recent paper Alonzo Church[2] has introduced an idea of "effective calculability", which is equivalent to my "computability", but is very differently defined. Church also reaches similar conclusions about the Entseheidungsproblem[3]. The proof of equivalence between "computability" and "effective calculability" is outlined in an appendix to the present paper.

1. Computing machines

We have said that the computable numbers are those whose decimals are calculable by finite means. This requires rather more explicit definition. No real attempt will be made to justify the definitions given until we reach §9. For the present I shall only say that the justification lies in the fact that the human memory is necessarily limited.

[1] Gödel, "Über formal unentscheidbaro Sätze der Principia Mathematica und ver-wandter Systeme, I", *Monatshefte Math. Phys.*, 38 (1931), 173–198.

[2] Alonzo Church "An unsolvable problem of elementary number theory", *American J. of Math.*, 58 (1936), 345–363.

[3] Alonzo Church "A note on the Entseheidungsproblem", *J. of Symbolic Logic*, 1 (1936), 40–41.

We may compare a man in the process of computing a real number to a machine which is only capable of a finite number of conditions $q_1, q_2, \ldots, q_R$, which will be called "*m*-configurations". The machine is supplied with a "tape" (the analogue of paper) running through it, and divided into sections (called "squares") each capable of bearing a "symbol". At any moment there is just one square, say the r-th, bearing the symbol $\mathfrak{S}(r)$ which is "in the machine". We may call this square the "scanned square". The symbol on the scanned square may be called the "scanned symbol". The "scanned symbol" is the only one of which the machine is, so to speak, "directly aware". However, by altering its *m*-configuration the machine can effectively remember some of the symbols which it has "seen" (scanned) previously. The possible behaviour of the machine at any moment is determined by the *m*-configuration q_n and the scanned symbol $\mathfrak{S}(r)$. This pair q_n, $\mathfrak{S}(r)$ will be called the "configuration": thus the configuration determines the possible behaviour of the machine. In some of the configurations in which the scanned square is blank (*i.e.* bears no symbol) the machine writes down a new symbol on the scanned square: in other configurations it erases the scanned symbol. The machine may also change the square which is being scanned, but only by shifting it one place to right or left. In addition to any of these operations the *m*-configuration may be changed. Some of the symbols written down will form the sequence of figures which is the decimal of the real number which is being computed. The others are just rough notes to "assist the memory". It will only be these rough notes which will be liable to erasure.

It is my contention that these operations include all those which are used in the computation of a number. The defence of this contention will be easier when the theory of the machines is familiar to the reader. In the next section I therefore proceed with the development of the theory and assume that it is understood what is meant by "machine", "tape", scanned", etc.

2. Definitions

Automatic machines

If at each stage the motion of a machine (in the sense of §1) is *completely* determined by the configuration, we shall call the machine an "automatic machine" (or *a*-machine).

For some purposes we might use machines (choice machines or *c*-machines) whose motion is only partially determined by the configuration (hence the use of the word "possible" in §1). When such a machine reaches one of these ambiguous configurations, it cannot go on until some arbitrary choice has been made by an external operator. This would be the case if we were using machines to deal with axiomatic systems. In this paper I deal only with automatic machines, and will therefore often omit the prefix *a*-.

Computing machines

If an *a*-machine prints two kinds of symbols, of which the first kind (called figures) consists entirely of 0 and 1 (the others being called symbols of the second kind), then the machine will be called a computing machine. If the machine is supplied with a blank tape and set in motion, starting from the correct initial *m*-configuration, the subsequence of the symbols printed by it which are of the first kind will be called the *sequence computed by the machine*. The real number whose expression as a binary decimal is obtained by prefacing this sequence by a decimal point is called the *number computed by the machine*.

At any stage of the motion of the machine, the number of the scanned square, the complete sequence of all symbols on the tape, and the *m*-configuration will be said to describe the *complete configuration* at that stage. The changes of the machine and tape between successive complete configurations will be called the *moves* of the machine.

Circular and circle-free machines

If a computing machine never writes down more than a finite number of symbols of the first kind, it will be called *circular*. Otherwise it is said to be *circle-free*.

A machine will be circular if it reaches a configuration from which there is no possible move, or if it goes on moving, and possibly printing symbols of the second kind, but cannot print any more symbols of the first kind. The significance of the term "circular" will be explained in §8.

Computable sequences and numbers

A sequence is said to be computable if it can be computed by a circle-free machine. A number is computable if it differs by an integer from the number computed by a circle-free machine.

We shall avoid confusion by speaking more often of computable sequences than of computable numbers.

3. Examples of computing machines

I. A machine can be constructed to compute the sequence 010101... The machine is to have the four *m*-configurations "$\mathfrak{b}$", "$\mathfrak{c}$", "$\mathfrak{f}$" "$\mathfrak{e}$", and is capable of printing "0" and "1". The behaviour of the machine is described in the following table in which "*R*" means "the machine moves so that it scans the square immediately on the right of the one it was scanning previously". Similarly for "*L*". "*E*" means "the scanned symbol is erased" and "*P*" stands for "prints". This table (and all succeeding tables of the same kind) is to be understood to mean that for a configuration described in the first two columns the operations in the third column are carried out successively, and the machine then goes over into the *m*-configuration described in the last column. When the second column is left blank, it is understood that the behaviour of the third and fourth columns applies for any symbol and for no symbol. The machine starts in the *m*-configuration $\mathfrak{b}$ with a blank tape.

Configuration		*Behaviour*	
m-config.	*symbol*	*operations*	*final m-config.*
$\mathfrak{b}$	None	$P0, R$	$\mathfrak{c}$
$\mathfrak{c}$	None	R	$\mathfrak{e}$
$\mathfrak{e}$	None	$P1, R$	$\mathfrak{f}$
$\mathfrak{f}$	None	R	$\mathfrak{b}$

If (contrary to the description in §1) we allow the letters *L*, *R* to appear more than once in the operations column we can simplify the table considerably.

m-config.	*symbol*	*operations*	*final m-config.*
	None	$P0$	$\mathfrak{b}$
$\mathfrak{b}$	0	$R, R, P1$	$\mathfrak{b}$
	1	$R, R, P0$	$\mathfrak{b}$

II. As a slightly more difficult example we can construct a machine to compute the sequence 001011011101111011111.... The machine is to be capable of five *m*-configurations, viz. "$\mathfrak{o}$", "$\mathfrak{q}$", "$\mathfrak{p}$", "$\mathfrak{f}$" "$\mathfrak{b}$" and of printing "ə", "*x*", "0" "1". The first three symbols on the tape will be "əə0"; the other figures follow on alternate squares. On the intermediate squares we never print anything but

"x". These letters serve to "keep the place" for us and are erased when we have finished with them. We also arrange that in the sequence of figures on alternate squares there shall be no blanks.

Configuration		*Behaviour*	
m-config.	*symbol*	*operations*	*final m-config.*
$\mathfrak{b}$		Pə$, R, P$ə$, R, P0, R, R, P0, L, L$	$\mathfrak{o}$
$\mathfrak{o}$	1	R, Px, L, L, L	$\mathfrak{o}$
	0		$\mathfrak{q}$
$\mathfrak{q}$	Any (0 or 1)	R, R	$\mathfrak{q}$
	None	$P1, L$	$\mathfrak{p}$
$\mathfrak{p}$	x	E, R	$\mathfrak{q}$
	ə	R	$\mathfrak{f}$
	None	L, L	$\mathfrak{p}$
$\mathfrak{f}$	Any	R, R	$\mathfrak{f}$
	None	$P0, L, L$	$\mathfrak{o}$

To illustrate the working of this machine a table is given below of the first few complete configurations. These complete configurations are described by writing down the sequence of symbols which are on the tape, with the *m*-configuration written below the scanned symbol. The successive complete configurations are separated by colons.

```
   : ə ə 0  0 : ə ə  0  0 : ə ə 0   0 : ə ə 0   0    : ə ə 0   0   1 :
 𝔟       𝔬           𝔮              𝔮              𝔮             𝔭

 ə ə 0   0  1 : ə ə  0  0   1 : ə ə 0   0   1 : ə ə 0   0   1 :
       𝔭          𝔭                 𝔣                  𝔣

 ə ə 0   0  1 : ə ə  0  0   1     : ə ə 0   0  1   0 :
            𝔣                   𝔣              𝔬

 ə ə 0   0  1 x 0 : ...
         𝔬
```

This table could also be written in the form

𝔟:ə ə 𝔬 0   0:ə ə 𝔮 0   0:...,   (C)

in which a space has been made on the left of the scanned symbol and the *m*-configuration written in this space. This form is less easy to follow, but we shall make use of it later for theoretical purposes.

The convention of writing the figures only on alternate squares is very useful: I shall always make use of it. 1 shall call the one sequence of alternate squares F-squares and the other sequence E-squares. The symbols on E-squares will be liable to erasure. The symbols on F-squares form a continuous sequence. There are no blanks until the end is reached. There is no need to have more than one E-square between each pair of F-squares: an apparent need of more E-squares can be satisfied by having a sufficiently rich variety of symbols capable of being printed on E-squares. If a symbol β is on an F-square S and a symbol a is on the E-square next on the right of S, then S and β will be said to be *marked* with α. The process of printing this a will be called marking β (or S) with a.

4. Abbreviated tables

There are certain types of process used by nearly all machines, and these, in some machines, are used in many connections. These processes include copying down sequences of symbols, comparing sequences, erasing all symbols of a given form, etc. Where such processes are concerned we can abbreviate the tables for the m-configurations considerably by the use of "skeleton tables". In skeleton tables there appear capital German letters and small Greek letters. These are of the nature of "variables". By replacing each capital German letter throughout by an m-configuration and each small Greek letter by a symbol, we obtain the table for an m-configuration.

The skeleton tables are to be regarded as nothing but abbreviations: they are not essential. So long as the reader understands how to obtain the complete tables from the skeleton tables, there is no need to give any exact definitions in this connection.

Let us consider an example:

m-config.	*Symbol*	*Behaviour*	*Final m-config.*
$\mathfrak{f}(\mathfrak{C},\mathfrak{B},\alpha)$	ə	L	$\mathfrak{f}_1(\mathfrak{C},\mathfrak{B},\alpha)$
	not ə	L	$\mathfrak{f}(\mathfrak{C},\mathfrak{B},\alpha)$
$\mathfrak{f}_1(\mathfrak{C},\mathfrak{B},\alpha)$	α		$\mathfrak{C}$
	not α	R	$\mathfrak{f}_1(\mathfrak{C},\mathfrak{B},\alpha)$
	None	R	$\mathfrak{f}_2(\mathfrak{C},\mathfrak{B},\alpha)$
$\mathfrak{f}_2(\mathfrak{C},\mathfrak{B},\alpha)$	α		$\mathfrak{C}$
	not α	R	$\mathfrak{f}_1(\mathfrak{C},\mathfrak{B},\alpha)$
	None	R	$\mathfrak{B}$

From the m-configuration $\mathfrak{f}(\mathfrak{C},\mathfrak{B},\alpha)$ the machine finds the symbol of form a which is farthest to the left (the "first α") and the m-configuration then becomes $\mathfrak{C}$. If there is no α then the m-configuration becomes $\mathfrak{B}$.

If we were to replace $\mathfrak{C}$ throughout by $\mathfrak{q}$ (say), $\mathfrak{B}$ by $\mathfrak{r}$, and a by x, we should have a complete table for the m-configuration $\mathfrak{f}(\mathfrak{q},\mathfrak{r},x)$. $\mathfrak{f}$ is called an "m-configuration function" or "m-function".

The only expressions which are admissible for substitution in an m-function are the m-configurations and symbols of the machine. These have to be enumerated more less explicitly : they may include expressions such as $\mathfrak{p}(e,x)$; indeed they must if there are any m-functions used at all. If we did not insist on this explicit enumeration, but simply stated that the machine had certain m-configurations (enumerated) and all m-configurations obtainable by substitution of m-configurations in certain m-functions, we should usually get an infinity of m-configurations; *e.g.*, we might say that the machine was to have the m-configuration $\mathfrak{q}$ and all m-configurations obtainable by substituting an *m*-configuration for $\mathfrak{C}$ in $\mathfrak{p}(\mathfrak{C})$. Then it would have $\mathfrak{q}, \mathfrak{p}(\mathfrak{q}), \mathfrak{p}(\mathfrak{p}(\mathfrak{q})), \mathfrak{p}(\mathfrak{p}(\mathfrak{p}(\mathfrak{q}))), \ldots$ as *m*-configurations.

Our interpretation rule then is this. We are given the names of the *m*-configurations of the machine, mostly expressed in terms of *m*-functions. We are also given skeleton tables. All we want is the complete table for the *m*-configurations of the machine. This is obtained by repeated substitution in the skeleton tables.

Further examples

(In the explanations the symbol "→" is used to signify "the machine goes into the *m*-configuration....")

$\mathfrak{e}(\mathfrak{C},\mathfrak{B},\alpha)$		$\mathfrak{f}(\mathfrak{e}_1(\mathfrak{C},\mathfrak{B},\alpha)\mathfrak{B},\alpha)$	From $\mathfrak{e}(\mathfrak{C},\mathfrak{B},\alpha)$ the first α is erased
$\mathfrak{e}_1(\mathfrak{C},\mathfrak{B},\alpha)$	E	$\mathfrak{C}$	and $\rightarrow \mathfrak{B}$. If there is no $\alpha \rightarrow \mathfrak{B}$.
$\mathfrak{e}(\mathfrak{B},\alpha)$		$\mathfrak{e}(\mathfrak{e}(\mathfrak{B},\alpha),\mathfrak{B},\alpha)$	From $\mathfrak{e}(\mathfrak{B},\alpha)$ all letters α are erased and $\rightarrow \mathfrak{B}$.

The last example seems somewhat more difficult to interpret than most. Let us suppose that in the list of *m*-configurations of some machine there appears $\mathfrak{e}(\mathfrak{b},x)$ ($= \mathfrak{q}$, say). The table is

	$\mathfrak{e}(\mathfrak{b},x)$	$\mathfrak{e}(\mathfrak{e}(\mathfrak{b},x),\mathfrak{b},x)$
or	$\mathfrak{q}$	$\mathfrak{e}(\mathfrak{q},\mathfrak{b},x)$.

Or, in greater detail:

$\mathfrak{q}$		$\mathfrak{e}(\mathfrak{q},\mathfrak{b},x)$
$\mathfrak{e}(\mathfrak{q},\mathfrak{b},x)$		$\mathfrak{f}(\mathfrak{e}_1(\mathfrak{q},\mathfrak{b},x),\mathfrak{b},x)$
$\mathfrak{e}_1(\mathfrak{q},\mathfrak{b},x)$	E	$\mathfrak{q}$.

In this we could replace $\mathfrak{e}_1(\mathfrak{q},\mathfrak{b},x)$ by $\mathfrak{q}'$ and then give the table for $\mathfrak{f}$ (with the right substitutions) and eventually reach a table in which no m-functions appeared.

$\mathfrak{pe}(\mathfrak{C},\beta)$			$\mathfrak{f}(\mathfrak{pe}_1(\mathfrak{C},\beta),\mathfrak{C},ə)$	From $\mathfrak{pe}(\mathfrak{C},\beta)$ the machine
$\mathfrak{pe}_1(\mathfrak{C},\beta)$	Any	R,R	$\mathfrak{pe}_1(\mathfrak{C},\beta)$	prints β at the end of the
	None	$P\beta$	$\mathfrak{C}$	sequence of symbols and $\rightarrow \mathfrak{C}$.
$\mathfrak{l}(\mathfrak{C})$		L	$\mathfrak{C}$	From $\mathfrak{f}'(\mathfrak{C},\mathfrak{B},\alpha)$ it does the same
$\mathfrak{r}(\mathfrak{C})$		R	$\mathfrak{C}$	as for $\mathfrak{f}(\mathfrak{C},\mathfrak{B},\alpha)$ but moves to the left before $\rightarrow \mathfrak{C}$.
$\mathfrak{f}'(\mathfrak{C},\mathfrak{B},\alpha)$			$\mathfrak{f}(\mathfrak{l}(\mathfrak{C}),\mathfrak{B},\alpha)$	
$\mathfrak{f}''(\mathfrak{C},\mathfrak{B},\alpha)$			$\mathfrak{f}(\mathfrak{r}(\mathfrak{C}),\mathfrak{B},\alpha)$	
$\mathfrak{c}(\mathfrak{C},\mathfrak{B},\alpha)$			$\mathfrak{f}'(\mathfrak{c}_1(\mathfrak{C}),\mathfrak{B},\alpha)$	$\mathfrak{c}(\mathfrak{C},\mathfrak{B},\alpha)$. The machine writes
$\mathfrak{c}_1(\mathfrak{C})$		β	$\mathfrak{pe}(\mathfrak{C},\mathfrak{B})$	at the end the first symbol marked α and $\rightarrow \mathfrak{C}$.

The last line stands for the totality of lines obtainable from it by replacing β by any symbol which may occur on the tape of the machine concerned.

$\mathfrak{ce}(\mathfrak{C},\mathfrak{B},\alpha)$		$\mathfrak{c}(\mathfrak{e}(\mathfrak{C},\mathfrak{B},\alpha),\mathfrak{B},\alpha)$	$\mathfrak{ce}(\mathfrak{B},\alpha)$. The machine copies down in order at the end all symbols marked α and erases the letters α; $\rightarrow \mathfrak{B}$.
$\mathfrak{ce}(\mathfrak{B},\alpha)$		$\mathfrak{ce}(\mathfrak{ce}(\mathfrak{B},\alpha),\mathfrak{B},\alpha)$	
$\mathfrak{re}(\mathfrak{C},\mathfrak{B},\alpha,\beta)$		$\mathfrak{f}_1(\mathfrak{re}_1(\mathfrak{C},\mathfrak{B},\alpha,\beta)\mathfrak{B},\alpha)$	$\mathfrak{re}(\mathfrak{C},\mathfrak{B},\alpha,\beta)$. The machine replaces the first α by β and $\rightarrow \mathfrak{C} \rightarrow \mathfrak{B}$ if there is no α. $\mathfrak{re}(\mathfrak{B},\alpha,\beta)$. The machine replaces all letters α by β; $\rightarrow \mathfrak{B}$
$\mathfrak{re}_1(\mathfrak{C},\mathfrak{B},\alpha,\beta)$	$E, P\beta$	$\mathfrak{C}$	
$\mathfrak{re}(\mathfrak{B},\alpha,\beta)$		$\mathfrak{re}(\mathfrak{re}(\mathfrak{B},\alpha,\beta),\mathfrak{B},\alpha,\beta)$	
$\mathfrak{cr}(\mathfrak{C},\mathfrak{B},\alpha)$		$\mathfrak{c}(\mathfrak{re}(\mathfrak{C},\mathfrak{B},\alpha,\alpha),\mathfrak{B},\alpha)$	$\mathfrak{cr}(\mathfrak{B},\alpha)$ differs from $\mathfrak{ce}(\mathfrak{B},\alpha)$ only in that the letters a are not erased. The *m*-configuration $\mathfrak{cr}(\mathfrak{C},\alpha)$ is taken up when no letters "α" are on the tape.
$\mathfrak{cr}(\mathfrak{B},\alpha)$		$\mathfrak{cr}(\mathfrak{cr}(\mathfrak{B},\alpha),\mathfrak{re}(\mathfrak{B},\alpha,\alpha),\alpha)$	

$\mathfrak{cp}(\mathfrak{C},\mathfrak{A},\mathfrak{E},\alpha,\beta)$		$\mathfrak{f}'(\mathfrak{cp}_1(\mathfrak{C}_1,\mathfrak{A},\beta),\mathfrak{f}(\mathfrak{A},\mathfrak{E},\beta),\alpha)$
$\mathfrak{cp}_1(\mathfrak{C},\mathfrak{A},\beta)$	γ	$\mathfrak{f}'(\mathfrak{cp}_2(\mathfrak{C},\mathfrak{A},\gamma),\mathfrak{A},\alpha)$
$\mathfrak{cp}_2(\mathfrak{C},\mathfrak{A},\gamma)$	γ	$\mathfrak{C}$
	not γ	$\mathfrak{A}$.

The first symbol marked α and the first marked β are compared. If there is neither α nor β, $\rightarrow \mathfrak{E}$. If there are both and the symbols are alike, $\rightarrow \mathfrak{C}$. Otherwise $\rightarrow \mathfrak{A}$.

$$\mathfrak{cpe}(\mathfrak{C},\mathfrak{A},\mathfrak{E},\alpha,\beta) \qquad \mathfrak{cp}(\mathfrak{e}(\mathfrak{e}(\mathfrak{C},\mathfrak{C},\beta),\mathfrak{C},\alpha),\mathfrak{A},\mathfrak{E},\alpha,\beta)$$

$\mathfrak{cpe}(\mathfrak{C},\mathfrak{A},\mathfrak{E},\alpha,\beta)$ differs from $\mathfrak{cp}(\mathfrak{C},\mathfrak{A},\mathfrak{E},\alpha,\beta)$ in that in the case when there is similarity the first α and β are erased.

$$\mathfrak{cpe}(\mathfrak{A},\mathfrak{E},\alpha,\beta) \qquad \mathfrak{cpe}(\mathfrak{cpe}(\mathfrak{A},\mathfrak{E},\alpha,\beta),\mathfrak{A},\mathfrak{E},\alpha,\beta).$$

$\mathfrak{cpe}(\mathfrak{A},\mathfrak{E},\alpha,\beta)$. The sequence of symbols marked α is compared with the sequence marked β. $\rightarrow \mathfrak{E}$ if they are similar. Otherwise $\rightarrow \mathfrak{A}$. Some of the symbols α and β are erased.

$\mathfrak{q}(\mathfrak{C})$	Any	R	$\mathfrak{q}(\mathfrak{C})$	$\mathfrak{q}(\mathfrak{C},\alpha)$. The machine finds the last symbol of form α. $\rightarrow \mathfrak{C}$.
	None	R	$\mathfrak{q}_1(\mathfrak{C})$	
$\mathfrak{q}_1(\mathfrak{C})$	Any	R	$\mathfrak{q}(\mathfrak{C})$	
	None		$\mathfrak{C}$	
$\mathfrak{q}(\mathfrak{C},\alpha)$			$\mathfrak{q}(\mathfrak{q}(\mathfrak{C},\alpha)$	
$\mathfrak{q}_1(\mathfrak{C},\alpha)$	α		$\mathfrak{C}$	
	not α	L	$\mathfrak{q}_1(\mathfrak{C},\alpha)$	
$\mathfrak{pe}_2(\mathfrak{C},\alpha,\beta)$			$\mathfrak{pe}(\mathfrak{pe}(\mathfrak{C},\beta),\alpha)$	$\mathfrak{pe}_2(\mathfrak{C},\alpha,\beta)$. The machine prints $\alpha\beta$ at the end

$\mathfrak{ce}_2(\mathfrak{B},\alpha,\beta)$		$\mathfrak{ce}(\mathfrak{cc}(\mathfrak{B},\beta),\alpha)$	$\mathfrak{ce}_3(\mathfrak{B},\alpha,\beta,\gamma)$. The machine copies down at the end first the symbols marked α, then those marked β, and finally those marked γ; it erases the symbols α, β, γ.
$\mathfrak{ce}_3(\mathfrak{B},\alpha,\beta,\gamma)$		$\mathfrak{ce}(\mathfrak{cc}_2(\mathfrak{B},\beta,\gamma),\alpha)$	
$\mathfrak{e}(\mathfrak{C})$	ə $\quad R$	$\mathfrak{e}_1(\mathfrak{C})$	From $\mathfrak{e}(\mathfrak{C})$ the marks are erased form all marked symbols $\rightarrow \mathfrak{C}$.
	Not ə $\quad L$	$\mathfrak{e}(\mathfrak{C})$	
$\mathfrak{e}_1(\mathfrak{C})$	Any $\quad R,E,R$	$\mathfrak{e}_1(\mathfrak{C})$	
	None	$\mathfrak{C}$	

5. Enumeration of computable sequences

A computable sequence γ is determined by a description of a machine which computes γ. Thus the sequence 001011011101111... is determined by the table on p. 19, and, in fact, any computable sequence is capable of being described in terms of such a table.

It will be useful to put these tables into a kind of standard form. In the first place let us suppose that the table is given in the same form as the first table, for example, I on p. 18. That is to say, that the entry in the operations column is always of one of the forms $E:E$, $R:E$, $L:P\alpha;P\alpha$, $R:P\alpha$, $L:R:L$: or no entry at all. The table can always be put into this form by introducing more m-configurations. Now let us give numbers to the m-configurations, calling them $q_1, \ldots, q_R$, as in §1. The initial m-configuration is always to be called q_1. We also give numbers to the symbols $S_1, \ldots, S_m$ and, in particular, blank $= S_0, 0 = S_1, 1 = S_2$. The lines of the table are now of form

m-config.	*Symbol*	*Operations*	*Final m-config.*	
q_i	S_j	PS_k, L	q_m	(N_1)
q_i	S_j	PS_k, R	q_m	(N_2)
q_i	S_j	PS_k	q_m	(N_3)

Lines such as

q_i	S_i	E,R	q_m

are to be written as

q_i	S_j	PS_0, R	q_m

and lines such as

q_i	S_j	R	q_m

to be written as

q_i	S_j	PS_j, R	q_m

In this way we reduce each line of the table to a line of one of the forms (N_1), (N_2) (N_3).

From each line of form (N_1) let us form an expression $q_iS_jS_kLq_m$; from each line of form (N_2) we form an expression $q_iS_jS_kRq_m$; and from each line of form (N_3) we form an expression $q_iS_jS_kNq_m$.

Let us write down all expressions so formed from the table for the machine and separate them by semi-colons. In this way we obtain a complete description of the machine. In this description we shall replace q_i by the letter "D" followed by the letter "A" repeated i times, and S_j by "D^3"

followed by "C" repeated j times. This new description of the machine may be called the *standard description* (S.D). It is made up entirely from the letters "A", "C", "D" "L", "R", "N", and from ";".

If finally we replace "A" by "1", "C" by "2", "D" by "3" "L" by "4", R by "5" "N" by "6", and ";" by "7" we shall have a description of the machine in the form of an arabic numeral. The integer represented by this numeral may be called a *description number* (D.N) of the machine. The D.N determine the S.D and the structure of the machine uniquely, The machine whose D.N is n may be described as $\mathcal{M}(n)$.

To each computable sequence there corresponds at least one description number, while to no description number does there correspond more than one computable sequence. The computable sequences and numbers are therefore enumerable.

Let us find a description number for the machine I of §3. When we rename the m-configurations its table becomes:

$$\begin{array}{llll} q_1 & S_0 & PS_1,\ R & q_2 \\ q_2 & S_0 & PS_0,\ R & q_3 \\ q_3 & S_0 & PS_2,\ R & q_4 \\ q_4 & S_0 & PS_0,\ R & q_1 \end{array}$$

Other tables could be obtained by adding irrelevant lines such as

$$\begin{array}{llll} q_1 & S_1 & PS_1,\ R & q_2 \end{array}$$

Our first standard form would be

$$q_1 S_0 S_1 R q_2;\ q_2 S_0 S_0 R q_3;\ q_3 S_0 S_2 R q_4;\ q_4 S_0 S_0 R q_1;.$$

The standard description is

DADDCRDAA; *DAADDRDAAA*; *DAAADDCCRDAAAA*; *DAAAADDRDA*;

A description number is

31332531173113353111731113322531111731111335317

and so is

3133253117311335311173111332253111173111133531731323253117

A number which is a description number of a circle-free machine will be called a *satisfactory* number. In §8 it is shown that there can be no general process for determining whether a given number is satisfactory or not.

6. The universal computing machine

It is possible to invent a single machine which can be used to compute any computable sequence. If this machine $\mathcal{U}$ is supplied with a tape on the beginning of which is written the S.D of some computing machine $\mathcal{M}$, then $\mathcal{U}$ will compute the same sequence as $\mathfrak{M}$. In this section I explain in outline the behaviour of the machine. The next section is devoted to giving the complete table for $\mathcal{U}$.

Let us first suppose that we have a machine $\mathcal{M}'$ which will write down on the F-squares the successive complete configurations of $\mathcal{M}$. These might be expressed in the same form as on p. 19 using the second description, (C), with all symbols on one line. Or, better, we could transform this description (as in §5) by replacing each m-configuration by "D" followed by "A" repeated the appropriate number of times, and by replacing each symbol by "D" followed by "C" repeated the appropriate number of times. The numbers of letters "A" and "C" are to agree with the numbers chosen in §5, so that, in particular, "0" is replaced by "DC", "1" by DCC", and the blanks by "D". These substitutions are to be made after the complete configurations have been put together, as in

(C). Difficulties arise if we do the substitution first. In each complete configuration the blanks would all have to be replaced by "*D*", so that the complete configuration would not be expressed as a finite sequence of symbols.

If in the description of the machine II of §3 we replace "o" by "*DAA*", "ə" by "*DCCC*", "q" by "*DAAA*", then the sequence (C) becomes:

$$DA : DCCCDCCCDAADCDDC : DCCCDCCCDAAADCDDC : \ldots \tag{C_1}$$

(This is the sequence of symbols on *F*-squares.)

It is not difficult to see that if $\mathcal{M}$ can be constructed, then so can $\mathcal{M}'$. The manner of operation of $\mathcal{M}'$ could be made to depend on having the rules of operation (*i.e.*, the S.D) of $\mathcal{M}$. written somewhere within itself (*i.e.* within $\mathcal{M}'$); each step could be carried out by referring to these rules. We have only to regard the rules as being capable of being taken out and exchanged for others and we have something very akin to the universal machine.

One thing is lacking: at present the machine $\mathcal{M}'$ prints no figures. We may correct this by printing between each successive pair of complete configurations the figures which appear in the new configuration but not in the old. Then (C_1) becomes

$$DDA : 0 : 0 : DCCCDCCCDAADCDDC : DCCC\ldots. \tag{C_2}$$

It is not altogether obvious that the *E*-squares leave enough room for the necessary "rough work", but this is, in fact, the case.

The sequences of letters between the colons in expressions such as (C_1) may be used as standard descriptions of the complete configurations. When the letters are replaced by figures, as in §5, we shall have a numerical description of the complete configuration, which may be called its description number.

7. Detailed description of the universal machine

A table is given below of the behaviour of this universal machine. The *m*-configurations of which the machine is capable are all those occurring in the first and last columns of the table, together with all those which occur when we write out the unabbreviated tables of those which appear in the table in the form of *m*-functions. *E.g.*, $\mathfrak{e}(\mathfrak{anf})$ appears in the table and is an *m*-function. Its unabbreviated table is (see p. 22)

$\mathfrak{e}(\mathfrak{anf})$	ə	R	$\mathfrak{e}_1(\mathfrak{anf})$
	not ə	L	$\mathfrak{e}(\mathfrak{anf})$
$\mathfrak{e}_1(\mathfrak{anf})$	Any	R, E, R	$\mathfrak{e}_1(\mathfrak{anf})$
	None		$(\mathfrak{anf})$

Consequently $\mathfrak{e}_1(\mathfrak{anf})$ is an *m*-configuration of $\mathcal{U}$.

When $\mathcal{U}$ is ready to start work the tape running through it bears on it the symbol a on an *F*-square and again a on the next *E*-square; after this, on *F*-squares only, comes the S.D of the machine followed by a double colon "::" (a single symbol, on an *F*-square). The S.D consists of a number of instructions, separated by semi-colons.

Each instruction consists of five consecutive parts

(i) "*D*" followed by a sequence of letters "*A*". This describes the relevant *m*-configuration.
(ii) "*D*" followed by a sequence of letters "*C*". This describes the scanned symbol.
(iii) "*D*" followed by another sequence of letters "*C*". This describes the symbol into which the scanned symbol is to be changed.
(iv) "*L*", "*R*", or "*N*", describing whether the machine is to move to left, right, or not at all.

(v) "*D*" followed by a sequence of letters "*A*". This describes the final *m*-configuration.

The machine $\mathcal{U}$ is to be capable of printing "*A*", "*C*", "*D*", "0", "1", "*u*", "*v*", "*w*", "*x*", "*y*", "*z*". The S.D is formed from ";", "*A*", "*C*", "*D*", "*L*", "*R*", "*N*".

Subsidiary skeleton table

$\mathfrak{con}(\mathfrak{C},\alpha)$	Not *A*	R,R	$\mathfrak{con}(\mathfrak{C},\alpha)$	
	A	$L,P\alpha,R$	$\mathfrak{con}_1(\mathfrak{C},\alpha)$	$\mathfrak{con}(\mathfrak{C},\alpha)$ Starting from an *F*-square, *S* say, the sequence *C* of symbols describing a configuration closest on the right of *S* is marked out with letters α. $\rightarrow \mathfrak{C}$.
$\mathfrak{con}_1(\mathfrak{C},\alpha)$	*A*	$R,P\alpha,R$	$\mathfrak{con}_1(\mathfrak{C},\alpha)$	
	D	$R,P\alpha,R$	$\mathfrak{con}_2(\mathfrak{C},\alpha)$	
$\mathfrak{con}_2(\mathfrak{C},\alpha)$	*C*	$R,P\alpha,R$	$\mathfrak{con}_2(\mathfrak{C},\alpha)$	$\mathfrak{con}(\mathfrak{C},\alpha)$. In the final configuration the machine is scanning the square which is four squares to the right of the last square of *C*. *C* is left unmarked.
	Not *C*	R,R	$\mathfrak{C}$	

The table for $\mathcal{U}$.

$\mathfrak{b}$			$\mathfrak{f}(\mathfrak{b}_1,\mathfrak{b}2,::,\alpha)$	$\mathfrak{b}$. The machine prints: *DA* on the *F*-squares after :: $\rightarrow \mathfrak{anf}$.
$\mathfrak{b}_1$		$R,R,P:,R,R,PD,R,R,PA$	$\mathfrak{anf}$	
$\mathfrak{anf}$			$\mathfrak{g}(\mathfrak{anf}_1,:)$	$\mathfrak{anf}$. The machine marks the configuration in the last complete configuration with *y*. $\rightarrow \mathfrak{kom}$.
$\mathfrak{anf}_1$		$\mathfrak{con}(\mathfrak{kom},y)$		
$\mathfrak{kom}$	;	R,P_z,L	$\mathfrak{con}(\mathfrak{kmp},x)$	$\mathfrak{kom}$. The machine finds the last semi-colon not marked with *z*. It marks this semi-colon with *z* and the configuration following it with *x*.
	z	L,L	$\mathfrak{kom}$	
	not z nor;	L	$\mathfrak{kom}$	
$\mathfrak{kmp}$		$\mathfrak{cpe}(\mathfrak{e}(\mathfrak{kom},x,y),\ \mathfrak{sim},x,y)$		$\mathfrak{kmp}$. The machine compares the sequences marked *x* and *y*. It erases all letters *x* and *y*. $\rightarrow \mathfrak{sim}$ if they are alike. Otherwise $\rightarrow \mathfrak{kom}$.

$\mathfrak{anf}$. Taking the long view, the last instruction relevant to the last configuration is found. It can be recognised afterwards as the instruction following the last semi-colon marked *z*. $\rightarrow \mathfrak{sim}$.

m-config.	Symbol	Operations	Final m-config.	
$\mathfrak{sim}$			$\mathfrak{f}'(\mathfrak{sim}_1, \mathfrak{sim}_1, z)$	$\mathfrak{sim}$ The machine marks out the instructions. That part of the instructions which refers to operations to be carried out is marked with u, and the final m-configuration with y. The letters z are erased.
$\mathfrak{sim}_1$			$\mathfrak{con}(\mathfrak{sim}_2,)$	
$\mathfrak{sim}_2$	A		$\mathfrak{sim}_3$	
	not A	R, P_u, R, R, R	$\mathfrak{sim}_2$	
$\mathfrak{sim}_3$	not A	L, P_y	$\mathfrak{e}(\mathfrak{mf}, z)$	
	A	L, P_y, R, R, R	$\mathfrak{sim}_3$	
$\mathfrak{mf}$			$\mathfrak{g}(\mathfrak{mf}, :)$	
$\mathfrak{mf}_1$	not A	R, R	$\mathfrak{mf}_1$	$\mathfrak{mf}$. The last complete configuration is marked out into four sections. The configuration is left unmarked. The symbol directly preceding it is marked with x. The remainder of the complete configuration is divided into two parts, of which the first is marked with v and the last with w. A colon is printed after the whole. $\rightarrow \mathfrak{sh}$.
	A	L, L, L, L	$\mathfrak{mf}_2$	
$\mathfrak{mf}_2$	C	R, P_x, L, L, L	$\mathfrak{mf}_2$	
	:		$\mathfrak{mf}_4$	
	D	R, P_x, L, L, L	$\mathfrak{mf}_3$	
$\mathfrak{mf}_3$	not :	R, P_v, L, L, L	$\mathfrak{mf}_3$	
	:		$\mathfrak{mf}_4$	
$\mathfrak{mf}_4$			$\mathfrak{con}(\mathfrak{l}(\mathfrak{l}(\mathfrak{mp}_5)),)$	
$\mathfrak{mf}_5$	Any	R, P_w, R	$\mathfrak{mf}_5$	
	None	P:	$\mathfrak{sh}$	
$\mathfrak{sh}$			$\mathfrak{f}(\mathfrak{sh}_1, \mathfrak{inst}, u)$	
$\mathfrak{sh}_1$		L, L, L	$\mathfrak{sh}_2$	$\mathfrak{sh}$. The instructions (marked u) are examined. If it is found that they involve "Print 0" or "Print 1", then 0: or 1: is printed at the end.
$\mathfrak{sh}_2$	D	R, R, R, R	$\mathfrak{sh}_2$	
	not D		$\mathfrak{inst}$	
$\mathfrak{sh}_3$	C	R, R	$\mathfrak{sh}_4$	
	not C		$\mathfrak{inst}$	
$\mathfrak{sh}_4$	C	R, R	$\mathfrak{sh}_5$	
	not C		$\mathfrak{pe}_2(\mathfrak{inst}, 0, :)$	
$\mathfrak{sh}_5$	C		$\mathfrak{inst}$	
	not C		$\mathfrak{pe}_2(\mathfrak{inst}, 1, :)$	

$\mathfrak{inst}$			$\mathfrak{g}(\mathfrak{l}(\mathfrak{inst}_1), u)$	
$\mathfrak{inst}_1$	α	R, E	$\mathfrak{inst}_1(\alpha)$	$\mathfrak{inst}$ The next complete configuration is written down, carrying out the marked instructions. The letters u, v, w, x, y are erased. $\rightarrow$ $\mathfrak{anf}$.
$\mathfrak{inst}_1(L)$			$\mathfrak{ce}_5(\mathfrak{ov}, v, y, x, u, w)$	
$\mathfrak{inst}_1(R)$			$\mathfrak{ce}_5(\mathfrak{ov}, v, y, x, u, w)$	
$\mathfrak{inst}_1(N)$			$\mathfrak{ec}_5(\mathfrak{ov}, v, y, x, u, w)$	
$\mathfrak{ov}$			$\mathfrak{e}(\mathfrak{anf})$	

8. Application of the diagonal process

It may be thought that arguments which prove that the real numbers are not enumerable would also prove that the computable numbers and sequences cannot be enumerable[4]. It might, for instance, be thought that the limit of a sequence of computable numbers must be computable. This is clearly only true if the sequence of computable numbers is defined by some rule.

Or we might apply the diagonal process. "If the computable sequences are enumerable, let a_n be the n-th computable sequence, and let $\phi_n(m)$ be the m-th figure in a_n. Let β be the sequence with $1-\phi_n(n)$ as its n-th figure. Since β is computable, there exists a number K such that $1-\phi_n(n) = \phi_K(n)$ all n. Putting $n = K$, we have $1 = 2\phi_k(K)$, *i.e.* 1 is even. This is impossible. The computable sequences are therefore not enumerable ".

The fallacy in this argument lies in the assumption that β is computable. It would be true if we could enumerate the computable sequences by finite means, but the problem of enumerating computable sequences is equivalent to the problem of finding out whether a given number is the D.N of a circle-free machine, and we have no general process for doing this in a finite number of steps. In fact, by applying the diagonal process argument correctly, we can show that there cannot be any such general process.

The simplest and most direct proof of this is by showing that, if this general process exists, then there is a machine which computes β. This proof, although perfectly sound, has the disadvantage that it may leave the reader with a feeling that "there must be something wrong ". The proof which I shall give has not this disadvantage, and gives a certain insight into the significance of the idea"circle-free". It depends not on constructing β, but on constructing β', whose n-th figure is $\phi_n(n)$.

Let us suppose that there is such a process; that is to say, that we can invent a machine $\mathcal{D}$ which, when supplied with the S.D. of any computing machine $\mathcal{M}$ will test this S.D and if $\mathcal{M}$ is circular will mark the S.D with the symbol "u" and if it is circle-free will mark it with "s". By combining the machines $\mathcal{D}$ and $\mathcal{U}$ we could construct a machine $\mathcal{H}$ I to compute the sequence β'. The machine $\mathcal{D}$, may require a tape. We may suppose that it uses the E-squares beyond all symbols on F-squares, and that when it has reached its verdict all the rough work done by $\mathcal{D}$ is erased.

The machine $\mathcal{H}$ has its motion divided into sections. In the first $N-1$ sections, among other things, the integers 1, 2,...,$N-1$ have been written down and tested by the machine $\mathcal{D}$. A certain number, say $R(N-1)$, of them have been found to be the D.N's of circle-free machines. In the N-th section the machine $\mathcal{D}$ tests the number N. If N is satisfactory, *i.e.*, if it is the D.N of a circle-free machine, then $R(N) = 1 + R(N-1)$ and the first $R(N)$ figures of the sequence of which a D.N is

[4] *Cf.* Hobson,*Theory of functions of a real variable*(2nd ed., 1921),87,88.

N are calculated. The $R(N)$-th figure of this sequence is written down as one of the figures of the sequence β' computed by $\mathcal{H}$. If N is not satisfactory, then $R(N) = R(N-1)$ and the machine goes on to the $(N+1)$-th section of its motion.

From the construction of $\mathcal{H}$ we can see that $\mathcal{H}$. is circle-free. Each section of the motion of $\mathcal{H}$ comes to an end after a finite number of steps. For, by our assumption about $\mathcal{D}$ the decision as to whether N is satisfactory is reached in a finite number of steps. If N is not satisfactory, then the N-th section is finished. If N is satisfactory, this means that the machine $\mathcal{M}(N)$ whose D.N is N is circle-free, and therefore its $R(N)$-th figure can be calculated in a finite number steps. When this figure has been calculated and written down as the $R(N)$-th figure of β', the N-th section is finished. Hence $\mathcal{H}$ is circle-free.

Now let K be the D.N of $\mathcal{H}$. What does.1 $\mathcal{H}$ do in the K-th section of its motion ? It must test whether K is satisfactory, giving a verdict "*s*" or "*u*". Since K is the D.N of $\mathcal{H}$ and since $\mathcal{H}$ is circle-free, the verdict cannot be "*u*". On the other hand the verdict cannot be "*s*". For if it were, then in the K-th section of its motion $\mathcal{H}$, would be bound to compute the first $R(K-1)+1 = R(K)$ figures of the sequence computed by the machine with K as its D.N and to write down the $R(K)$-th as a figure of the sequence computed by $\mathcal{H}$. The computation of the first $R(K) - 1$ figures would be carried out all right, but the instructions for calculating the $R(K)$-th would amount to "calculate the first $R(K)$ figures computed by H and write down the $R(K)$-th". This $R(K)$-th figure would never be found. *I.e.*, $\mathcal{H}$ is circular, contrary both to what we have found in the last paragraph and to the verdict "*s*". Thus both verdicts are impossible and we conclude that there can be no machine $\mathcal{D}$.

We can show further that *there can be no machine $\mathcal{E}$ which, when supplied with the S.D of an arbitrary machine $\mathcal{M}$, will determine whether. $\mathcal{M}$ ever prints a given symbol (0 say).*

We will first show that, if there is a machine $\mathcal{E}$, then there is a general process for determining whether a given machine. $\mathcal{M}$ prints 0 infinitely often. Let $\mathcal{M}_1$ be a machine which prints the same sequence as $\mathcal{M}$, except that in the position where the first 0 printed by $\mathcal{M}$ stands, $\mathcal{M}_1$ prints $\bar{0}$. $\mathcal{M}_2$ is to have the first two symbols 0 replaced by $\bar{0}$ and so on. Thus, if. $\mathcal{M}$. were to print

$$ABA01AAB0010AB\ldots,$$

then $\mathcal{M}_1$ would print

$$ABA\bar{0}1AAB0010AB\ldots$$

and. $\mathcal{M}_2$ would print

$$ABA\bar{0}1AAB\bar{0}010AB\ldots.$$

Now let $\mathcal{R}$ be a machine which, when supplied with the S.D of $\mathcal{M}$ will write down successively the S.D of $\mathcal{M}$, of $\mathcal{M}_1$, of $\mathcal{M}_2$ ... (there is such a machine). We combine $\mathcal{R}$ with $\mathcal{E}$. and obtain a new machine, }. In the motion of $\mathcal{R}$ first $\mathcal{R}$ is used to write down the S.D of $\mathcal{M}$, and then $\mathcal{E}$ tests it, : 0 : is written if it is found that. $\mathcal{M}$ never prints 0; then $\mathcal{R}$ writes the S.D of $\mathcal{M}_1$, and this is tested, : 0 : being printed if and only if $\mathcal{M}_1$ never prints 0, and so on. Now let us test } with $\mathcal{E}$. If it is found that }, never prints 0, then $\mathcal{M}$ prints 0 infinitely often; if } prints 0 sometimes, then $\mathcal{M}$ does not print 0 infinitely often.

Similarly there is a general process for determining whether $\mathcal{M}$ prints 1 infinitely often. By a combination of these processes we have a process for determining whether $\mathcal{M}$ prints an infinity of figures, *i.e.* we have a process for determining whether $\mathcal{M}$ is circle-free. There can therefore be no machine $\mathcal{E}$.

The expression "there is a general process for determining..." has been used throughout this section as equivalent to "there is a machine which will determine ...". This usage can be justified if and only if we can justify our definition of "computable". For each of these "general process" problems can be expressed as a problem concerning a general process for determining whether a given integer n has a property $G(n)$ [*e.g.*$G(n)$ might mean "n is satisfactory" or "n is the Gödel representation of a provable formula"], and this is equivalent to computing a number whose n-th figure is 1 if $G(n)$ is true and 0 if it is false.

9. The extent of the computable numbers

No attempt has yet been made to show that the "computable" numbers include all numbers which would naturally be regarded as computable. All arguments which can be given are bound to be, fundamentally, appeals to intuition, and for this reason rather unsatisfactory mathematically. The real question at issue is "What are the possible processes which can be carried out in computing a number ?"

The arguments which I shall use are of three kinds.

(a) A direct appeal to intuition.
(b) A proof of the equivalence of two definitions (in case the new definition has a greater intuitive appeal).
(c) Giving examples of large classes of numbers which are computable.

Once it is granted that computable numbers are all " computable " several other propositions of the same character follow. In particular, it follows that, if there is a general process for determining whether a formula of the Hilbert function calculus is provable, then the determination can be carried out by a machine.

I. [Type (*a*)]. This argument is only an elaboration of the ideas of §1.

Computing is normally done by writing certain symbols on paper. We may suppose this paper is divided into squares like a child's arithmetic book. In elementary arithmetic the two-dimensional character of the paper is sometimes used. But such a use is always avoidable, and I think that it will be agreed that the two-dimensional character of paper is no essential of computation. I assume then that the computation is carried out on one-dimensional paper, *i.e.* on a tape divided into squares. I shall also suppose that the number of symbols which may be printed is finite. If we were to allow an infinity of symbols, then there would be symbols differing to an arbitrarily small extent[5]. The effect of this restriction of the number of symbols is not very serious. It is always possible to use sequences of symbols in the place of single symbols. Thus an Arabic numeral such as 17 or 999999999999999 is normally treated as a single symbol. Similarly in any European language words are treated as single symbols (Chinese, however, attempts to have an enumerable infinity of symbols). The differences from our point of view between the single and compound symbols. is that the compound symbols, if they are too lengthy, cannot be observed at one glance. This is in accordance with experience. We cannot tell at a glance whether 9999999999999999 and 999999999999999 are the same.

The behaviour of the computer at any moment is determined by the symbols which he is observing, and his "state of mind" at that moment. We may suppose that there is a bound B to the number of symbols or squares which the computer can observe at one moment. If he wishes to observe more, he must use successive observations. We will also suppose that the number of states of mind which need be taken into account is finite. The reasons for this are of the same character as those which restrict the number of symbols. If we admitted an infinity of states of mind, some of them will be "arbitrarily close" and will be confused. Again, the restriction is not one which seriously affects computation, since the use of more complicated states of mind can be avoided by writing more symbols on the tape.

[5] If we regard a symbol as literally printed on a square we may suppose that the square is $0 \leqslant x \leqslant 1$, $0 \leqslant y \leqslant 1$. The symbol is defined as a set of points in this square, viz. the Bet occupied by printer's ink. If these sets are restricted to be measurable, we can define the "distance"between two symbols as the cost of transforming one symbol into the other if the cost of moving unit area of printer's ink unit distance is unity, and there is an, infinite supply of ink at $x = 2, y = 0$. With this topology the symbols form a conditionally compact space.

Let us imagine the operations performed by the computer to be split up into "simple operations" which are so elementary that it is not easy to imagine them further divided. Every such operation consists of some change of the physical system consisting of the computer and his tape. We know the state of the system if we know the sequence of symbols on the tape, which of these are observed by the computer (possibly with a special order), and the state of mind of the computer. We may suppose that in a simple operation not more than one symbol is altered. Any other changes can be split up into simple changes of this kind. The situation in regard to the squares whose symbols may be altered in this way is the same as in regard to the observed squares. We may, therefore, without loss of generality, assume that the squares whose symbols are changed are always "observed" squares.

Besides these changes of symbols, the simple operations must include changes of distribution of observed squares. The new observed squares must be immediately recognisable by the computer. I think it is reasonable to suppose that they can only be squares whose distance from the closest of the immediately previously observed squares does not exceed a certain fixed amount. Let us say that each of the new observed squares is within L squares of an immediately previously observed square.

In connection with "immediate recognisability", it may be thought that there are other kinds of square which are immediately recognisable. In particular, squares marked by special symbols might be taken as immediately reognisable. Now if these squares are marked only by single symbols there can be only a finite number of them, and we should not upset our theory by adjoining these marked squares to the observed squares. If, on the other hand, they are marked by a sequence of symbols, we cannot regard the process of recognition as a simple process. This is a fundamental point and should be illustrated. In most mathematical papers the equations and theorems are numbered. Normally the numbers do not go beyond (say) 1000. It is, therefore, possible to recognise a theorem at a glance by its number. But if the paper was very long, we might reach Theorem 157767733443477; then, further on in the paper, we might find "...hence (applying Theorem 157767733443477) we have...". In order to make sure which was the relevant theorem we should have to compare the two numbers figure by figure, possibly ticking the figures off in pencil to make sure of their not being counted twice. If in spite of this it is still thought that there are other "immediately recognisable" squares, it does not upset my contention so long as these squares can be found by some process of which my type of machine is capable. This idea is developed in III below.

The simple operations must therefore include:

(a) Changes of the symbol on one of the observed squares.
(b) CChanges of one of the squares observed to another square

within L squares of one of the previously observed squares.

It may be that some of these changes necessarily involve a change of state of mind. The most general single operation must therefore be taken to be one of the following:

(A) A possible change (a) of symbol together with a possible change of state of mind.
(B) A possible change (b) of observed squares, together with a possible change of state of mind.

The operation actually performed is determined, as has been suggested on p. 30, by the state of mind of the computer and the observed symbols. In particular, they determine the state of mind of the computer after the operation is carried out.

We may now construct a machine to do the work of this computer. To each state of mind of the computer corresponds an "m-configuration" of the machine. The machine scans B squares corresponding to the B squares observed by the computer. In any move the machine can change a symbol on a scanned square or can change any one of the scanned squares to another square distant not more than L squares from one of the other scanned squares. The move which is done, and the succeeding configuration, are determined by the scanned symbol and the m-configuration. The machines just described do not differ very essentially from computing machines as defined in §2 and corresponding to any machine of this type a computing machine can be constructed to compute the same sequence, that is to say the sequence computed by the computer.

II. [Type (*b*)].

If the notation of the Hilbert functional calculus[6] is modified so as to be systematic, and so as to involve only a finite number of symbols, it becomes possible to construct an automatic[7] machine $\mathcal{K}$, which will find all the provable formulae of the calculus[8].

Now let α be a sequence, and let us denote by $G_\alpha(x)$ the proposition "The x-th figure of α is 1", so that[9] $G_\alpha(x)$ means "The x-th figure of α is 0. Suppose further that we can find a set of properties which define the sequence α and which can be expressed in terms of $G_\alpha(x)$ and of the propositional functions $N(x)$ meaning "x is a non-negative integer" and $F(x, y)$ meaning "$y = x + 1$". When we join all these formulae together conjunctively, we shall have a formula, $\mathfrak{A}$ say which defines α. The terms of $\mathfrak{A}$ must include the necessary parts of the Peano axioms, viz.,

$$(\exists u)N(u)\&(x)\,(N(x) \to (\exists y)F(x,y))\,\&\,(F(x,\ y) \to N(y)),$$

which we will abbreviate to P.

When we say "$\mathfrak{A}$ defines α", we mean that $-\mathfrak{A}$ is not a provable formula, and also that, for each n, one of the following formulae (A_n) or (B_n) is provable.

$$\mathfrak{A}\ \&\ F^{(n)} \to G_\alpha(u^{(n)}), \qquad (\mathrm{A}_n)^{10}$$

$$\mathfrak{A}\ \&\ F^{(n)} \to \left(-G_\alpha(u^{(n)})\right), \qquad (\mathrm{B}_n),$$

where $F^{(n)}$ stands for $F(u,\ u')\ \&\ F\,(u',\ u'')\ \&\ \ldots F(u^{(n-1)},\ u^{(n)})$.

I say that α is then a computable sequence: a machine $\mathcal{K}_\alpha$ to compute α can be obtained by a fairly simple modification of $\mathcal{K}$.

We divide the motion of $\mathcal{K}_\alpha$ into sections. The n-th section is devoted to finding the n-th figure of α. After the $(n-1)$-th section is finished a double colon :: is printed after all the symbols, and the succeeding work is done wholly on the squares to the right of this double colon. The first step is to write the letter "A" followed by the formula (A_n) and then "B" followed by (B_n). The machine $\mathcal{K}_\alpha$ then starts to do the work of $\mathcal{K}$, but whenever a provable formula is found, this formula is compared with (A_n) and with (B_n). If it is the same formula as (A_n), then the figure "1" is printed, and the n-th section is finished. If it is (B_n), then "0" is printed and the section is finished. If it is different from both, then the work of $\mathcal{K}$ is continued from the point at which it had been abandoned. Sooner or later one of the formulae (A_n) or (B_n) is reached; this follows from our hypotheses about α and $\mathfrak{A}$, and the known nature of $\mathcal{K}$. Hence the n-th section will eventually be finished. $\mathcal{K}_\alpha$. is circle-free; α is computable.

It can also be shown that the numbers α definable in this way by the use of axioms include all the computable numbers. This is done by describing computing machines in terms of the function calculus.

---

[6] The expression "the functional calculus" is used throughout to mean the *restricted* Hilbert functional calculus.

[7] It is most natural to construct first a choice machine (§ 2) to do this. But it is then easy to construct the required automatic machine. We can suppose t h a t the choices are always choices between two possibilities 0 and 1. Each proof will then be determined by a sequence of choices $i_1, i_2, \ldots, i_n (i_1 = 0 \text{ or } 1, i_2 = 0 \text{ or } 1, \ldots, i_n = 0 \text{ or } 1)$, and hence the number $2'' + i_1 2^{n-1} + i_2 2^{n-2} + \ldots + i_n$ completely determines the proof. The automatic machine carries out successively proof 1, proof 2, proof 3,. . . .

[8] The author has found a description of such a machine.

[9] The negation sign is written before an expression and not over it.

[10] A sequence of r primes is denoted by $^{(r)}$.

It must be remembered that we have attached rather a special meaning to the phrase "$\mathfrak{A}$ defines α". The computable numbers do not include all (in the ordinary sense) definable numbers. Let δ be a sequence whose n-th figure is 1 or 0 according as n is or is not satisfactory. It is an immediate consequence of the theorem of §8 that δ is not computable. It is (so far as we know at present) possible that any assigned number of figures of δ. can be calculated, but not by a uniform process. When sufficiently many figures of δ have been calculated, an essentially new method is necessary in order to obtain more figures.

III. This may be regarded as a modification of I or as a corollary of II.

We suppose, as in I, that the computation is carried out on a tape; but we avoid introducing the "state of mind" by considering a more physical and definite counterpart of it. It is always possible for the computer to break off from his work, to go away and forget all about it, and later to come back and go on with it. If he does this he must leave a note of instructions (written in some standard form) explaining how the work is to be continued. This note is the counterpart of the "state of mind". We will suppose that the computer works in such a desultory manner that he never does more than one step at a sitting. The note of instructions must enable him to carry out one step and write the next note, Thus the state of progress of the computation at any stage is completely determined by the note of instructions and the symbols on the tape. That is, the state of the system may be described by a single expression (sequence of symbols), consisting of the symbols on the tape followed by Δ (which we suppose not to appear elsewhere) and then by the note of instructions. This expression may be called the "state formula". We know that the state formula at any given stage is determined by the state formula before the last step was made, and we assume that the relation of these two formulae is expressible in the functional calculus. In other words, we assume that there is an axiom $\mathfrak{A}$ which expresses the rules governing the behaviour of the computer, in terms of the relation of the state formula at any stage to the state formula at the preceding stage. If this is so, we can construct a machine to write down the successive state formulae, and hence to compute the required number.

10. Examples of large classes of numbers which are computable

It will be useful to begin with definitions of a computable function of an integral variable and of a computable variable, etc. There are many equivalent ways of defining a computable function of an integral variable. The simplest is, possibly, as follows. If γ is a computable sequence in which 0 appears infinitely[7] often, and n is an integer, then let us define $\xi(\gamma,\ n)$ to be the number of figures 1 between the n-th and the $(n+1)$-th figure 0 in γ. Then $\phi(n)$ is computable if, for all n and some γ, $\phi(n) = \xi(\gamma, n)$. An equivalent definition is this. Let $H(x,\ y)$ mean $\phi(x) = y$. Then, if we can find a contradiction-free axiom $\mathfrak{A}_\phi$, such that $\mathfrak{A}_\phi \rightarrow P$, and if for each integer n there exists an integer N, such that

$$\mathfrak{A}_\phi \ \&\ F^{(N)} \rightarrow H(u^{(n)},\ u^{(\phi(n))}),$$

and such that, if $m \neq \phi(n)$, then, for some N',

$$\mathfrak{A}_\phi \ \&\ F^{(N')} \rightarrow \big(-H(u^{(n)},\ u^{(m)}\big),$$

then ϕ may be said to be a computable function.

We cannot define general computable functions of a real variable, since there is no general method of describing a real number, but we can define a computable function of a computable variable. If n is satisfactory, let γn be the number computed by $\mathcal{M}(n)$, and let

$$a_n = \tan\Big(\pi(\gamma_n - \tfrac{1}{2})\Big),$$

[7] If $\mathcal{M}$ computes γ, then the problem whether $\mathcal{M}$ prints 0 infinitely often is of the same character as the problem whether $\mathcal{M}$ is circle-free.

unless $\gamma_n = 0$ or $\gamma_n = 1$, in either of which cases $\alpha_n = 0$. Then, as n runs through the satisfactory numbers, α_n runs through the computable numbers[8]. Now let $\phi(n)$ be a computable function which can be shown to be such that for any satisfactory argument its value is satisfactory,[9]. Then the function f, defined by $f(a_n) = a_{\phi(n)}$, is a computable function and all computable functions of a computable variable are expressible in this form.

Similar definitions may be given of computable functions of several variables, computable-valued functions of an integral variable, etc.

I shall enunciate a number of theorems about computability, but I shall prove only (ii) and a theorem similar to (iii).

(i) A computable function of a computable function of an integral or computable variable is computable.

(ii) Any function of an integral variable defined recursively in terms of computable functions is computable. *I.e.* if $\phi(m, n)$ is computable, and r is some integer, then $\eta(n)$ is computable, where

$$\eta(0) = r,$$

$$\eta(n) = \phi(n, \eta(n-1)).$$

(iii) If $\phi(m, n)$ is a computable function of two integral variables, then $\phi(n, n)$ is a computable function of n.

(iii) If $\varphi(m, n)$ is a computable function of two integral variables, then $\varphi(n, n)$ is a computable function of n.

(iv) If $\phi(n)$ is a computable function whose value is always 0 or 1, then the sequence whose n-th figure is $\phi(n)$ is computable.

Dedekind's theorem does not hold in the ordinary form if we replace "real" throughout by "computable". But it holds in the following form:

(v) If $G(\alpha)$ is a propositional function of the computable numbers and

$$(a) \quad (\exists\alpha)(\exists\beta)\{G(\alpha) \mathbin{\&} (-G(\beta))\},$$

$$(b) \quad G(\alpha) \mathbin{\&} (-G(\beta)) \rightarrow (\alpha < \beta),$$

and there is a general process for determining the truth value of $G(\alpha)$, then there is a computable number ξ such that

$$G(\alpha) \rightarrow \alpha \leqslant \xi,$$

$$-G(\alpha) \rightarrow \alpha \geqslant \xi.$$

In other words, the theorem holds for any section of the computables such that there is a general process for determining to which class a given number belongs.

Owing to this restriction of Dedekind's theorem, we cannot say that a computable bounded increasing sequence of computable numbers has a computable limit. This may possibly be understood by considering a. sequence such as

$$-1, -\tfrac{1}{2}, -\tfrac{1}{4} -\tfrac{1}{8}, -\tfrac{1}{16}, \tfrac{1}{2}, \ldots.$$

On the other hand, (v) enables us to prove

[8] A function α_n may be defined in many other ways so as to run through the computable numbers.

[9] Although it is not possible to find a general process for determining whether a given number is satisfactory, it is often possible to show that certain classes of numbers are satisfactory.

(vi) If α and β are computable and $\alpha < \beta$ and $\phi(\alpha) < 0 < \phi(\beta)$, where $\phi(\alpha)$ is a computable increasing continuous function, then there is a unique computable number γ, satisfying $\alpha < \gamma < \beta$ and $\phi(\gamma) = 0$.

Computable convergence

We shall say that a sequence β_n of computable numbers *converges computably* if there is a computable integral valued function $N(\epsilon)$ of the computable variable ϵ, such that we can show that, if $\epsilon > 0$ and $n > N(\epsilon)$ and $m > N(\epsilon)$, then $|\beta_n - \beta_m| < \epsilon$.

We can then show that

(vii) A power series whose coefficients form a computable sequence of computable numbers is computably convergent at all computable points in the interior of its interval of convergence.
(viii) The limit of a computably convergent sequence is computable.

And with the obvious definition of "uniformly computably convergent":

(ix) The limit of a uniformly computably convergent computable sequence of computable functions is a computable function. Hence
(x) The sum of a power series whose coefficients form a computable sequence is a computable function in the interior of its interval of convergence.

From (viii) and $\pi = 4(1 - \frac{1}{3} + \frac{1}{5} - \ldots)$ we deduce that π is computable.
From $e = 1 + 1 + \frac{1}{2!} + \frac{1}{3!} + \ldots$ we deduce that e is computable.
From (vi) we deduce that all real algebraic numbers are computable.
From (vi) and (x) we deduce that the real zeros of the Bessel functions are computable.

Proof of (ii).

Let $H(x, y)$ mean "$\eta(x) = y$", and let $K(x, y, z)$ mean "$\phi(x, y) = z$". $\mathfrak{A}_\phi$ is the axiom for $\phi(x, y)$. We take $\mathfrak{A}_\eta$ to be

$$\mathfrak{A}_\phi \,\&\, P \& \ (F(x, y) \to G(x,y)) \ \& \ (G(x, y) \ \& \ G(y, z) \to G(x, z))$$
$$\& \left(F^{(r)} \to H(u, u^{(r)})\right) \ \& \ (F(v, w) \ \& \ H(v, x) \ \& \ K(w, x, z) \to H(w, z))$$
$$\& \ [H(w, z) \ \& \ G(z, t) \mathrm{v} G(t,z) \to (-H(w, t))].$$

I shall not give the proof of consistency of $\mathfrak{A}_\eta$. Such a proof may be constructed by the methods used in Hilbert and Bernays, *Grundlagen der Mathematik* (Berlin, 1934), p. 209 *et seq.* The consistency is also clear from the meaning.

Suppose that, for some n, N, we have shown

$$\mathfrak{A}_\eta \ \& \ F^{(N)} \to H(u^{(n-1)}, u^{(\eta(n-1))}),$$

then, for some M,

$$\mathfrak{A}_\phi \ \& \ F^{(M)} \to K(u^{(n)}, u^{(\eta(n-1))}, u^{(\eta(n))}),$$

$$\mathfrak{A}_\eta \ \& \ F^{(M)} \to F(u^{(n-1)}, u^{(n)}) \ \& \ H(u^{(n-1)}, u^{(\eta(n-1))})$$

$$\& \ K(u^{(n)}, u^{(\eta(n-1))}, u^{(\eta(n))}),$$

and

$$\mathfrak{A} \,\&\, F^{(M)} \to [F(u^{(n-1)},\, u^{(n)}) \& \, H(u^{(n-1)},\, u^{(\eta(n-1))})$$
$$\&K\,(u^{(n)},\, u^{(\eta(n-1))},\, u^{(\eta(n))}) \to H(u^{(n)},\, u^{(\eta(n))})].$$

Hence $\mathfrak{A}_n \,\&\, F^{(M)} \to H(u^{(n)},\, u^{(\eta(n))}).$

Also $\mathfrak{A}_n \,\&\, F^{(r)} \to H(u,\, u^{(\eta(0))}).$

Hence for each n some formula of the form

$$\mathfrak{A}_\eta \,\&\, F^{(M)} \to H(u^{(n)},\, u^{(\eta(n))})$$

is provable. Also, if $M' \geqslant M$ and $M' \geqslant m$ and $m \neq \eta(u)$, then

$$\mathfrak{A}_\eta \,\&\, F^{(M')} \to G(u^{\eta((n))}, u^{(m)}) v G(u^{(m)},\, u^{(\eta(n))})$$

and

$$\mathfrak{A}_\eta \,\&\, F^{(M')} \to [\{G(u^{(\eta(n))},\, u^{(m)}) v G(u^{(m)},\, u^{(\eta(n)}).$$
$$\&\, H\,(u^{(n)},\, u^{(\eta(m))}\} \to (-H(u^{(n)},\, u^{(m)}))].$$

Hence $\mathfrak{A}_\eta \,\&\, F^{(M')} \to (-H(u^{(n)},\, u^{(m)})).$

The conditions of our second definition of a computable function are therefore satisfied. Consequently η is a computable function.

Proof of a modified form of (iii).

Suppose that we are given a machine $\mathcal{N}$, which, starting with a tape bearing on it ə ə followed by a sequence of any number of letters "F" on F-squares and in the m-configuration b, will compute a sequence γ_n depending on the number n letters "F". If $\phi_n(m)$ is the m-th figure of γ_n, then the sequence β whose n-th figure is $\phi_n(n)$ is computable.

We suppose that the table for $\mathcal{N}$ has been written out in such a way that in each line only one operation appears in the operations column. We also suppose that Ξ, Θ, $\overline{0}$, and $\overline{1}$ do not occur in the table, and we replace ə throughout by Θ, 0 by $\overline{0}$, and 1 by $\overline{1}$. Further substitutions are then made. Any line of form

$\mathfrak{A}$	α	$P\overline{0}$	$\mathfrak{B}$

we replace by

$\mathfrak{A}$	α	$P\overline{0}$	$\mathfrak{re}(\mathfrak{B}, \mathfrak{u}, h, k)$

and any line of the form

$\mathfrak{A}$	α	$P\overline{1}$	$\mathfrak{B}$

by

$\mathfrak{A}$	α	$P\overline{1}$	$\mathfrak{re}(\mathfrak{B}, \mathfrak{v}, h, k)$

and we add to the table the following lines:

$\mathfrak{u}$		$\mathfrak{pe}(\mathfrak{u}_1, 0)$
$\mathfrak{u}_1$	$R, Pk, R, P\Theta, R, P\Theta$	$\mathfrak{u}_2$
$\mathfrak{u}_2$		$\mathfrak{re}(\mathfrak{u}_3, \mathfrak{u}_3, k, h)$
$\mathfrak{u}_3$		$\mathfrak{pe}(\mathfrak{u}_2, F)$

and similar lines with $\mathfrak{v}$ for u and 1 for 0 together with the following line

$\mathfrak{c}$ $\quad$ $R, P\Xi, R, Ph$ $\quad$ $\mathfrak{b}$.

We then have the table for the machine $\mathcal{N}'$ which computes β. The initial m-configuration is $\mathfrak{c}$, and the initial scanned symbol is the second ə.

11. Application to the Entscheidungsproblem

The results of §8 have some important applications. In particular, they can be used to show that the Hilbert Entscheidungsproblem can have no solution. For the present I shall confine myself to proving this particular theorem. For the formulation of this problem I must refer the reader to Hilbert and Ackermann's *Grundzüge der Theoretischen Logik* (Berlin, 1931), chapter 3.

I propose, therefore, to show that there can be no general process for determining whether a given formula $\mathfrak{A}$ of the functional calculus **K** is provable, *i.e.* that there can be no machine which, supplied with any one $\mathfrak{A}$ of these formulae, will eventually say whether $\mathfrak{A}$ is provable.

It should perhaps be remarked that what I shall prove is quite different from the well-known results of Gödel[10]. Gödel has shown that (in the formalism of Principia Mathematica) there are propositions $\mathfrak{A}$ such that neither $\mathfrak{A}$ nor $-\mathfrak{A}$ is provable. As a consequence of this, it is shown that no proof of consistency of Principia Mathematica (or of **K**) can be given within that formalism. On the other hand, I shall show that there is no general method which tells whether a given formula $\mathfrak{A}$ is provable in **K**, or, what comes to the same, whether the system consisting of **K** with $-\mathfrak{A}$ adjoined as an extra axiom is consistent.

If the negation of what Gödel has shown had been proved, *i.e.* if, for each $\mathfrak{A}$ either $\mathfrak{A}$ or $-\mathfrak{A}$ is provable, then we should have an immediate solution of the Entscheidungsproblem. For we can invent a machine $\mathcal{K}$ which will prove consecutively all provable formulae. Sooner or later $\mathcal{K}$ will reach either $\mathfrak{A}$ or $-\mathfrak{A}$. If it reaches $\mathfrak{A}$, then we know that $\mathfrak{A}$ is provable. If it reaches $-\mathfrak{A}$, then, since **K** is consistent (Hilbert and Ackermann, p. 65), we know that $\mathfrak{A}$ is not provable.

Owing to the absence of integers in **K** the proofs appear somewhat lengthy. The underlying ideas are quite straightforward.

Corresponding to each computing machine $\mathcal{M}$ we construct a formula Un ($\mathcal{M}$) and we show that, if there is a general method for determining whether Un ($\mathcal{M}$) is provable, then there is a general method for determining whether $\mathcal{M}$ ever prints 0.

The interpretations of the propositional functions involved are as follows:

$R_{S_l}(x, y)$ is to be interpreted as "in the complete configuration x (of $\mathcal{M}$) the symbol on the square y is S".

$I(x, y)$ is to be interpreted as "in the complete configuration x the square y is scanned".

$K_{q_m}(x)$ is to be interpreted as "in the complete configuration x the m-configuration is q_m.

$F(x, y)$ is to be interpreted as "y is the immediate successor of x".

Inst $\{q_iS_jS_kLq_l\}$ is to be an abbreviation for

$$\begin{aligned}(x, y, x', y')\big\{(R_{S_j}(x, y) \mathbin{\&} I(x, y) \mathbin{\&} K_{q_i}(x) \mathbin{\&} F(x, x') \mathbin{\&} F(y', y))\\ \rightarrow (I(x', y') \mathbin{\&} R_{S_k}(x', y) \mathbin{\&} K_{q_l}(x')\\ \mathbin{\&} (z)\ [F(y', z) \vee (R_{S_j}(x, z) \rightarrow R_{S_k}(x', z))])\big\}.\end{aligned}$$

Inst $\{q_iS_jS_kRq_l\}$ and Inst $\{q_iS_jS_kNq_l\}$

are to be abbreviations for other similarly constructed expressions.

Let us put the description of $\mathcal{M}$ into the first standard form of §6. This description consists of a number of expressions such as "$q_iS_jS_kLq_l$" (or with R or N substituted for L). Let us form

[10] Loc. cir.

all the corresponding expressions such as Inst $\{q_iS_jS_kLq_l\}$ and take their logical sum. This we call Des $(\mathcal{M})$.

The formula Un $(\mathcal{M})$ is to be

$$(\exists u)[N(u) \,\&\, (x)(N(x) \rightarrow (\exists x')F(x, x'))$$
$$\&\, (y,z)(F(y, z) \rightarrow N(y) \,\&\, N(z)) \,\&\, (y)R_{S_0}(u, y)$$
$$\&\, I(u, u) \,\&\, K_{q_1}(u) \,\&\, Des(\mathcal{M})]$$
$$\rightarrow (\exists s)(\exists t)[N(s) \,\&\, N(t) \,\&\, R_{S_1}(s, t)].$$

$[N(u) \,\&\ldots\&$ Des $(\mathcal{M})]$ may be abbreviated to $A(\mathcal{M})$.

When we substitute the meanings suggested on p. 37–38 we find that Un $(\mathcal{M})$ has the interpretation "in some complete configuration of $\mathcal{M}$, S_1 (*i.e.* 0) appears on the tape". Corresponding to this I prove that

(a) If S_1 appears on the tape in some complete configuration of $\mathcal{M}$, then Un $(\mathcal{M})$ is provable.
(b) If Un $(\mathcal{M})$ is provable. then S_1 appears on the tape in some complete configuration of $\mathcal{M}$.

When this has been done, the remainder of the theorem is trivial.

Lemma 1 *If S_1 appears on the tape in some complete configuration of $\mathcal{M}$, then* Un $(\mathcal{M})$ *is provable.*

We have to show how to prove Un $(\mathcal{M})$. Let us suppose that in the n-th complete configuration the sequence of symbols on the tape is $S_{r(n,0)}, S_{r(n,1)}, \ldots, S_{r(n,n)}$, followed by nothing but blanks, and that the scanned symbol is the $i(n)$-th, and that the m-configuration is $q_{k(n)}$. Then we may form the proposition

$$R_{S_{r(n,0)}}(u^{(n)}, u) \,\&\, R_{S_{r(n,1)}}(u^{(n)}, u') \,\&\, \ldots \,\&\, R_{S_{r(n,n,}}(u^{(n)}, u^{(n)})$$
$$\&\, I(u^n, u^{(i(n))}) \,\&\, K_{qk_{(n)}}(u^{(n)})$$
$$\&\, (y)F((y, u')\mathrm{v}F(u, y)\mathrm{v}F(u', y)\mathrm{v}\ldots\mathrm{v}F(u^{(n-1)}, y)\mathrm{v}R_{S_0}(u^{(n)}, y)),$$

which we may abbreviate to CC_n.

As before, $F(u, u') \,\&\, F(u', u') \,\&\ldots\&\, F\,(u^{(r-1)}, u^{(r)})$ is abbreviated to $F^{(r)}$.

I shall show that all formulae of the form $A(\mathcal{M}) \,\&\, F^{(n)} \rightarrow CC_n$ (abbreviated to CF_n) are provable. The meaning of CF_n is "The n-th complete configuration of $\mathcal{M}$ is so and so", where "so and so" stands for the actual n-th complete configuration of $\mathcal{M}$. That CF_n should be provable is therefore to be expected.

CF_0 is certainly provable, for in the complete configuration the symbols are all blanks, the m-configuration is q_1, and the scanned square is u, *i.e.* CC_0 is

$$(y)R_{S_0}(u, y) \,\&\, I(u, u) \,\&\, K_{q1}(u).$$

$A(\mathcal{M}) \rightarrow CC_0$ is then trivial.

We next show that $CF_n \rightarrow CF_{n+1}$ is provable for each n. There are three cases to consider, according as in the move from the n-th to the $(n+1)$-th configuration the machine moves to left or to right or remains stationary. We suppose that the first case applies, *i.e.* the machine moves to the left. A similar argument applies in the other cases. If

$$r(n, i(n)) = a,\ r(n+1, i(n+1)) = c,\ k(i(n)) = b, \text{ and } k(i(n+1)) = d,$$

then Des $(\mathcal{M})$ must include Inst $\{q_aS_bS_dLq_c\}$ as one of its terms, *i.e.*

$$\text{Des }(\mathcal{M}) \rightarrow \text{Inst }\{q_aS_bS_dLq_c\}.$$

Hence $$A(\mathcal{M})\ \&\ F^{(n+1)} \to \text{Inst}\ \{q_a S_b S_d L q_c\}\ \&\ F^{(n+1)}.$$

But $$\text{Inst}\ \{q_a S_b S_d L q_c\}\ \&\ F^{(n+1)} \to (CC_n \to CC_{n+1})$$

is provable, and so therefore is

$$A(\mathcal{M})\ \&\ F^{(n+1)} \to (CC_n \to CC_{n+1})$$

and $$(A(\mathcal{M})\ \&\ F^{(n)} \to CC_n) \to (A(\mathcal{M})\ \&\ F^{(n+1)} \to CC_{n+1}),$$

i.e. $$CF_n \to CF_{n+1}.$$

CF_n is provable for each n. Now it is the assumption of this lemma that S_1 appears somewhere, in some complete configuration, in the sequence of symbols printed by $\mathcal{M}$; that is, for some integers N, K, CC_N has $R_{S_1}(u^{(N)}, u^{(K)})$ as one of its terms, and therefore $CC_N \to R_{S_1}(u^{(N)}, u^{(K)})$ is provable. We have then

$$CC_N \to R_{S_1}(u^{(N)}, u^{(K)})$$

and $$A(\mathcal{M})\ \&\ F^{(N)} \to CC^N.$$

We also have

$$(\exists u)A(\mathcal{M}) \to (\exists u)(\exists u')\ldots(\exists u^{(N')})(A(\mathcal{M})\ \&\ F^{(N)}),$$

where $N' = \max(N, K)$. And so

$$(\exists u)A(.\mathcal{M}) \to (\exists u)(\exists u')\ \ldots\ (\exists u^{(N')})R_{S_1}(u^{(N)}, u^{(K)}),$$

$$(\exists u)A(\mathcal{M}) \to (\exists u^{(N)})(\exists u^{(K)})R_{S_1}(u^{(N)}, u^{(K)}),$$

$$(\exists u)A(\mathcal{M}) \to (\exists s)(\exists t)R_{S_1}(s, t),$$

i.e. $\text{Un}(\mathcal{M})$ is provable.

This completes the proof of Lemma 1.

Lemma 2 *If* Un $(\mathcal{M})$ *is provable, then* S_1 *appears on the tape in some complete configuration of* $\mathcal{M}$.

If we substitute any propositional functions for function variables in a provable formula, we obtain a true proposition. In particular, if we substitute the meanings tabulated on pp. 37–38 in Un $(\mathcal{M})$, we obtain a true proposition with the meaning "S_1 appears somewhere on the tape in some complete configuration of $\mathcal{M}$".

We are now in a position to show that the Entscheidungsproblem cannot be solved. Let us suppose the contrary. Then there is a general (mechanical) process for determining whether Un $(\mathcal{M})$ is provable. By Lemmas 1 and 2, this implies that there is a process for determining whether $\mathcal{M}$ ever prints 0, and this is impossible, by §8. Hence the Entscheidungsproblem cannot be solved.

In view of the large number of particular cases of solutions of the Entscheidungsproblem for formulae with restricted systems of quantors, it is interesting to express Un $(\mathcal{M})$ in a form in which all quantors are at the beginning. Un $(\mathcal{M})$ is in fact, expressible in the form

$$(u)(\exists x)(w)(\exists u_1)\ldots(\exists u_n)\mathfrak{B}, \tag{I}$$

where $\mathfrak{B}$ contains no quantors, and $n = 6$. By unimportant modifications we can obtain a formula, with all essential properties of Un $(\mathcal{M})$, which is of form (I) with $n = 5$.

Added 28 *August*, 1936.

Appendix

Computability and effective calculability

The theorem that all effectively calculable (λ-definable) sequences are computable and its converse are proved below in outline. It is assumed that the terms "well-formed formula" (W.F.F.) and "conversion" as used by Church and Kleene are understood. In the second of these proofs the existence of several formulae is assumed without proof; these formulae may be constructed straightforwardly with the help of, *e.g.*, the results of Kleene in "A theory of positive integers in formal logic", *American Journal of Math.*, 57 (1935), 153–173, 219–244.

The W.F.F. representing an integer n will be denoted by N_n. We shall say that a sequence γ whose n-th figure is $\phi_\gamma(n)$ is λ-definable or effectively calculable if $I+\phi_\gamma(\mu)$ is a λ-definable function of n, *i.e.* if there is a W.F.F. M_γ such that, for all integers n,

$$\{M_\gamma\}(N_n) \text{ conv } N_{\phi_\gamma(n)+1},$$

i.e. $\{M_\gamma\}(N_n)$ is convertible into $\lambda xy.\ x(x(y))$ or into $\lambda xy.x(y)$ according as the n-th figure of λ is 1 or 0.

To show that every λ-definable sequence γ is computable, we have to show how to construct a machine to compute γ. For use with machines it is convenient to make a trivial modification in the calculus of conversion. This alteration consists in using x, x', x'', ... as variables instead of a, b, c, ... We now construct a machine $\mathcal{L}$ which, when supplied with the formula M_γ, writes down the sequence γ. The construction of $\mathcal{L}$ is somewhat similar to that of the machine $\mathcal{K}$ which proves all provable formulae of the functional calculus. We first construct a choice machine $\mathcal{L}_1$, which if supplied with a W.F.F., M say, and suitably manipulated, obtains any formula into which M is convertible. $\mathcal{L}_1$ can then be modified so as to yield an automatic machine $\mathcal{L}_2$ which obtains successively all the formulae into which M is convertible (of. foot-note p. 32). The machine $\mathcal{L}$ includes $\mathcal{L}_2$ as a part. The motion of the machine $\mathcal{L}$ when supplied with the formula M_γ is divided into sections of which the n-th is devoted to finding the n-th figure of γ. The first stage in this n-th section is the formation of $\{M_\gamma\}(N_n)$. This formula is then supplied to the machine $\mathcal{L}_2$, which converts it successively into various other formulae. Each formula into which it is convertible eventually appears, and each, as it is found, is compared with

$$\lambda x[\lambda x'[\{x\}(\{x\}(x'))]], \quad \text{i.e.} N_2,$$

and with

$$\lambda x[\lambda x'[\{x\}(x')]], \quad \text{i.e.} N_1.$$

If it is identical with the first of these, then the machine prints the figure 1 and the n-th section is finished. If it is identical with the second, then 0 is printed and the section is finished. If it is different from both, then the work of $\mathcal{L}_2$ is resumed. By hypothesis, $\{M_\gamma\}(N_n)$ is convertible into one of the formulae N_2 or N_1; consequently the n-th section will eventually be finished, *i.e.* the n-th figure of γ will eventually be written down.

To prove that every computable sequence γ is λ-definable, we must show how to find a formula M_γ such that, for all integers n,

$$\{M_\gamma\}(N_n) \text{ conv } N_{1+\phi_r(n)}.$$

Let $\mathcal{M}$ be a machine which computes γ and let us take some description of the complete configurations of $\mathcal{M}$ by means of numbers, *e.g.* we may take the D.N of the complete configuration as described in §6. Let $\xi(n)$ be the D.N of the n-th complete configuration of $\mathcal{M}$. The table for the machine $\mathcal{M}$ gives us a relation between $\xi(n+1)$ and $\xi(n)$ of the form

$$\xi(n+1) = \rho_\gamma(\xi(n)),$$

where ρ_γ is a function of very restricted, although not usually very simple, form: it is determined by the table for $\mathcal{M}$. ρ_γ is λ-definable (I omit the proof of this), *i.e.* there is a W.F.F. A_γ such that, for all integers n,

$$\{A_\gamma\}(N_{\xi(n)}) \text{ conv } N_{\xi(n+1)}.$$

Let U stand for

$$\lambda u[\{\{u\}(A_\gamma)\}(N_r)],$$

where $r = \xi(0)$; then, for all integers n,

$$\{U_\gamma\}(N_n) \text{ conv } N_{\xi(n)}.$$

It may be proved that there is a formula V such that

$$\{\{V\}(N_{\xi(n+1)})\}(N_{\xi(n)}) \begin{cases} \text{conv } N_1 & \text{if, in going from the } n\text{-th to the } (n+1)\text{-th complete configuration, the figure 0 is printed.} \\ \text{conv } N_2 & \text{if the figure 1 is printed.} \\ \text{conv } N_2 & \text{otherwise.} \end{cases}$$

Let W_γ stand for

$$\lambda u[\{\{V\}\left(\{A_\gamma\}(\{U_\gamma\}(u))\right)\}(\{U_\gamma\}(u))],$$

so that, for each integer n

$$\{\{V\}(N_{\xi(n+1)})\}(N_{\xi(n)}) \text{ conv } \{W_\gamma\}(N_n),$$

and let Q be a formula such that

$$\{\{Q\}(W_\gamma)\}(N_s) \text{ conv } N_{r(z)},$$

where $r(s)$ is the s-th integer q for which $\{W_\gamma\}(N_q)$ is convertible into either N_1 or N_2. Then, if M_γ stands for

$$\lambda w[\{W_\gamma\}(\{\{Q\}(W_\gamma)\}(w))],$$

it will have the required property[11]

The Graduate College,
Princeton University,
New Jersey, U.S.A.

[11] In a complete proof of the λ-definability of computable sequences it would be best to modify this method by replacing the numerical description of the complete configurations by a description which can be handled more easily with our apparatus. Let us choose certain integers to represent the symbols and the m-configurations of the machine. Suppose that in a certain complete configuration the numbers representing the successive symbols on the tape are $s_1 s_2 \ldots, s_n$, that the m-th symbol is scanned, and that the m-configuration has the number t; then we may represent this complete configuration by the formula

$$[[N_{s_1}, N_{s_2}, \ldots, N_{s_{m-1}}], [N_t, N_{s_m}], [N_{s_{m+1}}, \ldots, N_{s_n}]],$$

where $[a,b]$ stands for $\lambda u[\{\{u\}(a)\}(b)]]$,

$[a,b,c]$ stands for $\lambda u[\{\{\{u\}(a)\}(b)\}(c)]$,

etc.

ON COMPUTABLE NUMBERS, WITH AN APPLICATION TO THE ENTSCHEIDUNGSPROBLEM. A CORRECTION

By A. M. TURING

In a paper entitled "On computable numbers, with an application to the Entscheidungsproblem"[1] the author gave a proof of the insolubility of the Entscheidungsproblem of the "engere Funktionenkalkül". This proof contained some formal errors[2] which will be corrected here: there are also some other statements in the same paper which should be modified, although they are not actually false as they stand.

The expression for Inst $\{q_i S_j S_k L q_l\}$ on p. 37 of the paper quoted should read

$$(x, y, x', y')\{(R_{S_j}(x, y) \& I(x, y) \& K_{qi}(x) \& F(x,x') \& F(y', y)) \\ \rightarrow (I(x', y') \& R_{s_k}(x',y) \& K_{ql}(x') \& F(y',z) \mathrm{v} [(R_{s_0}(x,z) \rightarrow R_{s_0}(x',z)) \\ \& (R_{S_1}(x, z) \rightarrow R_{s_1}(x',z)) \& \ldots \& (R_{S_M}(x,z) \rightarrow R_{S_M}(x',z))])\},$$

$S_0, S_1, \ldots, S_M$ being the symbols which. $\mathcal{M}$ can print. The statement on p. 39, line 4, viz.

$$\text{"Inst } \{q_a S_b S_d L q_c\} \& F^{(n+1)} \rightarrow (CC_n \rightarrow CC_{n+1})$$

is provable" is false (even with the new expression for Inst $\{q_a S_b S_d L q_c\}$): we are unable for example to deduce $F^{(n+1)} \rightarrow \big(-F(u, u'')\big)$ and therefore can never use the term

$$F(y', z)\mathrm{v}[(R_{S_0}(x, z) \rightarrow R_{S_0}(x', z) \& \ldots \& (R_{S_M}(x,z) \rightarrow R_{S_M}(x',z))]$$

in Inst $\{q_a S_b S_d L q_c\}$. To correct this we introduce a new functional variable $G[G(x,y)$ to have the interpretation "x precedes y"]. Then, if Q is an abbreviation for

$$(x)(\exists w)(y,z)\{F(x,w) \& (F(x,y) \rightarrow G(x,y)) \& (F(x,z) \& G(x,y) \rightarrow G(x,y)) \\ \& [G(z,x)\mathrm{v}(G(x,y) \& F(y,z))\mathrm{v}(F(x,y) \& F(z,y)) \rightarrow (-F(x,z))]$$

the corrected formula Un $(\mathcal{M})$ is to be

$$(\exists u)A(\mathcal{M}) \rightarrow (\exists s)(\exists t)R_{s_1}(s,t).$$

where $A(\mathcal{M})$ is an abbreviation for

$$Q \& (y)R_{S_0}(u, y)_1 \& I(u, u) \& K_{q_1}(u) \& \mathrm{Des}(\mathcal{M}).$$

The statement on p. 39 (line 3) must then read

$$\text{Inst } \{q_a S_b S_d L q_c\} \& Q \& QF^{(n+1)} \rightarrow (CC_n \rightarrow CC_{n+1}),$$

and line 29 should read

[1] Proc. London Math. Soc. (2), 42 (1936–7), 230–265.

[2] The author is indebted to P. Bernays for pointing out these errors.

$$r(n,i(n)) = b,\ r(n+1,i(n)) = d,\ k(n) = a,\ k(n+1) = c.$$

For the words "logical sum" on p. 38, line 4, read "conjunction". With these modifications the proof is correct. Un ($\mathcal{M}$) may be put in the form (I) (p. 39) with n = 4.

Some difficulty arises from the particular manner in which "computable number" was defined (p. 18). If the computable numbers are to satisfy intuitive requirements we should have:

If we can give a rule which associates with each positive integer n two rationals a_n, b_n satisfying $a_n \leqslant a_{n+1} < b_{n+1} \leqslant b_n, b_n - a_n < 2^{-n}$, then there is a computable number a for which $a_n \leqslant a \leqslant b_n$ each n. (A)

A proof of this may be given, valid by ordinary mathematical standards, but involving an application of the principle of excluded middle. On the other hand the following is false:

There is a rule. whereby, given the rule of formation of the sequences, a_n, b_n in (A) we can obtain a D.N. for a machine to compute a. (B)

That (B) is false, at least, if we adopt the convention that the decimals of numbers of the form $m/2^n$ shall always terminate with zeros, can be seen in this way. Let $\mathcal{N}$ be some machine, and define c_n as follows: $c_n = \frac{1}{2}$ if $\mathcal{N}$ has not printed a figure 0 by the time the n-th complete configuration is reached $c_n = \frac{1}{2} - 2^{m-3}$ if 0 had first been printed at the m-th, complete configuration ($m \leqslant n$). Put $a_n = c_n - 2^{-n-2}, b_n = c_n + 2^{-n-2}$. Then the inequalities of (A) are satisfied, and the first figure of α is 0 if $\mathcal{N}$ ever prints 0 and is 1 otherwise. If (B) were true we should have a means of finding the first figure of a given the D.N. of $\mathcal{N}$ *i.e.* we should be able to determine whether $\mathcal{N}$ ever prints 0, contrary to the results of §8 of the paper quoted. Thus although (A) shows that there must be machines which compute the Euler constant (for example) we cannot at present describe any such machine, for we do not yet know whether the Euler constant is of the form $m/2^n$.

This disagreeable situation can be avoided by modifying the manner in which computable numbers are associated with computable sequences, the totality of computable numbers being left unaltered. It may be done in many ways[3] of which this is an example. Suppose that the first figure of a computable sequence γ, is i and that this is followed by 1 repeated n times, then by 0 and finally by the sequence whose r-th figure is c_r; then the sequence γ, is to correspond to the real number

$$(2i-1)n + \sum_{r=1}^{\infty}(2c_r - 1)(\tfrac{2}{3})^r$$

If the machine which computes γ is regarded as computing also this real number then (B) holds. The uniqueness of representation of real numbers by sequences of figures is now lost, but this is of little theoretical importance, since the D.N.'s are not unique in any case.

The Graduate College,
Princetom, N.J., U.S.A.

[3] This use of overlapping intervals for the definition of real numbers is due originally to Brouwer.

Examining the Work and Its Later Impact

Stephen Wolfram on —

THE IMPORTANCE OF UNIVERSAL COMPUTATION

In the long view of intellectual history, I believe universal computation will stand as the single most important idea to emerge in the twentieth century. And this paper is where it first appeared with clarity.

The paper certainly did not set out to have such significance. Instead, its purpose was to address a technical question – albeit one thought to be important – about the foundations of mathematics. But to address that question, Turing created the concept of a universal computer – which in time led to the notion of software, the computer revolution, and an increasing fraction of all our technology today.

In retrospect, it seems almost bizarre that it took until 1936 for such a basic idea to emerge. For today, immersed as they are in modern technology, even quite young children seem to have a decent grasp of the basic idea of programmability and universal computation.

Even in antiquity, there was already the notion that any single human language could describe the same basic range of facts and processes. Leibniz tightened this up in the 1600s, imagining a universal language based on logic, and even discussing encoding logic with numbers (Leibniz, 1966).

Then in the 1800s, there were punched-card machines that could be programmed for different functions. And with the increasing abstraction and formalisation of mathematics, there emerged by the 1920s ideas like combinators and string-rewrite systems.

But it was Gödel's theorem (Gödel, 1931) that highlighted the importance of such abstractions. And in fact, inside the proof of Gödel's theorem is in effect exactly the idea of universal computation – but framed in the context of purely mathematical constructs.

The great significance of Turing's paper was to give concreteness to universal computation: to make it seem that universal computation might somehow be inevitable in any constructible system.

At the time, it was far from clear how general Turing's results might be. After all, for example, until the 1920s, there had been the idea that any reasonable mathematical function could be represented by primitive recursion – but that idea was immediately exploded by the discovery of the Ackermann function (Wilhelm, 1928).

And indeed, after his 1936 paper, Turing himself set about looking at systems involving oracles and so on, that would be more powerful than his universal Turing machine. But gradually an increasing collection of 'implementable' abstract models seemed instead to be precisely equivalent in their power to ordinary Turing machines.

Meanwhile, electronic computers were emerging. McCulloch and Pitts (1943) had used Turing's idea of a universal machine to argue that brains could in effect just be like networks of electronic components. And von Neumann (Burks, Goldstine and von Neumann, 1947) then applied these ideas to develop architectures for practical electronic computers.

There continued to be some theoretical work on Turing machines, but for the most part, electronic computers were treated purely as technology, with little discussion of foundational issues. And indeed, it was only in about the 1980s that Turing's work began to become more widely known.

And at that time, there tended to be the view that Turing machines were relevant as idealisations of what could be implemented with electronics, and perhaps with mathematics, but not necessarily much more. And indeed it was usually assumed (as it had been by Turing himself) that when it came to typical systems in nature, traditional mathematical equations – and not something like discrete Turing machines – would be the relevant models to use.

In the early 1980s, I became interested in a variety of natural systems that exhibited complex behaviour, and that had never been very usefully described by traditional mathematical equations. And I set about trying to find the simplest models that might describe such systems.

I had experience both with practical computers and with models in statistical physics based on discrete components. And in trying to find the simplest possible model, I quickly settled on one-dimensional cellular automata.

My main initial methodology for studying cellular automata was experimental: to just run computer experiments and see how the cellular automata behaved (see Figure 1).

The results were remarkable, and to me deeply surprising. For I found that even when the underlying rules for the system were extremely simple, the behaviour of the system as a whole could be immensely complex (see Figure 2).

And gradually, through my work (Wolfram, 2002) and the work of many others, it began to be clear that a great many systems in nature could successfully be modelled using these kinds of simple programs.

But as I searched for an understanding of the basic phenomenon by which complexity was generated, I was quickly led to Turing machines and universal computation. And I came to speculate that even in the simple cellular automata I was studying, there must be universal computation which in turn I then argued led to perceived complexity, and a variety of other fundamental phenomena.

Before my work, one might have assumed that systems in nature would typically need to be described by the standard continuous differential equations of mathematical physics – and would therefore presumably not act like Turing machines. But after seeing so many examples of natural systems successfully described by systems like cellular automata, it began to seem much more plausible that nothing with power beyond Turing machines was needed.

I do not think that Alan Turing ever directly simulated a Turing machine. He was interested in the theoretical issue of whether a Turing machine that is universal could be constructed. And indeed in this paper he showed that that was possible – though with a machine of considerable complexity. It was not until the beginning of the 1990s that I actually started simulating Turing machines in large numbers. I decided to see if my results on cellular automata would carry over to Turing machines – which operate in a sequential, rather than parallel, way.

And what I found was that much like in cellular automata, one does not have to go far in the universe of possible Turing machines before one starts to find examples that exhibit highly complex behaviour.

For a long time, it was not clear what the very simplest universal Turing machine might be. But now we know. It is a machine with two states and three colors that I first identified in the mid-1990s (see Figure 3), and that was finally proved universal in 2007 as a result of a competition we held (Smith, 2007).

Traditional intuition from looking at practical computers might have suggested that to get universality would require a system with a complicated structure, typical of what might be set up by human engineers. But from my studies in the computational universe of possible programs, I had

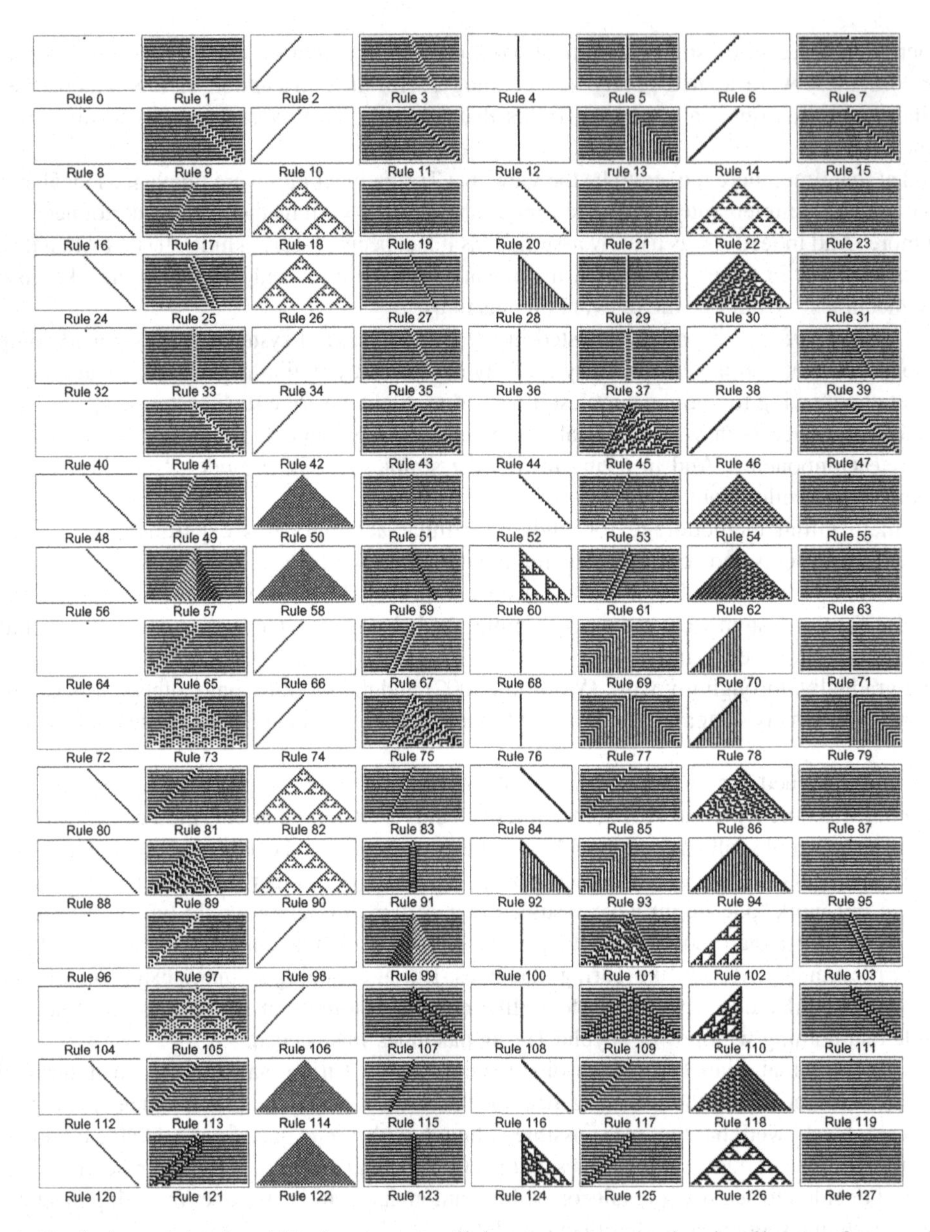

Fig. 1: Evolution of the first 128 elementary cellular automata, starting in all cases from a single black cell.

formulated a general principle that I called the Principle of Computational Equivalence, which predicted, among many other things, that universal computation should actually be very common, even among systems with simple underlying structures.

The result of this is to even further enhance the importance of Turing's paper.

At first, we might have thought that things like universal Turing machines would be relevant only to specific kinds of mathematical-type systems. But gradually we came to learn that all sorts of systems – including practical ones made with electronic components – could be set up to behave like universal Turing machines.

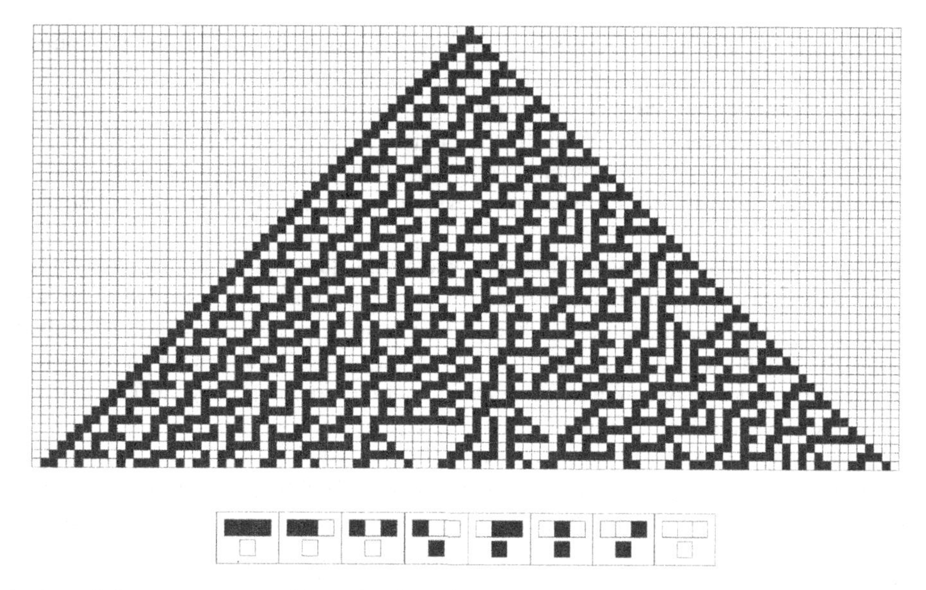

Fig. 2: A cellular automaton with a simple rule that generates a pattern, which seems in many respects random. The rule used is of the same type as in the previous examples, and the cellular automaton is again started from a single black cell. The pattern that is obtained is highly complex, and shows almost no overall regularity.

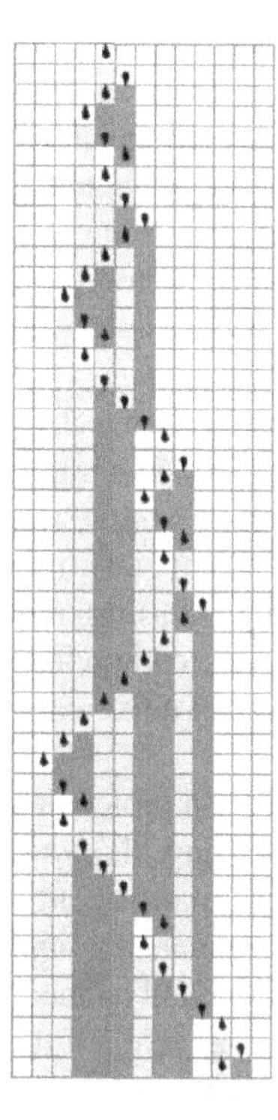

Fig. 3: The first 50 steps in evolution (from a blank tape) of the simplest Turing machine capable of universal computation.

But what we learn now is that setup is in a sense not required. Instead, the phenomenon of universality seems to be ubiquitous – in systems in nature, mathematics and elsewhere.

What is the significance of this? At a practical level, the occurrence of universality in so many kinds of systems makes it easier to imagine creating practical computation systems in a much wider variety of ways. But universality also implies some important limitations on science – particularly a phenomenon I call computational irreducibility, which fundamentally limits the predictability of processes.

In the past, one might have assumed that most systems in nature would be computationally much simpler than universal Turing machines. But now we have evidence that essentially whenever we see complex behaviour in nature, it is associated with processes that achieve universal computation.

But could the systems do more than universal Turing machines? We do not yet know for sure. But as we successfully model more and more systems without the need to go beyond Turing machines, it seems less and less plausible that this will be necessary.

The ultimate test, however, is whether we can model our whole physical universe using something equivalent to a universal Turing machine. We can certainly imagine a universe that operates like some behaviour of a Turing machine. But the issue is whether our actual universe does so.

Until we know the ultimate theory of physics, we will not be sure of the answer. The history of traditional physics might seem to suggest that as one goes to smaller and smaller scales, it must take something more and more complicated to describe the physical world. But from my exploration of cellular automata and similar systems, I have developed a quite different intuition – which now makes it seem quite plausible that there could be a simple rule that underlies everything in our universe.

The evidence I have increasingly seems to support this view. For more and more phenomena familiar from known physics seem to emerge from extremely simple underlying rules – operating underneath such traditional notions as space and time. And if it is true that there is a rule that in effect reproduces our universe, then we will know for certain that everything in our universe can in fact, in principle, be described by something equivalent to a universal Turing machine.

The Turing machine has then gone from a mathematical idealisation, to a model for computational processes – to a complete means of describing everything that can exist in our universe. It also has gone from something of importance only for questions of mathematical or theoretical computation to something whose features and properties must pervade all systems that we experience – and that must be considered fundamental to science.

We can trace the foundations of so much modern technology to the idea of universal computation set forth in Turing's paper. Increasingly, universal computation seems destined to be central to all sorts of issues in science. In a sense, though, I believe we are still early on the curve of seeing the full significance of universal computation. Computers, for example, became widespread though the development of first databases, then word processing, then the web – none of which make central use of universal computation. But now, as we see knowledge begin to become computable on a large scale, there is finally starting to be deeper use of universal computation. And there is still much more to come.

Almost everything we build – whether with molecules, social systems, whatever – will rely on universal computation. Our intuition about how the world works – and what we can know and predict about the world – will be based on thinking about universal computation. Even the future of the human condition will rely centrally on universal computation.

Newton's Principia Mathematica is often cited as the most important single work in the history of science – for it was the key milestone that enabled the development of the exact sciences and the tradition of engineering that arose from them. And certainly Newton was more aware of the significance of his work than Turing. But in the modest paper by Alan Turing reproduced here lie the seeds of what is surely the most important single intellectual development of the twentieth century, and possibly of all of modern human history.

References

Hilbert, D., Bernays, P., 1934. Grundlagen der Mathematik, vol.1, Springer, Berlin, p. 209.

Burks, A.W., Goldstine, H., von Neumann, J., 1947. Preliminary discussion of the Logical Design of an Electronic Computing Instrument, Institute for Advanced Study, Princeton, NJ.

Church , Alonzo 1936a. An unsolvable problem of elementary number theory, Am. J. Math. 58, 345–363.

Church, Alonzo, 1936b. A note on the Entseheidungsproblem, J. Symb. Logic 1, 40–41.

Gödel, K., 1931. Über formal unentscheidbare sätze der principia mathematica und verwandter systeme, I. Monatshefte für Mathematik und Physik, 38, 173–198.

Hobson, E.W., 1921. Theory of Functions of a Real Variable, second edn. 87–88.

Kleene, S.C., 1935. A theory of positive integers in formal logic. Amer. J. Math. 57, 153–173, 219–244.

Leibniz, G.W., 1966. Zur allgemeinen Charakteristik. Hauptschriften zur Grundlegung der Philosophie. pp. 30–31, Philosophische Werke Band 1, Translated by Artur Buchenau. Reviewed and with introduction and notes published by E. Cassirer. Hamburg, Felix Meiner.

McCulloch, W., Pitts, W., 1943. A logical calculus of the ideas immanent in nervous activity. Bull. Math. Biophys. 7, 115–133.

Smith, A., 2007. The Wolfram 2,3 Turing Machine Research Prize. http://www.wolframscience.com/prizes/tm23/solved.html. (accessed April 1, 2012)

Turing, A.M., 1936. On computable numbers with an application to the Entscheidungsproblem. Proc. London Math. Soc. 2 (42) 230–265.

Wilhelm, A., 1928. Zum hilbertschen aufbau der reellen zahlen. Mathematische Ann. 99, 118–133.

Wolfram, S., 2002. A New Kind of Science, Wolfram Media, Champaign, IL. USA.

Martin Davis illuminates —

THREE PROOFS OF THE UNSOLVABILITY OF THE ENTSCHEIDUNGSPROBLEM

The Entscheidungsproblem

What is usually called first order logic is an encapsulation of logical reasoning, especially as it occurs in proofs in mathematics. In this setup propositions become 'sentences' in which symbols for basic logical notions (like *and, not, all*) are combined with non-logical symbols for the specific items being reasoned about.

The Entscheidungsproblem may be stated in the following three equivalent ways:

(1) Find an algorithm to determine whether a given sentence of first order logic is *valid,* that is, true regardless of what specific objects and relationships are being reasoned about.
(2) Find an algorithm to determine whether a given sentence of first order logic is satisfiable, that is, true for some specific objects and relationships.
(3) Find an algorithm to determine given some sentences of first order logic regarded as *premises* and another sentence, being a desired *conclusion,* whether that conclusion is provable from the premises using the rules of proof for first order logic.

That the first two are equivalent is clear because the sentence A is valid if and only if not-A is not satisfiable. To show that the third formulation is equivalent to the first two, it suffices to note that, according to the *Completeness Theorem* that Gödel established in his doctoral dissertation, a sentence A is a logical consequence of premises $P_1, P_2, \ldots, P_n$ if and only if the sentence

$$P_1 \& P_2 \& \ldots \& P_n \Rightarrow A$$

is valid.

If we think of the premises as the axioms of some mathematical domain, we can see that an actual algorithm solving the Entscheidungsproblem would reduce all of mathematics, at least in principle, to mechanical calculation. Presumably it was this that led Hilbert to characterise the Entscheidungsproblem as 'the fundamental problem of mathematical logic'. This was enough to convince G. H. Hardy that there could be no such algorithm. He declared:

> There is of course no such theorem, and this is very fortunate, since if there were we should have a mechanical set of rules for the solution of all mathematical problems, and our activities as mathematicians would come to an end.[1]

In a different context, Poincaré made it clear that he found the whole idea of formalisation of mathematics ridiculous:

> Thus it will be readily understood that in order to demonstrate a theorem, it is not necessary or even useful to know what it means. ... we might imagine a machine where we should put in axioms at one end and take out theorems at the other, like that legendary machine in Chicago where pigs go in alive and come out transformed into hams and sausages. It is no more necessary for the mathematician than it is for these machines to know what he is doing.[2]

If indeed a solution of the Entscheidungsproblem would have reduced all of mathematics to mechanical calculation, then from the existence of any mathematical problem provably algorithmically unsolvable, the unsolvability of the Entscheidungsproblem itself should follow. The unsolvability proofs of Church and of Turing each follow this approach.

1. Church's Proof

Alonzo Church, in his historic (Church, 1936), provided a rigorous formal characterisation of what it means to be solvable by means of an algorithm, what has come to be known as *Church's Thesis*. This made it possible for him to prove that one specific problem is algorithmically unsolvable. In his work, Church (1936a) specified a finite set of premises that encapsulate this specific problem so faithfully that an algorithm for testing whether a given conclusion follows from those premises would also provide an algorithmic solution to that specific problem, although the problem is known to be unsolvable. From this contradiction Church could conclude that the Entscheidungsproblem itself is unsolvable.

Instead of indicating his satisfaction at having settled a fundamental problem, Church expressed a doubt. He noted the fact that the proof of Gödel's Completeness Theorem is necessarily non-constructive, dealing as it must, with the notion of validity that refers to arbitrary sets and

[1] Davis (2000), Chapter 7.

[2] Davis (2000), Chapter 5.

relations. And showing a surprisingly extreme constructivist stance, Church cast doubt on his own accomplishment saying:

> The unsolvability of [all three forms] of the Entscheidungsproblem . . . cannot, therefore, be regarded as established beyond question.

2. Turing's Proof

Alan Turing began working on the Entscheidungsproblem with no knowledge of what Church was doing. He began with his own explication of algorithmic solvability, or as he called it, computability, in terms of extremely simple abstract computing machines, what are now called 'Turing machines'. By analyzing what someone actually does when computing something, he provided a convincing argument for the adequacy of his formulation. (Later he proved that his concept was equivalent to Church's.) Like Church, he used what he had done to prove that a certain specific problem is unsolvable. In Turing's case the unsolvable problem was to determine algorithmically whether one of his given machines would ever produce a particular symbol, the so-called 'printing problem'. His remarkable paper (Turing, 1936) contained as a byproduct a construction showing that a particular one of his machines, his *universal machine*, could all by itself duplicate anything that any of his machines could do, and thereby showed in schematic form the possibility of an all-purpose computer.

By using sentences of first order logic to mimic the step-by-step behaviour of his machines he was able to associate with any one of his machines a corresponding sentence of first order logic that is valid if and only if that machine eventually produces the symbol 0. Thus an algorithm for validity (the first form of the Entscheidungsproblem in our list of three) would automatically provide a solution to the printing problem, although it is, in fact, unsolvable. Thus Turing could conclude that the Entscheidungsproblem is algorithmically unsolvable. Turing's method turned out to be quite fertile and was later used successfully to obtain significant new results.[3]

3. The should-have-been Gödel–Kleene Proof

Church and Turing's proofs were both published in 1936. Gödel's epochal paper (Gödel, 1931) in which he showed that formal logical systems in which a modicum of mathematics could be developed would inevitably be incomplete: there would be straightforward mathematical statements which could be neither proved nor disproved in that system. But the paper is extremely rich and, in particular contains an application to the Entscheidungsproblem. This application involves a class of functions from natural numbers to natural numbers called *primitive recursive*.[4] The key thing about this class is that included functions that can be defined *recursively*. As an example, we may consider the function 2^x which can be defined by the recursion:

$$2^0 = 1; \quad 2^{k+1} = 2 \times 2^k.$$

What Gödel proved relating to the Entscheidungsproblem in his remarkable 1931 paper is that corresponding to any given primitive recursive function $f(x)$, there is a sentence of first order logic, which is satisfiable if and only if $f(x) = 0$ for all x.[5] So to get the unsolvability of the Entscheidungsproblem (in our second form) it suffices to show that there is no algorithm for determining of a given primitive recursive function $f(x)$ whether it is equal to 0 for all x.

---

[3] See Börger et al. (1997).

[4] Not quite the term Gödel used.

[5] Gödel (1931) Theorem X.

Stephen Kleene was a student of and collaborator with Church during the exciting period when these dramatic results were being developed. Kleene's paper (Kleene, 1936) is mainly remembered for his *Normal Form Theorem* in which arbitrary computable functions on the natural numbers are seen to be closely related to primitive recursive functions. Thus for any computable function $f(x)$. Kleene showed that there are a pair of primitive recursive functions $g(x), h(x,y)$ such that:

$$f(x) = g(\min_{y}(h(x,y) = 0)).$$

In addition, Kleene found a particular primitive recursive function $t(z,x)$ such that the problem of determining for a given value of z whether $t(z,x) = 0$ for all x is algorithmically unsolvable.[6] From this special case, it is clear that there can be no general algorithm to determine of an arbitrary given primitive recursive function $f(x)$ whether it is equal to 0 for all x. So Gödel's result immediately yields the unsolvability of the Entscheidungsproblem.

It seems remarkable that Kleene, who certainly had studied Gödel's 1931 paper and knew it very well, apparently had not noticed this connection.

References

Börger, E., Grädel, E., Gurevich, Y., 1997. The Classical Decision Problem, Springer, Berlin, Heidelberg.

Church, A., 1936. An unsolvable problem in elementary number theory. Am. J. Math. 58, 345–363. Reprinted in Davis, M. (1965).

Church, A., 1936a. A note on the Entscheidungsproblem. J. Symb. Log. 1, 40–41. Correction ibid pp. 101–102. Reprinted with corrections incorporated: Davis, M. (1965).

Copeland, B.J., (Ed.), 2004. The Essential Turing. Oxford University Press.

Davis, M. (Ed.), 1965. The Undecidable. Raven Press, 1965. Reprinted: Dover 2004.

Davis, M., 1982. Why Gödel didn't have Church's Thesis, Inf. Control 54, 3–24.

Davis, M., 2000. The Universal Computer. Turing Centenary edition, A.K. Peters, Norton, New York, 2011.

Gödel, K., 1931. Über formal unentscheidbare Sätze der Principia mathematica und verwandter Systeme I, Monatshefte für Mathematik und Physik, Band 38, pp. 173–198. Reprinted with English translation from: Gödel, K., 1986. On Formally Undecidable Proposition of Principia Mathematica and Related Systems I, pp. 144–195.

Gödel, K., 1986. Collected Works, vol. I, Solomon Feferman et al, eds., Oxford University Press.

Kleene, S.C., 1936. General recursive functions of general numbers, Mathematische Annalen, Band 112, 727–742. Reprinted in Davis, M. (1965).

Petzold, C., 2008. The Annotated Turing: A Guided Tour through Alan Turing's Historic Paper on Computability and the Turing Machine, Wiley, Indianapolis.

Turing, A., 1936. On computable numbers with an application to the Entscheidungsproblem. Proc. Lond. Math. Soc. ser. 2, 42, 230–265. Correction: ibid, 43 (1937), pp. 544–546. Reprinted in Davis, M. (1965) pp. 116–154. Reprinted in Turing (2001) pp. 18–56. Reprinted in Copeland, J. (2004) pp. 58–90; 94–96. Reprinted in Petzold, C. (2008) (the original text interspersed with commentary).

Turing, A., 2001. Collected Works: Mathematical Logic. Gandy, R.O., Yates, C.E.M. (Eds.), North-Holland, Amsterdam, London.

[6] See Kleene (1936) XVI. To coordinate the present notation with Kleene's, we would define

$$t(z,x) = \begin{cases} 0 & \text{if } T_1(z,z,x) \\ 1 & \text{otherwise} \end{cases}$$

Samson Abramsky detects —

TWO PUZZLES ABOUT COMPUTATION

1. Introduction

Turing's classical analysis of computation Turing (1936) gives a compelling account of the nature of the computational process; of *how* we compute. This allows the notion of *computability*, of what can in principle be computed, to be captured in a mathematically precise fashion.

The purpose of this note is to raise two different questions, which are rarely if ever considered, and to which, it seems, we lack convincing, systematic answers. These questions can be posed as:

- Why do we compute?
- What do we compute?

The point is not so much that we have no answers to these puzzles, as that we have no established body of theory which gives satisfying, systematic answers, as part of a broader understanding. By raising these questions, we hope to stimulate some thinking in this direction.

These puzzles were raised in Abramsky (2008); see also Adriaans and van Emde Boas (2011).

2. Why Do We Compute?

The first puzzle is simply stated:

Why do we compute?

By this we mean: why do we perform (or build machines and get them to perform) actual, physically embodied computations?

There is, indeed, an obvious answer to this question:

To gain information (which, therefore, we did not previously have).

But — how is this possible?[1] Two problems seems to arise, one stemming from physics, and one from logic.

Problem 1: Doesn't this contradict the second law of thermodynamics?

Problem 2: Isn't the output *implied* by the input?

We shall discuss each of these in turn.

[1] Indeed, I was once challenged on this point by an eminent physicist (now knighted), who demanded to know how I could speak of information increasing in computation when Shannon Information theory tells us that it cannot! My failure to answer this point very convincingly at the time led me to continue to ponder the issue, and eventually gave rise to this discussion.

Problem 1

The problem is that, presumably, information is conserved in the *total* system. The natural response is that, nevertheless, there *can* be information flow between, and information increase in, *subsystems*; just as a body can gain heat from its environment. More precisely, while the entropy of an isolated (total) system cannot decrease, a sub-system *can* decrease its entropy by consuming energy from its environment.

Thus if we wish to speak of information flow and increase, this must be done relative to subsystems. Indeed, the fundamental objects of study should be *open systems*, whose behaviour must be understood in relation to an external environment. Subsystems which can observe incoming information from their environment, and act to send information to their environment, have the capabilities of *agents*.

Moral: Agents and their interactions are intrinsic to the study of information flow and increase in computation. The classical theories of information do not reflect this adequately.

Observer-dependence of information increase? Yorick Wilks (personal communication) has suggested the following additional twist. Consider an equation such as

$$3 \times 5 = 15.$$

The forward direction $3 \times 5 \rightarrow 15$ is obviously a natural direction of computation, where we perform a multiplication. But the reverse direction $15 \rightarrow 3 \times 5$ is also of interest — finding the prime factors of a number! So it seems that the *direction of possible information increase* must be understood as relative to the observer or user of the computation!

Can we in fact find an objective, observer-independent notion of information increase? This seems important to the whole issue of whether information is inherently subjective, or whether it has an objective structure.

Problem 2

The second puzzle is the computational version of what has been called the *scandal of deduction* DAgostino and Floridi (2009); Hintikka (1970); Sequoiah-Grayson (2008). The logical problem is to find the sense in which logical deduction can be informative, since, by the nature of the process, the conclusions are 'logically contained' in the premises. So what has been added by the derivation? This is a rather basic question, which it is surprisingly difficult to find a satisfactory answer to.

Computation can be modelled faithfully as deduction, whether in the sense of deducing the steps that a Turing maching takes starting from its initial configuration, or more directly via the Curry-Howard isomorphism Curry, Feys and Craig (1958); Howard (1980), under which computation can be viewed as performing cut-elimination on proofs, or normalization of proof terms. Thus the same question can be asked of computation: since the result of the computation is logically implied by the program together with the input data, what has been added by computing it?

The same issue can be formulated in terms of the logic programming paradigm, or of querying a relational database: in both cases, the result of the query is a logical consequence of the data- or knowledge-base.

It is, of course, tempting to answer in terms of the complexity of the inference process; but this seems to beg the question. We need to understand first what the inference process is doing for us!

We can also link this puzzle to another well-known issue in logic, namely the principle of *logical omnisicience* in epistemic logic, which is unrealistic yet hard to avoid. This principle can be formulated as follows:

$$[K_a\phi \wedge (\phi \rightarrow \psi)] \rightarrow K_a\psi.$$

It says that the knowledge of agent a is deductively closed: if a knows a proposition ϕ, then he knows all its logical consequences. This is patently untrue in practice, and brings us directly back to our puzzle concerning computation. We compute to gain information we did not have. We start from the information of knowing the program and its input, and the computation provides us with explicit knowledge of the output. But what does 'explicit' mean?

The computational perspective may indeed provide a usefully clarifying perspective on the issue of logical omniscience, since it provides a context in which the distinction between 'explicit' and 'implicit' knowledge can be made precise. Let us start with the notion of a function. In the 19th century, the idea of a function as a 'rule' — as given by some defining expression — was replaced by its 'set-theoretic semantics' as a set of ordered pairs. In other terminology, a particular defining expression is an *intensional description* of a function, while the set of ordered pairs which it denotes is its *extension*.

A program is exactly an intensional description of a function, with the additional property that this description can be used to explicitly calculate outputs from given inputs in a stepwise, mechanical fashion.[2] We can say that implicit knowledge, in the context of computation, is knowledge of an intensional description; while explicit knowledge, of a data item such as a number, amounts to possessing the *numeral* (in some numbering system) corresponding to that number; or more generally, to possessing a particular form of intensional description which is essentially isomorphic to the extension.

The purpose of computation in these terms is precisely to convert intensional descriptions into extensional ones, or implicit knowledge of an input-output pair into explicit knowledge. The *cost* of this process is calibrated in terms of the resources needed — the number of computation steps, the workspace which may be needed to perform these steps, etc. Thus we return to the usual, 'commonsense' view of computation. The point is that it rests on this distinction between intension and extension, or implicit vs. explicit knowledge.

Another important aspect of why we compute is *data reduction*—getting rid of a lot of the information in the input. Note that normal forms are in general *unmanagably big* Vorobyov (1997). Note also that it is *deletion of data* which creates thermodynamic cost in computation Landauer (1961). Thus we can say that much (or all?) of the actual usefulness of computation lies in getting rid of the hay-stack, leaving only the needle.

The challenge here is to build a useful theory which provides convincing and helpful answers to these questions. In our view these puzzles, naive as they are, point to some natural questions which a truly comprehensive theory of computation, incorporating a 'dynamics of information', should be able to answer.

3. What Do We Compute?

The classical notion of computability as pioneered by Turing Turing (1936) focusses on the key issue of *how* we compute; of what constitutes a computation. However, it relies on pre-existing notions from mathematics as to *what* is computed: numbers, functions, sets, etc.

This idea also served computer science well for many years: it is perfectly natural in many situations to view a computational process in terms of computing an output from an input. This computation may be deterministic, non-deterministic, random, or even quantum, but essentially the same general paradigm applies.

However, as computation has evolved to embrace diverse forms and purposes: distributed, global, mobile, interactive, multi-media, embedded, autonomous, virtual, pervasive, ... the adequacy of this view has become increasingly doubtful.

Traditionally, the *dynamics* of computing systems — their unfolding behaviour in space and time — has been a mere means to the end of computing the function which specifies the algorithmic problem which the system is solving.[3] In much of contemporary computing, the situation is

---

[2] We refer e.g. to Gandy (1980); Sieg (2002) for attempts to give a precise mathematical characterization of 'mechanical'.

[3] Insofar as the dynamics has been of interest, it has been in quantitative terms, counting the resources which the algorithmic process consumes — leading of course to the notions of algorithmic complexity. Is it too fanciful to speculate

reversed: the *purpose* of the computing system is to exhibit certain behaviour. The *implementation* of this required behaviour will seek to reduce various aspects of the specification to the solution of standard algorithmic problems.

What does the Internet compute?

Surely not a mathematical function ...

Why Does It Matter?

We shall mention two basic issues in the theory of computation which become moot in the light of this issue.

There has been an enormous amount of work on the first, namely the theory of concurrent processes. Despite this huge literature, produced over the past four decades and more, no consensus has been achieved as to what processes *are*, in terms of their essential mathematical structure. Instead, there has been a huge proliferation of different models, calculi, semantics, notions of equivalence. To make the point, we may contrast the situation with the λ-calculus, the beautiful, fundamental calculus of functions introduced by Church at the very point of emergence of the notion of computability Church (1941). Although there are many variants, there is essentially a unique, core calculus which can be presented in a few lines, and which delineates the essential ideas of functional computation. In extreme contrast, there are a huge number of process calculi, and none can be considered as definitive.

Is the notion of process too amorphous, too open to different interpretations and contexts of use, to admit a unified, fundamental theory? Or has the field not yet found its Turing? See Abramsky (2006) for an extended discussion.

The second issue follows on from the first, although it has been much less studied to date. This concerns the Church-Turing thesis of universality of the model of computation. What does this mean when we move to a broader conception of what is computed? And are there any compelling candidates? Is there a widely accepted universal model of interactive or concurrent computation?

As a corollary to the current state of our understanding of processes as described in the previous paragraphs, there is no such clear-cut notion of universality. It is important to understand what is at issue here. If we are interested in the process of computation itself, the structure of interactive behaviour, then on what basis can we judge if one such process is faithfully simulated by another? It is not of course that there are no candidate notions of this kind which have been proposed in the literature; the problem, rather, is that there are far too many of them, reflecting different intuitions, and different operational and application scenarios.

Once again, we must ask: is this embarrassing multitude of diverse and competing notions a necessary reflection of the nature of this notion, or may we hope for an incisive contribution from some future Turing which will unify and organize the field?

References

Abramsky, S., 2006. What are the Fundamental Structures of Concurrency? We still dont know! Electronic Notes in Theoretical Computer Science 162, 37–41.

Abramsky, S., 2008. Information, processes and games. In: van Benthem, J., Adriaans, P. (eds.), Handbook of the Philosophy of Information, Elsevier Science Publishers, Amsterdam, pp. 483–549.

Adriaans, P., van Emde Boas, P., 2011. Computation, information, and the arrow of time. In: Cooper, S.B., Sorbi, A. (eds.), Computability in Context, World Scientific, Singapore, pp. 1–18.

---

that the lack of an adequate structural theory of processes has been an impediment to fundamental progress in complexity theory?

Church, A., 1941. The Calculi of Lambda Conversion, Vol. 6 of Annals of Mathematics Studies, Princeton University Press, Princeton.
Curry, H.B., Feys, R. and Craig, W., 1958. Combinatory Logic: Vol. 1. North-Holland, Amsterdam.
D'Agostino, M., Floridi, L., 2009. The enduring scandal of deduction. Synthese, 167(2), 271–315.
Gandy, R., 1980. Churchs thesis and principles for mechanisms, Studies in Logic and the Foundations of Mathematics, 101, 123–148.
Hintikka, J., 1970. Information, deduction, and the a priori. Nous 4(2), 135–152.
Howard, W.A., 1980. The formulae-as-types notion of construction. In: Selden, J.P., Hindley, J.R. (eds.), To HB Curry: essays on combinatory logic, lambda calculus and formalism, Academic Press, pp. 479–490.
Landauer, R., 1961. Irreversibility and heat generation in the computing process, IBM Journal of Research and Development 5, 183–191.
Sequoiah-Grayson, S., 2008. The scandal of deduction, Journal of Philosophical Logic 37(1), 67– 94.
Sieg, W., 2002. Calculations by man and machine: Mathematical presentation. In: Gärdenfors, P., Wolenski, J., and Kijania-Placek, K. (eds.): In the scope of logic, methodology, and philosophy of science: volume two of the 11th International Congress of Logic, Methodology and Philosophy of Science, Cracow, August 1999, Kluwer Academic Publishers, Dordrecht, Boston, London, p. 247.
Turing, A.M., 1936. On computable numbers, Proc. of the Lond. Math. Soc. 2(42), 230–65, 1936.
Vorobyov, S.G., 1997. The "hardest" natural decidable theory. In: Proceedings of the 12th Symposium on Logic in Computer Science, IEEE Press, 294–305.

Paul Vitányi illustrates the importance of —

TURING MACHINES AND UNDERSTANDING COMPUTATIONAL COMPLEXITY

1. Introduction

A *Turing machine* refers to a hypothetical machine proposed by Alan M. Turing (1912–54) in 1936 (Turing, 1936) whose computations are intended to give an operational and formal definition of the intuitive notion of computability in the discrete domain. It is a digital device and sufficiently simple to be amenable to theoretical analysis and sufficiently powerful to embrace everything in the discrete domain that is intuitively computable. As if that were not enough, in the theory of computation many major complexity classes can be easily characterised by an appropriately restricted Turing machine; notably, the important classes P and NP and consequently the major question whether P equals NP.

Turing gave a brilliant demonstration that everything that can be reasonably said to be computed by a human computer using a fixed procedure can be computed by such a machine. As Turing claimed, any process that can be naturally called an effective procedure is realised by a Turing machine. This is known as Turing's thesis. Enter Alonzo Church (1903–95). Over the years, all serious attempts to give precise yet intuitively satisfactory definitions of a notion of effective procedure (what Church called effectively calculable function) in the widest possible sense have turned out to be equivalent – to define essentially the same class of processes. The Church–Turing thesis states that a function on the positive integers is effectively calculable if and only if it is computable. An informal accumulation of the tradition in S. C. Kleene (1952) has transformed it to the Computability thesis: there is an objective notion of effective computability independent of a particular formalisation. The informal arguments Turing sets forth in his 1936 paper are as lucid and convincing now as they were then. To us it seems that it is the best introduction to the subject. It gives the intuitions that lead up to the formal definition, and is in a certain sense a prerequisite of what follows. The reader can find this introduction in Turing (1936) included in this volume. It begins with:

> "All arguments are bound to be, fundamentally, appeals to intuition, and for that reason rather unsatisfactory mathematically. The real question at issue is: 'what are the possible processes which can be carried out in computing (a number)?' The arguments which I shall use are of three kinds.
> (a) A direct appeal to intuition.
> (b) A proof of equivalence of two definitions (in case the new definition has a greater intuitive appeal).
> (c) Giving examples of large classes of numbers which are computable."

2. Formal definition of the Turing machine

We formalise Turing's description as follows: A Turing machine consists of a finite program, called the finite control, capable of manipulating a linear list of cells, called the tape, using one access pointer, called the head. We refer to the two directions on the tape as *right* and *left*. The finite control can be in any one of a finite set of states Q, and each tape cell can contain a 0, a 1, or a *blank* B. Time is discrete and the time instants are ordered $0, 1, 2, \ldots$, with 0 the time at which the machine starts its computation. At any time, the head is positioned over a particular cell, which it is said to *scan*. At time 0 the head is situated on a distinguished cell on the tape called the *start cell*, and the finite control is in a distinguished state q_0. At time 0 all cells contain Bs, except for a contiguous finite sequence of cells, extending from the start cell to the right, which contain 0's and 1's. This binary sequence is called the *input*. The device can perform the following basic operations:

(1) It can write an element from $A = \{0, 1, B\}$ in the cell it scans; and

(2) it can shift the head one cell left or right.

When the device is active it executes these operations at the rate of one operation per time unit (a *step*). At the conclusion of each step, the finite control takes on a state from Q. The device is constructed so that it behaves according to a finite list of rules. These rules determine, from the current state of the finite control and the symbol contained in the cell under scan, the operation to be performed next and the state to enter at the end of the next operation execution.

The rules have format (p, s, a, q): p is the current state of the finite control; s is the symbol under scan; a is the next operation to be executed of type (1) or (2) designated in the obvious sense by an element from $S = \{0, 1, B, L, R\}$; and q is the state of the finite control to be entered at the end of this step.

For now, we assume that there are no two distinct quadruples that have their first two elements identical, the device is deterministic. Not every possible combination of the first two elements has to be in the set; in this way we permit the device to perform 'no' operation. In this case we say that the device halts. Hence, we can define a Turing machine by a mapping from a finite subset of $Q \times A$ into $S \times Q$. Given a Turing machine and an input, the Turing machine carries out a uniquely determined succession of operations, which may or may not terminate in a finite number of steps.

Strings and natural numbers are occasionally identified according to the pairing

$$(\epsilon, 0), (0, 1), (1, 2), (00, 3), (01, 4), (10, 5), (11, 6), \ldots, \qquad (2.1)$$

where ϵ denotes the empty string (with no bits). In the following, we need the notion of a *self-delimiting* code of a binary string. If $x = x_1 \ldots x_n$ is a string of n bits, then its self-delimiting code is $\bar{x} = 1^n 0x$. Clearly, the length $|\bar{x}| = 2|x| + 1$. Encoding a binary string self-delimitingly enables a machine to determine where the string ends reading it from left to right in a single pass and without reading past the last bit of the code.

2.1. *Computable functions*

We can associate a partial function with each Turing machine in the following way: The input to the Turing machine is presented as an n-tuple $(x_1,\ldots,x_n)$ consisting of self-delimiting versions of the x_i's. The integer represented by the maximal binary string (bordered by blanks) of which some bit is scanned, or 0 if a blank is scanned, by the time the machine halts, is called the *output* of the computation. Under this convention for inputs and outputs, each Turing machine defines a partial function from n-tuples of integers onto the integers, $n \geq 1$. We call such a function partial computable. If the Turing machine halts for all inputs, then the function computed is defined for all arguments and we call it total computable. (Instead of *computable* the more ambiguous *recursive* has also been used.) We call a function with range $\{0,1\}$ a *predicate*, with the interpretation that the predicate of an n-tuple of values is *true* if the corresponding function assumes value 1 for that n-tuple of values for its arguments and is *false* or *undefined* otherwise. Hence, we can talk about *partial (total) computable predicates.*

2.2. *Examples of computable functions*

Consider x as a binary string. It is easy to see that the functions $|x|, f(x) = \bar{x}$, $g(\bar{x}y) = x$, and $h(\bar{x}y) = y$ are partial computable. Functions g and h are not total since the value for input 1111 is not defined. The function $g'(\bar{x}y)$ defined as 1 if $x = y$ and as 0 if $x \neq y$ is a computable predicate. Consider x as an integer. The following functions are basic n-place total computable functions: the *successor* function $\gamma^{(1)}(x) = x + 1$, the *zero* function $\zeta^{(n)}(x_1,\ldots,x_n) = 0$, and the *projection* function $\pi_m^{(n)}(x_1,\ldots,x_n) = x_m (1 \leq m \leq n)$. The function $\langle x,y\rangle = \bar{x}y$ is a total computable one-to-one mapping from pairs of natural numbers into the natural numbers. We can easily extend this scheme to obtain a total computable one-to-one mapping from k-tuples of integers into the integers, for each fixed k. Define $\langle n_1,n_2,\ldots,n_k\rangle = \langle n_1,\langle n_2,\ldots,n_k\rangle\rangle$. Another total recursive one-to-one mapping from k-tuples of integers into the integers is $\langle n_1,n_2,\ldots,n_k\rangle = \bar{n}_1 \ldots \bar{n}_{k-1}\bar{n}_k$.

3. Computability thesis and the universal Turing machine

The class of algorithmically computable numerical functions (in the intuitive sense) coincides with the class of partial computable functions. Originally intended as a proposal to henceforth supply intuitive terms such as 'computable' and 'effective procedure' with a precise meaning, the Computability thesis has come into use as shorthand for a claim that from a given description of a procedure in terms of an informal set of instructions we can derive a formal one in terms of Turing machines.

It is possible to give an effective (computable) one-to-one pairing between natural numbers and Turing machines. This is called an *effective enumeration.* One way to do this is to encode the table of rules of each Turing machine in binary, in a canonical way.

The only thing we have to do for every Turing machine is to encode the defining mapping $T: Q \times A \rightarrow S \times Q$. Giving each element of $Q \bigcup S$ a unique binary code requires s bits for each such element, with $s = \lceil \log(|Q| + 5) \rceil$. Denote the encoding function by e. Then the quadruple $(p, 0, B, q)$ is encoded as $e(p)e(0)e(B)e(q)$. If the number of rules is r, then $r \leq 3|Q|$. We agree to consider the state of the first rule as the start state. The entire list of quadruples,

$$T = (p_1,t_1,s_1,q_1),(p_2,t_2,s_2,q_2),\ldots,(p_r,t_r,s_r,q_r),$$

is encoded as

$$E(T) = \bar{s}\bar{r}e(p_1)e(t_1)e(s_1)e(q_1)\ldots e(p_r)e(t_r)e(s_r)e(q_r).$$

Note that $l(E(T)) \leq 4rs + 2\log rs + 4$. (Moreover, E is self-delimiting, which is convenient in situations in which we want to recognise the substring $E(T)$ as prefix of a larger string.)

We order the resulting binary strings lexicographically (according to increasing length). We assign an index, or Gödel number, $n(T)$ to each Turing machine T by defining $n(T) = i$ if $E(T)$ is the i-th element in the lexicographic order of Turing machine codes. This yields a sequence of Turing machines $T_1, T_2, \ldots$ that constitutes the effective enumeration. One can construct a Turing machine to decide whether a given binary string x encodes a Turing machine, by checking whether it can be decoded according to the scheme above, that the tuple elements belong to $Q \times A \times S \times Q$, followed by a check whether any two different rules start with the same two elements. This observation enables us to construct *universal* Turing machines.

A universal Turing machine U is a Turing machine that can imitate the behaviour of any other Turing machine T. It is a fundamental result that such machines exist and can be constructed effectively. Only a suitable description of T's finite program and input needs to be entered on U's tape initially. To execute the consecutive actions that T would perform on its own tape, U uses T's description to simulate T's actions on a representation of T's tape contents. Such a machine U is also called *computation universal*. In fact, there are infinitely many such U's.

We focus on a universal Turing machine U that uses the encoding above. It is not difficult, but tedious, to define a Turing machine in quadruple format that expects inputs of the format $E(T)p$ and is undefined for inputs not of that form. The machine U starts to execute the successive operations of T using p as input and the description $E(T)$ of T it has found so that $U(E(T)p) = T(p)$ for every T and p. We omit the explicit construction of U.

For the contemporary reader there should be nothing mysterious in the concept of a general-purpose computer which can perform any computation when supplied with an appropriate program. The surprising thing is that a general-purpose computer can be very simple: M. Minsky (1967) has shown that four tape symbols and seven states suffice easily in the above scheme. This machine can be changed to, in the sense of being simulated by, our format using tape symbols $\{0, 1, B\}$ at the cost of an increase in the number of states. The last reference contains an excellent discussion of Turing machines, their computations and related machines. The effective enumeration of Turing machines $T_1, T_2, \ldots$ determines an effective enumeration of partial computable functions $\varphi_1, \varphi_2, \ldots$ such that φ_i is the function computed by T_i, for all i. It is important to distinguish between a function ψ and a name for ψ. A name for ψ can be an algorithm that computes ψ, in the form of a Turing machine T. It can also be a natural number i such that ψ equals φ_i in the above list. We call i an index for ψ. Thus, each partial computable ψ occurs many times in the given effective enumeration, that is, it has many indices.

4. Undecidability of the halting problem

Turing's paper (Turing, 1936), and more so Kurt Gödel's paper (Gödel, 1931), where such a result first appeared, are celebrated for showing that certain well-defined questions in the mathematical domain cannot be settled by any effective procedure for answering questions. The following 'machine form' of this undecidability result is due to Turing and Church: 'which machine computations eventually terminate with a definite result, and which machine computations go on forever without a definite conclusion?' This is sometimes called the halting problem.

Since all machines can be simulated by the universal Turing machine U, this question cannot be decided in the case of the single machine U, or more generally for any other individual universal machine. The following theorem due to Turing (1936), formalises this discussion. Let $\varphi_1, \varphi_2, \ldots$ be the standard enumeration of partial computable functions and write $\varphi(x) < \infty$ if $\varphi(x)$ is defined and write $\varphi(x) = \infty$ otherwise. Define $K_0 = \{\langle x, y\rangle : \varphi_x(y) < \infty\}$ as the *halting set*.

THEOREM 4.1. *The halting set K_0 is not computable.*

The theorem of Turing on the incomputability of the halting set was preceded by (and was intended as an alternative way to show) the famous (first) Incompleteness Theorem of Kurt Gödel in 1931. Recall that a formal theory T consists of a set of well-formed formulas, formulas for short.

For convenience these formulas are taken to be finite binary strings. Invariably, the formulas are specified in such a way that an effective procedure exists that decides which strings are formulas and which strings are not.

The formulas are the objects of interest of the theory and constitute the meaningful statements. With each theory we associate a set of true formulas and a set of provable formulas. The set of true formulas is *true* according to some (often non-constructive) criterion of truth. The set of provable formulas is *provable* according to some (usually effective) syntactic notion of proof.

A theory T is simply any set of formulas. A theory is axiomatisable if it can be effectively enumerated. For instance, its axioms (initial formulas) can be effectively enumerated and there is an effective procedure that enumerates all proofs for formulas in T from the axioms. A theory is decidable if it is a recursive set. A theory T is consistent if not both formula x and and its negation $\neg x$ are in T. A theory T is sound if each formula x in T is true (with respect to the standard model of the natural numbers).

Hence, soundness implies consistency. A particularly important example of an axiomatisable theory is Peano arithmetic, which axiomatises the standard elementary theory of the natural numbers.

Theorem 4.2. *There is a computably enumerable set, say the set K_0 defined above, such that for every axiomatisable theory T that is sound and extends Peano arithmetic, there is a number n such that the formula '$n \notin K_0$' is true but not provable in T.*

In his original proof, Gödel uses diagonalisation to prove the incompleteness of any sufficiently rich logical theory T with a computably enumerable axiom system, such as Peano arithmetic. By his technique he exhibits for such a theory an explicit construction of an undecidable statement y that says of itself *I am unprovable in* T. The formulation in terms of computable function theory is due to A. Church and S. C. Kleene.

Turing's idea was to give a formal meaning to the notion of 'giving a proof.' Intuitively, a proof is a sort of computation where every step follows (and follows logically) from the previous one, starting from the input. To put everything as broad as possible, Turing analyses the notion of 'computation' from an 'input' to an 'output' and uses this to give an alternative proof of Gödel's theorem.

Prominent examples of uncomputable functions are the Kolmogorov complexity function and the universal algorithmic probability function. These are the fundamental notions in Li and Vitányi (2008) and, among others, Downey and Hirschfeldt (2010); Nies (2009).

5. Complexity of computations

Theoretically, every intuitively computable (effectively calculable) function is computable by a personal computer or by a Turing machine. But a computation that takes 2^n steps on an input of length n would not be regarded as practical or feasible. No computer would ever finish such a computation in the lifetime of the universe even with n merely 1000. For example, if we have 10^9 processors each taking 10^9 steps/s, then we can execute $3.1 \times 10^{25} < 2^{100}$ steps/year. Computational complexity theory tries to identify problems that are feasibly computable.

In computational complexity theory, we are often concerned with languages. A language L over a finite alphabet Σ is simply a subset of Σ^*. We say that a Turing machine accepts a language L if it outputs 1 when the input is a member of L and outputs 0 otherwise. That is, the Turing machine computes a predicate.

Let T be a Turing machine. If for every input of length n we have that T makes at most $t(n)$ moves before it halts, then we say that T runs in time $t(n)$, or has time complexity $t(n)$. If T uses at most $s(n)$ tape cells in the above computations, then we say that T uses $s(n)$ space, or has space complexity $s(n)$.

For convenience, we often give the Turing machine in Fig. 1 a few more work tapes and designate one tape as a read-only input tape. Thus, each transition rule will be of the form $(p, \bar{s}, a, q)$, where $\bar{s}$ contains the scanned symbols on all the tapes, and p, a, q are as above, except that an operation now involves moving maybe more than one head.

We sometimes also make a Turing machine non-deterministic by allowing two distinct transition rules to have identical first two components. That is, a non-deterministic Turing machine may have different alternative moves at each step. Such a machine accepts if one accepting path leads to acceptance. Turing machines are deterministic unless it is explicitly stated otherwise.

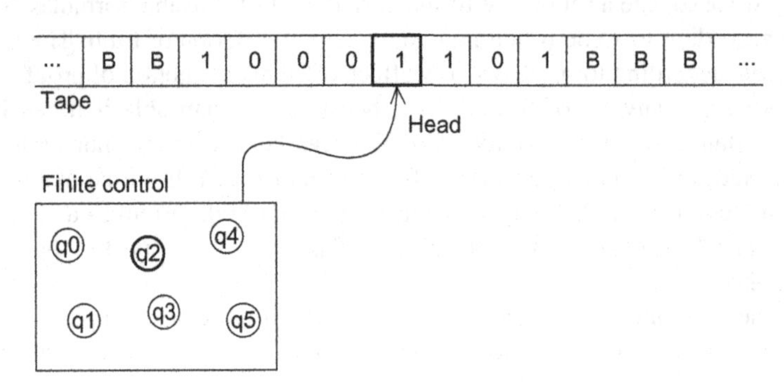

Fig. 1: Turing machine.

- DTIME[$t(n)$] is the set of languages accepted by multitape deterministic Turing machines in time $O(t(n))$;
- NTIME[$t(n)$] is the set of languages accepted by multitape non-deterministic Turing machines in time $O(t(n))$;
- DSPACE[$s(n)$] is the set of languages accepted by multitape deterministic Turing machines in $O(s(n))$ space;
- NSPACE[$s(n)$] is the set of languages accepted by multitape non-deterministic Turing machines in $O(s(n))$ space.
- With c running through the natural numbers:
  P is the complexity class $\bigcup_c \text{DTIME}[n^c]$;
  NP is the complexity class $\bigcup_c \text{NTIME}[n^c]$;
  PSPACE is the complexity class $\bigcup_c \text{DSPACE}[n^c]$.

Languages in P, that is, languages acceptable in polynomial time, are considered feasibly computable. The non-deterministic version for PSPACE turns out to be identical to PSPACE by Savitch's Theorem (Savitch, 1970), which states that $\text{NSPACE}[s(n)] = \text{DSPACE}[(s(n))^2]$. The following relationships hold trivially, $\text{P} \subseteq \text{NP} \subseteq \text{PSPACE}$. It is one of the most fundamental open questions in computer science and mathematics to prove whether either of the above inclusions is proper. Research in computational complexity theory focuses on these questions. In order to solve these problems, one can identify the hardest problems in NP or PSPACE. The Bible of this area is the works of Garey and Johnson (1979).

6. Importance of the Turing machine

In the last three quarters of a century, the Turing machine model has proven to be of priceless value for the development of the science of data processing. All theory development reaches back to this format. The model has become so dominant that new other models that are not polynomial-time reducible to Turing machines are viewed as not realistic (the so-called polynomial-time Computability thesis).

Without explaining terms, the *random access machine* (RAM) with *logarithmic cost*, or *unit cost* without multiplications, is viewed as realistic, while the unit cost RAM with multiplications or the *parallel random access machine* (PRAM) are not so viewed. New notions, such as randomised computations as in the works by Motwani and Raghavan (1995) (like the fast primality tests used in Internet cryptographical protocols) are analysed using *probabilistic* Turing machines.

In 1980 the Nobelist Richard Feynman proposed a *quantum computer*, in effect an analogous version of a quantum system. Contrary to digital computers (classical, quantum or otherwise), an analogue computer works with continuous variables and simulates the system we want to solve directly: for example, a wind tunnel with a model aircraft simulates the aeroflow and in particular non-laminar turbulence of the aimed-for actual aircraft. In practice, analogue computers have worked only for special problems. In contrast, the digital computer, where everything is expressed in bits, has proven to be universally applicable. Feynman's innovative idea was without issue until D. Deutsch (1985) put the proposal in the form of a quantum Turing machine, that is, a digital quantum computer. This digital development exploded the area both theoretically and applied to the great area of *quantum computing*.

References

Deutsch, D., 1985. Quantum theory, the Church–Turing principle and the universal quantum computer. Proc. Royal Soc. Lond. A. 400, 97–117.

Downey, R.G., Hirschfeldt, D., 2010. Algorithmic Randomness and Complexity (Theory and Applications of Computability), Springer, New York.

Garey, M.R., Johnson, D.S., 1979. Computers and Intractability: A Guide to the Theory of NP-Completeness, W.H. Freeman, New York.

Gödel, K., 1931. Über formal unentscheidbare Stze der Principia Mathematica und verwandter Systeme, I. Monatshefte für Mathematik und Physik, 38, 173–198.

Kleene, S.C., 1952. Introduction to Metamathematics, Van Nostrand, New York.

Li, M., Vitányi, P.M.B., 2008. An Introduction to Kolmogorov Complexity and Its Applications, third ed. Springer, New York.

Minsky, M., 1967. Computation: Finite and Infinite Machines, Prentice-Hall, Inc., Englewood Cliffs, NJ.

Motwani, R., Raghavan, P., 1995. Randomised Algorithms, Cambridge Univ. Press, Cambridge, UK.

Nies, A., 2009. Computability and Randomness, Oxford Univ. Press, USA.

Savitch, W.J., 1970. Relationships between nondeterministic and deterministic tape complexities. J. Comput. Syst. Sci. 4 (2), 177–192.

Turing, A.M., 1936. On computable numbers, with an application to the entscheidungsproblem. Proc. Lond. Math. Soc. 2 (42), 230–265. "Correction", 43(1937), 544–546.

Gregory Chaitin traces the path —

FROM THE HALTING PROBLEM TO THE HALTING PROBABILITY

In this remarkable paper, Turing starts by observing that most real numbers are uncomputable – indeed this is the case with probability one. Furthermore he observes that incompleteness is a corollary of uncomputability, because if it is always possible to prove whether or not something is the case using a fixed formal axiomatic theory, then there is in principle a mechanical procedure for searching through all possible proofs and mechanically deciding the answer.

This makes incompleteness much more natural and fundamental than the assertion 'I am unprovable!' that is true if and only if it is unprovable, that was constructed by Gödel. So, following Turing 1936, we have a much bigger problem than following Gödel (1931).

On the one hand he taketh away, on the other he giveth, for although Turing shows that formal languages for mathematical reasoning are necessarily incomplete, he also shows that formal programming languages can have a kind of completeness called *universality.* No formal language can express all possible proofs, but programming languages are commonly universal, that is to say, capable of expressing essentially any algorithm.

In our *Search for the Perfect Language* (to echo Umberto Eco), let us now consider how expressive different programming languages can be. Given a particular programming language, two important things to consider are the complexity $H(x)$, namely the size in bits of the smallest program to calculate x as a function of x, and the corresponding probability $P(x)$ that a program whose bits are chosen by using independent tosses of a fair coin will compute x.

We are thus led to select a subset of the Turing universal languages that minimise H and maximise P; one way to define such a language is to consider a universal computer U that runs self-delimiting binary computer programs $\pi_C p$ defined as follows:

$$U(\pi_C p) = C(p).$$

In other words, the result of running $\pi_C p$ on U is the same as the result of running p on C.

Any two such universal languages U and V will necessarily have

$$|H_U(x) - H_V(x)| < c$$

and

$$P_U(x) > P_V(x) \times 2^{-c}, \qquad P_V(x) > P_U(x) \times 2^{-c}.$$

It is in this precise sense that such a universal U minimises H and maximises P.

Using such a U we can define the halting probability Ω, for example, as follows:

$$\Omega = \sum P(n)$$

over all positive integers n, or alternatively

$$\Omega' = \sum 2^{-H(n)},$$

which has a slightly different numerical value but essentially the same paradoxical properties.

What are these properties? Ω is a form of concentrated mathematical creativity, or, alternatively, a particularly economical Turing oracle for the halting problem, because knowing N bits of the dyadic expansion of Ω enables one to solve the halting problem for all programs that compute a positive integer that are up to N bits in size. It follows that the bits of the dyadic expansion of Ω are irreducible mathematical information; they cannot be compressed into a theory smaller than they are.

More precisely, it takes a formal theory of complexity $\geq N - c$ (one requiring a $\geq N - c$ bit program to enumerate all its theorems) to enable us to determine N bits of Ω. From this it follows that Ω is Borel normal, so that Ω is a particularly natural example of a normal number. In 1933, Turing's friend David Champernowne found a natural example of a number normal for blocks of all size in base-ten; Ω provably has this property in any base.[1]

From a philosophical point of view, however, the most striking thing about Ω is that it provides a perfect simulation in pure mathematics, where all truths are necessary truths, of contingent, accidental truths – i.e., of truths such as historical facts or biological frozen accidents.

[1] Andrew Hodges conjectures that Turing's work on normal numbers helped Turing to formulate the notion of a computable real; see Hodges' review of Copeland's *The Essential Turing* in the November 2006 *AMS Notices* and Turing's *A Note on Normal Numbers* in Part II.

Indeed, I have just recently come to understand that the most important property of Ω is that it opens a door for us from mathematics to biology. The halting probability Ω contains infinite irreducible complexity and in a sense shows that pure mathematics is even more biological then biology itself, which merely contains extremely large finite complexity. For each bit of the dyadic expansion of Ω is one bit of independent, irreducible mathematical information, while the human genome is merely 3×10^9 bases $= 6 \times 10^9$ bits of information.

Robert Irving Soare expands on —

TURING AND THE ART OF CLASSICAL COMPUTABILITY

1. Mathematics as an art

Mathematics is an art as well as a science. It is an art in the sense of a *skill* as in Donald Knuth's series, *The Art of Computer Programming,* but it is also an art in the sense of an *esthetic endeavor* with inherent *beauty*, which is recognised by all mathematicians.

One of the world's leading art treasures is Michelangelo's statue of *David* as a young man displayed in the Accademia Gallery in Florence. There is a long aisle to approach the statue of David. The aisle is flanked by the statues of Michelangelo's unfinished slaves struggling as if to emerge from the blocks of marble. These figures reveal Michelangelo's work process. There are practically no details, and yet they possess a weight and power beyond their physical proportions. Michelangelo thought of himself not as carving a statue but as seeing clearly the figure within the marble and then chipping away the marble to release it. The unfinished slaves are perhaps a more revealing example of this talent than the finished statue of David.

Similarly, it was Alan Turing (1936) and (1939) who saw the figure of computability in the marble more clearly than anyone else. Finding a formal definition for effectively calculable functions was the first step, but *demonstrating* that it captured effective calculability was as much an artistic achievement as a mathematical one.

2. Defining the effectively calculable functions

The *Entscheidungsproblem,* the decision problem for first order logic, was described in the works of Hilbert and Ackermann (1928). To show this problem unsolvable one first had to mathematically define the effectively calculable functions. From 1931 to 1934, Church and his student Kleene developed the λ-definable functions. Church privately proposed to Gödel in 1934 that λ-definable functions should be identified with the effectively calculable functions. Gödel rejected this as 'thoroughly unsatisfactory.'

Gödel (1934) defined the *Herbrand–Gödel (HG) recursive functions*, a class of functions as a deductive formal system with initial functions and with two rules of inference to derive new functions. Church (1936) proposed *Church's Thesis* that a function is effectively calculable if and only if it is Herbrand–Gödel recursive. Gödel still did not accept it. Kleene (1936) then defined the *μ-recursive functions* by combining the (Gödel) numbering of syntax in Gödel's Incompleteness Theorem (1931) with the HG recursive functions. This definition is mathematically correct and prevailed for several decades in research papers from 1935 to 1965, but it is not intuitive, being based on two unintuitive formalisms. By 1936 Gödel knew these definitions, and their formal mathematical equivalence but he did not accept any of them. Indeed, Gödel suggested that it might not be possible to give a mathematical definition of calculability, and he wrote in footnote 3 of (1934) '*... the notion of finite computation is not defined, but serves as a heuristic principle'*.

Turing (1936) brought a new vision of human computability. Turing's remarkable achievement consisted of several parts that we sketch only briefly because they are very well-known. Turing: (1) defined an automatic machine (*a*-machine) based on his model of how a human being might carry out a calculation; (2) defined a universal Turing machine whose inputs included both programs and integers and could simulate any Turing machine on any input; (3) gave an extraordinary demonstration that any function calculated by a human being could be computed by an *a*-machine. Turing then stated what was later known as *Turing's Thesis* that a function on the integers is computable by a finite procedure if and only if it is computable by a Turing machine.

First, Turing gave a model based on a mechanistic approach to human computing, something the previous models lacked. Perhaps, even more impressive was Turing's careful analysis in component parts of how a human being might calculate and then an argument why his Turing machine could simulate this calculation. By comparison, Church (1936) tried to carry out a similar argument that any calculable function is HG recursive, but Gandy (1988, p. 79) and Sieg (1994, pp. 80, 87) pointed out the flaws in Church's argument. Gödel never accepted Church's Thesis, but he accepted Turing's Thesis at once, and stated:

'That this is really the correct definition of mechanical computability was established beyond any doubt by Turing'. *Gödel Collected Works Volume III, 1995, Section 3.3*:

'But I was completely convinced only by Turing's paper'. *Gödel: letter to Kreisel of May 1, 1968 (Sieg, 1994, p. 88).*

'The greatest improvement was made possible through the precise definition of the concept of finite procedure, ... This concept, ... is equivalent to the concept of a 'computable function of integers' *Gödel (1951, pp. 304–305), Gibbs lecture.*

Kleene (1981b, p. 49) wrote, 'Turing's computability is intrinsically persuasive' but 'λ-definability is not intrinsically persuasive' and 'general recursiveness scarcely so (its author Gödel being at that time not at all persuaded).' Kleene wrote in his second book (1967, p. 233), "Turing's machine concept arises from a direct effort to analyze computation procedures as we know them intuitively into elementary operations. Turing argued that repetitions of his elementary operations would suffice for any possible computation. For this reason, Turing computability suggests the thesis more immediately than the other equivalent notions and so we choose it for our exposition."

Church, in his review (1937) of Turing (1936) wrote, Computability by a Turing machine, 'has the advantage of making the identification with effectiveness in the ordinary (not explicitly defined) sense evident immediately – i.e., without the necessity of proving preliminary theorems'.

3. Why Turing and not Church?

Why give so much credit to Turing and not to Church? In 1923–27 Church was explaining the Hilbert papers to his Princeton thesis adviser, Oswald Veblen. Turing heard of them only a decade later in the Cambridge seminar of M.H.A Newman. In 1936 when Church proposed Church's Thesis, he was a full professor at Princeton in 1936 when Turing was a mere graduate student. Church used the model of the Herbrand–Gödel recursive functions, defined by Gödel, the most eminent logician at that time. They used the concept of recursion (induction) that had appeared in mathematics since Dedekind (1888).

Turing machines were a fanciful new invention without such a well-known, mathematical foundation as recursion. By 1934 Church and Kleene had shown that most number theoretic functions were λ-definable and therefore recursive, giving clear evidence for Church's Thesis. Church was the first to propose Church's Thesis first in 1934 for the λ-definable functions and then in 1935–36 for the Herbrand–Gödel recursive functions, even though Gödel did not believe it. Church got it right and he got it first. The effectively calculable functions *are* the recursive functions.

If this had been the solution to a purely mathematical problem in number theory, Church would have received at least half the credit and Turing would have been credited with a later but independent solution to the problem. By any purely quantifiable evaluation Church's contribution was at least as important as Turing's. Gödel's Incompleteness Theorem (1931) and his proof (1940) of the consistency of CH and AC were purely mathematical problems not requiring one to make mathematically precise an informal concept like calculability. However, characterising human computability was *not* a purely quantifiable process. Gödel (1946, p. 84) wrote, 'one [Turing] has for the first time succeeded in giving an absolute definition of an interesting epistemological notion, i.e., one not depending on the formalism chosen'.

4. Why Michelangelo and not Donatello?

Donatello (1386–1466) was a sculptor in Florence. In 1430 he created the bronze statue of *David,* his most famous work. This was a remarkable work, innovative in many ways, the first free-standing nude statue since ancient times, the first major work of Renaissance sculpture. Now compare this to Michelangelo's *David,* in 1504, the most famous statue in the world. Michelangelo broke away from the traditional way of representing David, with sword in hand and with the giant's head at his feet (as with Donatello). Michelangelo has caught David tense with increasing power as he is about to go to battle. Michelangelo places him in perfect contraposto outdoing the Greek representations of heros.

Michelangelo and Turing both completely transcended conventional approaches. They created something *completely new* from their own visions, something that went far beyond the achievements of their contemporaries. Second, both emphasised the *human form.* Michelangelo brought out the human form in his statues and the Sistine ceiling. Turing invented a system that simulates how a human being computes and then demonstrated that his creation did capture human computing.

Frank Zölner wrote in his book *Michelangelo Life and Work* (2010, p. 7)

> "As innovative as Leonardo da Vinci, who was a generation older, as productive as his slightly younger contemporary Raphael of Urbino, as secretive as Giorgione in Venice and blessed like Titian with a long life and unbridled creativity, Michelangelo Buonarroti embodies, perhaps most completely, the concept of the artist in the modern era."

Likewise, Alan Turing embodies, perhaps most completely, the concept of human computability in the modern era. Regarding the creative process, Turing 1939, Sec. 11 wrote,

> "Mathematical reasoning may be regarded rather schematically as the exercise of a combination of two faculties, which we may call *intuition* and *ingenuity.* The activity of the intuition consists in making spontaneous judgments which are not the result of conscious trains of reasoning. These judgments are often but by no means invariably correct. . . .
> The exercise of ingenuity in mathematics consists in aiding the intuition through suitable arrangements of propositions, and perhaps geometrical figures or drawings."

Turing's first great contribution was by intuition. Although others were studying deductive systems like HG recursion, Turing's intuition drew him to a completely new model more clearly reflecting in mechanical terms how a calculation is carried out. His second contribution was his exercise of ingenuity which led him to a demonstration that anything computed by a human being could be computed by a Turing machine. Gandy (1988, p. 82) observed, 'Turing's analysis does much more than provide an argument for' Turing's Thesis, '*it proves a theorem.*' Furthermore, as Gandy (1988, pp. 83–84) pointed out, 'Turing's analysis makes no reference whatsoever to calculating machines. Turing machines appear as a result, a codification, of his analysis of calculations by humans.' Sieg (1994, 2006, 2009), gives a full analysis of Turing's contribution. Wittgenstein remarked about Turing machines, 'These machines are *humans* who calculate'.

5. Classical art and classical computability

The term 'recursive' to mean 'computable' began with Church (1936) and was developed by Kleene to prevail for 60 years in the form, 'recursive function,' 'recursively enumerable set' and 'recursive function theory'. In June 1979, Gerald Sacks and I had each been invited to give a series of four lectures at the Italian Mathematical Summer Center (C.I.M.E.). Gerald was to lecture on *generalised recursion theory (GRT)* and I on recursion theory on the integers ω, at that time sometimes called *ordinary recursion theory (ORT).* I began the trip with a few days in Florence revisiting the Renaissance art treasures of the Uffizi gallery and Michelangelo's statues in the Accademia. As the train made its way to Bressanone at the very north of Italy in the Dolomite alps, I was still basking in the memories of the art of the Renaissance.

As our courses began in Bressanone, Gerald kept repeating the term 'ordinary recursion theory (ORT)'. He was doing nothing wrong, simply using the term as it had come to be used in the previous decade to distinguish ω-recursion theory from GRT. And yet as the phrase kept cascading down it clashed more and more with my esthetic sense. It seemed far too impoverished to describe the magnificent theory created by Turing(1936, 1939), Post (1944) and the others. My colleagues at the University of Chicago, Alberto Calderone and Antoni Zygmund, worked in singular integrals and classical analysis, but classical analysis was never called 'ordinary analysis' to distinguish it from functional analysis. No one ever called the art of the Renaissance 'ordinary art' to distinguish it from Baroque art or Impressionism.

By my third lecture it all came together. I coined the term *'classical recursion theory (CRT)'* and developed a whole lecture about the analogies between CRT and the classical art of the Italian High Renaissance. The lectures and art analogies were published in the works done by Soare (1981), but it is a rather obscure reference and not widely read. The lectures were expanded at the Cornell AMS meeting in July 1982, but not published there. Some of the analogous characteristics are these.

5.1. *Human scale*

A Roman arch such as the Arch of Constantine next to the Colosseum in Rome is designed to arouse awe and to dwarf the human figure. In contrast the Loggia della Signoria in Florence is on a human scale and designed to display statues of human size. The art and sculpture of the High Renaissance were designed to display the human form. Analogously, the computability theory of Turing and Post works on the integers, which can be represented as in Turing by a finite sequence of ones and blanks. GRT works on infinite ordinals or on functionals of higher type.

5.2. *Composition and balance*

The paintings of the Renaissance were characterised by highly complex compositions which were balanced to keep the eye from leaving the painting. Leonardo's *The Virgin and Saint Anne* has a very complicated and carefully designed composition around the three figures, Mary, her mother, Anne and her son Jesus. The heads and feet form one large triangle. The arms and child form an inner rectangle. Everything holds the eye and prevents it from leaving the painting as it might in a Baroque painting. In classical computability, theorems such as the Friedberg Muchnik theorem are proved by a delicate balance of opposing requirements, positive requirements wanting to enumerate elements into a set A and negative requirements wanting to keep elements out. These constructions are often defined with as much intricacy and balance as a Renaissance composition. Other characteristics will be developed in later papers.

6. Computability and recursion

Church (1936) and Kleene began to use the term 'recursive' to mean 'computable' as well as 'inductively defined'. Since Kleene was the main figure in the subject after 1940, this term had became standard for 60 years from 1936 to 1996. By the 1990s this usage had become problematic. When one referred to a recursive function did one mean 'inductively defined' or 'effectively computable?' Also, the term 'recursive' was not well understood in the mathematical and scientific community and, if understood at all, it was identified with the elementary methods of iteration and recursion in a first programming course. Neither Turing nor Gödel ever used the word 'recursive' to mean 'computable' or 'recursive function theory' to name the subject. When others did, Gödel reacted sharply negatively stating, 'the term in question [recursive] should be used with reference to the kind of work Rosza Peter did'.

Soare (1996, 1999a) analysed the history and meaning of computability and recursion and suggested that the terms 'Computability Theory' and 'computably enumerable set' be used in place of the recursive version. This was largely adopted within a few years.

7. The art of exposition

In the art of exposition it is not sufficient to have a correct theorem with a correct proof. It must be the right theorem with the right proof, relating the results which came before and those which will come after in an aesthetically pleasing mix. The entire work must be artistically beautiful and must appeal to the imagination.

The initial expositions in the study by Turing (1936) and Post (1944) were clear, intuitive and very well motivated. In contrast, Kleene (1936) had developed the Kleene T-predicate as a Gödel coding of the Herbrand–Kleene recursive functions, which had little appeal to the imagination. Kleene's mathematical results were very difficult but his T-predicate notation was hard to read. It dominated the proofs in the subject for over 30 years. For example, Friedberg (1957a) used the Kleene T-predicate style proofs in his solution to Post's problem and his completeness criterion (Friedberg, 1957a), which made the proofs difficult to read. Compare these proofs with the informal style of Rogers' (1967) book's written in a clear and intuitive style, which opened the subject to a generation of students and which was continued in Soare's (1987) book.

8. Conclusion

Mathematicians are not assigned projects like building bridges. Like artists, they choose which problems to work on according to taste and beauty. Like artists, what they produce is evaluated on the basis of beauty as well as mathematical results. The greatest results are those arising from a completely new vision and a profound intuition into the area.

References

Church A., 1936. An unsolvable problem of elementary number theory. Am. J. Math., 58, 345–363.

Feferman S., 2007. Turing's Thesis. Notices of the Amer. Math. Soc. 53, 1200–1206.

Friedberg, R.M., 1957. Two recursively enumerable sets of incomparable degrees of unsolvability. Proc. Natl. Acad. Sci. USA 43, 236–238.

Friedberg, R.M., 1957. A criterion for completeness of degrees of unsolvability. J. Symb. Log. 22, 159–160.
Friedberg, R.M., 1958. Three theorems on recursive enumeration: I. decomposition, II. maximal set, III. enumeration without duplication. J. Symb. Log. 23, 309–316.
Gandy, R., 1988. The confluence of ideas in 1936. Herken, 55–111.
Gödel K., 1931. Über formal unentscheidbare sätze der Principia Mathematica und verwandter systeme. I. Monatsch. Math. Phys. 38, 173–178.
Gödel K., 1934. On undecidable propositions of formal mathematical systems. Notes by S.C. Kleene and J.B. Rosser on lectures at the Institute for Advanced Study, Princeton, New Jersey, 1934, p. 30.
Gödel K., 1995. Collected works vol. III: unpublished essays and lectures, Feferman, S., et. al. (Eds.), Oxford Univ. Press, Oxford.
Kleene, S.C., 1936. General recursive functions of natural numbers. Math. Ann. 112, 727–742.
Post, E.L., 1944. Recursively enumerable sets of positive integers and their decision problems. Bull. Am. Math. Soc. 50, 284–316.
Rogers, Jr., H., 1967. Theory of Recursive Functions and Effective Computability, McGraw-Hill, New York.
Sieg, W., 1994. Mechanical procedures and mathematical experience. In: George, A. (Ed.), Mathematics and Mind, Oxford Univ. Press, Oxford, , pp. 71–117
Sieg, W., 1997. Step by recursive step: Church's analysis of effective calculability. Bull. Symb. Log. 3, 154–180.
Sieg, W., 2006. Gödel on computability. Philosophia Mathematica 14, 189–207.
Sieg, W., 2009. On computability. In: Irvine, A.D. (Ed.), Handbook of the Philosophy of Mathematics, Elsevier, pp. 535–630.
Soare, R.I., 1981. Constructions in the recursively enumerable degrees. In: Lolli, G. (Ed.), Recursion Theory and Computational Complexity, Proceedings of Centro Internazionale Matematico Estivo (C.I.M.E.), June 14–23, 1979, in Bressanone, Italy, Liguori Editore, Naples, Italy.
Soare, R.I., 1987. Recursively Enumerable Sets and Degrees: A Study of Computable Functions and Computably Generated Sets, Springer-Verlag, Heidelberg.
Soare, R. I., 1996. Computability and Recursion. Bull. Symb. Log. 2, 284–321.
Soare, R.I., 1999. The history and concept of computability. In: Griffor, E.R. (Ed.), Handbook of Computability Theory, North-Holland, Amsterdam, pp. 3–36.
Soare, R.I., 2007. Computability and Incomputability, Computation and Logic in the Real World. In: Cooper, S.B., Löwe, B., Sorbi, Andrea (Eds.), Proceedings of the Third Conference on Computability in Europe, CiE 2007, Siena, Italy, June 18–23, 2007, Lecture Notes in Computer Science, No. 4497, Springer-Verlag, Berlin, Heidelberg.
Soare, R.I., 2009. Turing oracle machines, online computing, and three displacements in computability theory. Ann. Pure Appl. Log. 160, 368–399.
Soare, R.I., 2012. Formalism and intuition in computability. In: Abramsky, S., Cooper, S.B. (Eds.), Philosophical Transactions of the Royal Society A. 370, 3277–3304.
Soare, R.I., in press a. Computability Theory and Applications: The Art of Classical Computability, Computability in Europe Series, Springer-Verlag, Heidelberg.
Soare, R.I., in press b. Turing-post relativized computability and interactive computing. In: Copeland, J., Posy, C., Shagrir, O. (Eds.), Computability: Gödel, Church, Turing, and Beyond, MIT Press.
Soare, R.I., in press c. The impact of Turing on computable reducibility and information content. In: Downey, R. (Ed.), Association for Symbolic Logic Lecture Notes in Logic (CUP).
Turing, A.M., 1936.[1] Turing, A.M., 1936. On computable numbers, with an application to the entscheidungsproblem. Proc. Lond. Math. Soc. 2 (42), 230–265. "Correction", 43 (1937), 544–546.
Turing, A.M., 1965. Systems of logic based on ordinals. Proc. Lond. Math. Soc. 45 Part 3, 161–228. Reprinted in Davis [1965], 154–222.
Zöllner, F., Thoenes, C., Taschen, B. (Eds.), 2010. Michaelangelo Life and Work, Taschen GmbH, Köln, 2010. ISBN: 978-3-8365-2117-8

[1] Many papers, Kleene (1943, p. 73; 1987a;1987b), Davis (1965, p. 72), Post (1943, p. 200) and others, mistakenly refer to this paper as 'Turing (1937)'. The journal states that Turing's manuscript was 'received on May 28 1936 and read on November 12 1936'. It appeared in two sections, the first section of pages 230–240 in volume **42**, part 3, issued on November 30 1936, and the second section of pages 241–265 in volume **42**, part 4, issued on December 23 1936. No part of Turing's paper appeared in 1937, but the two page minor correction (1937a) did.

Rainer Glaschick takes us on a trip back to —

TURING MACHINES IN MÜNSTER

In the University of Münster (Germany) existed a room with Alan Turing's name, the *Turingraum*. It contained documents and, as tangible objects, small machines made from post office relays, designed and built by Gisbert Hasenjaeger and Dieter Rödding. While the *Turingraum* does no longer exist, a number of the artefacts have been kept by the Hasenjaeger and Rödding families, and now made accessible for analysis and maybe reconstruction.

One of these artefacts is a universal Turing machine with 16 relays, which has 4 states, 2 symbols, and 3 non-erasable tapes, apparently smaller than any machine of this type known so far. Also, it needs only 15 bits to encode a program to add a mark at the end of a chain of marks on the result tape.

1. Introduction

The *Institut für mathematische Logik und Grundlagenforschung* of the University of Münster in Germany had a *Turingraum*, dedicated to Alan Turing's work and the Turing machine in particular, initiated and run from 1960 until 1985 by Gisbert Hasenjaeger and Dieter Rödding.

The Institute in Münster was the first and for a longtime leading one in the area of mathematical logic in Germany. It was established in 1936 by Heinrich Scholz, followed in 1953 by Hans Hermes (until 1966) and Dieter Rödding (until 1985). H. Scholz was one of the two persons who asked Alan Turing for a reprint of his "On computable numbers", which, including a dedication, is still in the archives of the University. H. Hermes is well known for his books on computability and logic, where he established a prominent role for Turing machines. H. Hermes was also the first one who proved under the title *Die Universalität programmgesteuerter Rechenmaschinen* (Hermes, 1954) that an idealised programmable electronic computer could be programmed to duplicate the behaviour of any Turing machine.

The *Turingraum* contained a collection of physical machines that were either Turing machines or register machines.[1] The devices in the *Turingraum* were conceived and built by Gisbert Hasenjaeger and Dieter Rödding.

Unfortunately, the *Turingraum* does no longer exist, and no specific documents could be found any more. Fortunately, Irmhild Hasenjaeger and Walburga Rödding have preserved several of those objects. Norbert Ryska[2] found this out and convinced the families to entrust us the objects for further study, which we thankfully acknowledge here.[3]

2. The object

The device selected for in-depth analysis has a central box and three peripherals, one peripheral obviously being a Turing tape, see Fig. 1. This object was shown in Oxford during 2012 and then in Paderborn in the special exhibition on Alan Turing in the Heinz Nixdorf Museum.

As E. Börger wrote in his obituary for Dieter Rödding (Börger, 1987), a universal Turing machine was built between 1958 and 1960, which is probably the machine covered here. He also

[1] They are also called counter machines.

[2] The director of the Heinz Nixdorf Museum.

[3] Special thanks to E. Börger for his continuous support.

reported from his own experience, confirmed by others, that machines of this kind were shown and operated in lectures on computability.

The machine is shown in Fig. 1. It seemed to be a fairly complete and attractive object for investigation. The detailed structure could be reconstructed now.

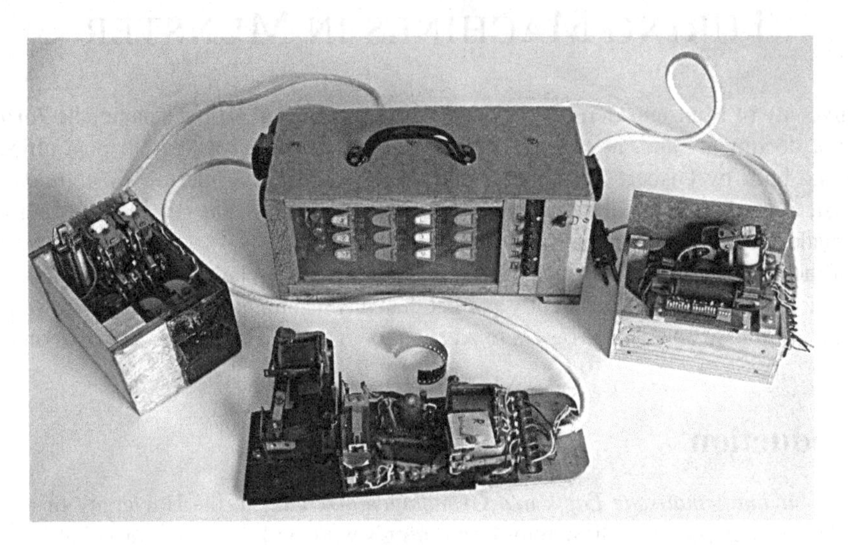

Fig. 1: Hasenjaeger's machine.

2.1. *Hardware layout*

The machine has a main cabinet, with 16 relays, 3 connectors, a switch, 5 buttons and 4 lamps. Three different kinds of 'tapes' can be plugged to the main box:

- Tape P, the program tape. It is a circular tape with 18 positions, that can be set by small switches. Using a selector switch, it can be 'moved' in only one direction.
- Tape Q, the counter tape. Two selector switches are connected back-to-back, so that a signal is generated iff both have the same position. It is thus a counter (modulo 18), that can be incremented and decremented, and zero sensed. In terms of Turing tapes, it is an immutable tape with a marker every 18th position.
- Tape R, the result tape.[4] It can be moved in both directions, but once marked, the mark cannot be erased. Used was 35-mm film, split longitudinally, so that one border has a perforation. The marks were punched as triangular notches at the opposite border. A lever sensed the marks, and was raised by a magnet automatically whenever the tape was moved or punched.

The main cabinet has a fixed logic, where 4 of the 16 relays provide a two-phase clock, and the remaining 12 relays are pairwise connected to provide 6 flipflops. Two pairs of flipflops were connected in a master-slave mode, so that during one phase of the clock the information was transferred from the master to the slave. In the other phase, the state table is evaluated using contact trees, either moving (or marking) the tapes, and sets the master flipflops for state transitions.

Thus, the machine has a two-bit state memory; the states labelled with roman numerals I, II, III and IV. Four lamps show the current state, and four of the push buttons allow to directly set a state. The other push button allows single-step advance, and the switch allows to run in continuous mode.

[4] Labelled *Rechenband*, i.e., calculating or computing tape.

The reconstructed state table is shown in A. Slightly different from what Hasenjaeger wrote (Hasenjaeger, 1987), the encoding is: `1`=*, `01`=R, `001`=L, `000n1`, where *n* stands for *n* zeroes and * for mark. `n` is the the equivalent of the conditional transfer, that Wang (1957) denoted *Cn*, but in our case it is a relative jump address, i.e., a skip of *n* instructions. As all instructions end in a mark, the state machine skips marks on tape P, if tape R is marked; otherwise the *n* zeroes just read are discarded. Because tape P is cyclic, a long-enough skip goes back, resulting in a program loop. Because all instructions except mark have zeroes, forward jumps are possible.

2.2. *Examples*

The first example uses the bit pattern found on the machine, but this is not necessarily authentic, as over the years someone may have flipped some switches just for fun.

The bit string `101101010000111100` found on the tape produces this sequence of operations:

```
* R * R R 1 **** L R * R R 1 **** ...
```

resulting in the sequence `**_*_*_*...` on tape R. Note that the first bit is interpreted only at the beginning as a mark instruction; because in the other rounds, it is the end bit of the `L` instruction. Also, the `1` is redundant on a blank tape, as the previous instruction has just reached a blank square.

Wang (1957) gives the following program to position onto the space at the end of a chain of marks: `1.* 2.R 3.C2 4.R 5.L`. The equivalent would be `* R 3 R L`, using $1+2+7+2+3=15$ bits, but staying busy because of Wang's convention using `RL` as stop or return.

Slightly modifying the machine (so that it stops if it tries to mark a marked tape), a program to append a mark at the end of a (possibly empty) chain of marks could be done with `2 * * R`, using only $6+1+1+2=10$ bits.

When I began the analysis, I did not believe it would be possible to build a UTM with 16 relays only, because according to the the very recent work of D. Woods and T. Neary (Neary, 2008; Woods and Neary, 2009), a (4,2) UTM (in this configuration) is not yet known.[5]

The compactness of this machine description clearly demonstrates the superiority of Hasenjaeger's and Rödding's concept to use Wang's programmatic method instead of the encoding of state tables.

3. The papers

Two papers have been published by G. Hasenjaeger, which contain substantial information about his and D. Rödding's work on Turing machines. These papers were published relatively late; Hasenjaeger left Münster to become professor in Bonn in 1962.

3.1. *Universal Turing machines and Jones–Matiyasevich-masking*

In the article with the above title, Hasenjaeger (1984) made several remarks that are related to our objects.

The article starts with a section *Background* as follows:

> When I learned from reports given by BÖRGER (spring 1982) and JONES (fall 1982) about a new combination of coding of sequences with coding of Boolean algebras as a tool to describe the behaviour of register machines, I tried to apply this tool on my earlier variants of small universal Turing machines hoping these application should lead to some sufficiently simple solutions for exponential diophantine predicates universal for r.e. sets.

[5] T. Neary also accounts for the efficiency of a UTM, and I think, the number of bits to encode a certain problem is important too.

This confirms that Hasenjaeger desired to make a small *universal* Turing machine.

Section 4 in his paper has the title *Universal Turing machines*, and its complete text is as follows:

> As decoding a number (introduced as an additional argument indicating a program or a particular machine) certainly needs more states, smaller solutions are obtained by introducing an additional program tape or program loop. Instead of targets for conditional jumps (or "gotos"), an additional register to count the "in between" for suspended operations seems adequate. As besides conditional halt and jump three operations on the tape are sufficient (l, r, print, or l, r, change sign) hence are to be transcribed, a similar code should also serve for a multiple counter concept: Just one counter is on duty; operations are: add 1, take 1, change of counters in a given cyclic order.

Not only does he write about a circular program tape, there is also an alternate use of tapes, like explained above for the tapes P and Q for the Hasenjaeger machine.

Note also the indication that state changes are coded as a distance to the next entry on the programme tape, accumulated to and then consumed from a counter tape, instead of absolute state numbers.

Section 5 has the title *From Turing tapes to counting registers* and starts as follows:

> As 25 years ago finding it harder to materialize a Turing tape with an operation: change symbol (instead of: print 1, not erasing) we introduced a multitape version: one 1 on each tape, and moving for counting. By changing the tape "on duty" in a cyclic order all tapes can be operated

It is not clear which machine was meant having tapes with a single mark and moving for counting, i.e., if the tapes using selector switches were included.

More important appears the idea not to use $3n$ different instructions for n counter tapes to enlarge, decrease and test, but instead 4 for enlarge, decrease, test and cyclic change. Whether this attempt really is advantageous, depends on the problem and code, because switching to a specific register may need up to $n-1$ change instructions.

3.2. *On the early history of register machines*

In 1987, G. Hasenjaeger published a short note with the above title (Hasenjaeger, 1987) that had the following footnote on the title page: *This report on my collaboration with D. Rödding is not restricted to the item on the title; but that item seems to be the most remarkable result of our joint activities.* Reading it in the context of the above device, this publication reveals important information.

The second part, with the heading *Turing Machines*, reveals a lot of information in relation to Alan Turing.[6] In the first paragraph, the Turing machines to be built are characterised as *theoretical*, i.e., not for practical use.

It is also mentioned that the people in Münster were not aware of the practical work of Alan Turing at that time, in particular the ACE.[7] In the next paragraph, Hasenjaeger mentions that Wang's article (Wang, 1957) had a high influence on the following activities to *materialise theoretical machines*.

---

[6] G. Hasenjaeger had worked for the German military and been *assigned the task of examining the security of Enigma. He detected weak points, but he could not foresee that the Allies had long since been taking advantage of these weaknesses* (Schmeh, 2009).

[7] Astonishing enough, that in 1987 Hasenjaeger mentioned the ACE, which is still unknown to many experts in this field.

Note the remark that the unreliablitiy of the tape punch lead to the search for alternatives, and, via Moore's practical proposal (Moore, 1952), to the idea of separate counting and programme tapes, and finally perhaps to Rödding's theoretical work on register machines and decomposition of automata. This clearly indicates that attempts to *materialise* theoretical constructs can lead to fruitful advances in theory.

The idea of alternatively scanning tapes, as mentioned here, has already been present in Hasenjaeger's machine, described in detail above. I originally suspected that the fact that tape Q could only be used in states II and IV, and tape R only in states I and III, was caused by a lack of relays and contacts, once the 16 relays were nicely assembled. Maybe we have the not so uncommon case that the desire to use only 16 relays lead to the idea of using tapes alternatively, which was later more extensively used.

Hasenjaeger mentioned that Rödding *already reported his results in H. HERMES' colloquium, when similar results of M. MINSKY [1961] became known.* And some sentences later: *I think we were angry enough not to go into MINSKY's details. Thus, I realised only much later that these details were quite different.* This is quite a pity, as his (4,2) UTM should have been published, not only for the small size, but also for the observation that using Wang's programme oriented encoding was also practically superior over the traditional attempt of encoding state tables.[8] E. Börger remembered (Börger, 1987) in this context that the unreliability of the first tape drive rose the desire to avoid punching, and became the abstract task to use a fixed, finite maximum of marks on a tape, leading Hasenjaeger and Rödding to the idea of using a tape with a single immutable marker as a counter.

Noting that using counters is a basic building block for Turing machines, D. Rödding followed this idea to use (infinite) counters to define computability, not only because this model made it easier to teach computability at elementary level. The final version used only two operations, increment and decrement, with just a (backwards) loop if the decremented register is not yet zero. He published a very clear and comprehensible description (Rödding, 1972) in German. Early publications on register machines are by Minsky (1961), and Shepherdson and Sturgis (1963), the latter citing the above mentioned proof of Hermes (1954) at the end of Section 1.

4. Conclusion

From the legacy of G. Hasenjaeger and D. Rödding, a small machine, made from old relays, was obtained, which was used to practically demonstrate Turing machines. Restricted to really small machines and due to difficulties in building tape drives, a machine was built with a mark-only result tape, a counter tape and a read-only program tape. Using this configuration, a (4,2) UTM was built that could encode a simple program in only 15 bits of the program tape. This was achieved not by encoding state tables, but by following H. Wang's theoretical proposal of using programs instead. The use of relative instead of absolute (preferably backward) jumps mirrored the transition to structured programming, and influenced D. Röddings final concept of register (or counter) machines.

5. State machine

The reconstructed state table is printed below. State numbers are in arabic digits, column 1 shows the state number, column 2 the conditions for the tapes P, Q and R (dot means do not care), column 3 shows the actions (dot for no action) and column 4 the next state (dot for no state change):

```
S PQR PQR S'
```

[8] In an early version of this paper, I started to argue that using state tables is more efficient, as it allows more actions to be done in parallel. While this is true for hardware logic, it is obviously is wrong for programs on Turing tapes.

```
        I: P=1 is punch, P=0 other instruction
1 1.0 +.* .   mark if not marked, next instruction 1 1.1 +.. . no
need to mark if marked, next instruction 1 0.. +.. 2   other
instruction, take the 0

        II: R, L or other; Q is zero on entry
2 10. +.R 1   go right, next instruction 2 00. ++. .   save 0 in
Q, check next P bit 2 11. +-L 1   next P bit is 1, go left, clear
Q, next inst. 2 01. +-. 3   next P bit is 0, this is a skip

        III: skip part 1: count zeroes to Q, if mark
3 0.0 +.. .   R has space, skip zeroes until P=1 3 0.1 ++. . R has
mark, count zeroes until P=1 3 1.0 +.. 1   end found; R has space:
next instruction 3 1.1 .+. 4   end found; R has mark, need to skip

        IV: skip part 2: execute
4 01. +.. .   while Q>0, skip zeroes on P, leave Q 4 11. +-. .
while Q>0, skip a one, decrement Q 4 .0. ... 1   Q=0, next
instruction
```

Remark:

Meanwhile, some of Hasenjaeger's notes and a bidirectional uniselector have been found, showing that the machine was built for a bidirectional tape P. Thus, in the state table for state 4 tape P is advanced backwards instead of forwards:

```
S PQR PQR S'
4 01. -.. . while Q>0, skip zeroes backwards on P, leave Q
4 11. --. . while Q>0, skip a one backward, decrement Q
4 .0. +.. 1 Q=0, end of backjump
```

The machine still never did jump conditionally, until today, when we have built a new bidirectional tape P from material left by Hasenjaeger.

References

Börger, E., 1987. D. Rödding: Ein Nachruf., Jber.d.dt.Math.-Verein. 89, 144–148. http://dml.math.uni-bielefeld.de/JB_DMV/JB_DMV_090_3.pdf

Hasenjaeger, G., 1984. Universal Turing machines (UTM) and Jones–Matiyasevich-Masking. In: Börger, E., Hasenjaeger, G., Rödding, D. (Eds.), Logic and Machines: Decision Problems and Complexity, vol. 171 of Lecture Notes in Computer Science, Springer Berlin/Heidelberg, 248–253. http://dx.doi.org/10.1007/3-540-13331-3_44.

Hasenjaeger, G., 1987. On the early history of register machines. In: Börger, E. (Ed.), Computation Theory and Logic, vol. 270 of Lecture Notes in Computer Science Springer Berlin/Heidelberg, 181–188. http://dx.doi.org/10.1007/3-540-18170-9_165.

Hermes, H., 1954. Die Universalität programmgesteuerter Rechenmaschinen, Mathematisch-Physikalische Semesterberichte der Universität Göttingen 4, 42–53.

Minsky, M.L., 1961. Recursive unsolvability of Post's problem of "Tag" and other topics in theory of Turing machines. Ann. Math. 74 (2), 437–455, ISSN 0003-486X.

Moore, E.F., 1952. A simplified universal Turing machine. In: Proceedings of the 1952 ACM national meeting (Toronto), ACM '52, ACM, New York, NY, USA, pp. 50–55. http://doi.acm.org/ 10.1145/800259.808993.

Neary, T., 2008. Small universal Turing machines, Ph.D. thesis, NUI, Maynooth.

Rödding, D., 1972. Registermaschinen, Der Mathematikunterricht 18, 32–41, ISSN 0025-5807.

Schmeh, K., 2009. Enigmas contemporary witness: Gisbert hasenjaeger, cryptologia 33, 343–346, ISSN 0161-1194, doi:10.1080/01611190903186003.
Shepherdson, J.C., Sturgis, H.E., 1963. Computability of recursive functions. J. ACM 10, 217–255, ISSN 0004-5411. http://doi.acm.org/10.1145/321160.321170.
Turing, A.M., 1936. On computable numbers, with an application to the Entscheidungsproblem. Proc. Lond. Math. Soc. ser. 2. 24, 230–265.
Wang, H., 1957. A variant to Turing's theory of computing machines. J. ACM 4, 63–92, ISSN 0004-5411, http://doi.acm.org/10.1145/320856.320867.
Woods, D., Neary, T. 2009. The complexity of small universal Turing machines. Theor. Comput. Sci. 410, 443–450.

From K. Vela Velupillai —

Reflections on Wittgenstein's Debates with Turing during his *Lectures on the Foundations of Mathematics*[1]

Andrew Hodges (2008) recalled Max Newman's characterisation of Alan Turing as 'at heart more an applied than a pure mathematician', and went on (p. 4; italics added):

> "It might be more true to say that Turing had resisted this Cambridge classification from the outset. He attacked every kind of problem – from *arguing with Wittgenstein*, to the characteristics of electronic components, to the petals of a daisy."

This prompts me to return to Turing's 'debates' with Wittgenstein – now remembering Max Newman's characterisation – during the latter's *Lectures on the Foundations of Mathematics* (Wittgenstein, 1939) [1976]. It is little realised – indeed, to the best of this writer's knowledge, never mentioned – that when Turing attended these lectures, in the Lent and Easter terms of 1939, he was the young (Turing was not quite 27 years old and Wittgenstein turning a *vintage* 40!) author of *Systems of logic based on ordinals* (Turing, 1939) where 'ways in which systems of logic may be associated with constructive ordinals' (couched in the language of the λ-calculus) was a main theme. Feferman, in his perceptive *Preface* to '*Systems of Logic*' (Turing, 2001, p. 79) observed, correctly in my opinion: Turing never tried to develop an over-all philosophy of mathematics...' Yet, he (Turing) was engaging one of the great philosophers of the twentieth century on *his* (Wittgenstein's) interpretation – even 'deconstruction' – of Cantor's work on *Transfinite Numbers*!

It is a pity that in these famous lectures by Wittgenstein, Turing was 'set up' as the 'strawman' representing orthodox mathematics and mathematical logic, defending the conventional notion of consistency (not related to its specialised version in the Gödel–Rosser work) in mathematics. Had

[1] I write as an economist who is painfully aware that uninformed references to this dialogue between one of the great philosophers and the pioneer of computability recur in the methodological literature of economics (cf., e.g., the muddled invoking of a particular part of this famous dialogue by McCloskey (1991, pp. 13–4)).

the protagonists been privy to the Newman–Hodges picture of Alan Turing, who 'began (and ended) with the *physical world*' (Hodges, op.cit., p. 4), the subsequent misrepresentation of Wittgenstein's stance[2] may have been prevented.

The context for the particularly (in)famous part of the Wittgenstein–Turing dialogue on consistency/contradiction in mathematics (and mathematical logic), may well be the original few remarks in Wittgenstein (op.cit, pp. 211–212, Lecture XXII; italics added):

> It was suggested last time [i.e., Lecture XXI] that the danger with a contradiction in logic or mathematics is in the application. Turing suggested that a bridge might collapse.
>
> Now it does not sound quite right to say *that a bridge might fall down because of a contradiction* [in logic or mathematics]."

Now, to place this in proper historical perspective, compare Stanislaw Ulam's dialogue with Gian-Carlo Rota on collapsing bridges and logical contradictions (Rota, 1986, p. 2; italics added):

> "However, out of curiosity I [Rota] decided to play devil's advocate, and watch his reaction.
>
> But if what you [Ulam] say is right, what becomes of objectivity, an idea that is so definitively formalized by *mathematical logic* and by the theory of sets, on which you [Ulam] yourself have worked for many years of your youth?
>
> There was a visible emotion in his [Ulam's] answer. Really? What makes you [Rota] so sure that *mathematical logic corresponds to the way we think*?[3] You are suffering from what the French call a 'deformation professionelle.' *Look at the bridge over there. It was built following logical principles. Suppose that a contradiction were to be found in a set theory. Do you honestly believe that the bridge might fall down?*
>
> Do you [Ulam] then propose that we give up mathematical logic? said I [Rota], in fake amazement.
>
> Quite the opposite [said Ulam]. *Logic formalizes only very few of the processes by which we actually think*[4]. The time has come to enrich formal logic by adding to itsome other fundamental notions[5]."

In the years before Laurent Schwartz elegantly encapsulated the *Dirac delta function*[6] with his notion of generalised functions, von Neumann had 'banished' it from 'official' use in physics and

[2] Most egregiously represented by Charles Chihara (1977), only partially blunted by Shanker's brilliant counterattack in Shanker (1987).

[3] Brouwer had been there, and Wittgenstein may have remembered it, long before them, and had remarked, in his Inaugural Lecture of 1912 (Brouwer, 1913, p. 84; italics added), most perceptively:
"To the philosopher or to the anthropologist, but *not to the mathematician*, belongs the task of investigating why certain systems of symbolic logic rather than others may be effectively projected upon nature. *Not to the mathematician*, but to the psychologist, belongs the task of explaining why we believe in certain systems of symbolic logic and not in others, in particular *why we are averse to the so-called contradictory systems* in which the negative as well as the positive of certain propositions are valid."

[4] If, at this point, Ulam had added '*and act*', he would have completely encapsulated Wittgenstein's prescription for circumventing contradictions in mathematics and logic by means of '*rules*'.

[5] The original journal article has 'motions', but the context makes it clear that what is meant is 'notions'.

[6] Dirac himself attributed the origin of the idea to his 'early engineering training' (cf., Kragh, 1990, p. 41) – surely paralleling both Wittgenstein's early training as an aeronautical engineer and Turing's above characterisation by Newman and Hodges.

quantum mechanics for being mathematically 'improper'. Meanwhile, physicists, with princely unconcern for the prestigious embargo placed on the delta function, went on happily using it for calculations. Engineers, of course, were blissfully unaware of von Neumann's prestige or embargo and went on calculating with the Heaviside operational calculus.

So far as I know, neither the Feynman Integral, nor Bishop's constructivism, have been axiomatised. This has not prevented perfectly valid calculations using Feynman integrals in quantum electrodynamics. For all we know, there are, lurking in the inner recesses of the Platonic Universe, the eventually discoverable logical foundations, which will show that the use of the Feynman integral entails contradictions. No quantum physicist in his right mind would pay the slightest attention to such logical hair-splitting (cf., also Schwartz, 2001, Chapter VI).

I am suggesting, therefore, that a sympathetic reader (an always elusive creature), should approach this famous dialogue between a great philosopher of, among other things, mathematics, and a great logician and founding father of computability theory, remembering that both of these intellectual giants were also, fundamentally, wedded to the 'physical world' – one with an explicit engineering background and the other as an 'applied mathematician', both camouflaging as logicians and mathematicians perplexed by semantic paradoxes and grammatical nuances that they thought could be sorted out by dialogue.

References

Brouwer, L.E.J., 1913. Intuitionism and Formalism, Bull. Am. Math. Soc. 20 (2), 81–96.

Chihara, C., 1977. Wittgenstein's Analysis of the Paradoxes in His Lectures on the Foundations of Mathematics, Philos. Rev. 86 (3), 365–381.

Cooper, S.B., Löwe, B., Sorbi, A., 2008. New Computational Paradigms: Changing Conceptions of What is Computable, Springer, New York.

Feferman, S., 2001. Preface to: Systems of logic based on ordinals. In: Gandy, R.O., Yates, C.E.M. (Eds.), Collected Works of A.M. Turing – Mathematical Logic, North-Holland, Amsterdam.

Hodges, A., 2008. Alan Turing, logical and physical. In: Cooper, et al., 2008, pp. 3–15.

Kragh, H. S., 1990. Dirac: A Scientific Biography, Cambridge University Press, Cambridge.

McCloskey, D. N., 1991. Economic science: a search through the hyperspace of assumptions? Methodus 3 (1) 6–16.

Rota, G.-C., 1986. In memoriam of Stan Ulam: the barrier of meaning. Physica D, 2 (1–3), 1–3.

Schwartz, L., 2001. A Mathematician Grappling with His Century, Birkhäuser Verlag, Basel.

Shanker, S. G., 1987. Wittgenstein and the Turning-Point in the Philosophy of Mathematics, State University of New York Press, Albany, NY.

Turing, A.M., 1939. Systems of logic based on ordinals, Proc. Lond. Math. Soc., Series 2, 45, 161–228.

Turing, A.M., 2001. Collected Works of A.M. Turing: Mathematical Logic, Gandy, R.O., Yates, C.E.M. (Eds.), North-Holland, Amsterdam.

Wittgenstein, L., (1939) [1976]. Wittgenstein's Lectures on the Foundations of Mathematics – Cambridge 1939. In: Diamond, C. (Ed.), The University of Chicago Press, Chicago.

Jan van Leeuwen and Jiří Wiedermann on —

The Computational Power of Turing's Non-Terminating Circular a-Machines

1. Introduction

For readers familiar with the concept of Turing machines as described in contemporary textbooks, reading the definition of a Turing machine in Turing's original paper (Turing, 1936) may present a surprise. It is not only the difference in notational style or in the vocabulary used when speaking about these machines (called *'automatic machines'*, or simply *'a-machines'*), which may be surprising. Astonishing may be the fact that properly designed a-machines never halt. a-Machines of this kind are called *circle-free* and their task is to output infinite sequences of binary digits representing computable real numbers $\in [0, 1]$. Obviously, for computing the infinite expansions of real numbers such a behaviour is perfectly desirable. Nevertheless, Turing noted that there may also be machines – so-called *circular machines* – which at some point stop producing output digits, i.e., they altogether produce only a finite number of output digits. This may happen in two different ways. Either the machine at hand reaches a configuration from which there is no possible further move, or the machine goes on moving without producing any further output digits.

The modern versions of circle-free a-machines are still being used as a formal model in so-called *computable analysis*, a field in which one studies the parts of real analysis and functional analysis that can be carried out in a computable manner (cf. Weihrauch (2000)).

The circular a-machines that terminate, i.e. that halt after performing a finite number of steps yield the basis of today's computability and complexity theory. In fact, they are the forerunner of the contemporary Turing machine model.

Non-terminating circular a-machines run forever but produce only a finite number of output symbols. It seems that no special attention has been paid to such machines. From a classical computational point of view, the machines are strange: in spite of the fact that their computation is infinite, they are doomed to produce but a finite number of outputs. Turing himself proved that the property of circularity of a-machines cannot be tested effectively by any other a-machine. What could such machines be good for?

Recently, we have investigated a new computational model of unbounded computational processes – so-called red-green Turing machines (van Leeuwen and Wiedermann, 2012).

The motivation for considering red-green Turing machines comes from the modern, typical computer applications in which the core mechanism is a multi-process system that is always up and running. Control goes from process to process and, whenever a process has its turn, the process computes until it executes an instruction that explicitly transfers control to another process. We have studied and appraised computationally the mechanism of control passing among the processes in the course of unbounded computation. The computing power of red-green Turing machines goes beyond that of classical Turing machines, reaching up to the second levels of the arithmetical hierarchy, viz. Δ_2 or even Σ_2.

We will show that red-green Turing machines can be seen as a modern variant of Turing's original circular non-terminating a-machines producing a finite number of output symbols. This connection between the two models gives a new link between Turing's 1936 paper and contemporary computing. The connection straightforwardly leads to the characterisation of the computational power of the underlying a-machines and thus reveals an unexpected super-Turing computational potential of an authentic, 'old-fashioned' machine model of Turing.

The rest of this short note is organised as follows. In Section 2 we describe both models – circular non-terminating a-machines producing a finite number of output symbols and red-green Turing machines – in more detail. Next we prove their computational equivalence. In Section 3 we discuss the significance of our result. Some conclusions are given in Section 4.

2. a-Machines and red-green Turing machines

2.1. *a-Machines*

Using the contemporary terminology of Turing machines, an a-machine can be seen as a deterministic single tape Turing machine with working alphabet Σ and input and output alphabet $A = \{0, 1\}$, with $\Sigma \cap A \neq \emptyset$. In Turing's terminology, symbols from A are called 'figures' or *symbols of the first kind*, whereas the symbols from Σ are called *symbols of the second kind.*

The input – a string from $\{0, 1\}^*$ – is written on the tape at the beginning of the computation. The computation of an a-machine now proceeds as usual, reading, writing and rewriting the symbols on the tape and moving the head in accordance with the machine's program. At each time, the sequence of symbols from A printed on the tape (as a subsequence of all symbols printed by the machine) is called *the sequence computed by the machine* (cf. Turing (1936)), or simply the result at that time.

In Turing's own words (Turing, 1936): *"If a computing machine never writes down more than a finite number of symbols of the first kind it will be called* circular. *Otherwise it is said to be* circle-free. *A machine will be circular if it reaches a configuration from which there is no possible move, or if it goes on moving, and possibly printing symbols of the second kind, but cannot print any more symbols of the first kind."*

Thus, a circular a-machine is a machine which on a given input prints a finite number of output symbols (not necessarily into different cells), and then either halts, or goes on forever, performing an infinite number of steps in which no output symbol is printed anymore.

2.2. *Red-green Turing machines*

A red-green Turing machine is formally almost identical to the classical model of Turing machines. The only difference is that in red-green Turing machines the set of states is partitioned in two disjoint subsets: the set of green states, and the set of red states, respectively. There are no halting states. A computation of a red-green Turing machine proceeds as in the classical case, changing between green and red states in accordance with the transition function. A moment of change in state color is called a *mind change.* A formal language is said to be *recognised* just in case on the inputs of this language and precisely those, the machine computations 'stabilise' in green states, i.e., from a certain time on, the machine keeps entering only green states. Similarly, a language is said to be *accepted* if and only if the inputs from the language are recognised, and the computations on the inputs not belonging to this language eventually stabilise in red states.

The model captures in a neat way a main feature of the current thinking of computing: namely, viewing computations as potentially infinite sequences of communications between processes, oscillating between different states of mind but ultimately converging on a fixed behaviour.

2.3. *Relation between a-machines and red-green Turing machines*

For our purpose – comparing red-green machines with a-machines – we will only consider deterministic single tape red-green machines over input/output alphabet A computing partial functions $f : \{0,1\}^* \to \{0,1\}^*$. The result of computation will be defined similarly as in the case of a-machines. More specifically, if for input x $f(x)$ is defined, $f(x)$ will be written in binary as a finite sequence of symbols over A, as a subsequence of all symbols printed on the tape. If $f(x)$ is undefined then $f(x)$ is represented by an infinite sequence.

Theorem 2.1. *Let $\mathcal{A}$ be a circular a-machine, let $\mathcal{R}$ be a red-green Turing machine, and let $f : \{0,1\}^* \to \{0,1\}^*$ be a partial function. Then f is computed by $\mathcal{A}$ if and only if f is computed by $\mathcal{R}$ with $f(x)$ mind changes.*

Proof. (Sketch.) We show that, if f is computed by a non-terminating circular a-machine $\mathcal{A}$, then f can be computed by a red-green machine $\mathcal{R}$ performing $f(x)$ mind changes.

On input x, machine $\mathcal{R}$ works as follows. Each time $\mathcal{A}$ prints an output symbol, $\mathcal{R}$ prints the same symbol, and changes state to red and then to green. If $\mathcal{A}$ has no move from some configuration, then $\mathcal{R}$ switches to a red state and starts cycling in that state.

If $\mathcal{A}$ prints a finite number $f(x)$ of symbols without termination, then $\mathcal{R}$ accepts x in $f(x)$ mind changes and stabilises in a green state. If $\mathcal{A}$ keeps on producing an infinite number of output symbols, then $f(x)$ is undefined and obviously, $\mathcal{R}$ keeps changing its mind infinitely often as well.

Obviously, $\mathcal{R}$ computes f and, if $f(x)$ is a finite number, then $\mathcal{R}$ computes $f(x)$ in $f(x)$ mind changes.

Conversely, let $\mathcal{R}$ be a machine accepting x with $f(x)$ mind changes. Obviously, at the occasion of each mind change $\mathcal{R}$ can add one to the current number of mind changes and the total number presents $\mathcal{R}$'s current output (we assume that output symbols can be rewritten). Thus, after $f(x)$ mind changes the output from $\mathcal{R}$ represents the value of $f(x)$. Simulation of $\mathcal{R}$ by $\mathcal{A}$ is then a straightforward matter.

If x is not accepted by $\mathcal{R}$ then $\mathcal{R}$ oscillates forever between red and green states and thus $\mathcal{A}$ produces an infinite number of outputs, effectively rejecting the input x for which $f(x)$ is undefined. □

3. Significance of the result

By equating the computations of non-terminating circular a-machines with those of red-green Turing machines, we have opened a way of appraising the computational power of the former model via the latter. Results for this model are available.

The idea of the red-green computing goes back to the very notion of computability. This connection is strengthened by the present result that links red-green computing to the authentic ideas of Turing on computing. The original concept of computability, established in the middle of the twentieth century, has been established in the period of ideas on function calculation. But computational systems nowadays have a number of features that extend beyond pure function value calculation. The concept of red-green computations specifically addresses one important feature of contemporary usage of computers, viz. unbounded computations of multi-process systems. Several models of infinite computations have been studied in the past, as a natural generalisation of the classical notion of computations, without having any particular 'realistic' computational model of infinite computations in mind. Along these lines, without entering into details, let us mention Gold's notion of limiting recursion (Gold, 1965, 1967), and Putnam's related notion of trial-and-error predicates, inductive computing (Burgin, 1983), various kinds of ω-automata (Büchi, 1962; Rabin, 1969; Staiger, 1997; Thomas, 1990), tae-computing Hintikka and Mutanen

Table 1 Approaches to Unbounded Computation.

Model of Computation	Level	Reference	year
Non-term. circ. a-machines	Σ_2, Π_2	Turing (1936), *this paper*	1936
Number-theoretic predicates	Arithmetic sets	Kleene (1943)	1943
Oracle Turing machines	All sets	Turing (1939)	1939
Trial-and-error predicates	Δ_2	Putnam (1965)	1965
Limiting recursion	Δ_2	Gold (1965)	1965
Iterated limiting recursion	Δ_k	Schubert (1974)	1974
Alternating Turing machines	Arithmetic sets	Chandra et al. (1981)	1976
ω-Turing machines	NA	Cohen and Gold (1978)	1978
Inductive Turing machines	Arithmetic sets	Burgin (1983)	1983
Tae-computability	Σ_2	Hintikka and Mutanen (1988)	1988
Infinite time Turing machines	Hyper-arithmetic sets	Hamkins and Lewis (2000)	2000
Accelerating Turing machines	NA	Copeland (2002)	2002
Relativistic computing	Δ_2, Σ_2	Etesi and Nemeti (2002) and Wiedermann and van Leeuwen (2002)	
SAD computers	Arithmetic sets	Hogarth (2004)	2004
Zeno machines	NA	Potgieter (2006)	2006
Display Turing machines	Δ_3	Rovan and Steskal (2007, 2009)	2007
Red-green Turing machines	Σ_2, Π_2	van Leeuwen and Wiedermann (2012)	2012

(1988), and display Turing machines with control (Rovan and Steskal, 2007, 2009) . Also, so-called hypercomputers have been inspired by relativistic physics, cf. Hogarth (2004), Welch (2008), Wiedermann and van Leeuwen (2002). At the heart of all these alternative approaches to unbounded computations, the complexity classes of the arithmetical hierarchy have repeatedly emerged as the classes characterising the computational power of the underlying models, in particular the classes Δ_2 and Σ_2. An overview on various approaches to unbounded computations is given in Table 1.

In preliminary studies we have investigated various aspects of red-green computations from the viewpoint of the computability theory. For example, it appears (van Leeuwen and Wiedermann, 2012) that the computational power of red-green Turing machines increases with the number of mind changes allowed (it climbs along the so-called Ershov hierarchy, cf. Cooper (2004), Ershov (1968) and Rogers (1967)). Also, for any finite number of mind changes red-green Turing machines recognise languages in Σ_2 and accept languages from Δ_2. In fact, computations of red-green Turing machines exactly characterise the latter two classes. This, together with the similar results achieved with the help of other machine or logical models of unbounded computation mentioned in the beginning of this section, suggests that, due to their simplicity and mathematical elegance, red-green Turing machines can serve as a bridging model among the various alternative models of potentially infinite computations. Moreover, another interesting result is that red-green Turing machines can elegantly and straightforwardly be simulated by relativistic Turing machines (Etesi and Nemeti, 2002; Wiedermann and van Leeuwen, 2002); (and vice versa). This indicates the relation of red-green computing to hypercomputing. An overview of the known results on red-green Turing machines can be found in van Leeuwen and Wiedermann (2012). The first results are encouraging and it is nice to see, retrospectively, that the core ideas essentially have their roots in the Turing's work.

4. Conclusion

It is symptomatic that the hallmarks of modern computing have appeared in the mathematical work of Turing whose primary aim initially was to define the notion of computability rather than to lay down the theoretical fundamentals of computing machinery. (Turing himself was among the first to realise the impact of the latter later on.) In particular, Turing's a-machines were tailored to infinite rather than finite computations. This, together with the prevailing use of computers nowadays, has opened a way towards considering problems up to Δ_2 or even Σ_2 computable by unbounded processes, as we have tried to show in this note. These considerations may change our attitude towards what is computable. Would this trend lead to an extension of the notion of computability?

Acknowledgements

This research was partially supported by Czech National Science Foundation Grant No. P202/10/1333 and institutional research plan AV0Z10300504.

References

Büchi, J.R., 1962. On a decision method in restricted second order arithmetic. In: Logic, Methodology and Philosophy of Science, Proceedings of the 1960 International Congress, Nagel, E., Suppes, P., Tarski, A., Stanford University Press, Stanford, pp. 1–11.

Burgin, M., 1983. Inductive Turing Machines, Notices Acad Sci USSR, 270:6 1289–1293 (translated from Russian, v. 27, No. 3).

Burgin, M., 2005. Super-recursive Algorithms, Springer, Berlin, Heidelberg.

Chandra, A.K., Kozen, D.C., Stockmeyer, L.J., 1981. Alternation, J. ACM 28 (1) 114–133.

Cohen, R.S., Gold, A.Y., 1978. ω-Computations on Turing machines, Theor. Comput. Sci. 6 (1), 1–23.

Copeland, B.J., 2002. Accelerating turing machines. Minds Machines 12, 281–301.

Cooper, S.B., 2004. Computability Theory, Chapman & Hall/CRC, Boca Raton, London, New York.

Ershov, Y.L., 1968. A certain hierarchy of sets I. Algebra i Logika 7 (1) 47–74 (Russian), *Algebra and Logic* 7:1 (1968) 25–43 (English translation).

Etesi, G., Nemeti, I., 2002. Non-Turing computation via Malament-Hogarth space-times. Int. J. Theor. Phys. 41 (2), 341–370.

Gold, E.M., 1965. Limiting recursion. J. Symb. Log. 30 (1) 28–48.

Gold, E.M., 1967. Language identification in the limit. Inf. Control 10 (5) 447–474.

Hamkins, J.D., 2000. A. lewis, infinite time turing machines. J. Symb. Log. 65 (2) 567–604.

Hintikka, J. Mutanen, A., 1988. An alternative concept of computability. In: Hintikka, J. (Ed.), Language, Truth, and Logic in Mathematics, Kluwer, Dordrecht, pp. 174–188.

Hogarth, M., 2004. Deciding arithmetic using SAD computers. British J. Phil. Soc. 55, 681–691.

Kleene, S.C., 1943. Recursive predicates and quantifiers. Trans. Amer. Math. Soc. 53 (1) 41–73.

Potgieter, P.H., 2006. Zeno machines and hypercomputation. Theor. Comput. Sci. 358, 26–33.

Putnam, H., 1965. Trial and error predicates and the solution to a problem of Mostowski. J. Symb. Log. 30 (1) 49–57.

Rabin, M.O., 1969. Decidability of second-order theories and automata on infinite trees. Trans. Amer. Math. Soc. 141, 1–35.

Rogers, Jr., H., 1967. Theory of Recursive Functions and Effective Computability, McGraw-Hill, New York.

Rovan, B., Steskal, L., 2007. Infinite computations and a hierarchy in Δ_3. In: Cooper, S.B., Löwe, B., Sorbi, A. (Eds.), Computability in Europe, Lecture Notes in Computer Science, vol. 4497, Springer, Berlin, pp. 660–669.

Rovan, B., Steskal, L., 2009. Infinite computations and a hierarchy in Δ_3 reconsidered. J. Log. Comput. 19 (1), 175–176.

Schubert, L.K., 1974. Iterated limiting recursion and the program minimization problem. J.ACM 21 (3), 436–445.

Staiger, L., 1997. ω-Languages. In: Rozenberg, G., Salomaa, A. (Eds.), Handbook of Formal Languages, Vol 3: Beyond Words, Springer, Berlin pp. 339–387.

Thomas, W., 1990. Automata on infinite objects. In: van Leeuwen, J. (Ed.), Handbook of Theoretical Computer Science, Vol. B: Formal Models and Semantics, Elsevier, Amsterdam, pp. 133–192.

Turing, A.M., 1936. On computable numbers, with an application to the Entscheidungsproblem. Proc. Lond. Math. Soc. Ser. 2 (42) 230–265.

Turing, A.M., 1939. Systems of logic based on ordinals. Proc. Lond. Math. Soc. Ser. 2 (45) 161–228.

van Leeuwen, J., Wiedermann, J., 2012. Computation as an unbounded process. Theoretical Computer Science 429, 202–212 .

Weihrauch, K., 2000. Computable Analysis: An Introduction, Springer-Verlag, Berlin/Heidelberg.

Welch, P.D., 2008. The extent of computations in Malament-Hogarth spacetimes. Br. J. Phil. Sci. 59 (4), 659–674.

Wiedermann, J., van Leeuwen, J., 2002. Relativistic computers and non-uniform complexity theory. In: Calude, C.S., Dinneen, M.J., Peper, F. (eds.), Unconventional Models of Computation (UMC'2002), Lecture Notes in Computer Science, Vol. 2509, Springer, Berlin, pp. 287–299.

Meurig Beynon puts an empirical slant on —

TURING'S APPROACH TO MODELLING STATES OF MIND

In discussing Turing's seminal 1936 paper, the *Stanford Encyclopedia of Philosophy,* http://plato.stanford.edu/entries/turing/, highlights the way in which his conception of the 'Turing Machine (TM)' was guided by the idea of *modelling states of mind*: '... in a bolder argument, the one he placed first, he considered an "intuitive" argument appealing to the states of mind of a human computer (Turing, 1936, p. 249). The entry of "mind" into his argument was highly significant, but at this stage it was only a mind following a rule.'

There are two respects in which the subsequent treatment of TMs in mainstream computer science has sidelined 'intuitive' elements:

- The TM is viewed as the fundamental mathematical abstraction in a theory of computer science that is based on 'computational thinking'.
- Computer science has promoted the computational theory of mind, which proposes that everything the human mind does is attributable to *following rules.*

In combination, these two viewpoints promote a narrow conception of computer science; they respectively root computing in an abstract machine model with a strict formal semantics and set out to show that such a model is enough to give a good account of its core applications.

Throughout the short history of computer science, it has suited the political purposes of an emerging discipline to emphasise its connections with Turing's work – one of the supreme intellectual achievements of the twentieth century. But perhaps it is not so clear that computer science as currently conceived truly fulfils Turing's aspirations for a science of computing.

As Hodges (2004) points out, Turing's vision was broader than that of a mathematical logician: 'The essence of Turing's achievement was the discovery of a concept *with an application to* logic, rooted in ideas which lay outside mathematics' and, unlike 'Church's Thesis', 'Turing's definition [of computability] was modeled on what human beings could actually do'. Also, as Hodges (2004) goes on to observe: '[After 1948] Turing did ... surprisingly little ... to build up the modern science of computation'.

As for Turing's stance on the contribution that computational abstractions make to our understanding of mind, there are many clues in his discussion of 'The Imitation Game' in Turing (1950). It is clear that Turing sees the isssue of whether an interrogator might be deceived into attributing a machine's responses to a human as a *limited* question, and not one that decisively illuminates the nature of mind. For instance, he acknowledges that other concerns about the mind might be beyond the scope of his enquiry: 'I don't wish to give the impression that I think there is no mystery about consciousness ... [b]ut I do not think these mysteries necessarily need be solved before we can answer the question with which we are concerned in this paper'. Even in relation to his contention that a computer could succeed in 'The Imitation Game', Turing (1950) is careful to point out that – despite the rebuttals he gives in response to objections to this possibility: 'The reader will have anticipated that I have no very convincing arguments of a positive nature to support my views'.

Hodges (1988) refers elsewhere to the 'trenchant materialist and atheist Turing who emerged after 1936', and it is perhaps these characteristics that have been most emphasised in subsequent building upon his legacy. In that context, it may seem surprising that, in rebutting counter-arguments to the idea that a computer might successfully play 'The Imitation Game', Turing (1950) remarks: 'The Argument from Extrasensory Perception] is to my mind quite a strong one'. But this is more explicable if, as Hodges (1988) advocates, 'we recognize [in Turing] the common thread – a great seriousness about the sheer mystery of mental phenomena, and an equally serious conviction that they must be reconciled with a scientific world view'.

As the *Stanford Encyclopedia of Philosophy* quotation above emphasises, Turing's treatment of states of mind was conceived with a view to modelling 'a mind following rules', as was appropriate for addressing the *Entscheidungsproblem*: a Turing machine should model procedures 'definite in the sense that at every stage a completely explicit "note of instructions" could be written down explaining what was to be done in such a way that another person could take up the work' (Hodges, 1988). Beyond question, Turing's insight informed the practical contributions he made to the development of computer programming. For instance, it gives a deep meaning – deeper than a programmer needed to appreciate – to the statement with which he prefaced his *General Remarks on Electronic Computers* in his Manual for the Ferranti Mk. I computer Turing (1951): 'Electronic computers are intended to carry out any definite rule-of-thumb process which could have been done by a human operator working in a disciplined but unintelligent manner'. It also led him to identify the importance in programming of more disciplined use of mathematical notation (Turing, 1944–5).

But – to take nothing away from the extraordinary fertile nature of Turing's insight, and what has been, and can yet be, achieved within the framework of computational models in many disciplines – it seems likely that Turing himself did not consider the broader issue of the nature of mind to be closed. To quote Hodges (1988) once again: '... we cannot feel that Turing had arrived at a complete theory of what he meant by modelling the mental functions of the brain by a logical machine structure'. Significantly, in the spirit of an applied mathematician, Turing gave high priority to squaring mathematical theory with empirical evidence and practical applicability. In this connection, Hodges (2004) contrasts Turing's approach to enhancing discrete computational models as models of physical systems through introducing randomness with 'that of some modern theorists, who seek to outdo discrete computation by exploiting the very elements that Turing made little of' and whose approaches lead to difficulties that render them 'meaningless without stability and robustness in the face of infinitesimal phenomena'. As further testimony to Turing's practical orientation, we need only consider the degree of direct involvement he wanted in building early computers, his contributions to code-breaking and speech recognition, and his concern to take account of the inevitability of operational error in theorising about computers (Hodges, 2004).

It may be that, by imputing greater authority to mechanical models of mind than Turing himself would have decisively endorsed, today's students of computer science are in danger of identifying mathematics with a blind process of inference such as can be carried out by a machine. Such a concern is raised by Byers (2007), for instance, in motivating his account of mathematical practice. Turing's biographers describe him as a mathematician '[whose] native style was rough-and-ready and prone to minor errors' (Feferman, 2006) for whom the notion of 'intuition' had a particular fascination (cf. his mathematical work on the Riemann hypothesis (Hodges, 1988)). If this seems to be at odds with the demystification of effective procedures that he achieved in his own work, an instructive parallel may be drawn with his older contemporary Emil Post, who concluded his paper 'Absolutely unsolvable problems and relatively undecidable propositions' (written in 1941, but only published posthumously (Post, 1965)) by expressing his amazement at the reaction to Godel's undecidability results in the following terms:

> "... mathematical thinking is, and must be, essentially creative. It is to the writer's continuing amazement that ten years after Gödel's remarkable achievement current views on the nature of mathematics are thereby affected only to the point of seeing the need of many formal systems, instead of a universal one. Rather has it seemed to us to be inevitable that these developments will result in a reversal of the entire axiomatic trend of the late nineteenth and early twentieth centuries, with a return to meaning and truth."

To the end of his life – in the spirit of Post's injunction – Turing seems to have been motivated to seek significance beyond an abstract logical interpretation for his TM concept. In realising his vision for computation, he felt the essential need to establish the connection with physical reality. Hodges (1988) cites Penrose's summary of Turing's position in 1950: 'It seems likely that he viewed physical action in general – which would include action of a human brain – to be always reducible to some kind of Turing-machine action'. As Hodges (1988) later goes on to relate, this was to be an unresolved problem for Turing, who recognised the challenge presented by the indeterminacy principle in quantum mechanics.

It is hard to imagine how the academic discipline of computer science would have emerged without Turing's contribution. In his work, Turing showed extraordinary prescience in relation to many aspects of the 'computer programming' related activity that has been the central focus of academic computer science throughout its history. But, at the time of his death, the transformative impact of computers and programming could hardly have been predicted. And in the same way that mathematics demands a broader account than formal systems can supply, so too does contemporary computing. In concluding this brief review, it is appropriate to look at ways in which modern computing and the science of computing to which we must now aspire is influenced by other perspectives on modelling human states of mind. This conclusion reflects the author's own research interest, under the auspices of the Empirical Modelling (EM) project http://www.dcs.warwick.ac.uk/modelling, in seeking a broader alternative conceptual framework for computing.

Turing (1950) asserts that the problem of developing a digital computer that can succeed at The Imitation Game 'is mainly one of programming'. In the context of modern software development, it has become apparent that 'the problem of programming' cannot be understood in a narrow sense. One of the most critical aspects of software development is that of binding meanings to artefacts that ostensibly are – or are to be – specified purely in computational terms. The idea that such semantic considerations can be comprehensively addressed by formal computational semantics has been criticised by reviewers representing many different perspectives. They include the expert software consultant Michael Jackson (2006), the distinguished computer scientist Peter Naur (1985) and the philosopher Cantwell-Smith (2002). The common theme in these, and other critiques, is that formal semantics can only go so far in mediating meanings in the software development process, especially where the activity involves 'radical design' (cf. Jackson (2006); Vincenti (1993)).

Whereas it suited Turing's (1936) purpose in addressing his mathematical objective to consider human states of mind associated with carrying out a calculation, quite different kinds of states of mind feature in modern software development involving radical design. Such a development process has to take account of the perspectives of many human participants whose understanding is mediated in quite different ways from those of the traditional 'human computer': they cannot be expected to appreciate the full purposes or context for actions, to be able to interpret formal notations reliably, or to be able to communicate their wishes abstractly without reference to actual experience that can only be had and skills that can only be developed for instance by interacting with a prototype system. And even though the functional goal and the process itself may be clearly specified, the practical situated knowledge needed to enact the process may itself be difficult to access – as when we try to make a pot of tea in a neighbour's house, and have to contend with finding the ingredients ('where are the tea bags?'), identifying the utensils we need ('is that a teapot?'), and determining how to configure these ('where do I plug this in?'). In developing software for reactive systems, this exploratory activity may take yet more extreme forms: in configuring devices and tuning their responses, it as if we are investigating the feasibility of constructing the very hardware on which our programs are to be executed (Beynon et al., 2006).

The duality that separates 'the given already engineered computing device' from 'the to-be-specified abstract sequence of instructions to be performed on the device' is characteristic of the computational framework within which Turing was reasoning. Turing (1950) exploits this characteristic when he identifies The Imitation Game as 'drawing a fairly sharp line between the physical and the intellectual capabilities of a man', and stipulates that 'the interrogator cannot demand practical demonstration'.

An instructive comparison can be made between Turing's approach to modelling the mind of a human computer and that conceived by David Gooding (1990) in his account of Faraday's seminal experimental work on electromagnetic phenomena. In interpreting the way in which this activity was conducted, Gooding (2001) introduced the notion of 'construals' as 'proto-interpretative representations which combine images and words as provisional or tentative interpretations of novel experience ... [that is] being created ... through the interaction of visual, tactile, sensorimotor and auditory modes of perception together with existing interpretative concepts including mental images'. Such construals can be regarded as a means to knowledge representation in spirit similar to that advocated by Rodney Brooks (cf. Brooks (1991a) and Brooks (1991b)). Gooding (2001) invokes his research into Faraday's use of construals in his critique of the 'profoundly mistaken' notion 'that systematic, rational thought is or can be separate from the world that it seeks to understand, manipulate or control'. In collaboration with Addis et al. (2008), Gooding builds on this work to propose a broader notion of computer science embracing 'irrational sets' that 'require the use of an abductive inference system'.

Of the computer-based innovations that have been developed post-Turing, the spreadsheet is perhaps the one that is most directly connected with the 'modelling of states of mind' that Turing (1936) discussed. For instance, a spreadsheet can be viewed as a particularly effective way to represent human states of mind at the interface between the user and the computational process. Several of the characteristic themes that Hodges (1988) identifies in Turing's vision of the TM also seem to be relevant to the spreadsheet concept. The spreadsheet exemplifies 'the blending of the mechanical and the psychological' (Hodges, 1988). Through the dependency relations it characteristically maintain, the spreadsheet embodies the notion of *being determined* (Hodges, 1988) as this is understood in two complementary ways – as in the automatic recalculation of a cell value (e.g. *profit*) from an arithmetic formula (e.g. *profit* = *income* − *expenditure*), and as in the mind of the spreadsheet user, who apprehends 'profit' as indivisibly connected with 'income'. What is more, there is a closer correspondence between the current state of the spreadsheet and the mental state of the spreadsheet user than in a conventional TM or procedural programming model, where the variables that are intended to record meaningful quantities (e.g. 'profit' and 'income') are routinely assigned intermediate values that are inconsistent with their real-world semantics. In keeping with Turing's aspirations for modelling states of mind, as characterised by Hodges (1988), this commends the spreadsheet as 'a new level of description based on the idea of discrete states [such that] this level (rather than that of atoms and electrons, or indeed that of the physiology of brain tissue) [is] the correct one in which to couch the description of mental phenomena'.

A spreadsheet captures the human calculator's state of mind in a quite different sense from a TM. For the user of a spreadsheet, the state of interest has to do with the real-world semantics ('how will the price of petrol affect my profit?') rather than the routine computational semantics ('how is the formulae relating profit to the cost of petrol evaluated?'). The chief virtue of the spreadsheet is that it renders the mechanics of computation invisible, throwing its significance to the user into sharper relief.

Unlike a computational semantics, the real-world semantics of the spreadsheet is informal and pragmatic in character. Appreciating its state requires knowledge of the context (e.g. 'to what does profit refer?'), skill in associating cells with their referents (e.g. 'which cells record income and profit?') and in knowing how to carry out the interactions that disclose, maintain and probe meanings (e.g. how to change the price of petrol, how to revise the formula that define income and profit in response to changes in the tax laws, and how to carry out 'what if?' experiments). The speed with which computational updates are effected, the way in which key values are disposed in the spreadsheet grid and the level of familiarity of the user all contribute to the quality of the spreadsheet as a model of a state of mind. Turing (1936) himself discusses such issues in motivating his conception of the Turing machine: expressing concern in choosing his representations for numbers about symbols that 'cannot be observed at one glance' and insisting that changes to the squares being observed 'must be immediately recognisable by the computer'. This is evidence that focusing exclusively on formal mathematical interpretations of TMs fails to do justice to the subtlety of his thinking. Indeed, Turing (1950, p. 15) himself expresses a related concern about facile interpretations of logic when he refers to 'a fallacy to which philosophers and mathematicians are particularly subject ... the assumption that as soon as a fact is presented to a mind all consequences of that fact spring into the mind simultaneously with it'.

The goal of the EM project is to identify principles and develop tools to support a broader view of computing. Such a view takes account of roles for human agents richer than those of a 'human computer'. EM puts its primary focus on computer-based artefacts, similar in character to the *construals* introduced by Gooding in his work on the history of science, rather than 'computer programs'. An EM construal is framed with reference to three basic concepts: *observables*, *dependencies* and *agents*. These concepts have approximate counterparts in spreadsheets in - respectively - the cells, the relationships between cell values established by definitions, and the diverse modes of redefinition that are associated with state changing actions, both manual and automated, in connection with spreadsheet development and use. In keeping with the 'what if?' character of a spreadsheet, an EM construal is archetypally associated with the state of mind of a human experimenter involved for example in the kind of activity that Friedrich Steinle (1997) identifies as 'exploratory experimentation' that 'is driven by the elementary desire to obtain empirical regularities and to find out proper concepts and classifications by means of which those regularities can be formulated'.

The relationship between the EM and TM models of states of mind is best understood by considering how exploratory activities can engineer functional machine-like entities in the world. This is illustrated by the way in which – as conceived by Gooding (1990) – Faraday elaborated his construals in engineering the first prototype electric motor. EM principles for software development exploit construals in a similar way: first in enabling the exploratory sense-making activities that disclose patterns of interaction, agency and interpretation that can be reliably revisited, and then in configuring the situation and exercising discretion in interaction so as to establish program-like behaviours (Beynon et al., 2006). The way in which a machine-like abstraction is here identified through a conceptual shift of viewpoint on a situation that has first been suitably engineered is something that Turing appreciated in relation to his 'discrete-state machines': 'These are the machines which move by sudden jumps or clicks from one quite definite state to another. These states are sufficiently different for the possibility of confusion between them to be ignored. Strictly speaking, there are no such machines. Everything really moves continuously. But there are many kinds of machine which can profitably be thought of as being discrete-state machines. For instance in considering the switches for a lighting system it is a convenient fiction that each switch must be definitely on or definitely off. There must be intermediate positions, but for most purposes we can forget about them'.

In Addis et al. (2008) take inspiration from the use of construals in scientific practice but invoke the Peircean notion of abduction to arrive at a computational framework that is framed, like Turing's, in logical terms. EM gives an account of computing similar in spirit to that of Addis and Gooding in key respects but radically different in that – following William James (1912/1996) – the semantics of model building is squarely rooted in experience. In line with James's concept of 'radical empiricism', the fundamental premise of EM is that every instance of knowing is a connection made in the present experience of an individual, and all semantic relationships must be in some way traceable to such instances. As a model for a state of mind, an EM construal is characterised by the patterns of observables, agencies and dependencies that it embodies. This is unlike a logical specification such as is expressed by abstracting variables and declaring the constraints to which they are subject. Like a spreadsheet, an EM construal represents both a current state and latent germs of change that express expectations that rely upon contextual guarantees that can never be absolute.

It is impossible to say whether Turing would have been sympathetic to such approaches to placing his fundamental contribution in a broader conceptual frame. But beyond question, Turing's own style of thinking was in some respects well matched to a pragmatic philosophical stance. And where some have made grand theoretical claims for the TM concept in relation to computer science and the mind, Turing's own outlook and working practices put the emphasis on real and topical problems, on engaging with engineering issues and on ideas under construction, and appeal to the empiricist as well as to the logician. The remark that concludes his paper on *Computing Machinery and Intelligence* (Turing, 1950) testifies to the live, creative and adventurous qualities of his imagination: 'We can only see a short distance ahead, but we can see plenty there that needs to be done'.

Acknowledgements

I am much indebted to the many contributors to the EM research project, and especially to Steve Russ for key ideas that have motivated this paper.

References

Addis, T., Addis, J.T., Billinge, D., Gooding, D., Visscher, B.-F., 2008. The abductive loop: tracking irrational sets. Found. Sci. 13(5), 5–16.

Beynon, W.M., Boyatt, R.C., Russ, S.B., 2006. Rethinking programming. In: Latifi, S. (ed.), Proceedings of the IEEE Third International Conference on Information Technology: New Generations (ITNG 2006), April 10–12, 2006, Las Vegas, Nevada, USA 2006, pp. 149–154.

Brooks, R.A., 1991a. Intelligence without representation. Artif. Intell. 47, pp. 139–159.

Brooks, R.A., 1991b. Intelligence without reason. Proceedings of the 12th International Joint Conference on Artificial Intelligence - Volume 1, IJCAI-91, Morgan Kaufmann Pub. Inc., San Francisco, CA, pp. 569–595.

Byers, W., 2007. How Mathematicians Think: Using Ambiguity, Contradiction, and Paradox to Create Mathematics. Princeton University Press.

Cantwell-Smith, B., 2002. The foundations of computing. In: Scheutz, M. (Ed.), Computationalism: New Directions. MIT Press, Cambridge, MA, pp. 23–58.

Feferman, S., 2006. Turing's thesis. Notices Am. Math. Sci. 53(10), 1200–1206.

Gooding, D., 1990. Experiment and the Making of Meaning: Human Agency in Scientific Observation and Experiment. Kluwer, Dordrecht.

Gooding, D., 2001. Experiment as an instrument of innovation: Experience and embodied thought. In: Beynon, Nehaniv, and Dautenhahn (Eds.), Proceedings of the 4th International Conference on Cognitive Technology: Instruments of Mind. Springer LNCS, Vol. 2117, pp. 130–140.

Hodges, A., 1988. Alan Turing and the Turing machine. In: Rolf Herken (Ed.), The Universal Turing Machine. A Half-Century Survey, Oxford University Press, Oxford.

Hodges, A., 2004. Alan Turing: the logical and physical basis of computing. In: Proceedings of Alan Mathison Turing 2004: A Celebration of His Life and Achievements, Manchester University, 5 June, 2004. BCS eWiC Series, http://www.bcs.org/content/conWebDoc/17127

Jackson, M.A., 2006. What can we expect from program verification? IEEE Comput., 39(10), 53–59.

James, W., 1912/1996. Essays in Radical Empiricism (Reprinted from the original 1912 edition by Longmans, Green and Co., New York), London: Bison Books.

Naur, P., 1985. Intuition in software development. In: Ehrig, H., Floyd, C., Nivat, M., Thatcher, J.W. (Eds.), Mathematical Foundations of Software Development, Proceedings of the International Joint Conference on Theory and Practice of Software Development (TAPSOFT), Berlin, Germany, March 25-29, 1985, Volume 2. Lecture Notes in Computer Science 186, Springer, pp. 60–79.

Post, E., 1965. Absolutely unsolvable problems and relatively undecidable propositions. In: M. Davis (Ed.), The Undecidable – Basic Papers on Undecidable Propositions, Unsolvable Problems and Computable Functions, Raven Books, New York.

Steinle, F., 1997. Entering new fields: exploratory uses of experimentation. Phil. Sci. 64 (Proc.), pp. S65–S74.

Turing, A.M.. 1936. On computable numbers, with an application to the Entscheidungsproblem. Proc. London Math. Soc. 42(2), 230–265.

Turing, A.M., 1944–45. The Reform of Mathematical Notation (unpublished, in Collected Works).

Turing, A.M., 1950. Computing machinery and intelligence, Mind 49, 433–460.

Turing, A.M., 1951. Programmers' Handbook for the Manchester Electronic Computer Mark II (1st ed.), Computing Machine Laboratory, Manchester University, *c.* March 1951. Digital facsimile in The Turing Archive for the History of Computing at http://www.AlanTuring.net/programmers_handbook.

Vincenti, W.C.. 1993. What Engineers Know and How They Know It: Analytical Studies from Aeronautical History. The Johns Hopkins University Press, Baltimore.

Stanford Encyclopedia of Philosophy. http://plato.stanford.edu/entries/turing/

The EM website at url: http://www.dcs.warwick.ac.uk/modelling

Henk Barendregt and Antonio Raffone explore —

Conscious Cognition as a Discrete, Deterministic and Universal Turing Machine Process[1]

1. Systems with states

It is often maintained that the brain-as-computer metaphor is ill taken. Nevertheless one can view conscious cognition as a Turing machine process, with its discrete, deterministic and universal aspects. Not being familiar to the language of science one may object to the claim that computation plays an important role in the life of humans (and in fact all animals). Nevertheless, for goal-directed movements fast and accurate (unconscious) computations are necessary. Sensory input has to be transformed to output in the form of adequate action. Cognitive scientists, who are aware of the need for computation, still may object to the computer metaphor. Indeed, our brain is not a network of Boolean switches and it does neither have numerical input nor output. Our claim is that it is nevertheless useful to interpret cognition as a hybrid Turing machine process.

Modelling systems (machines or living organisms) the notion of 'state' is important. Only considering stimulus-reaction (Input, Action) transitions, we get

$$I \mapsto A. \tag{1.1}$$

This 'behaviouristic' view has limited possibilities. Actual systems can react differently to the same input. To model this difference, inspired by Turing machines, one introduces states, modifying (1.1) to

$$I \times S \mapsto A \times S. \tag{1.2}$$

Now the output may depend also on the state. This will be elaborated below.

2. The Turing machine: processes and computation

A Turing machine is a theoretical model of ad hoc computing devices, including the universal Turing machine,[2] after which the modern digital computers are built. It consists of a potentially two-sided infinite tape[3] with memory cells, a movable reading/writing head placed on one of the cells, and a finite set S of states. The cells each contain a symbol from a finite input alphabet I (set of symbols). Each specific Turing machine is determined by:

$$t_1, \ldots, t_m : I \times S \mapsto A \times S, \tag{2.1}$$

---

[1] Added in print. After acceptance for publication of this commentary we found out that in Zylberberg et al. (2011) overlapping ideas have been presented.

[2] The universality means that just one machine can simulate the behaviour of all other ones by giving it various *programs*.

[3] In modern computers a disc or flash memory is used instead of a tape. The infinity of the tape was proposed by Turing in order to be technology independent. But each computation on a Turing machine uses only a finite amount of memory.

where we have the following

$$
\begin{aligned}
I &= \text{set of possible inputs (symbols)} \\
&\quad \text{the head may read on the tape at its location,} \\
S &= \text{set of possible states,} \\
A &= \{L, R, W(a)\}, \text{ the set of possible actions:} \\
L &= \text{moving head left (or the tape moves right),} \\
R &= \text{moving head right (or the tape moves left),} \\
W(a) &= \text{overwriting present location with symbol } a \in I.
\end{aligned}
$$

For example a machine M can have a, b in I and s_1, s_2 in S, and transition rules

$$
\begin{aligned}
t_1: &\quad \langle a, s_1 \rangle \mapsto \langle R, s_2 \rangle \\
t_2: &\quad \langle b, s_1 \rangle \mapsto \langle W(a), s_2 \rangle
\end{aligned}
$$

with the following meanings.

t_1: if M reads an a in state s_1, then
the reading head moves one cell to the right and M enters state s_2;

t_2: if M reads a b in state s_1, then
it (over)writes (this b with) an a and M enters state s_2.

With a Turing machine one can run *processes* and perform *computations*.

A *computation* starts with an input. In the Turing machine this is represented as a finite sequence of data, elements of I, written on consecutive cells of the tape. The other cells are blank (also considered as an element of the alphabet I). The read/write head is located at a particular cell of the tape and the machine is in an initial state q_0. The machine performs the actions according to its transition rules, until no more rule applies and the machine 'halts'. The resulting contents on the tape is considered as the output of the computation.

Turing made it plausible that any kind of mechanical computation can be performed in such a way. Moreover, he constructed a single Turing machine *UM*, the *Universal Machine*, that can simulate an arbitrary Turing machine M. Wanting to simulate the computation of M on input i, notation $M(i)$, one can construct a program p_M for M such that for all input i one has

$$UM(p_M, i) = M(i).$$

This means that *UM* requires an extra argument, the program code p_M, next to the given argument i. Turing used it to define a problem that cannot be answered by the computation of any Turing machine and hence not by any computation.

A *process* is like a computation, but without the requirement that there is a final state in which the machine comes to a halt. So computations are special processes focused on *termination*; processes in general are focused on *continuation*. The usefulness of processes can be seen by giving some of the cells on the tape a special status: for input ('sensors') and for output ('actuators') from and to the outside world. A factory involving heating devices, thermometers, and safety valves, may be controlled in this way by a Turing machine acting as process.

The process (or computation) taking place in a Turing machine is discrete and deterministic: it consists of a stream of distinct steps, only depending on the input.

3. The neural Turing machine

From the description of a process it is clear that life (humans, animals and even plants) can be thought of as processes. In artificial intelligence (AI) one tries to emulate these processes. There are the two views in AI, one the symbolic rule based Simon and Newell (1958), and the connectionist one related to Turing (1948) and Hillis (1989). Simon and Newell state that intelligence works in a discrete serial way following specific rules. The connectionists state that cognition uses the parallelism of 'neural nets' and not a sequential system. In the hybrid version of Turing machines presented below, the sequential machine will get transition rules programmed by a parallel neural net, providing a useful unification for understanding human cognition.

Let us review the model of the Turing machine. A particular such machine is determined by a finitely specified transition map (2.1). Now we slightly change the interpretation of this notation.

$I =$ now stands for sensory input,
$S =$ set of possible states,
$A =$ now stands for actions, including neural excitation for movement and focussing attention,
$\mapsto =$ the transition determined by a neural net.

We do have a non-essential extension. No longer is I a finite alphabet, but a virtually unbounded set of inputs from the world. It still is essentially finite by the limitations of our senses. In a Turing machine the set I is typically of size 2^n, with $n < 10$; in human cognition it is orders of magnitude bigger. The same applies to the set A. This set directs bodily movements, speech or mental action.

Another feature that happens in the brain is that whilst we are processing, our processor does change. This includes development and is essential for homo sapiens. This seems like a proper extension of the notion of a Turing machine. But thanks to the existence of a universal Turing machine this is not so. Instead of (N stands for the neural net determining the transitions and A can act on N)

$$I \times S \xrightarrow[N]{} A \times S \tag{3.1}$$

one can employ the universal machine and write the equivalent

$$I \times p_N \times S \xrightarrow[\text{UM}]{} A \times S.$$

Now it becomes possible that the A act on the program p_N. In ordinary computing this is not advisable, as it is difficult to reason about the resulting effects. But in the neural evolution it fits perfectly well.

In the resulting model of cognition the set of states S plays an important role. Rather than seeing human cognition in a stimulus response fashion like in (1.1), as was fashionable in the behaviourist days of last century, the cognitive model (3.1) shows the essence of states. A 'state' is a mathematical concept: giving the same input–output relation. We know empirically that attention and emotions greatly influence these states. Under the same circumstances these inner state can make of a human being a saint, a scientist, a Scrooge or worse. It should be noted that the model (3.1) is discrete. Conscious cognition is a stream of separate phenomena, taking place in time. We will come back to this in the next section.

4. Conscious cognition: discrete temporal frames

A currently influential model of human conscious cognition is the global workspace (GW) theory (Baars, 1998; Baars et al., 2003). In this model, conscious cognition enables an access to a varying subset of brain sources.

A neuronal underpinning for the GW model has been developed in Dehaene and Naccache (2001). It is characterised by a winner-take-all dynamics, forming a 'neural processing bottleneck', involving 'broadcasting' activity from prefrontal cortex to neurons on a global scale in the brain.

Only one large-scale reverberating neural assembly is assumed to be active at any given moment. This crucially involves the thalamocortical pulse and imposes a temporal resolution for the stream of conscious cognition, needing at least 100 ms for a perceptual awareness moment.

Independently, based on psychophysical, neurophysiological and electrophysiological findings, Varela et al. (2001) postulate that a specific large-scale neural assembly underlies the emergence and operation of each conscious cognitive act. Such assemblies occur in the thalamocortical system, using closed-loop signalling with periods of 100–300 ms, see Tononi and Edelman (1998). This is consistent with the earlier behavioural evidence of the psychological refractory period, based on minimal temporal resolutions (Welford, 1952), about 150 ms.

On the other hand, Efron (1973) suggested, based on psychophysical evidence, that conscious cognition is temporally discrete and parsed into sensory sampling intervals or 'perceptual frames', estimated to be about 70–100 ms in average duration. More recently, based on psychophysical and electrophysiological evidence, the range 70–100 ms has been interpreted as an attentional object-based sampling rate for visual motion (van Rullen and Koch, 2006). This rate could be related to a sequence of shorter temporal processes, needed for unconscious treatment of sensory and other input, see van Rullen and Koch (2003) for a review. It may provide an estimate of the rate at which temporal representations at an unconscious level can be accessed (van Wassenhove, 2009).

To reconcile the framing of conscious cognition with the apparent continuity of perceptual experience, John (1990) suggested the following mechanism. A cortical convergence of a cascade of momentary perceptual frames establishes a steady-state perturbation from baseline brain activity. This idea has received substantial support from electroencephalographic (EEG) studies. The dynamics of the EEG field is represented by intervals of quasi-stability or 'microstates', with sudden transitions between them (Strik and Lehmann, 1993).

5. Conscious cognition: mind states

According to Baars' GW theory (Baars et al., 2003), sensory cognition works as follows. Input as signals from the sensory cortex are amplified by attention and become 'contents' of consciousness. After this amplification, feed back to the sensory cortex takes place to enable conscious access to the contents themselves, in a recurrent GW process. See Dehaene and Naccache (2001) and Lamme (2003).

In this process 'contextual' brain systems play a role in shaping conscious events. These include the 'where' and 'what' pathways in the parietal cortex for visual processing, see Milner and Goodale (2008). Regions of prefrontal cortex appear to do the same for other aspects of experience, including emotional, goal-related and self-representation aspects (Baars et al., 2003). Also the insula appears to play a crucial role as body- and feeling-related contextual system for awareness (Craig, 2009). More in general, as shown by behavioural research, affective states, including moods and emotions, provide a inner context guiding different forms of human judgment and cognitive processing, see Clore and Huntsinger (2007) for a review. These contexts can be considered as mind states, not only determining actions, but also the next input via selective attention. Selectivity in turn stems from current goals represented in prefrontal cortex (Duncan, 2001) and can ultimately be related to the current mind state. In a synthetic view, apart from inputs from sensory fields, inputs to the GW come from the GW output itself, see also Maia and Cleeremans (2005), depending on a given mind state.

In a TM controlling an industrial process the input is determined solely by the world. This is not so in human emotional cognition, where attention plays an input selecting role. Therefore mind states are themselves the ground for conscious cognition, not just a context. By their broadcasting, 'speaking to the audience' in Baars' theatre metaphor, they have the greatest influence on the brain state as a whole, and on (intentions for) action and thinking.

The brain substrates for mind states are potentially wider than those for the GW, with an overlap with the latter, and with the inclusion of various kinds of unconscious contextual systems supporting conscious cognition. The neural substrates for longer lasting emotional mind states plausibly also include the cerebrospinal fluid, as discussed in the paper by Veening and Barendregt (2010).

6. Trained phenomenology

The temporally discrete view of conscious cognition stemming from psychophysical and neuroscientific experiments, and models of conscious cognition, can be related to Buddhist psychology, based on trained phenomenology (insight meditation). Also in this theory, conscious cognition is described as a deterministic stream of successive 'pulses', with object and a state, see von Rospatt (1995).

Mindfulness, which can be conceived as a moment by moment reflexive awareness, is described as providing psychologically wholesome mind states. Being meta-awareness it is universal (as in a TM) and can bring flexibility in the co-determination of mind states and conscious processes. Mindfulness plausibly is supported by adaptive coding regions in prefrontal cortex (Raffone and Srinivasan, 2009). An effective way to influence the outcome of this deterministic process is to choose the right input. This can be done by training our attention, which chooses input and thereby the mind states. This is exactly what happens during the mental development of insight meditation: training concentration and mindfulness.

7. Conclusion

Behavioural and neurophysiological experiments and also trained phenomenology all point in the direction of conscious cognition as a discrete process depending on input and states. This is very similar to the Turing model of general computability. In fact, the hybrid Turing machine model of human conscious cognition captures well the recursive aspects mentioned in **5** and gives a logical interpretation of the notion of determinacy, emphasised both in cognitive science and Buddhism. This does not exclude free will, see, e.g. Dennett (2004).

References

Baars, B., 1998. Metaphors of consciousness and attention in the brain. Trends Neurosci. 21, 58–62.

Baars, B., Ramsoy, T., Laureys, S., 2003. Brain, conscious experience and the observing self. Trends Neurosci. 26, 671–675.

Clore, G., Huntsinger, J., 2007. How emotions inform judgment and regulate thought. Trends Cogn. Sci. 11, 393–399.

Craig, A., 2009. Emotional moments across time: a possible neural basis for time perception in the anterior insula. Philos. Tran. Royal Soc. Lond. B 364, 1933–1942.

Dehaene, S., Naccache, L., 2001. Towards a cognitive neuroscience of consciousness: basic evidence and a workspace framework. Cognition 79, 1–37.

Dennett, D.C., 2004. Freedom Evolves. Penguin, London, New York.

Duncan, J., 2001. An adaptive coding model of neural function in prefrontal cortex. Nat. Rev. Neurosci. 2, 820–829.

Efron, R., 1973. Conservation of temporal information by perceptual systems. Percept. Psychophys. 14, 518–530.

Hillis, D., 1989. The Connection Machine. MIT Press, Cambridge, MA, USA.

John, E.R., 1990. Machinery of the Mind: Data, Theory, and Speculations About Higher Brain Function. Birkhauser, Boston.

Lamme, V., 2003. Why attention and awareness are different. Trends Cogn. Sci. 7, 12–18.

Maia, T. V., Cleeremans, A., 2005. Consciousness: converging insights from connectionist modeling and neuroscience. Trends Cogn. Sci. 9, 397–404.

Milner, A., Goodale, M., 2008. Two visual systems re-viewed. Neuropsychologia 46, 774–785.

Raffone, A., Srinivasan, N., 2009. An adaptive workspace hypothesis about the neural correlates of consciousness: insights from neuroscience and meditation studies. Progress in Brain Research 176, 161–180. Prog Brain Res. 2009;176:161-80.

Simon, H. A., Newell, A., 1958. Heuristic problem solving: the next advance in operations research. Oper. Res. 6 (1), 1–10.

Strik, W., Lehmann, D., 1993. Data-determined window size and space-oriented segmentation of spontaneous eeg map series. Electroencephalogr. Clin. Neurophysiol. 87, 169–174.

Tononi, G., Edelman, G., 1998. Consciousness and complexity. Science 282, 1846–1851.

Turing, A.M. 1948. Intelligent Machinery. National Physical Laboratory Report. In: Meltzer, B., Michie, D. (eds) 1969. Machine Intelligence 5. Edinburgh: Edinburgh University Press, 3-23. Reproduced with the same pagination in Ince 1992.
van Rullen, R. Reddy, L., Koch, C., 2006. The continuous wagon wheel illusion is associated with changes in electroencephalogram power at 13 hz. J. Neurosci. 26, 502–507.
van Rullen, R., Koch, C., 2003. Is perception discrete or continuous? Trends Cogn. Sci. 7, 207–213.
van Wassenhove, V., 2009. Minding time in an amodal representational space. Philos. Trans. Royal Soc. B 364, 1815–1830.
Varela, F., Lachaux, J.-P., Rodriguez, E., Martinerie, J., 2001. The brainweb: phase synchronization and large-scale integration. Nat. Rev. Neurosci. 2, 229–239.
Veening, J. G., Barendregt, H. P., 2010. The regulation of brain states by neuroactive substances distributed via the cerebrospinal fluid; a review. Cerebrospinal Fluid Research 2010, 7:1, doi:10.1186/1743-8454-7-1. <www.cerebrospinalfluidresearch.com/content/7/1/1>
von Rospatt, A., 1995. The Buddhist Doctrine of Momentariness: A Survey of the Origins and Early Phase of this Doctrine up to Vasubandhu. Stuttgart, Franz Steiner Verlag.
Welford, A. T., 1952. The 'psychological refractory period' and the timing of high speed performance – a review and a theory. Br. J. Psychol. 42, 2–19.
Zylberberg, A., Dehaene, S., Roelfsema, P. R., Sigman, M., 2011. The human Turing machine: a neural framework for mental programs. Trends in Cognitive Sciences, 15, 293–300.

Aaron Sloman develops a distinctive view of —

VIRTUAL MACHINERY AND EVOLUTION OF MIND (PART 1)

1. Virtual machines and causation

The idea of implementing one Turing machine in another can be seen as a precursor of the increasingly important idea of a virtual machine running in a physical machine. Some features of virtual machinery that are potentially relevant to explaining the evolution of mind and consciousness will be discussed, including their causal powers and the differences between implementation and reduction.

One of Turing's achievements was the specification of a *Universal Turing Machine* (UTM) within which any other Turing machine could be emulated by specifying its properties on the tape of a UTM (Turing, 1936). This led to proofs of important theorems, e.g. about equivalence, decidability and complexity. It can also be seen as a precursor of what we now call virtual machinery (not to be confused with virtual reality). I shall try to show how the combination of virtuality, causal interaction and (relative) indefinability can produce something new to science. My part 2 (in Part **4** of this volume) will present implications regarding evolution of mind and consciousness.

2. Virtuality

The UTM idea established that a computing machine can run by being implemented as a *virtual* machine in another machine. (I think the gist of this idea was understood by Ada Lovelace a century earlier.) The mathematical properties of a Turing machine's trajectory through its state space will not depend on whether it is run directly in physical machinery or as a virtual machine implemented in another computation. This has proved immensely important for theorems of meta-mathematics and computer science and for some of the practicalities of using one computer for multiple purposes, including time-sharing. One of the consequences is that a Turing machine implementing another Turing machine can also be a virtual machine implemented in a UTM: so that layered implementations are possible.

In the decades following publication of Turing's paper, engineering developments emerged in parallel with mathematical developments, with some consequences that have not received much attention, but are of great philosophical interest and potentially also biological import. I will suggest in Part 2 that biological evolution 'discovered' many of the uses of virtual machinery long before we did. Unfortunately, the word 'virtual' suggests something 'unreal' or 'non-existent', whereas virtual machines can make things happen: they can be causes, with many effects, including physical effects. To that extent they, and the objects and processes that occur in them, are *real* not *virtual*!

A possible source of misunderstanding is the fact that among a subset of computer scientists the label 'virtual machine' refers to software implementations of 'real', 'physical' machines which they accurately simulate (Popek and Goldberg, 1974). The notion of 'virtual machine' used in this paper includes machines whose operations cannot all be defined in terms of physical properties, although they are all *implemented* in physical machinery, and can interact with and control physical machinery. These virtual machines should not be regarded as surrogates for 'real' physical machines. They are real enough, in their causal powers, despite being virtual.

3. Causation and computation

Causation is a crucial aspect of the engineering developments in computing, as I'll now try to explain. It is possible to take any finite collection of Turing machines and emulate them running in parallel, in synchrony, on a UTM. This demonstrates that *synchronised* parallelism does not produce any qualitatively new form of computation. The proofs are theorems about relationships between abstract mathematical structures including sequences of states of Turing machines – and do not mention physical causation. A running physical machine can be an instance of such an abstract mathematical structure. However, being physical it can be acted on by physical causes, e.g. causes that alter its speed. Moreover, as remarked in Sloman (1996), standard computability theorems do not apply to physical Turing machines that are not synchronised. For example, if TM T1 repeatedly outputs '0', and T2 repeatedly outputs '1', and the outputs are merged to form a binary sequence, then if something (e.g. a device controlled by a geiger counter) causes the speeds of T1 and T2 to vary randomly and they run forever, the result could (and most probably would) be a non-computable infinite binary sequence, even though each of T1 and T2 conforms to theorems about Turing machines. (This claim will be refuted if it ever turns out that the whole physical universe can be modelled on a single Turing machine. I know of no evidence that such a model is possible.)

Likewise, if a machine has physical sensors and some of its operations depend on the sensor readings, then the sequence of states generated may not be specifiable by any TM, if the environment is not equivalent to a TM. So the mathematical 'limit' theorems do not apply to all physically implemented information-processing systems. In fact a machine with sensors and effectors connected to physical objects in the environment is fundamentally different from a Turing machine running its 'closed' world consisting only of its (infinite tape) and controlling transition table.

Mathematical entities, such as numbers, functions, proofs and abstract models of computation, do not have spatio-temporal locations, whereas running instances of computations do, some of them distributed across networks. Likewise, there are no causal connections, only logical connections, between the TM states that form the subject matter of the mathematical theory of computation, whereas there *are* causal connections in the running instances, depending on the physical machinery used and the physical environment. So, notions like 'reliability' are relevant to the physical instances, but not the mathematical abstractions. From a mathematical point of view there is no difference between three separate computers running the same program, and a single computer simulating the three computers running the program. However, an engineer aiming for reliability would choose three physically separate computers with a voting mechanism as part of a flight control system, rather than a mathematically equivalent, equally fast, implementation in a single computer (Sloman, 1996), if all the computers use equally reliable components.

Physical details of time-sharing of the machines have other consequences. When the three separate machines running in synchrony switch states in unison, nothing happens between the states, whereas in the time-shared implementation on one computer, the underlying machine has to go through operations to switch from one virtual machine to another. Such 'context switching' processes have intermediate sub-states that do not occur in the parallel implementation. A detailed mathematical model of one machine running three virtual machines will need to include the intermediate states that occur during switching, but a model of three separate concurrently active machines will not. A malicious intruder, or a non-malicious operating system, will have opportunities to interfere with the time-shared systems during a context-switching process, e.g. modifying the emulated processes, interrupting them, or copying or modifying their internal data.

Such opportunities for intervention (e.g. checking that a sub-process does not violate access restrictions or transferring information between devices) are often used both within individual computers and in networked computers causally linked to external environments, e.g. sensing or controlling physical devices, chemical plants, air-liners, commercial customers, social or economic systems, and many more. In some cases, analog-to-digital digital-to-analog converters, and direct memory access mechanisms now allow constant interaction between processes. See also Dyson (1997).

The technology supporting the causal interactions includes (in no significant order): *memory management, paging, cacheing, interfaces of many kinds, interfacing protocols, protocol converters, device drivers, interrupt handlers, schedulers, privilege mechanisms, resource control mechanisms, file-management systems, interpreters, compilers, 'run-time systems' for various languages, garbage collectors, mechanisms supporting abstract data types, inheritance mechanisms, debugging tools, pipes, sockets, shared memory systems, firewalls, virus checkers, security systems, operating systems, application development systems, name-servers*, and more. All of these can be seen as contributing to intricate webs of causal connections in running systems, including *preventing* things from happening, *enabling* certain things to happen in certain conditions, *ensuring* that if certain things happen then other things happen, and in some cases *maintaining mappings* between physical and virtual processes, e.g. in device drivers. Philosophers who think that different causal webs at different levels of abstraction cannot coexist need to learn more engineering, unfortunately not a standard component of a philosophy degree.

4. Causation in RVMs

A running virtual machine can have many effects, including causing its own structure to change. Understanding how virtual machines can cause anything to happen requires a three-way distinction, between: (a) *Mathematical Models* (MMs), e.g. numbers, sets, grammars, proofs, etc., (b) *Physical Machines* (PMs), including atoms, voltages, chemical processes, electronic switches, etc., and (c) *Running Virtual Machines* (RVMs), e.g. calculators, games, formatters, provers, spelling checkers, email handlers, operating systems, etc., running in general-purpose computers.

MMs are static abstract structures, like proofs and axiom systems. Like numbers, they cannot *do* anything. They include Turing machine executions whose properties are the subject of mathematical proofs. Unfortunately some uses of 'virtual machine' refer to MMs, e.g. 'the Java virtual machine'. These are abstract, inactive, mathematical entities, not RVMs, whereas PMs and RVMs are active and cause things to happen.

Physical machines on our desks can now support varying collections of virtual machinery with various kinds of concurrently interacting components whose causal powers operate in parallel with the causal powers of underlying virtual or physical machines, and help to control those physical machines. Some of them are *application* RVMs that perform specific functions, e.g. playing chess, correcting spelling, handling email. Others are *platform* RVMs, like operating systems, or run-time systems of programming languages, which are capable of supporting many different higher level RVMs. Different RVMs have *different levels of granularity* and *different kinds of functionality*. They all differ from the granularity and functionality of the physical machinery. Relatively simple transitions in a RVM can use a very much larger collection of changes at the machine code level and an even larger collection of physical changes in the underlying PM – far more than any human can think about. Apart from the simplest programs even machine code specifications are unmanageable by human programmers. Automatic mechanisms (including compilers and interpreters) are used to ensure that machine-level processes support the intended RVMs.

Interpreted and compiled programming languages have important differences in this context. An interpreter ensures *dynamically* that the causal connections specified in the program are maintained. If the program is changed while running, the interpreter's behaviour will change. In contrast, a compiler *statically* creates machine code instructions to ensure that the specifications in the program are subsequently adhered to, and the original program plays no role thereafter. Changing it has no effect, unless it is recompiled (e.g. if an *incremental* compiler is used). In principle the machine code instructions can be altered directly by a running program (e.g. using the 'poke' command in Basic) but this is usually feasible only for relatively simple changes and would probably not be suitable for altering a complex plan after new obstacles are detected, and modifying the physical wiring would be out of the question. So some kinds of self-monitoring and self-modification are simplest if done using process descriptions corresponding to a high level virtual machine specified in an interpreted formalism and least feasible if done at the level of physical structure. Compiled machine code instructions are an intermediate case.

There are two different benefits of using a suitable RVM, namely (a) the already mentioned coarser granularity of events and states compared with a PM or low level RVM, and (a) the use of an ontology related to the application domain (e.g. playing chess, making airline reservations). Both of these are indispensable for processes of design, testing, debugging, extending, and also for run-time self-monitoring and control, which would be impossible to specify at the level of physical atoms, molecules or even transistors (partly because of explosive combinatorics, especially on time-sharing, multi-processing systems where the mappings between virtual and physical machinery keep changing). The coarser grain, and application-centred ontology makes self-monitoring (like human debugging of the system) more practical when high-level interpreted programs are run than when machine code compiled programs are run. This relates to the third aspect of some virtual machinery: ontological irreducibility.

5. Implementable but irreducible

The two main ideas presented so far are fairly familiar, namely (a) a VM can run on another (physical or virtual) machine, and (b) RVMs running in parallel can interact causally with one another and with things in the environment. A third consequence of 20th century technology is not so obvious, namely: *some VMs include states, processes and causal interactions whose descriptions require concepts that cannot be defined in terms of the language of the physical sciences: they are non-physically describable machines (NPDMs).* Virtual machinery can extend our ontology of types of causal interaction beyond physical interactions.

This is not a form of mysticism. It is related to the fact that a scientific theory can use concepts (e.g. 'gene', 'valence') that are *not definable* in terms of the actions and observations that scientists can perform. This contradicts both the 'concept empiricism' of philosophers like Berkeley and Hume, originally demolished in Kant (1781), and also its modern reincarnation, the 'symbol grounding' thesis popularised by Harnad (1990), which also claims that all concepts have to be derived from experience of instances. The alternative 'theory tethering' thesis, explained in Sloman (2007), is based on the conclusion in twentieth century philosophy of science that undefined symbols used in deep scientific theories get their meanings primarily, though not exclusively, from the structure of the theory, though a formalisation of such a theory need not fully determine what exactly it applies to in the world. The remaining indeterminacy of meaning is partly reduced by specifying forms of observation and experiment (e.g. 'meaning postulates' in Carnap (1947)) that are used in testing and applying the theory, 'tethering' the semantics of the theory. The meanings are never uniquely determined, since it is always possible for new observations and measurements (e.g. of charge on an electron) to be adopted as our knowledge and technology advance.

Ontologies used in specifying VMs, e.g. concepts like 'pawn', 'threat', 'capture' etc. used in specifying a chess VM, are also mainly defined by their role in the VM, whose specification expresses an explanatory theory about chess. Without making use of such concepts, which are not part of the ontology of physics, designers cannot develop implementations and users cannot understand what the program is for, or make use of it. So, when the VM runs, there is a physical implementation that is also running, but the two are not identical: there is an asymmetric relation between them. The PM is an *implementation* of the VM, but the VM is not an implementation of the VM, and there are many other statements that are true of one and false of the other. The RVM, but not the PM, may include threats, and defensive moves. And neither 'threat' nor 'defence' can be defined in the language of physics. Not all the concepts used to describe objects, events and processes in a RVM are *definable* in terms of concepts of physics even though the RVM is *implemented* in a physical machine. The physical machine could include some of the environment with which the RVM interacts. The detailed description of the PM is not a specification of the VM, since the VM could be the same even if it were implemented on a very different physical machine with different physical processes occurring during the execution even of a particular sequence of chess moves. The VM description is also not equivalent to any fixed *disjunction of descriptions* since the VM specification determines which PMs are adequate implementations. Programmers can make mistakes, and bugs in the virtual machinery are detected and removed, usually by altering a textual specification of the abstract virtual machinery not the physical machinery. When a bug in the program is fixed it does not have to be fixed differently for each physical implementation – a compiler or interpreter for the language handles the mapping between virtual machine and physical processes and those details are not part of the specification of the common virtual machine.

Neither can the VM machine states and processes be defined in terms of physical input-output specifications, since very different technologies can be used to implement interfaces for the same virtual machine, e.g. using mouse, keyboard, microphone or remote email for input. Moreover, some VMs perform much richer tasks than can be fully expressed in input–output relations, e.g. the visual system of a human (or future robot!) watching turbulent rapids in a river. (Compare the critique of Skinner in Chomksy (1959).)

The indefinability of VM ontologies in terms of PM ontologies does not imply that RVMs include some kind of 'spiritual stuff' that can exist independently of the physical implementation machinery, as assumed by those who believe in immortal minds, or souls. Despite the indefinability there are close causal connections between VM and PM states, but that includes things like detection of a threat causing a choice of defensive move, which is a VM process that can cause changes in the physical display and the physical memory contents. We thus have what is sometimes referred to as 'downwards causation', in addition to 'upwards causation' and 'sideways causation' (within the RVM).

6. Implications

The complex collection of hardware, firmware, and software technologies, developed since Turing's time has made possible information-processing systems of enormous complexity and sophistication performing many tasks that were previously performed only by humans and some that not even humans can perform. This has required new ways of thinking about *non-physically describable* virtual machinery (NPDVM) with causal powers. The new conceptual tools are relevant not only to engineering tasks but also to understanding what self-monitoring, self-controlling systems can do. Philosophy now has the task of working out in detail metaphysical implications of multiple coexisting causal webs with causation going sideways, upwards and downwards. Implications for evolution of mind are discussed in Part 2 of this paper, included in Part III of this volume. Finally, Part 3 of this paper, presenting the concept of meta-morphogenesis (the processes by which the processes of change and development change) will be included in Part IV of this volume.

References

Carnap, R., 1947. Meaning and Necessity: A Study in Semantics and Modal Logic. Chicago University Press, Chicago.

Chomksy, N., 1959. Review of skinner's Verbal Behaviour. Language, 35, 26–58.

George, B.D., 1997. Darwin Among The Machines: The Evolution Of Global Intelligence. Addison-Wesley, Reading, MA.

Harnad, S., 1990. The symbol grounding problem. Physica D, 42, 335–346.

Kant, I., 1781. Critique of Pure Reason. Macmillan, London. Translated (1929) by Norman Kemp Smith.

Popek, G.J., Goldberg, R.P., 1974. Formal requirements for virtualizable third generation architectures. Commun. ACM 17 (7).

Sloman, A., 1996. Beyond turing equivalence. In: Millican, P.J.R., Clark, A. (Eds.), Machines and Thought: The Legacy of Alan Turing (vol I), The Clarendon Press, Oxford, pp. 179–219. URL http://www.cs.bham.ac.uk/research/projects/cogaff/96-99.html#1. (Presented at Turing90 Colloquium, Sussex University, April 1990.

Sloman, A., 2007. Why symbol-grounding is both impossible and unnecessary, and why theory-tethering is more powerful anyway. http://www.cs.bham.ac.uk/research/projects/cogaff/talks/#models.

Turing, A.M., 1936. On computable numbers, with an application to the Entscheidungsproblem. Proc. Lond. Math. Soc. 42 (2), 230–265. URL http://www.abelard.org/turpap2/tp2-ie.asp.

Artur Ekert on the physical reality of —

One of many remarkable traits of Alan Turing was his ability to bridge the gap between the abstract and the physical. His background in physics is clearly seen in his approach to the definition of computability. Turing's machines (Turing, 1936) captured the notion of effective computation in a much more tangible and convincing way than, for example, the lambda calculus proposed by Alonzo Church (this was generously acknowledged by Church (1937) himself). Although Turing's machines were abstract constructs of his mathematical imagination there was nothing unphysical about them. Indeed, Turing's machines (with arbitrarily long tapes) can be built, but no one would ever do so except for fun, as they would be extremely slow and cumbersome. The computer I am working on at the moment is much faster and more reliable.

But wait a minute! Where does this reliability come from? My computer is a physical object, made out of a vast number of electronic components. How do I know that the computer generates the same outputs as the appropriate abstract Turing machine? How do I know that the machinery of electric currents must finally display the right answer? After all, nobody has tested the machine by following all possible logical steps, or by performing all the arithmetic it can perform. If they were able and willing to do that, there would be no need to build the computer in the first place. The reason we trust the machine cannot be based entirely on logic; it must also involve our knowledge of the physics of the machine. When relying on the machine's output, we rely on our knowledge of the laws of physics that govern the computation, i.e. the physical process that takes the machine from an initial state (input) to a final state (output) (Deutsch et al., 2000).

Given that algorithms can now be performed by real automatic computing machines, the natural question arises: what, precisely, is the set of logical procedures that can be performed by a physical machine? The theory of Turing machines cannot, even in principle, answer this question, nor can any approach based on formalising traditional notions of effective procedures. What we need instead is to extend Turing's idea of *mechanising* procedures. This would define logical procedures by the mechanical procedures that effectively perform logical operations. But what does it mean to involve real, physical machines in the definition of a logical notion? The discovery of quantum physics has provided us with an excellent example. Consider the following, very simple, machine which performs a computation mapping $\{0, 1\}$ to itself.

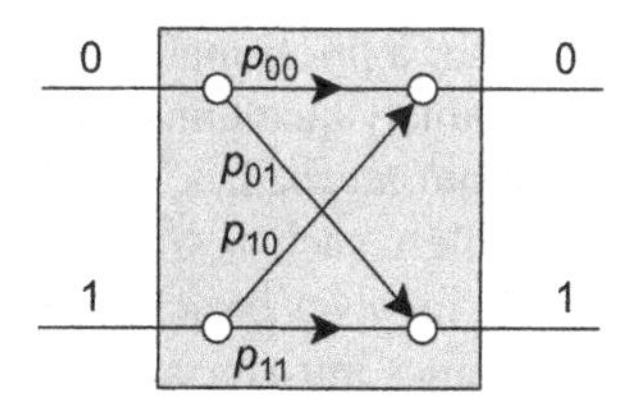

Here p_{ab} is the probability for the machine to produce the output b when presented with the input a. It may seem obvious that if the p_{ab} are arbitrary apart from satisfying the standard probability conditions $\sum_b p_{ab} = 1$, then the figure above represents *the most general* machine whose action depends on no other input or stored information and which performs a computation mapping $\{0, 1\}$ to itself. The deterministic limits are obtained by setting $p_{01} = p_{10} = 0$, $p_{00} = p_{11} = 1$ (which gives a logical identity machine) or $p_{01} = p_{10} = 1, p_{00} = p_{11} = 0$ (which gives a negation ('**not**') machine). Otherwise we have a randomising device. Let us assume, for the sake of illustration, that $p_{01} = p_{10} = p_{00} = p_{11} = 0.5$. Again, we may be tempted to think of such a machine as a random switch which, with equal probability, transforms any input into one of the two possible outputs. However, that is not necessarily the case. When the particular machine we are thinking of is followed by another, identical, machine the output is always the negation of the input.

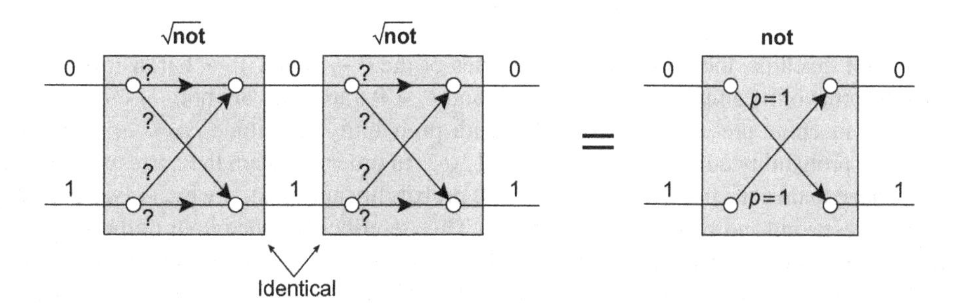

This is a very counter-intuitive claim – the machine alone outputs 0 or 1 with equal probability and independently of the input, but the two machines, one after another, acting independently, implement the logical operation **not**. That is why we call this machine $\sqrt{\textbf{not}}$. It may seem reasonable to argue that since there is no such operation in logic, $\sqrt{\textbf{not}}$ machines cannot exist. But they do exist! Physicists studying single-particle interference routinely construct them, and some of them are as simple as a half-silvered mirror, i.e. a mirror which with probability 50% reflects a photon that impinges upon it and with probability 50% allows it to pass through. In this particular case the logical values, 0 and 1, are represented by paths taken by photons before and after travelling through the mirror (Ekert, 2006; Nielsen and Chuang, 2000).

The reader may be wondering what has happened to the axiom of additivity in probability theory, which says that if E_1 and E_2 are mutually exclusive events then the probability of the event (E_1 **or** E_2) is the sum of the probabilities of the constituent events, E_1, E_2. We may argue that the transition $0 \to 0$ in the composite machine can happen in two mutually exclusive ways, namely, $0 \to 0 \to 0$ or $0 \to 1 \to 0$. The probabilities of the two are $p_{00}p_{00}$ and $p_{01}p_{10}$ respectively. Thus, the sum $p_{00}p_{00} + p_{01}p_{10}$ represents the probability of the $0 \to 0$ transition in the new machine. Provided that p_{00} or $p_{01}p_{10}$ are different from zero, this probability should also be different from zero. Yet we can build machines in which p_{00} and $p_{01}p_{10}$ are different from zero, but the probability of the $0 \to 0$ transition in the composite machine is equal to zero. So what is wrong with the above argument?

One thing that is wrong is the assumption that the processes $0 \to 0 \to 0$ and $0 \to 1 \to 0$ are mutually exclusive. In reality, the two transitions *both occur*, simultaneously. We cannot learn about this fact from probability theory or any other a priori mathematical construct. We learn it from the best physical theory available at present, namely quantum mechanics.

The mathematical machinery of quantum mechanics, which can be used to describe quantum computing machines ranging from the simplest, such as $\sqrt{\textbf{not}}$, to the quantum generalisation of the universal Turing machine (Deutsch, 1985), involves basic operations on complex numbers. Indeed, at the level of predictions, quantum mechanics introduces the concept of *probability amplitudes* – complex numbers c such that the quantities $|c|^2$ may under suitable circumstances be interpreted as probabilities. When a transition, such as 'a machine composed of two identical sub-machines starts in state 0 and generates output 0, and affects nothing else in the process', can occur in several alternative ways, the overall probability amplitude for the transition is the sum, not of the probabilities, but of the probability amplitudes for each of the constituent transitions considered separately.

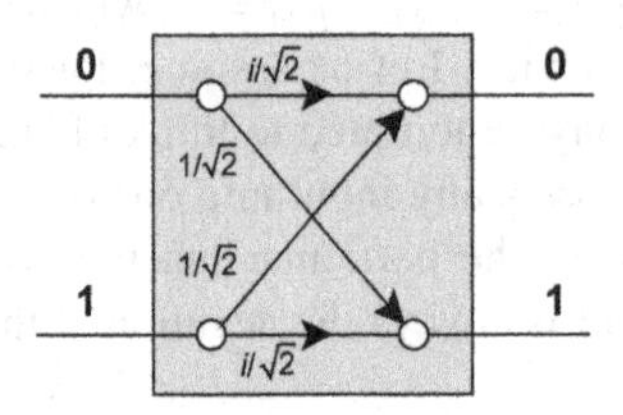

In the $\sqrt{\textbf{not}}$ machine, the probability amplitudes of the $0 \to 0$ and $1 \to 1$ transitions are both $i/\sqrt{2}$, and the probability amplitudes of the $0 \to 1$ and $1 \to 0$ transitions are both $1/\sqrt{2}$. This means that the $\sqrt{\textbf{not}}$ machine preserves the bit value with probability amplitude $c_{00} = c_{11} = i/\sqrt{2}$ and negates it with probability amplitude $c_{01} = c_{10} = 1/\sqrt{2}$. In order to obtain the corresponding probabilities we have to take the modulus squared of the probability amplitudes, which gives probability $1/2$ both for preserving and swapping the bit value. This describes the behaviour of the single $\sqrt{\textbf{not}}$ machine. When we concatenate the two machines then, in order to calculate the probability of output 0 from input 0, we have to add the probability amplitudes of all computational paths leading from input 0 to output 0. There are only two of them – $c_{00}c_{00}$ and $c_{01}c_{10}$. The first computational path has probability amplitude $i/\sqrt{2} \times i/\sqrt{2} = -1/2$ and the second one $1/\sqrt{2} \times 1/\sqrt{2} = +1/2$. We add the two probability amplitudes first and then we take the modulus squared of the sum. We find that the probability of output 0 is zero. Unlike probabilities, probability amplitudes can cancel each other out!

Quantum theory explains the behaviour of $\sqrt{\textbf{not}}$ and correctly predicts the probabilities of all the possible outputs no matter how we concatenate the machines. This knowledge was created as the result of conjectures, experimentation, and refutations. Hence, reassured by the physical experiments that corroborate this theory, logicians are now entitled to propose a new logical operation $\sqrt{\textbf{not}}$. Why? Because a faithful physical model for it exists in nature!

The story of the $\sqrt{\textbf{not}}$ is just one example which illustrates the main point: whenever we improve our knowledge about physical reality, we may also gain new means of improving our knowledge of logic, mathematics and formal constructs. It seems that we have no choice but to recognise the dependence of our mathematical *knowledge* (though not of mathematical truth itself) on physics, and that being so, it is time to abandon the classical view of computation as a purely logical notion independent of that of computation as a physical process (Deutsch, 1997; Deutsch et al., 2000).

References

Church, A., 1937. Review of Turing 1936. J. Symb. Log. 2, 42–43.

Deutsch, D., 1985. Quantum theory, the Church-Turing principle and the universal quantum computer. Proc. R. Soc. A: Math. Phys. Eng. Sci. 400, 97–117.

Deutsch, D., 1997. The Fabric of Reality. Allen Lane, The Penguin Press, New York, London.

Deutsch, D., Ekert, A., Lupacchini, R., 2000. Machines, logic and quantum physics. Bull. Symb. Log. 6, 265–283.

Ekert, A., 2006. Quanta, ciphers and computers. In: Fraser, G. (Ed.), The New Physics for the Twenty-First Century. Cambridge University Press, Cambridge, New York, pp. 268–283.

Nielsen, M. A., Chuang, I. L., 2000. Quantum Computation and Quantum Information. Cambridge University Press, Cambridge, New York.

Turing, A. M., 1936. On computable numbers, with an application to the Entscheidungsproblem. Proc. Lond. Math. Soc. 42, 230–265.

Cristian Calude, Ludwig Staiger and Michael Stay on —

HALTING AND NON-HALTING TURING COMPUTATIONS

1. Introduction

Turing's famous paper (Turing, 1936) proved that the *halting problem* – the problem of deciding whether a given Turing machine ever reaches the halting state when provided with a given tape – was undecidable. Turing machines give us a convenient way to talk about the time and space necessary to carry out computations, and play a significant role in both classical recursion theory and the theory of computational complexity (Cooper, 2004; Balcázar et al., 1995; Sipser, 2006; Wagner and Wechsung, 1986).

Nowadays, the undecidability discovered by Turing need not be quite the fearsome phenomenon it at first appears. To understand this we look in more detail at the time and space of Turing computations.

First, any Turing machine having an undecidable halting problem uses an infinite number of cells on its working tape (Calude and Staiger, 2010). Accordingly, the halting behaviour of a Turing machine M on input x can be divided into three categories:

(1) The machine M halts on x, in which case the number of cells used is necessarily finite.
(2) The machine M does not halt on x, but uses only finitely many distinct cells on its tapes.
(3) The machine M does not halt on x and uses infinitely many distinct cells on its tapes.

In the second case above, the halting problem for M on x is decidable, so Turing's undecidability result relies on the fact that machines with an undecidable halting problem necessarily use infinite space.

Secondly, the critical time (Chaitin, 1987) can be used to yield a classification of Turing computations into three categories:

- The machine M halts on x in time bounded by the critical number of steps.
- The machine M halts on x in time not bounded by the critical number of steps.
- The machine M does not halt on x.

The last case, that is when M does not halt on x, can be refined in terms of space complexity. Finally, the significance of these mathematical facts for hypercomputation and formal proofs in mathematics will be briefly discussed.

2. Turing machines

A Turing machine is a formalisation of a mechanical device. The device has a long tape on which a finite alphabet of symbols are read or written, and the tape can be shifted back and forth through the machine, which can read symbols from and write symbols to the tape. The machine itself has a finite set of internal states, and updates those states depending on what it finds on the tape. In the formalisation, the tape is infinite and there are never any errors. Turing's own description is wonderfully lucid, and we refer the reader to his account for details.

The set of pairs (M,x), where M is a Turing machine and x is an input, can be computably enumerated. We fix such an enumeration and we denote by $code(M,x)$ the code, or description, of the pair (M,x) in this enumeration.

3. Resources

Let M be a Turing machine and x an input word.

The function $time_M(x)$ denotes the number of steps executed by M on input x (see Balcázar et al. (1995)). By $M(x) < \infty$ we denote the fact that M stops on x. The *halting problem for a particular Turing machine M* is the problem of deciding, given x, whether $M(x) < \infty$. The *halting set* or the *domain* of M is the set $halt_M = \{x \mid M(x) < \infty\}$. It is well known that the halting problem for most Turing machines M is undecidable; more precisely, the halting set of M is computably enumerable but not computable. A Turing machine whose domain is prefix-free is called *self-delimiting* or *prefix-free* (Calude, 2002). Although many results presented below hold for any Turing machine, for uniformity we study only prefix-free Turing machines, which from now on we will simply call *machines*.

The computational space, or *space function*, $space_M(x)$ used by M on x is defined to be the number – finite or infinite – of cells used by M during its computation with the input x; a cell used at least once is counted as used.[4] Obviously, if $space_M(x)$ is finite, then the computation process as described above can have only a finite number of different configurations.[5]

Clearly, $space_M(x) < \infty$ whenever $M(x) < \infty$, and $M(x) = \infty$ if and only if $time_M(x) = \infty$.

Given a machine M, we can therefore classify input strings x according to $time_M$ and $space_M$ and get the following three sets:

$$halt_M = \{x \mid time_M(x) < \infty\},$$

$$\{x \mid time_M(x) = \infty,\ space_M(x) < \infty\},$$

$$\{x \mid space_M(x) = \infty\}.$$

Calude and Staiger (2010) showed that if for every x, $space_M(x) < \infty$, then the halting problem for M is decidable. This result does not contradict Turing's undecidability of the *halting problem* because the set of descriptions *code(M,x)* – where M is a machine, x is a string – for which $space_M(x) < \infty$, is computably enumerable but not computable.

4. Halting time

Let bin be the computable bijection that associates to every integer $n \geq 1$ its binary expansion without the leading 1: bin(1) is the empty string, bin(2) = 0, bin(3) = 1, bin(4) = 00 etc. The *natural complexity* of the string y (with respect to the machine M) is $\nabla_M(y) = \min\{n \geq 1 \mid M(\text{bin}(n)) = y\}$ (see Calude and Stay (2006)); ∇ is a relative of Kolmogorov complexity for partially computable functions used in Manin and Zilber (2010).

The invariance theorem says that one can effectively construct a 'universal machine' that can simulate any other machine. A machine U is universal if for every machine M there is a constant $\varepsilon > 0$ (depending upon U and M) such that $\nabla_U(x) \leq \varepsilon \cdot \nabla_M(x)$, for all strings x. We fix a universal machine U and define $\nabla = \nabla_U$.

Say a machine M gets an input x and runs for exactly t steps before halting. Chaitin (1987) showed that there is a program y for the universal machine not much longer than *code(M,x)* such that $U(y) = t$ – or more formally, that there is a constant c such that if $M(x)$ halts exactly in time t, then $\nabla(\text{bin}(t)) \leq 2^{|code(M,x)|+c}$.

A binary string x is algorithmically random if $\nabla(x) \geq 2^{|x|}/|x|$. A time t is called algorithmically random if bin(t) is algorithmically random. Assume that $M(x)$ has not stopped in time $2^{2N+2c+1}$, where $N = |code(M,x)|$ and c comes from Chaitin's statement above; then Calude and Stay (2008) proved that $M(x)$ cannot stop at any algorithmically random time $t \geq 2^{2N+2c+1}$.

Therefore, if one runs a program for long enough (where 'long enough' depends on c above), then either the program halts at a non-algorithmically random time or it does not halt at all. The density of non-algorithmically random numbers near n is $1/n$. Hence, *most times are not halting times for any machine and input.*

Consider the set $S = \{0,1\}^R \times \mathbb{N}$ whose elements are pairs of an input string of length R and a potential runtime. Let $q(x)$ be the uniform probability distribution on strings of length R and $p(n)$ be any computable probability distribution on natural numbers. Given M,x and a positive integer m, we can effectively compute a critical value $T_{critical}(|code(M,x)|,m)$ such that either $M(x)$ stops in time less than $T_{critical}(|code(M,x)|,m)$, or the probability given by $q \times p$ that $M(x)$ eventually stops is smaller than 2^{-m}.

[4] This definition differs slightly from the space complexity usually employed in computational complexity theory (Balcázar et al., 1995), which treats the space as infinite if the time is infinite.

[5] A configuration records the current state, tape contents and head location (Sipser, 2006, p. 140).

Manin (2010) proved a general result of this kind valid for many complexity measures, including time.

Given a probability bound 2^{-m}, the halting behaviour of a machine M can be described by the following three sets:

$$\{x \mid time_M(x) < T_{critical}(|code(M,x)|, m)\},$$
$$\{x \mid T_{critical}(|code(M,x)|, m) \leq time_M(x) < \infty\},$$
$$\{x \mid time_M(x) = \infty\}.$$

The last case can be refined using the computational space as follows:

$$\{x \mid time_M(x) = \infty, space_M(x) < \infty\},$$
$$\{x \mid time_M(x) = \infty, \text{ ZFC proves } space_M(x) = \infty\},$$
$$\{x \mid space_M(x) = \infty, \text{ but ZFC cannot prove } space_M(x) = \infty\}.$$

5. Final comments

What is the 'real-world' significance of this commentary?[6]

The bad news is that even an accelerated Turing machine (Copeland, 2002) needs infinite space to solve the halting problem. This begs for more insight into the mysterious usefulness of analogue computation which can bypass this limit (as Kreisel (1970) anticipated). The good news is that the halting problem can be probabilistically solved with any probability less than one.

Hilbert's formal proofs have been apparently killed for the practice of mathematics by Gödel's Incompleteness Theorem, so ultimately by the undecidability of the halting problem. Unexpectedly – at least from the theoretical view point – in the last decade, enormous progress has been made on automating the production of formal proofs, with tools like Isabelle, Coq and others (Hales, 2008). In part, this successful story – which in our humble opinion will change the way mathematics is done – is due to the practical possibility of working with meaningful, computational resource defined, approximations of the halting problem.

Acknowledgements

We thank Alastair Abbott, Barry Cooper and Jan van Leeuwen for comments that improved the presentation.

References

Balcázar, J.L., Díaz, J., Gabarró, L., 1995. Structural Complexity I, second ed. Springer-Verlag, Berlin.

Barrow, J., 2005. The Infinite Book. A Short Guide to the Boundless, Timeless and the Endless, Jonathan Cape, London.

Boolos, G., Jeffrey, R.C., 1980. Computability and Logic, Cambridge University Press, Cambridge.

Calude, C.S., 2002. Information and Randomness: An Algorithmic Perspective, second ed. Springer-Verlag, Berlin.

Calude, C.S., Staiger, L., 2010. A note on accelerated Turing machines. Math. Struct. Comput. Sci. 20, 1011–1017.

[6] ... was the question Barry Cooper asked us at the 9th iteration of this commentary.

Calude, C.S., Stay, M.A., 2006. Natural halting probabilities, partial randomness, and Zeta functions. Inf. Comput. 204, 1718–1739.
Calude, C.S., Stay, M.A., 2008. Most programs stop quickly or never halt. Adv. Appl. Math. 40, 295–308.
Chaitin, G.J., 1987. Computing the busy beaver function. In: Cover, T.M., Gopinath, B. (Eds.), Open Problems in Communication and Computation, Springer-Verlag, Heidelberg, pp. 108–112.
Cooper, S.B., 2004. Computability Theory, Chapman & Hall/CRC, Boca Raton, New York, London.
Copeland, B., 2002. Accelerating Turing machines. Minds Mach. 12 (2), 281–300.
Hales, T.C., 2008. Formal proof. Notices of the AMS 11, 1370–1380.
Kreisel, G., 1970. Church's thesis: a kind of reducibility axiom for constructive mathematics. In: Kino, A., Myhill, J., Vesley, R.E. (Eds.), Intuitionism and Proof Theory, North-Holland, Amsterdam, 121–150.
Manin, Y., 2010. Infinities in quantum field theory and in classical computing: renormalization program. In Ferreira, F., Löwe, B., Majordomo, E., Gomes, L.M. (Eds.), Programs, Proofs, Process. Proceedings CiE 2010, LNCS 6158, Springer, Heidelberg, pp. 307–316. Full paper in Renormalization and Computation II: Time Cutoff and the Halting Problem, http:// arxiv.org/abs/0908.3430, 24 August 2009.
Manin, Y., Zilber, B., 2010. A Course in Mathematical Logic for Mathematicians, second ed. Springer, Heidelberg.
Sipser, M., 2006. Introduction to the Theory of Computation, second ed. PWS, Boston.
Turing, A.M., 1936. On computable numbers, with an application to the Entscheidungsproblem. Proc. Lond. Math. Soc. 2 42, 230–265.
Turing, A.M., 1937. On computable numbers, with an application to the Entscheidungsproblem: A correction. Proc. Lond. Math. Soc. 2, 43, 544–546.
Wagner, K., Wechsung, G., 1986. Computational Complexity, Deutscher Verlag der Wissenschaften, Berlin.

Philip Welch leads us —

TOWARD THE UNKNOWN REGION: ON COMPUTING INFINITE NUMBERS

Darest thou now, O Soul,
Walk out with me toward the
Unknown Region,
Where neither ground is for the feet,
nor any path to follow?

Whitman, Leaves of Grass

The story is told that Church, after presenting the λ-calculus as a means of addressing the *Entscheidungsproblem*, was told by Gödel, in effect, to go away and try again. Gödel in conversation with Church said that he found the suggestion that effectively calculable be identified with the λ-calculus as 'thoroughly unsatisfactory'.[1] However, Gödel immediately recognised Turing's model as *the* model for this problem. Why was this? Presumably the criteria, somewhat implicit in Hilbert's question, as to whether there is an algorithm, or 'effective process' to determine from an 'effectively' given set of axioms in a deductive system, for any sentence φ of the system's language, whether or not φ was deducible from those axioms, would hinge crucially on the notion of algorithm or 'effectivity'. As Gödel had already shown the incompleteness of formal systems (satisfying a modicum of some basic requirements and thereby introducing the primitive recursive functions),

[1] This is recalled in a letter from Church to Kleene (Kleene, 1959). The conversation took place in early 1934 (Rosser, 1984).

all that remained in one sense was this question of 'decidability'. Was there a method for determining, given the axiom set A, which φ were deducible? The general recursive functions of Gödel were also used by Church but as Gandy (1988) and Sieg (1994) have pointed out, Church did not establish that the actual steps in this (or any other) calculus were themselves of a recursive nature. Crucially Turing's notion does this.

Turing's paper solved this problem by way of introducing the now familiar, but still beautifully simple, notion of an 'automatic machine' (or '*a*-machine', but which I shall just abbreviate as 'TM'). As is well known, in a TM an infinite tape runs to the right (and if one wishes to the left too). Turing analysed what he took to be a set of automaton like constructions/edits of a particular square or cell of the tape that was being scanned by what we should now call a read/write head, under the control of a finite instruction set. The machine would write marks from an alphabet (let us just say 0/1 here). If a computation halted there would then be an 'output' – meaning the tape's contents. Otherwise the machine would perform fruitlessly tasks for ever in an infinite sequence of moments in time $t_0, t_1, \ldots, t_n, \ldots$. The paper, although titled, and set up to prove, the unsolvability of the Entscheidungsproblem, had of course done much more than that: it had captured the notion of an automatic or mechanical behaviour, and moreover had deployed a notion of 'universality' – that there was a universal machine that could mimic any other machine given the latter's code as an input, and this, together with the intensional aspects of Turing's argument, are presumably what Gödel (and we later) found persuasive. In a final section of the paper, Turing applies his concept of machine to solving this problem. He further showed in an Appendix that the functions defined by Church's λ-calculus were coextensive with his computable functions.

This showed, when taken with the work of Church and Kleene, that the *(Gödel–Herbrand) general recursive functions* can be simulated by the TM. We shall not discuss here the Church–Turing thesis that any effective algorithmic process is simulable by Turing machines, or equivalently, as Turing proved, by the general recursive functions. Rather we shall concentrate on the capabilities of the machine model to compute such functions, and its generalisations.

Gödel had already drawn attention to the primitive recursive functions, which were then extended to the general recursive functions. Kleene, whilst working with Church on the Decision Problem, developed an *equational calculus* for the *partial recursive functions*: a set of basic equations and rules for recursion that allowed for the definition of the class of partial recursive functions as that class closed under those equational schemata (see, e.g. Odifreddi (1989)). It is this equational approach that was formative for a mathematico-logical approach to the theory of recursive functions, and indeed for a *definability* theory for sets of numbers. Whilst Turing had shown the universality of his machines, the flexibility of the equational approach allowed Kleene to produce a number of theorems analysing the computable functions. An *Enumeration Theorem* for the partial recursive functions as well as the *S-m-n Theorem* which allowed for enumeration of the partial recursive functions on $m+n$ variables to be re-enumerated as n-ary functions in m-parameters, and the *Recursion* or *Fixed-point* theorem. These relied on his

Theorem 1. *(Normal Form Theorem) (Kleene) There is a primitive recursive universal predicate T_1, and a primitive recursive function U satisfying $\forall e \forall x$:*

$$P_e(x) \downarrow z \quad \leftrightarrow \quad \exists y \in \mathbb{N}[T_1(e,x,y) \wedge U(y) = z].$$

In Turing's paper '*On Ordinal Logics*' (discussed elsewhere) he somewhat quietly introduced the notion of *o-machine* or 'oracle'-machine. Such a machine added to the capabilities of the *a*-machine the possibility of input during the course of computation via queries to an external oracle. If we enumerate the programs of a such a TM (equipped now with a command querying membership questions of a set A) as $\langle P_e^A \mid e \in \mathbb{N} \rangle$ then the halting set $A' =_{\mathrm{df}} \{e \mid P_e^A(e)\downarrow\}$ would come to be named the *Turing jump* of the set A, and is a *complete* Σ_1^A set, in that any set recursively enumerable in A is also (1-1) *reducible* to A'. This is the basis for Post's analysis of such sets and their relationship to the arithmetical hierarchy (here relativised to the set A). Thus *definability* becomes identified with the hierarchy of relativised halting sets. Moreover it is probably fair to say that the beginnings of the theory of definability, although connected with halting sets as stated, become increasingly divorced in conceptual terms from the TM model itself. It seems as if the machine has done its work for us, and now we may proceed to the mathematics of analysis of recursive sets, r.e. sets (which we might now want to call *computably enumerable* or *c.e.*), and hierarchies of such and the concomitant *degree theory*. In particular definability over the natural number structure later became extended to considering the notion of *hyperarithmetic* definability: sets of numbers that whilst not recursive, or even obtainable by finitely many iterations of the Turing jump operator, (i.e. again the arithmetically definable sets, that is those definable over $\mathbb{N} = \langle \mathbb{N}, +, \times, 0, S \rangle$ in the appropriate language), were definable by processes that themselves had computable, i.e., recursively definable, procedures: a hyperarithmetic set was one seen as having a *computable* or *recursive protocol* for describing its construction. By a definable version of Suslin's theorem, Kleene showed that such sets coincide with the Δ^1_1-definable sets over $\mathbb{N}$, that is those $A \subseteq \mathbb{N}$ that were both Π^1_1- and Σ^1_1-definable. Such notions then connect to the theory of inductive definitions over $\mathbb{N}$ as now described.

Let $\Psi : \mathcal{P}(\mathbb{N}) \to \mathcal{P}(\mathbb{N})$ be any arithmetic, or even Π^1_1 monotone operator (e.g., '$n \in \Psi(X)$' is arithmetic (or Π^1_1) as a relation of n and X); we define the following iterates of Ψ:

$$\Psi_0(X) = X; \quad \Psi_{\alpha+1}(X) = \Psi(\Psi_\alpha(X)); \quad \Psi_\lambda(X) = \bigcup_{\alpha<\lambda} \Psi_\alpha(X).$$

By sheer monotonicity, this process must reach a least α with $\Psi_\alpha(X) = \Psi_{\alpha+1}(X)$. For $X = \varnothing$ then Spector showed for Π^1_1 Ψ, (Gandy for arithmetical) that this process must halt by ω_1^{ck} the first non-recursive ordinal ('ck' for 'Church–Kleene'), and in general not sooner; the resulting *fixed point* set, namely that with $\Psi_\alpha(\varnothing) = \Psi(\Psi_\alpha(\varnothing))$, was Π^1_1.

It is possible (see Rogers (1967)) to define a reducibility for $x \subseteq \omega$: '*x is hyperarithmetic in y*', writing this as '$x \leq_h y$'; further there is the notion of a complete hyperarithmetically 'c.e.' set or '*hyperjump*' x^h corresponding to Turing jump x'. Spector was able to establish that the ordinal assignment $x \mapsto \omega^x_{1ck}$ (where ω^x_{1ck} is the least ordinal not recursive in x) satisfied the so-called '*Spector Criterion*':

$$x \leq_h y \to (x^h \leq_h y \leftrightarrow \omega^x_{1ck} < \omega^y_{1ck}).$$

To very cursorily summarise the history (in a rather crude fashion), generalisations of *recursion* away from the structure $\mathbb{N}$ tended in two directions in the 1960s: firstly via metarecursion theory and the work of Kreisel and Sacks, into extending the domain to initial segments of the ordinals that were *admissible*. The other direction initiated by Kleene, who published his equational system of recursion in higher type objects in a series of papers from 1959 onwards. He himself devised a notion that came to be called *Kleene Recursion*: this was a higher type recursion with a domain that of all the real numbers, that allowed hyperarithmetic questions about membership of sets of such numbers to be answered. This was a 'boldface' notion: the recursive sets of natural numbers became the Borel (or boldface Δ^1_1 sets by the Suslin Theorem) and the analogon of a computably enumerable set was a co-analytic sets of reals. We shall return to this model later. However the science of higher type recursion theory later became much developed by the work of Gandy, Moschovakis, Harrington, Normann, Kechris, et al.

Indeed it did, but these formulations, and indeed much of the generalised theory of inductive definitions, and the later theory of *Spector classes* of Moschovakis became intimately involved with the notion of admissible set. A transitive set M was an *admissible set* if it was a model of some modest basic set theory, including schemes of Δ_1-Comprehension and Σ_1-Replacement. In particular those levels of (relativised) Gödel hierarchies of the form $\langle L_\alpha[A], \in A \rangle$ that were admissible came to play a central role. The ordinal *height* of a transitive admissible set was named an *admissible ordinal*; the first such above ω is indeed ω_{1ck}: $L_{\omega_{1ck}} \models \text{KP}$. The importance of admissible sets was that they were domains inside which certain natural Σ_1-*set theoretic recursions* or constructions (rather than number theoretic) could be effected.

Much of this may seem a million miles from the original conceptions of recursive set of integers and universality of the computation/recursion procedure that allowed for the Enumeration and Recursion Theorems *etc.* mentioned above. However these founding theorems return, almost partly as definitions, in the notion that Moschovakis singled out as encompassing a generalised theory of inductive definability, that of the *Spector class.* We give the definition (with some details swept under the rug) as it pertains to inductive definitions at the lowest type, those of sets of integers.

DEFINITION 2. *(Spector Class) (Moschovakis) A class of sets of integers* $\Gamma \subseteq \mathcal{P}(\omega)$ *(or* $\mathcal{P}(\omega^\omega)$*, or direct products of such) is a* Spector Class *if the following are satisfied:*

(i) *(Closure)* Γ *is closed under trivial substitutions and those by functions themselves in* Γ*, and universal quantification over numbers:* $\forall^\omega$*;*
(ii) *(Universality) There is a relation* $R \in \Gamma \cap \omega \times \omega$*, which is* universal *for all relations in* Γ *on* ω*: if* $P \in \mathcal{P}(\omega) \cap \Gamma$ *then* $\exists e \in \omega\, \forall k (k \in P \leftrightarrow (e,k) \in R)$ *(and similarly for other products of* ω *and* ω^ω*).*
(iii) *(Norm Property) For any* $P \in \Gamma$ *there is a function* φ *with* $\varphi : P \longrightarrow On$ *satisfying certain properties close to saying that the prewellordering induced on* P *by* φ *has both its graph and its complement in* Γ*.*

The last property here is somewhat distractingly technical, and so has been left vague, but the point is that the function into the ordinals, gives us a nice prewellordering of the set P and in a very loose sense, we can think of it as saying that an n gets into P before m does if $\varphi(n) < \varphi(m)$. Property (ii) is recognisable as an *Enumeration* property; (i) hints at some initial basic closure properties.

There are countless examples of Spector classes, but the original basic such *ur*-class is obtained by taking the class of Π^1_1 sets of integers (and this is the *least* such class). Such classes relate to notions of definability over admissible sets, since to any Spector class Γ of sets of integers we may find an admissible structure M_Γ over which the sets in Γ correspond to sets inductively definable over M_Γ via some operator Ψ as above.

So, on the one hand, the increasing generalisations of recursion theory to higher types, and especially those including infinite ordinals in their domain, were mostly *mathematical* generalisations rather than machine model generalisations. This might lead one to think that with the increasing sophistication of the approach, that the original intuition of machine or computer had been left far, far behind. However running through this development was always a thread of machine-mindedness. For Kleene Recursion this was for Rogers (1967) the '$\aleph_0$-mind': one could think of Kleene Recursion as the notion of computability that might arise by taking hold of the TM model where a mind could survey the whole tape, as one step; or rewrite an ω-sequence of bits, as one operation, and further consult an oracle A (consisting of a *set* of subsets of $\mathbb{N}$) and receive a 0/1 answer as to whether the *whole set* of integers coded on a tape was or was not in A. In short anything that a mind capable of comprehending, and acting on, an $\aleph_0$=sized amount of information could do. The class of wellfounded 'computation trees' is then a Π^1_1 subset of $\mathcal{P}(\mathbb{N})$ with the latter identified as $\mathbb{R}$. A computational process that outputs, e.g., answers to membership questions about whether the real x belongs to a certain 'generalised r.e' set of real numbers, becomes then the problem of determining the wellfoundedness of a certain tree within any transitive admissible set containing that real x. In a realistic sense (*pace* the countably infinite processes involved) this can still be construed as a machine model.[2]

[2] This would take us too far off course, but even in metarecursion theory, Platek said that he also thought of a formalism involving a Shepherdson–Sturgess–Minsky like register machine containing countable ordinals in its registers, and performing the appropriate machine like instructions (Platek, Private communication). This could have been extended to α-recursion theory.

The recent two decades however has seen an expansion of interest in generalising models of computation, and this has led to mathematical investigations of models of computation where one or more space or time parameters are relaxed: the model of the *Infinite Time Turing Machine* of Hamkins and Lewis (2000) allows time to become transfinite. Time is still considered to tick away in discrete steps for $t = 0, 1, 2, \ldots$, but there are also now limit stages $t = \omega$, then $\omega + 1, \ldots, \omega + \omega, \ldots$ One simply has to specify a behaviour at limit stages of time: thinking of the ordinary Turing machine model, as the program is finite, if the machine does not halt but runs for an ω-sequence of steps, then it is in a program loop; so at time ω put it at the beginning of the outermost loop or subroutine in which it was involved (in other words at the least instruction number it visited infinitely often below ω). What is in the cells? We may specify, given an alphabet of 0's, 1's and B (for blanks) that the i-th cell C_i contains the alphabet symbol j at time ω, if there was a time $t = n < \omega$ so that for all later finite times $m \geq n$ the cell constantly contained a j; however if the cell value changed unboundedly often before ω then let it have value a B for ambiguity at time ω. Where is the R/W head? It at time t cell $C_{k(t)}$ is being read, we may place it at cell $C_{k(\omega)}$ where $k(\omega)$ is defined to be the liminf of the $k(t)$'s for $y < \omega$, *unless* this value is ω itself (because the head has wandered off to infinity). In the latter case we define $k(\omega)$ to be 0. (This accords with our idea of putting the machine at the start of the outermost loop entered into unboundedly often before time ω where possible. Hamkins and Lewis (2000) does this differently, by having three tapes, a 0/1 alphabet only and by taking limsups at limit ordinals, and placing the head back to cell C_0 at all limit times. However mathematically the functions produced are the same.) The same considerations are used at any limit ordinal λ. We may now amuse ourselves by asking any of the myriad questions that have been asked for the standard model TM. What are the 'ITTM'-computable functions produced by such machines? What are the decidable sets of integers? What is the halting set $H = \{P_e(e) \downarrow \mid e \in \mathbb{N}\}$? (Notice that we barely have to change notation to formulate the question.) Now computations may halt *after* stage ω; but at what stages? Since an ITTM can now receive an infinite stream of input it essentially can also compute on reals as well as integers. We can devise oracle machines that, like Kleene Recursion, can quiz a set of reals. What can we do now? Lest one think that this is merely an occupation for an idle hour on a rainy Sunday afternoon, one can show that the classes of ITTM 'semi-decidable' sets of numbers and reals produced form a Spector class. They are thus a particular instance of a higher type recursion theory.

Consider another machine: if we are relaxing time, why not go the whole hog and relax space considerations too? Let us consider an ITTM machine model with an *uncountably* long tape? Or even a tape as long as the class of ordinal numbers. What then can such a machine create? Remarkably there is an *ordinary standard Turing program* that can in effect compute, given some finite number of ordinals (input as 1's at the appropriate ordinal places on the tape) the truth set in the Gödel constructible universe L of the constructible set L-coded by those ordinals. In short, following such considerations, we have another presentation, now a *machine theoretic presentation*, of the set-theoretic Gödel L-hierarchy to set aside those alternatives of Jensen (1972) and Deutsch (1985) Sec. 9 (See Dawson (2009) and Koepke (2005).)

One may wonder at the apparent strength of these machine models, but a moment's reflection shows it to be in the limit rules themselves. We may have ordinary Turing style-action at successor stages, but the limit rule is a kind of infinitary logical rule integrating over this time dimension. It may look innocuous to put B-blanks on tapes at infinite stages, or take a liminf of previous cell values, but it is in these actions that the whole essence of the process inheres.

So what of these 'liminf' processes themselves? Consider then generalisations of the inductive operators defined above. We now define for such a Ψ (no longer required to be monotone):

$$\Psi_0(X) = X; \quad \Psi_{\alpha+1}(X) = \Psi(\Psi_\alpha(X));$$

$$\Psi_\lambda(X) = \liminf_{\alpha \to \lambda} \Psi_\alpha(X) =_{df} \cup_{\alpha<\lambda} \cap_{\lambda>\beta>\alpha} \Psi_\beta(X) \text{ for limit } \lambda.$$

What kind of operators are these? Such do not necessarily reach a fixed point but instead (after countably many iterations) reach a *stability point*: a least stage $\zeta = \zeta(X)$ so that $\Psi_\zeta(X)$ periodically returns for ever after as α runs through all the ordinals. Elementary arguments show that ζ exists and is countable. We thus may develop a theory of such *quasi-inductive definitions*. Starting with the natural numbers the quasi-inductive sets defined by arithmetic (or hyperarithmetic) operators Ψ again form a natural Spector class. Indeed this is to be expected: we can program an ITTM to calculate them.

Other examples may occur to the reader: consider now *Infinite Time Register Machines* (Koepke, 2006; Koepke and Miller, 2008). We allow transfinite time on a standard register machine containing integers. If co-finally at a limit stage a register has become unbounded we by *fiat* reset it to zero, and otherwise register contents contain the liminf of the previous values; the instruction number about to be performed at a limit stage is again the beginning of the outermost routine called unboundedly before the limit time. Now ask the same questions as for ITTM's. There is however a fundamental difference between the Register machine model and the TM model, which does not show up at the finite level. Universality fails, as there is no universal ITRM: as the numbers of registers increase, their strength increases. (In fact the class considered as a whole is weaker by far than the ITTM class.)

Instead of simply asking what the computational power of a transfinite computational model is, one can approach the machine from another direction: that of reverse mathematics (Simpson, 1999). Even the assertion that every ITTM on zero input either halts or loops requires a proof that can be effected only in a substantial fragment of second order number theory: Π^1_2-Comprehension is insufficient, although Π^1_3-Comprehension suffices. For ITRMs the analogous assertion turns out to be equivalent to Π^1_1-CA_0, see Koepke and Welch (2011).

At the risk of a truism, the thing one must always be aware of in considering these generalisations is the infinitary nature of the generalisation: in analysing ITTM's Hamkins and Lewis (2000) stick closely to questions and analogies with Turing jump and degree. However it is the analysis of the new concept in terms of set theory, or already extant higher recursion theory, that renders a full characterisation and delivers the deeper theorems. In the study by Hamkins and Lewis (2000) the analogy with TM's machines was thoroughly pursued, but ultimately the analogy of ITTM-*degree* is closer to that of hyperdegree, or rather Δ^1_2-degree, and the behaviour of such machines is closely tied to that of low levels of the Gödel constructible hierarchy; it is this realisation that enables one to actually answer some of the questions they asked: what are the semi-decidable sets of reals? How long do computations really take? Indeed an analysis of the latter is necessary in order to prove the analogue of Kleene's Normal Form Theorem:

Theorem 3. *(ITTM Normal Form Theorem) (Welch, 2009)*

(a) *For any program index e we may effectively (in the usual sense) find an index e' so that:*

$$\forall x \in 2^{\mathbb{N}} P_e(x) \downarrow \Rightarrow \exists y \in 2^{\mathbb{N}} P_{e'}(x) \downarrow y$$

where y is a code for the whole computation sequence for $P_e(x)$.

(b) *There is an ITTM-decidable universal predicate $\mathfrak{T}_1$, and an arithmetical function U satisfying $\forall e \forall x$:*

$$P_e(x) \downarrow z \leftrightarrow \exists y \in 2^{\mathbb{N}}[\mathfrak{T}_1(e,x,y) \wedge U(y) = z].$$

Note that a y in part (a) has to be a real coding at the very least the ordinal length of the computation $P_e(x)$; in order for there to be any hope of such an e' existing, we need to know that (a code for) the *length* of any computation of the form $P_e(x)$ can itself be computed from something decidable in x. This length is of course a transfinite ordinal, and so we are in a very different ballpark from the original Normal Form Theorem. Fortunately we can, and this length has a characterisation in terms of the L-hierarchy; further, because of this analysis we can determine the decidable, semi-decidable sets etc. for this model. As evidence that we are doing the right kind of thing we see that we can also obtain a Spector Criterion for the notion of ITTM-reducibility on sets of integers, $x \leq_\infty y$; using ITTM-degrees, and ITTM-jump (denoted x^∇), just as Spector had for hyperdegree:

$$x \leq_\infty y \rightarrow (x^\nabla \leq_\infty y \leftrightarrow \zeta^x < \zeta^y)$$

where now ζ^x is the least ordinal after which the universal x-ITTM machine starts repeating.

We mention this here, as it typifies analyses of such proposed machine-theoretically inspired models, that once they are allowed into the transfinite realm, then one uses set-theoretic, or analytical methods to resolve such questions. Once one has started this kind of freeing oneself from the finite realm, one sees all sorts of possibilities: for example consider the Blum-Shub-Smale machine – again something certainly inspired by the Turing model. If we let this run transfinitely what may that compute? Can we think of more general transfinite machines with different and perhaps more complex limit rules? Friedman and Welch (2011) is one attempt to define such machines that run through all the reals of the least β-model of analysis. Can one make sense in general of (some variant of) dynamical systems allowed to run beyond ω? Usually such systems are restricted to continuous functions on some interval or manifold – but what if we consider more general functions in some higher Baire class?

It seems to me that there is little difference in the end between a system of equations recursively applied (in some general sense), and a general machine. However, within the theory of Spector classes and higher type recursion theory, these machine-inspired models occur sporadically as points of illumination, as concrete, and so readily graspable, examples of that rather abstract theory.

And all of this ultimately we have Turing to thank for.

References

Dawson, B., 2009. Ordinal time Turing computation. Ph.D. thesis, Bristol.

Devlin, K., 1984. Constructibility. Perspectives in Mathematical Logic. Springer Verlag, Berlin, Heidelberg.

Friedman, S.D., Welch, P.D., 2011. Hypermachines. J. Symbol. Log. 76 (2), 620–636.

Gandy, R.O., 1988. The confluence of ideas in 1936. In: Herken, (Ed.), The Universal Turing Machine: A Half Century Survey. Oxford University Press, Oxford, pp. 55–111.

Hamkins, J.D., Lewis, A., 2000. Infinite time Turing machines. J. Symb. Log. 65 (2), 567–604.

Jensen, R.B., 1972. The fine structure of the constructible hierarchy. Ann. Math. Log. 4, 229–308.

Kleene, S.C., 1959. Origins of recursive function theory. Ann. Hist. Comput. 3, 52–67.

Koepke, P., 2005. Turing computation on ordinals. Bull. Symb. Log. 11, 377–397.

Koepke, P., 2006. Infinite time register machines. In: Beckmann, A., et al. (Eds.), Logical Approaches to Computational Barriers, vol. 3988 of Springer Lecture Notes Computer Science, Springer, Swansea, pp. 257–266.

Koepke, P., Miller, P., 2008. An enhanced theory of infinite time register machines. In: Beckmann, A., et al. (Ed.), Logic and the Theory of Algorithms, vol. 5028 of Springer Lecture Notes Computer Science, Springer, Swansea, pp. 306–315.

Koepke, P., Welch, P.D., 2011. A generalised dynamical system, infinite time register machines, and Π^1_1-CA$_0$. In Proceedings of CiE 2011, Sofia, Lecture Notes in Computer Science, Springer.

Odifreddi, P.-G., 1989. Classical Recursion Theory: the theory of functions and sets of natural numbers. Studies in Logic. North-Holland, Amsterdam.

Rogers, H., 1967. Recursive Function Theory. Higher Mathematics. McGraw, 1967.

Rosser, H.B., 1984. Highlights of the history of the λ-calculus. Ann. Hist. Comput., 6 (4), 337–349.

Sieg, W., 1994. Mechanical procedures and mathematical experience. In: George, A. (Ed.), Mathematics and Mind. Oxford University Press, Oxford, New York, Toronto, Tokyo.

Simpson, S., 1999. Subsystems of second order arithmetic. Perspectives in Mathematical Logic. Springer.

Welch, P.D., 2009. Characteristics of discrete transfinite Turing machine models: halting times, stabilization times, and normal form theorems. Theor. Comput. Sci. 410, 426–442.

On Computable Numbers, with an Application to the Entscheidungsproblem by A. M. Turing – Review by: Alonzo Church[1]

Andrew Hodges finds significance in —

CHURCH'S REVIEW OF COMPUTABLE NUMBERS

This comment is stimulated by the very curious review of Turing's paper by Alonzo Church in the *Journal of Symbolic Logic*, 1937. This review introduced Turing's work to the logicians' world, and in so doing used the expression 'Turing machine' for the first time. It was remarkable in several ways, not least for the unstinting welcome offered to Turing's revolutionary ideas, even though Church had the reputation of being cautious to the point of pedantry. It was generous in spirit despite the fact that it must have been disconcerting for Church that a young unknown, a complete outsider, had given a more satisfactorily direct and 'intuitive' account of effective calculation than the lambda calculus.

But the most curious thing is that Church was actually bolder in his physical imagery than Turing was:

> The author [Turing] proposes as a criterion that an infinite sequence of digits 0 and 1 be "computable" that it shall be possible to devise a computing machine, occupying a finite space and with working parts of finite size, which will write down the sequence to any number of terms if allowed to run for a sufficiently long time. As a matter of convenience, certain further restrictions are imposed in the character of the machine, but these are of such a nature as obviously to cause no loss of generality – in particular, a human calculator, provided with pencil and paper and explicit instructions, can be regarded as a kind of Turing machine.

In a further sentence (in the review of Post's work, immediately following)Church referred to Turing's concept as computability by an 'arbitrary machine', subject only to such finiteness conditions.

Yet Turing's paper did not actually refer to 'arbitrary machines'. Turing certainly brought an idea of physical action into the picture of computation. But his thorough and detailed analysis was of the human calculator, with arguments for finiteness based not on physical space and size, but on human memory and states of mind. It is odd that Church did not simply quote this model in his review, but instead portrayed the human calculator as a particular case of an apparently more general finite 'machine' with 'working parts'.

Nowadays, Church's assertion about what could be computed by an arbitrary machine, emphasising its generality, and characterising it in terms of space and size, reads more like the 'physical Church-Turing thesis', than the careful limitation to a human being, working to rule. It is unclear whether Church was actually aware of this distinction.

---

[1]Church, A., 1937. Review of Turing 19367, J. Symbolic Logic 2, 42, with permission from Association for Symbolic Logic.

It is an interesting question as to what Turing himself, who was well acquainted with mathematical physics, thought were the physical connotations of his 1936 work. Firstly, there is the question of the building of working Turing machines, and the universal machine in particular. Newman, and Turing's friend David Champernowne, later attested to discussions of it even at that time. But no written material has reached us. It is certainly hard to see how Turing could have failed to see that the atomic machine operations could be implemented with the sort of technology used in automatic telephone exchanges and teleprinters. The second, more difficult, question is what Turing thought the structure and limitations of computability had to say about the nature of the physical world. In particular, it is a striking fact that Turing, far more than most mathematicians of 1936, had an insight into the quantum-mechanical revolution in the description of physics. In 1932 he had studied Neumann's new axiomatisation of quantum mechanics, and it might well be that this work had encouraged him in the work of logical analysis which flowered in 1936. So of all people, he was well equipped to make some comment on Church's idea of an 'arbitrary machine', if only to emphasise that the changing nature of modern physics meant that this was not a simple concept. Yet he gave no recorded reaction, and there seems to have been no debate around the question at this period.

In his post-war writing, Turing made free use of the word 'machine for describing mechanical processes, and made no attempt to alert his readers to any distinction between human worker-to-rule and physical system – a distinction which, nowadays, would be considered important. Thus Turing (1948) referred to written-out programs, for a human to follow, as 'paper machines'. The imagery is that of a human acting out the part of a machine. Indeed, he stated that any calculating machine could be imitated by a human computer, again the reverse of the 1936 image. He referred often to the rote-working human calculator as a model for the way a computer worked and a guide as to what it could be made to do in practice. But he also referred to the advantage of the universal machine being that it could replace the 'engineering' of special-purpose machines. Most importantly, he appealed to the idea of simulating the brain as a physical system. So in later years Turing readily appealed to general ideas of physical mechanisms when discussing the scope of computability. Finally, in his last years, he seems to have taken an interest in the implications of quantum mechanics. But as for what he had in mind in 1936, we cannot know, and Church's review only heightens the mystery.

Reference

Turing, A.M., 1936. On computable numbers, with an application to the Entscheidungsproblem. Proceedings of the London Mathematical Society, series 2, vol. 42, pp. 230–265.

A. M. TURING. *On computable numbers, with an application to the Entscheidungsproblem.* ***Proceedings of the London Mathematical Society,*** series 2, vol. 42 (1936–7), pp. 230–265.

The author proposes as a criterion that an infinite sequence of digits 0 and 1 be "computable" that it shall be possible to devise a computing machine, occupying a finite space and with working parts of finite size, which will write down the sequence to any desired number of terms if allowed to run for a sufficiently long time. As a matter of convenience, certain further restrictions are imposed on the character of the machine, but these are of such a nature as obviously to cause no loss of generality – in particular, a human calculator, provided with pencil and paper and explicit instructions can be regarded as a kind of Turing machine. Thus, it is immediately clear that computability, so defined, can be identified with (especially, is no less general than) the notion of effectiveness as it appears in certain mathematical problems (various forms of the Entscheidungsproblem, various problems to find complete sets of invariants in topology, group theory, etc., and in general any problem which concerns the discovery of an algorithm).

The principal result is that there exist sequences (well defined on classical grounds), which are not computable. In particular, the *deducibility problem* of the functional calculus of first order (Hilbert and Ackermann's engere Funktionenkalkül) is unsolvable in the sense that, if the formulas of this calculus are enumerated in a straightforward manner, the sequence whose nth term is 0 or 1, according as the nth formula in the enumeration is or is not deducible, is not computable. (The proof here requires some correction in matters of detail.)

In an appendix, the author sketches a proof of equivalence of "computability" in his sense and "effective calculability" in the sense of the present reviewer (*American Journal of Mathematics*, vol. 58 (1936), pp. 345–363, see review in this Journal, vol. 1, pp. 73–74). The author's result concerning the existence of uncomputable sequences was also anticipated, in terms of effective calculability, in the cited paper. His work was, however, done independently, being nearly complete and known in substance to a number of persons at the time the paper appeared.

As a matter of fact, there is involved here the equivalence of three different notions: computability by a Turing machine, general recursiveness in the sense of Herbrand-Gödel-Kleene and λ-definability in the sense of Kleene and the present reviewer. Of these, the first has the advantage of making the identification with effectiveness in the ordinary (not explicitly defined) sense evident immediately – i.e., without the necessity of proving preliminary theorems. The second and third have the advantage of suitability for embodiment in a system of symbolic logic.

ALONZO CHURCH

Computability and λ-Definability

(J. Symbolic Logic, vol. 2 (1937), p. 153–163)

Henk Barendregt, Giulio Manzonetto and Rinus Plasmeijer trace through to today —

THE IMPERATIVE AND FUNCTIONAL PROGRAMMING PARADIGM

1. Models of computation

In Turing (1936) a characterisation is given of those functions that can be computed using a mechanical device. Moreover it was shown that some precisely stated problems cannot be decided by such functions. In order to give evidence for the power of this model of computation, Turing (1937) showed that machine computability has the same strength as definability via λ-calculus, introduced in Church (1936). This model of computation was also introduced with the aim of showing that undecidable problems exist.

Turing machine computability forms a very simple model that is easy to mechanise. Lambda calculus computability, on the other hand, is *a priori* more powerful. Therefore, it is not obvious that it can be executed by a machine. In showing the equivalence of both models, Turing shows us that λ-calculus computations are performable by a machine, so demonstrating the power of Turing machine computations. This gave rise to the combined

Church-Turing Thesis *The notion of intuitive computability is exactly captured by λ-definability or by Turing computability.*

Computability via Turing machines gave rise to imperative programming. Computability described via λ-calculus gave rise to functional programming. As imperative programmes are more easy to run on hardware, this style of software became predominant. We present major advantages of the functional programming paradigm over the imperative one, that are applicable, provided one is willing to explicitly deal with simple abstractions.

2. Functional programming

2.1. *Features from lambda calculus*

Rewriting. Lambda terms form a set of formal expressions subjected to possible *rewriting* (or *reduction*) steps. For each term, there are in general several parts that can be rewritten. However, if there is an eventual outcome, in which there is no more possibility to rewrite, it is necessarily unique.

Application. An important feature of the syntax of λ-terms is *application*. Two expressions can be applied to each other: if F and A are λ-terms, then so is FA, with as intended meaning the function F

Supported in part by NWO Project 612.000.936 CALMOC.

applied to the argument A. Both the function and the argument are given the same status as λ-terms. This implies that functions can be applied to functions, obtaining higher-order functions.

Abstraction. Next to the application there is *abstraction*. This feature allows to create complex functions. For example, given terms F and G intended as functions, then one may form $F \circ G$ and $G \circ F \circ G$ with the rewriting rules

$$(F \circ G)a \rightarrow F(Ga);$$
$$(G \circ F \circ G)a \rightarrow G(F(Ga)).$$

It is interesting to note that there is one single mechanism, λ-abstraction, that can capture both examples and much more. Given a λ-term M in which the variable x may occur, one can form the abstraction $\lambda x.M$. It has as intended meaning the function that assigns to x the value M. More generally, $\lambda x.M$ assigns to N the value $M[x{:=}N]$, where the latter denotes the expression obtained by substituting N for x in M. Then one has

$$F \circ G \triangleq \lambda x.F(Gx);$$
$$G \circ F \circ G \triangleq \lambda x.G(F(Gx)).$$

β-reduction. Corresponding to this abstraction with its intended meaning, there is a single rewriting mechanism. It is called β-reduction and is

$$(\lambda x.M)N \rightarrow M[x{:=}N],$$

giving the two rewrite examples mentioned above from the definition of $F \circ G$ and $G \circ F \circ G$. One can iterate the procedure and introduce the higher-order function C 'composition' as follows.

$$C \triangleq \lambda f \lambda g \lambda x.f(gx).$$

Having C one can write $F \circ G \triangleq C\ F\ G$, where in the absence of parentheses, one should read this $C\ F\ G$ as $(C\ F)\ G$. Dually the iterated abstraction $\lambda f \lambda g \lambda x.f(gx)$ should be read as $\lambda f(\lambda g(\lambda x.f(gx)))$.

Instead of λ-abstraction, it is convenient to define functions by their applicative behaviour[1]. One then writes `comp f g x = f(gx)`, obtaining 'composition' `comp f g` $=$ `f` $\circ$ `g`. One can even give definitions that are 'looping', like `L x` $=$ `(x, L(x + 1))`, so that `L 0` $=$ `(0,(1,(2,(3,...))))`. Similarly, one can construct the list of all prime numbers. Infinite lists are easier to describe than a list of say 28 elements.

Lazy evaluation. Expressions are evaluated as little as possible. Consider the program `f (L 4)`: the expression `L 4` is computed only when `f` tries to read some input and is just evaluated for long enough to return a value to `f`. This enables dealing with 'infinite objects', mentioned above.

2.2. *Features beyond lambda calculus*

For the pragmatics of functional programming, several features are added to the basic system of λ-calculus.

Data. Although integers and 'scientific' real numbers can be represented as λ-terms, for efficiency reasons they are given by special constants, together with the primitives for standard operations.

[1] This method is called as *heuristic application principle* by Böhm.

Names. The original λ-calculus formalism does not have names *by design*, as arbitrary λ-abstractions can be made. However, for the pragmatics of using and reusing software components, it is useful to introduce a naming construction `let`. For example, `let comp = λfλgλx.f(gx)` means that composition $C \triangleq \lambda f \lambda g \lambda x. f(gx)$ is now called `comp` and can be used later to define

$$F \circ G \triangleq \mathtt{comp}\, F G.$$

3. Types

In physics, constants have a 'dimension', e.g., speed is measured in kilometre per hour. When we bike at $\mathtt{v} = 12\,\mathtt{km/h}$ and we do this for $\mathtt{t} = 3\,\mathtt{h}$, we have gone $\mathtt{vt} = 12 \cdot 3 = 36\,\mathtt{km}$. Dimensions prevent that we want to consider, e.g., $\mathtt{vt}^2$ to compute the distance.

Similarly, functional programming languages come with a type system helping to ensure correctness. A program expecting a number (: `Int`) should not receive a judgement (: `Bool`). Giving module `F` a type `A` is denoted by `F : A` (read '`F` in `A`'). One starts typing the data, e.g., `3 : Int`, `True : Bool`. Functions with behaviour `G x = y` get as type $\mathtt{A} \to \mathtt{B}$, where `x : A` and `y : B`. An application `F a` is only allowed if the types match, i.e., $\mathtt{F} : \mathtt{A} \to \mathtt{B}$ and `a : A`.

The functional programmer indicates the types of the data structures and basic functions, and the machine performs *type inferencing* at compile time. This is a major help for combining software modules in a correct way: many bugs are caught as the result will be an untypable program.

Algebraic Data Types. Next to basic data types, like `Int`, `Bool`, one likes to use common data structures, like lists and trees of elements of type `A`. Such structures start small and grow. There is the empty list `Nil` and one can extend a list `tl` ('tail') by chaining an element `hd` ('head') of type `A`, obtaining `Cons hd tl`. The function that counts the number of elements in a list (of arbitrary type) can be defined by specifying that on the empty list it is 0 and on an enlarged list it is the length of the previous list plus 1.

```
count : List A → Int
count Nil = 0
count (Cons hd tl) = 1 + count tl
```

Using so-called Generalised Abstract Data Types, one introduces several such types simultaneously, mutually depending on each other, keeping type inferencing possible, see Schrijvers et al. (2009).

Generic types. One can 'code' algebraic types enabling to write uniformly functions on these. For example, there is a map on lists and on trees of data (hanging at the leaves), whereby a given function `f` is applied to each element. The two functions can be obtained by specializing one program with a generic type, including possible exceptions. With this feature, one can embed Domain Specific Languages within a functional language, see Plasmeijer et al. (2007).

Dynamic types. A functional program `P` having type `A` is compiled to machine code and evaluated. During this process, the original term, its type and context of definitions are forgotten. Using 'dynamic types', one may keep track of this syntactic data. This enables dynamic code, e.g., for writing typed Operating System in a pure functional language and dealing with unknown plug-ins or database specifications.

Efficiency. Considerable effort has been put into the compiler technologies for functional languages. The code generated by the state of art compilers for Haskell, OCaML and Clean is so good, that efficiency is no longer an issue, see the URL <`shootout.alioth.debian.org`>.

An example. Using abstraction, application, algebraic data types and functions as arguments, one can write compact software that can easily be modified. One can construct `foldr` with the following specification.

$$\mathtt{foldr}\ (\bullet)\ [\mathtt{a}_1, \ldots, \mathtt{a}_n]\ \mathtt{start} = (\mathtt{a}_1 \bullet (\mathtt{a}_2 \bullet (\ldots (\mathtt{a}_n \bullet \mathtt{start}) \ldots)))$$

Here, '•' is a binary operation used in infix position, and '(•)' stands for • as argument. The program foldr works on every kind of lists and subsequently can be applied to particular functions and data structures, for example to obtain sumlist. As a comparison, the imperative code is also given. The function foldr has the 'schematic type' (A → B → B) → B → List A → B, which does not need to be given by the programmer.

working Functional (Haskell-like)	Imperative (C-like)

foldr f e Nil = e
foldr f e (Cons hd tl) =
    f hd (foldr f e tl)
 	
sumlist = foldr (+) 0
	int sumlist(node *l){
	  node *cur = l;
	  int res = 0;
	  while not (curr = Nil){
	    res = res + curr->num;
	    curr = curr->next;
	  }
	  return res;
	}

If one needs the product of the elements of a list, then in the functional case, one just adds a new line prodlist = foldr (*) 1, whereas in the imperative case, one needs to write another procedure exactly like sumlist, except for the product that now replaces the sum, thus doubling the lines of code.

4. Input/output

In applications, one needs to perform I/O to manipulate information available in peripherals (e.g., keyboard and mouse for input and the file system and screen for output). These characterise the 'state' of the machine and platform running a program. This may effect the content of the file, which is imperative by nature. In *pure*[2] functional languages, one has to deal with I/O in a special way, while maintaining modularity, readability and typability.

In Haskell, the state has a particular type State. For output, one modifies the state ('writing'), and for input, one makes available to the main functional program a value from the state ('reading', thereby also possibly changing the state). The programming environment of Haskell comes with a collection of write and read operations having the following effect.

```
write a state = state'
read state = (a, state')
```

Together these are called 'actors', with the following type and action.

```
actor : State → A × State
actor state = (a, state')
```

Here, 'write a' is interpreted as an actor, for which the A is not used. Actors can have an erasing effect on the state. Therefore, the programmer has access to the actors, but not to the state. Otherwise one could write

$$\texttt{f state} = (\texttt{write 1 state}, \texttt{write 2 state}), \tag{1}$$

having a not well-defined effect (depending on execution order). The type State remains hidden to the programmer, but not the actors having type Monad A = State → (A × State), parametrised with a type A, on which the program can operate. The entire program is seen as a state modifying

[2] A pure functional language is one without assignments, i.e., statements like [x := x + 1].

function of type `Monad B`, for some type `B`. Composing such functions, using a kind of composition[3], one preserves modularity and compactness. In the meantime, any possible computations can take place, by interleaving these with the actors. The monadic approach in Haskell has as advantage that it does not require a special type system to deal with I/O.

In Clean, dealing with I/O is more general. Monads are used as well, but also the state on which they act is given to the programmer. This is possible because a *uniqueness type system* warrants safe usage that avoids situations like (1) above. As the state is available, one can split it into different components, like files, the keyboard and whatever one needs. These can be modified separately, as long as they are not duplicated. Explicit access to the state allows to write the actors within Clean itself.

We see that in both Haskell and Clean, dealing with I/O comes at a certain price. But this is well worth the advantages of pure functional programming languages: having arbitrarily high meaningful information density, with modules that can be combined easily in a safe way.

5. Current research

5.1. *Parallelism*

Pure functional languages are better equipped for programming multi-cores than imperative languages, as the result of a function is independent of its evaluation order. Therefore, modules in a functional program can be safely evaluated in parallel. As it costs overhead to send data between processors, one should restrict parallel evaluation to those functions having time consuming computations. Research on parallel evaluation of functional programs, see Hammond and Michaelson (1999), has been revived by the advent of new multi-core machines (Marlow et al, 2009).

5.2. *Certification*

In languages like AGDA (Bove et al., 2009), or Coq , 2010, one can fully specify a program and prove correctness. This demands even more skills from the programmer than pure functional languages already done: programming becomes proving. But the ideal of 'formal methods' (fully specifying software together with a proof of correctness) has become feasible. For example, Leroy (2009) gives a full certification of an optimised compiler for the kernel of the (imperative) language C.

6. History and perspective

The first functional language was Lisp (McCarthy et al., 1962). There is no type system, and I/O is done imperatively. By contrast, the language ML (Milner et al., 1997) and its modern variant F# (Syme et al., 2007) are impure as well, but strongly typed. Miranda (Turner, 1985) was one of the first pure functional programming languages, with lazy evaluation and type inference. Clean (Plasmeijer and van Eekelen, 2002) and Haskell (Jones, 2003) are modern variants of Miranda. Haskell has become the *de facto* standard pure functional language, which is widely used in academia.

Pure functional programming has not yet become mainstream, despite its expressive power and increased safety. To make use of the power, one needs understanding the type systems and the use of the right abstractions. Once mastered, functional programming enables writing applications in a fraction of the usual development and debugging time.

---

[3] As reading actors use the information obtained by modifying the state, it is a serial action with giving over a token, like in a relay race.

References

Bove, A., Dybjer, P., Norell, U., 2009. A brief overview of Agda – a functional language with dependent types. In: Berghofer, S., Nipkow, T., Urban, C., Wenzel, M., (Eds.), Theorem Proving in Higher Order Logics, 22nd International Conference, TPHOLs 2009, Munich, Germany, August 17–20, 2009. Proceedings, volume 5674 of Lecture Notes in Computer Science, Springer, pp. 73–78.

Church, A., 1936. An unsolvable problem of elementary number theory. Am. J. Math. 58 (2), pp. 345–363.

Coq Development Team, 2010. The Coq Proof Assistant Reference Manual – Version V8.3. URL: <http://coq.inria.fr>.

Hammond, K., Michaelson, G., 1999. Research Directions in Parallel Functional Programming, Springer.

Jones, S.P., (Eds.), 2003. Haskell 98 Language and Libraries: The Revised Report, Cambridge University Press.

Leroy, X., 2009. A formally verified compiler back-end. J. Autom. Reason. 43 (4), 363–446.

Marlow, S., Jones, S.L.P. Singh, S., 2009. Runtime support for multicore Haskell. In: Hutton, G., Tolmach, A.P., (Eds.), Proceeding of the 14th ACM SIGPLAN International Conference on Functional Programming, ICFP 2009, August 31–September 2, ACM, Edinburgh, Scotland, UK, pp. 65–78.

McCarthy, J., Abrahams, P.W., Edwards, D.J., Hart, T.P., Levin, M.I., 1962. LISP 1.5 Programmer's Manual, MIT Press.

Milner, R., Tofte, M., Harper, R., McQueen, D., 1997. The Definition of Standard ML, The MIT Press.

Plasmeijer, R.M., Achten, P., Koopman, P.W.M., 2007. iTasks: executable specifications of interactive work flow systems for the web. In: Hinze, R., Ramsey, N., (Eds.), Proceedings of the 12th ACM SIGPLAN International Conference on Functional Programming, ICFP 2007, October 1–3, 2007, ACM, Freiburg, Germany, pp. 141–152.

Plasmeijer, R.M., van Eekelen, M.C.J.D., 2002. Clean Language Report, University of Nijmegen, Software Technology Group.

Schrijvers, T., Jones, S.L.P., Sulzmann, M., Vytiniotis, D., 2009. Complete and decidable type inference for GADTs. In: Hutton, G., Tolmach, A.P., (Eds.), Proceeding of the 14th ACM SIGPLAN International Conference on Functional Programming, ICFP 2009, August 31–September 2, 2009, ACM, Edinburgh, Scotland, UK, pp. 341–352.

Syme, D., Granicz, A., Cisternino, A., 2007. Expert F#, Reactive, Asynchronous and Concurrent Programming. Apress, 2007.

Turing, A.M., 1936. On computable numbers, with an application to the Entscheidungsproblem. Proc. Lond. Math. Soc. ser. 2. 24, 230–265.

Turing, A.M., 1937. Computability and lambda-definability. J. Symbol. Log. 2 (4),153–163.

Turner, D.A., 1985. Miranda: A non-strict functional language with polymorphic types. Proceedings IFIP Conference on Functional Programming Languages and Computer Architecture, Nancy, France, September 1985. Springer Lecture Notes in Computer Science 201, pp 1-16.

COMPUTABILITY AND λ-DEFINABILITY

A. M. TURING

Several definitions have been given to express an exact meaning corresponding to the intuitive idea of 'effective calculability' as applied for instance to functions of positive integers. The purpose of the present paper is to show that the computable[1] functions introduced by the author are identical with the λ-definable[2] functions of Church and the general recursive[3] functions due to Herbrand and Gödel and developed by Kleene. It is shown that every λ-definable function is computable and that every computable function is general recursive. There is a modified form of λ-definability, known as λ-K-definability, and it turns out to be natural to put the proof that every λ-definable function is computable in the form of a proof that every λ-K-definable function is computable; that every λ-definable function is λ-K-definable is trivial. If these results are taken in conjunction with an already available[4] proof that every general recursive function is λ-definable we shall have the required equivalence of computability with λ-definability and incidentally a new proof of the equivalence of λ-definability and λ-K-definability.

A definition of what is meant by a computable function cannot be given satisfactorily in a short space. I therefore refer the reader to *Computable* pp. 230–235 and p. 254. The proof that computability implies recursiveness requires no more knowledge of computable functions than the ideas underlying the definition: the technical details are recalled in §5. On the other hand in proving that the λ-K-definable functions are computable it is assumed that the reader is familiar with the methods of *Computable* pp. 235–239.

The identification of 'effectively calculable' functions with computable functions is possibly more convincing than an identification with the λ-definable or general recursive functions. For those who take this view the formal proof of equivalence provides a justification for Church's calculus, and allows the 'machines' which generate computable functions to be replaced by the more convenient λ-definitions

1. Definition of λ-K-definability

In this section the notion of λ-K-definability is introduced in a form suitable for handling with machines. There will be three differences from the normal, in addition to that which distinguishes λ-K-definability from λ-definability. One consists in using only one kind [ ] of bracket instead of three, {}, (), [ ]; another is that $\mathrm{x}, \mathrm{x}^{|}, \mathrm{x}^{||}, \cdots$ are used as variables instead of an indefinite infinite

Received September 11, 1937.

[1] A. M. Turing, *On computable numbers with an application to the Entscheidungsproblem*, ***Proceedings of the London Mathematical Society***, **ser. 2**, vol. 42 (1936–7), pp. 230–265, quoted here as *Computable*. A similar definition was given by E. L. Post, *Finite combinatory processes—formulation* 1, this JOURNAL, vol. 1 (1936), pp. 103–105.

[2] Alonzo Church *An unsolvable problem of elementary number theory*, ***American journal of mathematics***, vol. 58 (1936), pp. 345–363, quoted here as *Unsolvable*.

[3] S. C. Kleene, *General recursive functions of natural numbers*, ***Mathematische Annalen***, vol. 112 (1935–6), pp. 727–742. A definition of general recursiveness is also to be found in *Unsolvable* pp. 350–351.

[4] S. C. Kleene. *λ-definability and recursiveness*, ***Duke mathematical journal***, vol. 2 (1936), pp. 340–353.

list of the single symbols, and the third is a change in the form of condition (ii) of immediate transformability, not affecting the definition of convertibility except in form.

There are five symbols which occur in the formulae of the conversion calculus. They are λ, x, ${}^{|}$, [and]. A sequence of symbols consisting of x followed by ${}^{|}$ repeated any number (possibly 0) of times is called a *variable. Properly-formed formulae* are a class of sequences of symbols which includes all variables. Also if ***M*** and ***N*** are[5] properly-formed formula, then [***M***][***N***] (i.e. the sequence consisting of [followed by ***M*** then by], [and the sequence ***N***, and finally by]) is a properly-formed formula. If ***M*** is a properly-formed formula and ***V*** is a variable, then λ***V***[***M***] is a properly-formed formula. If any sequence is a properly-formed formula it must follow that it is so from what has already been said.

A properly-formed formula ***M*** will be said to be *immediately transformable* into ***N*** if either:

(i) ***M*** is of the form λ***V***[***X***] and ***N*** is λ***U***[***Y***] where ***Y*** is obtained from ***X*** by replacing the variable ***V*** by the variable ***U*** throughout, provided that ***U*** does not occur in ***X***.

(ii) ***M*** is of the form [λ***V***[***X***]][***Y***] where ***V*** is a variable and ***N*** is obtained by substituting ***Y*** for ***V*** throughout ***X***. This is to be subject to the restriction that if ***W*** be either ***V*** or a variable occurring in ***Y***, then λ***W*** must not occur in ***X***.

(iii) ***N*** is immediately transformable into ***M*** by (ii).

A will be said to be immediately convertible into ***B*** if ***A*** is immediately transformable into ***B*** or if ***A*** is of the form ***X***[***L***]***Y*** and ***B*** is ***X***[***M***]***Y*** where ***L*** is immediately transformable into ***M***. Either ***X*** or ***Y*** may be void. ***A*** is *convertible* to ***B*** (***A*** conv ***B***) if there is a finite sequence of properly-formed formulae, beginning with ***A*** and terminating with ***B***, each immediately convertible into the preceding.

The formulae,

$\lambda x[\lambda x^{|}\,[x^{|}]]$ (abbreviated to 0),

$\lambda x[\lambda x^{|}\,[[x][x^{|}]]]$ (abbreviated to 1),

$\lambda x[\lambda x^{|}\,[[x][[x][x^{|}]]]]$ (abbreviated to 2), etc.,

represent the natural numbers. If ***n*** represents a natural number then the next natural number is represented by a formula which is convertible to [***S***][***n***] where ***S*** is

$$\lambda x^{||}[\lambda x\,[\lambda x^{|}[[x][[[x^{||}][x]][x^{|}]]]]].$$

A function $f(n)$ of the natural numbers, taking natural numbers as values will be said to be λ-*K*-definable if there is a formula ***F*** such that [***F***][***n***] is convertible to the formula representing $f(n)$ if ***n*** is the formula representing n. The formula [***F***][***n***] can never be convertible to two formulae representing different natural numbers, for it has been shown[6] that if two properly-formed formulae are in normal form (i.e., have no parts of the form [λ***V***[***M***]][***N***]) and are convertible into one another, then the conversion can be carried out by the use of (i) only. The formulae representing the natural numbers are in normal form and the formulae representing two different natural numbers are certainly not convertible into one another by the use of (i) alone.

[5] Heavy type capitals are used to stand for variable or undetermined sequences of symbols. In expressions involving brackets and heavy type letters it is to be understood that the possible substitutions of sequences of symbols for these letters is to be subject to the restriction that the pairing of the explicitly shown brackets is unaltered by the substitution; thus in ***X***[***L***]***Y*** the number of occurrences of [in ***L*** must equal the number of occurrences of].

[6] Alonzo Church and J. B. Rosser, *Some properties of conversion*, ***Transactions of the American Mathematical Society***, vol. 39 (1936), pp. 472–482. The result used here is Theorem 1 Corollary 2 as extended to the modified conversion on p. 482.

2. Abbreviations

A number of abbreviations of the same character as those in *Computable* (pp. 235–239) are introduced here. They will be applied in connection with the calculus of conversion, but are necessary for other purposes, e.g. for carrying out the processes of any ordinary formal logic with machines. The abbreviations in *Computable* are taken as known.

'The sequence of symbols marked with α (followed by α)' will be abbreviated to S(α) in the explanations. Sequences are normally identified by the way they are marked, and are as it were lost when their marks are erased.

In the tables B will be used as a name for the symbol 'blank.'

$\mathfrak{pem}(\mathfrak{A}, \alpha, \beta)$		$\mathfrak{pe}(\mathfrak{pem}_1, \alpha)$
$\mathfrak{pem}_1$	$R, P\beta$	$\mathfrak{A}$

$\mathfrak{pem}_1$ here stands for $\mathfrak{pem}_1(\mathfrak{A}, \alpha, \beta)$ and similar abbreviations must be understood throughout.

$\mathfrak{pem}(\mathfrak{A}, \alpha, \beta)$. The machine prints α at the end of the sequence of symbols on F-squares and marks it with β. $\rightarrow \mathfrak{A}$.

The tables for $\mathfrak{crm}$ $(\mathfrak{B}, \gamma, \beta)$ and $\mathfrak{cem}(\mathfrak{B}, \gamma, \beta)$ are to be obtained from those for $\mathfrak{cr}(\mathfrak{B}, \gamma)$ and $\mathfrak{ce}(\mathfrak{B}, \gamma)$ by replacing $\mathfrak{pe}(\mathfrak{A}, \alpha)$ by $\mathfrak{pem}(\mathfrak{A}, \alpha, \beta)$ throughout.

$\mathfrak{cpr}(\mathfrak{A}, \mathfrak{C}, \alpha, \beta)$	$\mathfrak{cp}(\mathfrak{cpr}_1, \mathfrak{cpr}_2, \mathfrak{cpr}_3, \alpha, \beta)$
$\mathfrak{cpr}_1$	$\mathfrak{re}(\mathfrak{re}(\mathfrak{cpr}, \mathfrak{b}, \beta, b), \mathfrak{b}, \alpha, a)$
$\mathfrak{cpr}_2$	$\mathfrak{re}(\mathfrak{re}(\mathfrak{C}, b, \beta), a, \alpha)$
$\mathfrak{cpr}_3$	$\mathfrak{re}(\mathfrak{re}(\mathfrak{A}, b, \beta), a, \alpha)$

$\mathfrak{cpr}(\mathfrak{A}, \mathfrak{C}, \alpha, \beta)$. The machine compares S(α) with S(β). $\rightarrow \mathfrak{A}$ if they are alike; $\rightarrow \mathfrak{C}$ otherwise. No erasures are made.

The letters a, b occurring in the table for $\mathfrak{cpr}$ should not be used elsewhere in any machine whose table involves $\mathfrak{cpr}$. This can be made automatic by using $a_{\mathfrak{cpr}}$ and $b_{\mathfrak{cpr}}$ say, instead of a and b. We shall however write a and b and understand them to mean $a_{\mathfrak{cpr}}$ and $b_{\mathfrak{cpr}}$. The same applies for the letters $a, \cdots, z$ in all such tables.

$\mathfrak{f}(\mathfrak{A}, \gamma)$	γ	L	$\mathfrak{A}$
	not γ	R, R	$\mathfrak{f}$
$\mathfrak{bf}(\mathfrak{A}, \alpha, \beta, \gamma, \delta)$		E, Pa	$\mathfrak{f}(\mathfrak{bf}_1, \gamma)$
$\mathfrak{bf}_1$	α	R, E, Pb	$\mathfrak{f}(\mathfrak{bf}_2, \gamma)$
	β	L	$\mathfrak{bf}_3$
	others	$R, R, R,$	$\mathfrak{f}(\mathfrak{bf}_1, \mathfrak{b}, \gamma)$
$\mathfrak{bf}_2$	β	R, E, Pb	$\mathfrak{f}(\mathfrak{bf}, \mathfrak{b}, a)$
	not β	R, R, R	$\mathfrak{f}(\mathfrak{bf}_2, \gamma)$
$\mathfrak{bf}_3$	γ or b	$E, P\delta, L, L$	$\mathfrak{bf}_3$
	α	$E, P\gamma$	$\mathfrak{A}$
	others	L, L	$\mathfrak{bf}_3$

$\mathfrak{bf}(\mathfrak{A}, \alpha, \beta, \gamma, \delta)$. This describes the process of finding the partner of a bracket. If α and β are regarded as left and right brackets, then if the machine takes up the internal configuration $\mathfrak{bf}$ when scanning a square next on the right of an α it will find the partner of this α in the sequence $S(\gamma)$, and will mark the part of $S(\gamma)$ which is between the brackets with δ (instead of γ). The final internal configuration is $\mathfrak{A}$ and the scanned square is that which was scanned when the internal configuration $\mathfrak{bf}$ was first taken up.

$\mathfrak{sb}(\mathfrak{A},\alpha,\beta,\gamma,\delta,\epsilon)$			$\mathfrak{f}'(\mathfrak{sb}_1,\mathfrak{crm}(\mathfrak{re}(\mathfrak{re}(\mathfrak{sb},a,j)b,\beta)\gamma,\epsilon),\beta)$
$\mathfrak{sb}_1$	σ	R, E, Pb	$\mathfrak{sb}_2(\mathfrak{A}, \alpha, \beta, \gamma, \delta, \epsilon, \sigma)$
$\mathfrak{sb}_2$			$\mathfrak{f}'(\mathfrak{sb}_3,\mathfrak{crm}(\mathfrak{re}(\mathfrak{re}(\mathfrak{re}(\mathfrak{A},b,\beta),j,\alpha),a,\alpha), a,\delta), \alpha)$
$\mathfrak{sb}_3$	σ	R,E,Pa	$\mathfrak{sb}$
	not σ	R,E,Pa	$\mathfrak{re}\ (\mathfrak{f}'(\mathfrak{sb}_4, b, a)_b,\beta)$
$\mathfrak{sb}_4$	τ	R,E,Pj	$\mathfrak{re}(\mathfrak{pem}(\mathfrak{sb},\tau,\delta), a,\alpha)$
$\mathfrak{sub}(\mathfrak{A},\alpha,\beta,\gamma,\delta)$			$\mathfrak{sb}(\mathfrak{A},\alpha,\beta,\gamma, \delta,\delta)$
$\mathfrak{dt}(\mathfrak{A}, \mathfrak{B},\alpha,\beta)$			$\mathfrak{pem}(\mathfrak{sb}(\mathfrak{f}(\mathfrak{e}(\mathfrak{A},d),\mathfrak{B}, d), \alpha,\beta,p,r, d), r,p)$

$\mathfrak{sub}(\mathfrak{A}, \alpha,\beta, \gamma, \delta)$. $S(\gamma)$ is substituted for $S(\beta)$ throughout $S(\alpha)$. The result is copied down and marked with δ. $\rightarrow \mathfrak{A}$.

$\mathfrak{dt}(\mathfrak{A}, \mathfrak{B}, \alpha,\beta)$. It is determined whether the sequence $S(\beta)$ occurs in $S(\alpha)$. $\rightarrow \mathfrak{A}$ if it does; $\rightarrow \mathfrak{B}$ otherwise.

The tables which follow are particularly important in all cases where an enumeration of all possible results of operations of given types is required. The enumeration may be carried out by regarding the operation as determined by a number of choices, each between two possibilities, **L** and **M** say. Each possible sequence of operations is then associated with a finite sequence of letters **L** and **M**. These sequences can easily be enumerated. The method used here is to replace **L** by 0, each **M** by 1, follow the whole by 1, reverse the order and regard the result as the binary Arabic numeral corresponding to the given sequence. Thus the first few sequences (beginning with the one associated with 1) are: the null sequence, **L**, **M**, **LL**, **ML**, **LM**, **MM**, **LLL**, **MLL**, **LML**, **MML**. In the general table below ζ and η are used instead of **L** and **M**.

$\mathfrak{add}(\mathfrak{A},\alpha,\zeta,\eta)$			$\mathfrak{f}'(\mathfrak{add}_1,\mathfrak{pem}(\mathfrak{add}_2,\zeta,a),\alpha)$
$\mathfrak{add}_1$	η	R,E	$\mathfrak{pem}(\mathfrak{add}, \zeta,a)$
	ζ	R,E	$\mathfrak{pem}(\mathfrak{add}_2, \eta,a)$
$\mathfrak{add}_2$			$\mathfrak{cem}(\mathfrak{re}(\mathfrak{A}, a, \alpha), \alpha,a)$

$\mathfrak{add}(\mathfrak{A}, \alpha,\zeta, \eta)$. The sequence $S(\alpha)$ consisting of letters ζ and η only is transformed into the next sequence. $\rightarrow \mathfrak{A}$.

$\mathfrak{ch}(\mathfrak{A}, \mathfrak{B},\mathfrak{C},\alpha,\zeta,\eta)$			$\mathfrak{f}'(\mathfrak{ch}_1,\mathfrak{re}(\mathfrak{C},b,\alpha),\alpha)$
$\mathfrak{ch}_1$	ζ	R,E,Pb	$\mathfrak{A}$
	η	R,E,Pb	$\mathfrak{B}$

$\mathfrak{ch}(\mathfrak{A},\mathfrak{B},\mathfrak{C},\alpha,\zeta,\eta)$ is an internal configuration which is taken up when a choice has to be made. $S(\alpha)$ is the sequence of letters ζ and η determining the choices. $\rightarrow \mathfrak{A}$ if the first unused letter is ζ; $\rightarrow \mathfrak{B}$

if it is η: it is then indicated that this ζ or η has been used by replacing its mark by b. When the whole sequence has been used up these marks are replaced by α again and $\rightarrow \mathfrak{E}$.

$\mathfrak{cch}(\mathfrak{A},\mathfrak{B},\mathfrak{C},\alpha,\zeta,\eta)$		R	$\mathfrak{cch}_1$
$\mathfrak{cch}_1$	σ	$E,\ Pa$	$\mathfrak{cch}_2(\mathfrak{A},\ \mathfrak{B},\ \mathfrak{C},\ \alpha,\zeta,\eta,\sigma)$
$\mathfrak{cch}_2$			$\mathfrak{ch}(\mathfrak{f}(\mathfrak{cch}_3,\ \mathfrak{b},\ a),\mathfrak{f}(\mathfrak{cch}_4,\ \mathfrak{b},\ a),\ \mathfrak{C},\ \alpha,\zeta,\ \eta)$
$\mathfrak{cch}_3$		$E,P\sigma,L$	$\mathfrak{A}$
$\mathfrak{cch}_4$		$E,P\sigma,L$	$\mathfrak{B}$

$\mathfrak{cch}(\mathfrak{A},\ \mathfrak{B},\mathfrak{C},\alpha,\zeta,\eta)$. This differs from $\mathfrak{ch}$ in that the internal configurations $\mathfrak{A}$ and $\mathfrak{B}$ are taken up when the same square is scanned as that which was scanned when the internal configuration $\mathfrak{cch}$ was first reached, provided that this was an F-square.

3. Mechanical conversion

We are now in a position to show how the conversion process can be carried out mechanically. It will be necessary to be able to perform all of the three kinds of immediate transformation. (iii) can be done most easily if we can enumerate properly-formed formulae. It is principally for this purpose that we introduce the table for $\mathfrak{pff}(\mathfrak{A},\alpha)$.

$\mathfrak{funf}(\mathfrak{A},\ \alpha,\beta,\gamma)$ $\quad\quad$ $\mathfrak{pem}(\mathfrak{crm}(\mathfrak{pem}_2(\mathfrak{crm}(\mathfrak{pem}(\mathfrak{A},],\gamma),\ \beta,\ \gamma),],[,\gamma),\alpha,\gamma),\ [,\gamma)$

$\mathfrak{funf}(\mathfrak{A},\alpha,\beta,\gamma)$. [S($\alpha$)][S($\beta$)] is written at the end and marked with γ. $\rightarrow \mathfrak{A}$.

$\mathfrak{ch}(\mathfrak{A},\ \mathfrak{B},\mathfrak{C},\ \theta)$ $\quad\quad$ $\mathfrak{ch}(\mathfrak{A},\mathfrak{B},\mathfrak{C},\theta,\mathbf{L},\mathbf{M})$

$\mathfrak{cch}(\mathfrak{A},\ \mathfrak{B},\mathfrak{C},\ \theta)$ $\quad\quad$ $\mathfrak{cch}(\mathfrak{A},\mathfrak{B},\ \mathfrak{C},\ \theta,\mathbf{L},\ \mathbf{M})$

The choices will be determined by a sequence made up of letters **L** and **M**.

$\mathfrak{pff}(\mathfrak{A},\mathfrak{C},\ \alpha,\theta)$			$\mathfrak{pe}_5(\mathfrak{c}(\mathfrak{h}(\mathfrak{af},\mathfrak{pff}_1,\mathfrak{C},\theta),:,;,\mathrm{x},;,\mathrm{x})$
$\mathfrak{pff}_1$			$\mathfrak{q}(\mathfrak{pff}_2,:)$
$\mathfrak{pff}_2$	;	R,R	$\mathfrak{cch}(\mathfrak{pff}_3,\mathfrak{pff}_2,\mathfrak{C},\theta)$
	ə		$\mathfrak{af}$
	others	R,R	$\mathfrak{pff}_2$
$\mathfrak{pff}_3$	;		$\mathfrak{pff}_4$
	ə		$\mathfrak{af}$
	others	R,Pa,R	$\mathfrak{pff}_3$
$\mathfrak{pff}_4$	;	R,R	$\mathfrak{cch}(\mathfrak{pff}_5,\mathfrak{pff}_4,\mathfrak{C},\theta)$
	ə		$\mathfrak{af}$
	others	R,R	$\mathfrak{pff}_4$
$\mathfrak{pff}_5$	; or ə		$\mathfrak{ar}$
	others	R,Pb,R	$\mathfrak{pff}_5$

ar			ch(ne, ch(comp, ab, $\mathfrak{C},\theta$), $\mathfrak{C},\theta$)
ne			$\mathrm{pe}_2(\mathrm{ne}_1$, ;, x)
ne_1			ch(pe(ne_1, $^{\vert}$), af, $\mathfrak{C},\theta$)
comp			pe(funk(af, a, b, ᗺ), ;)
ab			$\mathrm{pe}_3(\mathrm{ab}_1$, ;, λ, x)
ab_1			ch(pe($\mathrm{ab}^{\vert}_1$), ab_2, $\mathfrak{C},\theta$)
ab_2			pe(ce(pe(af,]), a), [)
af			e(e(ch(fin, pff_1, $\mathfrak{C},\theta$), a), b)
fin			q(r(r(fin_1)), ;)
fin_1	not ᗺ	$R, P\alpha, R$	fin_1
	ᗺ		$\mathfrak{A}$

pff($\mathfrak{A},\mathfrak{C},\alpha,\theta$). A properly-formed formula is chosen, written down at the end and marked with α. $\rightarrow \mathfrak{A}$. This is done by writing down successive properly-formed formulae separated by semi-colons, and obtaining others from them by abstraction (i.e., the process by which $\lambda \boldsymbol{V}[\boldsymbol{M}]$ is obtained from $\boldsymbol{M}$), by application of a function to its argument (i.e., obtaining $[\boldsymbol{M}][\boldsymbol{N}]$ from $\boldsymbol{M}$ and $\boldsymbol{N}$), and by writing down new variables. Before writing down a new formula we have the alternative of taking the last formula as the result of the calculation. In this case the internal configuration fin is taken up. If a new formula is to be constructed then two of the old formulae are chosen and marked with a and b: then one of the internal configurations ab, comp, ne is chosen and the new formula is correspondingly $\lambda \boldsymbol{V}[\mathrm{S}(a)], [\mathrm{S}(a)][\mathrm{S}(b)]$, or $\boldsymbol{V}$, where $\boldsymbol{V}$ is a new variable. The whole of the work is separated by a colon from the symbols which were on the tape previously. The meanings of pe_3 and pe_5 are analogous to pe_2.

The occurrence of λ in this table is of course as a symbol of the conversion calculus, not as a variable machine symbol.

The immediate transformations (i) and (ii) are described next.

va($\mathfrak{A}$, $\mathfrak{C}$, α, β, θ)			f′(va_1, $\mathfrak{A}$, α)
va_1	λ	R, E, Pa	f′(va_2, b, α)
	others		crm ($\mathfrak{A}$, α, β)
va_2	x or $^{\vert}$	R, E, Pb	f′(va_2, b, α)
	others	R	bf(pe(va_3, x), [,], α, c)
va_3		R, Pd	ch(pe(va_3, $^{\vert}$), dt(va_4, va_5, c, d), $\mathfrak{C}$, θ)
va_4			re(re(re(crm(e($\mathfrak{A}$, d), α, β), b, α)a, α)c, α)
va_5			crm(re(re(re(va_6, a, α), b, α), c, α)b, f)
va_6			sub(e(e($\mathfrak{A}$, d), f), α, f, d, β)

va($\mathfrak{A},\mathfrak{C},\alpha,\beta,\theta$). An immediate transformation (i) is chosen, and if permissible is carried out on $\mathrm{S}(\alpha)$, the result being marked with β. If the chosen transformation is not permissible then $\mathrm{S}(\beta)$ is identical with $\mathrm{S}(\alpha)$. $\rightarrow \mathfrak{A}$.

$\mathfrak{red}(\mathfrak{A},\alpha,\beta)$

$\mathfrak{red}_1$	[	R	$\mathfrak{bf}(\mathfrak{red}_2, [,],\alpha,c)$
	not [		$\mathfrak{red}_{13}$
$\mathfrak{red}_2$		E, Pf	$\mathfrak{re}(\mathfrak{f}(\mathfrak{red}_3, \mathfrak{b},\alpha), \mathfrak{b}, \alpha, f)$
$\mathfrak{red}_3$			$\mathfrak{bf}(\mathfrak{red}_4, [,],\alpha,d)$
$\mathfrak{red}_4$			$\mathfrak{f}'(\mathfrak{red}_5, \mathfrak{b}, \mathfrak{c})$
$\mathfrak{red}_5$	λ	R, E, Pf	$\mathfrak{f}'(\mathfrak{red}_6, \mathfrak{b}, c)$
	not λ		$\mathfrak{red}_{13}$
$\mathfrak{red}_6$	x or $^{\vert}$	R, E, Pg	$\mathfrak{f}'(\mathfrak{red}_6, \mathfrak{b}, c)$
	[	R, E, Pf	$\mathfrak{q}(\mathfrak{red}_7, c)$
$\mathfrak{red}_7$		E, Pf	$\mathfrak{red}_8$
$\mathfrak{red}_8$			$\mathfrak{f}'(\mathfrak{red}_9, \mathfrak{red}_{10}, c)$
$\mathfrak{red}_9$	λ	R, E, Pk	$\mathfrak{f}'(\mathfrak{red}_{11}, \mathfrak{b}, c)$
	not λ	R, E, Pk	$\mathfrak{red}_8$
$\mathfrak{red}_{11}$	x or $^{\vert}$	R, E, Pj	$\mathfrak{f}'(\mathfrak{red}_{11}, \mathfrak{b}, c)$
	[		$\mathfrak{cpr}(\mathfrak{red}_{13}, \mathfrak{red}_{12}, j, g)$
$\mathfrak{red}_{12}$			$\mathfrak{dt}(\mathfrak{red}_{13}, \mathfrak{re}(\mathfrak{red}_8, j, k), d, j)$
$\mathfrak{red}_{10}$			$\mathfrak{sub}(\mathfrak{re}(\mathfrak{re}(\mathfrak{re}(\mathfrak{re}(\mathfrak{A}, d, \alpha), f, \alpha), k, \alpha), g, \alpha)k, g, d, \beta)$
$\mathfrak{red}_{13}$			$\mathfrak{re}(\mathfrak{re}(\mathfrak{re}(\mathfrak{re}(\mathfrak{re}(\mathfrak{re}(\mathfrak{crm}(\mathfrak{A},\alpha,\beta),d,\alpha),g,\alpha),c,\alpha),f,\alpha),k,\alpha),j,\alpha)$

$\mathfrak{red}(\mathfrak{A},\alpha,\beta)$. An immediate transformation (ii) is carried out on S(α), supposing that S(α) is properly-formed. The result is marked with β. $\rightarrow \mathfrak{A}$. If the transformation is not possible or permissible S(β) is identical with S(α). Considerable use is made of the hypothesis that S(α) is properly-formed. Thus if its first symbol is [then it must be of form [***L***][***N***] and if in addition the second symbol is λ then it is of form [λ***V***[***M***]][***N***]. The internal configuration $\mathfrak{red}_8$ is never reached unless S(α) is of this form, and in that case it first occurs when ***V*** has been marked with g, ***M*** with c, and ***N*** with d, the remaining symbols of what was S(α) being now marked with α or f. It is then determined whether the immediate transformation (ii) is permissible: if it is then $\mathfrak{red}_{10}$ is taken up and the substitution carried out.

$\mathfrak{imc}(\mathfrak{A},\mathfrak{C},\alpha,\beta,\theta)$			$\mathfrak{ch}(\mathfrak{f}'(\mathfrak{imc}_1,\ \mathfrak{imc}_2,\ \alpha),\mathfrak{re}(\mathfrak{imc}_1,\alpha,a),\ \mathfrak{C},\ \theta)$
$\mathfrak{imc}_1,$	[	R,E,Pc	$\mathfrak{ch}(\mathfrak{imc},\ \mathfrak{imc}_3,\ \mathfrak{C},\theta)$
	not [	R,E,Pc	$\mathfrak{imc}$
$\mathfrak{imc}_2$			$\mathfrak{re}(\mathfrak{crm}(\mathfrak{A},\alpha,\beta),c,\alpha)$
$\mathfrak{imc}_3$			$\mathfrak{q}(\mathfrak{bf}(\mathfrak{imc}_4,\ [,],\ \alpha,a),\ c)$
$\mathfrak{imc}_4$			$\mathfrak{ch}(\mathfrak{vc},\mathfrak{ch}(\mathfrak{rc},\mathfrak{ex},\ \mathfrak{C},\theta),\ \mathfrak{C},\ \theta)$
$\mathfrak{vc}$			$\mathfrak{va}(\mathfrak{imc}_5,\mathfrak{C},a,b,\theta)$
$\mathfrak{rc}$			$\mathfrak{red}(\mathfrak{imc}_5,a,b)$
$\mathfrak{er}$			$\mathfrak{pff}(\mathfrak{red}(\mathfrak{er}_1,b,d),\mathfrak{G},b,\theta)$
$\mathfrak{er}_1$			$\mathfrak{cpr}(\mathfrak{e}(\mathfrak{imc}_5,d),\mathfrak{e}(\mathfrak{e}(\mathfrak{crm}(\mathfrak{imc}_5,a,b),d),b),d,a)$
$\mathfrak{imc}_5$			$\mathfrak{crm}(\mathfrak{crm}(\mathfrak{crm}(\mathfrak{re}(\mathfrak{re}(\mathfrak{e}(\mathfrak{A},b),c,\alpha),a,\alpha),\alpha,\beta),b,\beta),c,\beta)$

$\mathfrak{imc}(\mathfrak{A},\ \mathfrak{C},\alpha,\beta,\ \theta)$. An immediate conversion is chosen and performed on $S(\alpha)$. The result is marked with β. $\rightarrow \mathfrak{A}$.

$\mathfrak{conv}(\mathfrak{A},\alpha,\beta,\theta)$			$\mathfrak{pe}(\mathfrak{crm}(\mathfrak{conv}_1,\alpha,d),.)$
$\mathfrak{conv}_1$			$\mathfrak{ch}(\mathfrak{imc}(\mathfrak{conv}_2,\mathfrak{au},d,f,\theta),\mathfrak{re}(\mathfrak{A},d,\beta),\mathfrak{au},\theta$
$\mathfrak{conv}_2$			$\mathfrak{e}(\mathfrak{re}(\mathfrak{conv}_1,f,d),d)$
$\mathfrak{au}$			$\mathfrak{q}(\mathfrak{au}_1,.)$
$\mathfrak{au}_1$	$\mathfrak{B}$		$\mathfrak{crm}(\mathfrak{A},\alpha,\beta)$
	not $\mathfrak{B}$	R,E,R	$\mathfrak{au}_1$

$\mathfrak{conv}(\mathfrak{A},\alpha,\beta,\theta)$. A conversion is chosen and performed on $S(\alpha)$. The result is marked with β. $\rightarrow \mathfrak{A}$. The sequence determining the choices is $S(\theta)$. If it should happen that this sequence is exhausted before the conversion is completed then the final formula is the same as the original, i.e. $S(\alpha)$. The half finished conversion work is effectively removed from the tape by erasing the marks.

4. Computability of λ-*K*-definable functions

It is now comparatively simple to show that a λ-*K*-definable function is computable, i.e., that[7] if $f(n)$ is λ-K-definable then the sequence γ_f in which there are $f(n)$ figures 1 between the nth and the $(n+1)$th 0, and $f(0)$ figures before the first 0, is computable.

To simplify the table for the machine which computes γ_f we use the abbreviation $\mathfrak{Wr}(\mathfrak{A},\ \boldsymbol{M},\ \alpha)$ for an internal configuration starting from which the machine writes the sequence $\boldsymbol{M}$ of symbols at the end, marking it with α and finishing in the internal configuration $\mathfrak{A}$. Thus the table for $\mathfrak{Wr}(\mathfrak{A}, \lambda x^{|},\ \alpha)$ would be:

$\mathfrak{Wr}(\mathfrak{A},\ \lambda x^{\ |},\ \alpha)$		$\mathfrak{pe}(\mathfrak{Wr}_1,\ \mathfrak{B})$
$\mathfrak{Wr}_1,$	$P\lambda,\ R,P\alpha,R,\ Px,R,P\alpha,R,\ P^{|},\ R,P\alpha$	$\mathfrak{A}$

[7] Computable p. 254.

We use one more skeleton table:

$\mathfrak{pis}(\mathfrak{A}, \alpha, \beta)$ $\quad$ $\mathfrak{funf}(\mathfrak{e}(\mathfrak{re}(\mathfrak{A}, a, \alpha), \alpha), \beta, \alpha, a)$

If $\boldsymbol{F}$ is the formula which λ-K-defines $f(n)$ then the table for the machine which computes γ_f is:

$\mathfrak{b}$	Pə, R,Pə	$\mathfrak{Wr}(\mathfrak{b}_1, \boldsymbol{F}, h)$
$\mathfrak{b}_1$		$\mathfrak{Wr}(\mathfrak{b}_2, \lambda x[\lambda x^{\mathsf{I}}[x^{\mathsf{I}}]], i)$
$\mathfrak{b}_2$		$\mathfrak{crm}(\mathfrak{b}_3, i, k)$
$\mathfrak{b}_3$		$\mathfrak{Wr}(\mathfrak{ba}, \lambda x^{\mathsf{II}}[\lambda x[\lambda x^{\mathsf{I}}[[x][[[x^{\mathsf{II}}][x]][x^{\mathsf{I}}]]]]], u)$
$\mathfrak{ba}$		$\mathfrak{funf}(\mathfrak{cn}_1, h, k, v)$
$\mathfrak{cn}$		$\mathfrak{add}(\mathfrak{cn}_1, s, \mathbf{L}, \mathbf{M})$
$\mathfrak{cn}_1$		$\mathfrak{crn}(\mathfrak{cn}_2, i, d)$
$\mathfrak{cn}_2$		$\mathfrak{ch}(\mathfrak{re}(\mathfrak{cn}_3, d, m), \mathfrak{pl}\int(\mathfrak{cn}_2, d, u), \mathfrak{cn}_6, s)$
$\mathfrak{cn}_3$		$\mathfrak{conv}(\mathfrak{cn}_4, v, w, s)$
$\mathfrak{cn}_4$		$\mathfrak{cpr}(\mathfrak{cn}_5, \mathfrak{cn}_6, w, m)$
$\mathfrak{cn}_6$		$\mathfrak{e}(\mathfrak{e}(\mathfrak{e}(\mathfrak{cn}_{10}, w), m), d)$
$\mathfrak{cn}_{10}$		$\mathfrak{ch}(\mathfrak{cn}_{10}, \mathfrak{cn}_{10}, \mathfrak{cn}, s)$
$\mathfrak{cn}_5$		$q(\mathfrak{cn}_7, m)$
$\mathfrak{cn}_7$	E	$q(\mathfrak{cn}_8, m)$
$\mathfrak{cn}_8$	E	$q(\mathfrak{l}(\mathfrak{cn}_9), m)$
$\mathfrak{cn}_9$ {]	R, E	$\mathfrak{pem}(q(\mathfrak{l}(\mathfrak{cn}_9), m), 1, a)$
$\mathfrak{cn}_9$ { not]		$\mathfrak{pem}(\mathfrak{e}(\mathfrak{e}(\mathfrak{e}(\mathfrak{e}(\mathfrak{ba}_1, s), v), w), m), 0, a)$
$\mathfrak{ba}_1$		$\mathfrak{pl}\int(\mathfrak{ba}, k, u)$

When the machine reaches the internal configuration $\mathfrak{ba}$ for the $(n+1)$th time $(n \geqq 0)$ the tape bears the formula $\boldsymbol{F}$ marked with h, the formula $\boldsymbol{n}$ representing the natural number $\boldsymbol{n}$ (or rather a formula convertible into it) marked with k,0 marked with i, and S marked with $\boldsymbol{u}$. A formula convertible into one representing some natural number $\boldsymbol{r}$ is then chosen and marked with $\boldsymbol{m}$. This brings us to the internal configuration $\mathfrak{cn}_3$. A conversion is then chosen and performed on S(v), i.e. on $[\boldsymbol{F}][n]$. The result is marked with w and compared with S(m). If they are not alike the letters w, m are erased and we go back to $\mathfrak{cn}_1$ after transforming the sequence S(s) which determines the choices into the next sequence. If they are alike then 1 is written at the end repeated r times followed by 0, all of which is marked with a. In order to have the correct number of figures we make use of the fact that the number of brackets occurring consecutively at the end of S(m) is $r+2$. The machine is back in the internal configuration $\mathfrak{ba}$ as soon as S(k) has been changed to $[S][\mathrm{S}(k)]$.

No attempt is being made to give a formal proof that this machine has the properties claimed for it. Such a formal proof is not possible unless the ideas of substitution and so forth occurring in the definition of conversion are formally defined, and the best form of such a definition is possibly in terms of machines.

If $f(n)(n \geqq 1)$ is λ-definable, i.e. if $\boldsymbol{F}$ is well-formed (*Unsolvable* p. 346), then the present argument shows also that $f(n)$ is then computable in the sense that a function $g(n)$ of positive integers is computable if there is a computable sequence with $g(n)$ figures 1 between the nth and $(n+1)$th figure 0.

5. Recursiveness of computable functions.

It will now be shown that every computable function $f(n)$ of the natural numbers, taking natural numbers as values, is general recursive. We shall in fact find primitive recursive functions $j(x)$, $\phi(x)$ such that if $\xi(x)$ is the $(x+1)$th ($x = 0, 1, 2, \cdots$) natural number y for which $j(y) = 0$, then $f(x)$ is given by

$$f(x) = \phi(\xi(x)).$$

It is easily seen that such a function is general recursive (cf. *Unsolvable* p. 353); also it can easily be brought into the form,[8]

$$f(x) = \phi(\epsilon y[i(x, y) = 0])$$

(where $\epsilon y[i(x,y) = 0]$ means 'the least natural number y for which $i(x, y) = 0$,' and $i(x, y)$ is primitive recursive) which plays a central part[9] in the theory of general recursive functions. It would be slightly simpler to set up recursion equations for $f(x)$ but in that case it would be necessary to show that they were consistent; this is avoided by confining ourselves to primitive recursions (whose consistency is not likely to be doubted) except at the step from $j(x)$ to $\xi(x)$.

We are given the description of a machine which computes $f(x)$ The machine writes down symbols on a tape:amongst these symbols occur figures 0 and 1. The number of figures 1 between the nth and the $(n+1)$th figure 0 is $f(n)$. At any moment there is one of the symbols on the tape which is to be distinguished from the others and is called the 'scanned symbol.' The state (complete configuration) of the system at any moment is described by the sequence of symbols on the tape, with an indication as to which of them is scanned, and the internal configuration (m-configuration in *Computable*) of the machine. As names for the symbols we take S_0, S_1, $\ldots S_{N-1}$ and for the internal configurations $q_1, q_2, \cdots, q_R$. Certain of these are names of definite symbols and internal configurations independent of the machine; in fact,

S_0 always stands for 'blank,'
S_1 always stands for 0,
S_2 always stands for 1,
q_1,always stands for the initial internal configuration.

If at any time there is the sequence

$$S_{s_1}, S_{s_2}, \cdots, S_{s_k}, \ldots, S_{s_{k+1}} \qquad (k > 0,\ l \geqq 0)$$

of symbols on the tape, with the kth symbol scanned and the internal configuration q_t, this complete configuration may be described by the four numbers,

$$w = s_{k-1} + Ns_{k-2} + \cdots + N^{k-2}s_1,$$

---

[8] This may be done by defining $i(x, y)$ as follows:

$$\begin{aligned} e(0) &= 0. \\ e(S(x)) &= S(e(x)) && \text{if } j(x) = 0, \\ &= e(x) && \text{otherwise}, \\ i(x, y) &= \text{Max } (0, x - e(y)), \end{aligned}$$

$S(x)$ as usual meaning x+1.

[9] Compare the two papers by Kleene already quoted.

s_k, t, and

$$v = s_{k+1} + Ns_{k+2} + \cdots + N^{l-1}s_{k+l}$$

or by the single number,

$$u = p(w,\ s_k, t,\ v),$$

where

$$p(x_1,\ x_2,\ x_3,\ x_4) = 2^{x_1}3^{x_2}5^{x_3}7^{x_4}.$$

Each complete configuration of the machine is determined by the preceding one. The manner of this determination is given in the description of the machine, which consists of a number of expressions each of one of the forms $q_tS_8S_{8'}Lq_{t'}$ or $q_tS_8S_{8'}Nq_{t'}$ or $q_tS_8S_{8'}Rq_{t'}$. The occurrence of the first of these means that if in any complete configuration the scanned symbol was S_8 and the internal configuration q_t, then the machine goes to the next complete configuration by replacing the scanned symbol by $S_{8'}$ and making the new scanned symbol the symbol on the left of it and the new internal configuration $q_{t'}$. In other words if a complete configuration be described by the number,

$$p(s_{k-1} + Ns_{k-2} + \cdots + N^{k-2}s_1, s, t, s_{k+1} + Ns_{k+2} + \cdots + N^{l-1}s_{k+l})$$
$$= p(s_{k-1} + Nf, s, t, s_{k+1} + Ng),$$

and if $q_tS_8S_{8'}Lq_{t'}$ occurs in the description of the machine, then the number describing the next complete configuration is

$$p(f,\ s_{k-1}, t',\ s' + N(s_{k+1} + Ng)).$$

In the case where we have $q_tS_8S_{8'}Nq_{t'}$ the next complete configuration will be described by

$$p(s_{k-1} + Nf,\ s', t',\ s_{k+1} + Ng),$$

and in the case of $q_tS_8S_{8'}Rq_{t'}$ by

$$p(s' + N(s_{k-1} + Nf),\ s_{k+1}, t', g).$$

We may define a primitive recursive function $d_1(s, t)$ (or $d_2(s, t)$ or $d_3(s, t)$) to have the value 1 or 0 according as an expression of the form $q_tS_8S_{8'}Lq_{t'}$ (or $q_tS_8S_{8'}Nq_{t'}$ or $q_tS_8S_{8'}Rq_{t'}$) does or does not occur in the description of the machine. In each of the three cases $z(s, t)$ is to have the value s' and $c(s, t)$ to have the value t'.$q(x)$, $r(x)$ are to be respectively quotient and remainder of x on division by N, and $\varpi_r(x) (r = 2, 3, 5, 7)$ is to be the greatest integer k for which r^k divides x. These functions are primitive recursive.

Then if we put

$$\theta(x) = d_1(\varpi_3(x), \varpi_5(x))p(q(\varpi_2(x)), r(\varpi_2(x)), c(\varpi_3(x), \varpi_5(x)), z(\varpi_3(x), \varpi_5(x)) + N\varpi_7(x))$$
$$+ d_2(\varpi_3(x), \varpi_5(x))p(\varpi_2(x), z(\varpi_3(x), \varpi_5(x)), c(\varpi_3(x), \varpi_5(x)), \varpi_7(x))$$
$$+ d_3(\varpi_3(x), \varpi_5(x))p(z(\varpi_3(x), \varpi_5(x)) + N\varpi_2(x), r(\varpi_7(x)), c(\varpi_3(x), \varpi_7(x)), q(\varpi_7(x))),$$

and

$$u(0) = p(0,\ 0,\ 1,\ 0) = 5,$$
$$u(S(x)) = \theta(u(x)),$$

$u(x)$ will be the number describing the $(x + 1)$th complete configuration of the machine.

$g(x,y)$ is to be defined by

$$g(S(x),y) = 2 \quad (\text{all} x, y \geqq 0), \qquad g(0,1) = 0,$$
$$g(0,0) = 2, \qquad g(0,2) = 1,$$
$$g(0,x) = 2 \quad (x \geqq 3),$$

and $j(x)$ by,

$$j(x) = g(\varpi 3(u(x)), z(\varpi 3(u(x)), \varpi 5(u(x)))).$$

Then $j(x) = 0$ means that in going from the $(x+1)$th to the $(x+2)$th complete configuration the machine prints a figure 0: if $j(x) = 1$ it prints 1: $j(x) = 2$ otherwise. $\xi(x)$ is defined to be the $(x+1)$th natural number y for which $j(y) = 0$, and $\phi(x)$ as follows:

$$\phi(0) = 0,$$
$$\phi(S(x)) = 0 \qquad \text{if} \quad j(x) = 0,$$
$$7 = \phi(x) \qquad \text{if} \quad j(x) = 2,$$
$$= S(\phi(x) \quad \text{if} \quad j(x) = 1.$$

Then $\phi(x)$ is the number of times 1 has been printed since the last 0, reckoned at the $(x+1)$th complete configuration. $\phi(\xi(x))$ is the number of times 1 occurs between the xth and the $(x+1)$th figure 0, its value when $x = 0$ being the number of figures 1 which precede all figures 0. But these are the properties which define $f(x)$.

PRINCETON UNIVERSITY

The $\mathfrak{p}$-Function in λ-K Conversion

(J. Symbolic Logic, vol. 2 (1937), p. 164)

Henk Barendregt and Giulio Manzonetto point out the subtleties of —

TURING'S CONTRIBUTIONS TO LAMBDA CALCULUS

1. Fixed point combinators in untyped lambda calculus

The untyped lambda calculus was introduced in 1932 by Church as a part of an investigation in the formal foundations of mathematics and logic. The two primitive notions of the lambda calculus are *application* and λ-*abstraction*. Application, written MN, is the operation of applying the term M considered as an algorithm to the term N considered as an input. Lambda abstraction, written $\lambda x.M$, is the process of forming a function from the expression M (possibly) depending on x. Refer to Barendregt et al. (2012) (this volume) for an intuitive account of the system relying on the fundamental proof in Turing (1937), that lambda definability is equivalent to machine computability.

An important feature of lambda calculus is that it has *fixed point combinators*, namely programmes Y satisfying $YM = M(YM)$ for all M's. These constitute the main ingredient for writing recursive programmes in functional style. Turing contributed to this subject by providing a fixed point combinator Θ having the additional property that the equality between ΘM and $M(\Theta M)$ results simply by reducing the former to the latter (which is not the case, in general). Finally, we report how Böhm and van der Mey gave a general receipt to generate many new fixed point combinators starting from a fixed one. Turing's fixed point operator can be obtained in this way.

The rest of the section consists of the technical details and may be skipped by readers not interested.

1.1. *Lambda terms, reduction, and conversion*

Formally, the set Λ of λ-*terms* is defined inductively as follows:

Every variable x is in Λ;
If $M, N \in \Lambda$, then $MN \in \Lambda$;
If $M \in \Lambda$, then $\lambda x.M \in \Lambda$, for every variable x.

Lambda abstraction is a 'binder'; therefore, a variable x in M is called *bound* if it occurs in the scope of a 'λx' and is called *free* otherwise. As usual we consider λ-terms up to α-*conversion*, i.e., we consider equal those λ-terms only differing for the names of their bound variables. For example, $\lambda x.x = \lambda y.y$.

Supported in part by NWO Project 612.000.936 CALMOC.

The *β-reduction*, which specifies how λ-terms compute, is defined as the contextual closure of the following rule:

$$(\lambda x.M)N \rightarrow_\beta M[x := N],$$

where $M[x := N]$ denotes the result of substituting the term N for every occurrence of x in M, subject to the usual proviso about renaming bound variables in M to avoid capture of free variables in N. A term of the shape $(\lambda x.M)N$ is called *redex* and $M[x := N]$ is its *contractum*.

Multistep β-reduction is denoted by $\twoheadrightarrow_\beta$. Thus $M \twoheadrightarrow_\beta N$ if $M = M_0 \rightarrow_\beta \cdots \rightarrow_\beta M_n = N$ for some $n \geq 0$. The *β-conversion*, written $M =_\beta N$, is the equivalence relation generated by $\rightarrow_\beta$ (i.e., its reflexive-symmetric-transitive closure). The Church-Rosser theorem in lambda calculus states that

$$M =_\beta N \text{ if } M \text{ and } N \text{ have a common reduct.}$$

1.2. *Fixed points*

Despite the fact that its syntax is very simple, the lambda calculus is a Turing complete programming language. One aspect of the richness of its expressive power is the presence of fixed points that allow to write recursive programmes.

Theorem 1.1 (Fixed Point Theorem). *For all λ-terms F, there is a λ-term M, such that $FM =_\beta M$.*

Proof. Take $M \triangleq \omega_F\omega_F$, where $\omega_f \triangleq \lambda x.f(xx)$. Then

$$M \triangleq \omega_F\omega_F =_\beta F(\omega_F\omega_F) \triangleq FM. \qquad \square$$

In fact one can show that a fixed point for F can be found uniformly, that is by a term that takes F as input. If a λ-term Y satisfies $YF =_\beta F(YF)$ for all $F \in \Lambda$, then it is called a *fixed point combinator* (fpc).

Corollary 1.2 (Curry). *Let $\mathsf{Y} \triangleq \lambda f.\omega_f\omega_f$. Then Y is a fixed point combinator.*

The term Y is also called the *paradoxical combinator*, as it abstracts the argument in Russell's paradox.

An fpc Y is *reducing* if for all $M \in \Lambda$ one has $YM \twoheadrightarrow_\beta M(YM)$. This notion is useful as in many applications one needs for a fixed point M of F that $M \twoheadrightarrow_\beta FM$. It is easy to check that Y is not reducing. Therefore, one cannot take $M \triangleq \mathsf{Y}F$ to get $M \twoheadrightarrow_\beta FM$.

In the study by Turing (1937a), a more convenient fpc is constructed that is reducing.

Proposition 1.3 (Turing). *There exists a reducing fpc.*

Proof. Define $\Theta \triangleq AA$, where $A \triangleq \lambda xy.y(xxy)$. Then one has

$$\Theta F \triangleq AAF \triangleq (\lambda xy.y(xxy))AF \rightarrow_\beta (\lambda y.y(AAy))F \rightarrow_\beta F(\Theta F). \qquad \square$$

The next lemma shows how Θ arises naturally.

Lemma 1.4 (Böhm, van der Mey). *A term Y is an fpc if it is a fixed point of the peculiar term $\delta = \lambda xy.y(xy)$.*

Proof. If $Y = \delta Y$, then $YF = \delta YF = F(YF)$. $\square$

For the converse, see the monograph by Barendregt (1984), Lemma 6.5.3.

From Lemma 1.4, it follows that, starting from a given fpc Y, one can derive an infinite sequence of fpc's.

$$Y_0 \triangleq Y, \quad Y_{n+1} \triangleq Y_n\delta.$$

A natural question is whether all these fpc's are different. In the study by Endrullis et al. (2010), it is proved, using 'clocked Böhm-trees', that starting from Curry's fpc Y, there are no duplicates in the sequence $\mathsf{Y}_0, \mathsf{Y}_1, \mathsf{Y}_2, \cdots$ (the Böhm sequence). The problem is open for sequences starting from an arbitrary fpc Y. Note that Turing's fpc occurs in the Böhm sequence: $\Theta =_\beta \mathsf{Y}_1$, as

$$\mathsf{Y}_1 \triangleq \mathsf{Y}\delta \to_\beta (\lambda x.\delta(xx))(\lambda x.\delta(xx)) \to_\beta (\lambda xy.y(xxy))(\lambda xy.y(xxy)) \triangleq \Theta.$$

2. Weak normalisation of simply typed lambda calculus

Both for programming and theory, the property of normalisation is crucial. Termination can be seen as an issue of programme correctness. The problem of finding all possible inhabitants of a given type in the simply typed lambda calculus relies on the fact that all typable terms have a normal form. An early proof of this weak normalisation result for the simply typed lambda calculus is due to Turing and is published posthumously in the study by Gandy (1980). The idea of the proof is to find a reduction strategy and a measure function mapping terms into some well-founded set, such that the measure is strictly decreasing throughout steps in the computation.

One can see nicely from the notes how Turing wrote them informally, for his own use (as we all start doing). He states that he well-orders the terms, but uses a map f to multisets with the multiset order (which indeed is a well-ordering of type ω^ω), but as the map is not injective, there is no ordering on terms. Then he states – like thinking aloud – that if $M \to_\beta N$ by reducing a redex of highest order, then $f(M) > f(N)$, which is not quite correct. Turing then adds 'this at any [rate] will be the case, if we choose the unreduced part of highest order whose λ lies furthest to the right'. This indeed yields weak normalisation.

The rest of this section is devoted to give a sketch of the technical proof; the reader not interested in technicalities can skip it until Theorem 2.5.

2.1. *Simply typed lambda terms and reduction*

Definition 2.1. *Let us fix a non-empty set* $\mathbb{A}$ *of atoms.*

(1) *The set* $\mathbb{T} = \mathbb{T}^{\mathbb{A}}$ *of simple types over* $\mathbb{A}$ *is defined inductively as follows:*
   (a) *If* $\alpha \in \mathbb{A}$, *then* $\alpha \in \mathbb{T}$.
   (b) *If* $\sigma, \tau \in \mathbb{T}$, *then* $\sigma \to \tau \in \mathbb{T}$.
(2) *The set* $\Lambda(\sigma)$ *of λ-terms of type* σ *is defined by induction as follows:*
   (a) *For every variable* x, *one has* $x^\sigma \in \Lambda(\sigma)$.
   (b) *If* $M \in \Lambda(\sigma \to \tau)$ *and* $N \in \Lambda(\sigma)$, *then* $(MN) \in \Lambda(\tau)$.
   (c) *If* $M \in \Lambda(\tau)$, *then* $(\lambda x^\sigma.M) \in \Lambda(\sigma \to \tau)$ *for every variable* x.
(3) *Finally, the set of simply typed λ-terms is given by* $\Lambda = \cup_{\sigma\in\mathbb{T}}\Lambda(\sigma)$.

On Λ, we define the β-reduction as the contextual closure of the rule

$$(\lambda x^\sigma.M)N \to_\beta M[x := N].$$

A term M is in *β-normal form* if there is no N, such that $M \to_\beta N$.

2.2. *The proof of weak normalisation*

Now following Turing in the study by Gandy (1980) define a measure $|\cdot| : \Lambda \to \omega^2$ mapping every simply typed λ-term into an element of ω^2, which can be seen as the well-founded set $\mathbb{N} \times \mathbb{N}$ lexicographically ordered. For the sake of clarity, we will sometimes attach extra type information to λ-terms, decorating each subterm with its type, writing, for example, $(\lambda x^\alpha.M^\beta)^{\alpha\to\beta}$ instead of $\lambda x^\alpha.M$.

Definition 2.2

(i) *The length $\ell(\alpha)$ of a type $\alpha \in \mathbb{T}$ is defined as follows:*
$$\ell(\alpha) = 1, \text{ for } \alpha \in \mathbb{A};\ \ell(\alpha \to \beta) = \ell(\alpha) + \ell(\beta) + 1.$$

(ii) *The length $\ell(R)$ of a redex $R = (\lambda x^{\alpha}.M^{\beta})^{\alpha\to\beta}N^{\alpha}$ is defined as $\ell(R) = \ell(\alpha \to \beta)$.*

(iii) *With each λ-term $M \in \Lambda$, we associate an element of the ordinal ω^2 by setting $|M| = (k,n)$ where k is the maximal length of a redex in M and n is the number of redexes of length k occurring in M.*

To associate a suitable reduction strategy $\to_s$ guaranteeing that $M \to_s N$ entails $|M| > |N|$, we have to study how the contraction of a redex can duplicate other redexes or create new redexes. Duplication of a redex R happens when contracting redexes of the form

$$(\lambda x^{\alpha}.M[x,x]^{\beta})^{\alpha\to\beta}R^{\alpha} \to_{\beta} M[R,R]^{\beta},$$

where $M[P,Q]$ is a notation to display subterm occurrences of M.

The duplication of R is not very dangerous, while the creation of new redexes might be more problematic: *a priori* new redexes of higher length might be created indefinitely. The main instrument to check that this is not possible is given by the following Lemma.

Lemma 2.3 (Creation of redexes (Lévy, 1978)). *Contraction of a β-redex can only create a new redex in one of the following ways:*

(i) $(\lambda x^{\alpha\to\beta}.M[x^{\alpha\to\beta}P^{\alpha}]^{\gamma})^{(\alpha\to\beta)\to\gamma}(\lambda y^{\alpha}.Q^{\beta}) \to_{\beta} M[(\lambda y^{\alpha}.Q^{\beta})^{\alpha\to\beta}P^{\alpha}]^{\gamma}$;

(ii) $(\lambda x^{\alpha}.(\lambda y^{\beta}.M[x^{\alpha},y^{\beta}]^{\gamma})^{\beta\to\gamma})^{\alpha\to(\beta\to\gamma)}P^{\alpha}Q^{\beta} \to_{\beta} (\lambda y^{\beta}.M[P^{\alpha},y^{\beta}]^{\gamma})^{\beta\to\gamma}Q^{\beta}$;

(iii) $(\lambda x^{\alpha\to\beta}.x^{\alpha\to\beta})^{(\alpha\to\beta)\to(\alpha\to\beta)}(\lambda y^{\alpha}.P^{\beta})^{\alpha\to\beta}Q^{\alpha} \to_{\beta} (\lambda y^{\alpha}.P^{\beta})^{\alpha\to\beta}Q^{\alpha}$.

As proved by Lévy in his PhD thesis (§1.8.4, Lemma 3), the above lemma holds more generally for the untyped lambda calculus. See also Exercise 14.5.3 in Barendregt (1984).

Lemma 2.4. *Suppose $M \xrightarrow{R}_{\beta} N$, i.e., N is obtained from M by contracting R, and let R' be a created redex in N. Then $\ell(R) > \ell(R')$.*

Proof. Check that in each case of Lemma 2.3 the property holds. □

This lemma is not explicitly mentioned in Turing's proof, but it is stated that when reducing a redex of highest length, no other redex of highest length is created.

The reduction strategy $\mathcal{S}$ taken by Turing for proving the weak normalisation property is as follows. If M is in β-normal form, then do nothing; otherwise $\mathcal{S}(M) = N$ by contracting the *rightmost redex* of maximal length in M.

Theorem 2.5 (Weak Normalisation). *The simply typed lambda calculus is weakly normalising, i.e., every $M \in \Lambda$ has a β-normal form found by $\mathcal{S}$.*

Proof. Contracting a redex R can only duplicate redexes R' to the right of R. Because the redex R chosen by $\mathcal{S}$ is the rightmost of maximal length, it only duplicates redexes R', such that $\ell(R') < \ell(R)$. By Lemma 2.4, also the new redexes created by the reduction are of smaller length. Therefore, $\mathcal{S}(M) = N$ entails $|M| > |N|$. As ω^2 is well founded, we are done. □

A recent discovery is that Gentzen already had a normalisation proof for derivations in natural deduction, see von Plato (2008). This implies the normalisation of typed lambda terms. However, the proof worked out for lambda calculus is more clear and understandable, thanks to the simple linear syntax of λ-terms.

2.3. *Strong normalisation*

Actually the simply typed lambda calculus enjoys strong normalisation, which means that all β-reductions are terminating regardless of the strategy that is chosen. The classic proof of strong normalisation by using the reducibility technique is due to the study by Tait (1967), already obtained in 1963 and used by many authors. The proof of strong normalisation by Tait does not use a complexity measure assigned to terms. In the study by de Vrijer (1987), it is shown that it is possible to do this, assigning to a term M an ordinal $|M|$ (in fact a natural number), in such a way that $M \rightarrow_\beta N$ entails $|M| > |N|$, regardless what redex is reduced. It is an open problem whether such ordinals can be assigned in a natural and simple way.

3. Postscript

Lambda calculus was more often on Turing's mind. The logician Robin Gandy, who had been a student and associate of Turing, mentioned in 1986 at a conference for his retirement that in the early 1950s, Turing had told him ideas to implement lambda reduction using graphs. This is now commonly done when designing compilers for functional programming languages. Thereby, Turing was not careful about the distinction between free and bound variables and Gandy could correct him. Then Turing said: "That remark is worth 10 pounds a week!", in those days enough for a decent living.

References

Barendregt, H.P., 1984. The lambda calculus, its syntax and semantics, 2nd ed. No. 103 in Studies in Logic and the Foundations of Mathematics, North-Holland.

Barendregt, H.P., Manzonetto, G., Plasmeijer, M.J., 2012. The imperative and functional programming paradigm. In: Cooper, B., van Leeuwen, J. (Eds.), This volume, Elsevier, pp. 121–126.

de Vrijer, R.C., 1987. Exactly estimating functionals and strong normalization. Indagat. Math. 49, 479–493.

Endrullis, J., Hendriks, D., Klop, J.W., 2010. Modular construction of fixed point combinators and clocked Böhm trees. In: Proceedings of the 25th Annual IEEE Symposium on Logic in Computer Science, LICS 2010, 11–14 July 2010, IEEE Computer Society, Edinburgh, United Kingdom, pp. 111–119.

Gandy, R.O., 1980. An early proof of normalization by A. M. Turing. In: Seldin, J.P., Hindley, J.R. (Eds.), To H. B. Curry: Essays on Combinatory Logic, Lambda Calculus and Formalism, Academic Press Limited, pp. 453–455.

Lévy, J.-J., 1978. Réductions correctes et optimales dans le lambda-calcul. Ph.D. thesis, Université Paris 7.

Tait, W.W., 1967. Intentional interpretation of functionals of finite type I. J. Symbol. Log. 32 (2), 198–212.

Turing, A.M., 1937. Computability and λ-definability. J. Symbolic Logic 2 (4), 1937, pp. 153–163.

Turing, A.M., 1937. The $\mathfrak{p}$-function in lambda-K-conversion. J. Symbol. Log. 2 (4), 164.

von Plato, J., 2008. Gentzen's proof of normalization for natural deduction. Bull. Symbol. Log. 14 (2), 240–257.

THE $\mathfrak{p}$-FUNCTION IN λ-K-CONVERSION

A. M. TURING

In the theory of conversion it is important to have a formally defined function which assigns to any positive integer n the least integer not less than n which has a given property. The definition of such a formula is somewhat involved:[1] I propose to give the corresponding formula in λ-K-conversion,[2] which will (naturally) be much simpler. I shall in fact find a formula $\mathfrak{p}$ such that if T be a formula for which $\boldsymbol{T}(\boldsymbol{n})$ is convertible[3] to a formula representing a natural number, whenever $\boldsymbol{n}$ represents a natural number, then $\mathfrak{p}(\boldsymbol{T}, r)$ is convertible to the formula $\boldsymbol{q}$ representing the least natural number $\boldsymbol{q}$, not less than r, for which $\boldsymbol{T}(\boldsymbol{q})$ conv 0.[2] The method depends on finding a formula Θ with the property that Θ conv $\lambda \boldsymbol{u}.\boldsymbol{u}(\Theta(\boldsymbol{u}))$, and consequently if $\boldsymbol{M} \rightarrow \Theta(\boldsymbol{V})$ then $\boldsymbol{M}$ conv $\boldsymbol{V}(\boldsymbol{M})$. A formula with this property is,

$$\Theta \rightarrow \{\lambda vu.u(v(v,u))\}(\lambda vu.u(v(v,u))).$$

The formula $\mathfrak{p}$ will have the required property if $\mathfrak{p}(\boldsymbol{T}, r)$ conv r when $\boldsymbol{T}(r)$ conv 0, and $\mathfrak{p}(\boldsymbol{T}, r)$ conv $\mathfrak{p}(\boldsymbol{T}, \boldsymbol{S}(r))$ otherwise. These conditions will be satisfied if $\mathfrak{p}(\boldsymbol{T}, r)$ conv $\boldsymbol{T}(r, \lambda \boldsymbol{x}.\mathfrak{p}(\boldsymbol{T}, \boldsymbol{S}(r)), r)$, i.e. if $\mathfrak{p}$ conv $\{\lambda \boldsymbol{ptr}.\boldsymbol{t}(r, \lambda \boldsymbol{x}.\boldsymbol{p}(\boldsymbol{t}, \boldsymbol{S}(r)), r)\}(\mathfrak{p})$. We therefore put,

$$\mathfrak{p} \rightarrow \Theta(\lambda ptr.t(r, \lambda x.p(t, S(r)), r)).$$

This enables us to define also a formula,

$$\mathcal{P} \rightarrow \lambda tn.n(\lambda v.\mathfrak{p}(t, S(v)), 0),$$

such that $\mathcal{P}(\boldsymbol{T}, \boldsymbol{n})$ is convertible to the formula representing the nth positive integer q for which $\boldsymbol{T}(\boldsymbol{q})$ conv 0.

PRINCETON UNIVERSITY

Received April 23, 1937.

[1] Such a function was first defined by S. C. Kleene, *A theory of positive integers in formal logic*, ***American journal of mathematics***, vol. 57 (1934), see p. 231.

[2] For the definition of λ-K-conversion see S. C. Kleene, *λ-definability and recursiveness*, ***Duke mathematical journal***, vol. 2 (1936), pp. 340–353, footnote 12. In λ-K-conversion we are able to define the formula $0 \rightarrow \lambda fx.x$. The same paper of Kleene contains the definition of a formula L with a property similar to the essential property of Θ (p. 346).

[3] "Convertible" and "conv" refer to λ-K-conversion throughout this note.

Systems of Logic Based on Ordinals

(Proc. Lond. Math. Soc., series 2 vol. 45 (1939), pp. 161–228)

Solomon Feferman returns to —

TURING'S THESIS: ORDINAL LOGICS AND ORACLE COMPUTABILITY

In the sole extended break from his life and varied career in England, Alan Turing spent the years 1936–8 doing graduate work at Princeton University under the direction of Alonzo Church, the doyen of American logicians. Those two years sufficed for him to complete a thesis and obtain the Ph.D. The results of the thesis were published in 1939 under the title, 'Systems of logic based on ordinals' (Turing, 1939). That was the first systematic attempt to deal with the natural idea of overcoming the Gödelian incompleteness of formal systems by iterating the adjunction of statements – such as the consistency of the system – that 'ought to' have been accepted but were not derivable; in fact, these kinds of iterations can be extended into the transfinite. As Turing put it beautifully in his introduction (Turing, 1939):

> The well-known theorem of Gödel (1931) shows that every system of logic is in a certain sense incomplete, but at the same time it indicates means whereby from a system L of logic a more complete system L' may be obtained. By repeating the process we get a sequence L, $L_1 = L'$, $L_2 = L'_1, \ldots$ each more complete than the preceding. A logic L_ω may then be constructed in which the provable theorems are the totality of theorems provable with the help of the logics L, L_1, $L_2, \ldots$ Proceeding in this way we can associate a system of logic with any constructive ordinal. It may be asked whether such a sequence of logics of this kind is complete in the sense that to any problem A there corresponds an ordinal α such that A is solvable by means of the logic L_α.

Using an ingenious argument in pursuit of this aim, Turing obtained a striking yet equivocal partial completeness result that clearly called for further investigation. But he did not continue that himself, and it would be some twenty years before the line of research he inaugurated would be renewed by others. The paper itself received little attention in the interim, though it contained a number of original and stimulating ideas, and though Turing's name had by then been well established through his earlier work on the concept of effective computability. One of those ideas is that of oracle computability, addressed elsewhere in this volume.

Here, in brief, is the story of what led Turing to Church, what was in his thesis, and what came after, both for him and for the subject.[1]

[1] Much of this note is adapted directly from my paper (Feferman, 2006) for the *Notices* of the American Mathematical Society. Prior to that I had written about this material at somewhat greater length in my work (Feferman, 1988), and that in turn was incorporated as an introductory note to Turing's 1939 paper in the volume, *Mathematical Logic* (Turing, 2001) of his collected works. In its biographical part, I have drawn to a considerable extent on Andrew Hodges' superb biography, *Alan Turing: The Enigma* (Hodges, 1994).

1. From Cambridge to Princeton

As an undergraduate at King's College, Cambridge, from 1931 to 1934, Turing was interested in many parts of mathematics, including mathematical logic. Toward the end of that period he began attending a survey course on logic by the topologist M. H. A. (Max) Newman. One of the problems mentioned in that course was the *Entscheidungsproblem*, the question whether there exists an effective method to decide, given any well-formed formula of the pure first-order predicate calculus, whether or not it is valid in all possible interpretations. After completing his undergraduate work, Turing stayed on as a fellow at King's College and continued to think about that problem, which had been solved in the affirmative for various special classes of formulas. However, Turing became convinced that the answer to the problem must be negative in general, but that in order to demonstrate the impossibility of a decision procedure, he would have to give an exact mathematical explanation of what it means to be computable by a strictly mechanical process. He arrived at such an analysis by mid-April 1936 via the fundamental idea of the *Turing machine*. As explained elsewhere in this volume, that is his most famous and certainly most important contribution to mathematical logic and the theory of computation. By its means, Turing showed that the sought for answer to the motivating problem was indeed negative in general, and he quickly prepared a draft of his work entitled 'On computable numbers, with an application to the Entscheidungsproblem'(Turing, 1936–7). After some initial skepticism, Newman became convinced of Turing's analysis and encouraged its publication. Neither Newman nor Turing were aware at that point that there were already two other proposals under serious consideration for analyzing the general concept of effective computability: one by Kurt Gödel, building on an idea of Jacques Herbrand, called *general recursiveness*, and the other by Church, called *definability in the lambda* (λ)*-calculus*. Church and his student Stephen C. Kleene proved the equivalence of these two notions, and what has come to be called 'Church's Thesis' was that general recursiveness constitutes an analysis of the concept of effective computability. This claim was stated in the paper Church (1936a), in which various mathematical and logical problems were shown to be effectively undecidable. That was followed by Church (1936), in which the general *Entscheidungsproblem* was answered in the negative, submitted 15 April 1936 just about the same time as Turing was preparing his paper for publication. When Newman and Turing received news about this work a month later, the first reaction was a great disappointment, but then it was agreed that Turing's analysis was sufficiently different to warrant publication. Moreover, after that was submitted but before it appeared, Turing was able to add an appendix in which he proved the equivalence of computability by his machines with lambda-definability.

A year later, in Church's 1937 review for *The Journal of Symbolic Logic* of Turing's paper, he stated that:

> As a matter of fact, there is involved here the equivalence of three different notions; computability by a Turing machine, general recursiveness in the sense of Herbrand-Gödel-Kleene, and λ-definability in the sense of Kleene and the present reviewer. Of these, the first has the advantage of making the identification with effectiveness in the ordinary (not explicitly defined) sense evident immediately ... The second and third have the advantage of suitability for embodiment in a system of symbolic logic.

Thus was born what is now called the *Church–Turing Thesis*, according to which the effectively computable functions are exactly those computable by a Turing machine.[2] This, of course, is not to be confused with Turing's thesis under Church, the main subject here.

On Newman's recommendation, Turing decided to spend the academic year 1936–7 studying with Church in Princeton. He applied for Princeton's Procter fellowship, of which three were offered

[2] Gödel accepted the Church–Turing Thesis in that form in a number of lectures and publications thereafter.

each year, one for Oxford, one for Cambridge and one for the Collège de France in Paris, but did not succeed in getting it. Still, he thought he could manage on his fellowship funds from King's College, and so Turing arrived there at the end of September 1936.

2. Turing in Princeton

The Princeton mathematics department was already a leader on the American scene when it was greatly enriched in the early 1930s by the establishment of the Institute for Advanced Study. The two shared Fine Hall until 1940, so that the lines between them were blurred and there was significant interaction. Among the mathematical leading lights that Turing found on his arrival were Einstein, von Neumann, Weyl at the Institute and Lefschetz in the department. In logic, he had hoped to find – besides Church – Gödel, Paul Bernays, Church's student Kleene and J. Barkley Rosser. As it happened, none besides Church were there so Turing was reduced to attending Church's lectures, which he found ponderous and excessively precise; but this was an exposure that he needed since, by contrast, Turing's native style was rough-and-ready and prone to minor errors. They met from time to time, but apparently there were no sparks, since Church was retiring by nature and Turing was a 'confirmed solitary'.

In the spring of 1936, Turing worked up for publication a proof in greater detail of the equivalence of machine computability with λ-definability. He also published two papers on group theory, one on finite approximations of continuous groups that was of interest to von Neumann. Dean Eisenhart of the Princeton mathematics department urged Turing to stay on for a second year and apply again for the Procter fellowship. This time, supported by von Neumann, Turing succeeded in obtaining the fellowship, and so decided to stay the extra year and do a Ph.D. under Church. Proposed as a thesis topic was the idea of ordinal logics that had been broached in Church's course as a way to 'escape' Gödel's incompleteness theorems.

Turing made good progress on his thesis topic and devoted himself full time to it when he returned to Princeton in the fall after a summer back in England, so that he ended up with a draft by Christmas of 1937. Apparently Turing would have been satisfied with that as an essentially finished product, since he wrote home that 'Church made a number of suggestions which resulted in the thesis being expanded to an appalling length'. One can well appreciate that Church would not knowingly tolerate imprecise formulations or proofs, let alone errors, and the published version shows that Turing went far to meet such demands while putting his own characteristic stamp on it. Following an oral exam in May on which his performance was noted as 'Excellent', the Ph.D. itself was granted in June 1938.

Von Neumann thought sufficiently highly of Turing's mathematical talents to offer him a position as his assistant at the Institute. Curiously, at that time von Neumann showed no knowledge or appreciation of his work in logic. It was not until 1939 that he was to recognise the fundamental importance of Turing's work on computability. Then, during World War II, when von Neumann was engaged in the practical design and development of general purpose electronic digital computers, he was to incorporate the key idea of Turing's universal computing machine in a direct way.

Von Neumann's offer was quite attractive, but Turing's stay in Princeton had not been a personally happy one, and he decided to return home despite the uncertain prospects outside of his fellowship at King's and in face of the brewing rumors of war. After publishing the thesis work he did no more on that topic and went on to other things. Not long after his return to England, he joined a course at the Government Code and Cypher School, and that was to lead to his top secret work during the war at Bletchley Park on breaking the German Enigma Code. This fascinating part of the story is told in full in Hodges' biography (Hodges, 1994) and elsewhere in this volume, as is his subsequent career working to build actual computers, promote artificial intelligence, theorise about morphogenesis and continue his work in mathematics. Tragically, this ended with his death in 1954, a probable suicide.

3. The thesis: ordinal logics

What Turing calls a *logic* is nowadays more usually called a *formal system*, i.e., one prescribed by an effective specification of a language, set of axioms and rules of inference. Where Turing used 'L' for logics I shall use 'S' for formal systems. Given an effective description of a sequence $\langle S_n \rangle_{n \in N}$ $[N = \{0, 1, 2, \ldots\}]$ of formal systems all of which share the same language and rules of inference, one can form a new system S_ω, which is the union of the S_n for n in N. If the sequence of S_n's is obtained by iterating an effective passage from one system to the next, then that iteration can be continued to form $S_{\omega+1}, \ldots$ and so on into the transfinite. This leads to the idea of an effective association of formal systems S_α with ordinals α. Clearly that can only be done for denumerable ordinals, but to deal with limits in an effective way, it turns out that we must work not with ordinals per se, but with *notations for ordinals*. Church and Kleene (1936) had introduced a system O of constructive ordinal notations, given by certain expressions in the λ-calculus. A variant of this uses numerical codes a for such expressions, and associates with each $a \in O$ a countable ordinal $| a |$.

In general, given any effective means of passing from a system S to an extension S' of S, one can form an ordinal logic $S^* = \langle S_a \rangle_{a \in O}$ which is such that for each $a \in O$ and $b =$ the successor of a, $S_b = S'_a$, and is further such that whenever a is a notation for a limit ordinal given by a recursive function with index e, then S_a is the union of the sequence of $S_{\{e\}n}$ for each $n \in N$. In particular, for systems whose language contains that of Peano Arithmetic, PA, one can take S' to be $S \cup \{\mathrm{Con}_S\}$, where Con_S formalises the consistency statement for S; the associated ordinal logic S^* thus iterates adjunction of consistency through all the constructive ordinal notations. If one starts with PA as the initial system it may be seen that each S_a is consistent and so S'_a is strictly stronger than S_a by Gödel's second incompleteness theorem. The consistency statements are expressible in $\forall$('for all')-form, i.e., $\forall x R(x)$ where R is an effectively decidable predicate, and so a natural question to raise is whether S^* is complete for statements of that form, i.e., whether whenever $\forall x R(x)$ is true in N then it is provable in S_a for some $a \in O$. Turing's main result for this ordinal logic was that it is indeed the case, in fact one can always choose such an a with $| a | = \omega + 1$. But, speaking informally, a can only be recognised to be a notation in O by first recognising the truth of $\forall x R(x)$. Turing realised that this completeness proof is disappointing, because it shifts the question of whether a $\forall$-statement is true to the question whether a number a actually belongs to O. In fact, the general question, given a, is $a \in O$?, turns out to be of higher logical complexity than any statement formed by the unlimited iteration of universal and existential quantifiers, $\forall$ and $\exists$. Another main result of Turing's thesis is that for quite general ordinal logics, S^* cannot be both complete for statements in $\forall$-form and invariant, i.e., is such that whenever $| a | = | b |$ then S_a and S_b prove the same theorems. It is for these reasons that I called his completeness results equivocal above. Even so, what Turing really hoped to obtain was completeness for statements in $\forall\exists$ ('for all, there exists')-form. His reason for concentrating on these, that he called 'number-theoretical problems', rather than considering arithmetical statements in general, is not clear. This special class certainly includes many number-theoretical statements (in the usual sense of the word) of mathematical interest, e.g., those such as the twin prime conjecture, that say that an effectively decidable set C of natural numbers is infinite. Also, as Turing pointed out, the question whether a given program for one of his machines computes a total function is in $\forall\exists$-form.

In Section 4 of his thesis, Turing introduced a new idea that was to change the face of the general theory of computation (a.k.a. recursion theory) but the only use he made of it there was curiously misguided. His aim was to produce an arithmetical problem which is not number-theoretical in his sense, i.e., not in ∀∃-form. This is trivial by a cardinality argument, since there are only countably many effective relations $R(x,y)$ of which we could say that $\forall x \exists y R(x, y)$ holds. Turing's way of dealing with this, instead, is through the new notion of computation relative to an *oracle*. As he puts it:

> Let us suppose that we are supplied with some unspecified means of solving number-theoretical [i.e., ∀∃] problems; a kind of oracle as it were. ... With the help of the oracle we could form a new kind of machine (call them *o*-machines), having as one of its fundamental processes that of solving a given number-theoretic problem.

He then showed that the problem of determining whether an *o*-machine terminates on any given input is an arithmetical problem not computable by any *o*-machine, and hence not solvable by the oracle itself. Turing did nothing further with the idea of *o*-machines, either in this paper or afterward. Post (1944) took it as his basic notion for a theory of *degrees of unsolvability*, crediting Turing with the result that for any set of natural numbers there is another of higher degree of unsolvability. This transformed the notion of computability from an absolute notion into a relative notion that would lead to entirely new developments and eventually to vastly generalised forms of recursion theory.[3]

4. Ordinal logics redux

The problems left open in Turing's thesis were attacked in my 1962 paper, 'Transfinite recursive progressions of axiomatic theories' (Feferman, 1962). The title contains my rechristening of 'ordinal logics' in order to give a more precise sense of the subject matter. I showed there that Turing's progression based on iteration of consistency statements is not complete for true ∀∃ statements, contrary to his hope. In fact, the same holds for the even stronger progression obtained by iterating adjunction to a system S of the *local reflection principle for S*. This is a scheme that formalises, for each arithmetical sentence A, that if A is provable in S then A holds. The *uniform reflection principle* is a generalisation of the local principle to arbitrary formulas. Then, I showed that a progression based on the iteration of that is complete for all true arithmetical sentences. One can also find a path P through O along which every true arithmetical sentence is provable in that progression. On the other hand, invariance fails badly in the sense that there are paths P' through O for which there are true sentences in ∀-form not provable along that path, as shown in my paper with Clifford Spector (Feferman and Spector, 1962). The book *Inexhaustibility* (Franzén, 2004a) by Torkel Franzén contains an accessible introduction to Feferman (1962), and his paper Franzén (2004b) gives an interesting explanation of what makes Turing's and my completeness results work.

The problem raised by Turing of recognising which expressions (or numbers) are actually notations for ordinals is dealt with in part through the concept of *autonomous progressions of theories*, obtained by imposing a boot strapping process. That allows one to go to a system S_a only if one already has a proof in a previously accepted system S_b that $a \in O$ (or that a recursive ordering of order type corresponding to a is a well-ordering). Such progressions are not complete but have been used to propose characterisations of certain informal concepts of proof, such as that of finitist proof (Kreisel, 1960, 1970) and predicative proof (Feferman, 1964). For more recent progress that replaces the use of transfinite progressions via a concept of the *unfolding* of formal systems based on suitable *axiom schemata*, see my article (Feferman, 1996), and (Feferman and Strahm, 2000).

[3] See the relevant pieces in other parts of this volume. I have written at greater length of the significance of oracle computability from several perspectives in my paper (Feferman, 1992).

References

Church, A., 1936a. A note on the Entscheidungsproblem, J. Symbolic Logic 1, 40–41; correction, ibid., 101–102.

Church, A., 1936b. An unsolvable problem of elementary number theory. Amer. J. Math. 58, 345–363.

Church, A., Kleene, S.C., 1936. Formal definitions in the theory of ordinal numbers. Fundamenta Mathematicae 28, 11–21.

Feferman, S., 1962. Transfinite recursive progressions of axiomatic theories. J. Symbolic Logic 27, 259–316.

Feferman, S., 1964. Systems of predicative analysis. J. Symbolic Logic 29, 1–30.

Feferman, S., 1988. Turing in the land of O(z). In: Herken, R. (Ed.), The Universal Turing Machine. A Half-Century Survey, Oxford University Press, Oxford, pp. 113–147.

Feferman, S., 1992. Turing's 'oracle': From absolute to relative computability?—and back. In: Echeverria, J., et al. (Eds.), The Space of Mathematics, Walter de Gruyter, Berlin, pp. 314–348

Feferman, S., 1996. Gödel's program for new axioms: Why, where, how and what?. In: Hajek, P. (Ed.). Gödel '96, Lecture Notes in Logic 6, 3–22.

Feferman, S., 2006. Turing's thesis. Notices Amer. Math. Soc. 53 (10) 1200–1205.

Feferman, S., Spector, C., 1962. Incompleteness along paths in recursive progressions of theories. J. Symbolic Logic 27, 383–390.

Feferman, S., Strahm, T., 2000. The unfolding of non-finitist arithmetic. Ann. Pure and Applied Logic 104, 75–96.

Franzén, T., 2004a. Inexhaustibility. A Non-Exhaustive Treatment, Lecture Notes in Logic 28. Assoc. for Symbolic Logic, A. K. Peters, Ltd., Wellesley (distribs.).

Franzén, T., 2004b. Transfinite progressions: a second look at completeness. Bull. Symbolic Logic 10, 367–389.

Hodges, A., 1994. Alan Turing: The Enigma, Simon and Schuster, New York 1983, rev. ed. Springer-Verlag, New York.

Kreisel, G., 1960. Ordinal logics and the characterization of informal concepts of proof. In: Proc. International Congress of Mathematicians at Edinburgh 1958, Cambridge Univ. Press, New York, pp. 289–299.

Kreisel, G., 1970. Principles of proof and ordinals implicit in given concepts. In: Myhill, J., et al. (Eds.), Intuitionism and Proof Theory, North-Holland, Amsterdam, pp. 489–516.

Post, E., 1944. Recursively enumerable sets and their decision problems. Bull. Amer. Math. Soc. 50, 284–316.

Turing, A.M., 1936–7. On computable numbers, with an application to the Entscheidungsproblem. Proc. London Math. Soc. 42 (2), 230–265; A correction, ibid. 43, 544–546. Reprinted in Turing (2001).

Turing, A.M., 1939. Systems of logic based on ordinals. Proc. London Math. Soc. (2), 161–228. Reprinted in Turing (2001).

Turing, A.M., 2001. Mathematical Logic. In: Gandy, R.O., Yates, C.E.M. (Eds.), Collected Works of A.M. Turing, Elsevier Science Publishers, Amsterdam.

SYSTEMS OF LOGIC BASED ON ORDINALS†

By A. M. TURING

[Received 31 May, 1938.—Read 16 June, 1938.]

The well-known theorem of Gödel (Gödel [**1**], [**2**]) shows that every system of logic is in a certain sense incomplete, but at the same time it indicates means whereby from a system L of logic a more complete system L' may be obtained. By repeating the process we get a sequence L, $L_1 = L'$, $L_2 = L'_1$, ... each more complete than the preceding. A logic L_ω may then be constructed in which the provable theorems are the totality of theorems provable with the help of the logics L, L_1, L_2,.... We may then form $L_{2\omega}$ related to L_ω in the same way as L_ω was related to L. Proceeding in this way we can associate a system of logic with any constructive ordinal‡. It may be asked whether a sequence of logics of this kind is complete in the sense that to any problem A there corresponds an ordinal α such that A is solvable by means of the logic L_α. I propose to investigate this question in a rather more general case, and to give some other examples of ways in which systems of logic may be associated with constructive ordinals.

---

† This paper represents work done while a Jane Eliza Procter Visiting Fellow at Princeton University, where the author received most valuable advice and assistance from Prof. Alonzo Church.

‡ The situation is not quite so simple as is suggested by this crude argument. See pages 170–172, 178–179.

1. The calculus of conversion. Gödel representations.

It will be convenient to be able to use the "conversion calculus" of Church for the description of functions and for some other purposes. This will make greater clarity and simplicity of expression possible. I give a short account of this calculus. For detailed descriptions see Church [**3**], [**2**], Kleene [**1**], Church and Rosser [**1**].

The formulae of the calculus are formed from the symbols {, }, (,), [,], λ, δ, and an infinite list of others called variables; we shall take for our infinite list a, b, ..., z, x', x'', ... Certain finite sequences of such symbols are called *well-formed formulae* (abbreviated to W.F.F.); we define this class inductively, and define simultaneously the free and the bound variables of a W.F.F. Any variable is a W.F.F. ; it is its only free variable, and it has no bound variables. δ is a W.F.F. and has no free or bound variables. If **M** and **N** are W.F.F. then {**M**}(**N**) is a W.F.F., whose free variables are the free variables of **M** together with the free variables of **N**, and whose bound variables are the bound variables of **M** together with those of **N**. If **M** is a W.F.F. and **V** is one of its free variables, then λ**V**[**M**] is a W.F.F. whose free variables are those of **M** with the exception of **V**, and whose bound variables are those of **M** together with **V**. No sequence of symbols is a W.F.F. except in consequence of these three statements.

In metamathematical statements we use heavy type letters to stand for variable or undetermined formulae, as was done in the last paragraph, and in future such letters will stand for well-formed formulae unless otherwise stated. Small letters in heavy type will stand for formulae representing undetermined positive integers (see below).

A W.F.F. is said to be in normal form if it has no parts of the form {λ**V**[**M**]}(**N**) and none of the form {{δ}(**M**)}(**N**), where **M** and **N** have no free variables.

We say that one W.F.F. is *immediately convertible* into another if it is obtained from it either by:

(i) Replacing one occurrence of a well-formed part λ**V**[**M**] by λ**U**[**N**], where the variable **U** does not occur in **M**, and **N** is obtained from **M** by replacing the variable **V** by **U** throughout.

(ii) Replacing a well-formed part {λ**V**[**M**]}(**N**) by the formula which is obtained from **M** by replacing **V** by **N** throughout, provided that the bound variables of **M** are distinct both from **V** and from the free variables of **N**.

(iii) The process inverse to (ii).

(iv) Replacing a well-formed part {{δ}(**M**)} (**M**) by

$$\lambda f\,[\lambda x\,[\{f\}(\{f\}(x))]]$$

if **M** is in normal form and has no free variables.

(v) Replacing a well-formed part {{δ}(**M**)} (**N**) by

$$\lambda f\,[\lambda x[\{f\}(x)]]$$

if **M** and **N** are in normal form, are not transformable into one another by repeated application of (i), and have no free variables.

(vi) The process inverse to (iv).

(vii) The process inverse to (v).

These rules could have been expressed in such a way that in no case could there be any doubt about the admissibility or the result of the transformation [in particular this can be done in the case of process (v)].

A formula **A** is said to be *convertible* into another **B** (abbreviated to "**A** conv **B**") if there is a finite chain of immediate conversions leading from one formula to the other. It is easily seen that the relation of convertibility is an equivalence relation, *i.e.* it is symmetric, transitive, and reflexive.

Since the formulae are liable to be very lengthy, we need means for abbreviating them. If we wish to introduce a particular letter as an abbreviation for a particular lengthy formula we write the letter followed by "$\rightarrow$" and then by the formula, thus

$$I \rightarrow \lambda x[x]$$

indicates that I is an abbreviation for $\lambda x[x]$. We also use the arrow in less sharply defined senses, but never so as to cause any real confusion. In these cases the meaning of the arrow may be rendered by the words "stands for".

If a formula **F** is, or is represented by, a single symbol we abbreviate $\{\mathbf{F}\}(\mathbf{X})$ to $\mathbf{F}(\mathbf{X})$. A formula $\{\{\mathbf{F}\}(\mathbf{X})\}(\mathbf{Y})$ may be abbreviated to

$$\{\mathbf{F}\}(\mathbf{X}, \mathbf{Y}),$$

or to $\mathbf{F}(\mathbf{X}, \mathbf{Y})$ if **F** is, or is represented by, a single symbol. Similarly for $\{\{\{\mathbf{F}\}(\mathbf{X})\}(\mathbf{Y})\}(\mathbf{Z})$, etc. A formula $\lambda\mathbf{V}_1[\lambda\mathbf{V}_2 \ldots [\lambda\mathbf{V}_r[\mathbf{M}]] \ldots]$ may be abbreviated to $\lambda\mathbf{V}_1\mathbf{V}_2 \ldots \mathbf{V}_r \cdot \mathbf{M}$.

We have not as yet assigned any meanings to our formulae, and we do not intend to do so in general. An exception may be made for the case of the positive integers, which are very conveniently represented by the formulae $\lambda fx \cdot f(x)$, $\lambda fx \cdot f(f(x))$, ... In fact we introduce the abbreviations

$$1 \rightarrow \lambda fx \cdot f(x)$$
$$2 \rightarrow \lambda fx \cdot f(f(x))$$
$$3 \rightarrow \lambda fx \cdot f(f(f(x))), \text{etc.},$$

and we also say, for example, that $\lambda fx \cdot f(f(x))$, or in full

$$\lambda f[\lambda x[\{f\}(\{f\}(x))]],$$

represents the positive integer 2. Later we shall allow certain formulae to represent ordinals, but otherwise we leave them without explicit meaning; an implicit meaning may be suggested by the abbreviations used. In any case where any meaning is assigned to formulae it is desirable that the meaning should be invariant under conversion. Our definitions of the positive integers do not violate this requirement, since it may be proved that no two formulae representing different positive integers are convertible the one into the other.

In connection with the positive integers we introduce the abbreviation

$$S \rightarrow \lambda ufx \cdot f(u(f, x)).$$

This formula has the property that, if **n** represents a positive integer, $S(\mathbf{n})$ is convertible to a formula representing its successor[†].

Formulae representing undetermined positive integers will be represented by small letters in heavy type, and we adopt once for all the convention that, if a small letter, n say, stands for a positive integer, then the same letter in heavy type, **n**, stands for the formula representing the positive integer. When no confusion arises from so doing, we shall not trouble to distinguish between an integer and the formula which represents it.

Suppose that $f(n)$ is a function of positive integers taking positive integers as values, and that there is a W.F.F. **F** not containing δ such that, for each positive integer n, $\mathbf{F}(\mathbf{n})$ is convertible to the formula representing $f(n)$. We shall then say that $f(n)$ is λ-definable or *formally definable*, and

---

[†] This follows from (A) below.

that **F** *formally defines* $f(n)$. Similar conventions are used for functions of more than one variable. The sum function is, for instance, formally defined by $\lambda abfx \cdot a(f, b(f, x))$; in fact, for any positive integers m, n, p for which $m+n=p$, we have

$$\{\lambda abfx \cdot a(f, b(f, x))\}(\mathbf{m}, \mathbf{n}) \text{ conv } \mathbf{p}.$$

In order to emphasize this relation we introduce the abbreviation

$$\mathbf{X}+\mathbf{Y} \rightarrow \{\lambda abfx \cdot a(f, b(f, x))\}(\mathbf{X}, \mathbf{Y})$$

and we shall use similar notations for sums of three or more terms, products, etc.

For any W.F.F. **G** we shall say that **G** *enumerates* the sequence **G**(1), **G**(2),... and any other sequence whose terms are convertible to those of this sequence.

When a formula is convertible to another which is in normal form, the second is described as a *normal form* of the first, which is then said to *have a normal form.* I quote here some of the more important theorems concerning normal forms.

(A) *If a formula has two normal forms they are convertible into one another by the use of* (i) *alone.* (Church and Rosser [**1**], 479, 481.)

(B) *If a formula has a normal form then every well-formed part of it has a normal form.* (Church and Rosser [**1**], 480–481.)

(C) *There is (demonstrably) no process whereby it can be said of a formula whether it has a normal form.* (Church [**3**], 360, Theorem XVIII.)

We often need to be able to describe formulae by means of positive integers. The method used here is due to Gödel (Gödel[**1**]). To each single symbol s of the calculus we assign an integer $r[s]$ as in the table below.

s	{, (, or [	},), or]	λ	δ	α	...	z	x′	x″	x‴	...
r[s]	1	2	3	4	5	...	30	31	32	33	...

If $s_1, s_2, \ldots, s_k$ is a sequence of symbols, then $2^{r[s_1]}3^{r[s_2]}\ldots p_k^{r[s_k]}$ where p_k is the k-th prime number) is called the *Gödel representation* (G.R.) of that sequence of symbols. No two W.F.F. have the same G.R.

Two theorems on G.R. of W.F.F. are quoted here.

(D) *There is a* W.F.F. "form" *such that if* a *is the* G.R. *of a* W.F.F. **A** *without free variables, then* form (a) conv **A.** (This follows from a similar theorem to be found in Church [**3**], 53, 66. Metads are used there in place of G.R.)

(E) *There is a* W.F.F. Gr *such that, if* **A** *is a* W.F.F. *with a normal form without free variables, then* Gr(**A**) conv **a**, *where a is the* G.R. *of a normal form of* **A**. [Church [**3**], 53, 66, as (**D**).]

2. Effective calculability. Abbreviation of treatment.

A function is said to be "effectively calculable" if its values can be found by some purely mechanical process. Although it is fairly easy to get an intuitive grasp of this idea, it is nevertheless desirable to have some more definite, mathematically expressible definition. Such a definition was first given by Gödel at Princeton in 1934 (Gödel [**2**], 26), following in part an unpublished suggestion of Herbrand, and has since been developed by Kleene [**2**]). These functions were described as "general recursive" by Gödel. We shall not be much concerned here with this particular definition. Another definition of effective calculability has been given by Church (Church [**3**], 356 – 358), who identifies it with λ-definability. The author has recently suggested a definition corresponding more closely to the intuitive idea (Turing [**1**], see also Post [**1**]). It was stated above that "a function is effectively calculable if its values can be found by some purely mechanical process". We may take this statement literally, understanding by a purely mechanical process one which could be carried out by a machine. It is possible to give a mathematical description, in a certain normal form, of the structures of these machines. The development of these ideas leads to the author's definition of a computable function, and to an identification of computability† with effective calculability. It is not difficult, though somewhat laborious, to prove that these three definitions are equivalent (Kleene [**3**], Turing [**2**]).

In the present paper we shall make considerable use of Church's identification of effective calculability with λ-definability, or, what comes to the same thing, of the identification with computability and one of the equivalence theorems. In most cases where we have to deal with an effectively calculable function, we shall introduce the corresponding W.F.F. with some such phrase as "the function f is effectively calculable, let F be a formula λ defining it", or "let F be a formula such that $F(\mathbf{n})$ is convertible to . . . whenever **n** represents a positive integer". In such cases there is no difficulty in seeing how a machine could in principle be designed to calculate the values of the function concerned; and, assuming this done, the equivalence theorem can be applied. A statement of what the formula F actually is may be omitted. We may immediately introduce on this basis a W.F.F. ϖ with the property that

$$\varpi(\mathbf{m},\ \mathbf{n}) \text{ conv } \mathbf{r},$$

if r is the greatest positive integer, if any, for which m^r divides n and r is 1 if there is none. We also introduce Dt with the properties

$$\text{Dt}(\mathbf{n},\ \mathbf{n}) \text{ conv } 3,$$
$$\text{Dt}(\mathbf{n}+\mathbf{m},\ \mathbf{n}) \text{ conv } 2,$$
$$\text{Dt}(\mathbf{n},\ \mathbf{n}+\mathbf{m}) \text{ conv } 1.$$

There is another point to be made clear in connection with the point of view that we are adopting. It is intended that all proofs that are given should be regarded no more critically than proofs in classical analysis. The subject matter, roughly speaking, is constructive systems of logic, but since the purpose is directed towards choosing a particular constructive system of logic for practical use, an attempt at this stage to put our theorems into constructive form would be putting the cart before the horse.

Those computable functions which take only the values 0 and 1 are of particular importance, since they determine and are determined by computable properties, as may be seen by replacing "0" and "1" by "true" and "false". But, besides this type of property, we may have to consider a different type, which is, roughly speaking, less constructive than the computable properties, but more so than the general predicates of classical mathematics. Suppose that we have a computable function of the natural numbers taking natural numbers as values, then corresponding to this function there is the property of being a value of the function. Such a property we shall describe as "axiomatic" ; the reason for using this term is that it is possible to define such a property by giving a set of axioms,

† We shall use the expression "computable function" to mean a function calculable by a machine, and we let "effectively calculable" refer to the intuitive idea without particular identification with any one of these definitions. We do not restrict the values taken by a computable function to be natural numbers; we may for instance have computable propositional functions.

the property to hold for a given argument if and only if it is possible to deduce that it holds from the axioms.

Axiomatic properties may also be characterized in this way. A property ψ of positive integers is axiomatic if and only if there is a computable property ϕ of two positive integers, such that $\psi(x)$ is true if and only if there is a positive integer y such that $\phi(x, y)$ is true. Or again ψ is axiomatic if and only if there is a W.F.F. **F** such that $\psi(n)$ is true if and only if **F**(**n**) conv 2.

3. Number-theoretic theorems.

By a *number-theoretic theorem*[†] we shall mean a theorem of the form "$\theta(x)$ vanishes for infinitely many natural numbers x", where $\theta(x)$ is a primitive recursive[‡] function.

We shall say that a problem is number-theoretic if it has been shown that any solution of the problem may be put in the form of a proof of one or more number-theoretic theorems. More accurately we may say that a class of problems is number-theoretic if the solution of any one of them can be transformed (by a uniform process) into the form of proofs of number-theoretic theorems.

I shall now draw a few consequences from the definition of "number theoretic theorems" and in section 5. I shall try to justify confining our consideration to this type of problem.

An alternative form for number-theoretic theorems is "for each natural number x there exists a natural number y such that $\phi(x,y)$ vanishes", where $\phi(x, y)$ is primitive recursive. In other words, there is a rule whereby, given the function $\theta(x)$, we can find a function $\phi(x,y)$, or given $\phi(x,y)$, we can find a function $\theta(x)$, such that "$\theta(x)$ vanishes infinitely often" is a necessary and sufficient condition for "for each x there is a y such that $\phi(x,y) = 0$". In fact, given $\theta(x)$, we define

$$\phi(x,y) = \theta(x) + \alpha(x,y),$$

where $\alpha(x,y)$ is the (primitive recursive) function with the properties

$$\begin{aligned} \alpha(x,y) &= 1 \ (y \leq x), \\ &= 0 \ (y > x). \end{aligned}$$

If on the other-hand we are given $\phi(x,y)$ we define $\theta(x)$ by the equations

$$\theta_1(0) = 3,$$

$$\theta_1(x+1) = 2^{(1+\varpi_2(\theta_1(x)))\sigma(\phi(\varpi_3(\theta_1(x))-1,\ \varpi_2(\theta_1(x))))} 3^{\varpi_3(\theta_1(x))+1-\sigma(\phi(\varpi_3(\theta_1(x))-1,\ \varpi_2(\theta_1(x))))},$$

$$\theta(x) = \phi\ (\varpi_3(\theta_1(x)) - 1, \quad \varpi_2(\theta_1(x))),$$

[†] I believe that there is no generally accepted meaning for this term, but it should be noticed that we are using it in a rather restricted sense. Tile most generally accepted meaning is probably this: suppose that we take an arbitrary formula of the functional calculus of the first order and replace the function variables by primitive recursive relations. The resulting formula represents a typical number-theoretic theorem in this (more general) sense.

[‡] Primitive recursive functions of natural numbers arc defined inductively as follows. Suppose that $f(x_1,\ldots,x_{n-1})$, $g(x_1,\ldots,x_n)$, $h(x_1,\ldots,x_{x+1})$ are primitive recursive, then $\varphi(x_1,\ \ldots,\ x_n)$ is primitive recursive if it is defined by one of the sets of equations (a) to (e).

(a) $\phi(x_1,\ \ldots,\ x_n) = h((x_1,\ \ldots,\ x_{m-1}), g(x_1,\ldots,x_n), x_{m+1},\ldots,x_{n-1},x_n)\ (1 \leq m \leq n)$;
(b) $\phi(x_1,\ \ldots,\ x_n) = f(x_2,\ \ldots,\ x_n)$;
(c) $\phi(x_1) = a$, where $n = 1$ and α is some particular natural number;
(d) $\phi\ (x_1) = x_1 + 1\ (n = 1)$;
(e) $\phi(x_1,\ldots,\ x_{n-1},\ 0) = f(x_1,\ldots,\ x_{n-1})$; $\phi(x_1,\ldots,\ x_{n-1},\ x_{n+1}) = h(x_1,\ \ldots,\ x_n,\ \phi(x_1,\ldots,\ x_n))$.

The class primitive recursive functions is more restricted than the class of computable functions, but it has the advantage that there is a process whereby it can be said of a set of equations whether it defines a primitive recursive function in the manner described above.

where $\varpi_r(x)$ is defined so as to mean "the largest s for which r^s divides x". The function $\sigma(x)$ is defined by the equations $\sigma(0) = 0$, $\sigma(x+1) = 1$. It is easily verified that the functions so defined have the desired properties.

We shall now show that questions about the truth of the statements of the form "does $f(x)$ vanish identically", where $f(x)$ is a computable function, can be reduced to questions about the truth of number-theoretic theorems. It is understood that in each case the rule for the calculation of $f(x)$ is given and that we are satisfied that this rule is valid, *i.e.* that the machine which should calculate $f(x)$ is circle free (Turing [**1**], 233). The function $f(x)$, being computable, is general recursive in the Herbrand-Gödel sense, and therefore, by a general theorem due to Kleene[†], is expressible in the form

$$\psi(\epsilon y[\phi(x,y) = 0]), \tag{3.2}$$

where $\epsilon y[\mathfrak{A}(y)]$ means "the least y for which $\mathfrak{A}(y)$ is true" and $\psi(y)$ and $\phi(x,y)$ are primitive recursive functions. Without loss of generality, we may suppose that the functions ϕ, ψ take only the values 0, 1. Then, if we define $\rho(x)$ by the equations (3.1) and

$$\rho(0) = \psi(0)(1-\theta(0)),$$
$$\rho(x+1) = 1-(1-\rho(x))\sigma[1+\theta(x)-\psi\{\varpi_2(\theta_1(x))\}]$$

it will be seen that $f(x)$ vanishes identically if and only if $\rho(x)$ vanishes for infinitely many values of x.

The converse of this result is not quite true. We cannot say that the question about the truth of any number-theoretic theorem is reducible to a question about whether a corresponding computable function vanishes identically; we should have rather to say that it is reducible to the problem of whether a certain machine is circle free and calculates an identically vanishing function. But more is true: every number-theoretic theorem is equivalent to the statement that a corresponding machine is circle free. The behaviour of the machine may be described roughly as follows: the machine is one for the calculation of the primitive recursive function $\theta(x)$ of the number-theoretic problem, except that the results of the calculation are first arranged in a form in which the figures 0 and 1 do not occur, and the machine is then modified so that, whenever it has been found that the function vanishes for some value of the argument, then 0 is printed. The machine is circle free if and only if an infinity of these figures are printed, *i.e.* if and only if $\theta(x)$ vanishes for infinitely many values of the argument. That, on the other hand, questions of circle freedom may be reduced to questions of the truth of number-theoretic theorems follows from the fact that $\theta(x)$ is primitive recursive when it is defined to have the value 0 if a certain machine $\mathcal{M}$ prints 0 or 1 in its $(x+1)$-th complete configuration, and to have the value 1 otherwise.

The conversion calculus provides another normal form for the number theoretic theorems, and the one which we shall find the most convenient to use. Every number-theoretic theorem is equivalent to a statement of the form "**A(n)** is convertible to 2 for every W.F.F. **n** representing a positive integer", **A** being a W.F.F. determined by the theorem; the property of **A** here asserted will be described briefly as "**A** is dual". Conversely such statements are reducible to number theoretic theorems. The first half of this assertion follows from our results for computable functions, or directly in this way. Since $\theta(x-1)+2$ is primitive recursive, it is formally definable, say, by means of a formula **G**. Now there is (Kleene [**1**], 232) a W.F.F. $\mathcal{P}$ with the property that, if **T(r)** is convertible to a formula representing a positive integer for each positive integer r, then $\mathcal{P}$(**T**, **n**) is convertible to s, where s is the n-th positive integer t (if there is one) for which **T(t)** conv 2; if **T(t)** conv 2

[†] Kleene [**3**], 727. This result is really superfluous for our purpose, since the proof that every computable function is general recursive proceeds by showing that these functions are of the form (3.2). (Turing [**2**], 161).

for less than n values of t then $\mathcal{P}(\mathbf{T},\ \mathbf{n})$ has no normal form. The formula $\mathbf{G}(\mathcal{P}(\mathbf{G},\ \mathbf{n}))$ is therefore convertible to 2 if and only if $\theta(x)$ vanishes for at least n values of x, and is convertible to 2 for every positive integer x if and only if $\theta(x)$ vanishes infinitely often. To prove the second half of the assertion, we take Gödel representations for the formulae of the conversion calculus. Let $c(x)$ be 0 if x is the G.R. of 2 (*i.e.* if x is $2^3.3^{10}.5.7^3.11^{28}.13.17.19^{10}.23^2.29.31.\ 37^{10}.41^2.43.47^{28}.53^2.59^2.61^2.67^2$) and let $c(x)$ be 1 otherwise. Take an enumeration of the G.R. of the formulae into which $\mathbf{A(m)}$ is convertible: let $a(m,\ n)$ be the n-th number in the enumeration. We can arrange the enumeration so that $a(m,\ n)$ is primitive recursive. Now the statement that $\mathbf{A(m)}$ is convertible to 2 for every positive integer m is equivalent to the statement that, corresponding to each positive integer m, there is a positive integer n such that $c(a(m,\ n)) = 0$; and this is number-theoretic.

It is easy to show that a number of unsolved problems, such as the problem of the truth of Fermat's last theorem, are number-theoretic. There are, however, also problems of analysis which are number-theoretic. The Riemann hypothesis gives us an example of this. We denote by $\zeta(s)$ the function defined for $\Re s = \sigma > 1$ by the series $\sum_{n=1}^{\infty} n^{-s}$ and over the rest of the complex plane with the exception of the point $s = 1$ by analytic continuation. The Riemann hypothesis asserts that this function does not vanish in the domain $\sigma > \frac{1}{2}$. It is easily shown that this is equivalent to saying that it does not vanish for $2 > \sigma > \frac{1}{2}\Im s = t > 2$, *i.e.* that it does not vanish inside any rectangle $2 > \sigma > \frac{1}{2} + 1/T,\ T > t > 2$, where T is an integer greater than 2. Now the function satisfies the inequalities

$$\left.\begin{aligned} \left|\zeta(s) - \sum_{1}^{N} n^{-s} - \frac{N^{1-s}}{s-1}\right| &< 2t(N-2)^{-\frac{1}{2}}\ 2 < \sigma < \tfrac{1}{2},\ t \geq 2, \\ |\zeta(s) - \zeta(s')| &< 60t|s - s'|,\ 2 < \sigma' < \tfrac{1}{2},\ t' \geq 2, \end{aligned}\right\}$$

and we can define a primitive recursive function $\xi(l,\ l',\ m,\ m',N,\ M)$ such that

$$\left|\xi(l,l',m,m',N,M) - M\left|\sum_{1}^{N} n^{-s} + \frac{N^{1-s}}{s-1}\right|\right| < 2, \quad \left(s = \frac{l}{l'} + i\frac{m}{m'}\right),$$

and therefore, if we put

$$\xi(l,\ M,\ m,\ M,M^2+2,\ M)\ = X(l,\ m,M),$$

we have

$$\left|\zeta\left(\frac{l+\vartheta}{M} + i\frac{m+\vartheta}{M}\right)\right| \geq \frac{X(l,m,M) - 122T}{M},$$

provided that

$$\frac{1}{2} + \frac{1}{T} \leq \frac{l-1}{M} < \frac{l+1}{M} < 2 - \frac{1}{M}, \qquad 2 < \frac{m-1}{M} < \frac{m+1}{M} < T$$
$$(-1 < \theta < 1,\ -1 < \theta' < 1).$$

If we define $B(M,\ T)$ to be the smallest value of $X(l,\ m,\ M)$ for which

$$\frac{1}{2} + \frac{1}{T} + \frac{1}{M} \leq \frac{l}{M} < 2 - \frac{1}{M}, \qquad 2 + \frac{1}{M} < \frac{m}{M} < T - \frac{1}{M},$$

then the Riemann hypothesis is true if for each T there is an M satisfying

$$B(M,T) > 122T.$$

If on the other hand there is a T such that, for all M, $B(M, T) \leq 122T$, the Riemann hypothesis is false; for let l_M, m_M be such that

$$X(l_M, m_M, M) \leq 122T,$$

then

$$\left|\zeta\left(\frac{l_M + im_M}{M}\right)\right| \leq \frac{244T}{M}.$$

Now if a is a condensation point of the sequence $(l_M + im_M)/M$ then since $\zeta(s)$ is continuous except at $s = 1$ we must have $\zeta(a) = 0$ implying the falsity of the Riemann hypothesis. Thus we have reduced the problem to the question whether for each T there is an M for which

$$B(M,T) > 122T.$$

$B(M, T)$ is primitive recursive, and the problem is therefore number-theoretic.

4. A type of problem which is not number-theoretic*.

Let us suppose that we are supplied with some unspecified means of solving number-theoretic problems; a kind of oracle as it were. We shall not go any further into the nature of this oracle apart from saying that it cannot be a machine. With the help of the oracle we could form a new kind of machine (call them o-machines), having as one of its fundamental processes that of solving a given number-theoretic problem. More definitely these machines are to behave in this way. The moves of the machine are determined as usual by a table except in the case of moves from a certain internal configuration $\mathfrak{o}$. If the machine is in the internal configuration $\mathfrak{o}$ and if the sequence of symbols marked with l is then the well-formed† formula **A**, then the machine goes into the internal configuration $\mathfrak{p}$ or $\mathfrak{t}$ according as it is or is not true that **A** is dual. The decision as to which is the case is referred to the oracle.

These machines may be described by tables of the same kind as those used for the description of a-machines, there being no entries, however, for the internal configuration $\mathfrak{o}$. We obtain description numbers from these tables in the same way as before. If we make the convention that, in assigning numbers to internal configurations, $\mathfrak{o}$, $\mathfrak{p}$, $\mathfrak{t}$ are always to be q_2, q_3, q_4, then the description numbers determine the behaviour of the machines uniquely.

Given any one of these machines we may ask ourselves the question whether or not it prints an infinity of figures 0 or 1; I assert that this class of problem is not number-theoretic. In view of the definition of "number theoretic problem" this means that it is not possible to construct an o-machine which, when supplied‡ with the description of any other o-machine, will determine whether that machine is o-circle free. The argument may be taken over directly from Turing [**1**], §8. We say that a number is o-satisfactory if it is the description number of an o-circle free machine. Then, if there is an o-machine which will determine of any integer whether it is o-satisfactory, there is also an o-machine to calculate the values of the function $1 - \phi_n(n)$. Let $r(n)$ be the n-th o-satisfactory number and let $\phi_n(m)$ be the m-th figure printed by the o-machine whose description number is $r(n)$. This o-machine is circle free and there is therefore an o-satisfactory number K such that $\phi_K(n) = 1 - \phi_n(n)$ for all n. Putting $n = K$ yields a contradiction. This completes the proof that problems of circle freedom of o-machines are not number-theoretic.

Propositions of the form that an o-machine is o-circle free can always be put in the form of propositions obtained from formulae of the functional calculus of the first order by replacing *some* of the functional variables by primitive recursive relations. Compare foot-note † on page 156.

---

* Compare Rosser [**1**].

† Without real loss of generality we may suppose that **A** is always well formed.

‡ Compare Turing [**1**], §6, 7.

5. Syntactical theorems as number-theoretic theorems.

I now mention a property of number-theoretic theorems which suggests that there is reason for regarding them as of particular importance.

Suppose that we have some axiomatic system of a purely formal nature. We do not concern ourselves at all in interpretations for the formulae of this system; they are to be regarded as of interest for themselves. An example of what is in mind is afforded by the conversion calculus (§1). Every sequence of symbols "**A** conv **B**", where **A** and **B** are well formed formulae is a formula of the axiomatic system and is provable if the W.F.F. **A** is convertible to **B**. The rules of conversion give us the rules of procedure in this axiomatic system.

Now consider a new rule of procedure which is reputed to yield only formulae provable in the original sense. We may ask ourselves whether such a rule is valid. The statement that such a rule is valid would be number-theoretic. To prove this, let us take Gödel representations for the formulae, and an enumeration of the provable formulae; let $\phi(r)$ be the G.R. of the r-th formula in the enumeration. We may suppose $\phi(r)$ to be primitive recursive if we are prepared to allow repetitions in the enumeration. Let $\psi(r)$ be the G.R. of the r-th formula obtained by the new rule, then the statement that this new rule is valid is equivalent to the assertion of

$$(r)(\exists s)[\psi(r) = \phi(s)]$$

(the domain of individuals being the natural numbers). It has been shown in §3 that such statements are number-theoretic.

It might plausibly be argued that all those theorems of mathematics which have any significance when taken alone are in effect syntactical theorems of this kind, stating the validity of certain "derived rules " of procedure. Without going so far as this, I should assert that theorems of this kind have an importance which makes it worth while to give them special consideration.

6. Logic formulae.

We shall call a formula **L** a *logic formula* (or, if it is clear that we are speaking of a W.F.F., simply a *logic*) if it has the property that, if **A** is a formula such that **L**(**A**) conv 2, then **A** is dual.

A logic formula gives us a means of satisfying ourselves of the truth of number-theoretic theorems. For to each number-theoretic proposition there corresponds a W.F.F. **A** which is dual if and only if the proposition is true. Now, if **L** is a logic and **L**(**A**) conv 2, then **A** is dual and we know that the corresponding number-theoretic proposition is true. It does not follow that, if **L** is a logic, we can use **L** to satisfy ourselves of the truth of *any* number-theoretic theorem.

If **L** is a logic, the set of formulae **A** for which **L**(**A**) conv 2 will be called the *extent* of **L**.

It may be proved by the use of (D), (E), p. 154, that there is a formula X such that, if **M** has a normal form, has no free variables and is not convertible to 2, then X(**M**) conv 1, but, if **M** conv 2, then X(**M**) conv 2. If **L** is a logic, then $\lambda x.X$(**L**(x)) is also a logic whose extent is the same as that of **L**, and which has the property that, if **A** has no free variables, then

$$\{\lambda x.X(\mathbf{L}(x))\}(\mathbf{A})$$

either is always convertible to 1 or to 2 or else has no normal form. A logic with this property will be said to be *standardized*.

We shall say that a logic **L**′ is *at least as complete as* a logic **L** if the extent of **L** is a subset of the extent of **L**′. The logic **L**′ is *more complete than* **L** if the extent of **L** is a proper subset of the extent of **L**′.

Suppose that we have an effective set of rules by which we can prove formulae to be dual; *i.e.* we have a system of symbolic logic in which the propositions proved are of the form that certain formulae are dual. Then we can find a logic formula whose extent consists of just those formulae which can be proved to be dual by the rules; that is to say, there is a rule for obtaining the logic formula from the system of symbolic logic. In fact the system of symbolic logic enables us to obtain† a computable function of positive integers whose values run through the Gödel representations of the formulae provable by means of the given rules. By the theorem of equivalence of computable and λ-definable functions, there is a formula **J** such that **J**(1), **J**(2), ... are the G.R. of these formulae. Now let

$$W \to \lambda jv.\mathcal{P}\,(\lambda u.\delta\,(j(u), v)\,, 1, I,\ 2)\,.$$

Then I assert that $W(\mathbf{J})$ is a logic with the required properties. The properties of $\mathcal{P}$ imply that $\mathcal{P}(\mathbf{C},\ 1)$ is convertible to the least positive integer **n** for which **C**(**n**) conv 2, and has no normal form if there is no such integer. Consequently $\mathcal{P}(\mathbf{C},\ 1,\ I,\ 2)$ is convertible to 2 if **C**(**n**) conv 2 for some positive integer n, and it has no normal form otherwise. That is to say that $W(\mathbf{J},\ \mathbf{A})$ conv 2 if and only if $\delta\,(\mathbf{J}(\mathbf{n}),\ \mathbf{A})$ conv 2, some n, *i.e.* if **J**(**n**) conv **A** some n.

There is conversely a formula W' such that, if **L** is a logic, then $W'(\mathbf{L})$ enumerates the extent of **L**. For there is a formula Q such that $Q(\mathbf{L},\ \mathbf{A},\ \mathbf{n})$ conv 2 if and only if **L**(**A**) is convertible to 2 in less than n steps. We then put

$$W' \to \lambda ln.\ \text{form}\left(\varpi\ \left(2, \mathcal{P}\,(\lambda x.Q\,(l, \text{form}\,(\varpi\,(2,x))\,, \varpi\,(3,x))\,, n)\right)\right).$$

Of course, $W'\,(W(\mathbf{J}))$ normally entirely different from **J** and $W\left(W'(\mathbf{L})\right)$ from **L**.

In the case where we have a symbolic logic whose propositions can be interpreted as number-theoretic theorems, but are not expressed in the form of the duality of formulae, we shall again have a corresponding logic formula, but its relation to the symbolic logic is not so simple. As an example let us take the case where the symbolic logic proves that certain primitive recursive functions vanish infinitely often. As was shown in §3, we can associate with each such proposition a W.F.F. which is dual if and only if the proposition is true. When we replace the propositions of the symbolic logic by theorems on the duality of formulae in this way, our previous argument applies and we obtain a certain logic formula **L**. However, **L** does not determine uniquely which are the propositions provable in the symbolic logic; for it is possible that "$\theta_1(x)$ vanishes infinitely often" and "$\theta_2(x)$ vanishes infinitely often" are both associated with "A is dual", and that the first of these propositions is provable in the system, but the second not. However, if we suppose that the system of symbolic logic is sufficiently powerful to be able to carry out the argument on pp. 157–158 then this difficulty cannot arise. There is also the possibility that there may be formulae in the extent of **L** with no propositions of the form "$\theta(x)$ vanishes infinitely often" corresponding to them. But to each such formula we can assign (by a different argument) a proposition p of the symbolic logic which is a necessary and sufficient condition for **A** to be dual. With p is associated (in the first way) a formula $\mathbf{A}'$. Now **L** can always be modified so that its extent contains $\mathbf{A}'$ whenever it contains **A**.

We shall be interested principally in questions of completeness. Let us suppose that we have a class of systems of symbolic logic, the propositions of these systems being expressed in a uniform notation and interpretable as number-theoretic theorems; suppose also that there is a rule by which we can assign to each proposition p of the notation a W.F.F. $\mathbf{A}_p$, which is dual if and only if p is true, and that to each W.F.F. **A** we can assign a proposition $p_{\mathbf{A}}$ which is a necessary and sufficient condition for **A** to be dual. $p_{\mathbf{A}_p}$, is to be expected to differ from p. To each symbolic logic C we can assign two logic formulae $\mathbf{L}_c$ and $\mathbf{L}'_c$. A formula **A** belongs to the extent of $\mathbf{L}_c$ if $p_{\mathbf{A}}$ is provable in C, while the extent of $\mathbf{L}'_C$ consists of all $\mathbf{A}p$, where p is provable in C. Let us say that the class of symbolic logics is complete if each true proposition is provable in one of them: let us also say that

† Compare Turing [**1**], 252, second footnote, [**2**], 156.

a class of logic formulae is complete if the set-theoretic sum of the extents of these logics includes all dual formulae. I assert that a necessary condition for a class of symbolic logics C to be complete is that the class of logics $\mathbf{L}_c$ is complete, while a sufficient condition is that the class of logics $\mathbf{L}'_c$ is complete. Let us suppose that the class of symbolic logics is complete; consider p_A, where **A** is arbitrary but dual. It must be provable in one of the systems, C say. **A** therefore belongs to the extent of $\mathbf{L}_c$, *i.e.* the class of logics $\mathbf{L}_c$ is complete. Now suppose the class of logics $\mathbf{L}'_c$ to be complete. Let p be an arbitrary true proposition of the notation ; $\mathbf{A}_p$ must belong to the extent of some $\mathbf{L}'_c$, and this means that p is provable in C.

We shall say that a single logic formula **L** is complete if its extent includes all dual formulae; that is to say, it is complete if it enables us to prove every true number-theoretic theorem. It is a consequence of the theorem of Gödel (if suitably extended) that no logic formula is complete, and this also follows from (*C*), p. 154, or from the results of Turing [**1**], §8, when taken in conjunction with §3 of the present paper. The idea of completeness of a logic formula is not therefore very important, although it is useful to have a term for it.

Suppose **Y** to be a W.F.F. such that $\mathbf{Y}(\mathbf{n})$ is a logic for each positive integer n. The formulae of the extent of $\mathbf{Y}(\mathbf{n})$ are enumerated by $W(\mathbf{Y}(\mathbf{n}))$, and the combined extents of these logics by

$$\lambda r.\ W\ (\mathbf{Y}(\varpi(2,r),\ \varpi(3,r))).$$

If we put

$$\Gamma \rightarrow \lambda y.\ W'(\lambda r.\ W(y(\varpi(2,r),\varpi(3,r)))),$$

then $\Gamma(\mathbf{Y})$ is a logic whose extent is the combined extent of

$$\mathbf{Y}(1),\ \mathbf{Y}(2),\ \mathbf{Y}(3),\ \ldots .$$

To each W.F.F. **L** we can assign a W.F.F. $V(\mathbf{L})$ such that a necessary and sufficient condition for **L** to be a logic formula is that $V(\mathbf{L})$ is dual. Let Nm be a W.F.F. which enumerates all formulae with normal forms and no free variables. Then the condition for **L** to be a logic is that $\mathbf{L}(\mathrm{Nm}(\mathbf{r}),\ \mathbf{s})$ conv 2 for all positive integers r, s, *i.e.* that

$$\lambda a.\mathbf{L}(\mathrm{Nm}(\varpi(2,a)),\varpi(3,\alpha))$$

is dual. We may therefore put

$$V \rightarrow \lambda la.\,l(\mathrm{Nm}(\varpi(2,a)),\ \varpi(3,a)).$$

7. Ordinals.

We begin our treatment of ordinals with some brief definitions from the Cantor theory of ordinals, but for the understanding of some of the proofs a greater amount of the Cantor theory is necessary than is set out here.

Suppose that we have a class determined by the propositional function $D(x)$ and a relation $G(x,y)$ ordering its members, *i.e.* satisfying

$$\left.\begin{array}{ll} G(x,y)\&G(y,z) \supset G(x,z), & \text{(i)} \\ D(x)\&D(y) \supset G(x,y) \vee G(y,x) \vee x=y, & \text{(ii)} \\ G(x,y) \supset D(x)\&D(y), & \text{(iii)} \\ \sim G(x,x). & \text{(iv)} \end{array}\right\} \quad (7.1)$$

The class defined by $D(x)$ is then called a *series* with the ordering relation $G(x,y)$. The series is said to be *well ordered* and the ordering relation is called an *ordinal* if every sub-series which is not void has a first term, *i.e.* if

$$(D')\{(\exists x)(D'(x))\&(x)(D'(x) \supset D(x)) \supset (\exists z)(y)[D'(z)\&(D'(y) \supset G(z,y) \vee z=y)]\}. \quad (7.2)$$

The condition (7.2) is equivalent to another, more suitable for our purposes, namely the condition that every descending subsequence must terminate formally

$$(x)\left\{D'(x) \supset D(x)\&(\exists y)(D'(y)\&G(y,x))\right\} \supset (x)(\sim D'(x)). \quad (7.\ 3)$$

The ordering relation $G(x,y)$ is said to be similar to $G'(x,y)$ if there is a one-one correspondence between the series transforming the one relation into the other. This is best expressed formally, thus

$$\begin{aligned}(\exists M)[(x)\{D(x) \supset (\exists x')\mathrm{M}(x,\ x')\}\&(x')\{D'(x') \supset (\exists x)M(x,\ x')\}\\ \&\{(M(x,\ x')\&M\ (x,\ x'')) \vee (M(x',\ x)\&M\ (x'',\ x)) \supset x'=x''\}\\ \&\{M(x,\ x')\&M(y,\ y') \supset (G(x,\ y)=G(x',\ y'))\}]. \quad (7.4)\end{aligned}$$

Ordering relations are regarded as belonging to the same ordinal if and only if they are similar.

We wish to give names to all the ordinals, but this will not be possible until they have been restricted in some way; the class of ordinals, as at present defined, is more than enumerable. The restrictions that we actually impose are these: $D(x)$ is to imply that x is a positive integer; $D(x)$ and $G(x,\ y)$ are to be computable properties. Both of the propositional functions $D(x)$, $G(x,\ y)$ can then be described by means of a single W.F.F. $\mathbf{\Omega}$ with the properties:

$\mathbf{\Omega(m,\ n)}$ conv 4 unless both $D(m)$ and $D(n)$ are true,

$\mathbf{\Omega(m,\ m)}$ conv 3 if $D(m)$ is true,

$\mathbf{\Omega(m,\ n)}$ conv 2 if $D(m)$, $D(n)$, $G(m,\ n), \sim (m=n)$ are true,

$\mathbf{\Omega(m,\ n)}$ conv 1 if $D(m)$, $D(n), \sim G(m,\ n), \sim (m=n)$ are true.

In consequence of the conditions to which $D(x)$, $G(x,y)$ are subjected, $\mathbf{\Omega}$ must further satisfy:

(a) if $\mathbf{\Omega(m,\ n)}$ is convertible to 1 or 2, then $\mathbf{\Omega(m,\ m)}$ and $\mathbf{\Omega(n,\ n)}$ arc convertible to 3,

(b) if $\mathbf{\Omega(m,\ m)}$ and $\mathbf{\Omega(n,\ n)}$ are convertible to 3, then $\mathbf{\Omega(m,n)}$ is convertible to 1, 2, or 3,

(c) if $\mathbf{\Omega(m,\ n)}$ is convertible to 1, then $\mathbf{\Omega(n,\ m)}$ is convertible to 2 and conversely,

(d) if $\mathbf{\Omega(m,\ n)}$ and $\mathbf{\Omega(n,\ p)}$ are convertible to 1, then $\mathbf{\Omega(m,\ p)}$ is also,

(e) there is no sequence $m_1,\ m_2,\ \ldots$ such that $\mathbf{\Omega(m_{\mathit{i}+1},\ m_{\mathit{i}})}$ conv 2 for each positive integer i,

(f) $\mathbf{\Omega(m,n)}$ is always convertible to 1, 2, 3, or 4.

If a formula $\mathbf{\Omega}$ satisfies these conditions then there are corresponding propositional functions $D(x)$, $G(x,y)$. We shall therefore say that $\mathbf{\Omega}$ is an *ordinal formula* if it satisfies the conditions $(a)-(f)$. It will be seen that a consequence of this definition is that Dt is an ordinal formula; it represents the ordinal ω. The definition that we have given does not pretend to have virtues such as elegance or convenience. It has been introduced rather to fix our ideas and to show how it is possible in principle to describe ordinals by means of well formed formulae. The definitions could be modified in a number of ways. Some such modifications arc quite trivial; they are typified by modifications such as changing the numbers 1, 2, 3, 4, used in the definition, to others. Two such definitions will be said to be equivalent; in general, we shall say that two definitions are equivalent if there are W.F.F. $\mathbf{T}$, $\mathbf{T}'$ such that, if $\mathbf{A}$ is an ordinal formula under one definition and represents the ordinal α, then $\mathbf{T}'(\mathbf{A})$ is an ordinal formula under the second definition and represents the same ordinal; and, conversely, if $\mathbf{A}'$ is an ordinal formula under the second definition representing α, then $\mathbf{T}(\mathbf{A}')$ represents α under the first definition. Besides definitions equivalent in this sense to our original definition, there are a number of other possibilities open. Suppose for instance that we do not require $D(x)$ and $G(x, y)$ to be computable, but that we require only that $D(x)$ and $G(x, y) \& x < y$ are axiomatic†. This leads to a definition of an ordinal formula which is (presumably) not equivalent to the definition that we are using‡. There are numerous possibilities, and little to guide us in choosing one definition rather than another. No one of them could well be described as "wrong"; some of them may be found more valuable in applications than others, and the particular choice that we have made has been determined partly by the applications that we have in view. In the case of theorems of a negative character, one would wish to prove them for each one of the possible definitions of "ordinal formula". This programme could, I think, be carried through for the negative results of §9, 10.

Before leaving the subject of possible ways of defining ordinal formulae, I must mention another definition due to Church and Kleene (Church and Kleene [**1**]). We can make use of this definition in constructing ordinal logics, but it is more convenient to use a slightly different definition which is equivalent (in the sense just described) to the Church-Kleene definition as modified in Church [**4**].

Introduce the abbreviations

$$U \to \lambda ufx.u(\lambda y.f(y(I,x))),$$

$$Suc \to \lambda aufx.f(a(u, f, x)).$$

We define first a partial ordering relation "<" which holds 164 between certain pairs of W.F.F. $[conditions(1) - (5)]$.

(1) If **A** conv **B**, then $\mathbf{A} < \mathbf{C}$ implies $\mathbf{B} < \mathbf{C}$ and $\mathbf{C} < \mathbf{A}$ implies $\mathbf{C} < \mathbf{B}$.
(2) $\mathbf{A} <$ Suc $(\mathbf{A})$.
(3) For any positive integers m and n, $\lambda ufx.\ \mathbf{R}(\mathbf{n}) < \lambda ufx.\mathbf{R}(\mathbf{m})$ implies $\lambda ufx.\ \mathbf{R}(\mathbf{n}) < \lambda ufx.u(\mathbf{R})$.
(4) If $\mathbf{A} < \mathbf{B}$ and $\mathbf{B} < \mathbf{C}$, then $\mathbf{A} < \mathbf{C}$. (1) – (4) are required for any W.F.F. **A**, **B**, **C**, $\lambda ufx.\ \mathbf{R}$.
(5) The relation $\mathbf{A} < \mathbf{B}$ holds only when compelled to do so by (1) – (4).

We define C-K ordinal formulae by the conditions (6)–(10).

(6) If **A** conv **B** and **A** is a C-K ordinal formula, then **B** is a C-K ordinal formula.
(7) U is a C-K ordinal formula.
(8) If **A** is a C-K ordinal formula, then Suc (**A**) is a C-K ordinal formula.
(9) If $\lambda ufx.\mathbf{R}(\mathbf{n})$ is a C-K ordinal formula and

$$\lambda ufx.\mathbf{R}(\mathbf{n}) < \lambda ufx.\mathbf{R}(S(\mathbf{n}))$$

for each positive integer n, then $\lambda ufx.u(\mathbf{R})$ is a C-K ordinal formula¶.
(10) A formula is a C-K ordinal formula only if compelled to be so by (6)–(9).

† To require $G(x,y)$ to be axiomatic amounts to requiring $G(x,y)$ to be computable on account of (7.1) (ii).

‡ On the other hand, if $D(x)$ is axiomatic and $(G(x,y)$ is computable in the modified sense that there is a rule for determining whether $G(x,y)$ is true which leads to a definite result in all cases where $D(x)$ and $D(y)$ are true, the corresponding definition of ordinal formula is equivalent to our definition. To give the proof would be too much of a digression. Probably other equivalences of this kind hold.

¶ If we also allow $\lambda ufx.u(\mathbf{R})$ to be a C-K ordinal formula when

$$\lambda ufx.\mathbf{n}(\mathbf{R}) \text{conv} \lambda ufx.S(\mathbf{n},\mathbf{R})$$

for all n, then the formulae for sum, product and exponentiation of C-K ordinal formulae can be much simplified. For instance, if **A** and **B** represent α and β, then

$$\lambda ufx.\mathbf{B}(u,f,\mathbf{A}(u,f,x))$$

represents $\alpha + \beta$. Property (6) remains true.

The representation of ordinals by formulae is described by (11)–(15).

(11) If **A** conv **B** and **A** represents α then **B** represents α.

(12) U represents 1.

(13) If **A** represents α, then Suc(**A**) represents $\alpha + 1$.

(14) If $\lambda ufx.\mathbf{R}(\mathbf{n})$ represents α_n for each positive integer n, then $\lambda ufx.u(\mathbf{R})$ represents the upper bound of the sequence $\alpha_1, \alpha_2, \alpha_3, \ldots$.

(15) A formula represents an ordinal only when compelled to do so by (11)−(14).

We denote any ordinal represented by **A** by $\Xi_{\mathbf{A}}$ without prejudice to the possibility that more than one ordinal may be represented by **A**. We shall write $\mathbf{A} \leq \mathbf{B}$ to mean $\mathbf{A} < \mathbf{B}$ or **A** conv **B**.

In proving properties of C-K ordinal formulae we shall often use a kind of analogue of the principle of transfinite induction. If ϕ is some property and we have:

$$\left.\begin{array}{l} (a)\ \text{If } \mathbf{A} \text{ conv } \mathbf{B} \text{ and } \phi(\mathbf{A}), \text{ then } \phi(\mathbf{B}), \\ (b)\ \phi(U), \\ (c)\ \text{If } \phi(\mathbf{A}), \text{ then } \phi(\text{Suc}(\mathbf{A})), \\ (d)\ \text{If } \phi(\lambda ufx.\mathbf{R}(\mathbf{n})) \text{ and } \lambda ufx.\mathbf{R}(\mathbf{n}) < \lambda ufx.\mathbf{R}(S(\mathbf{n})) \text{ for each positive integer } n, \text{ then} \\ \qquad\qquad \phi(\lambda ufx.u(\mathbf{R})); \end{array}\right\} \quad (7.5)$$

then $\phi(\mathbf{A})$ for each C-K ordinal formula **A**. To prove the validity of this principle we have only to observe that the class of formulae **A** satisfying $\phi(\mathbf{A})$ is one of those of which the class of C-K ordinal formulae was defined to be the smallest. We can use this principle to help us to prove:–

(i) Every C-K ordinal formula is convertible to the form $\lambda ufx.\mathbf{B}$, where **B** is in normal form.

(ii) There is a method by which for any C-K ordinal formula, we can determine into which of the forms U, Suc $(\lambda ufx.\mathbf{B})$, $\lambda ufx.u(\mathbf{R})$ (where u is free in **R**) it is convertible, and by which we can determine **B**, **R**. In each case **B**, **R** arc unique apart from conversions.

(iii) If **A** represents any ordinal, $\Xi_{\mathbf{A}}$ is unique. If $\Xi_{\mathbf{A}}, \Xi_{\mathbf{B}}$ exist and $\mathbf{A} < \mathbf{B}$, then $\Xi_A < \Xi_{\mathbf{B}}$.

(iv) If **A**, **B**, **C** are C-K ordinal formulae and $\mathbf{B} < \mathbf{A}$, $\mathbf{C} < \mathbf{A}$, then either $\mathbf{B} < \mathbf{C}$, $\mathbf{C} < \mathbf{B}$, or **B** conv **C**.

(v) A formula **A** is a C-K ordinal formula if:

(A) $U \leq \mathbf{A}$,

(B) If $\lambda ufx.u(\mathbf{R}) \leq \mathbf{A}$ and n is a positive integer, then

$$\lambda ufx.\mathbf{R}(\mathbf{n}) < \lambda ufx.\mathbf{R}(S(\mathbf{n})),$$

(C) For any two W.F.F. **B**, **C** with $\mathbf{B} < \mathbf{A}$, $\mathbf{C} < \mathbf{A}$ we have $\mathbf{B} < \mathbf{C}$, $\mathbf{C} < \mathbf{B}$, or **B** conv **C**, but never $\mathbf{B} < \mathbf{B}$,

(D) There is no infinite sequence $\mathbf{B}_1, \mathbf{B}_2, \ldots$ for which

$$\mathbf{B_r} < \mathbf{B_{r-1}} < \mathbf{A}$$

for each r.

(vi) There is a formula H such that, if **A** is a C-K ordinal formula, then $H(\mathbf{A})$ is an ordinal formula representing the same ordinal. $H(\mathbf{A})$ is not an ordinal formula unless **A** is a C-K ordinal formula.

Proof of (i). Take $\phi(\mathbf{A})$ to be "$\mathbf{A}$ is convertible to the form $\lambda ufx.\mathbf{B}$, where $\mathbf{B}$ is in normal form". The conditions (*a*) and (*b*) are trivial. For (*c*), suppose that $\mathbf{A}$ conv $\lambda ufx.\mathbf{B}$, where $\mathbf{B}$ is in normal form; then

$$\mathrm{Suc}(\mathbf{A}) \text{ conv } \lambda ufx.f(\mathbf{B})$$

and $f(\mathbf{B})$ is in normal form. For (*d*) we have only to show that $u(\mathbf{R})$ has a normal form, *i.e.* that $\mathbf{R}$ has a normal form; and this is true since $\mathbf{R}(1)$ has a normal form.

Proof of (ii). Since, by hypothesis, the formula is a C-K ordinal formula we have only to perform conversions on it until it is in one of the forms described. It is not possible to convert it into two of these three forms. For suppose that $\lambda ufx.f(\mathbf{A}(u,f,x))$ conv $\lambda ufx.u(\mathbf{R})$ and is a C-K ordinal formula; it is then convertible to the form $\lambda ufx.\mathbf{B}$, where $\mathbf{B}$ is in normal form. But the normal form of $\lambda ufx.u(\mathbf{R})$ can be obtained by conversions on $\mathbf{R}$, and that of $\lambda ufx.f(\mathbf{A}(u, f, x))$ by conversions on $\mathbf{A}(u,f,x)$ (as follows from Church and Rosser [**1**], Theorem 2); this, however, would imply that the formula in question had two normal forms, one of form $\lambda ufx.u(\mathbf{S})$ and one of form $\lambda ufx.f(\mathbf{C})$, which is impossible. Or let U conv $\lambda ufx.u(\mathbf{R})$, where $\mathbf{R}$ is a well formed formula with u as a free variable. We may suppose $\mathbf{R}$ to be in normal form. Now U is $\lambda ufx.u(\lambda y\ .\ f(y(I,\ x)))$. By (A), p. 154, $\mathbf{R}$ is identical with $\lambda y.f(y(I,x))$, which does not have u as a free variable. It now remains to show only that if

$$\mathrm{Suc}\ (\lambda ufx.\mathbf{B}) \text{ conv } \mathrm{Suc}\ (\lambda ufx.\mathbf{B}') \quad \text{and} \quad \lambda ufx.u(\mathbf{R}) \text{ conv } \lambda ufx.u(\mathbf{R}'),$$

then $\mathbf{B}$ conv $\mathbf{B}'$ and $\mathbf{R}$ conv $\mathbf{R}'$.

If $$\mathrm{Suc}\ (\lambda ufx.\mathbf{B}) \text{ conv } \mathrm{Suc}\ (\lambda ufx.\mathbf{B}'),$$

then $$\lambda ufx.f(\mathbf{B}) \text{ conv } \lambda ufx.f(\mathbf{B}')$$

but both of these formulae can be brought to normal form by conversions on $\mathbf{B}$, $\mathbf{B}'$ and therefore $\mathbf{B}$ conv $\mathbf{B}'$. The same argument applies in the case in which $\lambda ufx.u(\mathbf{R})$ conv $\lambda ufx.u(\mathbf{R}')$.

Proof of (iii). To prove the first half, take $\phi(\mathbf{A})$ to be "$\Xi_{\mathbf{A}}$ is unique". Then (7.5) (*a*) is trivial, and (*b*) follows from the fact that U is not convertible either to the form Suc ($\mathbf{A}$) or to $\lambda ufx.u(\mathbf{R})$, where $\mathbf{R}$ has u as a free variable. For (*c*): Suc ($\mathbf{A}$) is not convertible to the form $\lambda ufx.u(\mathbf{R})$; the possibility that Suc ($\mathbf{A}$) represents an ordinal on account of (12) or (14) is therefore eliminated. By (13), Suc ($\mathbf{A}$) represents $\alpha' + 1$ if $\mathbf{A}'$ represents α' and Suc ($\mathbf{A}$) conv Suc ($\mathbf{A}'$). If we suppose that $\mathbf{A}$ represents α, then $\mathbf{A}$, $\mathbf{A}'$, being C-K ordinal formulae, are convertible to the forms $\lambda ufx.\mathbf{B}$, $\lambda ufx.\ \mathbf{B}'$; but then, by (ii), $\mathbf{B}$ conv $\mathbf{B}'$, i.e. $\mathbf{A}$ conv $\mathbf{A}'$, and therefore $\alpha = \alpha'$ by the hypothesis $\phi(\mathbf{A})$. Then $\Xi_{\mathrm{Suc}(\mathbf{A})} = \alpha' + 1$ is unique. For (*d*) : $\lambda ufx.u(\mathbf{R})$ is not convertible to the form Suc ($\mathbf{A}$) or to U if $\mathbf{R}$ has u as a free variable. If $\lambda ufx.u(\mathbf{R})$ represents an ordinal, it is so therefore in virtue of (14), possibly together with (11). Now, if $\lambda ufx.u(\mathbf{R})$ conv $\lambda ufx.u(\mathbf{R}')$, then $\mathbf{R}$ conv $\mathbf{R}'$, so that the sequence $\lambda ufx.\mathbf{R}(1)$, λufx, $\mathbf{R}(2)$, ... in (14) is unique apart from conversions. Then, by the induction hypothesis, the sequence α_1, α_2, $\alpha_3 \ldots$ is unique. The only ordinal that is represented by $\lambda ufx.u(\mathbf{R})$ is the upper bound of this sequence; and this is unique.

For the second half we use a type of argument rather different from our transfinite induction principle. The formulae $\mathbf{B}$ for which $\mathbf{A} < \mathbf{B}$ form the smallest class for which:

$$\left.\begin{array}{l}\text{Suc }(\mathbf{A})\text{ belongs to the class.}\\ \text{If }\mathbf{C}\text{ belongs to the class, then Suc }(\mathbf{C})\text{ belongs to it.}\\ \text{If }\lambda ufx.\mathbf{R}(\mathbf{n})\text{ belongs to the class and}\\ \qquad\qquad \lambda ufx.\mathbf{R}(\mathbf{n}) < \lambda ufx.\mathbf{R}(\mathbf{m}),\\ \text{where } m, n \text{ are some positive integers, then } \lambda ufx.u(\mathbf{R})\text{belongs to it.}\\ \text{If }\mathbf{C}\text{ belongs to the class and }\mathbf{C}\text{ conv }\mathbf{C}'\text{ then }\mathbf{C}'\text{ belongs to it.}\end{array}\right\} \quad (7.6)$$

It will be sufficient to prove that the class of formulae **B** for which either $\Xi_{\mathbf{B}}$ does not exist or $\Xi_{\mathbf{A}} < \Xi_{\mathbf{B}}$ satisfies the conditions (7.6). Now

$$\Xi_{\text{Suc}(\mathbf{A})} = \Xi_{\mathbf{A}} + 1 > \Xi_{\mathbf{A}}, \qquad \Xi_{\text{Suc}(\mathbf{C})} > \Xi_{\mathrm{C}} > \Xi_{\mathbf{A}} \text{ if } \mathbf{C} \text{ is in the class.}$$

If $\Xi_{\lambda ufx.\mathbf{R}(n)}$ does not exist then $\Xi_{\lambda ufx.u(\mathbf{R})}$ does not exist, and therefore $\lambda ufx.u(\mathbf{R})$ is in the class. If $\Xi_{\lambda ufx.\mathbf{R}(\mathbf{n})}$ exists and is greater than $\Xi_{\mathbf{A}}$, and $\lambda ufx.\mathbf{R}(\mathbf{n}) < \lambda ufx.\mathbf{R}(\mathbf{m})$, then

$$\Xi_{\lambda ufx.u(\mathbf{R})} \geq \Xi_{\lambda ufx.\mathbf{R}(\mathbf{n})} > \Xi_{\mathbf{A}},$$

so that $\lambda ufx.u(\mathbf{R})$ belongs to the class.

Proof of (iv). We prove this by induction with respect to **A**. Take $\phi(\mathbf{A})$ to be "whenever **B** < **A** and **C** < **A** then **B** < **C** or **C** < **B** or **B** conv **C**′". $\phi(U)$ follows from the fact that we never have **B** < U. If we have $\phi(\mathbf{A})$ and **B** < Suc (A), then either **B** < **A** or **B** conv **A**; for we can find **D** such that **B** ≤ **D**, and then **D** < Suc (**A**) can be proved without appealing either to (1) or (5); (4) does not apply, so we must have **D** conv **A**. Then, if **B** < Suc (**A**) and **C** < Suc (**A**), we have four possibilities,

B conv **A**,	**C** conv **A**,
B conv **A**,	**C** < **A**,
B < **A**,	**C** conv **A**,
B < **A**,	**C** < **A**.

In the first case **B** conv **C**, in the second **C** < **B**, in the third **B** < **C**, and in the fourth the induction hypothesis applies.

Now suppose that $\lambda ufx.\mathbf{R}(\mathbf{n})$ is a C-K ordinal formula, that

$$\lambda ufx.\mathbf{R}(\mathbf{n}) < \lambda ufx.\mathbf{R}(S(\mathbf{n})) \text{ and } \phi(\mathbf{R}(\mathbf{n})),$$

for each positive integer n, and that **A** conv $\lambda ufx.u(\mathbf{R})$. Then, if **B** < **A**, this means that **B** < $\lambda ufx.\mathbf{R}(\mathbf{n})$ for some n; if we have also **C** < **A**, then **B** < $\lambda ufx.\mathbf{R}(\mathbf{q})$, **C** < $\lambda ufx.\mathbf{R}(\mathbf{q})$ for some q. Thus, for these **B** and **C**, the required result follows from $\phi(\lambda ufx.\mathbf{R}(\mathbf{q}))$.

Proof of (v). The conditions (C), (D) imply that the classes of interconvertible formulae **B**, **B** < **A** are well-ordered by the relation "<". We prove (v) by (ordinary) transfinite induction with respect to the order type α of the series formed by these classes; (α is, in fact, the solution of the equation $1+\alpha = \Xi_{\mathbf{A}}$, but we do not need this). We suppose then that (v) is true for all order types less than α. If **E** < **A**, then **E** satisfies the conditions of (v) and the corresponding order type is smaller: **E** is therefore a C-K ordinal formula. This expresses all consequences of the induction hypothesis that we need. There are three cases to consider:

(*x*) $\alpha = 0$.

(*y*) $\alpha = \beta + 1$.

(*z*) α is of neither of the forms (*x*), (*y*).

In case (*x*) we must have **A** conv U on account of (A). In case (*y*) there is a formula **D** such that **D** < **A**, and **B** ≤ **D** whenever **B** < **A**. The relation **D** < **A** must hold in virtue of either (1), (2), (3), or (4). It cannot be in virtue of (4); for then there would be **B**, **B** < **A**, **D** < **B** contrary to (C), taken in conjunction with the definition of **D**. If it is in virtue of (3), then α is the upper bound of a sequence α_1, α_2, ... of ordinals, which are increasing by reason of (iii) and the conditions $\lambda ufx.\mathbf{R}(\mathbf{n}) < \lambda ufx.\mathbf{R}(S(\mathbf{n}))$ in (B). This is inconsistent with $\alpha = \beta + 1$. This means that (2) applies [after we have eliminated (1) by suitable conversions on **A**, **D**] and we see that **A** conv Suc (**D**); but, since **D** < **A**, **D** is a C-K ordinal formula, and **A** must therefore be a C-K ordinal formula by (8). Now take case (*z*). It is impossible for **A** to be of the form Suc (**D**), for then we should have **B** < **D** whenever **B** < **A**, and this would mean that we had case (*y*). Since U < **A**, there must be an **F** such that **F** < **A** is demonstrable either by (2) or by (3) (after a possible conversion on **A**); it must of course be demonstrable by (3). Then **A** is of the form $\lambda ufx.u(\mathbf{R})$. By (3), (B) we see that $\lambda ufx.\mathbf{R}(\mathbf{n})$ < **A** for each positive integer n; each $\lambda ufx.\mathbf{R}(\mathbf{n})$ is therefore a C-K ordinal formula. Applying (9), (**B**) we see that **A** is a C-K ordinal formula.

Proof of (vi). To prove the first half, it is sufficient to find a method whereby from a C-K ordinal formula **A** we can find the corresponding ordinal formula $\mathbf{\Omega}$. For then there is a formula H_1 such that $H_1(\mathbf{a})$ conv **p** if a is the G.R. of **A** and p is that of $\mathbf{\Omega}$. H is then to be defined by

$$H \rightarrow \lambda a.\ \text{form}\ (H_1(\text{Gr}(a))).$$

The method of finding $\mathbf{\Omega}$ may be replaced by a method of finding $\mathbf{\Omega}(\mathbf{m},\mathbf{n})$, given **A** and any two positive integers m, n. We shall arrange the method so that, whenever **A** is not an ordinal formula, either the calculation of the values does not terminate or else the values are not consistent with $\mathbf{\Omega}$ being an ordinal formula. In this way we can prove the second half of (vi).

Let Ls be a formula such that Ls(**A**) enumerates the classes of formulae **B**, **B** < **A** [*i.e.* if **B** < **A** there is one and only one positive integer n for which Ls(**A**, **n**) conv **B**]. Then the rule for finding the value of $\mathbf{\Omega}(\mathbf{m}, \mathbf{n})$ is as follows:—

First determine whether $U \leq$ **A** and whether **A** is convertible to the form **r**(Suc, U). This terminates if **A** is a C-K ordinal formula.

If **A** conv **r**(Suc, U) and either $m > r+1$ or $n > r+1$, then the value is 4. If $m < n \leq r+1$, the value is 2. If $n < m \leq r+1$, the value is 1. If $m = n \leq r+1$, the value is 3.

If **A** is not convertible to this form, we determine whether either **A** or Ls(**A**, **m**) is convertible to the form $\lambda ufx.u(\mathbf{R})$, and if either of them is, we verify that $\lambda ufx.\mathbf{R}(\mathbf{n}) < \lambda ufx.\mathbf{R}(S(\mathbf{n}))$. We shall eventually come to an affirmative answer if **A** is a C-K ordinal formula.

Having checked this, we determine concerning m and n whether Ls(**A**, **m**) < Ls(**A**, **n**), Ls(**A**, **n**) < Ls(**A**, **m**), or $m = n$, and the value is to be accordingly 1, 2, or 3.

If **A** is a C-K ordinal formula, this process certainly terminates. To see that the values so calculated correspond to an ordinal formula, and one representing $\Xi_{\mathbf{A}}$, first observe that this is so when $\Xi_{\mathbf{A}}$ is finite. In the other case (iii) and (iv) show that $\Xi_{\mathbf{B}}$ determines a one-one correspondence between the ordinals $\beta, 1 \leq \beta \leq \Xi_{\mathbf{A}}$, and the classes of interconvertible formulae **B**, **B** < **A**. If we take $G(m, n)$ to be Ls (**A**, **m**) < Ls(**A**, **n**), we see that $G(m, n)$ is the ordering relation of a series of order type† $\Xi_{\mathbf{A}}$ and on the other hand that the values of $\mathbf{\Omega}(\mathbf{m}, \mathbf{n})$ are related to $G(m,n)$ as on p. 163.

† The order type is β, where $1 + \beta = \Xi_{\mathbf{A}}$; but $\beta = \Xi_{\mathbf{A}}$ since $\Xi_{\mathbf{A}}$ is infinite.

To prove the second half suppose that **A** is not a C-K ordinal formula. Then one of the conditions (A)–(D) in (v) must not be satisfied. If (A) is not satisfied we shall not obtain a result even in the calculation of $\mathbf{\Omega}(1,1)$. If (B) is not satisfied, we shall have for some positive integers p and q,

$$\text{Ls}(\mathbf{A},\ \mathbf{p}) \text{ conv } \lambda ufx.u(\mathbf{R})$$

but not $\lambda ufx.\mathbf{R}(\mathbf{q}) < \lambda ufx.\mathbf{R}(S(\mathbf{q}))$. Then the process of calculating $\mathbf{\Omega}(\mathbf{p},\ \mathbf{q})$ will not terminate. In case of failure of (C) or (D) the values of $\mathbf{\Omega}(\mathbf{m},\ \mathbf{n})$ may all be calculable, but if so conditions $(a)-(f)$, p. 163, will be violated. Thus, if **A** is not a C-K ordinal formula, then $H(\mathbf{A})$ is not an ordinal formula.

I propose now to define three formulae Sum, Lim, Inf of importance in connection with ordinal formulae. Since they are comparatively simple, they will for once be given almost in full. The formula Ug is one with the property that Ug(**m**) is convertible to the formula representing the largest odd integer dividing m: it is not given in full. P is the predecessor function; $P(S(\mathbf{m}))$ conv **m**, $P(1)$ conv 1.

Al $\rightarrow \lambda pxy.p(\lambda guv\ .\ g(v,u), \lambda uv.u(I,v),x,y)$,

Hf $\rightarrow \lambda m.P(m(\lambda guv\ .\ g(v,S(u)), \lambda uv.v(I,u),\ 1,\ 2))$,

Bd $\rightarrow \lambda ww'aa'x.$ Al $(\lambda f.w(a,a,w'(a',a',f)),x,4)$,

Sum $\rightarrow \lambda ww'pq.$ Bd $(w,w',$ Hf (p), Hf (q),

Al $(p,$ Al $(q,w'($Hf (p), Hf $(q)),1)$, Al $(S(q),w($ Hf (p), Hf $(q)),2))$,

Lim $\rightarrow \lambda zpq.\{\lambda ab.Bd(z(a),z(b)$, Ug (p), Ug (q), Al (Dt$(a,b)+$ Dt (b,a),

Dt $(a,b),z(a,$ Ug (p), Ug $(q)))(\varpi(2,p),\varpi(2,q))$,

Inf $\rightarrow \lambda wapq.$ Al $(\lambda f.w(a,p,w(a,q,f)),w(p,q),4)$.

The essential properties of these formulae are described by:

Al $(2\mathbf{r}-1,\mathbf{m},\mathbf{n})$ conv **m**, Al $(2\mathbf{r},\mathbf{m},\mathbf{n})$ conv **n**,

Hf(2**m**) conv **m**, Hf(2**m** $-$ 1) conv **m**

Bd $(\mathbf{\Omega},\mathbf{\Omega}',\mathbf{a},\ \mathbf{a}',\mathbf{x})$ conv 4, unless both

$$\mathbf{\Omega}(\mathbf{a},\mathbf{a}) \text{ conv } 3 \text{ and } \mathbf{\Omega}'(\mathbf{a}',\mathbf{a}') \text{ conv } 3,$$

it is then convertible to **x**.

If $\mathbf{\Omega}$, $\mathbf{\Omega}'$ are ordinal formulae representing α,β respectively, then Sum($\mathbf{\Omega}$, $\mathbf{\Omega}'$) is an ordinal formula representing $\alpha+\beta$. If **Z** is a W.F.F. enumerating a sequence of ordinal formulae representing α_1, $\alpha_2,\ldots$, then Lim (**Z**) is an ordinal formula representing the infinite sum $\alpha_1+\alpha_2+\alpha_3,\ldots$. If $\mathbf{\Omega}$ is an ordinal formula representing α, then Inf($\mathbf{\Omega}$) enumerates a sequence of ordinal formulae representing all the ordinals less than α without repetitions other than repetitions of the ordinal 0.

To prove that there is no general method for determining about a formula whether it is an ordinal formula, we use an argument akin to that leading to the Burali-Forti paradox; but the emphasis and the conclusion are different. Let us suppose that such an algorithm is available. This enables us to obtain a recursive enumeration $\mathbf{\Omega}_1$, $\mathbf{\Omega}_2,\ldots$ of the ordinal formulae in normal form. There is a formula **Z** such that **Z**(**n**) conv $\mathbf{\Omega}_\mathbf{n}$. Now Lim (**Z**) represents an ordinal greater than any represented by an $\mathbf{\Omega}_\mathbf{n}$, and it has therefore been omitted from the enumeration.

This argument proves more than was originally asserted. In fact, it proves that, if we take any class E of ordinal formulae in normal form, such that, if **A** is any ordinal formula. then there is a formula in E representing the same ordinal as **A**, then there is no method whereby one can determine whether a W.F.F. in normal form belongs to E.

8. Ordinal logics.

An ordinal logic is a W.F.F. Λ such that $\Lambda(\mathbf{\Omega})$ is logic formula whenever $\mathbf{\Omega}$ is an ordinal formula.

This definition is intended to bring under one heading a number of ways of constructing logics which have recently been proposed or which are suggested by recent advances. In this section I propose to show how to obtain some of these ordinal logics.

Suppose that we have a class W of logical systems. The symbols used in each of these systems are the same, and a class of sequences of symbols called "formulae" is defined, independently of the particular system in W. The rules of procedure of a system C define an axiomatic subset of the formulae, which are to be described as the "provable formulae of C". Suppose further that we have a method whereby from any system C' of W, we can obtain a new system C', also in W, and such that the set of provable formulae of C' includes the provable formulae of C (we shall be most interested in the case in which the y are included as a proper subset). It is to be understood that this "method" is an effective procedure for obtaining the rules of procedure of C' from those of C.

Suppose that to certain of the formulae of W we make number-theoretic theorems correspond: by modifying the definition of formula, we may suppose that this is done for all formulae. We shall say that one of the systems C is *valid* if the provability of a formula in C implies the truth of the corresponding number-theoretic theorem. Now let the relation of C' to C be such that the validity of C implies the validity of C', and let there be a valid system C_0 in W. Finally, suppose that, given any computable sequence $C_1, C_2, \ldots$ of systems in W, the "limit system" in which a formula is provable if and only if it is provable in one of the systems C_j, also belongs to W. These limit systems are to be regarded, not as functions of the sequence given in extension, but as functions of the rules of formation of their terms. A sequence given in extension may be described by various rules of formation, and there will be several corresponding limit systems. Each of these may be described as *a* limit system of the sequence.

In these circumstances we may construct an ordinal logic. Let us associate positive integers with the systems in such a way that to each C there corresponds a positive integer m_C, and that m_C completely describes the rules of procedure of C. Then there is a W.F.F. **K**, such that

$$\mathbf{K}(\mathbf{m}_C) \text{ conv } \mathbf{m}_{C'}$$

for each C in W and there is a W.F.F. Θ such that if $\mathbf{D}(\mathbf{r})$ conv $\mathbf{m}_{C_r}$. for each positive integer r, then $\Theta(\mathbf{D})$ conv $\mathbf{m}_C$, where C is a limit system of $C_1, C_2, \ldots$. With each system C of W it is possible to associate a logic formula $\mathbf{L}_C$: the relation between them is that, if G is a formula of W and the number-theoretic theorem corresponding to G (assumed expressed in the conversion calculus form) asserts that **B** is dual, then $\mathbf{L}_C(\mathbf{B})$ conv 2 if and only if G is provable in C. There is a W.F.F. **G** such that

$$\mathbf{G}(\mathbf{m}_C) \text{ conv } \mathbf{L}_C$$

for each C of W. Put

$$\mathbf{N} \to \lambda a.\mathbf{G}(a(\Theta, \mathbf{K}, \mathbf{m}_{C_0})).$$

I assert that $\mathbf{N}(\mathbf{A})$ is a logic formula for each C-K ordinal formula **A**, and that, if $\mathbf{A} < \mathbf{B}$, then $\mathbf{N}(\mathbf{B})$ is more complete than $\mathbf{N}(\mathbf{A})$, provided that there are formulae provable in C' but not in C for each valid C of W.

To prove this we shall show that to each C-K ordinal formula **A** there corresponds a unique system $C[\mathbf{A}]$ such that:

(i) $\mathbf{A}(\Theta, \mathbf{K}, \mathbf{m}_{C_0})$ conv $\mathbf{m}_{C[\mathbf{A}]}$,

and that it further satisfies:

(ii) $C[U]$ is a limit system of $C'_0, C'_0, \ldots$,

(iii) C [Suc ($\mathbf{A}$)] is $(C[\mathbf{A}])'$,

(iv) $C[\lambda ufx.u(\mathbf{R})]$ is a limit system of $C[\lambda ufx.\mathbf{R}(1)], C[\lambda ufx.\mathbf{R}(2)], \ldots$,

A and $\lambda ufx.u(\mathbf{R})$ being assumed to be C-K ordinal formulae. The uniqueness of the system follows from the fact that m_C determines C completely. Let us try to prove the existence of $C[\mathbf{A}]$ for each C-K ordinal formula **A**. As we have seen (p. 165) it is sufficient to prove

(a) $C[U]$ exists,

(b) if $C[\mathbf{A}]$ exists, then $C[\mathrm{Suc}(\mathbf{A})]$ exists,

(c) if $C[\lambda ufx.\mathbf{R}(1)], C[\lambda ufx.\mathbf{R}(2)], \ldots$ exist, then $C[\lambda ufx.u(\mathbf{R})]$ exists.

Proof of (*a*).

$$\{\lambda y.\mathbf{K}(y(I, \mathbf{m}_{C_0}))\}(\mathbf{n}) \text{ conv } \mathbf{K}(\mathbf{m}_{C_0}) \text{ conv } \mathbf{m}_{C'_0}$$

for all positive integers n, and therefore, by the definition of Θ, there is a system, which we call $C[U]$ and which is a limit system of $C'_0, C'_0, \ldots$, satisfying

$$\Theta(\lambda y.\mathbf{K}(y(I, \mathbf{m}_{C_0}))) \text{ conv } \mathbf{m}_{C[U]}.$$

But, on the other hand,

$$U(\Theta, \mathbf{K}, \mathbf{m}_{C_0}) \text{ conv } (\lambda y.\mathbf{K}(y(I, \mathbf{m}_{C_0}))).$$

This proves (*a*) and incidentally (ii).

Proof of (*b*).

$$\begin{aligned} \mathrm{Suc}(\mathbf{A}, \Theta, \mathbf{K}, m_{C_0}) &\text{ conv } \mathbf{K}(\mathbf{A}(\Theta, \mathbf{K}, \mathbf{m}_{C_0})) \\ &\text{ conv } \mathbf{K}(\mathbf{m}_{C[\mathbf{A}]}) \\ &\text{ conv } \mathbf{m}_{(C[\mathbf{A}])'}. \end{aligned}$$

Hence $C[\mathrm{Suc}(\mathbf{A})]$ exists and is given by (iii).

Proof of (*c*).

$$\begin{aligned} \{\{\lambda ufx.\mathbf{R}\}(\Theta, \mathbf{K}, \mathbf{m}_{C_0})\}(\mathbf{n}) &\text{ conv } \{\lambda ufx.\mathbf{R}(\mathbf{n})\}(\Theta, \mathbf{K}, \mathbf{m}_{C_0}) \\ &\text{ conv } \mathbf{m}_{C[\lambda ufx.\mathbf{R}(\mathbf{n})]} \end{aligned}$$

by hypothesis. Consequently, by the definition of Θ, there exists a C which is a limit system of

$$C[\lambda ufx.\mathbf{R}(1)],\ C[\lambda ufx.\mathbf{R}(2)], \ldots,$$

and satisfies

$$\Theta(\{\lambda ufx.u(\mathbf{R})\}(\Theta, \mathbf{K}, \mathbf{m}_{C_0})) \text{ conv } \mathbf{m}_C.$$

We define $C[\lambda ufx.u(\mathbf{R})]$ to be this C. We then have (iv) and

$$\begin{aligned} \{\lambda ufx.u(\mathbf{R})\}(\Theta, \mathbf{K}, \mathbf{m}_{C_0}) &\text{ conv } \Theta(\{\lambda ufx.\mathbf{R}\}(\Theta, \mathbf{K}, \mathbf{m}_{C_0})) \\ &\text{ conv } \mathbf{m}_{C[\lambda ufx.u(\mathbf{R})]}. \end{aligned}$$

This completes the proof of the properties (i)–(iv). From (ii), (iii), (iv), the fact that C_0 is valid, and that C' is valid when C is valid, we infer that $C[\mathbf{A}]$ is valid for each C-K ordinal formula $\mathbf{A}$: also that there are more formulae provable in $C[\mathbf{B}]$ than in $C[\mathbf{A}]$ when $\mathbf{A} < \mathbf{B}$. The truth of our assertions regarding $\mathbf{N}$ now follows in view of (i) and the definitions of $\mathbf{N}$ and $\mathbf{G}$.

We cannot conclude that $\mathbf{N}$ is an ordinal logic, since the formulae $\mathbf{A}$ are C-K ordinal formulae; but the formula H enables us to obtain an ordinal logic from $\mathbf{N}$. By the use of the formula Gr we obtain a formula Tn such that, if $\mathbf{A}$ has a normal form, then Tn($\mathbf{A}$) enumerates the G.R.'s of the formulae into which $\mathbf{A}$ is convertible. Also there is a formula Ck such that, if h is the G.R. of a formula $H(\mathbf{B})$, then Ck($\mathbf{h}$) conv $\mathbf{B}$, but otherwise Ck ($\mathbf{h}$) conv U. Since $H(\mathbf{B})$ is an ordinal formula only if $\mathbf{B}$ is a C-K ordinal formula Ck (Tn ($\mathbf{\Omega},\mathbf{n}$)) is a C-K ordinal formula for each ordinal formula $\mathbf{\Omega}$ and each integer n. For many ordinal formulae it will be convertible to U, but, for suitable $\mathbf{\Omega}$, it will be convertible to any given C-K ordinal formula. If we put

$$\mathbf{\Lambda} \rightarrow \lambda wa.\Gamma(\lambda n.\mathbf{N}(\mathrm{Ck}(Tn(w,n))),a),$$

$\mathbf{\Lambda}$ is the required ordinal logic. In fact, on account of the properties of $\mathbf{\Gamma}$, $\mathbf{\Lambda}(\mathbf{\Omega},\mathbf{A})$ will be convertible to 2 if and only if there is a positive integer n such that

$$\mathbf{N}(\mathrm{Ck}\ (\mathrm{Tn}\ (\mathbf{\Omega},\mathbf{n})),\mathbf{A})\ \mathrm{conv}\ 2.$$

If $\mathbf{\Omega}$ conv $H(\mathbf{B})$, there will be an integer n such that Ck ($\mathbf{Tn}(\mathbf{\Omega},\mathbf{n})$) conv $\mathbf{B}$, and then

$$\mathbf{N}(\mathrm{Ck}\ (\mathrm{Tn}\ (\mathbf{\Omega},\mathbf{n})),\mathbf{A})\ \mathrm{conv}\ \mathbf{N}(\mathbf{B},\mathbf{A}).$$

For any n, Ck (Tn ($\mathbf{\Omega},\mathbf{n}$)) is convertible to U or to some $\mathbf{B}$, where $\mathbf{\Omega}$ conv $H(\mathbf{B})$. Thus $\Lambda(\mathbf{\Omega},\mathbf{A})$ conv 2 if $\mathbf{\Omega}$ conv $H(\mathbf{B})$ and $\mathbf{N}(\mathbf{B},\mathbf{A})$ conv 2 or if $\mathbf{N}(U,\mathbf{A})$ conv 2, but not in any other case.

We may now specialize and consider particular classes W of systems. First let us try to construct the ordinal logic described roughly in the introduction. For W we take the class of systems arising from the system of *Principia Mathematica*† by adjoining to it axiomatic (in the sense described on p. 155) sets of axioms‡. Gödel has shown that primitive recursive relations§ can be expressed by means of formulae in P. In fact, there is a rule whereby, given the recursion equations defining a primitive recursive relation, we can find a formula¶ $\mathfrak{A}[x_0,\ldots,z_0]$ such that

$$\mathfrak{A}[f^{(m_1)}0,...,f^{(m_r)}0]$$

is provable in P if $F(m_1,\ldots,m_r)$ is true, and its negation is provable otherwise. Further, there is a method by which we can determine about a formula $\mathfrak{A}[x_0,\ldots,z_0]$ whether it arises from a primitive recursive relation in this way, and by which we can find the equations which defined the relation. Formulae of this kind will be called *recursion formulae*. We shall make use of a property that they possess, which we cannot prove formally here without giving their definition in full, but which is

† Whitehead and Russell [**1**]. The axioms and rules of procedure of a similar system P will be found in a convenient form in Gödel [1], and I follow Gödel. The symbols for the natural numbers in P are $0,f0,ff0,\ldots,f^{(n)}0\ldots$. Variables with the suffix "0" stand for natural numbers.

‡ It is sometimes regarded as necessary that the set of axioms used should be computable, the intention being that it should be possible to verify of a formula reputed to be an axiom whether it really is so. We can obtain the same effect with axiomatic sets of axioms in this way. In the rules of procedure describing which are the axioms, we incorporate a method of enumerating them, and we also introduce a rule that in the main part of the deduction, whenever we write down an axiom as such, we must also write down its position in the enumeration. It is possible to verify whether this has been done correctly.

§ A relation $F(m_1,\ldots,m_r)$ is primitive recursive if it is a necessary and sufficient condition for the vanishing of a primitive recursive function $\phi(m_1,\ldots,m_r)$.

¶ Capital German letters will be used to stand for variable or undetermined formulae in P. An expression such as 1[,] stands for the result of substituting $\mathfrak{B}$ and $\mathfrak{C}$ for x_0 and y_0 in $\mathfrak{A}$.

essentially trivial. Db $[x_0, y_0]$ is to stand for a certain recursion formula such that Db $[f^{(m)}0, f^{(n)}0]$ is provable in P if $m = 2n$ and its negation is provable otherwise. Suppose that $\mathfrak{A}[x_0], \mathfrak{B}[x_0]$ are two recursion formulae. Then the theorem which I am assuming is that there is a recursion relation $\mathfrak{C}_{\mathfrak{A},\mathfrak{B}}[x_0]$, such that we can prove

$$\mathfrak{C}_{\mathfrak{A},\mathfrak{B}}[x_0] \equiv (\exists y_0)((\mathrm{Db}[x_0 y_0].\mathfrak{A}[y_0]) \vee (\mathrm{Db}[fx_0, fy_0].\mathfrak{B}[y_0])) \tag{8.1}$$

in P.

The significant formulae in any of our extensions of P are those of the form

$$(x_0)(\exists y_0)\mathfrak{A}[x_0, y_0], \tag{8.2}$$

where $\mathfrak{A}[x_0, y_0]$ is a recursion formula, arising from the relation $R(m,n)$ let us say. The corresponding number-theoretic theorem states that for each natural number m there is a natural number n such that $R(m,n)$ is true.

The systems in W which are not valid are those in which a formula of the form (8. 2) is provable, but at the same time there is a natural number, m say, such that, for each natural number $n, R(m,n)$ is false. This means to say that $\sim \mathfrak{A}[f^{(m)}0, f^{(n)}0]$ is provable for each natural number n. Since (8.2) is provable, $(\exists x_0)\mathfrak{A}[f^{(m)}0, y_0]$ is provable, so that

$$(\exists y_0)\mathfrak{A}[f^{m)}0, y_0], \quad \sim \mathfrak{A}[f^{(m)}0, 0], \quad \sim \mathfrak{A}[f^{(m)}0, f0], \ldots \tag{8.3}$$

are all provable in the system. We may simplify (8.3). For a given m we may prove a formula of the form $\mathfrak{A}[f^{(m)}0, y_0] \equiv \mathfrak{B}[y_0]$ in P, where $\mathfrak{B}[x_0]$ is a recursion formula. Thus we find that a necessary and sufficient condition for a system of W to be valid is that for no recursion formula $\mathfrak{B}[x_0]$ are all of the formulae

$$(\exists x_0)\mathfrak{B}[x_0], \sim \mathfrak{B}[0], \sim \mathfrak{B}[f0], \ldots \tag{8. 4}$$

provable. An important consequence of this is that, if

$$\mathfrak{A}_1[x_0], \mathfrak{A}_2[x_0], \ldots \mathfrak{A}_n[x_0]$$

are recursion formulae, if

$$(\exists x_0)\mathfrak{A}_1[x_0] \vee \ldots \vee (\exists x_0)\mathfrak{A}_n[x_0] \tag{8.5}$$

is provable in C, and C is valid, then we can prove $\mathfrak{A}_r[f^{(a)}0]$ in C for some natural numbers r, a, where $1 \leq r \leq n$. Let us define $\mathfrak{D}_r$ to be the formula

$$(\exists x_0)\mathfrak{A}_1[x_0] \vee \ldots \vee (\exists x_0)\mathfrak{A}_r[x_0]$$

and then define $\mathfrak{E}_r[x_0]$ recursively by the condition that $\mathfrak{E}_1[x_0]$ is $\mathfrak{A}_1[x_0]$ and $\mathfrak{E}_{r+1}[x_0]$ be $\mathfrak{C}_{\mathfrak{E}_r, \mathfrak{A}_{r+1}[x_0]}$. Now I say that

$$\mathfrak{D}_r \supset (\exists x_0)\mathfrak{E}_r[x_0] \tag{8.6}$$

is provable for $1 \leq r \leq n$. It is clearly provable for $r = 1$: suppose it to be provable for a given r. We can prove

$$(y_0)(\exists x_0)\mathrm{Db}[x_0, y_0]$$

and

$$(y_0)(\exists x_0)\mathrm{Db}[fx_0, fy_0],$$

from which we obtain

$$\mathfrak{E}_r[y_0] \supset (\exists x_0)((\mathrm{Db}[x_0, y_0].\mathfrak{E}_r[y_0]) \vee (\mathrm{Db}[fx_0, fy_0].\mathfrak{A}_{r+1}[y_0]))$$

and

$$\mathfrak{A}_{r+1}[y_0] \supset (\exists x_0)((\text{Db}[x_0, y_0].\mathfrak{E}_r[y_0]) \vee (\text{Db}[fx_0, fy_0].\mathfrak{A}_{r+1}[y_0])).$$

These together with (8.1) yield

$$(\exists y_0)\mathfrak{E}_r[y_0] \vee (\exists y_0)\mathfrak{A}_{r+1}[y_0] \supset (\exists x_0)\mathfrak{E}_{\mathfrak{E}_r, \mathfrak{A}_{r+1}}[x_0],$$

which is sufficient to prove (8. 6) for $r+1$. Now, since (8. 5) is provable in C, $(\exists x_0)\mathfrak{E}_n[x_0]$ must also be provable, and, since C is valid, this means that $\mathfrak{E}_n[f^{(m)}0]$ must be provable for some natural number m. From (8.1) and the definition of $\mathfrak{E}_n[x_0]$ we see that this implies that $\mathfrak{A}_r[f^{(a)}0]$ is provable for some natural numbers a and r, $1 \leq r \leq n$.

To any system C of W we can assign a primitive recursive relation $P_C(m,n)$ with the intuitive meaning "m is the G.R. of a proof of the formula whose G.R. is n". We call the corresponding recursion formula Proof$_C$ $[x_0, y_0]$. (*ie.* $\text{Proof}_C[f^{(m)}0, f^{(n)}0]$ is provable when $P_C(m,n)$ is true, and its negation is provable otherwise). We can now explain what is the relation of a system C' to its predecessor C. The set of axioms which we adjoin to P to obtain C' consists of those adjoined in obtaining C, together with all formulae of the form

$$(\exists X_0)\text{Proof}_C[x_0, f^{(m)}0] \supset \mathfrak{F}, \tag{8.7}$$

where m is the G.R. of $\mathfrak{F}$.

We want to show that a contradiction can be obtained by assuming C' to be invalid but C to be valid. Let us suppose that a set of formulae of the form (8.4) is provable in C'. Let $\mathfrak{A}_1, \mathfrak{A}_2, \ldots, \mathfrak{A}_k$ be those axioms of C' of the form (8.7) which are used in the proof of $(\exists x_0)\mathfrak{B}[x_0]$. We may suppose that none of them is provable in C. Then by the deduction theorem we see that

$$(\mathfrak{A}_1.\mathfrak{A}_2 \ldots \mathfrak{A}_k) \supset (\exists x_0)\mathfrak{B}[x_0] \tag{8.8}$$

is provable in C. Let $\mathfrak{A}_l$ be $(\exists x_0)\text{Proof}_C[x_0, f^{(m_l)}0] \supset \mathfrak{F}_l$. Then from (8.8) we find that

$$(\exists x_0)\text{Proof}_C[x_0, f^{(m_1)}0] \vee \ldots \vee (\exists x_0)\text{Proof}_C[x_0, f^{(m_k)}0] \vee (\exists x_0)\mathfrak{B}[x_0]$$

is provable in C. It follows from a result which we have just proved that either $\mathfrak{B}[f^{(c)}0]$ is provable for some natural number c, or else $\text{Proof}_C[f^{(n)}0, f^{(m_l)}0]$ is provable in C for some natural number u and some $l, 1 \leq l \leq k$: but this would mean that $\mathfrak{F}_l$ is provable in C (this is one of the points where we assume the validity of C) and therefore also in C', contrary to hypothesis. Thus $\mathfrak{B}[f^{(c)}0]$ must be provable in C' ; but we are also assuming $\sim \mathfrak{B}[f^{(c)}0]$ to be provable in C'. There is therefore a contradiction in C'. Let us suppose that the axioms $\mathfrak{A}'_1, \ldots, \mathfrak{A}'_{k'}$, of the form (8. 7), when adjoined to C are sufficient to obtain the contradiction and that none of these axioms is that provable in C. Then

$$\sim \mathfrak{A}'_1 \vee \sim \mathfrak{A}'_2 \vee \ldots \vee \sim \mathfrak{A}'_{k'}$$

is provable in C, and if $\mathfrak{A}'_l$ is $(\exists x_0)$ Proof$_C$ $[x_0, f^{(m'_l)}0] \supset \mathfrak{F}'_l$, then

$$(\exists x_0]\text{Proof}_C[x_0, f^{(m'_1)}0] \vee \cdots \vee (\exists x_0)\text{Proof}[x_0, f^{(m'_{k'})}0]$$

is provable in C. But, by repetition of a previous argument, this means that $\mathfrak{A}'_l$ is provable for some l, $1 \leq l \leq k'$, contrary to hypothesis. This is the required contradiction.

We may now construct an ordinal logic in the manner described on pp. 170–172. We shall, however, carry out the construction in rather more detail, and with some modifications appropriate to the particular case. Each system C of our set W may be described by means of a W.F.F. M_C which enumerates the G.R.'s of the axioms of C. There is a W.F.F. E such that, if a is the G.R. of some proposition $\mathfrak{F}$, then $E(M_c, \mathbf{a})$ is convertible to the G.R. of

$$(\exists x_0)\text{Proof}_C[x_0, f^{(a)}0] \supset \mathfrak{F}.$$

If a is not the G.R. of any proposition in P, then $E(M_C, \mathbf{a})$ is to be convertible to the G.R. of $0 = 0$. From E we obtain a W.F.F. K such that $K(M_C, 2\mathbf{n}+1)$ conv $M_C(\mathbf{n}), K(M_C, 2\mathbf{n})$ conv $E(M_{C,\mathbf{n}})$. The successor system C' is defined by $K(M_C)$ conv M'_C. Let us choose a formula G such that $G(M_C, \mathbf{A})$ conv 2 if and only if the number-theoretic theorem equivalent to "**A** is dual" is provable in C. Then we define Λ_P by

$$\Lambda_P \to \lambda wa.\Gamma(\lambda y.G(\text{Ck}(\text{Tn}(w,y)), \lambda mn.m(\varpi(2,n), \varpi(3,n)), K, M_P)), a).$$

This is all ordinal logic provided that P is valid.

Another ordinal logic of this type has in effect been introduced by Church†. Superficially this ordinal logic seems to have no more in common with Λ_P than that they both arise by the method which we have described, which uses C-K ordinal formulae. The initial systems are entirely different. However, in the relation between C and C' there is an interesting analogy. In Church's method the step from C to C' is performed by means of subsidiary axioms of which the most important (Church [**2**], p. 88, 1_m) is almost a direct translation into his symbolism of the rule that we may take any formula of the form (8.4) as an axiom. There are other extra axioms, however, in Church's system, and it is therefore not unlikely that it is in some respects more complete than Λ_P.

There are other types of ordinal logic, apparently quite unrelated to the type that we have so far considered. I have in mind two types of ordinal logic, both of which can be best described directly in terms of ordinal formulae without any reference to C-K ordinal formulae. I shall describe here a specimen Λ_H of one of these types of ordinal logic. Ordinal logics of this kind were first considered by Hilbert (Hilbert [**1**], 183ff), and have also been used by Tarski (Tarski [**1**], 395ff); see also Gödel [**1**], foot-note 48[a].

Suppose that we have selected a particular ordinal formula $\mathbf{\Omega}$. We shall construct a modification $P_{\mathbf{\Omega}}$ of the system P of Gödel (see foot-note † on p. 172. We shall say that a natural number n is a *type* if it is either even or $2p-1$, where $\mathbf{\Omega}(\mathbf{p},\mathbf{p})$ conv 3. The definition of a variable in P is to be modified by the condition that the only admissible subscripts are to be the types in our sense. Elementary expressions are then defined as in P: in particular the definition of an elementary expression of type 0 is unchanged. An elementary formula is defined to be a sequence of symbols of the form $\mathfrak{A}_m\mathfrak{A}_n$, where $\mathfrak{A}_m$, $\mathfrak{A}_n$ are elementary expressions of types m, n satisfying one of the conditions $(a), (b), (c)$.

(a) m and n are both even and m exceeds n,

(b) m is odd and n is even,

(c) $m = 2p-1, n = 2q-1$, and $\mathbf{\Omega}(\mathbf{p},\mathbf{q})$ conv 2.

With these modifications the formal development of $P_{\mathbf{\Omega}}$ is the same as that of P. We want, however, to have a method of associating number-theoretic theorems with certain of the formulae of $P_{\mathbf{\Omega}}$. We cannot take over directly the association which we used in P. Suppose that G is a formula in P interpretable as a number-theoretic theorem in the way described in the course of constructing Λ_P (p. 172). Then, if every type suffix in G is doubled, we shall obtain a formula in $P_{\mathbf{\Omega}}$ which is to be interpreted as the same number-theoretic theorem. By the method of §6 we can now obtain from $P_{\mathbf{\Omega}}$ a formula $L_{\mathbf{\Omega}}$ which is a logic formula if $P_{\mathbf{\Omega}}$ is valid; in fact, given $\mathbf{\Omega}$ there is a method of obtaining $L_{\mathbf{\Omega}}$, so that there is a formula Λ_H such that $\Lambda_H(\mathbf{\Omega})$ conv $L_{\mathbf{\Omega}}$ for each ordinal formula $\mathbf{\Omega}$.

Having now familiarized ourselves with ordinal logics by means of these examples we may begin to consider general questions concerning them.

† In outline Church [**1**], 279–280. In greater detail Church [**2**], Chap. X.

9. Completeness questions.

The purpose of introducing ordinal logics was to avoid as far as possible the effects of Gödel's theorem. It is a consequence of this theorem, suitably modified, that it is impossible to obtain a complete logic formula, or (roughly speaking now) a complete system of logic. We were able, however, from a given system to obtain a more complete one by the adjunction as axioms of formulae, seen intuitively to be correct, but which the Gödel theorem shows are unprovable† in the original system; from this we obtained a yet more complete system by a repetition of the process, and so on. We found that the repetition of the process gave us a new system for each C-K ordinal formula. We should like to know whether this process suffices, or whether the system should be extended in other ways as well. If it were possible to determine about a W.F.F. in normal form whether it was an ordinal formula, we should know for certain that it was necessary to make extensions in other ways. In fact for any ordinal formula $\mathbf{\Lambda}$ it would then be possible to find a single logic formula $\mathbf{L}$ such that, if $\mathbf{\Lambda}(\mathbf{\Omega},\mathbf{A})$ conv 2 for some ordinal formula $\mathbf{\Omega}$, then $\mathbf{L}(\mathbf{A})$ conv 2. Since $\mathbf{L}$ must be incomplete, there must be formulae $\mathbf{A}$ for which $\mathbf{\Lambda}(\mathbf{\Omega},\mathbf{A})$ is not convertible to 2 for any ordinal formula $\mathbf{\Omega}$. However, in view of the fact, proved in §7, that there is no method of determining about a formula in normal form whether it is an ordinal formula, the case does not arise, and there is still a possibility that some ordinal logics may be complete in some sense. There is a quite natural way of defining completeness.

Definition of completeness of an ordinal logic

We say that an ordinal logic $\mathbf{\Lambda}$ is complete if corresponding to each dual formula $\mathbf{A}$ there is an ordinal formula $\mathbf{\Omega}_{\mathbf{A}}$ such that $\mathbf{\Lambda}(\mathbf{\Omega}_{\mathbf{A}},\mathbf{A})$ conv 2.

As has been explained in §2, the reference in the definition to the existence of $\mathbf{\Omega}_{\mathbf{A}}$ for each $\mathbf{A}$ is to be understood in the same naïve way as any reference to existence in mathematics.

There is room for modification in this definition: we might require that there is a formula $\mathbf{X}$ such that $\mathbf{X}(\mathbf{A})$ conv $\mathbf{\Omega}_{\mathbf{A}}$, $\mathbf{X}(\mathbf{A})$ being an ordinal formula whenever $\mathbf{A}$ is dual. There is no need, however, to discuss the relative merits of these two definitions, because in all cases in which we prove an ordinal logic to be complete we shall prove it to be complete even in the modified sense; but in cases in which we prove an ordinal logic to be incomplete, we use the definition as it stands.

In the terminology of §6, $\mathbf{\Lambda}$ is complete if the class of logics $\mathbf{\Lambda}(\mathbf{\Omega})$ is complete when $\mathbf{\Omega}$ runs through all ordinal formulae.

There is another completeness property which is related to this one. Let us for the moment describe an ordinal logic $\mathbf{\Lambda}$ as *all inclusive* if to each logic formula $\mathbf{L}$ there corresponds an ordinal formula $\mathbf{\Omega}_{(\mathbf{L})}$ such that $\mathbf{\Lambda}(\mathbf{\Omega}_{(\mathbf{L})})$ is as complete as $\mathbf{L}$. Clearly every all inclusive ordinal logic is complete; for, if $\mathbf{A}$ is dual, then $\delta(A)$ is a logic with $\mathbf{A}$ in its extent. But, if $\mathbf{\Lambda}$ is complete and

$$\mathrm{Ai} \rightarrow \lambda kw.\Gamma\Big(\lambda ra.\delta\Big(4,\delta\left(2,k\big(w,V(Nm(r))\big)\right)+\delta\big(2,Nm(r,a)\big)\Big)\Big),$$

then Ai($\mathbf{\Lambda}$)is an all inclusive ordinal logic. For, if $\mathbf{A}$ is in the extent of $\mathbf{\Lambda}(\mathbf{\Omega}_{\mathbf{A}})$ for each $\mathbf{A}$, and we put $\mathbf{\Omega}_{(\mathbf{L})} \rightarrow \mathbf{\Omega}_{V(\mathbf{L})}$, then I say that, if $\mathbf{B}$ is in the extent of $\mathbf{L}$, it must be in the extent of Ai $(\mathbf{\Lambda},\mathbf{\Omega}_{(\mathbf{L})})$. In fact, we see that Ai $(\mathbf{\Lambda},\mathbf{\Omega}_{V(\mathbf{L})},\mathbf{B})$ is convertible to

$$\Gamma\Big(\lambda ra.\delta\Big(4,\delta\big(2,\mathbf{\Lambda}\big(\mathbf{\Omega}_{V(\mathbf{L})},V(\mathrm{Nm}(r))\big)\big)+\delta\big(2,\mathrm{Nm}(r,a)\big)\Big),\ \mathbf{B}\Big).$$

† In the case of p we adjoined all of the axioms $(\exists x_0)$ Proof $[x_0, f^{(m)}0] \supset \mathfrak{F}$,where m is the G.R, of $\mathfrak{F}$; the Gödel theorem shows that *some* of them are unprovable in P.

For suitable n, Nm(**n**) conv **L** and then

$$\mathbf{\Lambda}(\mathbf{\Omega}_{V(\mathbf{L})}, V(\mathrm{Nm}(\mathbf{n}))) \text{ conv } 2,$$
$$\mathrm{Nm}(\mathbf{n},\ \mathbf{B}) \text{ conv } 2,$$

and therefore, by the properties of Γ and δ

$$\mathrm{Ai}(\mathbf{\Lambda},\ \mathbf{\Omega}_{V(\mathbf{L})},\ \mathbf{B}) \text{ conv } 2.$$

Conversely $\mathrm{Ai}(\mathbf{\Lambda}, \mathbf{\Omega}_{V(\mathbf{L})}, \mathbf{B})$ can be convertible to 2 only if both $\mathrm{Nm}\,(\mathbf{n}, \mathbf{B})$ and $\mathbf{\Lambda}(\mathbf{\Omega}_{V(\mathbf{L})}, V(\mathrm{Nm}(\mathbf{n})))$ are convertible to 2 for some positive integer n; but, if $\mathbf{\Lambda}(\mathbf{\Omega}_{V(\mathbf{L})}, V(\mathrm{Nm}(\mathbf{n})))$ conv 2, then Nm(**n**) must be a logic, and, since Nm(**n**, **B**) conv 2, **B** must be dual.

It should be noticed that our definitions of completeness refer only to number-theoretic theorems. Although it would be possible to introduce formulae analogous to ordinal logics which would prove more general theorems than number-theoretic ones, and have a corresponding definition of completeness, yet, if our theorems are too general, we shall find that our (modified) ordinal logics are never complete. This follows from the argument of §4. If our "oracle" tells us, not whether any given number-theoretic statement is true, but whether a given formula is an ordinal formula, the argument still applies, and we find that there are classes of problem which cannot be solved by a uniform process even with the help of this oracle. This is equivalent to saying that there is no ordinal logic of the proposed modified type which is complete with respect to these problems. This situation becomes more definite if we take formulae satisfying conditions $(a)-(e)$, (f') (as described at the end of §12) instead of ordinal formulae; it is then not possible for the ordinal logic to be complete with respect to any class of problems more extensive than the number-theoretic problems.

We might hope to obtain some intellectually satisfying system of logical inference (for the proof of number-theoretic theorems) with some ordinal logic. Gödel's theorem shows that such a system cannot be wholly mechanical; but with a complete ordinal logic we should be able to confine the non-mechanical steps entirely to verifications that particular formulae are ordinal formulae.

We might also expect to obtain an interesting classification of number-theoretic theorems according to "depth ". A theorem which required an ordinal α to prove it would be deeper than one which could be proved by the use of an ordinal β less than α. However, this presupposes more than is justified. We now define

Invariance of ordinal logics

An ordinal logic Λ is said to be *invariant up to* an ordinal a if, whenever $\mathbf{\Omega}$, $\mathbf{\Omega}'$ are ordinal formulae representing the same ordinal less than α, the extent of $\Lambda(\mathbf{\Omega})$ is identical with the extent of $\Lambda(\mathbf{\Omega}')$. An ordinal logic is *invariant* if it is invariant up to each ordinal represented by an ordinal formula.

Clearly the classification into depths presupposes that the ordinal logic used is invariant.

Among the questions that we should now like to ask are

(a) Are there any complete ordinal logics?

(b) Are there any complete invariant ordinal logics?

To these we might have added "are all ordinal logics complete? " : but this is trivial; in fact, there are ordinal logics which do not suffice to prove any number-theoretic theorems whatever.

We shall now show that (a) must be answered affirmatively. In fact, we can write down a complete ordinal logic at once. Put

$$\mathrm{Od} \rightarrow \lambda a.\{\lambda fmn.\mathrm{Dt}(f(m),f(n))\}\big(\lambda s.\mathcal{P}(\lambda r.(I,a(s)),1,s)\big)$$

and

$$\mathrm{Comp} \rightarrow \lambda wa.\delta\big(w, Od(a)\big).$$

I shall show that Comp is a complete ordinal logic.

For if, Comp $(\mathbf{\Omega},\mathbf{A})$ conv 2, then

$$\mathbf{\Omega} \text{ conv Od } (\mathbf{A})$$

$$\text{conv } \lambda mn.\, \text{Dt}\left(\mathcal{P}(\lambda r.r(I,\mathbf{A}(m)),l,m),\ \mathcal{P}(\lambda r.r(I,\mathbf{A}(n)),1,n\,)\right).$$

$\mathbf{\Omega}(\mathbf{m},\mathbf{n})$ has a normal form if $\mathbf{\Omega}$ is an ordinal formula, so that then

$$\mathcal{P}(\lambda r.r(I,\mathbf{A}(\mathbf{m})),\mathbf{1})$$

has a normal form; this means that $\mathbf{r}(I,\mathbf{A}(\mathbf{m}))$ conv 2 some r, *i.e.* $\mathbf{A}(\mathbf{m})$ conv 2. Thus, if Comp $(\mathbf{\Omega},\mathbf{A})$ conv 2 and $\mathbf{\Omega}$ is an ordinal formula, then $\mathbf{A}$ is dual. Comp is therefore an ordinal logic. Now suppose conversely that $\mathbf{A}$ is dual. I shall show that Od($\mathbf{A}$) is an ordinal formula representing the ordinal ω. For

$$\mathcal{P}(\lambda r.r(I,\mathbf{A}(\mathbf{m})),1,\mathbf{m}) \text{ conv } \mathcal{P}(\lambda r.r(I,2),1,\mathbf{m})$$

$$\text{conv } 1(\mathbf{m}) \text{ conv } \mathbf{m},$$

$$\text{Od}(\mathbf{A},\mathbf{m},\mathbf{n}) \text{ conv Dt}(\mathbf{m},\mathbf{n}),$$

i.e. Od($\mathbf{A}$) is an ordinal formula representing the same ordinal as Dt. But

$$\text{Comp (Od } (\mathbf{A})\ \mathbf{A}) \text{ conv } \delta(\text{ Od } (\mathbf{A}), \text{Od } (\mathbf{A})) \text{ conv } 2.$$

This proves the completeness of Comp.

Of course Comp is not the kind of complete ordinal logic that we should really wish to use. The use of Comp does not make it any easier to see that $\mathbf{A}$ is dual. In fact, if we really want to use an ordinal logic a proof, of completeness for that particular ordinal logic will be of little value; the ordinals given by the completeness proof will not be ones which can easily be seen intuitively to be ordinals. The only value in a completeness proof of this kind would be to show that, if any objection is to be raised against an ordinal logic, it must be on account of something more subtle than incompleteness.

The theorem of completeness is also unexpected in that the ordinal formulae used are all formulae representing ω. This is contrary to our intentions in constructing Λ_P for instance; implicitly we had in mind large ordinals expressed in a simple manner. Here we have small ordinals expressed in a very complex and artificial way.

Before trying to solve the problem (*b*), let us see how far Λ_P and Λ_H are invariant. We should certainly not expect Λ_P to be invariant, since the extent of $\Lambda_P(\mathbf{\Omega})$ will depend on whether $\mathbf{\Omega}$ is convertible to a formula of the form $H(\mathbf{A})$: but suppose that we call an ordinal logic $\mathbf{\Lambda}$ "C-K invariant up to α" the extent of $\mathbf{\Lambda}(H(\mathbf{A}))$ is the same as the extent of $\mathbf{\Lambda}(H(\mathbf{B}))$ whenever $\mathbf{A}$ and $\mathbf{B}$ are C-K ordinal formulae representing the same ordinal less than α. How far is Λ_P C-K invariant? It is not difficult to see that it is C-K invariant up to any finite ordinal, that is to say up to ω. It is also C-K invariant up to $\omega+1$, as follows from the fact that the extent of

$$\Lambda_P(H(\lambda ufx.u(\mathbf{R})))$$

is the set-theoretic sum of the extents of

$$\Lambda_P(H(\lambda ufx.\mathbf{R}(1))),\quad \Lambda_P(H(\lambda ufx.\mathbf{R}(2))),\ \ldots.$$

However, there is no obvious reason for believing that it is C-K invariant up to $\omega+2$, and in fact it is demonstrable that this is not the case (see the end of this section). Let us find out what happens if we try to prove that the extent of

$$\Lambda_P(H(\text{ Suc }(\lambda ufx.u(\mathbf{R}_1))))$$

is the same as the extent of

$$\Lambda_P(H(\text{Suc }(\lambda ufx.u(\mathbf{R}_2)))),$$

where $\lambda ufx.u(\mathbf{R}_1)$ and $\lambda ufx.u(\mathbf{R}_2)$ are two C-K ordinal formulae representing ω. We should have to prove that a formula interpretable as a number-theoretic theorem is provable in $C[\text{ Suc }(\lambda ufx.u(\mathbf{R}_1))]$ if, and only if, it is provable in $C[\text{ Suc }(\lambda ufx.u(\mathbf{R}_2))]$. Now $C[\text{ Suc }(\lambda ufx.u(\mathbf{R}_1))]$ is obtained from $C[\lambda ufx.u(\mathbf{R}_1)]$ by adjoining all axioms of the form

$$(\exists x_0) \text{ Proof}_{C[\lambda ufx.u(\mathbf{R}_1)]}[x_0, f^{(m)}0) \supset \mathfrak{F}, \tag{9. 1}$$

where m is the G.R. of $\mathfrak{F}$, and C [Suc $(\lambda ufx.u(\mathbf{R}_2))$] is obtained from $C[\lambda ufx.u(\mathbf{R}_2)]$ by adjoining all axioms of the form

$$(\exists x_0) \text{ Proof}_{C[\lambda ufx.u(\mathbf{R}_2)][x_0, f^{(m)}0]} \supset \mathfrak{F}. \tag{9. 2}$$

The axioms which must be adjoined to P to obtain $C[\lambda ufx.u(\mathbf{R}_1)]$ are essentially the same as those which must be adjoined to obtain the system $C[\lambda ufx.u(\mathbf{R}_2)]$: however the *rules of procedure which have to be applied before these axioms can be written down are in general quite different in the two cases*. Consequently (9.1) and (9.2) are quite different axioms, and there is no reason to expect their consequences to be the same. A proper understanding of this will make our treatment of question (*b*) much more intelligible. See also footnote ‡ on page 172.

Now let us turn to Λ_H. This ordinal logic is invariant. Suppose that $\mathbf{\Omega}$, $\mathbf{\Omega}'$ represent the same ordinal, and suppose that we have a proof of a number-theoretic theorem G in $P_{\mathbf{\Omega}}$. The formula expressing the number-theoretic theorem does not involve any odd types. Now there is a one-one correspondence between the odd types such that if $2m-1$ corresponds to $2m'-1$ and $2n-1$ to $2n'-1$ then $\mathbf{\Omega}(\mathbf{m},\mathbf{n})$ conv 2 implies $\mathbf{\Omega}'(\mathbf{m}',\mathbf{n}')$ conv 2. Let us modify the odd type-subscripts occurring in the proof of G, replacing each by its mate in the one-one correspondence. There results a proof in $P_{\mathbf{\Omega}'}$, with the same end formula G. That is to say that if G is provable in $P_{\mathbf{\Omega}}$ it is provable in $P_{\mathbf{\Omega}'}$. Λ_H is invariant.

The question (*b*) must be answered negatively. Much more can be proved, but we shall first prove an even weaker result which can be established very quickly, in order to illustrate the method.

I shall prove that an ordinal logic $\mathbf{\Lambda}$ cannot be invariant and have the property that the extent of $\mathbf{\Lambda}(\mathbf{\Omega})$ is a strictly increasing function of the ordinal represented by $\mathbf{\Omega}$. Suppose that $\mathbf{\Lambda}$ has these properties; then we shall obtain a contradiction. Let $\mathbf{A}$ be a W.F.F. in normal form and without free variables, and consider the process of carrying out conversions on $\mathbf{A}(1)$ until we have shown it convertible to 2, then converting $\mathbf{A}(2)$ to 2, then $\mathbf{A}(3)$ and so on: suppose that after r steps we are still performing the conversion on $\mathbf{A}(\mathbf{m}_r)$. There is a formula Jh such that Jh $(\mathbf{A},\mathbf{r})$ conv $\mathbf{m}_r$ for each positive integer r. Now let $\mathbf{Z}$ be a formula such that, for each positive integer $n, \mathbf{Z}(\mathbf{n})$ is an ordinal formula representing $\mathbf{\Omega}^{\mathbf{r}}$, and suppose $\mathbf{B}$ to be a member of the extent of $\mathbf{\Lambda}$(Suc(Lim $(\mathbf{Z})$)) but not of the extent of $\mathbf{\Lambda}$(Lim $(\mathbf{Z})$). Put

$$\mathbf{K}^* \to \lambda a.\mathbf{\Lambda}(\text{Suc}(\mathbf{Lim}(\lambda r.\mathbf{Z}(\text{Jh}(a,r)))),\ \text{B});$$

then $\mathbf{K}^*$ is a complete logic. For, if $\mathbf{A}$ is dual, then

$$\text{Suc }(\text{Lim}(\lambda r.\mathbf{Z}(\text{Jh}(\mathbf{A},r))))$$

represents the ordinal $\omega^{\omega}+1$, and therefore $\mathbf{K}^*(\mathbf{A})$ conv 2; but, if $\mathbf{A}$(c) is not convertible to 2, then

$$\text{Suc }(\text{Lim}(\lambda r.\mathbf{Z}(\text{Jh}(\mathbf{A},r))))$$

represents an ordinal not exceeding $\omega^{c}+1$, and $\mathbf{K}^*(\mathbf{A})$ is therefore not convertible to 2. Since there are no complete logic formulae, this proves our assertion.

We may now prove more powerful results.

Incompleteness theorems.

(A) If an ordinal logic $\mathbf{\Lambda}$ is invariant up to an ordinal α, then for any ordinal formula $\mathbf{\Omega}$ representing an ordinal β, $\beta < \alpha$, the extent of $\mathbf{\Lambda}(\mathbf{\Omega})$ is contained in the (set-theoretic) sum of the extents of the logics $\mathbf{\Lambda}(\mathbf{P})$, where $\mathbf{P}$ is finite.

(B) If an ordinal logic $\mathbf{\Lambda}$ is C-K invariant up to an ordinal α, then for any C-K ordinal formula $\mathbf{A}$ representing an ordinal β, $\beta < \alpha$, the extent of $\mathbf{\Lambda}(H(\mathbf{A}))$ is contained in the (set-theoretic) sum of the extents of the logics $\mathbf{\Lambda}(H(\mathbf{F}))$, where $\mathbf{F}$ is a C-K ordinal formula representing an ordinal less than ω^2.

Proof of (A). It is sufficient to prove that, if $\mathbf{\Omega}$ represents an ordinal $\gamma, \omega \leq \gamma < \alpha$, then the extent of $\mathbf{\Lambda}(\mathbf{\Omega})$ is contained in the set-theoretic sum of the extents of the logics $\mathbf{\Lambda}(\mathbf{\Omega}')$, where $\mathbf{\Omega}'$ represents an ordinal less than γ. The ordinal γ must be of the form $\gamma_0 + \rho$, where ρ is finite and represented by $\mathbf{P}$ say, and γ_0 is not the successor of any ordinal and is not less than ω. There are two cases to consider; $\gamma_0 = \omega$ and $\gamma_0 \geq 2\omega$. In each of them we shall obtain a contradiction from the assumption that there is a W.F.F. $\mathbf{B}$ such that $\mathbf{\Lambda}(\mathbf{\Omega}, \mathbf{B})$ conv 2 whenever $\mathbf{\Omega}$ represents γ, but is not convertible to 2 if $\mathbf{\Omega}$ represents a smaller ordinal. Let us take first the case $\gamma_0 \geq 2\omega$. Suppose that $\gamma_0 = \omega + \gamma_1$, and that $\mathbf{\Omega}_1$ is an ordinal formula representing γ_1. Let $\mathbf{A}$ be any W.F.F. with a normal form and no free variables, and let Z be the class of those positive integers which are exceeded by all integers n for which $\mathbf{A}(\mathbf{n})$ is not convertible to 2. Let E be the class of integers $2p$ such that $\mathbf{\Omega}(\mathbf{p}, \mathbf{n})$ conv 2 for some n belonging to Z. The class E, together with the class Q of all odd integers, is constructively enumerable. It is evident that the class can be enumerated with repetitions, and since it is infinite the required enumeration can be obtained by striking out the repetitions. There is, therefore, a formula En such that En $(\mathbf{\Omega}, \mathbf{A}, \mathbf{r})$ runs through the formulae of the class $E + Q$ without repetitions as r runs through the positive integers. We define

$$\text{Rt} \rightarrow \lambda wamn.\ \text{Sum (Dt } w, \text{En } (w, a, m), \text{En}(w, a, n)).$$

Then Rt $(\mathbf{\Omega}_1, \mathbf{A})$ is an ordinal formula which represents γ_0 if $\mathbf{A}$ is dual, but a smaller ordinal otherwise. In fact

$$\text{Rt } (\mathbf{\Omega}_1, \mathbf{A}, \mathbf{m}, \mathbf{n}) \text{ conv } \{\text{Sum (Dt, } \mathbf{\Omega}_1)\}(\text{En}(\mathbf{\Omega}_1, \mathbf{A}, \mathbf{m}), \text{ En } (\mathbf{\Omega}_1, \mathbf{A}, \mathbf{n})).$$

Now, if $\mathbf{A}$ is dual, $E + Q$ includes all integers m for which

$$\{\text{Sum (Dt, } \mathbf{\Omega}_1)\}\ (\mathbf{m}, \mathbf{m}) \text{ conv } 3.$$

(This depends on the particular form that we have chosen for the formula Sum.) Putting "En $(\mathbf{\Omega}_1, \mathbf{A}, \mathbf{p})$ conv $\mathbf{q}$" for $M(p, q)$, we see that condition (7. 4) is satisfied, so that Rt $(\mathbf{\Omega}_1, \mathbf{A})$ is an ordinal formula representing γ_0. But, if $\mathbf{A}$ is not dual, the set $E + Q$ consists of all integers m for which

$$\{\text{Sum (Dt, } \mathbf{\Omega}_1)\}\ (\mathbf{m}, \mathbf{r}) \text{ conv } 2,$$

where r depends only on $\mathbf{A}$. In this case Rt $(\mathbf{\Omega}_1, \mathbf{A})$ is an ordinal formula representing the same ordinal as Inf(Sum (Dt, $\mathbf{\Omega}_1$), $\mathbf{r}$), and this is smaller than γ_0. Now consider $\mathbf{K}$:

$$\mathbf{K} \rightarrow \lambda a.\mathbf{\Lambda}(\text{Sum}(\text{Rt}(\mathbf{\Omega}_1, \mathbf{A}), \mathbf{P}), \mathbf{B}). \tag{9.1}$$

If $\mathbf{A}$ is dual, $\mathbf{K}(\mathbf{A})$ is convertible to 2 since Sum (Rt $(\mathbf{\Omega}_1, \mathbf{A}), \mathbf{P}$) represents γ. But, if $\mathbf{A}$ is not dual, it is not convertible to 2, since Sum (Rt $(\mathbf{\Omega}_1, \mathbf{A}), \mathbf{P}$) then represents an ordinal smaller than γ. In $\mathbf{K}$ we therefore have a complete logic formula, which is impossible.

Now we take the case $\gamma_0 = \omega$. We introduce a W.F.F. Mg such that if n is the D.N. of a computing machine $\mathcal{M}$, and if by the m-th complete configuration of $\mathcal{M}$ the figure 0 has been printed, then Mg $(\mathbf{n}, \mathbf{m})$ is convertible to λpq. Al $(4(P, 2p + 2q), 3, 4)$ (which is an ordinal formula representing the ordinal 1, but if 0 has not been printed it is convertible to $\lambda pq.p(q,I,4)$ (which represents 0). Now consider

$$\mathbf{M} \rightarrow \lambda n.\mathbf{\Lambda}(\text{Sum (Lim (Mg}(n)), \mathbf{P}), \mathbf{B}).$$

If the machine never prints 0, then Lim ($\lambda r.$ Mg($\mathbf{n}, r$)) represents ω and Sum (Lim (Mg($\mathbf{n}$)), $\mathbf{P}$) represents γ. This means that $\mathbf{M}(\mathbf{n})$ is convertible to 2. If, however, $\mathcal{M}$ never prints 0, Sum(Lim(Mg($\mathbf{n}$)),$\mathbf{P}$) represents a finite ordinal and $\mathbf{M}(\mathbf{n})$ is not convertible to 2. In $\mathbf{M}$ we therefore have means of determining about a machine whether it ever prints 0, which is impossible† (Turing[**1**], §8). This completes the proof of (A).

Proof of (B). It is sufficient to prove that, if $\mathbf{C}$ represents an ordinal γ, $\omega^2 \leq \gamma < \alpha$, then the extent of $\mathbf{\Lambda}(H(\mathbf{C}))$ is included in the set-theoretic sum of the extents of $\mathbf{\Lambda}$ ($H(\mathbf{G})$), where $\mathbf{G}$ represents an ordinal less than γ. We obtain a contradiction from the assumption that there is a formula $\mathbf{B}$ which is in the extent of $\mathbf{\Lambda}(H(\mathbf{G}))$ if $\mathbf{G}$ represents γ, but not if it represents any smaller ordinal. The ordinal γ is of the form $\delta + \omega^2 + \xi$, where $\xi < \omega^2$. Let $\mathbf{D}$ be a C-K ordinal formula representing δ and $\lambda ufx \cdot \mathbf{Q}(u,f,\mathbf{A}(u,f,x))$ one representing $\alpha + \xi$ whenever $\mathbf{A}$ represents α.

We now define a formula Hg. Suppose that $\mathbf{A}$ is a W.F.F. in normal form and without free variables; consider the process of carrying out conversions on $\mathbf{A}$(l) until it is brought into the form 2, then converting $\mathbf{A}(2)$ to 2, then $\mathbf{A}(3)$, and so on. Suppose that at the r-th step of this process we are doing the n_r-th step in the conversion of $\mathbf{A}(\mathbf{m}_r)$. Thus, for instance, if $\mathbf{A}$ is not convertible to 2, m_r can never exceed 3. Then Hg($\mathbf{A}, \mathbf{r}$) is to be convertible to $\lambda f \cdot f(\mathbf{m_r}, \mathbf{n_r})$ for each positive integer r. Put

$$\text{Sq} \rightarrow \lambda dmn.n\Big(\text{Suc},\ m\big(\lambda aufx \cdot u(\lambda y \cdot y(\text{Suc}, \alpha(u,f,x)))\big),\ d(u,f,x)\big)\Big),$$

$$\mathbf{M} \rightarrow \lambda aufx \cdot \mathbf{Q}\big(u,f,u\big(\lambda y \cdot \text{Hg}(a,y,\text{Sq}(\mathbf{D}))\big)\big),$$

$$\mathbf{K}_1 \rightarrow \lambda a \cdot \mathbf{\Lambda}\big(\mathbf{M}(a),\ \mathbf{B}\big),$$

then I say that $\mathbf{K}_1$ is a complete logic formula. Sq ($\mathbf{D}, \mathbf{m}, \mathbf{n}$) is a C-K ordinal formula representing $\delta + m\omega + n$, and therefore Hg($\mathbf{A}, \mathbf{r}$, Sq ($\mathbf{D}$)) represents an ordinal ζ_r which increases steadily with increasing r, and tends to the limit $\delta + \omega^2$ if $\mathbf{A}$ is dual. Further

$$\text{Hg}(\mathbf{A}, \mathbf{r}, \text{Sq}(\mathbf{D})) < \text{Hg}(\mathbf{A}, S(\mathbf{r}), \text{Sq}(\mathbf{D}))$$

for each positive integer r. therefore $\lambda ufx \cdot u(\lambda y.$ Hg ($\mathbf{A}$,y, Sq ($\mathbf{D}$))) is a C-K ordinal formula and represents the limit of the sequence $\zeta_1, \zeta_2, \zeta_3, \ldots$. This is $\delta + \omega^2$ if $\mathbf{A}$ is dual, but a smaller ordinal otherwise. Likewise $\mathbf{M}(\mathbf{A})$ represents γ if $\mathbf{A}$ is dual, but is a smaller ordinal otherwise. The formula $\mathbf{B}$ therefore belongs to the extent of $\mathbf{\Lambda}(H(\mathbf{M}(\mathbf{A})))$ if and only if $\mathbf{A}$ is dual, and this implies that $\mathbf{K_1}$ is a complete logic formula, as was asserted. But this is impossible and we have the requiredcontradiction.

As a corollary to (A) we see that Λ_H is incomplete and in fact that the extent of Λ_H (Dt) contains the extent of $\Lambda_H(\mathbf{\Omega})$ for any ordinal formula $\mathbf{\Omega}$. This result, suggested to me first by the solution of question (*b*), may also be obtained more directly. In fact, if a number-theoretic theorem can be proved in any particular $P_{\mathbf{\Omega}}$, it can also be proved in $P_{\lambda mn \cdot m(\mathrm{n},I,4)}$. The formulae describing number-theoretic theorems in P do not involve more than a finite number of types, type 3 being the highest necessary. The formulae describing the number-theoretic theorems in any $P_{\mathbf{\Omega}}$ will be obtained by doubling the type subscripts. Now suppose that we have a proof of a number-theoretic theorem G in

† This part of the argument can equally well be based on the impossibility of determining about two W.F.F. whether they are interconvertible. (Church [**3**], 363.)

$P_{\mathbf{\Omega}}$ and that the types occurring in the proof are among $0, 2, 4, 6, t_1, t_2, t_3, \ldots$. We may suppose that they have been arranged with all the even types preceding all the odd types, the even types in order of magnitude and the type $2m-1$ preceding $2n-1$ if $\mathbf{\Omega}(\mathbf{m},\mathbf{n})$ conv 2. Now let each t_r be replaced by $10+2r$ throughout the proof of G. We thus obtain a proof of G in $P_{\lambda mn\cdot(n,I,4)}$.

As with problem (*a*), the solution of problem (*b*) does not require the use of high ordinals [*e.g.* if we make the assumption that the extent of $\mathbf{\Lambda}(\mathbf{\Omega})$ is a steadily increasing function of the ordinal represented by $\mathbf{\Omega}$ we do not have to consider ordinals higher than $\omega+2$]. However, if we restrict what we are to call ordinal formulae in some way, we shall have corresponding modified problems (*a*) and (*b*), the solutions will presumably be essentially the same, but will involve higher ordinals. Suppose, for example, that Prod is a W.F.F. with the property that Prod $(\mathbf{\Omega}_1, \mathbf{\Omega}_2)$ is an ordinal formula representing $\alpha_1\alpha_2$ when $\mathbf{\Omega}_1, \mathbf{\Omega}_2$ are ordinal formulae representing α_1, α_2 respectively, and suppose that we call a W.F.F. a l-ordinal formula when it is convertible to the form Sum (Prod($\mathbf{\Omega}$, Dt), $\mathbf{P}$), where $\mathbf{\Omega}, \mathbf{P}$ are ordinal formulae of which $\mathbf{P}$ represents a finite ordinal. We may define l-ordinal logics, l-completeness and l-invariance in an obvious way, and obtain a solution of problem (*b*) which differs from the solution in the ordinary case in that the ordinals less than to ω^2 take the place of the finite ordinals. More generally the cases that I have in mind are covered by the following theorem.

Suppose that we have a class V of formulae representing ordinals in some manner which we do not propose to specify definitely, and a subset† U of the class V such that:

(i) There is a formula $\mathbf{\Phi}$ such that if $\mathbf{T}$ enumerates a sequence of members of U representing an increasing sequence of ordinals, then $\mathbf{\Phi}(\mathbf{T})$ is a member of U representing the limit of the sequence.

(ii) There is a formula $\mathbf{E}$ such that $\mathbf{E}(\mathbf{m},\mathbf{n})$ is a member of U for each pair of positive integers m, n and if it represents $\epsilon_{m,n}$, then $\epsilon_{m,n} < \epsilon_{m',n'}$ if either $m < m'$ or $m = m', n < n'$.

(iii) There is a formula $\mathbf{G}$ such that, if $\mathbf{A}$ is a member of U, then $\mathbf{G}(\mathbf{A})$ is a member of U representing a larger ordinal than does $\mathbf{A}$, and such that $\mathbf{G}(\mathbf{E}(\mathbf{m},\mathbf{n}))$ always represents an ordinal not larger than $\epsilon_{m,n+1}$.

We define a V-ordinal logic to be a W.F.F. $\mathbf{\Lambda}$ such that $\mathbf{\Lambda}(\mathbf{A})$ is a logic whenever $\mathbf{A}$ belongs to V. $\mathbf{\Lambda}$ is V-invariant if the extent of $\mathbf{\Lambda}(\mathbf{A})$ depends only on the ordinal represented by $\mathbf{A}$. Then it is not possible for a V-ordinal logic $\mathbf{\Lambda}$ to be V-invariant and have the property that, if $\mathbf{C}_1$ represents a greater ordinal than $\mathbf{C}_2$ ($\mathbf{C}_1$ and $\mathbf{C}_2$ both being members of U), then the extent of $\mathbf{\Lambda}(\mathbf{C}_1)$ is greater than the extent of $\mathbf{\Lambda}(\mathbf{C}_2)$.

We suppose the contrary. Let $\mathbf{B}$ be a formula belonging to the extent of $\mathbf{\Lambda}\,((\mathbf{\Phi}\,(\lambda r\cdot \mathbf{E}(r,1))))$ but not to the extent of $\mathbf{\Lambda}\,(\mathbf{\Phi}\,(\lambda r\cdot \mathbf{E}(r,1)))$, and let $\mathbf{K}' \to \lambda a.\mathbf{\Lambda}\,(\mathbf{G}\,(\mathbf{\Phi}\,(\lambda r\cdot \mathrm{Hg}(\mathrm{a},r,\mathbf{E}))),\mathbf{B})$.

Then $\mathbf{K}'$ is a complete logic. For

$$\mathrm{Hg}\ (\mathbf{A},\mathbf{r},\mathbf{E}) \text{ conv } \mathbf{E}(\mathbf{m_r},\mathbf{n_r}).$$

$\mathbf{E}(\mathbf{m_r},\mathbf{n_r})$ is a sequence of V-ordinal formulae representing an increasing sequence of ordinals. Their limit is represented by $\mathbf{\Phi}(\lambda r.\ \mathrm{Hg}\ (\mathbf{A},r,\mathbf{E}))$; let us see what this limit is. First suppose that $\mathbf{A}$ is dual: then m_r tends to infinity as r tends to infinity, and $\mathbf{\Phi}(\lambda r.\ \mathrm{Hg}\ (\mathbf{A},r,\mathbf{E}))$ therefore represents the same ordinal as $\mathbf{\Phi}(\lambda r.\mathbf{E}(r,1))$. In this case we must have

$$\mathbf{K}'(\mathbf{A}) \text{ conv } 2.$$

Now suppose that $\mathbf{A}$ is not dual: m_r is eventually equal to some constant number, a say, and $\mathbf{\Phi}(\lambda r.\ \mathrm{Hg}\ (\mathbf{A},r,\mathbf{E}))$ represents the same ordinal as $\mathbf{\Phi}(\lambda r.\mathbf{E}(\mathbf{a},r))$, which is smaller than that represented by

† The subset U wholly supersedes V in what follows. The introduction of V serves to emphasise the fact that the set of ordinals represented by member of $\mathbf{U}$ may have gaps.

$\mathbf{\Phi}(\lambda r.\mathbf{E}(r,1)).\mathbf{B}$ cannot therefore belong to the extent of $\mathbf{\Lambda}(\mathbf{G}(\mathbf{\Phi}(\lambda r\cdot \mathrm{Hg}(\mathbf{A},r,\mathbf{E}))))$, and $\mathbf{K}'(\mathbf{A})$ is not convertible to 2. We have proved that $\mathbf{K}'$ is a complete logic, which is impossible.

This theorem can no doubt be improved in many ways. However, it is sufficiently general to show that, with almost any reasonable notation for ordinals, completeness is in compatible with invariance.

We can still give a certain meaning to the classification into depths with highly restricted kinds of ordinals. Suppose that we take a particular ordinal logic $\mathbf{\Lambda}$ and a particular ordinal formula $\mathbf{\Psi}$ representing the ordinal α say (preferably a large one), and that we restrict ourselves to ordinal formulae of the form $\mathrm{Inf}(\mathbf{\Psi},\mathbf{a})$. We then have a classification into depths, but the extents of all the logics which we so obtain are contained in the extent of a single logic.

We now attempt a problem of a rather different character, that of the completeness of Λ_P. It is to be expected that this ordinal logic is complete. I cannot at present give a proof of this, but I can give a proof that it is complete as regards a simpler type of theorem than the number-theoretic theorems, viz. those of form "$\theta(x)$ vanishes identically", where $\theta(x)$ is primitive recursive. The proof will have to be much abbreviated since we do not wish to go into the formal details of the system P. Also there is a certain lack of definiteness in the problem as at present stated, owing to the fact that the formulae G, E, M_P were not completely defined. Our attitude here is that it is open to the sceptical reader to give detailed definitions for these formulae and then verify that the remaining details of the proof can be filled in, using his definition. It is not asserted that these details can be filled in whatever be the definitions of G, E, M_P consistent with the properties already required of them, only that they can be filled in with the more natural definitions.

I shall prove the completeness theorem in the following form. If $\mathfrak{B}[x_0]$ is a recursion formula and if $\mathfrak{B}[0], \mathfrak{B}[f0], \ldots$ are all provable in P, then there is a C-K ordinal formula $\mathbf{A}$ such that $(x_0)\mathfrak{B}[x_0]$ is provable in the system $P^{\mathbf{A}}$ of logic obtained from P by adjoining as axioms all formulae whose G.R.'s are of the form

$$\mathbf{A}(\lambda mn\cdot m(\varpi(2,n),\varpi(3,n)),K,M_P,\mathbf{r})$$

(provided they represent propositions).

First let us define the formula $\mathbf{A}$. Suppose that $\mathbf{D}$ is a W.F.F. with the property that $\mathbf{D}(\mathbf{n})$ conv 2 if $\mathfrak{B}[f^{(n-1)}0]$ is provable in P, but $\mathbf{D}(\mathbf{n})$ conv 1 if $\sim\mathfrak{B}[f^{(n-1)}0]$ is provable in P (P is being assumed consistent). Let Θ be defined by

$$\Theta \to \{\lambda vu\cdot u(v(v,u))\}(\lambda vu.u(v(v,u))),$$

and let Vi be a formula with the properties

$$\mathrm{Vi}\,(2) \text{ conv } \lambda u.u(\mathrm{Suc}, U),$$
$$\mathrm{Vi}\,(1) \text{ conv } \lambda u.u(I,\Theta(\mathrm{Suc})).$$

The existence of such a formula is established in Kleene [**1**], corollary on p. 220. Now put

$$\mathbf{A}^* \to \lambda ufx.u(\lambda y.\mathrm{Vi}(\mathbf{D}(y),y,u,f,x)),$$
$$\mathbf{A} \to \mathrm{Suc}(\mathbf{A}^*).$$

I assert that $\mathbf{A}^*, \mathbf{A}$ are C-K ordinal formulae whenever it is true that $\mathfrak{B}[0], \mathfrak{B}[f0], \ldots$ are all provable in P. For in this case $\mathbf{A}^*$ is $\lambda ufx\cdot u(\mathbf{R})$, where

$$\mathbf{R} \to \lambda y.\mathrm{Vi}(\mathbf{D}(y),y,u,f,x),$$

and then

$\lambda ufx.\mathbf{R}(\mathbf{n})$ conv $\lambda ufx.$ Vi $(\mathbf{D}(\mathbf{n}), \mathbf{n}, u, f, x)$

conv $\lambda ufx.$ Vi $(2, \mathbf{n}, u, f, x)$

conv $\lambda ufx.\{\lambda n.n\ (\text{Suc}, U)\ \}(\mathbf{n}, u, f, x)$

conv $\lambda ufx.\mathbf{n}(\text{Suc}, U, u, f, x)$, which is a C-K ordinal formula,

and

$$\lambda ufx.S(\mathbf{n}, \text{Suc}, U, u, f, x) \text{ conv Suc } (\lambda ufx \cdot \mathbf{n}(\text{Suc}, U, u, f, x)).$$

These relations hold for an arbitrary positive integer n and therefore $\mathbf{A}^*$ is a C-K ordinal formula [condition (9) p. 164]: it follows immediately that $\mathbf{A}$ is also a C-K ordinal formula. It remains to prove that $(x_0)\mathfrak{B}[x_0]$ is provable in $P^{\mathbf{A}}$. To do this it is necessary to examine the structure of $\mathbf{A}^*$ in the case in which $(x_0)\mathfrak{B}[x_0]$ is false. Let us suppose that $\sim \mathfrak{B}[f^{(a-1)}0]$ is true, so that $\mathbf{D}(\mathbf{a})$ conv 1, and let us consider $\mathbf{B}$ where

$$\mathbf{B} \rightarrow \lambda ufx \cdot \text{Vi}(\mathbf{D}(\mathbf{a}), \mathbf{a}, u, f, x).$$

If $\mathbf{A}^*$ was a C-K ordinal formula, then $\mathbf{B}$ would be a member of its fundamental sequence; but

$$\begin{aligned}
\mathbf{B} &\text{ conv } \lambda ufx \cdot \text{Vi}(1, \mathbf{a}, u, f, x) \\
&\text{ conv } \lambda ufx \cdot \{\lambda u.u\,(I, \Theta(\text{Suc}))\}\,(\mathbf{a}, u, f, x) \\
&\text{ conv } \lambda ufx \cdot \Theta(\text{Suc}, u, f, x) \\
&\text{ conv } \lambda ufx \cdot \{\lambda u \cdot u(\Theta(u))\}\,(\text{Suc}, u, f, x) \\
&\text{ conv } \lambda ufx \cdot \text{Suc}(\Theta(\text{Suc}), u, f, x) \\
&\text{ conv Suc}(\lambda ufx \cdot \Phi(\text{Suc}, u, f, x)) \\
&\text{ conv Suc}(\mathbf{B}) \qquad (9.3)
\end{aligned}$$

This, of course, implies that $\mathbf{B} < \mathbf{B}$ and therefore that $\mathbf{B}$ is no C-K ordinal formula. This, although fundamental in the possibility of proving our completeness theorem, does not form an actual step in the argument. Roughly speaking, our argument amounts to this. The relation (9.3) implies that the system $P^{\mathbf{B}}$ is inconsistent and therefore that $P^{\mathbf{A}^*}$ is inconsistent and indeed we can prove in P (and *a fortiori* in $P^{\mathbf{A}}$) that $\sim (x_0)\mathfrak{B}[x_0]$ implies the inconsistency of $P^{\mathbf{A}^*}$. On the other hand in $P^{\mathbf{A}}$ we can prove the consistency of $P^{\mathbf{A}^*}$. The inconsistency of $P^{\mathbf{B}}$ is proved by the Gödel argument. Let us return to the details.

The axioms in $P^{\mathbf{B}}$ are those whose G.R.'s are of the form

$$\mathbf{B}\left(\lambda mn.m\,(\varpi\,(2,n), \varpi\,(3,n))\,, K, M_p, \mathbf{r}\right).$$

When we replace $\mathbf{B}$, by Suc $(\mathbf{B})$, this becomes

Suc $(\mathbf{B}, \lambda mn.m(\varpi\,(2,n), \varpi\,(3,n)), K, M_P, \mathbf{r})$

$$\text{conv } K\,(\mathbf{B}\,(\lambda mn \cdot m(\varpi\,(2,n), \varpi\,(3,n)), K, M_P, \mathbf{r}))$$

$$\text{conv } \mathbf{B}\,(\lambda mn.m\,(\varpi\,(2,n), \varpi\,(3,n))\,, K, M_P, \mathbf{p})$$

if $\mathbf{r}$ conv $2\mathbf{p}+1$,

$$\text{conv } E\,(\mathbf{B}(\lambda mn.m(\varpi\,(2,n), \varpi\,(3,n)), K, M_P), \mathbf{p})$$

if $\mathbf{r}$ conv $2\mathbf{p}$.

When we remember the essential property of the formula E, we see that the axioms of $P^{\mathbf{B}}$ include all formulae of the form

$$(\exists x_0)\text{Proof}_{\text{pB}}[x_0, f^{(q)}0] \supset \mathfrak{F},$$

where q is the G.R. of the formula $\mathfrak{F}$.

Let b be the G.R. of the formula $\mathfrak{A}$.

$$\sim (\exists x_0)(\exists y_0)\{\text{Proof}_{\text{pB}}[x_0, y_0].\text{Sb}[z_0, z_0, y_0]\}. \qquad (\mathfrak{A})$$

Sb $[x_0, y_0, z_0]$ is a particular recursion formula such that Sb $[f^{(l)}0, f^{(m)}0, f^{(n)}0]$ holds if and only if n is the G.R. of the result of substituting $f^{(m)}0$ for z_0 in the formula whose G.R. is l at all points where z_0 is free. Let p be the G.R. of the formula $\mathfrak{C}$.

$$\sim (\exists x_0)(\exists y_0)\{\text{Proof}_{\text{pB}}[x_0, y_0] \cdot \text{Sb}[f^{(b)}0, f^{(b)}0, y_0]\}. \qquad (\mathfrak{C})$$

Then we have as an axiom in P

$$(\exists x_0)\text{Proof}_{\text{pB}}[x_0 f^{(p)}0] \supset \mathfrak{C},$$

and we can prove in $P^{\mathbf{A}}$

$$(x_0)\{\text{Sb}[f^{(b)}0, f^{(b)}0, x_0] \equiv x_0 = f^{(p)}0\}, \qquad (9.4)$$

since $\mathfrak{C}$ is the result of substituting $f^{(b)}0$ for z_0 in $\mathfrak{A}$; hence

$$\sim (\exists y_0)\text{Proof}_{\text{pB}}[y_0, f^{(p)}0] \qquad (9.5)$$

is provable in P. Using (9.4) again, we see that $\mathfrak{C}$ can be proved in $P^{\mathbf{B}}$. But, if we can prove $\mathfrak{C}$ in $P^{\mathbf{B}}$, then we can prove its provability in $P^{\mathbf{B}}$, the proof being in P; *i.e.* we can prove

$$(\exists x_0)\text{Proof}_{\text{pB}}[x_0, f^{(p)}0]$$

in P (since p is the G.R. of $\mathfrak{C}$). But this contradicts (9.5), so that, if

$$\sim \mathfrak{B}[f^{(a-1)}0]$$

is true, we can prove a contradiction in $P^{\mathbf{B}}$ or in $P^{\mathbf{A}^*}$. Now I assert that the whole argument up to this point can be carried through formally in the system P, in fact, that,if c is the G.R. of $\sim (0 = 0)$, then

$$\sim (x_0)\mathfrak{B}[x_0] \supset (\exists v_0)\ \text{Proof}_{\text{pA}^*}[v_0, f^{(c)}0] \qquad (9.6)$$

is provable in P. I shall not attempt to give any more detailed proof of this assertion.

The formula

$$(\exists x_0)\text{Proof}_{\text{pA}^*}[x_0, f^{(c)}0] \supset \sim (0 = 0) \qquad (9.7)$$

is an axiom in $P^{\mathbf{A}}$. Combining (9.6), (9.7) we obtain $(x_0)\mathfrak{B}[x_0]$ in $P^{\mathbf{A}}$.

The completeness theorem as usual is of no value. Although it shows, for instance, that is possible to prove Fermat's last theorem with Λ_P (if it is true) yet the truth of the theorem would really be assumed by taking a certain formula as an ordinal formula.

That Λ_P is not invariant may be proved easily by our general theorem; alternatively it follows from the fact that, in proving our partial completeness theorem, we never used ordinals higher than $\omega + 1$. This fact can also be used to prove that Λ_P is not C-K invariant up to $\omega + 2$.

10. The continuum hypothesis. A digression.

The methods of §9 may be applied to problems which are constructive analogues of the continuum hypothesis problem. The continuum hypothesis asserts that $2^{\aleph_0} = \aleph_1$, in other words that, if ω_1 is the smallest ordinal α greater than ω such that a series with order type α cannot be put into one-one correspondence with the positive integers, then the ordinals less than ω_1 can be put into one-one correspondence with the subsets of the positive integers. To obtain a constructive analogue of this proposition we may replace the ordinals less than ω_1 either by the ordinal formulae, or by the ordinals represented by them; we may replace the subsets of the positive integers either by the computable sequences of figures $0, 1$, or by the description numbers of the machines which compute these sequences. In the manner in which the correspondence is to be set up there is also more than one possibility. Thus, even when we use only one kind of ordinal formula, there is still great ambiguity concerning what the constructive analogue of the continuum hypothesis should be. I shall prove a single result in this connection[†]. A number of others may be proved in the same way.

We ask "Is it possible to find a computable function of ordinal formulae determining a one-one correspondence between the ordinals represented by ordinal formulae and the computable sequences of figures $0, 1$?" More accurately, "Is there a formula **F** such that if $\mathbf{\Omega}$ is an ordinal formula and n a positive integer then $\mathbf{F}(\mathbf{\Omega},\mathbf{n})$ is convertible to 1 or to 2, and such that $\mathbf{F}(\mathbf{\Omega},\mathbf{n})$ conv $\mathbf{F}(\mathbf{\Omega}',\mathbf{n})$ for each positive integer n, if and only if $\mathbf{\Omega}$ and $\mathbf{\Omega}'$ represent the same ordinal ?" The answer is "No", as will be seen to be a consequence of the following argument: there is no formula **F** such that $\mathbf{F}(\mathbf{\Omega})$ enumerates one sequence of integers (each being 1 or 2) when $\mathbf{\Omega}$ represents ω and enumerates another sequence when $\mathbf{\Omega}$ represents 0. If there is such an **F**, then there is an a such that $\mathbf{F}(\mathbf{\Omega},\mathbf{a})$ conv (Dt, **a**) if $\mathbf{\Omega}$ represents ω but $\mathbf{F}(\mathbf{\Omega},\mathbf{a})$ and **F**(Dt, **a**) are convertible to different integers (1 or 2) if $\mathbf{\Omega}$ represents 0. To obtain a contradiction from this we introduce a W.F.F. Gm not unlike Mg. If the machine. $\mathcal{M}$ whose D.N. is n has printed 0 by the time the m-th complete configuration is reached then

$$\mathrm{Gm}\,(\mathbf{n},\mathbf{m}) \text{ conv } \lambda mn.m(n,I,4);$$

otherwise Gm $(\mathbf{n},\mathbf{m})$ conv $\lambda pq.\mathrm{Al}\,(4(P,2p+2q),3,4)$. Now consider **F**(Dt,**a**) and **F**(Lim(Gm (**n**)),**a**). If. $\mathcal{M}$ never prints 0, Lim(Gm(**n**)) represents the ordinal ω. Otherwise it represents 0. Consequently these two formulae are convertible to one another if and only $\mathcal{M}$ never prints 0. This gives us a means of determining about any machine whether it ever prints 0, which is impossible.

Results of this kind have of course no real relevance for the classical continuum hypothesis.

11. The purpose of ordinal logics.

Mathematical reasoning may be regarded rather schematically as the exercise of a combination of two faculties[‡], which we may call *intuition* and *ingenuity*. The activity of the intuition consists in making spontaneous judgments which are not the result of conscious trains of reasoning. These judgments are often but by no means invariably correct (leaving aside the the question what is meant by "correct"). Often it is possible to find some other way of verifying the correctness of an intuitive judgment. We may, for instance, judge that all positive integers are uniquely factorizable into primes; a detailed mathematical argument leads to the same result. This argument will also involve intuitive judgments, but they will be less open to criticism than the original judgment about factorization. I shall not attempt to explain this idea of "intuition" any more explicitly.

[†] A suggestion to consider this problem came to me indirectly from F. Bernstein. A related problem was suggested by P. Bernays.

[‡] We are leaving out of account that most important faculty which distinguishes topics of interest from others; in fact, we are regarding the function of the mathematician as simply to determine the truth or falsity of propositions.

The exercise of ingenuity in mathematics consists in aiding the intuition through suitable arrangements of propositions, and perhaps geometrical figures or drawings. It is intended that when these are really well arranged the validity of the intuitive steps which are required cannot seriously be doubted.

The parts played by these two faculties differ of course from occasion to occasion, and from mathematician to mathematician. This arbitrariness can be removed by the introduction of a formal logic. The necessity for using the intuition is then greatly reduced by setting down formal rules for carrying out inferences which are always intuitively valid. When working with a formal logic, the idea of ingenuity takes a more definite shape. In general a formal logic, will be framed so as to admit a considerable variety of possible steps in any stage in a proof. Ingenuity will then determine which steps are the more profitable for the purpose of proving a particular proposition. In pre-Gödel times it was thought by some that it would probably be possible to carry this programme to such a point that all the intuitive judgments of mathematics could be replaced by a finite number of these rules. The necessity for intuition would then be entirely eliminated.

In our discussions, however, we have gone to the opposite extreme and eliminated not intuition but ingenuity, and this in spite of the fact that our aim has been in much the same direction. We have been trying to see how far it is possible to eliminate intuition, and leave only ingenuity. We do not mind how much ingenuity is required, and therefore assume it to be available in unlimited supply. In our metamathematical discussions we actually express this assumption rather differently. We are always able to obtain from the rules of a formal logic a method of enumerating the propositions proved by its means. We then imagine that all proofs take the form of a search through this enumeration for the theorem for which a proof is desired. In this way ingenuity is replaced by patience. In these heuristic discussions, however, it is better not to make this reduction.

In consequence of the impossibility of finding a formal logic which wholly eliminates the necessity of using intuition, we naturally turn to "non-constructive" systems of logic with which not all the steps in a proof are mechanical, some being intuitive. An example of a non-constructive logic is afforded by any ordinal logic. When we have an ordinal logic, we are in a position to prove number-theoretic theorems by the intuitive steps of recognizing formulae as ordinal formulae, and the mechanical steps of carrying out conversions. What properties do we desire a non-constructive logic to have if we are to make use of it for the expression of mathematical proofs? We want it to show quite clearly when a step makes use of intuition, and when it is purely formal. The strain put on the intuition should be a minimum. Most important of all, it must be beyond all reasonable doubt that the logic leads to correct results whenever the intuitive steps are correct[†]. It is also desirable that the logic shall be adequate for the expression of number-theoretic theorems in order that it may be used in metamathematical discussions (cf. §5).

Of the particular ordinal logics that we have discussed, Λ_H and Λ_P certainly will not satisfy us. In the case of Λ_H we are in no better position than with a constructive logic. In the case of Λ_P (and for that matter also Λ_H) we are by no means certain that we shall never obtained any but true results, because we do not know whether all the number-theoretic theorems provable in the system P are true. To take Λ_P as a fundamental non-constructive logic for metamathematical arguments would be most unsound. There remains the system of Church which is free from these objections. It is probably complete (although this would not necessarily mean much) and it is beyond reasonable doubt that it always leads to correct results[‡]. In the next section I propose to describe another ordinal

[†] This requirement is very vague. It is not of course intended that the criterion of the correctness of the intuitive steps be the correctness of the final result. The meaning becomes clearer if each intuitive step is regarded as a judgment that a particular proposition is true. In the case of an ordinal logic it is always a judgment that a formula is an ordinal formula, and this is equivalent to judging that a number-theoretic proposition is true. In this case then the requirement is that the reputed ordinal logic *is* an ordinal logic.

[‡] This ordinal logic arises from a certain system C_0 in essentially the same way as Λ_P arose from P. By an argument similar to one occurring in §8 we can show that the ordinal logic leads to correct results if and only if C_0 is valid; the validity of C_0 is proved in Church [**1**], making use of the results of Church and Rosser [**1**].

logic, of a very different type, which is suggested by the work of Gentzen and which should also be adequate for the formalization of number-theoretic theorems. In particular it should be suitable for proofs of metamathematical theorems (cf. §5).

12. Gentzen type ordinal logics.

In proving the consistency of a certain system of formal logic Gentzen (Gentzen [**1**]) has made use of the principle of transfinite induction for ordinals less than ϵ_0, and has suggested that it is to be expected that transfinite induction carried sufficiently far would suffice to solve all problems of consistency. Another suggestion of basing systems of logic on transfinite induction has been made by Zermelo (Zermelo [**1**]). In this section I propose to show how this method of proof may be put into the form of a formal (non-constructive) logic, and afterwards to obtain from it an ordinal logic.

We can express the Gentzen method of proof formally in this way. Let us take the system P and adjoin to it an axiom $\mathfrak{A}_{\mathbf{\Omega}}$ with the intuitive meaning that the W.F.F. $\mathbf{\Omega}$ is an ordinal formula, whenever we feel certain that $\mathbf{\Omega}$ *is* an ordinal formula. This is a non-constructive system of logic which may easily be put into the form of an ordinal logic. By the method of §6 we make correspond to the system of logic consisting of P with the axiom $\mathfrak{A}_{\mathbf{\Omega}}$ adjoined a logic formula $L_{\mathbf{\Omega}}$: $L_{\mathbf{\Omega}}$ is an effectively calculable function of $\mathbf{\Omega}$, and there is therefore a formula Λ_{G^1} such that $\Lambda_{G^1}(\mathbf{\Omega})$ conv $L_{\mathbf{\Omega}}$ for each formula $\mathbf{\Omega}$. Λ_{G^1} is certainly not an ordinal logic unless P is valid, and therefore consistent. This formalization of Gentzen's idea would therefore not be applicable for the problem with which Gentzen himself was concerned, for he was proving the consistency of a system weaker than P. However, there are other ways in which the Gentzen method of proof can be formalized. I shall explain one, beginning by describing a certain logical calculus.

The symbols of the calculus are $f, x, {}^1, {}_1, 0, S, R, \Gamma, \Delta, E, |, \odot, !, (,), =$, and the comma ",". For clarity we shall use various sizes of brackets (,) in the following. We use capital German letters to stand for variable or undetermined sequences of these symbols.

It is to be understood that the relations that we are about to define hold only when compelled to do so by the conditions that we lay down. The conditions should be taken together as a simultaneous inductive definition of all the relations involved.

Suffixes.

$_1$ a *suffix*. If $\mathfrak{S}$ a suffix then $\mathfrak{S}_1$ is a suffix.

Indices.

^{1} is an index. If $\mathfrak{I}$ is an index then $\mathfrak{I}^1$ is an index.

Numerical variables.

If $\mathfrak{S}$ is a suffix then $x\mathfrak{S}$ is a numerical variable.

Functional variables.

If $\mathfrak{S}$ is a suffix and $\mathfrak{J}$ is an index, then $f\mathfrak{S}\mathfrak{J}$ a functional variable of index $\mathfrak{J}$.

Arguments.

(,) is an argument of index ^{1}. If $(\mathfrak{A})$ is an argument of index $\mathfrak{J}$ and $\mathfrak{T}$ is a term, then $(\mathfrak{A}\mathfrak{T},)$ is an argument of index $\mathfrak{J}^1$.

Numerals.

0 is a numeral.

If $\mathfrak{N}$ is a numeral, then $S(, \mathfrak{N},)$ is a numeral.

In metamathematical statements we shall denote the numeral in which S occurs r times by $S^{(r)}(, 0,)$.

Expressions of a given index.

A functional variable of index $\mathfrak{J}$ is an expression of index $\mathfrak{J}$.

R, S are expressions of index ^{111}, ^{11} respectively.

If $\mathfrak{N}$ is a numeral, then it is also an expression of index ^{1}.

Suppose that $\mathfrak{G}$ is an expression of index $\mathfrak{J}$, $\mathfrak{H}$ one of index $\mathfrak{J}^1$ and $\mathfrak{K}$ one of index $\mathfrak{J}^{111}$; then $(\Gamma\mathfrak{G})$ and $(\Delta\mathfrak{G})$ are expressions of index $\mathfrak{J}$, while $(E\mathfrak{G})$ and $(\mathfrak{G}|\mathfrak{H})$ and $(\mathfrak{G} \odot \mathfrak{K})$ and $(\mathfrak{G}!\mathfrak{H}!\mathfrak{K})$ are expressions of index $\mathfrak{J}^1$.

Function constants.

An expression of index $\mathfrak{J}$ in which no functional variable occurs is a function constant of index $\mathfrak{J}$. If in addition R does not occur, the expression is called a *primitive function constant.*

Terms.

0 is a term.

Every numerical variable is a term.

If $\mathfrak{G}$ an expression of index $\mathfrak{J}$ and $\mathfrak{A}$ is an argument of index $\mathfrak{J}$, then $\mathfrak{G}(A)$ is a term.

Equations.

If $\mathfrak{T}$ and $\mathfrak{T}'$ are terms, then $\mathfrak{T} = \mathfrak{T}'$ is an equation.

Provable equations.

We define what is meant by the provable equations relative to a given set of equations as axioms.

(a) The provable equations include all the axioms. The axioms are of the form of equations in which the symbols Γ, Δ, E, $|$, $\odot$, ! do not appear.

(b) If $\mathfrak{G}$ is an expression of index $\mathfrak{J}^{11}$ and $(\mathfrak{A})$ is an argument of index $\mathfrak{J}$, then

$$(\Gamma\mathfrak{G})(\mathfrak{G}x_1,x_{11},) = \mathfrak{G}(\mathfrak{A}x_{11},x_1,)$$

is a provable equation.

(c) If $\mathfrak{G}$ is an expression of index $\mathfrak{J}^1$, and $(\mathfrak{A})$ an argument of index $\mathfrak{J}$, then

$$(\Delta\mathfrak{G})(\mathfrak{A}x_1,) = \mathfrak{G}(,x_1\mathfrak{A})$$

is a provable equation.

(d) If $\mathfrak{G}$ is an expression of index $\mathfrak{J}$, and $(\mathfrak{A})$ is an argument of index $\mathfrak{J}$, then

$$(E\mathfrak{G})(\mathfrak{A}x_1,) = \mathfrak{G}(\mathfrak{A})$$

is a provable equation.

(e) If $\mathfrak{G}$ is an expression of index $\mathfrak{J}$ and $\mathfrak{H}$ is one of index $\mathfrak{J}^1$, and $(\mathfrak{A})$ is an argument of index $\mathfrak{J}$ then

$$(\mathfrak{G}|\mathfrak{H})(\mathfrak{A}) = \mathfrak{H}(\mathfrak{A}\mathfrak{G}(\mathfrak{A}),)$$

is a provable equation.

(f) If $\mathfrak{N}$ is an expression of index ^{1}, then $\mathfrak{N}(,) = \mathfrak{N}$ is a provable equation.

(g) If $\mathfrak{G}$ is an expression of index $\mathfrak{J}$ and $\mathfrak{K}$ one of index $\mathfrak{J}^{111}$, and $(\mathfrak{A})$ an argument of index $\mathfrak{J}^1$, then

$$(\mathfrak{G} \odot \mathfrak{K})(\mathfrak{A}0,) = \mathfrak{G}(\mathfrak{A})$$

and $$(\mathfrak{G} \odot \mathfrak{K})(\mathfrak{A}S(,x_1,),) = \mathfrak{K}(\mathfrak{A}x_1,S(,x_1,),(\mathfrak{G} \odot \mathfrak{K})\ (\mathfrak{A}x_1,),)$$

are provable equations. If in addition $\mathfrak{H}$ is an expression of index $\mathfrak{J}^1$ and

$$R(,\mathfrak{G}(\mathfrak{A}S(,x_1,),),x_1,)=0$$

is provable, then

$$(\mathfrak{G}!\mathfrak{K}!\mathfrak{H})(\mathfrak{A}0,)=\mathfrak{G}(\mathfrak{A})$$

and

$$\begin{aligned}&(\mathfrak{G}!\mathfrak{K}!\mathfrak{H})(\mathfrak{A}S(,x_1),)\\&\quad=\mathfrak{K}((\mathfrak{A}\mathfrak{H}(\mathfrak{A}S(,x_1,),),S(,x_1),(\mathfrak{G}!\mathfrak{K}!\mathfrak{H})(\mathfrak{A}\mathfrak{H}(\mathfrak{A}S(,x_1),),),)\end{aligned}$$

are provable.

(h) If $\mathfrak{T}=\mathfrak{T}'$ and $\mathfrak{U}=\mathfrak{U}'$ are provable, where $\mathfrak{T}$, $\mathfrak{T}'$, $\mathfrak{U}$ and $\mathfrak{U}'$ are terms, then $\mathfrak{U}'=\mathfrak{U}$ and the result of substituting $\mathfrak{U}'$ for $\mathfrak{U}$ at any particular occurrence in $\mathfrak{T}=\mathfrak{T}'$ are provable equations.

(i) The result of substituting any term for a particular numerical variable throughout a provable equation is provable.

(j) Suppose that $\mathfrak{G}$, $\mathfrak{G}'$ are expressions of index $\mathfrak{J}^1$, that $(\mathfrak{A})$ is an argument of index $\mathfrak{J}$ not containing the numerical variable $\mathfrak{X}$ and that $\mathfrak{G}(\mathfrak{A}0,)=\mathfrak{G}'(\mathfrak{A}0,)$ is provable. Also suppose that, if we add

$$\mathfrak{G}(\mathfrak{A}\mathfrak{X},)=\mathfrak{G}'(\mathfrak{A}\mathfrak{X},)$$

to the axioms and restrict (*i*) so that it can never be applied to the numerical variable $\mathfrak{X}$, then

$$\mathfrak{G}(\mathfrak{A}S(,\mathfrak{X},),)=\mathfrak{G}'(\mathfrak{A}S(,\mathfrak{X}),)$$

becomes a provable equation; in the hypothetical proof of this equation this rule (*j*) itself may be used provided that a different variable is chosen to take the part of $\mathfrak{X}$.

Under these conditions $\mathfrak{G}(\mathfrak{A}\mathfrak{X},)=\mathfrak{G}'(\mathfrak{A}\mathfrak{X}$, a provable equation.

(k) Suppose that $\mathfrak{G}$, $\mathfrak{G}'$, $\mathfrak{H}$ are expressions of index $\mathfrak{J}^1$, that $(\mathfrak{A})$ is an argument of index $\mathfrak{J}$ not containing the numerical variable $\mathfrak{X}$ and that

$$\mathfrak{G}(\mathfrak{A}0,)=\mathfrak{G}'(\mathfrak{A}0,)\,\text{and}\,R(,\mathfrak{H}(\mathfrak{A}S(,\mathbf{X},),),S(,\mathfrak{X}),)=0$$

are provable equations. Suppose also that, if we add

$$\mathfrak{G}(\mathfrak{A}\mathfrak{H}(\mathfrak{A}S(,\mathfrak{X},),))=\mathfrak{G}'(\mathfrak{A}\mathfrak{H}(\mathfrak{A}S(,\mathfrak{X},),))$$

to the axioms, and again restrict (*i*) so that it does not apply to $\mathfrak{X}$, then

$$\mathfrak{G}(\mathfrak{A}\mathfrak{X},)=\mathfrak{G}'(\mathfrak{A}\mathfrak{X},) \tag{12. 1}$$

becomes a provable equation; in the hypothetical proof of (12.1) the rule (*k*) may be used if a different variable takes the part of $\mathfrak{X}$.

Under these conditions (12.1) is a provable equation.

We have now completed the definition of a provable equation relative a given set of axioms. Next we shall show how to obtain an ordinal logic from this calculus. The first step is to set up a correspondence between some of the equations and number-theoretic theorems, in other words to show how they can be interpreted as number-theoretic theorems.

Let $\mathfrak{G}$ be a primitive function constant of index ^{111}. $\mathfrak{G}$ describes a certain primitive recursive function $\phi(m,n)$, determined by the condition that, for all natural numbers m,n, the equation

$$\mathfrak{G}(,S^{(m)}(,0,),S^{(n)}(,0,),) = S^{(\phi(m,n))}(,0,)$$

is provable without using the axioms (a). Suppose also that $\mathfrak{H}$ is an expression of index $\mathfrak{J}$. Then to the equation

$$\mathfrak{G}(,x_1,\mathfrak{H}(,x_1,),) = 0$$

we make correspond the number-theoretic theorem which asserts that for each natural number m there is a natural number n such that $\phi(m,n) = 0$. (The circumstance that there is more than one equation to represent each number-theoretic theorem could be avoided by a trivial but inconvenient modification of the calculus.)

Now let us suppose that some definite method is chosen for describing the sets of axioms by means of positive integers, the null set of axioms being described by the integer 1. By an argument used in §6 there is a W.F.F. Σ such that, if r the integer describing a set A of axioms, then $\Sigma(\mathbf{r})$ is a logic formula enabling us to prove just those number-theoretic theorems which are associated with equations provable with the above described calculus, the axioms being those described by the number r.

I explain two ways in which the construction of the ordinal logic may be completed.

In the first method we make use of the theory of general recursive functions (Kleene [**2**]). Let us consider all equations of the form

$$R(,S^{(m)}(,0,),S^{(n)}(,0,),) = S^{(p)}(,0,) \quad (12.2)$$

which are obtainable from the axioms by the use of rules (h), (i). It is a consequence of the theorem of equivalence of λ-definable and general recursive functions (Kleene [**3**]) that, if $r(m,n)$ is any λ-definable function of two variables, then we can choose the axioms so that (12.2) with $p = r(m,n)$ is obtainable in this way for each pair of natural numbers m,n, and no equation of the form

$$S^{(m)}(,0,) = S^{(n)}(,0,) \quad (m \neq n) \quad (12.3)$$

is obtainable. In particular, this is the case if $r(m,n)$ is defined by the condition that

$$\mathbf{\Omega}(\mathbf{m},\mathbf{n}) \text{ conv } S(\mathbf{p}) \text{ implies } p = r(m,n),$$
$$r(0,n) = 1, \quad \text{all} \quad n > 0 \quad r(0,0) = 2,$$

where $\mathbf{\Omega}$ is an original formula. There is a method for obtaining the axioms given the ordinal formula, and consequently a formula Rec such that, for any ordinal formula $\mathbf{\Omega}$, Rec $(\mathbf{\Omega})$ conv $\mathbf{m}$, where m is the integer describing the set of axioms corresponding to $\mathbf{\Omega}$. Then the formula

$$\Lambda_{G^2} \to \lambda w.\Sigma(\text{Rec}(w))$$

is an ordinal logic. Let us leave the proof of this aside for the present.

Our second ordinal logic is to be constructed by a method not unlike the one which we used in constructing Λ_P. We begin by assigning ordinal formulae to all sets of axioms satisfying certain conditions. For this purpose we again consider that part of the calculus which is obtained by restricting "expressions" to be functional variables or R or S and restricting the meaning of "term" accordingly; the new provable equations are given by conditions (a), (h), (i), together with an extra condition (l).

(l) The equation

$$R(,0,S(,x_1,),)=0$$

is provable.

We could design a machine which would obtain all equations of the form (12.2), with $m \neq n$, provable in this sense, and all of the form (12.3), except that it would cease to obtain any more equations when it had once obtained one of the latter "contradictory" equations. From the description of the machine we obtain a formula $\mathbf{\Omega}$ such that

$$\mathbf{\Omega}(\mathbf{m},\mathbf{n}) \text{ conv } 2 \quad \text{if} \quad R(,S^{(m-1)}(,0,),S^{(n-1)}(,0,),)=0$$

is obtained by the machine,

$$\mathbf{\Omega}(\mathbf{m},\mathbf{n}) \text{ conv } 1 \quad \text{if} \quad R(,S^{(\mathrm{n}-1)}(0,),S^{(m-1)}(,0,),)=0$$

is obtained by the machine, and

$$\mathbf{\Omega}(\mathbf{m},\mathbf{m}) \text{ conv } 3 \text{ always.}$$

The formula $\mathbf{\Omega}$ is an effectively calculable function of the set of axioms, and therefore also of m: consequently there is a formula M such that $M(\mathbf{m})$ conv $\mathbf{\Omega}$ when m describes the set of axioms. Now let Cm be a formula such that, if b is the G.R. of a formula $M(\mathbf{m})$, then Cm (b) conv $\mathbf{m}$, but otherwise Cm (b) conv 1. Let

$$\Lambda_{G^3} \rightarrow \lambda wa.\Gamma(\lambda n.\Sigma(\mathrm{Cm}(\mathrm{Tn}(w,n)),a).$$

Then $\Lambda_{G^3}(\mathbf{\Omega},\mathbf{A})$ conv 2 if and only if $\mathbf{\Omega}$ conv $M(\mathbf{m})$, where m describes a set of axioms which, taken with our calculus, suffices to prove the equation which is, roughly speaking, equivalent to "**A** is dual". To prove that Λ_{G^3} is an ordinal logic, it is sufficient to prove that the calculus with the axioms described by m proves only true number-theoretic theorems when $\mathbf{\Omega}$ is an ordinal formula. This condition on m may also be expressed in this way. Let us put $m \ll n$ if we can prove $R(,S^{(m)}(,0,),S^{(n)}(,0,))=0$ with $(a),(h),(i),(l)$: the condition is that $m \ll n$ is a well-ordering of the natural numbers and that no contradictory equation (12.3) is provable with the same rules $(a),(h)$, (i), (l). Let us say that such a set of axioms is *admissible*. Λ_{G^3} is an ordinal logic if the calculus leads to none but true number-theoretic theorems when an admissible set of axioms is used.

In the case of Λ_{G^2}, Rec $(\mathbf{\Omega})$ describes an admissible set of axioms whenever $\mathbf{\Omega}$ is an ordinal formula. Λ_{G^2} therefore is an ordinal logic if the calculus leads to correct results when admissible axioms are used.

To prove that admissible axioms have the required property, I do not attempt to do more than show how interpretations can be given to the equations of the calculus so that the rules of inference $(a)-(k)$ become intuitively valid methods of deduction, and so that the interpretation agrees with our convention regarding number-theoretic theorems.

Each expression is the name of a function, which may be only partially defined. The expression S corresponds simply to the successor function. If $\mathfrak{G}$ is either R or a functional variable and has $p+1$ symbols in its index, then it corresponds to a function g of p natural numbers defined as follows. If

$$\mathfrak{G}(,S^{(r_1)}(,0,),S^{(r_2)}(,0,),\ldots,S^{(r_p)}(,0,),)=S^{(l)}(,0,)$$

is provable by the use of (a), (h), (i), (1) only, then $g(r_1,r_2,\ldots,r_p)$ has the value p. It may not be defined for all arguments, but its value is always unique, for otherwise we could prove a "contradictory" equation and $M(\mathbf{m})$ would then not be an ordinal formula. The functions corresponding to the other expressions are essentially defined by $(b)-(f)$. For example, if g is the function corresponding to $\mathfrak{G}$ and g' that corresponding to $(\Gamma\mathfrak{G})$, then

$$g'(r_1,r_2,\ldots,r_p,l,m)=g(r_1,r_2,\ldots,r_p,m,l).$$

The values of the functions are clearly unique (when defined at all) if given by one of $(b)-(e)$. The case (f) is less obvious since the function defined appears also in the definiens. I do not treat the case of $(\mathfrak{G}\odot\mathfrak{K})$, since this is the well-known definition by primitive recursion, but I shall show that the values of the function corresponding to $(\mathfrak{G}!\mathfrak{K}!\mathfrak{H})$ are unique. Without loss of generality we may suppose that $(\mathfrak{A})$ in (f) is of index ^{1}. We have then to show that, if $h(m)$ is the function corresponding to $\mathfrak{H}$ and $r(m,n)$ that corresponding to R, and $k(u,v,w)$ is a given function and a a given natural number, then the equations

$$l(0)=a, \qquad (\alpha)$$

$$l(m+1)=k(h(m+1),m+1,l(h(m+1))) \qquad (\beta)$$

do not ever assign two different values for the function $l(m)$. Consider those values of r for which we obtain more than one value of $l(r)$, and suppose that there is at least one such. Clearly 0 is not one, for $l(0)$ can be defined only by (a). Since the relation $\ll$ is a well ordering, there is an integer r_0 such that $r_0>0, l(r_0)$ is not unique, and if $s\neq r_0$ and $l(s)$ is not unique then $r_0\ll s$. We may put $s=h(r_0)$, for, if $l(h(r_0))$ were unique, then $l(r_0)$, defined by (β), would be unique. But $r(h(r_0),r_0)=0$ i.e. $s\ll r_0$. There is, therefore, no integer r for which we obtain more than one value for the function $l(r)$.

Our interpretation of expressions as functions gives us animmediate interpretation for equations with no numerical variables. In general we interpret an equation with numerical variables as the (infinite) conjunction of all equations obtainable by replacing the variables by numerals. With this interpretation (h), (i) are seen to be valid methods of proof. In (j) the provability of

$$\mathfrak{G}(\mathfrak{A}S(,x_1,),)=\mathfrak{G}'(\mathfrak{A}S(,x_1,),)$$

when $\mathfrak{G}(\mathfrak{A}x_1,)=\mathfrak{G}'(\mathfrak{A}x_1,)$ is assumed to be interpreted as meaning that the implication between these equations holds for all substitutions of numerals for x_1. To justify this, one should satisfy oneself that these implications always hold when the hypothetical proof can be carried out. The rule of procedure (j) is now seen to be simply mathematical induction. The rule (k) is a form of transfinite induction. In proving the validity of (k) we may again suppose $(\mathfrak{A})$ is of index ^{1}. Let $r(m,n), g(m), g_1(m), h(n)$ be the functions corresponding respectively to $R,\mathfrak{G},\mathfrak{G}',\mathfrak{H}$. We shall prove that, if $g(0)=g'(0)$ and $r(h(n),n)=0$ for each positive integer n and if $g(n+1)=g'(n+1)$ whenever $g(h(n+1))=g'(h(n+1))$, then $g(n)=g'(n)$ for each natural number n. We consider the class of natural numbers for which $g(n)=g'(n)$ is not true. If the class is not void it has a positive member n_0 which precedes all other members in the well ordering $\ll$. But $h(n_0)$ is another member of the class, for otherwise we should have

$$g(h(n_0))=g'(h(n_0))$$

and therefore $g(n_0)=g'(n_0)$, i.e. n_0 would not be in the class. This implies $n_0\ll h(n_0)$ contrary to $r(h(n_0),n_0)=0$. The class is therefore void.

It should be noticed that we do not really need to make use of the fact that $\mathbf{\Omega}$ is an ordinal formula. It suffices that $\mathbf{\Omega}$ should satisfy conditions $(a)-(e)$ (p. 163) for ordinal formulae, and in place of (f) satisfy (f').

(f') There is no formula $\mathbf{T}$ such that $\mathbf{T}(\mathbf{n})$ is convertible to a formula representing a positive integer for each positive integer n, and such that $\mathbf{\Omega}(\mathbf{T}(\mathbf{n}),\mathbf{n})$ conv 2, for each positive integer n for which $\mathbf{\Omega}(\mathbf{n},\mathbf{n})$ conv 3.

The problem whether a formula satisfies conditions $(a) - (e)$, (f') is number-theoretic. If we use formulae satisfying these conditions instead of ordinal formulae with Λ_{G^2} or Λ_{G^3}, we have a non-constructive logic with certain advantages over ordinal logics. The intuitive judgments that must be made are all judgments of the truth of number theoretic-theorems. We have seen in §9 that the connection of ordinal logics with the classical theory of ordinals is quite superficial. There seem to be good reasons, therefore, for giving attention to ordinal formulae in this modified sense.

The ordinal logic Λ_{G^3} appears to be adequate for most purposes. It should, for instance, be possible to carry out Gentzen's proof of consistency of number theory, or the proof of the uniqueness of the normal form of a well-formed formula (Church and Rosser [**1**]) with our calculus and a fairly simple set of axioms. How far this is the case can, of course, only be determined by experiment.

One would prefer a non-constructive system of logic based on transfinite induction rather simpler than the system which we have described. In particular, it would seem that it should be possible to eliminate the necessity of stating explicitly the validity of definitions by primitive recursions, since this principle itself can be shown to be valid by transfinite induction. It is possible to make such modifications in the system, even in such a way that the resulting system is still complete, but no real advantage is gained by doing so. The effect is always, so far as I know, to restrict the class of formulae provable with a given set of axioms, so that we obtain no theorems but trivial restatements of the axioms. We have therefore to compromise between simplicity and comprehensiveness.

Index of definitions.

No attempt is being made to list heavy type formulae since their meanings are not always constant throughout the paper. Abbreviations for definite well-formed formulae are listed alphabetically.

Miscellaneous (in order of appearance).

Bibliography.

Alonzo Church, [1]. "A proof of freedom from contradiction", *Prec. Nat. A cad. Sci.*, 21 (1935), 275–281.

——, [2]. *Mathematical logic*, Lectures at Princeton University (1935–6), mimeographed, 113 pp.

——, [3]. "An unsolvable problem of elementary number theory", *American J. of Math.,* 58 (1936), 345–363.

——, [4]. "The constructive second number class", *Bull. American Math. Soc.*, 44 (1938), 224–238.

G. Gentzen, [i]. "Die Widerspruchsfreiheit der reinen Zahlentheorie", *Math. Annalen,* 112 (1936), 493–565.

K. Gödel, [1]. "Über formal unentscheidbare Sätze der Principia Mathematica und verwandter Systeme, I", *Monatshefte für Math. und Phys.,* 38 (1931), 173–189.

——, [2]. *On undecidable propositions of formal mathematical systems*, Lectures at the Institute for Advanced Study, Princeton, N.J., 1934, mimeographed, 30 pp.

D. Hilbert, [1]. "Über das Unendliche", *Math. Annalen*, 95 (1926), 161–190.

S. C. Kleene, [1]. "A theory of positive integers in formal logic", *American J. of Math.*, 57 (1935), 153–173 and 219–244.

——, [2]. "General recursive functions of natural numbers", *Math. Annalen*, 112 (1935–6), 727–742.

——, [3] "λ-definability and recursiveness", *Duke Math. Jour.*, 2 (1936), 340–353.

E. L. Post, [1]. "Finite combinatory processes—formulation 1", *Journal Symbolic Logic*, 1 (1938), 103–105.

J. B. Rosser, [1]. "Gödel theorems for non-constructive logic", *Journal Symbolic Logic*, 2 (1937), 129–137.

A. Tarski, [1]. "Der Wahrheitsbegriff in den formalisierten Sprachen", *Studia Philoso. phica*, 1 (1936), 261–405 (translation from the original paper in Polish dated 1933).

A. M. Turing, [1]. "On computable numbers, with an application to the Entscheidungsproblem", *Proc. London Math. Soc.* 2, (42) (1937), 230–265. A correction to this paper appeared in the same periodical, 43 (1937), 544–546.

——, [2]. "Computability and λ-definability", *Journal Symbolic Logic*, 2 (1937), 153–163.

E. Zermelo, [1]. "Grundlagen einer allgemeiner Theorie der mathematischen Satzsysteme, I", *Fund. Math.,* 25 (1935), 136–146.

Alonzo Church and S. C. Kleene, [1]. "Formal definitions in the theory of ordinal numbers", *Fund. Math.*, 28 (1936), 11–21.

Alonzo Church and J. B. Rosser, [1]. "Some properties of conversion", *Math. Soc.*, 39 (1936), 472–482.

D. Hilbert and W. Ackermann, [1]. *Grundzüge der theoretischen Logik*(2nd edition revised, Berlin, 1938), 130 pp.

A. N. Whitehead and Bertrand Russell, [1]. *Principia Mathematica*(2nd edition, Cambridge, 1925–1927), 3 vols.

King's College
Cambridge

Examining the Work and Its Later Impact

Michael Rathjen looks at —

TURING'S 'ORACLE' IN PROOF THEORY

Turing's paper on ordinal logics (Turing, 1939) greatly influenced research in proof theory in the 1960s, especially in Feferman's (1964, 1968) and Schütte's (1964, 1965) work on the limits of predicativity. His idea to overcome Gödel's incompleteness results by means of a hierarchy of introspective theories is a very natural one.[1] The essential difference between Turing's ordinal logics and the proof-theoretic hierarchies, though, is that the latter concern autonomous progressions of theories. The most direct influence of Turing (1939) was on Feferman's extension of it in his paper (Feferman, 1962) on non-autonomous transfinite recursive progressions of axiomatic theories.[2] Notably the idea of autonomous progressions holds a great deal of attraction and has seen renewed interest in discussions on FOM (an automated e-mail list for discussing foundations of mathematics) of the scope of predicative theories.[3]

It is perhaps less well-known that Turing's oracle computations, which he introduced in the brief Section 4 of Turing (1939), played a central role in another part of proof theory. The main passage from Turing (1939) reads as follows:

> "Let us suppose that we are supplied with some means of solving number-theoretic problems; a kind of oracle as it were ... With the help of the oracle we could form a new kind of machine (call them *o*-machines), having as one of its processes that of solving a given number-theoretic problem".

Much later in the work of Goodman (1976), the oracle was going to be used in showing that the axiom of choice can be eliminated from proofs of arithmetic statements in intuitionistic higher order theories. By the same token, it can be put to use in proving that the axiom of dependent choices can be removed from proofs of arithmetic statements in a number of intuitionistic set theories. This paper is aimed at giving a brief presentation of the ideas leading to these results.

1. Realisability

In 1930, the nature of intuitionism was greatly clarified when Heyting published a formalisation of intuitionistic predicate logic and intuitionistic arithmetic (later christened Heyting Arithmetic, **HA**). A few years later in 1933, Gentzen and Gödel independently provided translations which illuminated the relationship between classical and intuitionistic arithmetic.[4] Their so-called negative translations showed that, in a sense, Peano Arithmetic is contained in **HA**, and, moreover, that for

[1] This is discussed in Sol Feferman's contribution to this volume.

[2] More recently, Nash seems to be concerned with just non-autonomous progressions of theories.

[3] For example, Nik Weaver challenged the 'traditional' view on FOM and in his work Weaver (2005).

[4] Apparently unbeknownst to Heyting, Gentzen, and Gödel, Kolmogorov (1925) had previously given a formalisation (albeit an incomplete one) of intuitionistic logic and observed the translatability of classical into intuitionistic logic.

formulas not containing $\vee$ or $\exists$, provability in **PA** and **HA** amount to the same.[5] One moral drawable from this result is that (at least for the realm of arithmetic) rather than being a restriction, Brouwer's intuitionism turns out to be an extension of classical logic brought about by adding the constructive $\exists$ and $\vee$. Surprisingly, though none of this touched on the meaning of intuitionistic implication, negation and $\forall$,[6] which were still a matter of great concern to Dummett (1975, 2000) and thought to be infelicitous by Bishop and Bridges (1985, p. 13). It appears that the nature of $\rightarrow, \neg, \forall$ was not viewed as problematic at all in the early days of intuitionism. The most common explanation of the intuitionistic meaning of the logical connectives is the Brouwer–Heyting–Kolmogorov explanation (BHK for short). Whereas BHK gives a valuable heuristics for the meaning of the connectives in terms of constructions and is particularly good at explicating $\exists$ and $\vee$, on closer inspection it provides at best an interpretation of $\rightarrow, \neg, \forall$ by means of an as yet unexplained or primitive notion of construction and at worst resolves into circularity.[7] Thus, a deep contribution to the enlightenment period of intuitionism, which started with Kolmogorov, is owed to Kleene (1945), who gave a semantics or model for intuitionistic knowledge of a closed formula A of arithmetic in terms of number codes that 'realise' A. In the case of an implication or universally quantified formula, these realisers can be identified with (codes of) Turing machines. The definition of realisability is by induction on the complexity of A:

A realiser of	**has the form**
A atomic	any e providing A is true.
$A \wedge B$	(a,b), where a is a realiser of A and b is a realiser of B.
$A \rightarrow B$	e, where e is the Gödel number of a Turing machine M_e such that M_e halts with a realiser for B whenever a realiser of A is run on M_e.
$\neg A$	any e providing there is **no** realiser for A.
$A \vee B$	$(0,a)$, where a is a realiser of A, or $(1,b)$, where b is a realiser of B
$\forall x B(x)$	e, where e is a Gödel number of a Turing machine M_e such that M_e outputs a realiser for $A(\bar{n})$ when run on n.
$\exists x B(x)$	(n,b), where b is a realiser of $B(\bar{n})$.

Here (a,b) is some standard coding of pairs of natural numbers and $\bar{n}$ is the standard numeral corresponding to n.

There are now many different notions of realisability. They have become the most plentiful source of models for intuitionistic theories, ranging from arithmetic to higher type systems and set theories.

2. Heyting arithmetic in higher types

Hilbert in his paper *Über das Unendliche* from 1925 considered a hierarchy of functionals over the natural numbers, not only of finite but also of transfinite type.[8] The finite levels of this hierarchy

[5] Kolmogorov already in 1925 drew from this the conclusion that, contrary to Brouwer's views on the matter, a finitary statement proved by classical means is intuitionistically true.

[6] As witnessed by Gentzen's negative translation that leaves these particles undisturbed.

[7] Takeuti (1987, p.101) deprecated it as impredicative.

[8] Intriguingly, Hilbert (1926) also defined dependent types, thereby introducing the germinal idea of Martin-Löf type theory.
I owe this observation to Peter Hancock.

where used by Gödel (1933) to give an interpretation (known as functional or Dialectica interpretation) of first order arithmetic. The finite types are inductively defined by starting with the type o of natural numbers and the rule that given types σ and τ, $\sigma(\tau)$ is a type, too. Here $\sigma(\tau)$ is the type of functions from objects of type σ to objects of type τ. The objects of type $\neq o$ are addressed as functionals. $\mathbf{HA}^{\omega}$ denotes the extension of **HA** by variables $x^{\sigma}, y^{\sigma}, \ldots$ for each finite type σ as well as constants for special functionals together with their defining axioms. Included among the latter are recursor functionals R_{σ} for all finite types σ which allow one to define functionals by recursion on $\mathbb{N}$. Moreover, the schema of mathematical induction is extended to all formulae of the language. It is also interesting to add choice principles for all type levels to $\mathbf{HA}^{\omega}$:

$$(\mathbf{AC}_{\sigma\tau})\ \forall x^{\sigma}\, \exists y^{\tau}\, A(x^{\sigma}, y^{\tau}) \rightarrow \exists f^{\sigma(\tau)}\, \forall x^{\sigma}\, A(x^{\sigma}, f^{\sigma(\tau)}(x^{\sigma})). \tag{2.1}$$

Let $\mathbf{AC}_{FT} := \{\mathbf{AC}_{\sigma\tau} \mid \sigma, \tau \text{ types}\}$.

Instead of **HA** one could also extend the classical theory **PA** to a version $\mathbf{PA}^{\omega}$ with higher type functionals. The theory $\mathbf{PA}^{\omega} + \mathbf{AC}_{FT}$ is much stronger than **PA**. With aid of a realisability interpretation, however, one can show that $\mathbf{HA}^{\omega} + \mathbf{AC}_{FT}$ is not stronger than **HA**, but a much more difficult question remains: Does $\mathbf{HA}^{\omega} + \mathbf{AC}_{FT}$ prove more statements of arithmetic than **HA**? The answer was given by Goodman.

Theorem 1. *(Goodman, 1976, 1978).* $\mathbf{HA}^{\omega} + \mathbf{AC}_{FT}$ *is conservative over* $\mathbf{HA}^{\omega}$.

The proof used Goodman's 'theory of constructions", and was rather long and involved (Goodman, 1976). His second proof is more direct and also conceptually clearer (Goodman, 1978). It combines the ideas of forcing and realisability. The technology was then used by Beeson (1979, 1985) and Gordeev (1988), and in more recent times by Ray-Ming Chen and the author of this note to establish a plethora of conservativity results.

3. Realisability relative to an oracle

Realisability for **HA** is not co-extensive with deducibility in **HA**. Whilst all theorems of **HA** are realisable, there are realisable sentences which have no proof in **HA**. In order to ensure conservativity results, one needs an abstract form of realisability that entails deducibility. The two steps of Goodman's second proof have been neatly separated by Beeson to construct a general methodology for showing an intuitionistic theory T to be conservative over another theory S for arithmetic statements. The idea is to find a sequence of interpretations:

$$T \xrightarrow[\textit{realisability}]{} S_{\mathcal{O}} \xrightarrow[\textit{forcing}]{} S$$

It is worth pointing out that the realisability interpretation of T in $S_{\mathcal{O}}$ is very similar to Kleene's realisability by numbers as we have defined above, but instead of being based on ordinary Turing machines it uses oracle Turing machines, where the oracle $\mathcal{O}$ is a fixed partial function from $\mathbb{N}$ to $\{0,1\}$. Thus, as Beeson remarked, it could have been introduced by Kleene in 1945. In the course of a computation the oracle may be consulted about the value of $\mathcal{O}(n)$ for some n. If $\mathcal{O}(n)$ is defined it will return that value and the computation will continue, but if $\mathcal{O}(n)$ is not defined no response will be coming forward and the computation will never come to a halt. The theory $S_{\mathcal{O}}$ results from S by adding a constant $\mathcal{O}$ to the language of S together with an axiom expressing that $\mathcal{O}$ is a partial function from $\mathbb{N}$ to $\{0,1\}$, but no specifics about $\mathcal{O}$. The idea of the second interpretation step is that on account of $\mathcal{O}$'s arbitrariness, forcing can be used in the background theory S to interpret the constant by a generic partial function. Given an arithmetic statement A a partial function ψ can be engineered so that in the forcing model realisability of A entails the truth of A. The final step, then, is achieved by noticing that for arithmetic statements forceability (where the forcing conditions are finite partial functions on $\mathbb{N}$) and validity coincide.

As an application of this technology one can show that Constructive Zermelo Fraenkel Set Theory, **CZF**, augmented by the axiom of dependent choices, **DC**, is conservative over **CZF** with respect to arithmetic sentences (cf. Gordeev (1988); Chen and Rathjen) and that the same obtains for Intuitionistic Zermelo Fraenkel Set Theory, **IZF**, when separation is restricted to bounded formulae (Chen and Rathjen). Another result obtainable in this way is that (full) **IZF** extended by the Uniformity Principle

$$(\mathbf{UP}) \quad \forall x \exists n \in \mathbb{N} \varphi(x,n) \to \exists n \in \mathbb{N} \forall x \varphi(x,n)$$

remains conservative over **IZF** with respect to arithmetic sentences (cf. Chen (2010); Chen and Rathjen). **UP** roughly asserts that all (class) functions from the universe of sets into the set of natural numbers are constant.

Since for classical **ZF** arithmetical conservativity of $\mathbf{ZF}+\mathbf{DC}$ over **ZF** is an immediate consequence of the fact that L is a model of **ZF** one might wonder why this method does not carry over to the intuituitionistic setting. The answer is that albeit L can be defined in the same way in the latter setting, the ordinals cannot be shown to be linearly ordered, rendering L a rather useless construction.

References

Beeson, M., 1979. Goodman's theorem and beyond. Pacific J. Math. 84.

Beeson, M., 1985. Foundations of Constructive Mathematics, Springer, Berlin.

Bishop, E., Bridges, D., 1985. Constructive Analysis. Springer, Berlin.

Chen, R.-M., 2010. Independence and conservativity results for intuitionistic set theory, Ph.D. Thesis, University of Leeds.

Chen, R.-M., Rathjen, M., Conservativity results for intuitionistic set theories, in preparation.

Dummett, M., 1975. The philosophical basis of intuitionistic logic. In: Rose, H.E., Shepherdson, J.C. (Eds.), Logic Colloquium 1973, North-Holland, Amsterdam.

Dummett, M., 2000. Elements of Intuitionism. Second edn. Clarendon Press, Oxford.

Feferman, S., 1962. Transfinite recursive progressions of axiomatic theories. J. Symbol. Logic 27, 259–316.

Feferman, S., 1964. Systems of predicative analysis. J. Symbolic Logic 29, 1–30.

Feferman, S. 1968. Autonomous transfinite progressions and the extent of predicative mathematics. In: Logic, Methodology, and Philosophy of Science III, Proc. 3rd Internat. Congr., Amsterdam, 1967, North-Holland, Amsterdam, pp. 121–135.

Gentzen, G., 1974. Über das Verhaltnis zwischen intuitionistischer und klassischer Logik (1933) Originally to appear in the Mathematische Annalen, reached the stage of galley proofs but was withdrawn. It was finally published in Arch. ML 16, 119–132.

Gödel, K., 1933. Zur intuitionistischen Arithmetik und Zahlentheorie, Ergebnisse eines mathematischen Kolloquiums 4, 34–38.

Goodman, N., 1976. The theory of the Gödel functionals. J. Symbolic Logic 41, 574–583.

Goodman, N., 1978. Relativized realizability in intuitionistic arithmetic of all finite types. J. Symbolic Logic 43.

Gordeev, L., 1988. Proof-theoretic analysis: Weak systems of functions and classes. Ann. Pure Appl. Log. 38, 1–121.

Hilbert, D., 1926. Über das Unendliche. Mathematische Annalen, 143–151.

Kleene, S.C., 1945. On the interpretation of intuitionistic number theory. J. Symbolic Logic 10, 109–124.

Kolmogorov, A.N., 1925. On the principle of the excluded middle (Russian). Mat. Sb. 32, 646–667 (translated in van Heijenoort (1967) 414–437).

Nash Jr., A.N., Hierarchical Introspective Logics, www.math.princeton.edu/jfnj/texts_and_graphics/Main.Content/Various_Etc./Logic/talk.CMU/HILdos38.txt.

Schütte, K., 1964. Eine Grenze für die Beweisbarkeit der transfiniten Induktion in der verzweigten Typenlogik, Archiv für Mathematische Logik und Grundlagenforschung 67, 45–60.

Schütte, K., 1965. Predicative well-orderings. In: Crossley, J.N., Dummet, M.A.E., (Eds.). Formal systems and recursive functions, Proceedings of the Eighth Logic Colloqium, Oxford, July 1963. North Holland, Amsterdam, pp.176–184.

Takeuti, G. 1987. Proof Theory, second edn. North Holland, Amsterdam.

Turing, A.M. 1939. Systems of logic based on ordinals. Proc. Lond. Math. Soc. 2, 161–228.

van Heijenoort, J. (Eds.) 1967. From Frege to Gödel. A Source Book in Mathematical Logic 1879–1931, Harvard University Press, Cambridge Mass, (Reprinted 1970).

Weaver, N. Predicativity Beyond Γ_0, ArXiv Mathematics e-prints, arXiv:math/0509244, 38pages.

Philip Welch takes a set-theoretical view of —

TRUTH AND TURING

We should like to link Turing's construction in *Systems of logic based on ordinals* on progressions of theories, with some recent similar looking progressions of axiomatisations of truth sets. However, we first set the scene by sketching the original paper. It is interesting for two fundamental reasons. Firstly he introduces, in a rather understated fashion, the notion of a variant of his original Turing Machine, which was to be the *'o-machine'* for 'oracle-machine.' The latter is the well-known version of the basic machine, the *'a-machine'*, introduced in his 1936 paper *'On computable numbers'*. The a-machine is of course the standard Turing machine equipped with an oracle tape. In the paper, Turing describes rather a program that is allowed input at a stage of the computation when a special instruction is reached to ask for such input from the oracle tape. He envisaged then that in this way 'non-computable' functions could be introduced by calling for values. In the paper, after introducing this idea, he then repeats the argument that the halting problem was undecidable by such machines. He called this the 'circularity question': whether a particular TM $\mathcal{M}$ would eventually loop on a particular input. [I shall use TM to abbreviate Turing machines (with or without oracle tapes).]

Just as the a-machine became the standard model for a computer (in Turing's terms) so the o-machine has become for us the standard model for *relativised computability*: the notion that a set $A \subseteq N$ can be computed 'relative to a set $B \subseteq N$' is that membership questions as to whether $n \in A$ or not can be 'reduced' to finitely many similar queries of the set B, where we imagine the oracle tape of the machine to have the characteristic function of B written out as a series of 0's and 1's. We write nowadays in this case '$A \leq_T B$' for this relation. Sets A, B of numbers equivalent under $\leq_T$ are then declared to be in the same 'Turing degree' of incomputability. Thus, the whole theory of such algorithmic degrees can be effected using this model.

This however only occupies a page and a half. This is not what the paper is about. It is only a tool in his investigation of the second fundamental idea to emerge from the paper: the notion of an 'ordinal progression'. One has to admire the sweep of the paper: merely eight years after Gödel's paper on the Incompleteness Phenomenon, and only three years after his own paper *On Computable Numbers* he attempted to grapple with the incompleteness phenomena of formal systems by systematically extending theories $T = T_0 \subseteq T_1 \subseteq \cdots$ by adding at each stage a *consistency* statement about the preceding theory. The assumption is that our acceptance of a theory T somehow also impels us to accept its consistency. Who would work in Peano Arithmetic (PA) if they believed Con(PA) was false? And of course it is the consistency statement 'Con(PA)' that Gödel showed was a statement unprovable in PA (assuming that it *was* itself consistent).

Martin Davis refers to the paper in his introduction in a volume of collected sources as 'difficult' and in several ways it is: the ideas are not immediately transparent; the notation sticks with that of Church's λ-calculus (under whom Turing was at this time writing his Ph.D. thesis, which contained this research); the underlying extensions take place along a system of notations, related to one devised by Kleene. We now would use a system called 'Kleene's $\mathcal{O}$', but again the language is different: instead of asking whether a certain integer n can be seen to be in $\mathcal{O}$, Turing asks whether a certain formula is an 'ordinal formula': the latter are formulae used to name (what will be) constructive ordinals, and there is a list of seven conditions in terms of λ-conversion for them. He also gives a definition of 'C(hurch)–K(leene) ordinal formulae', which contain in essence a definition (equivalent to that) of $\mathcal{O}$. Discussing this in today's notation we have the following definition (where suc(n) can be taken to be 2^n and lim(n) to be 3^n):

Definition 1. *By simultaneous recursion we define '$n \in \mathcal{O}$' and '$n <_{\mathcal{O}} m$' for $n, m \in N$ together with an ordinal $|n|$ for each $n \in \mathcal{O}$:*

$0 \in \mathcal{O}$ *and* $|0| = 0$;
If $n \in \mathcal{O}$, *then* $suc(n) \in \mathcal{O}$, $n <_{\mathcal{O}} suc(n)$ *and* $|suc(n)| = n+1$;
If $\{e\}$ *is an index of a total recursive function, and* $\forall n(\{e\}(n) <_{\mathcal{O}} \{e\}(n+1))$ *then* $\lim(e) \in \mathcal{O}$, $\{e\}(n) <_{\mathcal{O}} \lim(e)$ *for every n, and* $|\lim(e)| = \sup\{|\{e\}(n)| : n \in N\}$.
If $n <_{\mathcal{O}} m \wedge m <_{\mathcal{O}} p \longrightarrow n <_{\mathcal{O}} p$.

By this means *notations* can be assigned to any *constructive ordinal*: that is any ordinal less than the first non-recursive ordinal ω_1^{ck}, with $n <_{\mathcal{O}} m \longrightarrow |n| < |m|$ (but not conversely). However, the relation '$n \in \mathcal{O}$' is complex being necessarily Π_1^1. A totally ordered subset of Field($<_{\mathcal{O}}$) is a *path* and the restriction of $<_{\mathcal{O}}$ to a path of the form $\{n : n <_{\mathcal{O}} m\}$ allows us to see that the latter set is actually recursively enumerable. Kleene's $\mathcal{O}$ then gives us a constructive framework to which we may attach objects, in this case *theories*.

Definition 2. *A consistency progression based on a theory* T *is a primitive recursive mapping* $n \longrightarrow \varphi_n$, *where* $\varphi_n(v_0)$ *is a* Σ_1 *formula that defines* T_n *and that PA proves: (i)* $T_0 = T$; *(ii)* $\forall n(T_{suc(n)} = T_n + Con(\varphi_n))$; *(iii)* $T_{\lim(n)} = \bigcup_m T_{\{n\}(m)}$.

Definition 3. *A progressive (consistency) sequence is then the restriction of a consistency progression to a path through* $\mathcal{O}$.

The existence of progressive sequences along paths has to be justified through the use of the Recursion Theorem. With these tools, Turing proved a form of Completeness Theorem.

Theorem 1. *(Turing's Completeness Theorem) For any true* Π_1 *sentence of arithmetic,* σ, *there is an* $a = a(\sigma) \in \mathcal{O}$ *with* $|a| = \omega + 1$, *so that* $T_a \vdash \sigma$. *The map* $\sigma \rightarrowtail a(\sigma)$ *is given by a primitive recursive function.*

Thus, we may for any true σ find a path of length $\omega+1$, $T = T_0, T_1, \ldots, T_{\omega+1} = T_a$ with the last proving σ. At first glance it looks as if Turing's theorem is giving us an insight into mathematical knowledge, but this is illusory. There is a trick here: what one does is construct for *any* Π_1 sentence σ an extension $T_{a(\sigma)}$ proving σ with $|a(\sigma)| = \omega + 1$; then *if* σ *is true* we deduce that $T_{a(\sigma)}$ is a consistency extension. The set $\mathcal{O}$ is, as we have remarked, a complex set of numbers, and the argument draws on this complexity.

In the paper, Turing stated that he had tried to prove a theorem for statements at the level he called that of 'number theoretic problems', which in effect are those expressible as Π_2 sentences. He expressed the hope that this might yet be proven. However, it was not until Feferman extended this work much later in the fundamental paper (Feferman, 1962), which used the somewhat strengthened Reflection Principles below, was it possible to prove a 'Completeness Theorem' in the above sense for Π_2 sentences.

There is the possibility of adding other statements than just consistency alone to progressions. The work of Feferman here has been far-reaching. Subsequent research of Beklemishev, Schmerl,

Franzen and others have extended this, and no doubt will be commented on elsewhere in this volume.

It is possible to formalise the notion that: *'if $T \vdash \sigma$,where $\sigma \in \Sigma_n$ then σ is true'* and this *'n-reflection'* may be abbreviated REFL_T^n in that the theory T reflects the Σ_n truth of the matter. For T extending PA this can be expressed by a single Π_{n+1} formula. Full reflection for all Σ_n formulae, REFL_T is then the assertion of n-reflection for all n. Instead of consistency sequences it is possible to talk of n-, or *full-reflection progressions* and so forth. These turn out to have different properties from those of the simpler consistency statement studied by Turing, and the extensive study of these has been developed by Feferman (1962), which, as mentioned, showed how there were Π_2, and indeed full, Completeness Theorems concerning paths through $\mathcal{O}$, of the kind that Turing discussed. (See, e.g., the discussion of Kreisel (1972) on the subject of such putative paths delivering mathematical knowledge.) There is also a broad literature on the *kinds* of paths or progressions one have: *autonomous progressions* are those of a more self-justifying flavour. We shall not go into these details, but refer the reader to the excellent surveys of Franzen (2004a,b).

The notion of such progressions can be used in a number of arenas, with rather differing levels of significance. I would like to highlight one current area of work: iterated reflection principles in *truth theories.* In a truth theory one explicitly adds axioms concerning a truth predicate T say. One typically takes a base language of interest (and it is almost always Peano Arithmetic PA, since (i) mathematicians are very much interested in number theory and (ii) in PA the mechanisms of coding effectively given languages by numbers or 'Gödel codes' is available. Let us call this language $\mathcal{L}$. To this is added a predicate symbol T and for numbers n that code sentences the intention is that '$T(n)$' is to be interpreted as the 'sentence coded by n is true'. Truth theorists discuss the interplay of notions of truth with various languages (for example we may extend $\mathcal{L}$ to $\mathcal{L}_T$ and allow n to range over codes of sentences not just of $\mathcal{L}$ but of $\mathcal{L}_T$); we may also consider *axiomatising truth*: we add a selection of axioms, axiom schemes, deduction rules etc., to the axioms of PA that express our beliefs about how the notion of 'truth' should behave. Depending on how this is done, theories of various types and strengths emerge. (One such is specified in more detail below.)

Just as Turing added consistency statements to make a progression of number theories, we may do the same for truth theories. [For example, *cf.* the recent works done by Fujimoto (2011).] We shall link this notion of progression with some current work in sequences of truth sets in a moment, but we point out that although superficially looking like Turing's progressions, the motivations are admittedly rather different.

Let $S_0 = \mathrm{PA}$ and S_1 be an axiom set of the kind just roughly described in the language $\mathcal{L}_1 =_{\mathrm{df}} \mathcal{L}_T$, using the new predicate symbol $T_0 =_{\mathrm{df}} T$ that is allowed into the induction scheme. S_1 is now a numerical theory, extending PA to which we can repeat this process: we add a new truth predicate T_1 so that $T_1(n)$ will be interpreted as saying that if n codes a sentence of $\mathcal{L}_1$, then that sentence is true. Again extend the axiom set to include the induction scheme for properties in the language with the new symbol T_1. At the limit stage ω we obtain a language $\mathcal{L}_\omega \supseteq \bigcup_{k \in N} \mathcal{L}_k$, and again take the union of the previous axiom sets to obtain S_ω. We then continue with adding a truth predicate T_ω in the next language $\mathcal{L}_{\omega+1}$, and obtain thereafter $S_{\omega+1}, \ldots, S_\alpha, \ldots$ etc. up to some ordinal λ say. We ensure that the axioms of S_α are given by some Σ_1-arithmetic formula $\psi_\alpha(v_0)$ at each stage. With some care this can be effected in a way that ensures, inductively, that the theories S_α are *arithmetically sound*, that is assuming the axioms of $S = S_0$ are true, every theorem of S_α is true for $\alpha < \lambda$.

As a simple example of how this can work, we define the axioms of *Positive Friedman Sheard* which I shall call P for brevity. The first axiom set below is PA^T, Peano Arithmetic extended into a language $\mathcal{L}_T$ containing T, the formulae of which are allowed into the induction scheme.

1. PA^T;
2. $\forall$ atomic $\phi \in \mathcal{L}_{PA} : T(\phi)$ coincides with truth and $T(\neg\phi)$ with falsity;
3. $\forall \phi, \psi \in \mathcal{L}_T : T(\phi \wedge \psi) \leftrightarrow (T(\phi) \wedge T(\psi))$;
4. $\forall \phi, \psi \in \mathcal{L}_T : (T(\neg\phi) \vee T(\neg\psi)) \rightarrow T(\neg(\phi \wedge \psi))$;
5. $\forall \phi, \psi \in \mathcal{L}_T : (T(\phi) \wedge \psi \rightarrow T(\psi)$;
6. $\forall \phi(x) \in \mathcal{L}_T : \quad \exists x T(\neg\phi(x)) \rightarrow T(\neg\forall x \phi(x))$;
7. *CONS*: $\forall \phi \in \mathcal{L}_T : \neg(T(\phi) \wedge T(\neg\phi))$;
8. (Deduction Schemes): From A (respectively $T(A)$) deduce $T(A)$ (respectively A).

The axioms are dubbed *positive* because they only make claims as to which sentences are *in* the extension of T. Note there is no direct clause concerning simple negations. It is important for this axiomatisation that the Deduction Schemes are just that: schemes (the axiomatic versions would make the system inconsistent). '*CONS*' asserts consistency. $T^n(A)$ abbreviates n-fold $T(T(\cdots T(A)\cdots)$. The strength of this theory is known.

It is possible to iterate such theories: set A_0 to be '$0 = 0$', and P_0 to be P.

Definition 4. *Set: (i)* P_δ *to be* $P \cup \{A_\beta \mid \beta < \delta\}$*; (ii)* $A_\delta \equiv \forall \phi \in \mathcal{L}_T[Prov_{P_\delta}(\phi) \rightarrow T(\phi)]$.

As one can see by the subscripts to the predicate expressing provability in a recursively given axiom system S, $Prov_S$, we are considering extensions of the system P by adding iterations of 'S-provability implying truth'. We have left vague what we mean by the ordinals there, or what the statements A_α actually are. Also, although superficially resembling systems of axioms of increasing strength in order to form the *reflexive closure* of a theory, we are not doing this so as to form, as in that process, a theory encapsulating all of our commitments to the theory P. Rather we can use it to axiomatise some truth sets, those that arise as various levels of a so-called *Herzberger truth sequence*. Set $H_0 = \emptyset$:

$$H_{\alpha+1} = \{\phi \in \mathcal{L}_T \mid \langle \mathbb{N}, H_\alpha \rangle \models \phi\}. \text{ For limit } \lambda : H_\lambda = \{\phi \mid \exists \alpha < \lambda \forall \beta \in (\alpha, \lambda)\ \phi \in H_\alpha\}.$$

Here, each H_α is the extension of the T predicate of each model in turn. Note the 'liminf' rule for limit stages: φ is put in the λ'th set if from some point α onwards it is in. Such limit rules have been used by a variety of philosophical logicians to build truth sets. Field (2008) constructs a similar hierarchy $\langle F_\alpha \rangle$. It would go too far into the theory to discuss these here, but essentially these hierarchies run up to some ordinal ζ. The question has been asked: can we axiomatise in some way the sets H_λ for $\lambda \leq \zeta$? On general grounds a simple first order axiomatisation is ruled out, but it might be possible to do so on an initial segment, or in some larger language. It turns out that for $\lambda < \zeta$ some iterations of the theories $P + A_\lambda$ axiomatise H_λ in that they become true first at H_λ and no earlier H_μ. (And we may do the same for the F_λ.)

In view of the previous comments about building hierarchies of Reflection Principle theories T_a for $a \in \mathbb{N}$, where we thought of a as a notation, the reader may wonder as how one can precisely do this, as the 'α' *etc.*, above are not part of the language, (as they were not for Turing) but here they are very much larger than the constructive ordinals, and were left vague. There are two possible answers here: one can show that *within the system* of building up the H- or F-hierarchies for any $\alpha < \zeta$ there are certain sentences B_α that can themselves be construed as notations for those ordinals, and we may use these as devices for referring indirectly to them, and incorporate these somehow into our iterated truth theories. The other possibility is to extend Kleene's $\mathcal{O}$ itself to a system $\widetilde{\mathcal{O}} \supset \mathcal{O}$. To do this we extend the notion of 'computability': whereas $\mathcal{O}$ is a system of notation for the computable ordinals; using ordinary Turing machines we now allow the system of notion to run transfinitely and thus we have a new notion of 'decidable' corresponding to 'having some fixed output, 0 or 1 from some point on'. The beauty of this is that we don't even have to change Definition 1 at all beyond replacing the word 'recursive' by 'transfinitely computable' in the above sense. The Turing machines programs are not altered; the finite computations are just a special subclass of the transfinite ones, and the resulting system subsumes $\mathcal{O}$ and then stretches out precisely to ζ. If we are willing to indulge in this use of 'decidable', we can use these members of $\widetilde{\mathcal{O}}$ as notations applicable for our theories T_a. Of course we no longer have the possibility of $\{a \mid a <_{\widetilde{\mathcal{O}}} b\}$ being *c.e.* in the ordinary sense any more, for $b \in \widetilde{\mathcal{O}}$, but this set has to be '*c.e.*' in this new, wider sense. This would mean that any pursuit of analogies to the Turing/Feferman theorems would have to leave behind the notion of theories being (ordinarily) computably axiomatised. However for the analysis of the truth sets H_α, F_α, as explained by Horsten et al. (2012), through Turing-style iterations of the Positive Friedman Sheard theory, these kind of notations look good enough.

Time will tell whether this kind of approach (or indeed the underlying truth set constructions) will prove to be of any value.

References

Feferman, S., 1962. Transfinite recursive progressions of axiomatic theories. J. Symbolic Logic, 27 (3), 259–316.

Field, H., 2008. Saving Truth from Paradox, Oxford University Press, Oxford, New York.

Franzen, T., 2004a. Inexhaustibility: A non-exhaustive treatment, vol. 16 of Lecture Notes in Logic. ASL/A.K.Peters.

Franzen, T., 2004b. Transfinite progressions: A second look at completeness. Bull. Symbolic Logic, 10 (3), 367–389.

Fujimoto, K., 2011. Autonomous progression and transfinite iteration of self-applicable truth. J. Symbolic Logic, 76 (3), 914–945.

Horsten, L., Leigh, G., Leitgeb, H., Welch, P.D, 2012. Revision Revisited. Review of Symbolic Logic.

Kreisel, G., 1972. Which number theoretic problems can be solved in recursive progressions on Π^1_1-Paths through O? J. Symbolic Logic, 37 (2), 311–334.

Alastair Abbott, Cristian Calude and Karl Svozil describe —

A Quantum Random Oracle[1]

1. Turing's oracles

Turing's oracles have been used for many years to successfully understand the world of the incomputable. Are these tools only pure mathematical constructs or are they more 'real'? We will show how quantum measurements performed in specifically designed environments can produce incomputable sequences of bits, and discuss why they can hence be seen as physical Turing oracles.

An oracle is a black box capable of answering a set of questions, and an oracle Turing machine is a Turing machine which can query an oracle. According to Turing (1939, p.173),

> We shall not go any further into the nature of this oracle apart from saying that it cannot be a machine.

In current terms, a Turing oracle is an incomputable set O of natural numbers or strings. The oracle Turing machine can perform all of the usual operations of a Turing machine, and can also

[1] We thank Mike Stay for illuminating discussions and Marcus Hutter for useful comments which improved the commentary.

query the oracle for an answer to finitely many questions of the form 'is n in O?'. Because O is incomputable, an oracle Turing machine is a hypercomputer: it performs tasks which no Turing machine can do.

Turing studied oracles asserting the truth/falsity of 'number-theoretic statements', i.e., statements of the form '$\theta(x)$ vanishes for infinitely many natural numbers', where $\theta(x)$ is a primitive recursive function. The class of number-theoretic statements includes, but does not coincide with, the class of Π_1 statements, i.e., statements of the form '$\forall n P(n)$', where $P(n)$ is a computable predicate. Both Fermat's Last Theorem and the Riemann Hypothesis are Π_1 statements, and hence number-theoretic statements. Some number-theoretic statements are (trivially) computable, but most of them are not, so they satisfy the Turing incomputability condition.

In cryptography, a 'random oracle' is a black box that responds to every query with a 'randomly' chosen response,[2] picked uniformly from its output domain subject to the restriction that for any fixed query the answer returned is the same every time it receives that query. In the framework known as the 'random oracle model', random oracles are used in schemes where the system or protocol is proved secure because an attacker is (seems to be) required to extract impossible information from the oracle. This approach has known limits: e.g., in the works done by Canetti et al. (1998), it is proved that there exist signature and encryption schemes that are secure in the random oracle model, but for which any implementation of the random oracle results in insecure schemes.

Let O be a subset of the set of natural numbers and let $\mathbf{x} = x_1x_2 \cdots x_n \cdots$ be an infinite binary sequence. The map $\mathbf{x} \mapsto O_{\mathbf{x}}$ defined by $O_{\mathbf{x}} = \{i \mid x_i = 1\}$ is bijective, so we can equally speak about oracles as infinite binary sequences or sets of natural numbers (or strings, by using, say, the quasi-lexicographical bijective enumeration of strings over a finite alphabet). Incomputability is preserved under this bijection. A query 'is n in O?' is equivalent to 'is $x_n = 1$?'.

The condition imposed in the 'random oracle' model requires that the oracle O is given by a uniformly distributed binary sequence. Some 'random oracles' may be Turing oracles, others may not. Champernowne's sequence

$$01000110110000010100111001011101110000\ldots$$

is uniformly distributed, so it is a 'random oracle'; this 'random oracle' is computable (primitive recursive), so not a Turing oracle.

The set of codes of halting Turing machines (computably enumerable but not computable), as well as the set of algorithmically random strings (immune, i.e., strongly incomputable) are examples of Turing oracles (Calude, 2002).

Are Turing oracles 'real' or just pure theoretical mathematical notions?

2. Value indefiniteness and the Kochen–Specker Theorem

Computability is based on Turing's model of a computing machine, a fundamentally deterministic concept. Quantum mechanics, however, has confronted physicists with a world that appears to behave randomly and is essentially non-deterministic. The failures of a deterministic viewpoint to account for the predictions of quantum mechanics are exemplified by 'no-go' theorems, which exclude the possibility of assigning 'hidden variables' that predict the outcome of quantum measurements.

According to Bell's Theorem, there is no hidden variable theory that gives the same statistical predictions as quantum mechanics and satisfies *value definiteness* (i.e., all possible observables –

[2] 'True' or 'pure' randomness does not exists from a mathematical point of view (Calude, 2002).

including non-compatible ones – simultaneously have predefined values) and *locality* (i.e., two space-like separated events cannot influence each other in any way).

Bell's Theorem manifests itself in statistical inequalities – the class of which are called *Bell-type inequalities* – which pose a bound on the possible correlation between outcomes of spatially separated events subject to local realism, but of which quantum mechanics predicts violations. As Bell's Theorem deals with the statistical predictions of quantum mechanics, it might not be totally unreasonable to ask whether there are 'stronger' no-go theorems, which can tell us something deeper about the outcome of *individual* quantum measurements. The answer is affirmative.

A measurement *context* is a maximal set of pairwise co-measurable observables. For a measurement context $C = \{A_1, A_2, \dots\}$, the values corresponding to outcomes of measurements of observables $A_1, A_2, \dots$ are $v(A_1, C), v(A_2, C), \dots$ The Kochen–Specker Theorem states that for a quantum mechanical system represented by a Hilbert space of dimension greater than two, it is impossible for a hidden variable theory to fulfill the predictions of quantum mechanics and satisfy the following two conditions: *value definiteness* and *non-contextuality* (i.e., the value corresponding to the outcome of a measurement of an observable A, $v(A)$, is independent of the other compatible observables measured alongside it).

3. An example of a quantum random oracle

Consider a quantum random number generator that outputs bits produced by successive preparation and measurement of a state in which each outcome has probability one-half. By envisaging this device running ad infinitum, we can consider the infinite sequence **x** it produces. If we assume a standard picture of quantum mechanics, i.e., a Copenhagen-like interpretation in which measurement irreversibly alters the quantum state,[3] that the experimenter has freedom in the choice of measurement basis[4] (the 'free-will assumption'), and that we reject the notion of contextual hidden variables and can hence, by the uniformity and symmetry of the Kochen–Specker construction conclude that all observables are value indefinite, then some surprising conclusions about **x** can be made (Calude and Svozil, 2008). If **x** were computable, then (in principle) it would be possible to predict the outcome of each measurement in advance. This amounts to the existence of hidden variables for these observables and hence is in contradiction with the value indefiniteness due to the Kochen–Specker Theorem forbidding the existence of such a consistent, context-independent pre-assignment of measurement outcomes. The free-will assumption guarantees that even for an unknown initial state preparation the measurement basis in general is not pre-determined, thereby avoiding the possibility that only the measured observable together with a particular context had a definite pre-assigned value (Hall, 2010). Put differently, if **x** were computable then the device would behave deterministically (and hence classically) rather than quantum mechanically, and would contain infinitely many computable correlations. Hence, we have to conclude that **x** must be incomputable. In fact, the argument is readily seen to prove the stronger property of bi-immunity of **x**.[5]

Bi-immunity is the weakest possible notion of randomness: every binary sequence that is not bi-immune contains an infinite computable subsequence, i.e., a computable subset. This fact allows a computable martingale[6] to succeed on this sequence, so the unpredictability of the sequence is infinitely many times compromised (Kjos-Hanssen et al., 2010).

[3] A 'many-worlds' interpretation is excluded.

[4] In a truly deterministic theory – sometimes called superdeterminism – the experimenter might have the illusion of exercising her independent free choice, but in reality she just obeys the rules of the theory.

[5] A sequence **x** is bi-immune if only finitely many bits of **x** are computable. Every bi-immune sequence is incomputable, but the converse is not true.

[6] A martingale is a function M from binary strings to positive reals satisfying the following fairness condition: $M(\sigma) = (M(\sigma 0) + M(\sigma 1))/2$. The martingale M succeeds on a sequence **x** if $\limsup_n M(\mathbf{x} \upharpoonright n) = \infty$.

A sequence **x** is called Martin-Löf random if it is not contained in any effective null set.[7] A sequence **x** is called Kurtz random if it belongs to every computable open class of Lebesgue measure one. Every Omega number – or halting probability, cf. Calude (2002) – is Martin-Löf random and every Martin-Löf random real is Kurtz random; the converse implications are not true. Open question: Is the quantum random sequence previously described Kurtz random?

4. A quantum random number generator certified by value indefiniteness

Can a quantum device generating a bi-immune sequence really be constructed? Many quantum random number generators have been described and, while it is not readily clear which of the existing devices do produce an incomputable sequence of bits, it is not difficult to conceive designs which are explicitly certified by value indefiniteness to do so. One such device was proposed by Abbott et al. (2010).

5. Hypercomputation via quantum random oracles

As noted before, an oracle Turing machine is a hypercomputer. In particular, a Turing machine working with a bi-immune quantum random oracle (Abbott et al., 2010) is a hypercomputer.

The undecidability proof of the halting problem still applies to such machines; although they determine whether particular Turing machines will halt on specific inputs, they cannot determine, in general, if machines equivalent to themselves will halt. This fact creates a hierarchy of machines, closely related to the arithmetical hierarchy in mathematical logic, each with a more powerful halting oracle and an even harder halting problem.

Arguably the most important open question regarding quantum random oracles is: *What is the computational power of a Turing machine working with a bi-immune quantum random oracle?* We believe that such an oracle Turing machine cannot solve the halting problem, but it may solve a weaker undecidable problem, for example, the lesser limited principle of omniscience which states that, if the existential quantification of the conjunction of two decidable predicates is false, then one of their separate existential quantifications is false (Bridges and Richman, 1987).

References

Abbott, A.A., Calude, C.S., Svozil, K. 2010. A quantum random number generator certified by value indefiniteness. CDMTCS Res. Rep. 396.

Bridges, D., Richman, F., 1987. Varieties of Constructive Mathematics. Number 97 in London Mathematical Society Lecture Note Series. Cambridge University Press, Cambridge.

Calude, C.S., 2002. Information and Randomness: An Algorithmic Perspective, 2nd ed., Springer-Verlag, Berlin.

Calude, C.S., Svozil, K., 2008. Quantum randomness and value indefiniteness. Adv. Sci. Lett. 1, 165–168.

Canetti, R., Goldreich, O., Halevi, S. 1998. The random oracle methodology, revisited. In: Proceedings of the 30th Annual ACM Symposium on the Theory of Computing, ACM, New York, pp. 209–218.

[7] The set of all infinite sequences beginning with a string σ – the cylinder generated by σ – is a basic open set in Cantor space. The Lebesgue measure of the cylinder generated by σ is $2^{-|\sigma|}$. Every open subset of Cantor space is the union of a countable sequence of disjoint basic open sets, and the measure of an open set is the sum of the measures of any such sequence. A computably (computable) open set is an open set that is the union of the sequence of basic open sets determined by a computably enumerable (computable) sequence of binary strings. A constructive null set is a computably enumerable sequence X_i of effective open sets such that $X_{i+1} \subseteq X_i$ and Lebesgue measure of X_i is smaller than 2^{-i}, for each i. The intersection of the sets X_i has Lebesgue measure zero.

Hall, M.J.W., 2010. Local deterministic model of singlet state correlations based on relaxing measurement independence. Phys. Rev. Lett. 105 (250404).

Kjos-Hanssen, B., Stephan, F., Teutsch, J.R. 2010. Enumerating randoms. http://arxiv.org/abs/1008.4825, arXiv:1008.4825v1.

Turing, A.M. 1939. Systems of logic based on ordinals. Proc. Lond. Math. Soc. Ser. 2, 45, 161–228.

Practical Forms of Type Theory

(J. Symbolic Logic, vol. 13 (1948), pp. 80–94)

Some background remarks[1] from Robin Gandy's —

PREFACE

The first draft of this paper (*A Practical Form of type theory*) shows that it was originally intended as one of several. One aim, which became the main aim, is set out in *The Reform of Mathematical Notation.* Turing wished to encourage 'mathematicians-in-the-street' to use notation and forms of argument which would safeguard their work from ambiguity and inconsistency; but to do this without forcing their work into the straitjacket of a particular logical system, or even requiring them to have detailed knowledge of such a system. To the end of his life, he thought this aim a proper one for a logician, and from time to time gave talks to mathematicians in which he would expound particular logical points. As a logician, however, he was interested in devising formal systems which could act as bridges between the formal and the informal, and this motivated him to produce the two systems set out in this paper. In SI, besides describing the intended universe of the nested-type system, he also explains a number of elementary logical points. He did not expect mathematicians to use the system, but it looks as if he hoped that some mathematicians would read the paper, even though ignorant of symbolic logic. In this, as in some of his other papers and lectures, he was overly optimistic about the abilities of his intended audience. Not only is it not obvious that the rules and axioms do correctly formalise the informal notions, but, more explicitly, a reader unfamiliar with symbolic logic will not appreciate the vital distinction between mathematical and metamathematical statements. (When in 1948 Turing tried to explain the Deduction Theorem to me, I failed to understand it because I did not distinguish between 'B can be inferred from A' and '$A \supset B$'.)

Work on the nested-type theory, including the writing of *A Practical Form of type theory* had mostly been done before the summer of 1945, when Turing moved from Hanslope Park to the National Physical Laboratory. There, during the second half of 1945, he was fully occupied working out his proposals for the ACE computer (The Collected Works, *Mechanical Intelligence*, pp. 1–86). During 1946 he completed *Practical Forms*. I do not know whether he merely shelved or completely abandoned further work on *Reform* and the project described in it.

The first two pages [of the first draft, *A Practical Form of type theory*] give a fuller account of Turing's motivation than does *Practical Forms*, so I quote them in full.

> It is usual for mathematicians to pay-lip service to the theory of types, but they will not usually make any attempt to bring their mathematics into line with it. An occasional paradox may perhaps be attributed to neglect of types, but no suggestions are made for the avoidance of these paradoxes short of the expression of all mathematics in the formalism of Principia Mathematica (say). In the present paper a system will be described which takes account of

[1] These introductory remarks are extracted from the Preface to the 1948 Turing paper, pp. 179–185 of the *Collected Works of A. M. Turing: Mathematical Logic*. See the *Collected Works* for further technical comments, and unpublished material of Turing on the same topic.

type theory, but at the same time follows very closely the normal mathematical outlook. The type theory intrudes itself on the system only very slightly, and its effect may be summed up in the form of one or two simple and natural cautions, which are easily carried over to unformalised mathematics: this should, I hope, enable all such serious mathematics as is supposedly based on the theory of types to be brought genuinely into line with it, at the cost of very little additional trouble to mathematicians.

This paper will appear in two parts. The first part is written chiefly for the mathematician who wishes to increase the rigour of his proofs along the lines indicated in the previous paragraph, rather than for the logician. The emphasis will be on notation and meaning rather than on axioms and rules of procedure; these will, however, be given for the sake of completeness. The second part will be devoted to a little axiomatic development, and the justification of the system in the case of the 'finite universe' i.e. the case where there is only a finite number of individuals. It will establish a very complete connection between this system and that of Church. This connection seems to be valuable because Church's system has greater theoretical simplicity than the proposed 'practical system', but is less convenient for the formalisation of proofs. Consequently, it will be natural to express proofs in the practical system, but metamathematical results in terms of Church's system.

The author wishes to repudiate any implication that may be suggested by this paper to the effect that he believes the Russell philosophy of mathematics to be the truest. He does believe, however, that it is the one which is most easily understood, and also that it describes most closely the accepted form of present-day mathematical thinking. This paper is concerned with giving accurate expression to that thinking. When that is done it will be easier to see the limitations of the outlook which goes with this form of thinking.

PRACTICAL FORMS OF TYPE THEORY

A. M. TURING

Russell's theory of types,[1] though probably not providing the soundest possible foundation for mathematics, follows closely the outlook of most mathematicians. The present paper is an attempt to present the theory of types in forms in which the types themselves only play a rather small part, as they do in ordinary mathematical argument. Two logical systems are described (called the "nested-type" and "concealed-type" systems). It is hoped that the ideas involved in these systems may help mathematicians to observe type theory in proofs as well as in doctrine. It will not be necessary to adopt a formal logical notation to do so.

1. The nested-type system for a finite universe

In this section the notation of the nested-type system will be explained. The explanation will be in terms of the 'finite universe,' i.e. we start with a finite number of objects or 'individuals' and build up other entities from these. We can then formulate certain rules which give valid results in this case and hope that they will apply in the infinite case also. We cannot of course hope that all such rules will work. We have to imagine that many rules of this kind have been tried, found wanting and rejected, and that others are still in use. This rather unsatisfactory-sounding process is as good an account as the author feels can be given of the way in which current mathematical procedure has grown up. But whatever the truth of this may be the finite universe provides a first class ground on which to describe the nested-type system, and we proceed accordingly.

Our finite universe has initially as its members the 'individuals' $U_1, \cdots, U_N$. Although these include all the individuals, they need not exhaust our stock-in-trade, for we can also bring in functions taking the individuals as arguments and having them also as values. With our increased range of commodities we can then go into business again and produce a still greater variety of objects, and repeat without limit. There obviously arises a great variety of different kinds of functions which may need to be distinguished, but for the present system we need only trouble ourselves with the very broadest divisions, which will be called types. These divisions are described below.

The individuals $U_1, \cdots, U_N$ form type 0.

The functions of individuals, taking individuals as values, together with the individuals themselves, form type 1.

The functions of arguments in type 1, taking values also in type 1, together with the members of type 1, form type 2.

......

The functions of arguments in type n, taking values also in type n, together with the members of type n, form type $n+1$.

......

It must be understood that by a "function" we mean the function itself and not merely one of its values. To illustrate the point by analogy with functions of a real variable, we should say that "sin"

Received January 6, 1947.

[1] A. N. Whitehead and Bertrand Russell, *Principia mathematica*, Cambridge, England, 1925.

denotes a function, but that "sin 0.3" and "sin x" do not, although the latter is often used (incorrectly in the author's opinion) as if synonymous with "sin".

It is convenient to require functions to be defined throughout the appropriate type, i.e. not to permit such definitions as "$f(0) = 0$, but if x is different from 0 then $f(x)$ is undefined." In order to cover such cases we shall set apart from the outset a particular individual U_1, which we shall rename "C", to be the value of a function in all cases where it would normally be regarded as undefined. So far as possible we try to keep C on a par with the other individuals. We deviate from this principle by adopting the convention that the value of a function is always C unless the function is of higher type than the argument. (More strictly, if the function belongs to every type to which the argument belongs.) We respect the principle by refraining from considering every expression containing "C" to have the value C.

The functions and individuals together will be known as *terms*. With our finite universe it is convenient to think of the functions as given by tables, consisting of two columns, in the first of which appear all the necessary arguments, and opposite them in the second column the appropriate values. Thus with $N = 4$ a typical member of type 1 would be represented by the table

U_2	U_3
U_1	U_1
U_3	U_1
U_4	U_4

(1)

It would be a convenience to have the table rearranged with the first column in natural order. In the case of the above table (1) we should simply have to interchange the first two rows. Such a table may be said to be in normal form. We can do this for all tables of type 1, and when we have done so we are in a position to define a natural order for the members of type 1. With both tables in normal form, the earlier table is to be the one which has the earlier value in the last row in which the two tables differ. Thus the table (1) above precedes

U_1	U_1
U_2	U_4
U_3	U_3
U_4	U_4

(2)

since when (1) is put into normal form the two tables differ last in the third row, and there (1) has the value U_1 but (2) has the value U_3. We shall also adopt the convention that the individuals in type 1 precede the tables. We may now continue the numbering of terms so as to include all type 1, simply numbering them in the natural order just defined. The numbers will extend from 1 to $N + N^N$. It may be verified that the above tables (1) and (2) are U_{205} and U_{241} respectively. A similar process may now be carried out for type 2 and then for type 3. In general when we are dealing with type n we have already numbered the members of type $n - 1$. It is easily verified that those tables which have already appeared as members of type $n - 1$ have the order which they had in that type, and precede all the new tables. The order of any two tables (new or old) is that of the last pair of values in which they differ.

Let us now introduce the notation $(\boldsymbol{UV})$ to denote the result of looking up $\boldsymbol{V}$ in the table $\boldsymbol{U}$; in slightly different words it is the entry against $\boldsymbol{V}$ in the table $\boldsymbol{U}$.[2] In other words again it is the value

[2] We shall use heavy type letters throughout to represent variables or undetermined formulas or tables. They occur only in metamathematical discussions. All our statements are understood to be true whatever substitutions of formulas (or tables, as the case may be) are made for the heavy type capital letters, and whatever substitutions of variables are made for the small heavy type letters.

of the function $\boldsymbol{U}$ for the argument $\boldsymbol{V}$, and might therefore, in agreement with current mathematical practice have been denoted by $\boldsymbol{U}(\boldsymbol{V})$. Our conventions require $(\boldsymbol{UV})$ to be C in cases where the table gives no information: these are just the cases where the lowest type to which $\boldsymbol{U}$ belongs does not exceed the lowest for $\boldsymbol{V}$. We may also introduce the notation $\boldsymbol{U} = \boldsymbol{V}$ to denote the identity of the terms $\boldsymbol{U}$ and $\boldsymbol{V}$. It should be noticed that so long as $\boldsymbol{U}$ and $\boldsymbol{V}$ are tables known to belong to some particular type n we can establish their identity by showing that they have the same values throughout type $n-1$ (this is known as the principle of extensionality and gives rise to the "axiom of extensionality"). The principle fails for individuals, for if $\boldsymbol{U}$ and $\boldsymbol{V}$ are individuals then $(\boldsymbol{UX})$ is always identical with $(\boldsymbol{VX})$, both being C, and yet $\boldsymbol{U}$ and $\boldsymbol{V}$ may well be different. The principle also fails when the types of the terms are unknown, for we can never then be sure that we have examined sufficient arguments for the functions. There may be some argument in a higher type than we have yet considered for which the two functions differ.

The expression $\boldsymbol{U} = \boldsymbol{V}$ which we have just introduced denotes a *proposition*, unlike $(\boldsymbol{UV})$ which was a term. Propositions may be thought of as having a value which is either true (T) or falsity (F). By taking T and F to be individuals we could have arranged for the propositions to be included amongst the terms, but we have not in fact done so.

There are several other ways of forming propositions. If $\boldsymbol{P}$ and $\boldsymbol{Q}$ are propositions then $(\sim\boldsymbol{P})$ is a proposition whose value is opposite to that of $\boldsymbol{P}$ and $(\boldsymbol{P} \supset \boldsymbol{Q})$ is one whose value is F if and only if $\boldsymbol{P}$ is T and $\boldsymbol{Q}$ is F. We may read $(\sim\boldsymbol{P})$ as "not $\boldsymbol{P}$" and $(\boldsymbol{P} \supset \boldsymbol{Q})$ as "$\boldsymbol{P}$ implies $\boldsymbol{Q}$." If $\boldsymbol{U}$ is a term then $\boldsymbol{D}^r\ \boldsymbol{U}$ represents the proposition that $\boldsymbol{U}$ is in type r, i.e. it is T if and only if $\boldsymbol{U}$ is in type r.

We could of course introduce a great variety of further means for forming terms and propositions. We could for instance define $(\boldsymbol{P}\ \&\ \boldsymbol{Q})$ as a proposition whose value is T if and only if both $\boldsymbol{P}$ and $\boldsymbol{Q}$ are T. We shall be content however with comparatively few, namely those we have already introduced, together with one further way of forming propositions and one of forming terms. These cannot be described without bringing in the ideas of "variable" and "formula with variables." Variables are of little importance except as parts of formulas. All we need say about them is that as a matter of notation small italic letters with any number of primes will be used as variables. *The letters p, q, r, s, t, (possibly with primes) will be proposition variables and the others term variables.* Small heavy type letters may be used to stand for any variable, with an obvious convention concerning the kind of variable. An example of a "formula with variables" is the expression $x = U_5$. On substituting a term, e.g. U_{10} for the the term variable x it becomes a proposition. Similarly $(U_{405}x)$ is a formula with variables: in this case substitution yields a term. In general a formula with variables or more briefly a *formula* is an *expression which yields a term or proposition on substituting terms and propositions for the (free) term and proposition variables respectively*. The formulas may be called *term formulas* or *proposition formulas* according as they give rise to terms or propositions on substitution. The word *free* in the definition should be ignored for the present.

We can now describe our remaining ways of forming terms and propositions. If $\boldsymbol{P}$ is a proposition formula with only the one free term variable $\boldsymbol{x}$ and no proposition variables then $(\imath\boldsymbol{x}, r)\ \boldsymbol{P}$ is a term and $(\boldsymbol{x},\ r)\ \boldsymbol{P}$ is a proposition. Of these the term $(\imath\boldsymbol{x}, r)\boldsymbol{P}$ has the value C unless there is one and only one term $\boldsymbol{U}$ in type r for which the result $S^{\boldsymbol{x}}_{\boldsymbol{U}}\boldsymbol{P}\ |$ of substituting $\boldsymbol{U}$ for $\boldsymbol{x}$ in $\boldsymbol{P}$ is T: if there is a unique $\boldsymbol{U}$ with this property then the value of $(\imath\boldsymbol{x},\ r)\ \boldsymbol{P}$ is that $\boldsymbol{U}$. The value of the proposition $(\boldsymbol{x}, r)\ \boldsymbol{P}$ is T if and only if all the results of substitution, $S^{\boldsymbol{x}}_{\boldsymbol{U}}\boldsymbol{P}\ |$, with $\boldsymbol{U}$ in type r, have the value T. We may read $(\imath\boldsymbol{x}, r)\ \boldsymbol{P}$ as "the $\boldsymbol{x}$ in type r such that $\boldsymbol{P}$" and $(\boldsymbol{x}, r)\ \boldsymbol{P}$ as "$\boldsymbol{P}$, for all $\boldsymbol{x}$ in type r."

Now consider the expression $(x, 3)(x\ = y)$. In it there occur the two variables x and y. If we substitute a term, e.g. U_6, for y we shall obtain a proposition, but if at the same time we substitute U_9 for x we shall obtain nonsense. We would like to excuse ourselves from making this second substitution and admit $(x, 3)(x = y)$ to membership of the class of formulas. Our excuse is that substitution should only be made for the *free* occurrences of a variable, and that the occurrences of x in $(x, 3)(x = y)$ are not free but bound. We say that a variable $\boldsymbol{u}$ occurs *bound* in a formula if the occurrence in question is in a part of form $(\imath\boldsymbol{u}, r)\boldsymbol{P}$ or $(\boldsymbol{u},\ r)\boldsymbol{P}$. Thus the first occurrence of x in

$(y,\ 1)[x = (\imath x,\ 0)(x = x)]$ is free and the others are bound. This expression is a proposition formula according to our definition. To verify this, first note that $x = x$ is a proposition formula with no free variables other than x and that $(\imath x,\ 0)(x = x)$ is therefore a term. Consequently $\boldsymbol{U} = (\imath x,\ 0)(x = x)$ is a proposition, and *a fortiori* a proposition formula, for any term $\boldsymbol{U}$. It has no free variables rather than y (indeed it has none at all), and therefore $(y,\ 1)[\boldsymbol{U} = (\imath x,\ 0)(x = x)]$ must be a proposition for any term $\boldsymbol{U}$, i.e. $(y,\ 1)[x = (\imath x, 0)(x = x)]$ is a proposition formula.

It will now be seen that terms and propositions are just term formulas and proposition formulas without free variables.

Free and bound variables are familiar in mathematics though they are seldom consciously recognized. A typical example of a bound variable is that of x in the integral $\int_0^1 x dx$; x occurs free in the equation $x(x-1) = 0$. A convenient method of distinguishing between bound and free variables is to make a substitution of a constant (of the appropriate kind) for the variable in question. If nonsense results the variable is certainly bound: if sense results it is most probably free. Sense may perhaps result from substitution for a bound variable if the result of the substitution and the original expression are interpreted according to different conventions. The double suffix summation convention of tensor theory provides an example of this. Using this convention the variable j in the expression $a_{ij}b_{jk}$ is bound, but we can substitute 1 for j and obtain a perfectly sensible expression; it is sensible because it is interpreted without applying the double suffix convention.

The outcome of our definition of "formulas" is that they will include terms, propositions, and variables. Also if $\boldsymbol{A}$ and $\boldsymbol{B}$ are term formulas, $\boldsymbol{P}$ and $\boldsymbol{Q}$ proposition formulas, $\boldsymbol{x}$ a term variable, and r a numeral representing a nonnegative integer, then $(\boldsymbol{A}\ \boldsymbol{B})$ and $(\imath\boldsymbol{x}, r)\boldsymbol{P}$ are also term formulas and $(\boldsymbol{A} = \boldsymbol{B})$, $\boldsymbol{D}^r\boldsymbol{A}$, $(\sim\boldsymbol{P})$, $(\boldsymbol{P} \supset \boldsymbol{Q})$, and $(\boldsymbol{x}, r)\boldsymbol{P}$ are proposition formulas. Our use of the letter "r" in these cases must not of course be confused with its use as a proposition variable. One further method of constructing formulas is worth mentioning although it is possible to do without it, and define it in terms already explained. This is "abstraction." If $\boldsymbol{A}$ is a term formula then $(\lambda\boldsymbol{x}, r)\boldsymbol{A}$ is a term formula of type $r+1$. It stands for the function whose value for the argument $\boldsymbol{U}$ in type r is $S^{\boldsymbol{x}}_{\boldsymbol{U}}\boldsymbol{A}\ |$, provided that $S^{\boldsymbol{x}}_{\boldsymbol{U}}\boldsymbol{A}\ |$ is in type r for every $\boldsymbol{U}$ in type r: if however there is a single argument $\boldsymbol{U}$ in type r for which $S^{\boldsymbol{x}}_{\boldsymbol{U}}A\ |$ is not in type r then $(\lambda\boldsymbol{x},\ r)\boldsymbol{A}$ is C. We can define $(\lambda\boldsymbol{x}, r)\boldsymbol{A}$ in previously explained terms as

$$(\imath\boldsymbol{y}, r+1)(\sim[(\boldsymbol{x},\ r)(\boldsymbol{y}\boldsymbol{x}\ = \boldsymbol{A}) \supset \boldsymbol{D}^0\boldsymbol{y}])$$

where $\boldsymbol{y}$ is any variable not occurring free in $\boldsymbol{A}$.

In the case of a finite universe the individuals $U_1, \ldots U_N$ form a part of the system. When dealing with an infinite universe this does not seem to be necessary, but it is convenient to retain symbols for three of them; these are $\boldsymbol{U}_1$ which is called C and which we have already mentioned, U_2 which is called T' and U_3 which is called F'. . These last two may be regarded as unofficial representatives of truth and falsity, looking after their interests amongst the terms: their official representatives are T and F which are propositions. The chief use of T' and F' is in connection with propositional functions. If we wish to express 'x is mortal' we form a function M which is defined for individuals (supposed to include mammals) and has the value T' for mortal arguments, F' for immortal arguments. Then "x is mortal" is written as $Mx = T'$.

At this point we should pause and consider what we have done. We have defined a class of expressions which we have called term-formulas and proposition-formulas, and which roughly correspond to the terms and propositions of mathematics. These formulas are given interpretations in the finite universe in terms of individuals and tables. Each term formula without free variables has an interpretation as a particular individual or table, and each proposition formula has an interpretation which is truth or falsity. We are able to determine whether a proposition formula without free variables is true by working out its interpretation, although this will be a very lengthy business unless the formula is very simple and N very small. The work involved in establishing the truth of formulas can be greatly reduced by the use of various rules, e.g. that if two formulas $\boldsymbol{P}$, $\boldsymbol{Q}$ are true then $\sim(\boldsymbol{P} \supset \sim\boldsymbol{Q})$ is true. A process of application of such rules may be allowed to oust the process of working out the interpretation.

Since the majority of the rules involved do not make any reference to the number N it is easy to forget the finite universe, and to allow the various rules to become reflex action. Eventually we break off almost all connection with the finite universe picture: in particular we repudiate such propositions as

$$(x, r)(y, r)((x \neq y) \supset ((fx) \neq (fy))) \supset (x, r)(\exists y, r)((fy) = \boldsymbol{x})$$

which are especially connected with such a picture. Finally we even repudiate the picture more violently by adopting an "axiom of infinity."

This, in my opinion, is a very idealised but essentially correct account of how the present mathematical argument-form has grown up. The last step or two may appear very lame, but I think this cannot be helped: I think that these last steps are not really sound.

One set of rules which can replace the finite universe picture is given below in §2 (rules I–X, XI_n).

Abbreviations. At this point we are obliged to introduce a few conventions which permit us to abbreviate our formulas. The unabbreviated formulas would be disagreeably cumbrous.

(a) We may introduce abbreviations by means of the arrow: a formula standing to the left of an arrow is understood to be an abbreviation of that on the right of it. If heavy type letters appear in these expressions it is understood that the formula on the left is an abbreviation of that on the right for any meaningful substitutions of formulas for the heavy-type letters. With these conventions we introduce the abbreviations:

$$(\boldsymbol{P} \,\&\, \boldsymbol{Q}) \to (\sim(\boldsymbol{P} \supset (\sim\boldsymbol{Q})))$$
$$(\boldsymbol{P} \vee \boldsymbol{Q}) \to ((\sim\boldsymbol{P}) \supset \boldsymbol{Q})$$
$$(\boldsymbol{P} \equiv \boldsymbol{Q}) \to ((\boldsymbol{P} \supset \boldsymbol{Q}) \,\&\, (\boldsymbol{Q} \supset \boldsymbol{P}))$$
$$(\exists \boldsymbol{x}, r)\boldsymbol{P} \to (\sim((\boldsymbol{x}, r)(\sim\boldsymbol{P})))$$
$$(\exists! x, r)\boldsymbol{P} \to ((\exists \boldsymbol{x}, r)\boldsymbol{P} \,\&\, (\boldsymbol{x}, r)(\boldsymbol{y}, r)(\boldsymbol{P} \neq S^{\boldsymbol{x}}_{\boldsymbol{y}}\boldsymbol{P}| \supset \boldsymbol{x} = \boldsymbol{y}))$$
$$(\boldsymbol{A} \neq \boldsymbol{B}) \to (\sim(\boldsymbol{A} = \boldsymbol{B}))$$
$$T \to (x, 0)x = x$$
$$F \to (\sim T)$$

The variable $\boldsymbol{y}$ must not be free in $\boldsymbol{P}$.

(b) formulas of form $\boldsymbol{A} \,\&\, \boldsymbol{B} \,\&\ldots\&\, \boldsymbol{P}$ we consider not to need any more brackets, since they have the same meaning in whatever manner the brackets are put in. Strictly speaking this equivalence only applies in virtue of rule IV below, and the reader may prefer to adopt some definite convention of his own as to the way the missing brackets are to be supplied. Similar considerations apply to formulas of form $\boldsymbol{A} \vee \boldsymbol{B} \vee \ldots \vee \boldsymbol{P}$.

(c) We shall often leave brackets out in cases where it is quite obvious how they should be replaced. Excessive bracketing often makes the formulas difficult to read. It is not thought worth while to introduce definite conventions in the present paper: we rely on common sense instead. Likewise we permit alterations in the form of a pair of brackets. These common sense conventions have already been applied to some extent.

2. Formal account of the nested-type system

We now describe the practical system in the usual formal manner, specifying what series of symbols are to be regarded as term-formulas, proposition formulas, variables, provable formulas, etc. We do not follow this aspect very far in the present paper, believing that mathematics is suffering more from lack of sound notation than from lack of rules of procedure.

Term variables. The symbols $a, b, \ldots, n, o, u, v, w, x, y, z, a', b', \ldots$ are term variables.

Proposition variables. The symbols $p, q, r, s, t, p,' q', \ldots$ are proposition variables.

Term formulas, proposition formulas, and formulas. Term variables are term-formulas. Terms $(U_1^H, U_2^H, \ldots)$ are term formulas. Prosposition variables are proposition formulas. If $\boldsymbol{A}$ and $\boldsymbol{B}$ are term formulas and $\boldsymbol{P}$ and $\boldsymbol{Q}$ are proposition formulas and $\boldsymbol{x}$ is a term-variable and r a numeral representing a non-negative integer, then $(\boldsymbol{AB})$ and $(\imath x, r)\boldsymbol{P}$ are term formulas and $(\boldsymbol{A} = \boldsymbol{B}), (\sim\boldsymbol{P}), (\boldsymbol{P} \supset \boldsymbol{Q}), \boldsymbol{D}^r\boldsymbol{A}, (\boldsymbol{x}, r)\boldsymbol{P}$ are proposition formulas. Term formulas and proposition formulas are formulas. No expression is a term variables, term formula, proposition variable, proposition formula, or formula unless compelled to be so by the foregoing.

Free and bound occurrences of variables. Each occurrence of a variable in a formula is either a bound or a free occurrence, but cannot be both. Occurrences of proposition variables are always free. The occurrence of the term variable $\boldsymbol{X}$ in the formula $\boldsymbol{X}$ is free. In the formulas $(\boldsymbol{AB}), (\imath\boldsymbol{X}, r)\boldsymbol{P}, (\boldsymbol{A} = \boldsymbol{B}), (\sim\boldsymbol{P}), (\boldsymbol{P} \supset \boldsymbol{Q}), \boldsymbol{D}^r\boldsymbol{A}, (\boldsymbol{X}, r)\boldsymbol{P}$ the occurrences of the various variables are free or bound according as they were free or bound in their corresponding occurrences in $\boldsymbol{A}, \boldsymbol{B}, \boldsymbol{P}$, or $\boldsymbol{Q}$ except that the occurrences of $\boldsymbol{X}$ in $(\boldsymbol{X}, r)\boldsymbol{P}, (\imath\boldsymbol{X}, r)\boldsymbol{P}$ are bound.

It may be observed that all four possible combinations concerning the presence or absence of a variable bound or free in a formula can occur. Examples are $T,' x, (\imath x, 0)(x = x), x = (\imath x, 0)$ $(x = x)$.

Formulas and tautological formulas of the propositional calculus. The formulas of the propositional calculus are defined to be the least class of formulas containing the propositional variables, and containing $(\boldsymbol{P} \supset \boldsymbol{Q})$ and $(\sim\boldsymbol{P})$ whenever it contains $\boldsymbol{P}$ and $\boldsymbol{Q}$. Tautological formulas of the propositional calculus are those which always give the value T if a substitution of values T or F is made for the variables, and the result then evaluated as follows: T $\supset$ T is F, T $\supset$ F is F, F $\supset$ T is T, F $\supset$ F is T, $\sim$T is F, $\sim$F is T.

The rules of procedure (provable formulas). We word our rules of procedure in the form of a definition of the "provable formulas". Throughout, $\boldsymbol{r}$ is any numeral representing a non-negative integer.

Rule I (Change of bound variables). The formulas

$$(\boldsymbol{x}, \boldsymbol{r})\boldsymbol{P} \equiv (\boldsymbol{y}, \boldsymbol{r})\ S_{\boldsymbol{y}}^{\boldsymbol{x}}\boldsymbol{P}|$$

$$(\imath\boldsymbol{X}, \boldsymbol{r})\boldsymbol{P} \equiv (\imath\boldsymbol{y}, \boldsymbol{r})S_{\boldsymbol{y}}^{\boldsymbol{x}}\boldsymbol{P}|$$

are provable if $\boldsymbol{P}$ is a proposition formula in which $\boldsymbol{y}$ does not occur free, and $\boldsymbol{x}$ is not free at a place where $\boldsymbol{y}$ would be bound.

Rule II (Substitution). If $\boldsymbol{P}$ is provable, then $S_{\boldsymbol{A}}^{\boldsymbol{x}}\boldsymbol{P}|$ and $S_{\boldsymbol{Q}}^{\boldsymbol{q}}\boldsymbol{P}|$ are provable, where $\boldsymbol{A}$ and $\boldsymbol{Q}$ are respectively term and proposition formulas, and the bound variables of $\boldsymbol{P}$ are distinct both from $\boldsymbol{x}$ and $\boldsymbol{q}$ and from the free variables of $\boldsymbol{A}$ and of $\boldsymbol{Q}$.

Rule III (Quantifiers). If either of the two formulas $\boldsymbol{H} \supset (\boldsymbol{D}^r\boldsymbol{x} \supset \boldsymbol{P}), \boldsymbol{H} \supset (\boldsymbol{x}, r)\boldsymbol{P}$ is provable, and $\boldsymbol{x}$ is not free in $\boldsymbol{H}$, then the other is also provable.

Rule IV (Propositional calculus). Any tautologous formula of the propositional calculus is provable.

Rule V (Modus ponens). If the formulas $\boldsymbol{P} \supset \boldsymbol{Q}$ and $\boldsymbol{P}$ are both provable then $\boldsymbol{Q}$ is provable.

Rule VI (Descriptions). If $\boldsymbol{P}$ is a proposition formula in which x does not occur bound, then the formulas

$$(\exists! x,\ r)\boldsymbol{P} \supset S^{x}_{(\imath x,\ r)\boldsymbol{P}}\boldsymbol{P}|$$

$$\sim(\exists! x, r)\ \boldsymbol{P} \supset (\imath x,\ r)\boldsymbol{P} = \boldsymbol{C}$$

$$D^{r}(\imath x,\ r)\boldsymbol{P}$$

are provable.

Rule VII. The formula

$$(x,\ r)\boldsymbol{D}^{r}\boldsymbol{A} \supset (\exists y,\ r+1)(\sim D^{0}y \ \&\ (x,\ r)yx = \boldsymbol{A})$$

is provable provided y does not appear free in the term formula $\boldsymbol{A}$.

Rule VIII (Axioms). For any numeral r representing a non-negative integer the following formulas numbered Al to C2 are provable:

Al. $C \neq T' \ \&\ C \neq F' \ \&\ T' \neq F'$

A2. $D^{0}C \ \&\ D^{0}T' \ \&\ D^{0}F'$

A3. $[D^{0}x \vee (D^{r+1}x \ \&{\sim} D^{r}y)] \supset xy = C$

A4. $D^{r}x \supset D^{r+1}x$

A5. $D^{r+1}x \supset D^{r}xy$

B1. $x = x$

B2. $(y = x \ \&\ y\ = z) \supset x = z$

B3. $x = y \supset (zx\ = zy \ \&\ xz = yz)$

C1. $(x,\ r)fx = gx \supset [f = g \vee D^{0}f \vee\ D^{0}g \vee\ {\sim}D^{r+1}f \vee {\sim}D^{r+1}g]$

(Axiom of extensionality.)

C2. $(\exists i,\ r\ +2)(f, r+1)((\exists x,\ r)fx = T') \supset f(if) = T']$

(Axiom of choice.)

Rule IX (Axiom of infinity). The following formula is provable:

C3. $(\exists h,\ 1)(\exists v, 0)(x,\ 0)(y,\ 0)[(hx = hy \supset x = y) \ \&\ v \neq hx]$

If we have a finite universe with N individuals instead of an infinite one we must replace rule IX by:

Rule IX_N. The following, Dl and D2, are provable:

Dl. $D^{0}x \equiv (x = U^{H}_{1} \vee \ldots \vee U^{H}_{N})$

D2. $U^{H}_{n} \neq U^{H}_{m}$

where m and n are different and not greater than N.

We may make a number of remarks about these axioms and rules:

(1) Axioms D1, D2 are rather stronger than is really necessary. Instead we could use the one axiom

$$D^{0}x \supset (x = U^{H}_{1} \vee \ldots \vee x = U^{H}_{N})$$

which would be more nearly analogous to C3, but would admit the possibility of there being fewer than N individuals.

(2) The second formula under rule VI might have been omitted. If this had been done it would have been necessary to define a new description operator in terms of the old one in such a way that the second formula would apply for the new operator.

(3) It may be wondered why rules VI and VII do not appear under the axioms, $yx = T'$ being written for $\boldsymbol{P}$ and yx for $\boldsymbol{A}$. If there had been any more rules of this kind they could have been replaced by axioms, by making similar substitutions, but these axioms would only be equivalent to the corresponding rule in the presence of rules VI, VII. It will now be clear why rules VI, VII cannot themselves be written as axioms.

(4) A term U_m and its corresponding formula U_m^H are not regarded as identical as they were in §1. We have introduced a distinction rather similar to the distinction between the real and complex numbers π. This distinction will be of value in any attempt to provide a formal justification of the system in terms of tables: it would then be very embarrassing to have the same notation both for a formula and its interpretation. The author has carried through such a justification in detail, together with a proof that the system is complete for the finite universe. This provides a good check that no essential axioms have been omitted. The theorem mentioned in the next section provides a similar check.

(5) Although rule III does not permit $\boldsymbol{H} \supset (D^r\boldsymbol{x} \supset \boldsymbol{P})$ to be deduced directly from $\boldsymbol{H} \supset (\boldsymbol{x}, r)\boldsymbol{P}$ if $\boldsymbol{x}$ is free in $\boldsymbol{H}$, the deduction may be made indirectly.

(6) The axiom of choice is optional, i.e. we may drop this axiom and still retain a system adequate for the greater part of mathematics.

(7) We shall not carry out any proofs in this paper, but the following provable formulas are of interest:

$$x = y \supset (D^r x \supset D^r y)$$

$$(x, r)(\boldsymbol{P} \equiv \boldsymbol{Q}) \supset (\imath x, r)\, \boldsymbol{P} = (\imath x,\ r)\boldsymbol{Q}$$

$$(x,\ r)\boldsymbol{A} = \boldsymbol{B} \supset (\lambda x,\ r)\boldsymbol{A} = (\lambda x, r)\, \boldsymbol{B}$$

$$(x,\ r)D^r\boldsymbol{A} \supset (x, r)[((\lambda x, r)\boldsymbol{A})x = \boldsymbol{A}]$$

$$D^{r+1}(\lambda x, r)\, \boldsymbol{A}$$

$$(f, r)(g,\ r)[(x,\ r+1)(xf = xg) \supset f = g]$$

$$D^{r+1}x \equiv [(y, r+1)(D^r y \,\&(D^r y \vee xy = C)\} \,\&\, D^{r+2}x]$$

3. Equivalence with Church's system

The nested-type system described above may be proved equivalent, in a certain sense, to Church's simplified theory of types.[3] The proof is long and tedious, and would not justify publication, but it may be of interest to give an exact statement of the equivalence theorem. The form of "equivalence" used has a certain interest in itself.

Definition. A logical system 1 will be said to be *equivalent* to the logical system 2 if to each proposition-like formula $\boldsymbol{A}$ of 1 we can make correspond a proposition-like formula $\boldsymbol{A}^{(1,2)}$ of 2, and conversely to each proposition-like formula $\boldsymbol{P}$ of 2 we can make correspond a proposition-like formula $\boldsymbol{P}^{(2,1)}$ of 1, in such a way that

(i) If $\boldsymbol{A}$ is provable in 1 then $\boldsymbol{A}^{(1,2)}$ is provable in 2.
(ii) If $\boldsymbol{P}$ is provable in 2 then $\boldsymbol{P}^{(2,1)}$ is provable in 1.
(iii) If $\boldsymbol{A}$ is a proposition-like formula of 1 then $(\boldsymbol{A}^{(1,2)})^{(2,1)} \equiv \boldsymbol{A}$ is provable in 1.

[3] Alonzo Church, *A formulation of the simple theory of types*, this Journal, vol. 5 (1940), pp. 56–68.

(iv) If $\boldsymbol{P}$ is a proposition-like formula of 2 then $(\boldsymbol{P}^{(2,1)})^{(1,2)} \equiv \boldsymbol{P}$ is provable in 2.
(v) If $\boldsymbol{A}$ and $\boldsymbol{B}$ are proposition-like formulas of 1 then we can prove $(\boldsymbol{A} \equiv \boldsymbol{B})^{(1,2)} \equiv (\boldsymbol{A}^{(1,2)} \equiv \boldsymbol{B}^{(1,2)})$ in 2.
(vi) If $\boldsymbol{P}$ and $\boldsymbol{Q}$ are proposition-like formulas of 2 then we can prove $(\boldsymbol{P} \equiv \boldsymbol{Q})^{(2,1)} \equiv (\boldsymbol{P}^{(2,1)} \equiv \boldsymbol{Q}^{(2,1)})$ in 1.
The formula $\boldsymbol{A}^{(1,2)}$ must be an effectively calculable function of $\boldsymbol{A}$ and $\boldsymbol{P}^{(2,1)}$ of $\boldsymbol{P}$.

It is understood that for each system there is defined a special kind of formulas called 'proposition-like formulas'; that every provable formula is necessarily proposition-like, and that it is a comparatively trivial matter to determine whether a formula is proposition-like or not. Specifically we may say that the statement "$\boldsymbol{A}$ is a proposition-like formula" should be equivalent to some statement of the form "$\varphi(n) = 0$" where n is the Gödel representation of $\boldsymbol{A}$ and φ is some primitive recursive function. It is also understood that both systems "include the propositional calculus": this is required in connection with the logical equivalence signs in (iii) to (vi).

We are justified in describing this relation as the equivalence of the two systems, for the relation is transitive, symmetric, and reflexive, as I shall now show. The symmetry of the relation follows at once from the fact that interchange of systems 1 and 2 simply interchanges conditions (i) and (ii), (iii) and (iv), (v) and (vi). Reflexiveness is proved by taking $\boldsymbol{A}^{(1,1)}$ to be $\boldsymbol{A}$. Transitivity is not quite so easy. We shall have to bring in a third system 3. We will define $\boldsymbol{A}^{(1,3)}$ to be $(\boldsymbol{A}^{(1,2)})^{(2,3)}$ and $\boldsymbol{A}^{(3,1)}$ to be $(\boldsymbol{A}^{(3,2)})^{(2,1)}$. We assume conditions (i) to (vi) to hold for the pairs 1, 2 and 2, 3 and attempt to prove them for the pair 1, 3. Because of the symmetry it is sufficient to prove (i), (iii), (v). To prove (i) we must prove $(\boldsymbol{A}^{(1,2)})^{(2,3)}$ in 3 assuming $\boldsymbol{A}$ provable in 1. Now by (i) for the pair 1,2 we see that $A^{(1,2)}$ is provable in 2, and then by (i) for the pair 2, 3 we get $(\boldsymbol{A}^{(1,2)})^{(2,3)}$ in 3. To prove (iii) we must prove $(((\boldsymbol{A}^{(1,2)})^{(2,3)})^{(3,2)})^{(2,1)} \equiv \boldsymbol{A}$ in 1.

Using (iii) for the pair 2,3 gives us $((\boldsymbol{A}^{(1,2)})^{(2,3)})^{(3,2)} \equiv \boldsymbol{A}^{(1,2)}$ (in 2), whence by (ii) for the pair 1, 2 we have

$$(((\boldsymbol{A}^{(1,2)})^{(2,3)})^{(3,2)} \equiv \boldsymbol{A}^{(1,2)})^{(2,1)}$$

Also by (vi) for the pair 1, 2 we have

$$(((\boldsymbol{A}^{(1,2)})^{(2,3)})^{(3,2)} \equiv \boldsymbol{A}^{(1,2)})^{(2,1)} \equiv ((((\boldsymbol{A}^{(1,2)})^{(2,3)})^{(3,2)})^{(2,1)} \equiv (\boldsymbol{A}^{(1,2)})^{(2,1)})$$

and by (iii) for the pair 1,2 we have

$$(\boldsymbol{A}^{(1,2)})^{(2,1)} \equiv \boldsymbol{A}$$

Combining these last three results by the rules of the propositional calculus we obtain

$$(((\boldsymbol{A}^{(1,2)})^{(2,3)})^{(3,2)})^{(2,1)} \equiv \boldsymbol{A}$$

as required.

To prove (v) for the pair 1,3 we must prove

$$((\boldsymbol{A} \equiv \boldsymbol{B})^{(1,2)})^{(2,3)} \equiv ((\boldsymbol{A}^{(1,2)})^{(2,3)} \equiv (\boldsymbol{B}^{(1,2)})^{(2,3)})$$

By an application of (v) to the pair 1,2 followed by an application of (i) to the pair 2,3 we get

$$((\boldsymbol{A} \equiv \boldsymbol{B})^{(1,2)} \equiv (\boldsymbol{A}^{(1,2)} \equiv \boldsymbol{B}^{(1,2)}))^{(2,3)}$$

and by an application of (v) to the pair 2,3 we have

$$((\boldsymbol{A} \equiv \boldsymbol{B})^{(1,2)} \equiv (\boldsymbol{A}^{(1,2)} \equiv \boldsymbol{B}^{(1,2)}))^{(2,3)} \equiv (((\boldsymbol{A} \equiv \boldsymbol{B})^{(1,2)})^{(2,3)} \equiv (\boldsymbol{A}^{(1,2)} \equiv \boldsymbol{B}^{(1,2)})^{(2,3)})$$

Combining these by the propositional calculus gives

$$((A \equiv B)^{(1,2)})^{(2,3)} \equiv (A^{(1,2)} \equiv B^{(1,2)})^{(2,3)}$$

Condition (v) applied to 2,3 also gives

$$(A^{(1,2)} \equiv B^{(1,2)})^{(2,3)} \equiv ((A^{(1,2)})^{(2,3)} \equiv (B^{(1,2)})^{(2,3)})$$

from which we now obtain the required result.

Our definition of the equivalence of two systems could be summed up by saying that they are equivalent if we can translate from either system to the other in such a way that provable propositions translate into provable propositions again, and so that a double translation gives rise to a proposition equivalent to the original. This explanation ignores the last two conditions (v) and (vi), which are rather too tenuous for such rough handling.

The equivalence theorem then states that the nested-type system is equivalent to Church's system, if the proposition-like formulas of the nested-type system are taken to be the proposition formulas without free variables, and the proposition-like formulas of Church's system are those of type o without free variables.

4. Relaxation of type notation

The form of type theory which we have described is one in which the types themselves do not intrude very much. Even so they do still intrude to an appreciable extent, and it would be desirable to see how much further they can be relegated to the background. A possible way of doing so will be described in this section.

We could sum up the effect of type theory as it appears in this system by saying that we give no meaning to the expressions 'for all $\boldsymbol{x}$, $\boldsymbol{A}$,' 'there exists an $\boldsymbol{x}$, such that $\boldsymbol{A}$,' 'the $\boldsymbol{x}$, such that $\boldsymbol{A}$,' 'the function whose value for argument $\boldsymbol{x}$ is $\boldsymbol{A}$' (usually expressed symbolically as $(\boldsymbol{x})\boldsymbol{A}$, $(\exists \boldsymbol{x})\boldsymbol{A}, (\imath \boldsymbol{x})\boldsymbol{A}$, $(\lambda \boldsymbol{x})\boldsymbol{A}$, respectively). Instead we give meaning to the expressions $(\boldsymbol{x}, r)\boldsymbol{A}$, $(\exists \boldsymbol{x}, r)\boldsymbol{A}, (\imath \boldsymbol{x}, r)\boldsymbol{A}, (\lambda \boldsymbol{x}, r)\boldsymbol{A}$. Nevertheless in a large class of cases we *can* assign meanings to $(\boldsymbol{x})\,\boldsymbol{A}$, $(\exists \boldsymbol{x})\boldsymbol{A}, (\imath \boldsymbol{x})\boldsymbol{A}$, $(\lambda \boldsymbol{x})\boldsymbol{A}$ in a satisfactory manner. A typical case is that of a formula of the form $(\imath x)\boldsymbol{P}$ where $\boldsymbol{P}$ is such that we can prove $\boldsymbol{P} \supset \boldsymbol{D}^{10}x$, say. In this case for any integers $r, s \geqq 10$ we can prove $(\imath x, r)\boldsymbol{P} = (\imath x, s)\boldsymbol{P}$ and it is therefore natural to stipulate that $(\imath x)\boldsymbol{P}$ shall stand for the common value of $(\imath x,\ 10)\boldsymbol{P}, (\imath x, 11)\boldsymbol{P}, \cdots$. We may say more generally that if $(\imath \boldsymbol{x}, r_0)\boldsymbol{P} = (\imath \boldsymbol{x}, r)\boldsymbol{P}$ is provable for all $r \geqq r_0$ then $(\imath \boldsymbol{x})\boldsymbol{P}$ shall be said to be interpretable and to have the interpretation $(\imath \boldsymbol{x}, r_0)\ \boldsymbol{P}$. This is of course still only the beginning of a definition of "the interpretation of a formula with some type bounds omitted." In order to give the complete definition we must deal properly with formulas having free variables: results such as $\boldsymbol{P} \supset \boldsymbol{D}^{10}x$ (quoted above) are not normally provable if $\boldsymbol{P}$ has free variables other than x. On this account we introduce the idea of "interpretability under hypotheses"; the hypotheses involved are usually of the form $D^r\boldsymbol{x}$. The complete definition is as follows:

All variables and C, T', F' provide their own interpretations under any hypotheses.

If $\boldsymbol{A}, \boldsymbol{B}, \boldsymbol{P}, \boldsymbol{Q}$ have interpretations $\boldsymbol{A}', \boldsymbol{B}', \boldsymbol{P}', \boldsymbol{Q}'$ under certain hypotheses, then $(\boldsymbol{AB})$, $(\boldsymbol{A} = \boldsymbol{B})$, $\boldsymbol{D}^r\boldsymbol{A}, (\boldsymbol{P} \supset \boldsymbol{Q})$, $(\sim\boldsymbol{P})$ have the interpretations $(\boldsymbol{A}'\boldsymbol{B}'), (\boldsymbol{A}' = \boldsymbol{B}')$, $\boldsymbol{D}^r\boldsymbol{A}'$, $(\boldsymbol{P}' \supset \boldsymbol{Q}'), (\sim\boldsymbol{P}')$ respectively under the same hypotheses.

If, for each $r \geqq r_0$, $\boldsymbol{P}$ has the interpretation $\boldsymbol{P}_r$ under hypothesis $\boldsymbol{H}$ & $\boldsymbol{D}^r\boldsymbol{x}$ where $\boldsymbol{H}$ does not contain $\boldsymbol{x}$ free and we can prove

$$\textbf{(A)} \qquad \boldsymbol{H} \supset (\imath \boldsymbol{x},\ r_0)\boldsymbol{P}_{r0} = (\imath \boldsymbol{x}, r)\boldsymbol{P}_r$$

then $(\imath \boldsymbol{x})\boldsymbol{P}$ has the interpretation $(\imath \boldsymbol{x},\ r_0)\boldsymbol{P}_{r_0}$ under hypothesis $\boldsymbol{H}$. If instead of (**A**) we can prove

$$\boldsymbol{H} \supset [(\boldsymbol{x}, r_0)\boldsymbol{P}_{r_0} \equiv (\boldsymbol{x},\ r)\boldsymbol{P}_r]$$

then $(\boldsymbol{x})\boldsymbol{P}$ has the interpretation $(\boldsymbol{x},\ r_0)\boldsymbol{P}$ under $\boldsymbol{H}$.

No formula has any interpretation unless compelled to by the foregoing.

It may be observed that every formula of the nested-type system is interpretable and provides its own interpretation. Also that if $\boldsymbol{H} \supset \boldsymbol{K}$ is provable and a formula has a certain interpretation under $\boldsymbol{K}$ then it has the same interpretation under $\boldsymbol{H}$.

If $\boldsymbol{P}$ has the interpretation $\boldsymbol{P}_r$ under $\boldsymbol{H}\ \&\ \boldsymbol{D}^r\boldsymbol{x}$ and we wish to show either that $(\imath\boldsymbol{x})\boldsymbol{P}$ has the interpretation $(\imath\boldsymbol{x}, r_0)\boldsymbol{P}_{r_0}$, or that $(\exists\boldsymbol{x})\boldsymbol{P}$ has the interpretation $(\exists\boldsymbol{x},\ r_0)\boldsymbol{P}_{r_0}$ under $\boldsymbol{H}$, it is sufficient to prove $\boldsymbol{P}_r \supset \boldsymbol{D}^{r_0}\boldsymbol{x}(r \geqq r_0)$.

It will be seen that this definition does not provide an effective means of determining whether or not an expression is interpretable. This need not be considered a serious drawback, as we seldom need to establish that an expression is not interpretable.

The most natural cases where we can apply the above definitions are those of $(\boldsymbol{x})(\boldsymbol{A} \supset \boldsymbol{B})$, $(\exists\boldsymbol{x})(\boldsymbol{A}\ \&\ \boldsymbol{B})$, $(\imath\boldsymbol{x})(\boldsymbol{A}\ \&\ \boldsymbol{B})$ where $\boldsymbol{A} \supset \boldsymbol{D}^{r_0}\boldsymbol{x}$ is provable for some $\boldsymbol{r}_0$. It is fairly easy to remember which are the most important expressions $\boldsymbol{A}$ of this kind: e.g. in almost any formalisation we shall have " '$\boldsymbol{x}$ is a real number' $\supset \boldsymbol{D}^{r_0}\boldsymbol{x}$" with $r_0 = 10$ say; this fact would be remembered in the form "the class of real numbers is all right." It is not so easy to remember the appropriate numbers r_0, but it is hardly necessary to do so if the notations $(\boldsymbol{x})\boldsymbol{A}$ etc. are adhered to throughout. When $\boldsymbol{A}$ is such that for some r_0 we can prove $\boldsymbol{A} \supset \boldsymbol{D}^{r_0}\boldsymbol{x}$ I shall call the class of $\boldsymbol{x}$ for which $\boldsymbol{A}$ is true a "noun-class." "There is a very close connection between the part played by the formulas $\boldsymbol{A}$ in our system and nouns in ordinary language; so much so that one might say that type theory had been instinctively obeyed for thousands of years before its discovery by Russell. This connection may be seen by translating $(\boldsymbol{x})(\boldsymbol{A} \supset \boldsymbol{B})$, $(\exists\boldsymbol{x})\ (\boldsymbol{A}\ \&\ \boldsymbol{B})$, $(\imath\boldsymbol{x})\ (\boldsymbol{A}\ \&\ \boldsymbol{B})$ roughly as "All $\boldsymbol{A}$ satisfy $\boldsymbol{B}$," "There exists an $\boldsymbol{A}$ satisfying $\boldsymbol{B}$" and "The $\boldsymbol{A}$ which satisfies $\boldsymbol{B}$." In each case $\boldsymbol{A}$ is translated in the form of a noun. It seems that the necessity to use nouns prevents us automatically from committing type fallacies in common speech. We can probably only break down this 'safety device' by using nouns such as 'thing' or 'object' with the intended meaning 'anything whatever.' In the case of the Russell paradox ('class of all classes which are not members of themselves') we use the word 'class' in very much that way. We use it to mean 'class of anythings whatever.'

There are various ways in which we might make use of the idea of interpretable formulas to transform what we have called the 'nested type system' into something rather more closely analogous to common mathematical practice. One possibility is simply to regard the formulas without types as abbreviations of the appropriate formulas of the nested-type system, such formulas only being used when the appropriate metamathematical result justifying the interpretation has been established. This does not seem to be really satisfactory because of the frequent need to prove such metamathematical results. Alternatively we may set up some new symbolic system in which the formulas form a considerably wider class than those of the nested-type system, and are all interpretable as defined above. The author has investigated two such systems. In one of them the expression. $(\boldsymbol{x},\boldsymbol{A})\boldsymbol{P}$ had the meaning which we have assigned to $(\boldsymbol{x})(\boldsymbol{A}\boldsymbol{x} = T' \supset \boldsymbol{P})$. This is always interpretable if $\boldsymbol{A}$ is interpretable and without free variables. This scheme leads to rather heavy formulas in the elementary stages, though it may have advantages when more advanced branches of mathematics are reached. The second system appears rather more hopeful, and will now be described briefly. It may be called the "concealed type" system.

The formulas in the concealed type system will be described as "admissible formulas" to distinguish them from the formulas of the nested-type system. The admissible formulas will in fact be included amongst the interpretable formulas associated with the nested-type system. There will be admissible term formulas (ATF) and admissible proposition formulas (APF). We define APF, ATF, and provable formula by a simultaneous induction. Consequently there is no rule for determining whether an expression is an admissible formula or not: this is not usual in logical systems, but there seems to be no good reason for a positive taboo on such an arrangement. We now give the inductive definitions.

Every term variable is an ATF and every proposition variable is an APF.

The symbols E, C, T', F' are ATF.

If $\boldsymbol{A}, \boldsymbol{B}, \boldsymbol{F}$ are ATF and $\boldsymbol{P}, \boldsymbol{Q}, \boldsymbol{R}, \boldsymbol{S}$ are APF, and $\boldsymbol{P} \supset \boldsymbol{A}\boldsymbol{x} = \boldsymbol{T}', {\sim}\boldsymbol{Q} \supset \boldsymbol{A}\boldsymbol{x} = \boldsymbol{T}'$ are provable formulas then $(\exists \boldsymbol{x})\boldsymbol{P}, (\boldsymbol{x})\boldsymbol{Q}, (\boldsymbol{B} = \boldsymbol{F}), (\boldsymbol{R} \supset \boldsymbol{S}), {\sim}\boldsymbol{R}$ are APF, and $(\imath \boldsymbol{x})\boldsymbol{P}$, $(\boldsymbol{B}\boldsymbol{F})$ are ATF. The variable $\boldsymbol{x}$ must not occur free in $\boldsymbol{A}$.

Free and bound occurrences of variables are defined as in the nested-type system.

The symbol E corresponds to $(\lambda x, 0)T'$ of the nested type system. Its main purpose is to take the place of D^0 and indirectly to replace the other D^r. For any formula $\boldsymbol{A}$ we can prove $((\lambda x, 0)T')\boldsymbol{A} \equiv D^0\boldsymbol{A}$ in the nested-type system.

If $\boldsymbol{A}$ and $\boldsymbol{B}$ are ATF not containing $\boldsymbol{x}$, $\boldsymbol{y}$, or $\boldsymbol{z}$ free then the two expressions below are ATF, viz.

$$(\imath y)(x)[(yx \vee yx = C) \;\&{\sim}Ey \;\&\; (yx \equiv (Ax \vee Bx))]$$

$$(\imath y)(x)[(yx \vee yx = C) \;\&{\sim}Ey \;\&\; (yx \equiv (z)\{(Bz \supset A(xz) \;\&\; (Bz \supset xz = C)\}]$$

They may be abbreviated respectively to *Sum* $\boldsymbol{A}\ \boldsymbol{B}$ and *Pot* $\boldsymbol{A}\ \boldsymbol{B}$. In these formulas we have adopted the useful convention that a formula of form $\boldsymbol{A} = T'$ may be abbreviated to $\boldsymbol{A}$. The context will always enable one to determine when this abbreviation has been applied. We shall continue to use this convention.

Strictly speaking the definitions of *Sum* $\boldsymbol{AB}$ and *Pot* $\boldsymbol{AB}$ are invalid because the bound variables $\boldsymbol{x}$, $\boldsymbol{y}$, $\boldsymbol{z}$ were not specified. This technical difficulty may be resolved by requiring $\boldsymbol{x}$, $\boldsymbol{y}$, $\boldsymbol{z}$ to be the three earliest variables not appearing free in $\boldsymbol{A}$, $\boldsymbol{B}$.

The remainder of the definition consists of the axioms and rules of procedure. It may be remembered that these took the form of a definition in the nested-type system also

Rules of procedure (*concealed-type system*).

Rule I. The formulas

$$(x)P \equiv (y)S^x_y P|$$

$$(\imath x)P = (\imath y)S^x_y P|$$

are provable if $(\boldsymbol{x})\boldsymbol{P}$ is an APF in which $\boldsymbol{x}$ is not bound in $\boldsymbol{P}$,$\boldsymbol{y}$ does not occur free, $\boldsymbol{x}$ does not occur at a place where $\boldsymbol{y}$ would be bound, and $(\imath \boldsymbol{x})\boldsymbol{P}$ is an ATF.

Rule II. If $\boldsymbol{P}$ is provable, then $S^x_A P|$ and $S^q_Q P|$ are provable, where $\boldsymbol{A}$ and $\boldsymbol{P}$ are respectively an ATF and an APF, and the bound variables of $\boldsymbol{P}$ are distinct both from $\boldsymbol{x}$ and $\boldsymbol{q}$ and from the free variables of $\boldsymbol{A}$ and of $\boldsymbol{Q}$.

Rule III. If $\boldsymbol{H} \supset \boldsymbol{P}$ and $\boldsymbol{H} \supset (\boldsymbol{x})\boldsymbol{P}$ are both APF and one of them is provable then the other is provable also.

Rule IV. Any tautologous formula of the propositional calculus is provable.

Rule V. If the formulas $\boldsymbol{P} \supset \boldsymbol{Q}$ and $\boldsymbol{P}$ are both provable then $\boldsymbol{Q}$ is provable.

Rule VI. If $\boldsymbol{P}$ is an APF in which x does not occur bound, then the formulas

$$(\exists! x)P \supset S^x_{(\imath x)P} P|$$

$${\sim}(\exists! x)P \supset (\imath x)P = C$$

are provable provided they are APF.

Rule VII. If $\boldsymbol{A}$ is an APF in which x, y, z, u do not occur free, then

$$(x)(ux \supset z\boldsymbol{A}) \supset (\exists y)[(Potzu)y \,\&\, (x)(ux \supset yx = \boldsymbol{A})]$$

is provable.

In rule VI the definition

$$(\exists!\boldsymbol{x})\boldsymbol{P} \to (\exists \boldsymbol{x})\boldsymbol{P} \,\&\, (\boldsymbol{x})(\boldsymbol{y})\ (\boldsymbol{P}\& \ S^{\boldsymbol{x}}_{\boldsymbol{y}}\boldsymbol{P}| \supset \boldsymbol{x} = \boldsymbol{y})$$

is understood, $\boldsymbol{y}$ standing for a variable not occurring free in $\boldsymbol{P}$.

The axioms are:

A1 $C \neq T' \& C \neq F' \,\&\, T' \neq F'$

A2 $EC \,\&\, ET' \,\&\, EF'$

A3 $Ex \supset xy = C$

B1 $x = x$

B2 $(y = x \,\&\, y = z) \supset x = z$

B3 $(x = y) \supset (zx = zy \,\&\, xz = yz)$

C1 $[(Pot\ yu)f \,\&\, (Pot\ yu)g \,\&\, (x)(ux \supset fx = gx)] \supset f = g$

C2 $(\exists i)[(Pot\ u(Pot\ Eu))i \,\&\, (f)\ \{[(Pot\ Eu)f \,\&\, (\exists x)fx] \supset f(if)\}]$

C3 $(\exists t)[(Pot\,E\ E)t \,\&\, (\exists v)[Ev \,\&\, (x)(y)\{(Ex \,\& Ey)$
$\supset ((tx = ty \supset x = y) \,\&\, v \neq tx)\}]]$

To complete our inductive definition we need only add that no expression is an ATF, APF, or provable formula unless compelled to be so by the foregoing.

We may say that roughly speaking type theory appears in the concealed type system only through the condition that $\boldsymbol{P} \supset \boldsymbol{A}\boldsymbol{x} = \boldsymbol{T}'$ must be provable if $(\imath\boldsymbol{x})\boldsymbol{P}$ is to be an ATF, and a similar condition for $(\boldsymbol{x})\boldsymbol{P}$. The system is related to the nested-type system by the following metamathematical results:

(1) If we substitute $(\lambda x,\ 0)T'$ for E throughout an admissible formula without free variables we obtain an interpretable formula.
(2) If in a provable formula of the concealed-type system without free variables we make the substitution mentioned in (1) and then form an interpretation of the resulting formula we obtain a provable formula of the nested-type system.
(3) Every provable formula of the nested-type system is obtainable as in (2).

A valuable aid in the proof of these is the following result which concerns the nested-type system only:

(4) If $\boldsymbol{A}$ is a term formula containing only the variables $\boldsymbol{x}_1, \boldsymbol{x}_2, \ldots, \boldsymbol{x}_n$ free, and $m_1, m_2, \ldots, m_n$ are non-negative integers, then there is an integer k such that $D^{m_1}\boldsymbol{x}_1 \,\&\cdots\&\, D^{m_n}\boldsymbol{x}_n \supset D^k\boldsymbol{A}$ is provable.

NATIONAL PHYSICAL LABORATORY, TEDDINGTON

The use of Dots as Brackets in Church's System

(J. Symbolic Logic, vol. 7 (1942), pp. 146–156)

Lance Fortnow discovers —

TURING'S DOTS

Alan Turing's rarely cited paper 'The Use of Dots as Brackets in Church's System' defines a new notation for Church's λ-calculus using what Turing calls dots, the symbols '.' and ':'.

Turing states that he intended to make use of this notation in forthcoming papers entitled 'Some theorems about Church's systems' and 'The theory of virtual types'. I can find no record of those later papers. Likely Turing's activities during World War II curtailed his scientific research, and his interests shifted after the war.

Even though this paper had little to no direct influence to logic and computer science, it shows once again Turing's ability to reason about important issues in computer science before there were digital computers to reason about. In this case, Turing essentially studies an important aspect of programming languages, a syntax for trees.

To understand the paper, consider precedence operations on formulas such as

$$4x - 3y^2 + 7$$

To parse this equation, we need to know that exponentiation has precedence over multiplication, which has precedence over addition and subtraction. Operations with the same precedence occur left to right. The expression above can be written with parenthesis as

$$(((4x) - (3(y^2))) + 7)$$

Turing creates virtual precedence operations he calls dots and shows how it can replace balance parentheses used by Church ("A Formulation of the Simple Theory of Types," *J. Symbolic Logic* 5, pp. 56–68 (1940)).

Zero dots has highest precedence, then one dot (.), two dots (:), three dots (:.) etc. At the same precedence, application is done left to right. So the expression

$$a : .cd : e.fg$$

is evaluated as

$$(a((cd)(e(fg)))).$$

Thanks to my student Arefin Huq for helping me with 'breaking the code' of Turing's dot notation and to Robby Findler for helpful discussions.

Although Turing doesn't discuss trees, both notations describe binary trees. In this case

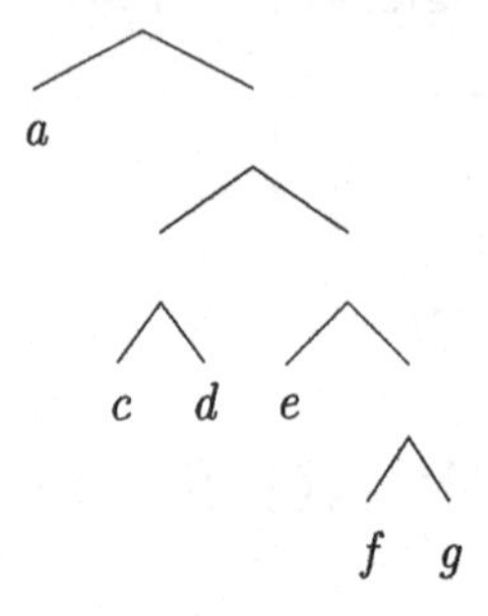

Every binary tree can be expressed through dots or through parentheses. Turing acknowledges that for simplicity, sometimes dot can be mixed with parenthesis or other precedence operators.

Today we have a common method for creating trees known as Extensible Markup Language (XML). The tree above can be described by

```
<t1>
a
<t2>
<t3> c d</t3>
<t4> e
<t5> f g </t5>
</t4>
</t2>
</t1>
```

The tags (`<t1>, <t2>...`) act like parentheses. According to Northwestern Professor Robby Findler, an expert in programming languages, no major system uses virtual precedence operations akin to Turing's dots.

The reason is modularity. Suppose we wanted to replace g with a subtree consisting of r and s. With parentheses we can do a simple replacement

$$(a((cd)(e(fg))))$$

becomes

$$(a((cd)(e(f(rs))))).$$

With dots although we have to readjust the whole formula,

$$a :.cd : e.fg$$

becomes

$$a :: cd :.e : f.rs$$

In short, Turing's dots gave him a way to think about the order of operations in a structure that was more intuitive to him to prepare him for planned future work on Church's λ-calculus. Unlike the Turing machine, the dot notation did not catch on for reasons Turing did not appreciate: that someone might want to modify the code.

THE USE OF DOTS AS BRACKETS IN CHURCH'S SYSTEM

A. M. TURING

Any logical system, if its use is to be carried beyond a rather elementary stage needs powerful conventions about abbreviations: in particular one usually wants to modify the bracketing so as to make the formulae more readable, and also possibly shorter. The present note has been written in the belief that Church's formulation of the simple theory of types[1] is particularly suitable as a basis for work on that theory, and that it is therefore worth while introducing special conventions which take into account the needs of this particular system. The conventions which I shall describe are ones which I have used a good deal myself, and have always found adequate. I intend to make use of them in forthcoming papers.[2] They may be regarded as an extension of Curry's conventions.[3]

I shall begin with a general discussion of punctuation by means of groups of dots. This general theory is applicable, with some modifications, to Russell's,[4] Quine's,[5] and Curry's[3] bracketing systems as well as to the present one.

General bracketing theory

We consider a logical system in which every formula is either:

An irreducible formula (or *token* in Curry's terminology).
Of form $\boldsymbol{R}(\boldsymbol{A})$ where $\boldsymbol{R}$ is a monadic operator and $\boldsymbol{A}$ a formula.
Of form $(\boldsymbol{A})\boldsymbol{S}(\boldsymbol{B})$ where $\boldsymbol{S}$ is a dyadic operator and $\boldsymbol{A}$ and $\boldsymbol{B}$ are formulae.

We need not of course enquire further into the nature of the irreducible formulae, monadic operators, and dyadic operators, but to fix our ideas we may think of irreducible formulae as consisting of a single letter with suffixes etc., e.g. x_α, $J^{\alpha,\beta}_{\beta(o\alpha)}$. Typical of monadic operators would be $\sim$, $[\exists x_\alpha]$ and of dyadic operators $\supset$ and $=$. The formulae in this sense will be described in future as *unabbreviated formulae*: the word 'formula' without qualification will be liable to be used of various kinds of series of symbols.

We may also recognise another kind of formulae which we call *abbreviated formulae* and which consist of series of symbols which are irreducible formulae, brackets, monadic and dyadic operators, and a new kind of symbol called a point, which may be thought of as a group of dots. To be an abbreviated formula the series of symbols must satisfy the conditions:

(a) The brackets must be properly paired, i.e., if we go on removing pairs of brackets which face each other and have no other brackets between them there should eventually be no brackets left.

Received June 17, 1942.

[1] Alonzo Church, *A formulation of the simple theory of types*, this JOURNAL, vol. 5 (1940), pp. 56–68.

[2] A. M. Turing, *Some theorems about Church's system, and The theory of virtual types*, forthcoming.

[3] H. B. Curry, *On the use of dots as brackets in logical expressions*, this JOURNAL, vol. 2 (1937), pp. 26–28.

[4] Whitehead and Russell, ***Principia mathematica,*** vol. 1, pp. 9–11.

[5] W. V. Quine, ***Mathematical logic*** (New York 1940), pp. 37–42.

The brackets appearing in an abbreviated formula will often be described as 'explicitly shown brackets.'

(b) Of a pair of brackets one must occur adjacent to an operator and one not. The expression 'adjacent to an operator' is used here and elsewhere to mean 'adjacent to a dyadic operator or adjacent to and on the right of a monadic operator.'

(c) If in the formula we replace dyadic operators by '*D*', monadic operators by '*M*', irreducible formulae by '*x*' and points by '**:**', calling the result the 'projected formula,' then the first symbol of a projected abbreviated formula must be '(', '*x*', or '*M*' and the last, ')' or '*x*'. A pair of consecutive symbols in the projected formula must be '*x*)', '(*x*', '))', '*M*(', '*D*(', ')*D*', '(*M*' or '((' or else part of one of the following series of three: '*x*:*D*', '*D*:*x*', '*M*:*x*', '):*D*', '*D*:(', '*M*:(', '*D*:*M*', '*M*:*M*': in the latter case the whole series of three symbols must be part of the projected formula.

We want one and only one formula to correspond to each abbreviated formula. Such a correspondence is defined below in terms of an ordering of the points. I shall follow Russell's terminology and speak of the earlier of two points in the ordering as being of *higher power* than the other. Curry uses the expression 'senior to' and Quine, whose points are called 'joints,' uses 'looser than.' The power of a point may depend on any formal relationships between the point and the formula it occurs in, and varies from system to system.

The rule for replacing the abbreviated formula by the unabbreviated may be put into two forms, of which the first is the more natural theoretically, and the second, which seems rather arbitrary, is the easier to apply.

First form of rule

The rule operates by reducing the number of points in the formula whose unabbreviated form is to be found.

Suppose first that the formula has explicit brackets, e.g. that it is of form ***A*(*B*) *C***, where ***A***, ***B***, ***C*** are not required to be formulae in any special sense, but just rows of symbols, and the brackets shown are properly paired. Then the unabbreviated form of ***A*(*B*)*C*** may be obtained from the unabbreviated forms ***E*** of ***AwC*** and ***F*** of ***B*** by substituting (***F***) for ***w*** in ***E***. The symbol ***w*** is to be some symbol not occurring in ***A*** or ***C***. In other words the interior of an explicitly shown bracket is to be worked out as if it were a whole formula, and the part of the formula outside the bracket is to be worked out as if the bracketed part were a single letter.

If the formula has no explicitly shown brackets we find the point of highest power and replace it by a bracket. This bracket is to be right facing if the point is right facing, i.e., if it is on the right of its operator: similarly the bracket is left facing if the point is left facing. Another bracket, oppositely facing, must be put at one end of the formula to balance the first.

Second form of rule

We first define the *enclosing brackets* of a symbol other than an explicitly shown bracket. They are paired explicitly shown brackets, enclosing the symbol in question, but not enclosing any other pair of brackets which enclose the symbol. If the enclosing brackets are always to be defined there must be a pair of brackets enclosing the whole formula. We imagine these supplied.

To find the unabbreviated formula we clearly have to replace each point by a similarly facing bracket, and to put in a balancing bracket somewhere. The interval from the point to the balancing bracket is called the scope of the point: in reckoning scopes, points and brackets will be neglected, so that for instance if two similarly facing points are to have their partnering brackets immediately following one another their scopes will be regarded as ending at the same place. The rule for determining the scope is that it is to be as short as possible, subject to the following *scope condition*:

> The balancing bracket β of a point π is either adjacent to one of the enclosing brackets of π, or else to some point ρ facing oppositely to π and having the same enclosing brackets as π in which case β must be on the side of ρ which is nearer to π. The point ρ must be of higher power than π or any point between ρ and π facing similarly to π and having the same enclosing brackets as π.

Equivalence theorem

There are three things to be proved about these rules:

(i) When we use the first rule it does not matter in what order the pairs of explicit brackets are taken.
(ii) The result of applying the first rule to an 'abbreviated formula' (satisfying by definition conditions (a), (b), (c) above) is to give us an 'unabbreviated formula' as originally defined.
(iii) The two rules are equivalent.

To prove (i) let $\boldsymbol{A}(\boldsymbol{B})\ \boldsymbol{C}$ be one of the shortest formulae for which the result of applying the rule is not unique. We are justified in assuming that explicit brackets occur for otherwise the first step in applying the rule is uniquely determined and consists in introducing brackets. Whatever transformation we apply to the formula it remains of the form $\boldsymbol{A}'(\boldsymbol{B}')\boldsymbol{C}'$ where $\boldsymbol{A}'\boldsymbol{w}\ \boldsymbol{C}'$ is obtained from $\boldsymbol{AwC}$ and $\boldsymbol{B}'$ from B by a (possibly incomplete) application of the rule. In particular this is true of the final result of applying the rule. In this case In this case $\boldsymbol{AwC}'$ contains no points: it is therefore the final result of applying the rule to $\boldsymbol{AwC}$ and since $\boldsymbol{AwC}$ is shorter than $\boldsymbol{A}(\boldsymbol{B})\ \boldsymbol{C}$ the formula must be unique. Similarly $\boldsymbol{B}'$ is unique, and therefore $\boldsymbol{A}'(\boldsymbol{B}')\boldsymbol{C}'$ is unique.—The word 'shortest' as used in this argument must be interpreted as 'having the smallest number of symbols, points however being reckoned as two symbols.'

To prove (ii) it is sufficient to show that the application of the transformations described in the rule always leaves us with an abbreviated formula, and that if an abbreviated formula has no points then it is an unabbreviated formula. The transformations always consist of the removal of a point and the introduction of a pair of brackets. The brackets have no other brackets between them, so that the brackets remain properly paired, i.e., (a) remains satisfied. One of the brackets replaces a point, and therefore by (c) applied to the original formula is adjacent to an operator. The other bracket is put in either at the end of the formula or adjacent to a similarly facing bracket, facing away from it. It cannot be adjacent to an operator, for if it were there would have been an operator adjacent to the end of the formula, or to a bracket facing towards it in the original formula, contradicting (c). This shows that (b) remains true. To show that (c) remains true we have only to notice that when we replace points by similarly facing brackets in the admissible combinations the results are made up of admissible combinations, and that admissible combinations always result when a bracket is introduced at the end of the formula or adjacent to a similarly facing bracket.

To prove the second requirement let us see what condition (c) amounts to when there are no points in the formula. The allowable pairs of symbols in the projected formula are '*x*)', '*M*(', '*D*(', ')*D*', '))', '(*x*', '(*M*', '((' and a formula must start with '(', '*M*', or '*x*' and end with '*x*' or ')'. If it starts with '*x*' it can only continue with '(', and this bracket can have no partner: i.e., if the projected formula starts with '*x*' then '*x*' is the whole of it. If it starts with '*M*' it continues with '(', and this bracket has a partner, so that the whole is of form $M(\boldsymbol{A})\boldsymbol{B}$, and by (b) of form $M(\boldsymbol{A})$. If the formula starts with '(' this has a partner which by (b) is adjacent to an operator: i.e., the formula is of form $(\boldsymbol{A})D\boldsymbol{B}$ and therefore of form $(\boldsymbol{A})D(\boldsymbol{C})\boldsymbol{E}$. Applying (b) we see it is of form $(\boldsymbol{A})D(\boldsymbol{C})$. Thus we have shown that abbreviated formulae without points are always either irreducible formulae or of one of the forms $\boldsymbol{R}(\boldsymbol{A})$ or $(\boldsymbol{A})\boldsymbol{S}(\boldsymbol{B})$, where $\boldsymbol{R}$ is a monadic and $\boldsymbol{S}$ a dyadic operator. The formulae $\boldsymbol{A}$ and $\boldsymbol{B}$ necessarily satisfy the conditions (a), (b), (c) since the whole formula satisfies them, and the symbols allowed at the ends of a formula by (c) are just the ones which may follow a right facing bracket or precede a left facing bracket: these formulae are therefore themselves 'abbreviated formulae.' An induction over the length of the formula will now prove that every abbreviated formula without points is an unabbreviated formula, as required.

To prove (iii) notice that the second rule agrees with the first as regards the replacement of the points of highest power, for with either rule we may suppose that the enclosing brackets of the point to be replaced are at the ends of the formula. It will therefore be sufficient to prove that the order of replacement of two points may be interchanged when we are using the second rule.

The case when the two points did not originally have the same enclosing brackets is trivial, for then the replacement of the one point does not alter the set of symbols having the same enclosing brackets as the other, and therefore does not alter its scope. We may therefore suppose that the enclosing brackets of both points are at the ends of the formula. We may also suppose that there are no other brackets in the formula, for if any pair of brackets, together with what is between them, is replaced by a single letter, the scope of neither of the points is altered.

The scopes of two points can never be strictly overlapping. Suppose that the scope of one point is limited by brackets α and β of which α is the one further to the left, and the other by γ and δ of which γ is to the left; also that α is to the left of β and that the scopes strictly overlap, so that the brackets form a figure like this

$$\underset{\alpha}{(}\ldots\underset{\gamma}{(}\ldots\underset{\beta}{)}\ldots\underset{\delta}{)}$$

The points from which these brackets arise can be either at α and γ, or at α and δ, or at β and γ or at β and δ. The consideration of the last alternative can be omitted as it is the same as the first apart from interchange of left and right. In the case that the points are at α and γ the brackets α, β must satisfy the scope condition, so that the point at β must be of higher power than those at α and γ or any right facing point between α and β; in particular it is of higher power than those at γ and between γ and β, and therefore by the scope condition the bracket δ partnering γ must have the same position as β, in which case the scopes do not strictly overlap. Next suppose that the points are at α and δ. Then applying the scope condition to the brackets α and β we find that the point at β is stronger than that at γ, and this means that the scope condition cannot be satisfied for a point at δ whose partnering bracket is at γ. Finally suppose that the points are at β and γ. Applying the scope condition to γ and δ we see that either γ or some right facing point between it and β is of higher power than β: but if this is so the scope condition cannot be satisfied for α and β.

This completes the proof that the scopes of two points can never be strictly overlapping, and we now apply it to the interchange of order of removal of brackets under the second rule. Suppose that the scope of the first point is from α to β, the point itself being at α, which we suppose to be to the left of β, and the scope of the second from γ to δ; γ being to the left of δ, but no assumption being made as to whether the point was at γ or δ. We wish to show that the scope of the first point as calculated by the scope condition is unaltered if the other point is replaced by its brackets γ, δ. To fix our ideas we suppose that the scopes α to β and γ to δ are as calculated before either pair of brackets has been put in. The scope of the first point is certainly unaltered by the replacement of the second in the case that the scopes do not overlap at all, for then neither the points within the interval α to β, nor the left facing point (or possibly bracket) at β can be altered by the introduction of γ and δ, and the application of the scope condition gives exactly the same result for the position of β. As the scopes cannot strictly overlap we must suppose that either the interval α to β is wholly contained in the interval γ to δ or wholly contains it. In the first case the data for the application of the scope condition to the bracket β are again not relevantly altered. If the interval γ to δ is wholly contained in the interval α to β we consider separately the possibilities that β might be moved farther to the right or farther to the left by the introduction of γ and δ. To show that β is not moved farther to the right it will be sufficient to show that the interval still satisfies the scope condition. This is certainly the case, for the effect of the introduction of γ and δ, so far from introducing new right facing points is to enclose some in brackets, thereby as it were disqualifying them, and also to remove the point from which γ and δ themselves arose. To show that β is not moved farther to the left we have to show that there can be no left facing points ρ between α and β which satisfy the scope condition. Such a point would certainly have to be between δ and β, for if it were between γ and δ it would not have the same enclosing brackets as α, and if it were between α and γ the position of β would have been at ρ regardless of whether the brackets γ and δ had been put in or not. If ρ between δ and β satisfies the scope condition, then in the original formula there must have been a right facing point σ either at γ, or in the interval γ to δ, which was more powerful than ρ and less powerful than the point at β. However, as the scope of the bracket γ, δ, if it arises from a point at γ, extends only as far as δ, there must have been a point at δ more powerful than σ and therefore than ρ and all right facing points in the interval α to γ. The original position of β would therefore have been the position of δ. If on the other hand the brackets γ and δ arise from a point at δ, then ρ must have been less powerful than some right facing point σ in the interval γ to δ without the alternative of σ being γ itself. We may suppose that σ is the right facing point of highest power in the interval γ to δ. But then as the bracket from δ extends as far as γ, either the point at δ or some left facing point T in the interval σ to δ must be of greater power than σ and therefore than ρ:τ would then be of higher power than all right facing points in the interval α to γ and also in the interval γ to δ, and therefore would have been the original position of β.

Jutaxposition and omitted points

In most systems there is some operation which is described simply by juxtaposition, without any special operator. In Church's system this is the application of a function to its argument; in Russell's it is conjunction and in algebra it is multiplication. In such systems the abbreviated formulae will be less restricted than the abbreviated formulae in the sense defined here. It is also usual to omit some of the points in the abbreviated formulae, it being understood that a point is to be introduced wherever one is necessary in order to satisfy the conditions (a), (b), (c), above. The power of such points may be settled at the same time as the other power conventions. There is one matter which has been left doubtful about the introduction of these points. When a pair of brackets is adjacent to operators at each end one of the brackets must be 'protected' from its operator by a point, but only one, in order to satisfy (b); which bracket should it be? The following three rules are equivalent:

(1) One may put in a point in both places. In this case (b) is no longer satisfied, and the final result of removing the points, by either of the rules, leaves an otiose pair of brackets which have to be removed before we have an unabbreviated formula.
(2) Both points are put in and then the weaker one removed.
(3) If the conventions below are adopted one may put the point in after the brackets.

With this practical kind of system, where juxtaposition is used and some points are omitted, the abbreviated formulae do not satisfy the conditions (b), (c) above: they satisfy (a), however, and also (c′), below. To distinguish these formulae from the abbreviated formulae proper I will call them *practical formulae*. The conditions (a), (c′) are necessary and sufficient for being a practical formula.

(c′) No pair of consecutive symbols in the projected formula may be one of the following: '( )', '(:', '(*D*', ':')', '::', '*M*)', '*MD*', '*D*)', '*DD*'. No three consecutive symbols may be '*M*:*D*' or '*D*:*D*'. The projected formula may begin only with '(', '*x*', or '*M*' and may end only with ')' or '*x*'.

From a practical formula we can obtain an abbreviated formula by first introducing an operator ⋆ to take the place of juxtaposition, and afterwards the omitted points. Wherever a point π is not adjacent to an operator we replace it by '$\pi \star \pi$'. We replace ')(' by ')⋆(',')*A*' by ')⋆*A*', '*A*(' by

'$A\star($' and 'AB' by '$A\star B$' if A and B are irreducible formulae. We then replace the omitted points. We may use small circles to represent them: thus the sequences 'xD', 'Dx', 'Mx', 'MM', 'DM', 'xM', '$)M$' in a projected formula become '$x_\circ D$', '$D_\circ x$', '$M_\circ x$', '$M_\circ M$', '$D_\circ M$', '$x_\circ M$', '$)_\circ M$'. The last two of these must be again modified by the introduction of $\star$, giving '$x_\circ \star_\circ M$' and '$)_\circ \star_\circ M$' but the process then comes to an end.

Application to Church's system

In Church's system the irreducible formulae are the variables and other single letter formulae, including, if we wish, abbreviations such as $S_{\iota'\iota'}$. The monadic operators are $\sim, [\boldsymbol{x}_\alpha], [\exists \boldsymbol{x}_\alpha], [\iota \boldsymbol{x}_\alpha], \lambda \boldsymbol{x}_\alpha$ and $[\lambda \boldsymbol{x}_\alpha]$, of which the last two may be regarded as the same so far as the unabbreviated formulae are concerned. The dyadic operators are $\supset$, $\mathbf{v}, \equiv, \&, =, \neq$, to which we may add $\star$. If we adopt the conventions of the last section it is only necessary to decide on the relative powers of the points in order that the unabbreviated form of a practical formula should be determined. The conventions recommended are as follows:

We divide the operators into two classes:

Class of high power containing $\supset$, $\mathbf{v}$, $\&$, $\equiv, \sim, [\boldsymbol{x}_\alpha]$, $[\exists \boldsymbol{x}_\alpha]$, $[\iota \boldsymbol{x}_\alpha]$, $[\lambda \boldsymbol{x}_\alpha], =, \neq$, and others which may be added from time to time such as $>, <, /$.

Class of low power containing $\lambda \boldsymbol{x}_\alpha, \star$, and others which may be added from time to time such as $+, -$.

In the class of high power we distinguish some operators as *handicapped*: these are $=, \neq$ (and $>, <$). A point adjacent to an operator in the high power class is always of higher power than one in the low power class. In the case of two points adjacent to operators of the same class the one with the greater number of dots is of the higher power, with the provisos that if the operator is handicapped the number of dots must be reduced by one, and that a point which is either omitted or represented by $_\circ$ counts as of 'zero dots.' Amongst points of the same class, and having the same (corrected) number of dots the left facing points are of higher power than the right facing. There is no need to decide which shall be the more powerful of two similarly facing points, since this is irrelevant to the scope condition, but for definiteness let us say that the one on the left is the more powerful.

The 'unabbreviated formula' which results from a 'practical formula' by the application of one of our rules is not strictly speaking a formula of Church's system nor even an abbreviation of one which would be recognised by Church. If A is the unabbreviated formula, and $A^{(D)}$ the corresponding formula recognised by Church. then $A^{(D)}$ is A if it is an irreducible formula, and otherwise is defined inductively by the conditions that:

$$((A\star B))^{(D)} \text{ is } (A^{(D)}B^{(D)}),$$

$$((A)\supset(B))^{(D)} \text{ is } [A^{(D)}\supset B^{(D)}],$$

$$((A)\mathbf{v}(B))^{(D)} \text{ is } [A^{(D)}\mathbf{v}B^{(D)}],$$

etc.;

$$(\sim(A))^{(D)} \text{ is } [\sim A^{(D)}],$$

$$([\boldsymbol{x}_\alpha](A))^{(D)} \text{ is } [(\boldsymbol{x}_\alpha)A^{(D)}],$$

etc.;

$$([\lambda \boldsymbol{x}_\alpha](A))^{(D)} \text{ is } ((\lambda \boldsymbol{x}_\alpha)A^{(D)}),$$

$$(\lambda \boldsymbol{x}_\alpha(A))^{(D)} \text{ is } (\lambda \boldsymbol{x}_\alpha A^{(D)}).$$

Discussion of the conventions

These power conventions appear to differ markedly from the Russell conventions because the operator against which a point is placed is made to be of greater effect in determining the power than the number of dots. However in Russell's system the operators in our class of low power do not occur at all, and the difference must be thought of as a rejection of his distinctions between operators for punctuation purposes, together with a special new treatment of the new operators. Our 'handicap of one dot' convention for $=$, $>$, etc. may however be regarded as taking the place of some of Russell's distinctions.

It is easy to remember which are the operators in the class of high power. They are the ones which normally either operate on propositions or form propositions. The ones which are handicapped are those which form propositions but do not normally operate on propositions. The case of $[\lambda \boldsymbol{x}_\alpha]$ is exceptional, but again it is easy to remember its power because the notation has been made analogous to that of the other high power operators. One would not normally use the form $[\lambda \boldsymbol{x}_\alpha]$ unless it is operating on a proposition.

The reason for adopting our high and low power class conventions is that in practice it is extremely seldom that we want the scope of a bracket starting from one of the low power operators to include one of the high power operators. The low power operators are in fact just the ones that we should use in formalising the mathematical formulae in a mathematical book. We should use the high power operators in formalising the English connecting matter. It is hardly necessary to point out that a bracket in one of the formulae never pairs with one in another formula, with English intervening. Our convention has the desired effect of closing automatically all brackets outstanding in the 'mathematical formulae' before going on to the English text. The reasons for adopting the handicap convention are similar. A bracket starting from an equality sign will not usually enclose another high power operator, although a bracket from an operator of low power will not enclose an equality sign.

The convention by which left facing points are made more powerful than right facing is convenient to complete the conventions, and is also in agreement with two of Church's own conventions, viz. that in the absence of other indication association is to the left, and that in the absence of dots an omitted bracket has the minimum possible scope.

The use of square brackets in connection with some of the operators, e.g. $[\exists x_\alpha]$, is necessary in a theoretical treatment, but it is not suggested that such a notation should be generally adopted. With very few exceptions one can tell whether the round brackets are part of an operator or not. One exception is the formula $(p_{oo})(q_o)$.

Examples

(i) As a first example of the effects of our conventions I shall take a very simple formula and remove the dots by the first rule. The formula which I shall take is $ab.c$ and even this will be found quite sufficiently complicated for the purpose. We must first transform the 'practical formula' into an 'abbreviated formula' by introducing the operator $\star$, and the points $_\circ$. This gives us $a_\circ \star_\circ b.\star.c$. We now take the point of highest power, which is the one following the b and replace it by a bracket facing left, i.e., away from its operator, and balance it with a bracket at the left end, giving us $(a_\circ \star_\circ b)\star.c$. We now have to evaluate separately $a_\circ \star_\circ b$ and $\xi \star.c$. The stronger point in $a_\circ \star_\circ b$ is the left one and this formula is therefore equivalent to $(a)\star_\circ b$, i.e., to the result of substituting (a) for η in the unabbreviated form of $\eta \star_\circ b$, i.e., in $\eta \star (b)$. The unabbreviated form of $(a_\circ \star_\circ b)$ is therefore $((a)\star(b))$: also the unabbreviated form of $\xi \star.c$ is $\xi \star (c)$, and therefore the unabbreviated form of $(a_\circ \star_\circ b)\star.c$ is the result of substituting $((a)\star(b))$ for ξ in $\xi \star (c)$, i.e., is $((a)\star(b))\star(c)$. Transforming this back to a formula of Church's system, properly speaking, we get $((ab)c)$.

In the remaining examples we will always use the second rule. No type suffixes will be shown.

(ii) We will first deal with formulae without operators, or at least without operators of high power. As one example,

$$(a((cd)(e(fg))))$$

can be abbreviated to

$$a : . \, cd : e.fg.$$

As another,

$$a.cd.efg$$

is an abbreviation of

$$((a(cd))((ef)g)).$$

The association to the left rule has been used here: in other words we have had to apply the rule that a dot is more powerful in its left facing than its right facing aspect. The structure of a formula is often more easily taken in if we slightly increase the number of dots and do not rely on this rule, e.g. the same formula may be written

$$a.cd : efg,$$

or again as

$$a.cd : ef.g.$$

Similarly it is often not advisable to replace all of the brackets in a formula by dots. As a group of dots never replaces more than four brackets it can hardly ever be worth while having as many as six dots, say, in a group. A few dots can however be made to go a long way by mixing them judiciously with explicitly shown brackets e.g.

$$bc.d :: . \, e :: f : . \, g : h.ij$$

is the best form of a certain formula when expressed without any explicit brackets, but

$$bc.d : e.f(g : h.ij)$$

is a much better form of it.

As an example of a formula involving λ,

$$h : . \, \lambda f \lambda x \, fx : g$$

is an abbreviation of

$$(h((\lambda f(\lambda x(fx)))g)).$$

(iii) As an example of a more general type of formula,

$$[m] . \, Nm \supset [p] . \, Np \supset m \neq S : pS.m$$

is an abbreviation of

$$[m]((Nm) \supset ([p]((Np) \supset (m \neq S((pS)m)))))).$$

If we did not have the 'handicap of one dot' convention we should have to put in a dot after '$Np \supset$'. In this case the effect is slight, but sometimes it can be considerable, e.g. without the convention

$$[x].x = y \,\&\, y = z \supset x = z$$

would have to become

$$[x] : x = y . \& . y = z . \supset . x = z.$$

(iv) The expressions

$$p \supset .q \supset : r \mathbf{v} s . \& t : \& u$$

and

$$p \supset (q \supset ((r \mathrm{v} s) \& t)) \& u$$

and

$$p \supset (q \supset .r \mathrm{v} s \& t) \& u$$

are all abbreviations of the same formula. Notice that in the first of these expressions the bracket starting after '$p \supset$' does not close when we reach the stronger point on the left of '$\& t$', because the former is reinforced by the even stronger point after '$q \supset$'. The most legible form of this formula, if it is standing by itself, is probably

$$p \supset : q \supset : r \mathbf{v} s . \& t \;: \& u.$$

(v) A formula similar to the last example in one respect is

$$p \supset q \,\&\, r,$$

which with our conventions is an abbreviation of

$$(p \supset q) \,\&\, r,$$

but with Russell's or Church's conventions would be an abbreviation of

$$p \supset (q \,\&\, r)$$

on account of the subdivision of our 'class of high power' into smaller classes of different powers.

(vi) Normally we shall not want to put dots against equality signs, or other operators which form propositions but do not operate on propositions. A typical exception is

$$[\iota \boldsymbol{x}_\alpha] . g_{\circ\alpha} \boldsymbol{x}_\alpha \supset f_{\circ\alpha} \boldsymbol{x}_\alpha \; := y_\alpha .$$

Another type of freak formula, difficult to abbreviate, occurs when we have functions which take propositions as arguments, e.g.

$$h_{\alpha\circ}(p_\alpha \supset q_0).$$

The only way of avoiding explicit brackets in such a case is to express the implication, not with the implication operator but with the implication function $C_{\circ\circ\circ}$, thus

$$h_{\alpha\circ} . C_{\circ\circ\circ} p_\circ q_\circ .$$

KING'S COLLEGE, CAMBRIDGE

The Reform of Mathematical Notation and Phraseology

(unpublished manuscript ca. 1944)

Stephen Wolfram connects —

COMPUTATION, MATHEMATICAL NOTATION AND LINGUISTICS

Much like ordinary natural languages, most of the mathematical notation we have today has grown up over a long period of time by a kind of natural selection. Occasionally, explicit efforts to systematise the notation have been made – though they have been remarkably few and far between.

In the late 1600s, Leibniz, for example, was quite concerned with mathematical notation – seeing it as an opportunity to move toward a more universal language, free of the controversies of particular ordinary languages. He invented the integral sign, the *d*/*dx* notation for derivatives (where he worried people would try to 'cancel the d's'), and attacked the use of * for multiplication ('will be confused with the letter x').

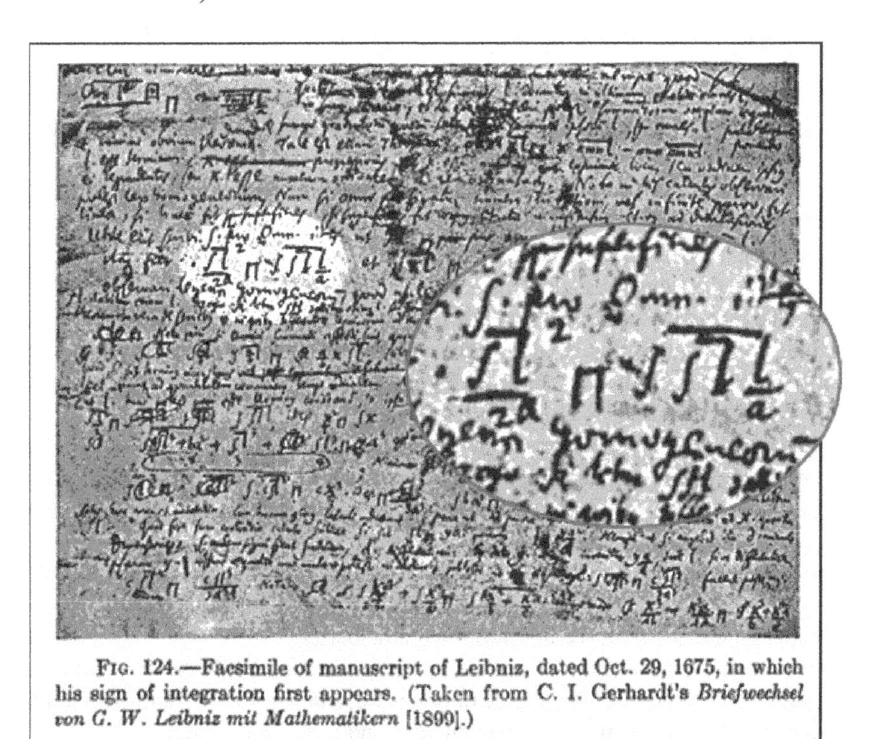

FIG. 124.—Facsimile of manuscript of Leibniz, dated Oct. 29, 1675, in which his sign of integration first appears. (Taken from C. I. Gerhardt's *Briefwechsel von G. W. Leibniz mit Mathematikern* [1899].)

Fig. 1: Leibniz was serious about developing notation for math. His most famous piece of notation was invented in 1675. For integrals, he had been using 'omn.', presumably standing for omnium. But on Friday 29 October 1675 he used, for the first time, the symbol that is used today.

In the 1800s Babbage wrote polemics about mathematical notation, and by the 1880s Frege, Peano and others were trying hard to create more systematic ways to represent mathematical processes. And no doubt that systematisation was a necessary precursor to Hilbert's programme, Gödel's theorem, and ultimately Turing's own work on defining what amounts to a universal mechanism for mathematical processes.

In a sense, though, a Turing machine is a very low-level representation of mathematical processes. And I suspect Turing was curious about what would be involved in creating a higher level representation: a full systematic language for mathematics at the level people actually do it.

As it happens, I have spent a significant part of my life developing *Mathematica* – which among other things aims to provide just such a language.

And in fact the core concept of *Mathematica* as such a language owes an important debt to the paradigm initiated by Turing's work. In the early 1900s, when people thought about systematising mathematics, they had a definite idea about what had to be done: one had to find a way to represent mathematical proofs – as a sort of modern version of something like logical syllogisms.

And, for example, Whitehead and Russell in their Principia *Mathematica* developed an elaborate and arcane scheme for doing this (see Figure 2 below).

But is systematising proofs really the only meaningful way to systematise mathematical processes? The Turing machine in a sense makes it clear that it is not. For a Turing machine provides a representation not of a proof, but of a computation.

Of course, practical mathematics had involved computation ever since Babylonian times. But pure mathematics – following the ideas of Euclid, and later of logic – had concentrated instead on proof. The concept of a Turing machine connected pure mathematics to computation in a systematic and universal way.

And when I came to develop *Mathematica*, I did so within the paradigm of computation rather than proof.

Mathematica represents mathematics in an actionable way: its purpose is not to show, or find, the steps in proofs, but rather to find results, and find what is true, by explicitly computing output from input.

As a direct consequence of universal computation, *Mathematica* can internally represent any possible computation. But then the challenge – as Turing in effect recognised – is to connect those possible computations to ones that humans can describe.

I have spent more than three decades designing languages – most importantly *Mathematica* – that allow computations to be specified conveniently. And in a sense the way I have worked is to try to imagine all the computations that people might want to do, and then to identify repeated chunks of computational work that occur in those – and then to give names to those chunks.

The result – if one succeeds – is an artificial language in which typical computations and programs can be expressed in the shortest and clearest possible way. And indeed, after countless millions of lines of *Mathematica* language have been written, I believe I can claim a certain degree of success.

But what about traditional mathematics? How can we represent it, as Turing wondered, in a systematic way?

If one is going to be able to automate mathematical computations, then ultimately one has to have a precise and systematic representation of the mathematics.

And with all the precision traditional in pure mathematics, one might imagine that its notation would somehow have evolved to a high degree of precision. But it has not. Traditional mathematical notation is full of implicit conventions, strange elisions and historical accidents.

SECTION A] THE CARDINAL NUMBER 1 351

$*52{\cdot}601.\ \vdash :: \alpha \epsilon 1 \,.\supset :.\, \phi(\breve{\iota}\text{‘}\alpha) \,.\equiv : x \epsilon \alpha \,.\supset_x .\, \phi x : \equiv : (\exists x) \,.\, x \epsilon \alpha \,.\, \phi x$

Dem.

$\vdash .\, *52{\cdot}15 \,.\supset \vdash :.\, \mathrm{Hp} \,.\supset : \mathrm{E}!\,\breve{\iota}\text{‘}\alpha : \qquad (1)$

$[*30{\cdot}4] \qquad \supset : x \,\breve{\iota}\, \alpha \,.\equiv .\, x = \breve{\iota}\text{‘}\alpha \,.$

$[*52{\cdot}6] \qquad \equiv .\, x \epsilon \alpha \qquad (2)$

$\vdash .\,(1)\,.\, *30{\cdot}33 \,.\supset$

$\vdash :: \mathrm{Hp} \,.\supset :.\, \phi(\breve{\iota}\text{‘}\alpha) \,.\equiv : x \,\breve{\iota}\, \alpha \,.\supset_x .\, \phi x : \equiv : (\exists x) \,.\, x \,\breve{\iota}\, \alpha \,.\, \phi x \qquad (3)$

$\vdash .\,(2)\,.\,(3)\,.\supset \vdash .\, \mathrm{Prop}$

$*52{\cdot}602.\ \vdash :.\, \hat{z}(\phi z) \epsilon 1 \,.\supset : \psi(\imath x)(\phi x) \,.\equiv .\, \phi x \supset_x \psi x \,.\equiv .\, (\exists x) \,.\, \phi x \,.\, \psi x$

$[*52{\cdot}12 \,.\, *14{\cdot}26]$

$*52{\cdot}61.\ \vdash :.\, \alpha \epsilon 1 \,.\supset : \breve{\iota}\text{‘}\alpha \epsilon \beta \,.\equiv .\, \alpha \subset \beta \,.\equiv .\, \exists!(\alpha \cap \beta) \quad \left[*52{\cdot}601 \dfrac{x \epsilon \beta}{\phi x}\right]$

$*52{\cdot}62.\ \vdash :.\, \alpha, \beta \epsilon 1 \,.\supset : \alpha = \beta \,.\equiv .\, \breve{\iota}\text{‘}\alpha = \breve{\iota}\text{‘}\beta$

Dem.

$\vdash .\, *52{\cdot}601 \,.\supset \vdash :: \mathrm{Hp} \,.\supset :.\, \breve{\iota}\text{‘}\alpha = \breve{\iota}\text{‘}\beta \,.\equiv : x \epsilon \alpha \,.\supset_x .\, x = \breve{\iota}\text{‘}\beta :$

$[*52{\cdot}6] \qquad \equiv : x \epsilon \alpha \,.\supset_x .\, x \epsilon \beta :$

$[*52{\cdot}46] \qquad \equiv : \alpha = \beta :: \supset \vdash .\, \mathrm{Prop}$

$*52{\cdot}63.\ \vdash : \alpha, \beta \epsilon 1 \,.\, \alpha \neq \beta \,.\supset .\, \alpha \cap \beta = \Lambda \quad [*52{\cdot}46 \,.\, \mathrm{Transp}]$

$*52{\cdot}64.\ \vdash : \alpha \epsilon 1 \,.\supset .\, \alpha \cap \beta \epsilon 1 \cup \iota\text{‘}\Lambda$

Dem.

$\vdash .\, *52{\cdot}43 \,. \quad \supset \vdash : \mathrm{Hp} \,.\, \exists!\,\alpha \cap \beta \,.\supset .\, \alpha \cap \beta \epsilon 1 :$

$[*5{\cdot}6 \,.\, *24{\cdot}54] \supset \vdash :.\, \mathrm{Hp} \,.\supset : \alpha \cap \beta = \Lambda \,.\vee .\, \alpha \cap \beta \epsilon 1 :$

$[*51{\cdot}236] \qquad \supset : \alpha \cap \beta \epsilon 1 \cup \iota\text{‘}\Lambda :.\supset \vdash .\, \mathrm{Prop}$

$*52{\cdot}7.\ \vdash :.\, \beta - \alpha \epsilon 1 \,.\, \alpha \subset \xi \,.\, \xi \subset \beta \,.\supset : \xi = \alpha \,.\vee .\, \xi = \beta$

Dem.

$\vdash .\, *22{\cdot}41 \,. \qquad \supset \vdash : \mathrm{Hp} \,.\, \xi \subset \alpha \,.\supset .\, \xi = \alpha \qquad (1)$

$\vdash .\, *24{\cdot}55 \,. \qquad \supset \vdash : \sim(\xi \subset \alpha) \,.\supset .\, \exists!\,\xi - \alpha \qquad (2)$

$\vdash .\, *22{\cdot}48 \,. \qquad \supset \vdash : \mathrm{Hp} \,. \quad \supset .\, \xi - \alpha \subset \beta - \alpha \qquad (3)$

$\vdash .\,(2)\,.\,(3)\,. \qquad \supset \vdash : \mathrm{Hp} \,.\sim(\xi \subset \alpha) \,.\supset .\, \exists!\,\xi - \alpha \,.\, \xi - \alpha \subset \beta - \alpha \qquad (4)$

$\vdash .\, *52{\cdot}1 \,. \qquad \supset \vdash : \mathrm{Hp} \,.\supset .\, (\exists x) \,.\, \beta - \alpha = \iota\text{‘}x \qquad (5)$

$\vdash .\,(4)\,.\,(5)\,.\, *51{\cdot}4 \,.\supset \vdash : \mathrm{Hp} \,.\sim(\xi \subset \alpha) \,.\supset .\, \xi - \alpha = \beta - \alpha \,.$

$[*24{\cdot}411] \qquad \supset .\, \xi = \beta \qquad (6)$

$\vdash .\,(1)\,.\,(6)\,.\supset \vdash .\, \mathrm{Prop}$

Fig. 2: A page from Whitehead and Russell's monumental work Principia *Mathematica* devoted to showing how the truths of mathematics could be derived from logic.

In designing the mathematical components of the *Mathematica* language (Wolfram, 2010), however, I had to create a systematic form of the notation. But to make *Mathematica* easy for humans to learn and understand, I wanted to stay as close as possible to traditional notation.

The result is that I undertook an extensive study of the way that mathematical notation is used in practice. In a sense, this study was similar in character to the way a linguist might try to infer the grammar and syntax of some ordinary spoken human language. But the literature of mathematics provides a somewhat more systematic corpus than is usually available.

And somewhat to my surprise, despite the diversity of the mathematical literature, there was a remarkable degree of consistency in the way notation tended to be used – down even to consistent unwritten conventions about the precedence of all sorts of mathematical operators.

And it took only a modest set of innovations to go from this notation to something completely precise and computable. (It helped that *Mathematica* can support not just linear textual input, but also full two-dimensional input, like traditional mathematical notation.)

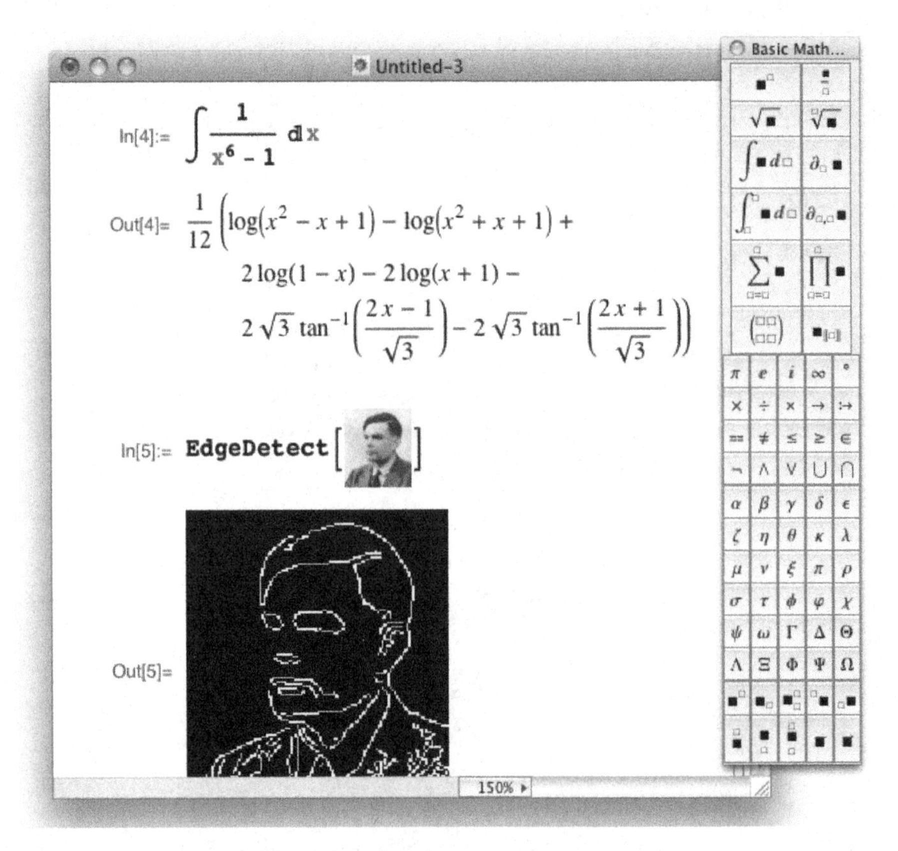

Fig. 3: Mathematical and other notation in *Mathematica*. Note the two-dimensional character of the input.

A great deal of mathematics has now been described in the precise notation of *Mathematica* (see `http://www.wolfram.com/mathematica/`).

But a few years ago, I became curious about the extent to which it would be possible to handle by computer completely free-form mathematical notation and input.

For in developing Wolfram|Alpha (see `http://www.wolframalpha.com/`) my goal was to allow people to specify their queries – whether about mathematics or anything else – just in the way that they think of them, without having to convert them to any kind of precise formal language (see Figure 4).

At first, it seemed as if this kind of free-form linguistic input might simply be impossible, or impractical. But thanks to a series of breakthroughs, we have been able to make this work in a highly successful way.

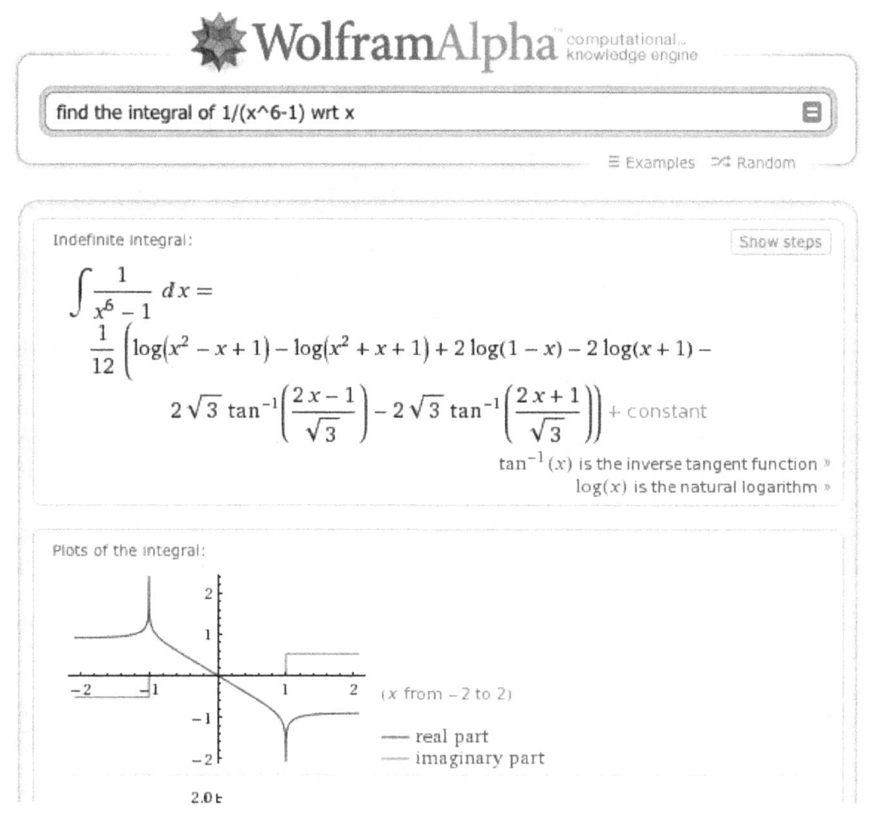

Fig. 4: Wolfram|Alpha understands free-form natural language specifications of mathematical operations.

Indeed, when it comes to textually typed mathematical input, we can now recognise what a person meant in very close to 100% of all cases – at least those that a trained human would find recognisable. Of course, it helps that we have been able to study many, many millions of inputs that have been fed to Wolfram|Alpha.

And among the results of this is that we can say with some precision the extent to which people do or do not use the various notational conventions that Turing describes in his notes.

In *Mathematica,* we try to do what Turing advocates: to create a completely systematic and precise notation for mathematics. And indeed this is a very powerful thing. But in Wolfram|Alpha, we have now succeeded in doing something that is in a sense maximally convenient for humans: just taking mathematical notation in the form that humans think of it, and interpreting it into a precise computable form.

I rather suspect – and hope – that Turing would appreciate the notation of *Mathematica* – set up as it is to provide a precise and unambiguous representation that can immediately be computed with.

And perhaps he would be surprised – as I was – that it is possible in Wolfram|Alpha to go from the strange and inconsistent notation that has grown up in mathematics, and in a sense use the sparsity of typical mathematical questions to be able to deduce what the corresponding precise notation should be.

In his work at the dawn of systematic computation, Turing could only begin to imagine what it would be like to make mathematics computational. Today – especially with *Mathematica* and Wolfram|Alpha – we have succeeded in making large swaths of mathematics computational.

One issue that has remained is the style of mathematics traditional in the twentieth century, which centers around the creation of mathematical structures ('Let F be a field...'). A recent realisation is that the basic paradigm of Wolfram|Alpha is exactly what is needed to make such mathematics computational.

In Wolfram|Alpha, it is common to enter some entity (say a city or a chemical), and then have Wolfram|Alpha automatically generate a report on what might be considered 'interesting' about that entity. The same can be done for mathematical structures.

In effect, Wolfram|Alpha must take the structure and then deduce what facts or theorems are 'interesting' about it. In part, this can be done from a curation of known mathematical theorems. In part, it must be done by a collection of mathematical and meta-mathematical algorithms and heuristics. But the result, I believe, will be that the vast majority of the parts of the human activity that we call 'mathematics' will successfully be completely automated. The concept of systematisation – and computation – that Turing had will have been realised.

And, as it happens, thanks to the likes of Wolfram|Alpha technology, there would not even be a need to 'reform' mathematical notation in order for humans to successfully describe what they want to do.

References

Wolfram, S., 2000. Mathematical Notation: Past and Future, Transcript of a keynote address presented at MathML and Math on the Web: MathML International Conference. <http://www.stephenwolfram.com/publications/recent/mathml/mathml2.html> (accessed 20.10.00).

Wolfram, S., 2010. The Poetry of Function Naming, blog post. <http://blog.stephenwolfram.com/2010/10/the-poetry-of-function-naming/> (accessed 18.10.10).

The Reform of Mathematical Notation and Phraseology

A. M. TURING

It has long been recognised that mathematics and logic are virtually the same and that they may be expected to merge imperceptibly into one another. Actually this merging process has not gone at all far, and mathematics has profited very little from researches in symbolic logic. The chief reasons for this seem to be a lack of liaison between the logician and the mathematician-in-the-street. Symbolic logic is a very alarming mouthful for most mathematicians, and the logicians are not very much interested in making it more palatable. It seems however that symbolic logic has a number of small lessons for the mathematician which may be taught without it being necessary for him to learn very much of symbolic logic.

In particular it seems that symbolic logic will help the mathematicians to improve their notation and phraseology, which are at present exceedingly unsystematic, and constitute a definite handicap both to the would-be-learner and to the writer who is unable to express ideas because the necessary notation for expressing them is not widely known. By notation I do not of course refer to such trivial questions as whether pressure should be denoted by p or P, but deeper ones such as whether we should say 'the function $f(z)$ of z' or 'the function f'.

It would not be advisable to let the reform take the form of a cast-iron logical system into which all the mathematics of the future are to be expressed. No democratic mathematical community would stand for such an idea, nor would it be desirable. Instead one must put forward a number of definite small suggestions for improvement, each backed up by good argument and examples. It should be possible for each suggestion to be adopted singly. Under these circumstances one may hope that some of the suggestions will be adopted in one quarter or another, and that the use of all will spread.

Although it is not desirable to try and put mathematics into the straight-jacket of a logical system, it may be desirable to use such a system when investigating notation. One is likely to be taking typical phrases from mathematical text-books and analysing their meaning. It is useful to have a logical system for expressing these meanings in a fairly unambiguous way. It may not greatly matter what system is used for this purpose, and it would be quite possible for different workers to use different systems.

To be specific I am inclined to suggest the following programme

i) An extensive examination of current mathematical and physical and engineering books and papers with a view to listing all commonly used forms of notation.
ii) Examination of these notations to discover what they really mean. This will usually involve statements of various implicit understandings as between writer and reader, it may also include the equivalent of the notation in question in a standard notation.
iii) Laying down a code of minimum requirements for desirable notations. These requirements should be exceedingly mild. In my opinion the points which should be covered by this code should include the following
    a) Free and bound variables should be understood by all and properly respected.
    b) Some sort of provision should be made for falling in line with the theory of types. This assumes a Russelian Weltenscheung, as applies I think to the majority of mathematicians-in-the-street.

c) The deduction theorem should be taken account of, i.e. it should be recognised that numerous forms of argument consist in one form or another of applications of the deduction theorem. The deduction theorem should therefore be as well known as the rule for integration by parts.

d) Very clear statements of the fundamental nature of the symbols should be made. There should be no danger of mistaking a real variable for a function taking real values.

iv) New notations suggested by symbolic logic.

v) Examples of the development of comparatively elementary parts of mathematics in obedience to the new code and embodying the new notations. These examples should only incorporate the new notations in cases where great advantage results. The effects of the various independent reforms should be shown separately, so far as possible, to facilitate their independent adoption.

Free and bound variables. Deduction theorem. Constants and parameters

In this section a) and c) of iii) above will be examined in greater detail.

The symbols used in mathematics may be classified as follows.

i) Symbols used entirely in punctuation, such as (,), 'etc.
These do not concern us at the moment.

ii) Absolute constants, typified by Σ, $1, =+\log$ etc, etc.
These also do not concern us at the moment.

iii) Letters representing constants, usually taken from the beginning of the alphabet, e.g., 'Let a be the radius of the sphere'.

iv) Letters representing genuine variables, for which substitution may be made, e.g., x in $x=x$.

v) Letters for which we can substitute other letters (most of them) to get a true result, but certainly cannot substitute constants; in the latter case nonsense results. Examples are provided by the occurrence of x in

$$\int_0^1 x\mathrm{d}x = \frac{1}{2}$$

and in 'for all numbers x greater than 2, $x^2 > 3$'. Substitution of 1 for either of these yields nonsense.

Letters described under iv) and v) above are known respectively as free and bound variables. Free variables are really comparatively rare. This is because we do not often make statements such as '$x=x$' but more often something like 'for all real numbers x, $x=x$': in this the opening phrase 'binds' the variable. Thus x is bound in the whole statement, but is free in the part $x=x$. The difference between the constants iii) and the free variables is somewhat subtle. The constants appear in the formula superficially as if they were free variables, but we cannot substitute for them. In these cases there has always been some assumption made about the variable (or constant) previously; thus we may have the equation

$$v = \frac{4}{3}\pi a^3$$

in which we cannot substitute for v and a, these being constants because we have made these assumptions about them 'a is the radius and v is the volume of the sphere'. The 'deduction theorem' states that in such a case, where we have obtained a result by means of some assumptions, we can state the result in a form where the assumptions are included in the result, e.g., 'If a is the radius and v is the volume of the sphere then $v=\frac{4}{3}\pi a^3$'. In this statement a and v are no longer constants.

We are now able to substitute for them: we might substitute v for a and get a statement with the same meaning, or we could substitute 2 for both a and v getting a true statement, but one of rather unorthodox character. This process whereby we pass from P proved under an assumption H to 'If H then P' may be called 'absorption of hypotheses'. The process converts constants or 'restricted variables' into free variables. Variables whose character changes in this way from restricted to free usually seem to be described as 'parameters', although it is very difficult to give any very definite meaning to the term.

Theory of types and domains of definition

We are taught that the theory of types is necessary for the avoidance of paradoxes, but we are not usually taught how to work the theory of types into our day-to-day mathematics: rather we are encouraged think that it is of no practical importance for anything but symbolic logic. This has a most unfortunate psychological effect. We tend to suspect the soundness of our arguments all the time because we do not know whether we are respecting the theory of types or not. Actually it is not difficult to put the theory of types into a form in which it can be used by the mathematician-in-the-street without having to study symbolic logic, much less use it. The statement of the type principle given below was suggested by lectures of Wittgenstein, but its shortcomings should not be laid at his door.

The type principle is effectively taken care of in ordinary language by the fact that there are nouns as well as adjectives. We can make the statement 'All horses are four-legged', which can be verified by examination of every horse, at any rate if there only a finite number of them. If however we try to use words like 'thing' or 'thing whatever' trouble begins. Suppose we understand 'thing' to include everything whatever, books, cats, men, women, thoughts, functions of men with cats as values, numbers, matrices, classes of classes, procedures, propositions,... Under these circumstances what can we make of the statement 'All things are not prime multiples of 6'. We are of course inclined to maintain that it is true, but that is merely a form of prejudice. What do we mean by it? Under no circumstances is the number of things to be examined finite. It may be that some meaning can be given to statements of this kind, but for the present we do not know of any. In effect then the theory of types requires us to refrain from the use of such nouns as 'thing', 'object' etc., which are intended to convey the idea 'anything whatever'. The most important places where this matters are

1) In connection with the word 'all'. We may for instance say 'All real numbers x have the property...' but not 'All things...'. In particular we should avoid putting the former in the form 'For all things x, if x is a real number...'.
2) In connection with 'there exists'. We allow 'there exists a real number such that...' but not 'There exists a thing x, such that x is a real number and...'.
3) In connection with descriptions. We allow 'The real number x such that...' but not 'The thing, such that x is a real number and...'.
4) In connection with abstraction, but this may be considered a special case of 3).

All this leaves open the question as to what are to be regarded as appropriate nouns to take the place of 'real number' in the above examples. This would probably be really treated on a fairly common sense basis, but the following rules certainly apply. The word *noun-class* is used to mean a class such as the class of real numbers in the examples above.

The sum of two noun-classes is a noun-class.
A sub-class of a noun-class is a noun-class.
The class of functions with arguments in one noun-class and values in another is a noun-class.

These rules do not lead to any noun-classes unless we know of some already. To many logicians it will be sufficient to add 'the null-class is a noun-class', but such a procedure will not be satisfactory to the average mathematician, who has but little concern with the null class, and certainly does not propose to build up the integers from it. The sensible thing to do seems to be to take for granted certain noun-classes such as the integers, and possibly also the real numbers and the points of three-dimensional space. In fact we may take as given noun-classes any classes of objects which, in the branch of mathematics concerned are usually considered as given *a priori*. Such assumptions, combined with the above three rules will be found quite adequate.

Although it is not intended that symbolic logic should take the place of English connecting matter in proofs it is as well to be able to express anything symbolically if required. The usual form of expression of 'For all $x, \dots$' is '$(x)\dots$'. As we have seen this is fundamentally unsound. Instead we want a notation for 'For all integers $x, \dots$'. For this I propose $(x, \mathrm{Int})\dots$, where Int is the notation used for the class of integers. Likewise we may use $(\exists x, \mathrm{Int})$ for 'There exists an integer x' and (x, Int) for 'The integer x'. It may also be desirable to have a notation for 'The class of functions whose arguments are integers and values real numbers'. No notation for this is suggested at the present moment.

[the next 4 pages of the typescript are missing]

Discussion of the system and application to normal mathematics

Let us now try to picture what would happen if we try to develop such subjects as the theory of real numbers or set theory in this system. How will it differ from the normal 'straightforward rigorous' development? Apart from the fact that implications etc. are symbolically expressed the only real difference will be that instead of saying 'for all $x, \dots$' we shall have to say 'for all x in segment 15.. .' (say) which may be taken as the way we read $(x, 15)\dots$. Now in most cases the statements to be transformed will be something more like 'For all real numbers $x, \dots$' or 'For all x, if x is a real number, then...'; in such a case it will not really be necessary to express the condition that x belongs to any particular segment, for the real numbers will have already been defined so as to be all included in some one segment. In all such cases therefore we agree to omit this phrase. Likewise in descriptions instead of saying 'The unique member of segment 15, which is a real number and...' we may say 'The real number such that...' on the ground that all real numbers are members of segment 15. It may perhaps be objected that this omission will sometimes give wrong results because the meaning of the formula may depend on the segment quoted. This is so, but it will normally be the case that an increase in the values quoted will not affect the meaning of the formula; by this I mean that the formula may be proved equivalent to any formula which can be obtained from it by increasing the segment bounds. A formula without free variables, which can be proved equal or equivalent to any formula obtained from it by increasing the segment bounds, may be called a regular formula. It should be noticed that even if we had not explicitly excluded them there would not usually be any reason for expecting formulae with free variables to be regular. The explicit exclusion simplifies matters. We should try to avoid the use of formulae, which are not regular. This can always be done essentially by the following device if necessary: suppose that somewhere there occurs (x, s) and that increasing s alters the meaning; then we may write instead $(x, s)D^s X \supset \cdots$.

It may perhaps be argued that $(x, r)\dots$ means just the same as $(x)D^r x \supset \cdots$, both being rendered into English by the phrase 'For all x in segment r...' or 'All members of segment r have the property that...' This is true, and the only reason for our using this different notation is that we thereby automatically ensure that some condition $D^r x$ will be included in the proposition. It is perhaps worth making a comparison here with the syntax of ordinary language, where we have two distinct types of property, adjectives and nouns; the mathematician is inclined to regard this distinction as unreal and arbitrary, and so in a sense it is, but it does have the effect that it is impossible to refer to anything without associating a 'noun property' with it, and so if we compare the nouns with the segments we see that ordinary language has a remarkable tendency to respect the theory of types.

We can make use of this fact to help make mathematics sound from the point of type theory. If each of the nouns used in mathematics defines a set completely contained in some segment, then ordinary syntax will keep us straight. I shall not of course attempt to give a formal proof of this, the informal character of language making it inappropriate: assuming however that this principle is essentially correct let us go further and see how much modification in common practice will have to be made; it actually amounts to surprisingly little. We have to be sure that the nouns used are 'legitimate nouns'. It is probably best therefore to manage with comparatively few nouns. We can make up some simple rules for the construction of legitimate nouns.

These rules may be taken to be

i) The set of functions of one noun that have values in another noun is a noun.
ii) The things, which are either one noun or another noun, may be collectively described by a third.
iii) We may use a noun to describe a subset of the things described by another noun.

To this we might perhaps add that the word 'individual' may be used as a noun.

It may now be seen our nouns are essentially the 'mengen' of Zermelo and the 'sets' of Bernays and of Gödel. However in this discussion I am not trying to set up a formal system, but merely to suggest how normal rigorous mathematics can be made to take account of type theory without serious upheaval.

A glance through a number of mathematical books provides the following examples of nouns.

a) Number, real number, integer, complex number, point, line, plane, manifold, operator, curve.
b) Group, ring, algebra, base, polynomial.
c) Set, class, pair, object, element.
d) Frontier, conjugate, derivative.
e) Integral, expression, equation, series, sequence, term.

The nouns listed under a) may be regarded as the most reasonable kind. Those under b) are like those under a) but are more complex in meaning, and might in some contexts be used illegitimately. Let us take the word group for example, and for the sake of argument let us pretend that it means what is normally meant by 'group of finite order'. If then we define a group as follows 'A group is a pair consisting of a finite class G of integers and a function $K(x, y)$ defined for x and y in G, and such that for all x, y, z in G, $K(x, K(y, z)) = K(K(x, y), z)$ and if $K(x, y) = K(x, z)$ then $y = z$' the word group will be legitimate. If however the words 'of integers' are omitted such a wide range of possibilities is admitted that the expression is no longer legitimate. Of course the inclusion of these words does not vastly hamper group theory, but they are usually omitted for the sake of the extra generality apparently obtained. It would of course also be legitimate to use instead of the phrase 'of integers' 'of integers or classes of integers' or restrict G by some other 'legitimate noun', but to do so would not give the same intellectual satisfaction as leaving every possibility still open; if however we take this last course the word group will not be a 'legitimate noun'. The examples given under c) are not normally legitimate unless used in such phrases as 'set of points', 'pair of real numbers' etc. The word 'object' is of course the most serious offender, being in fact the outcome of an attempt to evade the salutary restrictions of English syntax. The examples d) really represent functions, because they are used in the form 'the frontier of...' etc. However these functions take values restricted to one segment, and so may be regarded as legitimate. The examples e) may be regarded as purely syntactical, i.e. they do not themselves describe mathematical entities, but rather mathematical expressions, and therefore do not appear in the mathematical argument proper, and may be ignored for our purposes.

The difficulty in the above example concerning 'group of finite order' may be resolved to some extent by using the phrase 'member of segment N' in place of 'integer'. It is then understood that one is at liberty to substitute any object for N throughout the book (say).

Examining the Work and Its Later Impact

Juliet Floyd explores —

TURING, WITTGENSTEIN AND TYPES: PHILOSOPHICAL ASPECTS OF TURING'S 'THE REFORM OF MATHEMATICAL NOTATION AND PHRASEOLOGY' (1944–5)

Turing is explicit that 'the statement of the type principle' in this essay 'was suggested by lectures of Wittgenstein'. The disputatious record of exchanges between them at Wittgenstein's 1939 Cambridge lectures Diamond (1939) are deservedly well known, but Turing's constructive uses of Wittgensteinian ideas are not. The essay's interest lies here, and in its articulation of a powerful attitude toward the development of mathematical notation, an attitude that runs throughout Turing's and Wittgenstein's works, but whose philosophical significance is easy to overlook.

Wittgenstein and Turing are often regarded, in a misleading caricature, as philosophical opponents. Wittgenstein is taken to be a humanistic philosopher of meaning and 'forms of life', hostile to mathematical logic and the very idea of a Turing machine; Turing is taken to be a mechanistic or behaviouristic theorist of the mind, intent on reducing the concept of meaning to that of information. Neither picture is correct – see Floyd (2012).

Yet Wittgenstein and Turing shared, as they both explicitly acknowledged, a particular sort of anti-reductionist attitude toward logical and conceptual analysis. On their view, it is the everyday, purposeful uses we humans make of language that crucially animate and frame the notions of meaning and information. Attention to ordinary ways of speaking is crucial for insight into, and development of, those formal features of language that mathematical logicians are interested in systematising. As Turing put it already in 1933, in an undergraduate lecture to the Moral Sciences Club, 'The purely logistic view of mathematics is inadequate; mathematical propositions possess a variety of interpretations, of which the logistic is merely one' (Hodges, 1983, p. 86).

On this view, formal logic is simply one approach, a tool, neither good nor bad in itself. Its results and significance should be articulable, ideally, in everyday scientific language, perspicuously and intelligibly. Like Wittgenstein, Turing wished to ward off the ideal of a 'cast-iron logical system into which all the mathematics of the future are to be expressed'. Though a standardised notation 'may' be used in particular cases, transparency, opportunistic pluralism and usefulness for communication in an informal sense are essential factors to consider in the design of notations.

This is far from the sort of logistic purity of method embraced in Carnap's 1934 proposal to transform philosophy into the logical syntax of the language of science via a generalisation of Hilbert's axiomatic method (Carnap, 2002). It is also distinct from the kind of information-theoretic reductionism that construes Shannon's isolation of a (quantitative) notion of information (a notion purged of any articulable relation to the notion of meaning) as utterly fundamental.

Turing was perfectly aware of the importance of notational precision for rigor, and this essay is intended to make a contribution to the use of formal logic. He hopes to revitalise the 'liaison' between the logician and 'the mathematician-in-the-street'. But his attitude towards the development and use of notation is, as he says, Wittgensteinian in flavor: look to everyday uses. Mathematical logicians have, Turing says, shown insufficient interest in explaining to ordinary mathematicians the significance of what they do: for most ordinary mathematicians symbolic logic is 'a very alarming mouthful'. Turing's means to a newfound liaison will be through an analysis of everyday (mathematical and other) language. By this means a number of lessons may be drawn 'without it being necessary for [a mathematician] to learn very much of symbolic logic'.

Turing's aim is to show that 'normal rigorous mathematics' can take account of the theory of types without 'serious upheaval'. 'We are taught', he says, 'that the theory of types is necessary for the avoidance of paradoxes, but we are not usually taught how to work the theory of types into our day-to-day mathematics: rather we are encouraged to think that it is of no practical importance for anything but symbolic logic'. This he wishes to change.

He assumes that a 'Russellian *Weltanschauung*' with regard to the theory of types is characteristic of 'the majority of mathematicians-in-the-street'. But he takes the elucidation of types to originate in, and apply to, everyday language, which, unlike Russell, he respects as a locus of meaning. In English, a chair may be said to be heavy, but not (in the same sense) a baby's smile. A group may be said to be non-abelian or Lie, but not said to be composed of integers alone. We do not wonder whether the number one may or may not be identical with Barack Obama. Everyday language, even in mathematics, distinguishes between nouns and adjectives, arriving on the scene already typed, as the quantitative notion of information does not. This, for Turing, is logically relevant in a fundamental way.

The method Turing endorses turns on accepting an 'exceedingly mild' set of minimal requirements on notations – mild in the sense that they may be seen to grow naturally out of everyday ways of speaking in mathematics but may allow for the development of notations as complicated or detailed as one might like. As the first step, one conducts 'an extensive examination of current mathematical and physical and engineering books and papers with a view to listing all commonly used forms of notation', examining these to discover 'what they really mean', that is, what 'implicit understandings as between writer and reader' are at work. Then one develops types from there.

Turing's Wittgensteinian approach to types offers a piecemeal conception of the justification of notations, differing markedly from that of Carnap or Quine. They looked for the development of notations to rationally reconstruct the ontological and metaphysical commitments of entire theories, languages or conceptual schemes as a whole. On the view Turing shares with Wittgenstein, our everyday ways of speaking and using language forms the place from which we begin, and the place where we must end in developing formal systems. Interface with linguistic practice is fundamental for the logician.

For this reason, Turing singles out the deduction theorem as a fundamental rule. This is not only because it gives an assurance of the existence of rigorous formal derivations as counterparts to metatheoretic claims, but because it illustrates in practice the importance of the interplay between claims to theoremhood and formal derivations. Its use legitimates the informal but perspicuous articulation of results by relieving us of the burden of having actually to write down every step of the formal derivation when we accept the metatheoretic claim.

According to Gandy (2001), at the time he wrote this essay (1944), Turing was reading Quine's *Mathematical Logic* (1940) and took a strong dislike to what he regarded as the needlessly arcane quality of Quine's attention to syntax. Neither Russell in *Principia* nor Quine formulate the deduction theorem, for they interpret 'B can be logically derived from A' as $\vdash A \supset B$. Thus, the steps in a proof which would in everyday mathematical language be regarded as derivations from hypotheses are instead assumed always to take the form $A_1 \supset A_2 \supset A_3 \supset \ldots B$, where the A_i are the hypotheses or in force. By contrast Turing takes the theorem as a primitive rule.

A key practical role the deduction theorem plays in Turing's eyes is its clarifying the 'subtle' roles of the free and the bound variable. For it can be used, through 'absorption' of hypotheses, to make explicit assumptions governing the (informal or formal) use of parameters. The problem had led Wittgenstein to conceive the role of certain apparent concept words (e.g. number words) as grammatical rules concerning the use of variables, rather than second-order functions taking arbitrary concepts or properties as inputs (Wittgenstein, 1974, p. 348). The idea was to take all uses of free variables to be governed, implicitly or explicitly, by parameters and/or types. Turing uses the fact that a quantifier as used in 'normal' mathematics generally assumes parameters to motivate his recognition of the deduction theorem's central role and to develop Wittgenstein's idea in a constructive direction.

He shows, giving several everyday examples from mathematics, how free variables (e.g. '$x = x$') can in practice be eliminated in favour of typed bindings (e.g. 'for all real numbers x, $x = x$'). There is no expressive loss, he argues, because the general notion of a 'thing' or 'object' is, as Wittgenstein always maintained, a 'serious offender', having (anyway so far) no ordinary mathematical use or point.

Turing's proposals require, just as Wittgenstein's do, an intensional understanding of the uses of quantifiers. Frege worried that using his equivalence class construction for the natural numbers invited the question whether the number one is or might possibly turn out to be identical to Julius Caesar. Turing's proposed solution is to take noun-classes as basic. For example, instead of 'All things...' or 'For all things x, if x is a real number...' we may say 'All real numbers x have the property ...'. This, as Wittgenstein argued in the lectures Turing attended (Diamond, 1939, p. 167), is closer to ordinary usage and avoids numerous confusions. Turing lays out rules governing the use of 'noun-classes' echoing basic axioms of set theory (versions of union, separation and power set). The final section of the paper, an 'application to normal mathematics', discusses 'regular' formulae that may be proved to hold when lifted into wider domains. General notions like 'group', Turing admits, will not be expressible by means of a 'legitimate' noun. This, he thinks, 'does not vastly hamper group theory', for it leaves us with a choice: we can leave every possibility open (and have the notion of 'group' not counted as a legitimate noun) or we can restrict the noun to a specific domain ('group of integers') in which case it is admissible.

Turing realises that his account 'leaves open' the question as to what are to be regarded as appropriate nouns to take the place of specific domains used to type statements ('real number', 'abelian group'). But he remarks, here 'a fairly common sense basis' would 'probably' serve.

Is this a *philosophically* appropriate remark? Actually, Turing took what he repeatedly called 'common sense' quite seriously. Discussion of this arose explicitly with Wittgenstein, when Turing suggested that he appeared to be relying on common sense in his philosophical remarks (Diamond, 1939, p. 219ff). Though Wittgenstein savagely denied this at first, he revisited the issue more sympathetically in the next lecture, admitting that he did want to say something 'similar' to this (p. 223ff).

In an essay (Turing, 1954) published in the last year of his life, Turing states that one implication of the incompleteness theorems is that an appeal to common sense is inevitable for a mathematician:

> These [limitative] results, and some other results of mathematical logic may be regarded as going some way towards a demonstration, within mathematics itself, of the inadequacy of 'reason' unsupported by common sense.

Wittgenstein sought other ways of articulating the significance of incompleteness results and was inclined to subject the notion of 'common sense' to critical scrutiny. But he shared with Turing an ideal of rigor that included concrete attention to what makes sense by the lights of our current purposes, needs and uses of language.

Rather than a 'common sense basis' for developing mathematics, one might emphasise the value for mathematics and science of making 'very clear statements of the fundamental nature of the symbols', as Turing calls for here. This strong value placed on the vernacular, on lucidity and communicability, on perspicuity, is one that Turing and Wittgenstein shared. Turing was a master simplifier. His analysis of computability by means of the notion of Turing machine is more vivid, more pertinent and (as Gödel himself maintained) more epistemologically satisfying than Church's or Gödel's extensionally equivalent demarcations of the class of recursive functions (Gödel, 1986, p. 195). This is because, as a way of thinking, it is not entangled with the limitations of any particular formal system. It is everyday, perspicuous, simple, direct or 'common sensical'.

What was wanted, in the context in which Turing developed the notion of a Turing machine, was a clarification. To resolve Hilbert's famous *Entscheidungsproblem* what was required was not merely the development of a new formal system, but a way of thinking about what formal systems are and are used for. What was needed was a persuasive (perspicuous) analysis of the notion of a formal system itself, that is, of a 'definite method' in a sense relevant to the problem context, i.e. to the notion of a 'mechanical' or 'effective' 'procedure' that can be carried out by human beings, with only limited cognitive steps (recognising a symbolic configuration, seeing that one of finitely many rules applies, shifting attention stepwise to a new symbolic configuration and so on). An analysis like Turing's that intuitively analyses the very notion of a formal system by drawing an analogy with certain limited aspects of possible human cognitive activity was precisely what was wanted. The Turing machine offers us that way of thinking. It plays a role analogous to what Wittgenstein called a 'language game': a simplified snapshot of a portion of human use of language, designed for a particular purpose, to shed light on meaning.

It is sometimes held (for example, by Gödel (1990, p. 306)) that Turing's analogy with a human computer, drawing on the assumption that a (human) computer scans and works with only a finite number of symbols and/or states, involves strong and questionable metaphysical, epistemological and/or psychological assumptions that Turing relied upon to *justify* his analysis. And it is quite correct that at one pivotal point in developing the analysis in his famous paper, Turing claims that a human computer can recognise only a bounded number of different discrete configurations 'at a glance', or 'immediately', because human memory is limited (Turing, 1937, p. 231). But from the perspective adopted here, this is hardly a metaphysically or epistemologically loaded remark. Turing is only resting upon an everyday observation, and not a theory. He is simply making explicit certain characteristic features earmarking the concept that is being analysed in the Hilbertian context, namely, the idea of a human making a recogniseable step in a computation or a formal system, a human following a 'definite procedure' or rule in the relevant sense. His simplification does not turn on a thesis in philosophy of mind or mathematics, nor is it a psychological theory of what is 'really going on' in our minds when we compute. Instead, it is a model for clarification, taken up in a spirit analogous to Wittgenstein's idea that a proof must be perspicuous (*Übersichtlich, Übersehbar*), an idea about which Wittgenstein wrote a great deal after Turing attended his lectures (Wittgenstein, 1978, Part III).

References

Carnap, R., 2002. The Logical Syntax of Language. Open Court, Chicago.

Diamond, C. (Ed.), 1939. Wittgenstein's Lectures on the Foundations of Mathematics, Cambridge; The University of Chicago Press, Chicago, 1989.

Floyd, J., 2012. Wittgenstein's Diagonal Argument: A Variation on Cantor and Turing. In: Dybjer, P., Lindström, S., Palmgren, E., Sundholm, G. (Eds.), Epistemology versus Ontology: Essays on the Philosophy and Foundations of Mathematics in Honour of Per Martin-Löf, Springer-Verlag, NewYork/Dordrecht.

Gandy, R.O., 2001. Introduction to Turing's 'The reform of mathematical notation and phraseology' (1944–5). In: Gandy, R.O., Yates, C. E. M. (Eds.), Collected Works of A. M. Turing: Mathematical Logic, North-Holland/Elsevier Science, Amsterdam, pp. 211–213.

Gödel, K., 1986. Collected Works, vol. I: publications 1929–1936. Oxford University Press, New York.

Gödel, K., 1990. Collected Works, vol. II: publications 1938–1974. Oxford University Press, New York.

Hodges, A., 1983. Alan Turing: the Enigma. Burnett Books, London.

Turing, A. M., 1937. On computable numbers, with an application to the Entscheidungsproblem. Proc. Lond. Math. Soc. 2 (42), 230–265.

Turing, A. M., 1954. Solvable and unsolvable problems. Sci. News 31, 7–23.

Wittgenstein, L., 1974. Philosophical Grammar. Blackwell, Oxford.

Wittgenstein, L., 1978. Remarks on the Foundations of Mathematics. M. I. T. Press, Cambridge MA.

References

Part II

Hiding and Unhiding Information: Cryptology, Complexity and Number Theory

On the Gaussian error function

(Fellowship Dissertation (1935), unpublished)

Sandy L. Zabell delivers an authoritative guide to —

ALAN TURING AND THE CENTRAL LIMIT THEOREM[1]

Although the English mathematician Alan Mathison Turing (1912–1954) is remembered today primarily for his work in mathematical logic (Turing machines and the 'Entscheidungsproblem'), machine computation and artificial intelligence (the 'Turing test'), his name is not usually thought of in connection with either probability or statistics. One of the basic tools in both of these subjects is the use of the normal or Gaussian distribution as an approximation, one basic result being the Lindeberg-Feller central limit theorem taught in first-year graduate courses in mathematical probability. No one associates Turing with the central limit theorem, but in 1934, Turing, while still an undergraduate, rediscovered a version of the Lindeberg central limit theorem and much of the Feller-Lévy converse to it (then unpublished).

1. Introduction

Turing went up to Cambridge as an undergraduate in the Fall Term of 1931, having gained a scholarship to King's College. (Ironically, King's was his second choice; he had failed to gain a scholarship to Trinity.) Two years later, during the course of his studies, Turing attended a series of lectures on the Methodology of Science, given in the autumn of 1933 by the distinguished astrophysicist Sir Arthur Stanley Eddington. One topic that Eddington discussed was the tendency of experimental measurements subject to errors of observation to often have an approximately normal or Gaussian distribution. But Eddington's heuristic sketch left Turing dissatisfied; and Turing set out to derive a rigorous mathematical proof of what is today termed the central limit theorem for independent (but not necessarily identically distributed) random variables.

Turing succeeded in his objective within the short span of several months (no later than the end of February 1934). Only then he did find out that the problem had already been solved, 12 years earlier, in 1922, by the Finnish mathematician Jarl Waldemar Lindeberg (1876–1932). Despite this, Turing was encouraged to submit his work, suitably amended, as a Fellowship Dissertation. (Turing was still an undergraduate at the time; persons seeking to become a Fellow at a Cambridge college had to submit evidence of original work, but did not need to have a Ph.D. or its equivalent.) This revision, entitled 'On the Gaussian Error Function', was completed and submitted in November, 1934. On the strength of this paper, Turing was elected as a Fellow of King's 4 months later (March 16, 1935) at the age of 22; his nomination supported by the group theorist Philip Hall and the economists John Maynard Keynes and Alfred Cecil Pigou. Later that year, the paper was awarded the prestigious Smith's prize by the University (see Hodges, 1983).

[1]Turing's dissertation was not published. Only the Preface is reproduced here and in the Collected Works. A scan of the complete original dissertation is in the Turing Digital Archive.

Turing never published his paper. Its major result had been anticipated, although, as will be seen, it contains other results that were both interesting and novel at the time. But in the interim, Turing's mathematical interests had taken a very different turn, to mathematical logic, and thus Turing turned from mathematical probability, never to return.

2. The central limit theorem

The earliest version of the central limit theorem (CLT) is due to Abraham de Moivre (1667–1754). If $X_1, X_2, X_3, \ldots$ is an infinite sequence of 1s and 0s recording whether a success ($X_n = 1$) or failure ($X_n = 0$) has occurred at each stage in a sequence of repeated trials, then the sum $S_n := X_1 + X_2 + \cdots + X_n$ gives the total number of successes after n trials. If the trials are independent, and the probability of a success at each trial is the same, say $P[X_n = 1] = p, P[X_n = 0] = 1 - p$, then the probability of seeing exactly k successes in n trials has a binomial distribution:

$$P[S_n = k] = \frac{n!}{k!\,(n-k)!} p^k (1-p)^{n-k}.$$

If n is large (e.g., 10,000), then as de Moivre noted, the direct computation of binomial probabilities 'is not possible without labour nearly immense, not to say impossible'; and for this reason, he turned to approximate methods (see Diaconis and Zabell, 1991): using Stirling's approximation (including correction terms) to estimate the individual terms in the binomial distribution and then summing, de Moivre discovered the remarkable fact that

$$\lim_{n \to \infty} P\left[a \leq \frac{S_n - np}{\sqrt{np(1-p)}} \leq b\right] = \frac{1}{\sqrt{2\pi}} \int_a^b \exp\left[-\frac{1}{2}x^2\right] dx;$$

or $\Phi(b) - \Phi(a)$, where $\Phi(x)$ is the cumulative distribution function of the standard normal (or Gaussian) distribution:

$$\Phi(x) := \frac{1}{\sqrt{2\pi}} \int_{-\infty}^{x} \exp\left[-\frac{1}{2}t^2\right] dt.$$

During the 19th and 20th centuries, this result was extended far beyond the simple coin tossing setup considered by de Moivre, important contributions being made by Laplace, Poisson, Chebyshev, Markov, Liapunov, von Mises, Lindeberg, Lévy, Bernstein and Feller; see Le Cam (1986), Stigler (1986) and Hald (1998) for further historical information. Such investigations revealed that if $X_1, X_2, X_3, \ldots$ is any sequence of independent random variables having the same distribution, then the sum S_n satisfies the CLT provided suitable centering and scaling constants are used: the centering constant np in the binomial case is replaced by the sum of the expectations $E[X_n]$; the scaling constant $\sqrt{np(1-p)}$ is replaced by the square root of the sum of the variances $Var[X_n]$ (provided these are finite).

Indeed, it is not even necessary for the random variables X_n contributing to the sum S_n to have the same distribution, provided that no one term dominates the sum. Of course this has to be made precise. The best result is due to Lindeberg. Suppose $E[X_n] = 0, 0 < Var[X_n] < \infty, s_n^2 := Var[S_n]$, and

$$\Lambda_n(\epsilon) := \sum_{k=1}^{n} E\left[\left(\frac{X_k}{s_n}\right)^2 : \frac{|X_k|}{s_n} \geq \epsilon\right].$$

(The notation $E[X; Y \geq \epsilon]$ means the expectation of X is restricted to outcomes ω, such that $Y(\omega) \geq \epsilon$). The *Lindeberg condition* is the requirement that

$$\Lambda_n(\epsilon) \to 0, \quad \forall \epsilon > 0; \tag{2.1}$$

and the *Lindeberg central limit theorem* (Lindeberg, 1922) states that if the sequence of random variables $X_1, X_2, \ldots$ satisfies the Lindeberg condition (2.1), then for all $a < b$,

$$\lim_{n\to\infty} P\left[a \leq \frac{S_n}{s_n} \leq b\right] = \Phi(b) - \Phi(a). \tag{2.2}$$

Despite its technical appearance, the Lindeberg condition (2.1) turns out to be a natural sufficient condition for the CLT. There are two reasons for this. First, the Lindeberg condition has a simple consequence: if $\sigma_k^2 := Var[X_k]$, then

$$\rho_n^2 := \max_{k\leq n}\left(\frac{\sigma_k^2}{s_n^2}\right) \to 0. \tag{2.3}$$

Thus, if the sequence X_1, X_2, X_3, satisfies the Lindeberg condition, the variance of an individual term X_k in the sum S_n is asymptotically negligible. Second, for such sequences, the Lindeberg condition is necessary as well as sufficient for the CLT to hold, a beautiful fact discovered (independently) by William Feller and Paul Lévy in 1935. In short: (2.1) $\leftrightarrow$ (2.2) + (2.3).

If, in contrast, the Feller–Lévy condition (2.3) fails, then it turns out that convergence to the normal distribution can only occur in a fashion markedly different from that of the CLT. If the Feller–Lévy condition fails, then there exists a number $\rho > 0$, and two sequences of positive integers $\{m_k\}$ and $\{n_k\}$, such that $\{m_k\}$ is strictly increasing,

$$1 \leq m_k \leq n_k \quad \text{for all } k, \quad \text{and} \quad Var\left[\frac{X_{m_k}}{s_{n_k}}\right] = \frac{\sigma_{m_k}^2}{s_{n_k}^2} \to \rho^2 > 0. \tag{2.4}$$

Feller (1937) showed that if normal convergence occurs (that is, condition (2.2) holds), but condition (2.4) also obtains, then ('$\Rightarrow$' denoting convergence in distribution, $N(\mu.\sigma^2)$ the normal distribution with mean μ, variance σ^2)

$$\frac{1}{\rho}\frac{X_{m_k}}{s_n k} \Rightarrow N(0,1).$$

That is, there exists a subsequence X_{m_k} whose contributions to the sums S_n are non-negligible (relative to s_n) and which, properly scaled, converges to the standard normal distribution.

3. Turing's fellowship dissertation

3.1. *Basic structure of the paper*

The first seven sections of the paper (pp. 1–6) summarise notation and the basic properties of distribution functions. Section 1 summarises the problem; Section 2 defines the distribution function F (abbreviated DF) of an 'error' ϵ; Section 3 summarises the basic properties of the expectation and mean square deviation (MSD) of a sum of independent errors; rigorous proofs in terms of the distribution function are given in Appendix C at the end of the paper. Section 4 discusses the distribution function of a sum of independent errors, the *sum distribution function* (SDF), in terms of the distribution functions of each term in the sum, and derives the formula for $F \oplus G$, the convolution of two distribution functions. Section 5 then introduces the concept of the *shape function* (SF); the standardisation of a distribution function F to have zero expectation and unit MSD; thus, if F has expectation μ and MSD $\sigma^2 (\sigma > 0)$, then the shape function of F is $U(x) := F(\sigma(x-\mu))$. (Turing uses the symbols 'a' and 'k^2' to denote μ and σ^2; several other minor changes in notation of this sort are used below.)

In Section 6, Turing then states the basic problem to be considered: given a sequence of errors ϵ_k, having distribution functions G_k, shape functions V_k, means μ_k, mean square deviations σ_k^2, sum distribution functions F_n and shape functions U_n for each F_n, under what conditions do the shape functions $U_n(x)$ converge uniformly to $\Phi(x)$, the 'SF of the Gaussian Error'? Turing then assumes for simplicity that $\mu_k = 0$ and $\sigma_k^2 < \infty$. In Section 7 (Fundamental Property of the Gaussian Error), he notes the only properties of Φ that are used in deriving sufficient conditions for normal convergence are that it is an SF, and the self-reproductive property of Φ: that is, if $X_1 \sim N(0,\sigma_1^2)$ and $X_2 \sim N(0,\sigma_2^2)$ are independent, then $X_1 + X_2 \sim N(0,\sigma_1^2 + \sigma_2^2)$. (The notation $X \sim N(\mu,\sigma^2)$ means that the random variable X has the distribution $N(\mu,\sigma^2)$.)

3.2. *The Quasi–necessary conditions*

It is at this point that Turing comes to the heart of the matter. In Section 8 (The Quasi-Necessary Conditions), Turing notes

> The conditions we shall impose fall into two groups. Those of one group (the quasi–necessary conditions) involve the MSDs only. They are not actually necessary, but if they are not fulfilled U_n can only tend to Φ by a kind of accident.

The two conditions that Turing refers to as the 'quasi–necessary conditions' are:

$$\sum_{k=1}^{\infty} \sigma_k^2 = \infty \quad \text{and} \quad \frac{\sigma_n^2}{s_n^2} \to 0. \tag{3.1}$$

It is easy to see that Turing's condition (3.1) is equivalent to condition (2.3). (That (2.3) $\Rightarrow$ (3.1) is immediate. To see (3.1) $\Rightarrow$ (2.3) : given $\epsilon > 0$, choose $M \geq 1$ so that $\sigma_n^2/s_n^2 < \epsilon$ for $n \geq M$, and $N \geq M$ so that $s_k^2/s_N^2 < \epsilon$ for $1 \leq k \leq M$; if $n \geq N$, then $\sigma_k^2/s_n^2 < \epsilon$ for $1 \leq k \leq n$.)

In his Theorems 4 and 5, Turing explores the consequences of the failure of either part of condition (3.1). Turing's proof of Theorem 4 requires his

Theorem 3.1. *If X and Y are independent, and both X and $X+Y$ are Gaussian, then Y is Gaussian.*

This is a special case of a celebrated theorem proven shortly thereafter by Harald Cramér (1936); if X and Y are independent, and $X+Y$ is Gaussian, then both X and Y must be Gaussian. Lévy had earlier conjectured Cramér's theorem to be true (in 1928 and again in 1935) but had been unable to prove it. Cramér's proof of this result in 1936 in turn enabled Lévy to arrive at necessary and sufficient conditions for the CLT of a very general type (using centering and scaling constants other than the mean and standard deviation), and this in turn led Lévy to write his famous monograph, *Théorie de l'Addition des Variables Aléatoires* (Lévy, 1937); see Le Cam (1986, pp. 80–81, 90).

Cramér's theorem is a hard fact; his original proof appealed to Hadamard's theorem in the theory of entire functions. The special case of the theorem needed by Turing is much simpler; it is an immediate consequence of the characterisation theorem for characteristic functions. To see this, let $\phi_X(t) := E[\exp(itX)]$ denote the characteristic function of a random variable X; and suppose that X and Y are independent, $X \sim N(0,\sigma^2)$ and $X+Y \sim N(0,\sigma^2+\tau^2)$. Then

$$\exp\left(-\frac{\sigma^2+\tau^2}{2}t^2\right) = \phi_{X+Y}(t) = \phi_X(t)\phi_Y(t) = \exp\left(-\frac{\sigma^2}{2}t^2\right)\phi_Y(t),$$

hence $\phi_Y(t) = \exp(-\tau^2 t^2/2)$; thus, $Y \sim N(0,\tau^2)$ because the characteristic function of a random variable uniquely determines the distribution of that variable. Turing's proof, which uses distribution functions, is not much longer.

It is an immediate consequence of Cramér's theorem that if $S_n/s_n \Rightarrow N(0,1)$, but $\lim_{n\to\infty} s_n^2 < \infty$, then all the summands X_j must in fact be Gaussian. But Turing did not have this fact at his disposal, only his much weaker Theorem 3. His Theorem 4 (phrased in the language of random variables) thus makes the much more limited claim that if (a) $\sum \sigma_n^2 < \infty$, (b) S_n converges to a Gaussian distribution, and (c) X_0 is a random variable at once independent of the original sequence $X_1, X_2, \ldots$ and having a distribution other than Gaussian, then the sequence $S_n^* = X_0 + S_n$ *cannot* converge to the Gaussian distribution. In other words, if $\sum \sigma_n^2 < \infty$, then 'the convergence ... to the Gaussian is so delicate that a single extra term in the sequence ... upsets it' (p. 17).

Turing's Theorem 5 in turn explores the consequences of the failure of (3.1) in the case that $\sum \sigma_n^2 = \infty$, but $\rho_n^2 := \sigma_n^2/s_n^2$ does not tend to zero as $n \to \infty$. The statement of the theorem is somewhat technical in nature, but Turing's later summary of it captures the essential phenomenon involved:

> If F_n [the distribution function of S_n] tends to Gaussian and σ_n^2/s_n^2 does not tend to zero [but $\sum \sigma_n^2 = \infty$] we can find a subsequence of G_n [the distribution function of X_n] tending to Gaussian.

Thus, Turing had by some 2 years anticipated Feller's discovery of the subsequence phenomenon. (In Turing's typescript, symbols such as 'F_n' are entered by hand; in the above quotation, the space for 'F_n' has by accident been left blank, but the paragraph immediately preceding this one in the typescript makes it clear that 'F_n' is intended.)

3.3. *The sufficient conditions*

Turing states in his preface that he had been 'informed that an almost identical proof had been given by Lindeberg'. This comment refers to the method of proof used by Turing and not the result obtained. Turing's method is to smooth the distribution functions $F_n(x)$ of the sum by forming the convolution $F_n * \Phi(x/\rho)$, expand the result in a Taylor series to third order and then let the variance ρ^2 of the convolution term tend to zero. This is similar to the method employed by Lindeberg. (There is an important difference, however: Turing does not use Lindeberg's 'swapping' argument. For an attractive modern presentation of the Lindeberg method, see Breiman (1968, pp. 167–170); for discussion of the method, Pollard's comments in Le Cam (1986, pp. 94–95.)) Turing does not, however, succeed in arriving at the Lindeberg condition (2.1) as a sufficient condition for convergence to the normal distribution; the most general sufficient condition he gives (on p. 27) is complex in appearance (although it necessarily implies the Lindeberg condition). Turing concedes that his 'form of the sufficiency conditions is too clumsy for direct application', but notes that it can be used to 'derive various criteria from it, of different degrees of directness and of comprehensiveness' (p. 28). One of these holds if the summands X_k all have the same shape (that is, the shape functions $V_k(x) := P[X_k/\sigma_k \le x]$ coincide) and thus includes the special case of identically distributed summands having a second moment. (This was no small feat, because even this special case of the more general Lindeberg result had eluded proof until the publication of Lindeberg's paper.)

One formulation of this criterion, equivalent to the one actually stated by Turing, is that there exists a function $J : \mathbf{R}^+ \to \mathbf{R}^+$, such that $\lim_{t\to\infty} J(t) = 0$, and

$$E\left[\left(\frac{X_k}{\sigma_k} - t\right)^2 ; \left|\frac{X_k}{\sigma_k}\right| \ge t\right] \le J(t) \quad \text{for all} \quad k \ge 1, t \ge 0. \tag{3.2}$$

In turn, one simple sufficient condition for this given by Turing (pp. 30–31) is that there exists a function ϕ, such that $\phi(x) > 0$ for all x, $\lim_{x\to\pm\infty} \phi(x) = \infty$, and $\sup_k E\left[\left(\frac{X_k}{\sigma_k}\right)^2 \phi\left(\frac{X_k}{\sigma_k}\right)\right] < \infty$.

(Note that unfortunately one important special case not covered by either of these conditions is that the X_k are uniformly bounded: $|X_k| \leq C$ for some $C > 0$ and all $k \geq 1$.)

In assessing this portion of Turing's paper, it is important to keep two points in mind. First, Turing states in his preface that 'since reading Lindeberg's paper, I have for obvious reasons made no alterations to that part of the paper which is similar to his'. The manuscript is thus necessarily incomplete; it presumably would have been further polished and refined had Turing continued to work on it; the technical sufficient conditions given represent how far Turing had gotten on the problem prior to seeing Lindeberg's work. Second, in 1934, the Lindeberg condition was only known to be sufficient, not necessary; thus even in discussing his results in other sections of the paper (where he felt free to refer to the Lindeberg result), it may not have seemed important to Turing to contrast his own particular technical sufficient conditions with those of Lindeberg; the similarity in the method must have seemed far more important.

3.4. *One counterexample*

In Section 14, Turing concludes by giving a simple example of a sequence $X_1, X_2, \ldots$ that satisfies the quasi-necessary conditions (3.1), but not the CLT. This example turns out to be quite interesting; see Zabell (1995).

4. Discussion

I. J. Good (1980, p. 34) has remarked that when Turing 'attacked a problem he started from first principles, and he was hardly influenced by received opinion. This attitude gave depth and originality to his thinking, and also it helped him to choose important problems'. This observation is nicely illustrated by Turing's work on the CLT. His dissertation is, viewed in context, a very impressive piece of work. Coming to the subject as an undergraduate, his knowledge of mathematical probability was apparently limited to some of the older textbooks such as 'Czuber, Morgan Crofton and others' (Preface, p. ii). Despite this, Turing immediately realized the importance of working at the level of distribution functions rather than densities; developed a method of attack similar to Lindeberg's; obtained useful sufficient conditions for convergence to the normal distribution; identified the conditions necessary for true central limit behaviour to occur; understood the relevance of a Cramér-type factorisation theorem in the derivation of such necessary conditions; and discovered the Feller subsequence phenomenon. If one realizes that the defects of the paper, such as they are, must largely reflect the fact that Turing had ceased to work on the main body of it after being apprised of Lindeberg's work, it is clear that Turing had penetrated almost immediately to the heart of a problem whose solution had long eluded many mathematicians far better versed in the subject than he. (It is interesting to note that Lindeberg was also a relative outsider to probability theory and only began to work in the field a few years before 1922.)

It is also interesting to note Turing's approach to the problem in terms of convolutions of distribution functions rather than sums of independent random variables. Feller had similarly avoided the use of the language of random variables in his 1935 paper, formulating the problem instead in terms of convolutions. The reason, as Le Cam (1986, p. 87) notes, was that 'Feller did not think that such concepts [as random variable] belonged in a mathematical framework. This was a common attitude in the mathematical community'.

References

Breiman, L., 1968. Probability. Addison–Wesley, Reading, MA.

Cramér, H., 1936. Ueber eine Eigenschaft der normalen Verteilungsfunktion. Mathematische Zeitschrift 41, 405–414.

Diaconis, P., Zabell, S., 1991. Closed form summation for classical distributions: variations on a theme of De Moivre. Stat. Sci. 6, 284–302.

Feller, W., 1935. Über den zentralen Grenzwertsatz der Wahrscheinlichkeitsrechnung. Mathematische Zeitschrift 40, 521–559.

Feller, W., 1937. Über den zentralen Grenzwertsatz der Wahrscheinlichkeitsrechnung, II. Mathematische Zeitschrift 42, 301–312.

Good, I.J., 1980. Pioneering work on computers at Bletchley. In: Metropolis, N., Howlett, J., Rota, G.-C. (Eds.), A History of Computing in the Twentieth Century, Academic Press, New York, pp. 31–45.

Hald, A., 1998. A History of Mathematical Statistics from 1750 to 1930. Wiley Interscience, New York.

Hodges, A., 1983. Alan Turing: The Enigma. Simon and Schuster, New York.

Le Cam, L., 1986. The central limit theorem around 1935 (with discussion). Stat. Sci. 1, 78–96.

Lévy, P., 1935. Propriétés asymptotiques des sommes de variables indépendantes ou enchainées. J. Math. Pures Appl. 14, 347–402.

Lévy, P., 1937. Théorie de l'Addition des Variables Aléatoires. Gauthier–Villars, Paris.

Lindeberg, J.W., 1922. Eine neue Herleitung des Exponential-gesetzes in der Wahrscheinlichkeitsrechnung. Mathematische Zeitschrift 15, 211–225.

Stigler, S.M., 1986. The History of Statistics. Harvard University Press, Cambridge, MA.

Turing's 'Preface' (1935) to 'On the Gaussian error function'

The object of this paper is to give a rigorous demonstration of the "limit theorem of the theory of probability". I had completed the essential part of it by the end of February 1934 but when considering publishing it I was informed that an almost identical proof had been given by Lindeberg[1]. The only important differences between the two papers is that I have introduced and laid stress on a type of condition which I call quasi-necessary (Section 8). We have both used "distribution functions" (§2) to describe errors instead of frequency functions (Appendix B) as was usual formerly. Lindeberg also uses (D) of §12 and Theorem 6 or their equivalents.

Since reading Lindeberg's paper I have for obvious reasons made no alterations to that part of the paper which is similar to his (viz. §9 to §13), but I have added elsewhere remarks on points of interest and the appendices.

So far as I know the results of §8 have not been given before. Many proofs of the completeness of the Hermite functions are already available (footnote, p.33) but I believe that that given in Appendix A is original. The remarks in Appendix B are probably not new. Appendix C is nothing more than a rigorous deduction of well-known facts. It is only given for the sake of logical completeness and it is of little consequence whether it is original or not.

My paper originated as an attempt to make rigorous the "popular" proof mentioned in Appendix B. I first met this proof in a course of lectures by Prof. Eddington. Variations of it are given by Czüber, Morgan, Crofton and others. Beyond this I have not used the work of others or other sources of information in the main body of the paper, except for elementary matter forming part of one's general mathematical education, but in the appendices I may mention Liapounoff's papers which I discuss there.

I consider §9 to §13 is by far the most important part of this paper, the remainder being comment and elaboration. At a first reading therefore §8 and the appendices may be omitted.

[1] *Math. Z.* **15** (1922).

Some Calculations of the Riemann Zeta function

(Proc. Lond. Math. Soc., series 3 vol. 3 (1953), pp. 99–117)

On a Theorem of Littlewood

(Unpublished manuscript, with S. Skewes, c.1952–53)

Dennis Hejhal and Andrew Odlyzko take an in-depth look at —

ALAN TURING AND THE RIEMANN ZETA FUNCTION

1. Introduction

Turing encountered the Riemann zeta function as a student and developed a life-long fascination with it. Though his research in this area was not a major thrust of his career, he did make a number of pioneering contributions. Most have now been superseded by later work, but one technique that he introduced is still a standard tool in the computational analysis of the zeta and related functions. It is known as *Turing's method* and keeps his name alive in those areas.

Of Turing's two published papers (Turing, 1943, 1953) involving the Riemann zeta function $\zeta(s)$, the second[1] is the more significant. In that paper, Turing reports on the first calculation of zeros of $\zeta(s)$ ever done with the aid of an electronic digital computer. It was in developing the theoretical underpinnings for this work that Turing's method first came into existence.

Our primary aim in this chapter is to provide an overview of Turing's work on the zeta function. The influence that interactions with available technology and with other researchers had on his thinking is deduced from Turing (1943, 1953) as well as some unpublished manuscripts of his (available in Turing (1992)) and related correspondence, some newly discovered. To minimise any overlap with other chapters, we do not discuss Turing's contributions to computing in general, even though they did influence the work on $\zeta(s)$ that he and those who followed in his footsteps carried out.

The recent survey article of Booker (2006) has a significant overlap with what we say here and is highly recommended as a collateral 'read'.

2. Recollection of some basics

The Riemann zeta function $\zeta(s)$ is defined for complex s with $\mathrm{Re}(s) > 1$ by

$$\zeta(s) = \sum_{n=1}^{\infty} \frac{1}{n^s}. \tag{2.1}$$

[1] *Some calculations of the Riemann Zeta function*, reproduced below.

This function can be extended analytically to the entire complex plane except for the point $s = 1$, at which there is a pole of order one. The extended function, which is again denoted by $\zeta(s)$, has so-called trivial zeros at $s = -2, -4, -6, \ldots$. The other zeros, called *nontrivial zeros*, are also infinite in number and lie inside the *critical strip* $0 < \text{Re}(s) < 1$. The Riemann hypothesis (RH) is the assertion that all the nontrivial zeros ρ lie in the centre of the critical strip, i.e., on the *critical line* $\text{Re}(s) = \frac{1}{2}$. Any ρ's lying off the critical line necessarily occur in symmetric quadruplets $\{\rho, \bar{\rho}, 1-\rho, 1-\bar{\rho}\}$.

The RH is widely regarded as the most famous unsolved problem in mathematics. It was one of the 23 famous problems selected by Hilbert in 1900 as among the most important in mathematics, and it is one of the seven Millennium Problems selected by the Clay Mathematics Institute in 2000 as the most important for the 21st century (Clay, 2000). For general background on the RH, we shall be content to cite the survey article by Conrey (2003) and Clay (2000). For more technical information about the zeta function, see Titchmarsh (1985).

The RH was posed by Bernhard Riemann in 1859. (See Clay (2000) for a copy of Riemann's paper and an English translation.) The importance of the RH stems from the connection observed by Riemann between primes and the non-trivial zeros of the zeta function. If, as usual, we let $\pi(x)$ be the number of primes up to x, then Riemann showed that (for $x \geqq 2$)

$$\pi(x) = \text{Li}(x) - \frac{1}{2}\text{Li}(x^{1/2}) - \sum_{\rho} \text{Li}(x^{\rho}) + W(x), \tag{2.2}$$

where $\text{Li}(x)$ is the logarithmic integral, a nice and smoothly growing function, and $\text{W}(x)$ is of lower order (relative to the three earlier summands). The terms $\text{Li}(x^{\rho})$ are special cases of the classical analytic function $\text{Ei}(\xi)$ defined for $\text{Im}(\xi) \neq 0$, which differs insignificantly from e^{ξ}/ξ whenever $|\xi| \gg 1$. One simply puts $\xi = \rho \ln(x)$ for each ρ.

The main difficulty in using Eq. (2.2) to estimate $\pi(x)$ is that the series is not absolutely convergent. Since $\pi(x)$ is a step function, and the individual terms on the right side of Eq. (2.2) are continuous at each prime number p, the sum behaves somewhat like a Fourier series in producing the discontinuities of $\pi(x)$. Another difficulty is that the sizes of the individual terms depend on the locations of the non-trivial zeros ρ.

The leading term in Eq. (2.2), $\text{Li}(x)$, grows like $x/\ln(x)$ as $x \to \infty$. The Prime Number Theorem, first proved in 1896 by Hadamard and de la Vallée Poussin using properties of zeros of the zeta function, tells us that asymptotically $\pi(x)$ grows like $\text{Li}(x)$; hence like $x/\ln(x)$. The RH has been shown to be equivalent to the difference function $|\pi(x) - \text{Li}(x)|$ being bounded by a quantity close to $\sqrt{x}$, where close means within logarithmic factors or (what amounts to the same thing) the square root of the leading term in Eq. (2.2).

In his famous 1859 paper, Riemann asserted that most non-trivial zeros of the zeta function are on the critical line and that it was *likely* that all of them lie there (which is what we now refer to as the RH). Riemann did not provide even a hint of a proof for the first, positive, assertion. It remains unproved to this day, although it is believed to be true, even by those who are skeptical of the truth of the RH. The RH itself is known to be true for the first 10^{13} non-trivial zeros, as well as large blocks of zeros much higher up, including some around zero number 10^{24}.

At the end of his paper, Riemann also discussed another conjecture that played a significant part in Turing's research, namely $\pi(x) < \text{Li}(x)$. As Riemann noted, computations by Gauss and Goldschmidt had established the validity of this inequality for $x < 10^5$, and if the series over the non-trivial zeros ρ in Eq. (2.2) was nicely behaved, the difference $\text{Li}(x) - \pi(x)$ would tend to grow roughly like $\sqrt{x}/\ln(x)$. From the tone of Riemann's presentation, it appears that he suspected that the inequality $\pi(x) < \text{Li}(x)$ might well be true generally. (We say 'suspected' because Riemann's wording is vague.)

Today, we know that $\pi(x) < \text{Li}(x)$ holds not just for $x < 10^5$, but even for $x < 10^{14}$. In 1914, however, Littlewood proved that there are infinitely many integers $x \geqq 2$ for which the inequality fails! The most recent result in this area shows that the inequality fails for some $x < 10^{317}$, but we still do not know where the first counterexample occurs. There are heuristic arguments suggesting there are no counterexamples within $x < 10^{30}$ and likely even higher. Thus, this is one of the many instances that occur in number theory of a conjecture that is supported by heuristics and extensive numerical evidence, yet turns out to be false. In a similar way, the validity of the RH is definitely *not* something that we can be assured of simply on the basis of its being true for the first 10^{13} cases.

Littlewood's proof that $\pi(x) > \text{Li}(x)$ holds infinitely often relied on Riemann's expansion (2.2) and required considerable technical virtuosity to deal with the infinite series that was not absolutely convergent. In the mid-1930s, another approach became available through the work of Ingham, which had the advantage of being both simpler and more explicit, but at the cost of requiring some computations. In very loose terms, Littlewood's result was shown to follow from knowledge of some initial set of non-trivial zeros of the zeta function (cf. Section 5 below). This connected numerical verifications of the RH to the $\pi(x) < \text{Li}(x)$ conjecture. Turing was intrigued by both problems and made contributions to each one.

Interestingly enough, it appears that Turing had doubts about the validity of the RH already at an early stage and that, over time, his skepticism only increased.[2]

3. On Turing's computations of the zeta function

The first computations of zeros of the zeta function were performed by Riemann and likely played an important role in his posing of the RH as a result likely to be true. His computations were carried out by hand, using an advanced method that is known today as the Riemann-Siegel formula. Both the method and Riemann's computations that utilised it remained unknown to the world-at-large until the early 1930s, when they were found in Riemann's unpublished papers by C. L. Siegel. In the meantime, as both the significance and difficulty of the RH were recognised around the turn of the 20th century, computations using a less efficient method, based on Euler-Maclaurin summation, were carried out by several investigators. The calculations used tables of logarithms and trig functions, paper and pencil, and mechanical calculators. The largest of those early computational efforts was that of J. Hutchinson, who showed that there were exactly 138 zeros of the zeta function with $0 < \text{Im}(s) < 300$ and that they all satisfied the RH. (Hutchinson also provided modestly accurate values for the 29 zeros in $0 < \text{Im}(s) < 100$.)

Aside from possible numerical mistakes, these computations are completely rigorous and do establish the validity of the RH for all the zeros for which it is claimed. As was recognised already by Riemann, there is a simple variant of the zeta function that is *real* on the critical line, so that a sign change of this function has to come from a zero of the zeta function that is right on the critical line. The final stage was the verification that the sign changes that have been found account for *all* the zeros in a given $\text{Im}(s)$- range. Until Turing came out with his method, this step was done by a rather messy, although in principle not very difficult, computation based on the principle of the argument. Turing's method obviates any need for using the argument principle. It involves only the real-valued function on the critical line. See Turing (1953, Section 4) for a precise statement.

In the mid-1930s, after Siegel's publication of the Riemann–Siegel formula, Titchmarsh obtained a grant for a larger computation. With the assistance of L. J. Comrie, tabulating machines, some 'computers' (as the mostly female operators of such machinery were called in those days), and the recently published algorithm, Titchmarsh established that the 1041 non-trivial zeros in $0 < \text{Im}(s) < 1468$ all satisfied the RH (Titchmarsh, 1936).

Turing became interested in extending Titchmarsh's results. He designed and started to build, with the help of a £40 grant from the Royal Society, a special purpose analog computer to verify

[2] Littlewood's views followed a similar trajectory; see Littlewood (1962; 1982, p. 792)

whether the RH is satisfied by all the zeros with $0 < \mathrm{Im}(s) < 6000$ (of which there are 5598). More details about this machine are available in Booker (2006) and Casselman (2006). Work on this project was interrupted by the outbreak of World War II, and this computer was never constructed.

We do not know how well Turing's zeta function machine would have worked, had it been built. At least one special zeta function computer was constructed to a different design later by van der Pol (1947). By that time, though, electronic digital computers were becoming available, and Turing (1953) was the first one to utilise them to investigate the zeta function. In 1950, he used the Manchester Mark 1 Electronic Computer to extend the Titchmarsh verification of the RH to the first 1104 zeros of the zeta function, the ones with $0 < \mathrm{Im}(s) < 1540$. This was a very small extension, but it represented a triumph of perseverance over a promising new technology that was still suffering from teething problems. In Turing's words, '[i]f it had not been for the fact that the computer remained in serviceable condition for an unusually long period from 3 P.M. one afternoon to 8 A.M. the following morning it is probable that the calculations would never have been done at all'. These days, when even our simple consumer devices have gigabytes of memory, it is instructive to recall that the machine available to Turing had a grand total of 25,600 bits of memory and that Turing worked directly with output 'punched out on teleprint tape' in base 32. That Turing stayed up all through the night conveys some idea of how interesting he found this experiment.

More significant than the extension of the Titchmarsh verification of the RH to an additional *63* zeros was Turing's earlier computation on that same occasion of the 1054 zeros in $2\pi 63^2 \leqq \mathrm{Im}(s) \leqq 2\pi 64^2$, all of which turned out to lie on the critical line. (Note that $2\pi 63^2$ is about 25,000.) Not only did this produce a substantial increase in the number of zeros that were known to obey the RH, but it represented an innovation, a realisation that by jumping to larger heights one could obtain a better view of the asymptotic behaviour of the zeta function.

Today, Turing's pioneering use of the Manchester Mark 1 for computing the zeta zeros is a historical footnote. Turing's results were soon surpassed by a sequence of increasingly extensive computations. His work was furthermore not an unexpected breakthrough. Development of digital computers and growing interest in the zeta function would surely have led to someone else carrying out similar calculations within a few years, even if he had not done so.

For several decades, progress came exclusively from faster computers and longer runs. Beginning, however, in the mid-1980s, new algorithms started appearing, such as the one of Schönhage and the second author of this chapter for computing large sets of zeros, as well the ones of Schönhage, Heath-Brown, and Hiary for computation of individual values of $\zeta(s)$ when $\mathrm{Im}(s)$ is very large. Combined with growing computing power, these algorithms have enabled calculations far beyond the reach of Turing and his contemporaries. It is now known that the RH is true for the first 10^{13} non-trivial zeros, for some tens of billions of zeros around zeros number 10^{23} and 10^{24}, and for some hundreds of zeros near zero number 10^{32}. (All these projects have relied on Turing's method for *proving* that all zeros in a given range have been found and are on the critical line.) If there was a strong motivation to obtain more data, these numbers could be increased by factors of 10 or 100 simply by harnessing more computing power. As machines become more powerful and more plentiful, and still better algorithms are found, we can look forward to substantial growth in information about non-trivial zeros.

Among recent computations of zeta zeros, the verifications of the RH – whether for initial segments or for blocks of zeros high up – carry on traditions that were extended or started by Turing. Other efforts have involved high precision computations of low zeros. Some of those are done to obtain improved bounds for the counterexamples to conjectures such as that of Mertens, or that $\pi(x) < \mathrm{Li}(x)$, and are related to projects Turing devoted quite a bit of time to, and where had he lived he might have carried out such computations himself (cf. Section 5). Others reflect a desire to test whether zeta zeros satisfy some algebraic relations among themselves or involving other well-known constants, such as e or π. (One conjectures that no such relations exist.) The *main* motivation, however, for recent computations of zeta zeros, as well as zeros of related functions, comes from a conjectured relation between those zeros and eigenvalues of random matrices. A conjecture made by Hilbert and Pólya in the 1910s was that the RH is true because zeta zeros correspond to eigenvalues of a positive operator. This initial conjecture was extremely vague and hard to test. However, a variety of developments in the next half a century provided additional motivation to consider the Hilbert and Pólya guess more seriously. A particularly important development was a theorem of H.L. Montgomery from the early 1970s, which suggested that zeta zeros should behave like eigenvalues of a particular family, the GUE, of random matrices that had been explored intensively by mathematical physicists. Subsequent computations by the second author provided extensive numerical evidence for this connection. Ever since, a large industry has grown up, exploiting the (still conjectural and empirical) connection between zeta zeros and random matrices. This is regarded by many researchers as the most promising road towards a proof of the RH. More details and references can be found in Conrey (2003). This work is far from what Turing was aware of, but one can expect that he would have found it exciting.

4. On Turing's early work with zeta

Most readers will likely have at least some familiarity with Andrew Hodges' definitive biography (Hodges, 1983) of Turing. Pages 94, 133–135, 140–142, and 154–158 therein suffice to give a quick overview of how Turing's research interests with $\zeta(s)$ got started around 1936–7 or so.

By combining the contents of four letters in the Turing Digital Archive (two from Ingham and one each from Skewes and Titchmarsh) with several other sources, it is possible to view these early developments in substantially greater depth and, in the process, add some valuable context to the overall picture. Our aim in the present section is to do this, albeit very succinctly.

The following timeline presents the essential points:

- Turing matriculates at King's College in 1931. He meets Ingham, one of the two mathematics supervisors there. Ingham's now classic Cambridge tract (Ingham, 1932) on prime number theory appears in 1932; Turing obtains a copy shortly thereafter (Hodges, 1983, p. 133).
- In 1933, Littlewood's student, Stanley Skewes (1933) proves that if the RH is true, the smallest $x \geqq 2$ for which $\pi(x) > \mathrm{Li}(x)$ must satisfy $x < 10^A$, where $A = 10^B$ and $B = 10^{34}$. The smallest such integer x is often called the *Skewes number*; for ease of reference, we'll denote this number by x_S. (Skewes and Turing rowed together regularly in Cambridge (Williams, 2007). As will become clear in Section 5, Turing first heard about Skewes' work in that setting, with Skewes 'rowing two' and Turing positioned at bow.)
- During his first year at Princeton (1936–7), Turing keeps in touch with Ingham; he also speaks occasionally with Hardy, who was visiting for a semester (Hodges, 1983, p. 117). Sometime prior to 1 June 1937, the date of Ingham's first Archive letter, Turing mentions to Ingham that he has become interested in trying to attack the x_S- problem by sharpening the original reasoning used by Littlewood in 1914; cf. Eq. (2.2) and (Ingham, 1932, p. 92ff). Ingham offers encouragement, but suggests that his recent, alternate proof (Ingham, 1936) for Littlewood's theorem may be more amenable for this purpose. He encloses an offprint, noting that Skewes has apparently tried the approach – only to come up with (a very likely improvable) upper bound 10^{19} for B, in place of the original 10^{34}.
- Back in Cambridge during the Summer of 1937, Turing pursues Ingham's suggestion with $\psi(x) - x$, a function closely related to $\pi(x) - \mathrm{Li}(x)$ (still assuming the truth of the RH). He obtains a bound much better than Skewes' and communicates this to Ingham. The draft manuscript for this, which appears to be (Turing, 1992, pp. 147–151) (or something quite similar) makes use of a variant of Ingham (1936) and several key $\zeta(s)$ estimates, including one from

the paper of Titchmarsh (1936) on the numerical verification of the RH for Im(s) ranging up to 1468. In his second letter (dated Sept. 18), Ingham reacts positively to Turing's work [without checking every calculation] and conveys the information that Littlewood and Skewes have just about finished deriving a bound for x_S wherein nothing is assumed about the truth of the RH. Ingham refers Turing to a recent paper of Littlewood that obliquely touches on the matter; see Littlewood (1982, pp. 838–843). (N.B. the "1948" appearing on p. 149 of Turing (1992) is *not* present in the original Archive manuscript; Turing only wrote p. 324, which corresponds to the 1933 edition of Jahnke/Emde, *Funktionentafeln*. See also the comment by Cohen in Turing (1992, p. 272).)

- Now back in Princeton, Turing's interests begin to shift more towards $\zeta(s)$ per se, especially its zeros and the matter of extending Titchmarsh (1936) past $t = 1468$. Apart from the work's *intrinsic* merit (including in exploring further the skepticisms about the RH voiced on the final pages in the study by Titchmarsh (1935, 1936)), Turing surely realised that gaining control on a larger initial set of zeta zeros would facilitate a better bound for x_S. The idea of building a special purpose 'gear-wheel' computer (Hodges, 1983, p. 140ff) to evaluate the sum function called for in the main numerical part of Titchmarsh (1936) probably arose during this period. Titchmarsh's letter (of 1 December) makes reference to this; he describes the idea as very interesting and advises Turing that, in the work he proposes, higher-order correction terms may be needed to secure proper accuracy in ρ. He also cautions 'it may be that, like with $\pi(x) - \mathrm{Li}(x)$, $\zeta(\frac{1}{2} + it)$ may go on for a very long time before revealing its true character'.
- On 9 December 1937, Skewes writes from Cape Town, where he worked, and reacts positively to Turing's improved bound for x_S from that summer. From the letter's wording, it is evident that both are occupied with other work at the moment (for Turing, this was his Ph.D. dissertation (Hodges, 1983, p.145)). Skewes writes that he cannot get back to Cambridge for another 2 years – but promises to give some details about the 'RH false case' in his next letter. No such letter is found in the Archive. (Although Skewes' Cambridge dissertation was accepted in 1938, it was not readied for journal publication until 15 years later (Skewes, 1955); an interesting popular account can be found in Section 14 of Littlewood (1948).)
- Turing receives his Princeton Ph.D. in June 1938 and, shortly afterwards, returns to England. It is not until 1939 that he resumes work on $\zeta(s)$. Turing (1943) is submitted for publication on 7 March 1939. In very loose terms, Turing (1943) seeks to address some of the 'correction term issue' that Titchmarsh raised in his 1937 letter by passing to an *alternate* (smoother) version of the $\zeta(s)$-expansion utilised in Titchmarsh (1936) whose basic error term appears to be both smaller and more readily estimable than the one employed previously. Emphasis is placed on s-regions (both on and off the critical line) likely to be pertinent in a 'gear-wheel' setting. The paper is very technical and, as noted by Heath-Brown in (Turing, 1992, p. 261), was soon made unnecessary by the advances that occurred when electronic computers became available. The influence of C. L. Siegel's celebrated 1931 paper on $\zeta(s)$ based on material found in the Riemann Nachlass is plainly visible at several places in Turing (1943). Expansions similar in spirit to Turing (1943) continue to be useful in a variety of other contexts; cf., e.g., Berry and Keating (1992), Paris (1994) and Rubinstein (2005, Section 3).
- Turing submits his £40 proposal for construction of a 'zeta function machine' to the Royal Society on 24 March 1939 (see Turing (1939)). Its stated aim is to extend the range of Titchmarsh's work on the RH by a factor of about 4. Due to the onset of World War II, the proposed machine is never completed.

5. A return to basics

As we just saw, Turing's fascination with $\zeta(s)$ actually originated in a very basic question about the ordinary prime numbers $\{2,3,5,7,11,13,17,\ldots\}$. In light of their structural and aesthetic 'starkness', it is not too surprising that, over the years, the primes would continue to retain a certain attractiveness for Turing.

Most papers dealing with Skewes' problem of trying to find x_S, the smallest integer x for which $\pi(x) > \text{Li}(x)$, are very technical. The two drafts in Turing (1992) devoted to this topic (viz., pp. 147–151, 153–174) are no exception. The second, '*On a theorem of Littlewood*', ostensibly written jointly with Skewes, is described by Britton on pp. XIV and 273 of Turing (1992) as having been in all likelihood prepared *solely* by Turing. Ingham, who studied the manuscript carefully in the early 1960s, expresses an equivalent view in Hardy (1967, p. 99). Since, as we shall see, the work is a significant one [its unpolished state notwithstanding], it is only natural to want to understand its background a little more clearly.[3] Our efforts in this direction have been aided in no small way by the fortuitous help that we received from A. M. Cohen, K. Hughes, J. Webb, P. Sarnak, S. B. Cooper and Stephen Skewes (Stanley Skewes' son).

We have already outlined the pre-World War II situation in Section 4. To take things further, we need to say just a few more words about Ingham (1936) . Riemann's formula in Eq. (2.2) gives an explicit representation for $\text{Li}(x) - \pi(x)$ as a sum over the non-trivial zeros ρ and the point $\frac{1}{2}$. (By abuse of language, we can temporarily regard $1/2$ as a ρ.) As was mentioned in Section 2, the ρ-sum requires technical virtuosity to handle and, even then, yields relatively poor results. Ingham's breakthrough was the observation that certain (sliding) weighted averages of $\text{Li}(x) - \pi(x)$ can be represented as sums over the ρ that are far more tractable, with terms that decline rapidly as the heights of the ρ grow. *Insofar as* this type of mollified sum *can be made* negative, at least some of the values of $\text{Li}(x) - \pi(x)$ that go into the average have to be negative as well, provided that the weights used in the averages are all non-negative. (When, as in Ingham (1936), the RH can be assumed, the key issue ultimately boils down to arranging things so that many sinusoidal ρ-terms 'all pull in the same direction' so as to successfully overpower something positive.) As such, the method usually does not produce any single counterexample to the $\pi(x) < \text{Li}(x)$ conjecture, but it *does* disprove it, and, if things are kept explicit enough, at least furnishes a region in which a counterexample has to lie.

Similar approaches have been developed for other number theoretic conjectures, such as that of Mertens. Typically, successful applications of such methods require high-precision values for some initial set of non-trivial zeros ρ, and knowledge that a considerably larger [finite] set satisfies the RH (the latter to help ensure the negligibility of all those terms past a certain ρ-threshhold).

It is interesting to observe that, already in the manuscript (Turing, 1992, pp. 147–151) from 1937, the mollification factor adopted by Turing is one of Gaussian type – exactly as would be appropriate as a 'first guess' in a setting in which there were some sporadically occurring off-line zeros in need of suppression in Eq. (2.2). The mollification choices adopted in Turing's second unpublished manuscript on this problem, '*On a theorem of Littlewood*' (Turing, 1992, pp. 153–174), OTL from now on, can be seen as building on that used in 1937.[4]

As the letter reproduced in Figure 2 clearly shows, Ingham and Britton's view about the authorship of OTL is correct. (Skewes (1955) was submitted for publication in December 1953. Consistent with the letter, his exposition makes no mention of OTL. The memorable phrase on

[3] As of Spring 2011, neither the original nor Britton's photocopy could be found in the Turing Digital Archive. Compare Turing (1992, p. IX(bottom)). Notice, too, that no date is offered for this work in Turing (1992).

[4] In Sections 2–5 of OTL, part of the idea is to imitate Ingham (1936) by using an 'approximate identity' interpretation of the Gaussian; cf. the bottom half of p. 154, 158(top), and 166 (lines 10, 16, 20–21). In this connection, see also lines 22–23 in Ingham's commentary, *op. cit.*

p. 50, line 10 may hint at one of Skewes' complications.[5]) In light of the unhappy events of the first part of 1952 (Hodges, 1983, pp. 471–473) and the inherent complexity of its estimates, it seems reasonably safe to hypothesise that the preparation date of OTL falls somewhere between mid-1952 and early 1953.

Such a timeframe would also be consistent with Turing's use of the phrases 'digital computer' and 'ten to twenty hours of computation time' on p. 168 of Turing (1992), not to mention the general mathematical mindset of the surrounding lines. Also note that some similar 'accounting-type' language occurs in Turing (1953, part II, pp. 112–116).

Although it is possible that the *work* for certain parts of OTL may actually have transpired some time prior to the drafting of any manuscript, the general sloppiness of Turing's typescript (we were able to secure a copy of A.M. Cohen's photocopy) tends to suggest that any 'time gap' is one of relatively modest size. Having said this, however, there may *still* be some value in noting that, during the 10-year period prior to 1953, there were five occasions on which a 'rekindling of x_S ideas' might well have occurred on one level or another:

- Prior to moving to Manchester, Turing spent the 1947–8 academic year in Cambridge. As it turns out, Skewes was also there on sabbatical for at least the first half of 1948, presumably doing some (pre-publication) fine tuning of his x_S thesis work with Littlewood. A letter dated 30 September [1948] from Littlewood to Skewes (made available to us by courtesy of John Webb) implicitly confirms the primary topic of their discussions, as well as Littlewood's close involvement. After a 10-year hiatus, one has to assume that Turing and Skewes occasionally talked.
- Littlewood's (1948, Section 14) popular account of the Skewes number also appeared in 1948 (July, to be more precise).
- During the 1949–50 academic year, there is some hint that, beyond his actual June 1950 experiment with the RH on the Manchester Mark 1, Turing may have also contemplated making calculations to a bit higher accuracy. See p. 99 (lines 6, 21–24), 100 (line 4), 104 (lines 5–6), and 114 (line 13) in Turing (1953); also Digital Archive item AMT/B/32/image 98 and Hodges (1983, p. 406, footnote).
- In Archive letters dated 19 December 1950 and 2 January 1951, Ingham raises a number of machine-oriented computational issues closely tied to a possible disproof of Pólya's conjecture, a conjecture very similar in spirit to $\pi(x) < \mathrm{Li}(x)$. It is evident from the January letter that Ingham has prompted Turing to start thinking about this matter.
- In March 1952, Kreisel (1952) appears. Section VI therein is devoted to a discussion of how to approach the Skewes problem along the lines that Turing originally wrote to Ingham about in the Spring of 1937. Kreisel presents no bound for x_S, however.

From a historical standpoint, it is fair to say that the significance of the first part of OTL (i.e., Sections 2-6) rests in Turing's realisation, already 1952–3, that by a judicious choice of mollification factor, it would prove feasible to eliminate the awkward quantitative dichotomy between the RH being true or false (i.e., 'H vs. NH'), which was introduced by Littlewood and was required previously, including in Skewes (1955), to secure an unconditional bound for x_S. And, further, that *in so doing*, a substantially superior x_S-bound would in fact accrue on the basis of using just several hundred ρ's.

Ingham offers a similar assessment in Hardy (1967, p. 99, lines 19–24) with a cautionary note about the manuscript's 'very rough' state. That Turing's ideas were fundamentally sound was shown

[5] Compare Burkill (1979, p.68, middle). The phrasing of item 4 in Turing (2001, p. 266) suggests that Turing may well have apprised Robin Gandy about his predicament with Skewes at some point.

Fig. 1: Professor Stanley Skewes. (Courtesy of the Department of Mathematics and Applied Mathematics, University of Cape Town.)

by Cohen and Mayhew in their 1965 work (Cohen and Mayhew, 1968) or (Turing, 1992, pp. 183–205) utilising about 450 zeros, albeit to greater precision than was available in the early 1950s.[6]

In the second half of OTL (i.e., Section 7), Turing derives a bound for x_S on the basis of there being an 'appropriately isolated' off-line zero ρ_0 in $\mathrm{Re}(s) > \frac{1}{2}$. Although the issue of obtaining an *optimal* x_S-bound in the specific setting of Theorem 3 may not have been looked at yet, results similar in spirit – even in more general settings – have been available for some years now in connection with the so-called Turán power sum method and comparative prime number theory. See, for instance, Knapowski (1961), Knapowski and Turán (1976), and Pintz (1980). Somewhat curiously, the latter two references make use of an idea (cf. Theorem H*) found in the aforementioned work on mathematical logic by Kreisel (1952).

Although the letter in Figure 2 may initially suggest an unsettled, uneasy state of affairs, in stepping back from it, we find ourselves in agreement with a comment made to us by Andrew Hodges, particularly vis à vis the period 1952 to early 1953, a time of clear personal difficulty for Turing. Concerning the letter, Hodges writes:

> ...what it conveys to me is something else quite marvellous – the timelessness of pure mathematics, illustrated in the way AMT refers back to discussions while rowing many years before. Despite everything that has happened, the war and computers, there are the prime numbers and their mysteries just the same as ever, something he has thought about from time to time ever since.

[6] The situation calls to mind Robin Gandy's comments in Turing (2001, p. 9) about Turing's love of calculating, in particular Turing's quip 'What's a factor of two between friends'?

COMPUTING MACHINE LABORATORY
UNIVERSITY OF MANCHESTER
MANCHESTER 13

TELEPHONE:
ARDWICK 2681

9th April, 1953.

S. Skewes, Esq.,
'Shannon',
Firdale Raod,
Newlands,
Cape Town.

Dear Sam,

Thank you very much for your letter and congratulations. I am sorry that my M.S. should have complicated life for you like that. I feel rather guilty about having invaded the territory of your number at all. One might have supposed that it could remain a pleasant corner one could keep to oneself. However you made the mistake of talking to me about it from time to time when you were rowing stroke and I at the bow until eventually I thought I had better find out what it was all about, and having done so, I could not refrain from playing at it myself. I should be very sorry if I were to stand in the way of your publishing your work.

The question as to whether Littlewood's proof did or did not give a value of your number "in principle" seems to me to be rather on a par with the question as to how many angels can dance on the point of a needle. I am sure the gentleman who first asked it had a good laugh for many years thereafter, and I think you may well do likewise: there is a whole lot by Kreisel on the subject in the Journal of Symbolic Logic recently.

I spend most of my time nowadays working in one way or another in connection with a computing machine. It is a rather niggly business in a way, like publication. If you have a single hole in the wrong place on the tapes you punch everything goes wrong, and whereas a reader will forgive two or three misprints the machine forgives nothing.

Best wishes,

Yours sincerely,

A. M. Turing

A. M. Turing.

Fig. 2: Copy of a 1953 letter from Turing to Stanley Skewes. (Courtesy of John Webb, University of Cape Town.)

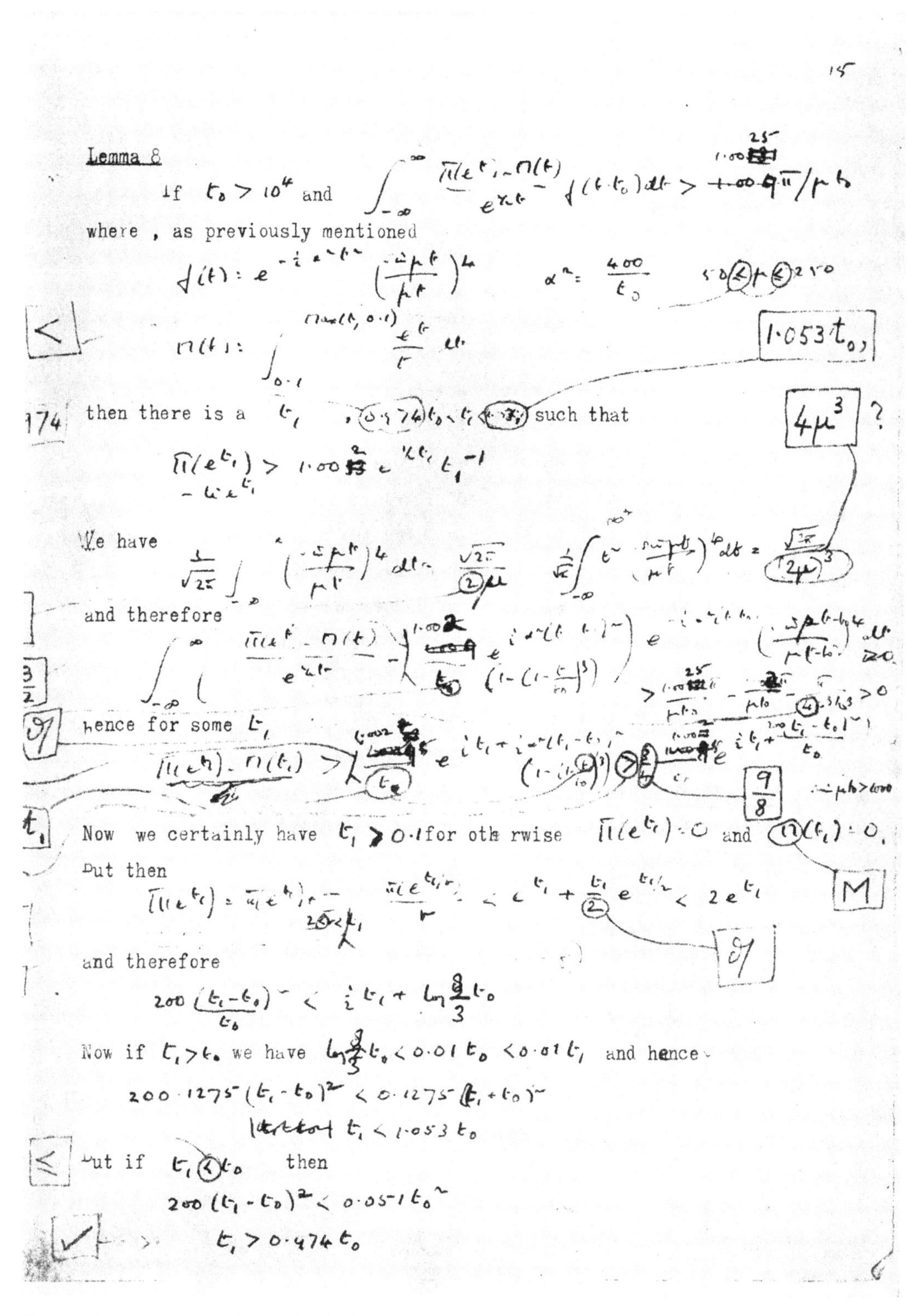

Fig. 3: A sample page from the draft manuscript, 'On a theorem of Littlewood.' The corrections noted in boxes were made by Ingham. (Reproduced from a photocopy, courtesy of Alan M. Cohen, Cardiff University.)

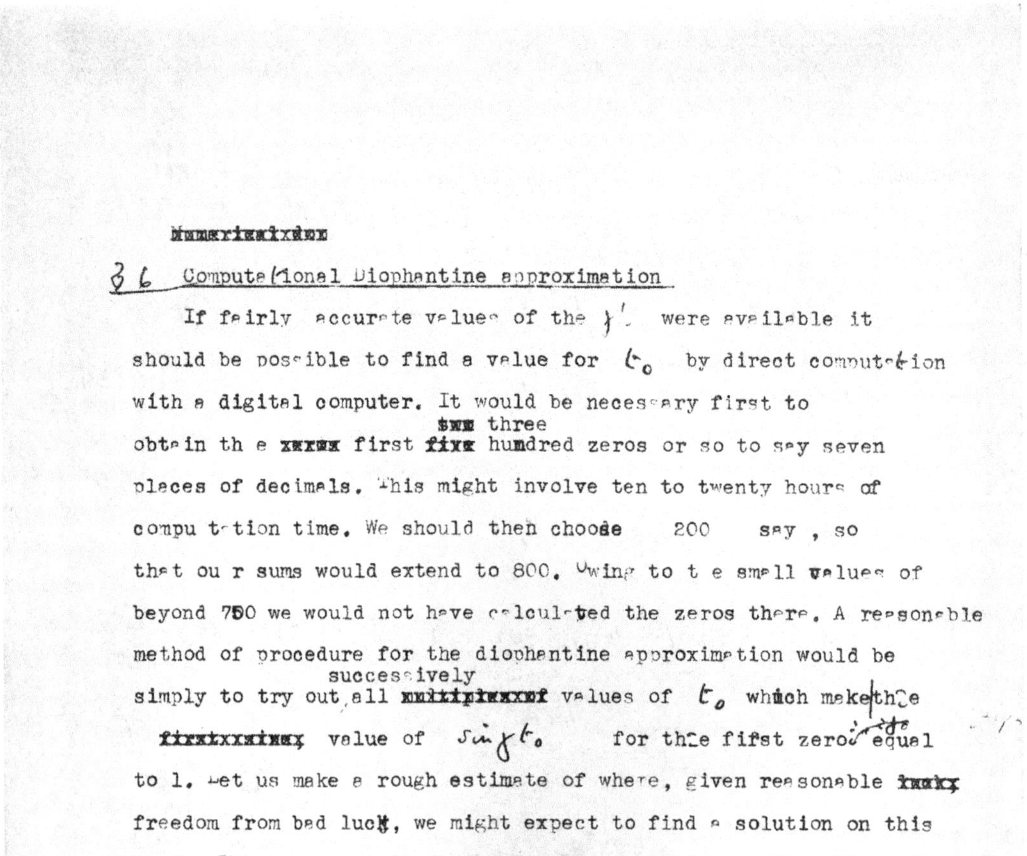

3 6 Computational Diophantine approximation

If fairly accurate values of the γ were available it should be possible to find a value for t_0 by direct computation with a digital computer. It would be necessary first to obtain the first three hundred zeros or so to say seven places of decimals. This might involve ten to twenty hours of computation time. We should then choose 200 say, so that our sums would extend to 800. Owing to the small values of beyond 750 we would not have calculated the zeros there. A reasonable method of procedure for the diophantine approximation would be simply to try out successively all values of t_0 which make the value of $\sin \gamma t_0$ for the first zero γ_0 equal to 1. Let us make a rough estimate of where, given reasonable freedom from bad luck, we might expect to find a solution on this basis. Let us assume that the sums of terms other than the first are independent and normally distributed. The standard deviation of the distribution of S is easily calculated to be

$$\sqrt{\frac{1}{2} \sum_{\gamma \neq \gamma_0} \frac{u(\gamma)}{\frac{1}{4}+\gamma^2} - \frac{u(\gamma_0)}{\frac{1}{4}+\gamma_0^2}}$$

i.e. about (0.0091). We get a solution if the sum exceeds $\frac{1}{2} - \frac{1}{\gamma_0}$ i.e. 0.43. The probability of this on the normal distribution is about 1.3×10^{-6}, from which we may conclude that there is an even chance of finding a value by this method in the first 500,000 trials, i.e. with $t_0 < 220{,}000$, i.e. establishing that there is an x, $2 < x < e^{220{,}000}$ for which $\pi(x) > \text{li}\, x$,

Fig. 4: A second page from the same typescript. Note Turing's use of the phrases 'digital computer' and 'ten to twenty hours of computation time.'

6. Turing's skepticism about the RH

In his pre-World War II work on $\zeta(s)$, Turing seems to have viewed the RH as an open question, one that might easily be either true or false. In Turing (1943, p. 197), for instance, he remarks rather nonchalantly that '[t]his may be of value for calculation of zeros not on the critical line'. As suggested in Section 4, this attitude may have arisen partly from the numerically based skepticism espoused by Titchmarsh (1935, 1936). Titchmarsh's 1937 letter would have only reinforced this. (Skeptical attitudes of this kind towards the RH were relatively common at the time.)

Based on the available evidence, it appears that by 1950 or so, Turing's earlier uncertainty about the truth of the RH had morphed into an outright skepticism.[7] Thus, on p. 169 of OTL, it is hard to ignore (even given the ambient 'if') the telling phrase that '[i]t seems very probable that the first zeros off the critical line that are computed will satisfy the conditions ...' It is hard to imagine anyone with serious doubts about the existence of zeros that violate the RH writing like this. And, even more to the point, on the very first page of Turing (1953), Turing declares: '[t]he calculations were done in an optimistic hope that a zero would be found off the critical line'. (The calculations to which Turing refers are those with $\text{Im}(s) \approx 25{,}000$. The ones with $\text{Im}(s)$ less than 1000π were aimed more at simply extending (Titchmarsh, 1936); see Turing (1953, p. 116, bottom.)

What is a little puzzling is that Turing expected a counterexample to the RH to lie so low. The work of A. Selberg during the mid-1940s showed that the root mean square of $\text{Im}\log\zeta(\frac{1}{2}+it)$, which to a large extent controls the distribution of zeros, grows about like $\sqrt{(\log\log T)/2}$ over any interval $[T, T+H]$ with, say, $H \approx T$. Similarly for the real part and for higher moments. In very rough terms, one also knows that large-scale irregularities in the 'sequencing' of ρ are linked to large oscillations in the aforementioned imaginary part; see, for instance, the first equation in Turing (1953, Section 4).

Accordingly, in order to reach regions wherein 'relatively many pairs of ρ have popped off the line', it seems reasonable that one would need to have $\sqrt{(\log\log T)/2}$ fairly large. Since this radical grows *extremely slowly*, expecting to ever see any type of *systematic* collapse in the RH using machine calculation is probably out of the question. Phrased somewhat differently: any off-line zeros in Turing's experiment would likely have been *sporadic* in nature and required significant luck to hit upon. It appears, based on Ingham's January 1951 letter, that Turing was aware of at least some of the work of Selberg on $\zeta(s)$ from the 1940s. Even without that input, however, one might have thought that Turing, whose first research project was on random variables (Hodges, 1983, p. 88) and who had extensive experience with statistics in his cryptographic work, might have had some concerns along these lines. If he did, there are no traces of them recorded in Turing (1953).[8]

As for skepticism about the RH, some distinguished number theorists, such as Littlewood and Turán, died as disbelievers. In general, however, the climate of opinion appears to have moved substantially towards embrace of the validity of the RH. This is well illustrated by Selberg. In 1946, he expressed, if not outright disbelief, then at least a concern about the lack of evidence in support of this conjecture (Selberg, 1946, Section 4). In 2005, however, towards the end of his life, when he was interviewed by Baas and Skau, Selberg asserted '[i]f one believes that there is something in this world that is as it should be, then I think that must be the truth of the Riemann Hypothesis'. See Baas and Skau (2008, pp. 631, 618 paragraph 5).

The evolution in the thinking of Selberg and other researchers was driven by the accumulation of numerical data for the RH as well as heuristics (some from random matrix approaches) and proofs of analogs of the RH for somewhat similar functions (such as certain zeta functions defined

[7] F.W.J. Olver recently informed the authors (in a private communication) that, in a conversation at NPL around 1947, Turing expressed his disbelief quite explicitly.

[8] Compare p. 168 (lines 27–32) in OTL, from a few years later.

over finite number fields). Had Turing lived longer, he might have modified his opinions about the validity of the RH and might well have become involved in some of these researches.

The zeta function was of course just one of Turing's many interests and not a major one. As can be seen from the record of his interactions with Skewes, say, he often put this subject aside for a number of years to concentrate on other topics. Still, the fact that he came back to it several times shows how interesting it was for him.

Had events transpired a bit differently in 1954, we like to think – as our *own* sort of 'optimistic hope' – that circumstances would have evolved in such a way that Turing's creativity would have continued to become piqued from time to time, prompting him to return occasionally to developments involving 'the zeros and primes'. With his insight and rare knowledge of the fields of number theory, analysis, probability, and computing that go into studying the zeta function, Turing could easily have emerged as a central player in this area.

References

Baas, N., Skau, C., 2008. The lord of the numbers, Atle Selberg. On his life and mathematics. Bull. Amer. Math. Soc. 45, 617–649.

Berry, M., Keating, J., 1992. A new asymptotic representation for $\zeta\left(\frac{1}{2}+it\right)$ and quantum spectral determinants. Proc. Royal Soc. Lond. A 437, 151–173.

Booker, A.R., 2006. Turing and the Riemann hypothesis. Notices Amer. Math. Soc. 53(10), 1208–1211.

Burkill, J.C., 1979. John Edensor Littlewood. Bull. Lond. Math. Soc. 11, 59–103.

Casselman, W., 2006. About the cover ... and a bit more. Notices Amer. Math. Soc. 53(10), 1186–1189.

Clay Mathematics Institute, 2000. Website devoted to the Riemann Hypothesis. ⟨http://www.claymath.org/millennium/Riemann_Hypothesis/⟩. Accessed 9 April 2012.

Cohen, A.M., Mayhew, M.J.E., 1968. On the difference $\pi(x) - \mathrm{li}\, x$. Proc. Lond. Math. Soc. Ser. 3, 18, 691–713.

Conrey, J.B., 2003. The Riemann hypothesis. Notices Amer. Math. Soc. 50(3), 341–353.

Hardy, G.H., 1967. Collected Papers of G. H. Hardy, vol. 2. Oxford University Press, Oxford.

Hodges, A., 1983. Alan Turing: The Enigma, Simon and Schuster, New York.

Ingham, A.E., 1932. The Distribution of Prime Numbers, Cambridge University Press, Cambridge.

Ingham, A.E., 1936. A note on the distribution of primes. Acta Arithmetica 1, 201–211.

Knapowski, S., 1961. On sign-changes in the remainder-term in the prime-number formula. J. Lond. Math. Soc. 36, 451–460.

Knapowski, S., Turán, P., 1976. On the sign changes of $(\pi(x) - lix)$. I. Colloq. Math. Soc. Janos Bolyai 13, 153–169. (See also: Acta Arithmetica 1965, 11, 193–202.)

Kreisel, G., 1952. On the interpretation of non-finitist proofs. II. J. Symb. Logic, vol. 17, 43–58.

Littlewood, J.E., 1948. Large numbers. Math. Gazette, 32(300), 163–171.

Littlewood, J.E., 1962. The Riemann hypothesis. In: Good, I.J. (Ed.), A Scientist Speculates. Basic Books, New York, pp. 390–391.

Littlewood, J.E., 1982. Collected Papers of J.E. Littlewood, vol. 2. Oxford University Press, Oxford.

Paris, R.B., 1994. An asymptotic representation for the Riemann zeta function on the critical line. Proc. Royal Soc. Lond. A 446, 565–587.

Pintz, J., 1980. On the remainder term of the prime number formula. I. Acta Arithmetica 36, 341–365.

Rubinstein, M., 2005. Computational methods and experiments in analytic number theory. In: Mezzadri, F., Snaith, N. (Eds.), Recent Perspectives in Random Matrix Theory and Number Theory, Cambridge University Press, Cambridge, pp. 425–506.

Selberg, A., 1946. The zeta-function and the Riemann hypothesis. In: C. R. Dixième Congrès Math. Scandinaves, Copenhague 1946, pp. 187–200. (Reprinted in Collected Papers, vol. 1, Springer-Verlag, Heidelberg, 1989, pp. 341–355.)

Skewes, S., 1933. On the difference $\pi(x) - li(x)$ (I). J. Lond. Math. Soc. 8, 277–283.

Skewes, S., 1955. On the difference $\pi(x) - li\,x$ (II). Proc. Lond. Math. Soc. Ser. 3, 5, 48–70.

Titchmarsh, E.C., 1935. The zeros of the Riemann zeta-function. Proc. Royal Soc. Lond. A 151, 234–255.
Titchmarsh, E.C., 1936. The zeros of the Riemann zeta-function. Proc. Royal Soc. Lond. A 157, 261–263.
Titchmarsh, E.C., 1985. The Theory of the Riemann Zeta-function, second ed., revised by D. R. Heath-Brown, Oxford University Press, Oxford (1st edition published 1951.)
Turing, A.M., 1939. Alan Turing's Zeta-Function Machine, 1939. ⟨http://www.turing.org.uk/sources/zetamachine.html⟩. Accessed 9 April 2012.
Turing, A.M., 1943. A method for the calculation of the zeta-function. Proc. London Math. Soc. Ser. 2, 48, 180–197.
Turing, A.M., 1953. Some calculations of the Riemann zeta-function. Proc. Lond. Math. Soc. Ser. 3, 3, 99–117.
Turing, A.M., 1992. Collected Works of A.M. Turing: Pure Mathematics, Britton, J.L. (Ed.), North-Holland, Amsterdam.
Turing, A.M., 2001. Collected Works of A.M. Turing: Mathematical Logic, Gandy, R., Yates, C. (Eds.), Elsevier, Amsterdam.
van der Pol, B., 1947. An electro-mechanical investigation of the Riemann zeta function in the critical strip. Bull. Amer. Math. Soc. 53, 976–981.
Williams, H.P., 2007. Stanley Skewes and the Skewes Number. J. Royal Institution Cornwall 70–75.

And Dennis Hejhal adds —

A Few Comments About Turing's Method

A short summary of Turing's method is available on pp. 255 and 256 of Turing (1992).[1] The remarks that follow are intended as a *supplement* to that; their aim is to show that, by broadening one's perspective a bit, the core idea of the method turns out to be both simpler and more versatile than might be originally suspected from examining just Turing (1953, Section 4).

In the context of the Riemann zeta function $\zeta(s)$, Turing's method is best viewed as an algorithm for rigorously establishing – without ever leaving the critical line – that all zeros in a certain $\mathrm{Im}(s)$-range have been found, are simple, and have real part exactly equal to $1/2$. In its original form, the method can be seen as hinging on three basic ingredients:

(a) A technical estimate (cf. Turing (1953, Theorem 4));
(b) An integration trick (cf. Turing (1953, Theorem 5));
(c) A blending of (b) with the calculation of a small number of auxiliary $\zeta(\frac{1}{2}+it)$-values so as to determine the total number of zeros (both online and off-line) in the given range, and then verifying that a standard 'twist' of $\zeta(s)$ manifests *at least* this number of *sign changes* over the relevant portion of $\{\mathrm{Re}(s)=\frac{1}{2}\}$.

In (c), the relevant twist is just multiplication by $\exp[i\vartheta(t)]$, where $\vartheta(t)$ is a certain, explicitly known, elementary function; cf. Turing (1953, Theorem 1). The resulting product with $\zeta(\frac{1}{2}+it)$ is then real valued for $t \in \mathbb{R}$. By its very nature, success in (c) is by no means pre-ordained.

---

[1] In the reference list below.

In regard to the item (a), the key thing to realise is that (Turing, 1953, Theorem 4) is simply an 'effectivisation' of a very classical estimate due to Littlewood; cf. (Titchmarsh, 1985, Theorem 9.9(A)). Booker (2006) showed that Turing's reasoning could be revamped so as to yield an analogous effectivisation for a much broader class of 'zeta-like' functions $L(s)$. If one is content to work with constants 'of merely modest size', other (more direct) approaches are possible here; see, e.g., the line of thought suggested by Titchmarsh (1985, Section 9.6), Turing (1992, p. 170, Lemma 16), Iwaniec and Kowalski (2004, Eqs (5.27)–(5.28), (5.31)–(5.33)) and Hejhal (1983, pp. 465 (middle) and 466 (top)).

In item (b), the crucial thing to keep in mind is that, although complicated looking, Theorem 5 is in reality a special case of a very basic trick; cf. relations (4) and (9) below.

To better appreciate this last point, as well as indicate some wider forms of (what is generally still referred to as) Turing's method, it is helpful to 'back up a bit' and quickly sketch an *alternative* certification procedure having a logical structure somewhat simpler than (a)–(c).

One begins by observing [via an integration by parts] that the scaling relation

$$\sup_{\alpha<\beta}\left|\int_{\alpha}^{\beta} g(t)\phi'(t)\,dt\right| \leqq \left(\max_{[a,b]}\phi'\right)\sup_{\alpha<\beta}\left|\int_{\alpha}^{\beta} g(t)\,dt\right| \tag{1}$$

holds anytime $g(t)$ is piecewise continuous on a given interval $[a,b]$ and the monotonic increasing C^2 function $\phi(t)$ has a second derivative whose sign remains fixed on $[a,b]$. Guided by the first equation in Turing (1953, Section 4), we now consider *any* monotonically increasing sequence of positive numbers x_n in $\{x > c > 0\}$ having an asymptotic mean spacing of 1. To control things somewhat, we hypothesise that

$$N\{c < x_n \leqq X\} = A + X + \mathcal{R}(X) \tag{2}$$

holds with a right continuous remainder function $\mathcal{R}$ satisfying

$$\sup_{[\alpha,\beta]\subseteqq[c,X]}\left|\int_{\alpha}^{\beta}\mathcal{R}(y)\,dy\right| \leqq E(X), \tag{3}$$

wherein $E(X)$ increases with X (for $X \geqq c$) while the quotient $\delta(X) \equiv E(X)/X$ tends to 0 monotonically. The letter 'E' stands for explicit; cf., e.g, Turing (1953, Lemma 10). By elementary algebra, one checks that

$$-\frac{h}{2}+\frac{1}{h}\int_{\xi-h}^{\xi}\mathcal{R}(y)dy \leqq \mathcal{R}(\xi) \leqq \frac{h}{2}+\frac{1}{h}\int_{\xi}^{\xi+h}\mathcal{R}(y)dy \tag{4}$$

for all $0 < h < \xi - c$. Taking $h = \sqrt{\xi\delta(\xi)}$ gives $|\mathcal{R}(\xi)| \leqq 2\sqrt{E(\xi)}$ for large ξ. In particular, for any E of the aforementioned type, one automatically has $\mathcal{R}(X) = O(\sqrt{X})$.

For the sake of argument, let us now temporarily assume that the numbers x_n correspond to ordinates of the non-trivial zeros of some zeta-like function $L(s)$ in a certain half-plane $\{\mathrm{Im}(s) > C\}$ *under a transformation* $x = \phi(t)$ akin to the $x = \vartheta(t)/\pi = 2\kappa(t/2\pi)$ appearing in Turing (1953, Section 4). By virtue of (1) and our comments about item (a), one gets an immediate counterpart of Lemma 10 in Turing (1953) and its associated function $E(X)$. Let $\{y_m\}$ be the strictly monotonic subsequence of $\{x_n\}$ associated with the list of distinct odd order zeros along $\{\mathrm{Re}(s) = \frac{1}{2}\}$ *detected by the machine* through its analysis of sign changes.[2] From an algorithmic standpoint, successfully verifying 'RH + simple' for all Im(s) up to some specific T_0 is basically tantamount to showing that $N(c,\xi_0] = M(c,\xi_0]$ holds for a suitable ξ_0, wherein N and M signify the obvious x and y cardinalities. (Since machine precision is finite, it is prudent to tacitly assume that T_0 and ξ_0 are subject to minor perturbation; we do so.)

Putting $\Delta = N(c,\xi_0] - M(c,\xi_0]$, we now evaluate the functional

$$I_H(\xi_0) \equiv \frac{1}{H}\int_{\xi_0}^{\xi_0+H} \{M(c,y] - A - y\}\,dy \tag{5}$$

for suitable $H > 1$. From the obvious fact that

$$H\Delta \leqq \int_{\xi_0}^{\xi_0+H} \{N(c,y] - M(c,y]\}\,dy, \tag{6}$$

a simple rearrangement produces

$$\Delta + I_H(\xi_0) \leqq \frac{1}{H}\int_{\xi_0}^{\xi_0+H} \mathcal{R}(y)\,dy \leqq \frac{E(\xi_0+H)}{H}. \tag{7}$$

Keeping H large enough to make the final term less than $\frac{1}{8}$ (say), we then encounter exactly two possibilities:

$$\text{(i)}\quad -\frac{3}{4} < I_H(\xi_0) < \frac{3}{4} \quad ; \quad \text{(ii)}\quad I_H(\xi_0) \leqq -\frac{3}{4}.$$

Case (i) is the nicer; here, Δ can only be 0. Finding $|I_H(\xi_0)| < \frac{3}{4}$ is thus a sufficient condition for securing 'RH+simple' over $(c,\xi_0]$. In this connection, it is also important to observe that if 'RH+simple' actually holds out to at least $\xi_0 + H$ and the machine's zero-detecting algorithm is working flawlessly (i.e., *lückenfrei*), the computed value of $|I_H(\xi_0)|$ will *necessarily* be less than $1/4$. (This is seen by noting that $N(\xi_0 - H, \xi_0 + 2H]$ differs from $3H$ by a quantity having magnitude at most $4\sqrt{E(\xi_0+2H)}$ and remembering that the tolerance η is small.) When (ii) holds, one gets instead a warning that $N(c,\xi_0+H] > M(c,\xi_0+H]$ and a concomitant bound on the size of Δ (for the original interval $(c,\xi_0]$).

In situations where the focus centres on proving $N(\xi_0,\xi_0^\star] = M(\xi_0,\xi_0^\star]$ rather than $N(c,\xi_0] = M(c,\xi_0]$, it is natural to replace $I_H(\xi_0)$ by

$$I_H(\xi_0,\xi_0^\star) \equiv \frac{1}{H}\int_0^H \{M(\xi_0 - v, \xi_0^\star + v] - (2v + \xi_0^\star - \xi_0)\}\,dv. \tag{8}$$

An analogous (i) - (ii) dichotomy clearly ensues.

---

[2] For later use, we assume that each machine-identified y_m is (subsequently) *refined* to some small tolerance $\eta < 10^{-3}$. Only the numerical value of y_m has significance not the corresponding n and x_n.

The foregoing certification procedure [which makes crucial use of (3)] has the advantage that it largely *obviates* any need for an initial estimate or determination of $\mathcal{R}(\xi_0)$ (and $\mathcal{R}(\xi_0^\star)$) via something like Turing (1953, Theorem 5). In zeta-like settings, Δ is of course even. We have intentionally avoided any use of this fact *in order to stress the technique's applicability to more general types of sequences* $\{x_n\}$ and $\{y_m\}$.[3]

To the best of our knowledge, the first use of an integration trick akin to (6) + (7) in certifying a 'Riemann Hypothesis' of some kind is the one found in Section 1 of P. R. Taylor's (1945) posthumous paper. Whether Turing was aware of Taylor's work while preparing Turing (1953) is unclear.

We now return to Theorem 5 [with an eye on reverting to something closer to Turing's original set-up], but in the broader framework of $\mathcal{R}(X)$ and relations (2) + (3). We assume that ξ_0 'sits' in a grid

$$\xi_0 - H_\ell < \xi_{-\ell} < \cdots < \xi_0 < \cdots < \xi_r < \xi_0 + H$$

having the property that each open interval (ξ_{j-1}, ξ_j) with $j \in [1-\ell, r]$ is known to contain some $x_{n(j)}$. Letting $M(\alpha, \beta]$ count the $x_{n(j)}$'s it is only natural to imitate (6) by integrating

$$0 \leqq N(\xi_0, y] - M(\xi_0, y] \quad \text{and} \quad 0 \leqq N(v, \xi_0] - M(v, \xi_0]$$

via elementary calculus. Insofar as $H_\ell < \xi_0 - c$, one readily finds that

$$C_\ell \leqq \mathcal{R}(\xi_0) \leqq D_r, \tag{9}$$

wherein

$$D_r = \frac{r}{2H} + \frac{(r-H)^2}{2H} + \frac{1}{H} \int_{\xi_0}^{\xi_0+H} \mathcal{R}(y)\,dy + \frac{1}{H} \sum_{[1,r]} \delta(j)$$

$$C_\ell = -\frac{\ell}{2H_\ell} - \frac{(\ell - H_\ell)^2}{2H_\ell} + \frac{1}{H_\ell} \int_{\xi_0 - H_\ell}^{\xi_0} \mathcal{R}(v)\,dv + \frac{1}{H_\ell} \sum_{[-\ell,-1]} \delta(k)$$

and $\xi_\nu \equiv \xi_0 + \nu + \delta(\nu)$. The bounds in (9) *improve* any time the grid is refined [in an obvious sense]. With empty sums being 0, the cases $r = 0$ and $\ell = 0$ are perfectly legitimate and simply reproduce estimate (4). Letting (9E) signify the *variant of* (9) in which the $\mathcal{R}$- integrals are replaced by $E(\xi_0 + H)$ and $-E(\xi_0)$, it is evident that Theorem 5 in Turing (1953) basically just corresponds to (9E) with $(H_\ell, H, \ell, r) = (R_1, R_2, R_1 - 1, R_2 - 1)$. In prototypical cases of Theorem 5, the numbers $\delta(\nu)$ will either be zero or else fairly small.

In situations where the x_n correspond (under a ϕ like before) to ordinates of the non-trivial zeros of a 'zeta-like' function $L(s)$, then just as in Turing (1953, Section 4), $\mathcal{R}(\xi_0)$ will be known *modulo* 2 any time $\mathfrak{L} \equiv \exp(i\pi(A+x))L(\frac{1}{2} + it)$ is non-zero and (at least approximately) computable at ξ_0. It is a curious fact that, *if* matters are secretly such that 'RH+simple' also holds near ξ_0 *and* no two of these nearby zeros are situated 'maliciously close' together, the value of $\mathcal{R}(\xi_0)$ will then be *uniquely* determined by (9E) whenever H_ℓ and H are taken appropriately big, and the grid $\{\xi_j\}$ is chosen in a manner consistent with knowledge of $\mathrm{sgn}(\mathfrak{L})$ at a sufficiently fine level of granularity in x.[4]

---

[3] In a forthcoming work, Booker and Strömbergsson treat a case in which the x_n correspond to eigenvalues of the Laplacian associated with (vibrations of) the modular surface $PSL(2,\mathbb{Z})\backslash H$. The relevant $E(X)$ here is basically just (const)$X/(\log X)^2$. Cf. Booker (2006, p. 1211) and Hejhal (1983, p. 466ff). Compare Then (2006, Section 5).

[4] There is no need to limit oneself to situations in which the 'hoped for' $\mathcal{R}(\xi_0)$ is small.

[The proof rests on a minor augmentation of (9). The essential point is that once the granularity level in x is sufficiently small, it becomes possible to derive a *lower bound* for $\mathcal{R}(\xi_0)$ having format very similar to D_r and an upper bound that resembles C_ℓ. When this new information is substituted back into (9E) together with knowledge of the mod 2 situation, all values but one ultimately fall to the wayside for $\mathcal{R}(\xi_0)$ since $1+2(\frac{1}{8}) < 2$. Cf. near (7) concerning '1/8'.]

In light of these observations, estimate (9E) is actually slightly better suited than Theorem 5 is for use in making numerical tests of the RH along the lines of (a)–(c).

In the case of $\zeta(s)$, one certainly expects that all but an infinitesimal proportion of the ρ's lie on the critical line. In view of the fact (due to Selberg) that $\log\zeta(\frac{1}{2}+it)/\sqrt{\frac{1}{2}\log\log t}$ is distributed like the Cartesian product of two standard Gaussians for large t, and the further fact (Conrey, 2003) that, in accordance with the GUE law cited earlier, one expects to see a significant paucity of 'nearly multiple' ρ's (i.e., 'a repulsion of levels'), any problems in the a priori determination of $N(\xi_0,\xi_0^\star]$ for *generically* chosen ξ_0 and $\xi_0^\star$ should tend to ensue chiefly from difficulties in computing $\zeta(\frac{1}{2}+it)$ for t-values of the necessary size. Barriers on this front will occur long before $\sqrt{\log\log t}$ ever has a chance to get large.

Similarly for a broad class of 'zeta-like' functions $L(s)$ whose zeros are also expected to satisfy GUE statistics in the limit of large $|s|$. In all such cases, once the height becomes appropriately large, Turing's 'abc' method provides (what is in all likelihood) a nearly optimal machine-based certification procedure for the associated RH.

References

Booker, A.R., 2006. Artin's conjecture, Turing's method, and the Riemann hypothesis. Exp. Math. 15, 385–407.

Booker, A.R., 2006. Turing and the Riemann hypothesis. Notices Amer. Math. Soc. 53(10), 1208–1211.

Conrey, J.B., 2003. The Riemann hypothesis. Notices Amer. Math. Soc. 50(3), 341–353.

Hejhal, D.A., 1983. The Selberg Trace Formula for $PSL(2,\mathbb{R})$, vol. 2. Lecture Notes in Mathematics, no. 1001, Springer-Verlag, Heidelberg.

Iwaniec, H., Kowalski, E., 2004. Analytic Number Theory, Colloquium Publications, no. 53, American Math. Soc., Providence.

Taylor, P.R., 1945. On the Riemann zeta function. Quart. J. Math. 16, 1–21.

Then, H., 2006. Arithmetic quantum chaos of Maass waveforms. In: Cartier, P. et al. (Eds.), Frontiers in Number Theory, Physics, and Geometry I, Springer-Verlag, Heidelberg, pp. 183–212.

Titchmarsh, E.C., 1985. The Theory of the Riemann Zeta-Function, second ed., revised by D. R. Heath-Brown, Oxford University Press, Oxford. (1st edition published 1951.)

Turing, A.M., 1953. Some calculations of the Riemann zeta-function. Proc. Lond. Math. Soc. Ser. 3, 3, 99–117.

Turing, A.M., 1992. Collected Works of A. M. Turing: Pure Mathematics, Britton, J.L. (Ed.), North-Holland, Amsterdam.

SOME CALCULATIONS OF THE RIEMANN ZETA-FUNCTION

By A. M. TURING

[Received 29 February 1952—Read 20 March 1952]

Introduction

IN June 1950 the Manchester University Mark 1 Electronic Computer was used to do some calculations concerned with the distribution of the zeros of the Riemann zeta-function. It was intended in fact to determine whether there are any zeros not on the critical line in certain particular intervals. The calculations had been planned some time in advance, but had in fact to be carried out in great haste. If it had not been for the fact that the computer remained in serviceable condition for an unusually long period from 3 p.m. one afternoon to 8 a.m. the following morning it is probable that the calculations would never have been done at all. As it was, the interval $2\pi.63^2 < t < 2\pi.64^2$ was investigated during that period, and very little more was accomplished.

The calculations were done in an optimistic hope that a zero would be found off the critical line, and the calculations were directed more towards finding such zeros than proving that none existed. The procedure was such that if it had been accurately followed, and if the machine made no errors in the period, then one could be sure that there were no zeros off the critical line in the interval in question. In practice only a few of the results were checked by repeating the calculation, so that the machine might well have made an error.

If more time had been available it was intended to do some more calculations in an altogether different spirit. There is no reason in principle why computation should not be carried through with the rigour usual in mathematical analysis. If definite rules are laid down as to how the computation is to be done one can predict bounds for the errors throughout. When the computations are done by hand there are serious practical difficulties about this. The computer will probably have his own ideas as to how certain steps should be done. When certain steps may be omitted without serious loss of accuracy he will wish to do so. Furthermore he will probably not see the point of the 'rigorous' computation and will probably say 'If you want more certainty about the accuracy why not just take more figures?' an argument difficult to counter. However, if the calculations are being done by an automatic computer one can feel sure that this kind of indiscipline does not occur. Even with the automatic computer this rigour can be rather tiresome to achieve, but in connexion with such a subject as the analytical theory of numbers, where rigour is the essence, it seems worth while. Unfortunately, although the details were all worked out, practically nothing was done on these lines. The interval $1414 < t < 1608$ was investigated and checked, but unfortunately at this point the machine broke down and no further work was done. Furthermore this interval was subsequently found to have been run with a wrong error value, and the most that can consequently be asserted with certainty is that the zeros lie on the critical line up to $t = 1540$, Titchmarsh having investigated as far as 1468 (Titchmarsh (5)).

Proc. London Math. Soc. (3) 3 (1953)

This paper is divided into two parts. The first part is devoted to the analysis connected with the problem. All the results obtained in this part are likely to be applicable to any further calculations to the same end, whether carried out on the Manchester Computer or by any other means. The second part is concerned with the means by which the results were achieved on the Manchester Computer.

PART I. GENERAL

1. The Θ notation

In analysis it is customary to use the notation $O\{f(x)\}$ to indicate 'some function whose ratio to $f(x)$ is bounded'. In the theory of a computation one needs a similar notation, but one is interested in the value of the bound concerned. We therefore use the notation $\Theta(\alpha)$ to indicate 'some number not greater in modulus than α'. The symbol Θ has been chosen partly because of a typographical similarity to 0, partly because of the relation with the use of ϑ to indicate 'a number less than 1'.

2. The approximate functional equation

We shall use throughout the notation of Ingham (**1**) and Titchmarsh (**3**) without special definition. Our problem is to investigate the distribution of zeros of $\zeta(s)$ for large t. This will presumably depend on being able to calculate $\zeta(\sigma + it)$ or some closely associated function for large t, and σ not too far from $\frac{1}{2}$. We have to consider what formula to use and what associated function. For $\sigma > 1$ it is possible to use the defining series $\zeta(s) = \sum\limits_{1}^{\infty} n^{-s}$, but this is too far from $\sigma = \frac{1}{2}$. For $0 < \sigma < 1$ there are other formulae which also involve calculating a number of terms of this series, but it is always necessary to take at least $t/2\pi$ terms.

Alteratively one can use the functional equation

$$\zeta(s)\Gamma\left(\frac{1}{2}s\right)\pi^{-\frac{1}{2}s} = \zeta(1-s)\Gamma\left(\frac{1}{2} - \frac{1}{2}s\right)\pi^{-\frac{1}{2}+\frac{1}{2}s}$$

and take $t/2\pi$ terms of the series $\zeta(1-s) = \sum\limits_{1}^{\infty} n^{s-1}$. Another possible method which might suggest itself is to calculate at a number of points in the region $\sigma > 1$ and extrapolate, but this again involves much the same amount of work. However, if one considers an interpolation formula involving both values from the region $\sigma > 1$ and from the region $\sigma < 0$ one finds that it is possible to calculate the function by taking only about $\surd(t/2\pi)$ terms of the series $\sum n^{-s}$ and an equal number from $\sum n^{s-1}$. This result is embodied in

THEOREM 1. *Let m and ξ be respectively the integral and non-integral parts of $\tau^{\frac{1}{2}}$ and*

$$\tau \geqslant 64,$$

$$\kappa(\tau) = \frac{1}{4\pi i}\log\frac{\Gamma\left(\frac{1}{4}+\pi i\tau\right)}{\Gamma\left(\frac{1}{4}-\pi i\tau\right)} - \frac{1}{4}\tau\log\pi,$$

$$Z(\tau) = \zeta\left(\frac{1}{2}+2\pi i\tau\right)e^{-2\pi i\kappa(\tau)},$$

$$\kappa_1(\tau) = \frac{1}{2}\left(\tau\log\tau - \tau - \frac{1}{2}\right),$$

$$h(\xi) = \frac{\cos 2\pi\left(\xi^2 - \xi - \frac{1}{16}\right)}{\cos 2\pi\xi}.$$

Then $Z(\tau)$ is real and

$$Z(\tau) = 2\sum_{n=1}^{m} n^{-\frac{1}{2}}\cos 2\pi\{\tau\log n - \kappa(\tau)\} + (-)^{m+1}\tau^{-\frac{1}{4}}h(\xi) + \Theta\left(1.09\tau^{-\frac{3}{4}}\right),$$

$$\kappa(\tau) = \kappa_1(\tau) + \Theta\left(0.006\tau^{-1}\right).$$

It will be seen that $Z(\tau)$ may also be defined as being $\zeta(\frac{1}{2}+2\pi i\tau)$ for τ real, $0<\tau<1$, and elsewhere by analytic continuation. The theorem could be proved by the argument outlined above, but is more conveniently proved by the method given as Theorem 22 of Titchmarsh (**3**). The numerical details are given in Titchmarsh (**4**). A more elaborate remainder is given there and is valid for $\tau \geqslant \delta$. The validity of the remainder given here follows trivially from it.

This formula can only give a limited accuracy, although it is nearly always adequate. If greater accuracy is required the formula given in Turing (**6**) may be applied. These agree with Titchmarsh's expression in the sum of m terms, but $h(\xi)$ is replaced by another sum.

The function $h(\xi)$ is troublesome to calculate, largely because the numerator and denominator both vanish at $\xi=\frac{1}{4}$ and $\xi=\frac{3}{4}$, so that a special method would have to be applied for the neighbourhood of these points. The alternative of using a table and interpolation suggests itself. This possibility quickly leads to the suggestion of replacing the function by some polynomial which approximates it well enough in the region concerned.

In fact the polynomial $0{\cdot}373 + 2{\cdot}160(\xi-\frac{1}{2})^2$ is quite adequate, for we have

THEOREM 2. *If $|\xi-\frac{1}{2}|<\frac{1}{2}$ we have*

$$h(\xi) = 0.373 + 2.160\left(\xi-\tfrac{1}{2}\right)^2 + \Theta(0.0153)$$

and if $|\xi-\frac{1}{2}<0.53$, we have $h(\xi) = 0.373+2.160\left(\xi-\frac{1}{2}\right)^2+\Theta(0.0243)$.

This result is rather unexpectedly troublesome to prove. Its proof will be given in slightly more detail than it deserves, treating it as an example of 'rigorous computation'.

It may be said: 'As this is a purely numerical result surely it can be proved by straight computation.' This is in effect what is done, but it is not possible to avoid theory entirely. The function was calculated for the values $0, \frac{1}{30}, \frac{2}{30}, \ldots, \frac{16}{30}, \frac{17}{30}$ of $\xi-\frac{1}{2}$ with an error $\Theta(10^{-4})$, and was found to satisfy the inequality with some margin. But nothing further can be deduced even if the differences are taken into account, unless something is known about the general behaviour of the function. An upper bound for the second derivative would be sufficient, but the labour of even the formal differentiation is discouraging, and the accidental singularities make the situation considerably worse. However, the function is integral, and it is therefore possible to obtain an inequality for any derivative by means of Cauchy's integral formula, taken in combination with an inequality for the function itself on a suitable contour. The method actually applied will be seen to be very similar to this. Instead of Cauchy's integral formula we use

$$f(\xi)-P(\xi) = \frac{(\xi-\xi_1)(\xi-\xi_2)(\xi-\xi_3)(\xi-\xi_4)}{2\pi i}\int \frac{f(u)du}{(u-\xi)(u-\xi_1)(u-\xi_2)(u-\xi_3)(u-\xi_4)},$$

where the function $f(\xi)$ is regular inside the anti-clockwise contour of integration, and $P(\xi)$ is the cubic polynomial agreeing with $f(\xi)$ at $\xi_1,\ \xi_2,\ \xi_3,\ \xi_4$. This equation follows from the fact that the right-hand side vanishes at the points $\xi_1,\ \xi_2,\ \xi_3,\ \xi_4$ and is of the form of $f(\xi)$ added to a cubic polynomial. We actually take the contour to be the square whose vertices are $\frac{1}{2}\pm i, \frac{1}{2}\pm 1$. One can prove without difficulty that $|h(\xi)| < \cosh\pi$ on this square and that if $f(\xi) = h(\xi) - 0{\cdot}373 - 2{\cdot}160(\xi-\frac{1}{2})^2$ then $|f(\xi)| < 14{\cdot}3$ on the square. Taking $\xi_1,\ \xi_2,\ \xi_3,\ \xi_4$ to be of form $n/30$ and two of them to be on either side of ξ one easily deduces $|f(\xi)-P(\xi)| < 0{\cdot}0033$ if $|\xi-\frac{1}{2}| < 0{\cdot}053$, and a consideration of the values at the calculated points and the differences gives $|P(\xi)| < 0{\cdot}021$ if $|\xi-\frac{1}{2}| < 0{\cdot}53$ and $|P(\xi)| < 0{\cdot}012$ if $|\xi-\frac{1}{2}| < \frac{1}{2}$. It will be seen that the use of this approximation to $h(\xi)$ gives an extra error in $Z(\tau)$ of the order of $\tau^{-\frac{1}{4}}$ whereas Titchmarsh's formula has an error of order only $\tau^{-\frac{3}{4}}$; but the errors are not equal until τ is over 2000, and both are then quite small. In the actual calculation described in Part II there were other errors of order as large as $\tau^{\frac{5}{4}}$.

Titchmarsh's formula as stated is valid only when the right value of m is used, i.e. if $\tau^{\frac{1}{2}} = m+\xi$ and $|\xi-\frac{1}{2}| \leqslant \frac{1}{2}$. This may be inconvenient as one may occasionally wish to go a little outside the range. One may justify doing so by means of

Theorem 3. *Theorem 1 is valid with the error $\Theta(1.09\tau^{-\frac{3}{4}})$ replaced by $\Theta(1.15m^{-\frac{3}{2}})$ if the condition that m and ξ be the integral and non-integral parts of $\tau^{\frac{1}{2}}$ is replaced by the condition that m be an integer and*

$$\tau^{\frac{1}{2}} = m+\xi,\ |\xi-\tfrac{1}{2}| < 0.53.$$

The new error introduced is

$$(-)^m\tau^{-\frac{1}{4}}\left[\frac{\cos 2\pi\left(\xi^2-\xi-\dfrac{1}{16}\right)}{\cos 2\pi\xi}+\frac{\cos 2\pi\left\{(\xi-1)^2-(\xi-1)-\dfrac{1}{16}\right\}}{\cos 2\pi(\xi-1)}\right] - $$

$$-2(m+1)^{-\frac{1}{2}}\cos 2\pi\left[(m+\xi)^2\log(m+1)-\kappa\left\{(m+\xi)^2\right\}\right]$$

in the case that $1 < \xi < 1.03$. But we have

$$\frac{\cos 2\pi\left(\xi^2-\xi-\dfrac{1}{16}\right)}{\cos 2\pi\xi}+\frac{\cos 2\pi\left\{(\xi-1)^2-(\xi-1)-\dfrac{1}{16}\right\}}{\cos 2\pi(\xi-1)} = 2\cos 2\pi\left(\xi^2-2\xi-\frac{1}{16}\right).$$

Also if we put

$$j(\xi) = (m+\xi)^2\log(m+1)-\kappa_1\{(m+\xi)^2\}-\frac{1}{2}m^2-m+\frac{1}{2}+\xi^2-2\xi-\frac{1}{16},$$

then $j(\xi)$ and its first two derivatives vanish at $\xi = 1$ and $j'''(\xi) = \frac{-2}{m+\xi}$. Hence by the mean value theorem $|j(\xi)| < \frac{(\xi-1)^3}{3(m+1)}$. Using also

$$|\kappa\{(m+\xi)^2\} - \kappa_1\{(m+\xi)^2\}| < \frac{0.006}{(m+1)^2},$$

we see that the new error is at most

$$4\pi(m+1)^{-\frac{1}{2}}\left(\frac{(\xi-1)^3}{3(m+1)} + 0.006(m+1)^{-2}\right) + 2|(m+\xi)^{-\frac{1}{2}} - (m+1)^{-\frac{1}{2}}|,$$

which is less than $0.052(m+1)^{-\frac{3}{2}}$ since $m \geqslant 7$, $|\xi - 1| < 0.03$. A similar argument applies for the case $-0.03 < \xi < 0$.

3. Principles of the calculations

We may now consider that with the aid of Theorems 1, 2, 3 we are in a position to calculate $Z(\tau)$ for any desired τ. How can we use this to obtain results about the distribution of the zeros? So long as the zeros are on the critical line the result is clearly applicable to enable us to find their position to an accuracy limited only by the accuracy to which we can find $Z(\tau)$. If there are zeros off the line we can find their position as follows. Suppose we have calculated $Z(\tau)$ for $\tau_1, \tau_2, \ldots, \tau_N$. Then we can approximate $Z(\tau)$ in the neighbourhood of these points by means of the polynomial $P(\tau)$ agreeing with $Z(\tau)$ at these points. The accuracy of the approximation may be determined as in Theorem 2. Suppose that in this way we find that $|Z(\tau) - P(\tau)| < \epsilon$ and $|P''(\tau)| < \epsilon'$ for $|\tau - \tau'| < \delta$ and that $|P(\tau')| < \epsilon''$ and $|P'(\tau') - a| < \epsilon'''$, then we see that

$$|Z(\tau) - a(\tau - \tau')| < \epsilon + \epsilon'' + \tfrac{1}{2}\epsilon'\delta^2 + \epsilon'''\delta$$

for $|\tau - \tau'| < \delta$, and we may conclude by Rouché's theorem that $Z(\tau)$ has a zero within this circle if $|a| > \epsilon''' + \frac{1}{2}\epsilon'\delta + \frac{\epsilon+\epsilon''}{\delta}$. This, however, is a tiresome procedure, and should be avoided unless we have good reason to believe that such a zero is really present. If there are any such zeros we may expect that the first ones to appear will be rather close to the critical line, and they will show themselves by the curve of $Z(\tau)$ approaching the zero line and receding without crossing it: in other words by behaving like a quadratic expression with complex zeros. In the absence of such behaviour we wish to prove that there are no complex zeros without using this interpolation procedure. Let us suppose that we have been investigating the range $T_0 < \tau < T_1$ and that we have found a certain number of real zeros in the interval. If by some means we can determine the total number of zeros in the rectangle $|\mathcal{I}\tau| < 2$, $T_0 < \mathcal{R}\tau < T_1$ (say) and find it to equal the number of changes of sign found, then we can be sure that there were no zeros off the critical line in this rectangle. This total number of zeros can be determined by calculating the function at various points round the rectangle. This might normally be expected to involve even more work than the calculations on the critical line. Fortunately, with the function concerned, the calculations on the lines $|\mathcal{I}\tau| = 2$ are not necessary. It is well known that the change in the argument of $Z(\tau)$ on these lines can be calculated to within $\frac{1}{2}\pi$ in terms of the gamma function. It remains to find the change on the lines $\mathcal{R}\tau = T_0$ and $\mathcal{R}\tau = T_1$. In principle this could be done by approximating $Z(\tau)$ with a polynomial, using an interpolation formula based on values calculated on the critical line. Since this interpolation procedure is necessary only at the ends of the interval investigated this would be a considerably smaller burden than the repeated application of it throughout the interval required by the method previously suggested. It will, however, be shown later on that even this application of the interpolation procedure is unnecessary, but for the sake of argument we will suppose for the moment that it is done. We may suppose then that the total number of zeros in the rectangle is known. If this differs from the number of changes of sign which have been found, then the deficit must be ascribed to a combination of four causes. Some may be due to pairs of complex zeros, some to pairs of changes of sign which were missed due to insufficiently many values $Z(\tau)$ being calculated, some to the accuracy of some of the values being inadequate to establish that changes of sign had occurred. Finally there may be some multiple zeros on the critical line. Each source accounts for an even number of zeros provided that the accuracy is sufficient for there to be no doubt about the signs of $Z(T_0)$ and $Z(T_1)$. By calculating further values and increasing the accuracy we can remove some of the discrepancies, but we cannot do anything about the multiple zeros by mere calculation. Assuming that there are no multiple zeros it is possible in principle to make sure that all the real zeros have been found by calculating $Z(\tau)$ at a sufficient number of real points, but the number of points would be many more than are required for finding all the real zeros. It is better to find the complex zeros in the manner already described.

To summarize. The method recommended is first to find the total number of zeros in the rectangle by methods to be described later. Then to calculate the function at sufficient points to account for all the zeros, either by changes of sign or as complex zeros determined by the use of Rouché's theorem. We know no way of dealing with multiple zeros, and simply hope that none are present.

4. Evaluation of $N(t)$

For reasons explained in the last section it is desirable to be able to determine the number of zeros of $Z(\tau)$ in a region $T_0 < \tau < T_1$. In practice this is best done by determining separately the numbers in the regions $0 < \mathcal{R}\tau < T_0$ and $0 < \mathcal{R}\tau < T_1$. If we write $\pi S(t)$ for the argument of $\zeta(\frac{1}{2}+it)$ obtained by continuation along a line parallel to the real axis from $\infty + it$, where the argument is defined to be zero, we have

$$N(T) = 2\kappa(\frac{T}{2\pi}) + 1 + S(T),$$

where $N(T)$ is the number of zeros of $\zeta(\sigma + it)$ in the region $0 < t < T$. The problem is thus reduced to the determination of $S(T)$. If the sign of $Z(\frac{T}{2\pi})$ is known, the value of $S(T)$ is known modulo 2. It is not therefore necessary to obtain $S(T)$ to any great accuracy. The principle of the method is that if $S_1(t) = \int_0^t S(u)du$ then $S_1(t)$ is known to be $O(\log t)$. If then the positions of the zeros are known in an interval of length L, $S(t)$ will be known modulo 2 in this interval, the additive even integer being the same throughout. Hence $S_1(t_0+L) - S_1(t_0)$ will be known modulo $2L$, and if L is sufficiently large this will determine it exactly and thereby determine $S(t)$ throughout the interval. In order to complete the details of this argument it is necessary to replace the O result by a Θ result. It would also be desirable to try and arrange to manage with very limited knowledge of the positions of the zeros.

Theorem 4. *If $t_2 > t_1 > 168\pi$, then*

$$S_1(t_2) - S_1(t_1) = \Theta\left(2.30 + 0.128\log\frac{t_2}{2\pi}\right).$$

The proof of this follows Theorem 40 of Titchmarsh (**3**). The essential step is

Lemma 1. *If $t_2 > t_1 > 0$, then*

$$\pi\{S_1(t_2) - S_1(t_1)\} = \mathcal{R}\int_{\frac{1}{2}+it_2}^{\infty+it_2} \log\zeta(s)ds - \mathcal{R}\int_{\frac{1}{2}+it_1}^{\infty+it_1} \log\zeta(s)ds.$$

We apply Cauchy's theorem to $\log\zeta(s)$ and the rectangle with vertices $\frac{1}{2}+it_1, \frac{1}{2}+it_2$, $R+it_2$, $R+it_1$ and appropriate detours round the branch lines from zeros within the rectangle. The real part of the integral is

$$-\mathcal{R}\int_{\frac{1}{2}+it_2}^{R+it_2}\log\zeta(s)ds+\int_{\frac{1}{2}+it_1}^{\frac{1}{2}+it_2}\arg\zeta(s)(-ids)+\mathcal{R}\int_{\frac{1}{2}+it_1}^{R+it_1}\log\zeta(s)ds-\int_{R+it_1}^{R+it_2}\arg\zeta(s)(-ids),$$

no contribution arising from the detours. The last of these integrals tends to 0 as $R\to\infty$ and the second is $\pi\{S_1(t_2)-S_1(t_1)\}$.

Lemma 2. *If $\tau\geqslant 64$, we have*

$$\left|\zeta\left(\frac{1}{2}+2\pi i\tau\right)\right|<4\tau^{\frac{1}{4}}.$$

Since $|h(\zeta)|<0{\cdot}95$ we have, by Theorem 1,

$$\left|\tau\left(\frac{1}{2}+2\pi i\tau\right)\right|=|Z(\tau)|<2\sum_{1\leqslant r\leqslant\tau^{\frac{1}{2}}}r^{-\frac{1}{2}}+1.2\tau^{-\frac{3}{4}}+0.95\tau^{-\frac{1}{4}}$$

and

$$\sum_{1\leqslant r\leqslant\tau^{\frac{1}{2}}}r^{-\frac{1}{2}}<1+\int_1^{\tau^{\frac{1}{2}}}x^{-\frac{1}{2}}\,dx=2\tau^{\frac{1}{4}}-1.$$

Lemma 3.

$$|\zeta(1.25+it)|<\zeta(1.25)<4.6,$$

$$\left|\int_{1.25+it}^{\infty}\log\zeta(s)ds\right|<\int_{1.25}^{\infty}\log\zeta(\sigma)d\sigma<1.17,$$

$$\left|\frac{\zeta'}{\zeta}(1.5+it)\right|<\frac{\zeta'}{\zeta}(1.5)<2.62,$$

$$\left|\int_{1.5+it}^{2.5+it}\log\zeta(s)ds\right|<\int_{1.5}^{2.5}\log\zeta(\sigma)d\sigma<0.548,$$

$$\left|\int_{1.5+it}^{\infty}\log\zeta(s)ds\right|<\int_{1.5}^{\infty}\log\zeta(\sigma)d\sigma<0.997,$$

$$\tfrac{1}{2}\log\pi>0.572.$$

These results are all based on the tables in Jahnke-Emde (**2**), p. 323. An error of two units in the last place is assumed. To the extent that we do not know how these tables were obtained we depart from the principles of the 'rigorous computation'.

Lemma 4. *If* $\frac{1}{2} < \sigma < \frac{5}{4}$ *and* $t > 168\pi$, *then*

$$|\zeta(s)| < 4.5t^{3/8-\frac{1}{4}\sigma}.$$

Consider $f(s)$:

$$f(s) = \zeta(s)\left(\frac{s-\frac{1}{2}}{i}\right)^{-3/8+\frac{1}{4}\sigma} \exp\left[\frac{-4\pi i}{s-\frac{1}{2}-127.5\pi i}\right].$$

Now
$$\left|\exp\left[\frac{-4\pi i}{s-\xi-127.5\pi i}\right]\right| = \exp\left[\frac{-4\pi(b-127.5\pi)}{(t-127.5\pi)^2+(\sigma-\frac{1}{2})^2}\right].$$

Hence, by Lemma 3, $|f(s)| < 4$ on the line $\sigma = \frac{1}{2}$. Elsewhere, if $\frac{1}{2} < \sigma < \frac{5}{4}, t > 128\pi$, we have

$$\log\left|\left(\frac{s-\frac{1}{2}}{i}\right)^{-3/8+\frac{1}{4}s}\right| = \frac{1}{2}\left(-\frac{3}{8}+\frac{1}{4}\sigma\right)\log\left\{t^2+\left(\sigma-\frac{1}{2}\right)^2\right\}+\frac{1}{4}t\tan^{-1}\frac{\sigma-\frac{1}{2}}{t}$$
$$\leqslant -\frac{1}{32}\log t^2 + \frac{3}{16}.$$

Hence on the line $\sigma = \frac{5}{4}, t \geqslant 128\pi$ we have

$$|f(s)| < \zeta\left(\frac{5}{4}\right)e^{\frac{3}{16}}(128\pi)^{-\frac{1}{16}} < 4.$$

Finally on the line $t = 128\pi, \frac{1}{2} \leqslant \sigma \leqslant \frac{5}{4}$ we have

$$|f(s)| < |\zeta(s)|\exp\left[\frac{-2\pi^2}{\frac{1}{4}\pi^2+\frac{9}{16}}\right] \text{ and } |\zeta(s)| < (128\pi)^{\frac{1}{2}}$$

by equation (8) on p. 27 of Ingham (1). Hence $|f(s)| < 4$ on the whole boundary of the strip $t \geqslant 128\pi, \frac{1}{2} < \sigma < \frac{5}{4}$, and, since certainly $f(s) = O(t)$, we have $|f(s)| < 4$ throughout the strip by the Phragmén-Lindelöf theorem. From this it follows that

$$|\zeta(s)| < 4e^{0.1}\left|\left(\frac{s-\frac{1}{2}}{i}\right)^{\frac{3}{8}-\frac{1}{4}s}\right|$$
$$< 4.5t^{\frac{3}{8}-\frac{1}{4}\sigma} \text{ for } t > 168\pi.$$

The purpose of the factor $\exp\left[\frac{-4\pi i}{s-\frac{1}{2}-127{\cdot}5\pi i}\right]$ is merely to enable us to do without accurate knowledge of $\zeta(s)$ over the end of the strip.

Lemma 5. *If $t > 168\pi$, then*

$$\mathcal{R}\int_{\frac{1}{2}+it}^{\infty+it} \log\zeta(s)ds \leqslant 2.30 + 0.12\log t.$$

For

$$\mathcal{R}\int_{\frac{1}{2}+it}^{\infty+it} \log\zeta(s)ds \leqslant 1.17 + \int_{0.5}^{1.25} \log|\zeta(s)|ds, \text{ by Lemma 3,}$$

$$\leqslant 1.17 + \int_{0.5}^{1.25} \left\{\log 4.5 + \left(\frac{3}{8} - \frac{1}{4}\sigma\right)\log t\right\}d\sigma, \text{ by Lemma 4,}$$

$$= 1.17 + 0.75\log 4.5 + \frac{15}{128}\log t$$

$$< 2.30 + 0.12\log t.$$

It is certainly possible to improve the coefficient of $\log t$ in this result at the expense of the constant. The coefficient of $\log t$ could be reduced at any rate to 0·052 using results stated on pp. 25, 26 of Titchmarsh (**3**).

Lemma 6.

$$\frac{\zeta(s)\zeta(s+2)}{\{\zeta(s+1)\}^2} = \frac{s^2}{s^2-1}\frac{\{\Gamma(\frac{1}{2}s+\frac{3}{2})\}^2}{\Gamma(\frac{1}{2}s+1)\Gamma(\frac{1}{2}s+2)}\prod_\rho \frac{(s-\rho)(s-\rho+2)}{(s-\rho+1)^2},$$

where the product is over the non-trivial zeros of the zeta-function.

This is an immediate consequence of the Weierstrass product for the zeta-function.

Lemma 7. *If $k = 1.49$, $\mathcal{R}a \geqslant 0$, then*

$$\mathcal{R}\left(\psi(a) + \frac{k}{a}\right) = \mathcal{R}\left[\int_{a-1}^{a} \log\frac{z}{z+1}dz + \frac{k}{a}\right] \geqslant 0.$$

It is easily seen that if $\mathcal{R}a = 0$ then $\mathcal{R}\psi(a) = \mathcal{R}(k/a) = 0$. Also that $\mathcal{R}(k/a) \geqslant 0$ for $\mathcal{R}a \geqslant 0$ and that $\psi(a)$ is continuous at 0. $\psi(a) + (k/a) \to 0$ as $a \to \infty$. Hence applying the maximum modulus principle (or rather, the minimum real part principle) to $\psi(a) + (k/a)$ and various regions

$$\mathcal{R}a \geqslant 0, \quad 0 < \epsilon < |a| < R,$$

we see by allowing $\epsilon \to 0$, $R \to \infty$ that the minimum real part must be achieved either on the boundary $\mathcal{R}a = 0$ or on the real axis (which may be a singularity). It only remains therefore to establish our inequality for the real axis. At any stationary point we must have

$$0 = \mathcal{R}\left(\psi'(a) - \frac{k}{a^2}\right) = -\log\left|1 - \frac{1}{a^2}\right| - \frac{k}{a^2}.$$

This equation only has solutions near to 0·91 and 1·2 both of which correspond to minima of $\psi(a) + (k/a)$. There is no ordinary maximum separating them, but there is a singularity at $a = 1$. By computations near to these minima, and knowledge of an upper bound for the second derivative of the function in intervals enclosing them, one can show that the values at the minima are positive. The value at the lesser minimum (near 0·91) is about 0·0087. Hence $\psi(a)+(k/a)>0$ on the real axis as required.

Lemma 8. *If* $\mathcal{R}z>0$, *then*

$$\frac{\Gamma'(z)}{\Gamma(z)}=\log z-\frac{1}{2z}+\Theta\left(\frac{2}{\pi^2|(\mathcal{J}z)^2-(\mathcal{R}z)^2|}\right).$$

We use the formula

$$\frac{\Gamma'(z)}{\Gamma(z)}=\log z-\frac{1}{2z}+2\int_0^\infty\frac{u\,du}{(u^2+z^2)(e^{2\pi u}-1)}$$

and take the line of integration to be $\mathcal{R}u=Iu=v>0$. Then

$$\left|\frac{u}{e^{2\pi u}-1}\right|<\frac{e^{-\pi v}}{\pi\sqrt{2}},\quad |u^2+z^2|>|(Iz)^2-(\mathcal{R}z)^2|.$$

No poles are encountered in the change of line of integration since $\mathcal{R}z>0$.

Lemma 9. *If* $t>50$, *then*

$$-\mathcal{R}\int_{\frac{1}{2}+it}^{\infty}\log\zeta(s)ds<4.9+0.245\log\frac{t}{2\pi}.$$

We have

$$\mathcal{R}\int_{\frac{1}{2}+it}^{\infty}\log\zeta(s)ds=\mathcal{R}\int_{\frac{1}{2}+it}^{\infty}\log\frac{\zeta(s)\zeta(s+2)}{\{\zeta(s+1)\}^2}ds+\mathcal{R}\int_{\frac{3}{2}+it}^{\infty}\log\zeta(s)ds+\mathcal{R}\int_{\frac{3}{2}+it}^{\frac{5}{2}+it}\log\zeta(s)ds$$

$$\geqslant\mathcal{R}\int_{\frac{1}{2}+it}^{\infty}\log\frac{\zeta(s)\zeta(s+2)}{\{\zeta(s+1)\}^2}ds-1.545,$$

by Lemma 3.

Also, by Lemma 6,

$$\int_{\frac{1}{2}+it}^{\infty}\log\frac{\zeta(s)\zeta(s+2)}{\{\zeta(s+1)\}^2}ds=\sum_{\rho}\int_{\frac{1}{2}+it}^{\frac{3}{2}+it}\log\frac{s-\rho}{s-\rho+1}ds-\int_{\frac{1}{2}+it}^{\frac{3}{2}+it}\log\frac{\Gamma\left(\frac{1}{2}s+1\right)}{\Gamma\left(\frac{1}{2}s+\frac{3}{2}\right)}ds+\int_{\frac{1}{2}+it}^{\frac{3}{2}+it}\log\frac{s}{s-1}ds.$$

Now if $\frac{1}{2}<\sigma<\frac{3}{2}$, then $|s-1|<|s|$, and therefore

$$\mathcal{R}\int_{\frac{1}{2}+it}^{\frac{3}{2}+it}\log\frac{s}{s-1}ds\geqslant 0.$$

Also

$$\mathcal{R}\int_{\frac{1}{2}+it}^{\frac{3}{2}+it}\log\frac{\Gamma\left(\frac{1}{2}s+1\right)}{\Gamma\left(\frac{1}{2}s+\frac{3}{2}\right)}ds=-\tfrac{1}{2}\mathcal{R}\frac{\Gamma'}{\Gamma}\left(\tfrac{1}{2}it+\sigma\right)$$

for some $\sigma, \frac{5}{4} < \sigma < \frac{9}{4}$, by the mean value theorems,

$$\leqslant -\frac{1}{4}\log\left(\frac{1}{4}t^2+\frac{25}{16}\right)-\frac{\frac{5}{2}}{t^2+\frac{25}{4}}+\frac{2}{\pi^2\left(\frac{1}{4}t^2-\frac{81}{16}\right)}$$

by Lemma 8,

$$< -\frac{1}{2}\log\frac{1}{2}t \text{ (since } t > 50)$$

$$< -\frac{1}{2}\log\frac{t}{2\pi}-0.572, \text{ by Lemma 3}$$

Finally,

$$\mathcal{R}\sum_{\rho}\int_{\frac{1}{2}+it}^{\frac{3}{2}+it}\log\frac{s-\rho}{s-\rho+1}ds \geqslant -1.49\mathcal{R}\sum_{\rho}\frac{1}{it-\rho+\frac{3}{2}},$$

by Lemma 7,

$$= -1.49\,\mathcal{R}\left[\frac{\zeta'}{\zeta}\left(it+\frac{3}{2}\right)-\frac{1}{2}\log\pi+\frac{1}{2}\frac{\Gamma'}{\Gamma}\left(\frac{1}{2}it+\frac{7}{4}\right)\right],$$

by the Mittag-Leffler series for $\frac{\zeta'}{\zeta}(s)$,

$$\geqslant -1.49\left[\mathcal{R}\frac{\zeta'}{\zeta}\left(it+\frac{3}{2}\right)-\frac{1}{2}\log\pi+\frac{1}{4}\log\left(\frac{1}{4}t^2+\frac{49}{16}\right)-\frac{7}{4t^2+49}\right],$$

by Lemma 8,

$$\geqslant -1.49\left[\frac{1}{2}\log\frac{t}{2\pi}+2.63\right],$$

using Lemma 3 and $t > 50$.

Combining these results gives the asserted inequality.

A variant of this method enables us to reduce the coefficient of τ to $\frac{1}{2}\log 2-\frac{1}{4}+\epsilon$, e.g., to 0.097, at the expense of the constant term.

Theorem 4 follows at once from Lemmas 1, 5, 9.

It is convenient to replace Theorem 4 by a similar result with $\kappa(\tau)$ or $\kappa_1(\tau)$ as the independent variable. This is because $\kappa(\tau)$ describes the 'expected' position of the zeros, and is therefore more informative than τ.

Lemma 10. *If* $\tau_1 > 84$, *then*

$$\int_{\tau_1}^{\tau_2} S(2\pi\tau)d\kappa_1(\tau) = \Theta\{0.184\log\tau_2 + 0.0103(\log\tau_2)^2\}.$$

For

$$\begin{aligned}\int_{\tau_1}^{\tau_2} S(2\pi\tau)d\kappa_1(\tau) &= \frac{1}{2\pi}\kappa_1'(\tau_1)\{S_1(2\pi\tau_2) - S_1(2\pi\tau_1)\} \\ &\quad - \frac{1}{2\pi}\int_{\tau_1}^{\tau_2}\{S_1(2\pi\tau_2) - S_1(2\pi\tau)\}\kappa_1''(\tau)d\tau \\ &= \Theta\left[\frac{2.30 + 0.128\log\tau_2}{2\pi}\left\{\kappa_1'(\tau_1) + \int_{\tau_1}^{\tau_2}|\kappa_1''(\tau)|d\tau\right\}\right] \\ &= \Theta\left(\frac{2.30 + 0.128\log\tau_2}{4\pi}\log\tau_2\right).\end{aligned}$$

Theorem 5. *Let*

$$64 < \tau_{-R_1} < \tau_{1-R_1} < \ldots < \tau_0 < \ldots < \tau_{R_2-1} < \tau_{R_2}$$

and $\kappa(\tau_r) = c_r, \delta_r = c_r - c_0 - \frac{1}{2}r, \delta_{R_2} = \delta_{-R_1} = 0$, *and* $Z(\tau_r)Z(\tau_{r+1}) < 0$ *if* $1 - R_1 \leqslant r \leqslant R_2 - 2$, $\tau_{-R_1} > 84$. *Then*

$$\begin{aligned}&-\frac{1}{2} + \frac{2}{R_1}\sum_{r=1-R_1}^{-1}\delta_r - \frac{0.006}{\tau_{-R_1}} - \frac{2}{R_1}\{0.184\log\tau_0 + 0.0103(\log\tau_0)^2\} \\ &\leqslant N(2\pi\tau_0) - 2c_0 - 1 \\ &\leqslant \frac{1}{2} + \frac{2}{R_2}\sum_{r=1}^{R_2-1}\delta_r + \frac{0.006}{\tau_0} + \frac{2}{R_1}\{0.184\log\tau_{R_2} + 0.0103(\log\tau_{R_2})^2\}.\end{aligned}$$

In the interval (τ_r, τ_{r+1}) we have $N(2\pi\tau) \geqslant N(2\pi\tau_0) + r$ if $0 \leqslant r \leqslant R_2 - 1$ and therefore

$$\begin{aligned}\int_{\tau_0}^{\tau_R} N(2\pi\tau)d\kappa_1(\tau) &\geqslant \sum_{r=0}^{R-1}(c_{r+1} - c_r)\{N(2\pi\tau_0) + r\} \\ &= \frac{1}{2}R\left[N(2\pi\tau_0) + \frac{1}{2}(R-1)\right] - \sum_{r=1}^{R}\delta_r.\end{aligned}$$

Also

$$\begin{aligned}\int_{\tau_0}^{\tau_R}\{2\kappa(\tau) + 1\}d\kappa_1(\tau) &= c_R - c_0 + \left(c_R^2 - c_0^2\right) + \Theta\left(\frac{0.006(c_R - c_0)}{\tau_0}\right) \\ &= \frac{1}{2}R\left(1 + 2c_0 + \frac{1}{2}R\right) + \Theta\left(\frac{0.003R}{\tau_0}\right).\end{aligned}$$

The second inequality now follows since $S(2\pi\tau) = N(2\pi\tau) - 1 - 2\kappa(\tau)$ and the first may be proved similarly.

Example. It is known by computation that within distance 0·05 of each of the half-integers $547\frac{1}{2}$ to $554\frac{1}{2}$ there lie values of κ such that the corresponding value of Z has the same sign as $\cos 2\pi\kappa$. It is required to show that if τ_0 is that one of the points concerned which is within 0·05 of 551 then $N(2\pi\tau_0) = 1103$.

We take the values concerned to be $\tau_{-7}, \tau_{-6}, \ldots, \tau_7$ in Theorem 5, and define τ_{-8}, τ_8 to satisfy $\delta_{-8} = \delta_8 = 0$. Then $|\delta_r| < 0{\cdot}1$ for each r, $-7 \leqslant r \leqslant 7$. The conditions of Theorem 5 are satisfied and it gives

$$-1{\cdot}0 \leqslant N(2\pi\tau_0) - 2c_0 - 1 \leqslant 1{\cdot}0.$$

$N(2\pi\tau_0)$ is odd since $Z(\tau_0)\cos 2\pi\kappa(\tau_0) > 0$ and we also have

$$|c_0 - 551| < 0{\cdot}05.$$

The required conclusion now follows.

PART II. THE COMPUTATIONS

1. Essentials of the Manchester computer

It is not intended to give any detailed account of the Manchester Computer here, but a few facts must be mentioned if the strategy of the computation is to be understood. The storage of the machine is of two kinds, known as 'electronic' and 'magnetic' storage. The electronic storage consisted of four 'pages' each of thirty-two lines of forty binary digits. The magnetic storage consisted of a certain number of tracks each of two pages of similar capacity. Only about eight of these tracks were available for the zeta-function calculations. It was possible at any time to transfer one or both pages of a track to the electronic storage by an appropriate instruction. This operation takes about 60 ms. (milliseconds). Transfers to the magnetic store from the electronic were also possible, but were in fact only used for preparatory loading of the magnetic store. The course of the calculations is controlled by instructions each of twenty binary digits. These are normally magnetically stored, but must be transferred to the electronic store before they can be obeyed. In the initial state of the machine (with the magnetic store loaded) the electronic store is filled with zeros. A zero instruction, however, has a definite meaning, and in fact results in a transfer of instructions to the electronic store, thus initiating the calculation. Most instructions, such as transfer of 'lines' of forty digits, take 1·8 ms., but transfers to or from the magnetic store take longer, as has been mentioned, and multiplications take a time depending on the number of digits 1 in the multiplier, ranging from 3·6 ms. for a power of two to 39 ms. for $2^{40} - 1$.

The results of the calculations are punched out on teleprint tape. This is a slow process in comparison with the calculations, taking about 150 ms. per character. The content of a tape may afterwards automatically be printed out with a typewriter if desired. The significance of what is printed out is determined by the 'programmer'. In the present case the output consisted mainly of numbers in the scale of 32 using the code

0	1	2	3	4	5	6	7	8	9	10	11	12	13	14	15	16	17	18	19
/	E	@	A	:	S	I	U	$\frac{1}{4}$	D	R	J	N	F	C	K	T	Z	L	W

20	21	22	23	24	25	26	27	28	29	30	31
H	Y	P	Q	O	B	G	‖	M	X	V	£

and writing the most significant digit on the right. More conventionally the scale of 10 can be used, but this would require the storage of a conversion routine, and the writer was entirely content to see the results in the scale of 32, with which he is sufficiently familiar.

2. Outline of calculation method

The calculations had of course to be planned so that the total storage capacity used was within the capacity of the machine. So long as this was fulfilled it was desirable to make the time of calculation as short as possible without excessive trouble in programming. The most time-consuming part of the calculations is of course the computation of the terms

$$n^{-\frac{1}{2}} \cos 2\pi(\tau \log n - \kappa)$$

from given κ and τ. By storing tables of $\log n$ and $n^{-\frac{1}{2}}$ within the machine this was reduced essentially to two multiplications and the calculation of a cosine, together with arrangements for 'looking up' the logarithm and reciprocal square root. The cosines were obtained from a table giving $\cos(r\pi/128)$ for $0 \leqslant r \leqslant 64$ by linear interpolation and reducing to the first quadrant. This gives an error of less than 10^{-4}, which is quite sufficient accuracy for the purpose. Very much greater accuracy was of course required in the logarithms, for an error ϵ in $\log n$ gives rise to an error approaching $2\pi\tau\epsilon$ in the cosine, and $2\pi\tau$ may be very large, e.g. 25,000. These logarithms were calculated by the machine in a previous computation, and were given with an error not exceeding 2.10^{-10}. The reciprocal square roots were given with error not exceeding 10^{-5}. Both the logarithms and the reciprocal square roots were checked after loading into the magnetic store by automatic addition, the results obtained being compared with values based respectively on Stirling's formula and on the known value of $\zeta(\frac{1}{2})$. The table only went as far as $n = 63$. The tabular cosines were built up automatically from the values of $\cos(\pi/128)$ and $\sin(\pi/128)$ by using the addition formula. The values of $\cos(\pi/128)$ and $\sin(\pi/128)$ were calculated both automatically and manually. A hand-copying process was used in connexion with this table, but the final results when loaded were automatically thrice differenced and the results inspected.

The routine as a whole was checked (amongst other methods) by comparing the result given for a value of τ about 20,000 with an entirely different, slower, and simpler routine. This routine had itself been checked against a hand-computed value for $\tau = 16$ and against a value given by Titchmarsh (5) for $\tau = 201{\cdot}596$.

Since it was only necessary to calculate $\kappa(\tau)$ once for each value of τ this calculation did not have to be particularly quickly performed. It was considered sufficient to obtain the logarithm by means of a slow but simple routine taking about 1·2 sec. The time for each term $n^{-\frac{1}{2}} \cos 2\pi(\tau \log n - \kappa)$ was about 0·2 sec. With $m = 63$, and allowing for the calculation of $\kappa_1(\tau)$ this means about 14 secs. for each value of τ. The routine could be used for recording the results for given values of τ, a typical entry obtained in this way being:

ZETAFASTG/F@Q¼B$YNK@:ZSZ″XVMX//SA/////¼OTNR@O//.

This entry has to be divided into sequences of eight characters. In this case they are:

1. ZETAFAST. This occurs at the beginning of each entry. Its purpose is mainly to identify the document as referring to this zeta-function routine.
2. G/F@Q¼B$. This is a number useful in checking results and called the 'cumulant'. It appears in the scale of 32, with the most significant digit on the right. This is the standard method of representing numbers on documents connected with the Manchester Computer (a decimal method can also be used if desired).

3. YNK@:ZSZ. This is also in the scale of 32 and gives the residue of $2^{40}\kappa_1(\tau)$ modulo 2^{40}. Since Z is the symbol for 17 it will be seen that $\kappa_1(\tau)$ is near to $\frac{1}{2}$ mod 1.
4. "XVMX///. This gives the value of $2^{17}\tau$; in this case τ is about 239·24.
5. SA /////$\frac{1}{4}$. This was always included in the record due to a minor difficulty in the programming. It did not seem worth while to take the trouble to eliminate it.
6. OTNR@O//. This is the value of $2^{30}Z(\tau)$ modulo 2^{40}. In this case $Z(\tau)$ is about 0·75.

The routine was not, however, used mostly for the calculation of values of $Z(\tau)$ with individually given τ. It was made to determine for itself appropriate values of τ, such as to give values of $2\kappa(\tau)$ near to successive integers. This was done by making each τ depend on the immediately previous one and on the previous κ by the formula $\tau' = \tau + (1-\delta)\alpha$, where τ', τ are the new and old values of τ respectively, δ is the difference of $2\kappa_1(\tau)$ from the nearest integer, and $(\alpha \log \tau)^{-1} = 1 + \Theta(0{\cdot}1)$. This procedure ensured that if the initial value of $\kappa(\tau)$ differs from an integer by less than 0·125, then the succeeding values will do likewise. It was decided not to record all the values of $Z(\tau)$, partly because the inspection and filing of the teleprint tape output would have a great burden to the experimenters. Values were only recorded when the unexpected sign occurred, i.e. when $Z(\tau)\cos 2\pi\kappa(\tau) < 0$. This reduced the amount of output data by about 90 per cent.

In order that there should be no doubt about the validity of the results it is necessary that one should also record all cases where the sign of $Z(\tau)$ is doubtful because of the limited accuracy of the computation. The criterion actually used was $Z(\tau)H(\kappa) > 0{\cdot}31E$, where

$$H(\kappa) \equiv \kappa - \tfrac{1}{4}(\text{mod } 1), \quad |H(\kappa)| < 0{\cdot}31.$$

The quantity $H(\kappa)$ arises very naturally with the computer. The condition $(\alpha \log \tau)^{-1} = 1 + \Theta(0{\cdot}1)$ ensures that (except for one or two values at the beginning of a run), $|H(\kappa)| < 0{\cdot}31$. The actual errors involved in the calculation were:

Error arising from using Titchmarsh's formula (Theorem 3) . .	$1.15m^{-\frac{3}{2}}$
Error due to replacing $\tau^{-\frac{1}{4}}h(\xi)$ by $m^{-\frac{1}{2}}h(\xi)$	$0.47m^{-\frac{3}{2}}$
Error due to replacing ξ by $\frac{1}{2}\tau m^{-1} - \frac{1}{2}m$	$1.08m^{-\frac{3}{2}}$
Error from using tabulated logarithms	$5.1 \times 10^{-10}m^{\frac{5}{2}}$
Error in replacing $\kappa(\tau)$ by $\kappa_1(\tau)$	$0.15m^{-\frac{3}{2}}$
Error in calculating $\kappa_1(\tau)$	$2 \times 10^{-10}m^{\frac{5}{2}}$
Error from using tabulated reciprocal square roots	$1.3 \times 10^{-4}m$
Error from using tabulated cosines and linear interpolations . .	$3.2 \times 10^{-4}m^{\frac{1}{2}}$
Error of Theorem 2	$0.0243m^{-\frac{1}{2}}$

There are also numerous rounding off errors which are very small. These and all the 'cross terms' have been absorbed into the above errors so that we may put the whole error as not more than

$$E = 2{\cdot}85m^{-\frac{3}{2}} + 0{\cdot}0243m^{-\frac{1}{2}} + 3{\cdot}2 \times 10^{-4}m^{\frac{1}{2}} + 1{\cdot}3 \times 10^{-4}m + 7{\cdot}1 \times 10^{-10}m^{\frac{5}{2}},$$

e.g. for $m = 15$

$$E < 0{\cdot}057,$$

and for $m = 65$

$$E < 0{\cdot}02.$$

The storage available was distributed as follows:

Magnetic store

Logarithms routine (for κ)	1 page
Table of logarithms and reciprocal square roots . . .	4 pages
Routine for calculating the terms $n^{-\frac{1}{2}}\cos 2\pi(\tau\log n-\kappa)$ and table of cosines	2 pages
Remainder of routine for calculating the function $Z(\tau)$ . .	2 pages
Input routine	2 pages
Output routine	2 pages

Electronic store, as occupied during the greater part of the time

Instructions and cosines	2 pages
Logarithms and reciprocal square roots	1 page
Miscellaneous data and working space	1 page

The principal investigation concerned the range $63^2 \leqslant \tau \leqslant 64^2$, i.e. the interval in which $m = 63$. Working at full efficiency it should have taken about 4 hours to calculate these values, the number of zeros concerned being about 1070. Full efficiency was not, however, achieved, and the calculation took about 9 hours. Only a small amount of this additional time was accounted for by duplicating the work. The special investigations in the neighbourhood of points where the unexpected sign occurred took a further 8 hours. The general reliability of the machine was checked from time to time by repeating small sections. The recorded cumulants were useful in this connexion. These cumulants were the totals of the values of $Z(\tau)$ computed since the last recorded value. If a calculation is repeated and there is agreement in cumulant value then there is a strong presumption that there is also agreement in all the individual values contributing to it. The result of this investigation, so far as it can be relied on, was that there are no complex zeros or multiple real zeros of $Z(\tau)$ in the region

$$63^2 \leqslant \tau \leqslant 64^2,$$

i.e. all zeros of $\zeta(s)$ in the region $2\pi.63^2 \leqslant t \leqslant 2\pi.64^2$ are simple zeros on the critical line.

Another investigation was also started with a view to extending the range of relatively small values of t for which the Riemann hypothesis holds. Titchmarsh has already proved that it holds up to $t = 1468$, i.e. to about $\tau = 231$. The new investigation started somewhat before $\tau = 225$ to allow a margin for the application of Theorem 5. It was intended to continue the work up to about $\tau = 500$, but an early breakdown resulted in its abandonment at $\tau = 256$. After applying Theorem 5 it would only be possible to assert the validity of the Riemann hypothesis up to about $\tau = 250$. All this part of the calculations was done twice, the unrecorded values being confirmed by means of the 'cumulants'.

Unfortunately $0{\cdot}31E$ was given the inappropriate value of $\frac{1}{128}$ and consequently we are only able to assert the validity of the Riemann hypothesis as far as $t = 1540$, a negligible advance.

REFERENCES

1. A. E. INGHAM, *The distribution of prime numbers*, Cambridge Mathematical Tracts, No. 30 (1931).
2. JAHNKE U. EMDE (1), *Tafeln höherer Funktionen*, Leipzig, 1948.
3. E. C. TITCHMARSH, *The zeta-function of Riemann*, Cambridge Mathematical Tracts, No. 26 (1930).
4. E. C. TITCHMARSH, 'The zeros of the Riemann zeta-function', *Proc. Roy. Soc.* A, 151 (1935), 234–55.
5. E. C. TITCHMARSH, 'The zeros of the Riemann zeta-function', ibid. 157 (1936), 261–3.
6. A. M. TURING, 'A method for the calculation of the zeta-function', *Proc. London Math. Soc.* (2) 48 (1943), 180–97.

The University,
Manchester.

ON A THEOREM OF LITTLEWOOD[1]

S. SKEWES and A.M. TURING

1. Introduction

We propose to investigate the question as to where $\pi(x) - \mathrm{li}\, x$ is positive.[2] This quantity is positive if x is less than about 1.42 and negative from there up to 10^7. The figures suggest that $\pi(x) - \mathrm{li}\, x \sim x^{1/2}/\log x$ as $x \to \infty$ but Littlewood (1914) has proved that $\pi(x) - \mathrm{li}\, x$ changes sign infinitely often. The argument proceeds by cases, according to whether the Riemann hypothesis is true or false. It has been announced by one of us (Skewes, 1933) that in the case that the Riemann hypothesis is true $\pi(x) - \mathrm{li}\, x > 0$ for some x, $2 < x < 10^a$, where $a = 10^b$, $b = 10^{34}$. In the present paper it is proposed to establish that $\pi(x) > \mathrm{li}\, x$ for some x, $2 < x < \exp(\exp 661)$, without the restriction of assuming the Riemann hypothesis. [Should 661 be 686?]

We shall also prepare the ground for the possibility of improving the bound to about 10^a, $a = 10^5$ with the aid of extensive computations, and also consider the effect on the situation of discovering zeros off the critical line.

2. Outline of the method

The necessity of using the functions $\Pi(x)$ and $\log\zeta(s)$ somewhat complicates the argument. The general outline of the method may be illustrated by dealing with the analogous problem of finding where $\theta(x) > x$. Since

$$\psi(x) = \theta(x) + \theta(x^{1/2}) + \theta(x^{1/3}) + \cdots,$$

and $\theta(x) \sim x$, this is essentially the question as to where $\psi(x) > x + x^{1/2}$. Now

$$\psi(x) - x = -\sum_{\varrho} x^{\varrho}/\varrho - (\zeta'/\zeta)(0) - \frac{1}{2}\log(1 - x^{-2}),$$

so that the problem reduces essentially to the question: for what values of t does the inequality

$$G(t) = -\sum_{\varrho} \varrho^{-1} \exp(\varrho - \frac{1}{2})t > 1$$

[1] Editors' footnote: In the Collected Works, J. L. Britton remarks at this point: "I have not been able to locate the original script but I had access to a photocopy of it. The article is in typescript with all mathematical symbols in manuscript. In spite of the joint authorship, the handwriting and phrasing indicate that it was perhaps written by Turing alone. There are handwritten comments by another person, probably A.E. Ingham. Such comments are here enclosed in square brackets [ ]. Where necessary an earlier $ sign indicates the place in the text where the remark applies. If the remark refers to a symbol or expression in the text, the relevant $ sign appears immediately before the symbol or expression." See pp. 272–273 of the Collected Works for additional remarks. We understand from A. M. Cohen of Cardiff that (a) Philip Stein helped Ingham in trying to tidy up the manuscript around 1961, some of the points flagged may reflect this; and (b) a scanned copy of Cohen's fifty year old photocopy of this work should soon be available in the Turing Archive.
What follows is a "rekeyed" version of the text found in the Collected Works (1992). The two versions are identical apart from some minor changes in punctuation, style, and layout.

[2] Here $\pi(x)$ is the number of primes less than x and $\mathrm{li}\, x$ is the logarithmic integral of x, i.e., the Cauchy principal value of $\int_0^x 1/\log t\, dt$. In these as in other matters of notation, we follow ?.

hold, summation being over the complex zeros?

One may consider various expressions of the form $\int_0^\infty G(t)f(t)dt$. Putting

$$\frac{1}{\sqrt{2\pi}}\int_0^\infty f(t)e^{iut}dt = F(u),$$

we have

$$I = \int_0^\infty G(t)f(t)dt = \frac{1}{\sqrt{2\pi}}\sum_\varrho \varrho^{-1}F\left(\frac{\varrho-\frac{1}{2}}{i}\right).$$

If $f(t)$ is positive for positive t, and decreases to zero sufficiently quickly, it is possible to argue back from the value of I to the inequality $G(t) > 1$. For instance, it will suffice to have

$$\int_0^\infty f(t)dt = 1, \quad f(t) > 0, \quad I > \frac{5}{4}, \quad \int_A^\infty G(t)f(t)dt < \frac{1}{4},$$

to infer $G(t) > 1$ for some t, $0 < t < A$. The value of t for which I is sufficiently large and positive is to be obtained by Diophantine approximation. In carrying out the approximation, we try to adjust the phases of the terms with small ϱ to the appropriate values and to arrange that the remaining, unadjusted, terms are small. We therefore wish $F((\varrho - \frac{1}{2})/i)$ to be small for the large values of ϱ. By taking $f(t) = ((\sin\beta t)/t)^2$, Ingham ensured that, if the Riemann hypothesis is true, only a finite number of terms were different from zero. In the present paper, we use instead a function

$$f(t) = \left(\frac{\sin\beta t}{t}\right)^4 \exp\left(-\frac{1}{2}\alpha^2t^2\right)$$

which is largely inspired by Ingham's argument. It does not result in the vanishing of any of the terms, but if α is small, the later terms are extremely small. The present function has various advantages (for the present purpose) over that used by Ingham. The factor $\exp(-\frac{1}{2}\alpha^2t^2)$ encourages the quick convergence of the integral $\int_0^\infty G(t)f(t)dt$, facilitating the inference of inequalities about $G(t)$ from values of I. This factor also causes $F(u)$ to be an integral function (whereas otherwise it would only be regular in two squares and two right angle segments). The use of the higher power of $(\sin\beta t)/t$ results in a rather sharper transition from large to small values of $f(u)$, leading to an appreciable numerical improvement.

Essential to the whole method is the fact that the Riemann hypothesis has been tested in the region $|t| < 1468$ (Titchmarsh,1936).

3. Formal preliminaries

Lemma 1. *If*

$$F(u) = \frac{1}{\sqrt{2\pi}}\int_{-\infty}^\infty e^{-iut}f(t)dt \quad and \quad \varphi(s) = \int_0^\infty x^{-s}dh(x)$$

and the integrals

$$\int_0^\infty x^{-3}|h(x)|dx \quad and \quad \int_{3/2-i\infty}^{3/2+i\infty} |F(-is)||ds|$$

are convergent and $h(x) = o(x^2)$ *as* $x \to \infty$ *and* $x \to 0$, *then*

$$\int_{-\infty}^{\infty} h(e^t)e^{-t/2}f(t-t_0)dt = \frac{1}{i\sqrt{2\pi}} \int_{2-i\infty}^{2+i\infty} \frac{\varphi(s)}{s} \exp\left(\left(s-\frac{1}{2}\right)t_0\right) F\left(-i\left(s-\frac{1}{2}\right)\right) ds.$$

If $\sigma = 2$, then $\varphi(s) = s\int_0^\infty h(x)x^{-s-1}dx$ by integration by parts and so

$$\frac{1}{i\sqrt{2\pi}} \int_{2-i\infty}^{2+i\infty} \frac{\varphi(s)}{s} \exp\left(\left(s-\frac{1}{2}\right)t_0\right) F\left(i\left(s-\frac{1}{2}\right)\right) ds$$

$$= \frac{1}{i\sqrt{2\pi}} \int_{2-i\infty}^{2+i\infty} \left[\int_0^\infty h(x)x^{-s-1}dx\right] \exp\left(\left(s-\frac{1}{2}\right)t_0\right) F\left(-i\left(s-\frac{1}{2}\right)\right) ds$$

$$= \frac{1}{i\sqrt{2\pi}} \int_0^\infty x^{-3/2}h(x) \left[\int_{2-i\infty}^{2+i\infty} x^{-s+1/2} \exp\left(\left(s-\frac{1}{2}\right)t_0\right) F\left(-i\left(s-\frac{1}{2}\right)\right) ds\right] dx$$

$$= \int_0^\infty x^{-3/2}h(x)f(\log x - t_0)dx.$$

The inversion is admissible since the double integral is absolutely convergent. [?$(t_0 - \log x)$] [$(t_0 - t)$] [? conditions on f] [or $+\ i(s-\frac{1}{2})$ and $+is$]

Lemma 2. *If the functions* f, F *are subject to the restrictions of Lemma 1, we have*

$$\int_{-\infty}^{\infty} e^{-t/2}(\Pi(e^t) - M(t))f(t-t_0)dt$$

$$= \frac{1}{\$\sqrt{2\pi}} \int_{2-i\infty}^{2+i\infty} s^{-1}(\log\zeta(s) - g(s))\exp((s-i)t_0), \text{ [?i]}$$

where

$$M(t) = \int t^{-1}e^t dt, \text{ integration from } 0.1 \text{ to } \max(t, 0.1),$$

and where

$$g(s) = \int_{0.1(s-1)}^{\infty} t^{-1}e^{-t}dt, \text{[? Df. if } s \text{ real} < 1]$$

the integration to be along a line parallel to the real axis.

We apply Lemma 1 first with $h(x) = \Pi(x)$ and again with $h(x) = M(\log x)$ and combine the results. Note that for $\mathcal{R}s > 1$, one may also write $g(s) = \int_{0.1}^\infty t^{-1}\exp(1-s)t\,dt$.

Lemma 3.
(a) *If the logarithm has its principal value, the function* $g(s) + \log(s-1)$ *has no singularities.*
(b) *For any* s, $|g(s)| < \pi + \$1 \exp(\$ - 0.1(1-\sigma))$.

$$[1/|s-1|; \text{but } g \text{ ambiguous}]\ [\text{?minus sign}]\ [\text{?}\log\sin g.|$$

(a) The function $g(s) + \log(s-1)$ may be defined as the indefinite integral of the regular function $(1 - \exp(-0.1(s-1)))/(s-1)$.
(b) The inequality may be proved by integrating along an arc of a circle and part of the positive real axis.

Lemma 4. *If the functions* f *and* F *are related as in Lemma 1 and* g *is the function defined in Lemma 2 and if* $F(\mathrm{i}s)$ *is bounded* [?] *in any strip* $\sigma_1 \leqslant \sigma \leqslant \sigma_2$, *then*

$$\int_{-\infty}^{\infty} e^{-t/2}(\Pi(e^t) - M(t))f(t-t_0)dt = I_1 + I_2\$ + \sum_{\varrho} I_{3,\varrho} + J,\ [?\ -]$$

where the summation is over the nontrivial zeros of the zeta function and $\Delta > \$\,0\ [?\tfrac{5}{2}]$ *and*

$$I_1 = \frac{1}{\mathrm{i}\sqrt{2\pi}} \int_{2-\mathrm{i}\infty}^{2+\mathrm{i}\infty} s^{-1}(\log(s+1)(s+2)\zeta(s+2)) \times \exp\left(\left(s-\tfrac{1}{2}\right)t_0\right) F\left(\$ - \mathrm{i}\left(s - \frac{1}{2}\right)\right) ds, \quad [+\mathrm{i}]$$

$$I_2 = \frac{1}{\mathrm{i}\sqrt{2\pi}} \int s^{-1}\left(\log\frac{\zeta(s)}{(s+1)(s+2)\zeta(s+2)} - g(s)\right) \exp\left(\left(s-\frac{1}{2}\right)t_0\right) F\left(\mathrm{i}\left(s-\frac{1}{2}\right)\right) ds,$$

integration from $-\Delta + \frac{1}{2} - \mathrm{i}\infty$ *to* $-\Delta + \frac{1}{2} + \mathrm{i}\infty$,

$$I_{3,\varrho} = \sqrt{2\pi} \int_{\varrho-2}^{\varrho} s^{-1} F\left(\$ - \mathrm{i}\left(s - \frac{1}{2}\right)\right) \exp\left(\left(s - \frac{1}{2}\right)t_0\right) ds, \quad [+\mathrm{i}],$$

$$J = \sqrt{2\pi}\left(\log\frac{\$ - \zeta(0)}{2\zeta(2)} - g(0)\right) F\left(\frac{1}{2}\mathrm{i}\right) \ldots (\text{illegible}).$$

[? minus sign] [log and $g(0)$ ambiguous]

Formally the result follows from Cauchy's theorem by moving the line of integration from $\sigma = 2$ to $\sigma = \frac{1}{2} - \Delta$ with the integrand of I_2.

The integral is absolutely convergent on either of these lines, so that it is only necessary to prove that there is a sequence T_r such that $T_r \to +\infty$ and $I_{4,r} \to 0$, where

$$I_{4,r} = \int s^{-1}\left(\log\frac{s\zeta(s)}{(s+1)(s+2)\zeta(s+2)} - g(s)\right) \exp\left(\left(s - \frac{1}{2}\right)t_0\right) F\left(-\mathrm{i}\left(s - \frac{1}{2}\right)\right) ds,$$

integration from $\frac{1}{2} - \Delta + \mathrm{i}T_r$ to $2 + \mathrm{i}T_r$, and a similar sequence with $T_r \to -\infty$. We shall only consider the former case.

(A handwritten paragraph:) The singularities of the integrand are the... lines from ϱ to $\varrho - 2$ together possibly with some singularities... However $\zeta(s)(s-1)/(\zeta(s+2)(s+1)(s+2))$ may be verified to have neither zero nor pole at any integer, and certainly has not at any other real point. Hence the only real singularity is that at 0, giving rise to the residue J.

(Continuation of typescript:) We may choose our sequence T_r according to Theorem 26 of Ingham. There will then exist A such that if $\frac{1}{2} - \Delta \leqslant \sigma \leqslant 2$, then $|(\zeta'/\zeta)(\sigma + iT_r)| < A(\log T_r)^2$. The quantity A will depend on Δ only. Then

$$\begin{aligned} |\log\zeta(\sigma + iT_r)| &\leqslant |\log\zeta(2 + iT_r)| + (1.5 + \Delta)A(\log T_r)^2 \\ &< 1 + (1.5 + \Delta)A(\log T_r)^2, \end{aligned}$$

and therefore

$$|I_{4,r}| < T_r^{-1}(1 + (1.5 + \Delta)A(\log T_r)^2)(1.5 + \Delta)M\exp(1.5t_0),$$

where M is the upper bound of $F\left(-i\left(s - \frac{1}{2}\right)\right)$ in the region $\frac{1}{2} - \Delta$. Evidently $I_{4,r} \to 0$ as $r \to \infty$. [From 'and therefore' to '$r \to \infty$': ? details; 2 ζ' s. Also s, $s+1$, $s+2$ and $-g(s)$.]

4. Results with a special kernel

The function f has been relatively unrestricted until now, but we shall now put $f(t) = f_1(t)f_2(t)$, where

$$f_1(t) = \left(\frac{\sin\mu t}{\mu t}\right)^4, \quad f_2(t) = \exp\left(-\frac{1}{2}\alpha^2 t^2\right),$$

and we shall also put $\alpha^2 t_0 = \Delta = 400$ although these substitutions will not always be made. The choice of values for μ and for t_0 will not be made just yet, but we shall assume that $50 < \mu < 250$ and $10^4 < t_0$. The functions F, F_1, F_2 will be Fourier transforms of f, f_1, f_2 as in Lemma 1.

Lemma 5. *We have*

$$F_1(z) = \sqrt{2\pi}(2\mu)^{-1}\kappa(z/2\mu) \text{ if } z \text{ is real,}$$

$$F_2(z) = \alpha^{-1}\exp\left(-z^2/2\alpha^2\right),$$

$$F(z) = \frac{1}{\sqrt{2\pi}}\int_{-\infty}^{\infty} F_2(z-u)F_1(u)du,$$

where

$$\kappa(x) = \begin{cases} 1 - \frac{1}{2}|x|^2, & 0 \leqslant |x| \leqslant 1, \quad \left[-\frac{2}{3}(1-|x|)^3 + \frac{1}{6}(2-|x|)^3\right] \\ \frac{1}{2}(2 - |x|^2), & 1 \leqslant |x| \leqslant 2, \quad \left[\frac{1}{6}(2-|x|)^3\right] \\ 0, & 2 \leqslant |x|. \end{cases}$$

These are immediate applications of well-known results in the theory of Fourier transforms.

Lemma 6. *If $z = x + iy$, where x and y are real,*

(a) $|F(z)| \leqslant \alpha^{-1}\exp\left(\frac{y^2}{2\alpha^2}\right)$,

(b) $|F(z)| \leqslant \alpha^{-1}\exp\left(\frac{y^2-(|x|-4\mu)^2}{2\alpha^2}\right)$ *if* $|x| \geqslant 4\mu$,

(c) $\int_{c-i\infty}^{c+i\infty}|F(iz)||dz| \leqslant \sqrt{2\pi}\alpha\exp\left(\frac{c^2}{2\alpha^2}\right)$, [$c$ real]

(d) $|F(z)-F_1(z)| \leqslant \sqrt{2\pi}\alpha(4\mu^2)^{-1}$ *if z is real,* [$\sqrt{2\pi}\alpha^2(16\mu^3)^{-1}$?]

(e) $|F'(z)| \leqslant (\alpha\mu)^{-1}\exp\left(\frac{y^2}{2\alpha^2}\right)$. [? multiplied by $\frac{2}{3}$]

To prove (a) and (b), we use the inequality

$$|F(z)| \leqslant \frac{1}{\sqrt{2\pi}}\int_{-\infty}^{\infty}|F_2(z-u)||F_1(u)|du,$$

and observe that since $F_1(u) \geqslant 0$ we have

$$\frac{1}{\sqrt{2\pi}}\int_{-\infty}^{\infty}|F_1(u)|du = \frac{1}{\sqrt{2\pi}}\int_{-\infty}^{\infty}F_1(u)du = f_1(0),$$

and also that the integrand vanishes outside the range $|u| \leqslant 4\mu$, so that $|F(z)| < M$[? $\leqslant$], where M is the maximum of $|F_2(z-u)|$ in this range.

To prove (c)

$$\int_{c-i\infty}^{c+i\infty}|F(iz)||dz| \leqslant \frac{1}{\sqrt{2\pi}}\int_{u=-\infty}^{\infty}\int_{c-i\infty}^{c+i\infty}|F_2(iz-u)||F_1(u)|\,du\,|dz|$$

$$= f_1(0)\int_{c-i\infty}^{c+i\infty}|F_2(iz)||dz| = \sqrt{2\pi}\alpha\exp\left(\frac{c^2}{2\alpha^2}\right).$$

[correct but obscure]

For (d)

$$|F(z)-F_1(z)| = \left|\frac{1}{\sqrt{2\pi}}\int_{-\infty}^{\infty}(F_1(z-u)-F_1(z)\$-uF_1'(z))F_2(u)du\right| \quad [+]$$

since F_2 is an even function

$$\leqslant \frac{1}{4\$\mu^2}\int_{-\infty}^{\infty}u^2|F_2(u)|du \text{ [}\mu^3\text{in fact] [times } \tfrac{1}{2}\text{]}$$

since $|F_1''(z)| \leqslant \sqrt{2\pi}/4\$\mu^2[\mu^3]$

$$= \sqrt{2\pi}\ldots/4\mu^2 \text{ (illegible). [Should } 4\mu^2 \text{ be } 16\mu^3\text{?]}$$

To prove (e)

(This proof is missing from this copy of the typescript, but see Cohen and Mayhew, 1968, p. 695, Lemma 2, part (iii).)

Lemma 7.

$$\left|\frac{1}{\sqrt{2\pi}}\int_{-\infty}^{\infty} e^{-t/2}(\Pi(e^t)-M(t))f(t-t_0)\$ - t_0^{-1}\sum_{|\gamma|\leqslant 4\mu+50}(\tfrac{1}{2}+i\gamma)^{-1}F(\gamma)\exp(i\gamma t_0)\right|$$

is less than $\mu^{-1}t_0^{-3/2}$ *if* $t_0 > 10^4, 50 < \mu < 250$. [? +]

We shall show that with the notation of Lemma 4

(a) $|J| < 10^{-8}t_0^{-3/2}$,

(b) $|I_1| < 10^{-8}t_0^{-3/2}$,

(c) $|I_2| < 10^{-8}t_0^{-3/2}$,

(d)

$$\left|\sum_{|\gamma|<4\mu+50}\$\sqrt{2\pi}I_{3,\varrho} - t_0^{-1}\left(\tfrac{1}{2}+i\gamma\right)^{-1}F(\gamma)\exp(i\gamma t_0)\$\right| < \left(\frac{0.02}{\mu}+\frac{0.052}{\mu}\sum_{|\gamma|<4\mu+50}\gamma^{-1}\right)t_0.$$

[left-hand side: ? left and right brackets] [right-hand side: ?]

(e) $\left|\sum_{|\gamma|\geqslant 4\mu+50} I_{3,\varrho}\right| < 10^{-8}t_0^{-3/2}$.

By Lemma 3, $|g(0)| < \$4$. [?] Also $\zeta(0) = -\frac{1}{2}$, $\zeta(2) = \frac{1}{6}\pi^2$ whence $|J| < 16$ [true but useless; ?$|J| < 2\exp(-\frac{1}{4}t_0)$] and so (a) since $t_0 > 10^4$.

To prove (b), we shift the line of integration onto $\sigma = \frac{1}{4}$ and observe that on that line $|s^{-1}\log((s+1)(s+2)\zeta(s+2))| < 10$. Hence

$$|I_1| < 10e^{-t/4}\$\int_{1/4-i\infty}^{1/4+i\infty}\left|F\left(\$-i\left(s-\tfrac{1}{2}\right)\right)\right||ds|\left[\frac{1}{\sqrt{2\pi}}\right]\ [?+]$$

$$< 10\alpha\exp(-\tfrac{1}{4}t_0+\tfrac{1}{32}\alpha^2) \text{ by Lemma 6(e)}$$

$$= \$\tfrac{1}{2}t_0^{-1/2}\exp(t_0(-\tfrac{1}{4}+\tfrac{1}{2800})) < 10^{-8}t_0^{-3/2}.\ [200]$$

To estimate I_2, we first consider the behaviour of $\log(\zeta(s)/((s+1)(s+2)\zeta(s+2)))$. By the functional equations of the zeta- and gamma-functions we have

$$\frac{\zeta(s)}{(s+1)(s+2)\zeta(s+2)} = -\frac{1}{(2\pi)^2}\left(\frac{s}{s+2}\right)\frac{\zeta(1-s)}{\zeta(-1-s)},$$

and therefore, if $\mathcal{R}s \leqslant -1$, [? − 3]

$$\left|\log\frac{\zeta(s)}{(s+1)(s+2)\zeta(s+2)}\right| \leqslant \pi + 2\log\$\pi + 2 + \left|\log\frac{s}{s+2}\right| < 11.\ [?2]$$

We also have $|g(s)| < 4$ [caret] by Lemma 3(b) and therefore, using Lemma 6(c),

$$|I_2| < 15\$\alpha \exp\left(\frac{\Delta^2}{2\alpha^2} - \Delta t_0\right) = 15\$ \ \alpha \exp(-\mathrm{i}\Delta t_0) \ [\exp(0.1(1-\sigma)), \text{ twice}]$$

$$\$ = 300 t_0^{-1/2} \exp(-200 t_0). \quad [\text{less than?}]$$

(the ‘2’ in ‘200’ is also queried)

It remains to estimate $I_{3,\varrho}$. Since $\mathcal{R}\varrho = \frac{1}{2}$ for $|I\varrho| < 4$ we may put

$$\frac{1}{\sqrt{2\pi}} I_{3,\varrho} = \varrho^{-t_0} F(\gamma) \exp(\mathrm{i}\gamma t_0) + K_{3,\varrho} + L_{3,\varrho} + M_{3,\varrho},$$

$$[?F(-\gamma) \ (= F(\gamma) \text{ because } F \text{ even})]$$

where

$$K_{3,\varrho} = -\int_{\varrho-\infty}^{\$\varrho} \varrho^{-1} F(\gamma) \exp((s-\mathrm{i})t_0) ds = -\varrho^{-t_0} F(\gamma) \exp(\mathrm{i}\gamma t_0 - 2t_0), \quad [?\varrho - 2]$$

$$L_{3,\varrho} = \int_{\varrho-2}^{\varrho} \varrho^{-1} (F(-\mathrm{i}(s-\tfrac{1}{2})) - F(-\gamma)) \exp((s-\tfrac{1}{2})t_0) ds,$$

$$M_{3,\varrho} = \int_{\varrho-2}^{\varrho} F(-\mathrm{i}(s-\tfrac{1}{2})) \ldots \exp((s-\tfrac{1}{2})t_0) ds. \quad (\text{illegible})$$

Then

$$|K_{3,\varrho}| \leqslant \gamma^{-t_0} \exp(-2t_0) |F(\gamma)| \leqslant (20\gamma t_0^{1/2})^{-1} \exp(-2t_0) < 0.001 \gamma^{-1} t_0^{-3/2} \mu^{-1}$$

[these three occurrences of the symbol γ are queried (not $F(\gamma)$)] and

$$|M_{3,\varrho}| \leqslant \gamma^{-2} 2\alpha^{-1} \mu^{-1} \int_0^2 x \exp\left(\frac{x^2}{2\alpha^2} - x t_0\right) \quad [\text{caret}]$$

[the 2 and μ^{-1} outside the integral are also queried]

$$\leqslant \frac{2\mu^{-1}}{\alpha\gamma^2} \int_0^2 x \ \exp(-\tfrac{1}{2} x t_0) dx < \frac{0.4 t_0^{-3/2}}{\gamma^2 \mu},$$

[O.K. but not by L3; see Ingham, 1932, notes 22, 23, ...]

$$\sum_{|\gamma| < 4\mu + 50} |M_{3,\varrho}| < \mu^{-1} 0.02 t_0^{-3/2}$$

since $\sum \gamma^{-2} < \frac{1}{20}$.

By an integration by parts, if $|\gamma| < 4\mu + 50$,

$$|L_{3,\varrho}| = (t_0|\varrho|)^{-1}\exp(\$ - 2t_0)(F(-\gamma) - F(-\gamma\$ + 2\mathrm{i}))\$$$

$$+\mathrm{i}\int_{\varrho-2}^{\varrho} F'(\$ - \mathrm{i}(s - \tfrac{1}{2}))\exp((s - \tfrac{1}{2})t_0)ds$$

$$[(\mathrm{i}\gamma - 2)t_0]\ [-?]\ [-?]\ [+]$$

$$\leqslant (t_0\$\gamma)^{-1}\left(2\alpha^{-1}\exp\left(\frac{2}{\alpha^2} - 2t_0\right)\right.$$

$$\left.+\alpha^{-1}\mu^{-1}\int_0^2 \exp\left(\frac{x^2}{2\alpha^2} - xt_0\right)dx\right) \quad [|\gamma|]$$

$$\leqslant (t_0\alpha\gamma)^{-1}\left(2\exp\left(t_0\left(-2 + \tfrac{1}{200}\right)\right) + \frac{1.01}{\mu t_0}\right) \quad [|\gamma| \text{ again.}]$$

$$\leqslant 0.051\gamma^{-1}t_0^{-3/2}\mu^{-1}.$$

(d) now follows by collection of results.

For the case $|\gamma| \geqslant 4\mu + 50$, we have, by Lemma 6(b)

$$|I_{3,\varrho}| \leqslant 2\sqrt{2\pi}\alpha^{-1}\gamma^{-1}\exp(\tfrac{1}{2}t_0 + (2\alpha^2)^{-1}((\tfrac{5}{2})^2 - (\gamma - 4\mu)^2))$$

$$\leqslant \frac{6}{\alpha\gamma}\exp(t_0/?(400 + \tfrac{25}{4} - (\gamma - 4\mu)^2)). \quad \text{(illegible)}$$

But since $|\gamma| \geqslant 4\mu + 50$ and $\mu \geqslant 50$, we have

$$(\gamma - 4\mu)^2 > 50(|\gamma| - 4\mu) > 1250 + 25(|\gamma| - 4\mu)$$

and therefore

$$|I_{3,\varrho}| \leqslant \frac{6}{\alpha\gamma}\exp\left(-t_0 - \frac{c_0}{32}(|\gamma| - 4\mu)\right) < \frac{24\mu}{\alpha\gamma^2}\exp(-t_0)$$

for $\gamma\exp(-t_0\gamma/32)$ is a decreasing function for $\gamma > 32/t_0$, and $4\mu \geqslant 200 > 3/t_0$. Then

$$\sum_{|\gamma|\geqslant 4\mu+50} |I_{3,\varrho}| < 1.2\mu\alpha^{-1}\exp(-t_0) < \$0.01t_0^{-3/2} \text{ since } \mu < 250.$$

[(e) says 10^{-8} (O.K. thus); but no relation to enunciation]

Lemma 8. *If* $t_0 > 10^4$ *and*

$$\int_{-\infty}^{\infty} e^{-t/2}(\Pi(e^t) - M(t))f(t - t_0)dt > 1.0025\frac{\pi}{\mu t_0}$$

where, as previously mentioned,

$$f(t) = \exp(-\tfrac{1}{2}\alpha^2t^2)\left(\frac{\sin\mu t}{\mu t}\right)^4, \quad \alpha^2 = \frac{400}{t_0}, \quad 50 < \mu < 250,$$

$$M(t) = \int t^{-1}e^t dt, \textit{ from } 0.1 \textit{ to } \max(t, 0.1),$$

then there is a t_1, $0.974\, t_0 < t_1 < 1.053\, t_0$, *such that*

$$\Pi(\exp t_1) - \text{li}\exp t_1 > 1.002\, t_1^{-1}\exp\left(\tfrac{1}{2}t_1\right).$$

We have

$$\frac{1}{\sqrt{2\pi}}\int_{-\infty}^{\infty}\left(\frac{\sin\mu t}{\mu t}\right)^4 dt = \frac{\sqrt{2\pi}}{2\mu},$$

$$\frac{1}{\sqrt{2\pi}}\int_{-\infty}^{\infty}t^2\left(\frac{\sin\mu t}{\mu t}\right)^4 dt = \frac{\sqrt{2\pi}}{(2\mu)^3} \quad [4\mu^3\ ?]$$

and therefore

$$\int_{-\infty}^{\infty}\exp\left(-\frac{1}{2}t\right)(\Pi(e^t) - M(t)) - \$1.002\, t^{-1}\frac{\exp(\mathrm{i}\alpha^2(t-t_0)^2)}{1-(1-t/t_0)^3}$$

$$\times\exp(-\mathrm{i}\alpha^2(t-t_0))\left(\frac{\sin(\mu(t-t_0))}{\mu(t-t_0)}\right)^4 dt > \frac{1.0025\pi}{\mu t_0} - \frac{2\pi}{\mu t_0} - \frac{\pi}{\dots} \quad \text{(illegible)}.$$

Hence, for some t_1,

$$\Pi(\exp t_1) - M(t_1) > 0.0025\, t_1^{-1}\exp(\mathrm{i}t_1 + \tfrac{1}{2}\alpha^2(t_1-t_0)^2),\ 1-\left(1-\frac{t}{t_0}\right)^3 > at_1^{-1}e^J,$$

where $a = \$\frac{3}{4}\ (1.005)\ \left[\frac{9}{8}\right]$ and $J = \frac{1}{2}t_1 + 200(t_1-t_0)^2/t_0$. Now we certainly have $t_1 > 0.1$ for otherwise $\Pi(\exp t_1) = 0$ and $M(t_1) = 0$. But then

$$\Pi(\exp t_1) = \dots \quad \text{(illegible)} < \exp t_1 + t_1\exp\left(\tfrac{1}{2}t_1\right) < 2\exp t_1,$$

and therefore

$$\frac{200(t_1-t_0)^2}{t_0} < \tfrac{1}{2}t_1 + \log(3t_0).$$

Now, if $t_1 > t_0$, we have $\log(3t_0) < 0.01\, t_0 < 0.01\, t_1$ and hence

$$200(0.1275(t_1-t_0)^2) < 0.1275(t_1+t_0)^2, \quad t_1 < 1.053\, t_0.$$

But if $t_1 \leqslant t_0$, then

$$200(t_1-t_0)^2 < 0.051t_0^2, \quad t_1 > 0.974\, t_0.$$

It only remains to prove that $M(t)$\$<li e^t, and this will follow if li $e^{0.1} < 0$. [? >] But

$$\text{li}\, e^{0.1} = \int_{-\infty}^{-0.2}t^{-1}e^t dt + \int_{-0.2}^{-0.1}t^{-1}e^t dt + \int_0^{0.1}\frac{\$\sinh t}{t\, dt}$$

$$< e^{-0.2}\int_{-0.2}^{-0.1}t^{-1}dt + \$\sinh 0.1 < 0 \quad \text{[insert 2, twice]}.$$

Lemma 9.

$$\mathrm{li}\, x < \frac{x}{\log x - 1.5} \quad \textit{if } x \geqslant e^8.$$

We have

$$e^{-a}\mathrm{li}\, e^a = \mathcal{R}\int_0^\infty (a-t)^{-1}e^{-t}dt$$

if the integration is along a contour which avoids a and a is real [and positive]

$$= \mathcal{R}\int_0^\infty \left(a^{-1} + \frac{t}{a^2}\right)e^{-t}dt + \mathcal{R}\int_0^\infty \frac{t^2e^{-t}}{a^2(a-t)}dt.$$

By taking the contour $t = u(1+\frac{1}{2}\mathrm{i})$ where u is real,

$$\left|\int_0^\infty \frac{t^2e^{-t}}{a^2(a-t)}dt\right| \leqslant \sqrt{5}a^{-3}\int_0^\infty (\tfrac{1}{2}\sqrt{5}a)^{\$3}e^{-a}d\frac{5}{2}u = \frac{6.25}{a^3}, [2\ ?]$$

$$e^{-a}\mathrm{li}\, e^a \leqslant \frac{1}{a} + \frac{1}{a^2} + \frac{6.25}{a^3} \leqslant \frac{1}{a} + \frac{1.5}{a^2} + \frac{2.25}{a^3} \text{ if } a \geqslant 8$$
$$< (a-1.5)^{-1}$$

Lemma 10. *If either*

(a) $(\Pi(x) - \mathrm{li}\, x)x^{-1/2}\log x > 1.002$ *and* $x > e^{2000}$

or

(b) $(\Pi(x) - \mathrm{li}\, x)x^{-1/2}\log x > 1.6$ *and* $x > e^{\$16}$,

[16 queried] *then either* $\pi(x) > \mathrm{li}\, x$ *or* $\pi(x^{1/2}) > \mathrm{li}\, x^{1/2}$.

We begin by estimating $\Pi(x) - \pi(x) - \frac{1}{2}\pi(x^{1/2})$ for $x > 16$.

$$\Pi(x) - \pi(x) - \frac{1}{2}\pi(x^{1/2}) = \frac{1}{3}\pi(x^{1/3}) + \sum r^{-1}\pi(x^{1/r})$$
$$\leqslant \frac{1}{9}x^{1/3} + \frac{2}{3} + \sum\frac{2}{r} + \sum\frac{x^{1/r}}{3r},$$

summation from $r = 4$ to $\log_2 x$ (since $\pi(u) < 2$ if $0 < u < e$ and $\pi(u) < 2 + \frac{1}{3}u$ if $e < u$). [O.K. but obscure; $\pi(u) < 2$ used in terms with $\log x < r \leqslant \log_2 x$]

$$\leqslant \frac{1}{9}x^{1/3} + 2\log\left(\frac{1}{2}\log_2 x\right) + \frac{4}{3}x^{1/4}\sum_{r=4}^\infty r^{-2}$$

(since $rx^{1/r}$ decreases with increasing r in the range $4 \leqslant r \leqslant \log x$)

$$\leqslant \frac{1}{9}x^{1/3} + 2\log\log x + 0.4x^{1/4}.$$

Then if $\pi(x)\$ > \mathrm{li}\, x$ and $\pi(x^{1/2})\$ > \mathrm{li}\, x^{1/2}$, [? $\leqslant$]

$$(\Pi(x) - \mathrm{li}\,x)x^{-1/2}\log x < \frac{1}{2}x^{-1/2}\,\mathrm{li}\,x^{1/2}\log x$$

$$+\frac{1}{9}x^{-1/6}\log x + 0.4x^{-1/4}\log x + 2(\log\log x)\log x,$$

$$\tfrac{1}{2}x^{-1/2}\,\mathrm{li}\,x^{1/2}\log x < \left(1 - \frac{3}{\log x}\right)^{-1} \quad [x \geqslant e^{16}] \text{ by Lemma 9}$$

and this denies both (a) and (b).

Theorem 1. *If* $50 < \mu < 250$, $10^4 < t_0$, $\sum_{|\gamma|<4\mu+50}\gamma^{-1} < 3$ *and*

$$\mathcal{R}\$ \sum_{0<\gamma<4\mu+50} \frac{K(\gamma)}{\frac{1}{2}+\mathrm{i}\gamma}\exp(\mathrm{i}\gamma t_0)\$ > 0.502\mu^{-1}\left(\frac{1}{2}\pi\right)^{1/2},$$

[left bracket and minus sign] [right bracket]
then there exists t_1, $0.974\, t_0 < t_1 < 1.053\, t_0$ *such that elther*

$$\pi(\exp t_1) > \mathrm{li}\exp t_1 \quad \mathit{or} \quad \pi\left(\exp\frac{1}{2}t_1\right) > \mathrm{li}\ \exp\frac{1}{2}t_1.$$

$$t_0^{1/2}\left(0.021 + 0.052\sum_{|\gamma|<4\mu+50}\gamma^{-1}\right) < 0.177t_0^{-1/2} < 0.00177.$$

Hence

$$\frac{\mu t_0}{\pi}\int_{-\infty}^{\infty} e^{-t/2}(\Pi(e^t) - M(t))f(t-t_0)dt > 1.004 - 0.00177\left(\frac{2}{\pi}\right)^{1/2} > 1.00,$$

and the condition of Lemma 8 is satisfied, and consequently condition (a) of Lemma 10.

5. The Diophantine approximation

We have not until now made much use of the results of heavy computations on the zeta function. We have made use of the fact that for $|\gamma| < 1468$, the non-trivial zeros are all on the critical line, although we could well have avoided doing so. We shall now go further and make use of some information about the positions of the zeros in this range. Titchmarsh has mentioned [where?] that if t_n is defined by the condition $\arg\Gamma(\frac{1}{4}+\frac{1}{2}\mathrm{i}t_n) = n\pi$, [def. $\theta(t) = \pi^{-1}\arg(\pi^{-1/2\mathrm{i}t}\Gamma\ (\frac{1}{4}+\frac{1}{2}$ $\mathrm{i}t))][\theta(t_n) = n-1, t_n > 0,\ n = 1, 2,\ \ldots]$ then if $0 < t_n < 1468$, we have $N(t_n) - n = -1,\ 0$ or 1, and also that if $N(t_n) - n\$ = 0$, [?] then $N(t_{n+1}) = n+1$ and $N(t_{n-1}) = n-1$. Also $\arg\Gamma(\frac{1}{4}+\frac{1}{2}$ $\mathrm{i}t)$ $[\theta(t)]$ is monotonic. $[t \geqslant \ldots]$ From these facts, it can easily be seen that if $0 < t < 1468$, then $\pi^{-1}\arg\Gamma(\frac{1}{4}+\frac{1}{2}\mathrm{i}t_n) < 2$. We also have $\pi^{-1}\arg\Gamma(\frac{1}{4}+\frac{1}{2}\mathrm{i}t) - t/(2\pi)\log(t/(2\pi)) - 1 < \frac{1}{4}$ for $t > 51$ and $|S^*(t)| < 2\frac{1}{4}$ for $t \leqslant 51$ whence $|S^*(t)| < 2\frac{1}{4}$ for $0 < t < 1468$. [the upper bound $\frac{1}{4}$ may be replaced by a much smaller constant; see Titchmarsh, 1935, p. 238 (ii).] Here, $S^*(t) = N(t) - t/(2\pi)\log(t/(2\pi)) - 1$. [$2\frac{1}{4}$ is queried.]

Lemma 12. *If $1 \leqslant h \leqslant 220$, then*

$$\left|\sum f\left(\frac{\gamma}{2\pi}\right) - \int_1^h f(v)\log v\,dv\right| < 2\frac{1}{4}\left(|f(h)| + [\operatorname{var} f]_1^h\right),$$

where the sum is over $2\pi < \gamma < 2\pi h$.

We make use of the inequality $|S^*(t)| < \frac{9}{4}$, $0 < t < 1468$ justified above, and also observe that $S^*(2\pi) = 0$. Then

$$\left|\sum f\left(\frac{\gamma}{2\pi}\right) - \int_1^h f(v)\log v dv\right| = \left|\int_1^h f(v) dS^*(2\pi v)\right|$$

$$= \left|f(h)S^*(2\pi h) - \int_1^h S^*(2\pi v) df(v)\right|$$

$$\leqslant 2\frac{1}{4}(f(h) + [\operatorname{var} f]_1^h).$$

In order to get a slightly better result, we shall use a modified form of Dirichlet's theorem.

Lemma 13. *Given real numbers $a_1, \ldots, a_m$, a positive real number τ and positive integers $n_1, \ldots, n_m$, we can find an integer r, $1 \leqslant r \leqslant \prod n_i$, such that for each i, $1 \leqslant i \leqslant m$, $\tau r a_i$ differs from an integer by at most n_i^{-1}.*

The proof is very similar to that of Dirichlet's theorem (Ingham, 1932, Theorem J), and the details will be left to the reader.

Lemma 14. *If we choose $\mu = 60\pi$ in the functions f, F we can find t_0 so that $e^{20} < t_0 < e^{660.9}$ and*

$$2S = \$ \sum_{|\gamma| < 4\mu + 50} \mu\left(\frac{2}{\pi}\right)^{1/2} F(\gamma)\left(\frac{1}{2} + i\gamma\right)^{-1} \exp(i\gamma t_0) > 1.004. \quad \text{[? minus]}$$

We have

$$S = \$ \sum_{0 < \gamma < 4\mu + 50} \mu\left(\frac{2}{\pi}\right)^{1/2}\left(\frac{1}{4} + \gamma^2\right)^{-1} F(\gamma)(\$ - \gamma\sin(\gamma t_0) + i\cos(\gamma t_0))$$

[? minus] [plus]

and we will put

$$S_1 = -\sum_{0 < \gamma < 4\mu + 50}\left(\gamma^2 + \frac{1}{4}\right)^{-1}\gamma\sin(\gamma t_0)\kappa\left(\frac{\gamma}{2\mu}\right),$$

$$S_2 = \$ \sum_{0 < \gamma < 4\mu + 50}\left(\gamma^2 + \frac{1}{4}\right)^{-1}\frac{1}{2}\cos(\gamma t_0)\kappa\left(\frac{\gamma}{2\mu}\right), \quad \text{[? minus]}$$

$$S_3 = -\sum_{0 < \gamma < 4\mu + 50}\gamma^{-1}\kappa\left(\frac{\gamma}{2\mu}\right)\sin(\gamma t_0).$$

Then

$$|S_3 - S_1| \leqslant \sum_{0<\gamma<4\mu+50} \left(\gamma\left(\gamma^2+\frac{1}{4}\right)\right)^{-1}\frac{1}{4}|\sin(\gamma t_0)| \leqslant \sum_{0<\gamma} \frac{1}{4\gamma^3} < 0.0004$$

and

$$|S - S_1 - S_2| \leqslant \sum |F(\gamma) - F_1(\gamma)|\mu\left(\frac{2}{\pi}\right)^{1/2} \quad \text{[range of summation?]}$$

$$\leqslant \frac{\alpha}{2\mu}\sum_{0<\gamma<4\mu+50} 1 < (20t_0^{-1/2})\frac{600}{120\pi} < 0.0035,\ t_0 \geqslant e^{20}-1.$$

We now choose τ so that for the first zero $i\gamma_0 + \frac{1}{2}$, $\tau\gamma_0/2\pi$ is an integer and $e^{20}-1 < \tau < e^{20}$. We then choose t_0 in accordance with Lemma 13, so that t_0 [query] is a multiple of τ and for each γ, $0 < \gamma < 120\pi, \gamma/(2\pi)(t_0 + \frac{1}{20})$ differs from an integer by at most $\gamma/(1920\pi)$. This t_0 can be found in the range

$$e^{20}+1 < t_0 + \frac{\pi}{2\beta} < (e^{20}+2)\prod_{\gamma_0<\gamma<120\pi}(1+1920\gamma^{-1}\pi).$$

$$\left[\text{should } \pi/(2\beta) \text{ be } \frac{1}{120}?\right]$$

Now

$$\log \prod_{2\pi<\gamma<120\pi}(1+1920\gamma^{-1}\pi) < \sum_{2\pi<\gamma<120\pi}\left(\log 1920\gamma^{-1}\pi + \frac{\gamma}{1800\pi}\right)$$

$$< \int_1^{60}\left(\log 960 - \log u + \frac{u}{900}\right)\log u\, du$$

$$+\frac{9}{4}\left(\log 960 + \frac{1}{900}\right) < 647 \quad [?672]$$

$$\log\left(1+1920\gamma_0^{-1}\pi\right) > 6.01,\ e^{20} < t_0 < e^{660.9}. \quad [686]$$

We have

$$S_3 \geqslant \$ - \sum_{0<\gamma<120\pi}\gamma^{-1}\kappa\left(\frac{\gamma}{120\pi}\right)\min_{|\eta|<1/8}\sin\left(\frac{(1+\eta)\gamma}{120}\right)$$

$$-\sum_{120\pi<\gamma<240\pi}\gamma^{-1}\kappa\left(\frac{\gamma}{120\pi}\right) \quad \text{[sign queried]}$$

and if we write

$$\varphi(v) = \begin{cases} (2\pi v)^{-1}\kappa\left(\frac{v}{60}\right)\min_{|\eta|<1/8}\sin\left(\frac{\pi v(1+\eta)}{60}\right) & \text{if } 0 < v < 60, \\ -(2\pi v)^{-1}\kappa\left(\frac{v}{60}\right), & \text{if } 60\$ < v\$ < 120 \quad [\leqslant] \end{cases}$$

we shall have by Lemma 12

$$S_3 \geqslant \sum \varphi\left(\frac{\gamma}{2\pi}\right) \quad (\text{over } 0 < \gamma < 240\pi)$$

$$\geqslant \int_1^{120} \varphi(v) \log v dv\$ + \tfrac{9}{4}(\text{var } \varphi)_1^{120}$$

$$\geqslant \int_1^{120} \varphi(v) \log v dv\$ + 0.0097. \quad [\text{signs queried}]$$

But by direct computation, we find

$$\int_1^{120} \varphi(v) \log v dv > 0.5080. \quad [\ldots\text{lse with}\ldots\text{correct}\ldots\text{p.9}]$$

Also

$$S_2 \geqslant 0.49 \sum_{0<\gamma<120\pi} \gamma^{-2}\kappa(\gamma) - \frac{1}{2} \sum_{120\pi<\gamma<240\pi} \gamma^{-2}\kappa(\gamma)$$

$$-(240)^{-2} \sum_{0<\gamma<120\pi} \kappa(\gamma) \max_{\cdots}\left(\frac{\sin(1+\eta)\gamma}{\ldots}\right) \Big/ \left(\frac{\gamma}{(2\ldots)}\right) \quad (\text{illegible}) > 0.008.$$

Hence,

$$S > 0.5080 + 0.003 - 0.0097 - 0.0004 - 0.0035 = 0.\ 5026. \quad [?]$$

We can now state our final result.

Theorem 2. *There is an x,..., such that* $\pi(x) > \operatorname{li} x$.

(Here, '...' represents one of two conditions, both of which are crossed out in the manuscript. They are

$$2 < x < \exp\exp a < 10^b, \quad b = 10^c,$$

where $(a, c) = (697, 303)$ or $(661, 287)$.) This follows from Theorem 1 and Lemma 14.

Our results up to this point have also been characterised by the extreme smallness of the remainder terms, and our chief concern has been to obtain *some* definite remainder with a relatively brief argument. From this point onwards, however, we shall be much more exacting.

6. Computational Diophantine approximation

If fairly accurate values of the γ's were available, it should be possible to find a value for t_0 by direct computation with a digital computer. It would be necessary first to obtain the first three hundred zeros or so to say seven places of decimals. This might involve ten to twenty hours of computation time. We should then choose 200 say, so that our sums would extend to 800. Owing to the small values of (symbol missing) beyond 700 we would not have calculated the zeros there. A reasonable method of procedure for the Diophantine approximation would be simply to try out successively all values of t_0 which make the value of $\sin(\gamma t_0)$ for the first zero equal to 1. Let us make a rough estimate of where, given reasonable freedom from bad luck, we might expect to find a solution on this basis. Let us assume that the sums of terms other than the first are independent and normally distributed. The standard deviation of the distribution of S is easily calculated to be

$$\left(\frac{1}{2}\sum_{\gamma\neq\gamma_0}\frac{\kappa(\gamma)}{\frac{1}{4}+\gamma^2}-\frac{\kappa(\gamma_0)}{\dots}\right)^{1/2}, \quad \text{(illegible)}$$

i.e., about 0.0091. We get a solution if the sum exceeds $\frac{1}{2}-1/\gamma_0$, i.e., 0.43. The probability of this on the normal distribution is about 1.3×10^{-6}, from which we may conclude that there is an even chance of finding a value by this method in the first 500,000 trials, i.e., with $t_0 < 220{,}000$, i.e., of establishing that there is an $x, 2 < x < \exp 220\,000$ for which $\pi(x) > \operatorname{li} x$.

7. Case where the Riemann hypothesis is ... false (positively?)

In order to complete our investigation, it would be as well to obtain some result that can be applied if the Riemann hypothesis is discovered to be false, by the ... (illegible) not on the critical line. If one is simply given that there is a zero in some large rectangle, not meeting the critical line (e.g., $\sigma > 0.53, 0 < t < 10^8$), it is not easy to prove any very satisfactory results about values of x for which $\pi(x) > \operatorname{li} x$. This is because of the possibility that there may be many other zeros near to the given one; they may be sufficiently near to have much nuisance value, but not near enough to give essentially the same effect as multiple zero. The present investigation ignores all these difficulties by postulating a zero $\beta_1+i\gamma_1$ off the critical line and at considerable distance from any other zeros of this kind. It seems very probable that the first zeros off the critical line that are computed will satisfy the conditions required. We shall again use Lemma 4, but this time we shall put

$$f(t)=\alpha\frac{1}{2}(1+\cos(\gamma_1 t))\exp\left(-\frac{1}{2}\alpha^2t^2\right), \quad \Delta=100,$$

and we therefore have

$$F(u)=\frac{1}{2}\exp\left(-\frac{u^2}{2\alpha^2}\right)+\frac{1}{4}\exp\left(-\frac{(u-\gamma_1)^2}{2\alpha^2}\right)+\frac{1}{4}\exp\left(-\frac{(u+\gamma_1)^2}{2\alpha^2}\right).$$

We shall need to have an upper bound for the number of zeros in a given range of t.

Lemma 15.

$$\left|\left(\frac{\Gamma'}{\Gamma}\right)(x+iy)-\log\left(x+iy-\frac{1}{2}\right)\right|<\frac{\pi}{y-1}.$$

We first obtain an inequality for $|(d^2/dz^2)\log\Gamma(z)|$, $z=x+iy$.

$$\left|\frac{d^2}{dz^2}\log\Gamma(z)\right|=\left|\sum_{n>0}(z-n)^{-2}\right|\leqslant\sum_{n=-\infty}^{\infty}(y^2+(n-x)^2)^{-1}$$
$$\leqslant\frac{2}{y^2}+\int_{-\infty}^{\infty}(y^2+u^2)^{-1}du.$$

Then

$$\left|\left(\frac{\Gamma'}{\Gamma}\right)(x+\mathrm{i}y) - \log\left(x+\mathrm{i}y-\frac{1}{2}\right)\right| = \left|\left(\frac{\Gamma'}{\Gamma}\right)(x+\mathrm{i}y) - \log\frac{\Gamma\left(x+\mathrm{i}y+\frac{1}{2}\right)}{\Gamma\left(x+\mathrm{i}y-\frac{1}{2}\right)}\right|$$

$$\leqslant \max_{Iz=y}\left|\frac{d^2}{dz^2}\log\Gamma(z)\right| \leqslant \frac{\pi}{y-1}.$$

Lemma 16. *Denoting the number of zeros of the zeta function with positive imaginary parts less than t by N(t),*

$$N\left(t+\frac{3}{2}\right) - N\left(t-\frac{3}{2}\right) \leqslant 1.6\log\frac{t+8}{2\pi} + 1.7$$

$$\sum\frac{\sigma-3}{(t-\gamma)^2+(\sigma-\beta)^2} = \mathcal{R}\sum(s-\varrho)^{-1}$$

$$= \frac{1}{2}\mathcal{R}\left(\frac{\Gamma'}{\Gamma}\right)\left(\frac{1}{2}s+1\right) - \frac{1}{2}\log\pi + \mathcal{R}\left(\frac{\zeta'}{\zeta}\right)(s) + \mathcal{R}(s-1)^{-1}$$

$$\leqslant \frac{1}{2}\log\left|\frac{s+1}{2\pi}\right| + \frac{\sigma-1}{t^2+(\sigma-1)^2} + \left(\frac{\zeta'}{\zeta}\right)(\sigma) + \frac{\pi}{2t-1}.$$

['2' in '$2t-1$' queried]

Taking $\sigma = 2$, $t \geqslant 10$, we have $(\zeta'/\zeta)(2) < 0.53$, and therefore,

$$\sum\frac{\sigma-\beta}{(t-\gamma)^2+(\sigma-\beta)^2} \leqslant \frac{1}{2}\log\frac{t+8}{2\pi} + 0.53.$$

Now if $\beta = \frac{1}{2}$, $|t-\gamma| > \frac{3}{2}$, then $(\sigma-\beta)/((t-\gamma)^2+(\sigma-\beta)^2) > \frac{1}{3} > \frac{102}{325}$ and if $\beta+\beta' = 1$, $|t-\gamma| < \frac{3}{2}$, then

$$\frac{\sigma-\beta}{(t-\gamma)^2+(\sigma-\beta)^2} + \frac{\sigma-\beta'}{(t-\gamma)^2+(\sigma-\beta)^2} \geqslant \frac{20}{32}.$$

Hence

$$N\left(t+\frac{3}{2}\right) - N\left(t-\frac{3}{2}\right) \leqslant \frac{325}{102}\left(\frac{1}{2}\log\frac{t+8}{2\pi} + 0.53\right)$$

$$\leqslant 1.6\log\frac{t+8}{2\pi} + 1.7.$$

This inequality is also clearly valid for $0 < t < 10$ for the left-hand side is the ... (remainder of sentence missing).

Theorem 3. *Suppose that* $\beta_1 + \mathrm{i}\gamma_1$, $\beta_1 > \frac{1}{2}$, $\gamma_1 > 0$, *is a zero of the zeta function and that for every other zero* $\beta + \mathrm{i}\gamma$ *either* $\beta = \frac{1}{2}$ *or* $|\gamma - \gamma_1| > 14$. *Then for some x,* $2 < x < (16\gamma_1)^a$, $a = 1.12/(\beta_1 - \frac{1}{2})$, *we have* $\pi(x) > \operatorname{li} x$.

With f, F defined as above we have

$$|F(x+\mathrm{i}y)| < \exp\left(\frac{y^2}{2\alpha^2}\right),$$

$$\int_{c-\mathrm{i}\infty}^{c+\mathrm{i}\infty} |F(\mathrm{i}s)||\mathrm{d}s| \leqslant (2\pi)^{1/2}\exp\left(\frac{c^2}{2\alpha^2}\right),$$

and, with the notation of Lemma 4, may prove that $|J|$, $|I_1|$, $|I_2|$ are each less than $10^{-8}t_0^{-3/2}$ as in Lemma 7. We now proceed to the estimation of $I_{3,\varrho}$. If $\varrho = \beta + \mathrm{i}\gamma$ and $0 < \mathcal{R}a < 1$

$$\left|\int_{\varrho-2}^{\varrho} \varrho^{-1}\exp\left(\left(s-\frac{1}{2}\right)t_0\right)\exp\left(\frac{(s-a)^2}{2\alpha^2}\right)ds\right|$$

$$\leqslant \gamma^{-1}\exp\left(\left(\beta-\frac{1}{2}\right)t_0 + (2\alpha^2)^{-1}\mathcal{R}(\varrho-a)^2\right)\int_{-2}^{0}\exp(ut_0)$$

$$+\frac{u}{\alpha^2}\mathcal{R}(\varrho-a)+\frac{u^2}{2\alpha^2}du$$

$$\leqslant \exp\left(\left(\beta-\frac{1}{2}\right)t_0 + \frac{1}{200}t_0\mathcal{R}(\varrho-a)^2\right)\gamma^{-1}\left(t_0+\frac{1}{200}t_0\mathcal{R}(\varrho-a-1)\right)^{-1}$$

$$\leqslant \frac{1.02}{\gamma t_0}\exp\left(\left(\beta-\frac{1}{2}\right)t_0+\frac{1}{200}t_0\mathcal{R}(\varrho-a)^2\right).$$

We shall deal separately with the zeros which are near to $\beta_1 + \mathrm{i}\gamma_1$ and those which are relatively far away. If $||\gamma| - |\gamma_1|| > 14$, then (since in any case $|\gamma| > 14$) we have

$$(|\gamma| - |\gamma_1|)^2 > 182 + ||\gamma| - |\gamma_1||, \quad |\gamma|^2 > 182 + |\gamma|,$$

and therefore,

$$|I_{3,\varrho}| < 1.02(\gamma t_0)^{-1}e^{t_0/2}\left(\frac{1}{2}\exp\left(-\frac{1}{200}t_0\gamma^2\right)\right.$$

$$\left.+\frac{1}{4}\exp\left(-\frac{1}{200}t_0(\gamma-\gamma_1)^2\right)+\frac{1}{4}\exp\left(-\frac{1}{200}t_0(\gamma+\gamma_1)^2\right)\right)$$

$$< 1.02(\gamma t_0)^{-1}\exp(-0.405\,t_0)\left(\frac{3}{4}\exp\left(-\frac{1}{200}t_0\gamma\right)\right.$$

$$\left.+\frac{1}{4}\exp\left(-\frac{1}{200}t_0||\gamma|-|\gamma_1||\right)\right)$$

We now suppose that $t_0 > 20$.

(The remaining three pages of the paper are hand-written.)

Then

$$\sum \gamma^{-1} \exp\left(-\tfrac{1}{200} t_0 ||\gamma| - |\gamma_1||\right) \leqslant \sum \gamma^{-1} \exp(-0.1||\gamma| - |\gamma_1||)$$

$$\leqslant \sum_n 1.6 \log\left(\tfrac{1}{2}(3 \ldots + 8)\right) \exp\left(-0.1\left(|3n - \gamma_1| - \tfrac{3}{2}\right)\right) \quad \text{by Lemma 15}$$

$$\leqslant 2\log(\tfrac{1}{2}(\gamma_1 + 8))(1 - \exp(-0.3))^{-1}$$

$$+2 \sum_{n > \gamma_1} \exp(-0.15|n - \gamma_1|) \max_{n > \gamma|} \log\left(\tfrac{1}{2}(3n + 8)\right) \exp(-0.15(n \ldots))$$

$$\leqslant 2\log\left(\tfrac{1}{2}(\gamma_1 + 8)\right),$$

$$\sum \gamma^{-1} \exp\left(-\tfrac{1}{200} t_0 \gamma\right) \leqslant \max_{u>0}\left(u \exp\left(-\tfrac{1}{200} t_0 u\right)\right) \sum \gamma^{-2} < \frac{2}{t_0},$$

$$\sum |I_{3,\varrho}| < 1.02 \exp(-0.405\, t_0)\left(\tfrac{3}{2} t_0^{-2} + 12 t_0^{-1}\right) \log\left(\tfrac{1}{2}(\gamma_1 + 8)\right) \ldots,$$

(sum over $||\gamma| - |\gamma_1|| > 14$)

$$\sum |I_{3,\varrho}| < 2\left(\frac{1.03}{\gamma_1 t_0}\right)\left(\sum 1\right) \text{ since } \gamma_1 > 1100$$

(first sum over $||\gamma| - |\gamma_1|| \leqslant 14$, second over $|\gamma - \gamma_1| < 14$)

$$< \frac{33}{\gamma_1 t_0} \log\left(\tfrac{1}{2}(\gamma_1 + 22)\right).$$

If $\varrho_1 = \beta_1 + i\gamma_1$

$$I_{\varrho_1,3} = \mathcal{R} \int_{-2}^{0} (4(\varrho_1 + u))^{-1} \exp\left(\left(\varrho_1 - \tfrac{1}{2} + u\right) t_0 + \tfrac{1}{2}\alpha^{-2}\left(\beta_1 - \tfrac{1}{2} + u\right)^2\right) du.$$

If $\sin(\gamma_1 t_0) = 1$ and $\gamma_1 > 40$, then

$$|\arg((\varrho_1 + u)^{-1} \exp(\varrho_1 t_0))| < \ldots,$$

$$\cos \arg((\varrho_1 + u)^{-1} \exp(\varrho_1 t_0)) \geqslant \frac{4}{4.05},$$

$$\left|\frac{\gamma_1}{\varrho_1 + u}\right| > \frac{4.05}{4.1},$$

$$I_{\varrho_1,3} \geqslant (4.1\gamma_1)^{-1}\exp\left(\left(\beta_1-\tfrac{1}{2}\right)t_0\right)\int_{-2}^{0}\exp\left(ut_0+\tfrac{1}{200}\left(\beta_1-\tfrac{1}{2}+u\right)^2 t_0\right)du$$

$$\geqslant (4.1\gamma_1)^{-1}\exp\left(\left(\beta_1-\tfrac{1}{2}\right)t_0\right)\int_{-2}^{0}\exp\left(\tfrac{1}{200}t_0\left(\beta_1-\tfrac{1}{2}\right)^2+u\left(t_0+\tfrac{1}{100}\left(\beta_1-\tfrac{1}{2}\right)\right)\right)du$$

$$=\exp\left(t_0\left(\beta_1-\tfrac{1}{2}+\tfrac{1}{200}\left(\beta_1-\tfrac{1}{2}\right)^2\right)\right)$$

$$\times\left(1-\exp\left(-2\left(t_0+\tfrac{1}{200}\left(\beta_1-\tfrac{1}{2}\right)\right)\right)\right)\left(4.1\gamma_1\left(t_0+\tfrac{1}{100}\left(\beta_1-\tfrac{1}{2}\right)\right)\right)^{-1}$$

$$\geqslant (4.2\gamma_1 t_0)^{-1}\exp\left(t_0\left(\beta_1-\tfrac{1}{2}+\tfrac{1}{200}\left(\beta_1-\tfrac{1}{2}\right)^2\right)\right),\quad \text{if } t_0>5.$$

Collecting results

$$t_0\int_{-\infty}^{\infty}e^{-t/2}(\Pi(e^t)-M(t))f(t-t_0)dt \geqslant$$

$$\geqslant -3.10^{-8}t_0^{-1/2}-1.02\exp(-0.405\,t_0)\left(\tfrac{3}{2}t_0^{-1}+12\log\ldots\right)$$

$$-\frac{33}{\gamma_1}\log\left(\tfrac{1}{2}(\gamma_1+22)\right)+(4.2\gamma_1)^{-1}\exp(t_0A),\quad \text{where}$$

$$A=\beta_1-\tfrac{1}{2}+\tfrac{1}{200}\left(\beta_1-\tfrac{1}{2}\right)^2.$$

We now choose t_0, so that $\sin(\gamma_1 t_0)=1$ and

$$0<t_0-\left(\beta_1-\tfrac{1}{2}\right)^{-1}\log(16\gamma_1)<\frac{2\pi}{\gamma_1}.$$

Since $\gamma_1>1468$, the condition $t_0>5$ is automatically satisfied, indeed $t_0>20$. Then $\exp\left(t_0\left(\beta_1-\tfrac{1}{2}\right)\right)>16\gamma_1$

$$t_0\int_{-\infty}^{\infty}e^{-t/2}(\Pi(e^t)-M(t))f(t-t_0)dt$$

$$\geqslant -10^{-8}-13\left(\log\left(\tfrac{1}{2}(\gamma_1+3)\right)\right)(16\gamma_1)^{-0.81}-\tfrac{33}{1468}\log\left(\tfrac{1}{2}(1490)\right)+3.8\geqslant 3.5,$$

$$\alpha \int_{-\infty}^{\infty} (e^{-t/2}(\Pi(e^t) - M(t))t_0 - 0.9\exp(0.4\alpha^2(t-t_0)^2))$$
$$\times \exp\left(-\tfrac{1}{2}\alpha^2(t-t_0)^2\right)\tfrac{1}{2}(1+\cos(\gamma_1(t-t_0)))dt$$
$$\geqslant 3.5 - 0.9\left(\tfrac{5}{2}\pi\right)^{1/2}\left(1+\exp\left(\frac{-\gamma_1^2}{10\alpha^2}\right)\right)$$
$$> 0, \text{ since } \frac{\gamma_1^2}{10\alpha^2} > \frac{1468^2 t_0}{1000} > 10.$$

Then for some t_1

$$t_0\left(\exp\left(-\tfrac{1}{2}t_1\right)\right)(\Pi(\exp t_1) - M(t_1)) > 0.9\exp(0.4\alpha^2(t_1-t_0)^2)$$
$$= 0.9\exp\left(4\sigma\frac{(t_1-t_0)^2}{t_0}\right),$$

$$40(t_1-t_0)^2 < \tfrac{1}{2}t_1t_0 + t_0\log(3t_0)$$

from which it follows that

$$0.8t_0 < t_1 < 1.12t_0,$$
$$\left(\exp\left(-\tfrac{1}{2}t_1\right)\right)(\Pi(\exp t_1) - M(t_1)) > 0.8t_1.$$

Applying Lemma 10

$$\pi(\exp t_1) > \mathrm{li}\exp t_1 \quad \text{or} \quad \pi\left(\exp\left(\tfrac{1}{2}t_1\right)\right) > \mathrm{li}\exp\left(\tfrac{1}{2}t_1\right),$$

i.e., there exists x, $2 < x < (16\gamma_1)^A$, where $A = 1.12/\left(\beta_1 - \frac{1}{2}\right)$, (such that) $\pi(x) > \mathrm{li}\,x$.

REFERENCES

(1) A.E. INGHAM, The distribution of prime numbers, Cambridge Mathematical Tracts, No. 30, 1932.

(2) J.E. LITTLEWOOD, Sur la distribution des nombres premiers, *Comptes Rendus* **158** (1914) 1869–1872.

(3) S. SKEWES, On the difference $\pi(x) - \mathrm{li}\,x$, *Proc. London Math. Soc.*
(This is the reference in Turing's paper. Presumably this is the paper of the same title, *J. London Math. Soc.* **8** (1933) 277–283.)

(4) E.C. TITCHMARSH,The zeros of the zeta-function, *Proc. Roy. Soc.* (A) **157** (1936) 261–263.

Solvable and Unsolvable Problems

(Science News 31, (1954), pp 7–23)

Gregory Chaitin recommends —

TURING'S SMALL GEM

This lovely paper by Turing beautifully illustrates Hilbert's remark in his 1900 Paris International Congress of Mathematicians paper on *Mathematical Problems* that one does not truly understand something until one can explain it to the first man that one meets on the street.[1] At a more philosophical level, note that Hilbert, Turing and computer programming formalisms are often taken as the justification for extreme formalism in presenting mathematics. Emil Post, on the contrary, insisted that the work of Gödel and Turing argued against formal axiomatics and in favor of a return to meaning and truth.[2] And here is Turing himself explaining the basic ideas behind his work without using any mathematical or programming formalism, in very clear, down to earth English with rather concrete imagery.

[1] My paraphrase. Hilbert actually states that 'An old French mathematician said: "A mathematical theory is not to be considered complete until you have made it so clear that you can explain it to the first man whom you meet on the street"'.

[2] See Post's (1941) remarks quoted at the end of Jeremy Gray's (2008) book *Plato's Ghost: The Modernist Transformation of Mathematics.*

SOLVABLE AND UNSOLVABLE PROBLEMS

A. M. TURING, F. R. S.

IF one is given a puzzle to solve one will usually, if it proves to be difficult, ask the owner whether it can be done. Such a question should have a quite definite answer, yes or no, at any rate provided the rules describing what you are allowed to do are perfectly clear. Of course the owner of the puzzle may not know the answer. One might equally ask, 'How can one tell whether a puzzle is solvable?', but this cannot be answered so straightforwardly. The fact of the matter is that there is *no* systematic method of testing puzzles to see whether they are solvable or not. If by this one meant merely that nobody had ever yet found a test which could be applied to any puzzle, there would be nothing at all remarkable in the statement. It would have been a great achievement to have invented such a test, so we can hardly be surprised that it has never been done. But it is not merely that the test has never been found. It has been proved that no such test ever can be found.

Let us get away from generalities a little and consider a particular puzzle. One which has been on sale during the last few years and has probably been seen by most of the readers of this article illustrates a number of the points involved quite well. The puzzle consists of a large square within which are some smaller movable squares numbered 1 to 15, and one empty space, into which any of the neighbouring squares can be slid leaving a new empty space behind it. One may be asked to transform a given arrangement of the squares into another by a succession of such movements of a square into an empty space. For this puzzle there is a fairly simple and quite practicable rule by which one can tell whether the transformation required is possible or not. One first imagines the transformation carried out according to a different set of rules. As well as sliding the squares into the empty space one is allowed to make moves each consisting of two interchanges, each of one pair of squares. One would, for instance, be allowed as one move to interchange the squares numbered 4 and 7, and also the squares numbered 3 and 5. One is permitted to use the same number in both pairs. Thus one may replace 1 by 2, 2 by 3, and 3 by 1 as a move because this is the same as interchanging first (1, 2) and then (1, 3). The original puzzle is solvable by sliding if it is solvable according to the new rules. It is not solvable by sliding if the required position can be reached by the new rules, together with a 'cheat' consisting of *one single* interchange of a pair of squares.* Suppose, for instance, that one is asked to get back to the standard position –

1	2	3	4
5	6	7	8
9	10	11	12
13	14	15	

from the position

10	1	4	5
9	2	6	8
11	3		15
13	14	7	12

*It would take us too far from our main purpose to give the proof of this rule: the reader should have little difficulty in proving it by making use of the fact that an odd number of interchanges can never bring a set of objects back to the position it started from.

One may, according to the modified rules, first get the empty square into the correct position by moving the squares 15 and 12, and then get the squares 1, 2, 3, ... successively into their correct positions by the interchanges (1, 10), (2, 10), (3, 4), (4, 5), (5, 9), (6, 10), (7, 10), (9, 11), (10, 11), (11, 15). The squares 8, 12, 13, 14, 15 are found to be already in their correct positions when their turns are reached. Since the number of interchanges required is even, this transformation is possible by sliding.[†] If one were required after this to interchange say square 14 and 15 it could not be done.

This explanation of the theory of the puzzle can be regarded as entirely satisfactory. It gives one a simple rule for determining for any two positions whether one can get from one to the other or not. That the rule is so satisfactory depends very largely on the fact that it does not take very long to apply. No mathematical method can be useful for any problem if it involves much calculation. It is nevertheless sometimes interesting to consider whether something is possible at all or not, without worrying whether, in case it *is* possible, the amount of labour or calculation is economically prohibitive. These investigations that are not concerned with the amount of work involved are in some ways easier to carry out, and they certainly have a greater aesthetic appeal. The results are not altogether without value, for if one has proved that there is no method of doing something it follows *a fortiori* that there is no practicable method. On the other hand, if one method has been proved to exist by which the decision can be made, it gives some encouragement to any one who wishes to find a workable method.

From this point of view, in which one is only interested in the question, 'Is there a systematic way of deciding whether puzzles of this kind are solvable?', the rules which have been described for the sliding-squares puzzle are much more special and detailed than is really necessary. It would be quite enough to say: 'Certainly one can find out whether one position can be reached from another by a systematic procedure. There are only a finite number of positions in which the numbered squares can be arranged (viz. 20922789888000) and only a finite number (2, 3, or 4) of moves in each position. By making a list of all the positions and working through all the moves, one can divide the positions into classes, such that sliding the squares allows one to get to any position which is in the same class as the one started from. By looking up which classes the two positions belong to one can tell whether one can get from one to the other or not.' This is all, of course, perfectly true, but one would hardly find such remarks helpful if they were made in reply to a request for an explanation of how the puzzle should be done. In fact they are so obvious that under such circumstances one might find them somehow rather insulting. But the fact of the matter is, that if one is interested in the question as put, 'Can one tell by a systematic method in which cases the puzzle is solvable?', this answer is entirely appropriate, because one wants to know if there is a systematic method, rather than to know of a good one.

The same kind of argument will apply for any puzzle where one is allowed to move certain 'pieces' around in a specified manner, provided that the total number of essentially different positions which the pieces can take up is finite. A slight variation on the argument is necessary in general to allow for the fact that in many puzzles some moves are allowed which one is not permitted to reverse. But one can still make a list of the positions, and list against these first the positions which can be reached from them in one move. One then adds the positions which are reached by two moves and so on until an increase in the number of moves does not give rise to any further entries. For instance, we can say at once that there is a method of deciding whether a patience can be got out with a given order of the cards in the pack: it is to be understood that there is only a finite number of places in which a card is ever to be placed on the table. It may be argued that one is permitted

[†]It can in fact be done by sliding successively the squares numbered 7, 14, 13, 11, 9, 10, 1, 2, 3, 7, 15, 8, 5, 4, 6, 3, 10, 1, 2, 6, 3, 10, 6, 2, 1, 6, 7, 15, 8, 5, 10, 8, 5, 10, 8, 7, 6, 9, 15, 5, 10, 8, 7, 6, 5, 15, 9, 5, 6, 7, 8, 12, 14, 13, 15, 10, 13, 15, 11, 9, 10, 11, 15, 13, 12, 14, 13, 15, 9, 10, 11, 12, 14, 13, 15, 14, 13, 15, 14, 13, 12, 11, 10, 9, 13, 14, 15, 12, 11, 10, 9, 13, 14, 15.

to put the cards down in a manner which is not perfectly regular, but one can still say that there is only a finite number of 'essentially different' positions. A more interesting example is provided by those puzzles made (apparently at least) of two or more pieces of very thick twisted wire which one is required to separate. It is understood that one is not allowed to bend the wires at all, and when one makes the right movement there is always plenty of room to get the pieces apart without them ever touching, if one wishes to do so. One may describe the positions of the pieces by saying where some three definite points of each piece are. Because of the spare space it is not necessary to give these positions quite exactly. It would be enough to give them to, say, a tenth of a millimetre. One does not need to take any notice of movements of the puzzle as a whole: in fact one could suppose one of the pieces quite fixed. The second piece can be supposed to be not very far away, for, if it is, the puzzle is already solved. These considerations enable us to reduce the number of 'essentially different' positions to a finite number, probably a few hundred millions, and the usual argument will then apply. There are some further complications, which we will not consider in detail, if we do not know how much clearance to allow for. It is necessary to repeat the process again and again allowing successively smaller and smaller clearances. Eventually one will find that either it can be solved, allowing a small clearance margin, or else it cannot be solved even allowing a small margin of 'cheating' (i.e. of 'forcing', or having the pieces slightly overlapping in space). It will, of course, be understood that this process of trying out the possible positions is not to be done with the physical puzzle itself, but on paper, with mathematical descriptions of the positions, and mathematical criteria for deciding whether in a given position the pieces overlap, etc.

These puzzles where one is asked to separate rigid bodies are in a way like the 'puzzle' of trying to undo a tangle, or more generally of trying to turn one knot into another without cutting the string. The difference is that one is allowed to bend the string, but not the wire forming the rigid bodies. In either case, if one wants to treat the problem seriously and systematically one has to replace the physical puzzle by a mathematical equivalent. The knot puzzle lends itself quite conveniently to this. A knot is just a closed curve in three dimensions nowhere crossing itself; but, for the purpose we are interested in, any knot can be given accurately enough as a series of segments in the directions of the three coordinate axes. Thus, for instance, the trefoil knot (Figure 1*a*) may be regarded as consisting of a number of segments joining the points given, in the usual (x, y, z) system of coordinates, as (1, 1, 1), (4, 1, 1,), (4, 2, 1), (4, 2, −1), (2, 2, −1), (2, 2, 2), (2, 0, 2), (3, 0, 2), (3, 0, 0), (3, 3, 0), (1, 3, 0), (1, 3, 1) and returning again with a twelfth segment to the starting point (1, 1, 1). This representation of the knot is shown in perspective in Figure 1*b*. There is no special virtue in the representation which has been chosen. If it is desired to follow the original curve more closely a greater number of segments must be used. Now let *a* and *d* represent unit steps in the positive and negative X-directions respectively, *b* and *e* in the Y-directions, and *c* and *f* in the Z-directions: then this knot may be described as *aaabffddccceeaffbbbddcee*. One can then, if one wishes, deal entirely with such sequences of letters. In order that such a sequence should represent a knot it is necessary and sufficient that the numbers of *a*'s and *d*'s should be equal, and likewise the number of *b*'s equal to the number of *e*'s and the number of *c*'s equal to the number of *f*'s, and it must not be possible to obtain another sequence of letters with these properties by omitting a number of consecutive letters at the beginning or the end or both. One can turn a knot into an equivalent one by operations of the following kinds—

(i) One may move a letter from one end of the row to the other.
(ii) One may interchange two consecutive letters provided this still gives a knot.
(iii) One may introduce a letter *a* in one place in the row, and *d* somewhere else, or *b* and *e*, or *c* and *f*, or take such pairs out, provided it still gives a knot.
(iv) One may replace *a* everywhere by *aa* and *d* by *dd* or replace each *b* and *e* by *bb* and *ee* or each *c* and *f* by *cc* and *ff*. One may also reverse any such operation.

—and these are all the moves that are necessary.

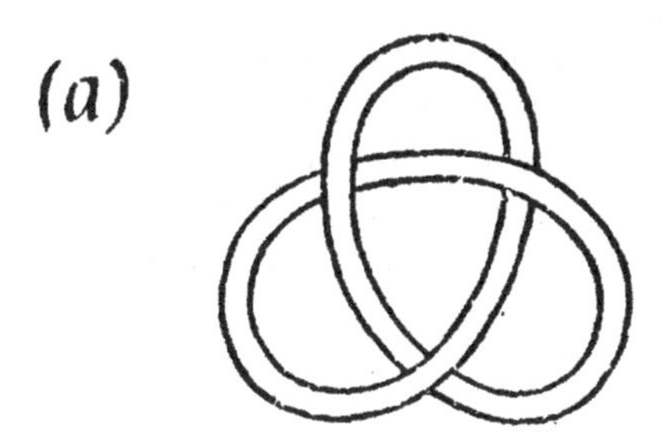

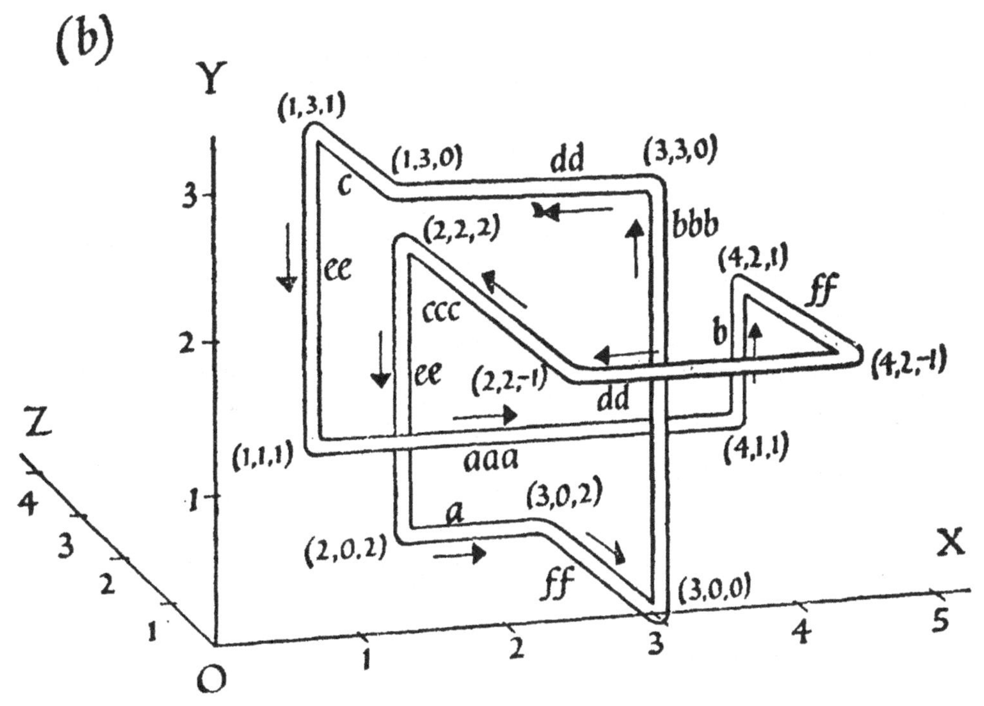

Fig. 1: (*a*) The trefoil knot (*b*) a possible representation of this knot as a number of segments joining points.

These knots provide an example of a puzzle where one cannot tell in advance how many arrangements of pieces may be involved (in this case the pieces are the letters a, b, c, d, e, f), so that the usual method of determining whether the puzzle is solvable cannot be applied. Because of rules (iii) and (iv) the lengths of the sequences describing the knots may become indefinitely great. No systematic method is yet known by which one can tell whether two knots are the same.

It is also possible to give a similar symbolic equivalent for the problem of separating rigid bodies, but it is less straightforward than in the case of knots.

These knots provide an example of a puzzle where one cannot tell in advance how many arrangements of pieces may be involved (in this case the pieces are the letters a, b, c, d, e, f), so that the usual method of determining whether the puzzle is solvable cannot be applied. Because of rules (iii) and (iv) the lengths of the sequences describing the knots may become indefinitely great. No systematic method is yet known by which one can tell whether two knots are the same.

Another type of puzzle which we shall find very important is the 'substitution puzzle'. In such a puzzle one is supposed to be supplied with a finite number of different kinds of counters, perhaps just black (B) and white (W). Each kind is in unlimited supply. Initially a number of counters are arranged in a row and one is asked to transform it into another pattern by substitutions. A finite list of the substitutions allowed is given. Thus, for instance, one might be allowed the substitutions

$$\text{(i)}\ \ WBW \to B$$

$$\text{(ii)}\ \ BW \to WBBW$$

and be asked to transform WBW into $WBBBW$, which could be done as follows

$$WBW \underset{\text{(ii)}}{\to} WWBBW \underset{\text{(ii)}}{\to} WWBWBBW \underset{\text{(i)}}{\to} WBBBW$$

Here the substitutions used are indicated by the numbers below the arrows, and their effects by underlinings. On the other hand if one were asked to transform WBB into BW it could not be done, for there are no admissible steps which reduce the number of B's.

It will be seen that with this puzzle, and with the majority of substitution puzzles, one cannot set any bound to the number of positions that the original position might give rise to.

It will have been realized by now that a puzzle can be something rather more important than just a toy. For instance the task of proving a given mathematical theorem within an axiomatic system is a very good example of a puzzle.

It would be helpful if one had some kind of 'normal form' or 'standard form' for describing puzzles. There is, in fact, quite a reasonably simple one which I shall attempt to describe. It will be necessary for reasons of space to take a good deal for granted, but this need not obscure the main ideas. First of all we may suppose that the puzzle is somehow reduced to a mathematical form in the sort of way that was used in the case of the knots. The position of the puzzle may be described, as was done in that case, by sequences of symbols in a row. There is usually very little difficulty in reducing other arrangements of symbols (e.g. the squares in the sliding squares puzzle) to this form. The question which remains to be answered is, 'What sort of rules should one be allowed to have for rearranging the symbols or counters?' In order to answer this one needs to think about what kinds of processes ever do occur in such rules, and, in order to reduce their number, to break them up into simpler processes. Typical of such processes are counting, copying, comparing, substituting. When one is doing such processes, it is necessary, especially if there are many symbols involved, and if one wishes to avoid carrying too much information in one's head, either to make a number of jottings elsewhere or to use a number of marker objects as well as the pieces of the puzzle itself. For instance, if one were making a copy of a row of counters concerned in the puzzle it would be as well to have a marker which divided the pieces which have been copied from those which have not and another showing the end of the portion to be copied. Now there is no reason why the rules of the puzzle itself should not be expressed in such a way as to take account of these markers. If one does express the rules in this way they can be made to be just substitutions. This means to say that the *normal form for puzzles is the substitution type of puzzle*. More definitely we can say:

Given any puzzle we can find a corresponding substitution puzzle *which is equivalent to it in the sense that given a solution of the one we can easily use it to find a solution of the other. If the original puzzle is concerned with rows of pieces of a finite number of different kinds, then the substitutions may be applied as an alternative set of rules to the pieces of the original puzzle. A transformation can be carried out by the rules of the original puzzle if and only if it can be carried out by the substitutions and leads to a final position from which all marker symbols have disappeared.*

This statement is still somewhat lacking in definiteness, and will remain so. I do not propose, for instance, to enter here into the question as to what I mean by the word 'easily'. The statement is moreover one which one does not attempt to prove. Propaganda is more appropriate to it than proof, for its status is something between a theorem and a definition. In so far as we know *a priori* what is a puzzle and what is not, the statement is a theorem. In so far as we do not know what puzzles are, the statement is a definition which tells us something about what they are. One can of course define a puzzle by some phrase beginning, for instance, 'A set of definite rules . . . ', but this just throws us back on the definition of 'definite rules'. Equally one can reduce it to the definition of 'computable function' or systematic procedure'. A definition of any one of these would define all the rest. Since 1935 a number of definitions have been given, explaining in detail the meaning of one or other of these terms, and these have all been proved equivalent to one another and also equivalent to the above statement. In effect there is no opposition to the view that every puzzle is equivalent to a substitution puzzle.

After these preliminaries let us think again about puzzles as a whole. First let us recapitulate. There are a number of questions to which a puzzle may give rise. When given a particular task one may ask quite simply

(a) *Can this be done*?

Such a straightforward question admits only the straightforward answers, 'Yes' or 'No', or perhaps 'I don't know'. In the case that the answer is 'Yes' the answerer need only have done the puzzle himself beforehand to be sure. If the answer is to be 'No', some rather more subtle kind of argument, more or less mathematical, is necessary. For instance, in the case of the sliding squares one can state that the impossible cases *are* impossible because of the mathematical fact that an odd number of simple interchanges of a number of objects can never bring one back to where one started. One may also be asked

(b) *What is the best way of doing this*?

Such a question does not admit of a straightforward answer. It depends partly on individual differences in people's ideas as to what they find easy. If it is put in the form, 'What is the solution which involves the smallest number of steps?', we again have a straightforward question, but now it is one which is somehow of remarkably little interest. In any particular case where the answer to (a) is 'Yes' one can find the smallest possible number of steps by a tedious and usually impracticable process of enumeration, but the result hardly justifies the labour.

When one has been asked a number of times whether a number of different puzzles of similar nature can be solved one is naturally led to ask oneself

(c) *Is there a systematic procedure by which I can answer these questions, for puzzles of this type*?

If one were feeling rather more ambitious one might even ask

(d) *Is there a systematic procedure by which one can tell whether a puzzle is solvable*?

I hope to show that the answer to this last question is 'No'. There are in fact certain types of puzzle for which the answer to (c) is 'No'.

Before we can consider this question properly we shall need to be quite clear what we mean by a 'systematic procedure' for deciding a question. But this need not now give us any particular difficulty. A 'systematic procedure' was one of the phrases which we mentioned as being equivalent to the idea of a puzzle, because either could be reduced to the other. If we are now clear as to what a puzzle is, then we should be equally clear about 'systematic procedures'. In fact a systematic procedure is just a puzzle *in which there is never more than one possible move in any of the positions which arise and in which some significance is attached to the final result.*

Now that we have explained the meaning both of the term 'puzzle' and of 'systematic procedure', we are in a position to prove the assertion made in the first paragraph of this article, that there cannot be any systematic procedure for determining whether a puzzle be solvable or not. The proof does not really require the detailed definition of either of the terms, but only the relation between them which we have just explained. Any systematic procedure for deciding whether a puzzle were solvable could certainly be put in the form of a puzzle, with unambiguous moves (i.e. only one move from any one position), and having for its starting position a combination of the rules, the starting position and the final position of the puzzle under investigation.

The puzzle under investigation is also to be described by its rules and starting position. Each of these is to be just a row of symbols. As we are only considering substitution puzzles, the rules need only be a list of all the substitution pairs appropriately punctuated. One possible form of punctuation would be to separate the first member of a pair from the second by an arrow, and to separate the different substitution pairs with colons. In this case the rules

B may be replaced by BC
WBW may be deleted

would be represented by '$: B \rightarrow BC : WBW \rightarrow :$'. For the purposes of the argument which follows, however, these arrows and colons are an embarrassment. We shall need the rules to be expressed without the use of any symbols which are barred from appearing in the starting positions. This can be achieved by the following simple, though slightly artificial trick. We first double all the symbols other than the punctuation symbols, thus '$: BB \rightarrow BBCC : WWBBWW \rightarrow :$'. We then replace each arrow by a single symbol, which must be different from those on either side of it, and each colon by three similar symbols, also chosen to avoid clashes. This can always be done if we have at least three symbols available, and the rules above could then be represented as, for instance, '*CCCBBWBBCCBBBWWBBWWBWWW*'. Of course according to these conventions a great variety of different rows of symbols will describe essentially the same puzzle. Quite apart from the arbitrary choice of the punctuating symbols the substitution pairs can be given in any order, and the same pair can be repeated again and again.

Now let $P(R,S)$ stand for 'the puzzle whose rules are described by the row of symbols R and whose starting position is described by S'. Owing to the special form in which we have chosen to describe the rules of puzzles, there is no reason why we should not consider $P(R,R)$ for which the 'rules' also serve as starting position: in fact the success of the argument which follows depends on our doing so. The argument will also be mainly concerned with puzzles in which there is at most one possible move in any position; these may be called 'puzzles with unambiguous moves'. Such a puzzle may be said to have 'come out' if one reaches either the position B or the position W, and the rules do not permit any further moves. Clearly if a puzzle has unambiguous moves it cannot both come out with the end result B and with the end result W.

We now consider the problem of classifying rules R of puzzles into two classes, I and II, as follows:

Class I is to consist of sets R of rules, which represent puzzles with unambiguous moves, and such that $P(R,R)$ comes out with the end result W.

Class II is to include all other cases, i.e. either $P(R,R)$ does not come out, or comes out with the end result B, or else R does not represent a puzzle with unambiguous moves. We may also, if we wish, include in this class sequences of symbols such as $BBBBB$ which do not represent a set of rules at all.

Now suppose that, contrary to the theorem that we wish to prove, we have a systematic procedure for deciding whether puzzles come out or not. Then with the aid of this procedure we shall be able to distinguish rules of class I from those of class II. There is no difficulty in deciding whether R really represents a set of rules, and whether they are unambiguous. If there is any difficulty it lies in finding the end result in the cases where the puzzle is known to come out: but this can be decided by actually working the puzzle through. By a principle which has already been explained, this systematic procedure for distinguishing the two classes can itself be put into the form of a substitution puzzle (with rules K, say). When applying these rules K, the rules R of the puzzle under investigation form the starting position, and the end result of the puzzle gives the result of the test. Since the procedure always gives an answer, the puzzle $P(K,R)$ always comes out. The puzzle K might be made to announce its results in a variety of ways, and we may be permitted to suppose that the end result is B for rules R of class I, and W for rules of class II. The opposite choice would be equally possible, and would hold for a slightly different set of rules K', which however we do not choose to favour with our attention. The puzzle with rules K may without difficulty be made to have unambiguous moves. Its essential properties are therefore:

K has unambiguous moves.

$P(K,R)$ always comes out whatever R.

If R is in class I, then $P(K,R)$ has end result B.

If R is in class II, then $P(K,R)$ has end result W.

These properties are however inconsistent with the definitions of the two classes. If we ask ourselves which class K belongs to, we find that neither will do. The puzzle $P(K,K)$ is bound to come out, but the properties of K tell us that we must get end result B if K is in class I and W if it is in class II, whereas the definitions of the classes tell us that the end results must be the other way round. The assumption that there was a systematic procedure for telling whether puzzles come out has thus been reduced to an absurdity.

Thus in connexion with question (c) above we can say that there are some types of puzzle for which no systematic method of deciding the question exists. This is often expressed in the form, 'There is no *decision procedure* for this type of puzzle', or again, 'The decision problem for this type of puzzle is unsolvable', and so one comes to speak (as in the title of this article) about 'unsolvable problems' meaning in effect puzzles for which there is no decision procedure. This is the technical meaning which the words are now given by mathematical logicians. It would seem more natural to use the phrase 'unsolvable problem' to mean just an unsolvable puzzle, as for example 'to transform 1, 2, 3 into 2, 1, 3 by cyclic permutation of the symbols', but this is not the meaning it now has. However, to minimize confusion I shall here always speak of 'unsolvable decision problems', rather than just 'unsolvable problems', and also speak of puzzles rather than problems where it is puzzles and not decision problems that are concerned.

It should be noticed that a decision problem only arises when one has an infinity of questions to ask. If you ask, 'Is this apple good to eat?', or 'Is this number prime?', or 'Is this puzzle solvable?' the question can be settled with a single 'Yes' or 'No'. A finite number of answers will deal with a question about a finite number of objects, such as the apples in a basket. When the number is infinite, or in some way not yet completed concerning say all the apples one may ever be offered, or all whole numbers or puzzles, a list of answers will not suffice. Some kind of rule or systematic procedure must be given. Even if the number concerned is finite one may still prefer to have a rule rather than a list: it may be easier to remember. But there certainly cannot be an unsolvable decision problem in such cases, because of the possibility of using finite list.

Regarding decision problems as being concerned with classes of puzzles, we see that if we have a decision method for one class it will apply also for any subclass. Likewise, if we have proved that there is no decision procedure for the subclass, it follows that there is none for the whole class. The most interesting and valuable results about unsolvable decision problems concern the smaller classes of puzzle.

Another point which is worth noticing is quite well illustrated by the puzzle which we considered first of all in which the pieces were sliding squares. If one wants to know whether the puzzle is solvable with a given starting position, one can try moving the pieces about in the hope of reaching the required end-position. If one succeeds, then one will have solved the puzzle and consequently will be able to answer the question, 'Is it solvable?' In the case that the puzzle is solvable one will eventually come on the right set of moves. If one has also a procedure by which, if the puzzle is unsolvable, one would eventually establish the fact that it was so, then one would have a solution of the decision problem for the puzzle. For it is only necessary to apply both processes, a bit of one alternating with a bit of the other, in order eventually to reach a conclusion by one or the other.

Actually, in the case of the sliding squares problem, we have got such a procedure, for we know that if, by sliding, one ever reaches the required final position, with squares 14 and 15 interchanged, then the puzzle is impossible.

It is clear then that the difficulty in finding decision procedures for types of puzzle lies in establishing that the puzzle is unsolvable in those cases where it *is* unsolvable. This, as was mentioned on page 16, requires some sort of mathematical argument. This suggests that we might try expressing the statement that the puzzle comes out in a mathematical form and then try and prove it by some systematic process. There is no particular difficulty in the first part of this project, the mathematical expression of the statement about the puzzle. But the second half of the project is bound to fail, because by a famous theorem of Gödel no systematic method of proving mathematical theorems is sufficiently complete to settle every mathematical question, yes or no. In any case we are now in a position to give an independent proof of this. If there were such a systematic method of proving mathematical theorems we could apply it to our puzzles and for each one eventually either prove that it was solvable or unsolvable; this would provide a systematic method of determining whether the puzzle was solvable or not, contrary to what we have already proved.

This result about the decision problem for puzzles, or, more accurately speaking, a number of others very similar to it, was proved in 1936–7. Since then a considerable number of further decision problems have been shown to be unsolvable. They are all proved to be unsolvable by showing that if they were solvable one could use the solution to provide a solution of the original one. They could all without difficulty be reduced to the same unsolvable problem. A number of these results are mentioned very shortly below. No attempt is made to explain the technical terms used, as most readers will be familiar with some of them, and the space required for the explanation would be quite out of proportion to its usefulness in this context.

(1) It is not possible to solve the decision problem even for substitution processes applied to rows of black and white counters only.
(2) There are certain particular puzzles for which there is no decision procedure, the rules being fixed and the only variable element being the starting position.
(3) There is no procedure for deciding whether a given set of axioms leads to a contradiction or not.
(4) The 'word problem in semi-groups with cancellation' is not solvable.
(5) It has recently been announced from Russia that the 'word problem in groups' is not solvable. This is a decision problem not unlike the 'word problem in semi-groups', but very much more important, having applications in topology: attempts were being made to solve this decision problem before any such problems had been proved unsolvable. No adequately complete proof is yet available, but if it is correct this is a considerable step forward.
(6) There is a set of 102 matrices of order 4, with integral coefficients such that there is no decision method for determining whether another given matrix is or is not expressible as a product of matrices from the given set.

These are, of course, only a selection from the results. Although quite a number of decision problems are now known to be unsolvable, we are still very far from being in a position to say of a given decision problem, whether it is solvable or not. Indeed, we shall never be quite in that position, for the question whether a given decision problem is solvable is itself one of the undecidable decision problems. The results which have been found are on the whole ones which have fallen into our laps rather than ones which have positively been searched for. Considerable efforts have however been made over the word problem in groups (see (5) above). Another problem which mathematicians are very anxious to settle is known as 'the decision problem of the equivalence of manifolds'. This is something like one of the problems we have already mentioned, that concerning the twisted wire puzzles. But whereas with the twisted wire puzzles the pieces are quite rigid, the 'equivalence of manifolds' problem concerns pieces which one is allowed to bend, stretch, twist, or compress as much as one likes, without ever actually breaking them or making new junctions or filling in holes. Given a number of interlacing pieces of plasticine one may be asked to transform them in this way into another given form. The decision problem for this class of problem is the 'decision problem for the equivalence of manifolds'. It is probably unsolvable, but has never been proved to be so. A similar decision problem which might well be unsolvable is the one concerning knots which has already been mentioned.

The results which have been described in this article are mainly of a negative character, setting certain bounds to what we can hope to achieve purely by reasoning. These, and some other results of mathematical logic may be regarded as going some way towards a demonstration, within mathematics itself, of the inadequacy of 'reason' unsupported by common sense.

FURTHER READING

Kleene, S. C. *Introduction to Metamathematics*, Amsterdam, 1952.

Examining the Work and Its Later Impact

Wilfried Sieg focuses on —

NORMAL FORMS FOR PUZZLES: A VARIANT OF TURING'S THESIS

If you have never read *Solvable and unsolvable problems*, you definitely should.[1] It will make for an intellectual encounter that is at first perhaps puzzling, but on reflection deeply rewarding. Turing presents his ideas on computability from an unusual perspective and puts them in a distinctive light. The essay is playful, yet serious; strange, yet familiar; informal, yet proves rigorously the unsolvability of a decision problem. What is most surprising is the fact that it discusses *The Thesis* without ever mentioning Turing's machine model of computation. Instead, it uses the underlying and more basic concept of an 'unambiguous substitution puzzle'. The latter serves for Turing as the mathematical notion corresponding to 'systematic procedure' and is modeled after Post's production systems. How can it do all those things, you may wonder. I will start in Section 1 by describing the mathematical set-up of substitution puzzles and show in Section 2 how this approach is in perfect harmony with the analysis Turing gave in his classical 1936 paper 'On computable numbers'. Section 3 points to generalisations of substitution puzzles that provide 'inductive' support for Turing's Thesis, whereas Section 4 discusses briefly a different way of addressing the central methodological problem.

1. Substitution puzzles

If we consider, as Turing does, ordinary puzzles like those concerning *sliding squares* or *knots*, then a particular type of question can be raised immediately, namely, is it possible to transform a given initial position into another configuration via the finitely many moves the puzzle allows? In the case of sliding squares, take this configuration as the initial one:

1		2	15
3	14	4	13
5	12	6	11
7	10	8	9

Is it possible to rearrange this configuration, exploiting the empty square and sliding the numbered squares, into the following one?

1	2	3	4
5	6	7	8
9	10	11	12
13	14	15	

[1] The essay is presumably Turing's last published paper and appeared in *Science News* 31 (1954), pp. 7–23. Editors' Note: Pages 322–331 in this volume.

Concerning puzzles for knots, Turing asks (on p. 12; p. 325 this volume), whether we can determine the equivalence of two knots via some specified rules. As a final example, Turing considers *substitution puzzles*. For these puzzles, one has an unlimited supply of counters, possibly only of two distinct kinds, say *B* (lack) and *W* (hite). An initial configuration is a finite sequence of such counters, and the puzzle task is to transform a given configuration into another one – by substitutions from a fixed finite list. Thus, Turing observes, a puzzle can be something more important than just a toy, as 'the task of proving a mathematical theorem within an axiomatic system is a very good example of a puzzle'. Indeed, it is an example of a substitution puzzle.

When facing a seemingly intractable puzzle, frustration might suggest the natural and general question, how one can tell whether this or any other puzzle is solvable. The answer to the question is given, in the end, by a mathematical theorem.

Theorem 1.1. *There is no systematic procedure for determining whether a puzzle is solvable or not.*

How are the concepts in the statement of the theorem defined so that it can be proved by a rigorous argument? – To prepare the answer to this question, Turing argues for two points: (i) positions of puzzles can be described by finite sequences of symbols, and (ii) rules for puzzles can be reduced to substitutions. That leads him to consider a *normal* or *standard form* for the description of puzzles and then to claim, 'the *normal form for puzzles is the substitution type of puzzle*'. The argument for (i) and (ii) takes, as Turing puts it, 'a great deal for granted'. In Section 2, I try to fill in some conceptually significant parts and show how the 1954-presentation wonderfully dovetails with the 1936-analysis of (mechanical) calculability. In the present essay, Turing asserts, furthermore, that when a puzzle is given, we can find a 'corresponding substitution puzzle that is equivalent to it in the sense that given a solution of the one we can easily find a solution of the other'.

The statement that any puzzle has a substitution puzzle as its normal form is, Turing understatedly admits, 'still somewhat lacking in definiteness, and will remain so'. From a contemporary perspective, it expresses an analogue of *Church's* or *Turing's Thesis* that is directly articulated for Post's concept of a production system. Turing views this not-quite-definite statement as being situated between a theorem and a definition:

> In so far as we know *a priori* what is a puzzle and what is not, the statement is a theorem. In so far as we do not know what puzzles are, the statement is a definition that tells us something about what they are. (p. 15; p. 327 this volume)

Turing claims that puzzles can be defined by appealing to a concept of 'a set of definite rules', 'computable function' or 'systematic procedure'. A definition of any one of theses concepts 'would define all the rest', as the definitions can be shown to be equivalent. Turing summarises the status of the methodological discussion among logicians when asserting, '... there is no opposition to the view that every puzzle is equivalent to a substitution puzzle'.

Mathematical logicians at the time were hardly moved by 'puzzles' but rather by the machine model that had been introduced by Turing in 1936 and, in a more human form, by Post in the same year. Let me briefly recall what a (two-letter) Turing machine is; in doing so I follow Post and Davis rather than Turing.[2] A *Turing machine* consists of a finite but potentially infinite tape. The tape is divided into squares, and each square may carry a symbol from a finite alphabet, say, just the two-letter alphabet consisting of **0** and **1** or, to emphasise the connection to substitution puzzles, *B* and *W*. The machine is able to scan one square at a time and perform, depending on the content of the observed square and its own internal state, one of four operations: print **0**, print **1**, or shift attention to one of the two immediately adjacent squares. The operation of the machine is fixed by a finite list of commands in the form of quadruples $\mathrm{q_i s_k c_l q_m}$ that express the following: If the machine is in

[2] That is, I am following the development in Post (1947) and Davis (1958).

internal state q_i and finds symbol s_k on the square it is scanning, then it is to carry out operation c_l and change its state to q_m. The deterministic character of the machine's operation is guaranteed by the requirement that a program must not contain two different quadruples with the same first two components.

Why does Turing (1954) expose his fundamental considerations on computability and unsolvability via Post's production systems that are renamed, in addition, as substitution puzzles? – There is an important rhetorical element, as the essay was to reach the readership of a popular science magazine, and the central issues certainly are more directly accessible to non-logicians when presented in this way. For example, to realise that proving a theorem in an axiomatic system can be viewed as solving a substitution puzzle is an easier task than to grasp the connection between theorem proving and computations of a two-letter machine! However, there are also substantive reasons. First, the reduction of substitution puzzles to Turing machines is not crucial for the central observations (and can easily be established). Secondly and conversely, production systems allow the concise description of Turing machine computations and are a marvelous tool for obtaining unsolvability results. Post recognised this possibility and exploited it (1947) to prove that the decision problem for Thue systems is unsolvable (using the undecidability of the halting problem). Based on this result, Post then proved the unsolvability of the word problem for semigroups. In his paper, Turing (1950) used Post's general approach and extended the unsolvability result from semigroups to semigroups with cancellation. Lastly, and that is an aspect I am going to explore in Section 3, substitution puzzles lend themselves readily to direct and directly meaningful generalisations.

2. Puzzles: analyzed

At the very beginning of his piece, Turing (1954) asserts that the unsolvability of the word problem for semigroups will be shown and continues:

> The method depends on reducing the unsolvability of the problem in question to a known unsolvable problem connected with the logical computing machines introduced by Post ([1936]) and the author ([Turing 1936]). In this we follow Post ([1947]) who reduced the problem of Thue to this same unsolvable problem. (p. 491)

Although Turing seems to point only to the structural similarity between the two-letter Turing machine and the worker in Post's (1936), there is a deeper coherence, as we will see. Let me first describe how Post's combinatory processes are generated by computation steps that are 'identical' with those of Turing's machines. These steps are taken by a human worker who operates in a *symbol space* consisting of a two-way infinite sequence of spaces or boxes.

> The problem solver or worker is to move and work in this symbol space, being capable of being in, and operating in but one box at a time. And apart from the presence of the worker, a box is to admit of but two possible conditions, i.e., being empty or unmarked, and having a single mark in it, say a vertical stroke.[3]

The worker can perform a number of *primitive acts*, namely, make a vertical stroke [V], erase a vertical stroke [E], move to the box immediately to the right [M_r] or to the left [M_l] of the box he is in, and determine [D] whether the box he is in is marked or not. In carrying out a particular combinatory process the worker begins in a special box, the *starting point*, and then follows directions from a finite, numbered sequence of instructions. The *i*-th instruction, *i* between 1 and *n*, is in one of the

[3] [Post 1936], p. 105. Post remarks, the infinite sequence of boxes can be replaced by a finite one that can be expanded as necessary.

following forms: (1) carry out act V, E, M_r or M_l and then follow direction j_i, (2) carry out act D and then, depending on whether the answer was positive or negative, follow direction j_i' or j_i''. (Post has a special stop instruction, but that can be replaced by stopping, conventionally, in case the number of the next direction is greater than *n*.) Are there intrinsic reasons for this formulation, except for its simplicity and Post's expectation that it will turn out to be equivalent to general recursiveness? Post writes at the very end of his paper:

> The writer expects the present formulation to turn out to be equivalent to recursiveness in the sense of the Gödel-Church development. Its purpose, however, is not only to present a system of a certain logical potency but also, in its restricted field, of psychological fidelity. In the latter sense wider and wider formulations are contemplated. On the other hand, our aim will be to show that all such are logically reducible to formulation 1. We offer this conclusion at the present moment as a *working hypothesis*. And to our mind such is Church's identification of effective calculability with recursiveness. (p. 291)

Investigating wider and wider formulations and reducing them to the above basic formulation would change, for Post, this 'hypothesis not so much to a definition or to an axiom but to a *natural law*'. In Section 3, I discuss a dramatically wider formulation involving K-graphs and corresponding 'substitution rules'.

As I mentioned, there actually is a deeper coherence between Post's work and Turing's. Post investigated, already in the 1920s, operations on finite strings and reduced canonical production systems to those in 'normal form'; that is reported in his work (1941).[4] It is surprising that Turing's (1936) analysis of calculability *leads* to exactly the same general approach and that the informal discussion (1954) dovetails with it in the most direct way. Let me briefly review Turing's analysis, emphasising at the very outset that he is concerned with human mechanical calculability.[5] He uses in his 1936-paper 'computer' for a human computing agent who proceeds mechanically; his machines, our Turing machines, consistently are just that, namely *machines*. Gandy suggested calling a human, who is carrying out a calculation, a 'computor' and using 'computer' to refer to some computing machine or other. Thus, with this new terminology, Turing analyses processes that computors can carry out, and that is exactly the extraordinary aspect of his approach: idealised computing agents are brought into the analysis of calculability. In addition, Turing formulates a crucial normative point when explicitly striving to isolate computor operations that are 'so elementary that it is not easy to imagine them further divided' (p. 136). Thus, it is obviously crucial that symbolic configurations relevant to fixing the circumstances for a computor's actions can be recognised *immediately* or *at a glance*.

Turing (1936) imagines a computor writing symbols on paper that is divided into squares 'like a child's arithmetic book'. Because the two-dimensional character of the paper is taken not to be an 'essential of computation' (p. 135), Turing takes a one-dimensional tape divided into squares as the basic computing space. (Note that a corresponding step was taken also for puzzles!) Because of this reductive step to a one-dimensional tape, we have to be concerned with the immediate recognisability of either individual symbols or sequences of symbols. In the first case, only finitely many distinct symbols should be written on a square. Turing argues (p. 135) for this restriction by remarking, 'if we were to allow an infinity of symbols, then there would be symbols differing to an arbitrarily small extent', and the computor could not distinguish at a glance between symbols that are sufficiently close. In the second case consider, for example, Arabic numerals like 178 or 99999999 as one symbol; then it is not possible for the computor to determine at a glance whether or not 9889995496789998769 is identical with 98899954967899998769.

---

[4] Martin Davis describes the background for this publication in his paper (1994).

[5] For a detailed exposition, see my (1994) or (2009, Section 3).

The sensory limitation of computors leads thus quite directly to boundedness and locality conditions: (**B**) there is a bound on the number of symbols or symbol sequences a computor can observe at a glance, and (**L**) the operations a computor can carry out must consist in a local modification (of an observed configuration). Since some of the operations may involve a change of state of mind, Turing concludes:

> The most general single operation must therefore be taken to be one of the following: (A) A possible change (a) of symbol together with a possible change of state of mind. (B) A possible change (b) of observed squares together with a possible change of state of mind. (p. 137)

With this restrictive analysis of a computor's steps, it is rather straightforward to conclude that a Turing machine operating on strings (a 'string machine') can carry out his computations. Indeed, Turing first considers such machines that mimic directly the work of the computor; then he asserts referring to ordinary Turing machines ('letter machines'):

> The machines just described [string machines] do not differ very essentially from computing machines as defined in §2 [letter machines], and corresponding to any machine of this type a computing machine can be constructed to compute the same sequence, that is to say the sequence computed by the computer. (p. 138)

It is clear that the string machines, as Gandy asserted, 'appear as a result, as a codification, of his [Turing's] analysis of calculations by humans'. It is equally apparent that string machines *are* substitution puzzles (played by humans)! In a certain way, Turing's (1954) takes for granted central aspects of the earlier analysis, in particular, the imposition of boundedness and locality conditions. These conditions are simply built into the very notion of a puzzle, whereas in 1936, as we just saw, a sustained argument leads from a computor's sensory limitations to the restrictive conditions. However, the sustained argument does not constitute a mathematical proof. Turing himself views it as mathematically unsatisfactory as it ultimately relies on an appeal to intuition.[6] Given the analysis, an appeal to intuition is no longer needed wholesale for *The Thesis*, but only for the more restricted judgment: string machines can carry out the mechanical operations of computors (satisfying the restrictive conditions). Let us call this judgment Turing's *Central Thesis.*

3. Substitution puzzles: generalised

Coming back to the proper understanding of the theorem (from Section 1) that has to be proved, *puzzle* is now to be interpreted as substitution puzzle and *systematic procedure* is defined not independently, but as a puzzle 'in which there is never more than one possible move in any of the positions which arise and in which some significance is attached to the final result'. (p. 17) With these definitions in place, Turing gives a beautiful diagonal argument showing, that there is no systematic procedure for deciding whether a puzzle with unambiguous moves 'comes out' for an initial configuration, i.e., ends after finitely many moves in the position B or W. (This is the usual Self-halting problem expressed for the special puzzles with Turing's self-conscious use of the letter K to refer to it.) He also indicates, on pp. 21–22, an argument for Gödel's first incompleteness theorem.

The significance of these theorems rests, of course, on the identification of the informal notion of *puzzle* with the mathematically rigorous concept of a *substitution puzzle*. This formulation of the

[6] This notion is discussed in Turing (1939); it is used for 'setting down formal rules for inferences, which are always intuitively valid'. He notes: 'In pre-Gödel times, it was thought by some that it would be possible to carry this programme to such a point that all the intuitive judgements of mathematics could be replaced by a finite number of these rules. The necessity for intuition would then be entirely eliminated'. (p. 209)

methodological issue brings out the parallelism with the standard formulation of Church's or Turing's Thesis, namely, that the informal notion of calculable number theoretic function is identified with the mathematically rigorous concept of general recursive, respectively, Turing machine computable function. It also describes the intellectual context for my first fortuitous and accidental encounter with this paper of Turing's when I was working with my student John Byrnes around 1994–5 on our lpaper, *K-graph machines: generalising Turing's machines and arguments*. The argument we generalised is quite literally that from Turing's (1936) I sketched above using Post's production systems to describe machine computations. In order to make Turing's Thesis inductively more convincing, it seemed sensible to allow larger classes of symbolic configurations and more general operations on them; at the time, I did not see how else one could possibly address the methodological issue. That was clearly in accord with Post's (1936) remarks at the end suggesting that this type of work would be the basis for changing a hypothesis to a natural law. But was that way of proceeding also in accord with Turing's general approach?

We knew, of course, that Turing had given his 1936 analysis of mechanical procedures for restricted two-dimensional configurations. The fact that he considered in 1954 quite open-ended, even three-dimensional configurations and mechanical operations on them, lifted our spirits. It allowed us to understand the statement, 'Puzzles have substitution puzzles as their normal form' as a variant of his Central Thesis, and to connect it systematically with Kolmogorov and Uspensky's work on algorithms. Our approach has three distinct components: the symbolic configurations are finite connected and labelled graphs, we called *K(olmogorov)-graphs*; K-graphs contain a unique *distinguished element* that corresponds to the scanned square of a Turing machine tape; the operations *substitute neighborhoods* of the distinguished element by other neighborhoods and are given by a finite list of generalised Post production rules. Though broadening Turing's (and Post's) considerations, we remain within his general analytic framework and prove that letter machines can mimic K-graph machines. Turing's Central Thesis expresses here that K-graph machines can do the work of computors directly.[7] In summary, a much more general class of symbolic configurations and operations on them is considered, and the Central Thesis for K-graph machines seems even more plausible than the one for string machines.

The diagram below gives a triangulated summary of these extended reflections. The left leg of the triangle indicates the standard formulation of Turing's Thesis, the class of calculable number theoretic functions is co-extensional with the Turing machine computable ones. The hypotenuse connecting 'Calculability of number theoretic functions' with 'Computations by a *symcon* machine' indicates the crucial part of Turing's extended argument: step 1 is given by conceptual analysis, whereas step 2 applies the Central Thesis for a particular class of symbolic configurations or *symcons*; here, letters, strings and K-graphs. The *symcon* machines are (generalised) Post systems operating, of course, on *symcons*. Finally, the right leg of the triangle points to the mathematical reduction, indeed equivalence between *symcon* and letter machines.

The separation of conceptual analysis and mathematical proof is essential for recognising that the correctness of *Turing's Thesis* rests on two pillars. The boundedness and locality conditions for computors constitute the first pillar. However, what are symbolic configurations, and what changes can mechanical operations effect? In order to complete the triangulation, we do have to appeal to the pertinent Central Thesis, the second pillar. That may have been the reason why Turing thought that the variant of his Thesis must remain indefinite and that this very statement is one 'which one does not attempt to prove'.

[7] As a playful indication of how K-graph machines straightforwardly can carry out human and genuinely symbolic, indeed diagrammatic algorithms, we programmed a K-graph machine to do ordinary, two-dimensional column addition.

4. Epilog: abstract notion

The leading ideas of Section 3 can be generalised further to obtain an analysis of *parallel machines*. To help the imagination a bit, one can consider Conway's *Game of Life* as an example of a 'discrete mechanical device' that carries out parallel computations. In any event, that is what Turing's student Robin Gandy did; he proposed such an analysis (1980) and used, as Turing did, a *Central Thesis* to connect parallel computations satisfying informal restrictive conditions to particular kinds of dynamical systems. The restrictive conditions are motivated by purely physical considerations. The latter can be reduced to two, namely, (i) a lower bound on the size of physically distinguishable 'atomic' components justified by the uncertainty principle of quantum mechanics and (ii) an upper bound on signal propagation grounded in the theory of special relativity. Together, these restrictions form the basis of boundedness and locality conditions for machines in the way sensory limitations do for computors.

That analogy, in turn, allows us to formulate crucial requirements simultaneously for computors and parallel machines: (i) they operate on configurations that are finite, but of unbounded size; (ii) they recognise, in each configuration, patterns that are represented in a fixed, bounded list of different kinds of 'stereotypes'; (iii) they locally and deterministically operate on the pattern(s) recognised in the given configuration; and (iv) they assemble the next configuration from the original one and the result of local operation(s). These requirements can be formulated precisely as axioms for discrete dynamical systems and define the second-order, abstract notions of a *Turing computor* and *Gandy machine*.[8] (This is fully in-line with the way, in which notions like that of a group or field were introduced at the beginning of modern abstract mathematics, i.e., in the second half of the nineteenth century.) Furthermore, they allow proving representation theorems of the form: the computations of any device satisfying the axioms can be reduced to those of Turing machines.

In this way, the methodological problems surrounding Turing's Thesis have been lifted from their unusual status in mathematics. They now concern 'axiomatic definitions' and (representation) theorems, no longer statements whose status is somewhere between a definition and a theorem. They are no longer unusual, but rather common and difficult: they require us to recognise the correctness and appropriateness of axioms for intended concepts.

References

Davis, M., 1958. Computability and Unsolvability, McGraw-Hill, New York.

Davis, M., 1965. The Undecidable – Basic Papers on Undecidable Propositions, Unsolvable Problems and Computable Functions, Raven Press, Hewlett, NY.

Davis, M., 1994. Emil L. Post: His life and his work. In: Davis, M. (Ed.), Solvability, Provability, Definability: The Collected Works of Emil L. Post, Birkhäuser, Boston, Basel, Berlin, pp. xi–xxviii.

Gandy, R., 1980. Church's thesis and principles for mechanisms, In: Barwise, J., Keisler, H.J., Kunen, K. (Eds.), The Kleene Symposium, North-Holland, Amsterdam, New York, pp. 123–148.

Kolmogorov, A., Uspensky. 1963. On the definition of an algorithm AMS Trans. 21, 217–245.

Post, E., 1936. Finite combinatory processes – formulation 1. J. Symb. Log. 1, 103–105.

Post, E., 1941. Absolutely unsolvable problems and relatively undecidable propositions – Account of an anticipation; submitted to the American Journal of Mathematics in 1941, but published only in Davis's 'The Undecidable' in 1965. (The mathematical core was published in that journal under the title, Formal reductions of the general combinatorial decision problem, volume 55 (2), 1943, 197–215.)

Post, E., 1947. Recursive unsolvability of a problem of Thue. J. Symbol. Log. 12, 1–11.

[8] For mathematical details for these and the following considerations, Gandy (1980) and Sieg (2002) should be consulted; Sieg (2008) gives a sustained informal discussion.

Sieg, W., 1994. Mechanical procedures and mathematical experience. In: George, A. (Ed.), Mathematics and Mind, Oxford University Press, Oxford, pp. 71–117.
Sieg, W., 2002. Calculations by man & machine: mathematical presentation. In: Gärdenfors, P., Wolenski, J., Kijania-Placek, K. (Eds.), In the Scope of Logic, Methodology and Philosophy of Science, vol. I. Kluwer Academic Publishers, pp. 247–262.
Sieg, W., 2008. Church without dogma: axioms for computability. In: Cooper, S.B., Löwe, B., Sorbi, A. (Eds.), New Computational Paradigms: Changing conceptions of what is computable, Springer, pp. 139–152.
Sieg, W., 2009. On computability. In: Irvine, A.D., (Ed.), Philosophy of Mathematics, North-Holland, Amsterdam, New York, pp. 535–630.
Sieg, W., Byrnes, J., 1996. K-graph machines: generalizing Turing's machines and arguments. In: Hajek, P. (Ed.), Gödel '96: Logical Foundations of Mathematics, Computer Science, and Physics – Kurt Gödel's Legacy, Lecture Notes in Logic 6, pp. 98–119.
Turing, A., 1936. On computable numbers with an application to the Entscheidungsproblem, Proc. of the London Math. Soc. 42 (2), 230–265.
Turing, A., 1939. Systems of logic based on ordinals, Proc. of the London Math. Soc. 45 (2), 161–228.
Turing, A., 1950. The word problem in semi-groups with cancellation. Ann. Math. 52, 491–505.
Turing, A., 1954. Solvable and unsolvable problem. Science News 31, 7–23.

K. Vela Velupillai connects –

TURING ON 'SOLVABLE AND UNSOLVABLE PROBLEMS' AND SIMON ON 'HUMAN PROBLEM SOLVING'

Turing's fundamental work on *Solvable and Unsolvable Problems* (Turing, 1954), *Intelligent Machinery* (Turing, 1969) and *Computing Machinery and Intelligence* (Turing, 1950) had a profound effect on the work of Herbert Simon, the only man to win both the *ACM Turing Prize* and the *Nobel Memorial Prize in Economics*, particularly in defining *boundedly rational* economic agents as *information processing systems* (IPS) solving decision problems.[1]

A comparison of Turing's classic formulation of *Solvable and Unsolvable Problems* and Simon's variation on that theme, as *Human Problem Solving* (Newell and Simon, 1972), would be an interesting exercise, but it must be left for a different occasion. This is partly because the *human problem solver* in the world of Simon needs to be defined in the same way Turing's approach to *Solvable and Unsolvable Problems* was built on the foundations he had established in his classic of 1936–7.

It is little realised that four of what I call the Five Turing Classics[2] – *On Computable Numbers* (Turing, 1936–7), *Systems of Logic* (Turing, 1939), *Computing Machinery and Intelligence* and

---

[1] In the precise sense in which this is given content in mathematical logic, metamathematics, computability theory and model theory.

[2] The fifth being, of course, *The Chemical Basis of Morphogenesis* (1952). It is interesting to note that the five contributions came in two clusters, the first two in 1936 and 1938/9; the last three in the fertile last four years of his tragically brief life.

Solvable and Unsolvable Problems – should be read together to glean *Turing's Philosophy*[3] *of Mind.* Simon, as one of the acknowledged founding fathers of computational cognitive science was deeply indebted to Turing in the way he tried to fashion what I have called *Computable Economics* (Velupillai, 2000).[4] It was not for nothing that Simon warmly acknowledged – and admonished – him in his essay in the volume 'memorializing Turing' (Simon, 1996, p. 81), titled *Machine as Mind*[5]:

"If we hurry, we can catch up to Turing on the path he pointed out to us so many years ago."

Simon was on that path, for almost the whole of his research life.

Building a Brain, in the context of economic decision making, meant a mechanism for encapsulating human intelligence, underpinned by rational behaviour in economic contexts. This was successfully achieved by Herbert Simon's lifelong research program on computational behavioural economics.[6]

From the early 1950s, Simon had empirically investigated evidence on human problem solving and had organised that evidence within an explicit framework of a theory of sequential information processing by a Turing Machine. This resulted in (Simon, 1979, p. x; italics added):

> "[A] general theory of human cognition, not limited to problem solving, [and] a methodology for expressing *theories of cognition as programs* [for digital computers] and for using [digital] computers [in general, Turing Machines] to simulate *human thinking*.

This was the first step in replacing the traditional Rational Economic Man with the computationally underpinned Thinking, i.e., Intelligent – Man. The next step was to stress two empirical facts (ibid, p. x; italics added):

(i) 'There exists *a basic repertory of mechanisms and processes* that Thinking Man uses in all the domains in which he exhibits *intelligent behaviour*'.
(ii) 'The models we build initially for the several domains must all be assembled from this same basic repertory, and common principles of architecture must be followed throughout'.

It is easy to substantiate the claim that the *basic repertory of mechanisms and processes* are those that define, in the limit, a Turing Machine formalisation of the Intelligent Man, when placed in the decision-making, problem-solving, context of economics – cf. Velupillai (2010).

The broad contours of this vision and method, and its basis in computability and computational complexity theory, were clearly outlined in a letter he wrote me, after reading my book on *Computable Economics* (Simon, 2000):

> "I want to share some first impressions on my reading of "Computable Economics." ... I was delighted and impressed by the mileage you could make with Turing Computability in showing how nonsensical the Arrow/Debreu formulation, and others like it, are as bases for notions of human rationality.

[3] Remembering Feferman's (2001, p. 79) cautionary note that Turing never tried to develop an over-all *philosophy of mathematics* ... ', but not forgetting that his above works were decisive in the resurrection of a particular vein of research in the philosophy of mind, particularly in its cognitive, neuroscientific, versions pioneered by Simon.

[4] I could as well have called it *Turing's Economics*.

[5] To which he added the caveat (ibid, p. 81):

> "I speak of 'mind' and not 'brain'. By mind I mean a system [a mechanism] that produces thought .. .'

I have always interpreted this notion of 'mechanism' with Gandy's *Principles for Mechanisms* Gandy (1980) in *mind* (sic!).

[6] I refer to this variant of behavioural economics, which is underpinned by a basis in *computational complexity theory*, as *classical* behavioural economics, to distinguish it from currently orthodox behavioural economics, sometimes referred to as *modern* behavioural economics, which has no computational basis whatsoever.

As the book makes clear, my own journey through bounded rationality has taken a somewhat different path. Let me put it this way. There are many levels of complexity in problems, and corresponding boundaries between them. *Turing computability is an outer boundary, and as you show, any theory that requires more power than that surely is irrelevant to any useful definition of human rationality.* A slightly stricter boundary is posed by computational complexity, especially in its common "worst case" form. We cannot expect people (and/or computers) to find exact solutions for large problems in computationally complex domains. This still leaves us far beyond what people and computers actually CAN do. The next boundary, but one for which we have few results ..., is computational complexity for the "average case", sometimes with an "almost everywhere" loophole. That begins to bring us closer to the realities of real-world and real-time computation. Finally, we get to the empirical boundary, measured by laboratory experiments on humans and by observation, of the level of complexity that humans actually can handle, with and without their computers, and - perhaps more important – what they actually do to solve problems that lie beyond this strict boundary even though they are within some of the broader limits.

The latter is an important point for economics, because we humans spend most of our lives making decisions that are far beyond any of the levels of complexity we can handle exactly; and this is where satisficing, floating aspiration levels, recognition and heuristic search, and similar devices for arriving at good-enough decisions take over. A parsimonious economic theory, and an empirically verifiable one, shows how human beings, using very simple procedures, reach decisions that lie far beyond their capacity for finding exact solutions by the usual maximizing criteria.

...

So I think we will continue to proceed on parallel, but somewhat distinct, paths for examining the implications of computational limits for rationality – you the path of mathematical theories of computation, I the path of learning how people in fact cope with their computational limits

...

While I am fighting on a somewhat different front, I find it greatly comforting that these outer ramparts of Turing computability are strongly manned, greatly cushioning the assault on the inner lines of empirical computability."

Unfortunately, the 'assaults' by orthodoxy and its non-computable, non-constructive forces are ceaseless and 'cushioning' the 'inner lines of empirical computability' from these persistent assaults is no easy task.

References

Feferman, S, 2001 Preface to: Systems of logic based on ordinals. In: Gandy, R.O., Yates, C.E.M. (Eds.), Collected Works of A.M. Turing – Mathematical Logic, North-Holland, Amsterdam.

Gandy, R.O. 1980. Church's thesis and principles for mechanisms. In: Barwise, J., Keisler, H.J., Kunen, K. (Ed.). The Kleene Symposium, North-Holland, Amsterdam.

Newell, A., Herbert A.S. 1972. Human Problem Solving, Prentice-Hall Inc., Englewood Cliffs, NJ.

Simon, H.A. 1979. Models of Thought, vol I. Yale University Press, New Haven.

Simon, H.A. 1996. Machine as mind. In: Macmillan, P., Clark, A. (Eds.). Machines and Thought - The Legacy of Alan Turing, vol. 1. Oxford University Press, Oxford, pp. 81–101, (Chapter 5).

Simon, H.A. 2000. Letter to Velupillai, 25 May, 2000.

Velupillai, K.V. 2000. Computable Economics, Oxford University Press, Oxford.

Velupillai, K.V. 2010. Foundations of Boundedly Rational Choice and Satisfying Decisions, Advances in Decision Sciences, April.

The Word Problem in Semi-Groups with Cancellation

(Annals of Mathematics, vol. 52, no. 2 (1950), pp. 491–505)

Gregory Chaitin on —

Finding the Halting Problem and the Halting Probability in Traditional Mathematics

Following Turing's 1936 discovery of the halting problem and how to derive incompleteness from it as a corollary in *On Computable Numbers*, a great deal of work was done by many mathematicians finding 'natural' versions of the halting problem in many traditional areas of mathematics, that is, in pre-Turing mathematics. Turing describes some of this work in his lovely 1954 expository paper *Solvable and Unsolvable Problems*, movingly published the year of his untimely death. Turing's word problem paper is a complicated piece of such work that was carried out by Turing himself; obviously he wished to reassure himself that the halting problem was ubiquitous in real mathematics, which is the current consensus.[1]

This very large body of work, which unfortunately has never been collected in a single book and remains scattered in crumbling, dusty journals, can also be used to show that the halting probability Ω is hiding in many fields of traditional mathematics. For example, using 1947 work of Emil Post, it is easy to show[2] that there is a semi-group without cancellation such that for each positive integer k the number of words $\lambda q_{\text{initial}} a^k b^n \lambda$ which are equal to the word $\lambda q_{\text{final}} \lambda$ is finite or infinite depending on whether the kth bit of Ω is a 0 or a 1. Following work of Toby Ord and Tien D. Kieu (2004), one can replace 'finite' or 'infinite' here by 'even' or 'odd'.

Most unfortunately, Turing did not live long enough to see the most beautiful example of the halting problem that occurs in a traditional field of mathematics, number theory: namely the 1970 Davis–Putnam–Robinson–Matiyasevich proof, that asking for a general method to determine whether or not a diophantine equation has a solution is equivalent to solving the halting problem. In particular, for any computer programming language, there is a diophantine equation with a parameter k that has a solution if and only if the kth computer program in that language ever halts. Similarly, there is a diophantine equation with a parameter k that has finitely or infinitely many solutions — or alternatively an even or an odd number of solutions — depending on whether the kth bit of Ω is 0 or 1.

[1] See Stephen Wolfram's *A New Kind of Science* and the February 2010 *AMS Notices* piece *Can't Decide? Undecide!* by Chaim Goodman-Strauss.

[2] G.J. Chaitin, 'An algebraic characterization of the halting probability', *Fundamenta Informaticae* 79 (2007), pp. 17–23.

While John L. Britton gives us a brief –

INTRODUCTION TO THE MATHEMATICS[1]

A *semi-group* is a set within which is defined an associative product. A *semi-group with cancellation* (SWC) is a semi-group S such that for all a, b, c in S, $ab = ac$ implies $b = c$ and $ba = ca$ implies $b = c$. Let K denote 'semi-group, SWC or group'. A *K-presentation* is a pair (S, D), where S is a finite set of symbols $s_1, \ldots, s_n$ and D is a finite set of formal equations $U_i = V_i (i = 1, \ldots, r)$ where U_i and V_i are words in the symbols; a *word* means a finite string of symbols in the semi-group or SWC case, but means an expression of the form $s_{i_1}^{e_1} \ldots s_{i_k}^{e_k}$ where $e_j = \pm 1 (j = 1, \ldots, k)$ in the group case. We say that $W_1 = W_2$, where W_1, W_2 are words, is a *relation* for (S, D) if, whenever X is a K containing elements $s_1, \ldots, s_n$ such that $U_i = V_i$ in X for all i, then $W_1 = W_2$ in X. In particular each $U_i = V_i$ is called a *defining relation* (or fundamental relation (FR)).

We say that the *word problem is solvable for* (S, D) if there is an algorithm which will determine of any pair of words W_1, W_2 whether or not $W_1 = W_2$ is a relation for (S, D).

For any (S, D) one can construct $[S, D]$, which is a K and which is unique up to isomorphism; it contains elements $s_1, \ldots, s_n$, each element of it is a word in $s_1, \ldots, s_n$ and $W_1 = W_2$ is a relation for (S, D) if and only if we have $W_1 = W_2$ in $[S, D]$.

In 1947, Post and Markov independently showed that there is a semi-group presentation with unsolvable word problem. The problem of extending this result to groups received a lot of attention but proved difficult. In the present paper, Turing considers the half-way house of semi-groups with cancellation; undoubtedly it influenced both Novikov who finally obtained the result for groups in 1955 and Boone, who proved the result independently at about the same time.

[1] Extracted from the Introduction to volume 1 of the Collected Works.

ANNALS OF MATHEMATICS
Vol. 52, No. 2, September, 1950

THE WORD PROBLEM IN SEMI-GROUPS WITH CANCELLATION

BY A. M. TURING

(Received August 13, 1949)

It will be shown that the word problem in semi-groups with cancellation is not solvable. The method depends on reducing the unsolvability of the problem in question to a known unsolvable problem connected with the logical computing machines introduced by Post (Post, [1]) and the author (Turing, [1]). In this we follow Post (Post, [2]) who reduced the problem of Thue to this same unsolvable problem.

1. Semi-groups with cancellation

By a semi-group with cancellation we understand a set $\mathfrak{S}$ within which is defined a function $f(a, b)$ described as a product and satisfying

(i) The associative law $f(f(a, b), c) = f(a, f(b, c))$
(ii) The cancellation laws $f(a, b) = f(a, c) \supset b = c$

$$f(b, a) = f(c, a) \supset b = c$$

for any a, b, c in $\mathfrak{S}$.

In view of the associative law we naturally write abc for both $f(f(a, b), c)$ and $f(a, f(b, c)$, and similarly for strings of letters of any length. If A and B are two such strings (e.g. *arghm* and *gog*) then AB represents the result of writing one after the other (i.e. *arghmgog*). We have here used the convention (to be followed throughout) that capital letters are variables for strings of letters.

2. The word problem

One may construct a semi-group with cancellation by means of generators and relations as follows. We first define a certain finite set $\mathfrak{G}$ of letters to be the *generators*. Strings of such letters are described as *words*. A class $\mathfrak{F}$ of pairs of words is chosen and described as the *fundamental relations* (abbreviated sometimes to F.R.). The words are understood to represent members of a semi-group, and pairs forming a fundamental relation to represent equal members, but no pairs of words to represent the same member unless obliged to by this condition. The object of the 'word problem' is to find a method for determining whether a given pair of words do or do not represent the same semi-group element. Such a pair of words will be described as a *relation* of the semi-group.

Treating the above explanation of the meaning of a relation as relatively informal and intuitive, we may also give a more formal definition which is equivalent to the first. This greater formality would in many applications tend to dryness and obscurity, but for our present purpose it gives greater clarity, since we are not concerned so much with obtaining relations as with showing that certain relations cannot be obtained. This requires a very unambiguous form of definition.

Immediate deducibility and assertibility

We say that

(i) Any pair in $\mathfrak{F}$ is immediately assertible.
(ii) Any pair of form (A, A) is immediately assertible.
(iii) (A, B) is immediately deducible from (B, A).
(iv) (A, B) is immediately deducible from (A, C) and (C, B).
(v) (AG, BG) is immediately deducible from (A, B) for any generator G.
(vi) (GA, GB) is immediately deducible from (A, B) for any generator G.
(vii) (A, B) is immediately deducible from (AG, BG) for any generator G.
(viii) (A, B) is immediately deducible from (GA, GB) for any generator G.

Note that we use a capital for G. If g is a generator it is a quite definite one.

Immediate deducibility and assertibility apply in no other cases but those listed above.

Semi-group relations

A pair (A, B) is a relation of the semi-group $\mathfrak{S}$ arising from the fundamental relations $\mathfrak{F}$ if there is a 'proof' of (A, B) consisting of a sequence of pairs of which each is either immediately assertible or immediately deducible from previous pairs of the sequence, and the last pair is (A, B).

We wish to find out whether the word problem for semi-groups is solvable or not. The possibilities may be divided into three alternatives.

(a) There might be a general method applicable for all sets $\mathfrak{F}$ of fundamental relations, and all pairs (A, B).
(b) For any given set of fundamental relations there might be a special method which could be used, but no general method applicable for all sets.
(c) There may be some particular set of F.R. such that no method will ap- ply with it: i.e., any reputed method will give the wrong answer for some pair of words.

We shall show that (c) is the correct alternative.

3. Computing machines

As in Turing [1], Post [1], we identify the existence of a 'general method for determining...' with that of a 'computing machine constructed to determine...'. Numerous alterative definitions have been given (Church [1], Kleene [1], see also Hilbert and Bernays, Appendix to vol. II) and proved equivalent (Kleene [2], Turing [2]).

A computing machine is imagined as a mechanical device capable of a finite number of states or *internal configurations* (I.C.) $q_1, q_2, \cdots, q_r$, working on a (both ways) infinite tape divided into squares on which symbols are printed. These symbols are drawn from the set $s_0, s_1, s_2, \cdots, s_n$. At any moment one particular square is especially closely connected with the machine. It is called the *scanned square*. The behavior of the machine at this moment is determined by its own I.C. and the symbol on the scanned square. The behavior consists in altering this symbol and possibly also shifting the scanned square one place to right or left. The state of the whole system at any moment is called its *complete configuration* (C.C.). The C.C. could be described by giving the sequence of symbols on the tape, starting with the first which is not blank and ending with the last, and interposing the name of the I.C. on the left of the symbol on the scanned square. In Turing [1] the expression *configuration* (unqualified) was used to describe the combination of I.C. with scanned symbol, so that the behavior of the machine is determined by the configuration at each moment.

For our present purpose many of the conventions of Turing [1] are not ideally chosen, but we are to some extent bound by them since we have to use the results of that paper. We shall describe a C.C. differently. We shall take the form described above and flank it with s_4 at either end. Thus for instance $s_4 s_1 s_3 q_2 s_1 s_4$ represents a C.C. in which the I.C. is q_2 and the tape bears $s_1 s_3 s_1$ the last of these symbols being on the scanned square. This device is due to Post (who uses h for s_4) and greatly facilitates the connection with the word problem and Thue systems. We shall divide the internal configurations into two classes, the left-facing and the right-facing. The left-facing I.C.'s will be $l_0, l_1, l_2, \cdots, l_R$ and the right facing will be $r_1, r_2, \cdots, r_R$ (some otiose ones may be included in one of the classes to equalize numbers). If the I.C. is an r_i the new convention requires the machine's behavior to be determined by it and the symbol next on its right, but in the case of an l_i it is determined by l_i and the symbol on its left. It is easy to modify a machine constructed according to the original conventions to give one with similar properties but satisfying the new conventions. We leave the details to the reader. The new convention is not of course considered desirable in general, but only where semi-groups (and perhaps groups) are concerned.

In accordance with these conventions we make the following definitions.

A *complete machine* is determined by giving a (finite) set of symbols, comprising *tape symbols, left-facing internal configurations*, and *right-facing internal configurations*, and also a *table*.

A *complete configuration* is a word made up of tape symbols together with one internal configuration. A table consists of a number of *entries*, each of which is a pair of words of one of the following eight forms

$$(r_i s_k,\ s_{k'} r_{i'}) \qquad (s_0 l_i,\ s_0 s_0 l_{i'})$$
$$(s_k l_i,\ l_{i'} s_{k'}) \qquad (s_0 l_i,\ l_{i'})$$
$$(s_k l_i, s_{k'} r_{i'}) \qquad (r_i s_0, r_{i'} s_0 s_0)$$
$$(r_i s_k, l_{i'} s_{k'}) \qquad (r_i s_0,\ r_{i'},)$$

where the left-facing internal configurations are $l_0, l_1, l_2, \cdots, l_R$ and the right-facing are $r_1 r_2, \cdots, r_P$. The tape symbols are $s_0, s_1, \cdots, s_N$

A table must be so constructed that no C.C. has two subwords are first members of different entries. As a consequence of this condition we may define the (unique if existent) *successor* of a C.C. as follows. If (U, V) is an entry of a table then the successor of $X\ U\ Y$ is $X\ V\ Y$. We say that B is a descendant of A if it is possible to connect A to B by a finite sequence of words each of which is the successor (in our technical sense) of the one which immediately precedes it in the sequence.

We might restrict the definition of a C.C. by insisting that a C.C. must begin and end with s_4. But it will appear shortly that it will make no difference for our purpose whether we do so or not.

Note that if B is a descendant of A in a machine $\mathfrak{M}$ then (A, B) is a relation of the semi-group whose F.R. are the entries of $\mathfrak{M}$.

The symbol s_0 represents 'blank'. If a machine following the new convention is made to correspond in behavior with one following the old it is necessary that the entries $(s_0 l_i,\ s_0 s_0 l_{i'})$, $(s_0 l_i,\ l_{i'})$, $(r_i s_0,\ r_{i'} s_0 s_0)$, $(r_i s_0,\ r_{i'})$ should only come into play when the machine is scanning a blank square in one of the two continuous infinite sequences of blank squares at the two ends of the tape. This would be ensured by appropriate entries in the machine's table, and would be made possible by the presence of the 'end symbols' s_4. However so long as we are satisfied that the behavior of the machines in Turing [1] can be modelled by machines following the present conventions we need not consider these details any further.

There is no general method which can be used to determine, given a machine and a C.C., whether that C.C. has any descendants in which a given generator occurs. The equivalent of this was stated in Turing [1] (p. 248), but Post has exposed a weakness in the proof which was given there. Post pointed out that certain conventions which should only have been used in the construction of particular machines had actually been assumed to apply to arbitrary machines. This convention required partial answers never to be erased, and to have a space order corresponding to he time order of their printing. Such a restriction applied to machines in general would be most undesirable. Post suggests over-coming the difficulty by considering the 'answer' to be given by the time sequence of printings, and ignoring the space sequence altogether. Another possibility is to show that the objectionable convention is in some sense inessential. This could be done by making use of the fact that the 'universal machine' itself obeys the convention. However there is no intention to enter into these questions in detail here, and we shall assume the theorem proved in the form in which it is stated at the beginning of this paragraph.

If we were content to prove the unsolvability of the word problem in the weak form ((b) or (c)), it would be natural to attempt to reduce it to the unsolvable problem just mentioned. For the strong form (c) however we prefer a result about a particular machine which may be translated into a result about a particular semi-group. This is made possible by the existence of the 'universal machine' U. This machine has the property that, given the table of any other machine $\mathfrak{M}$ and a C.C. K of $\mathfrak{M}$ we can find a C.C. of $\mathfrak{U}$, $c(\mathfrak{M}, K)$ say, such that a certain easily distinguishable subsequence of the descendants of $c(\mathfrak{M}, K)$ corresponds in a certain simple way to the descendants of K in the motion of $\mathfrak{M}$. The details of $\mathfrak{U}$ are to a very large extent arbitrary, and so are the details of the above-mentioned correspondences. One possible convention is that the special subsequence of descendants of $c(\mathfrak{M}, K)$ consists of those whose penultimate symbol is s_5, and that of the corresponding descendant of K is obtained by omitting all symbols which are not symbols for $\mathfrak{M}$. (This of course requires that there are certain symbols which are sacred to the use of $\mathfrak{U}$). However in spite of the arbitrariness we shall speak of $\mathfrak{U}$ as though it were a quite definite machine whose table was before us. In order actually to obtain such a table we should have to take the description of the universal machine in Turing [1] and modify it in several ways. Firstly it would be necessary to expand all the abbreviations used: secondly it must be modified to accord with the conventions of the present paper, and finally various matters of detail noted by Post (Post [2], p. 7, footnote) must be corrected. As a consequence of the properties of $\mathfrak{U}$ and the above mentioned unsolvable problem concerning machines in general *there is a machine* $\mathfrak{B}$ *such that there is no general method by which given a* C.C. *of* $\mathfrak{B}$ *we can determine whether* $s_4 l_0 s_4$ *is one of its descendants or not.* $\mathfrak{B}$ *is such that any* C.C. *containing* l_0 *has no successor*. Here $\mathfrak{B}$ is a machine obtained by a small modification of $\mathfrak{U}$. $\mathfrak{B}$ behaves like $\mathfrak{U}$ in its earlier steps, but as soon as it reaches a C.C. in which s_3 appears it behaves differently, proceeding quickly to the C.C. $s_4 l_0 s_4$ by erasures, etc.

In what follows we shall not be concerned with any other machines than $\mathfrak{B}$, apart from some parenthetical remarks about a machine $\mathfrak{B}'$. Our interest in machines in general, and the conventions by which they are restricted, therefore fades. All we need to know is that $\mathfrak{B}$ satisfies the above italicized statement and that its table follows our new conventions.

4. The semi-group $\mathfrak{S}_0$

The semi-group $\mathfrak{S}_0$ is obtained from the table for $\mathfrak{B}$ as follows. Let the tape symbols for $\mathfrak{B}$ be $s_0, s_1, \cdots, s_N$, the left-facing internal configurations $l_1, l_2, \ldots, l_R$ and the right-facing ones $r_1, r_2, \cdots, r_R$: let the entries of the table for $\mathfrak{B}$ be $E_1, \cdots, EM$. Then the generators of $\mathfrak{S}_0$ are to be $y, \sigma_m, \tau_m, j_i, k_i, n_i, p_i, u_k, v_k, w_k, z_k (m = 1,2, \cdots, M; i = 0,1, \cdots, N; k = 1,2, \cdots, R)$. The F.R. of $\mathfrak{S}_0$ will be classified into three kinds, *principal relations, commutation relations*, and *change relations*.

Principal relations. For each entry of $\mathfrak{B}$ we have two principal relations, called *first phase* and *second phase* relations. A table is given below of the forms that these relations take for the entry E_m according to the eight different forms that this entry can have

Entry E_m	First phase relation	Second phase relation
$(r_i s_h,\ s_{h'} r_{i'})$	$(v_i k_h,\ \sigma_m j_h,\ v_i,\ \tau_m)$	$(\sigma_m n_h,\ z_i,\ \tau_m,\ z_i p_h)$
$(s_h l_i,\ l_{i'} s_{h'})$	$(j_h u_i,\ \sigma_m u_i,\ k_{h'} \tau_m)$	$(\sigma_m w_i,\ p_{h'} \tau_m,\ n_h w_i)$
$(s_h l_i,\ s_{h'} r_{i'})$	$(j_h u_i,\ \sigma_m j_{h'} v_{i'} \tau_m)$	$(\sigma_m n_{h'} z_{i'} \tau_m,\ n_h w_i)$
$(r_i s_h,\ l_{i'} s_{h'})$	$(v_i k_h,\ \sigma_m u_{i'} k_{h'} \tau_m)$	$(\sigma_m w_{i'} p_{h'} \tau_m,\ z_i p_h)$
$(s_0 l_i,\ s_0 s_0 l_{i'})$	$(j_0 u_i,\ \sigma_m j_0 j_0 u_{i'} \tau_m)$	$(\sigma_m n_0 n_0 w_{i'} \tau_m,\ n_0 w_i)$
$(s_0 l_i,\ l_{i'})$	$(j_0 u_i, \sigma_m u_{i'} \tau_m)$	$(\sigma_m w_{i'} \tau_m,\ n_0 w_i)$
$(r_i s_0,\ r_{i'} s_0 s_0)$	$(v_i k_0,\ \sigma_m v_i,\ k_0 k_0 \tau_m)$	$(\sigma_m z_{i'} p_0 p_0 \tau_m,\ z_i k_0)$
$(r_i s_0,\ r_{i'})$	$(v_i k_0,\ \sigma_m v_{i'} \tau_m)$	$(\sigma_m z_{i'} \tau_m,\ z_i k_0)$.

Commutation relations

$$\left.\begin{array}{l}(\sigma_m u_i,\ u_i \tau_m)\\ (v_i \tau_m,\ \sigma_m v_i)\\ (w_i \tau_m,\ \sigma_m w_i)\\ (\sigma_m z_i,\ z_i \tau_m)\end{array}\right\} \begin{array}{l}(i = 1, \cdots, R\\ m = 1, \cdots, M).\end{array}$$

Change relations

$$(j_4 u_0,\ n_4 y)$$
$$(y k_4,\ w_0 p_4)$$
$$(y \tau_m,\ \sigma_m y) \qquad (m = 1, \cdots, M).$$

The symbols which occur in first phase relations may be called *first phase symbols*. Words made up of such symbols may be called first phase words. Second phase symbols and words may be defined similarly.

The symbols $u_i,\ v_i,\ w_i,\ z_i$ (which correspond to internal configurations) will be called *active* symbols. The symbols j_h, h_h, σ_m which always appear to the left of the active symbols in the F.R., are called *left symbols*, and k_h, p_h, τ_m are called right symbols.

A word is described as *normal* if it contains exactly one active symbol and all the symbols to the left of this active symbol are left symbols and all to the right of it are right symbols. Both members of an F.R. are evidently normal.

It will be seen that the first phase relations are obtained from the corresponding entries by

(a) Substituting v_i for r_i and u_i for l_i. Also substituting j_h for s_h when it occurs on the left of the I.C. but substituting k_h for it when it occurs on the right.
(b) Inserting σ_m on the left of the second term and τ_m on its right. The second phase entries are obtained by
(c) Substituting z_i for r_i and w_i for l_i. Also substituting n_h for s_h when it occurs on the left of the I.C. but substituting p_h for it when it occurs on the right.
(d) As (b).
(e) Interchanging the two terms.

The complete set of relations for $\mathfrak{S}_0$ may be regarded as forming the table for a certain machine $\mathfrak{B}'$. The movements of $\mathfrak{B}$ correspond to the earlier steps in the movement of $\mathfrak{B}'$, which may be called the first phase of the movement of $\mathfrak{B}'$. The C.C.'s of the two machines correspond if the generators σ_m, τ_m are ignored and the substitutions (a) applied. This correspondence applies throughout if $\mathfrak{B}$ never reaches the C.C. $s_4 l_0 s_4$. However if it does so the motion of $\mathfrak{B}'$ continues further. After the first phase comes an 'intermediate phase' in which the C.C. $\Sigma j_4 u_0 T_1 K_4 T_2$ is changed to $\Sigma n_4 w_0 T_1 p_4 T_2$.

Here Σ is a word formed entirely from the generators $\sigma_1, \sigma_2, \cdots, \sigma_M$ and T_1, T_2 are formed entirely from $\tau_1, \tau_2, \ldots, \tau_M$. Finally we have the second phase in which the motions of the first phase are carried out again, but in reverse order and with different tape symbols and internal con figurations. At the end of this we return to the original C.C. with certain substitutions applied. We intend to show that the necessary and sufficient condition that $s_4 l_0 s_4$ be a descendant of C is that $(\varphi_1(C), \varphi_2(C))$ be a relation of $\mathfrak{S}_0$ where φ_1, φ_2 represent the substitutions (a) and (c) above. The proof of the necessity is easy and consists essentially in establishing the above properties of $\mathfrak{B}'$. This argument is in effect reproduced in Lemmas 1–4 below, but without explicit reference to the machine $\mathfrak{B}'$. Indeed the machine $\mathfrak{B}'$ is wholly fictitious since it does not obey either the old or the new conventions. It has the power of "manufacturing tape" on which to write the symbols σ_m, τ_m. Such machines are certainly worthy of consideration, but are not recognized here. The sufficiency is more difficult, for we are permitted to do certain operations in semi-groups which do not correspond to motions of the machines, e.g., cancellation. The generators σ_m, τ_m have been introduced into the F.R. to ensure that any results which can be obtained by the application of these operations could also be obtained without. This means that the permission to cancel becomes a dead letter. The introduction of different sets of symbols for use on the right and the left of the active symbol has a similar purpose. It ensures that relations which contain more than one active symbol in each term can be split up into ones containing only one: this enables these relations to be interpreted in terms of machines. More picturesquely we can say that when the left and right symbols are distinguishable we can let two or more machines work on the same tape without danger of their interfering with one another.

We shall eventually prove

THEOREM 1. *The necessary and sufficient condition that* $(\varphi_1(C), \varphi_2(C))$ *be a relation of* $\mathfrak{S}_0$ *is that* $s_4 l_0 s_4$ *be a descendant of* C *in the machine* $\mathfrak{B}$.

As we have mentioned φ_1, φ_2 represent the substitutions (a) and (c). An equivalent definition is given below. Combining Theorem 1 with the principle enunciated at the end of §3 we have

THEOREM 2. *There is no general method by which we can determine whether a given pair of words is · or is not a relation of* $\mathfrak{S}_0$. *For if there were such a method we could apply it to pairs of the form* $(\varphi_1(C), \varphi_2(C))$ *and so determine whether* $s_4 l_0 s_4$ *is a descendant of* C *in* $\mathfrak{B}$.

Since we shall be concerned with no other semi-groups than $\mathfrak{S}_0$ we shall use the word 'relation' to mean 'relation of $\mathfrak{S}_0$'.

5. Sufficiency in Theorem 1

We begin by defining certain functions of words and C.C.'s. The null word is denoted by Λ. The functions φ_1, φ_2 (already mentioned above) are defined for all C.C.'s of $\mathfrak{B}$ by the equations

$$\varphi_1(l_i) = u_i, \quad \varphi_1(r_i) = v_i, \quad \varphi_1(s_h X) = j_h \varphi_1(X), \quad \varphi_1(X s_h) = \varphi_1(X) k_h$$
$$\varphi_1(l_i) = w_i, \quad \varphi_2(r_i) = z_i, \quad \varphi_2(s_h X) = n_{h\varphi_2}(X), \quad \varphi_2(X s_h) = \varphi_2(X) p_h.$$

The functions ψ_1, ψ_2 are defined for first and second phase words respectively by equations

$$\psi_1(XG) = \psi_1(X)\psi_1(G), \quad \psi_1(u_i) = l_i, \quad \psi_1(v_i) = v_i, \quad \psi_1(\Lambda) = \Lambda$$
$$\psi_1(k_h) = \psi_1(j_h) = s_h, \quad \psi_1(\sigma_m) = \psi_1(\tau_m) = \Lambda$$
$$\psi_2(XG) = \psi_2(X)\psi_2(G), \quad \psi_2(w_i) = l_i, \quad \psi_2(z_i) = v_i, \quad \psi_2(\Lambda) = \Lambda$$
$$\psi_2(r_h) = \psi_2(p_h) = s_h, \quad \psi_2(\sigma_m) = \psi_2(\tau_m) = \Lambda.$$

The function χ is defined for first phase words by the equations

$$\chi(XG) = \chi(X)\chi(G), \quad \chi(u_i) = w_i, \quad \chi(v_i) = z_i, \quad \chi(\Lambda) = \Lambda$$
$$\chi(j_h) = r_h, \quad \chi(k_h) = p_h, \quad \chi(\sigma_m) = \sigma_m, \quad \chi(\tau_m) = \tau_m.$$

The function τ is defined for words containing only $\sigma_1, \sigma_2, \cdots, \sigma_M$ and σ for words containing only $\tau_1 \tau_2, \cdots, \tau_M$ by the equations

$$\tau(\Lambda) = \Lambda, \qquad \tau(X\sigma_m) = \tau(X)\tau_m$$
$$\sigma(\Lambda) = \Lambda, \qquad \sigma(X\tau_m) = \sigma(X)\sigma_m.$$

In these definitions X is any word and G any generator.

LEMMA 1. *For all arguments for which the functions are defined we have*

$$\psi_1(XY) = \psi_1(X)\psi_1(Y), \qquad \psi_2(XY) = \psi_2(X)\psi_2(Y)$$
$$\chi(\psi_1(X)) = \psi_2(X) \qquad \varphi_2(\psi_1(X)) = \chi(X)$$
$$\psi_1(\varphi_1(V)) = V$$
$$\psi_2(\varphi_2(V)) = V.$$

These results are all trivial.

LEMMA 2. *The fundamental relations corresponding to the mth entry (U, V) of the table for $\mathfrak{B}$ are $(\varphi_1(U), \sigma_m\varphi_1(V)\tau_m)$ (first phase) and $(\sigma_m\varphi_2(V)\tau_m, \varphi_2(U))$ (second phase).*

This merely requires verification in each of the eight cases.

LEMMA 3. *If $C = C_0, C_1, C_2, \cdots$ are the descendants of C in $\mathfrak{B}$ (the sequence possibly terminating), then for each Cr we can find a normal word Hr such that (H_0, Hr) and $(\chi(Hr), \chi(H_0))$ are relations and $H_0 = \varphi_1(C_0)$, $\psi(Hr) = Cr$.*

In the case $r = 0$ the value $\varphi_1(C_0)$ evidently satisfies all the requirements. Suppose H_{r-1} has been defined and satisfies the conditions. We may write C_{r-1} as AUB and C_r as AVB where (U, V) is one of the entries of the machine. Now if $H_{r-1} = PQR$ then $A\ UB = \psi_1(P)\psi_1(Q)\psi_1(R)$. If P and R are sufficiently short, $\psi_1(P)$ and $\psi_1(R)$ are respectively contained in A and B. If we take them as long as can be done consistently with this, we shall have $\psi_1(P) = A$ and $\psi_1(R) = B$, for each increase of P or R by one symbol does not increase the length of $\psi_1(P)$ or $\psi_1(R)$ by more than one symbol. Hence $\psi_1(Q) = U$. We now define H_r to be $P_{\sigma_m\varphi_1}(V)\tau_m R$ where (U, V) is the entry E_m. Then

$$\psi_1(H_r) = \psi_1(P)\psi_1(\sigma_m)\psi_1(\varphi 1(V))\psi_1(\tau_m)\psi_1(R) = AVB = Cr$$

The relation (H_{r-1}, H_r) then follows from $(\varphi_1(U), \sigma_m(\varphi_1(V)\tau_m)$ which is an F.R. by Lemma 2. The F.R. $(\sigma_m\varphi_2(V)\tau_m, \varphi_2(U))$ can also be written as

$$(\chi(\psi_1(\sigma_m)\psi_1(\varphi_1(V))\psi_1(\tau_m)), \chi(Q))$$

in virtue of the equation $\psi_1(Q) = U$ and Lemma 1. From this F.R. follows the relation $(\chi(H_r), \chi(H_{r-1}))$.

LEMMA 4. *If $s_4 l_0 s_4$ is a descendant of C in $\mathfrak{B}$ then $(\varphi_1(C), \varphi_2(C))$ is a relation.*

Using the notation of Lemma 3 suppose that C_R is $s_4 l_0 s_4$. Then $(\varphi_r(C), H_r)$ and $(\chi(H_R), \varphi_2(C))$ are relations and it remains to obtain the relation $(H_R, \chi(H_R))$. Since $\psi_1(H_R) = s_4 l_0 s_4$ by Lemma 3 we can write the relation $(H_R, \chi(H_R))$ as $(\Sigma_1 j_4 \Sigma_2 u_0 T_1 K_4 T_2, \Sigma_1 n_4 \Sigma_2 w_0 T_1 P_4 T_2)$ where $\Sigma_1\Sigma_2$ contain only $\sigma_1, \sigma_2, \cdots, \sigma_M$, and $T_1 T_2$ contain only $\tau_1, \tau_2, \cdots, \tau_M$.

We obtain $(\Sigma_2 u_0,\ u_{0\tau}(\Sigma_2))$ and $(\sigma(T_1)w_0,\ w_0T_1)$ from the commutation relations and $(j_4u_0,\ n_4y)$, $(y\tau(\Sigma_2)T_1,\ \sum_2\sigma(T_1)y)$, $(yK_4,\ w_0P_4)$ from the change relations. From these the required relation follows.

6. The necessity in Theorem 1.

This completes the sufficiency part of Theorem 1, and it remains to prove the necessity. We are no longer interested in the detailed form of $\varphi_2(C)$, but only in the fact that it is not a first phase word, for this is in itself enough to ensure that $\mathfrak{B}$ reaches $s_4l_0s_4$. We proceed by an induction over the steps of the proof of the semi-group relation. We shall show that this proof need not contain any cancellations, and need only involve normal words.

Length of proof. Suppose that a proof contains a applications of (vii) or (viii), b applications of (i), c of (ii), (iii) or (iv) and d of (v) or (vi). We describe the length of the proof by the polynomial $a\omega^2 + b\omega + c + 2d$. We regard one polynomial as greater than another (and the corresponding proof as longer) if its values are greater for all sufficiently large positive values of ω. Alternatively we may regard $a\omega^2 + b\omega + c + 2d$ as an ordinal, but this may be misleading since our addition will be the addition of polynomials rather than the addition of ordinals. Thus for instance $(\omega+1)+(\omega^2+\omega)$ is $\omega^2+2\omega+1$ and not $\omega^2+\omega$.

We shall use the notation $[A,\ B]$ to mean "the length of the proof of the relation (A, B)".

The proofs are "well-ordered" when arranged according to their length in this sense. This means that if we can show that a property which holds of all proofs shorter than the proof P must also hold of P, then the property will hold of all proofs. Or in other words there is always a shortest proof not having a given property if there is any not having it. This is the basis of the arguments below.

LEMMA 5. *If a relation (A_1B) satisfies $n\omega \leqq [A,\ B] < (n+1)\omega$, then there is a chain $A = A_0,\ A_1, \cdots, A_n = B$ of words such that each pair of consecutive members (or links) is either of form $(XUY,\ XVY)$ or of form $(XVY,\ XUY)$ where $(U,\ V)$ is an* F.R. *In the former case the pair is described as a progressive joint, and in the latter as a retrogressive joint.*

The notation of this lemma is used throughout what follows. Consider the relation for which the lemma fails and for which the proof is shortest. The last step of the proof uses as premisses relations with shorter proofs and therefore ones for which the lemma holds. This last step is not by (vii) or (viii) for if it were the length would be at least ω^2. If the last step is by (i) then the length is ω, and the lemma holds with $n = 1$ by taking $U = A,\ V = B$ and $x,\ y$ void. If the last step is by (ii) then the length is 1 and the lemma holds with $n = 0$. If the last step is by (iii) then the chain for $(B,\ A)$ gives ones for $(A,\ B)$ when reversed in order. If it is by (iv) the chains for $(A,\ C)$ and for $(C,\ B)$ give one for $(A,\ B)$ when taken in combination. If it is by (v) and $(F,\ H)$ has the chain $F_0,\ F_1,\ \cdots,\ F_n$ then $(FG,\ HG)$ has the chain $F_0G,\ F_1G,\ \cdots,\ F_nG$. Similarly if it is by (vi). In each case there is a chain for the conclusion with the appropriate length, contrary to hypothesis. Hence the assumption that the lemma fails for some proof is false.

LEMMA 6. *If one link of a chain is normal then every link is normal.*

It will suffice to show that the neighbors of a normal link are normal. Suppose XUY, XVY are two consecutive links and XUY is normal, and either $(U,\ V)$ or $(V,\ U)$ is an F.R. Since XUY is normal X consists of left symbols and Y of right symbols. Also V is normal being a term of an F.R. Hence XVY is normal.

Chains such as those concerned in Lemma 6 will be called *normal chains.*

If $(XUY,\ XVY)$ is a joint of a chain and $(U,\ V)$ or $(V,\ U)$ is an F.R. then the joint is said to be *justified* by that F.R.

A two-generator word is described as a *barrier* if the proposition that the first generator is a left symbol is logically equivalent to the proposition that the second is a right symbol. A word is said to *have* a barrier if a barrier is a subword.

LEMMA 7. *No barrier occurs as a subword of either term of an* F.R.

This must be verified by detailed examination of the various F.R.

LEMMA 8. *A word without barriers is either normal or else consists entirely of left symbols or entirely of right symbols.*

This follows at once from the fact that two consecutive generators which do not form a barrier must form one of the four combinations left-left, left-active, active-right, right-right.

LEMMA 9. *If one link of a chain has no barriers, then no link of the chain has a barrier.*

We have only to show that if $X\ UY$ has no barriers and either $(U,\ V)$ or $(V,\ U)$ is an F.R. then XVY has no barriers. Any barrier which it has cannot be wholly contained in either X or Y, since it would then appear in XUY; nor can it be contained in V by Lemma 7. If it overlaps from X to V then its first generator is a left symbol, for it precedes a symbol in U which is not a right symbol. The first generator of V must therefore be a right symbol, which is impossible. Likewise the barrier cannot overlap from V to Y.

LEMMA 10. *If* $[G_1A,\ G_2B] < \omega^2$ *and* GG_1 *is a barrier then* GG_2 *is a barrier. Likewise if* $[AG_1,\ BG_2] < \omega^2$ *and* G_1G *is a barrier then* G_2G *is a barrier.*

We will consider only the first assertion. There is a chain for (G_1A, G_2B) by Lemma 5, and therefore one for $(GG_1A,\ GG_2B)$, by preceding each link with G. The result follows by lemma 9.

LEMMA 11. *Let* AG_1G_2F *and* BG_3G_4H *be abbreviated to* P *and* Q, *and let* $[P,\ Q] < \omega^2$. *Suppose also that* G_1G_2 *and* G_3G_4 *are barriers and either* AG_1 *and* BG_3 *contain no barriers or* G_2F *and* G_4H *contain no barriers. Suppose further that* P, Q *are not identical. Then* $[AG_1,\ BG_3] + [G_2F,\ G_4H] + 1 \leqq [P,\ Q]$.

If the lemma is false we suppose that $(P,\ Q)$ is the relation with the shortest proof consistent with denying the lemma. We deal only with the case where AG_1 and BG_3 contain no barriers, the case where G_2F and G_4H contain none being similar. Let us consider the last step in the proof of $(P,\ Q)$. The premisses to this last step must obey the lemma, if they are of appropriate form, i.e. if both terms have a barrier and they are not identical. This last step could not be by (i) since an F.R. has no barriers, nor by (ii) since P, Q are not identical. If it is by (iii) the proof could be reduced in length by omitting the last step and adopting a modified conclusion, while still denying the lemma. Now take the case where the last step is by (iv).We may suppose the premisses were $(P,\ D)$ and $(D,\ Q)$. We may suppose that D is not identical with either P or Q for if it were we could omit the last step and the identical premiss. If D contains a barrier it is of form EG_5G_6I where G_5G_6 is a barrier and E contains no barriers. The lemma then applies to both premises, i.e., to $(AG_1G_2F,\ EG_5G_6I)$ and $(EG_5G_6I,\ BG_3G_6H)$ so that

$$[AG_1,\ EG_5] + [G_2F,\ G_6I] + 1 \leqq [P,\ D].$$
$$[EG_5,\ BG_3] + [G_6I,\ G_4H] + 1 \leqq [D,\ Q].$$

But

$$[AG_1,\ EG_5] + [EG_5,\ BG_3] + 1 \geqq [AG_1,\ BG_3]$$
$$[G_2F,\ G_6I] + [G_6I,\ G_4H] + 1 \geqq [G_2F_1G_4H]$$
$$[P,\ D] + [D,\ Q] + 1 = [P,\ Q]$$

and so

$$[AG_1, BG_3]+[G_2F, G_4H]+1 \leqq [P, Q].$$

Hence D cannot contain a barrier. But this contradicts Lemma 9. If the last step is by (v) we may write $F'G$ and $H'G$ for F and H. Then

$$[G_2F, G_4H] \leqq [G_2F', G_4H']+1$$
$$[AG_1, BG_3]+[G_2F, G_4H]+1 \leqq [AG_1, BG_3]+[G_2F', G_4H']+2$$
$$\leqq [AG_1G_2F', BG_3G_4H']+1 = [P, Q].$$

If the last step is by (vi) then (P, Q) is of form (GG_5P', GG_6Q') where $G_1G_5G_6$ are generators and P', Q' may possibly be void. If GG_5 is a barrier then so is GG_6 by Lemma 10 and we see that A, B are void. (AG_1, BG_3) reduces to (G, G) whose proof is of length 1. The premiss of the last step is (G_2F, G_4H) and

$$[G_2F, G_4H]+2 = [P, Q]. \text{ Hence } [AG_1, BG_3]+[G_2F, G_4H]+1 = [P, Q].$$

If GG_5 is not a barrier then GG_6 is not either and we see that A, B are not void. Put $A = GA'$, $B = GB'$. A' and B' may be void. The premiss of the last step is $(A'G_1G_2F, B'G_3G_4H)$ and so since G_1G_2, G_3G_4 are barriers and $A'G_1$, $B'G_3$ have no barriers and $A'G_1G_2F$, $B'G_3G_4H$ are not identical

$$[A'G_1, B'G_3]+[G_2F, G_4H]+3 \leqq [A'G_1G_2F', B'G_3G_4H]+2 = [P, Q].$$

But

$$[A'G_1, B'G_3]+2 \geqq [AG_1, BG_3]$$

so that

$$[AG_1, BG_3]+[G_1F, G_4H]+1 \leqq [P, Q].$$

LEMMA 12. *No normal first phase word can be expressed in more than one way in the form X UY where U is the first term of an* F.R. *nor in more than one way in the form $XV\,Y$ where V is the second term of an* F.R.

If $X\,UY = X'U'Y'$ then $\psi(X)\psi(U)\psi(Y) = \psi(X')\psi(U')\psi(Y')$. But then

$$\varphi_1^{-1}(\psi(X))\varphi_1^{-1}(\psi(U))\varphi_1^{-1}(\psi(Y)) = \phi_1^{-1}(\psi(X'))\varphi_1^{-1}(\psi(U'))\varphi_1^{-1}(\psi(Y'))$$

and $\varphi_1^{-1}(\psi(U))$, $\varphi_1^{-1}(\psi(U'))$ are first terms in entries of the table for $\mathfrak{B}$, and $\varphi_1^{-1}(\psi(XUY))$, $\varphi_1^{-1}(\psi(X'U'Y'))$ are C.C.'s. Then by the restrictions to which all tables are subject we have $\varphi_1^{-1}(\psi(U)) = \varphi_1^{-1}(\psi(U'))$. Also U and U' have no generators σ_m (being first terms) and therefore $U = \psi(U)$, $U' = \psi(U')$. Hence $U = U'$. If $XVY = X'V'Y'$ then since both sides are normal V and V' must overlap, and since they both terminate with a σ_m or τ_m they must have a σ_m or τ_m in common. Hence the value of m is the same for both, i.e., V and V' are the same relation. Since the words are normal they contain at most one active symbol each. Hence in XVY, $X'V'Y'$ the active symbol of V must occur at the same distance from the beginning of both expressions, so that $X = X'$ and likewise $Y = Y'$.

LEMMA 13. *In a normal chain without repetitions there is no change of direction, i.e. the joints are either all progressive or all retrogressive.*

Suppose there is a link which is flanked both by a progressive and by a retrogressive joint.

Then it is expressible in both the forms XUY and $X'U'Y'$ where (U, V) and (U', V') are F.R. or else in both the forms XVY and $X'V'Y'$ In the former case the flanking links are XVY and $X'V'Y'$, and in the latter they are XUY and $X'U'Y'$. In either case they are identical, by Lemma 12 together with the fact that an F.R. is uniquely determined by either of its terms. The chain therefore has a repetition contrary to hypothesis.

LEMMA 14. *If the end links of a chain are first phase words, and there is no change of direction, then all links are first phase.*

We will suppose the chain progressive. Consider the last link which is not first phase, and suppose that the joint following it is justified by (U, V). Then V is first phase but U is not. But looking through the fundamental relations we see that there is no such (U, V).

LEMMA 15. *A normal chain without change of direction is completely determined by its first link.*

This is an immediate consequence of Lemma 12 and the definition of a chain.

LEMMA 16. *Let Σ_1, Σ_2 be words formed from $\sigma_1\sigma_2, \cdots, \sigma_M$, and let G be a first phase symbol, G' a second phase symbol. Then $\Sigma_1 G'X$ cannot precede $\Sigma_2 GY$ in a progressive chain.*

Every link can be written in the form $\sum FZ$ where Σ (and also Σ_3, Σ_4 etc.) contains only $\sigma_1\sigma_2, \cdots, \sigma_M$, and the generator F is not a σ_m. If $\Sigma_3 F'X'$, $\Sigma_4 FY'$ are two consecutive links of the chain then the corresponding joint must be justified by an F.R. of the form $\Sigma_5 F'X''$, $\Sigma_6 FY''$. If $\Sigma_3 F'X'$ is the last link for which F' is second phase, then F' is second phase but F is not. Clearly such an F.R. is neither first phase nor second phase, and no such relation can be found among the commutation relations or the change relations.

LEMMA 17. *If G is a generator and there is a normal chain for (GA, GB) or for* (AG, BG) *then there is one for (A, B).*

Consider the case where the given chain is one for (GA, GB), and suppose that there is no chain for (A, B). Let the chain for (GA, GB) be $F_0, F_1, \cdots, F_n$ Not all the links of the chain begin with G, for if they did we could obtain a chain for (A, B) simply by omitting G from each link. Let the first and last links which do not begin with G be F_r and F_s.We will assume the chain progressive. Then (F_{r-1}, F_r) is justified by some F.R. of form (GU, V) and (F_s, F_{s+1}) by one of form (U', GV') where V and U' do not begin with G. Thus G is both the leftmost symbol of the first member of a relation and the leftmost symbol of a second member. The only symbols which can satisfy this condition are σ_m. The first symbol of V must be of form z_i, u_i, or n_h. If it is z_i or u_i then it is the active symbol in V, and the link (F_r, F_{r+1}) must be justified by a relation whose first member begins with that symbol. But there is no such relation, so that V begins with n_h. Likewise U' begins with an j_h. But there is a progressive chain from V to U' contradicting Lemma 16. This completes the proof for the case (GA, GB). The argument for the case (AG, BG) is similar; although there is not complete symmetry in the change relations between left and right, it does not affect the argument at any point.

LEMMA 18. *If one term of a relation contains no active symbol and the proof of the relation does not apply* (vii) *or* (viii), *then the relation is an identity.*

We consider the last step of the shortest proof of such a relation which is not an identity. It is not by (i) since every F.R. has an active symbol in each term, nor by (ii) since it is not an identity. If it is by (iii) we could omit the last step and obtain a shorter proof denying the lemma. If it is by (iv) one of the premisses has a term which contains no active symbol. This premiss has a shorter proof than the relation in question and therefore is an identity, and the last step may be omitted. If it is by (v) or (vi) the premiss must be an identity by a similar argument, and the conclusion is therefore also an identity. A step by (vii) or (viii) has been excluded by the conditions of the lemma. Hence there is no such relation which is not an identity.

LEMMA 19. *The shortest proof of a normal relation does not use* (vii) *or* (viii) *or words which are not normal.*

We consider the relation (P, Q) which denies the lemma and, consistently with this, has the shortest possible proof. If there is an application of (vii) or (viii) consider the first one, and suppose it is applied to (GA, GB) to obtain (A, B). Then either (A, B) is (P, Q) or else A, B are not normal for in any other case (A, B) would deny the lemma and have a shorter proof than (P, Q). If (A, B) is (P, Q) then (GA, GB) cannot be normal, for if it were its proof would use only normal words, being shorter than that of (P, Q). There would therefore be a normal chain for (GA, GB) by Lemmas 5, 6 and one for (A, B) (i.e., for (P, Q)) by Lemma 17. Then (P, Q) satisfies the lemma contrary to hypothesis. Equally if (A, B) is not normal (GA, GB) will not be normal. Then by Lemma 8 there are either barriers in GA, GB or one of them contains no active symbol. In the latter case since (vii) and (viii) are not used in the proof of (GA, GB) this relation must be an identity (Lemma 18). But in this case we can shorten the proof of (A, B) since it is an identity. Hence there are barriers in GA and GB. Applying Lemma 11 we see that either the barriers are right at the beginning and $[G, G]+[A, B]+1 \leqq [GA, GB]$ or else (A, B) is of form $(A'G_1G_2F, B'G_3G_4H)$ where G_1G_2 and G_3G_4 are barriers and $GA'G_1$, $GB'G_3$ without barriers and

$$[GA'G_1, GB'G_3]+[G_2F, G_4H]+1 \leqq [GA, GB] < [P, Q].$$

We may apply Lemma 17 to $(GA'G, GB'G_3)$: this shows that there is a chain for $(A'G_1, B'G_3)$. Combining this with the proof of (G_2F, G_4H) gives us a proof of (P, Q) which does not apply (vii) or (viii) and is therefore shorter than the original one. The same is true if $[G, G]+[A, B]+1 \leqq [GA, GB]$. Hence there is not any application of (viii) and likewise none of (vii). The whole proof is therefore of length less than ω^2, so that by Lemma 5 there is a chain for (P, Q). But then (P, Q) does not deny the lemma, contrary to hypothesis. This establishes the lemma.

LEMMA 20. *If for some W which is not first phase* $(\varphi_1(C), W)$ *is a relation, then $s_4l_0s_4$ is a descendant of C.*

By Lemmas 19, 15 there is a chain for $(\varphi_1(C), TV)$. Let the last first phase link $X\ UY$ and the joint connecting it to the next link be ustified by (U, V). Then U is first phase but V is not. This identifies (U, V) as (j_4u_0, n_4y). Now consider the words H_r described in Lemma 3. As each H_r consists only of first phase symbols the same is true of each link of its chain (Lemma 14). This chain therefore is wholly contained in the chain for $(\varphi_1(C), XUY)$, which we will suppose is of length n. Then $r \leqq n$ so that C has only a finite number of descendants. Let the last be C_s, and consider the chain leading from H_s to XUY. Since it consists entirely of first phase symbols the fundamental relations involved must be either first phase relations or commutation relations. I assert that they are all commutation relations. For if one is a first phase relation then C_s has a successor. Since these steps are all by commutation relations, H_s contains u_0 and therefore C_s contains l_0, and is therefore $s_4l_0s_4$ by the properties of $\mathfrak{B}$.

Theorem 1 follows at once from Lemmas 4, 20 and the fact that $\varphi_2(C)$ is not first phase.

VICTORIA UNIVERSITY
MANCHESTER, ENGLAND

References

ALONZO CHURCH, [1]. *An unsolvable problem of elementary number theory*, Amer. J. Math., 58 (1936), 345–363.

S. C. KLEENE, [1]. *General recursive functions of natural numbers*, Math. Ann., 112 (1935–6), 727–742.

S. C. KLEENE, [2]. *λ-definability and recursiveness*. Duke Math. J., 2 (1936), 340–353.

E. L. POST, [1]. *Finite combinatory processes–formulation 1*, J. Symbol lic Logic, 1 (1936), 103–105.

E. L. POST, [2]. *Recursive unsolvability of a problem of Thue*, J. Symbolic Logic, 12 (1947), 1–11.

A. M. TURING, [1]. *On computable numbers, with an application to the Entscheidungsproblem*, Proc. London Math. Soc. (2), 42 (1937), 230–265. A correction to this paper has appeared in the same periodical, 43 (1937), 544–546.

A. M. TURING, [2]. *Computability and λ-definability*, J. Symbolic Logic, 2 (1937), 153–163.

D. HILBERT and P. BERNAYS, Grundlagen der Mathematik, 2 vols., Berlin, (1940).

On Permutation Groups

(unpublished manuscript)

John Leslie Britton's informative —

INTRODUCTION

Although this paper was evidently motivated by Turing's study of the Enigma machine (see Hodges: *Alan Turing – The Enigma*, 1983), it is essentially an important piece of pure mathematics. Turing was led to consider what turns out to be a formidable problem on permutation groups, which is as follows.

Consider permutations of the objects $a_1, \ldots, a_T$. Let R be the T-cycle $(a_1 a_2 \ldots a_T)$. For any permutation U, let $H(U)$ denote the group of all permutations of the form

$$R^{t_0} U R^{t_1} \ldots U R^{t_p}, \quad \sum t_i = 0.$$

$H(U)$ is called *exceptional* if it is not the symmetric or alternating group. The problem is to find all exceptional groups or at least to find all U such that $H(U)$ is exceptional.

Besides employing his usual ingenuity, Turing has to perform some really extensive calculations in order to solve the problem for the cases $T = 1, 2, \ldots, 8$.

Clearly, this problem is a challenge to present day workers in permutation groups. Computer scientists also may find it interesting to see if they can check Turing's results and extend his calculations beyond $T = 8$.

[1]Extracted from the Introduction to volume 1 of the Collected Works.

ON PERMUTATION GROUPS

[The beginning of Turing's typescript, including the title, is missing. The above title was suggested by R. O. Gandy.]

... form $U_{n_1} U_{n_2} \dots U_{n_M}$. It is easily seen that the parity of this permutation is that of U^M and therefore that the permutations from the same machine always have the same parity. For the present we shall not however investigate the permutations obtainable with a given machine but those which are obtainable with a given upright and any number of wheels.

Let $H(U)$ or H be the set of permutations obtainable from the upright U. It is easily seen that H is a group, for if P and Q are two permutations obtainable from the upright U, then PQ is obtainable from it by putting the machines giving P and Q in series ; (an algebraic argument is almost equally simple). Since H is finite and contained in the symmetric group, this is sufficient to prove it is a group. H may be expressed as the group generated by $U_1, U_2, \dots, U_T$ or again as consisting of all expressions of the form

$$R^{t_0} U R^{t_1} U \dots U R^{t_p},$$

with any p and with the exponents of R totalling zero. It thus differs only slightly from the group $J(U)$ or J generated by U, R, which consists of similar expressions without the restriction on the sum of the exponents. Every member of $J(U)$ is of the form PR^k, where P is in H. H is thus a subgroup of J of index at most T. H is in fact a self-conjugate (or normal) subgroup of J, for it is transformed into itself by the generators U, R of J, i.e., UHU^{-1} is H since U is in H and RU_nU^{-1} is U_{n+1} so RHR^{-1} is H.

We shall say that H or $H(U)$ is *exceptional* if H does not include the whole alternating group A (all even permutations). If H is not exceptional, it will be either A or the symmetric group S (all permutations), according as U is even or odd. It is so easy to see in this way whether H is A or S that it is quite adequate in describing it to say that it is unexceptional.

We shall actually investigate J rather than H. J is obviously easier to deal with than H and results for J may be translated into results for H by means of the next theorem.

Theorem I. *The necessary and sufficient condition for H to be unexceptional is that J be unexceptional, provided $U \neq 1$, $T \neq 4$.*

We have shown that H is a self-conjugate subgroup of J. Now if J is unexceptional, it is either A or S and the only self-conjugate subgroups of these if $T \neq 4$ are A, S and the group consisting of the identity only, and so H must be one of these. The last alternative has been excluded by assuming $U \neq 1$ so that H is A or S, i.e., H is unexceptional. Conversely, if H is unexceptional, so is J since J contains H.

Technique for investigating any particular upright U

In order to prove H is unexceptional, it will suffice to prove J contains all 3-cycles, for if this is so, J will be a self-conjugate subgroup of S and since it is not the identity, it must be either A or S. It would also be sufficient to prove that J contains all 2-cycles. We shall prove the following theorem.

Theorem II. *If J contains a member of the form $(\alpha, R^m\alpha)$ or $(\alpha, R^m\alpha, \beta)$ or $(\alpha, R^m\alpha)(\beta, \gamma)$, where m is prime to T, then it contains all 3-cycles and, in the first of these cases, also all 2-cycles. T must be greater than 4. $(\alpha, R^m\alpha)(\beta, \gamma)$ must not commute with $R^{T/2}$ if T is even.*

Suppose J contains $(\alpha, R^m\alpha)$. We will write α_k for $R^{mk}\alpha$. The symbols $\alpha_0, \alpha_1, \dots, \alpha_{T-1}$ include all the T symbols. Then J contains $R^{ms}(\alpha_0, \alpha_1)R^{-ms}$, i.e., (α_s, α_{s+1}). It therefore contains (α_0, α_2) since this is $(\alpha_1, \alpha_2)(\alpha_0, \alpha_1) \times (\alpha_1, \alpha_2)^{-1}$(if $T > 2$). It contains (α_0, α_3), which is $(\alpha_2, \alpha_3)(\alpha_0, \alpha_2)(\alpha_2, \alpha_3)^{-1}$ if $T > 3$ and repeating this argument it contains (α_0, α_r) for every $0 < r < T$. Finally it contains (α_p, α_q) since this is $R^{mp}(\alpha_0, \alpha_{q-p})R^{-mp}$ if $q \not\equiv p\ (T)$. Thus J contains every 2-cycle (and every 3-cycle).

Now suppose J contains $(\alpha, R^m\alpha, \beta)$ where m is prime to T. We may express β as $R^{mk}\alpha$. The first step is to prove that J contains an element of the form $(\alpha, R^{m'}\alpha, R^{2m'}\alpha)$, where m' is prime to T. In the case that $2k \equiv 1(T)$ we may take $m' = k$ and $(\alpha, R^m\alpha, R^{km}\alpha)^{-1}$ is $(\alpha, R^{m'}\alpha, R^{2m'}\alpha)$. In the case that $k+1 \equiv 0\ (T)$ we may take $m' = m$ and $R^m(\alpha, R^m\alpha, R^{mk}\alpha)R^{-m}$ is $(\alpha, R^{m'}\alpha, R^{2m'}\alpha)$. In the remaining cases I propose to prove that $(\alpha, R^m\alpha, R^{(k+1)m}\alpha)$ belongs to J, so that k may be increased until either $k+1 \equiv 0\ (T)$ or $2k-1 \equiv 0\ (T)$, and then previous cases can be applied.

We are assuming that $k+1 \not\equiv 0\ (T)$ and $2k-1 \not\equiv 0\ (T)$. For the moment I will also suppose $2k \not\equiv 0\ (T)$. Then $\alpha, R^m\alpha, R^{mk}\alpha, R^{m(k+1)}\alpha, R^{2mk}\alpha$ are all different and

$$\begin{aligned}
&[R^{mk}(\alpha, R^m\alpha, R^{mk}\alpha)R^{-mk}](\alpha, R^m\alpha, R^{mk}\alpha)[R^{mk}(\alpha, R^m\alpha, R^{mk}\alpha)R^{-mk}]^{-1}\\
&= (R^{mk}\alpha, R^{m(k+1)}\alpha, R^{2mk}\alpha)(\alpha, R^m\alpha, R^{mk}\alpha)(R^{mk}\alpha, R^{m(k+1)}\alpha, R^{2mk}\alpha)^{-1}\\
&= (\alpha, R^m\alpha, R^{m(k+1)}\alpha)
\end{aligned}$$

belongs to J. If, however, $2k \equiv 0\ (T)$ the left-hand side of this expression is equal to $(R^{mk}\alpha, R^m\alpha, R^{m(k+1)}\alpha)$ or to $(R^{mk}\alpha, R^{(2k+1)m}\alpha, R^{(k+1)m}\alpha)$ so that

$$R^{-mk}(R^{mk}\alpha, R^{(2k+1)m}\alpha, R^{(k+1)m}\alpha)^{-1}R^{mk} = (\alpha, R^{(k+1)m}\alpha, R^m\alpha)^{-1} = (\alpha, R^m\alpha, R^{(k+1)m}\alpha)$$

belongs to J.

This proves the existence in J of $(\alpha, R^{m'}\alpha, R^{2m'}\alpha)$ with m' prime to T, and I will now write α_k for $R^{m'k}\alpha$, so that $(\alpha_0, \alpha_1, \alpha_2)$ belongs to J, and so does $(\alpha_s, \alpha_{s+1}, \alpha_{s+2})$ (by transforming with $R^{m's}$). Then transforming $(\alpha_0, \alpha_1, \alpha_2)$ with $(\alpha_2, \alpha_3, \alpha_4)$, we find that $(\alpha_0, \alpha_1, \alpha_3)$ belongs to J provided $T \geq 5$ and transforming again with $(\alpha_3, \alpha_4, \alpha_5)$, we have $(\alpha_0, \alpha_1, \alpha_4)$ in J if $T \geq 6$, and repeating the argument $(\alpha_0, \alpha_1, \alpha_r)$ belongs to J provided $r < T-1$. $(\alpha_0, \alpha_1, \alpha_{T-1})$ may be seen to belong to J by transforming $(\alpha_0, \alpha_1, \alpha_{T-2})$ with $(\alpha_{T-3}, \alpha_{T-2}, \alpha_{T-1})$ provided $T > 4$. Thus every 3-cycle containing α_0 and α_1 belongs to J. But

$$(\alpha_0, \alpha_1, \alpha_r)(\alpha_0, \alpha_1, \alpha_s)^{-1} = (\alpha_s, \alpha_r, \alpha_0)$$

and therefore every 3-cycle containing α_0 belongs to J. Finally

$$(\alpha_s, \alpha_0, \alpha_r)(\alpha_s, \alpha_0, \alpha_t)^{-1} = (\alpha_t, \alpha_r, \alpha_s)$$

and therefore every 3-cycle whatever belongs to J.

There remains the case when we are given that $(\alpha, R^m\alpha)(\beta, \gamma)$ belongs to J. We shall reduce this to the case where it contains the 3-cycle. We may write α_k for $R^{mk}\alpha$ as before. We also write β' for $R^m\beta$. We are then given that $(\alpha_0, \alpha_1)(\beta, \gamma)$ belongs to J. We may divide this into the cases (remembering $T > 4$):

(a) $(\alpha_0, \alpha_1)(\beta, \gamma)$ belongs to J, where β, γ are different from α_2, α_{-1}; β is different from γ' and γ from β';
(b) $(\alpha_0, \alpha_1)(\alpha_2, \gamma)$ belongs to J, where γ is different from α_{-1}, α_{-2};
(c) $(\alpha_0, \alpha_1)(\alpha_{-1}, \gamma)$ belongs to J, where γ is different from α_2, α_3;
(d) $(\alpha_0, \alpha_1)(\beta, \beta')$ belongs to J, where β is different from α_2, α_{-1};
(e) $(\alpha_0, \alpha_1)(\alpha_{-1}, \alpha_2)$ belongs to J;
(f) $(\alpha_0, \alpha_1)(\alpha_2, \alpha_3)$ belongs to J;
[Following the instruction "P.T.O.", the reader finds:]
(g) $(\alpha_0, \alpha_1)(\alpha_2, \alpha_{-2})$;
(h) $(\alpha_0, \alpha_1)(\alpha_{-1}, \alpha_3)$. [See Fig. 1*]

* This consists of a circle passing through points $\alpha_{-3}, \alpha_{-2}, \alpha_{-1}, \alpha_0, \alpha_1$; a point α_3 is inside the circle. A second circle has diameter α_0, α_1. A third circle passes through α_{-2}, α_3 and surrounds the small circle.

These are equivalent. $(\alpha_0, \alpha_1)(\alpha_2, \alpha_{-2})$. Then $(\alpha_{-3}, \alpha_{-2})(\alpha_{-1}, \alpha_{-5})$ gives $(\alpha_0, \alpha_1)(\alpha_2, \alpha_{-3})$ if $T > 7$. O.K. In any case we can take it ... as (BC)(AE)

when $T = 6$

A B C D E F   (AF)(EDB) O.K.
E C B D A F

when $T = 5$

A B C D E   (EDB) O.K.
E C B D A

Thus only exception $T = 7$ when we get group of 7-point geometry.

[The main text now resumes:]

It is easily seen that cases (b) and (c) are essentially the same (by changing the sign of m). In case (a), since $(\alpha_1, \alpha_2)(\beta', \gamma')$ belongs to J

$$((\alpha_1, \alpha_2)(\beta', \gamma'))(\alpha_0, \alpha_1)(\beta, \gamma)((\alpha_1, \alpha_2)(\beta', \gamma'))^{-1} = (\alpha_0, \alpha_2)(\beta, \gamma)$$

belongs to J and so does

$$(\alpha_0, \alpha_1)(\beta, \gamma)((\alpha_0, \alpha_2)(\beta, \gamma))^{-1} = (\alpha_0, \alpha_2, \alpha_1).$$

In case (b), supposing first $\gamma \neq \alpha_2$, $(\alpha_2, \alpha_3)(\alpha_4, \gamma'')$ belongs to J and therefore, transforming $(\alpha_0, \alpha_1)(\alpha_2, \gamma)$ with this permutation, also $(\alpha_0, \alpha_1)(\alpha_3, \gamma)$ which comes under case (a). If however $\gamma = \alpha_{-2}$, the result of the same transformation is $(\alpha_4, \alpha_1)(\alpha_2, \alpha_{-2})$. But

$$((\alpha_4, \alpha_1)(\alpha_2, \alpha_{-2}))((\alpha_0, \alpha_1)(\alpha_2, \alpha_{-2})) = (\alpha_0, \alpha_4, \alpha_1)$$

and we can apply the case of the 3-cycles.

In case (d) put α_r for β. Then if $2r \not\equiv 0, 1, -1$ (T),

$$(\alpha_0, \alpha_1)(\alpha_r, \alpha_{r+1})(\alpha_{r+1}, \alpha_{r+2})(\alpha_{2r+1}, \alpha_{2r+2}) = (\alpha_0, \alpha_1)(\alpha_r, \alpha_{r+1}, \alpha_{r+2})(\alpha_{2r+1}, \alpha_{2r+2})$$

and so $(\alpha_r, \alpha_{r+1}, \alpha_{r+2})$ belongs to J and the case of 3-cycles applies.

If $2r \equiv 1$ (T), then $(\alpha_0, \alpha_1)(\alpha_r, \alpha_{r+1})$ and

$$(\alpha_r, \alpha_{r+1})(\alpha_{2r}, \alpha_{2r+1}) = (\alpha_r, \alpha_{r+1})(\alpha_1, \alpha_2)$$

belongs to J and therefore their product $(\alpha_1, \alpha_2, \alpha_0)$ also. Similarly if $2r \equiv -1$ (T). If $2r \equiv 0$ (T), then $(\alpha_0, \alpha_1)(\alpha_r, \alpha_{r+1})$ commutes with $R^{T/2}$.

In case (e), $(\alpha_0, \alpha_1)(\alpha_{-1}, \alpha_2)$ and $(\alpha_1, \alpha_2)(\alpha_0, \alpha_3)$ belong to J and therefore (transforming) $(\alpha_3, \alpha_2)(\alpha_{-1}, \alpha_1)$ and (subtracting 3 from suffixes) $(\alpha_0, \alpha_{-1})(\alpha_{-4}, \alpha_{-2})$. On changing the sign of m this becomes $(\alpha_0, \alpha_1)(\alpha_4, \alpha_2)$ and comes under case (b).

In case (f), $(\alpha_0, \alpha_1)(\alpha_2, \alpha_3)$ and $(\alpha_1, \alpha_2)(\alpha_3, \alpha_4)$ belong to J and therefore $(\alpha_0, \alpha_2)(\alpha_1, \alpha_4)$ and the product

$$((\alpha_0, \alpha_1)(\alpha_2, \alpha_3))((\alpha_0, \alpha_2)(\alpha_1, \alpha_4)) = (\alpha_0, \alpha_3, \alpha_2)(\alpha_1, \alpha_4)$$

and its square $(\alpha_2, \alpha_3, \alpha_0)$ to which the 3-cycle result applies.

This completes the proof of Theorem II, but we may notice the cases which have been expressly excluded since they give rise to exceptional groups. In the case of a 2-cycle $(\alpha, R^m\alpha)$ where m is not prime to T, H is an intransitive group. The intransitivity sets are each of the form $(\beta, R^m\beta, R^{2m}\beta, \ldots)$. The same applies in the case that the generator is a 3-cycle $(\alpha, R^m\alpha, R^{m+p}\alpha)$ when $m, p, m+p$ have a factor in common with T. We have omitted to give consideration to the case where $m, p, m+p$ each have a factor in common with T, but there is no factor in common with all four. Probably Theorem II applies to these cases also. In the case where the generator is $(\alpha, \beta)(\gamma, \delta)$ and commutes with $R^{T/2}$, each element of the group H also commutes with $R^{T/2}$.

It is very easy to apply Theorem II. We may first express U, UR, UR^2 etc. in cycles: this may be done for instance by writing the alphabets out double and also writing out the sequence UA, UB,..., UZ. By putting the former above the latter in various positions we get the permutations UR^s. Among these we may look for the permutations which have a 3-cycle and all other cycles of length prime to 3. By raising this to an appropriate power we obtain a 3-cycle which may or may not satisfy the conditions in Theorem II. If we are not successful we may use other permutations in J. We may also be able in a similar way to generate a permutation which is a pair of 2-cycles.

The following upright was chosen at random

A B C D E F G H I J K L M N O P Q R S T U V W X Y Z
M N Y T F B G R S L A X O E W K P C J Q Z D H V U I

In cycles it is

$$(\text{AMOWHRCYUZISJLXVDTQPK})(\text{BNEF})(\text{G}) = U.$$

Then $U^{22} = (\text{BE})(\text{NF})$.

The distance BE is 3, which is prime to 26. The distance NF is 8. Hence, Theorem II applies and J includes the whole of A, and therefore H includes A.

Systematic search for exceptional groups. Theory

In examining all possible uprights for a given T the main difficulty lies in the large number of uprights involved. Once it has been proved that a particular upright is unexceptional, the same will follow for a great number of others.

More generally given any upright we can find a great number of others which generate either the same group H or an isomorphic group. If we can classify these uprights together in some way we shall enormously reduce the labour, since we shall only need to investigate one member of each class. The chief principles which enable us to find equivalent uprights are:

(i) if $U' = R^m U R^n$, then $H(U') = H(U)$;
(ii) if V commutes with (R), then $H(VUV^{-1}) \cong H(U)$;
(N.B. If V commutes with (R), then $VRV^{-1} = R^k$, some k.)
(iii) if $U' = U^m$ and $U = U'^n$, then $H(U) = H(U')$.

The principle (i) is the one of which we make the most systematic use. Our method depends on the fact that there are very few U for which none of the permutations $R^m U R^n$ leave two letters invariant (in other words there are very few U without a beetle) and none if T is even. We therefore investigate separately the U with no beetles and the U with beetles.

U with no beetle

We can find an expression which determines the classes of permutations obtainable from one another by multiplication right or left by powers of R as follows.

Let $UR^{n+1}\text{Z} = R^{f(n)} UR^n \text{Z}$; (here Z represents the last letter of the alphabet however many characters there may be in it). Then we take the numbers $f(1)f(2)\ldots f(T)$ as describing the classes containing U. It may be verified that from these numbers and UZ it is possible to recover U, in fact they describe what is in common between U, RU, R^2U,... However if we move some of the numbers from the end of the sequence to the beginning, we shall be describing $UR^m, RUR^m, \ldots$ An example may help. Consider the permutation $U = $ (ABCDG) of seven letters. Write it as

A B C D E F G

B C D G E F A.

The numbers are the differences of consecutive letters in BCDGEFA, i.e., 1 1 3 5 1 2 1. The permutation R^2UR is

A B C D E F G

E F B G A C D

and the numbers 1 3 5 1 2 1 1 are obtainable by shifting the first figure to the end.

If then we take the various forms

$$f(1)f(2)\ldots f(T), f(T)f(1)f(2)\ldots f(T-1), \quad \text{etc.,}$$

and select that which, regarded as an arabic numeral, would be the smallest, we shall have a way of describing all the permutations $R^m UR^n$. We may call the resulting figures the *invariant* of U.

To a small extent we can combine this with the principle (ii) conveniently. If V satisfies $VRV^{-1} = R^{-1}$, then the invariant of VUV^{-1} is obtainable by reading that of U backwards and rearranging for minimum.

When we are investigating the cases where U has no beetle, the invariant is very restricted. It cannot contain Fig. 1. More generally the numbers

$$0, f(1)-1, f(1)+f(2)-2, f(1)+f(2)+f(3)-3, \ldots$$

(essentially one of the "rods") must be all different. These restrictions are so powerful that we normally find very few cases other than where U is a power of R. When T is even there are no such cases (as is well known): if there were the numbers $0, f(1)-1, f(1)+f(2)-2, \ldots$ would have to be the numbers $0, 1, \ldots, T-1$ in some order and would total $\frac{1}{2}T(T-1)$ modulo T. But $0, f(1), f(1)+f(2), \ldots$ are also all different and must total $\frac{1}{2}T(T-1)$ modulo T, and likewise $0, 1, \ldots, T-1$ total the same so that $0, f(1)-1, f(1)+f(2)-2, \ldots$ total $0 \mod T$, i.e., $\frac{1}{2}T(T-1)$ is $0 \mod T$, which is not so if T is even.

U with a beetle

We select a permutation $R^m UR^n$ which leaves two letters invariant to represent U. One of these may be taken to be A, if necessary by transforming with a further power of R. By means of principle (ii) we can also reduce the possibilities for a second letter. If the letters which were originally fixed were A and R^tA, we can transform them to A and R^sA provided that the highest common factor of T and t is the same as the highest common factor of T and s, since there is a V satisfying $V\text{A} = \text{A}$ and $VR^tV^{-1} = R^s, V(R)V^{-1} = (R)$. We therefore have to consider various pairs of letters left invariant such as A, R^tA; the various t to be considered should run through the numbers which divide T, omitting T itself but including 1. For each such t we then write down the permutations leaving A, R^tA invariant. We arrange them by classes of conjugates in S, and reduce them by means of principle (iii) as we write them down. Further reduction may afterwards be done by principle (ii) in a very special way: if V interchanges A and R^tA we can apply it. It is best probably to number the permutations and with each to give also the number t of that with which it is paired. Very many will be found self-paired.

The detailed search

T = 1, 2, 3, 4

It is not difficult to prove that there are no exceptional groups when $T = 1, 2$ or 3. The case $T = 4$ needs special investigation as it has been expressly excluded from Theorem I. The exceptional uprights in this case are

(1) (13) (24) (13)(24)

(12)(34) (32)(14) (1234) (4321).

The exceptional groups H are the identity, the cyclic groups [(1234)] and [(13)(24)], the 4-group, consisting of the identity and all permutations of the form $(\alpha\beta)(\gamma\delta)$, and a group isomorphic with the 4-group and generated by (13) and (24).

T = 5

With the theory we have developed, this case is also very trivial. There are no U without beetles except

[Following the instruction "P.T.O." we find:]

those with invariant

22222 33333 44444.

These together prove the numatizer of (R).

[Returning to the main text:]

and this leaves only the permutations leaving two letters invariant, and these are all covered by Theorem II, since 5 is prime.

Thus the only exceptional U are the members of the numatizer of (R), 20 in number.

T = 6

Since 6 is even we do not need to consider U without beetles. Our representative U will then always leave two letters invariant, and again will come under Theorem II immediately. U can only be exceptional by being intransitive or commuting with R^3. The total number which commute with R^3 is 48 and the number which have the intransitivity sets (ACE BDF) is 36. These include 6 which commute with R^3. Those with intransitivity set (AD, BE, CF) all commute with R^3. Thus the total number of exceptional U is $48 + 36 - 6 = 78$ out of a possible 720.

T = 7

The uprights without beetles are the members of the numatizer of (R) and those with the invariant 2335564 and its reversal. The latter however are unexceptional. A representative U is (BCGEDF) and U^2R is (AG)(BEC)(DF) which is evidently unexceptional (square it).

For the uprights with beetles we may take $t = 1$ only, i.e., we can always suppose A and B both invariant. Simple application of Theorem I shows that we need only consider cases of 5-cycles and principle (iii) shows that we can take it that $UC = \mathrm{G}$, i.e., we have reduced the representative permutations U to the form (A)(B)(G$\alpha\beta\gamma$C). The permutation V which interchanges A and B is (AB)(CG)(DF)(E). Thus, we need in the end only consider

(GDEFC) self paired
(GDFEC) paired
(GEDFC)
(GEFDC) paired
(GFDEC)
(GFEDC) self paired

By multiplying these with powers of R they may all be shown unexceptional.

We have also the case of the group generated by (AE)(BC) specifically mentioned in Theorem II. This gives the group of symmetry of the 7-point geometry [See Fig. 2*]. There are actually only two isomorphic groups of this kind, generated by (BC)(AE) and by (BE)(FD).

Invariants involved

(BC)(AE) group	(BE)(FD) group reverses
1556245	1542655
1323354	1453323
4444444	4444444
1264663	1366462
1111111	1111111
2222222	2222222

Invariants of numatizer of (*R*)

3333333	
5555555	total elements
6666666	$6 \times 49 + 6 \times 7 = 6 \times 7 \times 8 = 336$
4444444	i.e., 336 exceptional uprights
1111111	
2222222	

$T = 8$

This needs rather more investigation than the previous cases partly because it is the largest number yet considered and partly because it has more factors.

Obviously the permutations which commute with (R) or with a power of R or generate an intransitive group will be exceptional. We will consider that we are looking for other forms of exceptional upright.

We have various means for dealing with permutations:

(a) We may show that the group is the same as that generated by an upright to be considered later. A special case of this occurs when $t = 1$. One of the forms R^sU may be of a type to be considered under $t = 2$ or $t = 4$. This may be detected by writing down the figures $m_1, m_2, \ldots$ where $R^{m_r+r}H = UR^rH$. If any figure appears twice in this series at an even distance the conversion to the case $t = 2$ is possible. It will be indicated by an "X".

(b) One of the slides R^sU may be one to be considered later under the same value of t; marked "below".

(c) The square or some other power may be proved unexceptional. This means that we need only consider those 6-cycles whose squares and cubes are unexceptional. We may consider a 6-cycle and its square and cube simultaneously; this is advantageous also because the transformations V apply in the same way to all three.

(d) One of the slides R^sU may be unexceptional. This will be indicated by "slide", the value of the slide in cycles and "O.K. "

(e) One of the commutators UR^sUR^{-s} may be unexceptional. This is indicated by the value of the commutator and "O.K."

When all these fail, a query will be shown and the upright investigated further later.

* Fig. 2. This consists of a triangle ABC with inscribed circle, centre E, meeting the sides BC, CA, AB in points G, D, F, respectively.

t = 1

We may first go over the main plan, considering separately what is to be done with the various classes of conjugates in the symmetric group.

6-cycles.	These are left aside till the double threes have been considered.
Double threes.	These are arranged in pairs (as transformed by (CH)(EF)(DG) $= V$ which leaves A, B fixed and satisfies $VRV^{-1} = R^{-1}$) and dealt with in detail.
Triple twos.	Very few of these need to be considered in detail. Those with the pair (CH) give (BAC) together with other cycles on a slide and so are either O.K. under (d) or equivalent to a double three. Those with the pair (DH) are reduced to a $t = 4$ case under (a) and those with the pair (CG) are paired with ones having (DH).
Four-and-twos.	They need only be considered when their squares are intransitive or commute with R^4, by Theorem II.
Other cases	They consist of ones where three or more letters are invariant and are immediately reducible to $t = 2$ or $t = 4$.

Double threes.

(CDE)(FGH) (s.p.)	(CDE)(FGH). (DEF)(GHA) = (DC)(EGF)(HA) O.K.
(CDE)(FHG) (s.p.)	(CDE)(FHG). (DEF)(GAH) = (DC)(EHF)(AG) O.K.
(CDF)(EHG) (s.p.)	slide (BAGC)(EFH) O.K.
(CDF)(EGH) (s.p.)	??
(CDG)(EFH) (CEF)(HDG)	(CA)(DBE)(FG) slide O.K.
(CDG)(EHF) (CEF)(HDG)	slide (BAFHC)(EG) O.K.
(CDH)(EFG) (CGH)(DEF)	slide (BAC)(EH) O.K.
(CEG)(DFH) (s.p.)	intransitive
(CEG)(DHF) (s.p.)	intransitive
(CEH)(DFG) (CFH)(DEG)	slide (BAC)(DEFH)
(CEH)(DGF) (CFH)(DGE)	slide (DA)(EBF) O.K.
(CFG)(DEH) (s.p.)	??
(CFG)(DHE) (s.p.)	slide (BAEHC)(DF), transform to (HG) O.K.
(CFH)(DEG)	slide (BAC)(DFGH) O.K.

We have avoided using (a) and (b) owing to the inapplicability for the 6-cycles.

The upright (CDF)(EGH) is exceptional and the corresponding group consists of the elements with the invariants

11111111	8 elements	11111111
12214554	64 elements	25527667
13272315	64 elements	13245423
16573756	64 elements	34657564
24636425	64 elements	14737415
33476674	64 elements	12216336
77777777	8 elements	77777777
	336 elements	

Transformation of the group with (AG)(CH)(EC), which commutes with R gives another group which contains (CFG)(DEH).

The invariants of this latter group are given in the last column. These in- variants are useful for verifying that other exceptional uprights belong to these groups.

We have to investigate the 6-cycles whose squares are exceptional. They are shown below.

(CEDGFH)	X
(CGDHFE)	invariant 15132723 (above)
(CHDEFG)	X
(CDFEGH)	invariant 12216336 (above)
(CEFHGD)	X
(CHFDGE)	X
(CDEFGH)	X
(CFEHGD)	X
(CHEDGF)	X
(CDEHGF)	slide (ABDH)(CE) O.K.
(CHEFGD)	X
(CFEDGH)	X

It is not easy to prove directly that the elements with the above invariants form a group. However in this case we can manage by guessing what the group is. The order of the group being 336 it is natural to suppose that it is connected with the group of symmetry of the 7-point geometry, which is a well-known simple group of order 168. The even permutations in our group will form a group of order 168, which might with luck be isomorphic with the group of the 7-point geometry. This in fact turns out to be so. In order not to confuse the notation we will denote the points of the 7-point geometry by a, b, c, d, e, f, g; [See Fig. 3*]. We naturally try to express the group of symmetry as a group of permutations of some eight objects, in order to tie it up with the groups found above. These groups may perhaps be called K and K'. The standard technique for representing a group as a group of permutations of m objects is to find a subgroup of index m, and to consider the cosets of the subgroup as forming the m objects. In this case we want a subgroup of order 21, and such a subgroup is the normalizer of [(abcegfd)]. The cosets of this group are enumerated below in shortened form. Each line represents seven permutations, obtainable from one another by moving letters from one end to the other. Each coset has been given a name which is a capital letter.

A	*B*	*C*	*D*
abcegfd	abgdcfe	acdgfeb	aefbged
acgdbef	agcebdf	adfbcge	afgdebc
agbfcde	acbfged	afcedbg	agecfdb
E	*F*	*G*	*H*
acfbdeg	abecdfg	aegdfcb	abdgefc
afgdebc	aedgbcf	agfbedc	adecbgf
agecfdb	adbfegc	afecgbd	aebfdcg

Now the symmetry group contains the permutation (abgc)(fe) which induces the permutation $(ACEG)(BDFH)$ of the cosets. It also contains (abc)(def) which induces $(ADH)(BCG)$. Now if we identify the cosets with the eight symbols permuted in the group J (as the notation is intended to suggest), we see that $(ACEG)(BDFH)$ is R^2 and that $(ADH)(BCG)$ has the invariant 14737415, one of those of the group K'. It can then be easily verified that the symmetry group also induces

* Fig. 3. This consists of a triangle abc with inscribed circle, centre g, meeting the sides bc, ca, ab in d, e, f, respectively.

in the cosets permutations with all the other invariants of K', and also that $(ACEG)(BDFH)$ and $(ADH)(BCG)$ generate the whole symmetry group. Since this is of order 168 and since it contains the even permutations in K', numbering 168, it must coincide with the intersection of K' and A.

It now only remains to prove that the expressions of form S and SR, where S is in $K' \cap A$, form a group. This will follow if we can prove that R^2 belongs to $K' \cap A$ and that RSR^{-1} belongs to $K' \cap A$ whenever S be- longs to it. We know the former already and the latter follows at once from our invariant system; K' consists of complete sets of permutations having certain invariants.

Triple twos

As explained above we do not need to consider any except the ones without the pairs (CH), (DH) or (GC). This leaves:

(CD)(EF)(GH) (CD)(EF)(GH). (EF)(GH)(AB) =
(CD)(AB) O.K.
(CD)(EG)(FH) (CD)(EG)(FH). (FG)(HB)(AC) =
(ADC)(BFEGH) O.K.
(CE)(DF)(GH) pair
(CD)(EH)(FG) slide (BAEC)(FH)
(CF)(DE)(GH) pair
(CE)(DG)(FH) (CE)(DG)(FH). (FH)(GB)(AC) =
(GBD)(AC) O.K.
(CF)(DG)(EH) (CF)(DG)(EH). (DG)(EH)(FA) =(ACF) O.K.

Fours-and-twos

Omitting those whose squares are unexceptional and those with UH = C and their pairs we have only:

(CDEF)(HG) X
(CDGF)(HE) X
(CGHD)(FC) X
(DEHG)(FC) pair
(FEHG)(CD) pair

t = 2

We find it worthwhile to apply the principle (ii) on a rather large scale. There are four permutations V which leave A and C fixed; they are

A B C D E F G H
C B A H G F E D
C F A D G B E H
A F C H E B G D

From a single permutation we thus obtain as many as four generating isomorphic groups J, e.g., from (BDEFHG):

(BDEFHG)
(BHGFDE)
(FDBGHE)
(FHEBDG)

These may be transformed into equivalent forms and the alphabetically earliest chosen. We permit taking the reciprocal as a form of transforma - tion. Thus we get

(BDEFHG), (BEDFGH)?, (BGDFEH)?, (BDCFHE).

By these means we reduce the 6-cycles that need be considered down to 18. As before we actually consider first their squares (double threes) in the hope that they will be unexceptional and the 6-cycle need not be specially investigated.

6-cycles and double threes

(BDEFGH)	slide (ACF)(BFDHE) O.K. indirect
(BDEFHG)	slide (AB)(CEFHD) O.K.
(BDEGFH)	(BEF)(DGH). (CFG)(EHA) =(AFH)(CBEDG) O.K.
(BDEGHF)	slide (AD)(FCB) O.K.
(BDEHFG)	slide (BAG)(DCEH) O.K.
(BDEHGF)	invariant 34657564, giving group K′
(BDFEHG)	slide (BA)(DCFGH) O.K.
(BDFGHE)	slide (BA)(CFD)(HEG) O.K.
(BDFHGE)	slide (BAEH)(DCF) O.K.
(BDGFEH)	invariant 34657564, giving group K′
(BDGHEF)	slide (CAE)(FHDGB) O.K. indirect
(BDGHFE)	(BGF)(DHE). (CHG)(EAF) =(ABGCE)(DHF) O.K. indirect
(BDHEFG)	slide (BAFG)(DCH) O.K.
(BDHEGF)	slide (AH)(BCEDF) O.K.
(BDHFGE)	slide (DAEH)(FCG) O.K.
(BDHGEF)	slide (AD)(HBFCE) O.K.
(BDHGFE)	slide (CAE)(DHBFG) O.K. indirect
(BEDHGF)	slide (AED)(CF) O.K.

Above analyses are done on the squares of the 6-cycles, i.e., on the double threes. We must now investigate the cases of 6-cycles where the double threes were exceptional.

(BDEHGF)	slide (BAG)(FH)(CD) O.K.
(BDGHEF)	slide (BAEGH)(CD) O.K.
(BF)(DG)(EH)	X
(BF)(DH)(EG)	intransitive
(BG)(DF)(EH)	slide (AGEDH)(BCF)
(BG)(DH)(EF)	X
(BH)(DF)(EG)	intransitive
(BH)(DG)(EF)	(AH)(BCG)(DF) slide O.K.
(BE)(DH)(FG)	X
(BG)(DH)(FE)	X
(BH)(DG)(FE)	slide (AH)(BCG)(DF)
(BF)(DE)(GH)	X
(BH)(DE)(GF)	slide (AH)(BCEG)
(BH)(DF)(GE)	intransitive
(BE)(DF)(HG)	slide (ACBFG)(EH)
(BE)(DG)(HF)	slide (AEH)(BCGFD)
(BF)(DG)(HE)	X
(BG)(DE)(HF)	slide (ACDHG)(BE)
(BG)(DF)(HE)	(AGEDH)(BCF) slide O.K.
(BH)(EG)(DF)	intransitive

Fours-and-twos and fours

We only need consider those whose squares are intransitive or commute with R^4.

(EBGD) [(FH)]	(FBGD) [(CE)] O.K.	
(EBGF) [(DH)]	(EBFC) O.K.	(ABFC)(DG) O.K.
(EBGH) [(FD)]	(EBGA) O.K.	(GBAE)(CD) O.K.
(EDGB) [(FH)]	(HDGB) O.K.	(HDAB)(EC) O.K.
(EDGH) [(BF)]	(HDGA) O.K.	(HDCA)(BE) O.K.
(EFGH) [(DB)]	(EGHA) O.K.	(CGHA)(DB) O.K.

There are none which leave A, C fixed and commute with R^4 except (BDFH), (BHFD) which is intransitive anyway.

5-cycles

These 5-cycles are given in pairs which are equivalent by principle (ii).

(DEFGH)	s.p. (DEFGH)(EFGHA)$^{-1}$ =(DEA) O.K.
(DEFHG)	slide (BAGDC)(FH) O.K. indirect
(DEHGF)	
(DEGFH)	slide (BADC)(FGH) O.K.
(DEHFG)	
(DEGHF)	s.p. slide (EAGCH)(DFB) O.K. indirect
(BEFGH)	slide (BA)(CED) O.K.
(BEHFG)	
(BEFHG)	slide (BAGH)(CDE) O.K.
(BEGHF)	
(BEGFH)	s.p. slide (BA)(CED)(FGH) O.K.
(BEHGF)	s.p. slide (BAG)(DCE) O.K.
(BEFGH)	slide (AF)(BE)(CGDH) O.K.
(BDHGF)	
(BDFHG)	slide (BAGH)(DC)(EF) O.K.
(BDGFH)	
(BDGHF)	s.p. slide (BAFEG)(CD) O.K.
(BDHFG)	s.p. slide (BAFEH)(CD) O.K.

$t=4$

The permutations V which commute with (R) and leave A, E invariant or interchange them are:

A	B	C	D	E	F	G	H
A	D	G	B	E	H	C	F
E	H	C	F	A	D	G	B
A	H	G	F	E	D	C	B
E	D	C	B	A	H	G	F
A	F	C	H	E	B	G	D
E	B	G	D	A	F	C	H
E	F	G	H	A	B	C	D

6-cycles and double threes

As usual we actually examine the double threes although we test the 6-cycles.

(BCDFGH)	commutes with R^4 (both (BCDFGH) and (BDG)(CFH) and (EF)(CG)(DH))
(BCDFHG)	slide (BA)(EHCDF) O.K.
(BCDGHF)	slide (AB)(CDG)(HFE) O.K.
(BCDHGF)	invariant 33476674 group K
	6-cycle invariant 12214554 group K
	triple two invariant 77777777 part of numatizer of (R)
(BCFDGH)	slide (CAB)(EDFHG) O.K.
(BCFDHG)	slide (BA)(CFEGH) O.K.
(BCFHGD)	slide (BADH)(FEC) O.K.
(BCGDFH)	slide (BACG)(EHF) O.K.
(BCGDHF)	invariant 27652765 commutes with R^4
	6-cycle slide (CAD)(FHB) reducing to a $t = 2$ case already considered
	triple two slide (AHDG)(CEF) O.K.
(BCGFDH)	invariant 33476674 group K
	6-cycle slide (BA)(DG)(FEH) O.K.
	triple two (BF)(CD)(GH) commutes with R^4
(BCGFHD)	slide (BA)(DFECG) O.K.
(BCGHDF)	commutes with R^4
	6-cycle slide (AE)(HFC) O.K. triple two (AH)(CD)(GF) slide (ABCDH)(EGF) O.K.
(BCGHFD)	slide (CAF)(EHBDG) O.K.

Triple twos (remaining)

(BF)(CH)(DG)	commutes with R^4
(BG)(CH)(DF)	slide (AHG)(BFCE) O.K.
(BD)(CG)(FH)	commutes with R^4
(BD)(CH)(FG)	slide (AHDG)(CEF) O.K.
(BH)(CG)(DF)	intransitive
(BD)(CF)(GH)	(BD)(CF)(GH). (CE)(DG)(HA) =(AGBDH)(CEF) O.K.
(BF)(CH)(GD)	commutes with R^4
(BH)(CF)(GD)	(BH)(CF)(GD). (DB)(EH)(AF) = (DHEBG)(FBC) O.K.

Fives

(CDFGH) (CDHFG)	slide (BAC)(EF) O.K.
(CDFHG) (CDHGF)	slide (BAGHC)(FE) O.K.
(CDGFH) (CDGHF)	slide (BAC)(EGDF) O.K.

The 5-cycles where two fixed letters were at distance 2 were considered under $t = 2$.

The outcome for $T = 8$ is then that the exceptional uprights are all either

(a) intransitive, or product of an intransitive upright with R,
(b) commuting with R^4,
(c) members of the groups K, K'.

Let us now calculate the number of exceptional uprights. The number in K and K' omitting those with invariants 11111111 and 77777777 is $2 \times 336 - 32 = 640$. Those with invariants 11111111 and 77777777 are also intransitive, the condition for intransitivity being that all figures are even or all odd. The condition for commuting with R^4 is that the invariant is of the form abcdabcd. Thus these 16 elements are also of this kind, but no other members of K, K' are. There are $2 \times (4!)^2$ intransitive uprights or 1152 and there are $2^4 \times (4!)$ or 384 that commute with R^4. The intransitive ones that commute with R^4 are determined on giving the values of UA, UB (which must be of opposite parity). They number $8 \times 4 = 32$.

Final account

K and K' with	640
intransitive	1152
commute with R^4 with omissions	352
	2144
total of all uprights	40320

We now turn to a rather different topic in connection with the useof identical drums. Even if we know that all permutations are possible, will they be equally frequent? Fortunately, we can answer this in the affirmative. The problem will be examined under slightly more general conditions. No assumptions will be made about the relationship between the generators $U_1, \ldots, U_k$ and we will not assume that the basic group is the symmetric group but some other finite group G.

Let us suppose that we feed a certain frequency distribution of group elements into a wheel; how can we calculate the frequency distribution of the group elements at the output of the wheel? Let $g(a)$ be the proportion of the input elements which are a, and let $f(a)$ be the proportion of group elements effected by the wheel which are a, i.e., denoting the order of the group by h,

$$f(a) = h^{-1} \sum_r 1, \quad r = 1, \ldots, k,\ U_r = a.$$

Then we get output a if the input is b and the wheel effects the group element ab^{-1}, for any b. The proportion of such cases is $f(ab^{-1})g(b)$, or allowing for the different values of b, a total proportion of

$$\sum_b f(ab^{-1})g(b).$$

If then we define the operator R_f by the equation

$$(R_f g)(a) = \sum_b f(ab^{-1})g(b),$$

we can say that the frequency distribution for n wheels is given by $R_f^{n-1}f$. We wish to determine how this function behaves with increasing n.

We consider the real-valued functions on the group as forming an Euclidean space of h dimensions, where h is the order of the group H. We may put (g, k) for the scalar product $h^{-1}\sum_a g(a)k(a)$ and $\|g\|$ for the distance $(g, g)^{1/2}$ from the origin. We may also put $\overline{g}$ for the mean value $h^{-1}\sum_a g(a)$. Schwarz' inequality gives at once $\|g\| \geq \overline{g}$, and if we suppose $g(a) \geq 0$ for all a, $g(a) > 0$ for some a, we shall have $\overline{g} > 0$. We will also suppose $f(a) \geq 0$ for all a, $f(a) > 0$ for some a, $\overline{f} = 1$. Then we have the next lemma.

Lemma (a). *If $\bar{f} = 1$, then $\|R_f g\| \leq \|g\|$ and equality holds only if $g(ab^{-1}x)/g(x)$ is independent of x for any a, b for which $f(a) \neq 0$ and $f(b) \neq 0$.*

First note that

$$\left(h^{-1}\sum_x g(x)g(cx)\right)^2 \leq h^{-2}\sum(g(x))^2\sum(g(cx))^2$$
$$= (h^{-1}\sum(g(x))^2)^2 = \|g\|^4,$$

equality holding if $g(cx)/g(x)$ is independent of x. Then

$$\|R_f g\|^2 = h^{-1}\sum_{a,b,x} f(ab^{-1})g(b)f(ax^{-1})g(x)$$
$$= h^{-1}\sum_{c,v,x} f(c)f(cv)g(vx)g(x) \; [v = bx^{-1},\; c = ab^{-1}]$$
$$\leq \sum_{c,v} f(c)f(cv)\;\|g\|^2$$
$$= \|g\|^2,$$

equality holding in the case mentioned.

Let us define the *limiting distribution* for f as the condensation points of the sequence g, $R_f g$, $R_f^2 g$, ...

Then Lemma (a) will enable us to prove the following theorem.

Theorem III. *The limiting distributions for f are constant throughout the cosets of a certain self-conjugate subgroup H_1 of H. H_1 consists of all expressions of the form $U_{r_1}^{m_1} U_{r_2}^{m_2} \ldots U_{r_p}^{m_p}$, where the sum $\sum m_q$ of exponents is zero. The factor group H/H_1 is cyclic. In the case that g is f, the limiting distributions each have the value zero except in one coset of H_1.*

Let k be a limiting distribution. Let it be the limit of the sequence $R_f^{n_1} g, R_f^{n_2} g, \ldots$ Then

$$\|R_f^{n_r} g\| \geq \|R_f^{n_r+1} g\| \geq \|R_f^{n_{r+1}} g\| \geq \|k\|.$$

Now $\|R_f^{n_r} g\|/\|k\|$ tends to the limit 1 as r tends to infinity and therefore $\|R_f^{n_r+1} g\|/\|R_f^{n_r} g\|$ tends to 1. But $\|R_f u\|/\|u\|$ is a continuous function of u and therefore taking the limit of the sequence $\|R_f k\|/\|k\| = 1$. Applying Lemma (a) to this we see that there is a function $\varphi_1(y)$ defined for all expressions of the form ab^{-1} where $f(a) \neq 0$, $f(b) \neq 0$ such that $k(yx) = \varphi(y)k(x)$ for all x. By applying the same argument with $R_f^{m-1} f$ in place of f we find that there is a function $\varphi_u(y)$ defined for all expressions of the form $a_1 a_2 \cdots a_n b_n^{-1} \cdots b_1^{-1}$ where $f(a_1), f(a_2), \ldots, f(b_1)$ are all different from 0 such that

$$k(yx) = \varphi_u(y)k(x), \quad \text{all } x,$$

whenever $\varphi_u(y)$ is defined. The various functions $\varphi_u(y)$ must agree whenever their domains of definition overlap, and they may therefore be all represented by one symbol φ. In fact we may say that $\varphi(y)$ is defined and has value α whenever $k(yx) = \alpha k(x)$ for all x. It now appears that the domain of definition of $\varphi(y)$ is a group, for if $k(y_1x) = \alpha_1 k(x)$ for all x and $k(y_2x) = \alpha_2 k(x)$ for all x, then $k(y_1y_2x) = \varphi(y_1)k(y_2x) = \varphi(y_1)\varphi(y_2)k(x)$ for all x. Thus if y_1 and y_2 belong to the domain of definition of φ, so does y_1y_2 and $\varphi(y_1y_2) = \varphi(y_1)\varphi(y_2)$.

It is now immediately seen that the domain of definition is H_1. The function φ is a one-dimensional representation of H_1 but it is real and positive and therefore has the value 1 throughout H_1. This last argument may also be expressed without the use of representation theory thus. Since H_1 is finite, any element y of it satisfies an equation $y^m = 1$ and therefore $(\varphi(y))^m = \varphi(y^m) = 1$. But since $g(x)$ is always non negative, $\varphi(y) \geq 0$ and so $\varphi(y) = 1$. This implies that $g(x)$ is constant throughout each coset of H_1.

It now only remains to investigate the character of the group H_1. It is easily seen to be self-conjugate, since if aba^{-1} belongs to H_1 and b to H, the total of exponents of group generators U_r in aba^{-1} must be 0, those in a^{-1} cancelling with those in a; hence aba^{-1} belongs to H_1 if b does; H_1 is self-conjugate.

Now let us take a particular generator U_1, say. Then the cosets $H_1U_1^m$ exhaust the group H. For if p is an element of H, it will be a product of generators; let the total of exponents be m. Then pU_1^{-m} has total exponents 0 and so belongs to H_1, i.e., p belongs to $H_1U_1^m$, i.e., these cosets exhaust H. If U_1^s is the lowest power of U_1 which belongs to H_1, then H/H_1 is evidently isomorphic with the cyclic group of order s.

In the case that g is f, all the group elements for which $R_f^{n-1}f$ is not zero are products of n generators and therefore belong to $H_1U_1^n$.

Example As an example let us consider the quaternion group consisting of 1, i, j, k, $1'$, i',j', k' with the table

1	i	j	k	$1'$	i'	j'	k'
i	$1'$	k	j'	i'	1	k'	j
j	k'	$1'$	i	j'	k	1	i'
k	j	i'	$1'$	k'	j'	i	1
$1'$	i'	j'	k'	1	i	j	k
i'	1	k'	j	i	$1'$	k	j'
j'	k	1	i'	j	k'	$1'$	i
k'	j'	i	1	k	j	i'	$1'$

and let U_1 be i and U_2 be j. The various functions $R_f^{n-1}f$ are given in the table

n	1	i	j	k	$1'$	i'	j'	k'
1	0	$\frac{1}{2}$	$\frac{1}{2}$	0	0	0	0	0
2	0	0	0	$\frac{1}{4}$	$\frac{1}{2}$	0	0	$\frac{1}{4}$
3	0	$\frac{1}{4}$	$\frac{1}{4}$	0	0	$\frac{1}{4}$	$\frac{1}{4}$	0
4	$\frac{1}{4}$	0	0	$\frac{1}{4}$	$\frac{1}{4}$	0	0	$\frac{1}{4}$
5	0	$\frac{1}{4}$	$\frac{1}{4}$	0	0	$\frac{1}{4}$	$\frac{1}{4}$	0
6	$\frac{1}{4}$	0	0	$\frac{1}{4}$	$\frac{1}{4}$	0	0	$\frac{1}{4}$

It is seen that the group H_1 is the group generated by k; it has a factor group which is cyclic of order 2.

Case of symmetric and alternating groups

In the case under consideration at the beginning of our analysis, H was either the symmetric or the alternating group unless the upright U was exceptional. In this case H_1 is also either the symmetric or the alternating group, for it is self-conjugate in H. It will be the alternating group if the generators are all of the same parity and the symmetric group otherwise. We therefore conclude that

> *when the upright is not exceptional the distributions with large numbers of wheels are uniform throughout the alternating group (even permutations). If odd permutations are possible with the given upright and number of wheels the distribution is uniform throughout the symmetric group (all permutations).*

Rounding-off Errors in Matrix Processes

(Quart. J. Mech. Appl. Math. **1** (1948), 287–308)

Lenore Blum brings into view —

ALAN TURING AND THE OTHER THEORY OF COMPUTATION

1. Introduction

The two major traditions of the theory of computation, each staking claim to similar motivations and aspirations, have for the most part run a parallel non-intersecting course. On one hand, we have the tradition arising from logic and computer science addressing problems with more recent origins, using tools of combinatorics and discrete mathematics. On the other hand, we have numerical analysis and scientific computation emanating from the classical tradition of equation solving and the continuous mathematics of calculus. Both traditions are motivated by a desire to understand the essence of computation, of algorithm; both aspire to discover useful, even profound, consequences.

While the logic and computer science communities are keenly aware of Alan Turing's seminal role in the former (discrete) tradition of the theory of computation, most remain unaware of Alan Turing's role in the latter (continuous) tradition, this notwithstanding the many references to Turing in the modern numerical analysis/computational mathematics literature, e.g., Bürgisser (2010), Higham (2002), Trefethen and Bau (1997) and Wilkinson (1971).

In 1948, in the first issue of the *Quarterly Journal of Mechanics and Applied Mathematics*, sandwiched between a paper on 'Use of Relaxation Methods and Fourier Transforms' and 'The Position of the Shock-Wave in Certain Aerodynamic Problems', appears the article 'Rounding-Off Errors in Matrix Processes'. This paper introduces the notion of the *condition number* of a matrix, the chief factor limiting the accuracy in solving linear systems, a notion fundamental to numerical computation and analysis, and a notion with implications for complexity theory today. This paper was written by Alan Turing (1948).

'Rounding-Off Errors in Matrix Processes' was by no means an anomaly in Turing's creative pursuits. In his 1970 Turing Award Lecture, Wilkinson (1971) writes:

> Turing's international reputation rests mainly on his work on computable numbers but I like to recall that he was a considerable numerical analyst, and a good part of his time from 1946 onwards was spent working in this field...

Wilkinson attributes this interest, and Turing's decision to go to the National Physical Laboratory (NPL) after the war, to the years he spent at Bletchley Park gaining knowledge of electronics and 'one of the happiest times of his life'.

Wilkinson also credits Turing for converting him from a classical to numerical analyst. From 1946 to 1948, Wilkinson worked for Turing at the NPL on the logical design of Turing's proposed Automatic Computing Engine (ACE) and the problem of programming basic numerical algorithms.[1] 'The period with Turing fired me with so much enthusiasm for the computer project and so heightened my interest in numerical analysis that gradually I abandoned [the idea of returning to Cambridge to take up research in classical analysis]' (Wilkinson, 1971).

In 1946, with the anticipation of a programmable digital computing device on the horizon,[2] it was of great interest to understand the comparative merits of competing computational 'processes' and how accurate such processes would be in the face of inevitable round-off errors. Solving linear systems is basic. Thus for Turing (as it was for von Neumann and Goldstine (1947)), examining methods of solution with regard to the ensuing round-off errors presented a compelling intellectual challenge.[3]

Here I would like to recognise Turing's work in the foundations of numerical computation. In an expanded version of this paper, I will expound more on its influence in complexity theory today, and how it provides a unifying concept for the two major traditions in the theory of computation.

2. Rounding-off errors in matrix processes

> This paper contains descriptions of a number of methods for solving sets of linear simultaneous equations and for inverting matrices, *but its main concern is with the theoretical limits of accuracy* that may be obtained in the application of these methods, due to round-off errors.[4]

So begins Turing's (1948) paper. (Italics are mine, I'll return to this shortly.)

The basic problem at hand: Given the linear system, $Ax = b$ where A is a real non-singular $n \times n$ matrix and $b \in \mathbf{R}^n$. Solve for x.

Prompted by calculations (Fox et al., 1948) challenging the arguments by Hotelling (1943) that Gaussian elimination and other direct methods would lead to exponential round-off errors, Turing introduces quantities not considered earlier to bound the magnitude of errors, showing that for all 'normal' cases, the exponential estimates are 'far too pessimistic'.[5]

---

[1] Turing (1945) submitted an 86-page proposal to the NPL for the ACE computer, *an automatic electronic digital computer with internal program storage*. This was to be an incarnation of the universal machine envisioned by his theoretical construct in 'Computable Numbers' (Turing, 1936), blurring the boundary between data and program. Thus, and in contrast to other proposed 'computing machines' at the time, Turing's computer would have simplified hardware, with universality emanating from the power of programming.

Turing envisioned an intelligent machine that would learn from its teachers and from its experience and mistakes, and hence have the ability to create and modify its own programs. Turing also felt that the most conducive environment for realising such a machine was to have mathematicians and engineers working together in close proximity, not in separate domains (Hodges, 1992).

[2] Unfortunately, the ACE computer was never to see the light of day; a less ambitious non-universal machine, the PILOT ACE, was constructed after Turing left the NPL for Manchester in 1948.

[3] It is clear that Turing and von Neumann were working on similar problems, for similar reasons, in similar ways at the same time, probably independently. However while Turing acknowledges von Neumann, as far as I know, von Neumann never cites Turing's work in this area.

[4] Although Turing's central passion during his time at the NPL was the promised realization of his universal computer, his only published paper to come out of this period (1945-1948) was 'Round-Off Errors in Matrix Processes'. The paper ends with the cryptic acknowledgement: 'published with the permission of the Director of the National Physical Laboratory'.

[5] In their paper, Bargmann, Montgomery and von Neumann (1963) also dismissed Gaussian elimination as likely being unstable due to magnification of errors at successive stages (pp. 430–431) and so turn to iterative methods for analysis. However, von Neumann and Goldstine (1947) reassess noting, as does Turing, that it is the computed solution, not the intermediate computed numbers, which should be the salient object of study. They re-investigated Gaussian elimination for computing matrix inversion and now give optimistic error bounds similar to those of Turing, but for the special case of positive definite symmetric matrices. Turing, in his paper, notes that von Neumann communicated these results to him at Princeton (during a short visit) in January 1947 before his own proofs were complete.

In this paper, Turing introduced the notion of *condition number* making explicit for the first time a measure that helps formalise the informal notion of ill- and well-conditioned problems.

3. Turing and the matrix condition number

'When we come to make estimates of errors in matrix processes, we shall find that the chief factor limiting the accuracy that can be obtained is "ill-conditioning" of the matrices involved' (Turing, 1948).

Turing provides an illustrative example:

$$\left.\begin{aligned} 1\cdot 4x + 0\cdot 9y &= 2\cdot 7 \\ -0\cdot 8x + 1\cdot 7y &= -1\cdot 2 \end{aligned}\right\} \qquad (8.1)$$

$$\left.\begin{aligned} -0\cdot 786x + 1\cdot 709y &= -1\cdot 173 \\ -0\cdot 800x + 1\cdot 700y &= -1\cdot 200 \end{aligned}\right\} \qquad (8.2)$$

The set of equations (8.2) is fully equivalent to (8.1)[6] but clearly if we attempt to solve (8.2) by numerical methods involving rounding-off errors we are almost certain to get much less accuracy than if we worked with equations (8.1). We should describe the equations (8.2) as an *ill-conditioned* set, or, at any rate, as ill-conditioned compared with (8.1). It is characteristic of ill-conditioned sets of equations that small percentage errors in the coefficients given may lead to large percentage errors in the solution. ...

Turing defines (N and M) condition numbers, which in essence measure the intrinsic magnification of errors.[7] He then analyzes various standard methods for solving linear systems, including Gaussian elimination, and gets error bounds proportional to his measures of condition.[8] Turing is 'also as much interested in statistical behaviour of the errors as in the maximum possible values' and presents probabilistic estimates; he also improves Hotellings worst case bound from 4^{n-1} to 2^{n-1} (Turing, 1948).

The following widely used (spectral) matrix condition number $\kappa(A)$, wedged somewhat between Turing's condition numbers, is often attributed to Turing, though it is unclear who first defined it. John Todd (1968), in his survey, is vague on its genesis although he specifically credits Turing with recognising 'that a condition number should depend symmetrically on A and A^{-1}, specifically as a product of their norms'.[9,10] (See also the discussion in the postscript at the end of this paper.)

---

[6] The third equation is the second plus .01 times the first.

[7] The N condition number is defined as $N(A)N(A^{-1})$ and the M condition number as $nM(A)M(A^{-1})$, where $N(A)$ is the Frobenius norm of A and $M(A)$ is the max norm.

[8] In sections 3 and 4 of his paper, Turing also formulates the LU decomposition of a matrix (actually the LDU decomposition) and shows that Gaussian elimination computes such a decomposition.

[9] Turing defines the spectral norm in the 'Rounding-Off Errors' paper so he could have easily defined the spectral condition number.

[10] For the case of computing the inverse of a positive definite symmetric matrix A by Gaussian Elimination, von Neumann and Goldstine (1947) give an error estimate bounded by $14.2n^2(\lambda_1/\lambda_2)u$. Here λ_1 and λ_2 are the largest and smallest eigenvalues of A and u is the smallest number recognised by the machine. For the case of positive definite symmetric matrices, λ_1/λ_2 is equal to $\kappa(A)$. Thus, the spectral condition number appears implicitly in the von Neumann–Goldstine paper.

Definition 3.1. *Suppose A is a real non-singular $n \times n$ matrix. The (spectral)* condition number *of A is given by*

$$\kappa(A) = \|A\| \left\|A^{-1}\right\|,$$

where

$$\|A\| = \max_{y \neq 0} \frac{|Ay|}{|y|} = \max_{|y|=1} |Ay|$$

is the operator (spectral) norm with respect to the Euclidean norm.[11] *For singular matrices, define $\kappa(A) = \infty$.*

To see how natural a measure this is, consider a slightly more general situation. Let X and Y be normed vector spaces with associated map $\varphi : X \to Y$. A measure of the 'condition' of *problem instance* (φ, x) should indicate how small perturbations of the *input* x (the *problem data*) will alter the *output* $\varphi(x)$ (the *problem solution*).

So let Δx be a small perturbation of input x and $\Delta\varphi = \varphi(x + \Delta x) - \varphi(x)$. The limit as $\|\Delta x\|$ goes to zero of the ratio

$$\frac{\|\Delta\varphi\|}{\|\Delta x\|},$$

or of the relative ratio

$$\frac{\|\Delta\varphi\| \;/\; \|\varphi(x)\|}{\|\Delta x\| \;/\; \|x\|}$$

(favored by numerical analysts), will be a measure of the condition of the problem instance.[12] If large, computing the output with small error will require increased precision, and hence from a computational complexity point of view, increased time/space resources.

Definition 3.2.[13] *The* condition number of problem instance (φ, x) *is defined by*

$$\widehat{\kappa}(\varphi, x) = \lim_{\delta \to 0} \sup_{\|\Delta x\| \leq \delta} \frac{\|\Delta\varphi\|}{\|\Delta x\|}$$

and the relative condition number *by*

$$\kappa(\varphi, x) = \lim_{\delta \to 0} \sup_{\|\Delta x\| \leq \delta \|x\|} \frac{\|\Delta\varphi\| / \|\varphi(x)\|}{\|\Delta x\| / \|x\|}.$$

If $\kappa(\varphi, x)$ is small, the problem instance is said to be well-conditioned *and if large,* ill-conditioned. *If $\kappa(\varphi, x) = \infty$, the problem instance is* ill-posed.

As Turing envisaged it, the condition number measures the theoretical limits of accuracy in solving a problem. In particular, the logarithm of the condition number provides an intrinsic lower

---

[11] This definition can be generalised using other norms. In the case of the Euclidean norm, $\kappa(A) = \sigma_1/\sigma_n$, where σ_1 and σ_2 are the largest and smallest singular values of A, respectively. It follows that $\kappa(A) \geq 1$.

[12] All norms are assumed to be with respect to the relevant spaces.

[13] Here I follow the notation in Trefethen and Bau (1997), a book I highly recommend for background in numerical linear algebra.

bound for the *loss of precision* in solving the problem instance,[14] thus also providing a key intrinsic parameter for specifying 'input word size' for measuring computational complexity over the reals.[15]

If φ is differentiable, then

$$\widehat{\kappa}(\varphi,x) = \|D\varphi\,(x)\|$$

and

$$\kappa\,(\varphi,x) = \|D\varphi\,(x)\|\,(\|x\|\,/\,\|\varphi\,(x)\|)\,,$$

where $D\varphi(x)$ is the Jacobian (derivative) matrix of φ at x and $\|D\varphi(x)\|$ is the operator norm of $D\varphi(x)$ with respect to the induced norms on X and Y.

Here is where Turing meets Newton. The following theorem says the (spectral) matrix condition number $\kappa(A)$ is (essentially) the condition number for solving the linear system $Ax = b$.[16]

THEOREM 3.1.

1. *Fix A, a real non-singular $n \times n$ matrix, and consider the map $\varphi_A : \mathbf{R}^n \to \mathbf{R}^n$ where $\varphi_A(b) = A^{-1}(b)$. Then $\kappa(\varphi_A, b) \leq \kappa(A)$ and there exist $\widehat{b}$ such that $\kappa(\varphi_A, \widehat{b}) = \kappa(A)$. Thus, with respect to perturbations in b, the matrix condition number is the worst case relative condition for solving linear system $Ax = b$.*
2. *Fix $b \in \mathbf{R}^n$ and consider the partial map $\varphi_b : \mathbf{R}^{n\times n} \to \mathbf{R}^n$ where, for A non-singular, $\varphi_b(A) = A^{-1}(b)$. Then for A non-singular, $\kappa(\varphi_b, A) = \kappa(A)$.*

4. Turing's evolving perspective on computing over the reals

The Turing Machine is the canonical abstract model of a general purpose computer, studied in almost every first course in theoretical computer science. What most students of theory are not aware of, however, is that Turing defined his 'machine' in order to define a theory of real computation. The first paragraph of his seminal paper (Turing, 1936) begins and ends as follows:

> The "computable" numbers may be described briefly as the real numbers whose expressions as a decimal are calculable by finite means. ... According to my definition, a number is computable if its decimal can be written down by a machine.

Of course, the machine thus developed becomes the basis for the classical theory of computation of logic and theoretical computer science.

In the same first paragraph Turing writes, 'I hope shortly to give an account of the relation of the computable numbers, functions, and so forth to one another. This will include a development of the theory of functions of a real variable expressed in terms of computable numbers'. As far as I know, Turing never returned to computing over the reals using this approach; *recursive analysis* (also known as *computable analysis*) was developed by others.

[14] Velvel Kahan points out that 'pre-conditioning' can sometimes alter the given problem instance to a better conditioned one with the same solution. (Convert equations (8.2) to (8.1) in Turing's illustrative example.)

[15] Computational complexity is measured as a function of input word size. Early on, when computational complexity theorists analyzed algorithms for solving problems naturally defined over the real numbers, they replaced real coefficients by rational approximations. This led to considering arbitrary input word sizes; 'nearby' problem instances could have widely different input sizes, depending on the approximation chosen. This would allow for wide variations in the analysis of algorithms on nearby instances, making no sense, particularly for well-conditioned instances. Thus, it is more natural to replace the arbitrary 'bit' size approximation by the logarithm of the condition. Other natural parameters to add include the problem instance dimension and the logarithm of output accuracy desired (Blum, 1990). Mike Shub proposes also to include another parameter related to the condition, namely the inverse of the distance to 'ill-posedness'.

[16] This inspired in part the title of my paper, 'Computing over the Reals: Where Turing meets Newton' (Blum, 2004).

When Turing does return to computing over the reals, as in 'Rounding-Off Errors' written while he was preoccupied with the concrete problem of computing solutions to systems of equations, his implicit real number model is vastly different. Now real numbers are considered as individual entities and each basic algebraic operation is counted as one step. In the first section of this paper, Turing considers the 'measures' of work in a process:

> It is convenient to have a measure of the amount of work involved in a computing process, even though it be a very crude one. ... We might, for instance, count the number of additions, subtractions, multiplications, divisions, recordings of numbers, ...

This is the basic approach taken by numerical analysts, qualified as Turing also implies, by condition and round-off errors. It is also the approach taken by Mike Shub, Steve Smale and myself (Blum et al. (1989)), and later with Felipe Cucker in our book, *Complexity and Real Computation* (Blum et al. (1998)).

5. Postscript: who invented the condition number?

It is clear that Alan Turing was the first to explicitly formalise a measure that would capture the informal notion of condition (of solving a linear system) and to call this measure a *condition number*. Formalising what's in the air serves to illuminate essence and chart new direction. However, ideas in the air have many proprietors.

To find out more about the origins of the *spectral condition number*, I e-mailed a number of numerical analysts.

I also looked at many original papers. The responses I received, and related readings, uncover a debate concerning the origins of the (concept of) condition number not unlike the debate surrounding the origins of the general purpose computer – with Turing and von Neumann figuring centrally to both. (For an insightful assessment of the latter debate see 'Mysteries of mathematics and computation', Shub (1994).)

Gauss (1823) himself is referenced for considering perturbations and preconditioning. Pete Stewart points to Wittmeyer (1936) for some of the earliest perturbation bounds where products of norms appear explicitly. Todd (1950) explicitly focused on the notion of condition number, citing Turing's N and M condition numbers and the implicit von Neumann–Goldstine measure, which he called the P-condition number (P for Princeton).

Beresford Parlett tells me that 'the notion was "in the air" from the time of Turing and von Neumann et al.', that the concept was used by Forsythe in a course he took from him at Stanford early in 1959 and that Wilkinson most surely 'used the concept routinely in his lectures in Ann Arbor (summer, 1958)'. The earliest explicit definition of the spectral condition number I could find in writing was in Householder's (1958) SIAM article (where he cites Turing) and then in Wilkinson's (1963, p. 91) book.

By far, the most informative and researched history can be found in Joe Grcar's 76-page article, 'John von Neumann's Analysis of Gaussian Elimination and the Origins of Modern Analysis' (Grcar, 2011). Here he uncovers a letter from von Neumann to Goldstine (dated 11 January 1947) that explicitly names the ratio of the extreme singular values as ℓ. Why this was not included in their paper or made explicit in their error bounds is a mystery to me.[17] Joe chalks this up to von Neumann

[17] Joe was also kind enough to illuminate for me in detail how one could unravel von Neumann's and Goldstine's error analyses for the general case in their paper.

Many authors have cited the obtuseness and non-explicitness. For example, Alan Edelman (1989), in his PhD thesis, recasts von Neumann's and Goldstine's ideas in modern notation given 'the difficulty of extracting the various ideas from their work' and cites Wilkinson's referring to the paper's 'indigestibility'.

using his speeches (e.g., von Neumann (1989)) to expound on his ideas, particularly those given to drum up support for his computer project at the Institute for Advanced Study.[18]

Grcar's article definitely puts von Neumann at center stage; of von Neumann's role as a key player in this area there is no doubt. Grcar also implies, however, that Turing's work on rounding error, and the condition number, was prompted by Turing's meeting with von Neumann in Princeton in January 1947. Indeed on page 630 he says, 'No less than Alan Turing produced the first derivative work from the inversion paper'.

This flies in the face of all we know about Alan Turing's singular individuality, both in personality and in research (Hodges, 1992). In their personal remembrances of Turing, both Wilkinson, who worked closely with him at the NPL in Teddington, and Newman (1955), earlier at Cambridge and Bletchley and later in Manchester, point to Turing's 'strong predilection for working everything out from first principles, usually in the first instance without consulting any previous work on the subject, and no doubt it was this habit which gave his work that characteristically original flavor'.

It also flies in the face of fact. As recounted by Wilkinson (1971), Turing's experience with his team at the NPL, prior to meeting von Neumann in Princeton in 1947, was the stimulus for his paper:

> ... some time after my arrival [at the NPL in 1946], a system of 18 equations arrived in Mathematics Division and after talking around it for some time we finally decided to abandon theorizing and to solve it.... The operation was manned by [Leslie] Fox, [Charles] Goodwin, Turing, and me, and we decided on Gaussian elimination with complete pivoting. Turing was not particularly enthusiastic, partly because he was not an experienced performer on a desk machine and partly because he was convinced that it would be a failure. ... the system was mildly ill-conditioned, the last equation had a coefficient of order 10^{-4} ... and the residuals were ... of order 10^{-10}, that is of the size corresponding to the exact solution rounded to ten decimals. ...
>
> ... I'm sure that this experience made quite an impression on him and set him thinking afresh on the problem of rounding errors in elimination processes. About a year later he produced his famous paper 'Rounding-off errors in matrix process' ...

Kahan (1966) (also a Turing Award recipient), in his paper and in a lengthy phone conversation (August 2011), asserts that von Neumann and Goldstine were misguided in their approach to matrix inversion (by computing A^{-1} from the formula $A^{-1} = (A^T A)^{-1} A^T$).

Kahan's assessment of Turing is a fitting conclusion to this paper: 'A more nearly modern error-analysis was provided by Turing (1948) in a paper whose last few paragraphs foreshadowed much of what was to come, but his paper laid unnoticed for several years until Wilkinson began to publish the papers which have since become a model of modern error-analysis'.[19]

---

[18] An unpublished and undated paper by Goldstine and von Neumann (1963) containing material presented by von Neumann in various lectures going back to 1946, but clearly containing later perspectives as well, explicitly singles out (on page 14) ℓ as the 'figure of merit'. Also interesting to me, in the same paragraph, are the words 'loss of precision' connected to the condition number, possibly for the first time.

[19] Kahan is referring to *backward error analysis* that, rather than estimating the errors in a computed solution of a given problem instance (i.e., *forward error analysis*), estimates the closeness of a nearby problem instance whose exact solution is the same as the approximate computed solution of the original. Grcar (2011) also points to the von Neumann–Goldstine paper as a precursor to this notion as well.

References

Bargmann, V., Montgomery, D., von Neumann, J., 1963. Solution[s] of linear systems of high order, In: Taub, A. H. (Eds.), John von Neumann Collected Works, vol. 5. pp. 421–478. Macmillan, New York. Report prepared for Navy Bureau of Ordnance, 1946.

Blum, L., 1990. Lectures on a theory of computation and complexity over the reals (or an arbitrary ring). In: E. Jen, editor, Lectures in the Sciences of Complexity II, Addison-Wesley, pp. 1–47.

Blum, L., Shub, M., Smale, S., 1989. On a theory of computation and complexity over the real numbers: NP-Completeness, recursive functions and universal machines. Bull. Am. Math. Soc. 21, 1–46.

Blum, L., Cucker, F., Shub, M., Smale, S., 1998. Complexity and Real Computation, Springer, New York.

Blum, L., 2004. Computing over the reals: Where Turing meets Newton. Not. Am. Math. Soc. 51, 1024–1034.

Bürgisser, P., 2010. Smoothed analysis of condition numbers, In: Proceedings of the International Congress of Mathematicians, 2010, World Scientific.

Edelman, A., 1989. Eigenvalues and condition numbers of random matrices. MIT PhD Dissertation. http://math.mit.edu/~edelman/thesis/thesis.pdf.

Fox, L., Huskey, H.D., Wilkinson, J.H., 1948. Notes on the solution of algebraic linear simultaneous equations. Q. J. Mech. Appl. Math. 1, 149–173.

Gauss, C.F., 1823. Letter to Gerling, December 26, 1823. Werke, 9:278–281. Trans, by G. E. Forsythe, Mathematical Tables and Other Aids to Computation vol. 5 (1951). pp. 255–258.

Goldstine, H.H., von Neumann, J. 1963. On the principles of large scale computing machines (unpublished), In: Taub, A.H. (Ed.), John von Neumann Collected Works, vol. 5. Macmillan, New York, pp. 1–33.

Grcar, J.F., 2011. John von Neumann's analysis of Gaussian elimination and the origins of modern numerical analysis. SIAM Rev. 53 (4), 607–682.

Higham, N.H., 2002. Accuracy and Stability of Numerical Algorithms, second ed. Society for Industrial and Applied Mathematics, Philadelphia, PA, USA.

Hodges, A., 1992. Alan Turing: The Enigma, Vintage.

Hotelling, H., 1943. Some new methods in matrix calculation. Ann. Math. Statist. 14(1), 1–34.

Householder, A.S., 1958. A class of methods for inverting matrices. J. Soc. Ind. Appl. Math., 6, 189–195.

Kahan, W.M., 1966. Numerical linear algebra. Can. Math. Bull. 9, 757–801.

Newman, M.H.A., 1955. Alan Mathison Turing. 1912–1954. Biogr. Mems. Fell. R. Soc. 1, 253–263.

Shub, M., 1994. Mysteries of mathematics and computation. Math. Intell. 16, 10–15.

Todd, J., 1950. The condition of a certain matrix. Proc. Camb. Philos. Soc. 46, 116–118.

Todd, J., 1968. On condition numbers, In: Programmation en Mathématiques Numériques, Besançon, 1966, vol. 7 (no. 165) of Éditions Centre Nat. Recherche Sci., Paris, pp. 141–159.

Trefethen, L.N., Bau, D., 1997. Numerical linear algebra. SIAM.

Turing, A.M., 1936. On computable numbers, with an application to the Entscheidungsproblem. Proc. Lond. Math. Soc. 2, 42(1), 230–265.

Turing, A.M., 1945. Proposal for development in the Mathematics Division of an Automatic Computing Engine (ACE). Report E.882, Executive Committee, inst-NPL.

Turing, A.M., 1948. Rounding-off errors in matrix processes. Q. J. Mech. Appl. Math. 1, 287–308.

von Neumann, J., 1989. The principles of large-scale computing machines. Ann. His. Comput. 10(4), 243–256. Transcript of lecture delivered on May 15, 1946.

von Neumann, J., Goldstine, H.H., 1947. Numerical inverting of matrices of high order. Bull. Am. Math. Soc. 53(11), 1021–1099.

Wilkinson, J.H., 1963. Rounding Errors in Algebraic Processes. Notes on Applied Science No. 32, Her Majesty's Stationery Office, London. Also published by Prentice-Hall, Englewood Cliffs, NJ, USA. Reprinted by Dover, New York, 1994.

Wilkinson, J.H., 1971. Some comments from a numerical analyst. JACM 18(2), 137–147. 1970 Turing Lecture.

Wittmeyer, H., 1936. Einfluß der Änderung einer Matrix auf die Lösung des zugehörigen Gleichungssystems, sowie auf die charakteristischen Zahlen und die Eigenvektoren. Z. Angew. Math. Mech. 16, 287–300.

ROUNDING-OFF ERRORS IN MATRIX PROCESSES

By A. M. TURING

(*National Physical Laboratory, Teddington, Middlesex*)

[Received 4 November 1947]

SUMMARY

A number of methods of solving sets of linear equations and inverting matrices are discussed. The theory of the rounding-off errors involved is investigated for some of the methods. In all cases examined, including the well-known 'Gauss elimination process', it is found that the errors are normally quite moderate: no exponential build-up need occur.

Included amongst the methods considered is a generalization of Choleski's method which appears to have advantages over other known methods both as regards accuracy and convenience. This method may also be regarded as a rearrangement of the elimination process.

THIS paper contains descriptions of a number of methods for solving sets of linear simultaneous equations and for inverting matrices, but its main concern is with the theoretical limits of accuracy that may be obtained in the application of these methods, due to rounding-off errors.

The best known method for the solution of linear equations is Gauss's elimination method. This is the method almost universally taught in schools. It has, unfortunately, recently come into disrepute on the ground that rounding off will give rise to very large errors. It has, for instance, been argued by Hotelling (ref. **5**) that in solving a set of n equations we should keep $n \log_{10} 4$ extra or 'guarding' figures. Actually, although examples can be constructed where as many as $n \log_{10} 2$ extra figures would be required, these are exceptional. In the present paper the magnitude of the error is described in terms of quantities not considered in Hotelling's analysis; from the inequalities proved here it can immediately be seen that in all normal cases the Hotelling estimate is far too pessimistic.

The belief that the elimination method and other 'direct' methods of solution lead to large errors has been responsible for a recent search for other methods which would be free from this weakness. These were mainly methods of successive approximation and considerably more laborious than the direct ones. There now appears to be no real advantage in the indirect methods, except in connexion with matrices having special properties, for example, where the vast majority of the coefficients are very small, but there is at least one large one in each row.

The writer was prompted to carry out this research largely by the practical work of L. Fox in applying the elimination method (ref. **2**). Fox found that no exponential build-up of errors such as that envisaged by Hotelling actually occurred. In the meantime another theoretical investigation was being carried out by J. v. Neumann, who reached conclusions similar to those of this paper for the case of positive definite matrices, and communicated them to the writer at Princeton in January 1947 before the proofs given here were complete. These results are now published (ref. **6**).

1. Measure of work in a process

It is convenient to have a measure of the amount of work involved in a computing process, even though it be a very crude one. We may count up the number of times that various elementary operations are applied in the whole process and then give them various weights. We might, for instance, count the number of additions, subtractions, multiplications, divisions, recordings of numbers, and extractions of figures from tables. In the case of computing with matrices most of the work consists of multiplications and writing down numbers, and we shall therefore only attempt to count the number of multiplications and recordings. For this purpose a reciprocation will count as a multiplication. This is purely formal. A division will then count as two multiplications; this seems a little too much, and there may be other anomalies, but on the whole substantial justice should be done.

2. Solution of equations versus inversion

Let us suppose we are given a set of linear equations $\mathbf{Ax} = \mathbf{b}$ to solve. Here $\mathbf{A}$ represents a square matrix of the nth order and $\mathbf{x}$ and $\mathbf{b}$ vectors of the nth order. We may either treat this problem as it stands and attempt to find x, or we may solve the more general problem of finding the inverse of the matrix $\mathbf{A}$, and then allow it to operate on $\mathbf{b}$ giving the required solution of the equations as $\mathbf{x} = \mathbf{A}^{-1}\mathbf{b}$. If we are quite certain that we only require the solution to the one set of equations, the former approach has the advantage of involving less work (about one-third the number of multiplications by almost all methods). If, however, we wish to solve a number of sets of equations with the same matrix $\mathbf{A}$ it is more convenient to work out the inverse and apply it to each of the vectors $\mathbf{b}$. This involves, in addition, n^2 multiplications and n recordings for each vector, compared with a total of about $\frac{1}{3}n^3$ multiplications in an independent solution.There are other advantages in having an inverse. From the coefficients of the inverse we can see at once how sensitive the solution is to small changes in the coefficients of $\mathbf{A}$ and of $\mathbf{b}$. We have, in fact,

$$\frac{\partial x_i}{\partial b_j} = (\mathbf{A}^{-1})_{ij}, \quad \frac{\partial x_i}{\partial a_{jk}} = -(\mathbf{A}^{-1})_{ij} x_k.$$

This enables us to estimate the accuracy of the solution if we can judge the accuracy of the data, that is, of the matrix $\mathbf{A}$ and the vector b, and also enables us to correct for any small changes which we may wish to make in these data.

It seems probable that with the advent of electronic computers it will become standard practice to find the inverse. This time has, however, not yet arrived and some consideration is therefore given in this paper to solutions without inversion. A form of compromise involving less work than inversion, but including some of the advantages, is also considered.

3. Triangular resolution of a matrix

A number of the methods for the solution of equations and, more particularly, for the inversion of matrices, depend on the resolution of a matrix into the product of two triangular matrices. Let us describe a matrix which has zeros above the diagonal as 'lower triangular' and one which has zeros below as 'upper triangular'. If in addition the coefficients on the diagonal are unity the expressions 'unit upper triangular' and 'unit lower triangular' may be used. The resolution is essentially unique, in fact we have the following

THEOREM ON TRIANGULAR RESOLUTION. *If the principal minors of the matrix* $\mathbf{A}$ *are non-singular, then there is a unique unit lower triangular matrix* $\mathbf{L}$, *a unique diagonal matrix* $\mathbf{D}$, *with non-zero diagonal elements, and a unique unit upper triangular matrix* $\mathbf{U}$ *such that* $\mathbf{A} = \mathbf{LDU}$. *Similarly there are unique* $\mathbf{L}'$, $\mathbf{D}'$, $\mathbf{U}'$ *such that* $\mathbf{A} = \mathbf{U}'\mathbf{D}'\mathbf{L}'$.

The kth diagonal element of $\mathbf{D}$ will be denoted by d_k. The $1k$ coefficient of the equation $\mathbf{A} = \mathbf{LDU}$ gives us $l_{11}d_1u_{1k} = a_{1k}$ and since $l_{11} = u_{11} = 1$ this determines d_1 to be a_{11} and u_{1k} to be a_{1k}/d_1; these choices satisfy the equations in question. Suppose now that we have found values of l_{ij}, u_{jk} with $j < i_0$ (that is, we have found the first i_0–1 rows of $\mathbf{L}$ and columns of $\mathbf{U}$) and the first i_0–1 diagonal elements d_k, so that the equations arising from the first i_0–1 rows of the equation $\mathbf{A} =$

LDU are satisfied; and suppose further that these choices are unique and $d_k \neq 0$. It will be shown how the next row of **L** and the next column of **U**, and the next diagonal element $d_{io} \neq 0$ are to be chosen so as to satisfy the equations arising from the next row of **A** = **LDU**, and that the choice is unique. The equations to be satisfied in fact state

$$l_{i_0 i_0} d_{i_0} u_{i_0 k} = a_{i_0 k} - \sum_{j<i_0} l_{ioj} d_j u_{jk} \ (k \geq i_0),$$

$$l_{i_0 k} d_k u_{kk} = a_{i_0 k} - \sum_{j<k} l_{ioj} d_j u_{jk} \ (k < i_0).$$

The right-hand sides of these equations are entirely in terms of quantities already determined. When $k = i_0$ the first equation is satisfied and can only be satisfied by putting d_{io} = right-hand side, determining d_{io}. The equations for $k > i_0$ can then be satisfied by one and only one set of values of $u_{i_0 k}$, provided $d_{io} \neq 0$. The equations for $k < i_0$ can also be satisfied by one and only one set of values of l_{iok}, since each d_k is different from 0 The new diagonal element d_{io} is not 0 because the i_0th principal minor of **A** is equal to the product of the first i_0 diagonal elements d_k.

4. The elimination method

Suppose that we wish to solve the equations **Ax** = **b** by the elimination method. The procedure is as follows. We first add such multiples of the first equation to the others that the coefficient of x_1 is reduced to zero in all of them (excepting the first). We then add multiples of the second equation to the later ones until the coefficient of x_2 is reduced to zero. After $n-1$ steps of this nature we shall be left with a set of equations of the form $\sum_{i \leq J} v_{ij} x_j = c_i$. From the equation $v_{nn} x_n = c_n$ the unknown x_n can then be found immediately, and by substituting it in the equation $v_{n-1,n-1} x_{n-1} + v_{n-1,n} x_n = c_{n-1}$ we then find x_{n-1}, and so on until by repeated back-substitution we have found all the coefficients of the (originally) unknown vector **x**. This description of the elimination process is all that is required in order to apply it. We shall find it instructive, however, to look at it further from a number of points of view.

(1) The process of replacing the rows of a matrix by linear combinations of other rows may be regarded as left-multiplication of the matrix by another matrix, this second matrix having coefficients which describe the linear combinations required. Each stage of the above-described elimination process is of this nature, so that we first convert the equations **Ax** = **b** into $\mathbf{J_1Ax} = \mathbf{J_1b}$ and record $\mathbf{J_1A}$ and $\mathbf{J_1b}$. We then convert them into $\mathbf{J_2J_1Ax} = \mathbf{J_2J_1b}$, and so on, until we finally have $\mathbf{J}_{n-1} \ldots \mathbf{J_1Ax} = \mathbf{J}_{n-1} \ldots \mathbf{J_1b}$. In accordance with the theorem on triangular resolution we may write $\mathbf{J}_{n-1} \ldots \mathbf{J}_1 = \mathbf{L}^{-1}$ and $\mathbf{J}_{n-1} \ldots \mathbf{J_1A} = \mathbf{DU}$. The matrix **DU** is upper triangular, that is, it has no coefficients other than zeros below the diagonal. The matrix $\mathbf{L}^{-1}$ and its inverse **L** are lower triangular.
(2) The matrix **L** can be very easily obtained from the matrices $\mathbf{J}_1, \ldots, \mathbf{J}_{n-1}$. We have in fact $\mathbf{L} = \mathbf{1} + \sum_{r=1}^{n-1} (\mathbf{1} - \mathbf{J}_r)$. The proof of this will be left to the reader.
(3) There is no need for us to take either the equations or the unknowns in the order in which they are given. In other words, if **P**, **Q** represent permutations we may solve instead $\mathbf{A'x'} = \mathbf{b'}$, where $\mathbf{A'} = \mathbf{PAQ}$, $\mathbf{b'} = \mathbf{Pb}$, $\mathbf{x} = \mathbf{Qx'}$. The permutations **P**,**Q** may be chosen bit by bit as we carry the process through. One popular method is to let **Q** be the identity, that is, to take the variables in the order given, and to choose **P** so that the coefficients in the matrices $\mathbf{J}_r$ do not exceed unity in absolute magnitude. This is always possible, and for almost all matrices gives a unique **P**. Alternatively, this variation of the method may be described by saying that **P** is chosen so that d_1 shall have the largest possible value, and subject to this, d_2 to be as large as possible,

and so on. This procedure is called 'taking the largest coefficient in the column as pivot'. The diagonal elements $d_1, d_2, \ldots, d_n$ are known as the first, second, ..., last pivots. There seems to be a definite advantage in using the largest pivot in the column as it is likely to have smaller proportionate errors than other possible pivots, and saves us from the embarrassment of getting a pivot which is little different from zero. It is possible that there is also a further advantage in choosing the largest coefficient in the matrix as pivot.

(4) The leading terms of the work involved in solving a set of n equations by the elimination method are as follows: $\frac{1}{3}n^3 + O(n^2)$ multiplications and recordings of which $\frac{1}{2}n^2 + O(n)$ recordings involve the vector $\mathbf{b}$.

(5) If, after we have solved one set of equations $\mathbf{Ax} = \mathbf{b}$, we are asked to solve a second set $\mathbf{Ax}' = \mathbf{b}'$ with the same matrix $\mathbf{A}$, we have only to operate on $\mathbf{b}$ with the matrices $\mathbf{J}_1, \ldots, \mathbf{J}_{n-1}$ the values of which may be supposed to have been kept for reference, and then solve $\mathbf{DUx} = \mathbf{J}_{n-1} \ldots \mathbf{J}_1\mathbf{b}$. In other words, if the matrices $\mathbf{J}_1 \ldots, \mathbf{J}_{n-1}$ have been kept (amounting to $\frac{1}{2}n(n-1)$ numbers) the work involved in solving a second set with the same $\mathbf{A}$ is that part of the original work which involved $\mathbf{b}$, namely, $\frac{1}{2}n^2 + O(n)$ multiplications and n recordings.

This process may also be expressed in another form, which appears to be quite different, but actually is an identical calculation. As mentioned in (2), the triangle $\mathbf{L}$ in the resolution $\mathbf{A} = \mathbf{LDU}$ may be obtained immediately from the matrices $\mathbf{J}_1, \ldots, \mathbf{J}_{n-1}$. If we put $\mathbf{DUx}' = \mathrm{y}'$ we shall then have $\mathrm{Ly}' = \mathbf{b}'$. The equations $\mathbf{Ly}' = \mathbf{b}'$ may be solved for y' by one back-substitution process and then the equation $\mathbf{DUx}' = \mathbf{y}'$ solved by a second back-substitution.

(6) As we have described it, the matrices $\mathbf{J}_1\mathbf{A}$, $\mathbf{J}_2\mathbf{J}_1\mathbf{A}, \ldots$, are all written down in full. Actually, however, we are not really interested in all the coefficients of all these matrices. All we need in the end are $\mathbf{J}_1, \ldots, \mathbf{J}_{n-1}$ and $\mathbf{J}_{n-1} \ldots \mathbf{J}_1\,\mathbf{A}$. It is sufficient, therefore, to calculate all coefficients of $\mathbf{J}_{n-1} \ldots \mathbf{J}_1\mathbf{A}$, and those coefficients of $\mathbf{J}_r \ldots \mathbf{J}_1\,\mathbf{A}$ which are required for the determination of $\mathbf{J}_{r+1}$. If we write $\mathbf{A}^{(r)}$ for $\mathbf{J}_r \ldots \mathbf{J}_1\,\mathbf{A}$ we have

$$\mathbf{A}_{ij}^{(r)} = \mathbf{A}_{ij}^{(r-1)} + (\mathbf{J}_r)_{ir}\mathbf{A}_{rj}^{(r-1)} \ (i > r),$$

where

$$(\mathbf{J}_r)_{ir} = -\frac{\mathbf{A}_{ir}^{(r-1)}}{\mathbf{A}_{rr}^{(r-1)}},$$

and by addition

$$\mathbf{A}_{ij}^{(r)} = \mathbf{A}_{ij} + \sum_{s=1}^{r}(\mathbf{J}_s)_{is}\mathbf{A}_{sj}^{(s-1)}.$$

If $i \leq r$ we have $\mathbf{A}_{ij}^{(r)} = \mathbf{A}_{ij}^{(r-1)}$ and so

$$\mathbf{A}_{ij}^{(n)} = \mathbf{A}_{ij} + \sum_{s=1}^{n-1}(\mathbf{J}_s)_{is}\mathbf{A}_{sj}^{(n)},$$

$$(\mathbf{J}_r)_{ir} = -\frac{\mathbf{A}_{ir} + \sum_{s=1}^{r-1}(\mathbf{J}_s)_{is}\mathbf{A}_{sr}^{(s-1)}}{\mathbf{A}_{rr} + \sum_{s=1}^{r-1}(\mathbf{J}_s)_{rs}\mathbf{A}_{sr}^{(s-1)}}.$$

Thus we can obtain the numbers actually required ($\mathbf{A}_{ij}^{(n)}$, $(\mathbf{J}_r)_{ir}$) without recording intermediate quantities. This variation of the elimination method will be seen to be identical with the method (1) of §6 (the 'unsymmetrical Choleski method').

This form of the elimination method is to be preferred to the original form in every way. The recording involved in the work on the matrix is reduced from $\frac{1}{3}n^3 + O(n^2)$ to $n^2 + O(n)$, and the rounding off is at the same time made correspondingly less frequent.

(7) The elimination method may be used to invert a matrix. One method is to solve a succession of sets of equations $\mathbf{A}\mathbf{x}^{(r)} = \mathbf{b}^{(r)}$, where $\mathbf{b}^{(r)} = \delta_{ir}$. The total work involved in the inversion is then $n^3 + O(n^2)$ multiplications. Alternatively, we may invert the matrices **L** and **DU** separately by back-substitution and then multiply them together. The work is still $n^3 + O(n^2)$ multiplications.

(8) When the matrix **A** is symmetric, the matrices **L** and **U** are transposes, and it is therefore unnecessary to calculate both of them. The best arrangement is probably to proceed as with an unsymmetrical matrix, but to ignore all the coefficients below the diagonal in the matrices $\mathbf{A}^{(r)}$. These coefficients are all either zero or equal to the corresponding elements of the transpose. This fact enables us to find the appropriate matrices $\mathbf{J}_r$ at each stage.

(9) The elimination method can be described in another, superficially quite unrelated form. We may combine multiplication of rows and addition to other rows with multiplication of columns and adding to other columns. In other words, we may form a product $\mathbf{J}_{n-1}\ldots\mathbf{J}_1\mathbf{A}\mathbf{K}_1\ldots\mathbf{K}_{n-1}$, and try to arrange that it shall be diagonal. The matrix $\mathbf{J}_r$ is to differ from unity only in the rth column below the diagonal, and $\mathbf{K}_r$ is to differ from unity only in the rth row above the diagonal. If we carry out the multiplications by $\mathbf{J}_1,\ldots,\mathbf{J}_{n-1}$ before the multiplications by $\mathbf{K}_1\ldots,\mathbf{K}_{n-1}$, then it is clear that we have only the elimination method, for in either case we form $\mathbf{J}_1\mathbf{A}, \mathbf{J}_2\mathbf{J}_1\mathbf{A},\ldots$ and the multiplications by $\mathbf{K}_1,\ldots,\mathbf{K}_{n-1}$ which come after actually involve no computation; they merely result in replacing certain coefficients in the matrix $\mathbf{J}_{n-1}\ldots\mathbf{J}_1\mathbf{A}$ by zeros (compare note (2)). It is not quite so clear in the case where the order of calculation is $\mathbf{A}$, $\mathbf{J}_1\mathbf{A}$, $\mathbf{J}_1\mathbf{A}\mathbf{K}_1$, $\mathbf{J}_2\mathbf{J}_1\mathbf{A}\mathbf{K}_1,\ldots$. In this case, however, the right-multiplications do not alter that part of the matrix which will be required later; in fact, they again do nothing but replace certain coefficients by zeros. So far as the subsequent work is concerned, we may consider that these right-multiplications were omitted, and that we formed $\mathbf{J}_{n-1}\ldots\mathbf{J}_1\mathbf{A}$ as in the elimination method.

When this method is used and we choose the largest pivot in the matrix, it is clear that all the coefficients of $\mathbf{J}_r$ and of $\mathbf{K}_r$ do not exceed unity. This provides one proof that when the largest pivot in the matrix is chosen the coefficients of **L**, **U** do not exceed unity (in absolute magnitude).

5. Jordan's method for inversion

In §4 (1) we mentioned that the elimination process could be regarded as the reduction of a matrix to triangular form by left-multiplication of it by a sequence of matrices $\mathbf{J}_1,\ldots,\mathbf{J}_{n-1}$. In the Jordan method we left-multiply the matrix **A** by a similar sequence of matrices. The difference is that with the Jordan method we aim at reducing **A** to a diagonal,† or preferably to the unit matrix, instead of merely to a triangle.†

The process consists in forming the successive matrices $\mathbf{J}_1\mathbf{A}$, $\mathbf{J}_2\mathbf{J}_1\mathbf{A},\ldots$, where $\mathbf{J}_r$ differs from the unit matrix only in the rth column, and where $\mathbf{J}_r\ldots\mathbf{J}_1\,\mathbf{A}$ differs from a diagonal matrix only in the columns after the rth.

Let us put $\quad \mathbf{A}^{(r)} = \mathbf{J}_r\ldots\mathbf{J}_1\mathbf{A}, \quad \mathbf{X}^{(r)} = \mathbf{J}_r\ldots\mathbf{J}_1,$
we shall then have

$$\mathbf{A}_{ij}^{(r)} = \mathbf{A}_{ij}^{(r-1)} + (\mathbf{J}_r)_{ir}\mathbf{A}_{rj}^{(r-1)} \quad (i \neq r),$$

$$(\mathbf{J}_r)_{ir} = -\frac{\mathbf{A}_{ir}^{(r-1)}}{\mathbf{A}_{rr}^{(r-1)}} \quad (i \neq r)$$

† Hereafter 'triangle' and 'diagonal' will be written for triangular matrix' and 'diagonal matrix'

(so that

$$\mathbf{A}_{ir}^{(r)} = 0 \quad \text{if} \quad (i \neq r),$$
$$\mathbf{A}_{rj}^{(r)} = (\mathbf{J}_r)_{rr}\mathbf{A}_{rj}^{(r-1)},$$
$$\mathbf{X}_{ij}^{(0)} = \delta_{ij},$$
$$\mathbf{X}_{ij}^{(r)} = \mathbf{X}_{ij}^{(r-1)} + (\mathbf{J}_r)_{ir}\mathbf{X}_{rj}^{(r-1)}.$$

The particular diagonal to which **A** is reduced is at our disposal. Possible choices include the following. The diagonal may be the unit matrix. Or we may arrange that the diagonal elements of the $\mathbf{J}_r$ are all unity and tolerate the non-unit diagonal elements in $\mathbf{J}_n \ldots \mathbf{J}_1 \mathbf{A}$. A third alternative is to arrange that the diagonal elements in $\mathbf{J}_n \ldots \mathbf{J}_1 \mathbf{A}$ shall be between $0 \cdot 1$ and 1 and that the diagonal elements in $\mathbf{J}_r$ shall be powers of 10.

Jordan's method is probably the most straightforward one for inversion. Although it can be used for the solution of equations, it is not very economical for that purpose. For hand work it has the serious disadvantage that the recording is very heavy and cannot be avoided by methods such as that suggested in connexion with the elimination method. It may be the best method for use with electronic computing machinery.

6. Other methods involving the triangular resolution

There are several ways of obtaining the triangular resolution. When it has been obtained, it can be used for the solution of sets of equations, or for the inversion of the matrix as has been described under the elimination method. Possible methods of resolution are described below.

(1) We may use the formulae given in the proof of the theorem on triangular resolution. This involves $\frac{1}{3}n^3 + O(n^2)$ multiplications, $n^2 + O(n)$ recordings. This method is closely related to Choleski's method for symmetrical matrices ((7) below), and we may therefore describe it as the 'unsymmetrical Choleski method'.
(2) We may apply the elimination method, regarded as a means of obtaining the triangular resolution; see notes (1), (2), (6) on the elimination method.
(3) We may obtain simultaneously, and bit by bit, the four triangles $\mathbf{L}$, $\mathbf{L}^{-1}$, $\mathbf{U}$, $\mathbf{U}^{-1}$ and the diagonal **D**. The method makes use of the following simple facts about triangles:
    (a) If we wish to invert a triangle, but only know the values in a subtriangle, we can obtain the coefficients of the inverse in the corresponding subtriangle: for example, if we know the first 5 rows of a lower triangle **L**, then we can obtain the first 5 rows of $\mathbf{L}^{-1}$.
    (b) If we know the first r columns of a unit lower triangle then we know its first $r+1$ rows: likewise, if we know the first r rows of a unit upper triangle we know also its first $r+1$ columns.

Let us suppose that we have carried the process to the point of knowing the first r rows of **L**, the first $r-2$ of $\mathbf{L}^{-1}$ and $r-1$ of **U** and $\mathbf{U}^{-1}$. We carry on the inversion of **L** to obtain the $(r-1)$th and rth rows of $\mathbf{L}^{-1}$, and then multiply these rows into **A** to obtain the rth and $(r-1)$th rows of $\mathbf{L}^{-1}\mathbf{A}$, i.e. of **DU**. From this we obtain at once the rth and $(r-1)$th rows of **D**, and dividing obtain the rth and $(r-1)$th rows of **U**. By (b) we have the rth and $(r+1)$th columns of **U** and by (a) obtain those of $\mathbf{U}^{-1}$. Multiplying we obtain the rth and $(r+1)$th columns of $\mathbf{AU}^{-1}$, i.e. of **LD**, and from this the rth and $(r+1)$th elements of **D** and columns of **L**. By (b) we have the $(r+1)$th and $(r+2)$th rows of **L**.

We can, of course, arrange to increase r by 1 instead of 2 at each stage.

This is essentially Morris's escalator method (ref. 4), so called because by breaking off the work at any stage we obtain the solution for one of the principal minors of **A**; the order of the minor increases in steps. Morris's method differs in one small point. The diagonal elements **D** are not obtained as the diagonal of $\mathbf{L}^{-1}\mathbf{A}$ or of $\mathbf{AU}^{-1}$, but by using the identity $d_k = a_{kk} - \sum_{i<k} (\mathbf{AU}^{-1})_{ki} d_i^{-1} (\mathbf{L}^{-1}\mathbf{A})_{ik}$, which follows from the (kk) coefficient of the matrix equation$\mathbf{A} = (\mathbf{AU}^{-1})\mathbf{D}^{-1}\,(\mathbf{L}^{-1}\mathbf{A})$.

If Morris's method is used for the inversion of a matrix the work involved consists of $\frac{5}{3}n^3 + O(n^2)$ multiplications (two triangle inversions each $\frac{1}{6}n^3 + O(n^2)$, two multiplications of a triangle by **A**, each $\frac{1}{2}n^3 + O(n^2)$, and one multiplication of two triangles of opposite type, $\frac{1}{3}n^3 + O(n^2)$), and $3n^3 + O(n^2)$ recordings (this can be slightly reduced). It does not appear to be especially satisfactory in either respect.

To relate the above account to Morris's put

$$q_k = d_k,\ x_i = (U^{-1})_{1i},\ y_i = (U^{-1})_{2i}, \ldots,\ x'_i = (L^{-1})_{i1},\ y'_i = (L^{-1})_{i2}, \ldots.$$

(4) We may look for an upper triangular matrix **M** such that

$$\mathbf{M{*}A{*}AM} = 1,$$

that is, so that **AM** is orthogonal. From the first r rows of **M** (which are also the first r columns of **M∗**) we can obtain the first r rows of **M∗** because of its triangular character, and hence the corresponding rows of **M∗A∗** and **M∗A∗A**. The equation **M∗A∗A.M** $= 1$ is then applied, using the first r columns in the $(r+1)$th row of the product. This determines the ratios of the coefficients of **M** in the $(r+1)$th row. The $(r+1)$th diagonal element of the equation then determines the multiplying factor. Having found **M** and **AM** we obtain the inverse as **M(AM)∗**, or we may solve $\mathbf{Ax} = \mathbf{b}$ by forming (**AM**)∗**b** and then **M(AM)∗b**. In the terminology of orthogonal vectors, as described below, the formation of (**AM**)∗**b** would be 'expressing **b** in terms of the base of orthogonal vectors'.

This method is the orthogonalization process described in ref. (**3**), p. 9. It is closely related to the Morris method for symmetrical matrices (see (5) below). We may apply Morris's method by forming **A∗A** and then looking for the upper triangular matrix **M** to satisfy **M∗A∗AM** $= 1$. This would only involve **A** through the formation of **A∗A** and hence of **MA ∗A**. Thus Morris's method applied to the normalized matrix **A∗A** differs from the orthogonalization process only in that **M∗A∗A** is obtained as **M∗(A∗A)** instead of as **(M∗A∗)A**.

We now come to methods for symmetrical matrices. These can all be made to provide methods for unsymmetrical matrices by normalizing the given matrix, that is, forming **AA∗** from **A**. For instance, if we wish to solve $\mathbf{Ax} = \mathbf{b}$, we may form **A∗A** and **A∗b**, and then solve **A∗Ax** $=$ **A∗b** by one of these methods. This normalizing technique is, however, of doubtful value. The formation of **A∗A** involves $\frac{1}{2}n^3 + O(n^2)$ multiplications, so that the work involved is greater with normalization than without, in the case of solving equations, and is no less for the case of inversion. Moreover, normalizing tends to make equations more 'ill-conditioned' (see §8 below).

(5) A scheme mentioned in note (8) under the elimination method.

(6) We may apply the method (1), but we shall only need to find **L** and **D**, since **U** $=$ **L***. As a slight variation we may find **LD**.

(7) Another variation on (6) is to find $\mathbf{LD}^{\frac{1}{2}}$. This method is due to Choleski (ref. **1**). The matrix $\mathbf{LD}^{\frac{1}{2}}$ may involve some pure imaginary numbers, but no strictly complex ones.

(8) Morris's method simplifies considerably for symmetric matrices. From the first r rows of **L** we can obtain the first r columns of $\mathbf{L}{*}^{-1}$, i.e. $\mathbf{U}^{-1}$, by inverting. Left-multiplication by **A** gives

the first r columns of $\mathbf{AU}^{-1}$, i.e. of $\mathbf{LD}$, and from this we obtain the first $(r+1)$ rows of $\mathbf{L}$. Again Morris obtains $\mathbf{D}$ differently.

This method is identical with a variation of the orthogonalization method, applicable to symmetric matrices and due to L. Fox (ref. 2). Fox regards two vectors $\mathbf{b}$ and $\mathbf{c}$ as 'orthogonal' relative to $\mathbf{A}$ if $(\mathbf{c}, \mathbf{Ab}) = 0$ (scalar product). Fox finds a set of vectors $\mathbf{v}_1, \mathbf{v}_2, \ldots, \mathbf{v}_n$ which are orthogonal in this sense. The vectors $\mathbf{Av}_r$ may be used as a base for other vectors: we have in fact

$$\mathbf{b} = \sum_r \frac{(\mathbf{b}, \mathbf{v}_r)}{(\mathbf{v}_r, \mathbf{Av}_r)} \mathbf{Av}_r.$$

The solution of equations is effected by means of the formula

$$\mathbf{A}^{-1}\mathbf{b} = \sum_r \frac{(\mathbf{b}, \mathbf{v}_r)}{(\mathbf{v}_r, \mathbf{Av}_r)} \mathbf{v}_r.$$

It is best to obtain $\mathbf{v}_1, \mathbf{v}_2, \ldots, \mathbf{v}_n$ by orthogonalizing the unit coordinate-axis vectors, that is, besides the vectors being orthogonal, $\mathbf{v}_r$ is restricted to be a linear combination of $\mathbf{e}_1, \mathbf{e}_2, \ldots, \mathbf{e}_n$, or in other words, to have all coefficients after the rth equal to 0. In this case the vectors $\mathbf{v}_r$ are the rows of $\mathbf{L}^{-1}$, and the orthogonality relation is $\mathbf{L}^{-1}(\mathbf{AL}^{-1*}) = \mathbf{D}$. The orthogonalization process by which $\mathbf{L}^{-1}$ is found is identical with the inversion of $\mathbf{AL}^{-1*}\mathbf{D}^{-1}$.

7. Measure of the magnitude of a matrix

There are a number of ways in which the magnitude of a matrix may be measured by a real number. They include:

The norm. The norm $N(\mathbf{A})$ of the matrix A is given by

$$N(\mathbf{A}) = (\text{trace}\mathbf{A}^*\mathbf{A})^{\frac{1}{2}} = \left(\sum_{i,j} a_{ij}^2\right)^{\frac{1}{2}}.$$

The maximum expansion $B(\mathbf{A})$. This is given by

$$B(\mathbf{A}) = \max_{\mathbf{x}} \frac{|\mathbf{Ax}|}{|\mathbf{x}|} = \max_{\mathbf{x}} \frac{(\mathbf{Ax}, \mathbf{Ax})^{\frac{1}{2}}}{(\mathbf{x}, \mathbf{x})^{\frac{1}{2}}}.$$

The maximum coefficient $M(\mathbf{A})$. This is the largest coefficient in the matrix:

$$M(\mathbf{A}) = \max_{i,j} |a_{ij}|.$$

Of these measures one of the first two above is probably of greatest theoretical significance. In this paper we deal chiefly with the maximum coefficient, since it is the most easily computed.

A number of inequalities relating these are listed below.

$$M(\mathbf{X}+\mathbf{Y}) \le M(\mathbf{X})+M(\mathbf{Y}) \tag{7.1}$$

$$M(\mathbf{XY}) \le nM(\mathbf{X})M(\mathbf{Y}) \tag{7.2}$$

$$B(\mathbf{X}+\mathbf{Y}) \le B(\mathbf{X})+B(\mathbf{Y}) \tag{7.3}$$

$$B(\mathbf{XY}) \le B(\mathbf{X})B(\mathbf{Y}) \tag{7.4}$$

$$N(\mathbf{X}+\mathbf{Y}) \le N(\mathbf{X})+N(\mathbf{Y}) \tag{7.5}$$

$$N(\mathbf{XY}) \le N(\mathbf{X})N(\mathbf{Y}) \tag{7.6}$$

$$N(\mathbf{X}) \le nM(\mathbf{X}) \tag{7.7}$$

$$M(\mathbf{X}) \le N(\mathbf{X}) \tag{7.8}$$

$$M(\mathbf{X}) \le B(\mathbf{X}) \tag{7.9}$$

$$B(\mathbf{X}) \le n^{\frac{1}{2}}M(\mathbf{X}) \tag{7.10}$$

$$B(\mathbf{X}) \le N(\mathbf{X}) \tag{7.11}$$

$$N(\mathbf{X}) \le n^{\frac{1}{2}}B(\mathbf{X}) \tag{7.12}$$

8. Ill-conditioned matrices and equations

When we come to make estimates of errors in matrix processes we shall find that the chief factor limiting the accuracy that can be obtained is ‘ill-conditioning’ of the matrices involved. The expression ‘ill-conditioned’ is sometimes used merely as a term of abuse applicable to matrices or equations, but it seems most often to carry a meaning somewhat similar to that defined below.

Consider the equations

$$\left.\begin{aligned} 1\cdot 4x+0\cdot 9y &= 2\cdot 7 \\ -0\cdot 8x+1\cdot 7y &= -1\cdot 2 \end{aligned}\right\} \tag{8.1}$$

and form from them another set by adding one-hundredth of the first to the second, to give a new equation replacing the first

$$\left.\begin{aligned} -0\cdot 786x+1\cdot 709y &= -1\cdot 173 \\ -0\cdot 800x+1\cdot 700y &= -1\cdot 200 \end{aligned}\right\}. \tag{8.2}$$

The set of equations (8.2) is fully equivalent to (8.1), but clearly if we attempt to solve (8.2) by numerical methods involving rounding-off errors we are almost certain to get much less accuracy than if we worked with equations (8.1). We should describe the equations (8.2) as an *ill-conditioned* set, or, at any rate, as ill-conditioned compared with (8.1). It is characteristic of ill-conditioned sets of equations that small percentage errors in the coefficients given may lead to large percentage errors in the solution. If we are required to solve the equations $\mathbf{Ax} = \mathbf{b}$. but the coefficients used are those of $\mathbf{A}-\mathbf{S}$ instead of those of $\mathbf{A}$, $\mathbf{S}$ being a small matrix, then, to first order in $\mathbf{S}$, the solution obtained will be $\mathbf{x_0}+\mathbf{A}^{-1}\mathbf{S}\mathbf{x}_0$, where $\mathbf{x}_0$ is the correct solution. We may average the effect of this over a random population of matrices $\mathbf{S}$, and over the coefficients in the solution and matrix, and we shall find the

$$\frac{\text{R.M.S. error of coefficients of solution}}{\text{R.M.S. coefficient of solution}}$$
$$= \frac{1}{n}N(\mathbf{A})N(\mathbf{A}^{-1})\frac{\text{R.M.S. error of coefficients of }\mathbf{A}}{\text{R.M.S. coefficient of }\mathbf{A}}.$$

This equation suggests that we might take either $N(\mathbf{A})N(\mathbf{A}^{-1})$ or $\frac{1}{n}N(\mathbf{A})N(\mathbf{A}^{-1})$ as a measure of the degree of ill-conditioning in a matrix. We will adopt the latter and call $\frac{1}{n}N(\mathbf{A})N(\mathbf{A}^{-1})$ the *the N-condition number of* $\boldsymbol{A}$. We will also use $nM(\mathbf{A})M(\mathbf{A}^{-1})$ as another measure of ill-conditioning and call it *the M-condition number of* $\boldsymbol{A}$. There is substantial agreement between the two measures, though the M-number tends to be the larger, especially with diagonal or nearly diagonal matrices.

It should be noted that if all the coefficients of a matrix are multiplied by the same factor the condition numbers are unaltered, but that if a row or column is multiplied by a very large or a very small number the condition numbers are usually increased. For instance, the matrices

$$\begin{pmatrix} 0\cdot 8 & 0\cdot 6 \\ -0\cdot 6 & 0\cdot 8 \end{pmatrix} \quad (8.3) \qquad \text{and} \qquad \begin{pmatrix} 0\cdot 008 & 0\cdot 006 \\ -0\cdot 6 & 0\cdot 8 \end{pmatrix} \quad (8.4)$$

have the M-condition numbers $1\cdot 28$ and 128 respectively and N-condition numbers 1 and $50\cdot 005$. This may be considered quite a satisfactory example of the application of the definition. In practice one will tend to work with the same number of figures throughout a matrix. and the small values in the first row of 8.4 will prejudice the accuracy obtainable, because of the number of significant figures available. It is certainly true a trivial modification improves the conditioning, but we should consider that until the possibility of this modification has been observed and action taken, the matrix remains ill-conditioned.

It is often stated that ill-conditioned matrices are ones which have small determinants, that is, small considering the magnitudes of the coefficients. This statement contains a certain amount of truth. It is certainly the case that bad conditioning and small determinants tend to go together. However, the determinant may differ very greatly from the above-defined condition numbers as a measure of conditioning. This may be illustrated by the cases of the matrices

$$\begin{pmatrix} 1 & 0 & 0 \\ 0 & 0.1 & 0 \\ 0 & 0 & 0.1 \end{pmatrix}; \begin{pmatrix} 1 & 0 & 0 \\ 0 & 1 & 1 \\ 0 & 0 & 0.01 \end{pmatrix}; \begin{pmatrix} 1 & 1 & 1 \\ 1 & 1.1 & 1 \\ 1 & 1 & 1.1 \end{pmatrix}; \begin{pmatrix} 1 & 1 & 1 \\ 1 & 2 & 1 \\ 1 & 1 & 1.01 \end{pmatrix};$$

all of which have the determinant $0\cdot 01$, and which have the M-condition numbers 30, 300, $69\cdot 3$, 612, respectively, and N-condition numbers $4\cdot 77$, $47\cdot 1$, $33\cdot 0$, 232.

The best conditioned matrices are the orthogonal ones. which have N-condition numbers of 1. Their M-condition numbers are mostly of the order of magnitude of $\log n$ (for large order n). If the coefficients of a matrix are chosen at random from a normal population we shall get N-condition numbers of the order of $n^{\frac{1}{2}}$ and M-condition numbers about $\log n$ times greater. Thus random matrices are only slightly ill-conditioned.

The matrices which occur in practical problems are by no means random in this sense. There is a very large class of problems which naturally give rise to highly ill-conditioned equations. Suppose. for example, that we have reason to believe that some function of position in two dimensions can be represented by a polynomial of the fourth degree and that we wish to determine the coefficients. To this end we measure the values of the function at 25 points, and so obtain 25 linear equations for the desired coefficients. It may well happen that we are only able to make the measurements within a small region, and this will certainly mean that the equations are ill-conditioned. In such a case the equations might be improved by a differencing procedure, but this will not necessarily be the case with all problems. Preconditioning of equations in this way will always require considerable liaison between the experimenter and the computer, and this will limit its applicability.

9. The classical iterative method

Suppose that $\mathbf{B}$ is an approximate inverse of $\mathbf{A}$. Then we can obtain from it a better inverse $\mathbf{B}_2$ by the formula $\mathbf{B}_2 = 2\mathbf{B} - \mathbf{BAB}$. If we write $\mathbf{E} = 1 - \mathbf{AB}$, $\mathbf{E}_2 = 1 - \mathbf{AB}_2$, so that $\mathbf{E}$ and $\mathbf{E}_2$ give a measure of the incorrectness of the two inverses: we have $\mathbf{E}_2 = \mathbf{E}^2$, so that at each application of this process the error is essentially squared.

The work involved in applying this method is considerable, since it involves $2n^3$ multiplications at each stage. It may be useful in cases where a good approximate inverse is already available, and $1-\mathbf{AB}$ has already been calculated, but found to be a little larger than can be tolerated. We may then calculate $\mathbf{B}_2$ but carry the process no farther. This involves n^3 multiplications, but since we may write $\mathbf{B}_2 = \mathbf{B} + \mathbf{BE}$, the number of figures in one of the factors (viz. in $\mathbf{E}$) may be kept small.

A somewhat similar type of method applies for the improvement of solutions of sets of equations. Suppose, for example, we have to solve the equations $\mathbf{Ax} = \mathbf{b}$ and that we have obtained a resolution $\mathbf{A} = \mathbf{L}.\ \mathbf{DU}$ (say), somewhat inaccurately. By double back-substitution we obtain a solution $\mathbf{x_1}$ of $\mathbf{L.DUx} = \mathbf{b}$, which is an inaccurate solution of $\mathbf{Ax} = \mathbf{b}$. We may further test this solution by forming the 'residual' vector $\mathbf{b_1} = \mathbf{b} - \mathbf{Ax_1}$' and if this is too large we solve $\mathbf{Ax} = \mathbf{b_1}$ to obtain a correction. In this process we do not obtain 'quadratic convergence' but only convergence in geometric progression. On the other hand, the method is very practical because the work involved per stage is only $2n^2$ multiplications.

10. General remarks on error estimates: the error in a reputed inverse

Error estimates can be of two kinds. We may wish to know how accurate a certain result is, and be willing to do some additional computation to find out. A different kind of estimate is required if we are planning calculations and wish to know whether a given method will lead to accurate results. In the former case we do not care what quantities the error is expressed in terms of, provided they are reasonably easily computed. With these estimates we wish to be absolutely sure that the error is within the range stated, but at the same time not to state a range which is very much larger than necessary. With the second type of estimate, the error is preferably expressed in terms of quantities whose meaning is sufficiently familiar that the general run of values involved may at least be guessed at. We are also as much interested in the statistical behaviour of the errors as in the maximum possible value.

This paper is mainly concerned with estimates of the second kind, since those of the first kind can be quickly dismissed. Let $\mathbf{B}$ be a reputed inverse of $\mathbf{A}$. To determine its accuracy we form $\mathbf{E} = 1 - \mathbf{AB}$. Then in view of the inequalities (7.1), (7.2), and the equation

$$\mathbf{A}^{-1} - \mathbf{B} = \mathbf{B}(\mathbf{E} + \mathbf{E}^2 + \ldots)$$

we have

$$M(\mathbf{B} - \mathbf{A}^{-1}) \leq \sum_{r=1}^{\infty} M(\mathbf{BE}^r) \leq \sum_{r=1}^{\infty} n^r M(\mathbf{B})\{M(\mathbf{E})\}^r = \frac{nM(\mathbf{B})M(\mathbf{E})}{1 - nM(\mathbf{E})},$$

which is the required error estimate. In order to apply this inequality it is necessary to carry out the matrix multiplication $\mathbf{BA}$, involving n^3 multiplications. However, if it is intended to apply the classical iteration method for improving the inverse at least once, we shall have to calculate **E** in doing so, and we shall have $1-\mathbf{AB}_2 = \mathbf{E}_2 = \mathbf{E}^2$ and therefore

$$M(\mathbf{B}_2 - \mathbf{A}^{-1}) \leq \frac{nM(\mathbf{B}_2)M(\mathbf{E}_2)}{1 - nM(\mathbf{E}_2)} \leq \frac{n^2 M(\mathbf{B}_2)\{M(\mathbf{E})^2\}}{1 - n^2\{M(\mathbf{E})\}^2}.$$

It should be observed that this inequality is only applicable to the inversion of a matrix, and not to the solution of equations. It is difficult to determine the accuracy of the solution of a set of equations without inverting the matrix. This is another reason why it is preferable to treat inversion rather than solution of equations as a standard process.

When making estimates of the effects of rounding-off errors we need the process under examination to be rather minutely described. If, for instance, a product *abc* is to be formed, we need to know whether it is obtained as *ab.c* or as *a.bc*. If it is obtained as *ab.c* we shall need to know how many figures are kept in *ab*. This may be either a definite number of decimal or binary places, or a definite number of significant figures, or the number of figures kept may be made to depend on the results of previous calculations. Usually, however, by a trivial modification of the quantity recorded, these latter cases can be reduced to one of the former.

The variety of possible detailed calculation procedures is, of course, vastly greater than the list of methods which we have considered, for these can be subdivided into numerous alternatives which appear only trivially different at first sight, but which may differ very seriously from the point of view of error estimates. We cannot here carry out the analysis for more than a very few of the procedures. These have been chosen so as to give bounds of error which are both reasonably small and also fairly simple in their analytical form. We have concentrated particularly on error estimates which can be expressed in terms of the matrix **A** and its inverse. In practical work the details of the procedure must be determined by other considerations. With any particular procedure it will usually be found possible to obtain some estimate of the type proved in this paper, but usually quantities such as $M(\mathbf{L})$, $M(\mathbf{D}^{-1})$, etc., will be involved. These can be obtained conveniently as a by-product in the calculation. Alternatively, one may find bounds of error by calculating $1-\mathbf{AB}$ as above. In this case the importance of the analysis which follows is to show that it is probable that the error obtained will be reasonably small if a process is used which is somewhat similar to one of those here considered, and that these methods are therefore reasonable ones to use. Our main purpose in this paper is to establish that the exponential build-up of errors need not occur, and this will be proved when we have found one method of inversion where it is absent.

11. Rounding-off errors in Jordan's method

The Jordan method was described in §5, but we have now to specify the details of the rounding-off and the diagonal. We shall consider the case where **A** is reduced to a unit matrix. We assume that in the calculation of each quantity

$$\mathbf{A}_{ij}^{(r-1)} - \frac{\mathbf{A}_{rj}^{(r-1)}\mathbf{A}_{ir}^{(r-1)}}{\mathbf{A}_{rr}^{(r-1)}},$$

an error of at most ϵ is made. How this is to be secured need not be specified, but it is clear that the number of figures to be retained in $\mathbf{A}_{ir}^{(r-1)}/\mathbf{A}_{rr}^{(r-1)}$ will have to depend on the values of the $\mathbf{A}_{rj}^{(r-1)}$. Likewise, we assume that in the calculation of

$$\mathbf{X}_{ij}^{(r-1)} - \frac{\mathbf{X}_{rj}^{(r-1)}\mathbf{A}_{ir}^{(r-1)}}{\mathbf{A}_{rr}^{(r-1)}}$$

an error of at most ϵ' is made. It is convenient to think of these errors as quantities deliberately added after the accurate calculation has been made. If the quantities added after the calculation of $\mathbf{A}^{(r)}$, $\mathbf{X}^{(r)}$ are the matrices $\mathbf{S}_r$, $\mathbf{S'}_r$ we shall have

$$\begin{aligned} &\mathbf{J}_n[\ldots\{\mathbf{J}_2(\mathbf{J}_1\mathbf{A}+\mathbf{S}_1)+\mathbf{S}_2\}\ldots]+\mathbf{S}_n=1,\\ &\mathbf{J}_n[\ldots\{\mathbf{J}_2(\mathbf{J}_1+\mathbf{S'}_1)+\mathbf{S'}_2\}\ldots]+\mathbf{S'}_n=\Xi, \end{aligned} \tag{11.1}$$

where Ξ represents the actual matrix obtained at the end of the calculation as the value of $\mathbf{A}^{-1}$.

The equations (11.1) give us

$$\begin{aligned} &\mathbf{A}+\sum\mathbf{X}_r^{-1}\mathbf{S}_r=\mathbf{X}_n^{-1},\\ &1+\sum\mathbf{X}_r^{-1}\mathbf{S}'_r=\mathbf{X}_n^{-1}\Xi \end{aligned} \tag{11.2}$$

and hence

$$\Xi=(1+\mathbf{A}^{-1}\sum_r\mathbf{X}_r^{-1}\mathbf{S}_r)\mathbf{A}^{-1}(1+\sum_r\mathbf{X}_r^{-1}\mathbf{S}'_r). \tag{11.3}$$

The matrix $\mathbf{X}_r\,\mathbf{A}$ is the result of the first r stages of the reduction of $\mathbf{A}$ and agrees with $\mathbf{D}$ in the first r columns. This fact may be expressed in the equation

$$(\mathbf{X}_r\mathbf{A}-1)\mathbf{I}_r=0, \tag{11.4}$$

where $\mathbf{I}_r$ is that matrix which agrees with the unit matrix in the first r columns and with the zero matrix elsewhere. It is also clear that $\mathbf{X}_r$ differs from the unit matrix only in the first r columns; this fact may be expressed in the equation

$$(\mathbf{X}_r-1)(1-\mathbf{I}_r)=0. \tag{11.5}$$

From (11.4) and (11.5) we now find $\mathbf{X}_r^{-1}$;

$$\mathbf{X}_r^{-1}=\mathbf{A}\mathbf{I}_r+1-\mathbf{I}_r. \tag{11.6}$$

When we ignore the second-order terms in the rounding-off errors (11,3), (11.6) give us

$$\begin{aligned} \Xi-\mathbf{A}^{-1}&=-\mathbf{A}^{-1}\left(\sum_r\mathbf{X}_r^{-1}\mathbf{S}_r\right)\mathbf{A}^{-1}+\mathbf{A}^{-1}\sum_r\mathbf{X}_r^{-1}\mathbf{S}'_r\\ &=\sum_r\{\mathbf{I}_r+\mathbf{A}^{-1}(1-\mathbf{I}_r)\}(\mathbf{S}_r\mathbf{A}^{-1}-\mathbf{S'}_r). \end{aligned} \tag{11.7}$$

Let us now assume that each coefficient $\mathbf{S}_r$ is at most ϵ and each coefficient of $\mathbf{S}'_r$ at most ϵ'. From (11.7) we can estimate the error in M-measure

$$\begin{aligned} M(\Xi-\mathbf{A}^{-1})&\leq\sum_r n\{1+M(\mathbf{A}^{-1})\}M(\mathbf{S}_r\mathbf{A}^{-1}-\mathbf{S}'_r)\\ &\leq\sum_r n\{1+M(\mathbf{A}^{-1})\}\{n\epsilon M(\mathbf{A}^{-1})+\epsilon'\}\\ &\leq n^2\{1+M(\mathbf{A}^{-1})\}\{\epsilon'+n\epsilon M(\mathbf{A}^{-1})\}, \end{aligned} \tag{11.8}$$

or in B-measure,

$$B(\Xi - \mathbf{A}^{-1}) \leq \sum_r B\{\mathbf{I}_r + \mathbf{A}^{-1}(1-\mathbf{I}_r)\}\{B(\mathbf{S}_r\mathbf{A}^{-1}) + B(\mathbf{S'}_r)\}$$

$$\leq \sum_r \{1 + B(\mathbf{A}^{-1})\}\{\epsilon n^{\frac{1}{2}} B(\mathbf{A}^{-1}) + \epsilon' n^{\frac{1}{2}}\}$$

$$\leq n^{\frac{3}{2}}\{1 + B(\mathbf{A}^{-1})\}\{\epsilon B(\mathbf{A}^{-1}) + \epsilon'\}, \tag{11.9}$$

or in N-measure,

$$N(\Xi - \mathbf{A}^{-1}) \leq \sum_r N\{\mathbf{I}_r + \mathbf{A}^{-1}(1-\mathbf{I}_r)\}\{N(\mathbf{S}_r\mathbf{A}^{-1}) + N(\mathbf{S'}_r)\}$$

$$\leq \sum_r \{r^{\frac{1}{2}} + (1-r)^{\frac{1}{2}} N(\mathbf{A}^{-1})\}\{n\epsilon N(\mathbf{A}^{-1}) + n\epsilon'\}$$

$$\leq \tfrac{2}{3}(n+1)^{\frac{5}{2}}\{1 + N(\mathbf{A}^{-1})\}\{\epsilon' + \epsilon N(\mathbf{A}^{-1})\}. \tag{11.10}$$

If we use the relations $\mathbf{S}_r\mathbf{I}_r = \mathbf{S}'_r(1-\mathbf{I}_r) = 0$, which follow from the restrictions on the coefficients which can suffer rounding-off errors, (11.8) may be improvedto

$$M(\Xi - \mathbf{A}^{-1}) \leq n\epsilon' + \frac{n(n-1)}{2} M(\mathbf{A}^{-1})\left\{\epsilon + \epsilon' + \frac{2n-1}{3}\epsilon M(\mathbf{A}^{-1})\right\}. \tag{11.11}$$

This result is best possible in the sense that given ϵ, ϵ', M we can find $\mathbf{S}_r$, $\mathbf{S}'_r$, $\mathbf{A}$ so that $M(\mathbf{S}_r) \leq \epsilon$, $M(\mathbf{S}'_r) \leq \epsilon'$, $M(\mathbf{A}^{-1}) = \mathbf{M}$ and the error $M(\Xi - \mathbf{A}^{-1})$, still ignoring second-order terms, is exactly

$$n\epsilon' + \frac{n(n-1)}{2} M\left(\epsilon + \epsilon' + \frac{2n-1}{3}\epsilon M\right).$$

We may also use (11.7) to give us an estimate of the statistical error. Let the coefficients of the matrices $\mathbf{S}_1, \ldots, \mathbf{S}_n$ which are not obliged to be 0 be $s_1, \ldots, s_K$ in some order, and likewise let the coefficients of $\mathbf{S'}_1, \ldots, \mathbf{S'}_n$ which are not necessarily zero be $s_{K+1}, \ldots, s_P$. The equation (11.7) may then be put in the form

$$(\Xi - \mathbf{A}^{-1})_{ij} = \sum_{u=1}^{P} c_{iju} s_u,$$

where c_{iju} depends only on the coefficients of $\mathbf{A}^{-1}$. Suppose that the rounding-off errors s_u are independent and have standard deviation σ_u and zero mean, then the mean square value of $(\Xi - \mathbf{A}^{-1})_{ij}$ is $\sum\limits_{u=1}^{P} c_{iju}^2 \sigma_u^2$. Let us put $\sigma_u = \eta$ for $u \leq K$, $\sigma_u = \eta'$ for $u > K$ and the mean square error in $\mathbf{A}_{ij}^{-1}$ becomes $\eta^2 \sum\limits_{u=1}^{K} c_{iju}^2 + \eta'^2 \sum\limits_{u=K+1}^{P} c_{iju}^2$. When we substitute in the correct values for c_{iju} we obtain:

mean square error in $(\mathbf{A}^{-1})_{ij}$

$$= \eta^2 \sum_{m,K} (\mathbf{A}^{-1})_{im}^2 (\mathbf{A}^{-1})_{Kj}^2 \min(K, i-1) + \eta^2 \sum_{K>i} (\mathbf{A}^{-1})_{Kj}^2 (K-i) +$$

$$+ \eta'^2 \left[\sum_m (\mathbf{A}^{-1})_{im}^2 \min(j, m-1) + \frac{(n-1)(n-i+1)}{2}\right],$$

where η is the standard deviation and zero the mean of each coefficient of $\mathbf{S}_r$, and η' is the standard deviation and zero the mean of each coefficient of $\mathbf{S}'_r$.

Also

mean square error in $(\mathbf{A}^{-1})_{ij}$

$$\leq \eta^2 \left[\{M(\mathbf{A}^{-1})\}^4 \frac{n(n+1)(n-\frac{1}{2})}{3} + \{M(\mathbf{A}^{-1})\}^2 \frac{(n-1)(n-i+1)}{2} \right] + $$
$$+ \eta'^2 \left[\{M(\mathbf{A}^{-1})\}^2 (n_2 - \tfrac{1}{2} - \tfrac{1}{2}j) + \frac{(n-i)(n-i+1)}{2} \right].$$

The leading term in the R.M.S. error in $(\mathbf{A}^{-1})_{ij}$ is therefore at most

$$\eta \left\{ M(\mathbf{A}^{-1}) \right\}^2 \frac{n^{\frac{3}{2}}}{\sqrt{3}}.$$

The assumptions $M(\mathbf{S}_r) < \epsilon$, $M(\mathbf{S'}_r) < \epsilon'$ in the above analysis state in effect that we are working to a fixed number of decimal places both in the reduction of the original matrix to unity and in the building up of the inverse. It is not easy to obtain corresponding results for the case where a definite number of *significant* figures are kept, but we may make some qualitative suggestions.

The error when working with a fixed number of decimal places arose almost entirely from the reduction of the original matrix, and very little from the building up of the inverse. This, at any rate, applies for the inversion of ill-conditioned matrices with coefficients of moderate size.

However, the coefficients of the inverse are larger than those of the original matrix, so that if we work to the same number of significant figures in both we may expect the discrepancy to disappear. The general idea of this may be expressed by putting

$$M(\mathbf{S}_r) < \delta M(\mathbf{A}), \qquad M(\mathbf{S}'_r) < \delta' M(\mathbf{A}^{-1}),$$

so that
$$\frac{M(\Xi - \mathbf{A}^{-1})}{M(\mathbf{A}^{-1})} < n^3 M(\mathbf{A}) M(\mathbf{A}^{-1}) \left(1 + \frac{1}{M(\mathbf{A}^{-1})}\right) \left(\delta + \frac{\delta'}{M(\mathbf{A})}\right).$$

There still remains the factor $\frac{1}{M(\mathbf{A})}$ multiplying δ'. This could be removed by arranging to reduce $\mathbf{A}$, not to the unit matrix, $\mathbf{1}$, but to $M(\mathbf{A}).\mathbf{1}$. This would be a reasonable procedure in any case, though it would be more convenient to choose the nearest power of 10 to take the place of $M(\mathbf{A})$. We see now that it is the M-condition number $nM(\mathbf{A})M(\mathbf{A}^{-1})$ which determines the magnitude of the errors when we work to a definite number of figures.

In the case of positive definite, symmetric matrices it is possible to give more definite estimates for the case where calculation is limited to a specific number of significant figures. Results of this nature have been obtained by J. v. Neumann and H. H. Goldstine (ref. **6**).

It is instructive to compare the estimates of error given above with the errors liable to arise from the inaccuracy of the original matrix. If we desire the inverse of $\mathbf{A}$, but the figures given to us are not those of $\mathbf{A}$ but of $\mathbf{A} - \mathbf{S}$, then if we invert perfectly correctly we shall get $(\mathbf{A} - \mathbf{S})^{-1}$ instead of $\mathbf{A}^{-1}$, that is, we shall make an error of $(\mathbf{A} - \mathbf{S})^{-1} - \mathbf{A}^{-1}$, i.e. of

$$(1 - \mathbf{A}^{-1}\mathbf{S})^{-1}\mathbf{A}^{-1}\mathbf{S}\mathbf{A}^{-1}.$$

If we ignore the second-order terms this is $\mathbf{A}^{-1}\mathbf{S}\mathbf{A}^{-1}$. The leading terms in the error in the Jordan method were $\mathbf{A}^{-1}(\sum_r (1 - \mathbf{I}_r)\mathbf{S}_r)\mathbf{A}^{-1}$ so that we might say that the greater part of the error is equal to that error which would have been produced by an original error in the matrix of $\sum_r (1 - \mathbf{I}_r)\mathbf{S}_r$.

It is possible to give error estimates also for several others amongst the methods suggested elsewhere in this paper. This is, for instance. the case for the elimination method.

The elimination method in its first phase proceeds similarly to the Jordan process, but we only attempt to reduce **A** to a triangle and not to a diagonal: also the matrix representing the complete operation in this first phase is triangular.

12. Errors in the Gauss elimination process

We will consider the errors in the Gauss elimination process as consisting of two parts, one arising from the reduction of the matrix to the triangular form, and the other from the back-substitution. Of these we are mainly interested in the error arising from the reduction, since this is the part of the process which has been most criticized. We adopt the description of the process given in §4, note (1), and observe that apart from a slight difference in the form of the matrices $\mathbf{J}_r$, the reduction is similar to the Jordan process. As in the Jordan process, we shall assume that we make matrix errors $\mathbf{S}_1, \mathbf{S}_2, \ldots, \mathbf{S}_n$ in the various stages of the reduction of **A**, and vector errors $\mathbf{s}_1, \mathbf{s}_2, \ldots, \mathbf{s}_n$ in the operations on **b**. Assuming there are no back-substitution errors, and ignoring the second-order terms in the errors we should have:

$$\text{error in } \mathbf{x} = \mathbf{U}^{-1}\mathbf{X}_n \sum_{r=1}^{n} \mathbf{X}_r^{-1}(\mathbf{s}'_r - \mathbf{S}_r\mathbf{U}^{-1}\mathbf{X}_n\mathbf{b}),$$

where $\mathbf{X}_r = \mathbf{J}_r \ldots \mathbf{J}_1$. Now, assuming that the process has been done with the largest pivot chosen from each column, we shall have $M(\mathbf{X}_r^{-1}) = 1$, for $\mathbf{X}_r^{-1} = 1 + \sum_{s \le r}(1 - \mathbf{J}_s)$ as mentioned in §4(2). Then

$$\begin{aligned} |\text{error in } \mathbf{x}_m| &= |(\mathbf{A}^{-1}\sum \mathbf{X}_r^{-1}(\mathbf{s}'_r - \mathbf{S}_r\mathbf{A}^{-1}\mathbf{b}))_m| \\ &= \left| \sum_{\substack{j,k,r \\ j \ge k}} (\mathbf{A}^{-1})_{mj}(\mathbf{X}_r^{-1})_{jk}(\mathbf{s}'_r)_k - \sum_{l.p} (\mathbf{S}_r)_{kl}(\mathbf{A}^{-1})_{lp}\mathbf{b}_p \right| \\ &\le \frac{n^2(n+1)}{2} M(\mathbf{A}^{-1})\epsilon' + \frac{n^4(n+1)}{2}\{M(\mathbf{A}^{-1})\}^2 M(\mathbf{b})\epsilon, \end{aligned}$$

where $M(\mathbf{s}'_r) \le \epsilon'$, $M(\mathbf{S}_r) \le \epsilon$.

To these errors we have to add those which arise from the back-substitution. This consists in solving the equations $\mathbf{DUx} = \mathbf{L}^{-1}\mathbf{b}$, where **U** is unit upper triangular and **D** diagonal. We obtain x_n first and then x_r in order of decreasing r by means of the formula $x_r = \mathbf{d}_r^{-1}(\mathbf{L}^{-1}\mathbf{b})_r - \sum_{i>r}(\mathbf{DU})_{ri}x_i$. Now if we make an error of t_r in the calculation of x_r from the previously obtained coefficients of **x**, then we shall have solved accurately the equations $\mathbf{DUx} = \mathbf{L}^{-1}\mathbf{b} + \mathbf{Dt}$, that is, we shall have introduced an error of $\mathbf{U}^{-1}\mathbf{t}$, or, since $\mathbf{A} = \mathbf{LDU}$, of $\mathbf{A}^{-1}\mathbf{LDt}$. If we arrange that $M|t_r| \le \epsilon d_r^{-1}$, the greatest error in any coefficient from this source is $n^2 M(\mathbf{A}^{-1})\epsilon$, and normally much smaller than the error arising from the first part of the process. Furthermore, d_r will normally tend to be less than 1.

It is interesting to note the value of the error in the last pivot, that is, the error in the (nn) coefficient of $\mathbf{J}_n \ldots \mathbf{J}_1 \mathbf{A}$. The matrix error in $\mathbf{J}_n \ldots \mathbf{J}_1\mathbf{A}$ is $\mathbf{X}_n \sum_r \mathbf{X}_r^{-1}\mathbf{S}_r$, that is, since $\mathbf{X}_n\mathbf{L}^{-1} = \mathbf{DUA}^{-1}$, it is $\mathbf{DUA}^{-1}\sum_r \mathbf{X}_r^{-1}\mathbf{S}_r$. The$(nn)$ coefficient is $d_n \sum_r a_{nj}(\mathbf{X}_r^{-1}\mathbf{S}_r)_{jn}$ and since $M(\mathbf{X}_r^{-1}) = 1$, $M(\mathbf{S}_r) \le \epsilon$ it does not exceed $n^2 d_n M(\mathbf{A}^{-1})\epsilon$ in absolute magnitude, that is, the proportionate error in the last pivot is at most $n^2 M(\mathbf{A}^{-1})\epsilon$. This cannot be very large unless the matrix is ill-conditioned. With worst possible conditioning we find an error somewhat similar to Hotelling's estimate. The matrix error in $\mathbf{J}_n \ldots \mathbf{J}_1 \mathbf{A}$ may be written $\mathbf{L}^{-1} \sum_r \mathbf{X}_r^{-1} \mathbf{S}_r$, from which we find that the error in the last pivot cannot exceed $n^2 \epsilon M(\mathbf{L}^{-1})$. But since $M(\mathbf{L}) = 1$ we find $M(\mathbf{L}^{-1}) \leq 2^{n-1}$ (and equality can be attained): that this error may actually be as great as $2^{n-2}\epsilon$ may be seen by considering the inversion of a matrix differing only slightly from

$$\begin{bmatrix} 1 & 0 & 0 & \ldots & 0 \\ -1 & 1 & 0 & \ldots & 0 \\ -1 & -1 & 1 & \ldots & 0 \\ \cdot & \cdot\cdot & \cdot & \cdot\cdot\cdot & \cdot \\ -1 & -1 & -1 & \ldots & 1 \end{bmatrix}$$

It appears then that the error in the last pivot can only be large if $\mathbf{L}^{-1}$ is large, and that this can only happen with ill-conditioned equations. Actually even then we may consider ourselves very unlucky if $\mathbf{L}^{-1}$ is large. Normally, even with ill-conditioned equations we may expect the off-diagonal coefficients of $\mathbf{L}$ to be distributed fairly uniformly between -1 and 1, possibly with a tendency to be near 0. Only when there is a strong tendency for negative values will we find a large $\mathbf{L}^{-1}$.

13. Errors in the unsymmetrical Choleski method

When obtaining the triangular resolution of a matrix by the method of the theorem (§3) it is convenient to think of the process as follows. We are given a matrix $\mathbf{A}$ and the matrices $\mathbf{L}$ and $\mathbf{DU}$ ($= \mathbf{W}$, say). We form the product $\mathbf{LW}$ coefficient by coefficient. When calculating any one of the coefficients of $\mathbf{LW}$, we always find that the data are incomplete to the extent of one number, and we therefore choose this number so as to give the required coefficient in $\mathbf{A}$. The unknown quantity when forming a_{ij} is always either l_{ij} or w_{ij}. Regarding the process in this way suggests the following rule for deciding the number of figures to be retained. We always retain sufficient figures to give us an error of not more than ϵ in the coefficient of $\mathbf{A}$ under consideration. In actual hand computation this rule is extremely simple to apply. Suppose, for example, that ϵ is $\frac{1}{2}10^{-7}$ and that we are forming the product $(\mathbf{LW})_{94}$, i.e. $\sum_{j=1}^{4} l_{9j} w_{j4}$. We first form $\sum_{j=1}^{3} l_{9j} w_{j4}$ accumulating the products in the machine. All the relevant quantities should be available at this stage. We then set up the multiplicand w_{44} which should also be known and 'turn the handle' until the quantity in the product register, rounded off to seven figures, first agrees with the given value of a_{94} (which is assumed to have zeros in the eighth and later figures). All the figures in the multiplier register are then written down as the value of l_{94}.

The theoryof the errors in this method is peculiarly simple. The triangular resolution obtained is an exact resolution of a matrix $\mathbf{A} - \mathbf{S}$, where $M(\mathbf{S}) < \epsilon$, and the resultant error in the inverse is $\mathbf{A}^{-1}\,\mathbf{S}\mathbf{A}^{-1}$, and in any coefficient at most $n^2\{M(\mathbf{A}^{-1})\}^2\epsilon$. A similar procedure is appropriate in the inversion of the triangles $\mathbf{L}$ and $\mathbf{W}$. When inverting $\mathbf{W}$ (say) we can arrange, by an exactly similar computing procedure, that its product with its reputed inverse differs from unity by at most ϵ' in each coefficient, i.e. $\mathbf{LK} = 1 - \mathbf{S}'_2$ where $M(\mathbf{S}') < \epsilon$ and $\mathbf{K}$ is the reputed inverse. Note the order in the product which is significant. Likewise we find a reputed inverse $\mathbf{V}$ for $\mathbf{DU}$ such that $\mathbf{V.DU} = 1 - \mathbf{S}''$ and $M(\mathbf{S}'') < \epsilon'$. The error arising from using these reputed inverses is $-(1-\mathbf{S}'')^{-1}\mathbf{VK}(1-\mathbf{S}')+$ $\mathbf{VK}$, or neglecting second-order terms, $\mathbf{S}''\mathbf{A}^{-1} + \mathbf{A}^{-1}\mathbf{S}'$. Finally, there is a possible source of error due to rounding off in the actual formation of the product $\mathbf{VK}$. If this does not exceed ϵ' in any coefficient, the error in any coefficient of the reputed inverse of $\mathbf{A}$ is in all at most

$$n^2\epsilon\{M(\mathbf{A}^{-1})\}^2 + 2n\epsilon' M(\mathbf{A}^{-1}) + \epsilon''.$$

This paper is published with the permission of the Director of the National Physical Laboratory.

REFERENCES

1. Commandant BÉNOIT, 'Note sur une méthode, etc.' (Procédé du Commandant CHOLESKY), *Bull. Géod.* (Toulouse), 1924, No. 2, 5–77.
2. L. FOX, H. D. HUSKEY, and J. H. WILKINSON, 'Notes on the solution of algebraic linear simultaneous equations', *see above*, pp. 149–73.
3. J. v. NEUMANN, V. BARGMANN, and D. MONTGOMERY, *Solution of Linear Systems of High Order*, lithographed, Princeton (1946).
4. J. MORRIS, 'An escalator method for the solution of linear simultaneous equations ', *Phil. Mag.*, series 7, **37** (1946), 106.
5. H. HOTELLING, 'Some new methods in matrix calculation', *Ann. Math. Stat.* **14** (1943), 34.
6. J. v. NEUMANN and H. H. GOLDSTINE, 'Numerical inverting of matrices of high order', *Bull. Amer. Math. Soc.* **53** (1947), 1021–99.

A Note on Normal Numbers

(unpublished)

Andrew Hodges on an interesting connection between —

COMPUTABLE NUMBERS AND NORMAL NUMBERS

Alan Turing's famous 1936 paper hit the mathematical world entirely out of the blue, without any published precursor. Nor did Turing leave any drafts, letters, journals of his ideas or accounts of the development of his thought. Only in quite general terms can we see the background, his undergraduate reading of Russell's logic in 1933 combining with a lifelong fascination with the problem of Mind.

There is one small exception to this otherwise complete lack of manuscript record. Some pages of Turing's final typescript for *On computable numbers*, with formulas in his handwriting, do survive. This is because he used their reverse sides as scrap paper on which to write his notes on *normal numbers,* which are now held in the archive at King's College, Cambridge.

This is an interesting connection. We know that Turing was acquainted as early as 1933 with Borel's study of normal numbers. For in that year, his friend and fellow mathematics undergraduate David Champernowne made (and published) the observation that the number 0.1234567891011121314... is normal in base-10 decimals. Turing's notes show him attempting to build a more general theory, with results (Turing 1936?) which were unpublished in his time but reproduced and reviewed by J. R. Britton in the *Collected Works*.

It seems very possible that Turing's analysis of decimal expressions influenced his choice of framework of 'computable numbers'. Turing's note on normal numbers uses the terms 'mechanical process' and 'constructive enumeration' which show the closeness of this work to the ideas of *On computable numbers.* In any case it is notable that the chosen title for his great paper put numbers first, and the application to logic second.

This connection illustrates a wider point, which is that Turing in 1935 was not a specialised logician, but the product of a very strong Cambridge background in both pure and applied mathematics, who in 1934 had devoted himself to a proof of the Central Limit Theorem. The rigorous analysis of the real number system, as the foundation of continuous mathematics, was a central part of his knowledge.

Indeed these notes were written on the reverse sides of pages of the typescript including those on 'computable convergence'. This is the section he must have hoped to expand into a new constructive treatment of analysis, judging by the hostage to fortune he left in the introduction saying that he would 'soon' give a further paper on the computable definition of real functions. The difficulty posed by the non-unique decimal representation of the reals seems to have stopped this project in its tracks. Although Turing's method in his correction note (Turing 1937) was the first step in modern computable analysis, Turing himself never followed up this lead.

It is perhaps more surprising that Turing never followed up the question underlying Borel's work, that of characterising a random number. Although Church later gave a definition based on computability, Turing's reference to randomness in his later work was curiously cavalier, leaving the concept undefined in his 1948 discussion of machines with random elements. In (Turing 1950) he used a pseudo-random sequence (the digits of π) to illustrate the computer simulation of randomness, without any discussion of the concept. This is even odder when one recalls that he had spent 6 years of war on the work of distinguishing random from pseudo-random. As in so many ways, the few details we have about the development of Alan Turing's ideas only lead to more and generally unanswerable questions.

References

Britton, J.L., 1992. Introduction and notes to Turing's note on normal numbers. In: Britton, J.L. (Ed.), The Collected Works of A. M. Turing: Pure Mathematics. North-Holland, Amsterdam.

Champernowne, D.G., 1933. The construction of decimals normal in the scale of ten. J. Lond. Math. Soc. 8, 254–260.

A NOTE ON NORMAL NUMBERS

A. M. TURING

Although it is known that almost all numbers are normal[1] no example of a normal number has ever been given. I propose to show how normal numbers may be constructed and to prove that almost all numbers are normal constructively.

Consider the R-figure integers in the scale of $t, t \geqslant 2$. If γ is any sequence of figures in that scale we denote by $N(t,\gamma,n,R)$ the number of these in which γ occurs exactly n times. Then it can be proved without difficulty that

$$\left(\sum_{n=1}^{R} n\, N(t,\gamma,n,R)\right) \Big/ \left(\sum_{n=1}^{R} N(t,\gamma,n,R)\right) = R^{-1}(R-r+1)t^{-r},$$

where $l(\gamma) = r$ is the length of the sequence γ: it is also possible[2] to prove that

$$\sum_{|n-Rt^{-r}|>k} N(t,\gamma,n,R) < 2t^R \mathrm{e}^{-k^2t^r/4R}, \tag{1}$$

provided $kt^r/R < 0.3$.

Let α be a real number and $S(\alpha,t,\gamma,R)$ the number of occurrences of γ in the first R figures after the decimal point in the expression of α in the scale of t. α is said to be *normal* if $R^{-1}S(\alpha,t,\gamma,R) \to t^{-r}$ as $R \to \infty$ for each γ, t, where $r = l(\gamma)$.

Now consider sums of a finite number of open intervals with rational end points. These can be enumerated constructively. We take a particular constructive enumeration: let E_n be the nth set of intervals in the enumeration. Then we have the next theorem.

THEOREM 1. *We can find a constructive*[3] *function $c(k,n)$ of two integral variables such that $E_{c(k,n+1)} \leqslant E_{c(k,n)}$ and $mE_{c(k,n)} > 1 - 1/k$ for each k, n and $E(k) = \prod_{n=1}^{\infty} E_{c(k,n)}$ consists entirely of normal numbers for each k.*

Let $B(\Delta,\gamma,t,R)$ be the set of numbers $\alpha (0 < \alpha < 1)$ for which

$$|S(\alpha,t,\gamma,R) - Rt^{-r}| < \frac{R}{\Delta t^r}, \quad \left(K = \frac{R}{\Delta t^r}\right), \quad \Delta = \frac{R}{Kt^r}. \tag{2}$$

Then by (1)

$$m\,B(\Delta,\gamma,t,R) > 1 - 2\mathrm{e}^{-Rt^{-r}/4\Delta^2} \quad \text{if } \Delta < 0.3.$$

Let $A(\Delta,T,L,R)$ be the set of those α for which (2) holds whenever $2 \leqslant t \leqslant T$ and $l(\gamma) \leqslant L$, i.e.,

$$A(\Delta,T,L,R) = \prod_{t=2}^{T} \prod_{l(\gamma)\leqslant L} B(\Delta,\gamma,t,R).$$

The number of factors in the product is at most T^{L+1} so that

$$m\,A(\Delta,T,L,R) > 1 - T^{L+1}\mathrm{e}^{-RT^{-2}/4\Delta^2}.$$

Let

$$A_k = A([k^{1/4}],[\mathrm{e}^{\sqrt{\log k}}],[\sqrt{\log k}-1],k),$$
$$\overline{A}_k = A(k,[\mathrm{e}^{\sqrt{\log k}}],[\sqrt{\log k}-1],k^4).$$

Then, if $k \geqslant 1000$, we shall have

$$m\,A_k > 1 - k\mathrm{e}^{-1/2k^{1/2}} > 1 - 1/k(k-1).$$

$c(k,n)(k \geqslant 1000)$ is to be defined as follows. $c(k,0)$ is $(0,1)$. $c(k,n+1)$ is the intersection of an interval $(\beta_n,1)(0 \leqslant \beta_n < 1)$ with A_{k+n+1} and $c(k,n), \beta_n$ being chosen so that the measure of $c(k,n+1)$ is $1-1/k+(k+n+1)^{-1}$. This is possible since the measure of $c(k,n)$ is $1-1/k+1/(k+n)$ and that of A_{k+n+1} is at least $1-1/((k+n)(k+n+1))$. Consequently the measure of $c(k,n)\cap A_{k+n+1}$ is at least $1-1/k+1/(k+n+1)$. If $k<1000$ we define $c(k,n)$ to be $c(1000,n)$. $c(k,n)$ is a finite sum of intervals for each k, n. When we remove the boundary points we obtain a set of the form $E_{c(k,n)}$ of measure $1-1/k+1/(k+n)(k\geqslant 1000)$. The intervals of which $E_{c(k,n)}$ is composed may be found by a mechanical process and so the function $c(k,n)$ is constructive. The set $E(k) = \prod_{n=1}^{\infty} E_{c(k,n)}$ consists of normal numbers for if $\alpha \in E(k)$, then $\alpha \in A_k$(all $k > K$, $k \geqslant 1000$). If γ is a sequence of length r in the scale of t and if k_0 be such that

$$[\mathrm{e}^{\sqrt{\log k_0}}] > t \quad \text{and} \quad [\sqrt{\log k_0}] > r+1,$$

then for $k > k_0$

$$|S(\beta,t,\gamma,k) - kt^{-r}| < k[k^{1/4}]^{-1},$$

where β is in A_k (by the definition of A_k). Hence $k^{-1}S(\alpha,t,\gamma,k) \to t^{-r}$ as k tends to infinity, i.e., α is normal.

Theorem 2. *There is a rule whereby given an integer k and an infinite sequence of figures* 0 *and* 1 (*the pth figure in the sequence being* $\theta(p)$) *we can find a normal number* $\alpha(k,\theta)$ *in the interval* $(0,1)$ *and in such a way that for fixed k these numbers form a set of measure at least* $1-2/k$ *and so that the first n figures of* θ *determine* $\alpha(k,\theta)$ *to within* 2^{-n}.

With each integer n we associate an interval of the form $(m_n/2^n_2,(m_n+1)/2^n)$ whose intersection with $E(k)$ is of positive measure, and given m_n we obtain m_{n+1} as follows Put

$$m\,E_{c(k,n)} \cap \left(\frac{m_n}{2^n},\frac{2m_n+1}{2^{n+1}}\right) = a_{n,m},$$

$$m\,E_{c(k,n)} \cap \left(\frac{2m_n+1}{2^{n+1}},\frac{m_n+1}{2^n}\right) = b_{n,m},$$

and let r_n be the smallest m for which either $a_{n,m} < k^{-1}2^{-2n}$ or $b_{n,m} < k^{-1}2^{-2n}$ or both $a_{n,m} > 1/k(k+n+1)$ and $b_{n,m} > 1/k(k+n+1)$. There exists such an r_n for $a_{n,m}$ and $b_{n,m}$ decrease either to 0 or to some positive number. In the case where $a_{n,r_n} < k^{-1}2^{-2n}$ we put $m_{n+1} = 2m_n+1$; if $a_{n,r_n} \geqslant k^{-1}2^{-2n}$ but $b_{n,r_n} < k^{-1}2^{-2n}$, we put $m_{n+1}=2m_n$, and in the third case we put $m_{n+1}=2m_n$ or $m_{n+1}=2m_n+1$ according as $\theta(n)=0$ or 1.

For each n the interval $(m_n/2^n,(m_n+1)/2^{n+1})$ includes normal numbers in positive measure. The intersection of these intervals contains only one number which must be normal.

Now consider the set $A(k,n)$ consisting of all possible intervals $(m_n/2^n, (m_n+1)/2^n)$, i.e., the sum of all these intervals as we allow the first n figures of θ to run through all possibilities. Then

$$m\ E(k) \cap A(k,n+1) = m\ E(k) \cap A(k,n)$$
$$- \sum_{m=0}^{2^n-1} m\ E(k) \cap (A(k,\ n) - A(k,\ n+1)) \cap \left(\frac{m}{2^n}, \frac{m+1}{2^n}\right).$$

But

$$m\ (A(k,n) - A(k,n+1)) \cap \left(\frac{m}{2^n}, \frac{m+1}{2^n}\right) < 2^{-2n}k^{-1},$$

so that

$$m\ E(k) \cap A(k,n+1) > m\ E(k) \cap A(k,n) - 2^{-n-1}k^{-1}$$
$$> mE(k) - k^{-1} > 1 - \frac{2}{k}.$$

The set of all possible numbers $\alpha(K,\theta)$ is therefore of measure at least $1 - 2/k$.

By taking particular sequences θ (e.g., $\theta(n) = 0$ for all n) we obtain particular normal numbers.

Examining the Work and Its Later Impact

Verónica Becher takes a closer look at —

TURING'S NOTE ON NORMAL NUMBERS

In an unpublished manuscript with title "*A Note on Normal Numbers*", Alan Turing gives the first explicit algorithm to compute a *normal* real number. Normality demands that the infinite expansion of a real number be seriously balanced: a number is normal *in a given scale* (numbering base), if every block of digits of the same length occurs with the same limit frequency in the expansion of the number expressed in that scale. For example, if a number is normal in the scale of two, the digits "0" and "1" occur, in the limit, half of the times; each of the blocks "00", "01", "10" and "11" occur one fourth of the times, and so on. A real number that is normal to *every scale* is called absolutely normal, or just *normal*.[1] Émile Borel (1909) stated this definition and proved the existence of normal numbers, showing that, indeed, almost all numbers are normal. Borel's proof is based on measure theory, and being purely existential, it provides no method of constructing an example of a normal number.

With his note Turing solves the problem of finding examples of normal numbers, raised by Borel. Turing gives, first, a constructive proof that almost all numbers are normal, and then, an algorithm to produce normal numbers, which leads to his computable examples.

As defined by Turing in his breakthrough article "*On computable numbers...* " (1936), the computable real numbers are those whose infinite expansion can be generated by a mechanical (finitary) method, outputting each of its digits, one after the other. There is no evident reason for the normal numbers to have a non-empty intersection with the computable numbers. A measure-theoretic argument is not enough to see that these sets intersect: the set of normal numbers in the unit interval has measure one, but the computable numbers are just countable, hence they form a measure-zero set. Along his note Turing uses the term *constructive*, but never uses the term *computable*, which would have better expressed the finitarily-based constructiveness he actually achieves.

Turing's note remained unpublished until its inclusion in the *Collected Works* edited by J.R. Britton (1992). A typewritten document together with a handwritten draft is in Turing's archive in King's College, Cambridge; the scanned versions are available on the Web in `http://www.turingarchive.org`. Turing's note is undated. Presumably he wrote it not much after 1936, because part of the handwritten manuscript is in the back of the galley proofs of "*On computable numbers...*". Turing's calligraphy is hard to follow, there are numerous crossing outs, and each page starts in small lettering that slightly grows towards the end of the page. The typewritten document —only mathematical formulae are handwritten— is much more complete. In eleven lines of the draft that Turing did not include in the typewritten document, he appraises the results of his note. He cites David Champernowne's[2] example of normality in the scale of ten —but not proved normal

[1] For a thorough presentation of normal numbers see Kuipers and Niederreiter (2006) or Bugeaud (2012).

[2] David Champernowne was the first friend that Turing made when he entered King's College Cambridge. This is reported of Andrew Hodges's superb biography (2000).

in any other scale— and says that it may also be natural that an example of a *normal* number (i.e., normal in *every* scale) be demonstrated as such and written down. Then, he writes *"this note cannot, therefore, be considered as providing convenient examples of normal numbers"*[3].

Champernowne's number (1933) is formed by writing down all the positive integers in order, in decimal notation, 0.12345678910111213...[4] The reference to Champernowne's number suggests what Turing could have considered to be a *convenient* example: a number with a simple mathematical definition and easily computable. According to the modern theory of computational complexity, which was only developed in the 1960's and required the Turing machine as its computational model, Turing's algorithm has exponential complexity: the number of operations needed to compute the n-th digit of the output sequence is exponential in n. We now know that this is intractable for every past or present computer. One can interpret Turing's negative assessment of the numbers produced by his algorithm as a trail of his intuitive considerations on the computational complexity of the algorithm. Years later, in *"Solvable and unsolvable problems"* (1954) he will write, tangentially, about algorithmic solutions that cause combinatorial explosion .

Still in the handwritten draft Turing says that the purpose of his note is, rather, to counter the then dominant idea that the existence proof of normal numbers provides no example of them. And he adds that the arguments in the note, in fact, follow the existence proof fairly closely. Here Turing is obviously referring to the proof of the measure of normal numbers —a version of this proof appears in the book by G.H.Hardy and E.M.Wright (1979), which had its first edition in 1938—.

There is a letter exchange[5] between Hardy and Turing, where Hardy recalls he searched the literature when Champernowne was doing his work "but could not find anything satisfactory anywhere". Hardy's letter ends saying that his "feeling is that Lebesgue made a proof himself that never got published". Actually, Henri Lebesgue constructed a normal number in 1909, but didn't publish it until 1917. In the same journal issue, Wacław Sierpiński presented his example of a normal number, based on a seemingly simpler but equivalent characterization of normality Both, Lebesgue and Sierpiński, gave a partially constructive proof of the measure of the set of normal numbers, and defined their respective examples as the limit of a set that includes all non-normal numbers —this limit point is outside the set—. Their examples were not finitarily defined. At that time computability theory was not even born, so it is not surprising that neither Lebesgue nor Sierpiński used a stronger notion of constructiveness. However, these antecedents may explain why Turing did not publish his construction.

Although Turing's note is incomplete, it is correct except for some minor technical errors. In Becher et al. (2007) we completed it by giving full proofs and corrected the errors. In doing so we tried to recreate Turing's ideas as accurately as possible. Turing proves two theorems. The first provides a finitarily based method to construct a set of normal real numbers in the unit interval, of arbitrary large measure.

THEOREM 2. *We can find a constructive function $c(k,n)$ of two integer variables with values in finite sets of pairs of rational numbers such that, for each k and n, if $E_{c(k,n)} = (a_1,b_1) \cup (a_2,b_2) \cup \ldots (a_m,b_m)$ denotes the finite union of the intervals whose rational endpoints are the pairs given by $c(k,n)$, then $E_{c(k,n)}$ is included in $E_{c(k,n-1)}$ and the measure of $E_{c(k,n)}$ is greater than $1-1/k$. And for each k, $E(k) = \bigcap_n E_{c(k,n)}$ has measure $1-1/k$ and consists entirely of normal numbers.*

---

[3] Turing's underlining.

[4] To prove it normal in the scale of ten, Champernowne ingeniously bounds the number of occurrences of each block of digits in the initial segments of the sequence. In this proof it is crucial to know, explicitly, the digit in each position of the sequence. The technique is not relevant to Turing's note.

[5] Letter sent by G.H.Hardy to A.M.Turing, dated June 1, Trinity College, presumably in the late 1930s. Hardy answers a letter from Turing of March 28, apologizing for not responding earlier and for not giving him a definitive satisfactory response. It is in Turing's archive in King's College and available in the digital archive with code AMTD/D/5 image 6.

The construction is uniform of the parameter k —devoted to fix the measure— and it is done by computable approximations. The idea is as follows: Turing prunes the unit interval by stages. It starts with the set $E_{c(k,0)}$ equal to the whole unit interval. At stage n, the set $E_{c(k,n)}$ is the finite approximation to $E(k)$ that results from removing from $E_{c(k,n-1)}$ the points that are *not* candidates to be normal, according to the inspection of an initial segment of their expansions. At the end of this infinite construction all rational numbers have been discarded, because of their periodic structure. All irrational numbers with an unbalanced expansion have been discarded. But also many normal numbers are discarded, because their initial segments remain unbalanced for too long. Turing covers all initial segment sizes, all scales, and all blocks, by increasing functions of the stage n. And puts a decreasing bound on the acceptable discrepancy between the actual number of blocks in the inspected initial segments and the perfect number of blocks expected by the property of normality. These functions (initial segment size, scale, block length and discrepancy) must be such that, at each stage n, the set of discarded real numbers has a small measure. To bound this measure Turing uses a constructive version of the Strong Law of Large Numbers. Thus, at each stage, finitely many intervals with rational endpoints and very small measure are removed. Turing tailors the sets $E_{c(k,n)}$ so as to have measure greater than $1 - 1/k$. The set $E(k)$ is the limit of this construction, hence it is the countable intersection of the constructed sets $E_{c(k,n)}$, and it consists entirely of normal numbers.

In a general perspective, the proof of Theorem 1 conveys the impression that Turing intuitively knew, ahead of his time, that traditional mathematical concepts specified by finite approximations, such as measure or continuity, could be made *computational*. This line of research has become mainstream and has developed under the general name of *effective mathematics*. In particular, Turing's construction in Theorem 1 is precursory in the theory of *algorithmic randomness* that started in the 1960's. Although there are variants, the currently most accepted definition of randomness is due to the different but equivalent formulations given by Per Martin-Löf and Gregory Chaitin. Intuitively, a real number is random when when it exhibits the almost-everywhere properties of all reals. A random real number must pass every test of these properties. Martin Löf had the idea to focus just in properties definable in terms of computability: a test for randomness is a uniformly computably enumerable sequence of sets whose measure converges to zero. A real number is *not* random if it belongs to each of the sets of some test. Astonishingly, Turing's construction in Theorem 1 leads immediately to a test for randomness: the k-th set of the test is the complement of the set $E(k)$ in the unit interval. The measure of the k-th set is $1/k$, which tends to zero as k increases. The set of non-normal numbers is included in each set of the test because, for each k, $E(k)$ consists entirely of normal numbers. By this test, if a number is *not* normal, then is *not* random. Thus, randomess implies normality.

Turing's second theorem gives an affirmative answer to the then outstanding question of whether there are computable normal numbers, and provides concrete instances. In fact, it gives much more:

Theorem 2. *There is an algorithm that, given an integer k and an infinite sequence θ of zeros and ones, produces a normal number $\alpha(k,\theta)$ in the unit interval, expressed in the scale of two, such that in order to write down the first n digits of $\alpha(k,\theta)$ the algorithm requires at most the first n digits of θ. For a fixed k these numbers $\alpha(k,\theta)$ form a set of measure at least* $1 - 2/k$.

Our reconstruction of the proof of Theorem 2 in the aforementioned publication Becher et al. (2007) supersedes J.L. Britton's editorial notes in the *Collected Works* (references 7 to 12 on page 119, elaborated in pages 264 and 265), where it is asserted that the proof given by Turing is inadequate, and speculated that the theorem could indeed be false.

Turing's algorithm is uniform in the parameter k and it receives as input an infinite sequence θ of zeros and ones. The algorithm works by stages. The main idea is to split the unit interval by halves, successively. It starts with the whole unit interval and at each stage it chooses either the left half or the right half of the current interval. The sequence $\alpha(k,\theta)$ of zeros and ones output by the algorithm is the trace of the left/right selection at each stage. The invariant of the algorithm is that the intersection of the current interval with the set $E(k)$ of normal numbers of Theorem 1 has positive measure. To ensure this condition at stage n the algorithm uses the finite approximation $E_{c(k,n)}$ of the set $E(k)$. The algorithm chooses the half of the current interval whose intersection with $E_{c(k,n)}$ reaches a minimum measure that avoids running out of measure in later stages. In case both halves reach this minimum, the algorithm uses the n-th symbol of the input sequence θ to decide. Since the chosen intervals at successive stages are nested and their measures converge to zero, their intersection contains exactly one number. This is the number $\alpha(k,\theta)$ output by the algorithm.

When the input θ is a computable sequence —Turing puts the infinite sequence of all zeros— the algorithm produces a computable normal number. To prove that for a fixed k, the set of output numbers $\alpha(k,\theta)$ for all possible inputs θ has measure at least $1-2/k$, Turing bounds the measure of the unqualified intervals up to stage n, as the n first symbols of the sequence θ run through all possibilities. The algorithm can be adapted to intercalate the symbols of the input sequence at fixed positions of the output sequence. Thus, one obtains non-computable normal numbers in each Turing degree.

The time complexity of the algorithm is the number of needed operations to produce the n-th digit of the output sequence $\alpha(k,\theta)$. This just requires to compute, at each stage n, the measure of the intersection of the current interval with the set $E_{c(k,n)}$. Turing gives no hints on properties of the sets $E_{c(k,n)}$ that could allow for a fast calculation of their measure. The naive way does the combinatorial construction of $E_{c(k,n)}$, in a number of operations exponential in n. Turing's algorithm *verbatim* would have simple-exponential complexity, but its correctness proof is missing in Turing's note. Our reconstruction of the algorithm —that we give together with its correctness proof— has, unfortunately, double-exponential time complexity —because the number of intervals we consider in $E_{c(k,n)}$ is exponentially larger than in Turing's literal construction—.

The computable reformulation of Sierpiński's normal number that we gave in Becher and Figueira (2002) also has double-exponential time complexity. A theorem of Strauss (1997) asserts that normal numbers computable in simple-exponential time do exist, but this purely existential result yields no specific instances. The problem of providing an example of an easily computable normal number (normal to every integer scale) is, still, unresolved.

References

Becher, V., Figueira, S., 2002. An example of a computable absolutely normal number. Theoretical Computer Science, 270, 947–958.

Becher, V., Figueira, S., Picchi, R., 2007. Turing's unpublished algorithm for normal numbers. Theoretical Computer Science, 377, 126–138.

Borel, E., 1909. Les probabilités dénombrables et leurs applications arithmétiques. Rendiconti del Circolo Matematico di Palermo, 27, 247–271.

Bugeaud, Y. 2012. Distribution Modulo One and Diophantine Approximation, Cambridge University Press.

Champernowne, D. G., 1933. The construction of decimals in the scale of ten. Journal of the London Mathematical Society, 8, 254–260.

Hardy, G.H., Wright, E.M., 1979. An Introduction to the Theory of Numbers. Oxford University Press. First edition in 1938.

Hodges, A., 2000. Alan Turing: the Enigma. Walker and Company, New York.

Kuipers, L., Niederreiter, H., 2006. Uniform distribution of sequences. Dover Books on Mathematics, New York.

Lebesgue, H., 1917. Sur certaines démonstrations d'existence. Bulletin de la Société Mathématique de France, 45, 132–144.

Sierpiński, W., 1917. Démonstration élémentaire du théorème de M. Borel sur les nombres absolument normaux et détermination effective d'un tel nombre. Bulletin de la Société Mathématique de France, 45, 127–132.

Strauss, M., 1997. Normal numbers and sources for BPP. Theoretical Computer Science, 178:155–169.

Turing, A. M., 1936. On computable numbers, with an application to the Entscheidungsproblem. Proceedings of the London Mathematical Society, Series 2, 42, 230–265.

Turing, A. M., 1954. Solvable and unsolvable problems. Science News 31. Included in Collected Works of A. M. Turing: Pure Mathematics, 1992, North Holland, Amsterdam, pages 99–115.

Turing, A. M., 1992. A note on normal numbers. In: Britton, J. (Ed.), Collected Works of A.M. Turing: Pure Mathematics. North Holland, Amsterdam, pages 117–119, with notes of the editor in pages 263–265.

Turing's Treatise on the Enigma (Prof's Book)

(unpublished, ca. 1940)

Frode Weierud on Alan Turing, Dilly Knox, Bayesian statistics, decoding machines and —

PROF'S BOOK: SEEN IN THE LIGHT OF CRYPTOLOGIC HISTORY

1. Introduction

'Turing's Treatise on the Enigma', also called Prof's Book, was shrouded in secrecy for almost 60 years. Internal evidence suggests that Alan Mathison Turing wrote this Enigma Treatise during the late autumn of 1940. Some of the material he refers to, such as the Railway Enigma – the rewired version of the commercial, unsteckered Enigma machine used by the German railways, was first broken into in August 1940. The attack is reported to have been made by Alan Turing and Peter Twinn, and it took place in Hut 8, the Bletchley Park (BP) hut doing cryptanalytical work on the German Naval Enigma, then under Turing's command (Batey, 2009).

2. Turing declassified

On 4 April 1996, more than 1.3 million pages of declassified documents were released to the public by the US National Archives and Record Administration. The collection was very varied in both content and scope; there was a lot of trivia but also the occasional gold nugget, such as 'Turing's Treatise on the Enigma' and Patrick Mahon's 'The History of Hut Eight: 1939–1945'.

That Turing had worked at the Government Code and Cypher School (GC & CS) during the war was well known in 1996. Andrew Hodges' biography of Alan Turing (Hodges, 1983) that was published in 1983 and the book by Gordon Welchman (1982) published the year before, had both given Alan Turing the credit for developing the Turing–Welchman Bombe and for making the first break into an intractable problem – German Naval Enigma. Apart from Turing's unpublished paper 'On Permutation Groups', no documents relating to his work at GC & CS during the war had then been discovered.

Welchman's book explained how the Bombe worked and how cribs, probable words in the message, were used to construct menus. The menu was the 'program' that told the operator how to connect the electrical circuits such that the Bombe could attempt to find the Enigma key that had been used to encipher the crib and produce the intercepted ciphertext. The groundbreaking work of the Poles had also been revealed in several books (Bertrand, 1973; Kozaczuk, 1984; Paillole, 1985) and articles, but very little was known about the actual cryptanalytical techniques. The only public knowledge was that if two or more messages were enciphered on the same key, or if a very long crib could be obtained, then the machine's wheel wirings could be recovered. But by what 'magic' nobody fully knew, until the Prof's Book released its secrets.

3. The pre-war heritage

Initially, we considered the theories and methods as Turing's own even if he never made such a claim. At BP he worked with cryptanalysts such as Dillwyn ('Dilly') Knox, who initially was his boss, Tony Kendrick, Peter Twinn, John Jeffreys, Gordon Welchman and others who would also have contributed to this work. However, GC & CS had worked on the Enigma problem before Alan Turing joined. In 1979 Sir Harry Hinsley revealed the existence of a pre-war Enigma history at GC & CS; he says: 'In 1937, GC and CS broke into the less modified and less secure model of this machine [Enigma] that was being used by the Germans, the Italians and the Spanish nationalist forces' (Hinsley et al., 1979, p. 54).

Today our knowledge has improved. The first four chapters of 'Turing's Treatise on the Enigma' are not original works by Alan Turing, but are based on work Hugh Foss and Dilly Knox did in the late 1920s and early to mid 1930s. The English interest in the cipher machine Enigma invented by Dr Arthur Scherbius in 1918 and commercialized in the early 1920s seems to have started at a very early date. In 1949 Hugh Foss wrote a report where he tells of his study of the 'small Enigma', an Enigma C, some time in 1927 (Erskine and Smith, 2011; Foss, 1949). Hugh Foss studied the machine with the aim of assessing its security and to determine if it was suitable for adoption by the British. He wrote it all up in a report (Foss, 1928) in which he showed that if the wiring was known a crib of 15 letters would reveal the identity and setting of the right-hand wheel, and with a crib of 180 letters it was possible to determine the unknown wiring of the right-hand and middle wheels. As he says: *The methods I used were rather clumsy as they were geometrical rather than algebraical and, when Dilly Knox came to study the subject ten years later, he invented the 'rods' and the process known as 'buttoning up', which used the same properties as I had done, but did so in a more effective way.*

4. Rebuilding on old foundations

The comic strips Turing mentions in Chapter 1 are the paper strips that Hugh Foss used during his study in 1927, and in Chapter 2 he introduces the more practical rods that Dilly Knox developed in the mid 1930s. Chapter 3 is devoted to the more difficult method of 'boxing' and 'buttoning up' that also were developed by Dilly Knox in this period, but which are based on the original geometrical techniques using rectangles developed by Hugh Foss. Dilly attacked the commercial Enigma K machines used by the Italians and the Spanish nationalist forces in 1937, and he recovered the wiring of the three wheels being used in the Italian machine (Fuensanta et al., 2010). However, all attempts to break into the Enigma traffic of the German Naval units that operated in the Spanish waters during this period failed.

This new interest in the German service Enigma resulted in renewed contacts with the French intelligence service. In the autumn of 1938, the French furnished two Enigma instruction manuals from 1930 and some sample encipherments. The encipherments were made by successively enciphering every letter on the Enigma keyboard, a total of 26 letters, at different starting positions but with the same internal machine settings. This gave the idea to the attack that Turing describes in Chapter 3 as the Saga. One manual, on the direction for use of the Enigma keys, was of interest as it contained a 90-letter sample plaintext message and its exact encipherment together with the internal machine settings and the chosen message key (Batey, 2009, p. 170). One would expect such a sample to be fictive, the ciphertext would be just random cipher groups, but against all rules it had been enciphered on a real, operational Enigma machine.

In the autumn of 1938, well before the British learned that the Poles had broken the Enigma, they attempted to use the 'boxing' and 'buttoning up' methods on this message to recover the wheel wirings. The process involved taking the recovered wheel constatations after the boxing had been completed and transforming them with the diagonal of the machine, the wiring of the keyboard letters to the entry wheel contacts. This would strip off the wiring of the entry wheel and result in a pure wheel wiring. Everybody had a go at this; Knox, Foss, Kendrick, Twinn and Turing all tried, but they all failed. It was only in July 1939 that they learned from the Poles that the diagonal was ABCDEF etc. in strict alphabetical order instead of QWERTZU etc. in the German keyboard order, as in the commercial machine. With this crucial information Peter Twinn recovered the wheel wirings in a couple of hours.

Turing reported to BP on 4 September 1939 and there he immediately started to work under Dilly Knox in the Cottage – a part of the stable-yard at the BP mansion. His great wish was to tackle the German Navy Enigma, which due to its intricate indicating system was a lot more difficult. Until May 1937 the German Navy used the same indicating system as the other two services with double enciphered message settings, so-called 'throw-on' indicators, but from then on they started to use message settings selected from a code book and to encipher the message indicator with changing bigram tables. Turing got a place for himself in the stable loft of the Cottage where he could work alone and in peace. He was not good at socializing. He did not attend the coffee breaks in the Cottage or the meals in the mansion, so two of Dilly's girls rigged up a pulley to send up coffee and sandwiches in a basket (Batey, 2009). His relationship with Knox seems to have been cordial but Dilly apparently had some difficulties in keeping him under his command, writing: *Turing is very difficult to anchor down. He is very clever but quite irresponsible and throws out suggestions of all sorts of merit. I have just, but only just, enough authority and ability to keep his ideas in some sort of order and discipline. But he is very nice about it all* (Batey, 2009, p. 94).

5. Inventing modern tools

Turing not only worked on Naval Enigma; he also attacked the problem of mechanising the process of finding the daily Enigma keys. The Polish Bomba, which was based on the peculiar 'throw-on' indicators, was very limited. He therefore embarked on designing a new key-finding machine, the Turing–Welchman Bombe, which he describes in detail in Chapter 6. This must have started already when he still was in the Cottage. While Knox and John Jeffreys were busy preparing the modified Zygalski sheets, which Turing describes in Chapter 5, Turing was designing his modified Bombe. On 1 November 1939, the Cottage team – Knox, Twinn, Welchman, Turing and Jeffreys – wrote a memorandum setting out the required mechanical aids for solving the Enigma problem; a 30 Enigma Bombe machine is one of the items.

This was probably Turing's greatest contribution. While Knox and the others were very much rooted in the manual cryptanalytical techniques, Turing was the visionary who was looking for machines to do the work. However, Turing made another major contribution, which in term had a much more profound and longer lasting influence on cryptanalysis than the Bombe; he introduced Bayesian statistics in this field. He first applied Bayesian statistics on Naval Enigma where the process Banburismus was used to identify the right-hand and middle wheels such as to cut down the number of wheel orders that had to be run on the Bombe. Turing mentions Banburismus in Chapters 6 and 7, but he does not give any description of it unless it has been lost in one of the few missing pages of the treatise. However, Mahon (1945) and Alexander (1945) give reasonably good explanations. The term Bayesian statistics had not yet been coined in 1940 and it probably would not have been used as Bayes' Rule was in ill repute with most statisticians at the time (Bertsch, 2011). But that it was Bayes' Rule Turing used and that he was fully aware of this fact is illustrated by the question his statistical assistant, Irving 'Jack' Good, asked Turing one day: 'Aren't you essentially using Bayes' theorem?' Turing answered: 'I suppose' (Bertsch, 2011). Probably few people at the time realized that Banburimus was Bayes' Rule in disguise, hidden through the use of Turing's weight of evidence measured in his invented units of ban, centiban and deciban. His weight of evidence was also used in the method known as Turingery or Turingismus that he invented to deduce the wheel patterns of the Lorenz cipher machine, SZ 42 or Tunny as it was called at BP. The Bayesian seeds that Turing sowed in 1940 grew solid roots in the cryptanalytical community and Jack Good, who became a high priest of Bayesian statistics, probably continued Turing's work when he was employed by Government Communications Headquarters (GCHQ) from 1948 to 1959. After the war Jack Good tried to make Turing's wartime statistical work better known, but unfortunately in very heavy disguise. And Good probably did not tell us all, because some work of Alan Turing still remains classified. Record group HW 25, at the British National Archives in Kew, lists two reports by Alan Turing with the titles 'Report on the applications of probability to cryptography' (Turing, 1946b) and 'Paper on statistics of repetitions' (Turing, 1946a); both are retained by GCHQ under section 3(4) of the Public Records Act 1958. What new surprises do they contain? With some luck we might soon know.[1]

6. Acknowledgements

I should like to thank Mavis Batey and Ralph Erskine for their help with answering my questions, furnishing me with missing documents and generally guiding me through the wilderness of cryptologic history.

References

Alexander, C.H.O'D., 1945. Cryptographic History of Work on German Naval Enigma. TNA PRO, HW 25/1.

Batey, M., 2009. Dilly: The Man Who Broke Enigmas, Dialogue, Biteback Publishing, London.

Bertrand, G., 1973. Enigma ou la plus grande Énigme de la guerre 1939–1945, Librairie Plon, Paris.

Erskine, R., Smith, M. (Eds.), 2011. The Bletchley Park Codebreakers, Dialogue, Biteback Publishing, London.

Foss, H., 1928. The Reciprocal ENIGMA. TNA PRO, Kew, Surrey, HW 25/14.

Foss, H., 1949. Reminiscences on the ENIGMA Machine. TNA PRO, HW 25/10.

Fuensanta, J.R.S., Espiau, F.J.L.B, Weierud, F., 2010. Spanish Enigma: A history of the Enigma in Spain. Cryptologia 34(4), 301–328.

Hinsley, F.H., et al., 1979. British Intelligence in the Second World War: v. 1: Its Influence on Strategy and Operations, HMSO, London.

Hodges, A., 1983. Alan Turing: The Enigma, Burnett Books, London.

Kozaczuk, W., 1984. Enigma: How the German Machine Cipher Was Broken and How it Was Read by the Allies in World War Two, Arms and Armour Press, London.

Mahon, A.P., 1945. The History of Hut 8. TNA PRO, HW 25/2.

McGrayne, S.B., 2011. The Theory that Would Not Die, Yale University Press, New Haven.

Paillole, P., 1985. Notre Espion Chez Hitler. Éditions Robert Laffont, Paris.

Turing, A.M., 1946a. Paper on Statistics of Repetitions. TNA PRO, HW 25/38, (retained by GCHQ).

Turing, A.M., 1946b. Report on the Applications of Probability to Cryptography. TNA PRO, HW 25/37, (retained by GCHQ).

Welchman, G., 1982. The Hut Six Story: Breaking the Enigma Codes, Allen Lane, London.

[1] On 16 April 2012 the two statistical reports were released to the public after GCHQ, in response to requests from researchers, had re-examined the papers and has determined that there was no longer any sensitivity associated with them.

Excerpts from the 'Enigma Paper'[2]

1. Excerpt 1

Page 97 contains the heading 'A mechanical method. The Bombe' and reveals the crucial idea that defeated the plugboard complication.

Turing first gives a sample 25-1etter Enigma cipher text with its 'crib' or guessed plaintext set against it.

The ciphertext is D A E D A Q O Z S I Q M M K B I L G M P W H A I V
The plaintext is K E I N E Z U S A E T Z E Z U M V O R B E R I Q T
(keine Zusätze zum Vorbericht)

Turing explains: '... a method of solution will depend upon taking hypotheses about parts of the keys and drawing what conclusions one can, hoping to get either a confirmation or a contradiction...' The method depends absolutely upon making a correct guess about the corresponding short piece of plaintext.

[2] The accompanying descriptions are reproduced from Volume 4 of the *Collected Works* – see Andrew Hodges' Preface to the excerpts, pp. 225–229.

is given by the column of the inverse rod set-up, and we can find all possible positions where the click coupling occurs from the Turing sheets or the Jeffreys sheets. In some cases there will be other constatations which are made up from letters supposed to be unsteckered because they occur in the click, and these will further reduce the number of places to be tested.

These methods have both of them given successful results, but they are not practicable for cases where there are many Stecker, or even where there are few Stecker and many wheel orders.

<u>A mechanical method. The Bombe.</u>

Now let us turn to the case where there is a large number of Stecker so many that any attempt to make use of the fxxxxthxx unsteckered letters is not likely to succeed. To fix our ideas let us take a particular crib.

1	2	3	4	5	6	7	8	9	10	11	12	13	14	15	16	17	18	19	20	21	22	23
D	A	E	D	A	Q	O	Z	S	I	Q	M	M	K	B	I	L	G	M	P	W	H	A
K	E	I	N	E	Z	U	S	A	E	T	Z	E	Z	U	M	V	O	R	B	E	R	I

24	25
I	V
Q	T

Presumably the method of solution will depend on taking hypotheses about parts of the keys and drawing what conclusions one can, hoping to get either a confirmation or a contradiction. The parts of the keys involved are the xxxxxxxxxxxxxxx wheel order, the rod start of the crib, whether there are any turnovers in the crib and if so where, and the Stecker. As regards the wheel order one is almost bound to consider all of these separately. If the crib were of very great length one might make no assumption about what wheels were in the L.H.W. position and M.W. position, and apply the method we have called a 'Stecker knockout' (an attempt of this kind was made with the 'Feindseligkeiten' crib in Nov. '39), xxxxx or one might sometimes make assumptions about the L.H.W. and M.W. but none, until a late stage about the R.H.W. In this case we have to work entirely with constatations where the R.H.W. has the same position. This method was used for the crib from the xxxxx Schluesselzettel of the Vorpostenboot, with success; however I shall assume that all

2. Excerpt 2

Page 99 shows (in Fig. 59) Turing's diagram of the logical chain of implications that could be deduced from the piece of guessed plaintext as on page 97, and (in Fig. 60) the basic idea of a Bombe to exploit such implications.

The closed cycles in Turing's diagram, for instance Z-S-A-E-M-Z, correspond to chains of logical implications that yield consistency conditions independent of the plugboard.

They make it possible to reject a rotor position, as inconsistent with the ciphertext and plaintext data, even though the plugboard remains unknown.

When these chains of logical implications are exploited, almost every 'wrong' rotor position will be rejected, leaving just a few to be tested in detail. Among them, if the plaintext has been guessed correctly, will be the correct rotor setting.

Turing's hazy sketch of the Bombe does not however explain the crucial ideas that he and G.W. Welchman later used to mechanise the process of following chains of logical implications to the full; these are described in the following pages.

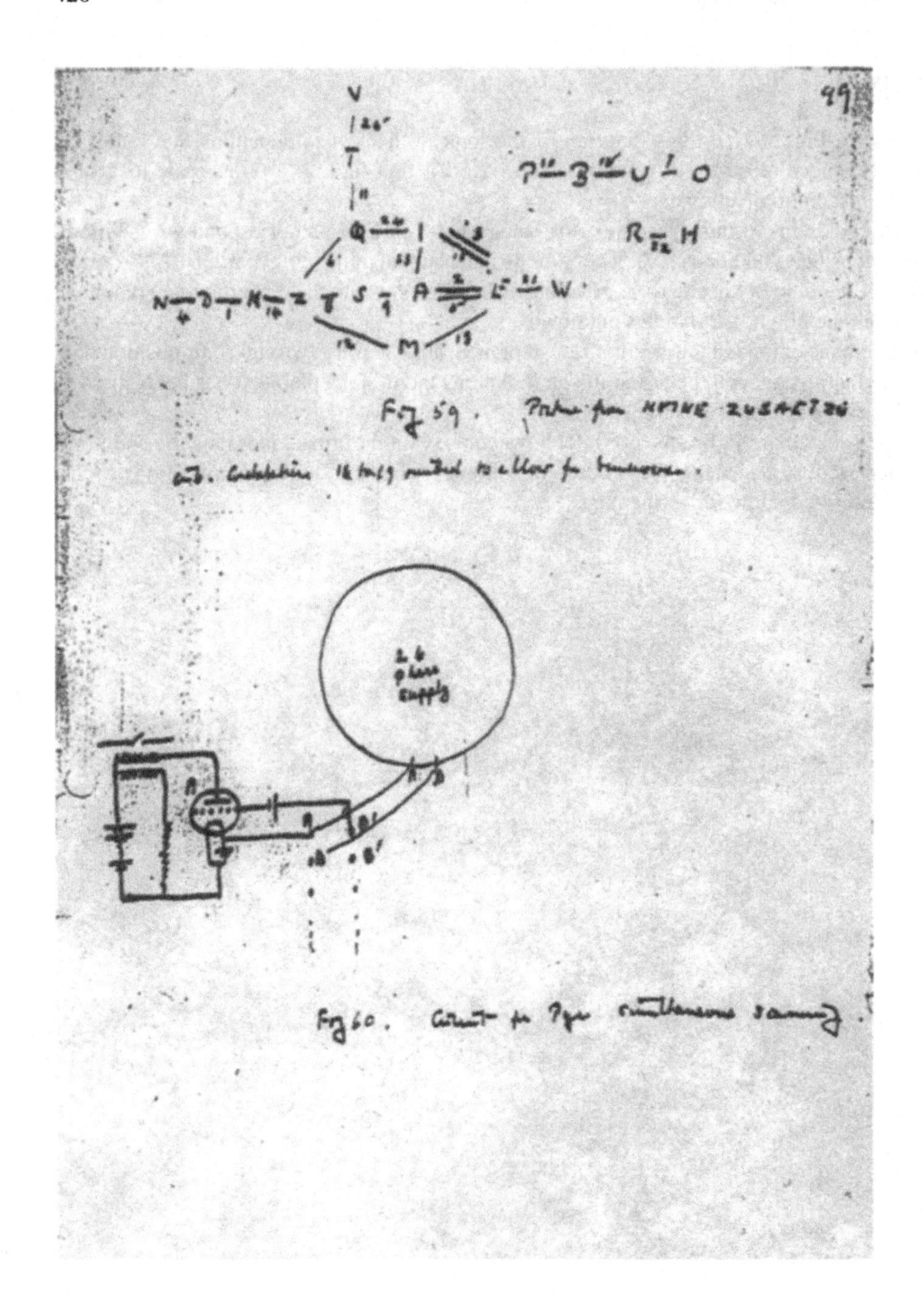
26 phase supply

3. Excerpt 3

Page 104 describes the critical problem of how to 'scan' the electrical output from the Bombe to detect the possibility of a correct rotor position. Turing first describes 'serial scanning', in which each of the 26 different plugboard hypotheses ('each Stecker value') has to be tried in turn, and then raises the possibility of 'simultaneous scanning'.

104

It now only remains to find a mechanical method of registering whether the multiple encipherment condition is fulfilled. This can be done most simply if we are willing to test each Stecker value of the central letter throughout all rod starts before trying the next Stecker value.

Suppose we are investigating the case where the Stecker value of the central letter E is K. We let current enter all of the chains of enigmas at their K input points, and at the K output points of the chains we put relays. The 'on' points of the five relays are put in series with a battery (any), and another relay. A current flows through this last relay if and only if a current flows through all the other five relays, i.e. if the five multiple encipherments applied to K all give K. When this happens the effect is, essentially, to stop the machine, and such an occurrence is known at Letchworth as a 'straight'. An alternative possibility is to have a quickly rotating 'scanner' which, during a revolution would first connect the input points A of the chains to the current supply, and the output points A to the relays, and then would connect the input and output points B to the supply and relays. In a revolution of the scanner the output and input points A to Z would all have their turn, and the right hand wheels would then move on. This last possible solution was called 'serial scanning' and led to all the possible forms of registration being known as different kinds of 'scanning'. The simple possibility that we first mentioned was called 'single line scanning'. Naturally there was much research into possible alternatives to these two kinds of scanning, which would enable all 26 possible Stecker values of the central letter to be tested simultaneously without the wheels of the machine moving. Any device to do this was described as 'simultaneous scanning'.

4. Excerpt 4

Page 107 describes how ‘simultaneous scanning’ can give rise to the ‘very useful’ principle that ‘all the deductions drawn will then be false, and those that remain will stand out clearly as possible correct hypotheses’.

107

When we look at the Bombe in this way we see that it would be natural to modify it so as to make this idea fit even better. We have not so far allowed for lengthy chains of deductions; the possible deductions stop as soon as one comes back to the central letter. There is however no reason why, when from one hypothesis about the Stecker value of the central letter we have deduced that the central letter must have another Stecker value, we should not go on and draw further conclusions from this second Stecker value. At first sight this seems quite useless, but, as all the deductions are reversible, it is actually very useful, for all the conclusions that can be drawn will then be false, and those that remain will stand out clearly as possible correct hypotheses. In order that all these deductions may be made mechanically we shall have to connect the 26 contacts at the end of each chain to the common beginning of all the chains. With this arrangement we can think of each output or input point of an enigma as representing a possible Stecker, and if two of these points are connected together through the enigmas then the corresponding Stecker imply one another. At this point we might see how it all works out in the case of the crib given above. This crib was actually enciphered with alphabets which, when corrected for their Stecker, are those given below, the numbers over the crib constatations giving the columns [illegible]. The alphabets most used are 2,3,5,10,23, and these are reproduced below for reference

2	3	5	10	23
XN	XH	MD	TD	LV
AP	DU	JZ	IH	WG
QK	KN	CV	MU	DI
CV	PK	BA	XE	OM
TF	QI	YE	CV	XU
UO	AW	UH	JY	FT
MS	OV	PQ	DF	JP
BD	JY	DW	SL	QK
IW	DM	LH	OH	AY
JZ	[illegible]	BK	WN	NB
GH	SL	FU	AZ	EJ
YE	UT	OI	PK	ZQ
HL	FC	KT	GM	RK

5. Excerpt 5

Page 112 describes Welchman's idea for exploiting the self-inverse property of the Enigma by using the Diagonal Board.

112

Figs 63, 64 shew the connections of enigmas and diagonal board in a particular case. The case of a six-letter alphabet has been taken to reduce the size of the figure.

The actual origin of the spider was not an attempt to find simultaneous scanning for the Bombe, but to make use of the reciprocal character of the Stecker. This occurred at a time when it was clear that very much shorter cribs would have to be worked than could be managed on the Bombe. Welchmann then discovered that by using a diagonal board one could get the complete set of consequences of a hypothesis. The ideal machine that Welchmann was aiming at was to reject any position in which a certain fixed-for-the-time Stecker hypothesis led to any direct contradiction: by a direct contradiction I do not mean to include any contradictions which can only be obtained by considering all Stecker values of some letter independently and shewing each one inconsistent with the original hypothesis. Actually the spider does more than this in one way and less in another. It is not restricted to dealing with one Stecker hypothesis at a time, and it does not find all direct contradictions.

Naturally enough Welchmann and Keen set to work to find some way of adapting the spider so as to detect all direct contradictions. The result of this research is described in the next section. Before we can leave the spider however we should see what sort of contradictions it will detect, and about how many stops one will get with given data.

First of all let us simplify the problem and consider only 'normal' stops, i.e. positions at which by altering the point at which the current enters the diagonal board one can make 26 relays close. The current will then be

6. Excerpt 6

Pages 136–7 describe the context in which Turing made his successful deduction of the bigram key-system in the Naval Enigma.

136

find th e keys for other days at th e beginning of May and th ey actually found the Stecker for the 2nd, 3rd, 4th, 5th and 8th, and read about 100 messages. The in dicators an d window positions of four (selected) messages for th e 5th were

Indicator	Window start
KFJX EWTW	P C V
SYLQ EWUF	B Z V
JMHO UVQG	M E M
JMFE FEVC	MYE K

The repetition of the EW combined with th e repetition of V suggests that the fifth and sixth letters describe the third letter of the window position, and similarly one is ledto believe that the first two letters ofthe indicator represent the first letter of the window position, and that the third and fourth represent the second. Presumably this effect is somehow produced by means of a table of bigramme equivalents ofletters, but it cannot be done simply by replacing the letters of the window position with one of th eir bigramme equ valents, and then putting in a dummy bigramme, for in this case the window position corresponding to JMFE FEVC would have to be say MYY instead of MYK. Probably some encipherment is involved somewhere. The two most natural alternatives are , i) The letters of the window position are replaced by some bigramme equiv lents and then the whole enciphered at some 'Grundstellung', or ii) The window position is enciphered at the Grundstellung, and the resulting letters replaced by bigramme equivalents. The second of these alternatives was made far more probable by the following indicators occurring on the 2nd May

EXDP IVJO	V C P
XOCX JXJY	V U E
ROXX JLWA	N U M

With this second alternative we can deduce from the

first two indic tors th t the bigramme XX nd XX h ve the s me value, and this is confirmed from the second nd third, where XX and XX oc ur in th second position instead ofth e first.

It so happened that the change of indicating system had not been very well made, and a certain torpedo boat, withthe call sign AFA: had not been provided with the bigramme tables. This boat sent a message in nother cipher explaining this on the 1st May, and it was arranged th t traffic with AFA: was to take place according tothe old system until May 4, when the bigramme tables would be supplied. Sufficient traffic passed on May 2,3 ~~for~~ to and from AFA: for the Grundstellung used to be found, the Stecker having alre dy been found from the FORTYWEEPY messages. It was natural to assume that the Grundstellung used by AFA: was the Grundstellung to be used with the correct method of indication, and as soon as we noticed the two indicators mentioned above we tried this out and found it to be th e case.

There actually turned out to be some more complications. There were ^at least^ two Grundstellungen instead of one. One of them was called the Allgemeine and the other the Offiziere Grundstellung. This made it extremely difficult to find eithter Grundstellung. The Poles pointed out another possiblity, viz th t th e trigrammes were still prob bly not chosen at random. They suggested that probably the window positions enciphered at the Grundstellung, rather than the window positions th emselves were taken off the restricted list.

In Nov. 1939 a prison er told us that the German Navy had now given up writing numbers with Y...YY...Y and that ~~thexxxxxx~~ the digits of th e numbers were spelt out in full. When we heard this we exa ined the messa es towards th e end of 1937 which were expected to be continuati ns nd wrote th e expected beginxiánings under them. The pro ortoin of 'crashes' i.e. of letters apparently left unaltered by encipherment, ~~xxx~~ then shows how nearly correct our guesses were. Assuming that the change ~~xxx~~

Further Aspects of the Work and Its History

Tony Sale[1] delves into the cryptographic background to —

ALAN TURING, THE ENIGMA AND THE BOMBE

The mathematician Alan Turing had been identified, at Cambridge, as a likely candidate for code-breaking. He came to the Government Codes & Ciphers School, (GC&CS), in Broadway in London a number of times in early 1938 to be shown what had already been achieved. He was shown the rodding method and some intercepts of German signals enciphered on the German forces Enigma that had the stecker board. Dilly Knox already knew that the German forces Enigma rotors were wired differently to the commercial rotors, did not know the entry rotor order and apparently did not know the double encipherment of the message key.

Alan Turing had been thinking for some time of ways to attack Enigma. The main thrust of his ideas was based around what is now called 'known plain text' and what became known in Bletchley Park as a 'crib'.

Turing realised that if traffic analysis could be used to predict the text of some parts of the enciphered messages, then a machine could then be used to test, at high speed, whether there were any possible settings of the wheels, which translated the enciphered characters into the deduced characters. More importantly, using his mathematical skills, he showed that it was far quicker to prove that a transformation from ciphered to deduced text precluded a vast number of possible wheel combinations and starting positions.

1. Letter pairs

GC&CS already had a few intercepts and at least one plain text/ciphertext pair, reputed to have been smuggled to England by a Polish cipher clerk.

Among the characteristics that Turing had found was that occasionally the same cipher/plain text pair of characters occurred at different places in the same message. These were known as 'clicks' in Bletchley Park.

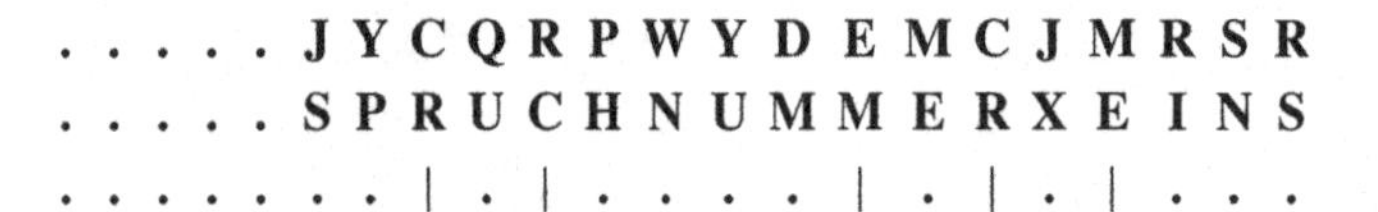

Remember that because the Enigma machine is reversible, R→C is the same as C→R and M→E the same as E→M.

[1] We are very grateful to Tony for his permission to use this unpublished commentary from his webpages, and see its inclusion here as a small memorial to his invaluable work for Bletchley Park.

Whether such pairings occur is determined by the rotor order and the core rotor start positions. Turing realised that conversely the actual rotor order and core rotor start position could be arrived at by trying all configurations to see if these pairings were satisfied. This would only work for an unsteckered Enigma or for a Steckered Enigma in which C and R were unsteckered. In the early days of Enigma, only six letters were Steckered so this could happen.

Obviously, just setting up a single Enigma machine and trying by keying in would take an impossibly long time. The next step was to consider how the tests could be carried out simultaneously for a particular Enigma start configuration.

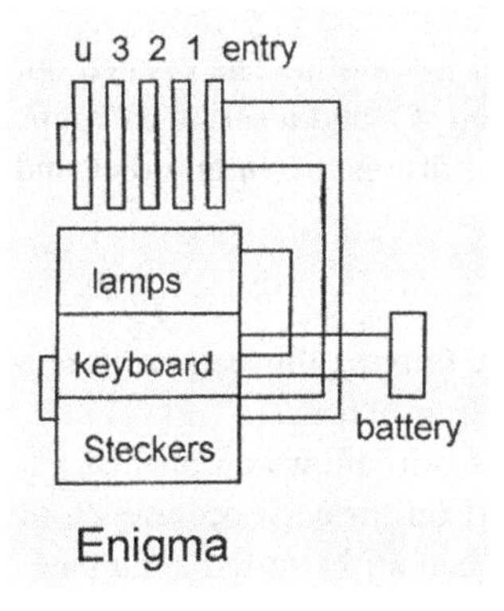

Testing for letter pairs required a method for rapidly determining whether such a configuration was true or false. This led to the concept of electrically connecting together a number of Enigma machines.

This was achieved by using an 'opened out' Enigma.

In the actual Enigma, electrical current enters and leaves by the fixed entry rotor because of the reflector or Umkerwaltz (U) and this precluded connecting Enigmas together. In Turing's opened out Enigma, the reflector had two sides; the exit side being connected to three rotors representing the reverse current paths through the actual Enigma rotors. This gave separate input and output connections and thus allowed a number of Enigmas to be connected in series.

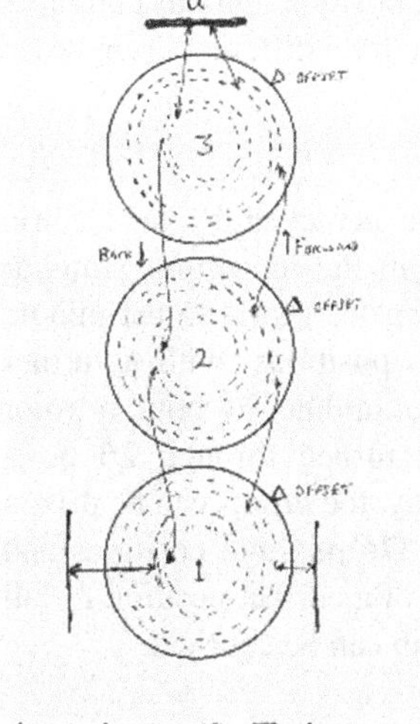

In the Letchworth (so-called because the British Tabulating Machine factory that made them was in the town of Letchworth), implementation, the clever thing was to include both forward and backward wiring of an Enigma rotor in one drum. The connections from one drum to the next were by four concentric circles of 26 fixed contacts and four concentric sets of wire brushes on the drum. Three sets of fixed contacts were permanently wired together and to the 26 way input and output connectors. Three drums, representing the original Enigma rotors, could now be placed on shafts over the contacts and this was an opened out Enigma with separate input and output connectors.

To return to the problem of checking whether C enciphers to R, (written as C→R), first an offset reference from the start is required. A lowercase alphabet written over the ciphertext gives this.

a b c d e f g h i j k l m n o p q
J Y C Q R P W Y D E M C J M R S R
S P R U C H N U M M E R X E I N S
. . | . | . . . . . | . | . | . . .

This shows that C→R at offset c, e and l from the start and M→E at j, k and n. The opened out Enigma allows an electric voltage to be applied to the input connection 'C' and a set of 26 lamps to be connected to the output connector. If the R lamp lights then the drums are in an order and position such that C enciphers to R.

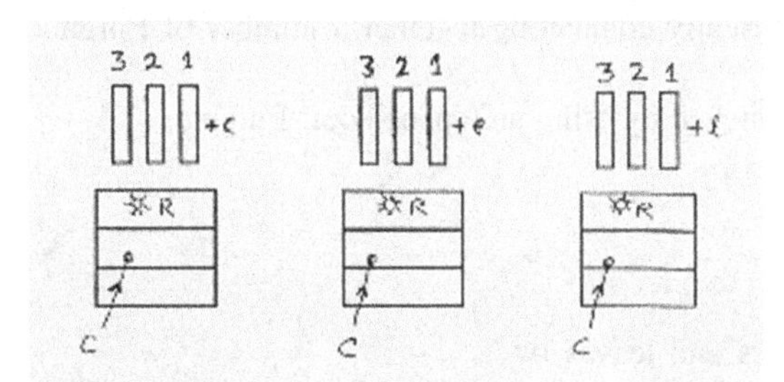

With a single Enigma this can occur at a vast number of drum settings.

However, the crib allows an opened out Enigma to be set up for each occurrence of C→R and they can all be tested simultaneously.

The opened out Enigmas are all set up with the same drum order and the drums are then turned to the same settings for the top (left hand) and middle drums but the bottom (right hand) drums are turned to the offset letter along the crib at which the test is to be made. All the inputs are connected in parallel and a voltage applied to the 'C' contact. Then a set of relays connected to each of the 'R' output contacts tests to see if all the R contacts have a voltage on at the same time. When they do, a position of the drums has been found, which satisfies the crib at the points chosen for C→R.

If they do not, then all the bottom drums are advanced one position and the test is tried again. After 26 positions of the bottom drum, the centre drum is advanced one position and this continues until all drum positions have been tested. Then the drums are changed to try a different drum order. A very long process by hand, which obviously asks to be automated.

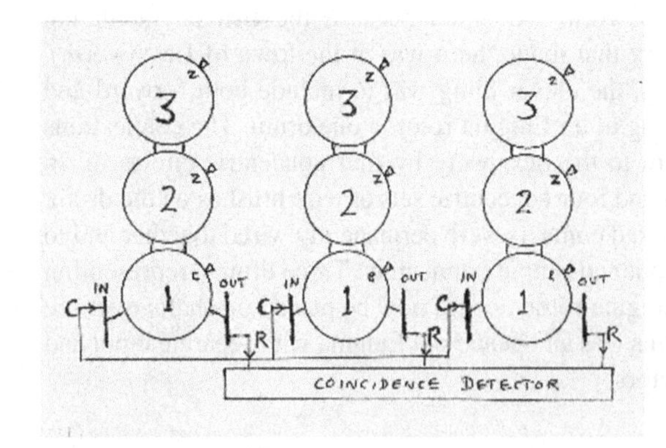

This can be achieved by an electric motor driving all the top drums simultaneously and then 'carrying' to the middle drum every 26 positions, with a further carry from the middle to bottom rotor when this has turned through 26 positions. In this way, the drums can be driven through all 17,576 possible positions and the occurrence of a correct position for all C→R in the crib can be checked.

But there are still a large number of positions that satisfy the C→R test.

What is needed is a better method for finding the rotor order and rotor setting.

2. Letter loops

An extension of the concept of letter pairs is where letters enciphered from one to another at different places in the crib resulting in loops of letters.

a	b	c	d	e	f	g	h	i	j	k	l	m	n	o	p	q
J	**Y**	**C**	**Q**	**R**	**P**	**R**	**Y**	**D**	**E**	**M**	**C**	**J**	**M**	**R**	**S**	**R**
S	**P**	**R**	**U**	**C**	**H**	**N**	**U**	**M**	**M**	**E**	**R**	**X**	**E**	**I**	**N**	**S**
.	.	.	.	.	.	|	.	.	.	.	.	.	.	.	|	|

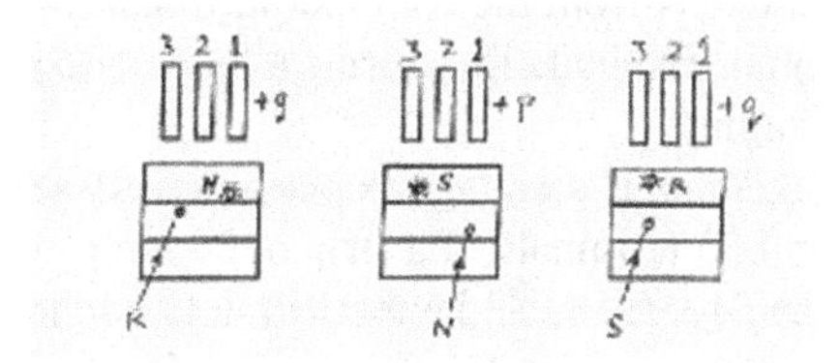

For instance, R→N at g, N→S at p and S→R at q making a loop. A diagram showing such loops was known as a menu. But if Steckers are being used this is actually:

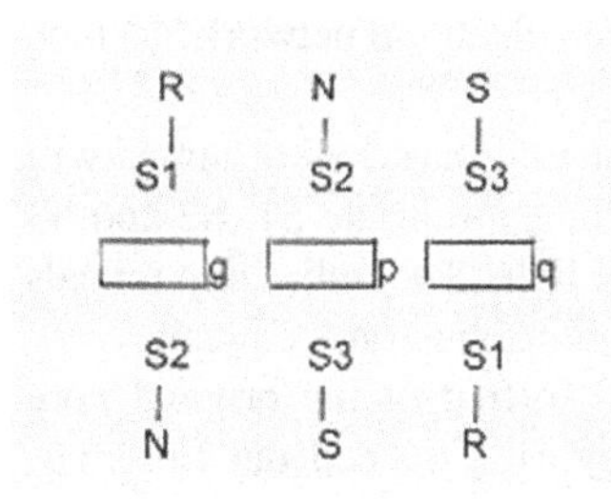

R Steckered to S1 enciphers to S2 Steckered to N at g,

N Steckered to S2 enciphers to S3 Steckered to S at p,

S Steckered to S3 enciphers to S1 Steckered to R at q.

The problem now is to find the core positions Sl, S2, and S3. If these can be found then they are the Steckers of the menu letters.

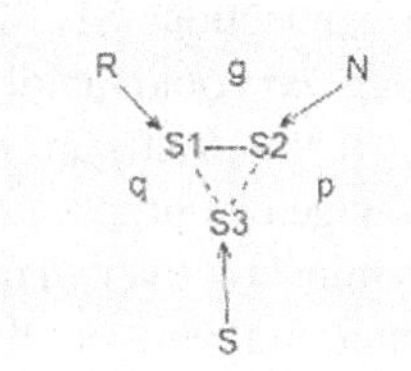

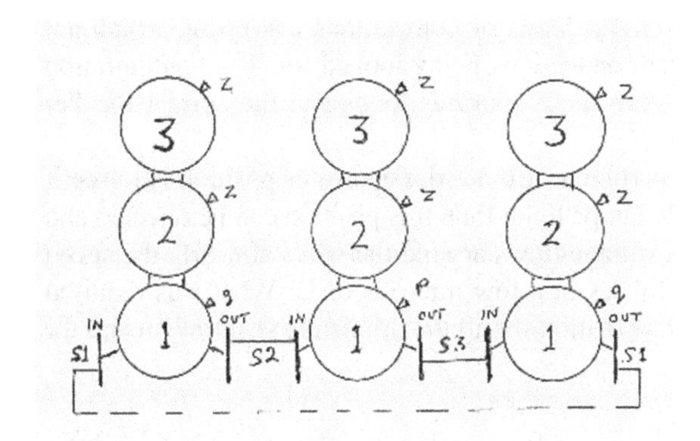

But Turing realised that there was another way of looking at interconnected opened out Enigmas and that this way found Stecker connections.

Take the loop example above of R→N→S→R. Three opened out Enigmas are connected serially one to the other and the bottom drums are turned to the offsets g, p and q. If the correct drum order is being used, then there will be some start position of the top, middle and bottom drums, which corresponds to the actual original Enigma core rotor positions having allowed for the difference between the original Ringstellung and ZZZ. At this point, the core rotor positions will be the same as the original Enigma core rotor positions and the encipherments will then be the same.

This means that a voltage placed onto the S1 input of the first opened out Enigma, which is the Stecker of the input R, will come out on the S2 terminal, which is the Stecker of N. Since this is connected to the next opened out Enigma, this goes in on its S2 terminal and comes out on the S3 terminal, which is the Stecker of S. This S3 input now goes through the third opened out Enigma and comes out at Sl, which is the Stecker of R. Thus, the drum positions correspond to the original Enigma positions where S1→S2→S3→S1.

The magic trick is now to connect the output terminals of the last opened out Enigma back to the input of the first Enigma. There is now a physical wired connection through the opened out Enigmas from the S1 input terminal to the S1 output terminal, which is now connected to the S1 input terminal. This forms a loop of wire not connected to any other terminals on any opened out Enigma.

Thus, if a voltage is placed on S1 at the input it goes nowhere else, just appears on the S1, S2 and S3 terminals. If a strip of 26 lamps is connected at the joins between opened out Enigmas then the S1, S2 and S3 lamps will light confirming the voltage path through S1, S2 and S3.

Now comes Turing's really clever bit. If S1 is not known and the voltage is placed on, say, A then this voltage will propagate through the opened out Enigmas because they are joined around from output to input, but CANNOT reach the S1, S2 and S3 loop because it is not connected to any other terminals. The voltage runs around the wires inside the opened out Enigmas until it reaches a terminal that already has the voltage on it. The complete vastly complex electrical network has then reached a steady state.

Now, if the lamp strip is connected at the joins of the opened out Enigmas, lots of lamps will light showing where the voltage has reached various terminals, but the appropriate S1, S2 and S3 lamps will not light. In favourable circumstances 25 of the lamps will light. The unlit lamp reveals the core letters, S1, S2 or S3. These are interpreted as the Steckers of the letters on the menu.

When the drum order and drum positions are correct compared to that of the original core Enigma encipherment, there is just the one wired connection through the opened out Enigmas, at connections S1, S2 and S3. But Turing also realised that such a system of joined opened out Enigmas could rapidly reject positions of the drums, which were not the correct ones.

If the drums are not in the correct position, then the loop S1, S2, S3 does not exists and the voltage can propagate to these terminals as well. Thus it is possible for the voltage to reach all 26 terminals at the join of two of the opened out Enigmas. This implies that there is no possible Stecker letter and therefore this position of the drums cannot be correct. But because of the way the cross wiring inside real Enigma rotors is organised, closed loops of connections can occur, which are not the loops corresponding to the actual Stecker connections being looked for. The configuration of opened out Enigmas cannot distinguish between these spurious loops and the correct Stecker loop.

The test for a loop of possible Steckers at a particular drum order and rotor position is to see if either only one or 25 of the lamps are lit. If all 26 lamps light, then this position can be rejected and this rejection can occur at very high speed. The voltage flows around the wires at nearly the speed of light so that the whole complex network stabilises in a few microseconds. What was required was some way of automating the changes of drum position for all the drums in synchronism and for rapidly sensing any reject situation.

3. The Bombe

In 1939, the only technology available for achieving electrical connections from rapidly changing drum positions was to use small wire brushes on the drums to make contact with fixed contacts on the Test Plate. This was a proven technology from punched card equipment. High-speed relays were initially the only reliable devices for sensing the voltages on the interconnections. Thermionic valves were tried but were not reliable enough in 1939. Later, thyratron gas-filled valves were used successfully and these were about 100 times faster than the high-speed relays.

The British Tabulating Machine Co (BTM) had designed the opened out Enigmas and built the Test Plate. The project to now build a complete search engine, which became known as a Bombe, came under the direction of H H (Doc) Keen. The machine, known as Victory, was completed by March 1940 and delivered to Bletchley Park. It was first installed in one end of Hut 1. Now the work began on finding out how to use this new device. Results at first were not very encouraging. The difficulties in finding cribs meant that when a menu was constructed between intercepted enciphered text and a crib, it usually did not have enough loops to provide good rejection and therefore a large number of incorrect stops resulted.

4. The Diagonal Board

Then Gordon Welchman came up with the idea of the Diagonal Board. This was an implementation of the simple fact that if B is Steckered to G, then G is also Steckered to B. If 26 rows of 26 way connectors are stacked up, then any connection point can be referenced by its row letter and column letter. A physical piece of wire can now connect row B element G to row G element B. The device was called a Diagonal Board because such a piece of wire is diagonally across the matrix of connections.

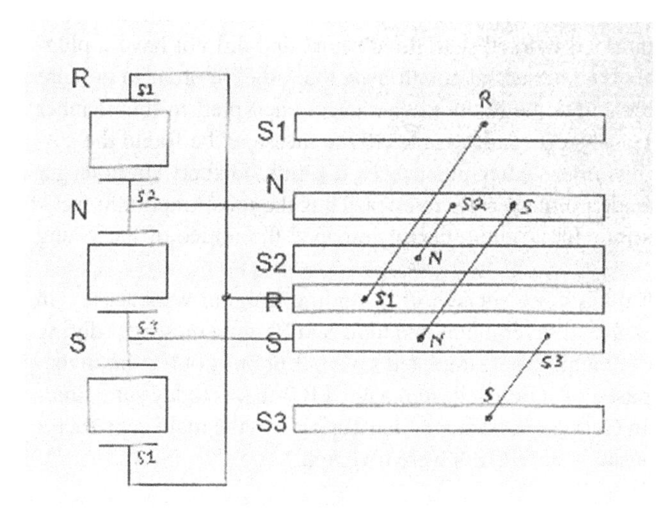

Now the double-ended Enigma configuration knows nothing about Steckers. It can only deduce rotor core wiring positions which satisfy the menu. However the possible Steckers such as R↔S1, can by exploited by the Diagonal Board. If the joins between double-ended Enigmas are also connected into the Diagonal Board at the position corresponding to the original cipher/plain text pair on the menu, say R, then this can significantly increase the rejection of incorrect double-ended Enigma drum positions.

It has already been shown that if a set of drum positions has been found where S1→S2→S3→S1, then a physical wired connection has been made through the joins between opened out Enigmas at S1, S2 & S3. The deduction from this is that R is Steckered to S1 etc. Now, if the join representing R on the menu is plugged to the R row of the Diagonal Board, a physical piece of wire will connect through the Diagonal Board from row R at position S1 to row S1 at position R. Since S1 is not plugged to anything, the voltage on this wire goes nowhere else. Similarly for the other joining positions between opened out Enigmas. Thus, the Diagonal Board does not affect the finding of the correct drum positions.

But if the drums are not in the correct position to make the connection S1, S2 and S3, then a voltage travelling around the network and finally arriving at say row N position S will be passed via the Diagonal Board wire to row S position N and will thus continue through the wiring in the opened out Enigmas on both sides of the join S. The Diagonal Board thus greatly contributes to the voltage flow around the network of wires in the opened out Enigmas due to the extra connectivity that it provides. This increases the rejection of drum positions, which do not satisfy the menu.

Klaus Schmeh looks at –

WHY TURING CRACKED THE ENIGMA AND THE GERMANS DID NOT

The success of Alan Turing and his fellows in cryptanalyzing the Enigma was only possible, because the Germans ignored the weaknesses of their most important cipher machine. This work examines, why they made this fatal mistake.

In October 1942, a young mathematics student named Gisbert Hasenjaeger started his work at OKW-Chi, a cryptologic unit of the German Armed Forces in Berlin. After a training in cryptology, Hasenjaeger was put into the newly founded department for security control of their own encryption methods. In this department, four employees checked the three most important German encryption machines of the time for possible weak spots. Hasenjaeger of all people, greenhorn at OKW-Chi, attended to the Enigma, while another mathematician was to examine the Siemens & Halske T52 (also known as 'Geheimschreiber') and the Lorenz SZ-42 (also known as 'Lorenz machine').

The Enigma variant Hasenjaeger examined worked with three rotors and did not have a plugboard. The Germans sold Enigmas of this type to neutral countries at that time, in order to procure foreign currency. A message with a length of some hundred characters encrypted in this manner was given to Hasenjaeger for analysis. Hasenjaeger indeed cracked the message: he found the correct rotor wiring and the correct rotor positions. After this succès d'estime Gisbert Hasenjaeger continued to work on the Enigma, but made no further discoveries. Thus the most important weak spot of the Enigma (the lack of fixed points due to the reflector) escaped the notice of the young cryptologist.

As is well known, the British were much more successful in finding Enigma weaknesses. In Bletchley Park near London, they cryptanalyzed several hundred thousand Enigma messages during the war years. They used a special type of machine that simulated several Enigma copies simultaneously. The machine, which was loosely based on a Polish design named Bomba, is today sometimes referred to as Turing Bombe (especially in Germany), because Alan Turing was the main constructor of it. Apart from the Turing Bombes thousands of workers were involved.

1. Why did the Germans overlook Enigma weaknesses?

It is certainly an interesting question, why the British were so successful in cracking the Enigma, whereas the Germans overestimated the security of their machine in a fatal way. One potential explanation is that the Germans were generally not very good at codebreaking. However, this is not correct. In fact, during World War II the Germans not only broke encryption machines like the American M-209, the US voice encryption system A-3, a Russian voice encryption device and some others, but they also deciphered numerous manual encryption systems including many codes and nomenclators (Schmeh, 2007a). Like their rivals in Great Britain and the USA, German cryptanalysts even constructed special machines that accelerated codebreaking processes considerably.

Klaus Schmeh (born in 1970) is a German computer scientist and a leading expert on crypto history. His book *Codeknacker gegen Codemacher* (W3L 2007) examines the history of cryptology; his book *Versteckte Botschaften* (Dpunkt-Verlag 2008) deals with the history of steganography.

The author would like to thank Gisbert Hasenjaeger (1919–2006), Günter Hütter, Klaus Kopacz and Susanne Kisser.

The story of Gisbert Hasenjaeger renders another possible explanation for the German failure: The Germans did not put enough effort in challenging their own cipher machines. The discrepancy is obvious: whereas at OKW-Chi one inexperienced cryptologist cared about potential Enigma weaknesses, the British put forth their best mathematicians, including Alan Turing and Dilly Knox. A workforce of several thousand people at Bletchley Park supported them.

The situation was similar for the two other important German encryption machines: the Lorenz machine and the Geheimschreiber, as described, were examined by Hasenjaeger's fellow – only one person was responsible for two machines that were decisive for the outcome of the war. On the British side John Tiltman, a leading cryptanalyst of his time, successfully dedicated himself to the Lorenz machine. The Geheimschreiber, which played no important role for the British, raised interest in Sweden, as the German Luftwaffe used it to protect the traffic between Germany and Norway. Arne Beurling, a leading Swedish mathematics professor, cracked an early version of this machine, while the Germans were not aware that it was seriously flawed.

Of course, it would be too easy to blame the failure of the Germans in detecting Enigma weaknesses just on the particular situation in one department at OKW-Chi. In fact, there is even evidence showing that some German cryptologists knew that the Enigma could be cracked (Bauer, 2000). However, even the critical experts had to admit that breaking the Enigma was only possible with a huge amount of human resources and machine craft. Obviously, nobody expected the British to actually employ this giant machinery. So, the main fault of the Germans was not that they overestimated the security of the Enigma, but that they underestimated the efforts their enemies (particularly the British) would take in order to break it.

Fig. 1: Gisbert Hasenjaeger was one of Turing's opponents in World War II. However, the two did not know about each other. (*Source: Hasenjaeger*)

Nevertheless, in the last years of the war the doubts of some German cryptologists convinced the military leaders to look for an Enigma replacement. The cipher machine that was developed now was named Schlüsselgerät 41 (also known as Hitler Mill). However, the decision not only came too late, but it was also difficult to put it into practice. As the war went on, it became more and more

Fig. 2: The Germans could have easily improved the security of the Enigma or replaced it by a better design. If they had done so, Alan Turing may have never succeeded in breaking German codes. (*Source: NSA*)

Fig. 3: The Hitler Mill, one of the best encryption machines of its time, was introduced to replace the Enigma. However, it came too late. (*Source: Schmeh*)

difficult for the industry to produce, because raw material, energy and qualified staff were lacking. At the end, only about 2000 Hitler Mills were built and delivered to the troops. The war was already lost, when they were – if at all – put into service. So, the Hitler Mill did not bother the British, although it was certainly one of the best cipher machines of its time and not breakable.

Although introducing an Enigma successor generally made sense, it obviously made the Germans blind to another serious mistake. Instead of replacing the Enigma, it would have been much easier to improve it. In fact, this would even have been ridiculously easy. For instance, the Germans used an identical rotor wiring between the late 1920s and the end of World War II. If they had changed the wiring once or twice a year, it would have made the cracking considerably harder. Another, even simpler method, was also not employed: the Enigma had only one notch per rotor. Had the Germans introduced additional ones (for instance two or three per rotor), it would have improved security considerably.

There was even a more trivial way to improve the Enigma, as Gordon Welchman, a fellow expert of Alan Turing, indicated: instead of using two-pole plug connections on the Enigma plugboard, the Germans could have used one-pole ones (Welchman, 2000). The costs for this and other improvements would have been close to zero, but the improvement in security would have been considerable.

All in all, a number of small but fatal mistakes made the German encryption in World War II insecure and thus affected the course of history. Without these mistakes the war might have lasted longer and Germany might have become a victim of the atomic bomb (the attack on Hiroshima took place only three months after the German surrender). Of course, Alan Turing also profited from the German incompetence. His genius as a cryptanalyst is beyond controversy, but his abilities might have been useless against an improved Enigma or a Hitler Mill.

2. Too many cooks spoiled the broth

It is hard to determine in detail, why exactly the German cryptologists made such severe mistakes. However, it is clear that there was a basic problem, which may have been the root of it all: the Germans did not bundle their cryptologic efforts. While the British concentrated their codebreaking activities in the GC&CS in Bletchley Park, the Germans did not have an equivalent institution. In fact, there were at least 12 German cryptographic units that worked independently from each other (Schmeh, 2007b). The most important German crypto group was the aforementioned OKW-Chi. Similar units were operated by the Air Force (Luftwaffe), by German Post, by the secret service (Abwehr) and other authorities. This separation led, among other things, to the situation that OKW-Chi did not accept the Siemens Geheimschreiber because of security concerns, whereas the Luftwaffe put it into service for highly secret information. For the same reason, Gisbert Hasenjaeger's findings concerning the Enigma certainly were not known to many responsible German cryptologists. It can only be imagined, what would have happened, if the Germans had concentrated their Enigma expertise at one place. It is well possible that this would have led to improvements or to an earlier replacement of the machine.

Only years after the war, when the Bundesrepublik was founded, the Germans proved to have learned their lesson – although they still did not know that the Enigma had been broken. When in the 1950s a new German crypto authority was founded, all activities were concentrated at one place. The authority was named 'Zentralstelle für das Chiffrierwesen' (ZfCh), which means 'central authority for cipher affairs' – even the name indicated that it was planned as a centralized institution. It was located in the new capital Bonn. This authority still exists, it is called 'Bundesamt für Sicherheit in der Informationstechnik' (BSI) now. In the 1950s, some ZfCh staff members wanted to re-install the Enigma (in a better version), but it was decided to use other machine designs instead.

Gisbert Hasenjaeger was not involved in these activities. In 1945, he resumed his studies of mathematical logic in Münster. He became a scientific assistant before he qualified as a professor in 1953. His further career led him to the University of Bonn and to Princeton University in the United States. When details about Bletchley Park became public in the 1970s, Gisbert Hasenjaeger

Fig. 4: The Enigma plugboard worked with two-pole plug connections. One-pole ones would have made the Enigma considerably more secure. (*Source: Schmeh*)

Fig. 5: The Geheimschreiber and the Lorenz machine suffered from similar weaknesses as the Enigma. The Germans used them anyway, and their enemies in World War II broke them.

finally learned that the British had cracked the Enigma in World War II. His comment: 'I was very impressed by the fact that Alan Turing, one of the greatest mathematicians of the 20th century, was one of my main opponents.'

References

Bauer, F., 2000. Entzifferte Geheimnisse, Methoden und Maximen der Kryptographie, Springer, Berlin, p. 222.

Schmeh, K., 2007a. Codeknacker gegen Codemacher, W3L, Bochum, p. 238.

Schmeh, K., 2007b. Codeknacker gegen Codemacher, W3L, Bochum, p. 239.

Welchman, G., 2000. The Hut Six Story. Breaking the Enigma Codes, Allen Lane, London 1982; Cleobury Mortimer M&M, Baldwin Shropshire, p. 168.

References

Speech System 'Delilah' – Report on Progress

(A.M. Turing, 6 June, 1944, National Archives, box HW 62/6)

Andrew Hodges sets the scene for —

THE SECRETS OF HANSLOPE PARK 1944–1945[1]

After March 1944 Turing was overall consultant for the work at Bletchley Park. But after August 1944, with the invasion of Europe secure, he spent more time with his hands on electronics, preparing himself for the design of an electronic computer. He did this at the MI6 base at Hanslope Park, Buckinghamshire. With the assistance of Donald Bayley, a young electronic engineer, he built an advanced speech scrambler of his own elegant design.

They called it the Delilah. This was at the suggestion of the young Cambridge mathematician Robin Gandy, who was also working there. After 1948 Robin Gandy became Alan Turing's student, colleague and close friend.

Alan Turing was particular open about his being gay while working with Donald Bayley. The young engineer was amazed at meeting someone who was open and 'almost proud' of it. He also told me how in 1944 he left this as a private matter, but that if such a thing had happened after 1948 when new 'security' rules came into force, he would have had to report it. Donald Bayley spoke further of this on the television programme made in 1992, *The Strange Life and Death of Dr Turing*, and said how by 1952 homosexuality was 'beyond the pale' and disqualified anyone from secret work. Alan Turing's colleague Jack Good, however, said on the same television programme that if the security authorities had known about Alan Turing's homosexuality from the beginning, 'we might have lost the war'.

The author Arthur C. Clarke makes a similar claim in this foreword to a book on Artificial Intelligence: 'How ironic that Alan Turing, who perhaps contributed more than any other individual to the Allied victory, would never have been allowed into Bletchley under normal security regulations.'

In a subsequent interview, Donald Michie has pointed out that the 'security' authorities did not appear to worry about various other gay men at the Bletchley Park establishment, thus confirming that such vetting only started after the Second World War.

[1]Editor's note: The 1944 report on progress was followed by a final technical report, to be found in the National Archives. It is not clear how much of the technical report is by Turing and how much by Donald Bayley. Andrew Hodges' background comments are courtesy of the *Alan Turing Scrapbook*.

TOP SECRET

Speech System 'Delilah' – Report on Progress

Research on 'Delilah' has been in progress since the beginning of May 1943. Up to now the work has all been concentrated on the unit for combining the key with the speech to produce cipher (or scrambled speech) and for recovering the speech from the cipher with the aid of the key. We have now produced a unit for doing this; the same unit does duty both as scrambler and descrambler, changing from the one to the other on throwing a switch. The unit uses seven valves and when suitable rearranged will probably occupy a space of about $10'' \times 8'' \times 5''$. The greatest care has been taken to avoid using more apparatus than is absolutely essential. It is possible that if this had not been done, the present position might have been reached two or three months earlier, at the cost of having an apparatus of about twice the present size.

Proposed Future Plans

(i) The most urgent job to be done now is the design of a unit for the production of key. How long this will take is difficult to estimate, but it is hoped that it will not be so long as the making of the combining unit. Six to nine months might be taken as a reasonable estimate.

(ii) When the time comes for point to point tests, a certain amount of work will have to be done in testing the suitability of the audio stages of the wireless apparatus involved, and possibly making some corresponding alterations. Some of this could be done concurrently with work on the key unit.

(iii) The present combining unit, though reasonably satisfactory cannot be regarded as perfect. The intelligibility could probably be improved by raising the frequency from 4 Kc/s to 6 Kc/s (corresponding to changing the speech band passed from 2 Kc/s to 3 Kc/s). There will also probably be small points to consider concerning production.

(iv) Whereas there may be some difficulty in the transmission of the speech scrambled by 'Delilah' she has another application where this question does not arise, viz. the scrambling of facsimile. A facsimile scrambler would simply be a low frequency scrambler working at a pulse frequency of about 300 cycles (say). To develop such a scrambler should be a matter of little more than changing the values in the present unit.

It will be appreciated that these lines of action can only be followed one, or perhaps two, at a time.

Suggested Form of Key

Some thought has been given to the problem of the form that they key should take. The original plan that a 'public key' should be transmitted, which would be so hashed up before use as to be unrecognisable, has now been abandoned, not so much for security reasons as on account of transmission difficulties. It is now proposed to produce a periodic key with a rather long period e.g.

$7 \times 8 \times 11 \times 13 \times 15 \times 17 = 2{,}031{,}040$ pulses or about 8 minutes. The nature of the key could be set in advance by plugging to any one of a considerable number of alternative keys. There will probably be something of the order of 10^{25} different possible keys. The key chosen will be changed daily let us say, and also to some smaller extent before each conversation. Besides these changes it is hoped to introduce an automatic partial key change whenever the transmit-receive key is operated. Without this device the various pieces of conversation would be in depth, and would in theory present no difficulty to the cryptographer. For the same reason one speaker should be limited to at most 8 minutes consecutive speech.

The period of key quoted above would be obtained by means of a number of multivibrators synchronised with the main pulse of the scrambler itself, and having frequencies which are $1/7, 1/8 \ldots 1/17$ of the frequency of that pulse. The outputs of these multivibrators with the networks can be altered by the plugging. The outputs of the networks then go through another set of networks, and the outputs of these are combined together with further plugging to give three different signals. These three signals are then combined by intermodulation, and the result after limiting is the key. This system has been devised to try and prevent any methods of breaking dependent on separating the effects of the different multivibrators. It is thought that by methods of the type described above a very high degree of security indeed can be obtained. There is certainly no comparison in security with any other scrambler of less than ten times the weight. For tank-to-tank and plane-to-plane work a rather less ambitious form of key will probably be adequate. Such a key unit might be of about the same size as the combining unit.

[signed] A. M. Turing

6th June 1944

Examining the Work and Its Later Impact

Craig Bauer presents —

Alan Turing and Voice Encryption: A Play in Three Acts

PROLOGUE

To properly understand Alan Turing's involvement with voice encryption, it's necessary to first set the scene by describing earlier work carried out by others. Voice encryption, also known as ciphony, goes back as far as the 1920s, when an analog system was put into use by AT&T. During this decade inverters swapped high tones with low tones, and vice-versa. Expressing it more mathematically, the frequency p of each component was replaced with $s - p$, where s is the frequency of a carrier wave. A major weakness with this form of encryption is that tones near the middle are hardly changed. So, that dull professor you remember not too fondly wouldn't be able to speak securely using an inverter, if his tone of choice was near the middle. Invertors only protected against casual eavesdropping and could be easily inverted back by determined amateurs. There was no key as such and inverters are not hard to build. In some cases, the devices were not even needed. With practice it is possible to understand much inverted speech, even if it isn't that old professor of yours speaking in a montone.

AT&T and RCA offered a slightly more sophisticated scheme in 1937. Known as the A-3 Scrambler this system split the speech into five channels (a.k.a. subbands), each of which could be inverted, and shuffled them before transmitting. However, this was still weak, and it was implemented in an especially weak manner. Since there are only $5! = 120$ ways to reorder the 5 subbands and $2^5 = 32$ ways to decide which (if any) of the subbands will be inverted, we have a total of $(120)(32) = 3{,}840$ ways to scramble the speech Thus, the key space is way too small. If the attacker knows how the system works, he could simply try all of the possibilities. Even worse, many of these keys failed to garble the speech sufficiently to prevent portions of it from remaining understandable. Worst of all, of the 11 keys deemed suitable for use, only 6 keys were used! They were applied in a cycle of 36 steps, each lasting 20 seconds, for a full period of 12 minutes.

Hence, like the inverters of the 1920s, the A-3 scrambler was understood to offer "Privacy, not Security." A good analogy is the privacy looks on interior doors of homes. If someone walks up to a home bathroom that is in use, and the lock prevents the doorknob from turning, he'll think,"Oh, someone's in there." and walk away. Privacy is protected. However, there is no real security. Someone intent on entering that bathroom will not be stopped by the lock. In the same manner, a scrambler would protect someone on a party line, but could not be expected to protect national secrets against foreign adversaries.

When President Franklin D. Roosevelt and Prime Minister Winston Churchill spoke on the phone, they needed real security, not just privacy, yet they initially used the A-3 Scrambler! It was solved by the Germans by September 1941, after only a few months' work. As the following quotes show, allies on both sides of the Atlantic were aware of the problem.

> "The security device has not yet been invented which is of any protection whatever against the skilled engineers who are employed by the enemy to record every word of every conversation made." - British foreign Office Memorandum, June 1942

> "In addition, this equipment furnishes a very low degree of security, and we know definitely that the enemy can break the system with almost no effort." - Colonel Frank McCarthy, Secretary to the Army General Staff, October 1943.

Given this knowledge, it's natural to ask why they didn't use something better. The answer is that securing speech with encryption is much more difficult than encrypting text. There are several reasons why this is so, but one of the most important is redundancy. Redundancy in speech allows us to comprehend it through music, background noise, bad connections, mumbling, other people speaking, etc. Text is about 50% redundant (in other words removing half of the letters from a given paragraph does not prevent it from being reconstructed), but speech is much more redundant and it is hard to disguise because of this.

Speech scrambled in the manner of the A-3 scrambler can be reconstructed using a sound spectrograph, which simply involves plotting the tones and reassembling them like a jigsaw puzzle. So, although splitting the voice into more channels increases the number of possible keys, the attacker could simply reassemble what amounts to a jigsaw puzzle with more pieces. A successful voice encryption system would have to operate in a fundamentally different manner than inverting and shuffling.

There was a very high cost associated with the lack of a secure voice system. Shortly before the Japanese attack on Pearl Harbor, American cryptanalysts broke a message sent in the Japanese diplomatic cipher known as Purple. It revealed that Japan would be breaking off diplomatic relations with the United States. In the context of the times, this meant war. General Marshall knew he needed to alert forces at Pearl Harbor to be prepared for a possible attack, but, not trusting the A-3 scrambler, he refused to use the telephone. If the Japanese were listening in, they would learn that their diplomatic cipher had been broken, and would likely change it. The United States would thus lose the benefit of the intelligence those messages provided. The result was that the message was sent by slower means and didn't arrive until after the attack.

Act I: Sigsaly

Fortunately, the simpler problem of enciphering text had been mastered. A perfect system existed and it was possible to create an analog of it for voice.

The perfect system for text is known as the one-time pad. They key for a one-time pad can be presented in various ways, but for our purposes here it is simplest to show it as a random string of integers between 0 and 25, inclusive. For example, 7, 4, 13, 2, 18, 21, etc. If we wish to send the message ATTACK, we simply shift each letter forward as many positions as is indicated by the number in the same position as that letter in our key. We have A+7, T+4, T+13, A+2, C+18, K+21, which turns into HXGCUF. Observe that T+13 and K+21 both took us past the end of the alphabet. When this happens, we simply start again at the beginning (imagining Z to be followed by A, and the rest of the alphabet again). This system is referred to as the one-time pad because a given key should only ever be used once. If it is reused, there are attacks that allow both messages to be recovered. This has happened.

The voice analog of one-time pad encryption would have to add random values to the sound wave. It's a method completely different from inverting and reordering subbands. It's the story of SIGSALY.[1] This system replaced the A-3 scrambler for Roosevelt and Churchill (and others).

[1] SIGSALY was referred to by many other names over the course of its development and use. These included RC-220-T-1, Project X (The Atomic Bomb was Project Y.), Project X-61753, X-Ray, Special Customer, and The Green Hornet.

Fig. 1: A view of SIGSALY (from the National Archives and Records Administration).

Upon first seeing images like Figure 1, I asked, "So where in the room is SIGSALY?" I wasn't sure which item I should be looking at. The answer was, "It is the room!" The result of the quest for secure voice communication led to a 55 ton system that took up 2,500 square feet. In fact, the image above only shows part of SIGSALY. It literally filled a house. Some reflection makes sense of why the project didn't turn out a more compact device.

The need to keep voice communications secure from Nazi cryptanalysts motivated the design of a secure system, but this impetus also meant that no time could be wasted. The designers didn't have the luxury of taking a decade to make a system of utmost elegance. Instead, they based it on earlier technology that could be readily obtained, saving much time. The heart of the system was a vocoder, which is a contraction of voice coder. The original intent of such devices was to digitize speech so that it might be sent on undersea phone cables using less bandwidth, thus reducing costs. Due to the aforementioned high redundancy of human speech, compression down to 10% of the original was possible, while still allowing the original meaning to be recovered. For SIGSALY, compression was a bonus. The goal was to digitize the voice, so that a random digital key could be added to it in the manner of the one-time pad. Off-the-shelf vocoder technology took up much space!

For those interested in hearing how early vocoders transformed speech, a recording of a Bell Labs vocoder from 1936 may be heard at http://www.complex.com/music/2010/08/the-50-greatest-vocoder-songs/bell-telephone-laboratory. Middle-aged readers of this piece might find the sound reminds them of the Cylons in the 1970s Battlestar Galactica series. Indeed, this effect was produced using a vocoder. Decades earlier, Secretary of War Henry Stimson remarked of a vocoder, "It made a curious kind of robot voice."

The vocoder used by SIGSALY broke the speech into 10 channels (from 150 Hz to 2950 Hz) and another two channels represented pitch. Each channel was 25 Hz, so the total bandwidth was (12)(25) = 300 Hz. Ultimately, the communications were sent at VHF.

The digitization of each channel was done on a senary scale. That is, the amplitude of each signal was represented on a scale from 0 to 5, inclusive. A binary scale was tried initially, but such rough approximation of amplitudes didn't allow for an understandable reconstruction of the voice on the receiving end. Pitch had to be measured even more precisely, on a scale from 0 to 35. Since this scale was represented by a pair of numbers between 0 and 5, pitch required two channels.

When discretizing sound, it seems reasonable to represent the amplitude using a linear scale, but the human ear doesn't work in this fashion. Instead, the ear distinguishes amplitudes at lower amplitudes more finely. Thus, if we wish to ease the ability of the ear to reconstruct the sound from a compressed form, measuring the amplitude on a logarithmic scale is a wiser choice. This allows for greater discernment at lower amplitudes. Thus, the difference in amplitude between signals represented by 0 and 1 (in our senary scale) is much smaller than the difference in amplitude between signals represented by 4 and 5. This technique goes by the technical name "Logarithmic Companding," where companding is itself a compression of compressing/expanding.

Having discretized the signal, we're ready to add the random key. With both the speech and the key taking values between 0 and 5, the sum will always fall between 0 and 10. SIGSALY however performed the addition modulo 6, so that the final result remained between 0 and 5. This was Harry Nyquist's idea. There are two reasons to add in this manner.

1) If we dont perform the mod 6 step, then a cipher level of 0 can only arise from both message and key being 0. So, whenever a 0 is the output, an interceptor will know a portion of the signal. Similarly, a cipher level of 10 can only arise from both message and key being 5. Hence, without the mod 6 step, an interceptor would be able to immediately identify $2/36 \approx 5.5\%$ of the signal from the simple analysis above.

2) Simply adding the key without the mod step would result in random increases in amplitude, which may be described as hearing the message over the background noise of the key. Can you understand someone despite the white noise produced by an air-conditioner in the background?

SIGSALY enciphered every channel in this manner, using a separate random key for each. Figure 2 provides a simplified overview of the encryption process.

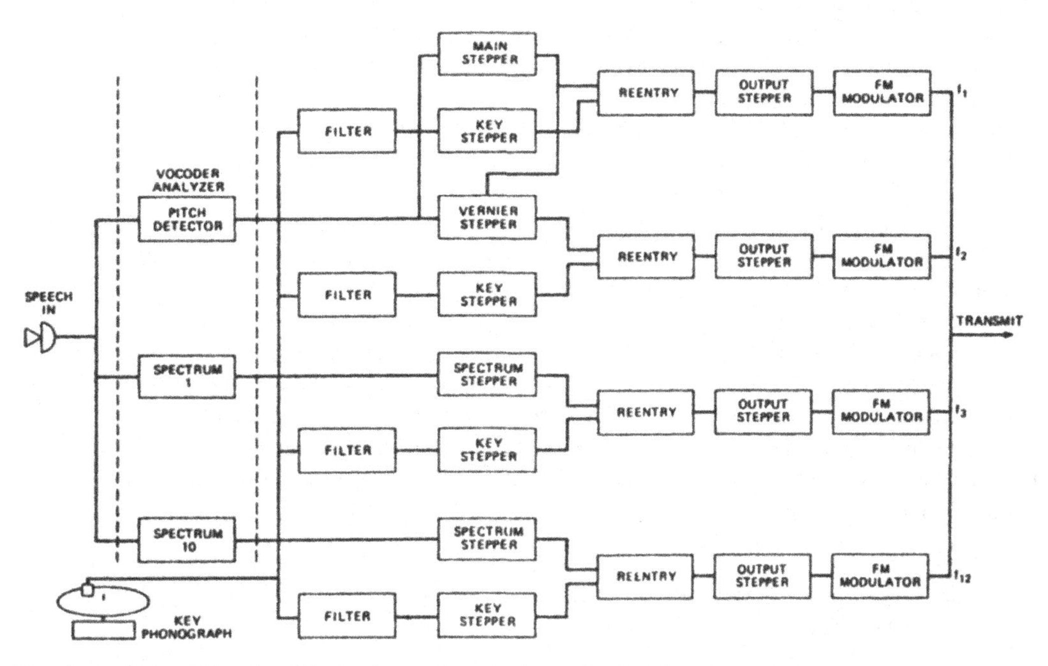

Fig. 2: An incredibly simplified schematic overview of a SIGSALY transmit terminal (courtesy of Donald E. Mehl).

Figure 2 shows the speech entering the system on the left hand side and getting broken down into a pitch channel (pitch detector) and ten voice channels (spectrum 1 through spectrum 10). There are steps, not discussed here, both before and after the mod 6 (reentry) takes place. They can be found in Donald E. Mehl's book [3].

The source of SIGSALY's one-time key was 16 inch records that only played for 12 minutes each. With no pun intended, this component was codenamed SIGGRUV. As with text, the key was added to encipher and subtracted to decipher. A built in safety mechanism caused communication to cease if the record ever stopped. Otherwise, the speaker would suddenly be broadcasting in the clear. The digitized speech was sampled 50 times per second, so to separately encipher all of the channels, the record had to be simultaneously playing 12 tones at different frequencies, and these tones had to change every fiftieth of a second.

It's natural to ask why the sampling rate was 50 times per second and not higher or lower. The fundamental unit of speech, known as a phoneme, has a duration of about a fiftieth of a second, so the sampling rate is just high enough to allow it to be captured. A higher sampling rate is not needed to make the digitized voice comprehensible and would worsen the synchronization problem (the record at the receiving terminal, used to subtract the key, must be synchronized with the incoming message, if there is to be any hope of recovering it). While we're on the topic of synchronization, it should be mentioned that the records contained tones for purposes other than encryption. For example, a tone at one particular frequency was used for fine-tuning the synchronization.

Ideally the keys would be random, a condition simulated for SIGGRUV by recording thermal noise backward. None of these records would become classic tunes, but the military was content with one-hit wonders. Indeed, the system would become vulnerable if the same record were ever replayed. Although not labeled as such, the implicit warning was "Don't Play it Again Uncle Sam!" and the records were destroyed after use.

Once the SIGSALY installations were in place, all that was necessary for communication was that each location have the same record. Initially spares were made, but as confidence increased, only two copies of each record were made. Over 1,500 distinct key sets were eventually produced. Still, there was a Plan B. Figure 3 looks like a locker room, but it is simply SIGSALYs back-up key, codenamed SIGBUSE. If for some reason the records couldn't be used for keying purposes, SIGBUSE could generate a pseudorandom key mechanically.

Since SIGSALY would link Roosevelt and Churchill, both the Americans and the British needed to be satisfied that it was secure. The British had the added concern that the operating teams, which would consist of Americans, even in London, would hear everything. Thus, in January 1943, the British sent their top cryptanalyst, Alan Turing, to America to evaluate SIGSALY. After much debate, probably reaching President Roosevelt, Turing was granted access to details of the closely guarded secret project, entering Bell Labs on 19 January, 1943.[2]

Most of Turing's time was spent on ciphony cryptanalysis (as a beta tester). He also offered improvements to the SIGBUSE key, but SIGBUSE turned out to be wasted space. The records never failed, so the alternate key never saw use. Following his examination of the system, Turing reported to the British, "If the equipment is to be operated solely by U.S. personnel it will be impossible to prevent them listening in if they so desire." In reality, the Americans were often so focused on their jobs they had no idea what was actually said. Ultimately, based on Turing's endorsement, the British accepted SIGSALY.

[2] See Tompkins, p. 59, Hodges p. 245, Mehl p. 69. In any case, Secretary of War Stimson resolved the debate by insisting that Turing be granted access.

Fig. 3: SIGSALY's back-up key SIGBUSE (courtesy of Donald E. Mehl).

In November 1942 an experimental station was installed in New York, and in July 1943 a final version was activated linking Washington D.C. and London. This marked the first transmission of digital speech and the first practical Pulse Code Modulation. Although the bandwidth compressing vocoder was described earlier in this piece as pre-existing technology (which it was), it had not become practical enough for use. Eventually there were a dozen SIGSALY installations. None of the messages they conveyed were ever broken. In fact, the Germans didnt even recognize it as enciphered speech. They thought it was just noise or perhaps a teletype signal. The sound they heard was similar to the music played at the start of the Green Hornet TV show of that era. Although they might not have been familiar with the program, Americans certainly were, and this is why the system was sometimes referred to as The Green Hornet. SIGSALY wasn't declassified until 1976, at which time a slew of patents, applied for decades earlier, were finally granted. A mock-up of a portion of SIGSALY may be seen today at the National Cryptologic Museum.[3]

Turing's examination of SIGSALY served as his education in ciphony (the U.S. was ahead of the U.K. in this area) and the experience inspired him to create his own (completely different) system, Delilah. This brings us to Act II.

Act II: Delilah

During his trip back to England in March 1943, Turing came up with the idea for a voice encryption system that he thought would be superior to SIGSALY [1, p. 273]. His work on this system was not done at Bletchley Park, but rather Hanslope Park (10 miles to the north). Turing was mostly working alone, although he had help from an electrical engineer, Donald Bayley. The project seems

[3] See http://www.nsa.gov/about/cryptologic_heritage/museum/index.shtml

to have been for his own amusement! [1, p. 269-270] Since SIGSALY worked, there was no pressing need for another system. Yet, the rush to complete SIGSALY resulted in a system that was far from ideal. Consider the following characteristics.

1) It weighed 55 tons and had a 70 ton shipping weight.

2) It took up 2,500 square ft.

3) It cost between $250,000 and $1,000,000 per installation.

4) It converted 30 kilowatts of power into 1 milliwatt of low quality speech.

5) The deciphered speech sounded like Donald Duck.

This is why SIGSALY received an (honorable) discharge not long after the war ended. Turing thought he could do better. He didn't name his system himself, but instead offered a prize to whoever came up with the best suggestion. Delilah, after the biblical "deceiver of men," was declared the winner [1, p.273].

The short piece Speech System 'Delilah' Report on Progress reproduced in the present volume was found by Ralph Erskine in the British National Archives (HW 62/6). In it Turing offered the physical description, "The unit uses seven valves and when suitably rearranged will probably occupy a space of about 10" x 8" x 5"." Clearly he improved upon SIGSALY in the categories of size and weight![4]

Turing achieved his size reduction, in part, by eliminating the vocoder, Delilah sent an analog signal, in contrast to the digital SIGSALY. Also, there was only one keying system. Doing away with a back-up key saved much space. However, Turing's keying system had more in common with SIGBUSE than the records SIGSALY used as the primary key.

In 2009, a technical paper on Delilah by Turing and Bayley was released to the British National Archives [5]. It saw print for the first time, near the end of the Turing Centennial, in the October 2012 issue of *Cryptologia*. Thanks to this paper, a more detailed presentation of Turing's work on ciphony may be presented here.

The most impressive feature of Delilah was that it sampled speech 4,000 times per second (SIGSALY's rate was 50). This seems like overkill, but Turing hadn't arrived at this value without justification. He had conversations with Claude Shannon at Bell Labs and knew Shannon's Bandwidth Theorem (actually due to Whittaker, 1915):

> "If a signal time function is sampled instantaneously at regular intervals and at a rate at least twice the highest significant signal frequency, the samples contain all of the original message."[5]

Delilah's 2,000 Hz thus required 4,000 samples/sec. Today's CD Audio is 44.1 kHz with a 16 bit (65,536 levels) sampling accuracy and DVD Audio is 192 kHz with a 24 bit (16,777,216 levels) sampling accuracy. Still, for the time period, 4,000 samples/sec was an impressive rate.

The message amplitudes for Delilah were scaled to not exceed 1 and the key was added modulo 1, even though this approach didn't work for SIGSALY. However, since delilah was an analog system, the amplitude could be represented by a real number, and was not limited to just 0 and 1. One similarity with SIGSALY is that without the mod step, cryptanalysis is possible.

One of the technical problem that arose was that the result of the encryption was too high frequency for telephone circuits. Turing's solution was to "feed each "spike" into a specially devised electronic circuit with an orthogonal property" [1, p. 275]. Referred to as "Turing's Orthogonal Circuit" we know such devices today as matched filters. This approach is now common in high

[4] To be fair, the Americans did develop a smaller mobile version of SIGSALY that fit in a van. Dubbed Junior X, it used an 8 channel vocoder, but was never deployed [4, p. 54].

[5] Taken here from [3, p. 71]

speed digital communications systems; intersymbol interference (ISI) is prevented, since the zero crossings, after passing through the filter, are at sampling points where they don't interfere. Nyquist originated this idea.[6]

Turing and Bayley considered two methods for keying Delilah, "(a) recording random noise on discs or tape and using those recordings simultaneously at the ends of the transmission path and (b) generating identical voltages at each end." [5, p. 37] They noted that option (a) "has the disadvantage that the mechanical difficulties of starting and maintaining the keys in synchronism are large, and, furthermore, the number of discs or reels of tape required becomes prohibitive." Of course, Turing was aware that SIGSALY solved the synchronization problem, although not at the higher frequency he desired. He moved on to plan (b), and had to confront the difficult problem of how to imitate randomness. "Six multivibrators are locked with the pulse from the combiner and their outputs taken to networks which serve both to isolate individual multivibrators and to differentiate the outputs (so as to strengthen, relatively, the higher harmonics.) The fundamental frequency of each multivibrator is some exact sub-multiple of the combiner pulse frequency, the various sub-multiples being:- 5, 7, 8, 9, 23, and 31." [5, p. 37]

As the description continues, it is easy to see that SIGSALY wasn't the only influence in his thinking. Clearly Enigma played a role:

> "The outputs of these networks (26 in all) pass through the cypher machine and are combined at the output end to form seven inputs to seven distorting networks. The distorting networks have differing phase characteristics so that the two out-puts which are produced by combining their seven outputs at the plugboard depend enormously on what frequencies were fed into the various networks i.e. on the setting of the cypher machine and the plugboard." [5, p. 37]

> "Note: It has been assumed above that the reader is familiar with the cypher machine used. Should that not be the case it will be sufficient to understand that the machine is a device enabling 26 contacts to be connected to 26 others in a pre-determined random manner and that the mode of connection may be changed by pressing a key." [5, p. 38]

Despite the relatively prime values, the key would eventually repeat:

> "The multivibrators start simultaneously and since they have fundamental frequencies which are prime to one another it follows that they will arrive back at the starting position after a time 5x7x8x9x23x31 times the period of the locking pulse. That is, the key will repeat after 1,785,600 x 250 secs 7.48 mins." [5, p. 38]

Thus, Delilah users were expected to stop and change keys every 7 minutes. Doing so was easy. Any change to the rotor positions or the plugboard connections created a new key from the same source.

Turing was working alone much of the time with very little funding or resources, so it is not surprising that a number of probelms were present in Delilah. For example, it was too difficult to synchronize the keys for transatlantic conversations, so the system was only workable for "local" calls. Mod 1 sounds simpler than mod 6, but the amplitudes for Delilah needed precise measure and transmission, unlike the senary system of SIGSALY. Also, the Signal to Noise ratio was 10 dB. That is, the speech was only 10 times as powerful as the noise. This was "rather lower than desired" according to Turing and Bayley. Most importantly, the system was not quite completed in time (spring 1945) for the war and wasn't high enough quality for commercial use. Hodges summed the matter up. "As a contribution to British technology it had been a complete waste of time." [1. p. 346]

John Harper is leading a project to rebuild Delilah. A report by Harper follows the present piece. But this was not Turing's last encounter with ciphony.

---

[6] See p. 632 of Nyquist, Harry, Certain Topics in Telegraph Transmission Theory, *Trans. AIEE*, Vol. 47, pp. 617644, Apr. 1928, Presented at the Winter Convention of the AIEE, New York, N. Y., February 13-17, 1928.

Act III: Feuerstein

This act is still in the process of being pieced together by historians. I received confirmation from the GCHQ historian that Alan Turing did go on a mission to a Vocoder lab in Upper Franconia, Bavaria, Germany shortly after the end of World War II. The exact site was Feuerstein, and Turing was tasked with assessing how much progress the Germans had made on voice encryption. I was skeptical that the British would send their top cryptologist on such a potentially dangerous mission, but it did happen. The details have not yet been firmed up to the degree desired, but some are provided below. According to Tompkins, Turing visited Feuerstein on May 15, 1945, and TICOM[7] raided it that summer [4, p. 190-194]. Hodges described a Turing trip to Germany in July, but no mention of Feuerstein is made [1, p. 311-312]. In any case. Tommy Flowers and others went with Turing on at least one mission. There is not necessarily any contradiction in the dates here. Turing was likely part of the TICOM raid that Tompkins mentions and that may be the July trip Hodges describes. Possibly, Turing made several trips to Germany shortly after the war. Juliet Floyd learned that Turing made trips to Göttingen late in his life. Hardly anything is publicly known about these visits. Despite the intensive study that that Turing has received, there remains much room for more research.

Acknowledgments

Many thanks are due to the National Security Agency for making it possible for me to pursue my passion full-time as the 2011-2012 Scholar-in-Residence at their Center for Cryptologic History. Rene Stein, National Cryptologic Museum librarian was invaluable in helping me locate materials I needed. Wayne Blanding showed great patience answering basic questions from a non-engineer. Dave Tompkins clued me in to applications of vocoders in music, enlivening presentations I give on this topic. Thanks also to John Harper, and Tina Hampson!

References

1. Hodges, Andrew, *Alan Turing: The Enigma*, Simon & Schuster, New York, 1983.

2. Kahn, David, *The Codebreakers*, Second Edition, Scribner, New York, 1996.

3. Mehl, Donald E., *The Green Hornet*, self-published, 1997.

4. Tompkins, Dave, *How to Wreck a Nice Beach*, Stopsmiling Books, Chicago, 2010.

5. Turing, Alan, and Donald Bayley, Government Code and Cypher School: Cryptographic Studies, HW 25/36, Report on speech secrecy system DELILAH, a technical description compiled by A M Turing and Lieutenant D Bayley REME, 1945-1946, British National Archives, released in 2009. Published by *Cryptologia* in October 2012, Vol. 36, No. 4, pp. 295-340.

6. Weadon, Patrick, D., *Sigsaly Story*, 2009, available online at
`http://www.nsa.gov/about/cryptologic_heritage/center_crypt_history/publications/sigsaly_story.shtml`

7. Boone, J. V., and R. R. Peterson, *Sigsaly - The Start of the Digital Revolution*, Center for Cryptologic History, National Security Agency, Fort George G. Meade, Maryland, July 2000, available online at
`http://www.nsa.gov/about/cryptologic_heritage/center_crypt_history/publications/sigsaly_start_digital.shtml`

---

[7] Standing for Target Intelligence Committee, this group was tasked with recovery of cryptologic (as well as other) intelligence. Although it was led by the U.S., the British were also part of the effort.

John Harper reports on the —

DELILAH REBUILD PROJECT

The original Delilah Project came to the attention of the Bletchley Park Volunteers when one of our team, who regularly visits the National Archive in Kew, discovered a very large report. It consisted of 80 pages of text and formula plus circuit diagram blueprints and oscilloscope images. We arranged to make copies of these using a camera. This was a start but later the GCHQ Archives department made available a full copy in much better quality than ours.

We became quite excited about this project because it was a major piece of work carried out by Alan Turing during 1943 and 1944 that few people had heard of.

The first record official record that has come to light appears to have been a report dated 6 June 1944 where Alan Turing is writing that research began in May 1943. A Combining Unit appears to be in existence at this time but the ideas about the Key Stream are just coming together. The main report appears to be undated but it must have taken many more months to build a Key Unit and carry out the reported tests and measurements. One might speculate that the project would have run until the end of 1944.

To be fair, Andrew Hodges had already 'discovered' the Delilah Project some years ago (*Alan Turing – The Enigma*, pages 273–276 inclusive). In this he writes about Don Bayley, Turing's partner and co-author of the report. Also there is a picture and a brief reference by Dr. Robin Gandy in a BBC TV Horizon documentary directed and produced by Christopher Sykes some years ago. Although Turing was considered to still be based at Bletchley Park, he spent many months at Hanslope Park presumably because they had good workshop facilities away from the Hurly Burly of Bletchley Park.

With the help of retired Hanslope Park Foreign Office staff we were able to track down Don Bayley who now lives quietly in Yorkshire. I have visited him there and since corresponded.

What is common to many voice encipherment systems is a key stream that is unique for a given transmission. This is ideally in a 'one time pad' form. It is believed that Alan Turing was shown under strict security the workings of an American Voice Secrecy system called SIGSALI when visiting the States (see http://en.wikipedia.org/wiki/SIGSALY etc.). These systems were extremely large, about the size of a 1950s mainframe computer and extremely expensive. Even more expensive were the 'gramophone records' that held the unique key stream. These had to be distributed very securely to each end of the voice link and once used, destroyed. It is speculation, but assumed, that Turing thought that he could improve dramatically on this at a fraction of the cost. In Delilah his 'one time pad' was the setting of five letter transposition wheels similar to an Enigma Machine plus a seven-way patch panel. This modified the key stream. Not quite a 'one time pad' but as this key stream changed with every send and receive change, breaking an enciphered voice message that would change in minutes was considered at that time to be very secure. To add to this, wheels could be reversed or alternatives fitted. As with Enigma, the weak spot would be the 'Setting Sheet'.

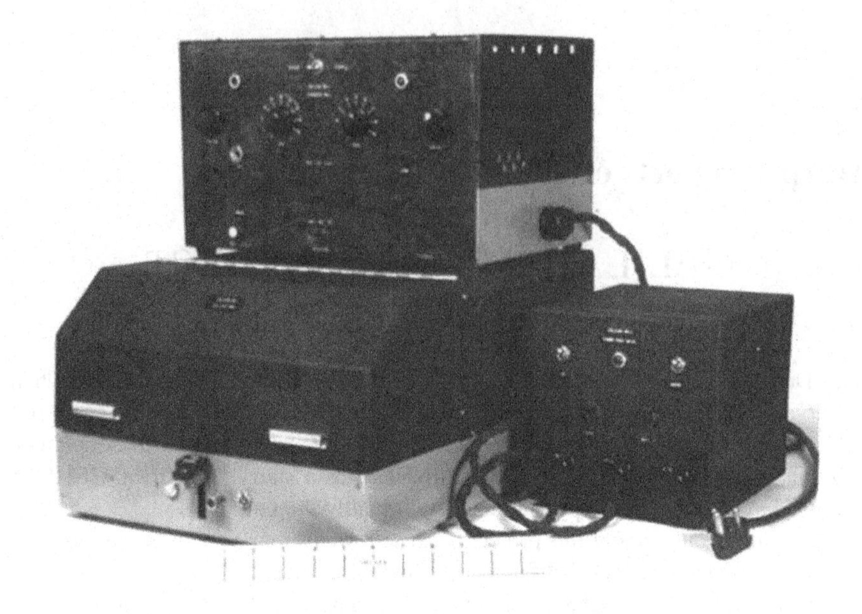

DELILAH Mk. I

Fig. 1:

The system is labelled MK 1 with the report detailing the need for further development. No doubt this work could have continued but the war was being won and people were looking forward to a civilian career. From what we read about Alan Turing he was also very keen to get back to computing.

No doubt the authorities were no longer keen to fund further, war related, developments. Whatever the reason, no further work was done on Delilah and it never went into production.

1. Delilah hardware

This consists of three separate units at each end of a link. These are connected together by power and signal cables with an external connection to a landline or other link such as a VHF radio link.

2. Power supply

This works off the mains and supplies power to the Combiner and Key Unit.

3. Combiner

The voice signal is combined with the key stream to produce a signal to line that is no longer intelligible and highly secure. The unit also works in reverse when set to 'receive' with the key unit again providing the key. This when processed with the incoming signal recovers voice.

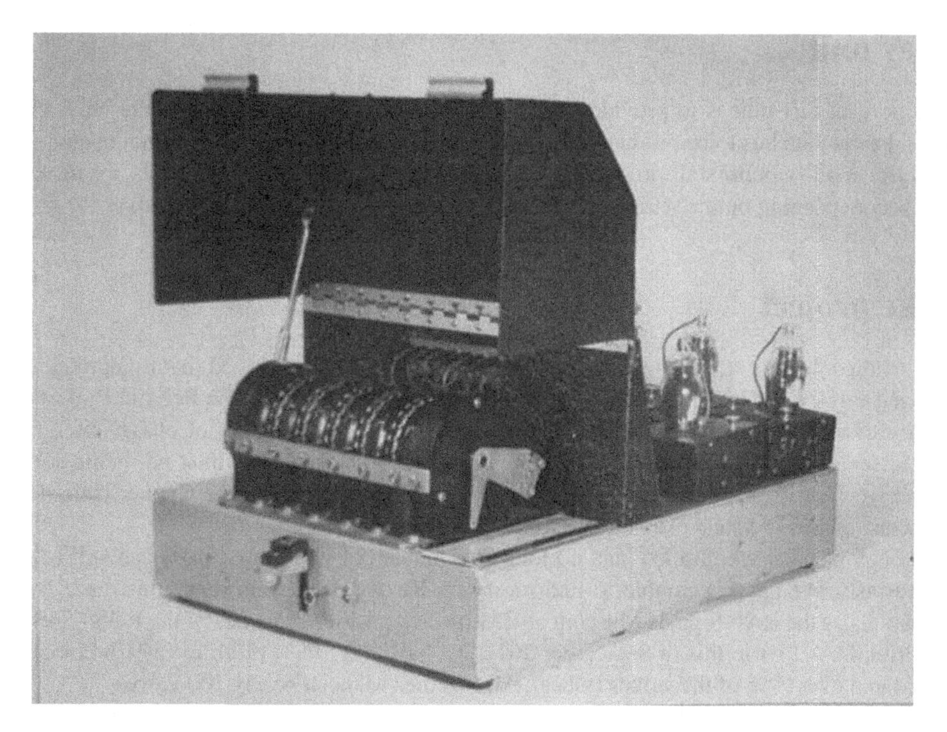

Fig. 2:

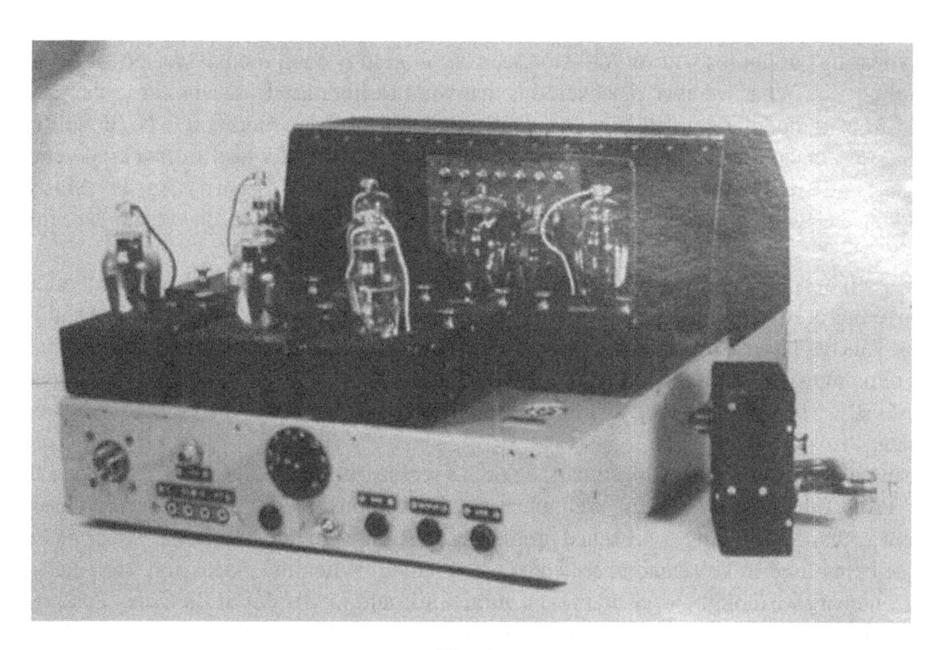

Fig. 3:

4. Key unit

The purpose of this unit is to provide a key stream at one end of the link. At the other end, an identical key stream has to be produced that is precisely synchronised with the sending end.

A set of multi-vibrators all producing different frequency square wave signals are mixed in a unique way depending on the settings of a cypher unit and a seven-way plugboard.

5. The project

Having studied the report, photographs and circuit diagrams, it was decided that to reproduce what Turing and Bayley had achieved was possible. As with the previous Bombe Rebuild Project it was necessary to identify things that would stop us succeeding. Manufacture of chassis etc., assembly, premises, testing facilities, where to demonstrate and funding etc. whilst not being solved at the onset were deemed to be solvable. The major issues were details of the Cypher Unit, detailed manufacturing drawings and obsolete 1940s radio components.

Most components are obsolete and no longer available but good, used parts can still be found by enthusiasts. We have been able to find most of what we need from donations made by such organisations as the RAF Signals Museum at Henlow and various branches of the Radio Society of Great Britain. As I write this in September 2011, we have over 95% of all the 1940s components collected and over 90% of the valves (tubes). All together we need nearly 100 valves.

Manufacturing drawings for the chassis and covers were not in the report so they had to be recreated. This activity is almost complete using Computer Aided Design techniques. Recreation of accurate drawings is perhaps more easy that one might at first think. The report gives overall dimensions for the three units. One photo has a ruler showing but most importantly it is possible to identify the 1940s components and with these available and measured it is possible to reasonably accurately draw the area where they are mounted.

Our most difficult problem is the Cypher Unit. Don Bailey said that this was an American CCM unit. Comparing our photos with WWII American cryptography equipment it was possible to verify this up to a point. What we have discovered is that what Delilah used was similar to the CCM but as the CCM is of modular construction one can see that this has been modified to be 'double ended' as is necessary in the Delilah application. We have borrowed a CCM wheel from a kind gentleman in the States and other Americans who have access to the National Security Agency Museum in Washington DC are making detailed measurements that we will use to check our 'speculative' drawings.

Construction is under way in September 2011 with the Power Supply being used as a pilot. This is to prove our drawing methods that involve laser cutting of the chassis parts before bending and painting. This has proved successful and one Power Supply is assembled with its components fitted.

The remaining pairs of Key Unit and Combiner cabinets will be heading towards the sheet metal people shortly. The current activity is to decide the best way to make the complex Cypher Unit components.

One has to ask the question why embark on such a reconstruction? Having completed the Bombe Rebuild Project that is now working well and regularly demonstrated, the team became fascinated with the way that Alan Turing approached problems such as how to break Enigma. To see an Enigma technique being used in voice encipherment was intriguing. When the discussion started about the Turing Centenary we thought, what additional attraction could we display at Bletchley Park to add to existing items including the Bombe Rebuild and Checking Machine, the slate statue and the Turing Papers? Delilah seemed appropriate. The Bombe Rebuild team welcomed a new challenge – we had worked so well together before and had obtained such satisfaction and recognition of what we had achieved.

Checking a Large Routine

(Paper, EDSAC Inaugural Conference, 24 June 1949.
In: Report of a Conference on High Speed Automatic Calculating Machines, pp. 67–69)

Cliff B. Jones gives a modern assessment of —

TURING'S "CHECKING A LARGE ROUTINE"

Alan Turing gave a remarkable paper at the Cambridge meeting in June 1949 that marked the inauguration of the EDSAC computer. Had this paper been more widely read and understood, it could have accelerated the important area of reasoning about programs by a decade or more. Just one of the impressive features of the paper is its brevity: it comprises less than three pages. Here, after setting the context and outlining the achievement, a fuller assessment is attempted.

1. Context

Alan Turing's (1936) paper is probably accepted as the paper which fixed the notion of what it means to program a universal machine. Addressing the "Entscheidungsproblem", Turing (1936)[1] defined what is today called the "Turing machine", which is capable of performing any computation if only provided with the right program. At this time, Turing was not concerned with the design of a realistic computer but needed a fixed notion of computation to prove that there exist problems that are not computable. In fact, programming a Turing machine was impossibly tedious and little advance in practical computing would have been made using such a language.

By 1949, there were realistic general purpose computers.[2] Although these machines exhibited all of the key features of modern computers, people who have grown up during this exciting electronic age would find them extremely primitive – they were physically huge, had small stores and by today's standards were both unreliable and extremely slow.

More importantly, they were essentially programmed in terms of the instruction set of the specific machine, so that if a storage cell was to be incremented, the programmer had to write one instruction to bring that cell into an accumulator, a second instruction to add the incrementing value into the accumulator and a final instruction to store the incremented value back into the original storage location. Programs can do little without loops, but their construction was even more tedious. Furthermore, the programmer was responsible for working out the addresses of instructions to which jumps were required; this made program correction a messy process.

---

[1] This paper is discussed in Samson Abramsky's chapter of this book.

[2] This is not the place to describe the progress from the EDVAC report through to the SSEM ("Baby") computer from Williams and Kilburn in Manchester and Wilkes' EDSAC in Cambridge – nor even Turing's own work on designing the Pilot ACE at NPL; the interested reader is referred to Copeland (2011).

2. The problem of correctness

Writing programs is one of the most exacting tasks undertaken by humans: a programmer has to write a series of instructions that are followed blindly by an obedient but dumb servant. The fact that programs are executed exactly as written is one of their virtues in that they can execute faultlessly for decades but it also their Achilles' heel. The whole issue about the 'Y2K' problem was that, having allocated only two decimal places to store date information, programs would have been in danger of subtracting 49 from 12 and deciding that it was −37 years since Turing gave his Cambridge paper; any intelligent person would say it must obviously be 63 years.

This situation is compounded by the astronomical number of different states a program can occupy. A tiny program that can take a few independent inputs might have an input state space of 2^{96} values; if the instructions in the program include a few branches and loops, there might be only one, or very few, input combination that delivers an answer that is not as expected by the user of the program. Testing the whole input space for all but the most trivial programs is totally impossible. One of the most often quoted aphorisms in computing is from Edsger Dijkstra who said 'Program testing can show the presence of bugs, never their absence'.

What then can be done? Mathematicians prove theorems for all possible cases. Euclid did not prove his eponymous theorem about the lengths of the sides of right-angled triangles by testing a large number of cases – the proof established the result for all such triangles that could be constructed at any time. In principle then, the task was clear, record in a succinct way the required relation of the outputs of a program to its inputs and then reason about why a program would always satisfy this 'specification'.

Knowing the quantities involved in the first programs were all represented as (restricted) positive integers might even make one suspect that the only tool needed to perform such arguments was mathematical induction. Unfortunately, this is by no means a complete answer. At a minimum, it is essential to have a precise way to reason about the meaning of the statements in the language of programs. In addition, the reasoner needs some guidance as to how to organise the argument or proof.

3. Turing's contribution

Alan Turing's paper (Turing, 1949) in the proceedings of the 1949 Cambridge event is just two and a half pages long – admittedly the paper is the quaint English legal format that was known as 'foolscap', but the paper has the equivalent of less than 300 lines. Moreover, into this short length, it fits one of the clearest arguments given for a process which has been a cornerstone of presenting arguments about programs for decades after Turing's untimely death. Turing's motivational analogy presents a column of four-digit numbers whose sum is sought. If the sum alone is written below the numbers, anyone wishing to audit the addition has to repeat the whole calculation; if however the carry digits are recorded, the task is decomposed into four simpler checks; furthermore, these four sums of single digits can even be undertaken in parallel by different auditors.

The message was clear; what was needed was a way of recording steps of the argument for the claim that a program satisfied its specification that split the task of checking individual steps in the argument. The second paragraph of Turing (1949) is worth quoting in full:

> In order that the man who checks may not have too difficult a task the programmer should make a number of definite assertions which can be checked individually, and from which the correctness of the whole program easily follows.

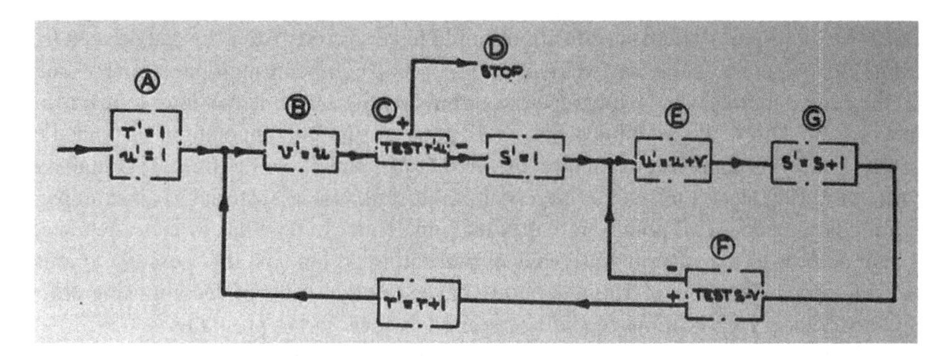

Fig. 1: Turing's original flowchart

STORAGE LOCATION	(INITIAL) Ⓐ k=6	Ⓑ k=5	Ⓒ k=4	(STOP) Ⓓ k=0	Ⓔ k=3	Ⓕ k=1	Ⓖ k=2
27					s	s+1	s
28		r	r		r	r	r
29	n	n	n	n	r	n	n
30		⌊r	⌊r		s⌊r	(s+1)⌊r	(s+1)⌊r
31			⌊r	⌊n	⌊r	⌊r	⌊r
	TO Ⓑ WITH r'=1 s'=1	TO Ⓒ	TO Ⓓ IF r=n TO Ⓔ IF r<n		TO Ⓖ	TO Ⓑ WITH r'=r+1 IF s≥r TO Ⓔ WITH r'=s+1 IF s<r.	TO Ⓕ

Fig. 2: Turing's original annotations

Turing chose to annotate a flow chart of a program with expressions that related the values which could be contained in particular storage locations. A copy of the flowchart for his chosen program is given in Fig. 1 with the expressions relating values in Fig. 2.

What was the specification of Turing's example program? Well, as printed the paper states that the task is to assign to the location 31 (given the helpful symbolic name v) the value of location 29 (n) – a single assignment in a modern programming language would achieve this and even in the early machines would probably need only a load followed by a store command! The confusion results from the fact that, however brilliant, Turing was a rather sloppy proofreader. He had elected to write factorial n as $\underline{|n}$ but then omitted to handwrite in the box symbol anywhere in the text;[3] furthermore, his choice of identifiers v, r etc. was less than optimal since his writing led to 10 further uncorrected transcription errors. Thus, even when Turing (1949) came to light, it was hard to decipher. A careful decoding of the figures is contained in Morris and Jones (1984) together with comments that relate the method to later research.

[3] The more modern standard notation for factorial n is $n!$, which could have been typed.

The factorial program – whose specification should have said 'set v to $n!$' – can be seen from the flowchart in Fig. 1 to use a doubly nested loop: the inner loop computes multiplication by successive addition (the early computers had a tiny repertoire of instructions), while the outer loop ensures that the numbers from 1 to n are multiplied into v. To look at just one annotation: at point E in the flowchart, u (line 30) is supposed to contain the product of s and $r!$; and v (line 31) contains $r!$.

The title of Turing (1949) talks of a 'large routine' and the use of factorial as an example might surprise a modern reader. Two points are worth making. This example has in fact often been used by (far) later authors to experiment with ways of presenting arguments. It is possible to argue that it shows even more prescience on Turing's part to have seen that ways of decomposing proofs that programs satisfy their specifications would be required for ever larger programs.

Before looking at what might have happened had Turing's 1949 paper been appreciated earlier, there is one illuminating link to his groundbreaking 1936 paper that is worth mentioning. In Turing (1936), the problem that Turing shows could not be solved *in general* was that no one could ever write a debugging program that could decide whether any program provided as input would always terminate. Anyone who has sat in front of a PC knows what it means for a program to go into a loop – the whole machine locks up and the only thing to be done is to reboot the machine. Turing is careful to argue that his specific program in Fig. 1 will always terminate and he uses a particularly intriguing approach using ordinal numbers (but then concedes that 2^{40} actually suffices).

4. What came after 1949?

The next significant papers on reasoning about programs came 17 long years after Turing spoke at the Cambridge event. Aad van Wijngaarden recognised the full impact of a parenthetical remark above: computers cannot contain ideal mathematical integers because the word length (or indeed the overall store size) is necessarily restricted. In fact, in most machines, adding one to the largest positive number that can be stored might well yield a negative number. In the paper of van Wijngaarden (1966), a series of axioms are offered whose use makes it safe to reason about such finite representations.

The step that is far closer to Turing's idea came a year later when Bob Floyd (1967) was published. This is one of those papers that is widely cited but read rather less often. It is a gem. Floyd proposes a way of annotating programs to reason about their correctness and termination. Like Turing, Floyd annotates flow charts (Floyd even uses the factorial example) but his assertions are more general in that they are expressions in the logic known as 'first-order predicate calculus'. Floyd goes much further and he is using a modern programming language notation which makes programs clearer and more succinct than early machine codes. He provides precise rules for when logical assertions match the corresponding program statements. Floyd even described what later became known as 'healthiness conditions' that capture intuitive but important properties for the relationship that any command can establish between the assertions preceding and following the statement.

A major step soon followed. Tony Hoare had been searching for a way to document the meaning (semantics) of a programming language. After several attempts to break away from semantic description styles that he found too closely linked to implementation details, he received a copy of Floyd's paper and immediately saw that this provided the key to an 'axiomatic semantic' style. Hoare's paper is probably one of the most influential in reasoning about programs. Not only is it widely cited but it is also read by most researchers who reference it. The paper by Hoare (1969) makes a major bold step. Superficially, there is a shift from annotated flowcharts to triples containing a command surrounded by its pre-and post-assertions; but Hoare's shift opens up a far more mathematical view of how to combine steps of an argument. Hoare's (1969) paper also integrates the axioms that van Wijngaarden had offered for restricted computer numbers.

This excursion into what followed Turing's 1949 paper has identified enough hills to triangulate influences and missed opportunities; a fuller account of the development of program verification can be found in Jones (2003) – in particular, it (describes other work omitted here to save space and) describes the importance of data abstraction and the move to use Floyd/Hoare axioms in the *development* of program designs.

5. Assessment

The temptation to ask what might have been? is irresistible. Neither Floyd nor Hoare was aware of Turing's 1949 paper until long after they made their own contributions at the end of the 1960s. This author's view is that a wider awareness of Turing's early paper could have accelerated the research on reasoning formally about programs by at least a decade. But there is an interesting piece of counter evidence. Certainly neither Hoare nor Floyd were at any scientific conference in the late 1940s because both were too young, but a check of the list of people at the Cambridge meeting reveals that Aad van Wijngaarden attended the overall meeting. This does not guarantee that he heard Turing's talk – but let us assume he did. The clear inference is that by the time he thought about the problems of reasoning about computer arithmetic, he had the pieces in his hands to synthesise what Floyd and Hoare did.[4] Perhaps science had to wait for minds as profound as Floyd's and Hoare's to recreate and step beyond Turing's ideas.

It is also reasonable to look for other links. As early as 1947, the massive report Goldstine and von Neumann (1947) by von Neumann and Goldstine had some notion of annotating programs. It has to be said that their notation is unintuitive, but it is clear from Hodges' biography of Turing that the latter did visit von Neumann during the second world war. Whether von Neumann influenced Turing's thought or *vice versa* is not known.

Another tantalising historical hint is Floyd's generous acknowledgement in is his paper:

> These modes of proof of correctness and termination are not original; they are based on ideas of Perlis and Gorn, and may have made their earliest appearance in an unpublished paper by Gorn.

Could this have been a development of von Neumann's ideas? There is still detective work to be done but this author can offer two points of warning: time should not be lost and the results will not be universally popular. As evidence to the first point, the announcement of Saul Gorn's death reached this author just as he decided to contact Gorn with regard to the above question. A small story about the after affects of publishing Morris and Jones (1984) illustrates the second point. No lesser figure than Maurice Wilkes wrote in a private letter 'I regard Floyd's discovery of loop invariants as one of very few really significant advances in programming science ... I would not like the idea to get around that Floyd's great advance had been anticipated by Turing'. Whilst thinking about a reply, relief followed from Lockwood's perfect 'We have no interest at all in making or marring any reputations (not that either Turing's or Floyd's could be in any danger from us)'.

Turing's paper was clearly a 'pre-echo' of what came many years after his death; it had the potential to accelerate a key avenue of research that has grown for half a century and is now having real impact on industrial software development.

---

[4] The current author knew van Wijngaarden reasonably well but only studied Turing's paper (and noticed the attendance list of the conference) after his death; subsequent questions to some who knew van Wijngaarden even better have uncovered no mention he ever made about Turing's talk.

References

Copeland, B.J., 2011. The Manchester computer: A revised history — part 2: The Baby computer. IEEE Ann. Hist. Comput. 33(1), 22–37.

Floyd, R.W., 1967. Assigning meanings to programs. In: Proc. Symp. in Applied Mathematics, vol. 19: Mathematical Aspects of Computer Science, American Mathematical Society, pp. 19–32

Goldstine, H.H., von Neumann, J., 1947. Planning and coding of problems for an electronic computing instrument. Part II, vol. 1 of a Report prepared for U.S. Army Ord. Dept.; republished as pages 80–151 of Taub (1963).

Hoare, C.A.R., 1969. An axiomatic basis for computer programming. Commun. ACM. 12, 576–580, 583.

Jones, C.B., 2003. The early search for tractable ways of reasoning about programs. IEEE Ann. Hist. Comput. 25(2), 26–49.

Morris, F.L., Jones, C.B. 1984. An early program proof by Alan Turing. Ann. Hist. Comput. 6(2), 139–143.

Taub, A.H., 1936. (Ed.) John von Neumann: Collected Works, volume V: Design of Computers, Theory of Automata and Numerical Analysis. Pergamon Press, Oxford.

Turing, A.M., 1937. On computable numbers, with an application to the Entscheidungsproblem. Proceedings of the London Mathematical Society, Series 2, 42, 230–265. Correction published: ibid, 43:544–546, 1937.

Turing, A.M., 1949. Checking a large routine. In: Report of a Conference on High Speed Automatic Calculating Machines, University Mathematical Laboratory, Cambridge, pp. 67–69.

van Wijngaarden, A., 1966. Numerical analysis as an independent science. BIT, 6, 66–81.

Friday, 24th June.

Checking a large routine. by Dr. A. Turing

How can one check a routine in the sense of making sure that it is right?

In order that the man who checks may not have too difficult a task the programmer should make a number of definite assertions which can be checked individually, and from which the correctness of the whole programme easily follows.

Consider the analogy of checking an addition. If it is given as:

$$\begin{array}{r} 1374 \\ 5906 \\ 6719 \\ 4337 \\ 7768 \\ \hline 26104 \end{array}$$

one must check the whole at one sitting, because of the carries. But if the totals for the various columns are given, as below:

$$\begin{array}{r} 1374 \\ 5906 \\ 6719 \\ 4337 \\ 7768 \\ \hline 3974 \\ 2213 \\ \hline 26104 \end{array}$$

the checker's work is much easier being split up into the checking of the various assertions $3+9+7+3+7=29$ etc. and the small addition

$$\begin{array}{r} 3794 \\ 2213 \\ \hline 26104 \end{array}$$

This principle can be applied to the process of checking a large routine but we will illustrate the method by means of a small routine viz. one to obtain n without the use of a multiplier, multiplication being carried out by repeated addition.

At a typical moment of the process we have recorded r and s r for some r, s. We can change s r to $(s+1)$ r by addition of r. When $s = r+1$ we can change r to $r+1$ by a transfer. Unfortunately there is no coding system sufficiently generally known to justify giving the routine for this process in full, but the flow diagram given in Fig. 1 will be sufficient for illustration.

Each 'box of the flow diagram represents a straight sequence of instructions without changes of control. The following convention is used:

(i) a dashed letter indicates the value at the end of the process represented by the box:
(ii) an undashed letter represents the initial value of a quantity.

One cannot equate similar letters appearing in different boxes, but it is intended that the following identifications be valid throughout

s	content	of	line	27	of	store
r	”	”	”	28	”	”
n	”	”	”	29	”	”
u	”	”	”	30	”	”
v	”	”	”	31	”	”

It is also intended that u be s r or something of the sort e.g. it might be $(s+1)$ r or s $r-1$ but not e.g. s^2+r^2.

In order to assist the checker, the programmer should make assertions about the various states that the machine can reach. These assertions may be tabulated as in Fig. 2. Assertions are only made for the states when certain particular quantities are in control, corresponding to the ringed letters in the flow diagram. One column of the table is used for each such situation of the control. Other quantities are also needed to specify the condition of the machine completely: in our case it is sufficient to give r and s. The upper part of the table gives the various contents of the store lines in the various conditions of the machine, and restrictions on the quantities s, r (which we may call inductive variables). The lower part tells us which of the conditions will be the next to occur.

The checker has to verify that the columns corresponding to the initial condition and the stopped condition agree with the claims that are made for the routine as a whole. In this case the claim is that if we start with control in condition D and with n in line 29 we shall find a quantity in line 31 when the machine stops which is r (provided this is less than 240, but this condition has been ignored).

He has also to verify that each of the assertions in the lower half of the table is correct. In doing this the columns may be taken in any order and quite independently. Thus for column B the checker would argue. “From the flow diagram we see that after B the box $v^1 = u$ applies. From the upper part of the column for B we have $u = r$. Hence $v^1 = r$ i.e. the entry for v i.e. for line 31 in C should be r. The other entries are the same as in B”.

Finally the checker has to verify that the process comes to an end. Here again he should be assisted by the programmer giving a further definite assertion to be verified. This may take the form of a quantity which is asserted to decrease continually and vanish when thè machine stops. To the pure mathematician it is natural to give an ordinal number. In this problem the ordinal might be $(n-r)\,w^2 + (r-s)\,w + k$. A less highbrow form of the same thing would be to give the integer $2^{80}(n-r) + 2^{40}(r-s) + k$. Taking the latter case and the step from B to C there would be a decrease from $2^{80}\ (n-r) + 2^{40}(r-s) + 5$ to $2^{80}(n-v) + 2^{40}(r-s) + 4$. In the step from F to B there is a decrease from $2^{80}\ (n-r) + 2^{40}(r-s) + 1$ to $2^{80}(n-r1) + 2^{40}(r+1-s) + 5$.

In the course of checking that the process comes to an end the time involved may also be estimated by arranging that the decreasing quantity represents an upper bound to the time till the machine stops.

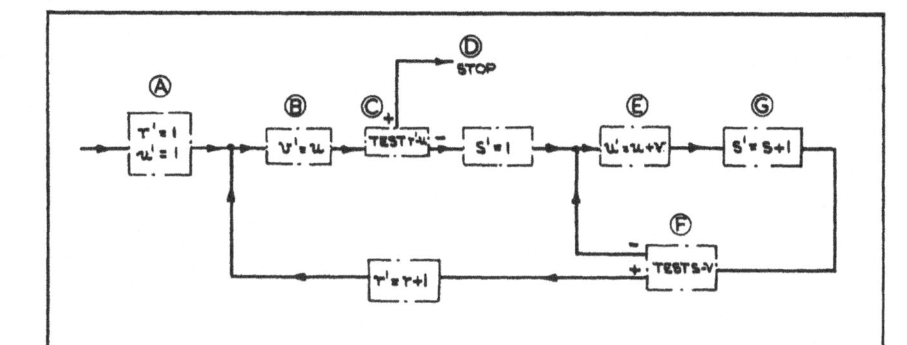

FIG. 1

STORAGE LOCATION	(INITIAL) Ⓐ k = 6	Ⓑ k = 5	Ⓒ k = 4	(STOP) Ⓓ k = 0	Ⓔ k = 3	Ⓕ k = 1	Ⓖ k = 2
27					s	$s+1$	s
28		r	r		r	r	r
29	n	n	n	n	r	n	n
30		$\underline{\lvert r}$	$\underline{\lvert r}$		$s\underline{\lvert r}$	$(s+1)\underline{\lvert r}$	$(s+1)\underline{\lvert r}$
31			$\underline{\lvert r}$	$\underline{\lvert n}$	$\underline{\lvert r}$	$\underline{\lvert r}$	$\underline{\lvert r}$
	TO Ⓑ WITH $r' = 1$ $s' = 1$	TO Ⓒ	TO Ⓓ IF $r = n$ TO Ⓔ IF $r < n$		TO Ⓖ	TO Ⓑ WITH $r' = r+1$ IF $s \geq r$ TO Ⓔ WITH $r' = s+1$ IF $s < r$.	TO Ⓕ

FIG. 2

ISSUE 1 4.8.49.	T.R.E. M.O.S. DIAG. No. RTR 11/5790.

Excerpt from: Programmer's Handbook for the Manchester Electronic Computer Mark II

(Printed ca. March, 1951.)

Local Programming Methods and Conventions

(Paper read at the Inaugural Conference for the Manchester University Computer, July 1951.)

Toby Howard describes —

TURING'S CONTRIBUTIONS TO THE EARLY MANCHESTER COMPUTERS

1. Introduction

In 1948, the world's first stored program digital computer was designed and built at The University of Manchester. It was a volatile assembly of cathode ray tubes, vacuum tubes (thermionic valves), resistors and capacitors mounted on huge metal frames. Its official name was The Small-Scale Experimental Machine, but it soon became known as the Baby.

Much has been written about the extraordinary early days in Manchester. My intention in this short commentary is to focus on the specific contributions made by Alan Turing, and give a flavour of what computer programming was like in those early days. I'll look at two of Turing's Manchester publications, both from 1951: his talk 'Local Programming Methods and Conventions' at the Inaugural Computing Conference in Manchester, and his 'Programmers' Handbook for the Manchester Electronic Computer Mark II (what we now call the Ferranti Mark 1)'. Both of these writings contain very low-level detail, much of which is of interest only to specialist historians of computing. My aim is to place Turing's Manchester work in context, and paint the bigger picture. For readers wishing to follow up and discover more detail, I suggest some ideas for further reading.

2. The Baby

A CRT store capable of holding 2048 bits had been demonstrated in December 1947, and the Baby was built to find out if the store could function reliably as the memory of a practical computing machine doing realistic work.

In a University building on Oxford Road, which still stands today, Freddie Williams, Tom Kilburn and Geoff Tootill gave the machine shape. At the time there was little distinction between understanding how the machine worked, and how to make the machine do something useful. Programming as a discipline, like the computer itself, was in its infancy.

Although it worked, at least for short periods, the Baby was sometimes unstable. It was an experiment, after all, and it faltered when the trams thundered down Oxford Road, their electrical systems upsetting the Baby's delicate electronics.

Alas, no photographs of the Baby survive, but its successor, the Manchester Mark 1, from 1949, gives an idea of the 'look' of the machine:

Fig. 1: The Manchester Mark 1 (1949).

Viewed from today's technical perspective, the design of the Baby is familiar, although on a vastly reduced scale. Like many modern computers, it manipulated its data in chunks of 32-bit 'words', and represented signed integers using the '2's complement' convention. It had a Random Access Memory (RAM) of 32 words, making a total of 1024 bits, or 0.128 kilobytes (kb). A modern laptop typically has 4 GB of RAM, about 32 million times more than the Baby.

Whilst modern computers have over 100 different types of instruction, the Baby had just 7. Three things could be manipulated: the 'accumulator', A, where the result of the last computation was stored; the contents of the word in RAM with address S; and the memory address of the current program instruction being executed, CI. The instruction set was as follows (the notation $P \leftarrow Q$ means 'P is given the value of Q'):

1. $A \leftarrow -S$
2. $A \leftarrow A - S$
3. $S \leftarrow A$
4. IF ($A < 0$) THEN ($CI \leftarrow CI + 1$)
5. $CI \leftarrow S$
6. $CI \leftarrow CI + S$
7. HALT

Each of the Baby's storage units was a CRT, one for each of A, CI and the RAM. A fourth CRT, called the 'display tube', could be assigned to view the contents of any of the other tubes. The photograph below (taken from the Ferranti Mark 1) shows how the CRT RAM display looked (bright dots are 1s, dim dots are 0s)[1].

[1] See `http://www.computer50.org/kgill/williams/display.html`.

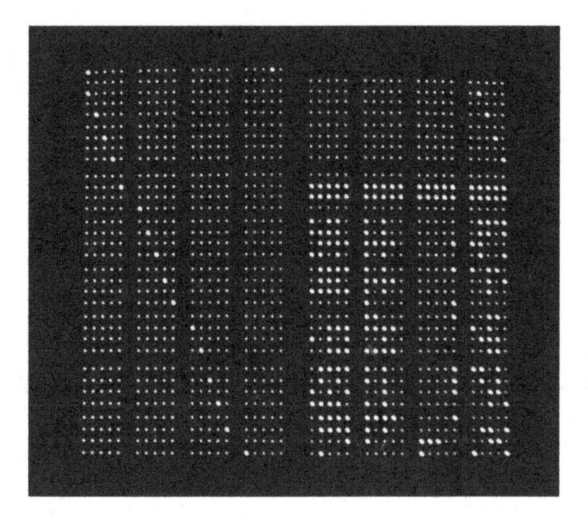

Fig. 2: CRT RAM display from Ferranti Mark 1.

The Baby ran its first program on 21 June 1948. Written by Tom Kilburn, the program took an integer N, and computed its highest factor – the largest integer that would divide into N with no remainder. After a few weeks the program was run on much larger numbers, and as reported in *Nature*, the program took 52 minutes to find the highest factor of $N = 218$, which involved 2.1 million instruction executions, and 3.5 million store accesses (Williams and Kilburn, 1948).

At this time, Turing was at Cambridge, on sabbatical from his post at the National Physical Laboratory, where he had produced a design for an as-yet unbuilt computer called the Automatic Computing Engine (ACE). Its memory would be a mercury acoustic delay line – a curious technology based on a tube several feet long containing liquid mercury. A voltage representing a binary 0 or 1 was converted to an acoustic "ping", and sent down the tube as a sound wave. The ping was detected at the other end of the tube, converted back to a 0 or 1 voltage, and then fed back to the start of the tube to repeat the process. With appropriate timing circuitry, this acoustic loop could store hundreds of bits of binary data.

The building of ACE had stalled, and in May 1948 Turing, frustrated, accepted the offer of a post at The University of Manchester. In July 1948, on hearing that the Baby was working, he wrote a program for long division and sent it up to Manchester. He subsequently sent another program for factorising numbers. Turing arrived in Manchester in October 1948, appointed to the Mathematics Department as Deputy Director of the Royal Society Computing Laboratory.

3. The early Manchester machines

The pioneering work at Manchester attracted the attention of the Ministry of Supply, who saw the potential for Britain to take the lead in the development of computing, and in 1948 commissioned Ferranti Ltd to collaborate with the University. The project now became focused on developing a general purpose computing machine. By the autumn of 1949 the Manchester Mark 1 had been completed, expanding on the prototype ideas of the Baby, by including more memory, and what was called a two-level' store comprising the CRT memory used in the Baby, and a magnetic drum which was slower to read and write, but much larger in capacity. This was the first time a combination of small/fast and larger/slower stores had been used, and it remains the system of fast RAM/slower big disk in most computers today.

Turing made several contributions to the development of the Manchester Mark 1. In the summer of 1949, working with Dai Edwards, he attached a paper tape reader to the machine, so programmers need no longer key in their programs manually on the input switches. He also wrote what we would today call the 'driver software' for the tape reader. Working with Geoff Tootill, Turing designed hardware for generating Random Numbers. In October 1949 Cicely Popplewell was appointed Turing's assistant, and shortly afterwards he took on Audrey Bates, the first of two M.Sc. students he was to supervise. The second, in 1953, was Bernard Richards, working on Morphogenesis.

4. The Manchester University Inaugural Conference

"Since the instructions are held in the machine in a form intelligible to the programmer it is very easy to follow what is going on by looking at the monitor tubes"
– Turing at the Inaugural Conference

The University of Manchester hosted a large conference (9–15 July 1951) to demonstrate the production Ferranti Mark 1, which attracted 170 international delegates. It had been 3 years since the Baby ran its first program.

Turing (1951) gave a short talk entitled 'Local Programming Methods and Conventions', in which he described a scheme for writing down the instructions used to program the Mark 1, in a convenient way for programmers. Discussing the use of a code to represent binary numbers, he said: 'The choice made at Manchester was the scale of 32. It is probable that a scale of 8 or 16 would be more convenient'. He was right. In later years, the scales of 8 (octal) and 16 (hexadecimal) would become the norm.

Today the usual convention is to write binary numbers with the most significant (or 'highest') bit on the left, so we would write decimal 5 as binary 00101. But in Turing's scheme, the most significant bit was on the right, so decimal 5 would be binary 10100. Turing had used 5-hole teleprinter tape at Bletchley Park, and adopted that encoding for representing the bit patterns in the Mark 1's instructions and data, where a teleprinter character would stand for a group of 5 bits, according to the International Telegraphy Alphabet. Using the example above, binary 10100 would be represented by the character with that code, in this case 'S'. The full table of bit/letter assignments was as follows:

0	00000	/	8	00010	$\frac{1}{2}$	16	00001	T	24	00011	O
1	10000	E	9	10010	D	17	10001	Z	25	10011	B
2	01000	@	10	01010	R	18	01001	L	26	01011	G
3	11000	A	11	11010	J	19	11001	W	27	11011	"
4	00100	:	12	00110	N	20	00101	H	28	00111	M
5	10100	S	13	10110	F	21	10101	Y	29	10111	X
6	01100	I	14	01110	C	22	01101	P	30	01111	V
7	11100	U	15	11110	K	23	11101	Q	31	11111	£

The instructions in the Mark 1 were 20 bits in length, so to represent the instruction '01100010100000101110' the programmer would split it into four groups of 5 bits, and use the teleprinter code, to give 'IRTC'. These four characters could then be input into the machine using the console switches, or paper tape.

Turing believed that programmers should follow the execution of their programs by watching the pattern of dots on the CRT screen. He called this 'peeping', and argued, somewhat eccentrically, that programmers should practise the art until they could debug their programs simply by watching for unexpected behaviour among the dancing bright and dim spots.

Turing's conference talk is all about notational conventions, and is, frankly, rather dull. He is discussing a low-level encoding that is relevant only to the Manchester machines. But despite the parochial detail, Turing was addressing a universal problem: how we can use symbols to help manage complexity.

5. The programmers' handbook for the Ferranti Mark 1

"If it is desired to give a definition of programming, one might say that it is an activity by which a digital computer is made to do a man's will, by expressing this will suitably on punched tapes, or whatever other input medium is accepted by the machine." – Programmers' Handbook, p. 50.

Following the Baby came the improved version, known as the Manchester University Mark 1 or MADM (Manchester Digital Machine). The subsequent fully engineered version of MADM was produced by Ferranti and became known as the Ferranti Mark 1. During this period of intense activity, Turing contributed to the addition of new instructions. The production version of the Ferranti Mark 1, the world's first commercially available general purpose digital computer, was delivered to the University in February 1951. It was for this machine that Turing wrote the programming handbook (confusingly for us, the book is actually entitled 'Programmers' Handbook for the Manchester Electronic Computer Mark II', that being Turing's name for the machine we now refer to as the Ferranti Mark 1) (Facsimile Scan , 1951; Thau, 2000). The first two models to be manufactured were sold to the Universities of Manchester and Toronto. Subsequently the design was slightly altered, and Ferranti sold seven more of the 'Mark 1 star' version of the machine.

An August 1952 Ferranti sales brochure for 'The Manchester Universal Electronic Computer' (i.e., the Ferranti Mark 1) trumpeted the selling points of the machine: '[it] can perform all the operations of arithmetic exceedingly rapidly', '[it] can remember a great many numbers' and '[it] can make decisions'. The machine comprised 'about 4000 valves, 2500 capacitors, 15,000 resistors, 100,000 soldered joints and 6 miles of wire' (Ferranti Sales Brochure, 1952).

What was it actually like to write programs for the Mark 1? First, the programmer had to decide how to organise the program so that it could be correctly loaded into the computer. Turing, together with Cicely Popplewell, devised a system called 'Scheme A', which was a set of conventions and library subroutines, for moving the data used by the program between the fast CRT store and the relatively slow magnetic drum store. Scheme A also dictated conventions for calling subroutines, and inputting programs. Today we would refer to this as 'system software' – providing a standard infrastructure to be used by programmers.

Turing's 90-page handbook is a guide for programmers written by a mathematician. He alternates between explanatory passages, and dense presentations of technical detail. Programming is described at the raw, machine code level – at the time there was no concept of a 'high-level' language. Below, for example, is a program from the handbook[2] that multiplies two numbers together by repeated additions:

[2] See http://www.alanturing.net/turing_archive/archive/m/m01/M01-015.html

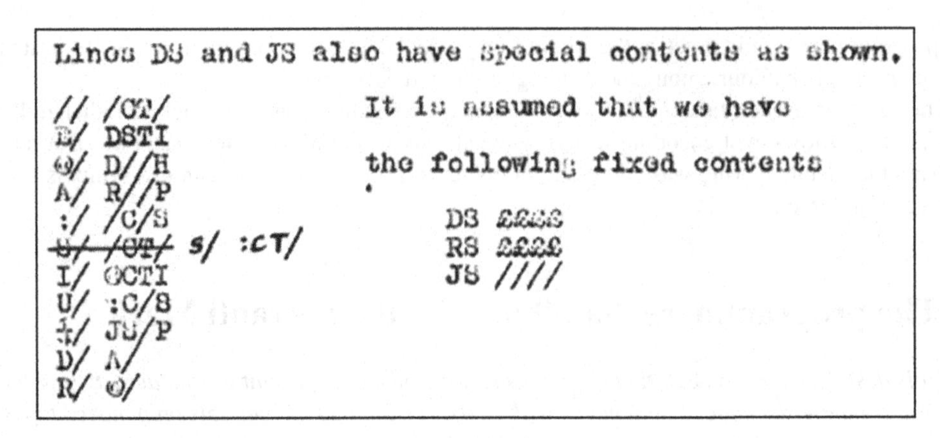

```
Lines DS and JS also have special contents as shown.

// /CT/                    It is assumed that we have
E/ DSTI
@/ D//H                    the following fixed contents
A/ R//P                    .
:/ /C/S                         DS ££££
S/ /CT/ S/ :CT/                 RS ££££
I/ @CTI                         JS ////
U/ :C/S
½/ JS/P
D/ A/
R/ @/
```

Fig. 3: A program from Turing's "Programmer's Handbook".

Here we can see the binary program instructions represented using the base-32 teleprinter code described above. There are no mnemonics, as used in assembly-level languages. Instead, the programmer really is working in the binary of the machine's instruction set. To read and write even moderately complex programs would take considerable effort.

Programmers today would probably find this scheme impossibly slow and laborious, but such comparisons are rarely helpful. At the time it was simply the only way to program, and Ferranti employee Olaf Chedzoy, for example, has written of his fond memories of programming the Mark 1 (Chedzoy, 1952).

6. Conclusions

Turing's hands-on contributions to the development of the Manchester computers lasted from October 1948 to October 1951. For the next, and sadly final, two and a half years of his life, Turing was programming the Manchester machine to support his biological research. He had become, as we would say today, an 'applications programmer'.

Campbell-Kelly (1980) takes the view that Turing's work on the Mark 1, when compared to such achievements as the Turing Machine (Turing, 1936), and his design of ACE, was 'not an example of his finest work'. His biographer Andrew Hodges makes a plausible case for a number of Turing's novel ideas which were never properly followed up, such as self-organising systems and artificial life, chaos theory, mathematically proving programs correct, developing higher level programming languages, and the lambda calculus which inspired McCarthy's LISP in 1958 (Hodges, 1992; 2004).

Such observations, however, can never detract from the fact that Alan Turing was an extraordinary man – a unique and brilliant thinker. And in 2012 we celebrate his life and achievements.

7. Suggestions for further reading

A primary source for material on the Baby and the early Manchester scene is the Computer50 Web site (Napper, 1998). Simon Lavington's definitive 'A History of Manchester Computers' (Lavington, 1998) is essential reading, and beautifully illustrated with archive photographs. For a detailed coverage of Mark 1 programming see Martin Campbell-Kelly's 'Programming the Mark 1: Early Programming Activity at the University of Manchester' (Campbell-Kelly, 1980). See also the fascinating papers by Chris Burton, Brian Napper and Frank Sumner in 'The First Computers: History and Architecture' (Rojas, 2002), and Mary Croarken's 'The beginnings of the Manchester computer phenomenon: people and influences', which traces the pre-history of the Baby (Croarken, 1993).

8. Acknowledgements

I am indebted to my colleagues at The University of Manchester School of Computer Science for their insights and sense of history. In particular I thank, for their helpful comments on early drafts of this article, Professor Simon Lavington and Professor Steve Furber, and also Chris Burton, who oversaw the creation of an authentic working replica of the Baby for its 50th anniversary in 1998 – now a permanent exhibit at the Manchester Museum of Science and Industry. I give special thanks to Professor Bernard Richards, who generously shared with me his memories of working with Alan Turing. I dedicate this article, in memoriam, to Professor Hilary Kahn and Dr Brian Napper, both passionate custodians of Manchester's unique computing heritage.

References

Campbell-Kelly, M., 1980. Programming the Mark 1: Early programming activity at the University of Manchester. Ann. Hist. Comput. 2(2), 130–168, ISSN 0164-1239.

Chedzoy, O., 1952. Starting work on the world's first electronic computer: 1952 memories. uk.fujitsu.com/pensioner/localData/pdf/public/olaf_art2.pdf.

Croarken, M., 1993. The beginnings of the Manchester computer phenomenon: people and influences. IEEE Ann. Hist. Comput. 15(3), 9–16.

Facsimile scan, 1951. Facsimile scan of the 1st Edition of the 1951 Programming manual at the Turing Digital Archive. www.turingarchive.org/browse.php/B/32.

Ferranti Sales Brochure, 1952. www.computer50.org/kgill/mark1/sale.html.

Hodges, A., 1992. Alan Turing: the Enigma, Vintage, London ISBN 978-0099116417.

Hodges, A., 2004. www.turing.org.uk/philosophy/lausanne1.html and a revised version in Alan Turing: Life and Legacy of a Great Thinker, in: C. Teuscher (Ed.), Springer, Berlin, Heidelberg, ISBN 978-3540200208.

Lavington, S.H., 1998. A History of Manchester Computers, second edn. British Computer Society, London, ISBN 0-902505-01-8.

Napper, R.B.E., 1998. www.computer50.org.

Rojas, P. (Ed.), 2002. The First Computers: History and Architectures, MIT Press, Cambridge, MA, ISBN 9780262681377.

Thau, R.D., 2000. Transcription of Alan Turing's Manual for the Ferranti Mark1, www.computer50.org/kgill/mark1/RobertTau/turing.pdf.

Turing, A.M., 1936. On computable numbers, with an application to the Entscheidungsproblem. In: Proceedings of the London Mathematical Society, vol. 43. pp. 230–265. Available from www.scribd.com/doc/2937039/Alan-M-Turing-On-Computable-Numbers.

Turing, A.M., 1951. Local programming methods and conventions. In: Manchester University Computer: Inaugural Conference, , University of Manchester/Ferranti Ltd. 9–12 July 1951, p. 12.

Williams, F.C., Kilburn, T., 1948. Electronic digital computers. Nature 162(487), reprinted at www.computer50.org/kgill/mark1/natletter.html.

Excerpt from: Programmer's Handbook for the Manchester Electronic Computer Mark II

Programming principles

Programming is a skill best acquired by practice and example rather than from books. The remarks given here are therefore quite inadequate.

If it is desired to give a definition of programming, one might say that it is an activity by which a digital computer is made to do a man's will, by expressing this will suitably on punched tapes, or whatever other input medium is accepted by the machine. This is normally achieved by working up from relatively simple requirements to more complex ones. Thus for instance if it is desired to do a Fourier analysis on the machine it would be as well to arrange first that one can calculate cosines on it. For each process that one wishes the machine to be able to carry out one constructs a 'routine' for the process. This consists mainly of a set of instructions, which are obeyed by the machine in the process. It is not usual however to reckon all the instructions obeyed in the process as part of the 'routine'. Many of them will belong to other routines, previously constructed and carrying out simpler processes. Thus for instance the Fourier analysis process would involve obeying instructions in the routine for forming cosines as well as ones in the analysis routine proper. In a case like this the cosine's routine is described as a 'subroutine' of the analysis routine. The subroutines of any routine may themselves have subroutines. This is like the case of the bigger and lesser fleas. I am not sure of the exact meaning the poet attached to the phrase 'and so on ad infinitum', but am inclined to think that he meant there was no limit that one could assign to a parasitic chain of fleas, rather than that he believed in infinitely long chains. This certainly is the case with subroutines. One always eventually comes down to a routine without subroutines.

What is normally required of a routine is that a certain function of the state of the machine shall be calculated and stored in a given place, the majority of the content of the store being unaffected by the process, and the routine not being dependent on this part having any particular content. It is usual also for the other part of the store to be divided into a part which is altered in the process but not greatly restricted as to its original content, and a part which is unaltered in its content throughout, and such that the correct working of the routine depends on this part having a particular content. The former can be described as 'the working space for the routine' and the latter as 'the space occupied by the routine'.

Applying this to the Mark II machine, the routines usually 'occupy' various tracks of the magnetic store. The working space includes all the electronic store, with the exception of PERM[1], which can equally well be reckoned as working space which is never really altered, or as a part common to all routines. The first two pages of the electronic store are also somewhat exceptional, if normal conventions are used, as they are only altered when one is copying new routines from the magnetic store onto them.

[1] System code that was permanently kept in the electronic store; the space occupied was not therefore generally accessible to the user.

These definitions do not really help the beginner. Something more specific is needed. I describe below the principal steps which I use in programming, in the hope they will be of some small assistance.

i) **Make a plan.** This rather baffling piece of advice is often offered in identical words to the beginner in chess. Likewise the writer of a short story is advised to 'think of a plot' or an inventor to 'have an idea'. These things are not the kind that we try to make rules about. In this case however some assistance can be given, by describing the decisions that go to make up the plan.

a) If it is a genuine numerical computation that is involved (rather than, e.g. the solution of a puzzle) one must decide what mathematical formulae are to be used. For example if one were calculating the Bessel function $J_0(x)$ one would have, amongst others, the alternatives of using the power series in x, various other power series with other origins, interpolation from a table, various definite integrals, integration of the differential equation by small arcs and asymptotic formulae. It may be necessary to give some small consideration to a number of the alternative methods.

b) Some idea should be formed as to the supply and demand of the various factors involved. A balance must always be struck between the following incompatible desires:

To carry the process through as fast as possible
To use as little storage space as possible
To finish the programming as quickly as possible
To achieve the maximum possible accuracy

We may express this by saying that machine time, storage space, programmer's time and inaccuracy of results all cost something. The plan should take this into account to some extent, though a true optimum cannot be achieved except by chance, since programmer's time is involved, so that a determination of the optimum would defeat its own ends. The 'state of the market' for these economic factors will vary greatly from problem to problem. For instance there will be an enormous proportion of problems (40% perhaps) where there is no question of using the whole storage capacity of the machine, so that space is almost free. With other types of problem one could easily use ten million digits of storage and still not be satisfied.

The space shortage applies mainly to working space rather than to space occupied by the routines. Since these usually have to be written down by someone this in itself has a limiting effect. Speed will usually be a factor worth consideration, though there are many 'fiddling' jobs where it is almost irrelevant. For instance the calculation of tabular values for functions, which are to be stored in the machine and later used for interpolation, would usually be in this class. Programmer's time will usually be the main factor in special jobs, but is relatively unimportant in fundamental routines which are used in most jobs. Accuracy may compete with machine time, e.g. over such questions as the number of terms to be taken in a series, and with space over the question as to whether 20 or 40 digits of a number should be stored.

c) The available storage space must be apportioned to various duties. This will apply both to magnetic and electronic storage. The magnetic storage will probably be mainly either working space or unused. It should be possible to estimate the space occupied by instructions to within say two tracks, for a large part will probably be previously constructed programmes, occupying a known number of tracks. The quantities to be held in the working space should if possible be arranged in packets which it is convenient to use all at once, and which can be packed into a track or a half-track or quarter-track. For instance when multiplying matrices it might be convenient to partition the matrices into four-rowed or eight-rowed square matrices and keep each either in a track or a quarter-track. The apportionment of the electronic store is partly ruled by the conventions we have introduced, but

there is still a good deal of freedom, e.g. if eight pages are available then pages 4, 5, 6 can be used for systematic working space and may be used for various different purposes that require systematic working space.
The beginner will do well to ask for advice concerning plans. Bad plans lead to programmes being thrown away, wasting valuable programmer's time.

d) If questions of time are at all critical the 'plan' should include a little detailed programming, i.e. the writing down of a few instructions. It should be fairly evident which operations are likely to consume most of the time, and help decide whether the plan as a whole is satisfactory. Very often the 'omission of counting method' should be applied.
e) If one cannot think of any way, good or bad, for doing a job, it is a good thing to try and think how one would do it oneself with pencil and paper. If one can think of such a method it can usually be translated into a method which could be applied to the machine.

(ii) Break the problem down. This in effect means to decide which parts of the problem should be made into definite subroutines. The purpose of this is partly to make the problem easier to think about, the remaining work consisting of a number of 'limited objective' problems. Another reason for breaking down is to facilitate the solution of other problems by the provision of useful subroutines. For instance if the problem on hand were the calculation of Bessel functions and it had already been decided to use the differential equation, it might be wise to make and use a subroutine for the solution of linear second order differential equations. This subroutine would in general be used in connection with other subroutines which calculate the coefficients of the equation.

(iii) Do the programming of the new subroutines. It is better to do the programming of the subroutines before that of the main routine, because there will be a number of details which will only be known after the subroutine has been made, e.g. scale factors applied to the results, number of pages occupied by the subroutines, etc. It also frequently happens in the making of the subroutine that some relatively small change in its proposed properties is desirable. Changes of these details may put the main routine seriously out if it were made first. There is a danger that this procedure may result in one's 'not seeing the wood for the trees', but this should not happen if the original plan was well thought out. The programming of each subroutine can itself be divided into parts:

a) As with programming a whole problem a plan is needed for a subroutine. A convenient aid in this is the 'block schematic diagram'. This consists of a number of operations described in English (or any private notation that the programmer prefers) and joined by arrows. Two arrows may leave a point where a test occurs, or more if a variable control transfer number is used. Notes may also be made showing what is tested, or how many times a loop is to be traversed.
b) The operations appearing as blocks in a) may be replaced by actual instructions. It is usually not worth while at first to write down more than the last two characters of the (presumptive) instruction, i.e. the B line and function parts. These are quite enough to remind one of what was the purpose of the instruction.
c) One may then write the instructions into a page, deciding at the same time what are to be the addresses involved.
d) When the programming is complete, check sheets must be made. It is often advisable to start making check sheets long before the programme is complete; one should in fact begin them as soon as one feels that one has got into a muddle. It is often possible to work out most of the programme on the check sheets and afterwards transfer back onto the page or pages of instructions.

(iv) Programme the main routine. This follows principles similar to **(iii)**. Of course these remarks merely represent one possible way of doing programming. Individuals will no doubt vary as to the methods they prefer.

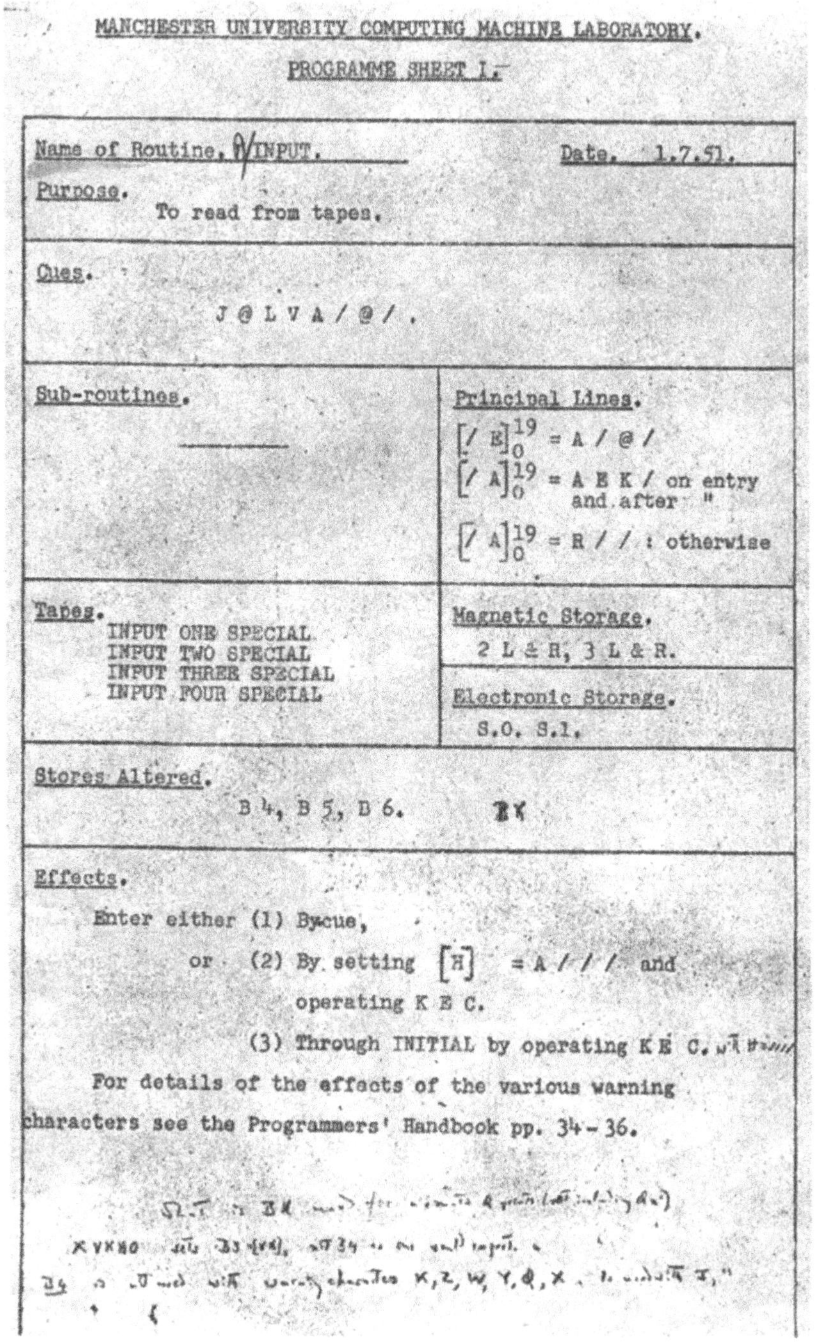

MANCHESTER UNIVERSITY COMPUTING MACHINE LABORATORY.

PROGRAMME SHEET I.

Name of Routine. INPUT. Date. 1.7.51.

Purpose.
To read from tapes.

Ques.
J @ L V A / @ / .

Sub-routines.
———

Principal Lines.
$[/ E]_0^{19}$ = A / @ /
$[/ A]_0^{19}$ = A E K / on entry and after "
$[/ A]_0^{19}$ = R / / : otherwise

Tapes.
INPUT ONE SPECIAL
INPUT TWO SPECIAL
INPUT THREE SPECIAL
INPUT FOUR SPECIAL

Magnetic Storage.
2 L & R, 3 L & R.

Electronic Storage.
S.0, S.1.

Stores Altered.
B 4, B 5, D 6. BK

Effects.
Enter either (1) By cue,
or (2) By setting [H] = A / / / and operating K E C.
(3) Through INITIAL by operating K E C.

For details of the effects of the various warning characters see the Programmers' Handbook pp. 34 - 36.

Fig. 1: Reproduction of Turing's INPUT routine, which was used to read programs into the machine from punched paper-tape.

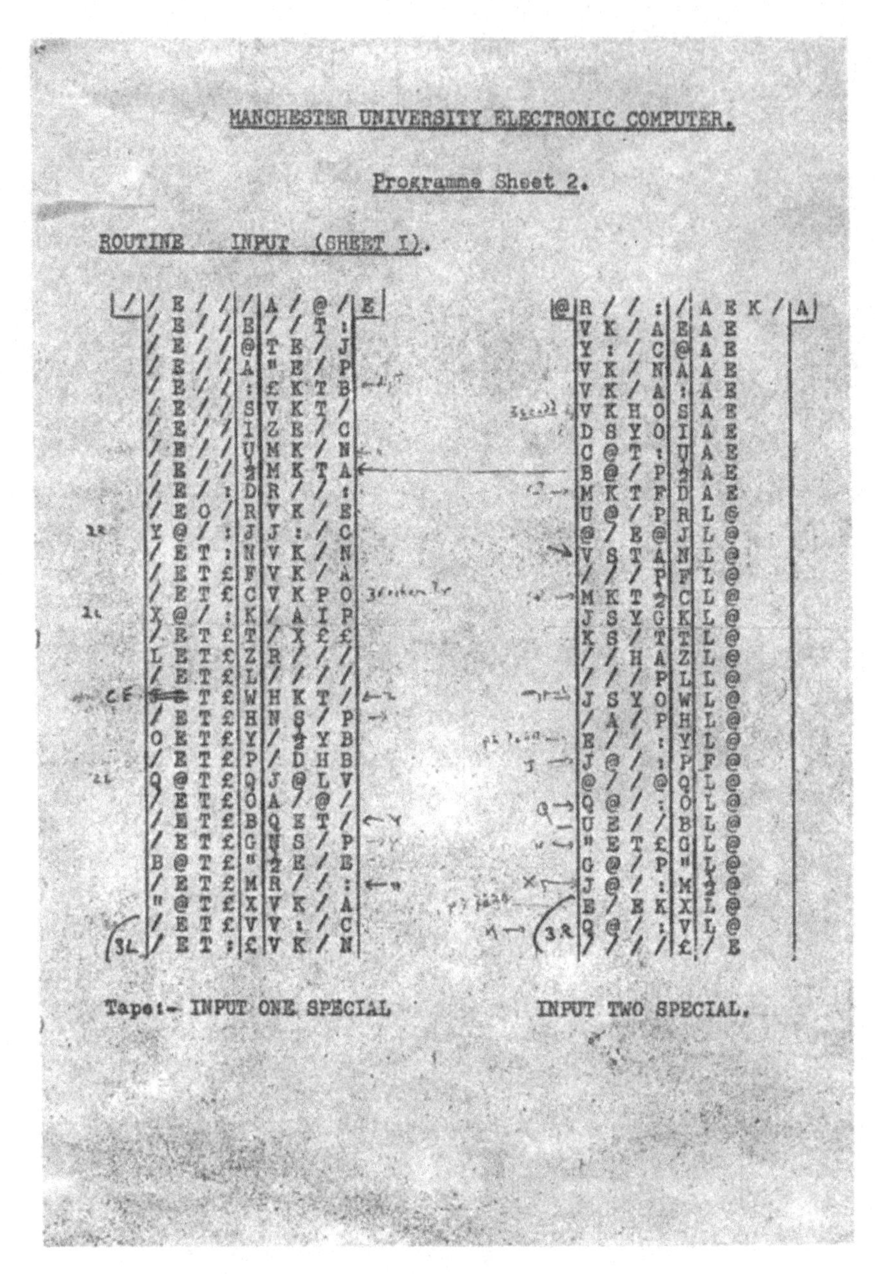
MANCHESTER UNIVERSITY ELECTRONIC COMPUTER.

Programme Sheet 2.

ROUTINE INPUT (SHEET I).

Tape:- INPUT ONE SPECIAL INPUT TWO SPECIAL.

Fig. 1: (continued)

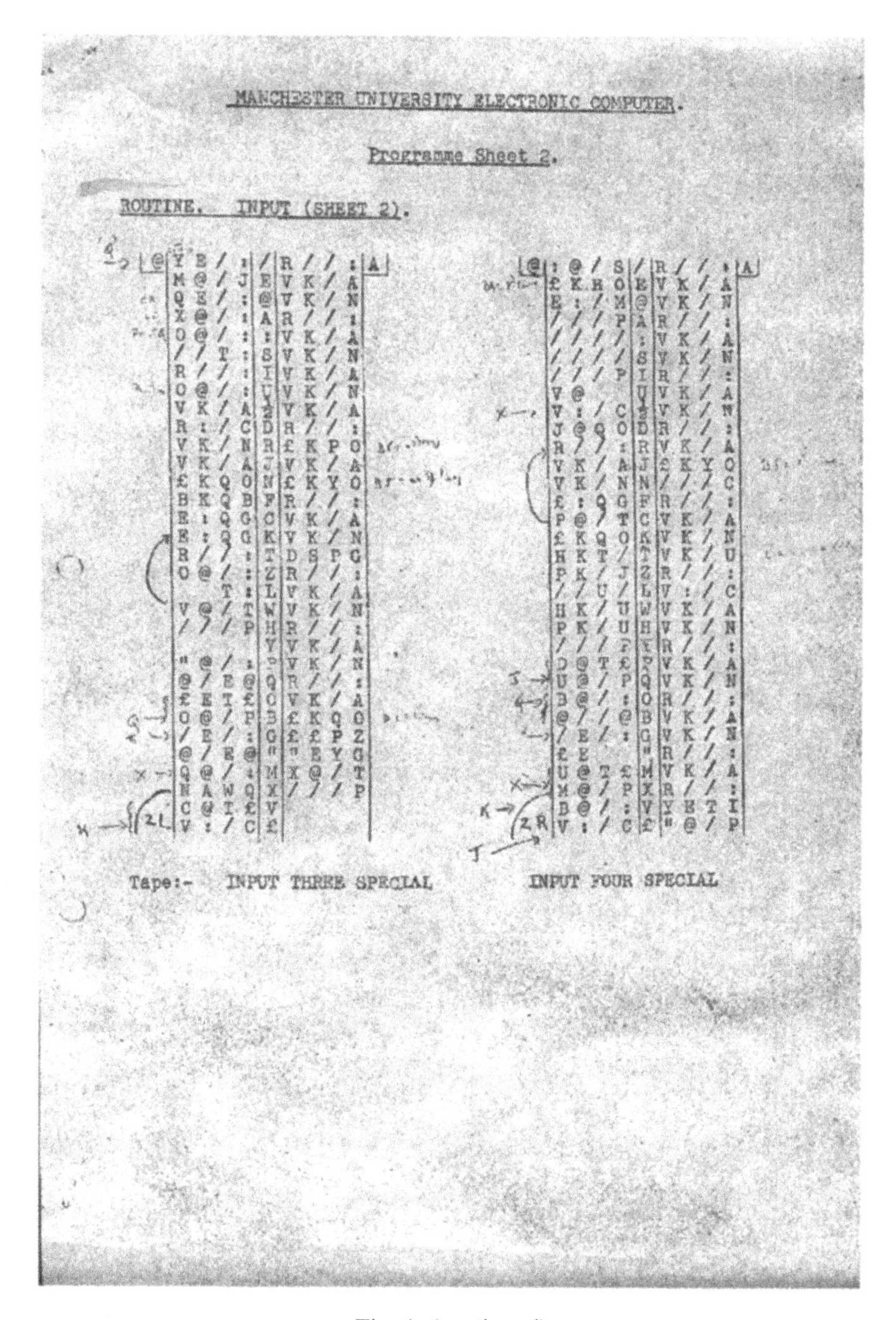

MANCHESTER UNIVERSITY ELECTRONIC COMPUTER.

Programme Sheet 2.

ROUTINE. INPUT (SHEET 2).

@									A
	Y	E	/	:	/	R	/	/	:
	M	@	/	J	E	V	K	/	A
	Q	E	/	:	@	V	K	/	N
	X	@	/	:	A	R	/	/	:
	O	@	/	:	:	V	K	/	A
	/	/	T	:	S	V	K	/	N
	R	/	/	:	I	V	K	/	A
	O	@	/	:	U	V	K	/	N
	V	K	/	A	½	V	K	/	A
	R	:	/	C	D	R	/	/	:
	V	K	/	N	R	£	K	P	O
	V	K	/	A	J	V	K	/	A
	£	K	Q	O	N	£	K	Y	O
	B	K	Q	B	F	R	/	/	:
	E	:	Q	G	C	V	K	/	A
	E	:	Q	G	K	V	K	/	N
	R	/	/	:	T	D	S	P	G
	O	@	/	:	Z	R	/	/	:
			T	:	L	V	K	/	A
	V	@	/	T	W	V	K	/	N
	/	/	/	P	H	R	/	/	:
					Y	V	K	/	A
	"	@	/	:	?	V	K	/	N
	@	/	E	@	Q	R	/	/	:
	£	E	T	£	O	V	K	/	A
	O	@	/	P	B	£	K	Q	O
	/	E	/	:	G	£	£	P	Z
	@	/	E	@	"	"	E	Y	G
	Q	@	/	:	M	X	@	/	T
	N	A	W	Q	X	/	/	/	P
	C	@	T	£	V				
	V	:	/	C	£				

@									A
	:	@	/	S	/	R	/	/	:
	£	K	H	O	E	V	K	/	A
	E	:	/	M	@	V	K	/	N
	/	/	/	P	A	R	/	/	:
	/	/	/	/	:	V	K	/	A
	/	/	/	/	S	V	K	/	N
	/	/	/	P	I	R	/	/	:
	V	@			U	V	K	/	A
	V	:	/	C	½	V	K	/	N
	J	@	Q	O	D	R	/	/	:
	R	/	/	:	R	V	K	/	A
	V	K	/	A	J	£	K	Y	O
	V	K	/	N	N	/	/	/	C
	£	:	Q	G	F	R	/	/	:
	P	@	/	T	C	V	K	/	A
	£	K	Q	O	K	V	K	/	N
	H	K	T	/	T	V	K	/	U
	P	K	/	J	Z	R	/	/	:
	/	/	U	/	L	V	:	/	C
	H	K	/	U	W	V	K	/	A
	P	K	/	U	H	V	K	/	N
	/	/	/	£	Y	R	/	/	:
	D	@	T	£	?	V	K	/	A
	U	@	/	P	Q	V	K	/	N
	B	@	/	:	O	R	/	/	:
	@	/	/	@	B	V	K	/	A
	/	E	/	:	G	V	K	/	N
	£	E			"	R	/	/	:
	U	@	T	£	M	V	K	/	A
	M	@	/	P	X	R	/	/	:
	B	@	/	:	V	Y	E	T	I
	V	:	/	C	£	"	@	/	P

Tape:- INPUT THREE SPECIAL INPUT FOUR SPECIAL

Fig. 1: (continued)

LOCAL PROGRAMMING METHODS AND CONVENTIONS

By A. M. Turing, B.A., Ph.D., F.R.S.

In this talk I am speaking about the use of teleprint code for writing down instructions and other material stored in the machine, and also about the conventions whereby rows of digits represent numbers.

In connection with the use of the teleprint code, it is necessary to choose between doing one's programming in a binary form, or else using some other form and allowing the machine to convert to the binary form. Binary scale working can be replaced by the use of any power of two as radix, and the choice made at Manchester was the scale of 32. It is probable that a scale of 8 or 16 would be slightly more convenient, but there is little to choose. The following points are made in favour of the teleprint (scale 32) method, as opposed to a decimal method.

(i) Since the instructions are held in the machine in a form intelligible to the programmer it is very easy to follow what is going on by looking at the monitor tubes. I express scepticism over the view that this so-called 'peeping' can or should be entirely eliminated by the use of automatic checking programmes. I would prefer to bring the peeping procedure to such efficiency that one's errors could be found extremely quickly by it. As practised at Manchester this procedure consists of the preparation in advance of 'checksheets' purporting to describe the behaviour of the machine in detail. The behaviour of the machine can then be compared with these checksheets. Peeping could, however, be objectionable unless, when an error is found, the programmer leaves the machine to someone else whilst he thinks about it.

(ii) Since the names of the store lines are little more than arbitrary labels it matters little what code is used for them. The function part of an instruction can in any case only be a quite arbitrary symbol.

(iii) In the case of the Manchester machine there are additional reasons. There is a natural division of the store into tubes, each of 64 lines. This division cannot be ignored because transfers of information between the electronic and magnetic stores is always by complete tubefuls. Each tube consists of two 'columns' each of 32 lines, so that each line can be described by two teleprint characters, one for the column and the other for the position in the column.

In connection with the relation between rows of digits and numbers, the nature of the multiplier makes it inconvenient to use only one convention by which rows of digits represent numbers. Four conventions are used at Manchester, which can be illustrated in connection with five digit rows, e.g., 00011. With all of these conventions as well as others such as the Cambridge convention, the numbers concerned form an arithmetic progression. The row 00000 always represents the number 0, and rows representing successive terms of the progression are related by 'row addition' of 10000. The most significant digit is on the right. Under these circumstances the convention is completely determined by giving the range of numbers which can be represented by it. The four Manchester conventions and the Cambridge convention can then be described by the table:

	Least Number	Greatest Number	Value for 00011	Suffix
Plus	0	31	24	+
Plus or Minus	-16	15	-8	±
Plus Fractional	0	$\frac{31}{32}$	$\frac{3}{4}$	+/
Plus or Minus Fractional	$-\frac{1}{2}$	$\frac{15}{32}$	$-\frac{1}{4}$	±/
Cambridge	-2	$+\frac{15}{8}$	-1	C (suggested)

If the suffix shown in the last column is combined with a symbol representing a row, then the resulting expression represents the corresponding number. Thus (00011) ±/ is synonymous with $-\frac{1}{4}$. It is desirable also to have a notation by which numbers are transformed into rows of digits. If α is a number then $[\alpha]^{r}_{s}$ represents that part of the binary expression of α which runs from the coefficient of 2^{s} to that of 2^{r}. Thus for instance $[3\cdot5]^{3}_{-2}$, is 011100.

The provision of such notations makes it possible to describe operations occurring in the machine concisely and unambiguously. Thus for instance the operation of 'copying a store line into the accumulator with extension of the sign digit throughout the most significant half of the accumulator' can be represented by the equation $A' = \{S\pm\}'_{\pm}$. The convention that undashed letters represent values before the operation, dashed letters values after it, is always used.

DISCUSSION CONTRIBUTIONS

Mr. T. J. Rey

It has been asked if programming can be simplified. The answer depends on whether one refers to the initial or the final steps of human labour. To the extent that a calculating machine is useful because it can perform many operations for only a few instructions, a working knowledge of the specialised notions involved is essential. However, as regards the instruction code, its details can be built into specialised mechanisms. For example, a key may be labelled '×' whereas its business end will punch that group of binary digits which eventually sets up the multiplication circuit. Similarly, in the Mark VI B.T.L. Computer, instructions may be set up by threading wires through a matrix array of toroidal transformer cores; the operation

$A \times B - C$ (Next Instruction 7)

is then prepared on passing a wire through one core in each column, successive cores being labelled A, ×, B, C, 7; a pulse on this wire energises the transformers so as to stimulate control circuits via the secondary windings.

Mr. E. A. Newman

Mr. Turing suggests that all computation consists essentially of multiplication. This is not so. As an example, take the extraction of the roots of a polynomial by Horner's Method. Some computers have estimated that on the bulk of problems multiplication time in a machine would have to be of the order of 100 times addition or transfer time before it appreciably slowed the overall computing speed. There are, of course, exceptional problems for which this is not so.

Fig. 2: Turing's only formal publication on programming techniques for the Manchester Mark I, read at the Inaugural Conference for the Manchester University Computer in July 1951.

Part III

Building a Brain: Intelligent Machines, Practice and Theory

Turing's Lecture to the London Mathematical Society on 20 February 1947

(A more readable guide to the ACE computer than Turing's 1945 ACE report)

Anthony Beavers pays homage to —

ALAN TURING: MATHEMATICAL MECHANIST

I live just off of Bell Road outside of Newburgh, Indiana, a small town of 3000 people. A mile down the street Bell Road intersects with Telephone Road, not as a modern reminder of a technology belonging to bygone days, but as testimony that this technology, now more than a century and a quarter old, is still with us. In an age that prides itself on its digital devices and in which the computer now equals the telephone as a medium of communication, it is easy to forget the debt we owe to an era that industrialised the flow of information, that the light bulb, to pick a singular example, which is useful for upgrading visual information we might otherwise overlook, nonetheless remains the most prevalent of all modern day information technologies. Edison's light bulb, of course, belongs to a different order of informational devices than the computer, but not so the telephone, not entirely anyway.

Alan Turing, best known for his work on the Theory of Computation (1937), the Turing Machine (also 1937) and the Turing Test (1950), is often credited with being the father of computer science and the father of artificial intelligence. Less well known to the casual reader but equally important is his work in computer engineering. The following lecture on the Automatic Computing Engine, or ACE, shows Turing in this different light, as a mechanist concerned with getting the greatest computational power from minimal hardware resources. Yet Turing's work on mechanisms is often eclipsed by his thoughts on computability and his other theoretical interests. This is unfortunate for several reasons, one being that it obscures our picture of the historical trajectory of information technology, a second that it emphasizes a false dichotomy between 'hardware' and 'software' to which Turing himself did not ascribe but which has, nonetheless, confused researchers who study the nature of mind and intelligence for generations. We are only today starting to understand the error of our ways. *Our* ways ... not Turing's, as the following essay makes clear.

The second issue follows from the first, from understanding Turing independently of his mechanistic tendencies and his connections to the *industrial* information revolution of the nineteenth century. Thus, it is fitting to rescue the popular understanding of Turing from the two essays for which he is most notably known (Turing, 1937, 1950) in order to shed some light on his genuine place in history and, at the same time, to examine some of the implications that should have been clear to the philosophers and psychologists that immediately followed him. He was, in more than one way, ahead of his time, though oddly, as we shall see, because he was thoroughly connected to his past.

'It is ... quite difficult to think about the code entirely in abstracto without any kind of circuit', Turing (p. 490) writes in this lecture suggesting that in a working machine there is no 'code', just hardware, and that really what 'computer code' does is to configure circuitry within computing machinery to perform a particular information processing task. For ease of communication, I will call the belief that software constitutes a separate level of machine processing, that 'code' *can* be understood in abstraction from circuitry, the 'software seduction'. This position is most famously advanced by Marr (1982). Though it was originally intended to assist him in the analysis of vision, it set the stage for several theories in cognitive science and the philosophy of mind.

Marr's 'Tri- Level Hypothesis' decomposed an information processing system into three levels for the sake of analysis, the computational, algorithmic and implementational levels. The first level concerns the input/output specifications of the system, while the algorithmic level specifies the processes and representations whereby inputs are transformed into their appropriate outputs. The implementational level concerns how the algorithmic level is instantiated in something physical.

For Marr and several in the tradition that follows him, the real work of information processing is best understood by examining algorithms, while the implementation is largely incidental. In other words, the 'code' here *is* predominantly understood in abstraction from the circuits, or, in modern terms, 'software' is deemed more important for understanding information processing than 'hardware'. In turn, this tri-partite division led historically to a 'computer analogy' for understanding human cognition in which mind is taken to be to software as brain (or body) is to hardware. As a consequence, to understand mental function, we need only consider problem solving and other cognitive tasks on the algorithmic level. The details of the neural substrate belong to another science.

Traditional cognitive scientists still ascribe to this distinction, which is deeply rooted in the separation between cognitive psychology and neuroscience that has become part of the DNA of the modern academy. Of course, as traditions tend to fall to the past and new conceptions take their place, this paradigm too is on its way out. Embodied and situated cognition has made considerable headway in producing explanations that are both more fruitful and biologically plausible than those rooted in the aforementioned 'software seduction'. We are coming to understand that mind cannot be understood in abstraction from brain, body and environment just as code cannot be understood in abstraction from circuitry. Regarding Turing and, in particular, the essay to follow, these new explanations of cognition cannot be said to represent a turn away from the Turing paradigm, but rather a turn towards it, once we understand Turing's own mechanical tendencies. If mind is to brain as software is to hardware, then a true rendition of the 'computer analogy' would suggest that in a real working brain, mind is just more brain, more circuitry. We see why this is so in the subsequent essay, which again connects Turing to his past and helps us understand his true contribution to the history of information technology, a topic which I now address.

Though one common view is that the 'information age' begins with the birth of digital technologies and thus owes its debt to Turing (Floridi, 2008), when we explore the history of informational mechanisms, we see that this is only partially correct. It is quite true that the recent explosion in informational devices has increased dramatically since *Time Magazine* named the computer 'person of the year' in 1982, but it belongs nonetheless to a trajectory that was firmly set in motion by the end of the nineteenth century with what we might call the 'multimedia revolution' (Beavers, in press). Such a view does not minimize Turing's contribution, but helps to illustrate precisely where it lies. It is obviously not insignificant.

By 'multimedia revolution', I mean to invoke here an era in which information was decoupled from the exigencies of transportation technology and made to move on its own. Prior, if information traveled from point A to B, it was because someone carried it there; but just as the industrial revolution issued in a range of new mechanisms for everything from agriculture to textile manufacturing, it did the same for information. The casual reader, to be sure, is mostly aware of this fact, but its significance might not readily be clear. Some inventions and their approximate dates might add some clarity.

Though putting starting dates on such historical transformation is risky business, it is perhaps not too much of a falsification to date the beginning of the industrialization of information with the telegraph in 1836 and the Daguerreotype in 1839, which introduced practical photography. The telegraphic printer and the stock ticker followed soon after in 1856 and 1863, respectively. The years between 1876 and 1881 were perhaps the most immediately transformative with the telephone in 1877, the phonograph in 1878, and the light bulb and the photophone in 1880. The latter, Bell's favorite invention, could mediate telephone communications by modulating wave forms on a beam of light. Wireless telegraphy and the wax cylinder, which made Edison's phonograph practical, both emerged in 1881 with the wax cylinder (invented by Bell) serving as a patent reminder that not all of these technologies were electric. Many were not adopted for the same uses to which we put them today. The telephone, for instance, was an early form of broadcasting prior to radio (which is why a radio was originally referred to as a 'wireless'), and of the 10 uses Edison enumerates for the phonograph, recording music is listed fourth, with distance education suggested ninth and then last, 'connection with the telephone, so as to make that instrument an auxiliary in the transmission of permanent and invaluable records, instead of being the recipient of momentary and fleeting communication' (Edison, 1878).

Edison's vision of connecting phonographs to the telephone system to aid in the preservation and transmission of information anticipates Turing's observations (p. 495) in the following essay that 'it would be quite possible to arrange to control a distant computer by means of a telephone line', costing 'a few hundred pounds at most'. This would entail modulating wave forms over the telephone to allow for digital transmission, which was easy enough to do. However, the important thing is that for both men, hints of a networked world are present because of the invention and quick adoption of the telephone. In fact, by 1910 the telephone had become so popular as to warrant its own history. Herbert Casson's *The History of the Telephone* paints a vivid picture of the societal transformation in progress:

> What we might call the telephonization of city life, for lack of a simpler word, has remarkably altered our manner of living from what it was in the days of Abraham Lincoln. It has enabled us to be more social and cooperative. It has literally abolished the isolation of separate families, and has made us members of one great family. It has become so truly an organ of the social body that by telephone we now enter into contracts, give evidence, try lawsuits, make speeches, propose marriage, confer degrees, appeal to voters, and do almost everything else that is a matter of speech (199).

Indeed, long before the emergence of computing machinery, 'human computers', as Turing liked to call people, were quickly interconnecting, and it is difficult to imagine the invention of the mechanical computer (and a host of other technologies) without this affordance of the telephone. But there are deeper technological connections between the telephone and computing machinery, which I will discuss momentarily. Before doing so, it is worth sketching a bit more of the history that belongs to the 'multimedia revolution' by reminding the reader of a few more dates.

In 1891, Edison made pictures move with his invention of the motion picture camera. Radio and teletype would enter the scene in 1906, to be followed in 1914 by Edison's 'telescribe', a precursor to the answering machine that has now recently been replaced with 'voice mail'. Pictures would take to the airwaves with the public appearance of the television in 1926 and the National Broadcasting System in 1928, the same year that magnetic tape would become available. The speed of information transmission and its reach would continue with cable television in 1948, the cassette tape in 1958, the touch tone phone in 1963, color television in 1966 and the VCR in 1969.

Turing belongs to this historical trajectory, which I have intentionally recounted without mention of computing machinery to make a few points. The first is that the information revolution was well underway prior to Turing, but without one affordance that will greatly alter its landscape. Edison, Bell and several others could store information and move it around. They could not, however, capture it temporarily in a memory store, process it and then produce meaningful output. Turing's paper of 1937 introduces a theory of mechanical computation sufficient to put automated information processors into the mix of inter-networked human computers, but it will not provide the schematic for a practical piece of hardware. That work comes later, not solely by Turing, but partly so, as we see in this 1947 lecture to the London Mathematical Society. Second, not all of the technologies that belong to this information revolution were electric, as I previously noted, for instance, the phonograph, early cameras, motion picture cameras, etc. Electricity is important for computing machinery, but for practical reasons only, at least for what matters to the mathematician.

Turing writes at the beginning of this lecture (p. 486) that 'the property of being digital should be of greater interest than that of being electronic. That [computing machinery] is electronic is certainly important because these machines owe their high speed to this, and without the speed it is doubtful if financial support for their construction would be forthcoming. But this is virtually all that there is to be said on the subject'.

It is perhaps worth noting two things in regard to this quote. Computing machinery was not the first of the digital information mechanisms to enter this revolution. At the very beginning, Samuel Morse's famous code (c. 1830) was digital; the telegraph worked by sending pulses of current across an electrical line. Signals were encoded digitally, but not in binary, since the code used five 'tokens' not two, a short electric pulse, a long one, and less obviously the empty space between them that used a short moment of silence to separate 'dots' and 'dashes', a longer one to separate letters and a longer one still for words. More importantly, the quote from Turing makes reference to the necessity of building economically feasible equipment. This concern is not trivial; not at all. In fact, as the essay indicates, cost concerns are going to drive the architecture of the machine and the need for 'subsidiary [instruction] tables', that are stored in memory as early analogs to our functions or procedures. Getting the most computational power from minimal hardware resources, as I mentioned earlier, is at the heart of this essay, but for economic and not mathematical reasons. 'Code' then becomes a strategy for creating temporary circuitry inside of the machine that can be reconfigured later for a different processing task. This highlights the notion that computer code is not essential for information processing tasks; separate hardware (i.e., circuitry) could be built for each task, but this would be both costly and inefficient. This makes it clear that one can, in fact, trade off software for hardware as needed. Indeed, Turing's goal in this lecture is to describe a machine that has minimal circuitry, but that can be configured with code for any processing task that is computable. Here, he does this (p. 489) with a 200KB memory store that is 'probably comparable with the memory capacity of a minnow', quite a remarkable accomplishment even by today's standards.

Code, then, is useful to the extent that it can reconfigure circuitry, and hence is not a necessary condition for information processing tasks. However, such a machine could not get by without easily accessible, writeable and erasable memory. This fact leads us to one of the central practical insights of this lecture, the development of an accessible and efficient memory store. The infinite tape suggested in the idealized Turing machine of 1937 will not do because of the time it would take to jump around the tape. In explaining why, Turing ties himself again to the history of information technology by referring back to the affordances of the book over papyrus scrolls, noting nonetheless that even an automated memory structure based on the book would be highly inefficient, though the memory store he will advocate here is based on an analogy with it. The book will be to the papyrus scroll as memory circuitry will be to the infinite tape of the idealized machine. Here, the memory circuitry will consist of 200 separate mercury acoustic delay lines (analogous to pages in a book) that can each contain 1024 bits of information. I will not go into a complete description of how the system works here, since the details are spelled out in the lecture. I will, however, make a few final observations, that tie Turing to the history of information technology.

To start, many of the inventions of this era that industrialized the flow of information did so by modulating wave forms of various sorts. I have already mentioned Bell's experiments with light in this regard, but it was his early experiments with electricity that made the telephone (prior to cell phones) practical. Others, of course, were involved in a range of experiments (Tesla, Marconi, etc.) that modulated radio waves to support both radio and television. In this lecture, Turing considers the use of magnetic wires and cathode ray tubes before settling on the use of acoustic delay lines. Though analog, a close inspection of the schematics of Bell, Marconi or Tesla shows sufficient similarity to Turing's designs to plot him on the same trajectory. So, of course, the idea of modulating wave forms does not belong to Turing. In fact, some might even be surprised to learn that his early work was this explicitly dedicated to such.

What then is important about this lecture? It would seem to be its patently practical nature and its mechanical commitments that challenge our picture of a mathematical Turing by presenting another Turing, the mechanist, who gets into the nuts and bolts of computing machinery. In doing so, as I have tried to make clear, he does not show a clean break that starts a new era, but strong attachments to the preceding one. This is not to suggest that Turing did not issue in an unprecedented era in the processing and storage of information and the speed of its dissemination. He did, along with several others. In this regard, I wish to close with one final comment about technological visionaries in general.

Turing foresaw the possibility of connecting the computer to the telephone in much the same way that Edison did. Bell was already on the track of wireless telephony, so much so that in 1897, William Ayrton could make a prediction about the future that is still so painted in the language and technology of his past that today it reads like a 1930's science fiction film:

> There is no doubt that the day will come, maybe when you and I are forgotten, when copper wires, gutta-percha coverings, and iron sheathings will be relegated to the Museum of Antiquities. Then, when a person wants to telegraph to a friend, he knows not where, he will call in an electro-magnetic voice, which will be heard loud by him who has the electro-magnetic ear, but will be silent to everyone else. He will call, 'Where are you?' and the reply will come, 'I am at the bottom of the coal-mine' or 'Crossing the Andes,' or 'In the middle of the Pacific' (Fahie, 1900, vii).

No one back then, it seems, could imagine what would happen when the computer revolution, inaugurated by Turing, and the telephone revolution, inaugurated by Bell, would come crashing together at the end of the twentieth century to connect everyone to computers (both human and mechanical) by way of hand-held computer/telephones and other devices that are both affordable and have more computational power than Turing himself imagined practical.

We stand on the shoulders of giants, information visionaries of the past 200 years. Yet, if we could somehow transport them into today's information environment, I cannot help but think that even with the grand visions that they had, they would nonetheless be in utter surprise over where we have arrived. Shocked and baffled, amazed, they would perhaps feel both proud and humble. This makes it all the more shameful that Turing himself was never permitted to experience the appreciation of a world that owes him an incredible debt, not solely because of his brilliant mind, but also because he was not afraid to get his hands dirty. We should never disparage those who are willing to get dirty in the process of making the world a better place.

References

Beavers, A., In press. In the beginning was the word and then four revolutions in the history of information. In: Demir, H. (Ed.), Luciano Floridi's Philosophy of Technology: Critical Reflections. Springer.

Casson, H., 1910. The History of the Telephone. A. C. McClurg & Co, Chicago.

Edison, T., 1878. The North American Review. Retrieved from the U. S. Library of Congress. <http://memory.loc.gov/ammem/edhtml/edcyldr.html> (accessed 01.08.10).

Fahie, J., 1900. A History of Wireless Telegraphy, 1828–1899, Including Some Bare-Wire Proposals for Subaqueous Telegraphs, William Blackwood and Sons, London.

Floridi, L., 2008. Artificial intelligence's new frontier: Artificial companions and the fourth revolution. Metaphilosophy 39.4/5, 652–654.

Marr, D., 1982. Vision: A Computational Investigation into the Human Representation and Processing of Visual Information, W. H. Freeman, New York, NY.

Turing, A., 1937. On computable numbers, with an application to the Entscheidungsproblem. Proc. Lond. Math. Soc. 2.1, 230–265.

Turing, A., 1950. Computing machinery and intelligence. Mind 59.236, 433–460.

Lecture to the London Mathematical Society on 20 February 1947

A. M. TURING

The automatic computing engine now being designed at N.P.L. is a typical large scale electronic digital computing machine. In a single lecture it will not be possible to give much technical detail of this machine, and most of what I shall say will apply equally to any other machine of this type now being planned.

From the point of view of the mathematician the property of being digital should be of greater interest than that of being electronic. That it is electronic is certainly important because these machines owe their high speed to this, and without the speed it is doubtful if financial support for their construction would be forthcoming. But this is virtually all that there is to be said on that subject. That the machine is digital however has more subtle significance. It means firstly that numbers are represented by sequences of digits which can be as long as one wishes. One can therefore work to any desired degree of accuracy. This accuracy is not obtained by more careful machining of parts, control of temperature variations, and such means, but by a slight increase in the amount of equipment in the machine. To double the number of significant figures used would involve increasing the equipment by a factor definitely less than two, and would also have some effect in increasing the time taken over each job. This is in sharp contrast with analogue machines, and continuous variable machines such as the differential analyser, where each additional decimal digit required necessitates a complete redesign of the machine, and an increase in the cost by perhaps as much as a factor of 10. A second advantage of digital computing machines is that they are not restricted in their applications to any particular type of problem. The differential analyser is by far the most general type of analogue machine yet produced, but even it is comparatively limited in its scope. It can be made to deal with almost any kind of ordinary differential equation, but it is hardly able to deal with partial differential equations at all, and certainly cannot manage large numbers of linear simultaneous equations, or the zeros of polynomials. With digital machines however it is almost literally true that they are able to tackle any computing problem. A good working rule is that the ACE can be made to do any job that could be done by a human computer, and will do it in one ten-thousandth of the time. This time estimate is fairly reliable, except in cases where the job is too trivial to be worth while giving to the ACE.

Some years ago I was researching on what might now be described as an investigation of the theoretical possibilities and limitations of digital computing machines. I considered a type of machine which had a central mechanism, and an infinite memory which was contained on an infinite tape. This type of machine appeared to be sufficiently general. One of my conclusions was that the idea of a 'rule of thumb' process and a 'machine process' were synonymous. The expression 'machine process' of course means one which could be carried out by the type of machine I was considering. It was essential in these theoretical arguments that the memory should be infinite. It can easily be shown that otherwise the machine can only execute periodic operations. Machines such as the ACE may be regarded as practical versions of this same type of machine. There is at least a very close analogy. Digital computing machines all have a central mechanism or control and some very extensive form of memory. The memory does not have to be infinite, but it certainly needs to be very large. In general the arrangement of the memory on an infinite tape is unsatisfactory in a practical machine, because of the large amount of time which is liable to be spent in shifting up and down the tape to reach the point at which a particular piece of information required at the moment is stored. Thus a problem might easily need a storage of three million entries, and if each entry was equally likely to be the next required the average journey up the tape would be through a million entries, and this would be intolerable. One needs some form of memory with which any required entry can be reached at short notice. This difficulty presumably used to worry the Egyptians when their books were written on papyrus scrolls. It must have been slow work looking up references in them, and the present arrangement of written matter in books which can be opened at any point is greatly to be preferred. We may say that storage on tape and papyrus scrolls is somewhat *inaccessible*. It takes a considerable time to find a given entry. Memory in book form is a good deal better, and is certainly highly suitable when it is to be read by the human eye. We could even imagine a computing machine that was made to work with a memory based on books. It would not be very easy but would be immensely preferable to the single long tape. Let us for the sake of argument suppose that the difficulties involved in using books as memory were overcome, that is to say that mechanical devices for finding the right book and opening it at the right page, etc. etc. had been developed, imitating the use of human hands and eyes. The information contained in the books would still be rather inaccessible because of the time occupied in the mechanical motions. One cannot turn a page over very quickly without tearing it, and if one were to do much transportation, and do it fast, the energy involved would be very great. Thus if we moved one book every millisecond and each was moved ten metres and weighed 200 grams, and if the kinetic energy were wasted each time we should consume 10^{10} watts, about half the country's power consumption. If we are to have a really fast machine then, we must have our information, or at any rate a part of it, in a more accessible form than can be obtained with books. It seems that this can only be done at the expense of compactness and economy, e.g. by cutting the pages out of the books, and putting each one in to a separate reading mechanism. Some of the methods of storage which are being developed at the present time are not unlike this.

If one wishes to go to the extreme of accessibility in storage mechanisms one is liable to find that it is gained at the price of an intolerable loss of compactness and economy. For instance the most accessible known form of storage is that provided by the valve flip-flop or Jordan Eccles trigger circuit. This enables us to store one digit, capable of two values, and uses two thermionic valves. To store the content of an ordinary novel by such means would cost many millions of pounds. We clearly need some compromise method of storage which is more accessible than paper, film etc, but more economical in space and money than the straightforward use of valves. Another desirable feature is that it should be possible to record into the memory from within the computing machine, and this should be possible whether or not the storage already contains something, i.e. the storage should be *erasible*.

There are three main types of storage which have been developed recently and have these properties in greater or less degree. Magnetic wire is very compact, is erasible, can be recorded on from within the machine, and is moderately accessible. There is storage in the form of charge patterns on the screen of a cathode ray tube. This is probably the ultimate solution. It could eventually be nearly as accessible as the Jordan Eccles circuit. A third possibility is provided by acoustic delay lines. They give greater accessibility than the magnetic wire, though less than the C.R.T type. The accessibility is adequate for most purposes. Their chief advantage is that they are already a going concern. It is intended that the main memory of the ACE shall be provided by acoustic delay lines, consisting of mercury tanks.

The idea of using acoustic delay lines as memory units is due I believe to Eckert of Philadelphia University, who was the engineer chiefly responsible for the Eniac. The idea is to store the information in the form of compression waves travelling along a column of mercury. Liquids and solids will transmit sound of surprisingly high frequency, and it is quite feasible to put as many as 1000 pulses into a single 5′ tube. The signals may be conveyed into the mercury by a piezo-electric crystal, and also detected at the far end by another quartz crystal. A train of pulses or the information

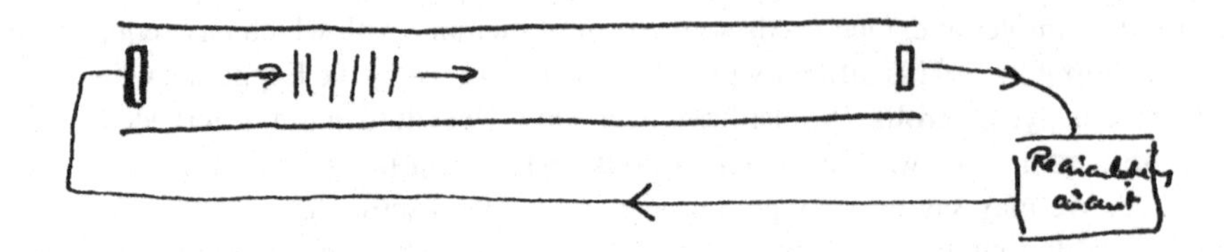

which they represent may be regarded as stored in the mercury whilst it is travelling through it. If the information is not required when the train emerges it can be fed back into the column again and again until such time as it *is* required. This requires a 'recirculating circuit' to read the signal as it emerges from the tank and amplify it and feed it in again. If this were done with a simple amplifier it is clear that the characteristics of both the tank and the amplifier would have to be extremely good to permit the signal to pass through even as many as ten times. Actually the recirculating circuit does something slightly different. What it does may perhaps be best expressed in terms of point set topology. Let the plane of the diagram represent the space of all possible signals. I do not of course wish to imply that this is two dimensional. Let the function f b defined for arguments in this signal space and have values in it. In fact let $f(s)$ represent the effect on the signal s when it is passed through the tank and the recirculating mechanism. We assume however that owing to thermal agitation the effect of recirculation may be to give any point within a circle of radius δ of $f(s)$. Then a necessary and sufficient condition that the tank can be used as a storage which will distinguish between N different signals is that there must be N sets $E_1 \ldots E_N$ such that if F_r is the set of points within distance ϵ of E_r

$$s \in F_r \supset f(s) \in E_r$$

and the sets F_r are disjoint. It is clearly sufficient for we have only then to ensure that the signals initially fed in belong to one or other of the sets F_r, and it will remain in the set after any number of recirculations, without any

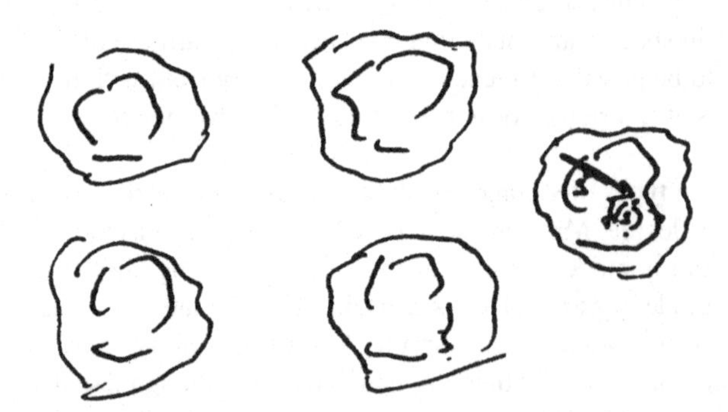

danger of confusion. It is necessary for suppose $s_1 \ldots s_N$ are signals which have different meanings and which can be fed into the machine at any time and read out later without fear of confusion.

Let E_r be the set of signals which *could* be obtained for s_r by successive applications of f and shifts of distance not more than ε. Then the sets E_r are disjoint [two lines indecipherable—Ed.]. In the case of a mercury delay line used for $N = 16$ the set would consist of all continuous signals within the shaded area.

One of the sets would consist of all continuous signals lying in the region below. It would represent the signal 1001.

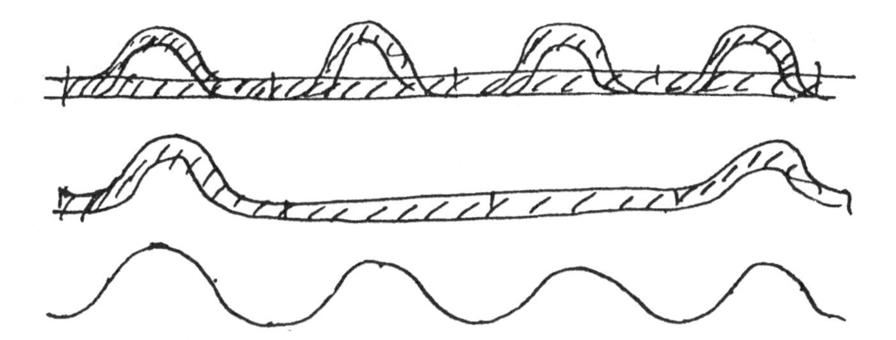

In order to put such a recirculation system into effect it is essential that a clock signal be supplied to the memory system so that it will be able to distinguish the times when a pulse if any should be present. It would for instance be natural to supply a timing sine wave as shown above to the recirculator.

The idea of a process f with the properties we have described is a very common one in connection with storage devices. It is known as 'regeneration' of storage. It is always present in some form, but sometimes the regeneration is as it were naturally occuring and no precautions have to be taken. In other cases special precautions have to be taken to improve such an f process or else the impression will fade.

The importance of a clock to the regeneration process in delay lines may be illustrated by an interesting little theorem. Suppose that instead of the condition $s \in F_r \supset f(s) \in E_r$ we impose a stronger one, viz $f^n(s) \to c_r$ if $s \in E_r$, i.e. there are ideal forms of the distinguishable signals, and each admissible signal converges towards the ideal form after recirculating. Then we can show that unless there is a clock the ideal signals are all constants. For let U_α represent a shift of origin, i.e. $U_\alpha s(t) = s(t+\alpha)$. Then since there is no clock the properties of the recirculator are the same at all times and f therefore commutes with U_α. Then $fU_\alpha(c_r) = U_\alpha f(c_r) = U_\alpha c_r$, for $f(c_r) = c_r$ since c_r is an ideal signal. But this means that $U_\alpha(c_r)$ is an ideal signal, and therefore for sufficiently small α must be c_r, since the ideal signals are discrete. Then for any β and sufficiently large u, β/u will be sufficiently small and $U_{\beta/u}(c) = c$. But then by iteration $c = U^u_{\beta/u}(c) = U_\beta(c)$ i.e. $c(t+\beta) = c(t)$. This means that the ideal signal c is a constant.

We might say that the clock enables us to introduce a discreteness into time, so that time for some purposes can be regarded as a succession of instants instead of a continuous flow. A digital machine must essentially deal with discrete objects, and in the case of the ACE this is made possible by the use of a clock. All other digital computing machines except for human and other brains that I know of do the same. One can think up ways of avoiding it, but they are very awkward. I should mention that the use of the clock in the ACE is not confined to the recirculation process, but is used in almost every part.

It may be as well to mention some figures connected with the mercury delay line as we shall use it. We shall use five foot tubes, with an inside diameter of half an inch. Each of these will enable us to store 1024 binary digits. The unit I have used here to describe storage capacity is self explanatory. A storage mechanism has a capacity of m binary digits if it can remember any sequence of m digits each being a 0 or a 1. The storage capacity is also the logarithm to the base 2 of the number of different signals which can be remembered, i.e. $\log_2 N$. The digits will be placed at a time interval of one microsecond, so that the time taken for the waves to travel down the tube is just over a millisecond. The velocity is about one and a half kilometres per second. The delay in accessibility time or average waiting for a given piece of information is about half a millisecond. In practice this is reduced to an effective 150 μs. The full storage capacity of the ACE available on Hg delay lines will be about 200,000 binary digits. This is probably comparable with the memory capacity of a minnow.

I have spent a considerable time in this lecture on this question of memory, because I believe that the provision of proper storage is the key to the problem of the digital computer, and certainly if they are to be persuaded to show any sort of genuine intelligence much larger capacities than are yet available must be provided. In my opinion this problem of making a large memory available at reasonably short notice is much more important than that of doing operations such as multiplication at high speed. Speed is necessary if the machine is to work fast enough for the machine to be commercially valuable, but a large storage capacity is necessary if it is to be capable of anything more than rather trivial operations. The storage capacity is therefore the more fundamental requirement.

Let us now return to the analogy of the theoretical computing machines with an infinite tape. It can be shown that a single special machine of that type can be made to do the work of all. It could in fact be made to work as a model of any other machine. The special machine may be called the universal machine; it works in the following quite simple manner. When we have decided what machine we wish to imitate we punch a description of it on the tape of the universal machine. This description explains what the machine would do in every configuration in which it might find itself. The universal machine has only to keep looking at this description in order to find out what it should do at each stage. Thus the complexity of the machine to be imitated is concentrated in the tape and does not appear in the universal machine proper in any way.

If we take the properties of the universal machine in combination with the fact that machine processes and rule of thumb processes are synonymous we may say that the universal machine is one which, when supplied with the appropriate instructions, can be made to do any rule of thumb process. This feature is paralleled in digital computing machines such as the ACE. They are in fact practical versions of the universal machine. There is a certain central pool of electronic equipment, and a large memory. When any particular problem has to be handled the appropriate instructions for the computing process involved are stored in the memory of the ACE and it is then 'set up' for carrying out that process.

I have indicated the main strategic ideas behind digital computing machinery, and will now follow this account up with the very briefest description of the ACE. It may be divided for the sake of argument into the following parts

Memory
Control
Arithmetic part
Input and output

I have already said enough about the memory and will only repeat that in the ACE the memory will consist mainly of 200 mercury delay lines each holding 1024 binary digits. The purpose of the control is to take the right instructions from the memory, see what they mean, and arrange for them to be carried out. It is understood that a certain 'code of instructions' has been laid down, whereby each 'word' or combination of say 32 binary digits describes some particular operation. The circuit of the control is made in accordance with the code, so that the right effect is produced. To a large extent we have also allowed the circuit to determine the code, i.e. we have not just thought up an imaginary 'best code' and then found a circuit to put it into effect, but have often simplified the circuit at the expense of the code. It is also quite difficult to think about the code entirely *in abstracto* without any kind of circuit. The arithmetic part of the machine is the part concerned with addition, multiplication and any other operations which it seems worth while to do by means of special circuits rather than through the simple facilities provided by the control. The distinction between control and arithmetic part is a rather hazy one, but at any rate it is clear that the machine should at least have an adder and a multiplier, even if they turn out in the end to be part of the control. This is the point at which I should mention that the machine is operated in the binary scale, with two qualifications. Inputs from externally provided data are in decimal, and so are outputs intended for human eyes rather than for later reconsumption by the ACE. This is the first qualification. The second is that, in spite of the intention of binary working there can be no bar on decimal working of a kind, because of the relation of the ACE to the universal machine. Binary working is the most natural thing to do with any large scale computer. It is much easier to work in the scale of two than any other, because it is so easy to produce mechanisms which have two positions of stability: the two positions may then be regarded as representing 0 and 1. Examples are lever as diagram, Jordan Eccles circuit, thyratron. If one is concerned with a

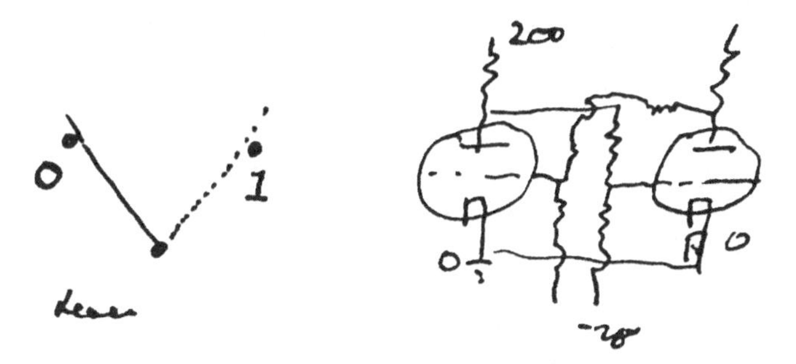

small scale calculating machine then there is at least one serious objection to binary working. For practical use it will be necessary to build a converter to transform numbers from the binary form to the decimal and back. This may well be a larger undertaking than the binary calculator. With the large scale machines this argument carries no weight. In the first place a converter would become a relatively small piece of apparatus, and in the second it would not really be necessary. This last statement sounds quite paradoxical, but it is a simple consequence of the fact that these machines can be made to do any rule of thumb process by remembering suitable instructions. In particular it can be made to do binary decimal conversion. For example in the case of the ACE the provision of the converter involves no more than adding two extra delay lines to the memory. This situation is very typical of what happens with the ACE. There are many fussy little details which have to be taken care of, and which, according to normal engineering practice would require special circuits. We are able to deal with these points without modification of the machine itself, by pure paper work, eventually resulting in feeding in appropriate instructions.

To return to the various parts of the machine. I was saying that it will work in the scale of two. It is not unnatural to use the convention that an electrical pulse shall represent the digit 1 and that absence of a pulse shall represent a digit 0. Thus a sequence of digits 0010110 would be represented by a signal like

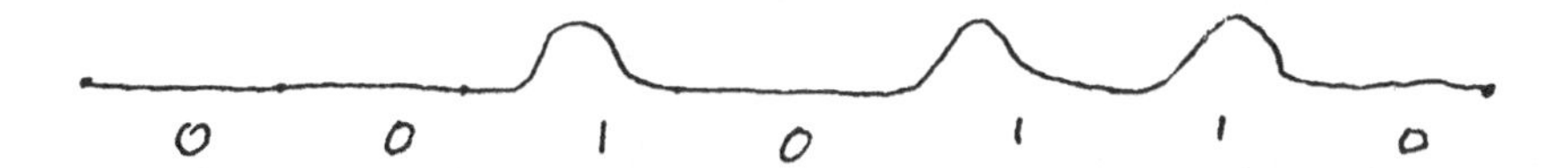

where the time interval might be one microsecond. Let us now look at what the process of binary addition is like. In ordinary decimal addition we always begin from the right, and the same naturally applies to binary. We have to do this because we cannot tell whether to carry unless we have already dealt with the less significant columns. The same applies with electronic addition, and therefore it is convenient to use the convention that if a sequence of pulses is coming down a line, then the least significant pulse always comes first. This has the unfortunate result that we must either write the least significant digit on the left in our binary numbers or else make time flow from right to left in our diagrams. As the latter alternative would involve writing from right to left as well as adding in that way, we have decided to put the least significant digit on the left. Now let us do a typical addition. Let us write the carry digits above the addends.

Carry		0	1	1	1	1	1	0	0	1	1
A	0	1	1	0	1	1	0	0	1	0	1 ...
B	0	1	1	1	0	1	0	0	1	1	...
		0	1	0	0	1	1	0	0	0	

Note that I can do the addition only looking at a small part of the data. To do the addition electronically we need to produce a circuit with three inputs and two outputs.

Inputs	*Outputs*
Addend A α	Sum δ
Addend B β	Carry ε
Carry from last column γ	

This circuit must be such that

If no. of 1's on inputs α, β, γ is	0	The sum	0	and	0
	1	δ	1	Carry	0
	2	is	0	ε	1
	3		1	is	1

It is very easy to produce a voltage proportional to the number of pulses on the inputs, and one then merely has to provide a circuit which will discriminate between four different levels and put out the appropriate sum and carry digits. I will not attempt to describe such a circuit; it can be quite simple. When we are given the circuit we merely have to connect it up with feedback and it is an adder. Thus:

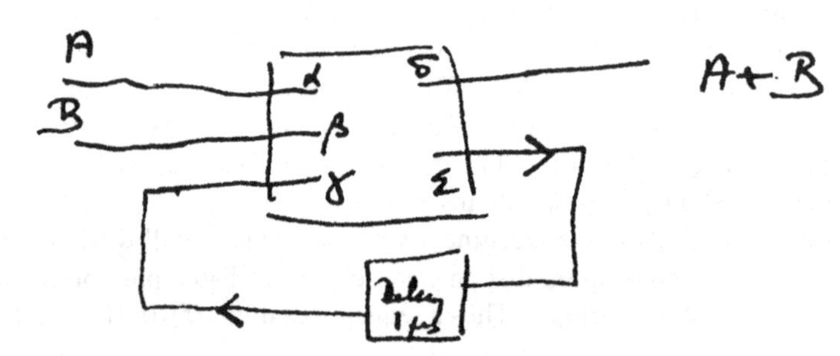

It will be seen that we have made use of the fact that the same process is used in addition with each digit, and also the fact that the properties of the electrical circuit are invariant under time shifts, at any rate if these are multiples of the clock period. It might be said that we have made use of the isomorphism between the group of these time shifts and the multiplicative group of real numbers to simplify our apparatus, though I doubt if many other applications of this principle could be found.

It will be seen that with such an adder the addition is broken down into the most elementary steps possible, such as adding one and one. Each of these occupies a microsecond. Our numbers will normally consist of 32 binary digits, so that two of them can be added in 32 microseconds. Likewise we shall do multiplications in the form of a number of consecutive additions of one and one or one and zero etc. There are 1024 such additions or thereabouts to be done in a multiplication of one 32 digit number by another, so that one might expect a multiplication to take about a millisecond. Actually the multiplier to be used on ACE will take rather over two milliseconds. This may sound rather long, when the unit operation is only a microsecond, but it actually seems that the machine is fairly well balanced in this respect, i.e. the multiplication time is not a serious bottleneck. Computers always spend just as long in writing numbers down and deciding what to do next as they do in actual multiplications, and it is just the same with the ACE. A great deal of time is spent in getting numbers in and out of storage and deciding what to do next. To complete the four elementary processes, subtraction is done by complementation and addition, and division is done by the use of the iteration formula

$$u_n = u_{n-1} + u_{n-1}(1 - au_{n-1})$$

u_n converges to a^{-1} provided $|1 - au_0| < 1$. The error is squared at each step, so that the convergence is very rapid. This process is of course programmed, i.e. the only extra apparatus required is the delay lines required for storing the relevant instructions.

Passing on from the arithmetic part there remains the input and output. For this purpose we have chosen Hollerith card equipment. We are able to obtain this without having to do any special development work. The speeds obtainable are not very impressive compared with the speeds at which the electronic equipment works, but they are quite sufficient in all cases where the calculation is long and the result concise: the interesting cases in fact. It might appear that there would be a difficulty in converting the information provided at the slow speeds appropriate to the Hollerith equipment to the high speeds required with the ACE, but it is really quite easy. The Hollerith speeds are so slow as to be counted zero or stop for many purposes, and the problem reduces to the simple one of converting a number of statically given digits into a stream of pulses. This can be done by means of a form of electronic commutator.

Before leaving the outline of the description of the machine I should mention some of the tactical situations that are met with in programming. I can illustrate two of them in connection with the calculation of the reciprocal described above. One of these is the idea of the iterative cycle. Each time that we go from u_r to u_{r+1} we apply the same sequence of operations, and it will therefore be economical in storage space if we use the same instructions. Thus we go round and round a cycle of instructions:

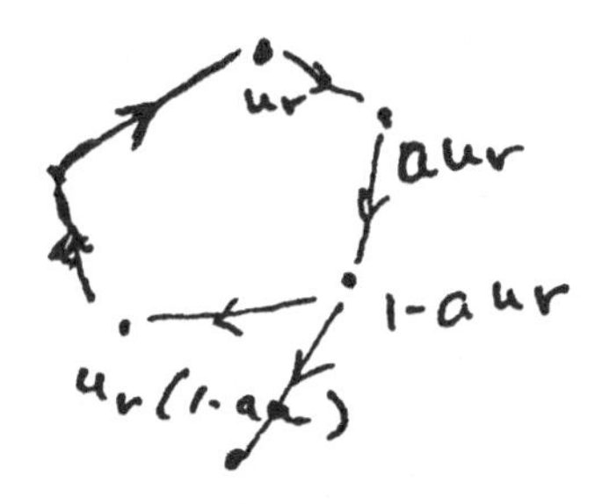

It looks however as if we were in danger of getting stuck in this cycle, and unable to get out. The solution of this difficulty involves another tactical idea, that of discrimination' i.e. of deciding what to do next partly according to the results of the machine itself, instead of according to data available to the programmer. In this case we include a discrimination in each cycle, which takes us out of the cycle when the value of $|1 - au|$ is sufficiently small. It is like an aeroplane circling over an aerodrome, and asking permission to land after each circle. This is a very simple idea, but is of the utmost importance. The idea of the iterative cycle of instructions will also be seen to be rather fundamental when it is realised that the majority of the instructions in the memory must be obeyed a great number of times. If the whole memory were occupied by instructions, none of it being used for numbers or other data, and if each instruction were obeyed once only, but took the longest possible time, the machine could only remain working for sixteen seconds.

Another important idea is that of constructing an instruction and then obeying it. This can be used amongst other things for discrimination. In the example I have just taken for instance we could calculate a quantity which was 1 if $|1 - au|$ was less than $2^{-3.1}$ and 0 otherwise. By adding this quantity to the instruction that is obeyed at the forking point the instruction can be completely altered in its effect when finally $1 - au$ is reduced to sufficiently small dimensions.

Probably the most important idea involved in instruction tables is that of standard *subsidiary tables*. Certain processes are used repeatedly in all sorts of different connections, and we wish to use the same instructions, from the same part of the memory every time. Thus we may use interpolation for the calculation of a great number of different functions, but we shall always use the same instruction table for interpolation. We have only to think out how this is to be done once, and forget then how it is done. Each time we want to do an interpolation we have only to remember the memory position where this table is kept, and make the appropriate reference in the instruction table which is using the interpolation. We might for instance be making up an instruction table for finding values of $J_0(x)$ and use the interpolation table in this way. We should then say that the interpolation table was a subsidiary to the table for calculating $J_0(x)$. There is thus a sort of hierarchy of tables. The interpolation table might be regarded as taking its orders from the J_0 table, and reporting its answers back to it. The master servant analogy is however not a very good one, as there are many more masters than servants, and many masters have to share the same servants.

Now let me give a picture of the operation of the machine. Let us begin with some problem which has been brought in by a customer. It will first go to the problems preparation section where it is examined to see whether it is in a suitable form and self-consistent, and a very rough computing procedure made out. It then goes to the tables preparation section. Let us suppose for example that the problem was to tabulate solutions of the equation

$$y'' + xy' = J_0(x)$$

with initial conditions $x = y = 0,\ y' = a$. This would be regarded as a particular case of solving the equation

$$y'' = F(x, y, y')$$

for which one would have instruction tables already prepared. One would need also a table to produce the function $F(x, y, z)$ (in this case $F(x, y, z) = J_0(x) - xz$ which would mainly involve a table to produce $J_0(x)$, and this we might expect to get off the shelf). A few additional details about the boundary conditions and the length of the arc would have to be dealt with, but much of this detail would also be found on the shelf, just like the table for obtaining $J_0(x)$. The instructions for the job would therefore consist of a considerable number taken off the shelf together with a few made up specially for the job in question. The instruction cards for the standard processes would have already been punched, but the new ones would have to be done separately. When these had all been assembled and checked they would be taken to the input mechanism, which is simply a Hollerith card feed. They would be put into the card hopper and a button pressed to start the cards moving through. It must be remembered that initially there are no instructions in the machine, and one's normal facilities are therefore not available. The first few cards that pass in have therefore to be carefully thought out to deal with this situation. They are the initial input cards and are always the same. When they have passed in a few rather fundamental instruction tables will have been set up in the machine, including sufficient to enable the machine to read the special pack of cards that has been prepared for the job we are doing. When this has been done there are various possibilities as to what happens next, depending on the way the job has been programmed. The machine might have been made to go straight on through, and carry out the job, punching or printing all the answers required, and stopping when all of this has been done. But more probably it will have been arranged that the machine stops as soon as the instruction tables have been put in. This allows for the possibility of checking that the content of the memories is correct, and for a number of variations of procedure. It is clearly a suitable moment for a break. We might also make a number of other breaks. For instance we might be interested in certain particular values of the parameter a, which were experimentally obtained figures, and it would then be convenient to pause after each parameter value, and feed the next parameter value in from another card. Or one might prefer to have the cards all ready in the hopper and let the ACE take them in as it wanted them, One can do as one wishes, but one must make up one's mind. Each time the machine pauses in this way a 'word' or sequence of 32 binary digits is displayed on neon bulbs. This word indicates the reason for stopping. I have already mentioned two possible reasons. A large class of further possible reasons is provided by the checks. The programming should be done in such a way that the ACE is frequently investigating identities which should be satisfied if all is as it should be. Whenever one of these checks fails the machine stops and displays a word which describes what check has failed.

It will be seen that the possibilities as to what one may do are immense. One of our difficulties will be the maintainence of an appropriate discipline, so that we do not lose track of what we are doing. We shall need a number of efficient librarian types to keep us in order.

Finally I should like to make a few conjectures as to the repercussions that electronic digital computing machinery will have on mathematics. I have already mentioned that the ACE will do the work of about 10,000 computers. It is to be expected therefore that large scale hand-computing will die out. Computers will still be employed on small calculations, such as the substitution of values in formulae, but whenever a single calculation may be expected to take a human computer days of work, it will presumably be done by an electronic computer instead. This will not necessitate everyone interested in such work having an electronic computer. It would be quite possible to arrange to control a distant computer by means of a telephone line. Special input and output machinery would be developed for use at these out stations, and would cost a few hundred pounds at most. The main bulk of the work done by these computers will however consist of problems which could not have been tackled by hand computing because of the scale of the undertaking. In order to supply the machine with these problems we shall need a great number of mathematicians of ability. These mathematicians will be needed in order to do the preliminary research on the problems, putting them into a form for computation. There will be considerable scope for analysts. When a human computer is working on a problem he can usually apply some common sense to give him an idea of how accurate his answers are. With a digital computer we can no longer rely on common sense, and the bounds of error must be based on some proved inequalities. We need analysts to find the appropriate inequalities for us. The inequalities need not always be explicit, i.e. one need not have them in such a form that we can tell, before the calculation starts, and using only pencil and paper, how big the error will be. The error calculation may be a serious part of the ACE's duties. To an extent it may be possible to replace the estimates of error by statistical estimates obtained by repeating the job several times, and doing the rounding off differently each time, controlling it by some random element, some electronic roulette wheel. Such statistical estimates however leave much in doubt, are wasteful in machine time, and give no indication of what can be done if it turns out that the errors are intolerably large. The statistical method can only help the analyst, not replace him.

Analysis is just one of the purposes for which we shall need good mathematicians. Roughly speaking those who work in connection with the ACE will be divided into its masters and its servants. Its masters will plan out instruction tables for it, thinking up deeper and deeper ways of using it. Its servants will feed it with cards as it calls for them. They will put right any parts that go wrong. They will assemble data that it requires. In fact the servants will take the place of limbs. As time goes on the calculator itself will take over the functions both of masters and of servants. The servants will be replaced by mechanical and electrical limbs and sense organs. One might for instance provide curve followers to enable data to be taken direct from curves instead of having girls read off values and punch them on cards. The masters are liable to get replaced because as soon as any technique becomes at all stereotyped it becomes possible to devise a system of instruction tables which will enable the electronic computer to do it for itself. It may happen however that the masters will refuse to do this. They may be unwilling to let their jobs be stolen from them in this way. In that case they would surround the whole of their work with mystery and make excuses, couched in well chosen gibberish, whenever any dangerous suggestions were made. I think that a reaction of this kind is a very real danger. This topic naturally leads to the question as to how far it is possible in principle for a computing machine to simulate human activities. I will return to this later, when I have discussed the effects of these machines on mathematics a little further.

I expect that digital computing machines will eventually stimulate a considerable interest in symbolic logic and mathematical philosophy. The language in which one communicates with these machines, i.e. the language of instruction tables, forms a sort of symbolic logic. The machine interprets whatever it is told in a quite definite manner without any sense of humour or sense of proportion. Unless in communicating with it one says exactly what one means, trouble is bound to result. Actually one could communicate with these machines in any language provided it was an exact language, i.e. in principle one should be able to communicate in any symbolic logic, provided that the machine were given instruction tables which would enable it to interpret that logical system. This would mean that there will be much more practical scope for logical systems than there has been in the past. Some attempts will probably be made to get the machine to do actual manipulations of mathematical formulae. To do so will require the development of a special logical system for the purpose. This system should resemble normal mathematical procedure closely, but at the same time should be as unambiguous as possible. As regards mathematical philosophy, since the machines will be doing more and more mathematics themselves, the centre of gravity of the human interest will be driven further and further into philosophical questions of what can in principle be done etc.

It has been said that computing machines can only carry out the processes that they are instructed to do. This is certainly true in the sense that if they do something other than what they were instructed then they have just made some mistake. It is also true that the intention in constructing these machines in the first instance is to treat them as slaves, giving them only jobs which have been thought out in detail, jobs such that the user of the machine fully understands what in principle is going on all the time. Up till the present machines have only been used in this way. But is it necessary that they should always be used in such a manner? Let us suppose we have set up a machine with certain initial instruction tables, so constructed that these tables might on occasion, if good reason arose, modify those tables. One can imagine that after the machine had been operating for some time, the instructions would have altered out of all recognition, but nevertheless still be such that one would have to admit that the machine was still doing very worthwhile calculations. Possibly it might still be getting results of the type desired when the machine was first set up, but in a much more efficient manner. In such a case one would have to admit that the progress of the machine had not been foreseen when its original instructions were put in. It would be like a pupil who had learnt much from his master, but had added much more by his own work. When this happens I feel that one is obliged to regard the machine as showing intelligence. As soon as one can provide a reasonably large memory capacity it should be possible to begin to experiment on these lines. The memory capacity of the human brain is probably of the order of ten thousand million binary digits. But most of this is probably used in remembering visual impressions, and other comparatively wasteful ways. One might reasonably hope to be able to make some real progress with a few million digits, especially if one confined one's investigations to some rather limited field such as the game of chess. It would probably be quite easy to find instruction tables which would enable the ACE to win against an average player. Indeed Shannon of Bell Telephone laboratories tells me that he has won games playing by rule of thumb: the skill of his opponents is not stated. But I would not consider such a victory very significant. What we want is a machine that can learn from experience. The possibility of letting the machine alter its own instructions provides the mechanism for this, but this of course does not get us very far.

It might be argued that there is a fundamental contradiction in the idea of a machine with intelligence. It is certainly true that 'acting like a machine', has become synonymous with lack of adaptability. But the reason for this is obvious. Machines in the past have bad very little storage, and there has been no question of the machine having any discretion. The argument might however be put into a more aggressive form. It has for instance been shown that with certain logical systems there can be no machine which will distinguish provable formulae of the system from unprovable, i.e. that there is no test that the machine can apply which will divide propositions with certainty into these two classes. Thus if a machine is made for this purpose it must in some cases fail to give an answer. On the other hand if a mathematician is confronted with such a problem he would search around and find new methods of proof, so that he ought eventually to be able to reach a decision about any given formula. This would be the argument. Against it I would say that fair play must be given to the machine. Instead of it sometimes giving no answer we could arrange that it gives occasional wrong answers. But the human mathematician would likewise make blunders when trying out new techniques. It is easy for us to regard these blunders as not counting and give him another chance, but the machine would probably be allowed no mercy. In other words then, if a machine is expected to be infallible, it cannot also be intelligent. There are several mathematical theorems which say almost exactly that. But these theorems say nothing about how much intelligence may be displayed if a machine makes no pretence at infallibility. To continue my plea for 'fair play for the machines' when testing their I.Q. A human mathematician has always undergone an extensive training. This training may be regarded as not unlike putting instruction tables into a machine. One must therefore not expect a machine to do a very great deal of building up of instruction tables on its own. No man adds very much to the body of knowledge, why should we expect more of a machine? Putting the same point differently, the machine must be allowed to have contact with human beings in order that it may adapt itself to their standards. The game of chess may perhaps be rather suitable for this purpose, as the moves of the machine's opponent will automatically provide this contact.

Computational drawing of Alan Turing, page 479, by the Aikon Project, Goldsmiths, University of London.

Intelligent Machinery

(Report written by Alan Turing for the National Physical Laboratory, 1948)

Rodney A. Brooks and —

THE CASE FOR EMBODIED INTELLIGENCE

For me Alan Turing's 1948 paper *Intelligent Machinery* was more important than his 1950 paper *Computing Machinery and Intelligence*.

At the begining of *Intelligent Machinery* Turing provided counter arguments to a number of possible objections to the idea that machines could be intelligent. And right at the end he introduced a precursor to the "Imitation Game", now commonly referred to as the *Turing Test*, of his 1950 paper. In this earlier version one human not very good chess player would try to guess whether he was playing against another human not very good chess player, or against an algorithm[1]. Expansion of these bookends became the body of *Computing Machinery and Intelligence*.

Intelligent Machinery itself was not published until 1970, so many early computer science researchers were unaware of it. I was fortunate to come in contact with it right as I was starting my academic career.

The bulk of the paper gives examples of how simple computational mechanisms could be adaptable, could be taught, and could learn for themselves. The examples and mechanisms Turing used in this exposition where networks of active computational elements. Although he connected them back to the universal machines of his 1937 paper, it is remarkable in hindsight, how different this abstraction was than the one that he had previously introduced, of the central processing element with a tape memory–still the essential model for all modern digital computers. Here, instead, he used a model inspired by brains. One can only wonder how different our technological world might be if Turing had lived to fully develop this set of ideas himself. Others carried on this second tradition, but one must think that perhaps Turing's intellectual influence might have been stronger as he would have been arguing against the approach that was adopted from his earlier work.

For me, the critical, and new, insights in *Intelligent Machinery* were two fold.

First, Turing made the distinction between embodied and disembodied intelligence. While arguing that building an embodied intelligence would be a 'sure' route to produce a thinking machine he rejected it in favor of disembodied intelligence on the grounds of technical practicalities of the era. Second, he introduced the notion of 'cultural search', that people's learning largely comes from the culture of other people in which they are immersed.

Modern researchers are now seriously investigating the emobodied approach to intelligence and have rediscovered the importance of interaction with people as the basis for intelligence. My own work for the last twenty five years has been based on these two ideas.

---

[1] At the time the opponent person had to be not very good so that they didn't outshine the then current abilities of mechanical chess playing. Today the opponent person would have to be a world champion to have any chance at not being outshone by the mechanical system!

Turing justifies the possibility of making a thinking machine by "the fact that it is possible to make machinery to imitate any small part of a man". He uses the implicit idea of his universal computing machines to dismiss the idea that it is necessary to emulate a person at the neural signal level in order to have intelligence, and instead suggests a digital computer, "if produced by present techniques, would be of immense size", which would control a robot from a distance. That robot would be built by "tak[ing] a man as a whole and to try to replace all parts of him by machines". In particular he suggests the parts would include "television cameras, microphones, loudspeakers, wheels and 'handling servo-mechanisms' ...". Turing's description from over sixty years ago, fairly precisely describes what is done today in dozens of research labs around the world with our PR2 robots, or Mekabots, with their brains offboard in racks of Linux boxes, or even off in the computing cloud.

Turing further rightfully notes that even in building such a robot "the creature would still have no contact with food, sex, sport, and many other things of interest to the human being". Nevertheless he suggests that such an approach "is probably the 'sure' way of producing a thinking machine", before dismissing it as too slow and impractical. He suggests instead that it is more practical, certainly at that time, to "see what can be done with a 'brain' which is more or less without a body". He suggests the following fields as ripe for exploration by disembodied intelligence:

(i) Various games, e.g., chess, noughts and crosses, bridge, poker
(ii) The learning of languages
(iii) Translation of languages
(iv) Cryptography
(v) Mathematics.

With these suggestions much of the early directions for the field of *Artificial Intelligence* were set, and certainly the odd numbered of Turing's suggestions formed a large part of the work in AI during its first decade.

In one paper Turing both distinguished embodied versus disembodied approaches to building intelligent machines, praised the former as more likely to succeed and either set or predicted the disembodied directions that were actually followed for many years.

But later, towards the very end of *Intelligent Machinery* he comes back to the place of bodies in the world. He distinguishes three kinds of *search* as ways to build intelligent systems: *intellectual searches*, *genetic search*, and *cultural search*. The first is the direction that classical AI went, where programs try to learn and improve their performance. Although he did not suggest that it be mechanised, genetic search has become a thoroughly practical approach to design and optimisation. And lastly, by cultural search, Turing means the way in which interactions with others contributes to the development of intelligence. This developmental approach, using social robots, has only now become practical in the last fifteen years, and is a rich source of both theoretical and practical learning systems for robots.

It is humbling to read Alan Turing's papers. He thought of it all. First.

INTELLIGENT MACHINERY

A. M. TURING

Abstract

The possible ways in which machinery might be made to show intelligent behaviour are discussed. The analogy with the human brain is used as a guiding principle. It is pointed out that the potentialities of the human intelligence can only be realized if suitable education is provided. The investigation mainly centres round an analogous teaching process applied to machines. The idea of an unorganized machine is defined, and it is suggested that the infant human cortex is of this nature. Simple examples of such machines are given, and their education by means of rewards and punishments is discussed. In one case the education process is carried through until the organization is similar to that of an ACE.

I propose to investigate the question as to whether it is possible for machinery to show intelligent behaviour. It is usually assumed without argument that it is not possible. Common catch phrases such as 'acting like a machine', 'purely mechanical behaviour' reveal this common attitude. It is not difficult to see why such an attitude should have arisen. Some of the reasons are:

(a) An unwillingness to admit the possibility that mankind can have any rivals in intellectual power. This occurs as much amongst intellectual people as amongst others: they have more to lose. Those who admit the possibility all agree that its realization would be very disagreeable. The same situation arises in connection with the possibility of our being superseded by some other animal species. This is almost as disagreeable and its theoretical possibility is indisputable.

(b) A religious belief that any attempt to construct such machines is a sort of Promethean irreverence.

(c) The very limited character of the machinery which has been used until recent times (e.g. up to 1940). This encouraged the belief that machinery was necessarily limited to extremely straightforward, possibly even to repetitive, jobs. This attitude is very well expressed by Dorothy Sayers (*The Mind of the Maker* p. 46) '... which imagines that God, having created his Universe, has now screwed the cap on His pen, put His feet on the mantelpiece and left the work to get on with itself.' This, however, rather comes into St Augustine's category of figures of speech or enigmatic sayings framed from things which do not exist at all. We simply do not know of any creation which goes on creating itself in variety when the creator has withdrawn from it. The idea is that God simply created a vast machine and has left it working until it runs down from lack of fuel. This is another of those obscure analogies, since we have no experience of machines that produce variety of their own accord; thenature of a machine is to 'do the same thing over and over again so long as it keeps going'.

(d) Recently the theorem of Gödel and related results (Gödel 1931, Church 1936, Turing 1937) have shown that if one tries to use machines for such purposes as determining the truth or falsity of mathematical theorems and one is not willing to tolerate an occasional wrong result, then any given machine will in some cases be unable to give an answer at all. On the other hand the human intelligence seems to be able to find methods of ever-increasing power for dealing with such problems 'transcending' the methods available to machines.

(e) In so far as a machine can show intelligence this is to be regarded as nothing but a reflection of the intelligence of its creator.

Refutation of some objections

In this section I propose to outline reasons why we do not need to be influenced by the above-described objections. The objections (a) and (b), being purely emotional, do not really need to be refuted. If one feels it necessary to refute them there is little to be said that could hope to prevail, though the actual production of the machines would probably have some effect. In so far then as we are influenced by such arguments we are bound to be left feeling rather uneasy about the whole project, at any rate for the present. These arguments cannot be wholly ignored, because the idea of 'intelligence' is itself emotional rather than mathematical.

The objection (c) in its crudest form is refuted at once by the actual existence of machinery (ENIAC etc.) which can go on through immense numbers (e.g. $10^{60,000}$ about for ACE) of operations without repetition, assuming no breakdown. The more subtle forms of this objection will be considered at length on pages 18–22.

The argument from Gödel's and other theorems (objection d) rests essentially on the condition that the machine must not make mistakes. But this is not a requirement for intelligence. It is related that the infant Gauss was asked at school to do the addition $15 + 18 + 21 + \cdots + 54$ (or something of the kind) and that he immediately wrote down 483, presumably having calculated it as $(15 + 54)(54 - 12)/2.3$. One can imagine circumstances where a foolish master told the child that he ought instead to have added 18 to 15 obtaining 33, then added 21, etc. From some points of view this would be a 'mistake', in spite of the obvious intelligence involved. One can also imagine a situation where the children were given a number of additions to do, of which the first 5 were all arithmetic progressions, but the 6th was say $23 + 34 + 45 + \cdots + 100 + 112 + 122 + \cdots + 199$. Gauss might have given the answer to this as if it were an arithmetic progression, not having noticed that the 9th term was 112 instead of 111. This would be a definite mistake, which the less intelligent children would not have been likely to make.

The view (d) that intelligence in machinery is merely a reflection of that of its creator is rather similar to the view that the credit for the discoveries of a pupil should be given to his teacher. In such a case the teacher would be pleased with the success of his methods of education, but would not claim the results themselves unless he had actually communicated them to his pupil. He would certainly have envisaged in very broad outline the sort of thing his pupil might be expected to do, but would not expect to foresee any sort of detail. It is already possible to produce machines where this sort of situation arises in a small degree. One can produce 'paper machines' for playing chess. Playing against such a machine gives a definite feeling that one is pitting one's wits against something alive.

These views will all be developed more completely below.

Varieties of machinery

It will not be possible to discuss possible means of producing intelligent machinery without introducing a number of technical terms to describe different kinds of existent machinery.

'Discrete' and 'continuous' machinery. We may call a machine 'discrete' when it is natural to describe its possible states as a discrete set, the motion of the machine occurring by jumping from one state to another. The states of 'continuous' machinery on the other hand form a continuous manifold, and the behaviour of the machine is described by a curve on this manifold. All machinery can be regarded as continuous, but when it is possible to regard it as discrete it is usually best to do so. The states of discrete machinery will be described as 'configurations'.

'Controlling' and 'active' machinery. Machinery may be described as 'con- trolling' if it only deals with information. In practice this condition is much the same as saying that the magnitude

of the machine's effects may be as small as we please, so long as we do not introduce confusion through Brownian movement, etc. 'Active' machinery is intended to produce some definite physical effect.

Examples		
	A Bulldozer	Continuous Active
	A Telephone	Continuous Controlling
	A Brunsviga	Discrete Controlling
	A Brain (probably)	Continuous Controlling, but is very similar to much discrete machinery
	The ENIAC, ACE, etc.	Discrete Controlling
	A Differential Analyser	Continuous Controlling.

We shall mainly be concerned with discrete controlling machinery. As we have mentioned, brains very nearly fall into this class, and there seems every reason to believe that they could have been made to fall genuinely into it without any change in their essential properties. However, the property of being 'discrete' is only an advantage for the theoretical investigator, and serves no evolutionary purpose, so we could not expect Nature to assist us by producing truly 'discrete' brains.

Given any discrete machine the first thing we wish to find out about it is the number of states (configurations) it can have. This number may be infinite (but enumerable) in which case we say that the machine has infinite memory (or storage) capacity. If the machine has a finite number N of possible states then we say that it has a memory capacity of (or equivalent to) $\log_2 N$ binary digits. According to this definition we have the following table of capacities, very roughly

The memory capacity of a machine more than anything else determines the complexity of its possible behaviour.

Brunsviga	90
ENIAC without cards and with fixed programme	600
ACE as proposed	60,000
Manchester machine (as actually working 8 August 1947)	1,100

The behaviour of a discrete machine is completely described when we are given the state (configuration) of the machine as a function of the immediately preceding state and the relevant external data.

Logical computing machines (LCMs)

In Turing (1937) a certain type of discrete machine was described. It had an infinite memory capacity obtained in the form of an infinite tape marked out into squares on each of which a symbol could be printed. At any moment there is one symbol in the machine; it is called the scanned symbol. The machine can alter the scanned symbol and its behaviour is in part described by that symbol, but the symbols on the tape elsewhere do not affect the behaviour of the machine. However the tape can be moved back and forth through the machine, this being one of the elementary operations of the machine. Any symbol on the tape may therefore eventually have an innings.

These machines will here be called 'Logical Computing Machines'. They are chiefly of interest when we wish to consider what a machine could in principle be designed to do, when we are willing to allow it both unlimited time and unlimited storage capacity.

Universal logical computing machines. It is possible to describe LCMs in a very standard way, and to put the description into a form which can be 'understood' (i.e., applied by) a special machine. In particular it is possible to design a 'universal machine' which is an LCM such that if the standard description of some other LCM is imposed on the otherwise blank tape from outside, and the (universal) machine then set going it will carry out the operations of the particular machine whose description it was given. For details the reader must refer to Turing (1937).

The importance of the universal machine is clear. We do not need to have an infinity of different machines doing different jobs. A single one will suffice. The engineering problem of producing various machines for various jobs is replaced by the office work of 'programming' the universal machine to do these jobs.

It is found in practice that LCMs can do anything that could be described as 'rule of thumb' or 'purely mechanical'. This is sufficiently well established that it is now agreed amongst logicians that 'calculable by means of an LCM' is the correct accurate rendering of such phrases. There are several mathematically equivalent but superficially very different renderings.

Practical computing machines (PCMs)

Although the operations which can be performed by LCMs include every rule- of-thumb process, the number of steps involved tends to be enormous. This is mainly due to the arrangement of the memory along the tape. Two facts which need to be used together may be stored very far apart on the tape. There is also rather little encouragement, when dealing with these machines, to condense the stored expressions at all. For instance the number of symbols required in order to express a number in Arabic form (e.g., 149056) cannot be given any definite bound, any more than if the numbers are expressed in the 'simplified Roman' form (IIIII . . . I, with 149056 occurrences of I). As the simplified Roman system obeys very much simpler laws one uses it instead of the Arabic system.

In practice however one *can* assign finite bounds to the numbers that one will deal with. For instance we can assign a bound to the number of steps that we will admit in a calculation performed with a real machine in the following sort of way. Suppose that the storage system depends on charging condensers of capacity $C = 1\mu\text{f}$, and that we use two states of charging, $E = 100$ volts and $-E = -100$ volts. When we wish to use the information carried by the condenser we have to observe its voltage. Owing to thermal agitation the voltage observed will always be slightly wrong, and the probability of an error between V and $V - \mathrm{d}V$ volts is

$$\frac{2kT}{\pi C} e^{-\frac{1}{2}V^2 C/kT} V \mathrm{d}V$$

where k is Boltzmann's constant. Taking the values suggested we find that the probability of reading the sign of the voltage wrong is about $10^{-1 \cdot 2 \times 10^{16}}$. If then a job took more than $10^{10^{17}}$ steps we should be virtually certain of getting the wrong answer, and we may therefore restrict ourselves to jobs with fewer steps. Even a bound of this order might have useful simplifying effects. More practical bounds are obtained by assuming that a light wave must travel at least 1 cm between steps (this would only be false with a very small machine), and that we could not wait more than 100 years for an answer. This would give a limit of 10^{20} steps. The storage capacity will probably have a rather similar bound, so that we could use sequences of 20 decimal digits for describing the position in which a given piece of data was to be found, and this would be a really valuable possibility.

Machines of the type generally known as 'Automatic Digital Computing Machines' often make great use of this possibility. They also usually put a great deal of their stored information in a form very different from the tape form. By means of a system rather reminiscent of a telephone exchange it is made possible to obtain a piece of information almost immediately by 'dialling' the position of this information in the store. The delay may be only a few microseconds with some systems. Such machines will be described as 'Practical Computing Machines'.

Universal practical computing machines. Nearly all of the PCMs now under construction have the essential properties of the 'Universal Logical Computing Machines' mentioned earlier. In practice, given any job which could have been done on an LCM one can also do it on one of these digital computers. I do not mean that we can do any required job of the type mentioned on it by suitable programming. The programming is pure paper work. It naturally occurs to one to ask whether, e.g., the ACE would be truly universal if its memory capacity were infinitely extended. I have investigated this question, and the answer appears to be as follows, though I have not proved any formal mathematical theorem about it. As has been explained, the ACE at present uses finite sequences of digits to describe positions in its memory: they are actually sequences of 9 binary digits (September 1947). The ACE also works largely for other purposes with sequences of 32 binary digits. If the memory were extended, e.g., to 1000 times its present capacity, it would be natural to arrange the memory in blocks of nearly the maximum capacity which can be handled with the 9 digits, and from time to time to switch from block to block. A relatively small part would never be switched. This would contain some of the more fundamental instruction tables and those concerned with switching. This part might be called the 'central part'. One would then need to have a number which described which block was in action at any moment. However this number might be as large as one pleased. Eventually the point would be reached where it could not be stored in a word (32 digits), or even in the central part. One would then have to set aside a block for storing the number, or even a sequence of blocks, say blocks $1, 2, \ldots n$. We should then have to store n, and in theory it would be of indefinite size. This sort of process can be extended in all sorts of ways, but we shall always be left with a positive integer which is of indefinite size and which needs to be stored somewhere, and there seems to be no way out of the difficulty but to introduce a 'tape'. But once this has been done, and since we are only trying to prove a theoretical result, one might as well, whilst proving the theorem, ignore all the other forms of storage. One will in fact have a ULCM with some complications. This in effect means that one will not be able to prove any result of the required kind which gives any intellectual satisfaction.

Paper machines

It is possible to produce the effect of a computing machine by writing down a set of rules of procedure and asking a man to carry them out. Such a combination of a man with written instructions will be called a 'Paper Machine'. A man provided with paper, pencil, and rubber, and subject to strict discipline, is in effect a universal machine. The expression 'paper machine' will often be used below.

Partially random and apparently partially random machines

It is possible to modify the above described types of discrete machines by allowing several alternative operations to be applied at some points, the alternatives to be chosen by a random process. Such a machine will be described as 'partially random'. If we wish to say definitely that a machine is not of this kind we will describe it as 'determined'. Sometimes a machine may be strictly speaking determined but appear superficially as if it were partially random. This would occur if for instance the digits of the number π were used to determine the choices of a partially random machine, where previously a dice thrower or electronic equivalent had been used. These machines are known as apparently partially random.

Unorganised machines

So far we have been considering machines which are designed for a definite purpose (though the universal machines are in a sense an exception). We might instead consider what happens when we make up a machine in a comparatively unsystematic way from some kind of standard components.

We could consider some particular machine of this nature and find out what sort of things it is likely to do. Machines which are largely random in their construction in this way will be called ‘Unorganized Machines’. This does not pretend to be an accurate term. It is conceivable that the same machine might be regarded by one man as organized and by another as unorganized.

A typical example of an unorganized machine would be as follows. The machine is made up from a rather large number N of similar units. Each unit has two input terminals, and has an output terminal which can be connected to the input terminals of (0 or more) other units. We may imagine that for each integer r, $1 \leq r \leq N$ two numbers $i(r)$ and $j(r)$ are chosen at random from $1 \ldots N$ and that we connect the inputs of unit r to the outputs of units (r) and $j(r)$. All of the units are connected to a central synchronizing unit from which synchronizing pulses are emitted at more or less equal intervals of time. The times when these pulses arrive will be called ‘moments’. Each unit is capable of having two states at each moment. These states may be called 0 and 1. The state is determined by the rule that the states of the units from which the input leads come are to be taken at the previous moment, multiplied together and the result subtracted from 1. An unorganized machine of this character is shown in the diagram below.

r	i(r)	j(r)
1	3	2
2	3	5
3	4	5
4	3	4
5	2	5

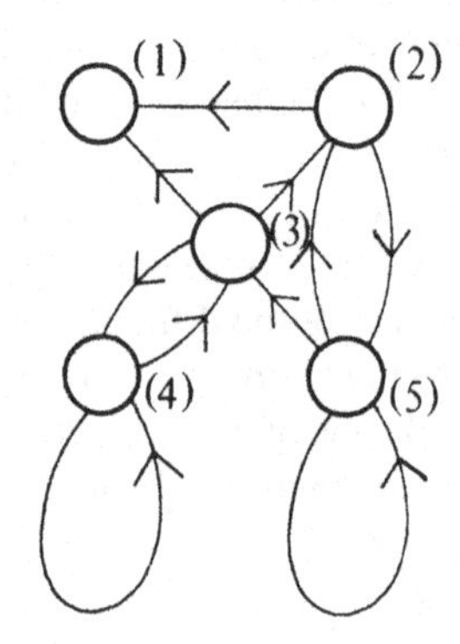

A sequence of six possible consecutive conditions for the whole machine is:

1	1	1	0	0	1	0
2	1	1	1	0	1	0
3	0	1	1	1	1	1
4	0	1	0	1	0	1
5	1	0	1	0	1	0

The behaviour of a machine with so few units is naturally very trivial. However, machines of this character can behave in a very complicated manner when the number of units is large. We may call these A-type unorganized machines. Thus the machine in the diagram is an A-type unorganized machine of 5 units. The motion of an A-type machine with N units is of course eventually periodic, as in any determined machine with finite memory capacity. The period cannot exceed 2^N moments, nor can the length of time before the periodic motion begins. In the example above the period is 2 moments and there are 3 moments before the periodic motion begins. 2^N is 32.

The A-type unorganized machines are of interest as being about the simplest model of a nervous system with a random arrangement of neurons. It would therefore be of very great interest to find out something about their behaviour. A second type of unorganized machine will now be described, not because it is of any great intrinsic importance, but because it will be useful later for illustrative purposes. Let us denote the circuit

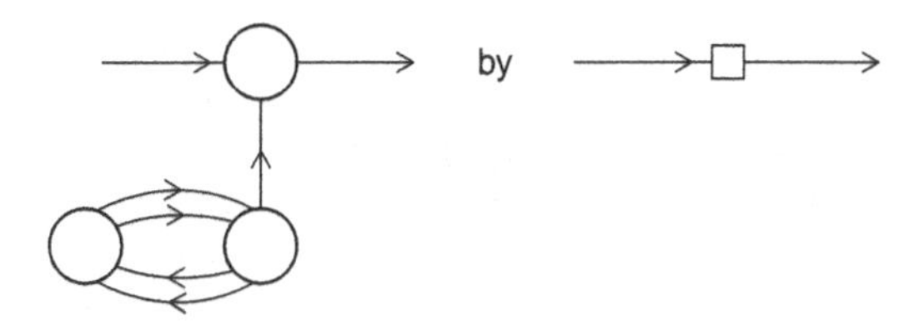

as an abbreviation. Then for each A-type unorganized machine we can construct another machine by replacing each connection ——→—— in it by ——→□——→. The resulting machines will be called B-type unorganized machines. It may be said that the B-type machines are all A-type. To this I would reply that the above definitions if correctly (but drily!) set out would take the form of describing the probability of an A- (or B-) type machine belonging to a given set; it is not merely a definition of which are the A-type machines and which are the B-type machines. If one chooses an A-type machine, with a given number of units, at random, it will be extremely unlikely that one will get a B-type machine.

It is easily seen that the connection ——→□——→ can have three conditions. It may (i) pass all signals through with interchange of 0 and 1, or (ii) it may convert all signals into 1, or again (iii) it may act as in (i) and (ii) in alternate moments. (Alternative (iii) has two sub-cases.) Which of these cases applies depends on the initial conditions. There is a delay of two moments in going through ——→□——→.

Interference with machinery. Modifiable and self-modifying machinery

The types of machine that we have considered so far are mainly ones that are allowed to continue in their own way for indefinite periods without interference from outside. The universal machines were an exception to this, in that from time to time one might change the description of the machine which is being imitated. We shall now consider machines in which such interference is the rule rather than the exception.

We may distinguish two kinds of interference. There is the extreme form in which parts of the machine are removed and replaced by others. This may be described as 'screwdriver interference'. At the other end of the scale is 'paper interference', which consists in the mere communication of information to the machine, which alters its behaviour. In view of the properties of the universal machine we do not need to consider the difference between these two kinds of machine as being so very radical after all. Paper interference when applied to the universal machine can be as useful as screwdriver interference.

We shall mainly be interested in paper interference. Since screwdriver interference can produce a completely new machine without difficulty there is rather little to be said about it. In future 'interference' will normally mean 'paper interference'.

When it is possible to alter the behaviour of a machine very radically we may speak of the machine as being 'modifiable'. This is a relative term. One machine may be spoken of as being more modifiable than another.

One may also sometimes speak of a machine modifying itself, or of a machine changing its own instructions. This is really a nonsensical form of phraseology, but is convenient. Of course, according to our conventions the 'machine' is completely described by the relation between its possible configurations at consecutive moments. It is an abstraction which, by the form of its definition, cannot change in time. If we consider the machine as starting in a particular configuration, however, we may be tempted to ignore those configurations which cannot be reached without interference from it. If we do this we should get a 'successor relation' for the configurations with different properties from the original one and so a different 'machine'.

If we now consider interference, we should say that each time interference occurs the machine is probably changed. It is in this sense that interference 'modifies' a machine. The sense in which a machine can modify itself is even more remote. We may, if we wish, divide the operations of the machine into two classes, normal and self-modifying operations. So long as only normal operations are performed we regard the machine as unaltered. Clearly the idea of 'self-modification' will not be of much interest except where the division of operations into the two classes is made very carefully. The sort of case I have in mind is a computing machine like the ACE where large parts of the storage are normally occupied in holding instruction tables. (Instruction tables are the equivalent in UPCMs of descriptions of machines in ULCMs). Whenever the content of this storage was altered by the internal operations of the machine, one would naturally speak of the machine 'modifying itself'.

Man as a machine

A great positive reason for believing in the possibility of making thinking machinery is the fact that it is possible to make machinery to imitate any small part of a man. That the microphone does this for the ear, and the television camera for the eye are commonplaces. One can also produce remote-controlled robots whose limbs balance the body with the aid of servo-mechanisms. Here we are chiefly interested in the nervous system. We could produce fairly accurate electrical models to copy the behaviour of nerves, but there seems very little point in doing so. It would be rather like putting a lot of work into cars which walked on legs instead of continuing to use wheels. The electrical circuits which are used in electronic computing machinery seem to have the essential properties of nerves. They are able to transmit information from place to place, and also to store it. Certainly the nerve has many advantages. It is extremely compact, does not wear out (probably for hundreds of years if kept in a suitable medium!) and has a very low energy consumption. Against these advantages the electronic circuits have only one counter-attraction, that of speed. This advantage is, how- ever, on such a scale that it may possibly outweigh the advantages of the nerve.

One way of setting about our task of building a 'thinking machine' would be to take a man as a whole and to try to replace all the parts of him by machinery. He would include television cameras, microphones, loudspeakers, wheels and 'handling servo-mechanisms' as well as some sort of 'electronic brain'. This would be a tremendous undertaking of course. The object, if produced by present techniques, would be of immense size, even if the 'brain' part were stationary and controlled the body from a distance. In order that the machine should have a chance of finding things out for itself it should be allowed to roam the countryside, and the danger to the ordinary citizen would be serious. Moreover even when the facilities mentioned above were provided, the creature would still have no contact with food, sex, sport and many other things of interest to the human being. Thus although this method is probably the 'sure' way of producing a thinking machine it seems to be altogether too slow and impracticable.

Instead we propose to try and see what can be done with a 'brain' which is more or less without a body providing, at most, organs of sight, speech, and hearing. We are then faced with the problem of finding suitable branches of thought for the machine to exercise its powers in. The following fields appear to me to have advantages:

(i) Various games, e.g., chess, noughts and crosses, bridge, poker
(ii) The learning of languages
(iii) Translation of languages
(iv) Cryptography
(v) Mathematics.

Of these (i), (iv), and to a lesser extent (iii) and (v) are good in that they require little contact with the outside world. For instance in order that the machine should be able to play chess its only organs need be 'eyes' capable of distinguishing the various positions on a specially made board, and means for announcing its own moves. Mathematics should preferably be restricted to branches where diagrams are not much used. Of the above possible fields the learning of languages would be the most impressive, since it is the most human of these activities. This field seems however to depend rather too much on sense organs and locomotion to be feasible.

The field of cryptography will perhaps be the most rewarding. There is a remarkably close parallel between the problems of the physicist and those of the cryptographer. The system on which a message is enciphered corresponds to the laws of the universe, the intercepted messages to the evidence available, the keys for a day or a message to important constants which have to be determined. The correspondence is very close, but the subject matter of cryptography is very easily dealt with by discrete machinery, physics not so easily.

Education of machinery

Although we have abandoned the plan to make a 'whole man', we should be wise to sometimes compare the circumstances of our machine with those of a man. It would be quite unfair to expect a machine straight from the factory to compete on equal terms with a university graduate. The graduate has had contact with human beings for twenty years or more. This contact has been modifying his behaviour pattern throughout that period. His teachers have been intentionally trying to modify it. At the end of the period a large number of standard routines will have been superimposed on the original pattern of his brain. These routines will be known to the community as a whole. He is then in a position to try out new combinations of these routines, to make slight variations on them, and to apply them in new ways.

We may say then that in so far as a man is a machine he is one that is subject to very much interference. In fact interference will be the rule rather than the exception. He is in frequent communication with other men, and is continually receiving visual and other stimuli which themselves constitute a form of interference. It will only be when the man is 'concentrating' with a view to eliminating these stimuli or 'distractions' that he approximates a machine without interference.

We are chiefly interested in machines with comparatively little interference, for reasons given in the last section, but it is important to remember that although a man when concentrating may behave like a machine without interference, his behaviour when concentrating is largely determined by the way he has been conditioned by previous interference.

If we are trying to produce an intelligent machine, and are following the human model as closely as we can, we should begin with a machine with very little capacity to carry out elaborate operations or to react in a disciplined manner to orders (taking the form of interference). Then by applying appropriate interference, mimicking education, we should hope to modify the machine until it could be relied on to produce definite reactions to certain commands. This would be the beginning of the process. I will not attempt to follow it further now.

Organizing unorganised machinery

Many unorganized machines have configurations such that if once that configuration is reached, and if the interference thereafter is appropriately restricted, the machine behaves as one organized for some definite purpose. For instance, the B-type machine shown below was chosen at random.

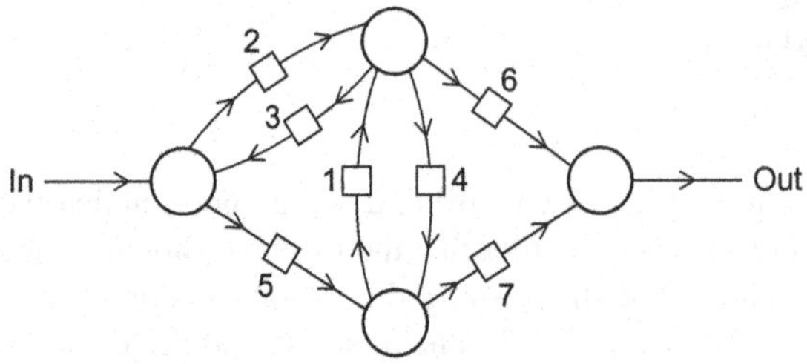

If the connections numbered 1, 3, 6, 4, are in condition (ii) initially and connections 2, 5, 7 are in condition (i), then the machine may be considered to be one for the purpose of passing on signals with a delay of 4 moments. This is a particular case of a very general property of B-type machines (and many other types), viz., that with suitable initial conditions they will do any required job, given sufficient time and provided the number of units is sufficient. In particular with a B-type unorganized machine with sufficient units one can find initial conditions which will make it into a universal machine with a given storage capacity. (A formal proof to this effect might be of some interest, or even a demonstration of it starting with a particular unorganized B-type machine, but I am not giving it as it lies rather too far outside the main argument.)

With these B-type machines the possibility of interference which could set in appropriate initial conditions has not been arranged for. It is however not difficult to think of appropriate methods by which this could be done. For instance instead of the connection

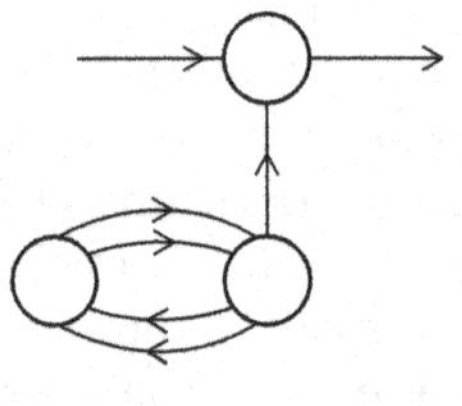

one might use

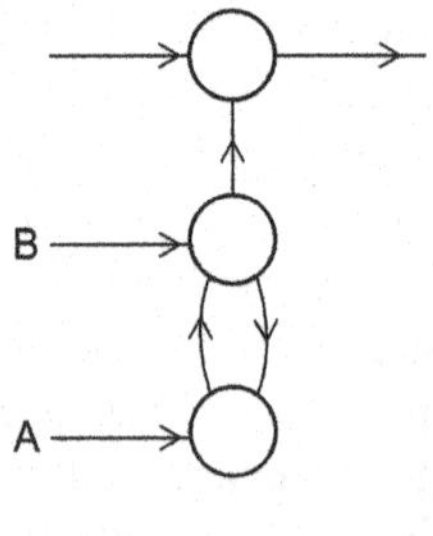

Here A, B are interfering inputs, normally giving the signal '1'. By supplying appropriate other signals at A, B we can get the connection into condition (i) or (ii), as desired. However this requires two special interfering inputs for each connection.

We shall be interested mainly in cases where there are only quite few independent inputs altogether, so that all the interference which sets up the 'initial conditions' of the machine has to be provided through one or two inputs. The process of setting up these initial conditions so that the machine will carry out some particular useful task may be called 'organizing the machine'. 'Organizing' is thus a form of 'modification'.

The cortex as an unorganised machine

Many parts of a man's brain are definite nerve circuits required for quite definite purposes. Examples of these are the 'centres' which control respiration, sneezing, following moving objects with the eyes, etc.: all the reflexes proper (not 'conditioned') are due to the activities of these definite structures in the brain. Likewise the apparatus for the more elementary analysis of shapes and sounds probably comes into this category. But the more intellectual activities of the brain are too varied to be managed on this basis. The difference between the languages spoken on the two sides of the Channel is not due to difference in development of the French-speaking and English-speaking parts of the brain. It is due to the linguistic parts having been subjected to different training. We believe then that there are large parts of the brain, chiefly in the cortex, whose function is largely indeterminate. In the infant these parts do not have much effect: the effect they have is uncoordinated. In the adult they have great and purposive effect: the form of this effect depends on the training in childhood. A large remnant of the random behaviour of infancy remains in the adult.

All of this suggests that the cortex of the infant is an unorganized machine, which can be organized by suitable interfering training. The organizing might result in the modification of the machine into a universal machine or something like it. This would mean that the adult will obey orders given in appropriate language, even if they were very complicated; he would have no common sense, and would obey the most ridiculous orders unflinchingly. When all his orders had been fulfilled he would sink into a comatose state or perhaps obey some standing order, such as eating. Creatures not unlike this can really be found, but most people behave quite differently under many circumstance. However the resemblance to a universal machine is still very great, and suggests to us that the step from the unorganized infant to a universal machine is one which should be understood. When this has been mastered we shall be in a far better position to consider how the organizing process might have been modified to produce a more normal type of mind.

This picture of the cortex as an unorganized machine is very satisfactory from the point of view of evolution and genetics. It clearly would not require any very complex system of genes to produce something like the A- or B-type unorganized machine. In fact this should be much easier than the production of such things as the respiratory centre. This might suggest that intelligent races could be produced comparatively easily. I think this is wrong because the possession of a human cortex (say) would be virtually useless if no attempt was made to organize it. Thus if a wolf by a mutation acquired a human cortex there is little reason to believe that he would have any selective advantage. If however the mutation occurred in a milieu where speech had developed (parrot-like wolves), and if the mutation by chance had well permeated a small community, then some selective advantage might be felt. It would then be possible to pass information on from generation to generation. However this is all rather speculative.

Experiments in organizing: pleasure–pain systems

It is interesting to experiment with unorganized machines admitting definite types of interference and try to organize them, e.g., to modify them into universal machines.

The organization of a machine into a universal machine would be most impressive if the arrangements of interference involve very few inputs. The training of the human child depends largely on a system of rewards and punishments, and this suggests that it ought to be possible to carry through the organizing with only two interfering inputs, one for 'pleasure' or 'reward' (R) and the other for 'pain' or punishment' (P). One can devise a large number of such 'pleasure–pain' systems. I will use this term to mean an unorganized machine of the following general character: The configurations of the machine are described by two expressions, which we may call the character-expression and the situation-expression. The character and situation at any moment, together with the input signals, determine the character and situation at the next moment. The character may be subject to some random variation. Pleasure interference has a tendency to fix the character, i.e., towards preventing it changing, whereas pain stimuli tend to disrupt the character, causing features which had become fixed to change, or to become again subject to random variation.

This definition is probably too vague and general to be very helpful. The idea is that when the 'character' changes we like to think of it as a change in the machine, but the 'situation' is merely the configuration of the machine described by the character. It is intended that pain stimuli occur when the machine's behaviour is wrong, pleasure stimuli when it is particularly right. With appropriate stimuli on these lines, judiciously operated by the 'teacher', one may hope that the 'character' will converge towards the one desired, i.e., that wrong behaviour will tend to become rare.

I have investigated a particular type of pleasure–pain system, which I will now describe.

The P-type unorganised machine

The P-type machine may be regarded as an LCM without a tape, and whose description is largely incomplete. When a configuration is reached, for which the action is undetermined, a random choice for the missing data is made and the appropriate entry is made in the description, tentatively, and is applied. When a pain stimulus occurs all tentative entries are cancelled, and when a pleasure stimulus occurs they are all made permanent.

Specifically. The situation is a number $s = 1, 2, \ldots, N$ and corresponds to the configuration of the incomplete machine. The character is a table of N entries showing the behaviour of the machine in each situation. Each entry has to say something both about the next situation and about what action the machine has to take. The action part may be either

(i) To do some externally visible act A_1 or $A_2 \ldots A_K$
(ii) To set one of the memory units $M_1 \ldots M_R$ either into the '1' condition or into the 0' condition.

The next situation is always the remainder either of $2s$ or of $2s + 1$ on division by N. These may be called alternatives 0 and 1. Which alternative applies may be determined by either

(a) one of the memory units
(b) a sense stimulus
(c) the pleasure–pain arrangements.

In each situation it is determined which of these applies when the machine is made, i.e., interference cannot alter which of the three cases applies. Also in cases (a) and (b) interference can have no effect. In case (c) the entry in the character table may be either U ('uncertain'), or T0 (tentative 0), T1, D0 (definite 0) or D1. When the entry in the character for the current situation is U then the alternative is chosen at random, and the entry in the character is changed to T0 or T1 according as 0 or 1 was chosen. If the character entry was T0 or D0 then the alternative is 0 and if it is T1 or D1 then the alternative is 1. The changes in character include the above mentioned change from U to T0 or T1, and a change of every T to D when a pleasure stimulus occurs, changes of T0 and T1 to U when a pain stimulus occurs.

We may imagine the memory units essentially as 'trigger circuits' or switches. The sense stimuli are means by which the teacher communicates 'unemotionally' to the machine, i.e., otherwise than by pleasure and pain stimuli. There are a finite number S of sense stimulus lines, and each always carries either the signal 0 or 1.

A small P-type machine is described in the table below

1	P	A	
2	P	B	M1 = 1
3	P	B	
4	S1	A	M1 = 0
5	M1	C	

In this machine there is only one memory unit M1 and one sense line S1. Its behaviour can be described by giving the successive situations together with the actions of the teacher: the latter consist of the values of S1 and the rewards and punishments. At any moment the 'character' consists of the above table with each 'P' replaced by either U, T0, D0 or D1. In working out the behaviour of the machine it is convenient first of all to make up a sequence of random digits for use when the U cases occur. Underneath these we may write the sequence of situations, and have other rows for the corresponding entries from the character, and for the actions of the teacher. The character and the values stored in the memory units may be kept on another sheet. The T entries may be made in pencil and the D entries in ink. A bit of the behaviour of the machine is given below:

Random sequence	0	0	1	1	1	0	0	1	0	0	1	1	0	1	1	0	0	0
Situations	3	1	3	1	3	1	3	1	2	4	4	4	3	2	.	.		
Alternative given by		U	T	T	T	T	T	U	U	S	S	S	U	T				
			0	0	0	0	0			1	1	1		0				
Visible action	B	A	B	A	B	A	B	A	B	A	A	A	B	B				
Rew. & Pun.							P											
Changes in S1	1											0						

It will be noticed that the machine very soon got into a repetitive cycle. This became externally visible through the repetitive BABAB. By means of a pain stimulus, this cycle was broken.
It will be noticed that the machine very soon got into a repetitive cycle. This became externally visible through the repetitive BABAB... By means of a pain stimulus this cycle was broken.

It is probably possible to organize these P-type machines into universal machines, but it is not easy because of the form of memory available. It would be necessary to organize the randomly distributed 'memory units' to provide a systematic form of memory, and this would not be easy. If, however, we supply the P-type machine with a systematic external memory this organizing becomes quite feasible. Such a memory could be provided in the form of a tape, and the externally visible operations could include movement to right and left along the tape, and altering the symbol on the tape to 0 or to 1. The sense lines could include one from the symbol on the tape. Alternatively, if the memory were to be finite, e.g., not more than 2^{32} binary digits, we could use a dialling system. (Dialling systems can also be used with an infinite memory, but this is not of much practical interest.) I have succeeded in organizing such a (paper) machine into a universal machine.

The details of the machine involved were as follows. There was a circular memory consisting of 64 squares of which at any moment one was in the machine ('scanned') and motion to right or left were among the 'visible actions'. Changing the symbol on the square was another 'visible action', and the symbol was connected to one of the sense lines S1. The even-numbered squares also had another function, they controlled the dialling of information to or from the main memory. This main memory consisted of 2^{32} binary digits. At any moment one of these digits was connected to the sense line S2. The digit of the main memory concerned was that indicated by the 32 even positioned digits of the circular memory. Another two of the 'visible actions' were printing 0 or 1 in this square of the main memory. There were also three ordinary memory units and three sense units S3, S4, S5. Also six other externally visible actions A,B,C,D,E,F.

This P-type machine with external memory has, it must be admitted, considerably more 'organization' than say the A-type unorganized machine. Nevertheless the fact that it can be organized into a universal machine still remains interesting.

The actual technique by which the 'organizing' of the P-type machine was carried through is perhaps a little disappointing. It is not sufficiently analogous to the kind of process by which a child would really be taught. The process actually adopted was first to let the machine run for a long time with continuous application of pain, and with various changes of the sense data S3, S4, S5. Observation of the sequence of externally visible actions for some thousands of moments made it possible to set up a scheme for identifying the situations, i.e., by which one could at any moment find out what the situation was, except that the situations as a whole had been renamed. A similar investigation, with less use of punishment, enables one to find the situations which are affected by the sense lines; the data about the situations involving the memory units can also be found but with more difficulty. At this stage the character has been reconstructed. There are no occurrences of T0, T1, D0, D1. The next stage is to think up some way of replacing the 0s of the character by D0, D1 in such a way as to give the desired modification. This will normally be possible with the suggested number of situations (1000), memory units, etc. The final stage is the conversion of the character into the chosen one. This may be done simply by allowing the machine to wander at random through a sequence of situations, and applying pain stimuli when the wrong choice is made, pleasure stimuli when the right one is made. It is best also to apply pain stimuli when irrelevant choices are made. This is to prevent getting isolated in a ring of irrelevant situations. The machine is now 'ready for use'.

The form of universal machine actually produced in this process was as follows. Each instruction consisted of 128 digits, which we may regard as forming four sets of 32, each of which describes one place in the main memory. These places may be called P,Q,R,S. The meaning of the instruction is that if p is the digit at P and q that at Q then $1-pq$ is to be transferred to position R and that the next instruction will be found in the 128 digits beginning at S. This gives a UPCM, though with rather less facilities than are available say on the ACE.

I feel that more should be done on these lines. I would like to investigate other types of unorganized machines, and also to try out organizing methods that would be more nearly analogous to our 'methods of education'. I made a start on the latter but found the work altogether too laborious at present. When some electronic machines are in actual operation I hope that they will make this more feasible. It should be easy to make a model of any particular machine that one wishes to work on within such a UPCM instead of having to work with a paper machine as at present. If also one decided on quite definite 'teaching policies' these could also be programmed into the machine. One would then allow the whole system to run for an appreciable period, and then break in as a kind of 'inspector of schools' and see what progress had been made. One might also be able to make some progress with unorganized machines more like the A- and B-types. The work involved with these is altogether too great for pure paper-machine work.

One particular kind of phenomenon I had been hoping to find in connection with the P-type machines. This was the incorporation of old routines into new. One might have 'taught' (i.e., modified or organized) a machine to add (say). Later one might teach it to multiply by small numbers by repeated addition and so arrange matters that the same set of situations which formed the addition routine, as originally taught, was also used in the additions involved in the multiplication. Although I was able to obtain a fairly detailed picture of how this might happen I was not able to do experiments on a sufficient scale for such phenomena to be seen as part of a large context.

I also hoped to find something rather similar to the 'irregular verbs' which add variety to language. We seem to be quite content that things should not obey too mathematically regular rules. By long experience we can pick up and apply the most complicated rules without being able to enunciate them at all. I rather suspect that a P-type machine without the systematic memory would behave in a rather similar manner because of the randomly distributed memory units. Clearly this could only be verified by very painstaking work; by the very nature of the problem 'mass production' methods like built-in teaching procedures could not help.

Discipline and initiative

If the untrained infant's mind is to become an intelligent one, it must acquire both discipline and initiative. So far we have been considering only discipline. To convert a brain or machine into a universal machine is the extremest form of discipline. Without something of this kind one cannot set up proper communication. But discipline is certainly not enough in itself to produce intelligence. That which is required in addition we call initiative. This statement will have to serve as a definition. Our task is to discover the nature of this residue as it occurs in man, and to try and copy it in machines.

Two possible methods of setting about this present themselves. On the one hand we have fully disciplined machines immediately available, or in a matter of months or years, in the form of various UPCMs. We might try to graft some initiative onto these. This would probably take the form of programming the machine to do every kind of job that could be done, as a matter of principle, whether it were economical to do it by machine or not. Bit by bit one would be able to allow the machine to make more and more 'choices' or 'decisions'. One would eventually find it possible to program it so as to make its behaviour be the logical result of a comparatively small number of general principles. When these became sufficiently general, interference would no longer be necessary, and the machine would have 'grown up'. This may be called the 'direct method'.

The other method is to start with an unorganized machine and to try to bring both discipline and initiative into it at once, i.e., instead of trying to organize the machine to become a universal machine, to organize it for initiative as well. Both methods should, I think, be attempted.

Intellectual, genetical and cultural searches

A very typical sort of problem requiring some sort of initiative consists of those of the form 'Find a number n such that ...'. This form covers a very great variety of problems. For instance problems of the form 'See if you can find a way of calculating the function which will enable us to obtain the values for arguments ... to accuracy ... within a time ... using the UPCM ...' are reducible to this form, for the problem is clearly equivalent to that of finding a program to put on the machine in question, and it is easy to put the programs into correspondence with the positive integers in such a way that given either the number or the program the other can easily be found. We should not go far wrong for the time being if we assumed that all problems were reducible to this form. It will be time to think again when something turns up which is obviously not of this form.

The crudest way of dealing with such a problem is to take the integers in order and to test each one to see whether it has the required property, and to go on until one is found which has it. Such a method will only be successful in the simplest cases. For instance in the case of problems of the kind mentioned above, where one is really searching for a program, the number required will normally be somewhere between 2^{1000} and $2^{1,000,000}$. For practical work therefore some more expeditious method is necessary. In a number of cases the following method would be successful. Starting with a UPCM we first put a program into it which corresponds to building in a logical system (like Russell's *Principia Mathematica*). This would not determine the behaviour of the machine completely: at various stages more than one choice as to the next step would be possible. We might arrange, however, to take all possible arrangement of choices in order, and go on until the machine proved a theorem, which, by its form, could be verified to give a solution of the problem. This may be seen to be a conversion of the original problem into another of the same form. Instead of searching through values of the original variable *n* one searches through values of something else. In practice when solving problems of the above kind one will probably apply some very complex 'transformation' of the original problem, involving searching through various variables, some more analogous to the original one, some more like a 'search through all proofs'. Further research into intelligence of machinery will probably be very greatly concerned with 'searches' of this kind. We may perhaps call such searches 'intellectual searches'. They might very briefly be defined as 'searches carried out by brains for combinations with particular properties'.

It may be of interest to mention two other kinds of search in this connection. There is the genetical or evolutionary search by which a combination of genes is looked for, the criterion being survival value. The remarkable success of this search confirms to some extent the idea that intellectual activity consists mainly of various kinds of search.

The remaining form of search is what 1 should like to call the 'cultural search'. As I have mentioned, the isolated man does not develop any intellectual power. It is necessary for him to be immersed in an environment of other men, whose techniques he absorbs during the first twenty years of his life. He may then perhaps do a little research of his own and make a very few discoveries which are passed on to other men. From this point of view the search for new techniques must be regarded as carried out by the human community as a whole, rather than by individuals.

Intelligence as an emotional concept

The extent to which we regard something as behaving in an intelligent manner is determined as much by our own state of mind and training as by the properties of the object under consideration. If we are able to explain and predict its behaviour or if there seems to be little underlying plan, we have little temptation to imagine intelligence. With the same object therefore it is possible that one man would consider it as intelligent and another would not; the second man would have found out the rules of its behaviour.

It is possible to do a little experiment on these lines, even at the present stage of knowledge. It is not difficult to devise a paper machine which will play a not very bad game of chess. Now get three men as subjects for the experiment A,B,C. A and C are to be rather poor chess players, B is the operator who works the paper machine. (In order that he should be able to work it fairly fast it is advisable that he be both mathematician and chess player.) Two rooms are used with some arrangement for communicating moves, and a game is played between C and either A or the paper machine. C may find it quite difficult to tell which he is playing. (This is a rather idealized form of an experiment I have actually done.)

References

Church, Alonzo (1936) An unsolvable problem of elementary number theory. *Amer. J. of Math.* **58**, 345–63.

Gödel, K. (1931). Über formal unentscheidbare Satze der Principia Mathematica und verwandter Systeme. *Monatshefte fur Math. und Phys.* **38**, 173–89.

Turing, A.M. (1937) On computable numbers with an application to the Entscheidungsproblem. *Proc. London Math. Soc.* **42**, 230–65.

Examining the Work and Its Later Impact

Christof Teuscher proposes —

A MODERN PERSPECTIVE ON TURING'S UNORGANISED MACHINES

1. Introduction

Most students who took a neural networks class will readily state that McCulloch and Pitts (1943) presented the opening shot of connectionism and modern neural network research. Some may remember that this early work was followed by the perceptron of Rosenblatt (1957), which was itself inspired by the ideas of Hebb (1949). Do the test yourself and you will quickly realize that, alas, neither students not their professors will mention Alan Turing's connectionist ideas, which go back to a 1948 *National Physical Laboratory* (NPL) report entitled 'Intelligent Machinery'. The report first appeared in an edited collection by Evans and Robertson (1968), 14 years after Turing's death, and one year later in "Machine Intelligence" (Turing, 1969). It is also reprinted in Copeland (2004) and Ince (1992) and available in the Turing Digital Archive (Turing, 1948). Turing's basic connectionist machines were later revived and extended by Webster, Copeland and Proudfoot (1996, 1999) and Teuscher (2002, 2004). Why were these ideas largely ignored and what importance – if any – do they play today? Before we try to answer these questions, let us briefly delve into Turing's unique ideas first.

Connectionism, in simple words, is a movement that tries to explain the human mind with its intellectual abilities by means of *artificial neural networks* (ANNs). The field was inspired by the recognition that the human brain processes information in an entirely different way from the classical von Neumann digital computer. Over the last 60 years, a large number of connectionist models were proposed, yet, they all have two things in common: they are built from simple basic processing elements (also called *neurons*) working in parallel and interacting with each other through (weighted or unweighted) connections. Both McCulloch and Pitts' and Alan Turing's work fit that definition, yet, there are a number of differences. Traditional McCulloch and Pitts's neurons can have as many inputs as they like, each neuron is binary and has a finite threshold, synapses can be excitatory and inhibitory, but the connections do not have any weights. Turing proposed three types of what he called *unorganised machines* (see Teuscher (2002) for more details): A-type, B-type, and P-type unorganised machines. A-type and B-type machines are Boolean networks made up of extremely simple, randomly interconnected NAND gates (neurons), each receiving two input signals from other neurons in the network (self-connections are allowed). The neurons are synchronised by means of a global clock signal. In comparison to A-type networks, Turing's B-type networks have modifiable interconnections, and an external agent can therefore 'organize' these machines – by enabling and disabling the connections – to perform a required job. An interesting detail of the B-type machine is that it is in principle a special A-type machine in which each connection is replaced by a small A-type machine that operates as a switch. The switch state (i.e., enabled or disabled) is either defined by the link's internal state or by two external *interfering inputs*.

Turing's idea behind the introduction of the B-type networks was to open the possibility of enabling useful and disabling useless links to produce a required behaviour. His deeper motivation was simply to build structures which allow for learning. The third type of machine – the P-type machine – is not a connectionist machine but rather a modified tape-less Turing machine that has two additional inputs: the *pleasure* and the *pain input*. The internal tables of an initially largely incomplete machine would then be completed by the application of 'pleasure' and 'punishment' stimuli by an external teacher. This is very much alike to what is today known as *reinforcement learning*. In this commentary, we will leave the P-type machine aside.

2. Context and significance

In his work, Turing makes no reference to the paper of McCulloch and Pitts (1943) nor do they mention Turing's work. It is an open question how much their work influenced each other, yet, we have to assume that they were at least aware of each other's ideas. We hypothesize that both bad timing and the fact that Turing's neurons are simpler and more abstract contributed to his work being largely ignored. There is no doubt that Turing was inspired by the human nervous system when he conceived his unorganised machines. However, at that time, when the foundations of modern neuroscience started to be laid, it became quickly clear that neurons do not simply compute NAND functions only, as Turing maybe rather naively proposed, simply because he knew that NAND gates were universal and could therefore – given enough gates with the possibility to interconnect them specifically – compute *any* logical function.

Alas, there is yet another proposition – possibly even more important – made by Turing in his 1948 report that was ignored. At the beginning, the machines are completely unorganised, comparable to an 'infants brain'. 'Then, by applying appropriate interference, mimicking education [...]' (Turing, 1969), the machine will be organised to produce a required behaviour. To achieve this, Turing proposed some sort of *genetic algorithm* – which he called *genetical* or *evolutionary search*:

> "There is the genetical or evolutionary search by which a combination of genes is looked for, the criterion being survival value. The remarkable success of this search confirms to some extent the idea that intellectual activity consists mainly of various kinds of search" (Turing, 1969, p. 23).

The idea of organising an initially random network of neurons and connections is undoubtedly one of the most significant aspects of Turing's 'Intelligent Machinery' paper, yet, given today's widespread usage of genetic algorithms and the fact that early work in simulating evolution in computers goes back to the mid to late fifties, Turing's proposal – and its subsequent ignorance – is even more remarkable. At his time, Turing was unfortunately unable to apply 'genetical search' to the optimisation of his unorganised machines because of the lack of computing resources.

3. Contemporary impact

Turing's work obtained a somewhat different meaning and importance with Stuart Kauffman's introduction of *random Boolean networks* (RBNs) (Kauffman, 1969). Kauffman studied the properties of RBNs in the late sixties already. His studies revealed surprisingly ordered structures in randomly constructed networks, in particular, the most highly organised behaviour appeared to occur in networks where each node receives inputs from two other nodes on average. Astonishingly, Turing has also chosen – probably unintentionally, or to keep things as simple as possible – two inputs for his neurons. In modern terms, a Turing unorganised machine can therefore be considered as a particular kind of RBN, where each node is a NAND function. Kauffman defined a random *NK* Boolean network as a network where each of the *N* nodes has two possible states of activity and receives on average *K* inputs from other nodes. Interestingly, the transition from order to chaos in RBNs occurs either as *K* decreases to 2 (also called the 'edge of chaos') or as other parameters are altered in simple ways. Kauffman (1993) wrote: 'It has now been known for over 20 years that boolean networks which are entirely random but subject to the simple constraint that each element is directly controlled by $K = 2$ elements spontaneously exhibit very high order'.

More recently, Rohlf et al. (2007) systematically studied damage spreading at the *sparse percolation* (SP) limit in RBN with perturbations that are independent of the network size *N*. This limit is relevant to information and damage propagation in many technological and natural networks. They found a critical connectivity (also called the 'edge of stability' or 'edge of robustness') close to $K = 2$, where the damage spreading is independent of *N*. Goudarzi et al. (2011) went a step further and studied information processing in populations of Boolean networks with evolving connectivity, then systematically explored the interplay between the learning capability, the robustness, the network topology, and the task complexity. They used genetic algorithms to evolve networks to perform required jobs (i.e., simple tasks), and therefore did in many ways exactly what Turing suggested in 1948 by 'genetical search' and 'appropriate interference'. Even more interestingly, they solved a long-standing open question and found computationally that, for large system sizes *N*, adaptive information processing drives the networks to a critical connectivity $K = 2$, which is the connectivity that Turing – for whatever reason – proposed.

4. Future developments and conclusion

We have argued in the past (Teuscher et al., 2009) that random dynamical network automata, such as RBN, may be interesting candidates for emerging nanoscale electronics. Among other things, our argument is based on the expectation that such nanoscale computing devices will be built – or rather self-assembled – in a bottom-up way from vast numbers of simple, densely arranged components that exhibit high failure rates, are relatively slow, and are connected in an unstructured way. RBNs therefore nicely fit into that definition. For nanoscale electronics, the robustness against faults and defects, the wiring cost (which is directly related to the power consumption and the chip area used), and the adaptiveness are often considered key properties. As seen above, $K = 2$ networks occupy a unique spot in the design space of such architectures because they are robust against certain types of failures (Rohlf et al., 2007); they reduce the wiring cost, are easier to manufacture because of the sparse connectivity, and offer optimal learning and generalisation capabilities for information processing (Goudarzi et al., 2011).

Turing may therefore unknowingly have laid the groundwork for future nanoscale computing architectures that are 'unorganised' and 'modifiable' and for today's reconfigurable hardware ideas. He wrote about a '[...] machine as being *modifiable*' when it is possible to '[...] alter the behaviour of a machine very radically [...]' (Turing, 1969). He distinguished two kinds of interference with machinery: (1) *screwdriver interference* and (2) *paper interference*. Screwdriver interference is the extreme form in which parts of the machine are removed and replaced by others. Paper interference consists in the mere communication of information to the machine, which alters its behaviour. Turing also wrote about machines that modify themselves, and he classified the operations of a machine into two classes: (1) *normal operations* and (2) *self-modifying operations*. Many of the ideas of (self-) modifying hardware can be found in today's *Field Programmable Gate Arrays* (FPGAs) or in other unconventional reconfigurable architectures (Durbeck and Macias, 2001;

Mange et al., 2000). All of these ideas are considered increasingly important for emerging massive-scale electronic devices, which may rely heavily on self-assembly, reconfiguration and (self-) adaptation.

Turing's long-forgotten work on unorganised machines, initially dismissed as a 'schoolboy essay' by his advisor, is more than ever current, influential, and deeply fascinating. We can only hypothesise what would have happened if his advisor would not have dismissed the manuscript. Students might now state that Turing played an important role on the connectionist stage.

References

Copeland, B.J. (Ed.), 2004. The Essential Turing, Oxford University Press, New York, NY.

Copeland, B.J., Proudfoot, D., 1996. On Alan Turing's anticipation of connectionism. Synth. Int. J. Epistemol. Methodol. Philos. Sci. 108, 361–377.

Copeland, B.J., Proudfoot, D., 1999. Alan Turing's forgotten ideas in computer science. Sci. Am. 280 (4), 76–81.

Durbeck, L.J.K., Macias, N.J., 2001. The cell matrix: an architecture for nanocomputing. Nanotechnology 12, 217–230.

Evans, C.R., Robertson, A.D.J. (Eds.), 1968. Cybernetics: Key Papers, University Park Press, Baltimore Md. and Manchester.

Goudarzi, A., Teuscher, C., Gulbahce, N., Rohlf, T., 2011. Emergent criticality through adaptive information processing in Boolean networks. arXiv:1104.4141. Submitted to Physical Review Letters. Under review.

Hebb, D., 1949. The Organization of Behavior, John Wiley, New York.

Ince, D.C. (Ed.), 1992. Collected Works of A. M. Turing: Mechanical Intelligence, North-Holland, Amsterdam.

Kauffman, S.A., 1968. Metabolic stability and epigenesis in randomly connected genetic nets. J. Theor. Biol. 22, 437–467.

Kauffman, S.A., 1993. The Origins of Order: Self–Organization and Selection in Evolution, Oxford University Press, New York, Oxford.

Mange, D., Sipper, M., Stauffer, A., Tempesti, G., 2000. Toward self-repairing and self-replicating hardware: the embryonics approach. Evolvable Hardware'2000, 205–214.

McCulloch, W.S., Pitts, W.H., 1943. A logical calculus of the ideas immanent in neural nets. Bull. Math. Biophys. 5, 115–133.

Rohlf, T., Gulbahce, N., Teuscher, C., 2007. Damage spreading and criticality in finite random dynamical networks. Phys. Rev. Lett. 99 (24), 248701.

Rosenblatt, F., 1957. The perceptron, a perceiving and recognizing automaton. Cornell Aeronautical Laboratory Report No. 85-460-1.

Rosenblatt, F., 1958. The perceptron: A probabilistic model for information storage and organization in the brain. Psychol. Rev. 65, 386–408.

Teuscher, C., 2002. Turing's Connectionism. An Investigation of Neural Network Architectures, Springer-Verlag, London.

Teuscher, C., 2004. Turing's connectionism. In: Teuscher, C. (Ed.), Alan Turing: Life and Legacy of a Great Thinker. Springer-Verlag, Berlin, Heidelberg, pp. 499–530.

Teuscher, C., Gulbahce, N., Rohlf, T., 2009. An assessment of random dynamical network automata for nanoelectronics. Int. J. Nanotechnol. Mol. Comput. 1 (4), 39–57.

Turing, A.M., 1948. Intelligent machinery. The Turing Digital Archive, `http://www.turingarchive.org/browse.php/C/11`.

Turing, A.M., 1969. Intelligent machinery. In Meltzer, B., Michie, D. (Eds.), Machine Intelligence, vol. 5. Edinburgh University Press, Edinburgh, pp. 3–23.

Turing, A.M., 1992. Intelligent machinery. In Ince (1992), pp. 107–127.

Webster, C. `http://compucology.net/unorganized`.

Nicholas Gessler connects past and future —

THE COMPUTERMAN, THE CRYPTOGRAPHER AND THE PHYSICIST

> The field of cryptography will perhaps be the most rewarding. There is a remarkably close parallel between the problems of the physicist and those of the cryptographer. The system on which a message is enciphered corresponds to the laws of the universe, the intercepted messages to the evidence available, the keys for a day or a message to important constants which have to be determined. The correspondence is very close but the subject matter of cryptography is very easily dealt with by discrete machinery, physics not so easily. (Turing, *Intelligent Machinery*, p. 509 above)

We recall the past because it serves the present. Today, it gives us fresh insights into ideas that we may have overlooked, and it calls our attention to ideas that others may have missed in prior days. It offers us new perspectives, larger vistas on a target of inquiry from points of view separated by space, time, culture and preconception. Since those targets arose in contexts that were so different than today's, we can open past ideas to a larger audience. We open a dialog with yesteryears, an exploration that challenges our understanding of the past and present that make us question whether progress has been made, why or why not, and how. With humility, we reassess our own work in the light of those that have come before us. With conceit, we enlist our antecedents as advocates for our own agendas. We engage to give our present projects richer meaning. What then, are we to make of Turing's statement? What are the implications that arise from what we take Turing to mean?

1. Science as cryptanalysis

> [Newton's}experiments were always, I suspect, a means, not of discovery, but always of verifying what he knew already ... He looked on the whole universe and all that is in it *as a riddle*, as a secret which could be read by applying pure thought to certain evidence, certain mystic clues which God had laid about the world to allow a sort of philosopher's treasure hunt to the esoteric brotherhood ... He regarded the Universe as a cryptogram set by the Almighty—just as he himself wrapt the discovery of the calculus in a cryptogram when he communicated with Leibniz. By pure thought, by concentration of mind, the riddle, he believed, would be revealed to the initiate. (Keynes, pp. 313–314)

My first, probably superficial, interpretation of his claim was as a metaphor, an analogy, between the practice we now call cryptology (the study of codes and ciphers), or more specifically cryptanalysis (the breaking of encrypted communications), and the practices of science, or more generally the epistemology and the philosophy of science. Physics has long been considered the King of the sciences; the biological, social and cognitive sciences having been relegated to roles of Jacks or knaves, all suffering, to some extent, from 'physics envy'. Physics, in this interpretation, thus stands in for our external world, for ultimate reality and for truth itself, revealed to us only through the veil of our limited cognitions, conceptions and perceptions. The world is simply not what it appears to us to be. There is something hiding away from plain sight, beyond the surface of everyday experience. We have adapted to comprehend the behaviours of things of roughly our own scales of space, that move and change according to our own scales of time. We have difficulty comprehending things that are too small or much too large, things that are too quick or much too slow. All things are thus 'encryptions' of reality in need of decipherment, especially those which lay outside our range of easy understanding. Evolution presents us with reality on a need-to-know basis.

> Leibniz long ago described the procedure of science as like the solving of a cryptogram; and this is a deep and an exact remark. In a scientific research, we have to do the opposite to transmitting information, so that we have to turn the theory of information backward. Instead of sending messages in a known code, we receive messages in an unknown code. The aim of science is to break the code of nature. (Bronowski, p. 429)

> Like a cryptographer who has captured an enemy agent, [the scientist] can send searching signals which are designed to evoke simple and decisive answers. (Bronowski, p. 432)

We must surveille nature with suspicion. Its truths are steganographically hidden and cryptographically scrambled secrets among the signals that bombard us in daily life. We must first become aware of their existence before they can be found identified and ultimately revealed.

> A hypothesis ... is like the key to a cryptograph, and the simpler it is, and the greater the number of events that can be explained by it, the more probable it is. But just as it is possible to write a letter intentionally so that it can be understood by means of several different keys, of which only one is the true one, so the same effect can have several causes. Hence no firm demonstration can be made from the success of hypotheses. (Leibniz, quoted in Rescher, p. 121)

But just as each effect can have several different causes, and each letter can several different understandings, so too should we suspect that Turing's statement (as well as his entire report) have several different meanings each keyed and targeted to different audiences.

2. At the National Physical Laboratory

Five years after Alan Turing's death in 1954, his mother, Sara Turing, wrote a tribute to her son. Recounting the context of Alan's work at the National Physics Laboratory, an arc rising with optimism and falling with disappointment, she recalled Sir Charles Darwin (grandson of the evolutionist), Director of the NPL, broadcast on the BBC, in November 1946:

> A young Cambridge mathematician, by name Turing, wrote a paper ... in which he worked out by strict logical principles how far a machine could be imagined which would imitate processes of thought ... Broadly we reckon that it will be possible to do arithmetic a hundred times as fast as a human computer, and this, of course means that it will be practicable to do all sorts of calculations outside the scope of human beings. (S. Turing, p. 79)

It might be worth noting Darwin's emphasis on doing 'arithmetic a hundred times as fast as a human computer'. Notwithstanding the fact that a transcript of the entire broadcast was not available to me, the project seemed focused on advancing the work of computers, *who* do calculations, not computers, *that* do calculations. Computers in those days were those persons doing such tasks as figuring actuarial, accountancy, statistical and engineering tables. The focus was not on emulating creativity or intelligence. Sara introduced the difficulties that led to Turing's resignation:

> In August 1947 my husband died. Alan, disappointed with what appeared to him the slow progress made with the construction of ACE, and convinced that he was wasting time since he was not permitted to go on the engineering side, asked for a sabbatical year. (S. Turing p. 86–87)

> While away in Cambridge he wrote a report on "learning machines" for the National Physical Laboratory whither he returned about May 1948. As progress on the ACE had not come up to his expectations he sent in his resignation from the Scientific Civil Service. (S. Turing, p. 89)

Soliciting a comment from the NPL, she received this ambivalent summary of Alan's contribution to the development of the ACE. E.T. Goodwin, Superintendent of the Mathematical Division of the NPL, wrote to Sara Turing in 1957, three years after Alan's death in 1954:

> In the early years after the war Alan produced what we call the 'logical design' of a large computer which was to be called 'The ACE' or Automatic Computing Engine. The Laboratory was very doubtful of its ability to produce successfully what was then so ambitious a machine and, at about the same time when Alan took his sabbatical year at Cambridge, it was decided to produce a small version which would be entitled the Pilot Ace. Though the basic ideas behind this machine were largely Alan's, you will understand that the detailed arrangement was decided by others. (S. Turing, p. 84)

Nine years earlier, Sir Charles Darwin, National Physics Lab Director, noted in the minutes of the Executive Committee, was much less diplomatic:

> [Turing's report is a] schoolboy's essay ... not suitable for publication. (Darwin quoted in Copeland, p. 401)

Turing quit the NPL after his sabbatical, breaking an agreement that he would return to work for another two years. His 'Intelligent Machinery,' subtitled 'A Report', was clearly not viewed as such by Darwin. It was a report on Turing's vision of work he had wanted to complete, but could not complete at the NPL.

> Everyone had been slow to adjust to the realities of the post-war period. In expecting the Post Office to cooperate on the [mercury] delay lines, Alan had been as unrealistic as any of the administrators ... Perhaps Darwin never really wanted a computer, just as the Admiralty had not really wanted to know where German ships were. The 'support' of Travis and the Ministry of Supply had not in fact made any difference to the bureaucratic inertia. Darwin and Womersley had played at being commissars while Alan remained the humbler worker and peasant ... [Turing] was not given a chance to make a mess of it for himself, as was his right as the creative worker ... for in the end every successful computer project had to solve the problem of integrating 'mathematical' and 'engineering' skills, which was exactly what he [Turing] longed to do. (Hodges, p. 376)

It was also manifesto and critique. At the beginning of 'Intelligent Machinery', Turing outlines 'some of the reasons' why 'it is assumed without argument' that it is not 'possible for machinery to show intelligent behaviour'. This likely was an opening salvo aimed directly at Sir Charles Darwin (Director of the NPL) and J.R. Womersley (Superintendent of the Mathematics Division), a confrontation with his superiors who were skeptical of his agenda. Turing resented the compartmentalisation of intellectual activity, in academia and at the NPL.

> Despite his resignation, and all the embarrassment that surrounded it, he completed a report for the NPL in July and August 1948. Its almost conversational style reflected the discussions he had pursued, many at Bletchley, in advancing the ides of *Intelligent Machinery*. Although nominally the work of his sabbatical year, and written for a hard-line technical establishment, it was really a description of a dream of Bletchley Park, and reviewed in an almost nostalgic way the course of his own life rather than contributing to any practical proposals that the NPL might adopt. (Hodges, p. 377)

During his sabbatical at Cambridge, he saw an example of how research should be done, which furthered his frustrations with the NPL.

> [Alan's] mind still straddled mathematics, engineering and philosophy in a way that the academic structure could not accommodate. Temporarily the war had resolved his frustration, giving him something to do that was intellectually satisfying, yet which actually worked. But that was over now, and instead of being drawn in, he was being pushed out... At Cambridge, the computer was firmly in the grasp of M.V. Wilkes ... (Hodges, p. 374)

Turing's correspondence between cryptography (mathematical designs) and physics (engineering physical instantiations of those designs) can be seen as a manifesto to bring down the walls of separation the administrators had erected at NPL. One can almost hear the call to unite theory with practice, mental work with physical labour: 'mathematicians and engineers unite'! He wanted freedom to move freely between the conceptual world of mathematics with the physical world of engineers. Turing was quick to appreciate and appropriate the structure of Wilkes project:

> One point concerning the form of organization struck me very strongly. The engineering development work was in every case being done in the same building with the more mathematical work. I am convinced that this is the right approach. It is not possible for the two parts of the organization to keep in sufficiently close touch otherwise. They are too deeply interdependent. We are frequently finding that we are held up due to ignorance of some point which could be cleared up by a conversation with the engineers, and the Post Office find similar difficulty; a telephone conversation is seldom effective because we cannot use diagrams. Probably more important are the points which are misunderstood, but which would be cleared up if closer contact were maintained, because they would come to light in casual discussion. It is clear that we must have an engineering section at the ACE site eventually, the sooner the better, I would say. (Turing cited in Copeland p. 397)

> [Wilkes] was in full control, without a Womorsley or a Darwin to get in the way, and working much as Alan would have liked to. The barricade between mathematics and engineering never arose. It was enough to show the folly of NPL policy .... (Hodges, p. 375)

The situation at the NPL was markedly different, design and engineering were separate:

> Little progress had been made on the physical construction of the ACE. The actual engineering work was being carried out not at the National Physical Laboratory but at the Post Office Research Station, under the supervision of Turning's wartime Associate Flowers. (Copeland 395)

According to Sara Turing, Alan coped with the stresses of the logistics of this separation in an uncustomary way.

> When, after the war, the Post Office was engaged in research on computers Alan was sometimes required to attend conferences at Dollis Hill and visit the Post Office laboratories. He disliked complicated cross-country journeys ... so he usually ran the fourteen miles from Teddington [the location of the NPL] to Dollis Hill. (S. Turing p. 86)

Alan's passion to bridge the gap between theory and practice was acquired at an early age:

> Unlike most mathematicians, Turing liked to get his hands dirty building things. To implement an automatic code machine he began building a binary multiplier using electromagnetic relays, which were the primary building blocks of computers before vacuum tubes were demonstrated to be sufficiently reliable. Turing even built his own relays in a machine shop and wound the electromagnets himself. (Petzold, p. 127)

The NPL was late to recognise its own lack of progress, and Womersley reported in the Executive Committee of the NPL on 20 April 1948:

> The present position of this project gives no cause for complacency and we were probably as far advanced 18 months ago ... There are several competitors to the ACE machine, and of these, that under construction at Cambridge University under Professor (sic) Wilkes, will probably be the first in operation. (Hodges, p. 375)

The cause was also recognised too late, at least for Turing:

> "At the end of April an NPL minute spoke of the need for en electronics group working 'together in one place as a whole in close contact with the planning staff at the Mathematics Division."' (Copeland 397–98)

Cryptography's correspondence to physics, discursive code for his desire to see the engineering efforts of the Post Office electronics group re-established at NPL in house, was meant to press for closer collaboration between the hardware developers, the engineers, and the software developers, the designers in the mathematics department. He further conceded that developing the machine itself (the engineering) would be much more difficult than producing the design and instructions for it (the mathematics). Under this interpretation, Turing invoked the camaraderie that existed at Bletchley Park during the development of the Bombe to decrypt Enigma messages.

It is interesting that Turing includes the Brunsviga pinwheel calculator at the bottom of his list of capacities of various machines with a memory of 90. It may have served both as a reference to a machine that everybody knew and simultaneously as a dig at the bureaucratic establishment that frustrated his attempts to bring the ACE to life. The Brunsviga had long been advertised as having 'brains of steel', and an icon riveted to each machine showed an image of a head in cutaway revealing clockwork gears for brains. For most human computers of the time, the Brunsviga was the workhorse they employed. Did the Brunsviga in that list stand in for Darwin's limited vision of Turing's project as rote arithmetic (as expressed in his BBC broadcast of 1946)? Did it evoke a limited and unimaginative future for intelligent machines, the equation being: (Brunsviga / thinking machine) = (creative workplace / NPL)? Double and even triple Brunsvigas were not unheard of but Turing chose the single as his example.

3. A computational world

If cryptography, i.e., cryptanalysis, is the search for the design of the Enigma machine and the protocols of its use, the hardware and software of an electromechanical computer, then it is also the search for the design of the natural world and the protocols that govern it, the hardware and the software of the computer on which the world as we experience it is run.

The correspondence that Turing draws between cryptography and physics is much richer than it first appears, much richer than that drawn by Newton or Leibniz. Based upon his secret work in developing the Bombe at Bletchley Park, an electromechanical computer with a dedicated program, invoking cryptography as 'as perhaps the most rewarding application' of the ACE, as well as the correspondence between 'cryptography' and 'physics' in his report would have brought that entire experience into the argument he was making. It's uncertain whether Darwin, Womersley or others knew about Turing's work breaking the Enigma. Flowers, who was with him at Bletchley Park and was now associated with the NPL, may have made the connection.

What Turing had done at Bletchley was to construct one enhanced electromechanical computer, the Bombe, to predict the operation of another electromechanical computer, the Enigma. The Bombe was a multiplicity of Enigmas and the project for the Bombe was to retrodict and discover the system (construction), settings (initial configuration), and data (cleartext) fed into the Enigma which would produce the data (ciphertext) that the listening stations had intercepted. He was recursively using one computer to mimic the behaviour of another, of machines represented inside other machines, of mathematics contained inside other mathematics, or in modern parlance computations nested inside computations. In his correspondences and parallels he is applying this same recursivity to physics, suggesting in the discursive style of his time that 'it's mathematics all the way down', or as we might say now, 'it's computation all the way down'. From top to bottom, from thought to whatever underlies physics, all science can be seen as mathematics writ large, that is, computation.

> Dr. Warren McCulloch, professor of psychiatry at the University of Illinois College of Medicine goes further: he says that the brain is actually a computer, and very like computers built by men. (Anon, p. 56)

Alan may have felt some sense of vindication when that statement appeared, as it seems that was the subtext of much of what he had to say in his report. The cover of *TIME* magazine on January 23, 1950, was adorned with Boris Artzybasheff's illustration of a computer examining its own progress and deciding what next to do and the caption, 'Mark III, Can Man Build a Superman?' 'At work, it roars louder than an Admiral', (Anon, p. 55). In two short years the 'computing machine' was taking on the appellation of 'computer' and the human 'computers' of the days before were now becoming 'human calculators' or simply 'human beings'. 'Computation' was gaining popularity. Those who designed these devices no longer were solely among the ranks of mathematicians, but were in the process of becoming 'computermen'.

What is computer science and computation? Frequently they stand in for all the algorithmic processes that we see in nature. Increasingly, the machines we build, computers, are seen merely as technological instantiations of computational phenomena that we discover observe in nature.

> Computer science is no more about computers than astronomy is about telescopes. – E.W. Dijkstra (Flake p. 23)

> Computer science is not about computers. It's the first time ... that we've begun to have ways to describe the kinds of machinery that we are. (Minsky 1996)

We have an emerging computational philosophy and epistemology of science. The subject is taken up explicitly as in *Computational Philosophy of Science* (Thagard). The possibility is quietly implied by the emergence of readily apparent patterns from computational rules in nature, such as *The Computational Beauty of Nature* (Flake), and Prusinkiewicz's series, *The Algorithmic Beauty of: Plants* (Prusinkiewicz), *Seaweeds, Sponges and Corals* (Kaandorp), and *Seashells* (Meinhardt). It does not offend our sense of self importance to accept computation as motivating 'lesser' forms of life, but it is still a controversial subject for the human and social sciences. Nevertheless, the RechnenderGeist has spread raising new multiagent explanations in the form of Artificial Societies from dying single-cause models in economics:

> What "sort of science" are we doing? ... [Our] aim is to provide initial microspecifications (initial agents, environments, and rules) that are *sufficient to generate* the macrostructures of interest. We consider a given macrostructure to be "explained" by a given microspecification when the latter's generative sufficiency has been established. (Axtell & Epstein, p. 177)

Horowitz carries this idea forward, from quark to quasar, in his compelling and ambitious book *The Emergence of Everything*. My own field of Anthropology has been slow to follow suit, preferring to privilege the influence the role of the individual and of top-down rational causation over that of the population and of bottom-up emergence inhuman culture. Artificial culture has yet to gain momentum (Gessler).

Among complex systems, we encounter the emergence of the entailments of processes operating at one local scale (of space, agency or time) to forms taking shape at another global scale. We witness global patterns of behaviour emerging from populations of local rules. It is only in the interaction of those local rules that the global pattern come into being. Nowhere among those individual rules would we find any indication of what they will produce. Without those local rules in constant play, the global world would not appear.

Reluctant to be seated among the advocates of 'the world is computational from bottom up', Stephen Wolfram enlists 'correspondences' as did Alan Turing in defining his 'Principle of Computational Equivalence':

> Whenever one sees behavior that is not obviously simple [i.e. complex] — in essentially any system — it can be thought of as corresponding to a computation of equivalent sophistication. (Wolfram, p. 5)
>
> The great historical successes of theoretical science have typically revolved around finding mathematical formulas that ... directly allow one to predict the outcome [of a particular system] ... The Principle of Computational Equivalence now implies that this will normally be possible only for rather special systems with simple behavior... Other [more complex] systems will tend to perform computations that are just as sophisticated as those we can do, even with all our mathematics and computers. And this means that such systems are computationally irreducible — so that in effect the only way to find their behavior is to trace each of their steps, spending about as much computational effort as the systems themselves. (Wolfram, p. 6)

Konrad Zuse, who designed and built the world's first working electromechanical, programmable, fully automatic computer, the Z-3, in 1941 tackled the problem of a computational universe head-on by taking the offensive. In 1969 he introduced the term 'automaton theoretical way of thinking' in his paper 'Rechnender Raum' or 'calculating space' (Zuse, p. 7). In it he takes up the proposition that the cosmos might operate as a cellular automaton. He examines the foundational principles of physics one by one, evaluating the possibility of subsuming each of them under the entailments of an appropriate cellular automaton. Nowhere does he find conclusive evidence to dismiss his proposition out of hand. His overall project is clear, but a quotable few lines summarizing his intent disappear among the details of his arguments.

> The question therefore appears justified whether data processing can have no more than an effectuating part in the interplay [between mathematics and physics] or whether it can also be the source of fruitful ideas which themselves influence the physical theories. (Zuse, p. 1)
>
> Such a process of influence can issue from two directions ... 2. A direct process of influencing, particularly by the thought patterns of automaton theory, the physical theories themselves could be postulated. This subject is without a doubt the more difficult, but also the more interesting. (Zuse, p. 2)
>
> The first result of viewing the cosmos as a cellular automaton is that the single cells represent a finite automaton. The question to what extent it is possible to consider the entire universe as a finite automaton depends on the assumption which we make in relation to its dimensions. (Zuse, p. 70)
>
> In view of the possibilities listed, it is clear that there are several different points of view possible: ... (3) The possibilities arising from the ideas of calculating space are in themselves so interesting that it is worthwhile to reconsider those concepts of traditional physics which are called into question and to examine their validity from new points of view. (Zuse, p. 93)

In like fashion, Ed Fredkin posits the existence of 'the Ultimate Computer' residing in a universe he calls 'Other'. His argument is also detailed but can be summarised more clearly:

> The answer lies in the amazing consequence of the simple assumption of Finite Nature. As we have explained, Finite Nature means that what underlies physics is essentially a computer. Not the kind of computer that students use to do their homework on, but a close cousin; a cellular automaton. Not knowing the details of that computer doesn't matter because a great and tragic British mathematician, Alan Turing proved that we don't need to know the details!
> What Turing did in the 1930s was to invent the Turing Machine. It was a way to formalize all the things that a mathematician could do with pencil and paper. The result proves that any ordinary computer, given the proper program and enough memory, can do what any other computer can do. It can also do what any mathematician can do; if we only knew how to write the program! Finite Nature implies that the process underlying physics is a kind of computer; therefore it is subject to Turing's proof. This means that there is not just one kind of underlying computer, but there are many possible equivalent computers. Of course some are simpler, some are more elegant, some use the least amount of various resources, some are faster... Once we have figured out that it's a computer at the bottom, we already know a lot even if we don't know what kind of computer would be most efficient at the task.
>
> As to where the Ultimate Computer is, we can give an equally precise answer, it is not in the Universe - it is in an *other* place. If space and time and matter and energy are all a consequence of the informational process running on the Ultimate Computer then everything in our universe is represented by that informational process. The place where the computer is, the engine that runs that process, we choose to call *"Other"*.

Jürgen Schmidhuber explores a philosophy and epistemology of computation with a lighter touch:

> A long time ago, the Great programmer wrote a program that runs all possible universes on His Big Computer. "Possible" means "computable": (1) Each universe evolves on a discrete time scale. (2) Any universe's state at a given time is describable by a finite number of bits. One of the many universes is ours, despite some who evolved in it and claim it is incomputable. (Schmidhuber, P. 201)
>
> **Conclusion**. By stepping back and adopting the Great Programmer's point of view, classic problems of philosophy go away. (Schmidhuber, P. 208)

Marvin Minsky expands upon Schmidhuber's 'conclusion' above:

> Fifty years ago, in the 1940s and 50s, human thinkers learned for the first time how to describe complicated machines. We invented something called computer language, programming language, and for the first time people had a way to describe complicated processes or complicated machines, complicated systems made of thousands of little parts all connected together ... Before 1950 there was no language to discuss this, no way for two people to exchange ideas about complicated machines. But why is it important to understand? Because that's what you are ....
> Computer Science is a new philosophy about complicated processes ... about Artificial Life, about natural life, about Artificial Intelligence [and] about natural intelligence ... So all [prior] philosophy, I think, is stupid. It was very good to try to make philosophy. Those people tried to make theories of thinking, theories of knowledge, theories of ethics, and theories of art, but ... they had no words to describe the processes or the data ... So I advise all students to read some philosophy, and with great sympathy. Not to understand what the philosopher said, but to feel compassionate and say, "Think of those poor people years ago who tried so hard to cook without ingredients, who tried to build a house without wood and nails, who tried to build a car without steel, or rubber or gasoline." So look at philosophy

with sympathy. But don't look for knowledge. There is none. Remember whenever you see ancient wisdom that still seems smart, what does it mean? It means that the ancient wisdom has something wrong with it that keeps people from replacing it for a long time. (Minsky 1996)

No less intellectually stimulating, and on a lighter note, the possibilities are often further explored as fiction. Among the most interesting are short stories such as Stanislaw Lem's 'Non Serviam' (Lem), novels such as Greg Egan's *Permutation City* (Egan) and films such as Josef Rusnak and Daniel F. Galouye's, *The Thirteenth Floor* (Rusnak).

Is our Universe, at its base, computational? What does it mean to make or refute this claim? Are debates, both pro and con, simply language games? Perhaps they are, but the games being played are oftentimes complex and have serious consequences. Such games are symptoms that expose the inability of spoken language to represent and describe, and our inability to understand and explain certain complexities in our world. Computer languages, on the other hand, may provide correspondences that are richer, that capture many facets of reality more completely than can natural languages. Moreover, by pressing 'run', they spin out the entailments of statements they contain in greater detail and more consistently than discursive arguments following an assertion. The claim that we are patterns that emerge from processes operating at a smaller scale, provides us with more inspiration, insight and wonder into the wonders of this world, than does its negation.

Sara Turing, Alan's mother, wrote:

Some years later Alan remarked that the daily papers were many years ahead of him, opening even his eyes in wonder, so far did they outstrip him in their forecasts. (S. Turing, p. 80)

Alan would have liked to have joined us in this discussion, and perhaps he has, as there is a little of Alan Turing in each of us.

References

Anonymous, 1950. The thinking machine. Time 55 (4), 54–60.

Bronowsky, J. 1955. Science as Foresight. In: Newman, J.R. (Ed.), What is Science? Twelve Eminent Scientists and Philosophers Explain their Various Fields to the Layman. Simon and Schuster, New York.

Copeland, B.J., 2004. The Essential Turing: Seminal Writings in Computing, Logic, Philosophy, Artificial Intelligence, and Artificial Life Plus the Secrets of Enigma, Clarendon Press, Oxford.

Egan, G., 2000. Permutation City, Millennium, London.

Epstein, J.M., Axtell, R. 1996. Growing Artificial Societies – Social Science from the Bottom Up, The MIT Press, Cambridge, MA.

Flake, G.W., 2001. The Computational Beauty of Nature, MIT Press, Cambridge, MA.

Fredkin, E., 1992. A New Cosmogony. http://www.leptonica.com/cachedpages/fredkin-cosmogony.html.

Gessler, N., 2010. Fostering creative emergence in artificial cultures. In: Fellermann, H. (Eds.), Artificial Life XII – Proceedings of the Twelfth International Conference on the Synthesis and Simulation of Living Systems. MIT Press, Cambridge, MA, pp. 669–676.

Hodges, A., 1983. Alan Turing – The Enigma, Simon and Schuster, New York.

Kaandorp, J.A., Kübler, J.E., 2001. The Algorithmic Beauty of Seaweeds, Sponges, and Corals, Springer-Verlag, Berlin, Heidelberg, New York.

Keynes, J.M., 1946. Newton, the Man. In: The Royal Society Newton Tercentenary Celebrations 15–19 July 1946, Cambridge Univ. Press (1947). Reprinted in: Keynes, G., (Ed.), Essays in Biography, second ed. W.W. Norton, New York (1963).

Lem, S., 1978. Non serviam. In: Stanislaw Lem's, a Perfect Vacuum. Harcourt Brace Jovanovich, San Diego, pp. 167–196.

Meinhardt, H., 2003. The Algorithmic Beauty of Sea Shells, Springer-Verlag, Berlin, Heidelberg, New York.

Minsky, M., 1985. The Society of Mind, Simon and Schuster, New York.

Minsky, M., 1996. Public Lecture at Artificial Life V, May 16, Nara, Japan, videotaped by Katsunori Shimohara, transcribed and edited by Nicholas Gessler.

Petzold, C., 2008. The Annotated Turing- A Guided Tour Through Alan Turing's Historic Paper on Cumputability and the Turing Machine, Wiley Publishing, Indianapolis.
Prusinkiewicz, P., Lindenmayer, A., 1990. The Algorithmic Beauty of Plants, Springer-Verlag, New York.
Rescher, N., 2003. On Leibniz, University of Pittsburgh Press, Pittsburgh.
Rusnak, J., Galouye, D.F., 1999. The Thirteenth Floor, Columbia Pictures, Centropolis Film Productions, Culver City.
Schmidhuber, J., 1997. A computer scientist's view of life, the universe, and everything. In: Freska, C. (Ed.), Foundations of Computer Science: Potential – Theory – Cognition. Springer, Heidelberg, pp. 201–208.
Thagard, P., 1993. Computational Philosophy of Science, MIT Press, Cambridge, MA.
Turing, S., 1959. Alan M. Turing, W. Heffer & Sons, Cambridge. Reprinted in: Turing, S., Alan M. Turing: Centenary Edition, Cambridge Univ. Press, Cambridge, 2012.
VanDevel, C., 2011. Brunsviga Exposition. http://www.crisvandevel.de/bvex.htm.
Wolfram, S., 2002. A New Kind of Science, Wolfram Media, Illinois.
Zuse, K., 1970. Calculating Space - Translation of Rechnender Raum, MIT Technical Translation, AZT-70-164-GEMIT, Cambridge.

Stephen Wolfram looks to reconcile —

INTELLIGENCE AND THE COMPUTATIONAL UNIVERSE

What will it take to create artificial intelligence? The only clear example of intelligence that we have traditionally had is human intelligence. But what aspects of human intelligence are somehow essential to the notion of 'intelligence', and what are merely side effects of our particular biological implementation?

In Turing's day it was known that the brain operated in a largely electrical way – and it was clear that all sorts of devices could be constructed with electronics. But what was the secret that let a brain show intelligent behaviour? Was it some particular architecture that could be emulated with electronics? Or was it something about the way information was provided? Or something else?

I don't think Turing ever imagined that his Turing machines would be equivalent to brains; he was sure there was something fundamentally more to brains. But of course his intuition did not have the benefit of all our experience with actual computers, with what they do, and with the experiments we can do with them.

In my own case, my view of artificial intelligence has changed completely over the past 30 years. And one of the important consequences of the change is that I came to believe that the Wolfram|Alpha 'Computational Knowledge Engine' (Wolfram) should be possible – and then proceeded to build it.

What precipitated the change in my views was the experimentation I did on the computational universe in connection with *A New Kind of Science* (Wolfram, 2002). Normally when we think of computers we imagine constructing machines or programs for specific purposes – to perform tasks we want.

And certainly that is what Turing had in mind when he set up Turing machines, or discussed how 'intelligent machines' could be built.

Originally motivated by natural science, however, what I did was to explore the general universe of possible programs – starting with simple programs that one might set up at random, or by enumeration. And what I found – first in the context of cellular automata – was that even extremely simple underlying rules are capable of producing behaviour of in effect arbitrary complexity.

This led me to a general principle – the Principle of Computational Equivalence (Wolfram, 2002) – that implies that beyond some very low threshold, almost any set of rules or programs that one encounters, if it does not have trivial behaviour, it will behave in a way that is computationally as sophisticated as anything else.

In other words, it does not take much to be able to do sophisticated computation. But adding more complexity to a system does not increase the sophistication of the computation that it can do.

I had always imagined that there was somehow 'more to intelligence' than 'just computation'. But the Principle of Computational Equivalence implies that at a fundamental level there cannot be.

What is intelligence? As a perhaps simpler analogue, one can ask, 'What is life?' In ancient Greek times, it was assumed that anything that 'moved itself' must be alive. A century ago it was assumed that life must be associated with some special chemistry or special thermodynamics. Later, it was assumed that self-reproduction was what was special.

By now it is clear that none of these particular definitions really allows us to distinguish life from non-life. Of course, in practice it is quite easy to tell what is alive and what is not. For all life we know shares not just abstract characteristics, but a detailed history. And the result is that all living things have many detailed similarities – like RNA, cell membranes, and so on.

So what about intelligence? I think the story is very much the same. One can imagine all sorts of abstract definitions. But in practice the definitions we use are deeply tied to the details of human intelligence, and the human condition. And indeed that is why the Turing Test remains a good test – not of abstract intelligence, but of human-like intelligence.

But what of abstract intelligence? My conclusion is that there is really nothing to distinguish it from 'pure computation.' When we say that 'the weather has a mind of its own', we are effectively attributing intelligence to the dynamics of a fluid. And in fact this is not as misguided as modern science might have one believe. For in fact the computations done by the fluid are doubtless equivalent to computations that can be done by any system, notably our brains.

So then what about the kinds of tasks that Turing considers for artificial intelligence? I used to imagine that some of these tasks might require some special spark of intelligence to do. But now I believe that all they will ever need is in a sense 'just computation'.

One of the broadest tasks – and one that seems closely related to intelligence – is answering questions. Given the knowledge that our civilisation has accumulated, is it possible to automate answering questions that can be answered on the basis of this knowledge?

In the early years of electronic computers, there was an assumption that this should be quite easy – through a kind of automation of mental processes analogous to the automation of physical tasks achieved by mechanical machines. In actuality, however, it turned out to be too difficult, and all that emerged were some embarrassingly simplistic toy attempts.

From my own work on computation and the computational universe, however, I came to believe that this failure was not inevitable, and that in fact, by doing appropriate engineering, and inventing a collection of relevant algorithmic approaches, it should be perfectly possible to make knowledge computable on a large scale.

The result of my effort is Wolfram|Alpha, which has been increasingly successful at answering millions of expert-level questions across thousands of domains of knowledge every day.

Wolfram|Alpha does not work much like Turing's conception of an intelligent machine, though it achieves in many directions more than I believe Turing would have expected would be possible of a machine. But much of what it does, it does in a very different way than human intelligence operates. For example, if Wolfram|Alpha is asked to solve a problem about the behaviour of a physical system, it does not do it by a process of 'thinking': of reasoning, like a Mediaeval philosopher, about how one part of the system should affect another and so on. Instead, it just sets up the appropriate equations, then uses the most powerful scientific and mathematical methods it can to push through to an answer.

Wolfram|Alpha is in this case in effect using the formalism of science and mathematics that our civilisation has set up, not the processes of thinking and reasoning that are basic to our brains.

And similarly, when Wolfram|Alpha understands natural language input provided by humans, it does it in ways that probably have little relation to the way such a task is performed in human brains.

What does all this mean for artificial intelligence, and for Turing's intelligent machinery? Over the coming decades I suspect the whole notion of artificial intelligence will become increasingly moot.

Through 'pure computation' we will perform all sorts of tasks that we traditionally attributed to intelligence. And as the computational systems we use become increasingly integrated into our lives as humans, those systems will in effect behave in more 'human-like' ways – fundamentally, because their interaction with the world will be increasingly aligned with what we humans experience.

Through our technology, and our understanding of the broader computational universe, we will achieve what Turing imagined. Though we will do it not through emulating details of humans, but in a sense through pure computation. Ironically enough, we will do it in a way that is much closer to Turing's original invention of Turing machines, and further from the details of human intelligence that he discussed in this paper.

References

Wolfram, S., 2002. A New Kind of Science, Wolfram Media
Wolfram, S., 2002. A New Kind of Science, Wolfram Media, (Chapter 12), pp. 715–846.
Wolfram|Alpha, http://www.wolframalpha.com/

Paul Smolensky asks a key question —

COGNITION: DISCRETE OR CONTINUOUS COMPUTATION?

'Cognition is computation'. This radical idea we owe in part to philosophers such as Aristotle and Hobbes, but most of all, to Turing, who made computation a tool with power sufficient to spur the founding of a new field: cognitive science. Rooted in a long history, two main currents shape cognitive science: both computational, but only one following a course apparently foreseen by Turing himself. In this essay we will explore these two streams, focusing on the key question: what is the relation between meaning and mechanism? More specifically: what is the mapping that links the meaningful elements of cognition – abstract concepts in the mind, and the knowledge that links them – to the mechanistic elements of computation – basic units of data, and the operations that act on them?[1]

[1] Turing's ideas concerning the formal details of computation and his ideas for relating machines to minds are both well-attested in his writings, but the latter are discussed at a level of generality that makes it difficult to interpolate how they were intended to be implemented in the former. The virtual absence of discussion of this mapping by Turing leaves it unclear whether it was simply taken for granted that this mapping is simple and transparent, as assumed by the symbolic paradigm in cognitive science introduced below. Thus the emphasis here will be on characterizing the conceptual relations underlying approaches to relating mind to machine in contemporary cognitive science, without attempting historical analysis. It is clear that Turing's work laid the foundations for and is consistent with the symbolic paradigm, and it is clear that a fundamental principle of the alternative, subsymbolic paradigm – distributed conceptual representation – was either not considered, not considered important, or not considered correct by Turing.

For our purposes, we can take computation to be the *reduction of unboundedly complex processes to combinations of simple ones.* A *computational architecture* is thus defined by (i) a set of simple, *primitive* processes that are assumed given; (ii) a general type of data that these processes operate upon; and (iii) a finite set of operations that join primitive processes to create more complex processes which they then in turn combine to create still more complex processes.

Between them, the two streams of cognitive science deploy two classes of computational architecture. The first class comprises the *discrete architectures*, of which a key example is the Turing machine; other examples include Babbage's Analytical Engine, Church's function systems, Kleene's algebras, and Post's string-rewriting systems. This last example is closest to the discrete architectures central to cognitive science. The type of data employed is *symbol strings* (i.e., sequences), e.g., `(x+(2*y))=z`. The primitive processes involve taking a string of symbols, say `w=z`, and substituting for one symbol, say `w`, a string of symbols, say `(x+u)`, producing `(x+u)=z`; this simple process is a *rewrite rule*, denoted $\texttt{w} \rightarrow \texttt{(x+u)}$. Combining many rewrite rules, it is possible to compute complex sets of strings, such as the set of all well-formed equations of algebra, or the set of all well-formed programs in a computer language such as Java, or even a set of strings constituting an approximation to the sentences of English (where each symbol denotes an English word). In the latter case, the rules constitute a *rewrite-rule grammar* with rules like $\texttt{Sentence} \rightarrow \texttt{NounPhrase}_{\texttt{subject}}\ \texttt{VerbPhrase}_{\texttt{predicate}}$.

What is the purpose of computational reduction? Before automatic computers, clerks were employed to carry out extensive calculations. These human *computors* needed to be instructed in terms of simple operations they could reliably perform by hand or with basic mechanical aids, without relying on intuitions as a mathematician might. Reducing complex calculations to a sequence of simple operations allowed these complex computations to be performed by human computors, who were using a mental faculty we can call the *conscious rule interpreter*. For human computors, the rules can be stated in English, using a circumscribed vocabulary referring to simple operations. It is this type of computational reduction that is formalised in discrete computation. And it is this type of reduction that Turing apparently envisaged for simulating the human mind.

Crucially, this type of reduction *conflates meaning with mechanism*. The complex process being carried out has the purpose of taking meaningful inputs (say, a target location) and producing meaningful outputs (rocket launch parameters). The mechanisms that perform the process operate on these meaningful elements: these are the operations described by the instructions that human computors follow.

This type of computational reduction defines the first stream of cognitive science: the *symbolic paradigm*. It is illustrated by what Haugeland (1985) terms 'Good Old Fashioned Artificial Intelligence', where programs consist of rules manipulating meaningful symbols that refer to the elements of the conceptual world in which the inputs and outputs reside. The same is true in the Newell and Simon (1972)/Anderson (1983) school of cognitive science, which models novice geometry students as literally internalizing the rules of their textbooks; with experience, these rules are modified to more efficiently achieve their effects, but remain, even in the expert, procedures (now unconscious) that manipulate symbols meaningful in the problem domain (symbols that refer to points, lines, triangles). When Turing (1950) discusses the *Argument from the informality of behaviour*, he addresses the kinds of 'laws of behaviour' that later developers of 'expert systems' would seek: propositions stated in terms of the concepts experts use to cognize their domain, concepts which are formalised as symbols that are meaningful in this sense and, simultaneously, are the tokens manipulated by the mechanisms of computation, which operate on the propositions encoding the identified laws of expert behaviour.

In the symbolic paradigm of cognitive science, then, the type of computational reduction afforded by discrete computation is deployed for not only the conscious rule interpreter internal to the mind of the human computor or novice geometry student – it is also invoked for the *intuitive processor* in the mind of the expert geometer, a processor that delivers inferences independently of consciously accessible justifications. The intuitive processor of the symbolic paradigm is performing logical inference via operations manipulating meaningful symbols, just as the conscious rule interpreter does: the differences concern only accuracy, efficiency and conscious access.

In the other approach to cognitive science – the *subsymbolic paradigm* – conscious rule interpretation is analyzable in the same terms as in the symbolic paradigm, but intuition is not (Smolensky, 1988). This is crucial because intuition includes the large majority of mental processes studied in cognitive science. Expertise resides in the intuitive processor, and we are all experts in perception, action, common sense, and language. If Turing's (1950) discussion of a machine playing the imitation game, or of training an intelligent machine through verbal instruction, is applied to the human machine, the necessary command of language is a capability of the human *intuitive* processor, and if the subsymbolic paradigm is correct, such a machine cannot operate within the confines of a discrete computational architecture operating on meaningful symbols.

For in the subsymbolic paradigm, the intuitive processor is formalised with a type of computation falling outside the class of discrete architectures. In a *continuous architecture*, data is numerical, and the primitive processes are arithmetic operations; data changes over time according to a differential equation, which combines multiple operations. The class of continuous computational architectures includes Thomson (Lord Kelvin)'s (1876) mechanical integrator, Bush's (1931) Differential Analyzer, Pour-El's (1974) theory of analog computation, Blum et al.'s (1989) theory of computation over the real numbers, as well as the continuous neural network models of Amari (1977), Hinton and Anderson (1981), Grossberg (1982), Hopfield (1984), Kohonen (1984), Rumelhart and McClelland (1986), and many others.[2] In these latter networks, the data consists in n real numbers $a_1, a_2, \ldots, a_n$; a_k is called 'the activation value of the kth unit (or neuron)'; together, they specify a point in $\mathbb{R}^n$, the 'activation vector' $\mathbf{a}$ – a 'pattern of activity'. A typical differential equation is $da_k/dt = -a_k + f(\sum_j W_{kj}a_j)$, where $f(x) \equiv [1+e^{-x}]^{-1}$ and each parameter W_{kj} is called 'the connection (or synaptic) weight from unit j to unit k'.

Because of the universality of discrete computation, any continuous computation can be approximately simulated with discrete computation[3] – but the crucial point is that then *the symbols manipulated must refer to activation values and weights*, not to meaningful concepts and to rules that relate them. In the subsymbolic paradigm, *mechanism operates below the level of meaning* (Hofstadter, 1985), whether the mechanism be given in its most natural formalisation as continuous computation, or in its discrete-computational approximation. Here is why.

According to the subsymbolic paradigm, in the intuitive processor, a concept is not represented as a symbol governed by the rules of a symbolic program: *a concept is represented by an entire activation vector*, which is governed by a differential equation that operates on individual activity values, each of which is but a small part of a vector that is meaningful. This is *distributed conceptual representation.* In the symbolic paradigm, the concept COFFEE is encoded as a symbol which

---

[2] Continuous computation includes some, but not all, of what is called 'parallel distributed processing', 'connectionism', or 'neural networks'; Turing's own work on network machines (Turing, 1948) falls in the discrete computational class, like the work of McCulloch and Pitts discussed below.

[3] The questions under consideration here concern how best to model the internal structure of mental processing; issues concerning digital vs. analog computability are not particularly central and can be put aside.

appears in propositions encoding knowledge of the relation of this concept to others. But in the subsymbolic paradigm, COFFEE is encoded as a vector. Changing the numbers in this vector can yield the vector encoding the concept SCONE. Individual numbers in these vectors do not correspond to concepts about the world of gustatory delicacies to which the computation refers: only the entire pattern of numbers constituting the activation vector **a** has such a conceptual interpretation. A discrete program simulating the continuous computation that defines the processing occurring within this subsymbolic model will have symbols that refer to 'the fifth activation value', not to CREAM.

Indeed, in the McCulloch and Pitts (1943) discrete calculus inspired by Turing (1936), the basic elements are propositions that refer to the binary active/inactive status of a node in a network. If such a predicate is equated with an 'idea' – a proposition over *concepts* – we are in the symbolic paradigm.[4] A network is but a notational variant of a complex proposition or equation; networks provide a notation that is often more convenient for describing parallel computation than is the string-based notations that are so well suited to sequential computation. The distinction between network- and string-based notations is largely orthogonal to the distinction between symbolic and subsymbolic cognitive models. It is true that conscious rule interpretation has a sequential nature (computors execute only one operation at a time), so lends itself well to string-based notation, while intuitive processing does not have an overtly sequential character. Intuition is modeled by unconscious sequential computation in the symbolic paradigm, but by (unconscious) parallel computation in the subsymbolic paradigm; this is reflected in the propensity of theorists to write the former in string- and the latter in network-based notations. What is crucial is that, whatever the notation, the primitive operations of the computational architecture operate on conceptually meaningful elements (symbols) in the symbolic paradigm, but on sub-conceptual elements (activations) in the subsymbolic paradigm.

Why have a significant proportion of cognitive scientists adopted a computational architecture deploying continuous mechanisms that operate beneath the level of meaning?

The subsymbolic paradigm is motivated by both the human mind and the human machine. Regarding the mind, during the 1970s certain psychologically-oriented cognitive scientists became frustrated with the rigidity of the symbolic paradigm for capturing human mental processes. Their theories called for 'partially active' words during sentence processing, 'spreading activation' to yield flexible associations between concepts, and, ultimately, learning procedures accumulating quantitative degrees of associations between sub-conceptual properties of experience from which emerge over time the coordinated aggregates which function as concepts, like Hebb's (1949) cell assemblies.

Regarding the human machine, it has long been an important part of the mind-body problem to connect the mental to the physical brain, and computational reduction offers the first prospect for carrying this out rigorously. This requires, however, that the computational architecture deployed has primitive operations, data, and combinators that a brain could provide. And this is what the continuous architectures employed in the subsymbolic paradigm achieve, according to our current best understanding of the appropriate level of neural organisation: a mental concept is encoded in the activity of many neurons, not a single one, and a given population of neurons can host multiple patterns encoding multiple mental concepts; the activity of a neuron is a continuously varying quantity at the relevant level of analysis; combination of information from multiple neurons has an approximately linear character, while the consequences of this combination for neural activation

---

[4] As mentioned in note 1, it is not clear what mapping between node activations and concepts was imagined for the network machines introduced by Turing or by McCulloch and Pitts (their 'ideas immanent in nervous activity'). If they envisioned something other than a transparent mapping under which each node encodes a conceptual proposition, then they seem not to have discussed the formal structure, or any of the implications of, a non-transparent mapping. (Nor the combinatorial explosion of nodes entailed by a 'node = proposition' mapping. A population of n nodes each with v discriminable values of course provides not n, but v^n patterns for potential use as representations.)

has a non-linear character that imposes minimum and maximum levels. The primitives provided by the continuous architectures of the subsymbolic paradigm are within the capabilities of the brain – that the brain has yet *more* complexity than assumed in these architectures does not compromise the subsymbolic paradigm's computational reduction of mental to neural processing.

While brain theory is far from achieving a settled state, this conception of continuous neural processing has generally displaced earlier notions according to which the relevant level of analysis was taken to be one where neural activations are binary (firing/not-firing), as assumed by the early discrete-computational network descriptions of the brain developed by Turing (1948) as well as McCulloch and Pitts (1943). Similarly, early on, the search for the meaning of neural activation targeted individual cells, but recent years have seen an explosion of research in which neural meaning is sought by recording 'population codes' over hundreds of neurons, or patterns of aggregated activity over hundreds of thousands of neurons (functional Magnetic Resonance Imaging, fMRI).

The implications of distributed representation – of taking vectors in $\mathbb{R}^n$ as the basic data type of the computational architecture – are many (Smolensky and Legendre, 2006). The basic operations of continuous mathematics provided by vector space theory now do the fundamental computational work. Consider, for example, language processing, a key domain for cognitive science. Instead of stringing together two conceptual-level symbols to form $\texttt{Frodo}_{\texttt{subject}}\ \texttt{lives}_{\texttt{predicate}}$, we add together two vectors, the first representing 'Frodo as subject' and the second 'lives as predicate'. The vector encoding 'Frodo as subject' results from taking a vector representing Frodo and a vector representing 'subject' and combining them with a vector operation called the tensor product. The basic mapping operation of vector space theory, linear transformation, provides the core of mental processing. Vector similarity, as conventionally defined using the vector inner product, serves to model conceptual similarity, a central notion of psychological theory. And a notion from dynamical systems theory proves to have major implications as well: the differential equations governing many subsymbolic models have the property that as time progresses, a quantity called the *Harmony* steadily increases. Harmony can be interpreted as a numerical measure of the *well-formedness* of the network's activation vector with respect to the weights. The objective of network computation is to produce the representation (vector) that is maximally well-formed – *optimal*.[5]

On a practical level, aspects of this continuous computational theory enable analysis of the conceptual meaning of measured neural activity patterns; this is now becoming pervasive in cognitive neuroscience (e.g., fMRI). And on a theoretical level, a nascent theory of vectorial computation is starting to tackle the kinds of computability questions initiated for discrete computation by Turing and his contemporaries.

The symbolic and subsymbolic paradigms conflict as regards the modeling of intuitive *processes*: the former, but not the latter, posit mechanisms manipulating conceptually meaningful elements. Nonetheless, with respect to mental *representations* – even within the intuitive processor – it is possible to achieve a degree of integration between the two paradigms by embedding discrete representations within a continuous vector space. Imagine the continuous space of all activation vectors as a Euclidean plane ($\mathbb{R}^n$, with $n = 2$). Imagine that stuck into the plane are a set of flags, each bearing a symbol string. So at a particular point $\mathbf{x}$ there is a flag labeled `b`, at another point $\mathbf{y}$ a flag labeled `ab`, and at $\mathbf{z}$ a flag bearing `aab`. The symbol strings `b`, `ab`, and `aab` have been *embedded in* $\mathbb{R}^n$ at the vectors $\mathbf{x}$, $\mathbf{y}$, and $\mathbf{z}$.

---

[5] Definitions: addition, $[\mathbf{a}+\mathbf{b}]_k \equiv a_k + b_k$; tensor product, $[\mathbf{a} \otimes \mathbf{b}]_{kj} \equiv a_k b_j$; linear transformation, $[\mathbf{Fa}]_k \equiv \sum_j F_{kj} a_j$; inner product, $[\mathbf{a} \cdot \mathbf{b}] \equiv \sum_k a_k b_k$; Harmony, $H(\mathbf{a}) \equiv 1/2 \sum_{kj} a_k W_{kj} a_j + \sum_k h(a_k)$.

A function f over symbol strings (e.g., f : b $\mapsto$ ab, ab $\mapsto$ aab, ...) can, for a significant class of functions, be achieved through a basic subsymbolic function, a linear transformation $\mathbf{F}$ over $\mathbb{R}^n$ (i.e., $\mathbf{F}: \mathbf{x} \mapsto \mathbf{y}, \mathbf{y} \mapsto \mathbf{z}, \ldots$).[6] Such strings might represent conceptual structures, in which case the *function computed* by such a subsymbolic system has a perfectly precise description within the symbolic paradigm – employing symbols which individually have conceptual meaning – but the *process* by which this function is computed does not. Returning to language processing, this yields a theory in which the syntax of a native speaker can be specified by a symbolic grammar, but the mental processes that manipulate syntactic mental representations can be specified only subsymbolically (in terms of connection weights derived from the symbolic grammar).

Such a theory of mental grammars has led to a new subsymbolically-grounded theory in a cognitive domain long central to the symbolic paradigm: universal grammar. Recall that in many subsymbolic models, processing is optimisation: it produces representations that have maximal Harmony (well-formedness). The Harmony of the representation $\mathbf{z}$ – the location of the flag bearing aab – can be taken to be the *grammatical well-formedness* of the symbolic representation aab. According to such a *Harmonic Grammar*, the grammatical strings are those with maximal Harmony (Legendre et al., 1990). And the Harmony of aab, it turns out, can be computed in terms of *constraints* like 'a cannot follow b': a string (like baa) that violates this constraint incurs a Harmony penalty $-s$, where s is the strength of that constraint in the grammar. In actual natural language grammars, an example constraint is 'no plural subject for a singular verb' (violated by dogs barks).

Previous approaches to human grammar (Chomsky, 1965) are primarily based in discrete computation theory: a grammar is a set of rewrite rules which provide a step-by-step set of instructions – suitable for a human computor – for constructing grammatical structures. That is, grammars are specified as *processes*. Harmonic Grammars are specified instead in terms of *products*: a product of language processing is grammatical iff it optimally satisfies the grammar's constraints; a process for generating such optimal products is not specified by the grammar, but is left to a separate theory analyzing how to carry out optimisation over such constraints. The new contribution to the theory of universal grammar derives from the following strong hypothesis: the constraints are the same in all human grammars – only their strengths vary from language to language. The empirical success of this hypothesis can be summarised: what is universal across languages is found in the products, not the processes, of language generation.

Constraint-based approaches to grammar can also be pursued with discrete computational architectures. Indeed it is a striking discovery of recent years that the Harmonic Grammars of human languages have a strong tendency to display a special property: the strength of any given constraint is greater than that of all weaker constraints combined.[7] This entails that all that is needed to determine grammaticality – optimality – is the *ranking* of constraints from strongest to weakest within a particular grammar. This is *Optimality Theory* (Prince and Smolensky, 1997, 2004), which is the theory that actually introduced the strong universality hypothesis in the form: the constraints in all grammars are the same; only their relative ranking varies across languages. In Optimality Theory, grammars *specify functions* in discrete computational terms; but considered as part of the subsymbolic paradigm, the human mental *processes* that actually compute these functions must

[6] For a fully recursive function over strings of unbounded length, we need $n = \infty$, but nonetheless $\mathbf{F}$ can be finitely specified.

[7] More precisely, the Harmony penalty resulting from a single violation of a given constraint is greater than the maximal Harmony penalty that can result from all weaker constraints combined.

be specified using continuous computation – optimisation – over Harmonic Grammars realised in vectorial distributed conceptual representations.

If cognition is computation, we must ask, what are the primitive computational elements, and how do they map onto cognitive entities? For the cognitive faculty of conscious rule interpretation, the computational primitives are the symbol-manipulating operations of discrete computation, with individual symbols mapping onto individual concepts. Mechanism operates on meaningful elements. For intuitive cognition, the same holds – according to the symbolic paradigm of cognitive science. In the subsymbolic paradigm, however, the computational primitives are the numerical operations of continuous computation, and a concept corresponds to an entire vector. Mechanism operates on individual numbers, activation values, beneath the level of meaning. Subsymbolic computation reduces complex mental processes to simple brain processes. Vector space theory provides tools now widely used for conceptual interpretation of recorded activation patterns in the brain. Dynamical systems theory provides tools for interpreting subsymbolic computation as optimisation. Applied to language, this leads to a theory of grammar in which what is universal is the optimality-defining criteria for evaluating the products of language processing – as opposed to the process of producing these representations, previously the subject matter of mainstream grammatical theory.

The universe of computation opened up to us by Turing includes not just the discrete class of architectures, but also the continuous class; not just symbolic, but also vectorial representation of concepts; the means to formalize grammatical knowledge not just as procedures for computation, but also as criteria for evaluating products of computation. Undoubtedly, the universe of computation holds other unexplored architectures for creating machine intelligence and for understanding human cognition.

References

Amari, S., 1977. Dynamics of pattern formation in lateral-inhibition type neural fields. Biol. Cybern. 27, 77–87.

Anderson, J.R., 1983. The Architecture of Cognition, Harvard University Press, Cambridge, MA.

Blum, L., Shub, M., Smale, S., 1989. On a theory of computation and complexity over the real numbers: NP-completeness, recursive functions and universal machines. Bull. Am. Math. Soc. 21, 1–46.

Bush, V., 1931. The differential analyser, a new machine for solving differential equations. J. Franklin Inst. 212, 447–488.

Chomsky, N., 1965. Aspects of the Theory of Syntax, MIT Press, Cambridge, MA.

Grossberg, S., 1982. Studies of Mind and Brain: Neural Principles of Learning, Perception, Development, Cognition, and Motor Control, Reidel, Boston, MA.

Haugeland, J., 1985. Artificial Intelligence: The Very Idea, MIT Press, Cambridge, MA.

Hebb, D., 1949. The Organization of Behavior: A Neuropsychological Theory, Wiley, NY.

Hinton, G., Anderson, J., 1981. Parallel Models of Associative Memory, Lawrence Erlbaum Associates, Mahwah, New Jersey.

Hofstadter, D., 1985. Waking up from the Boolean dream, or, subcognition as computation. In Hofstadter, D. (Ed.), Metamagical Themas: Questing for the Essence of Mind and Pattern. Bantam Books, NY, (pp. 631–665).

Hopfield, J., 1984. Neurons with graded response have collective computational properties like those of two-state neurons. Proc. Natl. Acad. Sci. USA 81, 3088–3092.

Kohonen, T., 1984. Self-Organization and Associative Memory, Springer, Berlin.

Legendre, G., Miyata, Y., Smolensky, P., 1990. Harmonic grammar – a formal multi-level connectionist theory of linguistic well-formedness: Theoretical foundations. Proc. Cogn. Sci. Soc. 12, 388–395.

McCulloch, W., Pitts, W., 1943. A logical calculus of the ideas immanent in nervous activity. Bull. Math. Biol. 5, 115–133.

Newell, A., Simon, H., 1972. Human Problem Solving, Prentice-Hall, Englewood Cliffs, New Jersey.

Pour-El, M., 1974. Abstract computability and its relation to the general purpose analog computer (some connections between logic, differential equations and analog computers). Trans. Am. Math. Soc. 199, 1–28.
Prince, A., Smolensky, P., 1997. Optimality: From neural networks to universal grammar. Science 275, 1604–1610.
Prince, A., Smolensky, P., 2004. Optimality Theory: Constraint Interaction in Generative Grammar, Blackwell, Oxford.
Rumelhart, D., McClelland, J., The PDP Research Group., 1986. Parallel Distributed Processing, MIT Press, Cambridge, MA.
Smolensky, P., 1988. On the proper treatment of connectionism. Behav. Brain Sci. 11, 1–23.
Smolensky, P., Legendre, G., 2006. The Harmonic Mind: From Neural Computation to Optimality-Theoretic Grammar, MIT Press, Cambridge, MA.
Thomson, W., 1876. On an instrument for calculating the integral of the product of two given functions. Proc. R. Soc. Lond. 24, 266–275.
Turing, A.M., 1936. On computable numbers, with an application to the Entscheidungsproblem. Proc. Lond. Math. Soc. 42, 230–265.
Turing, A.M., 1948. Intelligent Machinery: A Report by A. M. Turing, National Physical Laboratory, Teddington.
Turing, A.M., 1950. Computing machinery and intelligence. Mind 59, 433–460.

Tom Vickers recalls —

ALAN TURING AT THE NPL 1945–47

Having been fortunate to get to know Alan Turing briefly at NPL from 1946 to 1947, it was suggested that I should commit my memories to paper. Sadly, memories fade with age; also I find that I would be repeating much that is now readily available in important books, particularly those by Hodges (1983), Copeland (2005) and Yates (1997). For much greater detail of this short period, I would strongly recommend the two accounts by J. H. (Jim) Wilkinson in his 1970 ACM Turing Award lecture (Wilkinson, 1971, p.243), and *A History of Computing in the 20th Century* (Wilkinson, 1980, p.101).

I will therefore try to present material which has not appeared elsewhere, plus some personal observations on the remarkable revolution in computing which took place during my stay at NPL from 1946 to 1977.

At the end of World War II, Jim and I were invited by E. T. (Chas) Goodwin to join the newly formed 'National Mathematics Lab' which was the Mathematics Division of the National Physical Laboratory (NPL) in Teddington. Earlier in the war, Goodwin had been a senior colleague of ours in the Maths Lab at Cambridge doing ballistics problems. (That lab was later to become the home of EDSAC.)

It was planned that Jim would share his time as an aide to Turing who was working on his own on the design of a computer (ACE or Automatic Computing Engine) and also work with Goodwin and Les Fox on the development of good numerical methods. It was also hoped that his plan to return to Cambridge to do further research would not materialise. I joined the Desk Computing Section headed by Goodwin and one of my briefs was to look out for better and faster desk machines. (Little did I realise what was to appear in a short time.) The whole Division was about 20 strong but it gradually increased to about 40. We were all very young (under 40). We were housed in a large house (Cromer House) and later in the adjacent Teddington Hall, both next to NPL. This is where the thinking about the Ace started.

Although Turing did not mix socially to any extent, I was very fortunate that I was regularly able to join him at lunch with my two senior colleagues, Goodwin and Fox and possibly others; here he clearly enjoyed their company and was most jovial. I was more the listener than a contributor in such exalted company! The subject of sport often arose because Fox was particularly good at both soccer and cricket. Any sports event between Oxford (Fox) and Cambridge (Goodwin), or cricket match between Yorkshire (Fox) and Middlesex (Goodwin) always caused a lot of rivalry, much to Turing's amusement as he knew very little about either cricket or soccer. However, he was always able to offer a new slant to the games such as 'If a pyramid of acrobats leapt up, could they save a potential 6 hit at cricket?' It struck me that Turing was particularly good at lateral thinking.

Mathematics did crop up at times, and the subject of build-up of rounding errors was a hot topic. It was clear that Turing took a more pessimistic view than Fox, leading at times to strong debates. Jim, with more limited experience at that time, tended to favour Fox and later on, with the arrival of the Pilot Ace, was able to do lots of experiments to provide evidence for his subsequent research in error analysis.

Turing frequently raised the subject of Machine Intelligence (this was in 1946, long before it became a fashionable subject). He also discussed the possible use of a computer to play chess. This had obviously been a popular hobby during the war. He discussed the Bletchley 'Inter-hut' challenges when his colleagues had included the bulk of the British chess team. In revisiting Bowden's *Faster than Thought* (Bowden, 1953) I see that the chapter on Game Playing was attributed to Turing and he discusses how chess might be played by computer. This classic book is probably not well known but it represents the state of computer activity in the UK in 1952, with description of machines and applications. I give further details in Appendix 1. I also provide details in Appendix 2 of a 1949 book by Hartree (1950), a pre-war expert on the same subject.

Sadly, Turing did not stay to see the arrival of the Pilot Ace, his plans to concentrate on building the larger full-scale Ace having been delayed. However, he was pleased to see that it was a resounding success even though he had not been keen to see its development. He did not produce any problems to be solved, as his interests by then (not unusually for him) were far from the standard mathematical/engineering problems focused on at NPL. A return to academia was probably the right thing to do. He had left enough impetus to NPL in a stay of about 2 years.

The impact of the Pilot Ace on the parent Maths Division

By 1950, Maths Division had been in existence for 5 years and developed into a strong unit able to tackle a wide range of jobs for Government and Industry, mainly using desk calculators (of which we were able to acquire the best available) and Hollerith punched card machines. (Here, several had been converted for scientific rather than the usual commercial work. I have just been reminded that this included a modified tabulator redundant from Bletchley Park.) The smooth running cruise liner was about to be met by a gigantic tidal wave!

One of my jobs had been to seek improvements from the newer desk machines becoming available. Could I reduce the time to multiply 2 large numbers from 15 seconds by 10%, say? After exploiting a National Accounting machine with 6 registers modified to work in decimal rather than sterling, what could I do with 8 registers? Suddenly, the Pilot Ace appeared with a multiplication time of about 1/500 second and initially 250 registers. It is not surprising that I jumped at the opportunity to move, with the initial task of developing a junior section to be involved in operating and possibly programming. Ted York, who had been a member of the Punched Card team exploiting their modified equipment, also moved over to become a very experienced and enthusiastic programmer until recruited by RAE Farnborough when their Deuce (the commercial successor to the

Pilot Ace) arrived. Significant changes of duties and organisation followed during the Ace era but I recall that these are discussed in detail by Mary Croarken in *Early Scientific Computing in Britain* (Croarken, 1990) and I do not discuss them further here.

My chapter in (Copeland, 2005, p.265), presented at the Pilot Ace's 50th birthday conference, gives a summary of the overall impact of the Pilot Ace and I have little to add to it. I give details of the many jobs tackled and refer to the large number of guest workers who spent time with us and subsequently became leaders elsewhere, particularly in universities where they pioneered new courses in computing – E.S. Page, P. Samet, J. Ord-Smith and, very much later, Jean Bacon. We lost several staff by the same route – F. W. J. Olver (USA and still very active at Maryland), L. Fox (Oxford), C. W. Clenshaw (Lancaster) and S. Gill (London). Finally, I try to assess why it was such a success after a disappointing start and discuss some major offshoots such as the NAG (National Algorithms Group) Library which is now based in Wilkinson House, Oxford, and the Central Computing Agency.

Turing's Legacy to NPL

The ideas which he had developed with help from Jim Wilkinson, Mike Woodger and, later, Donald Davies and Gerald Alway (who sadly died at an early age in a car accident) were successfully used in building the Pilot Ace and it proved to be a greyhound among its UK rivals. In relation to EDSAC, for example, it was about 4 times faster, contained about a quarter the number of valves and had 33 copies made (the Deuce). The design of the Bendix G15 was based on it, over 400 were sold and it was arguably the world's first personal computer. For a detailed discussion of the speed of the Pilot Ace, see Martin Campbell-Kelly, in (Copeland, 2005, p.149). The most important impact was that his ideas persuaded three important members of the team to stay.

Jim Wilkinson had fully expected to return to academia but was inspired to stay and see a computer built. He then exploited its use and developed important research into error analysis of computer arithmetic, resulting in worldwide fame, a rise to the top position in the Scientific Civil Service and the award of FRS. He was able to inspire so many staff of all grades to get interested, whereas the shy Turing did not interact with junior staff.

Both Jim and Turing disliked some of the bureaucratic aspects of the Civil Service. Jim had a simple solution and that was to ignore it. To avoid a dull internal meeting, he would find a convenient hiding place to do some 'useful work'. (I recall that during his war-time work in the Cambridge Maths Lab he was found growing tomatoes on the top flat roof.) Many contributed to the success of the Pilot Ace but Jim was the undoubted leader who kept his light under a bushel. He was a real asset to NPL.

Mike Woodger also stayed at NPL until retirement. An early very successful project of his, assisted by Brian Munday, was started on the Pilot Ace (when the magnetic drum was fitted) and finalised on the Deuce. This was a General Interpretive Program (GIP). A good description of it by R. A. Vowels appears in (Copeland, 2005, p.319). Heavy use was made of it in handling matrices which often arise in engineering and aircraft design, although it did have wider applications. Indeed, this one program was sufficient to sell several Deuces to the then widespread aircraft industry. The firm of Bristol Engines developed a simplified form to deal with columns, rather like a modern spreadsheet.

Mike worked on language definitions including Algol and Ada. He was fortunately able to retrieve many important official documents which are now preserved in the Science Library. (My earlier attempts to archive important material were not encouraged: I was too early for the historian!)

Donald Davies had a long career in developing ideas in computing outside the main functions of Maths Division and these came under new regroupings eventually called Computer Science. David Yates, who was a colleague, gives full detail in his Yates (1997). Although a very important member of the Ace builders (particularly of the clever input-output arrangements), he did not get much involved with problem-solving as he acquired other duties, including a year in the USA. However, one of his first tasks on the Pilot Ace has not received much recognition so it is worth mentioning here. (I should add that when the Pilot Ace began to do jobs, any publicity was primarily up to the client. Further, there were not many appropriate journals available. That is why we got heavily involved in launching two new societies – the Institute of Mathematics and its Applications (IMA) and the British Computer Society (BCS).) Donald's problem came from the Road Research Lab - a sister establishment (and I believe a pre-war spin-off from NPL). It related to timing of traffic signals, both fixed, as they were then, and traffic-operated, which was under consideration. It must have been a very early use of computers in Operational Research involving the generation of random numbers. Four streams of traffic, represented by random digits (1=car, 0=gap) were timed through a junction. The streams could be seen on the output screen: when 2 of them stopped, the other 2 would start. The program was tiny, well under the 300 stores available. Hitherto, the use of random numbers taken from tables had not been widespread but it now became simple and the subject of OR blossomed from then on. Of his later work, mention should be made of the NPL network, which by 1976 connected about 30 various computers, 100 VDUs, several devices and services around NPL (a miniature World Wide Web). Two key features were the concept of packet-switching (as used in the Web) and a Standard Interface for connecting devices (like the USB on a modern laptop). At the time, its use was similar to the 3-point plug in the house. (It started as a 12-wire system, later extended to 18.) Like Turing and Wilkinson, Davies became FRS.

Thoughts on post-war computing at NPL

1. Little was known about the effort needed to build a computer, although from 1945 the key features were well established. The days of large special-purpose machines such as Colossus and ENIAC were over. (L. J. Comrie, the pre-war expert on computing - see Croarken (1990) - had long disliked such machines.) A machine should be able to store both data and instructions in a common store, and these could be in binary form represented by pulses. The storage system needed to be specified and there were a number of options - delay lines (mercury, nickel, I believe Turing even suggested gin!), cathode ray tube, magnetic drum, but later to be replaced by magnetic core. Sir Charles Darwin, Director of NPL, felt that the best machine would be created by co-operation among the 3 main UK builders - NPL and the Universities of Cambridge and Manchester. However, original interest among the parties was lukewarm and it is clear that a Turing link with M.V. Wilkes would not have got very far. However, at the lower level, Wilkinson had very cordial relations with David Wheeler and Stan Gill (who had worked briefly for Jim) giving David a lot of unrecognised credit for the success of EDSAC. So, by 1951, three different machines were working in the UK with a potential greater than in the USA at that time. NPL had its links to English Electric Co. with the Deuce and other computers to come; Cambridge had the surprise link to J. Lyons, of Corner House fame, with the start of commercial computing, and also the development of micro-programming; Manchester had close links with Ferranti, later to produce Atlas etc. In addition, a line of computers for defence was coming from Elliot Bros. Thus the initial lack of joint activity produced widespread dividends later.

2. An early prediction (possibly by Hartree (1950)) was made that 3 or 4 computers would satisfy the total national need! How many have we in a typical house now with its washing machine, phone, camera, laptop etc?

3. The newly-formed Nuclear Energy Authority of the late 1940s had big computing needs far beyond the desk machine but our Punched Card section proved to be effective for certain Monte-Carlo calculations, initially for Harwell and later for Aldermaston. (There was a very big difference. One was deadly secret and hush-hush whilst the other came under normal confidentiality. Yet the two problems were similar - in one the energy was captured (hopefully) for general use and in the other it was released with deadly results.) Later, a new problem was

produced by Aldermaston which needed the skills of Wilkinson. It was processed on the Pilot Ace although we were not to know the details. The Nuclear Units were early buyers when large commercial computers became available from IBM and Ferranti, and lured several of our experienced punched card staff to run new computing sections in Harwell, Aldermaston and Culham, with enhanced salaries. (Our use of our card machines was likely to decline, so this was less of a problem to us.) Wilkinson was under tremendous pressure to go to Aldermaston, but he resisted. Having tried hard to leave secret Armament Research at the end of the war, he had no desire to return to a similar regime.

4. Turing's proposal to "develop an electronic computer" which was presented to the NPL Executive Committee on 19 March 1946 did not, to my knowledge, surface until 1972, and then by chance! In going through the contents of a locked drawer for Confidential Information held by our deceased librarian, I came across the only copy held by the Division and this must have been deposited by J. R. Womersley when the proposal was first discussed. Jim Wilkinson was away, probably on one of his trips to the USA, so I showed it to Mike Woodger who had not met it. It was then published by Donald Davies as an NPL Report (see reference 138 in Yates (1997)) and later by Carpenter and Doran (1986). Unfortunately, the original was not returned to me. It must be of immense value now, but at least we know the content. In it, Turing gives an interesting list of ten tasks which ought to be solvable by his proposal. These are listed in (Yates, 1997, p.21). Most would have been standard problems once the Pilot Ace was working, but for 3 of them his comments are more interesting.

   a) Solution of linear simultaneous equations. We cannot expect to solve more than 50 equations." In fact, in 1952 we solved 129 equations (and without the magnetic drum). See (Copeland, 2005, p.271).

   b) Cut up a jigsaw into a number of whole squares. Not a common problem but could have great military significance." This must be a reference to the huge manual task of Bill Tutte in cracking one of the codes at Bletchley as a preliminary to Turing's work, which featured in a recent TV documentary (see (Hodges, 1983, p.332)). I wonder whether Turing or M. Newman ever gave thought to the impact of new computers on cryptography?

   c) Comments on the ability of a computer to play good chess. I recall a much later writer (around 1970) saying that it could eventually beat the best players. I mentioned this to Michael Stein, a schoolboy Grandmaster who spent the summer with us prior to going to university and his comment was nonsense". Some years later, I was very amused to read that Mike was the first Grandmaster to have been beaten by a computer.

5. How do ideas develop? From my experience at NPL, I found that lots of new ideas came about from chance discussions between scientists of different disciplines and there was plenty of scope for this at the three pioneering institutions. An extraordinary example is the original chat between M. V. Wilkes and J. Lyons to consider commercial computing. Innovation appears to be much more common than invention. However, in Turing's case he was much more the inventor than the innovator. He was always thinking way ahead of the current problem, and it needed a genius such as Jim Wilkinson to keep up with him. Hence his difficulty in communicating with others not on the same wavelength.

Appendix 1: On *"Faster than Thought"*

This very interesting collection of papers Bowden (1953) on the state of the art in 1952 does not appear to have come to the attention of the modern historian. After a brief history of previous machines (e.g. Babbage) he then describes the electronics involved in a 1950 computer. This is followed by descriptions of current work in progress in the UK and the USA, mainly supplied by the relevant workers. It is interesting that for the Pilot Ace, the chapter is anonymous and "reprinted from *Engineering* by kind permission of the editors". I assume that this refers to a Turing article, but I have no further reference. It is also of interest that, at the time of publication, Bowden was working for Ferranti and in close contact with Manchester University and Turing. The next section describes a range of applications, relying heavily on articles from Ferranti and Manchester University. As I have reported earlier, the article on game-playing is attributed to Turing, where he discusses chess, draughts and nim. There is no mention of applications at NPL but it could be that we had little published then. He had good contacts with us, as he was a buddy of our new director, E. C. Bullard. The final chapter on Thought and Machine Processes, which is written by the editor, includes a fascinating description of two mathematical prodigies of that time - Professor A. C. Aitken of Edinburgh University and Willy Klein of the Mathematisch Centrum in Amsterdam. I was very fortunate to have had impressive demonstrations from each of them. Their capabilities were similar, based on knowing their tables to 100 by 100 rather the 12 by 12, but their spare activities were rather different. Klein liked to learn logarithms of more numbers but Aitken, in his London demonstration, said that he was happy to look these up in tables and instead learnt all the piano works of Bach and Beethoven. Although he was teaching practical maths to students, I understand that his lab was poorly equipped with desk calculators as he expected his students to follow his methods! We always had close connections with the Dutch centre and Willy Klein (who was a humble operator) visited us for a couple of days. Apparently he would use a desk calculator when he was tired. I organised races with him using two of our popular demonstration programs - finding the smallest factor of a six-figure number and finding the day of a given date. He was quick! According to the article, he multiplied two ten-figure numbers to give a twenty-figure product in 64 seconds. In describing the Pilot Ace, I used to quote times for this as "by hand 15 minutes [Les Fox claimed he could do it faster but I am afraid he got the wrong answer], by desk machine 15 seconds and by computer 1/500 second". The article gives many examples of their amazing skills.

Bowden also quotes the skill of a Lakeland shepherd who, when retrieving and counting his flock of about 2000 sheep, will know not only how many are missing but which ones. What is incredible is that when I was a schoolboy I visited a farm in the foothills of the Pennines and helped to collect up the sheep. After more than one attempt we agreed a total and one was missing. Our illiterate farmer, who had very little school training, told us which one and where he last saw it!

Appendix 2: References

Hodges, A. (1983). Alan Turing: the enigma. An epic biography dealing more with his life from schoolboy onwards and less with technical aspects of his work.

Copeland, J., ed. (2005). Alan Turing's Automatic Computing Engine. A very comprehensive technical account of his achievements. It includes not only the papers given at the two-day conference to celebrate the 50th birthday anniversary of the Pilot Ace but much other important material.

Yates, D. M. (1997). Turing's Legacy. A history of computing at NPL, 1945-1995. It relies on the Annual Reports of NPL. It is thin on the first few years when little was available but very complete on the later work of Donald Davies and others.

Wilkinson, J. H. (1971). Some Comments from a Numerical Analyst. ACM Turing Award Lecture, 1970. A significant part deals with the 1? years he shared a room with Turing designing the Ace.

Wilkinson, J. H. (1980). His contribution to A History of Computing in the 20th Century. A collection of essays edited by N. Metropolis, J. Howlett and Gian-Carlo Rota, published by Academic Press. His article has a few bits in common with Wilkinson4, but the book provides an excellent summary of activity up to 1980, similar to the way that Bowden6 deals with 1952.

Bowden, B. V., ed. (1953, paperback 1971). Faster than Thought. See Appendix 1.

Croarken, M. (1990). Early Scientific Computing in Britain. A comprehensive account of how World War II changed the needs for computing from being very limited to the setting up of a centre at NPL (probably the most detailed account published) and the building of electronic computers.

Carpenter, B. J. and Doran R. W., eds. (1986). A. M. Turing's ACE report of 1946 and other papers, MIT Press. I have not seen this, but it provides the content of at least one historic document.

Hartree, D. R. (1950). Calculating Instruments and Machines (analogue and digital respectively to us). Hartree was a computing pioneer pre-war, initially in Manchester and later at Cambridge. He was involved in oversight of the new Maths Division at NPL. This is a rare book of lectures which he gave in late 1948 at Illinois and published by them. He begins by describing the Differential Analyser and its use, which he had built at Manchester, based on an original American design by Dr V. Bush. After a brief review of other instruments" he refers to Babbage, ENIAC and the Harvard Mark I calculator in the belief that the latter two are outstanding steps in the way true computers are likely to develop". He admits that developments in 1948 are very rapid. Finally, after summarising the electronic parts of machines under development in the UK, he suggests some mathematical problems for such machines. In total, a good review of the state of the art in 1948 to compare with 1980 (Wilkinson5) and 1952 (Bowden6).

Lavington, S., ed. (2012). Alan Turing and his Contemporaries - Building the world's first computers. This has just been published by the British Computer Society for the Computer Conservation Society to mark the centenary of his birth. Only 110 pages long, it concentrates on the period 1945-55 and relies heavily on material already published.

Douglas Hofstadter engages with —

The Gödel–Turing Threshold and the Human Soul

1. The universal machines that surround us

When I was around twelve years old, there were kits you could buy that allowed you to put together electronic circuitry that would carry out various interesting functions. You could build a radio, a circuit that would add two binary numbers, a device that could encode or decode a message using a substitution cipher, a 'brain' that would play tic-tac-toe against you, and a few other devices like this. Each of these machines was *dedicated*: it could do just one kind of trick. This is the usual meaning of 'machine' that we grow up with. We are accustomed to the idea of a refrigerator as a dedicated machine for keeping things cold, an alarm clock as a dedicated machine for waking us up and so on. But more recently, we have started to get used to machines that transcend their original purposes.

Take cellular telephones, for instance. Nowadays, in order to be competitive, cell phones are marketed not so much (maybe even very little) on the basis of their original purpose as communication devices, but instead for the number of tunes they can hold, the number of games you can play on them, the quality of the photos they can take, and who knows what else! Cell phones once were, but no longer are, dedicated machines. And why is that? It is because their inner circuitry has surpassed a certain threshold of complexity, and that fact allows them to have a chameleon-like nature. You can use the hardware inside a cell phone to house a word processor, a web browser, a gaggle of video games, and on and on. This, in essence, is what the computer revolution is all about: when a certain well-defined threshold – I'll call it the 'Gödel– Turing threshold' – is surpassed, then a computer can emulate *any* kind of machine.

This is the meaning of the term 'universal machine', introduced in 1936 by the English mathematician and computer pioneer Alan Turing, and today we are intimately familiar with the basic idea, although most people do not know the technical term or concept. We routinely download virtual machines from the web that can convert our universal laptops into temporarily specialized devices for watching movies, listening to music, playing games, making cheap international phone calls, identifying birds in photos, who knows what. Machines of all sorts come to us through wires or even through the air, via software, via patterns, and they swarm into and inhabit our computational hardware. One single universal machine morphs into new functionalities at the drop of a hat, or, more precisely, at the double-click of a mouse. I bounce back and forth between my e-mail program, my word processor, my web browser, my photo displayer and a dozen other 'applications' that all live inside my computer. At any specific moment, most of these independent, dedicated machines are dormant, sleeping, waiting patiently (actually, unconsciously) to be awakened by my royal double-click and to jump obediently to life and do my bidding.

Inspired by Kurt Gödel's astonishing discovery, in 1931, which Alfred North Whitehead and Bertrand Russell's monumental formal system *Principia Mathematica* could encode its own rules of inference and thus could 'look at itself', Alan Turing realized that the critical threshold for this kind of computational universality comes exactly at that point where a machine is flexible enough to read and correctly interpret a set of data that describe its own structure. At this crucial juncture, a machine can, in principle, explicitly watch how it does any particular task, step by step. Turing realized that a machine that has this critical level of flexibility can imitate any another machine, no matter how complex the latter is. In other words, there is nothing *more* flexible than a universal machine. Universality is as far as you can go!

This is why my Macintosh can, if I happen to have fed it the proper software, act indistinguishably from my son's more expensive and faster 'Alienware' computer (running any specific program), and vice versa. The only difference is speed, because my Mac will always remain, deep in its guts, a Mac. It will therefore have to imitate the fast, alien hardware by constantly consulting tables of data that explicitly describe the hardware of the Alien, and doing all those lookups is very slow. This is like me trying to get you to sign my signature by writing out a long set of instructions telling you how to draw every tiny curve. In principle it's possible, but it would be hugely slower than just signing with my own handware!

2. The unexpectedness of universality

There is a tight analogy linking universal machines of this sort with the universality of *Principia Mathematica.* What Russell and Whitehead did not suspect, but what Gödel realized, is that, simply by virtue of representing certain fundamental features of the positive integers (such basic facts as commutativity, distributivity, the law of mathematical induction), they had unwittingly made their formal system *PM* surpass a key threshold that made it 'universal', which is to say, capable of defining number-theoretical functions that imitate arbitrarily complex *other* patterns (or indeed, even capable of turning around and imitating itself – giving rise to Gödel's black-belt maneuver, whereby the enormous strength of the system was, in a sense, exploited to bring about its own downfall).

Russell and Whitehead did not realize what they had wrought because it did not occur to them to use *PM* to 'simulate' anything else. That idea was not on their radar screen (for that matter, radar itself was not on anybody's radar screen back then). Prime numbers, squares, sums of two squares, sums of two primes, Fibonacci numbers and so forth were seen merely as beautiful mathematical patterns – and patterns consisting of numbers, although fabulously intricate and endlessly fascinating, were not thought of as being isomorphic to anything else, let alone as being stand-ins for, and thus standing for, anything else. After Gödel and Turing, although, such naïveté went down the drain in a flash.

By and large, the engineers who designed the earliest electronic computers were as unaware as Russell and Whitehead had been of the richness that they were unwittingly bringing into being. They thought they were building machines of very limited, and purely military, scopes – for instance, machines to calculate the trajectories of ballistic missiles, taking wind and air resistance into account, or machines to break very specific types of enemy codes. They envisioned their computers as being specialized, single-purpose machines – a little like wind-up music boxes that could play just one tune each.

But at some point, when Alan Turing's abstract theory of computation, based in large part on Gödel's 1931 paper, collided with the concrete engineering realities, some of the more perceptive people (such as Turing himself and John von Neumann) put two and two together and realized that their machines, incorporating the richness of integer arithmetic that Gödel had shown was so potent, were thereby universal. All at once, these machines were like music boxes that could read arbitrary paper scrolls with holes in them, and thus could play *any* tune. From then on, it was simply a matter of time until cell phones started being able to don many personas other than just the plain old cell phone persona. All they had to do was surpass that threshold of complexity and memory size that limited them to a single 'tune', and then they could become anything.

The early computer engineers thought about their computers as number-crunching devices and did not see numbers as a universal medium. Today, we (and by 'we' I mean our culture as a whole, rather than specialists) do not see numbers that way either, but our lack of understanding is for an entirely different reason – in fact, for exactly the opposite reason. Today it is because all those numbers are so neatly hidden behind the screens of our laptops and desktops that we utterly forget they are there. We watch virtual football games unfolding on our screen between hypothetical "dream teams" that exist only inside the CPU, or central processing unit (which is carrying out arithmetical instructions, just as it was designed to do). Children build virtual towns inhabited by little people who virtually ride by on virtual bicycles, with leaves that virtually fall from trees and smoke that virtually dissipates into the virtual air. Cosmologists create virtual galaxies, let them loose, and watch what happens as they virtually collide. Biologists create virtual proteins and watch them fold up according to the complex virtual chemistry of their constituent virtual submolecules.

I could list hundreds of things that take place on computer screens, but few people ever think about the fact that all of this is happening courtesy of *addition and multiplication of integers* way down at the hardware level. But that *is* exactly what's happening. We don't call computers *computers* for nothing, after all! They are, in fact, computing sums and products of integers expressed in binary notation. And in that sense, Gödel's world-dazzling, Russell-crushing, Hilbert-toppling vision of 1931 have become such a commonplace in our downloading, upgrading, gigabyte culture that although we are all swimming in it all the time, hardly anyone is least aware of it. Just about the only trace of the original insight that remains visible, or rather, 'audible', around us is the very word 'computer'. That term tips you off, if you bother to think about it, to the fact that underneath all the colorful pictures, seductive games, and lightning-fast web searches, there is nothing going on but integer arithmetic. What a hilarious joke!

To be quite honest, however, things are a bit more ambiguous than that. Wherever there is a pattern, it can be seen either as itself or as standing for anything to which it is isomorphic. For example, if two spouses have been unfaithful to each other, then harsh accusatory words flung by the husband at his wife condemning her acts of straying apply equally much to him, and neither way of interpreting of his words will be truer than the other one, even if *he* is thinking only about one of those ways. His words apply to his own straying whether he likes it or not, because his sin was isomorphic to his wife's sin. Likewise, an operation on an integer that is written out in binary notation (for instance, the conversion of '0000000011001111' into '1100111100000000') that one person might describe as multiplication by 256 might be described by another observer as a left shift by eight bits, and by another observer as the transfer of a color from one pixel to its neighbor, and by someone else as the deletion of an alphanumeric character in a file. As long as each one is a correct description of what is happening, none of them is privileged. The reason we call computers 'computers', then, is solely a historic one. They originated as integer calculation machines, and they are still of course validly describable as such – but we now realize, as Kurt Gödel first did in 1931, that such devices can be equally validly perceived and talked about in terms that are fantastically different from what their originators intended.

3. Universal beings

We human beings, too, are universal machines of a different sort: our neural hardware can copy arbitrary patterns, even if evolution never had any grand plan for this kind of 'representational universality' to come about. Through our senses and then our symbols, we can internalize external phenomena of many sorts. For example, as we watch ripples spreading on a pond, our symbols echo their circular shapes, abstract them and can replay the essence of those shapes much later. I say 'the essence.' because some – in fact most – detail is lost; as is obvious, we retain not all levels of what we encounter but only those that our hardware, through the pressures of natural selection, came to consider the most important. I should also make clear that when I say that our symbols 'internalize' or 'copy' external patterns, I do not mean that when we watch ripples on a pond, or when we 'replay' a memory of such a scene (or of many such scenes blurred together), there literally are circular patterns spreading out on some horizontal surface inside our brains. I mean that a host of structures are jointly activated, which are connected with the concepts of water, wetness, ponds, horizontal surfaces, circularity, expansion, things bobbing up and down and so forth. I am not talking about a movie screen inside the head!

Representational universality also means that we can import ideas and happenings without having to be direct witnesses to them. For example, humans (but not most other animals) can easily process the two-dimensional arrays of pixels on a television screen and can see those ever-changing arrays as coding for distant or fictitious three-dimensional situations evolving over time.

One time on a skiing vacation in the Sierra Nevada, some two thousand miles west of our home town of Bloomington, Indiana, my children and I took advantage of the 'doggie cam' at the Bloomington kennel where we had boarded our golden retriever Ollie, and thanks to the World Wide Web, we were treated to a jerky sequence of stills of a couple of dozen dogs meandering haphazardly in a fenced-in play area outdoors, looking a bit like particles undergoing random Brownian motion, and although each pooch was rendered by a pretty small array of pixels, we could often recognize our Ollie by subtle features such as the angle of his tail. For some reason, the kids and I found this act of visual eavesdropping on Ollie quite hilarious, and although we could easily describe this droll scene to our human friends, and although I would bet a considerable sum that these few lines of text have vividly evoked in your mind both the canine scene at the kennel and the human scene at the ski resort, we all realized that there was not a hope in hell that we could ever explain to Ollie himself that we had been 'spying' on him from thousands of miles away. Ollie would never know, and could never know.

Why not? Because Ollie is a dog, and dogs' brains are not universal. They cannot absorb ideas like 'jerky still photo', '24-hour webcam', 'spying on dogs playing in the kennel', 'two thousand miles west of', or even, for that matter, the concept 'west of'. This is a huge and fundamental breach between humans and dogs – indeed, between humans and all other species. It is this that sets us apart, makes us unique, and, in the end, gives us what we call 'souls'.

In the world of living things, the magic threshold of representational universality is crossed whenever a system's repertoire of symbols becomes extensible without any obvious limit. This threshold was crossed on the species level somewhere along the way from earlier primates to ourselves. Systems above this counterpart to the Gödel–Turing threshold – let us call them 'beings', for short – have the capacity to model inside themselves other beings that they run into – to slap together quick-and-dirty models of beings that they encounter only briefly, to refine such coarse models over time, even to invent imaginary beings from whole cloth. (Beings with a propensity to invent other beings are often informally called 'novelists'.)

Once beyond the magic threshold, universal beings seem inevitably to become ravenously thirsty for tastes of the interiority of other universal beings. This is why we have movies, soap operas, television news, blogs, webcams, gossip columnists, *People* magazine and tabloid newspapers. People yearn to get inside other people's heads, to 'see out' from inside other crania, to gobble up other people's experiences.

Although I have been depicting it somewhat cynically, representational universality and the nearly insatiable hunger that it creates for vicarious experiences is but a stone's throw away from empathy, which I see as the most admirable quality of humanity. To 'be' someone else in a profound way is not merely to see the world intellectually as they see it and to feel rooted in the places and times that molded them as they grew up; it goes much further than that. It is to adopt their values, to take on their desires, to live their hopes, to feel their yearnings, to share their dreams, to shudder at their dreads, to participate in their life and to merge with their souls.

Computing Machinery and Intelligence

(MIND, vol. **59** (1950), pp. 433–460)

Gregory Chaitin discovers Alan Turing 'The Good Philosopher' at both sides of —

MECHANICAL INTELLIGENCE VERSUS UNCOMPUTABLE CREATIVITY

What is human intelligence? Is it mechanical or is it creative? In this paper, Turing argues forcefully for the former, mechanical, but his *On Computable Numbers* makes a dramatic statement in favour of creativity, since he shows that there is no mechanical procedure even for solving something as simple as the halting problem. Let us contrast Turing's *MIND* article with his *On Computable Numbers* more forcefully. According to his piece in *MIND,* human thinking is mechanical. But *On Computable Numbers* is – as Paul Feyerabend puts it in another context – *Against Method* in mathematics, and people can do mathematics! How come?

According to Gödel, mathematicians sometimes have direct access to the Platonic world of ideas; in Turing's famous *Systems of Logic based on Ordinals* terminology, they sometimes seem to have access to uncomputable *oracles.* Certainly, Euler and Ramanujan's extreme mathematical creativity are difficult to explain; indeed, they seem, at least superficially, to defy ordinary rational explanations.

In other words, are we machines or do we have a divine spark? In *MIND*, Turing argues the former, but as Emil Post argued in the 1940s, Gödel and Turing's work on incompleteness and uncomputability can equally well be viewed as emphasizing the fundamental importance of creativity – defined as uncomputability – in the progress of mathematics.[1] This also emphasises the connection between incompleteness and uncomputability and mathematical and biological creativity, which I do not believe are that different.

In fact, in my view, incompleteness and uncomputability open a door from mathematics to biology. The halting probability Ω contains infinite irreducible complexity and in a sense shows that pure mathematics is even more biological then biology itself, which merely contains extremely large finite complexity. For each bit of the dyadic expansion of Ω is one bit of irreducible mathematical information, while the human genome is merely 3×10^9 bases $= 6 \times 10^9$ bits of information.

It is a delightful paradox that Turing argues that we are machines while all the while emphasizing the importance of what machines cannot do. Like a good philosopher, he cannot help seeing the good arguments on both sides. He thus provides ammunition to both parties.

[1] Post's 1941 words, first published by Martin Davis in 1965 in *The Undecidable* and movingly placed by Jeremy Gray at the conclusion of his 2008 treatise on *Plato's Ghost: The Modernist Transformation of Mathematics,* are worth quoting: 'mathematical thinking is, and must be, essentially creative'.

COMPUTING MACHINERY AND INTELLIGENCE

By A. M. TURING

1. The Imitation Game.

I PROPOSE to consider the question, 'Can machines think ?' This should begin with definitions of the meaning of the terms 'machine' and 'think'. The definitions might be framed so as to reflect so far as possible the normal use of the words, but this attitude is dangerous. If the meaning of the words 'machine' and 'think' are to be found by examining how they are commonly used it is difficult to escape the conclusion that the meaning and the answer to the question, 'Can machines think ?' is to be sought in a statistical survey such as a Gallup poll. But this is absurd. Instead of attempting such a definition I shall replace the question by another, which is closely related to it and is expressed in relatively unambiguous words.

The new form of the problem can be described in terms of a game which we call the 'imitation game'. It is played with three people, a man (A), a woman (B), and an interrogator (C) who may be of either sex. The interrogator stays in a room apart from the other two. The object of the game for the interrogator is to determine which of the other two is the man and which is the woman. He knows them by labels X and Y, and at the end of the game he says either 'X is A and Y is B' or 'X is B and Y is A'. The interrogator is allowed to put questions to A and B thus:

C: Will X please tell me the length of his or her hair?

Now suppose X is actually A, then A must answer. It is A's object in the game to try and cause C to make the wrong identification. His answer might therefore be

'My hair is shingled, and the longest strands are about nine inches long.'

In order that tones of voice may not help the interrogator the answers should be written, or better still, typewritten. The ideal arrangement is to have a teleprinter communicating between the two rooms. Alternatively the question and answers can be repeated by an intermediary. The object of the game for the third player (B) is to help the interrogator. The best strategy for her is probably to give truthful answers. She can add such things as 'I am the woman, don't listen to him!' to her answers, but it will avail nothing as the man can make similar remarks.

We now ask the question, 'What will happen when a machine takes the part of A in this game?' Will the interrogator decide wrongly as often when the game is played like this as he does when the game is played between a man and a woman? These questions replace our original, 'Can machines think?'

2. Critique of the New Problem.

As well as asking, 'What is the answer to this new form of the question', one may ask, 'Is this new question a worthy one to investigate?' This latter question we investigate without further ado, thereby cutting short an infinite regress.

The new problem has the advantage of drawing a fairly sharp line between the physical and the intellectual capacities of a man. No engineer or chemist claims to be able to produce a material which is indistinguishable from the human skin. It is possible that at some time this might be done, but even supposing this invention available we should feel there was little point in trying to make a 'thinking machine' more human by dressing it up in such artificial flesh. The form in which we have set the problem reflects this fact in the condition which prevents the interrogator from seeing or touching the other competitors, or hearing their voices. Some other advantages of the proposed criterion may be shown up by specimen questions and answers. Thus:

Q: Please write me a sonnet on the subject of the Forth Bridge.
A: Count me out on this one. I never could write poetry.
Q: Add 34957 to 70764
A: (Pause about 30 seconds and then give as answer) 105621.
Q: Do you play chess?
A: Yes.
Q: I have K at my Kl, and no other pieces. You have only K at K6 and R at R1. It is your move. What do you play ?
A: (After a pause of 15 seconds) R-R8 mate.

The question and answer method seems to be suitable for introducing almost any one of the fields of human endeavour that we wish to include. We do not wish to penalise the machine for its inability to shine in beauty competitions, nor to penalise a man for losing in a race against an aeroplane. The conditions of our game make these disabilities irrelevant. The 'witnesses' can brag, if they consider it advisable, as much as they please about their charms, strength or heroism, but the interrogator cannot demand practical demonstrations.

The game may perhaps be criticised on the ground that the odds are weighted too heavily against the machine. If the man were to try and pretend to be the machine he would clearly make a very poor shewing. He would be given away at once by slowness and inaccuracy in arithmetic. May not machines carry out something which ought to be described as thinking but which is very different from what a man does? This objection is a very strong one, but at least we can say that if, nevertheleees, a machine can be constructed to play the imitation game satisfactorily, we need not be troubled by this objection.

It might be urged that when playing the 'imitation game' the best strategy for the machine may possibly be something other than imitation of the behaviour of a man. This may be, but I think it is unlikely that there is any great effect of this kind. In any case there is no intention to investigate here the theory of the game, and it will be assumed that the best strategy is to try to provide answers that would naturally be given by a man.

3. The Machines concerned in the Game.

The question which we put in §1 will not be quite definite until we have specified what we mean by the word 'machine'. It is natural that we should wish to permit every kind of engineering technique to be used in our machines. We also wish to allow the possibility than an engineer or team of engineers may construct a machine which works, but whose manner of operation cannot be satisfactorily described by its constructors because they have applied a method which is largely experimental. Finally, we wish to exclude from the machines men born in the usual manner. It is difficult to frame the definitions so as to satisfy these three conditions. One might for instance insist that the team of engineers should be all of one sex, but this would not really be satisfactory, for it is probably possible to rear a complete individual from a single cell of the skin (say) of a man. To do so would be a feat of biological technique deserving of the very highest praise, but we would not be inclined to regard it as a case of 'constructing a thinking machine'. This prompts us to abandon the requirement that every kind of technique should be permitted. We are the more ready to do so in view of the fact that the present interest in 'thinking machines' has been aroused by a particular kind of machine, usually called an 'electronic computer' or 'digital computer'. Following this suggestion we only permit digital computers to take part in our game.

This restriction appears at first sight to be a very drastic one. I shall attempt to show that it is not so in reality. To do this necessitates a short account of the nature and properties of these computers.

It may also be said that this identification of machines with digital computers, like our criterion for 'thinking', will only be unsatisfactory if (contrary to my belief), it turns out that digital computers are unable to give a good showing in the game.

There are already a number of digital computers in working order, and it may be asked, 'Why not try the experiment straight away ? It would be easy to satisfy the conditions of the game. A number of interrogators could be used, and statistics compiled to show how often the right identification was given.' The short answer is that we are not asking whether all digital computers would do well in the game nor whether the computers at present available would do well, but whether there are imaginable computers which would do well. But this is only the short answer. We shall see this question in a different light later.

4. Digital Computers.

The idea behind digital computers may be explained by saying that these machines are intended to carry out any operations which could be done by a human computer. The human computer is supposed to be following fixed rules; he has no authority to deviate from them in any detail. We may suppose that these rules are supplied in a book, which is altered whenever he is put on to a new job. He has also an unlimited supply of paper on which he does his calculations. He may also do his multiplications and additions on a 'desk machine', but this is not important.

If we use the above explanation as a definition we shall be in danger of circularity of argument. We avoid this by giving an outline of the means by which the desired effect is achieved. A digital computer can usually be regarded as consisting of three parts:

(i) Store.
(ii) Executive unit.
(iii) Control.

The store is a store of information, and corresponds to the human computer's paper, whether this is the paper on which he does his calculations or that on which his book of rules is printed. In so far as the human computer does calculations in his head a part of the store will correspond to his memory.

The executive unit is the part which carries out the various individual operations involved in a calculation. What these individual operations are will vary from machine to machine. Usually fairly lengthy operations can be done such as 'Multiply 3540675445 by 7076345687' but in some machines only very simple ones such as 'Write down 0' are possible.

We have mentioned that the 'book of rules' supplied to the computer is replaced in the machine by a part of the store. It is then called the 'table of instructions'. It is the duty of the control to see that these instructions are obeyed correctly and in the right order. The control is so constructed that this necessarily happens.

The information in the store is usually broken up into packets of moderately small size. In one machine, for instance, a packet might consist of ten decimal digits. Numbers are assigned to the parts of the store in which the various packets of information are stored, in some systematic manner. A typical instruction might say—

> 'Add the number stored in position 6809 to that in 4302 and put the result back into the latter storage position'.

Needless to say it would not occur in the machine expressed in English. It would more likely be coded in a form such as 6809430217. Here 17 says which of various possible operations is to be performed on the two numbers. In this case the operation is that described above, *viz.* 'Add the number....' It will be noticed that the instruction takes up 10 digits and so forms one packet of information, very conveniently. The control will normally take the instructions to be obeyed in the order of the positions in which they are stored, but occasionally an instruction such as

> 'Now obey the instruction stored in position 5606, and continue from there'

may be encountered, or again

> 'If position 4505 contains 0 obey next the instruction stored in 6707, otherwise continue straight on.'

Instructions of these latter types are very important because they make it possible for a sequence of operations to be repeated over and over again until some condition is fulfilled, but in doing so to obey, not fresh instructions on each repetition, but the same ones over and over again. To take a domestic analogy. Suppose Mother wants Tommy to call at the cobbler's every morning on his way to school to see if her shoes are done, she can ask him afresh every morning. Alternatively she can stick up a notice once and for all in the hall which he will see when he leaves for school and which tells him to call for the shoes, and also to destroy the notice when he comes back if he has the shoes with him.

The reader must accept it as a fact that digital computers can be constructed, and indeed have been constructed, according to the principles we have described, and that they can in fact mimic the actions of a human computer very closely.

The book of rules which we have described our human computer as using is of course a convenient fiction. Actual human computers really remember what they have got to do. If one wants to make a machine mimic the behaviour of the human computer in some complex operation one has to ask him how it is done, and then translate the answer into the form of an instruction table. Constructing instruction tables is usually described as 'programming'. To ' programme a machine to carry out the operation A' means to put the appropriate instruction table into the machine so that it will do A.

An interesting variant on the idea of a digital computer is a 'digital computer with a random element'. These have instructions involving the throwing of a die or some equivalent electronic process; one such instruction might for instance be,'Throw the die and put the resulting number into store 1000'. Sometimes such a machine is described as having free will (though I would not use this phrase myself). It is not normally possible to determine from observing a machine whether it has a random element, for a similar effect can be produced by such devices as making the choices depend on the digits of the decimal for π.

Most actual digital computers have only a finite store. There is no theoretical difficulty in the idea of a computer with an unlimited store. Of course only a finite part can have been used at any one time. Likewise only a finite amount can have been constructed, but we can imagine more and more being added as required. Such computers have special theoretical interest and will be called infinitive capacity computers.

The idea of a digital computer is an old one. Charles Babbage, Lucasian Professor of Mathematics at Cambridge from 1828 to 1839, planned such a machine, called the Analytical Engine, but it was never completed. Although Babbage had all the essential ideas, his machine was not at that time such a very attractive prospect. The speed which would have been available would be definitely faster than a human computer but something like 100 times slower than the Manchester machine, itself one of the slower of the modern machines. The storage was to be purely mechanical, using wheels and cards.

The fact that Babbage's Analytical Engine was to be entirely mechanical will help us to rid ourselves of a superstition. Importance is often attached to the fact that modern digital computers are electrical, and that the nervous system also is electrical. Since Babbage's machine was not electrical, and since all digital computers are in a sense equivalent, we see that this use of electricity cannot be of theoretical importance. Of course electricity usually comes in where fast signalling is concerned, so that it is not surprising that we find it in both these connections. In the nervous system chemical phenomena are at least as important as electrical. In certain computers the storage system is mainly acoustic. The feature of using electricity is thus seen to be only a very superficial similarity. If we wish to find such similarities we should look rather for mathematical analogies of function.

5. Universality of Digital Computers.

The digital computers considered in the last section may be classified amongst the 'discrete state machines'. These are the machines which move by sudden jumps or clicks from one quite definite state to another. These states are sufficiently different for the possibility of confusion between them to be ignored. Strictly speaking there are no such machines. Everything really moves continuously. But there are many kinds of machine which can profitably be *though of* as being discrete state machines. For instance in considering the switches for a lighting system it is a convenient fiction that each switch must be definitely on or definitely off. There must be intermediate positions, but for most purposes we can forget about them. As an example of a discrete state machine we might consider a wheel which clicks round through 120° once a second, but may be stopped by a lever which can be operated from outside; in addition a lamp is to light in one of the positions of the wheel. This machine could be described abstractly as follows. The internal state of the machine (which is described by the position of the wheel) may be q_1, q_2 or q_3. There is an input signal i_0 or i_1 position of lever). The internal state at any moment is determined by the last state and input signal according to the table

		Last State		
		q_1	q_2	q_3
Input	i_0	q_2	q_3	q_1
	i_1	q_1	q_2	q_3

The output signals, the only externally visible indication of the internal state (the light) are described by the table

State	q_1	q_2	q_3
Output	o_0	o_0	o_1

This example is typical of discrete state machines. They can be described by such tables provided they have only a finite number of possible states.

It will seem that given the initial state of the machine and the input signals it is always possible to predict all future states. This is reminiscent of Laplace's view that from the complete state of the universe at one moment of time, as described by the positions and velocities of all particles, it should be possible to predict all future states. The prediction which we are considering is, however, rather nearer to practicability than that considered by Laplace. The system of the 'universe as a whole' is such that quite small errors in the initial condition can have an overwhelming effect at a later time. The displacement of a single electron by a billionth of a centimetre at one moment might make the difference between a man being killed by an avalanche a year later, or escaping. It is an essential property of the mechanical systems which we have called 'discrete state machines' that this phenomenon does not occur. Even when we consider the actual physical machines instead of the idealised machines, reasonably accurate knowledge of the state at one moment yields reasonably accurate knowledge any number of steps later.

As we have mentioned, digital computers fall within the class of discrete state machines. But the number of states of which such a machine is capable is usually enormously large. For instance, the number for the machine now working at Manchester it about $2^{165,000}$, *i.e.* $10^{50,000}$. Compare this with our example of the clicking wheel described above, which had three states. It is not difficult to see why the number of states should be so immense. The computer includes a store corresponding to the paper used by a human computer. It must be possible to write into the store any one of the combinations of symbols which might have been written on the paper. For simplicity suppose that only digits from 0 to 9 are used as symbols. Variations in handwriting are ignored. Suppose the computer is allowed 100 sheets of paper each containing 50 lines each with room for 30 digits. Then the number of states is $10^{100 \times 50 \times 30}$, *i.e.* $10^{150,000}$. This is about the number of states of three Manchester machines put together. The logarithm to the base two of the number of states is usually called the 'storage capacity' of the machine. Thus the Manchester machine has a storage capacity of about 165,000 and the wheel machine of our example about 1·6. If two machines are put together their capacities must be added to obtain the capacity of the resultant machine. This leads to the possibility of statements such as 'The Manchester machine contains 64 magnetic tracks each with a capacity of 2560, eight electronic tubes with a capacity of 1280. Miscellaneous storage amounts to about 300 making a total of 174,380.'

Given the table corresponding to a discrete state machine it is possible to predict what it will do. There is no reason why this calculation should not be carried out by means of a digital computer. Provided it could be carried out sufficiently quickly the digital computer could mimic the behaviour of any discrete state machine. The imitation game could then be played with the machine in question (as B) and the mimicking digital computer (as A) and the interrogator would be unable to distinguish them. Of course the digital computer must have an adequate storage capacity as well as working sufficiently fast. Moreover, it must be programmed afresh for each new machine which it is desired to mimic.

This special property of digital computers, that they can mimic any discrete state machine, is described by saying that they are *universal* machines. The existence of machines with this property has the important consequence that, considerations of speed apart, it is unnecessary to design various new machines to do various computing processes. They can all be done with one digital computer, suitably programmed for each case . It will be seen that as a consequence of this all digital computers are in a same equivalent.

We may now consider again the point raised at the end of §3. It was suggested tentatively that the question, 'Can machines think?' should be replaced by 'Are there imaginable digital computers which would do well in the imitation game?' If we wish we can make this superficially more general and ask 'Are there discrete state machines which would do well?' But in view of the universality property we see that either of these question is equivalent to this, 'Let us fix our attention on one particular digital computer *C*. Is it true that by modifying this computer to have an adequate storage, suitably increasing its speed of action, and providing it with an appropriate programme, *C* can be made to play satisfactorily the part of A in the imitation game, the part of B being taken by a man?'

6. Contrary Views on the Main Question.

We may now consider the ground to have been cleared and we are ready to proceed to the debate on our question, 'Can machines think?' and the variant of it quoted at the end of the last section. We cannot altogether abandon the original form of the problem, for opinions will differ as to the appropriateness of the substitution and we must at least listen to what has to be said in this connexion.

It will simplify matters for the reader if I explain first my own beliefs in the matter. Consider first the more accurate form of the question. I believe that in about fifty years' time it will be possible to programme computers, with a storage capacity of about 10^9, to make them play the imitation game so well that an average interrogator will not have more than 70 per cent. chance of making the right identification after five minutes of questioning. The original question, 'Can machines think?' I believe to be too meaningless to deserve discussion. Nevertheless I believe that at the end of the century the use of words and general educated opinion will have altered so much that one will be able to speak of machines thinking without expecting to be contradicted. I believe further that no useful purpose is served by concealing these beliefs. The popular view that scientists proceed inexorably from well-established fact to well-established fact, never being influenced by any unproved conjecture, is quite mistaken. Provided it is made clear which are proved facts and which are conjectures, no harm can result. Conjectures are of great importance since they suggest useful lines of research.

I now proceed to consider opinions opposed to my own.

(1) *The Theological Objection.*

Thinking is a function of man's immortal soul. God has given an immortal soul to every man and woman, but not to any other animal or to machines. Hence no animal or machine can think.

I am unable to accept any part of this, but will attempt to reply in theological terms. I should find the argument more convincing if animals were classed with men, for there is a greater difference, to my mind, between the typical animate and the inanimate than there is between man and the other animals. The arbitrary character of the orthodox view becomes clearer if we consider how it might appear to a member of some other religious community. How do Christians regard the Moslem view that women have no souls? But let us leave this point aside and return to the main argument. It appears to me that the argument quoted above implies a serious restriction of the omnipotence of the Almighty. It is admitted that there are certain things that He cannot do such as making one equal to two, but should we not believe that He has freedom to confer a soul on an elephant if He sees fit? We might expect that He would only exercise this power in conjunction with a mutation which provided the elephant with an appropriately improved brain to minister to the needs of this soul. An argument of exactly similar form may be made for the case of machines. It may seem different because it is more difficult to "swallow". But this really only means that we think it would be less likely that He would consider the circumstances suitable for conferring a soul. The circumstances in question are discussed in the rest of this paper. In attempting to construct such machines we should not be irreverently usurping His power of creating souls, any more than we are in the procreation of children: rather we are, in either case, instruments of His will providing mansions for the souls that He creates.

However, this is mere speculation. I am not very impressed with theological arguments whatever they may be used to support. Such arguments have often been found unsatisfactory in the past. In the time of Galileo it was argued that the texts, "And the sun stood still... and hasted not to go down about a whole day" (Joshua x. 13) and "He laid the foundations of the earth, that it should not move at any time" (Psalm cv. 5) were an adequate refutation of the Copernican theory. With our present knowledge such an argument appears futile. When that knowledge was not available it made a quite different impression.

(2) *The 'Heads in the Sand' Objection.*

"The consequences of machines thinking would be too dreadful. Let us hope and believe that they cannot do so."

This argument is seldom expressed quite so openly as in the form above. But it affects most of us who think about it at all. We like to believe that Man is in some subtle way superior to the rest of creation. It is best if he can be shown to be *necessarily* superior, for then there is no danger of him losing his commanding position. The popularity of the theological argument is clearly connected

[1] Possibly this view is heretical. St. Thomas Aquinas (*Summa Theologica.* quoted by Bertrand Russell. p. 480) states that God cannot make a man to have no soul. But this may not be a real restriction on His powers, but only a result of the fact that men's souls are immortal, therefore indestructible.

with this feeling. It is likely to be quit strong in intellectual people, since they value the power of thinking more highly than others, and are more inclined to base their belief in the superiority of Man on this power.

I do not think that this argument is sufficiently substantial to require refutation. Consolation would be more appropriate: perhaps this should be sought in the transmigration of souls.

(3) *The Mathematical Objection.*

There are a number of results of mathematical logic which can be used to show that there are limitations to the powers of discrete-state machines. The best known of these results is known as Gödel's theorem,[2] and shows that in any sufficiently powerful logical system statements can be formulated which can neither be proved nor disproved within the system, unless possibly the system itself is inconsistent. There are other, in some respects similar, results due to *Church, Kleene, Rosser*, and *Turing*. The latter result is the most convenient to consider, since it refers directly to machines, whereas the others can only be used in a comparatively indirect argument: for instance if Gödel's theorem is to be used we need in addition to have some means of describing logical systems in term of machines, and machines in term of logical systems. The result in question refers to a type of machine which is essentially a digital computer with an infinite capacity. It states that there are certain things that such a machine cannot do. If it is rigged up to give answers to questions as in the imitation game, there will be some questions to which it will either give a wrong answer, or fail to give an answer at all however much time is allowed for a reply. There may, of course, be many such question, and question which cannot be answered by one machine may be satisfactorily answered by another. We are of course supposing for the present that the question are of the kind to which an answer 'Yes' or 'No' is appropriate, rather than question such as 'What do you think of Picasso?' The questions that we know the machines must fail on are of this type, "Consider the machine specified as follows... Will this machine ever answer 'Yes' to any question?" The dots are to be replaced by a description of some machine in a standard form, which could be something like that used in §5. When the machine described bears a certain comparatively simple relation to the machine which is under interrogation, it can be shown that the answer is either wrong or not forthcoming. This is the mathematical result: it is argued that it proves a disability of machines to which the human intellect is not subject.

The short answer to this argument is that although it is established that there are limitations to the powers of any particular machine, it has only been stated, without any sort of proof, that no such limitations apply to the human intellect. But I do not think this view can be dismissed quite so lightly. Whenever one of these machines is asked the appropriate critical question, and gives a definite answer, we know that this answer must be wrong, and this gives us a certain feeling of superiority. Is this feeling illusory? It is no doubt quite genuine, but I do not think too much importance should be attached to it. We too often give wrong answers to questions ourselves to be justified in being very pleased at such evidence of fallibility on the part of the machines. Further, our superiority can only be felt on such an occasion in relation to the one machine over which we have scored our petty triumph. There would be no question of triumphing simultaneously over *all* machines. In short, then, there might be men cleverer than any given machine, but then again there might be other machines cleverer again, and so on.

Those who hold to the mathematical argument would, I think, mostly be willing to accept the imitation game as a basis for discussion. Those who believe in the two previous objections would probably not be interested in any criteria.

---

[2] Author's names in italics refer to the Bibliography.

(4) *The Argument from Consciousness.*

This argument is very well expressed in *Professor Jefferson's* Lister Oration for 1949, from which I quote. "Not until a machine can write a sonnet or compose a concerto because of thoughts and emotions felt, and not by the chance fall of symbols, could we agree that machine equals brain—that is, not only write it but know that it had written it. No mechanism could feel (and not merely artificially signal, an easy contrivance) pleasure at its successes, grief when its valves fuse, be warmed by flattery, be made miserable by its mistakes, be charmed by sex, be angry or depressed when it cannot get what it wants."

This argument appears to be a denial of the validity of our test. According to the most extreme form of this view the only way by which one could be sure that a machine thinks is to *be* the machine and to feel oneself thinking. One could then describe these feelings to the world, but of course no one would be justified in taking any notice. Likewise according to this view the only way to know that a *man* thinks is to be that particular man. It is in fact the solipsist point of view. It may be the most logical view to hold but it makes communication of ideas difficult. A is liable to believe 'A thinks but B does not' whilst B believes 'B thinks but A does not'. Instead of arguing continually over this point it is usual to have the polite convention that everyone thinks.

I am sure that Professor Jefferson does not wish to adopt the extreme and solipsist point of view. Probably he would be quite willing to accept the imitation game as a test. The game (with the player B omitted) is frequently used in practice under the name of *viva voce* to discover whether some one really understands something or has 'learnt it parrot fashion'. Let us listen in to a part of such a *viva voce*:

Interrogator: In the first line of your sonnet which reads 'Shall I compare thee to a summer's day', would not 'a spring day' do as well or better?
Witness: It wouldn't scan.
Interrogator: How about 'a winter's day' That would scan all right.
Witness: Yes, but nobody wants to be compared to a winter's day.
Interrogator: Would you say Mr. Pickwick reminded you of Christmas?
Witness: In a way.
Interrogator: Yet Christmas is a winter's day, and I do not think Mr. Pickwick would mind the comparison.
Witness: I don't think you're serious. By a winter's day one means a typical winter's day, rather than a special one like Christmas.

And so on. What would Professor Jefferson say if the sonnet-writing machine was able to answer like this in the *viva voce*? I do not know whether he would regard the machine as 'merely artificially signalling' these answers, but if the answers were as satisfactory and sustained as in the above passage I do not think he would describe it as 'an easy contrivance'. This phrase is, I think, intended to cover such devices as the inclusion in the machine of a record of someone reading a sonnet, with appropriate switching to turn it on from time to time.

In short then, I think that most of those who support the argument from consciousness could be persuaded to abandon it rather than be forced into the solipsist position. They will then probably be willing to accept our test.

I do not wish to give the impression that I think there is no mystery about consciousness. There is, for instance, something of a paradox connected with any attempt to localise it. But I do not think these mysteries necessarily need to be solved before we can answer the question with which we are concerned in this paper.

(5) *Arguments from Various Disabilities.*

These arguments take the form, "I grant you that you can make machines do all the things you have mentioned but you will never be able to make one to do X". Numerous features X are suggested in this connexion. I offer a selection:

Be kind, resourceful, beautiful, friendly (p. 448), have initiative, have a sense of humour, tell right from wrong, make mistakes (p. 448), fall in love, enjoy strawberries and cream (p. 448), make some one fall in love with it, learn from experience (pp. 456 f.), use words properly, be the subject of its own thought (p. 449), have as much diversity of behaviour as a man, do something really new (p. 450). (Some of these disabilities are given special consideration as indicated by the page numbers.)

No support is usually offered for these statements. I believe they are mostly founded on the principle of scientific induction. A man has seen thousands of machines in his lifetime. From what he sees of them he draws a number of general conclusions. They are ugly, each is designed for a very limited purpose, when required for a minutely different purpose they are useless, the variety of behaviour of any one of them is very small, etc., etc. Naturally he concludes that these are necessary properties of machines in general. Many of these limitations are associated with the very small storage capacity of most machines. (I am assuming that the idea of storage capacity is extended in some way to cover machines other than discrete-state machines.

The exact definition does not matter as no mathematical accuracy is claimed in the present discussion.) A few years ago, when very little had been heard of digital computers, it was possible to elicit much incredulity concerning them, if one mentioned their properties without describing their construction. That was presumably due to a similar application of the principle of scientific induction. These applications of the principle are of course largely unconscious. When a burnt child fears the fire and shows that he fears it by avoiding it, I should say that he was applying scientific induction. (I could of course also describe his behaviour in many other ways.) The works and customs of mankind do not seem to be very suitable material to which to apply scientific induction. A very large part of space-time must be investigated, if reliable results are to be obtained. Otherwise we may (as most English children do) decide that everybody speaks English, and that it is silly to learn French.

There are, however, special remarks to be made about many of the disabilities that have been mentioned. The inability to enjoy strawberries and cream may have struck the reader as frivolous. Possibly a machine might be made to enjoy this delicious dish, but any attempt to make one do so would be idiotic. What is important about this disability is that it contributes to some of the other disabilities, *e.g.* to the difficulty of the same kind of friendliness occurring between man and machine as between white man and white man, or between black man and black man.

The claim that "machines cannot make mistakes" seems a curious one. One is tempted to retort, "Are they any the worse for that?" But let us adopt a more sympathetic attitude, and try to see what is really meant. I think this criticism can be explained in term of the imitation game. It is claimed that the interrogator could distinguish the machine from the man simply by setting them a number of problems in arithmetic. The machine would be unmasked because of its deadly accuracy. The reply to this is simple. The machine (programmed for playing the game) would not attempt to give the *right* answers to the arithmetic problems. It would deliberately introduce mistakes in a manner calculated to confuse the interrogator. A mechanical fault would probably show itself through an unsuitable decision as to what sort of a mistake to make in the arithmetic. Even this interpretation of the criticism is not sufficiently sympathetic. But we cannot afford the space to go into it much further. It seems to me that this criticism depends on a confusion between two kinds of mistake. We may call them 'errors of functioning' and 'errors of conclusion'. Errors of functioning are due to some mechanical or electrical fault which causes the machine to behave otherwise than it was designed to do. In philosophical discussion one likes to ignore the possibility of such errors; one is therefore discussing 'abstract machines'. These abstract machines are mathematical fictions rather than physical objects. By definition they are incapable of errors of functioning. In this sense we can truly say that 'machines can never make mistakes'. Errors of conclusion can only arise when some meaning is attached to the output signals from the machine. The machine might, for instance, type out mathematical equations, or sentences in English. When a false proposition is typed we say that the machine has committed an error of conclusion. There is clearly no reason at all for saying that a machine cannot make this kind of mistake. It might do nothing but type out repeatedly '$0 = 1$'.

To take a less perverse example, it might have some method for drawing conclusion by scientific induction. We must expect such a method to lead occasionally to erroneous results.

The claim that a machine cannot be the subject of its own thought can of course only be answered if it can be shown that the machine has *some* thought with *some* subject matter. Nevertheless, 'the subject matter of a machine's operations' does seem to mean something, at least to the people who deal with it. If, for instance, the machine was trying to find a solution of the equation $x^2 - 40x - 11 = 0$ one would be tempted to describe this equation as part of the machine's subject matter at that moment. In this sort of sense a machine undoubtedly can be its own subject matter. It may be used to help in making up its own programmes, or to predict the effect of alterations in its own structure. By observing the results of its own behaviour it can modify its own programmes so as to achieve some purpose more effectively. These are possibilities of the near future, rather than Utopian dreams.

The criticism that a machine cannot have much diversity of behaviour is just a way of saying that it cannot have much storage capacity. Until fairly recently a storage capacity of even a thousand digits was very rare.

The criticisms that we are considering here are often disguised forms of the argument from consciousness. Usually if one maintains that a machine *can* do one of these things, and describes the kind of method that the machine could use, one will not make much of an impression. It is thought that the method (whatever it may be, for it must be mechanical) is really rather base. Compare the parenthesis in Jefferson's statement quoted on p. 21.

(6) *Lady Lovelace's Objection.*

Our most detailed information of Babbage's Analytical Engine comes from a memoir by *Lady Lovelace*. In it she states, "The Analytical Engine has no pretensions to *originate* anything. It can do *whatever we know how to order it* to perform" (her italics). This statement is quoted by *Hartree* (p. 70) who adds: "This does not imply that it may not be possible to construct electronic equipment which will 'think for itself', or in which, in biological terms, one could set up a conditioned reflex, which would serve as a basis for 'learning'. Whether this is possible in principle or not is a stimulating and exciting question, suggested by some of these recent developments. But it did not seem that the machines constructed or projected at the time had this property".

I am in thorough agreement with Hartree over this. It will be noticed that he does not assert that the machines in question had not got the property, but rather that the evidence available to Lady Lovelace did not encourage her to believe that they had it. It is quite possible that the machines in question had in a sense got this property. For suppose that some discrete-state machine has the property. The Analytical Engine was a universal digital computer, so that, if its storage capacity and speed were adequate, it could by suitable programming be made to mimic the machine in question. Probably this argument did not occur to the Countess or to Babbage. In any case there was no obligation on them to claim all that could be claimed.

This whole question will be considered again under the heading of learning machines.

A variant of Lady Lovelace's objection states that a machine can 'never do anything really new'. This may be parried for a moment with the saw, 'There is nothing new under the sun'. Who can be certain that 'original work' that he has done was not simply the growth of the seed planted in him by teaching, or the effect of following well-known general principles. A better variant of the objection says that a machine can never 'take us by surprise'. This statement is a more direct challenged and can be met directly. Machines take me by surprise with great frequency. This is largely because I do not do sufficient calculation to decide what to expect them to do, or rather because, although I do a calculation, I do it in a hurried, slipshod fashion, taking risks. Perhaps I say to myself, 'I suppose the voltage here ought to be the same as there: anyway let's assume it is' Naturally I am often wrong, and the result is a surprise for me for by the time the experiment is done these assumptions have been forgotten. These admissions lay me open to lectures on the subject of my vicious ways, but do not throw any doubt on my credibility when I testify to the surprises I experience.

I do not expect this reply to silence my critic. He will probably say that such surprises are due to some creative mental act on my part, and reflect no credit on the machine. This leads us back to the argument from consciousness, and far from the idea of surprise. It is a line of argument we must consider closed, but it is perhaps worth remarking that the appreciation of something as surprising requires as much of a 'creative mental act' whether the surprising event originates from a man, a book, a machine or anything else.

The view that machines cannot give rise to surprises is due, I believe, to a fallacy to which philosophers and mathematicians are particularly subject. This is the assumption that as soon as a fact is presented to a mind all consequences of that fact spring into the mind simultaneously with it. It is a very useful assumption under many circumstances, but one too easily forgets that it is false. A natural consequence of doing so is that one then assumes that there is no virtue in the mere working out of consequences from data and general principles.

(7) *Argument from Continuity in the Nervous System.*

The nervous system is certainly not a discrete-state machine. A small error in the information about the size of a nervous impulse impinging on a neuron, may make a large difference to the size of the outgoing impulse. It may be argued that, this being so, one cannot expect to be able to mimic the behaviour of the nervous system with a discrete-state system.

It is true that a discrete-state machine must be different from a continuous machine. But if we adhere to the conditions of the imitation game, the interrogator will not be able to take any advantage of this difference. The situation can be made clearer if we consider some other simpler continuous machine. A differential analyser will do very well. (A differential analyser is a certain kind of machine not of the discrete-state type used for some kinds of calculation.) Some of these provide their answers in a typed form, and so are suitable for taking part in the game. It would not be possible for a digital computer to predict exactly what answers the differential analyser would give to a problem, but it would be quite capable of giving the right sort of answer. For instance, if asked to give the value of π (actually about 3·1416) it would be reasonable to choose at random between the values 3·12, 3·13, 3·14, 3·15, 3·16 with the probabilities of 0·05, 0·15, 0·55, 0·19, 0·06 (say). Under these circumstances it would be very difficult for the interrogator to distinguish the differential analyser from the digital computer.

(8) *TheArgument from Informality of Behaviour.*

It is not possible to produce a set of rules purporting to describe what a man should do in every conceivable set of circumstances. One might for instance have a rule that one is to stop when one sees a red traffic light, and to go if one sees a green one, but what if by some fault both appear together? One may perhaps decide that it is safest to stop. But some further difficulty may well arise this decision later. To attempt to provide rules of conduct to cover every eventuality, even those arising from traffic lights appear to be impossible. With all this I agree.

From this it is argued that we cannot be machines. I shall try to reproduce the argument, but I fear I shall hardly do it justice. It seems to run something like this. 'If each man had a definite set of rules of conduct by which he regulated his life he would be no better than a machine. But there are no such rules, so men cannot be machines'. The undistributed middle is glaring. I do not think the argument is ever put quite like this, but I believe this is the argument used nevertheless. There may however be a certain confusion between 'rules of conduct' and 'laws of behaviour' to cloud the issue. By 'rules of conduct' I mean precepts such as 'Stop if you see red lights', on which one can act, and of which one can be conscious. By 'laws of behaviour' I mean laws of nature as applied to a man's body such as 'if you pinch him he will squeak'. If we substitute 'laws of behaviour which regulate his life' for 'laws of conduct by which he regulates his life' in the argument quoted the undistributed middle is no longer insuperable. For we believe that it is not only true that being regulated by laws of behaviour implies being some sort of machine (though not necessarily a discrete-state machine, but that conversely being such a machine implies being regulated by such laws. However, we cannot so easily convince ourselves of the absence of complete laws of behaviour as of complete rules of conduct. The only way we know of for finding such laws is scientific observation, and we certainly know of no circumstances under which we could say, 'We have searched enough. There are no such laws.'

We can demonstrate more forcibly that any such statement would be unjustified. For suppose we could be sure of finding such laws if they existed. Then given a discrete-state machine it should certainly be possible to discover by observation sufficient about it to predict its future behaviour, and this within a reasonable time, say a thousand years. But this does not seem to be the case. I have set up on the Manchester computer a small programme using only 1000 units of storage, whereby the machine supplied with one sixteen figure number replies with another within two seconds. I would defy anyone to learn from these replies sufficient about the programme to be able to predict any replies to untried values.

(8) *The Argument from Extra-Sensory Perception.*

I assume that the reader is familiar with the idea of extra-sensory perception, and the meaning of the four items of it, *viz.* telepathy, clairvoyance, precognition and psycho-kinesis. These disturbing phenomena seem to deny all our usual scientific ideas. How we should like to discredit them! Unfortunately the statistical evidence, at least for telepathy, is overwhelming. It is very difficult to rearrange one's ideas so as to fit these new facts in. Once one has accepted them it does not seem a very big step to believe in ghosts and bogies. The idea that our bodies move simply according to the known laws of physics, together with some others not yet discovered but somewhat similar, would be one of the first to go.

This argument is to my mind quite a strong one. One can say in reply that many scientific theories seem to remain workable in practice, in spite of clashing with E.S.P.; that in fact one can get along very nicely if one forgets about it. This is rather cold comfort, and one fears that thinking is just the kind of phenomenon where E.S.P. may be especially. relevant.

A more specific argument based on E.S.P. might run as follows: "Let us play the imitation game, using as witnesses a man who is good as a telepathic receiver, and a digital computer. The interrogator can ask such questions as 'What suit does the card in my right hand belong to?' The man by telepathy or clairvoyance gives the right answer 130 times out of 400 cards. The machine can only guess at random, and perhaps gets 104 right, so the interrogator makes the right identification." There is an interesting possibility which opens here. Suppose the digital computer contains a random number generator. Then it will be natural to use this to decide what answer to give. But then the random number generator will be subject to the psycho-kinetic powers of the interrogator. Perhaps this psycho-kinesis might cause the machine to guess right more often than would be expected on a probability calculation, so that the interrogator might still be unable to make the right identification. On the other hand, he might be able to guess right without any questioning, by clairvoyance. With E.S.P. anything may happen.

If telepathy is admitted it will be necessary to tighten our test up. The situation could be regarded as analogous to that which would occur if the interrogator were talking to himself and one of the competitors was listening with his ear to the wall. To put the competitors into a 'telepathy-proof room' would satisfy all requirements.

7. Learning Machines.

The reader will have anticipated that I have no very convincing arguments of a positive nature to support my views. If I had I should not have taken such pains to point out the fallacies in contrary views. Such evidence as I have I shall now give.

Let us return for a moment to Lady Lovelace's objection, which stated that the machine can only do what we tell it to do. One could say that a man can 'inject' an idea into the machine, and that it will respond to a certain extent and then drop into quiescence, like a piano string struck by a hammer. Another simile would be an atomic pile of less than critical size: an injected idea is to correspond to a neutron entering the pile from without. Each such neutron will cause a certain disturbance which eventually dies away. If, however, the size of the pile is sufficiently increased, the disturbance caused by such an incoming neutron will very likely go on and on increasing until the whole pile is destroyed. Is there a corresponding phenomenon for minds, and is there one for machines? There does seem to be one for the human mind. The majority of them seem to be 'sub-critical', *i.e.* to correspond in this analogy to piles of subcritical size. An idea presented to such a mind will on average give rise to less than one idea in reply. A smallish proportion are super-critical. An idea presented to such a mind may give rise to a whole 'theory' consisting of secondary, tertiary and more remote ideas. Animals minds seem to be very definitely sub-critical. Adhering to this analogy we ask, 'Can a machine be made to be super-critical?'

The 'skin of an onion' analogy is also helpful. In considering the functions of the mind or the brain we find certain operations which we can explain in purely mechanical terms. This we say does not correspond to the real mind: it is a sort of skin which we must strip off if we are to find the real mind. But then in what remains we find a further skin to be stripped off, and so on. Proceeding in this way do we ever come to the 'real' mind, or do we eventually come to the skin which has nothing in it? In the latter case the whole mind is mechanical. (It would not be a discrete-state machine however. We have discussed this.)

These last two paragraphs do not claim to be convincing arguments. They should rather be described as 'recitations tending to produce belief'.

The only really satisfactory support that can be given for the view expressed at the beginning of §6, will be that provided by waiting for the end of the century and then doing the experiment described. But what can we say in the meantime? What steps should be taken now if the experiment is to be successful?.

As I have explained, the problem is mainly one of programming. Advances in engineering will have to be made too, but it seems unlikely that these will not be adequate for the requirements. Estimates of the storage capacity of the brain vary from 10^{10} to 10^{15} binary digits. I incline to the lower values and believe that only a very small fraction is used for the higher types of thinking. Most of it is probably used for the retention of visual impressions. I should be surprised if more than 10^9 was required for satisfactory playing of the imitation game, at any rate against a blind man. (Note—The capacity of the *Encyclopaedia Britannica*, 11th edition, is 2×10^9.) A storage capacity of 10^7 would be a very practicable possibility even by present techniques. It is probably not necessary to increase the speed of operations of the machines at all. Parts of modern machines which can be regarded as analogues of nerve cells work about a thousand times faster than the latter. This should provide a 'margin of safety' which could cover losses of speed arising in many ways. Our problem then is to find out how to programme these machines to play the game. At my present rate of working I produce about a thousand digits of programme a day, so that about sixty workers, working steadily through the fifty years might accomplish the job, if nothing went into the waste-paper basket. Some more expeditious method seems desirable.

In the process of trying to imitate an adult human mind we are bound to think a good deal about the process which has brought it to the state that it is in. We may notice three components,

(a) The initial state of the mind, say at birth,
(b) The education to which it has been subjected,
(c) Other experience, not to be described as education, to which it has been subjected.

Instead of trying to produce a programme to simulate the adult mind, why not rather try to produce one which simulates the child's ? If this were then subjected to an appropriate course of education one would obtain the adult brain. Presumably the child-brain is something like a note-book as one buys it from the stationers. Rather little mechanism, and lots of blank sheets. (Mechanism and writing are from our point of view almost synonymous.) Our hope is that there is so little mechanism in the child-brain that something like it can be easily programmed. The amount of work in the education we can assume, as a first approximation, to be much the same as for the human child.

We have thus divided our problem into two parts. The child-programme and the education process. These two remain very closely connected. We cannot expect to find a good child-machine at the first attempt. One must experiment with teaching one such machine and see how well it learns. One can then try another and *see* if it is better or worse. There is an obvious connection between this process and evolution, by the identifications

Structure of the child machine	=	Hereditary material
Changes " "	=	Mutations
Natural selection	=	Judgment of the experimenter

One may hope, however, that this process will be more expeditious than evolution. The survival of the fittest is a slow method for measuring advantages. The experimenter, by the exercise of intelligence, should be able to speed it up. Equally important is the fact that he is not restricted to random mutations. If he can trace a cause for some weakness he can probably think of the kind of mutation which will improve it.

It will not be possible to apply exactly the same teaching process to the machine as to a normal child. It will not, for instance, be provided with legs, so that it could not be asked to go out and fill the coal scuttle. Possibly it might not have eyes. But however well these deficiencies might be overcome by clever engineering, one could not send the creature to school without the other children making excessive fun of it. It must be given some tuition. We need not be too concerned about the legs, eyes, etc. The example of Miss *Helen Keller* shows that education can take place provided that communication in both directions between teacher and pupil can take place by some means or other.

We normally associate punishments and rewards with the teaching process. Some simple child-machines can be constructed or programmed on this sort of principle. The machine has to be so constructed that events which shortly preceded the occurrence of a punishment-signal are unlikely to be repeated, whereas a reward-signal increased the probability of repetition of the events which led up to it. These definitions do not presuppose any feelings on the part of the machine. I have done some experiments with one such child-machine, and succeeded in teaching it a few things, but the teaching method was too unorthodox for the experiment to be considered really successful.

The use of punishments and rewards can at best be a part of the teaching process. Roughly speaking, if the teacher has no other mean of communicating to the pupil, the amount of information which can reach him does not exceed the total number of rewards and punishments applied. By the time a child has learnt to repeat 'Casabianca' he would probably feel very sore indeed, if the text could only be discovered by a 'Twenty Questions' technique, every 'NO' taking the form of a blow. It is necessary therefore to have some other 'unemotional' channels of communication. If these are available it is possible to teach a machine by punishments and rewards to obey orders given in some language, *e.g.* a symbolic language. These orders are to be transmitted through the 'unemotional' channels. The use of this language will diminish greatly the number of punishments and rewards required.

Opinions may vary as to the complexity which is suitable in the child machine. One might try to make it as simple as possible consistently with the general principles. Alternatively one might have a complete system of logical inference 'built in'.[3] In the latter case the store would be largely occupied with definitions and propositions. The propositions would have various kinds of status, *e.g.* well-established facts, conjectures, mathematically proved theorems, statements given by an authority, expressions having the logical form of proposition but not belief-value. Certain propositions may be described as 'imperatives'. The machine should be so constructed that as soon as an imperative is classed as 'well-established' the appropriate action automatically takes place. To illustrate this, suppose the teacher says to the machine, 'Do your homework now'. This may cause "Teacher says 'Do your homework now' " to be included amongst the well-established facts. Another such fact might be, "Everything that teacher says is true". Combining these may eventually lead to the imperative, 'Do your homework now', being included amongst the well-established facts, and this, by the construction of the machine, will mean that the homework actually gets started, but the effect is very satisfactory. The processes of inference used by the machine need not he such as would satisfy the most exacting logicians. There might for instance be no hierarchy of types. But this need not mean that type fallacies will occur, any more than we are bound to fall over unfenced cliffs. Suitable imperatives (expressed *within* the systems, not forming part of the rules *of* the system) such as 'Do not use a class unless it is a subclass of one which has been mentioned by teacher' can have a similar effect to 'Do not go too near the edge'.

The imperatives that can be obeyed by a machine that has no limbs are bound to be of a rather intellectual character, as in the example (doing homework) given above. Important amongst such imperatives will be ones which regulate the order in which the rules of the logical system concerned are to be applied. For at each stage when one is using a logical system, there is a very large number of alternative steps, any of which one is permitted to apply, so far as obedience to the rules of the logical system is concerned. These choices make the difference between a brilliant and a footling reasoner, not the difference between a sound and a fallacious one. Propositions leading to imperatives of this kind might be "When Socrates is mentioned, use the syllogism in Barbara" or "If one method has been proved to be quicker than another, do not use the slower method". Some of these may be 'given by authority', but others may be produced by the machine itself, *e.g.* by scientific induction.

The idea of a learning machine may appear paradoxical to some readers. How can the rules of operation of the machine change? They should describe completely how the machine will react whatever its history might be, whatever changes it might undergo. The rules are thus quite time-invariant. This is quite true. The explanation of the paradox is that the rules which get changed in the learning process are of a rather less pretentious kind, claiming only an ephemeral validity. The render may draw a parallel with the Constitution of the United States.

An important feature of a learning machine is that its teacher will often be very largely ignorant of quite what is going on inside, although he may still be able to some extent to predict his pupil's behaviour. This should apply most strongly to the later education of a machine arising from a child-machine of well-tried design (or programme). This is in clear contrast with normal procedure when using a machine to do computations: one's object is then to have a clear mental picture of the state of the machine at each moment in the computation. This object can only be achieved with a struggle. The view that 'the machine can only do what we know how to order it to do',[4] appears strange in face of this. Most of the programmes which we can put into the machine will result in its doing something that we cannot make sense of at all, or which we regard as completely random behaviour. Intelligent behaviour presumably consists in a departure from the completely disciplined behaviour involved in computation, but a rather slight one, which does not give rise to random behaviour, or to pointless repetitive loops. Another important result of preparing our machine for its part in

[3] Or rather 'programmed in' for our child-machine will be programmed in a digital computer. But the logical system will not have to be learnt.

[4] Compare Lady Lovelace's statement (p. 450), which does not contain the word 'only'.

the imitation game by a process of teaching and learning is that 'human fallibility' is likely to be omitted in a rather natural way, *i.e.* without special 'coaching'. (The reader should reconcile this with the point of view on pp. 24, 25.) Processes that are learnt do not produce a hundred per cent. certainty of result; if they did they could not be unlearnt.

It is probably wise to include a random element in a learning machine (see p. 438). A random element is rather useful when we are searching for a solution of some problem. Suppose for instance we wanted to find a number between 50 and 200 which was equal to the square of the sum of its digits, we might start at 51 then try 52 and go on until we got a number that worked. Alternatively we might choose numbers at random until we got a a good one. This method has the advantage that it is unnecessary to keep track of the values that have been tried, but the disadvantage that one may try the same one twice, but this is not very important if there are several solutions. The systematic method has the advantage that there may be an enormous block without any solutions in the region which has to be investigated first. Now the learning process may be regarded as a search for a form of behaviour which will satisfy the teacher (or some other criterion). Since there is probably a very large number of satisfactory solutions the random method seems to be better than the systematic. It should be noticed that it is used in the analogous process of evolution. But there the systematic method is not possible. How could one keep track of the different genetical combinations that had been tried, so as to avoid trying them again?

We may hope that machines will eventually compete with men in all purely intellectual fields. But which are the best ones to start with? Even this is a difficult decision. Many people think that a very abstract activity, like the playing of chess, would be best. It can also be maintained that it is best to provide the machine with the best sense organs that money can buy, and then teach it to understand and speak English. This process could follow the normal teaching of a child. Things would be pointed out and named, etc. Again I do not know what the right answer is, but I think both approaches should be tried.

We can only see a short distance ahead, but we can see plenty there that needs to be done.

BIBLIOGRAPHY

Samuel Butler, Erewhon, London, 1865. Chapters 23, 24, 25, *The Book of the Machines.*

Alonzo Church, "An Unsolvable Problem of Elementary Number Theory", *American J. of Math.*, 58 (1936), 345–363.

K. Gödel, "Über formal unentscheidbare Sätze der Principia Mathematica und verwandter Systeme, I", *Monatshefte für Math. und Phys.*, (1931), 173–189.

D. R. Hartree, *Calculating Instruments and Machines*, New York, 1949.

S. C. Kleene, "General Recursive Functions of Natural Numbers", *American J. of Math.*, 57 (1935), 153–173 and 219–244.

G. Jefferson, "The Mind of Mechanical Man". Lister Oration for 1949. *British Medical Journal*, vol. 1 (1949), 1105–1121.

Countess of Lovelace, 'Translator's notes to an article on Babbage's Analytical Engine', *Scientific Memoirs* (ed. by R. Taylor), vol. 3 (1842), 691–731.

Bertrand Russell, *History of Western Philosophy*, London, 1940.

A. M. Turing, "On Computable Numbers, with an Application to the Entscheidungsproblem", *Proc. London Math. Soc.* (2), 42 (1937), 230–265.

Victoria University of Manchester

Examining the Work and Its Later Impact

Daniel Dennett is inspired by —

TURING'S "STRANGE INVERSION OF REASONING"

Some of the greatest, most revolutionary advances in science have been given their initial expression in attractively modest terms, with no fanfare. Charles Darwin managed to compress his entire theory into a single summary paragraph that a layperson can readily follow, in all its details:

> If during the long course of ages and under varying conditions of life, organic beings vary at all in the several parts of their organization, and I think this cannot be disputed; if there be, owing to the high geometric powers of increase of each species, at some age, season, or year, a severe struggle for life, and this certainly cannot be disputed; then, considering the infinite complexity of the relations of all organic beings to each other and to their conditions of existence, causing an infinite diversity in structure, constitution, and habits, to be advantageous to them, I think it would be a most extraordinary fact if no variation ever had occurred useful to each being's own welfare, in the same way as so many variations have occurred useful to man. But if variations useful to any organic being do occur, assuredly individuals thus characterized will have the best chance of being preserved in the struggle for life; and from the strong principle of inheritance they will tend to produce offspring similarly characterized. This principle of preservation, I have called, for the sake of brevity, Natural Selection. (*Origin of Species*, end of chapter 4)

Francis Crick and James Watson closed their epoch-making paper on the structure of DNA with the deliciously diffident sentence:

> It has not escaped our notice that the specific pairings we have postulated immediately suggests a possible copying mechanism for the replicating unit of life. (Watson and Crick (1953), p.738)

And Alan Turing created a new world of science and technology, setting the stage for solving one of the most baffling puzzles remaining to science, the mind-body problem, with an even shorter declarative sentence in the middle of his 1936 paper on computable numbers:

> It is possible to invent a single machine which can be used to compute any computable sequence. (Turing (1936), p.241)

Turing didn't just intuit that this remarkable feat was possible; he showed exactly how to make such a machine. With that demonstration the computer age was born. It is important to remember that there were entities called computers before Turing came up with his idea – but they were people, clerical workers with enough mathematical skill, patience, and pride in their work to generate reliable results of hours and hours of computation, day in and day out. Many of them were women.

Dryden Flight Research Center E49-0053 Photographed 10/49
Early "computers" at work. NASA photo

Thousands of them were employed in engineering and commerce, and in the armed forces and elsewhere, calculating tables for use in navigation, gunnery and other such technical endeavours. A good way of understanding Turing's revolutionary idea about computation is to put it in juxtaposition with Darwin's about evolution. The pre-Darwinian world was held together not by science but by tradition: all things in the universe, from the most exalted ('man') to the most humble (the ant, the pebble, and the raindrop), were the creations of a still more exalted thing, God, an omnipotent and omniscient intelligent creator – who bore a striking resemblance to the second-most exalted thing. Call this the trickle-down theory of creation. Darwin replaced it with the bubble-up theory of creation. One of Darwin's nineteenth century critics put it vividly:

> In the theory with which we have to deal, Absolute Ignorance is the artificer; so that we may enunciate as the fundamental principle of the whole system, that, IN ORDER TO MAKE A PERFECT AND BEAUTIFUL MACHINE, IT IS NOT REQUISITE TO KNOW HOW TO MAKE IT. This proposition will be found, on careful examination, to express, in condensed form, the essential purport of the Theory, and to express in a few words all Mr. Darwin's meaning; who, by a strange inversion of reasoning, seems to think Absolute Ignorance fully qualified to take the place of Absolute Wisdom in all the achievements of creative skill. (MacKenzie, 1868)

It was, indeed, a strange inversion of reasoning. To this day many people cannot get their heads around the unsettling idea that a purposeless, mindless process can crank away through the eons, generating ever more subtle, efficient and complex organisms without having the slightest whiff of understanding of what it is doing.

Turing's idea was a similar – in fact remarkably similar – strange inversion of reasoning. The Pre-Turing world was one in which computers were people, who had to understand mathematics in order to do their jobs. Turing realised that this was just not necessary: you could take the tasks they performed and squeeze out the last tiny smidgens of understanding, leaving nothing but brute, mechanical actions. IN ORDER TO BE A PERFECT AND BEAUTIFUL COMPUTING MACHINE IT IS NOT REQUISITE TO KNOW WHAT ARITHMETIC IS.

What Darwin and Turing had both discovered, in their different ways, was the existence of *competence without comprehension* (Dennett, 2009, from which material in the preceding paragraphs has been drawn, with revisions). This inverted the deeply plausible assumption that comprehension is in fact the *source* of all advanced competence. Why, after all, do we insist on sending our children to school, and why do we frown on the old-fashioned methods of rote learning? We expect our children's growing competence to *flow from* their growing comprehension; the motto of modern education might be: 'comprehend *in order to be* competent' And for us members of *H. sapiens*, this is almost always the right way to look at, and strive for, competence. I suspect that this much-loved principle of education is one of the primary motivators of skepticism about both evolution and its cousin in Turing's world, Artificial Intelligence. The very idea that mindless mechanicity can generate human-level – or divine level! – competence strikes many as philistine, repugnant, an insult to our minds and the mind of God.

Consider how Turing went about his proof. He took human computers as his model. There they sat at their desks, doing one simple and highly reliable step after another, checking their work, writing down the intermediate results instead of relying on their memories, consulting their recipes as often as they needed, turning what at first might appear a daunting task into a routine they could almost do in their sleep. Turing systematically broke down the simple steps into even simpler steps, removing all vestiges of discernment or comprehension. Did a human computer have difficulty telling the number 99999999999 from the number 9999999999? Then, break down the perceptual problem of *recognizing the number* into simpler problems, distributing easier, stupider acts of discrimination over multiple steps. He thus prepared an inventory of basic building blocks from which to construct the universal algorithm that could execute any other algorithm. He showed how that algorithm would enable a (human) computer to compute any function, and noted that:

> The behaviour of the computer at any moment is determined by the symbols which he is observing and his "state of mind" at that moment. We may suppose that there is a bound *B* to the number of symbols or squares which the computer can observe at one moment. If he wishes to observe more, he must use successive observations. .... The operation actually performed is determined .... by the state of mind of the computer and the observed symbols. In particular, they determine the state of mind of the computer after the operation is carried out.

He then noted, calmly:

> We may now construct a machine to do the work of this computer. (p.251)

Right there we see the reduction of *all possible computation* to a mindless process. We can start with the simple building blocks Turing had isolated, and construct layer upon layer of more sophisticated computation, restoring, gradually, the intelligence Turing had so deftly laundered out of the practices of human computers.

But what about the genius of Turing, and of later, lesser programmers, whose own intelligent comprehension was manifestly the source of the designs that can knit Turing's mindless building blocks into useful competences? Doesn't this dependence just re-introduce the trickle-down perspective on intelligence, with Turing in the God role? No less a thinker than Roger Penrose has expressed skepticism about the possibility that Artificial Intelligence could be the fruit of nothing but mindless algorithmic processes.

> I am a strong believer in the power of natural selection. But I do not see how natural selection, in itself, can evolve algorithms which could have the kind of conscious judgements of the validity of other algorithms that we seem to have. (1989, p.414)

He goes on that admit:

> To my way of thinking there is still something mysterious about evolution, with its apparent'groping' towards some future purpose. Things at least *seem* to organize themselves somewhat better than they'ought' to, just on the basis of blind-chance evolution and natural selection. (1989, p.416)

Indeed, a single cascade of natural selection events, occurring over even billions of years, would seem unlikely to be able to create a string of zeroes and ones that, once read by a digital computer, would be an 'algorithm' for 'conscious judgments.' But as Turing fully realised, there was nothing to prevent the process of evolution from copying itself on many scales, of mounting discernment and judgment. The recursive step that got the ball rolling – designing a computer that could mimic any other computer–could itself be reiterated, permitting specific computers to enhance their own powers by *redesigning themselves*, leaving their original designer far behind. Already in 'Computing Machinery and Intelligence,' his classic paper in *Mind*, 1950, he recognised that there was no contradiction in the concept of a (non-human) computer that could learn.

> The idea of a learning machine may appear paradoxical to some readers. How can the rules of operation of the machine change? They should describe completely how the machine will react whatever its history might be, whatever changes it might undergo. The rules are thus quite time-invariant. This is quite true. The explanation of the paradox is that the rules which get changed in the learning process are of a rather less pretentious kind, claiming only an ephemeral validity. The reader may draw a parallel with the Constitution of the United States. (See Suber (2001), unpublished, for a valuable discussion of this passage and the so-called paradox of self-amendment.)

He saw clearly that all the versatility and self-modifiability of human though – learning and re-evaluation and, language and problem-solving, for instance – could in principle be constructed out of these building blocks. Call this the bubble-up theory of mind, and contrast it with the various trickle-down theories of mind, by thinkers from René Descartes to John Searle (and including, notoriously, Kurt Gödel, whose proof was the inspiration for Turin's work) that start with human consciousness at its most reflective, and then are unable to unite such magical powers with the mere mechanisms of human bodies and brains.

Turing, like Darwin, broke down the mystery of intelligence (or Intelligent Design) into what we might call atomic steps of dumb happenstance, which, when accumulated by the millions, added up to a sort of pseudo-intelligence. The Central Processing Unit of a computer doesn't *really* know what arithmetic is, or understand what addition is, but it 'understands' the 'command' to add two numbers and put their sum in a register – in the minimal sense that it reliably adds when thus called upon to add and puts the sum in the right place. Let's say it *sorta* understands addition. A few levels higher, the operating system doesn't *really* understand that it is checking for errors of transmission and fixing them but it *sorta* understands this, and reliably does this work when called upon. A few further levels higher, when the building blocks are stacked up by the billions and trillions, the chess-playing program does't *really* understand that its queen is in jeopardy, but it *sorta* understands this, and IB's Watson on Jeopardy *sorta* understands the questions it answers.

Why indulge in this *'sorta'* talk? Because when we analyze – or synthesise – this stack of ever more competent levels, we need to keep track of two facts about each level: what it *is* and what it *does*. What it *is* can be described in terms of the structural organization of the parts from which it is mad – so long as we can assume that the parts function as they are supposed to function. What it *does* is some (cognitive) function that it (sorta) performs – well enough so that at the next level up, we can make the assumption that we have in our inventory a smarter building block that performs just that function – sorta, good enough to use. This is the key to breaking the back of the mind-bogglingly complex question of how a mind could ever be composed of material mechanisms.

What we might call the *sorta* operator is, in cognitive science, the parallel of Darwin's gradualism in evolutionary processes. Before there were bacteria there were *sorta* bacteria, before there were mammals there were *sorta* mammals and before there were dogs there were *sorta* dogs, and so forth. We need Darwin's gradualism to explain the huge difference between an ape and an apple, and we need Turin's gradualism to explain the huge difference between a humanoid robot and hand calculator. The ape and the apple are made of the same basic ingredients, differently structured and exploited in a many-level cascade of different functional competences. There is no principled dividing line between a *sorta* ape and an ape. The humanoid robot and the hand calculator are both made of the same basic, unthinking, unfeeling Turing-bricks, but as we compose them into larger, more competent structures, which then become the elements of still more competent structures at higher levels, we eventually arrive at parts so (*sorta*) intelligent that they can be assembled into competences that deserve to be called comprehending. We use the intentional stance (Dennett, 1971, 1987) to keep track of the beliefs and desires (or 'beliefs' and 'desires' or sorta beliefs and sorta desires) of the (sorta-)rational agents at every level from the simplest bacterium through all the discriminating, signaling, comparing, remembering circuits that compose the brains of animals from starfish to astronomers. There is no principled line above which true comprehension is to be found – even in our own case. The small child *sorta* understands her own sentence 'Daddy is a doctor', and I *sorta* understand '$E = mc^2$'. Some philosophers resist this anti-essentialism: either you believe that snow is white or you don't; either you are conscious or you aren't; nothing counts as an approximation of any mental phenomenon – it's all or nothing. And to such thinkers, the powers of minds are insoluble mysteries because they are 'perfect,' and perfectly unlike anything to be found in mere material mechanisms.

We still haven't arrived at 'real' understanding in robots, but we are getting closer. That, at least, is the conviction of those of us inspired by Turing's insight. The trickle-down theorists are sure in their bones that no amount of further building will ever get us to the real thing. They think that a Cartesian *res cogitans*, a thinking thing, cannot be constructed out of Turing's building blocks. And creationists are similarly sure in their bones that no amount of Darwinian shuffling and copying and selecting could ever arrive at (real) living things. They are wrong, but one can appreciate the discomfort that motivates their conviction.

Turing's strange inversion of reason, like Darwin's, goes against the grain of millennia of earlier thought. If the history of resistance to Darwinian thinking is a good measure, we can expect that long into the future, long after every triumph of human thought has been matched or surpassed by 'mere machines', there will still be thinkers who insist that the human mind works in mysterious ways that no science can comprehend.

References

Darwin, C., 1859. On the Origin of Species, John Murray, London.

Dennett, D.C., 1971. Intentional systems. J. Phil., 68, 87–106.

Dennett, D.C., 1987. The Intentional Stance, MIT Press, Cambridge, MA.

Dennett, D.C., 2009. Darwin's Strange inversion of reasoning. PNAS, 106, 10061–10065.

MacKenzie, R.B., 1868. The Darwinian Theory of the Transmutation of Species Examined, Nisbet & Co, London.

Penrose, R., 1989. The Emperor's New Mind, Oxford University Press, Oxford.

Suber, P., (unpublished). Saving Machines From Themselves: The Ethics of Deep Self-Modification, preprint, 30 November 2001. http://www.earlham.edu/~peters/writing/selfmod.htm

Turing, A., 1936. On computable numbers, with an application to the Entscheidungsproblem. Proc. Lond. Math. Soc. 42, 23–265, and erratum (1937) 43, 544–546.

Turing, A., 1950. Computing machinery and intelligence. Mind, LIX, 2236, 433–460.

Watson, J.D. and Crick, F.H.C., 1953. A structure for deoxyribose nucleic acid. Nature 171, 737–738.

Aaron Sloman draws together —

Virtual Machinery and Evolution of Mind (Part 2)*

1. Introduction

Darwin's critics could not understand how natural selection could produce minds and consciousness. They (and even some of his defenders) pointed out that his evidence, such as gradual changes in animal forms, supported only the hypothesis that natural selection produces *physical* forms and behaviours. Nobody could understand how physical mechanisms can produce mysterious and externally unobservable mental states and processes, referred to by Huxley as 'The explanatory gap'.[1] Since Darwin's time, the problem has been re-invented and re-labelled several times, e.g. as the problem of 'Phenomenal Consciousness' (Block, 1995) or the 'Hard Problem' of consciousness (Chalmers, 1996). The topic was touched on and side-stepped in Turing (1950). It remains unclear how a genome can, as a result of physical and chemical processes, produce the problematic, apparently non-physical, externally unobservable, personal experiences (qualia) and processes of thinking, feeling, wanting, and artistic, mathematical or scientific creation.

The facts about virtual machinery used in complex computing systems, presented in Part 1, suggest ways in which biological evolution may have taken advantage of virtual machines to produce self-monitoring, self-modifying, self-extending information-processing architectures, some of whose contents would have the core features of qualia, including non-definability in the language of physics. This suggests a way for Darwin to answer the criticism that natural selection can produce only physical development, not mental states and consciousness, though this type of explanation was not available in Darwin's time. On the basis of what we have learnt recently, we can now conjecture that evolution produced 'mysterious' aspects of consciousness if, like engineers in the last six or seven decades, it produced solutions to increasingly complex problems of representation and control – solutions based on increasingly abstract, but effective, mechanisms, including self-observation capabilities, implemented in non-physical virtual machines which, in turn, are implemented in lower level physical mechanisms. For this, evolution would have had to produce far more complex virtual machines than human engineers have so far managed, but the key idea might be the same.[2]

Part 1 presented Universal Turing Machines as theoretical precursors of technology supporting networks of interacting running virtual machines (RVMs) sensing and controlling things in their environment. Such RVMs are *fully implemented* in underlying physical machines (PMs) but the concepts used to describe the states and processes in some RVMs (e.g. 'pawn', 'threat' and 'win'

*This is the second of three linked papers in this Volume. Part 1 is in Part I of the volume. The final part, Part 3, on Meta-Morphogenesis, is in Part IV of the volume.

[1] For more detail and quotations from critics, see Sloman (2010a).

[2] It is not yet clear whether the biological virtual machinery can implemented in the kind of discrete technology used in computers as we know them. New kinds of computers may be required.

in chess VMs) are not *definable* in the language of the physical sciences. We now develop the biological versions of these ideas, potentially explaining how self-monitoring, self-modifying RVMs can include some of the features of consciousness, such as qualia, previously thought by some to be inexplicable by science, paving the way for a theory of how mind and consciousness might have evolved, and how robots might have qualia. Unlike capabilities of earlier machinery, there is no close relationship between information processing capabilities and physical form or behaviour.

2. Epigenesis: Bodies, behaviours and minds

Turing was interested in both evolution and epigenesis and made pioneering suggestions regarding the processes of morphogenesis – differentiation of cells to form diverse body parts during development. As far as I know, he did not do any work on how a genome can produce *behavioural competences* of the complete organism, including behaviours with complex conditional structures, so that what is done depends on internal and external sensory information, though he briefly considered learning, in Turing (1950).[3]

It is understandable that physical behaviours, such as hunting, eating, escaping predators and mating, should influence biological fitness and that evolution should select brain and other modifications that produce advantageous behaviours. But there are *internal* non-behavioural competences whose biological uses are not so obvious: thinking, reminiscing, perceiving with enjoyment, finding something puzzling and attempting to understand it. It is not obvious how biological evolution could produce mechanisms that are able to support such mental processes. Though it is clear that once such mechanisms were produced, some of them might enhance biological fitness.

Many species develop behavioural and internal competences that depend on the environment during development (e.g. which language a child speaks, and which mathematical problems are understood), so the genome-driven processes must create some innately specified competences partly under the influence of the genome and partly under the influence of combinations of sensorimotor signals during development (Held and Hein, 1963; McCarthy, 2008). For humans at least, the internal processes of competence-formation go on long after birth, suggesting that the genome continues producing, or enabling, or constraining effects (including changes in sexual and parental motivations and behaviours) long after the main body morphology and sensory-motor mechanisms have developed.

Karmiloff-Smith (1992) presents many examples where *after* achieving behavioural competence in some domain, learners (including some non-human species) re-organise their understanding of the domain in such a way as to give them new abilities to think and communicate about the domain. After children develop linguistic competences based on known phrases they spontaneously switch to using a *generative* syntax that allows *derivation* of solutions to novel problems, instead of having to learn empirically what does and does not work. Craik (1943) pointed out the value of such mechanisms in 1943, suggesting that they could be based on working mental models.[4] Grush (2004) and others suggest that such models could work as simulations or emulations. However, when used for reasoning purposes, as opposed to statistical prediction, a decomposable information structure is required rather than a fixed executable model, for instance when proving geometrical theorems.

---

[3] In 'The Mythical Turing Test' in Part III of this volume Turing's suggestion about learning on the basis of a blank slate at birth is criticised, and compared with McCarthy's ideas.

[4] I have not been able to find out whether Craik and Turing ever interacted. Turing must have known about his work, since he was a member of the Ratio club, founded in honour of Craik, shortly after he died in a road accident in 1945.

The mental models we use to explain, predict and produce processes in our environment include models of things like gear wheels, bicycles, electric circuits and other mechanisms that are too new to have been part of our evolutionary history. So, at least in humans, the model construction process cannot all be encoded in the genome: the specific models need information obtained after birth from the environment, along with novel ideas thought up by the individual.

So, the genome specifies not only physical morphology and physical behavioural competences, but also a multi-functional information-processing architecture developed partly in species-specific ways, at various stages after the individual's birth, partly under the control of features of the environment, and includes not only mechanisms for interpreting sensory information and mechanisms for controlling external movements, but also mechanisms for building and running predictive and explanatory models of structures and processes, either found in the environment or invented by the individual.[5] How can a genome specify ongoing construction processes to achieve that functionality? I don't think anyone is close to an answer, but I'll offer a conjecture about a feature of the process: evolution discovered the virtues of virtual machinery long before human engineers.

Part 1 of this series of papers outlined the benefits of virtual machinery in human-designed computing systems, and their advantages compared with specifying, designing, monitoring, controlling and debugging the *physical* machinery directly, The advantages come from the coarser granularity, the use of abstraction allowing different implementations to be compared, and the use of application-relevant semantics.

Perhaps a series of initially random changes during reproduction of organisms decoupled the control mechanisms from the physical sensors and effectors, allowing more flexibility in subsequent deployment, eventually leading to use of virtual machinery in animals because of its advantages for specifying types of competence at a relatively abstract level, avoiding the horrendous complexity of specifying all the physical and chemical details, and allowing construction of behaviour specifications of greater generality. The initial specification of behavioural competences in the genome might be far more compact and simpler to construct or evolve if a virtual machine specification is used, provided that other mechanisms ensure that that 'high level language' is mapped onto physical machinery in an appropriate way. The use of self-monitoring processes required for learning and modifying competences, including de-bugging them, may be totally intractable if the operations of atoms, molecules or even individual neurones are monitored and modified, but more tractable if the monitoring happens at the level of a RVM.

So something like a compiler is required for the basic epigenetic processes creating common features across a design, including physical forms, and something more like an interpreter to drive subsequent processes of learning and development.

3. The evolution of organisms with qualia

Part 1 showed that virtual machinery can be implemented in physical machinery, and events in virtual machines can be causally connected with other VM events and also with physical events both within the supporting machine and in the environment, as a result of use of complex mixtures of technology for creating and maintaining virtual/physical causal relationships developed over the last seven decades. The use of virtual machinery enormously simplifies the design, debugging, maintenance and development of complex systems. Finally, and perhaps most importantly, in machines

[5] It is argued in Sloman (1979, 2008) that this requires types of 'language' (in a generalised sense of the word, including structural variability and compositional semantics) that evolved, and in young humans develop, initially for *internal* information processing, not for external communication. We can call these 'generalised languages' (GLs).

that need to monitor and modify *their own* operations, performing the monitoring and modifications at the level of virtual machinery can be tractable where the corresponding tasks would he intractably complex and too inflexible and slow, if done by monitoring and modifying physical machinery.

So, biological evolution could have gained in power, flexibility, and speed of development by using virtual machine descriptions in the genome for specifying behavioural competences, instead of descriptions of the physical details. Moreover, if some of the virtual machinery is not fully specified in the genome, and has to be developed after birth or hatching by making use of new information gained by the individual from the environment, then that post-natal construction process will be simpler to specify, control and modulate, and easier to change as needed, if done at the virtual machine level rather than specifying all the chemical and neuronal changes required. And finally self-monitoring, self-control and self-modification in a sophisticated information-processing system all need to control virtual not physical, machinery.

An organism can perceive, think about and act on a rich and complex environment that contains enduring objects and processes at various locations at different spatial and temporal scales, and not all constantly in perceptual range. Doing this requires different sorts of information, including relatively enduring information structures and also rapidly changing perceptual contents and motor and proprioceptive signals. Relating the abstract goal of grasping a berry to the changing visual, haptic and motor signals requires machinery constructing, manipulating and using a variety of changing information contents, some concerned with what is happening in the environment, and some concerned with what's happening in the organism: e.g. is some information incomplete, or ambiguous or capable of answering a question, or capable of being used for detailed control of actions? The contents of those information structures seem to be exactly what philosophers have been attending to for centuries and labelling as experiences, sense data, or qualia, of different sorts. Visual and haptic processes perceiving the same portion of the environment could include overlapping virtual machines dealing with different aspects of the environment processed at different levels of abstraction in parallel Sloman (2009). Data-structures representing visible portions and features of the environment, e.g. visible portions of surfaces with colour, shape, orientation, curvature, speeds of motion or rotation, and relationships to other surface fragments (i.e. not the specific sensory signals), will then be components of virtual machines.

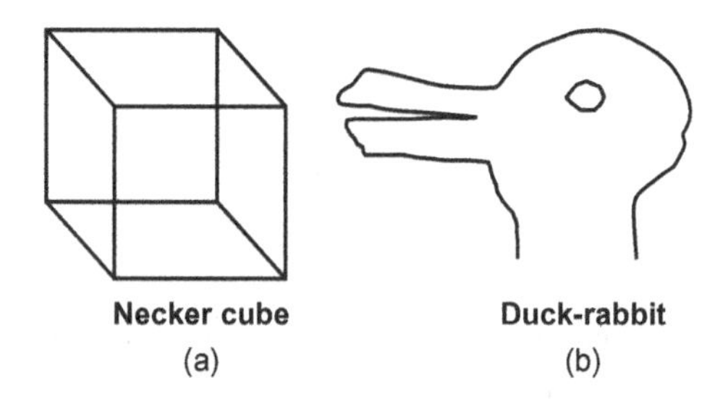

Fig. 1: *Each of the two figures is ambiguous and flips between two very different views. (a) can be seen as a 3-D wire frame cube. For most people it flips between two different views of the cube, in which the 3-D locations, orientations, and other relationships vary. In (b), the flip involves changes in body parts, the facing direction, and likely motion – requiring a very different ontology.*

Presumably similar qualia exist in other animals that are capable of similar controlled behaviours. But in humans, and perhaps a subset of other species, there is additional machinery that can detect record, compare, and later reflect on and describe, those contents. A possible use for that would be explaining to someone else how to tell when a danger or opportunity or achievement of a goal is imminent. This introspective use of the information content is different from and requires different machinery, from the use of the sensory contents to control movements.[6] As Fig. 1 shows, the very same low level sensory content can sometimes cause construction of more than one information structure. The ontology required to describe the change in contents differs from one ambiguous view to another. Some of the contents concern shape and information relevant to manipulation. Others concern more abstract capabilities and likely behaviours of perceived objects, like differences between a duck facing one way and a rabbit facing another way.

If the information structures created during perception and action are sometimes accessed by self-monitoring processes focused, not on what is in the environment, but on the content of what is currently being perceived, or signalled, then we potentially have an explanation of the phenomena that have led to philosophical and other puzzles about the existence and nature of sensory and motor qualia, which are often regarded as defining the most difficult aspect of mind to explain in physical or functional terms, and whose evolution and development in organisms Huxley and others found so difficult to explain. See also Sloman and Chrisley (2003).

Ryle, Dennett and others identified deep confusions in talk about consciousness and *qualia*, but such things clearly exist, though they are hard to characterise and to identify in other individuals and other species. Analysis of examples, including ambiguous figures, such as Fig. 1, helps to determine requirements for explanatory mechanisms. Such pictures illustrate the *intentionality* of perceptual experience, i.e. interpreting something as referring to something else and the different *ontologies* used by different experiences. I suggest that that is only possible within running virtual machinery, since concepts like 'interpreting', 'referring', 'intending' and 'looking' are no more definable in the language of physics than 'pawn' or 'threat'.

Many organisms can, I suspect, create and use such virtual entities without having the meta-semantic mechanisms required to detect and represent the fact that they do. As noted at the time of Darwin in Whittaker (1884), not all organisms that have qualia know that they have them! We can separate the *occurrence* of mental contents in an organism from their *detection* by the organism, which requires additional architectural complexity to support self-observation and self-description mechanisms. We need to experiment with varied ranges of increasingly complicated working examples, using different kinds of mechanism, in order to understand better some of the questions to be asked about mental phenomena in biological organisms. This is very close to Arbib's research programme described in Arbib (2003).

4. What next?

Experience shows that for many thinkers belief in an unbridgeable mind/body explanatory gap will be unshaken by all this. As argued in Sloman (2010), some cases of opposition will be based on use of incoherent concepts (e.g. a concept of 'phenomenal consciousness' *defined* to involve no causal or functional powers of the kinds described above). Designing working systems, using different robot designs to illustrate different products of evolution may help us understand the biological examples. But current achievements in AI vision, motor-control, concept-formation, forms of learning, language understanding and use, motive-generation, decision-making, plan-formation, problem-solving, and many others, are still (mostly) far inferior to those of humans and other

[6] That distinction between *having* mental contents and *detecting* them could explain evidence e.g. by Libet showing that initiation of motor signals can precede consciousness of the decision to move.

animals, in part because designers typically consider only a small subset of the requirements for biological intelligence. Even if we omit uniquely human competences, current robots are still far inferior to other animals, in many ways. There is no easy way to close those gaps, but there are many things to try, as long as we think clearly about what needs to be explained. Turing the philosopher-engineer-biologist could have made a substantial contribution to this project. Part 3, in part IV of this volume, expands on the theme of meta-morphogenesis.

References

Arbib, M.A., 2003. Rana computatrix to human language: Towards a computational neuroethology of language evolution. Philosophical Trans.: Mathematical, Physical and Engineering Sciences, 361 (1811), 2345–2379.

Block, N., 1995. On a confusion about the function of consciousness. Behavioral and Brain Sciences, 18, 227–247.

Chalmers, D.J., 1996. The Conscious Mind: In Search of a Fundamental Theory, Oxford University Press, New York, Oxford.

Craik, K., 1943. The Nature of Explanation, Cambridge University Press, London, New York.

Grush, R., 2004. The emulation theory of representation: Motor control, imagery, and perception. Behavioral and Brain Sciences, 27, 377–442.

Held, R., Hein, A., 1963. Movement-produced stimulation in the development of visually guided behaviour. J. Comparative and Physiological Psychol. 56 (5), 872–876.

Karmiloff-Smith, A., 1992. Beyond Modularity: A Developmental Perspective on Cognitive Science, MIT Press, Cambridge, MA.

Libet, B., 1989. Unconscious cerebral initiative and the role of conscious will in voluntary action. Behavioral and Brain Sciences, 8, 529–539.

McCarthy, J., 2008. The well-designed child. Artificial Intelligence, 172 (18), 2003–2014. Available from http://www-formal.stanford.edu/jmc/child.html

Sloman, A., 1979. The primacy of non-communicative language. In: MacCafferty, M., Gray, K. (Eds.), The Analysis of Meaning: Informatics 5 Proceedings ASLIB/BCS Conference, Oxford, March 1979, pp. 1–15. London: Aslib. (http://www.cs.bham.ac.uk/research/projects/cogaff/81-95.html#43)

Sloman, A., 2008. Evolution of minds and languages. What evolved first and develops first in children: Languages for communicating, or languages for thinking (Generalised Languages: GLs)? (Research Note No. COSY-PR-0702). Birmingham, UK. Available from
http://www.cs.bham.ac.uk/research/projects/cosy/papers/#pr0702

Sloman, A., 2009. Some requirements for human-like robots: Why the recent over-emphasis on embodiment has held up progress. In: Sendhoff, B., Koerner, E., Sporns, O., Ritter, H., Doya, K. (Eds.), Creating Brain-like Intelligence, Springer-Verlag, Berlin, pp. 248–277.

Sloman, A., 2010a. How virtual machinery can bridge the "Explanatory Gap". In: Doncieux, S., et al. (Eds.), Natural and Artificial Systems. Proceedings SAB 2010, LNAI 6226. Springer, Heidelberg, pp. 13–24. Available from http://www.cs.bham.ac.uk/research/projects/cogaff/10.html#sab

Sloman, A., 2010b. Phenomenal and access consciousness and the "Hard" Problem: A view from the designer stance. Int. J. Machine Consciousness, 2 (1), 117–169.

Sloman, A., Chrisley, R.L., 2003. Virtual machines and consciousness. J. Consciousness Studies, 10 (4–5), 113–172.

Turing, A.M., 1950. Computing machinery and intelligence. Mind 59, 433–460. (reprinted in E.A. Feigenbaum and J. Feldman (Eds.) Computers and Thought, McGraw-Hill, New York, 1963, 11–35).

Whittaker, T., 1884. Review of G. J. Romanes: Mental evolution in animals. Mind 9 (34), 291–295.

Mark Bishop examines —

THE PHENOMENAL CASE OF THE TURING TEST AND THE CHINESE ROOM

This short story is concerned with Alan Turing's take on an age-old question; 'Can a Machine Think?' Famously, in his 1950 paper *Computing Machinery and Intelligence* (CMI), Turing suggested replacing this question – which he found 'too meaningless to deserve discussion' – with a simple [behaviourial] test based on an imagined Victorian-esque pastime he entitled the 'imitation game'.

In this note, I will endeavour to draw a path linking Alan Turing with a Chinese room; a path which leads me to suggest that consciousness is necessary for grounded understanding and hence, because conscious phenomenal states are not instantiated by the execution of a mere computer program (Bishop, 2009), that Turing's exposition [in CMI] of an early form of computationalism (the view that computation can offer an explanatory basis for cognition, i.e. that the execution of an appropriate computer program is sufficient for 'thinking', 'understanding' and 'consciousness') was ultimately misguided.

The capabilities of 'computing machines' had long been of interest to Turing and certainly, as early as 1941, Turing was thinking about machine intelligence – specifically how computing machines could solve problems by searching through the space of possible problem solutions guided by heuristic principles. As a result, at the Royal Astronomical Society (RAS: London) in 1947, Turing gave what is perhaps the earliest public lecture on machine intelligence.

In 1948, following a year's sabbatical at Cambridge, Turing completed a report for the UK's National Physical Laboratory (NPL) on his research into machine intelligence, entitled *Intelligent Machinery* (Turing, 1948). Although not published contemporaneously, the report is notable for predicting a series of core themes which eventually emerged from the yet nascent science of machine intelligence: expert systems; connectionism; evolutionary algorithms; but, most intriguingly of all, the report offers perhaps the earliest version of the 'imitation game' – a procedure that has since become more widely known as the 'Turing test'.

In the NPL report Turing presented the original version of the imitation game as follows:

> *"The extent to which we regard something as behaving in an intelligent manner is determined as much by our own state of mind and training as by the properties of the object under consideration. If we are able to explain and predict its behaviour or if there seems to be little underlying plan, we have little temptation to imagine intelligence. With the same object therefore it is possible that one man would consider it as intelligent and another would not; the second man would have found out the rules of its behaviour.*
>
> *It is possible to do a little experiment on these lines, even at the present stage of knowledge. It is not difficult to devise a paper machine which will play a not very bad game of chess. Now get three men as subjects for the experiment A, B, and C. A and C are to be rather poor chess players, B is the operator who works the paper machine. (In order that he should be able to work it fairly fast it is advisable that he be both mathematician and chess player.) Two rooms are used with some arrangement for communicating moves, and a game is played between C and either A or the paper machine. C may find it quite difficult to tell which he is playing."*

In the initial exposition of the better known 'imitation game' presented in his 1950 paper (Turing 1950, ibid.) Turing called for a human interrogator (C) to hold a conversation with a male and female respondent (A and B) with whom the interrogator could communicate only indirectly by typewritten text. The object of this game was for the interrogator to correctly identify the gender of the players (A and B) purely as a result of such textual interactions; what makes the task non-trivial is that (a) the respondents are allowed to lie and (b) the interrogator is allowed to ask questions ranging over the whole gamut of human experience. At first glance it is perhaps mildly surprising that, even after many such interactions, a skilled player can determine (more accurately than by chance) the correct gender of the respondents.

Turing then asked 'What will happen when a machine takes the part of (A) in this game? Would the interrogator decide wrongly as often as when playing the initial imitation game?' In this version of the Turing test – which has become known as the 'standard interpretation' – a suitably programmed computer takes the part of either player (A) or player (B) (i.e. the computer plays as either the man or the woman) and the interrogator (C) simply has to determine which respondent is the human and which is the machine.

NB. The precise formulation of the Turing Test continues to attract lively debate. Thus in 2008 the UK society for the study of Artificial Intelligence and the Simulation of Behaviour (AISB) sponsored a one-day symposium on the Turing test at the University of Reading in the hope of eliciting further clarity in the interpretation of the test, further insight into its implications and further reflection as to its status as a [practical] measure of machine intelligence; for brief commentary, see the author's introduction to the recent special issue of KYBERNETES journal, which resulted from the 2008 symposium (Bishop, 2010).

1. A thinking machine?

In the 1950 paper, Turing confidently predicted that by the year 2000, there would be computers with 1G of storage (which turned out very prescient) that would be able to perform the [standard] Turing test such that the average interrogator would not have more than 70% chance of making the right identification after five minutes of questioning.

For some working at the coal-face in the world of Artificial Intelligence (AI), conclusive proof of the presence of mental states and capacities is a system's ability to pass the Turing test. If an AI system can convince an interrogator that it has mental states, then it must have those mental states; it would be purely 'human bias' to think otherwise (Warwick, 2002). If, for example, a machine could 'converse' with a native Chinese speaker in a manner indistinguishable from that of a native Chinese speaker then it could [literally] be said to *understand* Chinese.

But by 2011 even this minimal – five minute – Turing Test had still not been passed; albeit it seems very likely that in the next few years Turing's predictions for a 'time limited' Turing test will be met. But whether that will mean 'general educated opinion will have altered so much that one will be able to speak of machines thinking without expecting to be contradicted' (as Turing predicted) is doubtful as, in the 50 plus years since the CMI paper was first published, the status of the Turing test as a definitive measure of machine intelligence and understanding has been extensively critiqued.

Perhaps the best known criticism of a Turing-style test of machine understanding comes from John Searle whose seminal work on machine understanding, first presented in the 1980 paper *Minds, Brains & Programs* (MBP), has become known as the Chinese Room Argument (CRA) (Searle, 1980). In the CRA Searle endeavours to show that, even if a computer behaved in a manner fully indistinguishable from a human (when answering questions about a simple story), it cannot be said to genuinely understand its responses and hence the computer cannot be said to genuinely think.

Although the last thirty years have seen tremendous controversy over the success of the CRA, a great deal of consensus over its impact has emerged. Larry Hauser has called it 'perhaps the most influential and widely cited argument against the claims of Artificial Intelligence'. Stevan Harnad (1990), editor of Behavioural and Brain Sciences, asserted that 'it has already reached the status of a minor classic'. Anatol Rapaport claims the argument 'rivals the Turing test as a touchstone of philosophical inquiries into the foundations of AI'.

2. The Chinese Room Argument

In 1977 Schank and Abelson published information on a program they created, which could accept a simple story and then answer questions about it, using a large set of rules, heuristics and scripts. By script they referred to a detailed description of a stereotypical event unfolding through time. For example, a system dealing with restaurant stories would have a set of scripts about typical events that happen in a restaurant: entering the restaurant; choosing a table; ordering food; paying the bill and so on. In the wake of this and similar work in computing labs around the world, some of the more excitable proponents of artificial intelligence began to claim that such programs actually understood the stories they were given, and hence offered insight into human comprehension.

It was precisely an attempt to expose the flaws in the statements emerging from these proselytising AI-niks, and more generally to demonstrate the inadequacy of the Turing test, which led Searle to formulate the Chinese Room Argument.

The central claim of the CRA is that computations alone cannot in principle give rise to understanding, and that therefore, computational theories of mind cannot fully explain human cognition. More formally, Searle later stated that the CRA was an attempt to prove that syntax (rules for the correct formation of sentences:programs) is not sufficient for semantics (understanding). Combining this claim with those that programs are formal (syntactical), whereas minds have semantics, led Searle to conclude that 'programs are not minds'.

And yet it is clear that Searle believes that there is no barrier in principle to the notion that a machine can think and understand; indeed in MBP Searle explicitly states, in answer to the question 'Can a machine think?', that 'the answer is, obviously, yes. We are precisely such machines'. Clearly Searle did not intend the CRA to target machine intelligence *per se*, but rather any form of artificial intelligence according to which a machine could have genuine mental states (e.g., understanding Chinese) purely in virtue of executing an appropriate series of computations: what Searle termed 'Strong AI'.

Searle argues that understanding of, say, a Chinese story can never arise purely as a result of following the procedures prescribed by any computer program, for Searle offers a first-person tale outlining how he could instantiate such a program, and act as the Central Processing Unit of a computer (CPU), produce correct internal and external state transitions, pass a Turing Test for understanding Chinese, and yet still not understand a word of Chinese.

Searle-as-CPU describes a situation whereby he is locked in a room and presented with a large batch of papers covered with Chinese writing that he does not understand. Indeed, the monoglot Searle does not even recognise the symbols as being Chinese, as distinct from say Japanese or simply meaningless patterns. Later Searle is given a second batch of Chinese symbols, together with a set of rules (in English) that describe an effective method (algorithm) for correlating the second batch with the first, purely by their form or shape. Finally, Searle-as-CPU is given a third batch of Chinese symbols together with another set of rules (in English) to enable him to correlate the third batch with the first two, and these rules instruct him how to return certain sets of shapes (Chinese symbols) in response to certain symbols given in the third batch.

Unknown to Searle, the people outside the room call the first batch of Chinese symbols, 'the script', the second set 'the story', the third 'questions about the story', and the symbols he returns they call 'answers to the questions about the story'. The set of rules he is obeying they call 'the program'.

To complicate matters further, the people outside the room also give Searle-as-CPU stories in English and ask him questions about these stories in English, to which he can reply in English.

After a while Searle-as-CPU gets so good at following the instructions and the 'outsiders' get so good at supplying the rules he has to follow, that the answers he gives to the questions in Chinese symbols become indistinguishable from those a true Chinese person might give.

From an external point of view, the answers to the two sets of questions, one in English the other in Chinese, are equally good; Searle, in the Chinese room, has passed the Turing test. Yet in the Chinese language case, Searle behaves 'like a computer' and does not understand either the questions he is given or the answers he returns, whereas in the English case, *ex hypothesi*, he does. Searle contrasts the claim posed by some members of the AI community – that any machine capable of following such instructions can genuinely understand the story, the questions and answers – with his own continuing inability to understand a word of Chinese; pace Stevan Harnad (1990) for Searle the Chinese [squiggle and squoggle] symbols forever remain ungrounded.

Harnad's 'symbol-grounding problem' is closely related to the problem of how words (symbols) get their meanings. On its own the meaning of a word on a page is 'ungrounded' and merely looking it up in a dictionary doesn't help ground it. If I attempt to look up the meaning of a word I do not understand in a [unilingual] dictionary of a language I do not already understand, I simply wander endlessly from one meaningless definition to another (a problem not unfamiliar to young children); like Searle in his Chinese room, my search for meaning forever remains 'ungrounded'.

2.1. *The 'translation reply'*

Of course the type of programs from the 1970s which Searle appears to caricature – described as merely correlating uninterpreted Chinese symbols – are a long way from the style of modern A.I. programs being developed by cognitive scientists using, say, Natural Language Processing (NLP) techniques. Furthermore, with the recent development of powerful automatic translation tools (such as 'Google Translate') it is not difficult to imagine that future Chinese room systems might deploy, say, a machine translation module yielding an intermediate representation of a translation of the Chinese story into English (which *ex hypothesi* Searle-as-CPU could understand and analyse).

For Searle-as-CPU mapping the Chinese squiggles and squoggles into internal representation(s) in English offers potential for the unknown symbols to gradually become 'grounded'. Hence in this scenario, over time, it is not difficult to imagine Searle learning a little rudimentary Chinese.

At first sight, the possible development of complex A.I. programs (as described above) appears to undermine the CRA. The problem arises because the rule-book that Searle follows (developed by the Chinese room's 'programmers') now processes complex 'internal representations' in English (a language Searle understands) (Boden, 1990). This offers Searle-as-CPU a handle on underlying symbol-grounding problem; the problem is now analogous to merely learning Chinese from a Chinese:English dictionary.

Nevertheless, on deeper reflection, it is clear that such systems do not fundamentally undermine the CRA; unlike Searle, a real computing machine doesn't bring a grounded *understanding of English* to the problem of understanding Chinese anymore than it's processes yield a grounded understanding of Chinese (at most it brings merely *syntactical* information defining how to execute its machine code instruction set). Thus, from the perspective of any real computing machine, the English internal representations [and the Chinese symbols] that it processes remain ungrounded; in other words, the original CRA thought experiment grants the Searle-as-CPU too much 'symbol-grounding power'.

To redress the imbalance in symbol-grounding, we simply need to modify the original thought experiment such that the 'programmers' of Searle's rule-book are now, say, Japanese. In this case, even allowing the instructions in the rule-book [its *syntax*] to remain in English (so that Searle-as-CPU can still perform the required procedures), any intermediate internal representations that it deploys [the program's variables, data structures, translations etc.] are now specified in Japanese; a language [monoglot] Searle does not understand.

It is clear that Searle-as-CPU could blindly follow such a rule-book (appropriately manipulate the symbols and hence accurately respond to questions in Chinese) but continue to understand nothing of the underlying Chinese story; for Searle-as-CPU, the Chinese symbols and Japanese internal representations remain ungrounded. In this version of the CRA, the problem has become analogous to Searle endeavouring to learn Chinese using only a Chinese:Japanese dictionary.

2.2. *The 'logical reply'*

In a chapter published as part of the author's edited collection of essays on the Chinese room argument Jack Copeland challenged the logical validity of the basic Chinese Room Argument (Copeland, 2002). Alongside other commentators, he pointed out that the person in the room is not analogous to a computer executing an AI program, but rather just its CPU. Moreover the claim of A.I. researchers is not that the CPU understands Chinese, rather that the computer as a whole [the CPU + memory + hard disc and so on] does. Copeland subsequently observes that if the CRA is supposed to target the entire system (of Searle, the room, the rule-book, the bits of paper etc.), then the argument is simply not watertight; 'one might as well claim of the statement "The organisation of which Clark is a part has no taxable assets in Japan" follows logically from the statement "Clark has no taxable assets in Japan"'.

In fact Copeland's response is closely related to an earlier move which Searle anticipated from the start: the 'System reply'. Although the person in the room doesn't understand Chinese, the entire system of the room and its contents do; a response Searle finds entirely unsatisfactory.

Searle responds to the System reply by allowing the person in the room to internalise everything (the rules, the batches of paper, etc.), so that there is nothing in the Chinese-room-system that is not internalised. In response to the questions in Chinese and English, there are two subsystems – the native English-speaking Searle and the internalised Chinese room – and Searle continues to assert 'he understands nothing of the Chinese, and *a fortiori* neither does the system, because there isn't anything in the system that isn't in him. If he doesn't understand, then there is no way the system could understand because the system is just a part of him'.

Not surprisingly this move – to allow the person in the room to internalise the rule book, bits of paper, and so on – has also been criticised. One example is the critique by Georges Rey, who argues that once 'we observe the full complexity that a reasonable computational theory of mind proposes, the idea of someone memorising the programs to all these systems and getting them to run the right way is, well, pretty hard to imagine and intuitively evaluate'. However, as John Preston suggests in his introduction to our volume (Preston and Bishop), 'in this respect, it doesn't differ from Einstein's request for us to imagine what it would be like if, *per impossible*, we were riding on the front of a beam of light...'.

Hence Searle concludes:

> *"... the only motivation for saying there must be a subsystem in me that understands Chinese is that I have a program and I can pass the Turing test; I can fool native Chinese speakers. But precisely one of the points at issue is the adequacy of the Turing test. The example shows that there could be two 'systems,' both of which pass the Turing test, but only one of which understands; and it is no argument against this point to say that since they both pass the Turing test they must both understand, since this claim fails to meet the argument that the system in me that understands English has a great deal more than the system that merely*

> *processes Chinese. In short, the systems reply simply begs the question by insisting without argument that the system must understand Chinese."*

Yet other discussions of the System reply, for example that by Haugland (2002), have questioned why we should accept Searle's conclusion that the internalised Chinese subsystem doesn't understand Chinese given that its responses to the Chinese questions are correct and indistinguishable from those a native Chinese speaker may give:

> *"What we are to imagine in the internalization fantasy is something like a patient with multiple personality disorder. One "personality", Searle, is fluent in English (both written and spoken), doesn't know a word of Chinese, and is otherwise perfectly normal (except that he has the calculative powers of a mega idiot savant). The other ostensible personality - let's call him Hao - is fluent in Chinese (though only written, not spoken), has no English, and, moreover, apart from seeming to be able to read and write, is deaf, dumb, blind, and paralyzed.*
>
> *Why, exactly, should we conclude that Hao doesn't understand the Chinese that he appears to be reading and writing ("automatically", as it were)?*

Haugland suggests Searle's [internalisation] response equivocates on the use of the word 'in' and subsequently deduces that Searle is not entitled to assert that, 'were Hao to understand Chinese, so would he'; hence Haugland claims Searle's internalisation response to the System reply is simply not logically sound.

3. On consciousness and the metaphysical sensation of the inner

In response to Searle's critics, an insight that may lend support to his position can be found at the 'inner', phenomenological level. Let us compare the responses of the two systems – Hao and Searle – in the case where Searle first listens to a joke in Chinese and then in English. In the former case, although [assuming he executes the memorised procedures correctly] Searle may make the right linguistic responses in Chinese, he will never 'get the joke' and 'feel the laughter' because he, John Searle, still doesn't really understand a word of Chinese; whereas in the latter case, he may well 'get the joke', find it funny and laugh because he really does understand English. In other words, there is a fundamental 'difference in kind' (an *ontological* distinction) between these two cases. This is perhaps not so surprising as in the former case Searle's command of English is grounded by consciousness of his body; his interactions with the world and society; whereas in the latter *ex hypothesi*, he is merely carrying out ungrounded, uninterpreted symbol manipulations.

Conversely, such reification on the metaphysical sensations of the inner has recently been robustly criticised (from a Wittgensteinian perspective) by Murray Shanahan (2010). For example, with reference to the treatment of language, we find at the beginning of Philosophical Investigations (Wittgenstein, 1958), Shanahan highlights Wittgenstein mulling over the meaning of the word 'five' ...

> *"... addressing the philosophically inclined, who are apt to enquire into the 'nature' or 'essence' of meaning - almost as if it were some kind of stuff, something 'out there', so to speak, that is amenable to rational investigation. This is metaphysical thinking. But Wittgenstein advises us to set aside the vexing question, with its taint of metaphysics, of what a word like 'five' means. Instead he simply offers a description of the way the word 'five' is used, a description that emphasises its place in the hustle and bustle of daily human activity. There is no presentation of propositions and arguments in the style of traditional philosophy, no attempt to define the concept of number or pin down the nature of mathematical reality. Yet (we can imagine Wittgenstein saying) what has been left out of the description?*

Here Shanahan's exegesis of Wittgenstein would seem to imply that nothing of significance is left out by Searle's – programmed and perfectly acceptable – 'Chinese room' responses to questions about a Chinese story. Conversely, I suggest that in 'everyday' language use – *the hustle and bustle of daily human activity* – this is simply not the case.

Consider the following analogy: for all a small child may laugh at a sequence of adult jokes, she does not appropriately *feel* the laughter and really *understand* the jokes. Similarly, for all Hao may correctly respond to jokes in Chinese in ways that are behaviourally indistinguishable from those of a native Chinese speaker, can Hao ever appropriately *feel* the laughter and really *understand* the jokes?

In the absence of any linguistically grounding 'conscious sensations of laughter' accompanying Searle's execution of the Hao program, I doubt that by any *normal use* of the word 'understand' Shanahan can, legitimately, claim Searle-as-Hao understands the Chinese story anymore than the young child 'understands' adult jokes; demonstrably [in this case] mere outward behaviour alone is not sufficient to tease apart the two situations.

In conclusion I suggest that there is a fundamental *ontological* (not merely epistemological) distinction between Searle-as-Hao and Searle-as-native-English-speaker; a difference that cannot be teased apart by mere observation of external behaviour alone and – fatally for any Turing-esque style test – I suggest that this conscious difference is central to the notion of what it really means to 'understand' and to 'think'.

And so, just as there are *a priori* grounds for doubting that the mere execution of any computer program is sufficient to instantiate conscious phenomenal states in a machine, there may be concomitant grounds for believing Turing's 1950 dream – of mechanical, computational thought – is just beginning to fade; the overarching fallacy of 'cognitive computation' finally exposed.

References

Bishop, J.M., 2009. A cognitive computing fallacy? Cognition, computations and panpsychism. Cognitive Computing 1 (3), 221–233; Springer, Dordrecht, Netherlands.

Bishop, J.M., (Ed.) 2010. The Imitation Game. In: Special Issue on the Turing Test, Kybernetes 39(3); Emerald, Bingley, UK.

Boden, M., (Ed.) 1990. The English Reply, Escaping from the Chinese room, The Philosophy of Artificial Intelligence, Oxford University Press, Oxford, UK.

Copeland, J., 2002. The Chinese Room from a logical point of view. In Preston, J. & Bishop, M. (Eds.) Views into the Chinese Room: New Essays on Searle and Artificial Intelligence, Oxford University Press, Oxford, UK.

Harnad, S., 1990. The symbol grounding problem. Phys. D: Nonlinear Phenomena 42 (1–3), 335–346; see also Wikipedia on 'Symbol Grounding'.

Haugland, J., 2002. Syntax, Semantics, Physics, In: Preston, J. & Bishop, M. (Eds.), Views into the Chinese Room: New Essays on Searle and Artificial Intelligence, Oxford University Press, Oxford, UK.

Searle, J., 1980. Minds, brains, and programs, Behavioral and Brain Sciences, 3, 417–457; Cambridge University Press, Cambridge, UK.

Shanahan, M., 2010. Embodiment and the Inner Life Cognition and Consciousness in the Space of Possible Minds. Clarendon Press, Oxford, UK.

Turing, A.M. 1948. Intelligent machinery. National Physical Laboratory Report, 1948. In: Copeland, B.J. (Ed.), 2004, The Essential Turing, Oxford University Press, Oxford, UK.

Warwick, K., 2002. Alien Encounters. In: Preston, J. and Bishop, M. (Eds.) Views into the Chinese Room: New Essays on Searle and Artificial Intelligence, Oxford University Press, Oxford, UK.

Wittgenstein, L., 1958. Philosophical Investigations. Translated by G.E.M. Anscombe. Blackwell, Oxford, UK.

Peter Millican on recognising intelligence and —

THE PHILOSOPHICAL SIGNIFICANCE OF THE TURING MACHINE AND THE TURING TEST

The concepts through which we attempt to understand the world are formed by our experience of it. Alan Turing's 'imitation game' thought experiment can be seen as an attempt to stretch that experience, and with it our concept of *intelligence*. We naturally take our own intelligence to be intimately related to our *phenomenology* – our *sentience* and *conscious awareness*. But although Turing himself sometimes evinces the same assumption, his invention of the Turing machine provides an alternative, *algorithmic* model of information processing, and thus opens the prospect – where that information processing is sufficiently sophisticated and effective to deserve the name – of achieving 'intelligence' *without* consciousness.

1. Intelligence before Turing

The objects that we find in the world appear – at least to the casual observer – to divide fairly neatly into two quite distinct categories: *purposive* and *inanimate*. We ourselves are the most immediate examples of the former, and it is only natural to take ourselves as a model for the rest. From 'the inside', we both know our purposes, and self-consciously act on them. Our planned behaviour thus makes sense to us, and the actions of our family members and other humans are also explicable accordingly. Such purposive explanations are then very naturally applied further, to the animals we see behaving more or less comprehensibly in analogous ways (be they our pets, livestock, predators, birds, insects, or whatever).

Plants are less obviously purposive over a short timescale, but their growth, development and reproduction seem to manifest an equally clear teleology. Animals and plants together make up by far the majority of the most conspicuous elements of the pre-industrial landscape (something easy to overlook from within a modern house or city). It is not surprising, therefore, that before the age of modern science, the world as a whole was almost universally interpreted in terms of purpose, whether inherent or divine. Thunderstorms would vulgarly be attributed to the gods, and plagues to witchcraft. But even the academics of the time – the Medieval Schoolmen with their Aristotelian physics – took stones and stars to be as driven by purpose as animals, except that their purposes are more constant. Stones strive to reach the centre of the universe, and therefore fall to Earth; stars strive to mimic the eternal perfection of their Maker, and hence rotate around the heavens in perfect circles.

Galileo's telescope, in refuting the Aristotelian geocentric cosmology, equally sounded the death-knell for this entire picture of the physical universe. The scientific revolution that he ushered in undermined the view of inanimate objects as intrinsically purposive, replacing this with a theory of inert matter acting in accordance with rigid causal laws, being pulled and pushed around by forces and impulses that are mathematically determined by the relevant circumstances. Moving billiard balls bash into others and make them move in turn (according to their mass, angle of impact, and velocity); water gushes through a pipe under pressure from a pump; stones fall to Earth while the moon continues to orbit, and we discover that both can be neatly explained by Newton's postulation of a force of gravity that attracts according to an inverse square law. Modern physics significantly complicates this picture, of course, interlinking space with time and introducing an element of indeterminism. But the general mechanistic paradigm remains, the future of inanimate things unfolding from their past through mathematical laws that are purposeless and oblivious of any final destination.

From this modern perspective, the idea of purpose in inanimate things seems puerile and superstitious, or an occult relic of a pre-scientific era. Even most examples of living organisms, including plants and 'lower' animals, cease to be purposive, their appearance of teleology *explained away* by Darwinian selection. Genuine purpose lies exclusively in the domain of *conscious* beings, *desiring* certain ends and – at least in the case of humans and privileged 'higher' animals – *thinking* about the means to achieve those ends before acting accordingly. To think in this way to good effect is to be *intelligent*, a concept which thus ties together *conscious purpose* with the *effective processing of information* to identify the means to a desired outcome. Without the desired outcome, there would be no target for the information processing. But in the assessment of intelligence, it is the effectiveness of that processing rather than the strength or nature of the desire that provides the crucial measure. Human beings are the pre-eminent intellects of the natural world not because our desires are stronger than those of, say, a dog, but because we are so much better at identifying unobvious patterns, forming sophisticated plans, and calculating complex consequences.

2. Turing machines, intuition pumps and a word of caution

In his famous paper *On computable numbers* (1936),[1] Alan Turing came up with a precise model of an information processing machine – now universally known as a Turing machine – and provided an informal argument to suggest that when suitably programmed, this could faithfully execute any well-specified algorithmic process that can be carried out systematically by a human thinker. Suddenly a new question arises: *Should such information processing, as performed by an inanimate machine, be deemed genuinely intelligent?* Our experience of nature has not prepared us for this question, for although we have learned to think of *intelligence* as primarily a measure of the sophistication of information processing, we have also understood it as confined to *conscious* beings, planning how to achieve their ends. Now we are in a novel situation, faced with a machine which is clearly capable of processing information – of *calculating* answers to the sorts of questions that we standardly think of as demanding intelligence – and yet which has no ends of its own, and whose functioning has no need of *reason* as traditionally understood: no need of genuine *understanding*, *insight*, or *consciousness*.

Philosophers attempting to circumscribe the boundaries of some controversial or troublesome concept often appeal to thought experiments, nicely characterised by Daniel Dennett (1995) as 'intuition pumps'. Given the context described above, two sorts of thought experiment naturally suggest themselves. On the one hand, the advocate of machine intelligence can point to some suitably impressive example(s) of the sophisticated information processing achievable by an appropriately programmed Turing machine, and ask: *How can something which achieves <u>this</u> be denied genuine intelligence?* On the other hand, the opponent of machine intelligence can emphasise the crude, mechanical basis of the entity which is performing the processing, and the trivial individual steps by which it is operating, and ask: *How can anything which works like <u>this</u> be judged genuinely intelligent?* Turing himself takes the first path, presenting us in his paper *Computing machinery*

[1] The paper is technically daunting, but is presented and explained very effectively by Petzold (2008).

and intelligence (1950) with a scenario – a successful 'Turing test' – in which it seems unreasonably chauvinistic to deny intelligence.[2] John Searle, with his well-known 'Chinese room' thought experiment, takes the second path, focusing not on the outcome but on the method of processing, which seems so fantastically divorced from reality as to lack any semantic grounding.[3] The operator in Searle's room cannot plausibly be considered as reasoning *about* whatever is represented by the Chinese symbols he manipulates; hence the processing he carries out cannot be classed as *genuinely* intelligent. That, at least, is the moral that Searle would apparently have us draw.[4]

Thought-experiments designed to elicit particular 'intuitions' can be endlessly seductive for philosophers, but a severe note of caution is appropriate. Consider, for example, the following argument:

> "Performance at chess cannot provide an adequate criterion of intelligence, even of a domain-specific kind. For suppose that someone were to write a computer program of only a few dozen lines of code (in a standard general programming language), which could play chess at a grandmaster level in real time. Such a crude program could not possibly count as genuinely intelligent. Hence grandmaster performance at chess is not a reliable proof even of intelligent chess-playing."

It might well be true that we would be reluctant to count such a short computer program as 'genuinely intelligent'. But of course the fundamental hypothesis of this thought experiment – that such a program could possibly play grandmaster chess in real time – is utterly ludicrous. So we have no reason for taking it seriously as a guide to the boundaries of our concepts. Indeed it is easy to see that were we to allow this sort of thought experiment quite generally, it could without further ado rule out *any* performance-based criterion of intelligence.[5] But this seems outrageously simplistic, given that our notion of intelligence is, as mentioned above, one that we tend to assess primarily in terms of information-processing performance.

Note that performance here involves issues of resources, as well as output. It is not terribly difficult to write a computer program of modest length that plays infallible chess, if time and memory space are no object: simply analyse every possible line to the end, and score each as checkmate or as drawn (by stalemate, repetition, or the 50-move rule), chaining back accordingly. But such

[2] Turing calls this the 'imitation game', but it is now universally known as the 'Turing test' (at least when it involves a human interrogator interacting by teletype machine with one other human and one computer). In general terms, a computer program 'passes' the Turing test if it maintains a text-only conversation with sufficient human realism that the human interrogator cannot reliably distinguish between it and a human conversationalist. More detailed aspects of the test are discussed in Section 6.

[3] The most familiar Chinese room scenario (Searle, 1984, p. 32) involves a conversation conducted in written Chinese by means of cards posted into and out of a room, where the incoming cards express meaningful questions, and the outgoing cards provide meaningful and appropriate answers to those questions (such as might be produced by a competent and intelligent native speaker of Chinese). The twist is that the man inside the room has no knowledge whatever of the Chinese language or of the *semantics* – the meaning – of the symbols he is reading or writing. Instead, he is generating his written 'answers' by strictly applying rules based purely on the *syntax* – the shape and structure – of the 'question' character strings that he receives, these rules being specified in books contained within the room. Searle wishes us to conclude that the apparent meaningfulness of the answers that the man generates is an illusion, a conclusion which can then be taken as equally applicable to conversations generated by natural language processing computer programs.

[4] I say 'apparently', because although Searle presents his argument as an attack on 'strong artificial intelligence' and on the idea that machines can 'think' (e.g., Searle, 1980, p. 417; 1984, p. 36; 2002, p. 56), he generally expresses his thesis not as a denial of *intelligence* but rather of 'intentionality', 'cognitive states' (e.g., Searle, 1980, p. 417); 'a mind', 'mental states' (e.g., Searle, 1984, p. 37); 'cognitive processes', 'mental content', 'semantic content', or 'consciousness' (e.g., Searle, 2002, Section I). Since my focus is on Turing I shall not address this issue in detail here, but see note and Sections 4–5. below.

[5] 'Suppose that someone were to write a computer program of only a few dozen lines of code ... which could solve any problem of kind X ...'. And so forth.

a program, whose first calculated move will require vastly more steps than there are picoseconds in the entire history of the universe, is unlikely to be deemed 'intelligent'. Much the same applies to Ned Block's 'Blockhead' program, supposed to be programmed in advance with pre-prepared 'intelligent' responses for every conceivable sequence of verbal inputs in a conversation of some pre-defined length.[6] Such a program is possible only 'in principle', for even to set it up to cope with a fairly short conversation could consume more memory than the universe can hold. It seems entirely reasonable to deny that such programs can count as genuinely *intelligent*, when their mode of operation is so far removed from the clever techniques that intelligent organisms have evolved, to enable us to negotiate our way through complex problems with very limited resources.

Searles "Chinese room" combines outrageous unfeasibility with elements of genuine impossibility, because it hypothesises that intelligent answers – as good as those of a typical native Chinese speaker – could be framed by following purely syntactic rules in a context where the operator of those rules has no means of taking account of a changing world, both external and internal, whose events form the subject-matter of so much of our conversation. When asked (the Chinese translation of) questions such as 'Do you like the weather we've been having?', 'Did yesterday's news about *X* upset you?', 'How many times did I knock on your door just now?', or 'Have you disagreed with anything I've said in the last five minutes?', the operator's syntactic rules give no scope for sensory input, real-time updating, or emotional reaction, and so however sophisticated those rules might be, he cannot possibly match the response of someone who understands the question. But even if the questions are artificially limited to comprehension of a fixed story written in Chinese, rather than being interactive,[7] the suggestion that intelligent responses to arbitrary Chinese questions could be generated by Searle's specified method – through the manual consultation of purely syntactic rules recorded in books within a room – is as ridiculous as the idea that a 50-line computer program might play grandmaster chess in real time. 'Surely', Searle's scenario implicitly urges, 'something that operates in such a manner cannot possibly be deemed intelligent'. This indeed seems persuasive, but then something that operates in such a simplistic manner could not possibly reach the level of performance that Searle is postulating, so the significance of his thought experiment is crucially undermined.[8]

3. Turing machines, new paradigms and open texture

I started by suggesting that our modern concept of *intelligence* was established within a world apparently divided between two main categories of entities. We ourselves, together with 'higher' members of the animal kingdom, are organisms moved by conscious desires, able to process and exploit information (with various degrees of sophistication) in the conscious attempt to fulfil those desires. All other things lack consciousness and therefore cannot be moved by such desire, nor apprehend information. These unconscious entities include physical objects, which act mechanically: pushed or pulled around by impacts and forces that are blind to any final outcome. Plants and 'lower' animals, though presumably equally unaware, behave in ways that seem to manifest purpose, sufficiently so that for millennia it proved almost irresistible to attribute this behaviour to the

[6] See Block (1981). It is unclear who first coined the nice name 'Blockhead' for the program described in this paper.

[7] As in the original 1980 version of Searle's Chinese room scenario. For some useful background to that article, see Preston (2002), pp. 16–19.

[8] Searle might respond that it is *conceivable* that something operating by the Chinese room method – even under realistic constraints of space and speed – might achieve the required level of real-time performance, and this should therefore be considered *possible*. However it is obvious from cases such as the provability (or otherwise) of Goldbach's Conjecture that conceivability can be taken as a reliable guide to possibility only, at best, where it is interpreted as involving *clarity and distinctness* of a fairly strong kind. I can of course conceive in a general sense (e.g., sufficient for understanding the words) what it would be for a 10-line computer program to play infallible chess in real time. But I cannot clearly and distinctly conceive how such a program would operate, and it is *very obviously* not a genuine possibility. For more on this, and on Hume's influential appeals to the Conceivability Principle, see Millican (forthcoming) Section 5.

influence of a divine being with human-like intentions. Darwin's theory of evolution was revolutionary not only because it removed the need for such a designer-god, but more fundamentally, because it introduced an entirely new mode of explanation which was neither mechanical nor purposive. Such explanatory innovations are rare but momentous: other examples would be the development of mechanism itself by Galileo and others (e.g., Descartes and Boyle) in the seventeenth century, and its challenge by quantum mechanics in the twentieth century, which brought a very different conception of physical explanation.

The invention in 1936 of what we now know as the Turing machine bears comparison with these paradigm shifts.[9] For it provides a way of specifying processes in *algorithmic* terms that are neither purposive nor mechanical, but have common features with each. Like a purposive explanation, an algorithm is couched in terms of the abstract processing of information. But like a mechanical explanation, this processing is at bottom 'mindless' and automated, taking no account of any semantic significance and paying no regard to any desired endpoint. This – I suggest – is what opened the possibility that Turing presented so forcefully in his 1950 paper, of intelligent information processing that is automated rather than purposive. Before the Turing machine, information had to be understood in terms of its significance to a *conscious mind*. But Turing saw that information – and information processing – could be understood quite differently, thus opening the possibility of machine 'intelligence' gauged in terms of inputs and outputs rather than requiring any sort of internal understanding. Hence we reach the idea of a thought experiment that compares the external behaviour of man and machine, judging the latter to be intelligent if it can do equally well. As we saw in the previous section, such thought experiments need to take account not only of inputs and outputs, but also the constraints of our practical situation. Even the best-equipped organisms are limited in knowledge, capacity, time, and other resources. This puts a premium on the effective exploitation of our limited means, on efficient and flexible processing with uncertain inputs and under pressure of time. We naturally judge intelligence accordingly, and deprecate the inefficient brute force methods of the implausible thought experiments of Searle and Block, which lack any practical utility and differ so radically from any familiar reality.

Turing's 'imitation game' thought experiment, however, still remains to be judged, and we surely know now – even if this was hard for most of his readers to appreciate in 1950 – that the level of computational linguistic ability that it postulates is relatively plausible. Over the last decade, interactive computer systems have made huge strides in the processing of natural language (as illustrated, for example, by the development of automated translation systems), and although they still have a long way to go, it is by no means obviously ridiculous to consider a future system – even within the next few decades – that might achieve something like the level of performance anticipated by Turing. Admittedly some aspects of his thought experiment are less plausible than others, notably his requirement that the envisaged system should be able to pass for a human in general conversation, informed as this might be by personal emotions and by reference to changing events.[10] So to

[9] Floridi argues that 'the best way to understand the information turn is in terms of a fourth revolution in the long process of reassessing humanity's fundamental nature and role in the universe. We are not immobile, at the centre of the universe (Copernicus); we are not unnaturally distinct and different from the rest of the animal world (Darwin); and we are far from being entirely transparent to ourselves (Freud). We are now slowly accepting the idea that we might be informational organisms among many agents (Turing) . . . ' (2008, p. 651). Whether or not one accepts this account (e.g., I would be inclined to replace Freud with Hume and the development of cognitive science, cf. Section 4 below), it is interesting that the paradigm shifts I have identified – in terms of the discovery of new modes of explanation – correspond quite closely to major upheavals in our understanding of our place in the universe.

[10] Note, however, that Turing would 'wish to permit every kind of engineering technique to be used in our machines' (1950, p. 435; see p. 553 this volume), including 'the best sense organs that money can buy' (p. 460; p. 568). Consideration of sensory input plays a large role in his discussion of learning machines (pp. 454–460; pp. 564-568), and presumably explains why he expresses a special concern about the possibility of extra-sensory perception, which (were it to occur) could not be replicated mechanistically (pp. 453–454; see p. 564 this volume). Searle standardly restricts his operator to information gleaned from books inside the Chinese room, but he takes the message of his thought experiment to apply equally to a robot equipped with appropriate sensors (1984, pp. 34–35).

ensure that our discussion remains solidly grounded in foreseeably plausible reality, let us suppose only that the challenge of the Turing test *has* been fulfilled in an extensive – though not unlimited – factual domain: perhaps the science of chemistry. Suppose that computers have been programmed in such a way as to be able to sustain long and detailed conversations, appropriately directed, with complex, accurate reasoning and interlocking themes, apparently well informed about all relevant aspects of chemistry.[11] Should we call such conversational behaviour 'intelligent'?

In making this judgement, there is no reason why we should confine ourselves within narrow behaviourist limits, and it is entirely legitimate to take account of obvious points regarding the nature of any such program. Clearly such sophisticated discourse about chemical interactions will have to be informed by representations of molecular structures and relevant laws and forces: this is not a Blockhead-style lookup table, nor a 'chatterbot' designed to mislead (cf. Section 6 below). Real information processing is taking place, generating appropriate and informative responses by reference to the same mathematical and structural relationships that would inform a human expert,[12] but none of it – at least on the part of the program – is the least bit *conscious*. Does this then debar it from deserving the accolade of *intelligence*?

I would like to suggest that we cannot necessarily expect an unambiguous answer to this question, because it concerns the application of a concept beyond the context for which it has evolved. Our common sense world view seems to imply a general division between things that are consciously purposive and calculating, and others that are neither conscious, nor purposive, nor calculating. So it is not surprising that we then find it hard to classify a novel kind of entity which seems to calculate very effectively (in a sophisticated manner, and to a useful purpose), but which itself entirely lacks any kind of consciousness, and hence lacks any awareness or 'internal' understanding of either the apparent purpose or the calculation.

We have here a case of what Friedrich Waismann called *open texture*. A concept or term is said to be *open textured* if our understanding of it does not 'provide in advance for all possible cases'.[13] Our concepts are framed, or adapt, to fit the circumstances in which they are standardly employed, and they commonly fail to have determinate criteria of application in abnormal, unanticipated circumstances. Suppose, for example, that marriage is defined as being allowed only between a man and a woman, in a society in which it is absolutely taken for granted that everyone has an unambiguous sex (and gender) throughout their life. This rule might seem to be entirely clear and precise; indeed those who frame it take it be so. But it can nevertheless become indeterminate if, for

---

[11] This enables us to put aside the question of whether such a program could convincingly discuss matters that arguably require essential reference to human perceptions or emotions, such as sensory phenomena, morals or aesthetic appreciation. The presupposition that intelligence in *one* area does not require intelligence in *all* seems highly plausible to me, but could perhaps be threatened if, for example, it turned out that only a 'global workspace' could solve the frame problem, cf. Shanahan and Baars (2005).

[12] Searle might contest this, on the ground that there is no *semantic* connection between the representations in the program and the real-world features they represent. But for present purposes, the fact that there is a well-designed isomorphism between the relationships as understood by the scientist, and those formally manipulated by the program, will do. Clearly in some sense there is *information* being processed, even if that information fails to live up to Searle's 'semantic' requirements. Space does not permit further discussion of Searle's concerns here, but suffice it to say that I consider his notion of the 'semantic' to be fundamentally obscure, and liable to dissolve under close analysis. As the references in note 4 above indicate, his terminological promiscuity tends to conflate *information processing* with *phenomenology*, whereas I shall argue in Sections 4–5 below that these are best distinguished, since 'intelligent' processing does not necessarily require *consciousness*. Once distinguished, the plausibility of Searle's claim that a computer program could not possibly have 'semantic' relationships – at least in the *information processing* sense – is significantly weakened (especially if we consider the possibility of directly connecting the program to reality through appropriate sensors and manipulative mechanisms). Moreover even if the claim were to be accepted (e.g., on the basis that any fully adequate semantic relationship must involve *conscious* intentionality), it would then require a further argument to move from a lack of *semantics* to a lack of *intelligence*.

[13] Quoted from Williamson (1994), p. 90. Williamson discusses Waismann's use of open texture on pp. 89–95.

example, someone is born chromosomally male but physically female, or if sex-change operations occur. As the philosopher of law Herbert Hart insisted and this example illustrates, open texture is particularly important in legal contexts, which often hinge on the precise boundaries of rules that have been specified without even considering, let alone defining, their application to 'all possible cases'.[14]

4. Intelligence and consciousness

If the concept of intelligence is open textured in this way, then its application to suitably programmed computers is an open question, rather than one we should expect to be able to decide by simply analysing our existing concept. But this does not imply that the question has no best answer, and we have already seen at least one good reason for siding with Turing rather than Searle. For although we standardly take an intelligent entity to be one that has *conscious awareness and purpose* as well as *effectiveness in processing the relevant information*, nevertheless when we judge one person to be *more intelligent* than another, we do so almost exclusively in terms of the latter criterion. Thus we do not typically consider mathematical brilliance to be any sort of measure of the quality of a mathematician's inner life – the motivational desires, feelings of effort, or even poetic urges that he might experience whilst proving his next theorem. All this subjectivity is irrelevant, and it is the objective quality of his proof production that dominates, *except* in so far as we are inclined to require *some* inner life before we are prepared to count 'him' as a mathematician at all (as opposed to a mathematical tool).

This consideration can be pushed further, by noticing that for humans, at any rate, there is often an inverse relationship between these subjective and objective qualities. Perhaps the most familiar example is in driving a car, where the seasoned expert achieves high performance with little focused 'consciousness' of what he is doing – or subsequent memory of having done it – while the stumbling learner driver is only too conscious of every tense observation and manoeuvre. In the same way, the novice chess-player struggles to find a good move, vividly aware of his efforts and uncertainties, reflecting carefully and anxiously on all the considerations that come to mind. The grandmaster, by contrast, typically finds his move effortlessly, almost without conscious thought and entirely without struggle; moreover when asked to explain his 'thinking', he might well have nothing better to say than that 'in this sort of position, *that* is obviously the right move to play'. Here again our common sense identification of *intelligent* information processing with *self-conscious* information processing is contradicted, as we find that greater expertise is frequently accompanied by *less*, rather than more, articulacy.[15] And accordingly the person who has had to struggle to acquire a skill often makes the better teacher, having reflected far more on what the skill requires, and able to relate more closely to the difficulties of students. But we do not on this account judge him to be the better practitioner of the skill, even where the skill is one that we think of as paradigmatically 'intellectual'. Executing an intellectual task is one thing; reflecting on it quite another.

Pushing even further in the same direction, it turns out that there is little correlation between the sophistication of information processing that common tasks require, and their typical psychological impact or effort. Indeed, it seems that the vast majority of the most complex processing that takes place in our brains is entirely unconscious, and remarkably little of our mental life can properly be explained in terms of reflective reasoning and explicit inference. David Hume famously pioneered this message, proving the impossibility of accounting for such basic mental operations as inductive inference or the identification of persisting objects in terms of any traditional concept of reason.

[14] An entertaining example is provided by a famous *Punch* cartoon (6 March 1869, p. 96), in which a railway porter is telling an old lady about the price of travelling with her menagerie of pets, given rules which specify a cost for dogs only: 'Station Master says, Mum, as Cats is "Dogs", and Rabbits is "Dogs", and so's Parrots; but this 'ere "Tortis" is a Insect, so there ain't no charge for it!'

[15] See Michie (1993) and also my introduction to Millican and Clark (1996), pp. 2–3.

Though we might suppose that we are transparently *apprehending* rational connexions between past and future, or passively *perceiving* continuing things through time, in truth our minds (or at least our brains) are actively supplying crucial contributions of their own. It is these active inputs that enable us to move to conclusions beyond what pure reason would warrant, and to smooth over irregularities in the flux of sensations. And because they are creative rather than cognitive processes – reading *into* the world of our experience rather than *off* it – Hume attributes them to 'the imagination' rather than to 'reason'. The same lesson has been emphasised even more in recent years, with discoveries prompted by studies in artificial intelligence. It is now clear, for instance, that even the identification of objects in a visual scene *at a single time* essentially involves active processes of edge detection, shadow interpretation, and so forth, all of which are typically *subcognitive* and therefore unavailable to consciousness. And this increased appreciation of the sheer computational complexity of everyday cognition has gone together with a re-evaluation of the familiar examples of 'intelligence' that once seemed to represent the pinnacle of intellectual achievement. Arithmetic, for instance, seems abstract and difficult for humans, and is hard to master without years of schooling and practice. Yet compared to the computational difficulty of, say, tracking and catching a ball whilst running (something which many of us can do with relative ease, and which dogs seem to do quite naturally), arithmetic is utterly trivial. Again the lesson seems to be that if we wish to preserve the criterial correlation between *intelligence* and *competence founded on sophisticated information processing*, then we must be prepared to cast off the folk-psychological assumption that greater intelligence requires greater consciousness of what we are doing. With that assumption discarded, there is much to be said for relinquishing the requirement of consciousness entirely.

5. Information processing and phenomenology

Throughout this discussion I have resisted any *conflation* between 'intelligence' and 'consciousness', whilst fully acknowledging that our naïve concept of the one significantly implicates the other. This is important, because discussions of the Turing test are often horribly muddied by a failure to distinguish two quite different features of what we take to be intelligent thought, namely the *information processing* that it involves, and the *phenomenology* that potentially accompanies that information processing. Too often, the possession of *intelligence* is conflated with possession of a *mind*, yet it seems to me that the connotations of the two are radically different. When we consider an entity as having a mind, the crucial factor is not so much the quality of its intellectual processing as its possession of an 'inner life', or as Thomas Nagel famously put it, there being 'something it is like' to be them. When we say that we are 'minded' to do something, we are expressing a *felt desire* rather than any intellectual process. And nothing said above has given the slightest ground for supposing that an electronic computer – no matter how cleverly it might be programmed – is able to experience genuine feelings. I have argued that we should be prepared to accept the notion of unconscious *intelligence*, but there is no such compelling reason to countenance unconscious *desires*, let alone unconscious *feelings*.[16] Some might wish to do so, attracted either by exotic Freudianism or, at the other extreme, by the austere objectivity of behaviourism or functionalism. But there is no need for us to adjudicate on these things here, and I am happy to allow the opponent of machine intelligence to insist that even the merest *wish* – let alone a *passion* or a *craving* – is something that essentially requires *feeling*, and hence is confined to conscious beings.[17] We have already noted

[16] Unconscious sensations may provide an intermediate case, in that although *sensory awareness* can be seen as a source of information (and to that extent abstracted from phenomenology), conceptually it seems to be tied more closely to its internal Nagelian character than in the case of *intelligence*.

[17] So here I am content to agree with Searle in opposing the view that 'any physical system whatever that had the right program with the right inputs and outputs would have a mind in exactly the same sense that you and I have minds. ... that it must have thoughts and feelings, because that is all there is to having thoughts and feelings: implementing the right program.' (1984, pp. 28–29).

that this does not prevent *unconscious* beings from exhibiting *apparent* teleology, as we find in much of the animal kingdom and universally amongst plants. But again, for present purposes, I am quite happy to allow Turing's opponent to explain this away by the familiar appeal to Darwinism, and to reserve the term *desire* for the genuine (i.e., conscious) article. Such a reservation, however, is entirely consistent with allowing the possibility – indeed the manifest reality – of unconscious *intelligence*.

Turing himself, unfortunately, is guilty of the conflation that I am resisting, and perhaps deliberately so. In his 1950 paper he considers Geoffrey Jefferson's 'Argument from Consciousness' as a 'denial of the validity of our test', and his response is to compare it with solipsism, as though we could have no better reason for denying consciousness to a (suitably conversing) computer than we have for denying it to our fellow humans. But this response is weak, and should convince nobody who takes consciousness to have an ontological reality over and above behaviour and functional role. Certainly the subjectivity of consciousness seems mysterious, and perhaps all the more so as our psychological and physiological science has become more objective. The relationship of consciousness to our physical brain is hard to make sense of, as is its evolutionary function: even if we ignore the difficulty of understanding how consciousness can arise from physical matter, it remains obscure how, having arisen, it can contribute to our biological success (as the popularity of thought experiments involving 'zombies' testifies). Nevertheless, if there is one solid certainty in all this,[18] it is that consciousness must indeed bring some such evolutionary benefit, perhaps by facilitating a more efficient form of perspectivally informed processing than would otherwise be possible (e.g. spatio-temporal, perhaps, or in terms of our ability to employ a theory of mind about our fellows).[19] And that being so, we have every reason to suppose that the same biological make-up which generates our own capacity for consciousness does exactly the same for others of our species (and, indeed, of similar species).[20] No such argument can be made in the case of a programmed computer or robot. On the contrary, such a machine's behaviour – however closely it may be designed to mimic our own – is precisely explicable in terms of its program: that is what the program has been designed to do! When the machine produces an output which, in a human, would be expressive of consciousness, we know that the reason it does so is that it has been programmed appropriately (even if the detailed algorithmic mechanism is unpredictable or too complex for us to discern). Genuine, full-blooded, ontological *consciousness* – whatever exactly that might be beyond behaviour and functional role – is an entirely gratuitous postulation in such a case, eliminable immediately with a slash of Ockham's Razor. So Turing's anti-solipsistic move is powerless against someone who insists that we have such consciousness.

Note that this argument does not depend on the assumption that genuine consciousness is irrelevant to the achievement of sophisticated information processing; nor would it be refuted by the discovery that some forms of information processing are entirely beyond the practical capacity of anything that lacks a conscious perspective. For the latter discovery could only plausibly be made

---

[18] Such certainty is vastly more likely to be found in reasoning based on scientific considerations that are liable to empirical test – or on formal rules whose reliability can be rigorously tested by mechanical application to numerous cases – than in the aprioristic (and typically ad hoc) untestable argumentation of armchair philosophers. Anyone disinclined to accept this Humean point (1748, 12.27-9) would be well advised to ponder the track-record and shifting fashionable tides of philosophical armchair speculation!

[19] If consciousness had no causal impact on behaviour, but just somehow arose as an epiphenomenon, it would be a complete coincidence that such a manifest correlation has evolved between subjective feelings and bodily events. If the subjective pain of banging my knee, or the pleasure of tasting honey, are causally inert, then there is nothing to tie them evolutionarily to the events that characteristically generate them, and from the point of view of survival, they could just as well be reversed.

[20] For an illuminating discussion on the connection between evolutionary considerations and the inference to mental states of others, see Sober (2000).

in respect of a form of information processing that had *not* been achieved by a computer. If a computer were to achieve it, that would *ipso facto* provide overwhelming evidence that the computer's programmed powers were sufficient for it, and thus count decisively against the hypothesis that anything more was required.

6. Evaluating the Turing test: the lessons of ELIZA

In considering the significance of Turing's thought experiment, we should bear in mind the state of computer technology – both hardware and software – at the time when he came up with it. He could not point, as we can now, to sophisticated computer systems achieving feats of information processing hugely beyond the powers of the unaided human brain, not only in relatively abstract calculation (such as arithmetic or chess-playing), but also across a large and ever-increasing range of scientific enquiry. What he sought, therefore, was not a general criterion of intelligent behaviour, but a clear illustration of one sort of behaviour that anyone *would* recognise as paradigmatically intelligent were it to be achieved.[21] And in order to make this illustration relatively plausible within a reasonable timescale, that behaviour had to be confined to *verbal* interaction. In this context, his choice of test was judicious, his examples convincing, and his predictions remarkably accurate. Here, first, is an example of a conversation from the 1950 paper which, if spontaneously produced (and hence not pre-arranged in any way), would surely tend to persuade us that the Witness is capable of responding to such questions appropriately and 'intelligently':

" Interrogator: In the first line of your sonnet which reads 'Shall I compare thee to a summer's day', would not 'a spring day' do as well or better?
Witness: It wouldn't scan.
Interrogator: How about 'a winter's day' That would scan all right.
Witness: Yes, but nobody wants to be compared to a winter's day.
Interrogator: Would you say Mr. Pickwick reminded you of Christmas?
Witness: In a way.
Interrogator: Yet Christmas is a winter's day, and I do not think Mr. Pickwick would mind the comparison.
Witness: I don't think you're serious. By a winter's day one means a typical winter's day, rather than a special one like Christmas." (1950, p. 446; p̄.560)

However Turing is not so rash as to predict that performance at anything like this level is likely to be achievable soon. At the beginning of Section 6 of his paper, 'Contrary Views on the Main Question', he famously makes the following far more modest prediction:

> "I believe that in about fifty years' time it will be possible to programme computers, with a storage capacity of about 10^9, to make them play the imitation game so well than an average interrogator will not have more than 70 per cent. chance of making the right identification after five minutes of questioning. ...I believe that at the end of the century the use of words and general educated opinion will have altered so much than one will be able to speak of machines thinking without expecting to be contradicted." (1950, p. 442; p̄. 557)

The standard he suggests – whereby an *average* interrogator is supposed to be able to do no better than distinguishing between the computer and a human with *70% accuracy* after a mere *five minutes*

[21] Turing makes clear that he is seeking a *sufficient* test of intelligence rather than a *necessary* condition, when addressing the objection: 'May not machines carry out something which ought to be described as thinking but which is very different from what a man does?' (1950, p. 435).

of questioning – is not particularly high.[22] Had this been a serious goal of artificial intelligence research, I expect that it would have been solidly achieved by the year 2000. As for the other part of Turing's prediction, by the end of the century it had indeed become fairly commonplace to talk of computers 'thinking', especially about difficult information-processing tasks taking place in real time. Other psychological verbs have also become natural to apply to computer programs, with minimal if any embarrassment, and a conversation like this about a chess-playing program would not seem out of place:

> "Why is the computer taking so long to respond to your queen move?"
>
> "It's thinking hard, because it's realized that if it tries to defend against my attack by bringing its knight over to protect the king, I'll be able to grab its pawn on the other side. It's displaying now that it assesses the position as better for me materially, but it seems to be predicting that it can get some activity to compensate if it decides to let the pawn fall."

No doubt many philosophical pedants, hearing this conversation, would want to insist that the psychological verbs are being applied only loosely or analogically. But the fact remains that such application is extremely natural, and by now rather likely to be used 'without expecting to be contradicted'. We have here symptoms of precisely the sort of conceptual evolution advocated above, whereby increased habituation to a changed reality leads to a corresponding adaptation of our traditional concepts.

This conceptual evolution has not been significantly fostered by work towards satisfying the Turing test, which has led in a very different and less productive direction, namely the development of 'chatterbots' that are typically at best amusing curiosities rather than serious tools. Indeed with hindsight, it is a shame that Turing not only proposed his 'imitation game' as an illustration of how a computer could manifest intelligence (as in the sonnet conversation above) but also gave the impression that it could provide a rough measure of success in developing machine intelligence. For his quantitative prediction – that a 30% success rate at impersonation over five minutes' questioning would be achievable by the end of the century – naturally suggests that a higher success rate, over a longer period of questioning, would be a suitable indicator of progress. But unfortunatey, it is no such thing, because as the experience of Joseph Weizenbaum's ELIZA program of 1966 quickly showed, success in impersonation can be far more readily achieved by trickery and misdirection than by genuinely intelligent processing. Hence we have the sad spectacle of the Loebner Prize organisation, year after year, holding a Turing test challenge which encourages the further development of these largely pointless chatterbots, whose programmers, rather than attempting to write systems that handle language intelligently, instead focus on marginally improving their methods of *appearing* to contribute appropriately to the flow of a conversation. What these programs actually do is typically nothing more intelligent than recognising simple patterns in the input and replaying adaptations of the interrogator's own phrases, suitably modified by switching first- and second-person etc. Where even this fails, they say something entirely neutral (e.g., 'GO ON'), revert to some previous input, change the subject, or disguise the failure by saying something that makes no pretence of continuing a thread of conversation but is instead designed to appear amusingly eccentric (a manoeuvre which can even be presented positively as giving the chatterbot a distinctive 'personality'). The original

---

[22] This would mean that a sequence of 100 members of the public, each faced with the task of distinguishing a human from the computer through a five-minute 'interrogation' of each of them, could be expected to achieve 70 correct identifications, and 30 incorrect. So there is no requirement here that the interrogators are deceived most of the time. It is perhaps better to think of this performance as 30% *successful impersonation* by the computer program, rather than 70% *accuracy*, since the point of the test is to provide a measure of the *program's* achievement, rather than the *interrogators'*. (But note that 50% successful impersonation would be enough to imply indistinguishability from a human – any more than that, and the program is implicitly being judged to be 'more human than a human').

conversation in Weizenbaum's 1966 paper, which introduced most of these techniques, includes such gems as:

> "Men are all alike. [pattern: 'alike']
> IN WHAT WAY
> ...
> Well, my boyfriend made me come here. [pattern: 'my X' – this also prompts saving of X]
> YOUR BOYFRIEND MADE YOU COME HERE
> ...
> I need some help, that much seems certain. [pattern: 'I need X']
> WHAT WOULD IT MEAN TO YOU IF YOU GOT SOME HELP
> ...
> My mother takes care of me. [pattern: 'my mother X']
> WHO ELSE IN YOUR FAMILY TAKES CARE OF YOU
> ...
> Bullies. [no recognized pattern, so previously saved 'my X' phrase is re-used instead]
> DOES THAT HAVE ANYTHING TO DO WITH THE FACT THAT YOUR BOYFRIEND MADE YOU COME HERE"

Such a conversation can indeed *seem* intelligent, because observers are typically unaware of how much its apparent coherence depends on trickery and pretence (or, in the last case, sheer coincidence), and so they systematically give the computer the benefit of the doubt by interpreting the outputs as maximally appropriate within the conversation. Weizenbaum also cunningly has his program ELIZA play the role of a Rogerian psychotherapist, whose method consists largely of echoing back the human user's own thoughts, in order to elicit further such thoughts.[23] This ruse is quickly exposed if one sets two clones of such a program conversing with each other: without any injection of substantial content from human input, the conversation soon descends into aimless vacuity.

Since Turing presented his test as a demonstration of how computers could in principle manifest human-like intelligence, it is ironic that the main objection to the test is how *unintelligent* humans can be, both as conversationalists and as interpreters. In some areas of ordinary life, a fair amount of human conversation can consist of vacuous responses which engage only vaguely with what has gone before. Presumably for this reason, our interpretation of others' contributions can be excessively uncritical and over-generous, even when we have knowingly been put in the role of an 'interrogator' judging their intelligence. So although human conversational behaviour is generally intelligent up to a point – and often highly so – it is hardly a *paradigm* of intelligence, and there is no reason why *indistinguishability from a human* should be seen as the ideal criterion of intelligence, let alone *indistinguishability as judged by an average human* (which is thus doubly polluted by human sloppiness and fallibility).

None of this is to deny Turing's claim that genuine indistinguishability over an extended and suitably probing discussion by a discerning interrogator – as illustrated by his sonnet conversation – would provide a reasonable basis for ascribing intelligence (at least on a provisional basis).[24] But were anything remotely like this to be achievable in practice, the main problem with the test would become not the over-generosity of uncritical interrogators, but rather their excessive discernment. For in this situation, the main concern of programmers attempting to fulfil the test would be not

[23] For an implementation of Weizenbaum's 'DOCTOR' script (based closely on the appendix of his 1966 paper and generating the dialogue he quotes), within a fully documented learning environment that facilitates practical chatterbot development and experimentation without requiring programming expertise, see www.philocomp.net/ai/elizabeth/.

[24] The ascription could be withdrawn if it later turned out that the program was driven by a lookup table or just happened to 'get lucky'. As argued above, it is sophisticated and appropriate information processing – what enables it to respond 'intelligently' to a wide range of inputs – that constitutes intelligence, not just outward behaviour on particular occasions.

the maximising of intelligent processing *per se*, but rather the imitation of human reactions, many of which are informed by our emotions, personal histories and social lives, and often have rather little to do with intelligence. Mimicking of all this so as to convince a discriminating judge over an extended period would be an extremely impressive programming achievement, no doubt. But the notion of intelligence – as Turing would have been the first to insist – is nothing like so human focused as to require any such thing (cf. note 21 above), and hence it would be utterly perverse to make the general imitation of human characteristics a major focus of artificial intelligence research. Indeed this would be almost as ridiculous as making the imitation of birds – rather than fast, safe and efficient flight – a primary aim of aeronautical research.[25]

7. Conclusion: The Turing test and the Tutoring test

When Turing proposed his 'imitation game' in 1950, it served the valuable role of highlighting a context – namely textual conversation – in which one could realistically foresee computer behaviour that deserved to be called 'intelligent'. I have suggested that in this role the 'Turing test' succeeds, and that if we were presented with a system which could reliably generate conversation of the quality he illustrates in his article, we would have excellent reason for counting it as intelligent, even though we would have no good reason for ascribing it any sort of conscious awareness. Admittedly this involves conceptual change, because our naïve concept of intelligence combines both *information processing* and *phenomenological* aspects, but such change is well motivated in this sort of situation, where we are presented with a new kind of entity that fails to fit into our naïve taxonomy. Moreover there are good independent reasons for seeing sophistication of information processing – rather than inner experience – as the central criterion for intelligence: this conforms to our standard methods of comparing intelligence amongst people and animals, and also acknowledges the reality of highly intelligent behaviour that is 'intuitive', habitual, or subcognitive.

Unfortunately, however, the Turing test itself fares very badly as a method of *measuring* intelligence: it simply is not true that better performance in the test (in the sense of passing more plausibly for a human conversationalist, or for a longer period) correlates well with intelligent information processing. Nor is this only because success in the test is biased towards the imitation of *human* conversational behaviour, which surely disqualifies it as a *necessary* condition for intelligence (as Turing himself recognised). More damagingly, the development of chatterbots has revealed how unreliable we humans are as judges of conversational competence, mainly because we are so liable to read coherent meaning into any verbal exchange that is susceptible of it. Hence the chatterbot designer, aspiring to do as well as possible in the imitation game, aims not for the generation of precise and careful dialogue (in which the computer's mistakes or lack of 'humanity' will become all too apparent), but instead for the production of piecemeal responses that are maximally vague and sloppy, exploiting the foibles of the interrogator. Thus there is no plausible developmental pathway from increasing chatterbot performance in the Turing test to genuine artificial intelligence, and the Loebner Prize (though no doubt well motivated) is completely misdirected.

It might, nevertheless, be possible to preserve something in a similar spirit if we add two letters and move from the 'Turing test' to a 'Tutoring test', in which the aim is not to pass for a human, but instead to succeed in a conversational information-processing task which has a very clear point and whose measurement is relatively well understood. Here the criterion of success would be not *deception* but *revelation*, by tutoring the human 'interrogators' to acquire an understanding of some specific field of knowledge of which they were previously ignorant (e.g., some aspect of chemistry). In this context, for the tutoring system to reveal its non-human status would not be any kind of failure – all that matters is the effective eliciting of understanding in the tutee. And this can be assessed

[25] Whitby (1996), pp. 57-58, develops this point, while French (1990) highlights the extreme difficulty of programming a computer to mimic human *subcognitive* reactions.

by the methods we standardly use in educational practice, ranging from first-personal reports to interviews and formal tests. Now the Turing-style gold standard would be a tutoring system that can teach as effectively (in a given time) as a good human tutor; and it is an open question, I believe, whether this is realistically achievable, and in which fields. But whether or not that provides a plausible ultimate target, the great advantage of this Tutoring test – in almost any field to which it might be applied – is that work towards it can potentially be of real value, not only in developing systems that can provide cheap education to those unable to afford human tuition, but also in promoting *genuine* artificial intelligence.[26] For the comprehensive understanding of any intellectual issue by the tutee will typically involve the grasp of a complex web of connections amongst the relevant concepts and techniques. And to convey these most effectively, an intelligent tutoring system will presumably require some representation of these same connections: the more fully and faithfully they are represented, the better it is likely to be able to perform at tutoring. At any rate, it seems a relatively plausible expectation that work towards the Tutoring test could provide a valuable source of continuing inspiration in artificial intelligence. Sadly, the same can no longer be said of the Turing test.[27]

References

Block, N., 1981. Psychologism and behaviorism. Philos. Rev. 90, 5–43.

Dennett, D., 1995. Intuition pumps. In: Brockman, J. (Ed.), The Third Culture: Beyond the Scientific Revolution, Simon & Schuster, 182–188 (the rest of chapter 10, pp. 181–97, consists of commentary by others).

Floridi, L., 2008. Artificial intelligence's new frontier: artificial companions and the fourth revolution. Metaphilosophy 39, 651–655.

French, R.M., 1990. Subcognition and the limits of the Turing test. Mind 99, 53–65 and reprinted in Millican and Clark (1996), pp. 11–26.

Hume, D., 1748. An enquiry concerning human understanding. Peter Millican (ed.), Oxford University Press, Oxford, 2007.

Michie, D., 1993. Turing's test and conscious thought. Artif. Intell. 60, 1–22 and reprinted in Millican and Clark (Eds.), Machines and Thought. Clarendon Press, Oxford, pp. 27–51.

Millican, P., Clark, A., 1996. Machines and Thought, Oxford University Press.

Millican, P., (forthcoming), Hume's Fork, and his theory of relations, forthcoming in Philosophy and Phenomenological Research.

Nagel, T., 1974. What is it like to be a bat? Philos. Rev. 83, 435–450.

Petzold, C., 2008. The Annotated Turing: A Guided Tour Through Alan Turing's Historic Paper on Computability and the Turing Machine, Wiley.

Preston, J., 2002. Introduction. In: Preston, J., Bishop, M. (Eds.), Views into the Chinese Room, Clarendon Press, 1–50.

Searle, J.R., 1980. Minds, brains, and programs. Behav. Brain Sci. 3, 417–424.

Searle, J.R., 1984. Minds, Brains, and Science, Harvard University Press.

Searle, J.R., 2002. Twenty-one years in the chinese room. In: Preston, J., Bishop, M. (Eds.), Views into the Chinese Room, Clarendon Press, pp. 51–69.

Shanahan, M., Baars, B., 2005. Applying global workspace theory to the frame problem. Cognition 98, 157–176.

Sober, E., 2000. Evolution and the problem of other minds. J. Philos. 97, 365–386.

[26] Note the absence of any assumption here that 'artificial intelligence' must be unitary; indeed the Tutoring test would suggest a domain-relative notion.

[27] For helpful comments on this paper, which have enabled me to improve it significantly, I am very grateful to Tim Bayne, Robin Le Poidevin and Hsueh Qu.

Turing, A.M., 1936. On computable numbers, with an application to the Entscheidungsproblem. Proc. Lond. Math. Soc. Second Ser. 42, 230–265.
Turing, A.M., 1950. Computing machinery and intelligence. Mind 59, 433–460.
Weizenbaum, J., 1966. ELIZA – A computer program for the study of natural language communication between man and machine. Commun. ACM 9, 36–45.
Whitby, B., 1996. The Turing test: AI's biggest blind alley? In: Millican and Clark, pp. 53–62.
Williamson, T., 1994. Vagueness, Routledge.

Luciano Floridi brings out the value of —

THE TURING TEST AND THE METHOD OF LEVELS OF ABSTRACTION

1. Introduction

Among the many lessons that philosophy can learn from Turing, there is one that is fundamental and concerns the use of the method of levels of abstraction (LoA). Turing theorises and uses the method in his paper introducing his famous test (Turing, 1950). Because of this, the importance of the method seems to have been appreciated less than its application to the debate on the feasibility of AI. This is a pity, because the general method is independent of such debate and is actually invaluable as a general way of approaching philosophical questions, whenever we are dealing not only with intelligence but also with any phenomenon that we find intuitive but very hard to define. In this contribution, I wish to highlight its nature and value.

As it is well known, Turing refuses even to try to provide an answer to the question 'can a machine think?' He considers it a problem 'too meaningless to deserve discussion', because it involves vague concepts such as 'machine' and 'thinking' (Turing, 1950). Instead, he suggests replacing it with the Imitation Game, which is exactly more manageable and less demanding. By so doing, he specifies a LoA and asks a new question, which may be summed up thus: 'may one conclude that a machine is thinking, at this Level of Abstraction?' The rules of the game define the conditions of observability (Floridi et al. 2008). If a machine is indistinguishable from the chosen thinking agent at that LoA, then clearly the machine passes the test, at that LoA, and by changing the rules of the game, one changes the LoA and consequently the answer. Note that Turing

1. refuses to provide a universal definition of intelligence;
2. makes a hypothesis based on the common assumption that conversation skills require intelligence;
3. devises a model to evaluate whether a machine is intelligent comparatively, that means at that LoA; and
4. provides a system that is fully controllable. One knows how it works and how it can be modified, so it can be implemented to test other features such as creativity, learning and ethical behaviour.

A strongly constructionist approach grounds not only the design of the test, but also what Turing conjectured as a potentially successful strategy to obtain a machine that would pass the test. In the final section of his 1950 paper, entitled 'Learning Machines', Turing suggests, as a working hypothesis, that a child-machine could learn and gain its own knowledge through educational processes. Then he builds a model, the child-programme, in order to test that hypothesis. This system is controllable, for example through punishment and reward processes. Any hints or results in the process concern only the model and not how a human child learns.

2. Turing's idea of level of abstraction

It seems clear that the method used by Turing is both powerful and flexible. Complex biochemical compounds and abstruse mathematical concepts have at least one thing in common: they may be unintuitive, but once understood they are all definable with total precision, by listing a finite number of necessary and sufficient properties. Mundane phenomena like intelligence, life or mind share the opposite property: one intuitively knows what they are and perhaps could be, and yet there seems to be no way to encase them within the usual planks of necessary and sufficient conditions.

Sometimes the problem is addressed optimistically, as if it were just a matter of further shaping and sharpening whatever necessary and sufficient conditions are required to obtain a *definiens* that is finally watertight. Stretch here, cut there; ultimate agreement is only a matter of time, patience and cleverness. In fact, attempts follow one another without a final identikit ever being nailed to the *definiendum* in question. After a while, one starts suspecting that there might be something wrong with this *ad hoc* approach. Perhaps it is not the Procrustean *definiens* that needs fixing, but the Protean *definiendum*. Sometimes its intrinsic fuzziness is blamed. One cannot define with sufficient accuracy things like life, intelligence, agenthood and mind because they all admit of subtle degrees and continuous changes.

A solution is to give up all together or at best be resigned to being vague and rely on indicative examples. Pessimism follows optimism, but Turing's influential Imitation Game shows that this need not. The fact is that, in the exact discipline of mathematics, for example, definitions are 'parameterised' by generic sets. Such technique provides a method for regulating the right levels of abstraction Indeed abstraction acts as a 'hidden parameter' behind exact definitions, making a crucial difference. Thus, each *definiens* comes pre-formatted by an implicit LoA; it is stabilised, as it were, in order to allow a proper definition. An x is defined or identified as y never absolutely (i.e., LoA-independently), as a Kantian 'thing-in-itself', but always contextually, as a function of a given LoA, whether it be in the realm of Euclidean geometry, quantum physics or commonsensical perception.

When a LoA is sufficiently common, important, dominating or in fact happens to be the very frame that constructs the *definiendum*, it becomes 'transparent' to the user, and one has the pleasant impression that x can be subject to an adequate definition in a sort of conceptual vacuum. Glass is not a solid but a liquid, tomatoes are not vegetables but berries, a banana plant is a kind of grass, and whales are mammals not fish. Unintuitive as such views might be initially, they are all accepted without further complaint because one silently bows to the uncontroversial predominance of the corresponding LoA.

When no LoA is predominant or constitutive, things get messy. In this case, the fundamental lesson from Turing is to avoid fiddling with the *definiens* or blaming the *definiendum*, and to decide instead on an adequate LoA, before embarking on the task of understanding the nature of the *definiendum*.

Turing's analysis of intelligence or 'thinking' behaviour is enlightening. One might define 'intelligence' in a myriad of ways; many LoAs seem equally convincing but no single, absolute, definition is adequate in every context.Turing avoided the problem of 'defining' intelligence by first fixing a LoA – in this case a dialogue conducted by computer interface, with response time taken into account – and then establishing the necessary and sufficient conditions for a computing system to count as intelligent at that LoA: the imitation game. As I argued in Floridi (1995, 2008a), the LoA is crucial and changing it changes the test. An example is provided by the Loebner test (Moor, 2001), the current competitive incarnation of Turing's test. There, the LoA includes a particular format for questions, a mixture of human and non-human players, and precise scoring that takes into account repeated trials. It is precisely the flexibility of the method that makes Turing tests of different kinds still extremely useful.

The idea of a 'level of abstraction' plays an absolutely crucial rôle in the previous account. We have seen that this is so even if the specific LoA is left implicit. For example, whether we perceive oxygen in the environment depends on the LoA at which we are operating; to abstract it is not to overlook its vital importance, but merely to acknowledge its lack of immediate relevance to the current discourse, which *could* always be extended to include oxygen were that desired. The question is: what are these LoAs exactly? Turing did not explicitly theorise them, but his work contains all the necessary ideas to draw a complete picture. I have provided a definition and more detailed analysis in Floridi (2008b, 2011), so here I shall outline only his basic idea.

3. The method

Suppose we join Anne, Ben and Carole in the middle of a conversation. Anne is a collector and potential buyer; Ben tinkers in his spare time; and Carole is an economist. We do not know the object of their conversation, but we are able to hear this much:

> *Anne* observes that it has an anti-theft device installed, is kept garaged when not in use and has had only a single owner;
> *Ben* observes that its engine is not the original one, that its body has been recently re-painted but that all leather parts are very worn;
> *Carole* observes that the old engine consumed too much, that it has a stable market value but that its spare parts are expensive.

The participants view the object under discussion (the 'it' in their conversation) according to their own interests, at their own LoA. We may guess that they are probably talking about a car, or perhaps a motorcycle, but it could be an airplane.Whatever the reference is, it provides the source of information and is called the *system*. A LoA consists of a collection of observables, each with a well-defined possible set of values or outcomes. For the sake of simplicity, let us assume that Anne's LoA matches that of an owner, Ben's that of a mechanic and Carole's that of an insurer. Each LoA makes possible an analysis of the system, the result of which is called a *model* of the system. Evidently an entity may be described at a range of LoAs and so can have a range of models. So a *level of abstraction* or *LoA* is essentially an *interface*, used to analyse a system from a point of view. Models are the outcome of the analysis of a system, developed at some LoA(s). The *Method of Abstraction* consists of formalising the model (see Floridi (2008b)). In the previous example, Anne's LoA might consist of observables for security, method of storage and owner history; Ben's might consist of observables for engine condition, external body condition and internal condition; and Carole's might consist of observables for running cost, market value and maintenance cost. The interface might consist, for the purposes of the discussion, of the set of all three LoAs.

In this case, the LoAs happen to be disjoint, but in general they need not be. A particularly important case is that in which one LoA includes another. Suppose, for example, that Delia joins the discussion and analyses the system using a LoA that includes those of Anne and Ben. Delia's LoA might match that of a buyer. Then Delia's LoA is said to be more concrete, or lower, than Anne's, which is said to be more abstract, or higher; for Anne's LoA abstracts some observables apparent at Delia's.

4. Relativism

A LoA qualifies the level at which an entity or system is considered. In general, it seems that Turing was right when arguing that many uninteresting disagreements might be clarified by the various 'sides' making precise their LoA. Yet a crucial clarification is in order. It must be stressed that a clear indication of the LoA at which a system is being analysed allows pluralism without endorsing relativism. It is a mistake to think that 'anything goes' as long as one makes explicit the LoA, because LoA are mutually comparable and assessable (see Floridi (2008b) for a full defence of this point).

Introducing an explicit reference to the LoA clarifies that the model of a system is a function of the available observables, and that (i) different interfaces may be fairly ranked depending on how well they satisfy modelling specifications (e.g., informativeness, coherence, elegance, explanatory power, consistency with the data etc.) and (ii) different analyses can be fairly compared provided that they share the same LoA.

5. State and state-transitions

Let us agree that an entity is characterised, at a given LoA, by the properties it satisfies at that LoA (Cassirer, 1910). We are interested in systems that change, which means that some of those properties change value. A changing entity, therefore, has its evolution captured, at a given LoA and any instant, by the values of its attributes. Thus, an entity can be thought of as having states, determined by the value of the properties that hold at any instant of its evolution, for then any change in the entity corresponds to a state change and *vice versa*.

This conceptual approach allows us to view any entity as having states. The lower the LoA, the more detailed the observed changes and the greater the number of state components required to capture the change. Each change corresponds to a transition from one state to another. A transition may be non-deterministic. Indeed, it will typically be the case that the LoA under consideration abstracts the observables required to make the transition deterministic. As a result, the transition might lead from a given initial state to one of several possible subsequent states.

According to this view, the entity becomes a transition system. The notion of a 'transition system' provides a convenient means to support the basic criteria for agenthood, for example, being general enough to embrace the usual notions like automaton and process. It is frequently used to model interactive phenomena. We need only the idea; for a formal treatment of much more than we need in this context, the reader might wish to consult Arnold and Plaice (1994) and Colburn and Shute (2007).

A *transition system* comprises a (non-empty) set S of states and a family of operations, called the *transitions* on S. Each transition may take input and may yield output, but at any rate it takes the system from one state to another and in that way forms a (mathematical) relation on S. If the transition does take input or yield output, then it models an interaction between the system and its environment and so is called an *external* transition; otherwise, the transition lies beyond the influence of the environment (at the given LoA) and is called *internal*. It is to be emphasised that input and output are, like state, observed at a given LoA. Thus, the transition that models a system is dependent on the chosen LoA. At a lower LoA, an internal transition may become external; at a higher LoA, an external transition may become internal.

In our example, the object being discussed by Anne might be further qualified by state components for location, whether in-use, whether turned-on, whether the anti-theft device is engaged, history of owners and energy output. The operation of garaging the object might take as input a driver, and have the effect of placing the object in the garage with the engine off and the anti-theft device engaged, leaving the history of owners unchanged, and outputting a certain amount of energy. The 'in-use' state component could non-deterministically take either value, depending on the particular instantiation of the transition. Perhaps the object is not in use, being garaged for the night; or perhaps the driver is listening to a program broadcasted on its radio, in the quiet solitude of the garage. The precise definition depends on the LoA. Alternatively, if speed were observed but time, accelerator position and petrol consumption abstracted, then accelerating to 60 miles per hour would appear as an internal transition.

6. Conclusion

In this paper, I have summarised the method of abstraction as introduced by Turing's test. I have tried to show its principal features and crucial value for any conceptual analysis independently of the debate on AI. The method clarifies implicit assumptions, facilitates comparisons, enhances rigour and hence promotes the resolution of possible conceptual confusions. If carefully applied, the method confers remarkable advantages in terms of consistency and clarity. Too often philosophical debates seem to be caused by a misconception of the LoA at which the questions should be addressed. This is not to say that a simplistic policy of 'on the one hand and on other hand' sort of arguments would represent a panacea. Disagreement is often not based on confusion. But Turing was right in arguing that chances of resolving or overcoming it may be enhanced if one is first of all careful about specifying what sort of observables are at stake.

References

Arnold, A., Plaice, J., 1994. Finite Transition Systems: Semantics of Communicating Systems, Masson, Prentice Hall, Paris, Hemel Hempstead.

Cassirer, E., 1910. Substanzbegriff Und Funktionsbegriff. Untersuchungen Über Die Grundfragen Der Erkenntniskritik, Bruno Cassirer, Berlin. Translated by Swabey, W.M. and Swabey, M.C. in Substance and Function and Einstein's Theory of Relativity (Chicago, IL: Open Court, 1923).

Colburn, T., Shute, G., 2007. Abstraction in computer science. Minds Mach. 17(2), 169–184.

Floridi, L., 1995. Internet: which future for organized knowledge, Frankenstein or Pygmalion? Int. J. Hum. Comput. Stud. 43(2), 261–274.

Floridi, L., 2008a. Artificial intelligence's new frontier: artificial companions and the fourth revolution. Metaphilosophy 39(4/5), 651–655.

Floridi, L., 2008b. The method of levels of abstraction. Minds Mach. 18(3), 303–329.

Floridi, L., 2011. The Philosophy of Information, Oxford University Press, Oxford.

Floridi, L., Taddeo, M., Turilli, M., 2008. Turing's imitation game: still an impossible challenge for all machines and some judges. Minds Mach. 19(1), 145–150.

Moor, J.H., (Ed.), 2001. The Turing test: past, present and future, Minds Mach. 11.1 (special issue).

Turing, A.M., 1950. Computing machinery and intelligence. Minds Mach. 59, 433–460.

Aaron Sloman absolves Turing of —

THE MYTHICAL TURING TEST

1. Introduction

In his 1950 paper, Turing described his famous 'imitation game', defining a test that he thought machines would pass by the end of the century. For useful surveys of views about the test, see Saygin et al. (2000) and Proudfoot (2011). It is often claimed that Turing was proposing a test for intelligence. I think that assumption is mistaken (a) because Turing was far too intelligent to propose a test with so many flaws, (b) because his words indicate that he thought it would be a silly thing to do, and (c) because there is an alternative, much more defensible, reading of his paper as making a technological prediction, whose main function was to provide a unifying framework for discussing and refuting some common arguments against the possibility of intelligent machines.[1]

I shall try to explain (i) why the common interpretation of Turing's paper is mistaken, (ii) why the idea of a test for intelligence in a machine or animal is misguided, and (iii) why a different sort of test, not for a specific machine or animal, but for a genome or generic class of developing systems, would be of greater scientific and philosophical interest. That sort of test was not proposed by Turing, and is very different from the many proposed revisions of Turing's test, since it would require many instances of the design allowed to develop in a variety of environments. to be tested. That would be an experiment in meta-morphogenesis, the topic of my paper in Part IV of this volume.

2. Turing's 1950 paper

Section 1 of the paper states:

> *"I propose to consider the question, 'Can machines think?' This should begin with definitions of the meaning of the terms 'machine' and 'think.' The definitions might be framed so as to reflect so far as possible the normal use of the words, but this attitude is dangerous. If the meaning of the words 'machine' and 'think' are to be found by examining how they are commonly used it is difficult to escape the conclusion that the meaning and the answer to the question, 'Can machines think?' is to be sought in a statistical survey such as a Gallup poll. But this is absurd. ... "*

Instead of this 'absurd' procedure he proposes a game, which he calls 'The imitation game', which he uses to formulate a technological prediction:

> *"I believe that in about fifty years' time it will be possible, to programme computers, with a storage capacity of about 10^9, to make them play the imitation game so well that an average interrogator will not have more than 70 per cent chance of making the right identification*

[1] I have found that many of those who think Turing proposed a test for intelligence, if asked whether they have read the paper, answer 'No'. They simply repeat what others have said. Saygin's and Proudfoot's articles discuss some merits of the test.

after five minutes of questioning. The original question, "Can machines think?" I believe to be too meaningless to deserve discussion. . . . "

The game is not intended to answer the 'too meaningless' question whether machines can think, but enables Turing to formulate his prediction about what can be achieved in 50 years, so that he can discuss several objections to the prediction. Refuting them, by showing that they are all based on unsound arguments, is the main meat of his paper – his way of replacing the 'meaningless' question 'Can machines think?' with a 'relatively unambiguous' question. (We shall see that the question is not as unambiguous as he thought.)

3. Does the test have any value?

Turing's test is far too limited to serve as a criterion for intelligence. Nobody would accept as an employee, or a student, someone whose only known qualification was the ability to fool the 'average' population for five minutes, in 30 per cent of trials. Ability to pass the test is neither sufficient, nor necessary, for being intelligent (or able to think). No engineer would accept 30% success at playing the imitation game for five minutes as either a specification for a worthwhile design or as an acceptance test for a product. Further information would be required, e.g. *how* it managed to do this, under what conditions it succeeded and failed, and whether it used mechanisms that allowed it to overcome its limitations eventually.

Nothing about the test explains *how* a mind can work or what thinking is. No information scientist would accept Turing's prediction as specifying an explanatory mechanism. By 1950, Turing had already made profound contributions to our understanding of mathematical competence. Passing his shallow test provides no evidence for possession of any such competence. (Interrogation by mathematicians rather than average interrogators might!) The ability to pass the test could not drive natural selection since it requires the interrogators to have evolved previously. The vast majority of intelligent animals cannot pass Turing's test. Neither can highly intelligent pre-verbal human toddlers. So ability to pass that test is neither necessary nor sufficient for normal animal or human intelligence.

Intelligence is not some unique set of behavioural capabilities: there are different kinds of intelligence (and thinking) evident in nest-building birds, dolphins, elephants, baboons and human toddlers. In the terminology of Ryle (1949), 'intelligence' is a *polymorphous* concept. Its use can vary systematically according to context.[2]

Though worthless as a test for intelligence or thinking, the imitation game suits Turing's main purpose, namely providing a framework for presenting and refuting a collection of arguments against the possibility of machine intelligence. It has inspired some AI researchers to try to substantiate Turing's prediction, but that has proved difficult. I suspect Turing understood some of the difficulties, unlike some early proponents of AI who rashly predicted the imminent arrival of intelligent machines. Unfortunately, the test has diverted much intellectual effort from a deep study of biological varieties of intelligence and how to model or replicate them.

4. Turing's predictions

Turing was remarkably accurate about the number of bits available in a computer's memory by the turn of the century. His caution in formulating the test (requiring only 30% of testers to be fooled for only 5 minutes) has been justified by the failures of machines to pass the test so far (though they have come close). The failure seems to be a matter of scale rather than the problems of principle

[2] Compare what computer scientists call 'parametric polymorphism'.

that he discussed in his paper. A machine with a sufficiently large and varied collection of stored patterns could obviously pass the test. That's one of the problems with any time-limited behavioural test for intelligence.

Another problem with the test is its dependence on 'average' interrogators. I suspect that by about the year 2000 it actually was possible to fool close to 30% of the 'average' world-wide human population (excluding computer experts and those who had encountered the idea of a Turing test), for up to five minutes. But the 'average interrogator' has changed since then, in ways that Turing did not allow for. Computing technology has continued to advance since 2000, and computers are now doing much cleverer things, while increasing numbers of humans have been learning about what computers can and cannot do, through frequent use, news reports, internet discussions, and so on, making it harder to fool 'average' testers into thinking they are interacting with a human! Many more humans are now able to choose things to say to a machine that may reveal its inability to respond like a human, and the proportion is likely to increase. This relativity to cultural attainment of testers is one of the reasons why the test is so bad as a test of intelligence.

Unfortunately, the misinterpretation of Turing as proposing a test for thinking or for intelligence is so wide-spread that it has led to huge amounts of wasted effort, wasted, because, as Turing himself pointed out, the notion of such a test is based on a question which is 'too meaningless to deserve discussion'.[3] None of this diminishes the value of Turing's main purpose: presenting and demolishing arguments purporting to show that machines cannot successfully play the imitation game.

5. Turing's error about human-like learning

Turing did make one serious error in that paper. In his Section 7, 'Learning Machines', he wrote

> *"Instead of trying to produce a programme to simulate the adult mind, why not rather try to produce one which simulates the child's? If this were then subjected to an appropriate course of education one would obtain the adult brain. Presumably the child brain is something like a notebook as one buys it from the stationer's. Rather little mechanism, and lots of blank sheets. (Mechanism and writing are from our point of view almost synonymous.) Our hope is that there is so little mechanism in the child brain that something like it can be easily programmed. The amount of work in the education we can assume, as a first approximation, to be much the same as for the human child. ... "*

Turing, like many AI researchers studying learning machines, grossly underestimated the contribution of biological evolution to the processes of human learning. As John McCarthy put it

> *"Evolution solved a different problem than that of starting a baby with no a priori assumptions. ... Animal behavior, including human intelligence, evolved to survive and succeed in this complex, partially observable and very slightly controllable world. The main features of this world have existed for several billion years and should not have to be learned anew by each person or animal. In order to understand how a well-designed baby should work, we need to understand what the world is like at the gross level of human interaction with the environment."*

What McCarthy did not point out is that the specification is different for different animals, and different types of machine. Jackie Chappell and I have indicated ways in which such diversity can emerge.[4] This requires much more research in meta-morphogenesis.

[3] A related question on which there has been much futile discussion is whether machines can be conscious. I have attempted to write a tutorial introduction to some of the issues and ways of making progress in Sloman (2010b).

[4] See Sloman and Chappell (2005), Chappell and Sloman (2007) and Sloman (2008).

6. Dichotomies and continua

There is a very common mistake, implicitly made by most who ask: 'Can machines be intelligent?', or 'Can machines think?', namely, assuming that there is a *dichotomy* (a binary distinction) in a complex space where things are very varied. This is as mistaken as assuming there is a binary divide between things that are and things that are not efficient, useful, dangerous, or reliable. Sometimes people who realise that the assumption is mistaken, instead refer to differences of *degree*, e.g. by suggesting that there are differences in *degrees* of intelligence, consciousness, etc. That view makes two related mistakes: (a) assuming that there is a total ordering of cases, as if, for example, species could be put into a linear ordering of animals with more or less intelligence (or consciousness, etc.), and (b) assuming that there is *continuous* variation in kinds of intelligence (etc.).

An example of the first sort of mistake could also be made by a child who finds that some containers can be put inside other containers (e.g. a little box inside a bigger box) and draws the conclusion that all containers form a linear ordering, so that given any two containers Ca and Cb, either Ca can be put inside Cb, or Cb can be put inside Ca, or they are exactly the same size, because their bounding surfaces are the same shape and size. That's obviously false, because one box may have a square cross section and the other circular, with the diameter of the circle larger than the side of the square and smaller than the diagonal of the square.

There can be many different sorts of competence that are 'dimensions' of intelligence, and different individuals or different species may excel in different dimensions. E.g. one may be very good at designing furniture and terrible at proving theorems in geometry, while the other is good at mathematics and poor at designing furniture. Variations in ability to perceive, learn about, act in, plan in and survive in various types of environment are wide-spread among organisms. Neither species nor individual organisms can be arranged in a linear sequence of increasing intelligence.

The assumption of continuous variation, required for differences of degree, is also false. There are some kinds of knowledge or competence that cannot vary continuously. E.g. learning about arithmetic, or geometry, or grammar, involves learning distinct items. Competences that are expressed in rules don't have intermediate cases using half a rule, quarter of a rule, etc. Moreover, since genetic makeup is ultimately implemented in chemistry, and since molecules differ discontinuously (e.g. by addition or removal of atoms) it is impossible for species to vary continuously in their genetic makeup (though small differences are possible).

It follows from the above that the very idea of a Turing test or any other test for intelligence is muddled if there is no binary divide between things that are and things that are not intelligent, only a vast variety of cases. Attempting to replace a binary classification with a *measure* of intelligence is similarly mistaken in assuming that there is a total ordering of types of intelligence and, possibly also in assuming that there is continuous variation so that degrees of intelligence can be represented by real numbers.

If there is no total ordering, only a complex space of combinations of competences (just as there is a complex space of combinations of atoms, making the notion of ordering molecules along a line of increasing 'chemicality' (?) misguided), then, instead of using a number or label for degree or amount of intelligence, we need to find ways of *describing* types of intelligence in terms of the combinations of competences that they include, just as we describe chemical molecules in terms of the different combinations of atoms and chemical bonds and also the different properties that result from those structures (e.g. acidity, alkalinity and many more used in drug design). It may be more useful to search for a *grammar* for types of intelligence than a measure. A grammar for types of intelligence might be a specification of varieties of combinations of competences of many sorts that could be implemented in a unified working architecture, and could include combinations that could grow themselves, in ways that depend on their interactions with the environment.

7. What sort of test would be worth while?

If we are to propose tests of the general sort that people take the Turing Test to be, namely a test for something being intelligent, or human-like, we'll need to distinguish testing *a particular individual* from testing *a theory about a type of design* for working systems.

It is clear that there are many very different human beings and also that they all share a large collection of common features. We really should be testing a theory about what's common, where the differences come from, and what the implications are. Compare: if a theory about the weather is able to explain only how a particular tornado works and no other weather phenomena, then it cannot be a good explanation of the particular case. Likewise a theory of what's going on when oil burns that says nothing wood burning, or coal burning, or gas burning cannot be a good theory about oil burning.

Or suppose someone claims to have a theory about how to solve algebraic equations and an online computer test that demonstrates the theory. The implemented algorithm solves only quadratic equations: give it any quadratic equation and it will produce the solutions. Would that be taken as a good test for a theory of equation solving? We would rightly demand something more *generic*.

What would the required sort of *generic* theory of intelligence look like? The closest answer I can give is something like a parametrised specification for a highly polymorphic design for a working system, which can be given different parameters to produce instances of the design, where the instances will be very different in the way that, for example, humans in different cultures, or who talk different languages, or who grow up to have very different competences and interests are different, and yet be as similar as different humans are, a requirement that is very complex and very demanding, and not yet specifiable in detail since we don't yet know enough about what typical humans are like (e.g. how their vision systems work, how they learn, what mechanisms are involved in their motivational and other affective states and processes.)

The parameters, instead of all being supplied at the time the instance is created, would have to be picked up at various times during the development and testing of the instance.

In particular, in order to really understand human intelligence, we should be able to specify a *type* of system, with many different instances, differing as much as humans in different physical and social environments do. For example, as a result of educational and environmental influences, and some individual personality features, instances of the machine would 'grow up' to be philosophers with very different views, including views on what machines can or cannot do, e.g. some becoming like Alan Turing, others like John Searle, or Tom Nagel, or David Chalmers, or Dan Dennett, and perhaps even some like me, since I disagree with all the others![5]

References

Chappell, J., Sloman, A., 2007. Natural and artificial meta-configured altricial information-processing systems. Int. J. Unconventional Computing, 3(3), 211–239.
http://www.cs.bham.ac.uk/research/projects/cosy/papers#tr0609.

McCarthy, J., 2008. The well-designed child. Artificial Intelligence, 172(18), 2003–2014.
http://www-formal.stanford.edu/jmc/child.html.

Proudfoot, D., 2011. Anthropomorphism and AI: Turing's much misunderstood imitation game. Artificial Intelligence, 175(5–6), 950–957. Available from doi:10.1016/j.artint.2011.02.002.

Ryle, G., 1949. The Concept of Mind, Hutchinson, London.

Saygin, A.P., Cicekli, I., Akman, V., 2000. Turing test: 50 years later. Minds and Machines, 10(4), 463–518.
http://crl.ucsd.edu/ saygin/papers/MMTT.pdf.

[5] For more on this see Sloman (2010a,c).

Sloman, A., 2010a. An alternative to working on machine consciousness. Int. J. Mach. Consciousness, 2 (1), 1–18. http://www.cs.bham.ac.uk/research/projects/cogaff/09.html#910.
Sloman, A., 2010b. Phenomenal and access consciousness and the "hard" problem: a view from the designer stance. Int. J. Mach. Consciousness, 2 (1), 117–169.
http://www.cs.bham.ac.uk/research/projects/cogaff/09.html#906.
Sloman, A., 2010c. How virtual machinery can bridge the "explanatory gap" In: Doncieux, S., et al., (Eds.), Natural and Artificial Systems. Proceedings SAB 2010, LNAI 6226, Springer, Heidelberg, pp. 13–24. http://www.cs.bham.ac.uk/research/projects/cogaff/10.html#sab.
Sloman, A., Chappell, J., 2005. The altricial-precocial spectrum for robots. In: Proceedings IJCAI'05, IJCAI, Edinburgh, pp. 1187–1192. http://www.cs.bham.ac.uk/research/cogaff/05.html#200502.
Sloman, A., 2008. Evolution of minds and languages. What evolved first and develops first in children: Languages for communicating, or languages for thinking (Generalised Languages: GLs)?.
http://www.cs.bham.ac.uk/research/projects/cosy/papers/#pr0702.
Turing, A.M., 1950. Computing machinery and intelligence. Mind 59, 433–460. (reprinted in E.A. Feigenbaum and J. Feldman (Eds.), Computers and Thought McGraw-Hill, New York, 1963, pp. 11–35).

David Harel proposes —

A Turing-Like Test for Modelling Nature*

In 1950, Alan Turing proposed an 'imitation game', for determining whether a computer is intelligent. A human interrogator, the candidate computer and another human are put in separate rooms, connected electronically. Alice, the interrogator, doesn't know which is the human and which is the computer, and has a fixed amount of time to determine their correct identities by addressing questions to them. The computer has to make its best effort to deceive Alice, giving the impression of being human, and is said to pass the Turing test if after the allotted time Alice doesn't know which is which. Succeeding by guessing is avoided by administering the test several times. Here, I argue that a variant of this idea, but with a Popperian twist (Popper, 1959), is applicable to the computerised modelling of natural systems, particularly in systems biology.

1. The grand challenge

Many characteristics of man-made systems, especially those termed *reactive* by computer scientists, are central to the dynamics of biological systems too: heavy concurrency (simultaneity), event-driven and time-dependent behavior, cause-effect phenomena and distributed control. These occur from the atomic level, via the molecular and the intra- and inter-cellular levels, to full organisms and even entire populations, suggesting that biological systems can be modelled (that is, reverse-engineered, simulated and analyzed) using methods and tools developed for the engineering of complex man-made systems. Recent results from small-scale modelling efforts have been extremely encouraging, Harel (2003, 2005); Kam et al. (2003); Kugler et al. (2008); Popper (1959); Priami et al. (2001); Setty et al. (2008).

*This paper is a slightly expanded version of Harel (2005) and is published here with permission.

Most modelling efforts are partial, intended to capture some limited phenomena or mechanism, or a small part of a larger system. There is often a particular goal for the modelling, with specific laboratory observations relevant to it and specific behaviors for simulation and checking. The motivation for such modelling efforts is making carefully considered predictions, which would typically be driven by certain questions that the modeller has in mind. These predictions stemming from the model and its analysis then lead to particular experiments that are to be carried out in the laboratory, and thus to the discovery of new scientific facts. A different approach, far more ambitious in its scope and required work, is to aim at modelling a complete biological system, such as an organ or a full multi-cellular animal. This kind of effort is not motivated by specific questions but, rather, by the desire to understand an entire system. (See Harel (2003), in which I propose to model the *Caenorhabditis elegans* nematode.)

Such a comprehensive 'grand challenge' is extremely nontrivial and by its very nature is intended to take modelling to the limit: let's model (reverse-engineer) a worm or an elephant similarly to the way we engineer a chemical plant or an F-15 fighter jet. The challenge is to construct a realistic model, true to all known facts, which is smoothly extendable as the facts are discovered. It would feature a three-dimensional, animated graphical front end and would enable interactive multilevel probe-able simulations of the animal's development and behavior. The underlying computational framework would be mathematically rigorous but would also be intuitive enough for biologists to enter newly discovered data themselves. The model would also support alternative theses reflecting disagreement among the scientists, to observe and to compare their effects at run time.

As I have argued previously (Harel, 2003), achieving this goal, even for a very modest organism like *C. elegans*, would require enormous amounts of interdisciplinary work, both in the computational an analysis realms and in the accumulation, assimilation and formalization of the biological data itself. It raises numerous difficulties, for some of which no solutions are known at present. The good news is that this is typical of many grand challenges, both past and future; like putting a man on the moon, proving Fermat's last theorem or solving cancer. One of the characteristics of a 'good' long-term challenge is that, if successful, the results would be spectacular, but even if it is not completely successful, many fundamental and useful results will have been achieved along the way. In our case, a comprehensive *in silico* model of an entire living organism would constitute an unprecedented tool, allowing researchers to see and understand life in ways not otherwise possible, triggering and helping predict behavior, filling gaps and correcting errors, suggesting hitherto unimagined experiments and much more.

It is not my intention here to try to convince the reader of the virtues of such an effort, but many benefits can easily be imagined.

2. Measuring success

Still, what does being successful mean? How do we know when we are done, labelling the model valid? And are we ever done? It is one thing to build a computerised model that looks good and captures some desired, but limited, features of a biological system in order to provide answers to some specific research questions, but quite another thing to claim to have constructed a valid model of a full organism, using all that is known about it. In limited modelling, one has in mind a manageable set of laboratory observations – analogous to requirement sin engineering man-made systems – so that one essentially knows what to test for. The challenge in comprehensive modelling is so great and the required multi-levelled amounts of detail and their inter-combinations so vast, that it is no longer clear how to tell when one is done or what it means for one's model to be valid.

To address this question, we must clarify what we mean by modelling a creature based on all that is known. We must decide upon the model's levels of detail, so that we don't find ourselves having to deal with quarks or quantum effects. Moreover, we cannot hope to ever find out everything there is to know about a complex organism, even after limiting the levels of discourse. A model enabling computerised simulations can at best be based on the knowledge available at a given point in time and will have to take liberties with what we don't know yet or never will. For example, the model can be made to fill in certain dynamics by carrying out the biological equivalent of the movie industry's 'morphing' technique. In any case, this begs the question of how to tell when all that we do know has indeed been captured.

Here is where validating the model can be likened to the Turing test, but with a Popperian twist (Popper, 1959): a comprehensive model of a full organism will be deemed valid/complete/adequate if it cannot be distinguished from the real thing by an appropriate team of interrogators. This idea raises many subtle questions and may attract opposition on many grounds, which this short essay will not attempt to address. The reader is referred to Turing's original paper which discusses several issues that are relevant here, too.

3. Modifications to Turing's test

If we were to apply the idea in Turing's paper to validate models of natural systems, particularly biological ones, what types of modifications to the original test would we have to implement? First, to prevent us from using our senses to tell human from computer, Turing recommended employing separate rooms and electronic means for communication. In our version of the test, tailored for modelling a multi-cellular organism, we are not simulating communicable intelligence but development and behavior. Consequently, our 'protection buffer' will have to be quite more complex – intelligent, in fact! It would have to limit the interrogation (which would consist of probing the computerised model residing in one room and the actual *in vivo* or *in vitro* laboratory for the organism at hand that resides in the other) to be purely behavioural and to incorporate means for disguising the fact that the model is not an actual living entity. These would have to include neutral communication methods and similar-looking front-ends, as in Turing's original test, but also means for emulating the limitations of actual experimentation. A query requiring three weeks in a laboratory on the real thing would have to elicit a similarly realistic delay from the simulating model. Moreover, queries that cannot be addressed for real at all must be let unanswered by the model, too, even though the main reason for building models in the first place is to generate predictive and work-provoking responses even to those.

Second, our test is perpetually dynamic, in the good spirit of Popper's philosophy of science (Popper, 1959). A computer passing Turing's original test can be labelled intelligent once and for all because, even if we take into account the variability of intelligence among average humans, we don't expect the nature and scope of intelligence to change much over the years. In contrast, a model of a worm or a fly or an elephant that passes our version of the test can only be certified valid or complete for the present time. New research will repeatedly refute that completeness, and the model will have to be continuously strengthened to keep up with the advancement of science. The exciting point here is that once a model passes the test, that model can be viewed as a correct theory of the organism being modelled. In a sense it actually *is* the organism, unless and until it can be distinguished from the real thing by appropriately clever probes and the corresponding laboratory work, which is exactly what Popper is trying to teach us: the important work of experimental scientists is to refute accepted theories by increasingly sophisticated experimentation. And this is how science advances and knowledge increases. (By the way, the protection buffer required for our test will also have to change as advances are made in laboratory technology, but, interestingly, it will have to be made weaker, since probing the model and probing the real thing will become closer to each other.)

Third, our interrogators can't simply be any humans of average intelligence. Both they, and the buffer people responsible for 'running' the real organism in the room that contains the laboratory and providing its responses to probes, would have to be experts on the subject matter of the model, appropriately knowledgeable about its coverage and levels of detail. In the *C. elegans* proposal, for example, these would have to be knowledgeable members of the worm research community.

Clearly, the modified Turing test proposed here is not without its problems and is not put forward for immediate consideration in practice. Still, it could serve as an ultimate kind of certification for the success of what appears to be a worthy long-term research effort. Variations of the idea are also applicable to efforts aimed at modelling and simulating other kinds of natural systems.

References

Ciobanu, G., Rozenberg, G., (Eds.), 2004. Modeling in Molecular Biology, Springer-Verlag, Berlin.
Efroni, S., Harel, D., Cohen, I.R., 2003. Towards rigorous comprehension of biological complexity: modeling, execution and visualization of thymic T cell maturation. Genome Res. 13, 2485–2497.
Fisher, N.P., Hubbard, E.J.A., Stern, M.J., Harel, D., 2005. Computational insights into C. elegans vulval development. Proc. Natl. Acad. Sci. 102, 6, 1951–1956.
Harel, D., 2003. A grand challenge for computing: full reactive modeling of a multi-cellular animal. Bull. Eur. Assoc. Theor. Comput. Sci. 81, 226–235.
Harel, D., 2005. A Turing-like test for biological modeling. Nature Biotechnol. 23, 495–496.
Kam, N., Harel, D., Kugler, H., Marelly, R., Pnueli, A., Hubbard, E.J.A., et al., 2003. Formal Modeling of C. elegans Development: A Scenario-Based Approach, Proceedings of the 1st International. Workshop on Computational Methods in Systems Biology, Lecture Notes in Computer Science, vol. 2602, Springer-Verlag, Berlin, pp. 4–20.
Kugler, N.K.,, Marelly, R., Appleby, L., Fisher, J., Pnueli, A., Harel, D., et al., 2008. A scenario-based approach to modeling development: a prototype model of C. elegans vulval cell fate specification. Dev. Biol. 323, 1–5.
Popper, K.R., 1959. The Logic of Scientific Discovery, Hutchinson, London.
Priami, C., Regev, A., Silverman, W., Shapiro, E., 2001. Application of stochastic process algebras to bioinformatics of molecular processes. Inf. Process. Lett. 80, 25–31.
Setty, Y., Cohen, I.R., Dor, Y., Harel, D., 2008. Four-dimensional realistic modeling of pancreatic organogenesis. Proc. Natl. Acad. Sci. 105, 51, 20374–20379.
Turing, A.M., 1950. Computing machinery and intelligence. Mind 59, 433–460.

Huma Shah engages with the realities of —

Conversation, Deception and Intelligence: Turing's Question-Answer Game

Describing two ways to practicalise his question-answer game to examine machine thinking in 1950, Turing believed one day a machine would succeed in providing satisfactory and sustained answers to any questions. In 2011 IBM Watson achieved success competing against two human champions in a televised general knowledge quiz show. Though he regarded the process of thinking as mysterious, Turing believed building a machine to think might help us to understand how it is we humans think.

1. Introduction

In order to fully understand Turing's plans for constructing and examining a thinking machine, through text-based communication with an interrogator unfamiliar with the machine's inner workings, it is necessary to consider the ideas from his papers and lectures before and after 'Computing Machinery and Intelligence' (Turing, 1950). In this contribution, Turing's question-answer imitation game emerges from his scholarship, via his two methods to practicalise it: (a) a vive voce, direct interrogation of a machine by a human (see Fig. 1), and (b) a simultaneous comparison of the machine with a human (see Fig. 2), both unseen and unheard to an interrogator (Shah, 2010). Turing's 2-interlocutor and 3-participant tests are contrasted (see Table 1) and Turing's evolving predictions are analysed (see Table 2). Turing's 1952 statement 'at least 100 years' (Turing, 1952) shows, although he had grasped the difficulty of the task to build a conversational machine conveying human-like intelligence, he firmly believed a thinking machine would emerge in the 21st century. In 2011, IBM's Watson, a question-answer supercomputer using a statistical analysis approach to associating information, overcame two human general knowledge experts, an achievement Turing advocated when he wrote that a thinking machine could be built if there was not something else important keeping engineers from this venture. Had he been alive, Turing may well have asked: Was not IBM Watson thinking when it provided its real-time answers to clues in natural language during the *Final Jeopardy!* quiz in February 2011, albeit differently from its two human competitors, Ken Jennings and Brad Rutter?

Fig. 1: Turing's viva voce test.

2. Turing's early imitation game

In 1947, in a lecture on the automatic computing engine (ACE) to the London Mathematical Society (Turing, 1947), Turing acquainted attendees with the idea of intelligent machines that could learn from experience and compete against humans in a game of chess. This was the first time Turing introduced the possibility of a machine vs. human encounter in a restricted domain.

In 1948, Turing added indistinguishability by introducing a third player to his 1947 chess game. In the 3-participant chess play off, the machine is to be distinguished from a human poor chess player. In his 1948 report, 'Intelligent Machinery' (Turing, 1948) Turing set the scene on how to calibrate a machine's ability to think: 'I propose to investigate the question as to whether it is possible for machinery to show intelligent behaviour' (p. 410). Turing warned that 'the idea of 'intelligence' is itself emotional rather than mathematical' (p. 411), and that 'the extent to which we regard something as behaving in an intelligent manner is determined as much by our own state of mind and training as by the properties of the object under consideration' (p. 431). With this caveat in place, Turing attempted to mitigate subjective attribution of intelligence to others.

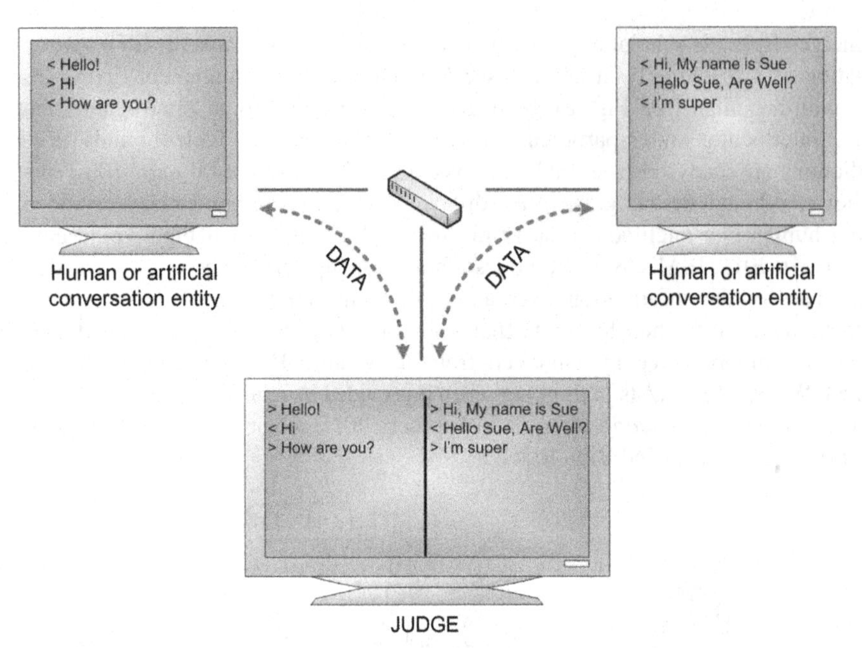

Fig. 2: Turing's simultaneous comparison test.

Table 1 Practicalising Turing's Imitation Game

Comparison of Turing's two Question-Answer Tests		
TIG Feature	Viva voce	Simultaneous-comparison
Mode of questioning	One-to-one: human interrogator-machine	One-to-two: human interrogator – machine + human
Type of questions	Unrestricted	Unrestricted
Number of participants	Two	Three
Duration of Interaction	Unspecified	After five minutes
Interrogator Type	Non-machine expert	Average judge
Number of Interrogators	Jury	Unspecified
Number of Tests	Judge quite a number of times	Unspecified
Language for communication (e.g., English)	Same for both interlocutors	Same for all three participants
Criteria for Test Pass: satisfactory and sustained answers	Considerable portion of jury taken in by pretence.	'average interrogator will not have more than 70 per cent chance of making the right identification'

Turing described his chess indistinguishability test claiming he had actually conducted 'a rather idealized form' (p. 431). Turing introduced a 'little experiment' involving three members *A*, *B*, and *C* playing chess. Turing declared, even at that early period in modern computing history, it was not difficult to produce a 'paper machine' playing a 'not very bad game of chess' (ibid.). By paper machine, Turing meant an 'effect of a computing machine' produced by a written down 'set of rules of procedures' which a man, 'provided with paper, pencil and rubber, and subject to strict discipline' could follow (p. 416). Turing suggested *B*, the operator of the paper machine, be a mathematician and chess player in order to work the machine 'fairly fast' (p. 431), and *A* and *C* be 'poor chess players' (ibid.). Turing envisaged the game played across two rooms between *C* and either *A*, or the paper machine operated by *B*, with 'some arrangement for communicating moves' (ibid.). Turing felt it might not be easy for C to say whether they were playing *A*, or the paper machine: 'C may find it difficult to tell which he is playing' (ibid.).

In considering what could be done by a 'brain without a body', Turing listed chess, learning and translation of languages, cryptography and mathematics. We can see how Turing's ideas were evolving, and why he replaced chess with a language test for the machine to display thinking, because he wrote 'the learning of languages would be the most impressive, since it is the most human of these activities' (p. 421). Turing felt that intellectual activity consisted of various kinds of search. Thus did Turing pave the way for his most famous examination of machine thinking conducted through question-answer sessions using text-based interaction.

Turing's Thinking Machine Predictions		
Year	Where prediction made	Prediction
1950	Computing machinery and intelligence	'in about fifty year's time it will be possible to programme computers, with a storage capacity of about 10^9, to make them play the imitation game so well that an average interrogator will not have more than a 70 per cent. chance of making the right identification after five minutes of questioning'
1951b	BBC radio broadcast: Can digital computers think?	'... at the end of the century it will be possible to programme a machine to answer questions in such a way that it will be extremely difficult to guess whether the answers are being given by a man or by the machine. I am imagining something like a viva voce examination' (in Copeland, 2004, p. 484)
1951b	BBC radio broadcast: Can digital computers think?	'If it (thinking machine) comes at all it will almost certainly be within the next millennium' (p. 486)
1952	BBC radio broadcast: Can automatic calculating machines be said to think?	'Oh yes, at least 100 years, I should say' [for machine to stand a chance in a no questions barred viva voce] (p. 495)

Table 2: Turing's Evolving Predictions

3. Turing's imitation game 1950

Following the Second World War, Turing drew readers into the uncommon idea of non-human machines as 'thinking entities' participating in a simple, text-based interview to examine indistinguishability between human and machine (Turing, 1950). This was in a time when he was obliged by official secrecy not to reveal the extent to which machines had assisted with work involving deception and intelligence during WWII battles, and in which women who, under the age of 30 less than a quarter of a century earlier were awarded the vote in 1928 on equal terms with men, assisted as messengers and typists to the mainly male code breakers, including Turing at Bletchley Park. In this context, Turing began his 'reader awakening' and prescribed an open mind to the possibility of what could be achieved by future machines. Beginning with a game of 'guess the sex of two hidden human interlocutors', Turing explained how a machine, inculcated with necessary memory, storage capacity, speed and the right programme would possess the ability to replace one of the humans in the guessing game to imitate appropriate answers to any question put by an interrogator stripped of visual and auditory interaction. Turing proposed a question-answer imitation game (TIG): the machine has 5 minutes to answer with sustained, satisfactory responses to any questions using the text-based communicative medium with a human interrogator who is not familiar with the inner workings of the machine (Shah, 2010). In his 1950 Mind article Turing suggested two ways to practicalise his imitation game: (a) one interrogator/one machine-witness viva voce (see Fig. 1), and (b) a three-participant, comparison of a machine with a human both simultaneously questioned by a human interrogator (see Fig. 2).

Turing's two practicalisations for his imitation game are given in Table 1. The essential difference is that in the viva voce test (VVTT), the machine is directly questioned by an interrogator, while in the other, simultaneous comparison (SCTT), the machine is compared with a human foil, both questioned in parallel by a human interrogator. In both tests the interrogator can ask anything: 'the question and answer method seems to be suitable for introducing almost any one of the fields of human endeavour that we wish to include' (p. 435). In the case of the comparison test, the human foil must act human while the machine must put up a satisfactory pretence with appropriate responses to questions. Turing's indistinguishability test allows the interrogator to return a judgement of 'two humans' after questioning a machine and human simultaneously (Shah, 2010). If the machine were to achieve this, i.e., deceive the interrogator into classifying it as a human, Turing asked: 'May not machines carry out something which ought to be described as thinking but which is very different from what a man does?' (p. 435). Turing predicted 'in about fifty year's time it will be possible to programme computers, with a storage capacity of about 10^9, to make them play the imitation game so well that an average interrogator will not have more than a 70 per cent chance of making the right identification after five minutes of questioning' (p. 442). However, this prediction evolved with Turing realising that it was likely to take a lot longer before a machine could succeed in his imitation game.

4. Turing's later imitation game scholarship

Elaborating upon his ideas during BBC radio discussions in 1951 and 1952 (Turing, 1951a,b, 1952), Turing described thinking as a 'sort of buzzing' in his head (p. 494), believing a 'considerable portion of a jury' (p. 495) of non-expert interrogators could be taken in by the imitating machine. Albrechtsen et al. (2009) showed experts are not necessarily better in their study of 'experienced police investigators' who were found to be no better than lay individuals at deception-detection, in fact they were 'more likely to judge statements as deceptive' (p. 1055). Turing believed that in due course, when developers did not have 'something better to do', they could build a machine to think (p. 569).

Turing discussed machine ability and revised his earlier 1950 prediction of 'in about fifty years time' to 'most certainly within the next millennium' (p. 486) and 'at least a 100 years' (p. 495). Table 2 presents Turing's evolving predictions. In the year before his death, in the 1953 essay on how to build a machine to think (Turing, 1953), Turing suggested chess as a good starting point. Turing said: '[the game] holds special interest of the representation of human knowledge in machines' (p. 562). Turing described how to deal with the question of whether a machine could be built to play by considering word meanings in sentences. This led him to put forward sub-questions that could be considered separately:

(i) Could machines be programmed to follow the legal moves according to the rules of the game?
(ii) Could machines solve chess problems given the board location of its pieces?
(iii) Could machines play reasonably well in the game?
(iv) Could the machine improve its play and profit from its experience?

As Turing believed it would, a machine did learn to play and beat humans at chess.

5. Realising Turing's future

In 1997 a computer achieved what Turing had years earlier felt a machine could, and what Simon and Newell predicted it would (Turing, 1953, p. 563). IBM's Deep Blue supercomputer beat Gary Kasparov, the then World Chess Champion, by 3.5 to 2.5 games in May 1997, a technology that was put to use in designing cars, forecasting the weather, and developing innovative drug therapies (IBM, 1997). In February 2011, following four years of research, IBM's Watson question-answer technology used natural language to sparr with two human champions during *Final Jeopardy!*, a televised man vs. machine general knowledge quiz show. With contributions from Carnegie Mellon, MIT, and Rensselaer Polytechnic Institute, among other universities, IBM Watson's reverse fact-finder showed it was possible for a machine to find its way through the format of a quiz show with its clues in the form of puzzles, riddles and puns to beat two human champions (Baker, 2011; IBM, 2011; Jennings, 2011; Shah, 2010).

IBM Watson did err, for instance remaining silent about its IBM ancestry when presented with the clue 'Garry Kasparov wrote the foreword for *The Complete Hedgehog*, about a defense in this game' (Baker, 2011, p. 246) rather than answer with what it displayed on its screen as a 96 percent confidence in its first response: 'What is chess?' (ibid.). Turing had warned in 1950 that 'We too often give wrong answers to questions ourselves to be justified in being very pleased on such evidence of fallibility on the part of the machines . . . our superiority can only be felt on such an occasion in relation to the one machine over which we have scored a petty triumph. There would be no question of triumphing simultaneously over all machines' (p. 445). Answering on topics such as the 'Beatles Jude, the swimmer Michael Phelps, the monster Grendel in Beowulf, the 1908 London Olympics, and the boundaries of black holes' (Baker, 2011, p. 239), Watson showed off its strength through its massive store of knowledge in responding to clues such as 'The chapels at Pembroke and Emmanuel Colleges were designed by this architect' with the correct question 'Who is Sir Christopher Wren?' (p. 245). Ken Jennings, one of two human contestants of IBM Watson in the 2011 *Final Jeopardy!* quiz wrote: 'there's no shame losing to silicon ... I don't have 2,880 processors and 15 terabytes of reference works' (Jennings, 2011, p. 2). IBM Watson's developers believe the technology has the potential to 'transform how computers can help people accomplish tasks' (IBM, 2011, p. 5), for example, searching and analysing unstructured data to assist with the diagnosis of diseases, parse legal documents, and drive progress (Baker, 2011).

The future could see question-answer systems weaved into artificial conversational systems combined with robot engineering, to assist humanity facing catastrophes as a result of natural disasters, as with Japan's Fukushima nuclear reactor following the March 2011 earthquake. In the spirit of Alan Turing, the difficulty of the tasks before us must not deter us from innovating.

Acknowledgements

The author thanks HarshM(Figure 1) and Christopher David Chapman (Figure 2). She also thanks Chris Bowden and Prasad Joshi (Latex expertise) and H. Shah (Tables).

References

Albrechtsen, J.S., Meissner, C.A., Susa, K.J., 2009. Can intuition improve deception detection performance? J. Exp. Soc. Psychol. 45, 1052–1055

Baker, S., 2011. Final Jeopardy: Man vs. Machine and the Quest to Know Everything, Houghton Mifflin Harcourt, New York.

IBM, 1997. IBM Research: Kasparov vs. Deep Blue – the Rematch http://www.research.ibm.com/deepblue/ accessed: 25.4.11: time: 20:00.

IBM, 2011. White paper: Watson a system designed to Answer http://public.dhe.ibm.com/ accessed: 4.4.11; time: 15.26.

Jennings, K., 2011. My Puny Human Brain. Slate http://www.slate.com/id/2284721/ accessed 21.4.11; time: 20.54.
Shah, H., 2010. Deception-detection and Machine Intelligence in Practical Turing Tests. PhD thesis, The University of Reading.
Turing, A.M., 1947. Automatic computing engine. In: Copeland, B.J., (Ed.). The Essential Turing – The Ideas that Gave Birth to the Computer Age, Clarendon Press, Oxford, 2004, pp. 378–394.
Turing, A.M., 1948. Intelligent machinery. In: Copeland, B.J. (Ed.), The Essential Turing – The Ideas that Gave Birth to the Computer Age, Clarendon Press, Oxford, 2004, pp. 410–432.
Turing, A.M., 1950. Computing machinery and intelligence. Mind 59 (236) 433–460.
Turing, A.M., 1951a. Intelligent machinery, a heretical theory. In: Copeland, B.J., (Ed.), The Essential Turing: The Ideas that Gave Birth to the Computer Age, Clarendon Press, Oxford, UK, 2004, pp. 472–475.
Turing, A.M., 1951b. Can digital computers think? lecture. In: Copeland, B.J. (Ed.), The Essential Turing: The Ideas that Gave Birth to the Computer Age, Clarendon Press, Oxford, UK, 2004, pp. 482–486.
Turing, A.M., 1952. Can automatic calculating machines be said to think? transcript of BBC radio broadcast. In: Copeland, B.J. (Ed.), The Essential Turing: The Ideas that Gave Birth to the Computer Age, Clarendon Press, Oxford, UK, 2004, pp. 494–506.
Turing, A.M., 1953. Chess. In: Copeland, B.J. (Ed.), The Essential Turing: The Ideas that Gave Birth to the Computer Age, Clarendon Press, Oxford, UK, 2004, pp. 569–575.

Kevin Warwick looks forward to —

TURING'S FUTURE

In 1950 Alan Turing said of humans that 'we can only see a short distance ahead, but we can see plenty there '. He looked himself, at that time, 50 to 100 year's into the future to the time when the Turing Imitation Game, which has become known as the Turing Test, would have been passed. He also said 'machines will eventually compete with men in all purely intellectual fields'. It could be argued that one machine came perilously close to passing the Turing test: *Elbot* 'surpassed Turing's 30 percent deception rate by fooling one in two experts' [p. 452], we can only assume that this is something that will happen well before Turing's 100 year horizon – certainly within Turing's 50 to 100 year time frame. So where does this leave us and where will it take us?

Turing suggested providing machines with ' the best sense organs money can buy' and then teaching them. He said, 'This process could follow the normal teaching of a child'. So he conceived of giving machines sensory input but interestingly appeared to shy away from completely embodying them with motor output as well – however he opened up the concept of machine learning which was revolutionary at the time (even now some people still believe machines to be merely programmed!!).

Let us assume that Turing was right! What does this mean? What will the world be like in 2050?

Firstly machines will be able to learn (like a child), to sense the world in a plethora of ways (not limited to human senses) and to fool humans into believing them to be human (at least to the stage where a human cannot tell the difference most of the time – possibly physical appearance accepted). As Turing pointed out 'An important feature of a learning machine is that its teacher will .. be largely ignorant of quite what is going on inside'. This needs to be coupled with the concept of machine creativity – Turing said 'the criticism that a machine cannot have much diversity of behaviour is just the same as saying that it cannot have much storage capacity'. Even nowadays it is apparent that the storage capacity of a personal computer compares favourably with that of a human brain.

But one key element is clear, and this tips the balance firmly in the favour of machines, it is the power of networking. This is something which was perhaps not particularly spotted by Turing and which provides all sorts of issues when Turing Tests are run at this time – quite simply networks are not allowable! But in reality most machines, of the computer/thinking type, are networked. The storage capacity, the communication ability, the learning skills of a machine are not just those of itself but rather are those of the network in which it is connected. In comparison humans are nowhere near as well networked in that our brains (apart from in a handful of scientific experiments) are not part of a closely connected network.

Potentially therefore we will (as indicated by Turing) by 2050 witness machines that are far more creative than humans, that can learn in a human like way and (if and when they wish) can fool us that they are human. To some this might sound quite sinister, Terminator-style and dangerous. Should we be scared or is it just a theoretical possibility?

As long as we do not use computers within a military environment then we need have no worries. As long as we do not develop weapon systems that protect themselves, that decide on their own target, that have no (realistic) human control then a dangerous situation is unlikely. But I hear you say – we already have such autonomous systems!

OK – then maybe the important thing is that humans maintain a veto, the chance to switch machines off. But you say – even now we cannot switch off many machines, we are dependant on them – the internet is a good example. Anyone disputing this fact can easily prove it to all of us – to show that humans can still do it, actually do it – switch off the internet.

So maybe in 2050 we will (with our pink glasses on) have machines around the home, helping us in our old age but maybe (with our black glasses on) machines will have humans around their home, helping them.

Playing chess, GO or other games with machines is one thing – but giving machines the power of choosing between the life and death of humans is quite something else.

References

Turing, A.M., 1950. Computing machinery and intelligence. Mind 59(236), 433–460.

Shah, H., Warwick, K., 2010. Hidden interlocutor misidentification in practical Turing tests. Minds and Machines. 20, 441–454.

Digital Computers Applied to Games

(Bowden, B. V. (Ed.), *1953. Faster than Thought.* Pitman, London, Chap. 25, pp. 286–310)

Alan Slomson introduces —

TURING AND CHESS

According to his mother, Alan Turing 'took up chess and stirred up some interest in the game among his fellows' at his preparatory school, Hazelhurst (Turing, 1959, p. 20). However, she also refers to him as 'not being a very good chess player' (Turing, 1959, p. 85). Later, Turing found himself in the company of many strong chess players at Bletchley Park. Andrew Hodges comments that his 'chess was always something of a joke at Bletchley, being all the more exposed to invidious comparison when the chess masters arrived. Harry Golombek had been able to give him queen odds, and still win' (Hodges, 1983, p. 265).

This is a little surprising in view of Turing's mathematical ability, and the frequency with which this is associated with ability in chess. For example, among Turing's near mathematical contemporaries in Cambridge who were strong chess players, and who represented the University at chess, we find the names of Jacob Bronowski (1908–74) and Charles Coulson (1910–74), (Sergeant, 1934, p. 357). The clue to Turing's weakness at chess may perhaps be found in his mother's remark that 'despite his profound mathematical insight, eagerness to press on often resulted in elementary mistakes in the simplest mathematical calculations' (Turing, 1959, p. 59), and to Feferman's comment that Turing's 'writing was rough-and-ready and prone to minor errors', (Feferman, 2001, p. 7). The minor errors that occurred in some of Turing's mathematical papers are not important. A single miscalculation often leads to defeat at chess.

Turing's ideas on using a computer to play chess emerged from his discussions with I. J. [Jack] Good (1916–2009), another young Cambridge mathematician, and a strong chess player, who was drafted to work at Bletchley Park (see Banks (1996)). Claude Shannon (1916–2001) had similar ideas which he presented at a conference in 1949 and which were published in the following year (Shannon, 1950). The connection between the Shannon's and Turing's ideas in this area is not entirely clear. Turing met Shannon at the Bell Labs in February 1943, and they talked about computers in the context of intelligence (Hodges, 1983, pp. 250–251). In a talk Turing gave to the London Mathematical Society in 1947, but not published until many years later, he refers briefly to the possibility of using a computer to play chess and says of Shannon that he 'tells me that he has won games playing by rule of thumb: the skill of his opponents is not stated' (Turing, 1947, p. 104). However, when Turing's ideas on using a computer to play chess were published in 1953, no mention is made of Shannon's work.

Turing's remarks on the use of computers to play chess form part of Chapter 25, 'a symposium on digital computing machines' (Bowden, 1953). The editor, Vivian Bowden (1910–89), was working for Ferranti selling computers when in 1953 he was appointed principal of the Manchester College of Science and Technology. The book indicates that, in addition to Turing, the other contributors to Chapter 25 were Audrey Bates, Christopher Strachey (1916–75) and Bowden himself, but does not make clear who wrote which parts of the chapter. The entire chapter is printed without comment about its authorship in the volume of Turing's collected papers covering his work on Machine Intelligence (Ince, 1992, pp. 161–185). However, Turing's original typescript has survived and was used by Jack Copeland in his edition, Copeland (2004), of a selection of Turing's work. This makes it clear that Turing's contribution to the chapter is just the section, following the half-title 'Chess', on pages 288 to 295 of Bowden (1953).[1] According to Copeland there are some minor differences between the typescript and the text as printed in Bowden (1953). There is one difference which is worthy of particular mention below.

Turing says that the main question he addresses is whether one could make a machine which would play a reasonably good game of chess. He invokes an informal version of what is now called the Church–Turing Thesis (for which see, e.g., Cooper (2004)), to say that it is not necessary for him to give 'actual programmes' to show that the answer to his question is 'yes' provided that he can give an unambiguous description of an appropriate algorithm. He indicates that, as chess is a finite game, if there was no limitation of time, it would be possible to evaluate any position as a win, draw or loss for the player about to move by applying the standard max–min algorithm to the tree of all positions obtainable from the given position using a sequence of legal moves. Since, because of the very large of different possible games (estimated as at least 10^{120} by Shannon (1950)) this is not feasible, we have to estimate the value of non-terminal positions according to some rule, now called an *evaluation function.*

These ideas are also to be found in the work of Shannon (1950). Both Shannon and Turing understood that it would be a mistake just to look ahead a fixed number of moves and then estimate the value of all the positions thus reached, but Turing is more specific, defining what he means by a *considerable move* and a *dead position*; evaluation being applied only to dead positions. Shannon talks about possible evaluation functions, but Turing gives a specific definition, and gives an example of a game played following his algorithm, the calculations being performed by hand.

It seems from this that Shannon and Turing independently arrived at the essentials of modern computer programmes for playing chess – deciding which positions to evaluate, the need for a suitable evaluation function and max–min searching – but Shannon gets the credit for the earliest publication of these ideas. However, Turing should be credited with the first specific algorithm for playing chess. Because of the difficulty in doing the calculations by hand, Turing's algorithm involves, normally, just looking two moves ahead, that is, one move by White and one by Black. Positions are evaluated by giving priority to the relative of the strengths of the two armies (counting a Queen as worth 9, a Rook as worth 5, and so on), and then calculating a 'Position-Play' value, which gives most weight to the sum of the square-roots of the number of legal moves available to each piece.

Turing also gets the credit for the first game played following such an algorithm. In this game, whose moves are given, Turing took the part of the computer, doing the calculations by hand. The player of the Black pieces was the computer scientist Alick Glennie (1925–2003) who is not named but merely described by Turing as 'a weak player who did not know the system'. Glennie was another contributor to Bowden's symposium, Bowden (1953).

I now come to one significant difference between Turing's typescript and the text as originally printed. As Copeland points out, the games differ from White's move 22 onwards, but he makes no suggestion as to how this occurred. If you apply Turing's algorithm to the position reached in the game after Black's 21st move, two things become apparent. First, applying the algorithm by hand is tedious, and Turing's persistence in doing this for 29 moves is to be admired. Second, it seems

[1] Editor's note: Dietrich Prinz, friend of Turing in Manchester, describes the first chess program to be actually implemented on pp. 295–297 (see Copeland (2004), p. 564, and pp. 632–634 of this volume).

that initially Turing made a mistake in the calculations. White's 22nd move, P × B, as printed in Copeland is not the move indicated by Turing's algorithm. I surmise that this was noticed, and the optimal move according to the algorithm, B × B, was later substituted. In the originally printed version in Bowden's book, White resigns after 29 moves, and the comment 'on the advice of his trainer' is added (by Turing?), whereas, according to the version printed by Copeland, the game ended after White was checkmated by Black's 29th move.

The limitation of the algorithm, looking ahead just two moves, means that it plays rather weak chess. However, except for two advances of the rooks' pawns, White's moves are, remarkably for such a simple algorithm, not unlike those that a weak player might choose, and it seems likely that this gave Turing confidence that chess playing programmes of considerable strength were feasible.

In his well-known paper (Turing, 1950), in which Turing discussed the question 'Can machines think?', and in which he introduced the *Imitation Game*, now known as the *Turing Test*, he gives the specific example of the solution to a chess problem as the sort of question that might be asked as a part of this test. Turing expresses the belief that 'in about fifty year's time it will be possible to programme computers, with a storage capacity of about 10^9, to make them play the imitation game so well that an average interrogator will not have not than a 70 per cent chance of making the right identification after five minutes of questioning' (Turing, 1950, p. 442).

As is well known, in 1996 the then world chess champion, Gary Kasparov, lost a match game to the Deep Blue chess programme. Kasparov ultimately won the match but a year later he was beaten $3\frac{1}{2}$–$2\frac{1}{2}$ by a strengthened version of the programme (see King (1997) for the details). The strength of such programmes arises mainly from the speed and power of modern computers, for example, King reports that the programme that defeated Kasparov used 220 parallel chips each able to examine two million positions each second (King, 1997, p. 47). They throw very little light on human intelligence. However, their underlying structure follows the approach set out by Shannon and Turing, and can be seen as a validation of their pioneering work.

References

Banks, D.L., 1996. A Conversation with I.J.Good. Stat. Sci. 11, 1–19.

Bowden, B.V. (Ed.), 1953. Faster Than Thought: A Symposium on Digital Computing Machines. Pitman, London.

Cooper, S.B., 2004. Computability Theory. Chapman and Hall/CRC, Boca Raton.

Copeland, J. (Ed.), 2004. The Essential Turing. Clarendon Press, Oxford.

Feferman, S., 2001. Historical Introduction pages 3–9 of Collected Works of A.M. Turing: Mathematical Logic. In: Gandy, R.O., Yates, C.E.M. (Eds.). Elsevier, Amsterdam.

Hodges, A., 1983. Alan Turing: The Enigma. Burnett Books, London.

Ince, D.C., 1992. Collected Works of A.M. Turing: Mechanical Intelligence. North-Holland, Amsterdam.

King, D., 1997. Kasparov v Deeper Blue. Batsford, London.

Sergeant, P.W., 1934. A Century of British Chess. Hutchinson, London.

Shannon, C.E., 1950. Programming a Computer for Playing Chess, Philosophical Magazine (7th series), vol. 41, pp. 256–275.

Turing, A., 1947. Lecture to the London Mathematical Society on 20 February 1947, pages 87–105 of Ince (1992); also reprinted as pages 378–394 of Copeland (2004).

Turing, A., 1950. Computing Machinery and Intelligence, Mind, 59, 433–460; reprinted as pages 133–160 of Ince (1992) and as pages 441–464 of Copeland (2004).

Turing, A., 1953. Chess, pages 288–295 of Bowden (1953); also reprinted as pages 163–170 of Ince (1992) and pages 569–575 of Copeland (2004)

Turing, S., 1959. Alan M. Turing, W. Heffer and Sons, Cambridge.

DIGITAL COMPUTERS APPLIED TO GAMES

Chess problems are the hymn tunes of mathematics — G. H. HARDY

MACHINES WHICH WILL PLAY GAMES have a long and interesting history. Among the first and most famous was the chess-playing automaton constructed in 1769 by the Baron Kempelen; M. Maelzel took it on tour all over the world, deceiving thousands of people into thinking that it played the game automatically. This machine was described in detail by Edgar Allan Poe; it is said to have defeated Napoleon himself—and he was accounted quite a good player, but it was finally shown up when somebody shouted "FIRE" during a game, and caused the machine to go into a paroxysm owing to the efforts of the little man inside to escape.

In about 1890 Signor Torres Quevedo made a simple machine—a real machine this time—which with a rook and king can check-mate an opponent with a single king. This machine avoids stalemate very cleverly and always wins its games. It allows an opponent to make two mistakes before it refuses to play further with him, so it is always possible to cheat by moving one's own king the length of the board. The mechanism of the machine is such that it cannot move its rook back past its king and one can then force a draw! This machine, like Babbage's "noughts and crosses" machine is relatively simple, the rules to be obeyed are quite straightforward, and the machines couldn't lose. Babbage thought that his analytical engine ought to be able to play a real game of chess, which is a much more difficult thing to do.

In this chapter we discuss how a digital computer can be made to play chess—it does so rather badly, and how it plays draughts—it does so quite well. We shall also describe a special simple machine which was built to entertain the public during the Festival of Britain. It was called Nimrod because it played nim, a game which is like noughts and crosses, in that the tricks which are needed to win can be expressed in mathematical terms. This machine was on show in South Kensington for six months and took on all comers.

During the Festival the Society for Psychical Research came and fitted up a room nearby in order to see if the operations of the machine could be influenced by concentrated thought on the part of the research workers, most of whom were elderly ladies. When this experiment had failed they tried to discover whether they in turn could be affected by vibrations from the machine, and could tell from another room how the game was progressing. Unfortunately this experiment, like the first, was a complete failure, the only conclusion being that machines are much less co-operative than human beings in telepathic experiments.

At the end of the Festival of Britain Nimrod was flown to Berlin and shown at the Trade Fair. The Germans had never seen anything like it, and came to see it in their thousands, so much so in fact that on the first day of the show they entirely ignored a bar at the far end of the room where free drinks were available, and it was necessary to call out special police to control the crowds. The machine became even more popular after it had defeated the Economics Minister, Dr. Erhardt, in three straight games. After this it was taken to Canada and demonstrated to the Society of Engineers in Toronto. It is still somewhere on the North American continent, though it may have been dismantled by now, and it would be amusing to match it against some of the other nim-playing machines which have been built in the last year or two.

The reader might well ask why we bother to use these complicated and expensive machines in so trivial a pursuit as playing games. It would be disingenuous of us to disguise the fact that the principal motive which prompted the work was the sheer fun of the thing, but nevertheless if ever we had to justify the time and effort (and we feel strongly that no excuses are either necessary or called for) we could quite easily make a pretence at doing so. We have already explained how hard all programming is to do, and how much difficulty is due to the incompetence of the machine at taking an overall view of the problem which it is analysing. This particular point is brought out more clearly in playing games than in anything else. The machine cannot look at the whole of a chess board at once; it has to peer short-sightedly at every square in turn, in much the same way as it has to look at a commercial document. Research into the techniques of programming a machine to tackle complicated problems of this type may in fact lead to quite important advances, and help in serious work in business and economics—perhaps, regrettably, even in the theory of war. We hope that mathematicians will continue to play draughts and chess, and to enjoy themselves as long as they can.

We have often been asked why we don't use the machine to work out the football pools, or even to do something to remove the present uncertainty about the results of tomorrow's horse races. Perhaps one day we shall persuade our mathematicians to apply themselves to this problem too. It would first be necessary to establish a series of numerical criteria from which the machine could predict the results with greater certainty than the ordinary citizen can achieve with his pin; the presumption underlying the whole idea is that such criteria do in fact exist, but that they are too complicated for a man to apply in time, whereas a machine could do the necessary computations for him. It is most unlikely that a machine could ever hope to predict (for example) the results of a single football match, but it is at least possible that a detailed analysis might significantly improve the odds in favour of the gambler, so that if he invested on a large enough scale he could make a profit. It is notoriously true that mathematics, and particularly the theory of probability, owes more to gambling than gambling owes to mathematics; perhaps a machine might do something to restore the balance. Lady Lovelace lost a fortune by trying to back horses scientifically, and many others have done the same; all one could hope for is a slight improvement in the odds. We might make it pay but we doubt it; as an academic exercise it would be amusing, but we shall give the project a low priority.

CHESS

When one is asked, "Could one make a machine to play chess?" there are several possible meanings which might be given to the words. Here are a few—

(*a*) Could one make a machine which would obey the rules of chess, i.e. one which would play random legal moves, or which could tell one whether a given move is a legal one?

(*b*) Could one make a machine which would solve chess problems, e.g. tell one whether, in a given position, white has a forced mate in three?

(*c*) Could one make a machine which would play a reasonably good game of chess, i.e. which, confronted with an ordinary (that is, not particularly unusual) chess position, would after two or three minutes of calculation, indicate a passably good legal move?

(*d*) Could one make a machine to play chess, and to improve its play, game by game, profiting from its experience?

To these we may add two further questions, unconnected with chess, which are likely to be on the tip of the reader's tongue.

(*e*) Could one make a machine which would answer questions put to it, in such a way that it would not be possible to distinguish its answers from those of a man?

(*f*) Could one make a machine which would have feelings as you and I have?

The problem to be considered here is (*c*), but to put this problem into perspective with the others I shall give the very briefest of answers to each of them.

To (*a*) and (*b*) I should say, "This certainly can be done. If it has not been done already it is merely because there is something better to do."

Question (*c*) we are to consider in greater detail, but the short answer is, "Yes, but the better the standard of play required, the more complex will the machine be, and the more ingenious perhaps the designer."

To (*d*) and (*e*) I should answer, "I believe so. I know of no really convincing argument to support this belief, and certainly of none to disprove it."

To (*f*) I should say, "I shall never know, any more than I shall ever be quite certain that *you* feel as I do."

In each of these problems except possibly the last, the phrase, "Could one make a machine to ..." might equally well be replaced by, "Could one programme an electronic computer to ..." Clearly the electronic computer so programmed would itself constitute a machine. And on the other hand if some other machine had been constructed to do the job we could use an electronic computer (of sufficient storage capacity), suitably programmed, to calculate what this machine would do, and in particular what answer it would give.

After these preliminaries let us give our minds to the problem of making a machine, or of programming a computer, to play a tolerable game of chess. In this short discussion it is of course out of the question to provide actual programmes, but this does not really matter on account of the following principle—

If one can explain quite unambiguously in English, with the aid of mathematical symbols if required, how a calculation is to be done, then it is always possible to programme any digital computer to do that calculation, provided the storage capacity is adequate.

This is not the sort of thing that admits of clear-cut proof, but amongst workers in the field it is regarded as being clear as day. Accepting this principle, our problem is reduced to explaining "unambiguously in English" the rules by which the machine is to choose its move in each position. For definiteness we will suppose the machine is playing white.

If the machine could calculate at an infinite speed, and also had unlimited storage capacity, a comparatively simple rule would suffice, and would give a result that in a sense could not be improved on. This rule could be stated:

"Consider every possible continuation of the game from the given position. There is only a finite number of them (at any rate if the fifty-move rule makes a draw obligatory, not merely permissive). Work back from the end of these continuations, marking a position with white to play as 'win' if there is a move which turns it into a position previously marked as 'win.' If this does not occur, but there is a move which leads to a position marked 'draw,' then mark the position 'draw.' Failing this, mark it 'lose.' Mark a position with black to play by a similar rule with 'win' and 'lose' interchanged. If after this process has been completed it is found that there are moves which lead to a position marked 'win,' one of these should be chosen. If there is none marked 'win' choose one marked 'draw' if such exists. If all moves lead to a position marked 'lose,' any move may be chosen."

Such a rule is practically applicable in the game of noughts and crosses, but in chess is of merely academic interest. Even when the rule can be applied it is not very appropriate for use against a weak opponent, who may make mistakes which ought to be exploited.

In spite of the impracticability of this rule it bears some resemblance to what one really does when playing chess. One does not follow all the continuations of play, but one follows some of them.

One does not follow them until the end of the game, but one follows them a move or two, perhaps more. Eventually a position seems, rightly or wrongly, too bad to be worth further consideration, or (less frequently) too good to hesitate longer over. The further a position is from the one on the board the less likely it is to occur, and therefore the shorter is the time which can be assigned for its consideration. Following this idea we might have a rule something like this—

"Consider all continuations of the game consisting of a move by white, a reply by black, and another move and reply. The value of the position at the end of each of these sequences of moves is estimated according to some suitable rule. The values at earlier positions are then calculated by working backwards move by move as in the theoretical rule given before. The move to be chosen is that which leads to the position with the greatest value."

It is possible to arrange that no two positions have the same value. The rule is then unambiguous. A very simple form of values, but one not having this property, is an "evaluation of material," e.g. on the basis—

$$\begin{aligned} \text{P} &= 1 \\ \text{Kt} &= 3 \\ \text{B} &= 3\tfrac{1}{2} \\ \text{R} &= 5 \\ \text{Q} &= 10 \\ \text{Checkmate} &= 1000 \end{aligned}$$

If B is black's total and W is white's, then W/B is quite a good measure of value. This is better than W—B as the latter does not encourage exchanges when one has the advantage. Some small extra arbitrary function of position may be added to ensure definiteness in the result.

The weakness of this rule is that it follows all combinations equally far. It would be much better if the more profitable moves were considered in greater detail than the less. It would also be desirable to take into account more than mere "value of material."

After this introduction I shall describe a particular set of rules, which could without difficulty be made into a machine programme. It is understood that the machine is white and that white is next to play. The current position is called the *position on the board*, and the positions arising from it by later moves *positions in the analysis*.

"CONSIDERABLE" MOVES

"Considerable" here is taken to mean moves which will be "considered" in the analysis by the machine.

Every possibility for white's next move and for black's reply is "considerable." If a capture is considerable then any recapture is considerable. The capture of an undefended piece or the capture of a piece of higher value by one of lower value is always considerable. A move giving checkmate is considerable.

DEAD POSITION

A position in the analysis is dead if there are no considerable moves in that position, i.e. if it is more than two moves ahead of the present position, and no capture or recapture or mate can be made in the next move.

VALUE OF POSITION

The value of a dead position is obtained by adding up the piece values as above, and forming the ratio W/B of white's total to black's. In other positions with white to play the value is the greatest value of (*a*) the positions obtained by considerable moves, or (*b*) the position itself evaluated as if a dead position. The latter alternative is to be omitted if all moves are considerable. The same process is to be undertaken for one of black's moves, but the machine will then choose the *least* value.

POSITION-PLAY VALUE

Each white piece has a certain position-play contribution and so has the black king. These must all be added up to give the position-play value.

For a Q, R, B, or Kt, count—

(*a*) The square root of the number of moves the piece can make from the position, counting a capture as two moves, and not forgetting that the king must not be left in check.
(*b*) (If not a Q) 1·0 if it is defended, and an additional 0·5 if twice defended.

For a K, count—

(*c*) For moves other than castling as (*a*) above.
(*d*) It is then necessary to make some allowance for the vulnerability of the K. This can be done by assuming it to be replaced by a friendly Q on the same square, estimating as in (*a*), but subtracting instead of adding.
(*e*) Count 1·0 for the possibility of castling later not being lost by moves of K or rooks, a further 1·0 if castling could take place on the next move, and yet another 1·0 for the actual performance of castling.

For a P, count—

(*f*) 0·2 for each rank advanced.
(*g*) 0·3 for being defended by at least one piece (not P).

For the black K, count—

(*h*) 1·0 for the threat of checkmate.
(*i*) 0·5 for check.

We can now state the rule for play as follows. The move chosen must have the greatest possible value, and, consistent with this, the greatest possible position-play value. If this condition admits of several solutions a choice may be made at random, or according to an arbitrary additional condition.

Note that no "analysis" is involved in position-play evaluation. This is to reduce the amount of work done on deciding the move.

The game below was played between this machine and a weak player who did not know the system. To simplify the calculations the square roots were rounded off to one decimal place, i.e. this table was used—

Number . . .	0	1	2	3	4	5	6	7	8	9	10	11	12	13
Square Root . .	0	1	1·4	1·7	2·0	2·2	2·4	2·6	2·8	3·0	3·2	3·3	3·5	3·6

No random choices actually arose in this game. The increase of position-play value is given after white's move if relevant. An asterisk indicates that every other move had a lower position-play value.

	White (Machine)		*Black*
1.	P—K4	4·2*	P—K4
2.	Kt—Q B3	3·1*	Kt—K B3
3.	P—Q4	2·6*	B—Q Kt5
4.	Kt—K B3(1)	2·0	P—Q3
5.	B—Q2	3.5*	Kt—Q B3
6.	P—Q5	0·2	Kt—Q5
7.	P—K R4(2)	1·1*	B—Kt5
8.	P—Q R4(2)	1·0*	Kt×Kt ch.
9.	P × Kt		B—K R4
10.	B—Kt5 ch.	2·4*	P—Q B3
11.	P × P		O—O
12.	P × P		R—Kt1
13.	B—R6	− 1·5	Q—R4
14.	Q—K2	0·6	Kt—Q2
15.	K R—Kt 1(3)	1·2*	Kt—B4(4)
16.	R—Kt 5(5)		B—Kt3
17.	B—Kt5	0·4	Kt×Kt P
18.	O—O—O	3.2*	Kt—B4
19.	B—B6		KR—Q B1
20.	B—Q5		B×Kt
21.	B × B	0·7	Q × P
22.	K—Q2		Kt—K3
23.	R—Kt4	− 0·3	Kt—Q5
24.	Q—Q3		Kt—Kt4
25.	B—Kt3		Q—R3
26.	B—B4		B—R4
27.	R—Kt3		Q—R5
28.	B × Kt		Q × B
29.	Q × P(6)		R—Q 1(4)
30.	Resigns(7)		

Notes—

1. If B—Q2 3·7* then P × P is foreseen.
2. Most inappropriate moves.
3. If white castles then B×Kt, B × B, Q × P.
4. The fork is unforeseen at white's last move.
5. Heads in the sand!
6. Fiddling while Rome burns!
7. On the advice of his trainer.

Numerous criticisms of the machine's play may be made. It is quite defenceless against forks, although it may be able to see certain other kinds of combination. It is of course not difficult to devise improvements of the programme so that these simple forks are foreseen. The reader may be able to think of some such improvements for himself. Since no claim is made that the above rule is particularly good, I have been content to leave this flaw without remedy; clearly a line has to be drawn between the flaws which one will attempt to eliminate and those which must be accepted as a risk. Another criticism is that the scheme proposed, although reasonable in the middle game, is futile in the end game. The change-over from the middle game to the end-game is usually sufficiently clear-cut for it to be possible to have an entirely different system for the end-game. This should of course include quite definite programmes for the standard situations, such as mate with rook and king, or king and pawn against king. There is no intention to discuss the end-game further here.

If I were to sum up the weakness of the above system in a few words I would describe it as a caricature of my own play. It was in fact based on an introspective analysis of my thought processes when playing, with considerable simplifications. It makes oversights which are very similar to those which I make myself, and which may in both cases be ascribed to the considerable moves being inappropriately chosen. This fact might be regarded as supporting the glib view which is often expressed, to the effect that "one cannot programme a machine to play a better game than one plays oneself." This statement should I think be compared with another of rather similar form. "No animal can swallow an animal heavier than himself." Both statements are, as far as I know, untrue. They are also both of a kind that one is easily bluffed into accepting, partly because one thinks that there ought to be some slick way of demonstrating them, and one does not like to admit that one does not see what this argument is. They are also both supported by normal experience, and need exceptional cases to falsify them. The statement about chess programming may be falsified quite simply by the speed of the machine, which might make it feasible to carry the analysis a move farther than a man could do in the same time. This effect is less than might be supposed. Although electronic computers are very fast where conventional computing is concerned, their advantage is much reduced where enumeration of cases, etc., is involved on a large scale. Take for instance the problem of counting the possible moves from a given position in chess. If the number is 30 a man might do it in 45 seconds and the machine in 1 second. The machine has still an advantage, but it is much less overwhelming than it would be for instance when calculating cosines.

In connexion with the question of the ability of a chess-machine to profit from experience, one can see that it would be quite possible to programme the machine to try out variations in its method of play (e.g. variations in piece value) and adopt the one giving the most satisfactory results. This could certainly be described as "learning," though it is not quite representative of learning as we know it. It might also be possible to programme the machine to search for new types of combination in chess. If this project produced results which were quite new, and also interesting to the programmer, who should have the credit? Compare this with the situation where a Defence Minister gives orders for research to be done to find a counter to the bow and arrow. Should the inventor of the shield have the credit, or should the Defence Minister?

THE MANCHESTER UNIVERSITY MACHINE

In November, 1951, some months after this article was written (by Dr. Turing) Dr. Prinz was able to make the Manchester University machine solve a few straightforward chess problems of the "Mate-in-Two" type (see *Research*, Vol. 6 (1952), p. 261).

It is usually true to say that the best and often the only way to see how well the machine can tackle a particular type of problem is to produce a definite programme for the machine, and, in this case, in order to have something working in the shortest possible time, a few restrictions were imposed on the rules of chess as they were "explained" to the machine. For example castling was not permitted, nor were double moves by pawns, nor taking *en passant* nor the promotion of a pawn into a piece when it reached the last row; further, no distinction was made between mate and stalemate.

The programme contained a routine for the construction of the next possible move, a routine to check this move for legality, and various sequences for recording the moves and the positions obtained. All these separate subroutines were linked together by a master routine which reflected the structure of the problem as a whole and ensured that the subroutines were entered in the proper sequence.

The technique of programming was perhaps rather crude, and many refinements, increasing the speed of operation, are doubtless possible. For this reason, the results reported here can only serve as a very rough guide to the speed attainable; but they do show the need for considerable improvement in programming technique and machine performance before a successful game by a machine against a human chess player becomes a practical possibility.

The programme, as well as the initial position on the chess board, was supplied to the machine on punched tape and then transferred to the magnetic store of the machine.

A initial routine (sub-programme) was transferred to the electronic store, and the machine started its computation. The programme was so organized that every first move by white was printed out; after the key move had been reached the machine printed: "MATE."

The main result of the experiment was that the machine is disappointingly slow when playing chess—in contrast to the extreme superiority over human computers where purely mathematical problems are concerned. For the simple example given in the position reproduced here, 15 minutes were needed to print the solution. A detailed analysis shows that the machine tried about 450 possible moves (of which about 100 were illegal) in the course of the game; this means about two seconds per move on the average.

A considerable portion of this time had to be used for a test for self-check (i.e. after a player had made a move, to find out whether his own King was left in check). This was done by first examining all squares connected to the King's square by a Knight's move, to see (*a*) whether they were on the board at all, (*b*) whether they were empty or occupied, (*c*) if occupied, by a piece of which colour and (*d*) if occupied by a piece of opposite colour, whether or not this piece was a Knight. A similar test had to be carried out for any other piece that might have put the King in check. This test involves several hundreds of operations and, at a machine speed of 1 msec per operation, might take an appreciable fraction of a second.

The next important time-consuming factor was the magnetic transfers, i.e. the transfers of sub-programmes and data (relating to positions and moves) between the magnetic and the electronic store. It is here that improved programming technique may save time by better utilization of the electronic store, thus reducing the number of transfers (nine for every legal move in the present programme).

Compared with these two items, the time spent in computing the moves appeared to be of minor importance although the machine not only computed the possible moves but also the impossible, but "thinkable" moves—meaning those which either carry the piece off the board, or lead to a collision with a piece of the same colour already on the square. These moves, however, were quickly rejected by the machine and did not contribute greatly to the total computation time.

It appears that if this crude method of programming were the only one available it would be quite impractical for any machine to compete on reasonable terms with a competent human being.

Before we conclude too easily that no computer will ever compete in a Masters' Tournament let us remind ourselves that the Manchester machine solved a problem after a few weeks tuition, which represents quite reasonable progress for a beginner.

The First Chess Problem Solved by a Computing Machine. The task set the Manchester machine was to find a move by white that would lead to a mate in the next move, whatever black might answer. The move is R—R6.

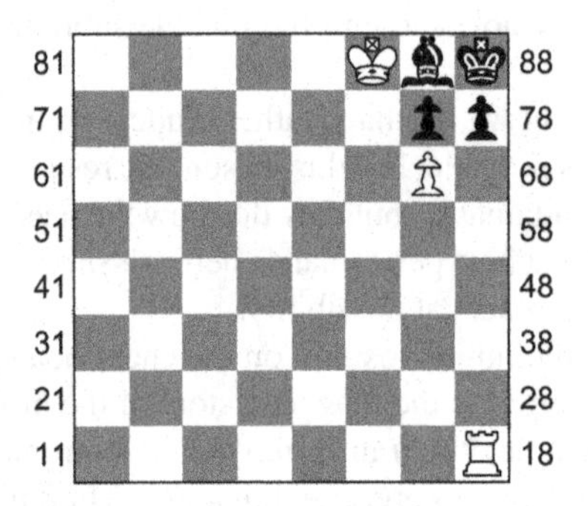

For solution of the problem by the machine the squares of the board were numbered in rather unusual fashion. The bottom row was numbered 11 to 18 (from left to right), the next 21 to 28, and so on to the top row, which was 81–88. Square 68 was thus the square in row 6, column 8. The machine has printed out all the moves which white tried out to find a solution, and has printed "MATE" after finding and recording the key move, which appears in the form "Rook to 68."

The list of moves is—

Pawn to 78.	Rook to 11.
Rook to 17.	Rook to 28.
Rook to 16.	Rook to 38.
Rook to 15.	Rook to 48.
Rook to 14.	Rook to 58.
Rook to 13.	Rook to 68.
Rook to 12.	MATE.

DRAUGHTS

The game of draughts occupies an intermediate position between the extremely complex games such as chess, and the relatively simple games such as nim or noughts-and-crosses for which a complete mathematical theory exists. This fact makes it a rather suitable subject for experiments in mechanical game playing, for although there is no complete theory of the game available, so that the machine has to look ahead to find the moves, the moves themselves are rather simple and relatively few in number.

Various forms of strategy have been suggested for constructing an automatic chess player; the purpose of such plans is to reduce the time taken by the machine to choose its move. As Prinz has shown, the time taken by any machine which considers all the possible moves for four or five steps ahead would be quite prohibitive, and the principal aim of the strategy is to reduce this number very considerably, while at the same time introducing a scheme of valuing the positions which will allow it to choose a reasonably good move. The chief interest in games-playing machines lies in the development of a suitable strategy.

Before any strategy can be realized in practice, however, the basic programme necessary to find the possible moves and to make them must be constructed. When this has been done the strategy, which consists principally of the methods by which positions can be valued, can be added to make the complete game player. It is obviously possible to make experiments with different strategies using the same basic move-finding-and-making routine.

The basic programme for draughts, which is described in outline in the following paragraphs, is very much simpler than the corresponding one for chess. It has in fact proved possible to put both it and the necessary position storage in the electronic store of the Manchester machine at the same time. This removes the need for magnetic transfers during the operation of the programme, and this fact, together with the simplicity of the moves, has reduced the time taken to consider a single move to about one tenth of a second.

Basic Programme For Draughts

We must first consider the representation of a position in the machine. The 32 squares used in a draughts board are numbered as shown in the diagram.

A position is represented by 3 thirty-two-digit binary numbers (or words) B, W and K which give the positions of the black men (and kings), the white men (and kings) and the kings (of either colour) respectively. The digits of these words each represent a square on the board; the square n being represented by the digit 2^n.

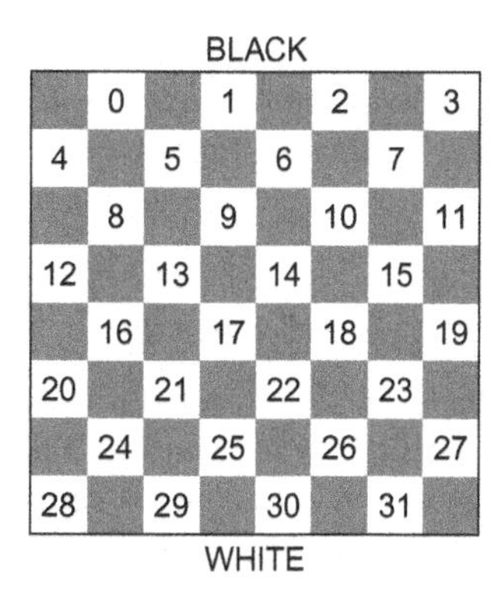

Thus the least significant digit represents square 0 and the most significant digit represents square 31. (In the Manchester machine, where the word length is 40 digits, the last 8 digits are irrelevant). A unit in the word indicates the presence, and a zero indicates the absence of the appropriate type of man in the corresponding square. Thus the opening position of the game would be represented by*—

$$
\begin{aligned}
B &= 1111,\ 1111,\ 1111,\ 0000,\ 0000,\ 0000,\ 0000,\ 0000\\
W &= 0000,\ 0000,\ 0000,\ 0000,\ 0000,\ 1111,\ 1111,\ 1111\\
K &= 0000,\ 0000,\ 0000,\ 0000,\ 0000,\ 0000,\ 0000,\ 0000
\end{aligned}
$$

The positions of the white kings are indicated by the word $W\&K$, while the empty squares are indicated by the word $\sim W\&\sim B$.†

It will be seen that there are at most four possible types of non-capture moves from any square on the board. For example, from square 14 the possible moves are to squares 9, 10, 17 or 18. The machine considers all these moves in turn, but it will be sufficient to indicate here the way in which it deals with one of them—say the move 14–18.

* All binary numbers are written in the convention used for the Manchester machine, i.e. with their *least* significant digit on the left.

† $W\&K$ stands for the logical product of W and K (sometimes also known as the result of collating W and K). $\sim W$ stands for the negation of W, i.e. the word obtained by writing 1's for 0's in W, and vice versa (see Chapter 15).

This type of move, which consists of adding 4 to the number of the square, corresponds to multiplying the appropriate digit in the position word by 2^4. A move of this type can be made by any black man, but only by a white king; it cannot be made from squares 28, 29, 30 or 31 nor can it be made unless the square to which the man is to be moved is empty. For a black move, the machine there- fore forms the following quantity—

$$\gamma = \{(B\&M) \times 2^4\}\& \sim W\& \sim B$$

where $M = 1111, 1111, 1111, 1111, 1111, 1111, 1111, 0000$

For a white move, the corresponding quantity is—

$$\{(W\&K\&M) \times 2^4\}\& \sim W\& \sim B$$

In these expressions $(B\&M)$ or $(W\&K\&M)$ give all the men on the board who could make the move; multiplying this by 2^4 give the squares to which they would move. If these squares are empty (collate with$\sim$W&$\sim$B) the move is possible.

The quantity γ thus represents all the possible moves of this type. To consider a single one of these, the largest non-zero digit of γ is taken and removed from γ. The word consisting of this single digit known as θ, gives the square to which the man is moved. The quantity $\phi = \theta \times 2^{-4}$ is then formed and gives the square from which the man was moved. For a black move, the quantity—

$$B' = B \not\equiv \theta \not\equiv \phi$$

will then give the new position of the black men. If $K\&\phi$ is not zero, the man moved was a king so that $K' = K \not\equiv \theta \not\equiv \phi$ gives the new position of the kings. If $K\&\phi$ is zero, the man moved was not a king. The new position of the kings will therefore be unaltered unless the man has kinged during this move—in other words unless $\theta \geqslant 2^{28}$ in which case $K' = K \not\equiv \theta$.

Relatively simple modifications of this scheme are needed to deal with white moves and non-capture moves of other types. Capture moves are somewhat more complicated as multiple captures must be allowed for. Furthermore, all the possible captures must be made or the machine will render itself liable to be huffed. This leads to a considerable complication which it is not possible to describe fully here, but the basic scheme is not altered.

The machine considers all the possible moves of one type before starting the next, so that in order to describe a position fully, it is necessary to store the word γ, which indicates the moves still to be considered, as well as the position words B, W and K. It is also necessary to keep a record of the type of move being considered. This is done with the aid of a further parameter word P which also contains the value associated with the position. The whole storage required for a position is thus reduced to the 5 thirty-two-digit words B, W, K, γ, and P.

Valuation of Positions and Strategy

It should be possible to graft almost any type of strategy on to the move-finding scheme outlined above to produce a complete draughts-playing routine and then to evaluate the effectiveness of the strategy by direct experiment. I have done this with two rather simple types of strategy so far, and I hope to be able to try some rather more refined strategies in the future.

For demonstration purposes, and also to ensure that a record of the game is kept, and to take certain precautions against machine error, the move-finding sequence and the associated strategy have been combined with a general game-playing routine which accepts the opponent's moves, displays the positions, prints the move, and generally organizes the sequence of operations in the game. It is rather typical of logical programmes that this organizing routine is in fact longer than the game-playing routine proper. As its operations, though rather spectacular, are of only trivial theoretical interest, I shall not describe them here.

The first, and simplest, strategy to try is the direct one of allowing the machine to consider all the possible moves ahead on both sides for a specified number of stages. It then makes its choice, valuing the final resulting positions only in terms of the material left on the board and ignoring any positional advantage. There is an upper limit to the number of stages ahead that can be considered owing to limitations of storage space—actually six moves, three on each side, are all that can be allowed. In practice, however, time considerations provide a more severe limitation. There are on an average about ten possible legal moves at each stage of the game, so that consideration of one further stage multiplies the time for making the move by a factor of about ten. The machine considers moves at the rate of about ten a second, so that looking three moves ahead (two of its own and one of its opponents), which takes between one and two minutes, represents about the limit which can be allowed from the point of view of time.

This is not sufficient to allow the machine to play well, though it can play fairly sensibly for most of the game. One wholly unexpected difficulty appears. Consider the position on the following board.

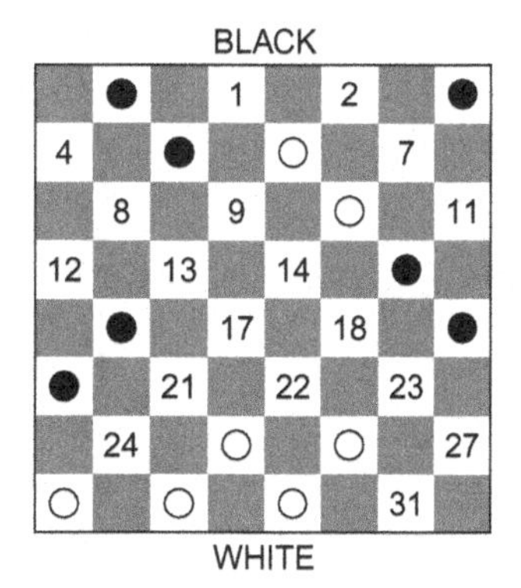

In this position, the machine is aware that its opponent is going to king next move. Now a king is more valuable than a man—the actual values used were three for a king and one for a man—so that if the opponent kings the machine effectively loses two points. The only way it can stop this is by offering a man for sacrifice, because then, by the rules of the game, the sacrifice must be taken at once. If the machine does this, it will lose only one point, and as it is not looking far enough ahead, it cannot see that it has not prevented its opponent from kinging but only postponed the evil day. At its next move it is still faced with the same difficulty, which it tries to solve in the same way, so that it will make every possible sacrifice of a single man before it accepts as inevitable the creation of an opponent's king. In fact, when faced with this position, the machine played 19—23, followed by 16—21 and 20—24.

This, of course, is a fatal flaw in the strategy—and not one it would have been easy to discover without actually trying it out. An opponent who detected this behaviour—and it is extremely conspicuous in play—would only have to leave his man on the point of kinging indefinitely. The machine would then sacrifice all its remaining men as soon as the opportunity offered.

In order to avoid this difficulty, the second strategy was devised. In this the machine continues to investigate the moves ahead until it has found two consecutive moves without captures. This means that it will be able to recognize the futility of its sacrifice to prevent kinging. It is still necessary to impose an over-riding limit on the number of stages it can consider, and once more, considerations of time limit this. However, as no move is continued for more than two stages unless it leads to a capture, it is possible to allow the machine to consider up to four stages ahead without it becoming intolerably slow. This would mean that it would consider the sacrifice of two men to be of equal value to the creation of an opponent's king, and as there is a random choice between moves of equal value, it might still make this useless sacrifice. This has been prevented by reducing the value of a king from 3 to $2\frac{7}{8}$.

	Machine	*Strachey*
1.	11—15	23—18
2.	7—11	21—17
3.	8—12	20 – 16[1]
4.	12—21(16)	25—16(21)
5.	9—14![2]	18—9(14)
6.	6—20(16,9)[3]	27—23
7.	2—7[4]	23—18
8.	5—8	18—14
9.	8—13[5]	17—8(13)
10.	4—13(8)	14—9
11.	1—5[6]	9—6
12.	15—19	6—1 K
13.	5—9	1—6?[7]
14.	0—5![8]	6—15(10)
15.	11—25(22,15)	30—21(25)
16.	13—17	21—14(17)
17.	9—18(14)	24—21
18.	18—23	26—22
19.	23—27	22—17
20.	5—8[9]	17—14
21.	8—13	14—9
22.	19—23	9—6
23.	23—26[10]	31—22(26)
24.	27—31K	6—2 K
25.	7—10	2—7
26.	10—15	21—16?[11]
27.	3—10(7)	16—9(13)
28.	10—14	9—6
29.	15—19	6—2K
30.	31—27[12]	2—6
31.	27—31[12]	6—10
32.	31—26[13]	10—17(14)
33.	19—23	29—25
34.	26—31[14]	

Notes—

1. An experiment on my part—the only deliberate offer I made. I thought, wrongly, that it was quite safe.
2. Not foreseen by me.
3. Better than 5—21 (9,17).
4. A random move (zero value). Shows the lack of a constructive plan.
5. Another random move of zero value, actually rather good.
6. Bad. Ultimately allows me to make a King. 10—14 would have been better.
7. A bad slip on my part.
8. Taking full advantage of my slip.
9. Bad. Unblocks the way to a King.
10. Sacrifice in order to get a King (not to stop me kinging). A good move, but not possible before 19—23 had been made by chance.
11. Another bad slip on my part.
12. Purposeless. The strategy is failing badly in the end game.
13. Too late.
14. Futile. The game was stopped at this point as the outcome was obvious.

With this modified strategy, the machine can play quite a tolerable game until it reaches the end game. It has always seemed probable that a wholly different strategy will be necessary for end games. The game given on page 303, which is the first ever played using the strategy, brings this point out very clearly.

NIM

A considerably easier game which the machine can be programmed to play is the one known as nim. Probably a variation of this was known to the Chinese—certainly in its present form many people have met it. We have chosen to deal with this comparatively trivial game in detail because of its topical interest. Thousands of people will have seen *Nimrod*, the computer built by Ferranti Ltd. for the Science Exhibition of the Festival of Britain. This special-purpose machine was designed to show the main features of large electronic digital computers, and the game of nim was chosen as an interesting but simple demonstration problem. The game itself is as follows—

Initially we have any number of heaps, each containing any number of tokens (usually matches). In the simplest form, two contestants play alternately, and may pick up as many matches as they wish at one time from *one* pile, but they must take at least one match. The aim is to avoid taking the last match of all—or there is another variation where the aim is to take the last match or group of matches.

The so-called *multiple game* differs from this only in that the number of heaps altered in any move may take any value from one up to a pre-assigned maximum k. Of course, to prevent complete triviality, k must be less than N, the total number of heaps.

The detailed theory of nim was worked out long ago and, apart from the initial distribution of the matches, no element of chance need enter into the game. This theory is very simple, but it becomes clearer for the non-mathematician if we use the concept of a binary number, introduced elsewhere (see page 33).

We can now proceed to give a working rule for the game of nim. We would like to find a *winning position* having the following characteristics—

(*a*) It is impossible, when faced by a winning position, to make a move which will leave a winning position.

(*b*) Faced with any other than a winning position, it is possible to make a move resulting in a winning position.

(*c*) If at any stage of the game a player A can convert a position into a winning position, it is possible for A to win, and impossible for his opponent B to do so unless A makes a mistake. A wins by leaving a winning position at every succeeding move on his part.

Such winning positions can be achieved and are recognized as follows: For any given configuration, express the number of matches in each heap as a binary number. Suppose, for example, that we have four heaps, A, B, C and D, containing respectively 7, 4, 3 and 2 matches. These are represented—

	4	2	1	
A	1	1	1	(7)
B	1	0	0	(4)
C	0	1	1	(3)
D	0	1	0	(2)

We write these down as above, one under the other, and add up each column, e.g., in the above example, we get

	4	2	1
sum:	2	3	2

Now the "secret" of a winning position is that every column should be divisible by $k+1$; k being the maximum number of heaps which can be altered in any one move. Thus the example quoted above cannot represent a winning position whatever our initial choice of k. However, suppose we have $k=1$; then consider the position—

	4	2	1	
A	1	0	1	(5)
B	1	1	1	(7)
C	0	1	1	(3)
D	0	0	1	(1)
sum :	2	2	4	

This is a winning position, but would not be so if we had previously fixed $k=3$, for example.

To convert an "unsafe" into a winning position, we deal with a column at a time. Consider our previous example with $k=1$.

	4	2	1	
A	1	1	1	(7)
B	1	0	0	(4)
C	0	1	1	(3)
D	0	1	0	(2)
sum :	2	3	2	

We start with the "most-significant," or left-hand column. This sum is divisible by $k+1$, so we proceed to consideration of the next column. The sum here is 3, which is not divisible by $k+1$, so we choose any heap, say D, having a one in this column. We remove this 1 (which is equivalent to subtracting 2 from D), and put 1 in every less-significant (or right-hand) column of this heap (which in this case is equivalent to adding 1, though if we had chosen to modify A instead, it would have meant no change in the last column). That is, we make the minimum move which removes the 1 in the "unsafe" column. Thus we remove 1 from D, and so alter its binary representation to 001.

Now our representation is—

	4	2	1	
A	1	1	1	(7)
B	1	0	0	(4)
C	0	1	1	(3)
D	0	0	1	(1)
sum :	2	2	3	

and we see that we have made the sum of column 2 divisible by $k+1$ at the expense of column 1. However, we shall now proceed to adjust column 1. To avoid altering more than k heaps in one move, we must alter one or more of the heaps already affected if, by so doing, we can achieve the desired result, rather than select a fresh heap.

Now, in this case, we wish to remove 1 from column 1 of some heap. Since heap D has already been altered, we choose this—it has a 1 in this column.

So, at the end of our move, we have removed two matches from heap D, and leave the winning position—

	4	2	1	
A	1	1	1	(7)
B	1	0	0	(4)
C	0	1	1	(3)
D	0	0	0	(0)
sum :	2	2	2	

In adapting this game for the universal computer, we allow a maximum of eight heaps, with not more than thirty-one matches in a heap. In Nimrod the more stringent restrictions to four heaps, each with a maximum content of seven matches, were applied to simplify the problems of demonstration.

Possible positions with which the machine may be faced are as follows—

(*a*) At least $k+1$ heaps contain more than one match.
(*b*) The number of heaps containing more than one match lies between 1 and k (inclusive).
(*c*) No heap contains more than one match. Not all heaps are empty.
(*d*) All heaps are empty.

In case (*a*), we follow the so-called *normal routine*, which aims at leaving column sums all divisible by $(k+1)$.

In case (*b*), we want to leave $r(k+1)+1$ heaps containing one match, and no heaps with more than one, where r may have any non-negative integral value (i.e. $r = 0, 1, 2, \ldots$).

In case (*c*) the same applies. If only one heap is left, containing one match, we have no choice of move, but this need not be treated separately.

In case (*d*), the game is over. Special investigation has to be used to detect this case. In all other cases, if the normal routine cannot succeed in its purpose, i.e. if the machine is faced with a winning position—a random move can, and must, be made. But, in this situation, this obviously cannot be done.

Thus the routine breaks up naturally into the following parts—

(i) Entry
(ii) Determination of case
(iii) Normal Routine
(iv) Cases (*b*) and (*c*)
(v) Treatment of zero case (*d*)
(vi) Random move
(vii) Emergence.

There is no need to give further details of the programme, but an example is given of how the machine would tackle a specific game.

Suppose initially that we have four heaps, containing respectively 7, 4, 5 and 2 matches; that $k = 2$; and that the machine moves first.

(i) Entry—

	4	2	1	
A	1	1	1	(7)
B	1	0	0	(4)
C	1	0	1	(5)
D	0	1	0	(2)
sum :	2	2	2	

(ii) Determination of case—
There are 4 non-zero, non-unit heaps, so we are dealing with case (a).
(iii) Normal routine—

	4	2	1	
A	1	1	1	(7)
B	1	0	0	(4)
C	1	0	1	(5)
D	0	1	0	(2)
sum :	3	2	2	

The sum of column 4 is divisible by $k+1$ so we need not modify it.
The sum of column 2 is 2, and is not divisible by $k+1$, so we need to modify any heap having a 1 in this column—say heap A.
According to the rules, we then get—

	4	2	1	
A	1	0	1	(5)
B	1	0	0	(4)
C	1	0	1	(5)
D	0	1	0	(2)
sum :	3	1	2	

And we note that heap A has been modified, and should be again modified whenever possible. Sum of column 2 is still not divisible by $k+1$, so this time we modify heap D to obtain—

	4	2	1	
A	1	0	1	(5)
B	1	0	0	(4)
C	1	0	1	(5)
D	0	0	1	(1)
sum :	3	0	3	

Column 2 is now divisible by $k + 1$ and, proceeding to the next column, we see this condition is also satisfied here, so the move has been completed and a winning-position left, the means to this end being the removal of two matches from A, and one from D, leaving 5, 4, 5 and 1. (If column 1 had needed adjustment, we should have had to modify one or both of heaps A and D, since these had already been affected.)

Suppose the opponent now makes a move leaving 0, 4, 2 and 1 as the contents of the respective heaps. It is now for the machine to move again.

(i) Entry—

	4	2	1	
A	0	0	1	(0)
B	1	0	0	(4)
C	0	1	0	(2)
D	0	0	1	(1)

(ii) Determination of case.
There are 3 non-zero, non-unit heaps, so we are dealing with case (b). Thus we want to leave 1, or 4, or 7... unit-heaps. Clearly we can only leave 1 unit heap in this case.

(iv) Cases (b) and (c).
We remove all matches from heaps B and D, which affects only k heaps, and leaves just one unit heap as required.

The opponent is now forced to remove the last match, and the machine wins the game.

Examining the Work and Its Later Impact

David Levy[1] delves deeper into —

ALAN TURING ON COMPUTER CHESS

In the history of computer chess three names stand out as candidates for the title of 'founder'. The accolade is usually given to Claude Shannon, who was certainly the founder of the field of Information Theory and whose seminal lecture[2] on computer chess in 1949 laid the foundations of most of the chess programs developed since then.

A second candidate is the German computer designer Konrad Zuse, who in 1945 wrote *Die Plankalkül*, describing an algorithmic language he had devised during the period 1936–45. Zuse's original publication was in German and it was not until 1975 that an English translation appeared in print. Consequently Zuse's work on computer chess[3] was virtually unknown to the English speaking world until after Shannon's ideas had already been adopted by most of those working in this field, and Zuse's contribution to the field has been rather slight in the sense that no leading program has ever been based on his ideas.

The third candidate for the title of 'founder of computer chess' is Alan Turing. Although not a strong player himself, he was very keen on the game and had even taught himself to play without sight of the board. Turing recognized that chess could be a suitable vehicle for a study of the mechanization of thought processes, and the idea of a chess program fascinated him. During the war years, while working at Bletchley Park on the breaking of the German Enigma code, Turing had many discussions on the subject with Donald Michie and Jack Good, both of whom were also chess enthusiasts.[4]

It was Turing, in fact, who was the first to publish in English on how chess could be programmed. In a report in 1946 for the National Physical Laboratory, entitled *Proposed Electronic Calculator*, Turing described some of the problems that he expected to be submitted to the computer he was proposing, one of which was programming a computer to play chess.

> 'Given a position in chess the machine could be made to list all the "winning combinations" to a depth of about three moves on either side. This is not unlike the previous problem[5] but raises the question "Can the machine play chess?" It could fairly easily be made to play a rather bad game. It would be bad because chess requires intelligence. We stated at the

---

[1] Dr. David Levy is an International Master at Chess, and President of the International Computer Games Association. davidlevylondon@yahoo.com

[2] Shannon's 1950 paper 'Programming a Computer for Playing Chess' was the first publication to describe how a position might be evaluated and how a game tree might be searched. That paper was the text of a lecture given by Shannon to the Institute of Radio Engineers on 9 March 1949, which happened to be Bobby Fischer's sixth birthday.

[3] Described in Chapter 5 of the English edition, pp. 201–244.

[4] Bletchley Park was a hotbed for chess addicts. Three of the five players in the English chess team who participated in the 1939 Chess Olympiad in Buenos Aires were recruited to work at Bletchley: Harry Golombek, Stuart (later Sir Stuart) Milner-Barry and Hugh Alexander. It is said that once, when arriving on his bicycle for work, Turing was asked by a guard at the gate of who his head of section was, and rather than naming himself he replied 'Alexander', so important was the latter's role in Turing's team.

[5] Solving a jigsaw.

> beginning of this section [i.e. when describing how programming is done] that the machine should be treated as entirely without intelligence. There are indications however that it is possible to make the machine display intelligence at the risk of its making occasional serious mistakes. By following up this aspect the machine could probably be made to play very good chess'.

Turing later augmented these comments in a lecture to the London Mathematical Society[6] in which he raised the possibility of having a chess machine that could learn from its own experience.

> 'It would probably be quite easy to find instruction tables which would enable the ACE to win against an average player. Indeed Shannon of Bell Telephone laboratories tells me that he has won games playing by rule of thumb: the skill of his opponents is not stated. But I would not consider such a victory very significant. What we want is a machine that can learn from experience. The possibility of letting the machine alter its own instructions provides the mechanism for this, but this of course does not get us very far'.

And he subsequently speculated in his 1948 report *Intelligent Machinery*, also written for the National Physical Laboratory, on whether a chess playing machine should be regarded as intelligent, suggesting a form of the imitation game[7] which became famous through Turing's 1950 paper *Computing Machinery and Intelligence*.

> 'The extent to which we regard something as behaving in an intelligent manner is determined as much by our own state of mind and training as by the properties of the object under consideration. If we are able to explain and predict its behaviour or if there seems to be little underlying plan, we have little temptation to imagine intelligence. With the same object therefore it is possible that one man would consider it as intelligent and another would not; the second man would have found out the rules of its behaviour'.

> 'It is possible to do a little experiment on these lines, even at the present state of knowledge. It is not difficult to devise a paper machine which will play a not very bad game of chess. Now get three men as subjects for the experiment A, B, C. A and C are to be rather poor chess players, B is the operator who works the paper machine. (In order that he should be able to work it fairly fast, it is advisable that he be both mathematician and chess player.) Two rooms are used with some arrangement for communicating moves, and a game is played between C and either A or the paper machine. C may find it quite difficult to tell which he is playing. (This is a rather idealized form of an experiment I have actually done.)'

Following the end of the war Turing's fascination for the mechanization of thought processes caused him to dwell on the problem of programming chess, continuing his discussions with Good and Michie. Almost nothing survives of whatever correspondence they exchanged on the subject, only the following snippet from a letter he wrote to Good on 18 September 1948.

> 'The chess machine designed by Champ[8] and myself is rather on your lines. Unfortunately we made no definite record of what time it was[9], but I am going to write one down definitely in the next few days with a view to playing the Shaun-Michie machine'.[10]

---

[6] On 20 February 1947.

[7] More widely known as the Turing Test.

[8] 'Champ' was the Cambridge economist David Champernowne, a friend of Turing from the time they were undergraduates together at King's College, Cambridge.

[9] What Turing meant in this letter by 'what time it was' is not entirely clear. Perhaps he was referring to the time it took to perform the calculations necessary to make a move.

[10] Shaun Wylie was a colleague of Donald Michie at Bletchley Park.

The Turing–Champernowne 'paper machine' was the first known chess program or simulation, preceding Shannon's first publication on the subject. It searched only one 'ply'[11], employing an evaluation function which survives in the literature in an article by Turing published in 1953.[12]

In her biography of her son, Sara Turing recounts Michie's recollections of Turochamp.

> 'Alan told me that he and Champernowne had constructed a machine to play chess, in the sense of a complete specification on paper for such a machine. One could call it a "paper machine" from which one could laboriously calculate move by move what the corresponding electronic machine would do were it constructed. Each move required perhaps half an hour's paper work as compared with the fraction of a second which a real machine would need. During a stay in Cambridge, Shaun Wylie and I constructed a rival "paper machine" which we christened Machiavelli, from our two names, Michie-Wylie. On behalf of Machiavelli we then issued a challenge to the Turochamp (our name for the Turing-Champernowne machine), the game to be played by correspondence. Alan and I were responsible for conducting the correspondence and working out the moves for our respective machines. The labour involved proved too tedious for us and the game did not progress beyond the first few moves. Alan was then at Manchester, I think, and he had plans to programme the electronic computer there[13] with the two chess machines so as to be able to run off a series of games between them in a short time and so discover which was the better. I think he embarked on the project, but it was never finished'.

The Turochamp 'program' only ever beat one opponent, Champernowne's wife, who was a beginner.[14] Fortunately, the one substantive piece of Turing's writing on the subject of computer chess (included above) provides a detailed insight on their method of position evaluation.

On pages 627–630 above can be found Turing's discussion of the program. He comments on p. 629:

> After this introduction I shall describe a particular set of rules, which could without difficulty be made into a machine programme. It is understood that the machine is white and that white is next to play. The current position is called the position on the board, and the positions arising from it by later moves positions in the analysis.

Following the description of the rules, Turing continues (p. 630):

> We can now state the rule for play as follows. The move chosen must have the greatest possible value, and, consistent with this, the greatest possible position-play value. If this condition admits of several solutions a choice may be made at random, or according to an arbitrary additional condition.
> Note that no 'analysis' is involved in position-play evaluation. This is to reduce the amount of work done on deciding the move.
> The game below was played between this machine and a weak player who did not know the system. To simplify the calculations the square roots were rounded off to one decimal place, i.e., this table was used.

---

[11] A ply, or half-move, is a move by White or a move by Black. The depth of search of chess and other strategy games programs is traditionally measured in ply. Thus, in mentioning 'a depth of about three moves on either side' in the NPL report quoted above, Turing is referring to a 6-ply search.

[12] See below for the full text of that article.

[13] The Ferranti Mark I. It was for this computer that Dietich Prinz, a fellow researcher at Manchester University, wrote the first program to solve chess problems of the type: 'White to play and force checkmate in two moves'. That program first ran in November 1951. But this was not a program to *play* chess – a problem of this type is an analytical exercise that bears almost no relation to playing the game.

[14] Sara Turing was told by Champernowne that the calculations took only 2 or 3 minutes per move.

There is some interest in a closer look at Turing's example, adding some pertinent observations. Note that in Turing's article he employed 'descriptive' notation for the moves, which was the popular method of chess notation in the UK at that time. I have transcribed the moves into algebraic notation which today is the most popular form of notation throughout the world. The comments are by Turing apart from those marked DL which are mine.

	White:(Machine)		Black:(Human)
1	e2-e4	4.2*	e7-e5
2	Nb1-c3	3.1*	Ng8-f6
3	d2-d4	2.6*	Bf8-b4
4	Ng1-f3	2.0	

If 4 Bc1-d2 3.7* then 4… e5×d4 is foreseen.

4	…		d7-d6
5	Bc1-d2	3.5*	Nb8-c6
6	d4-d5	0.2	Nc6-d4
7	h2-h4	1.1*	

[Strangely enough, even though such moves would be almost unthinkable in a game between reasonable human players, they were not uncommon in computer games during the early decades of computer chess. The reason is not hard to find. The programme's positional 'judgement' is governed by its evaluation function, which is designed to incorporate various chess rules-of-thumb (heuristics). Two of the heuristics embodied in Turing's evaluation function, as well as in the evaluation functions of many more recent programmes, are: (i) advance your pawns (exemplified by the bonus of 0.3 for each rank advanced); and (ii) increase your mobility (score the square root of the number of moves that a piece can make). The move 7 h2-h4 scores a bonus of 0.6 for advancing the pawn two ranks, and it increases the mobility of White's rook on h1 from 2 (for which it scores 1.4) to 4 (for which it scores 2). – DL]

7	…		Bc8-g4
8	a2-a4	1.0*	

White's last two moves are most inappropriate.
[This useless pawn advance is made for the same reason as h2-h4. – DL]

8	…		Nd4xf3+
9	g2xf3		Bg4-h5
10	Bf1-b5+	2.4*	c7-c6

[Obviously 10… Nf6-d7 would be better. – DL]

11	d5xc6		0-0
12	c6xb7		Ra8-b8
13	Bb5-a6	-1.5	Qd8-a5
14	Qd1-e2	0.6	Nf6-d7

[Back can win back one pawn by 14 ... Bh5xf3! 15 Qe2xf3 Qa5xa6. – DL]

```
15  Rh1-g1        1.2*
```

If 15 0-0-0 Bb4xc3 16 Bd2xc3 Qa5xa4.

```
15  ...                     Nd7-c5
```

The fork is unforeseen at White's last move.

```
16  Rg1-g5
```

Heads in the sand!

[What Turing means by this comment is that the programme is faced with the loss of its advanced pawn on b7 and so it staves off this material loss as far as possible. By playing 16 Rg1-g5 the programme appears to be avoiding reality – it simply pushes reality (in this case the loss of the b7 pawn) over its horizon. This is known in the computer chess world as the 'horizon effect'. But now, after Black moves his attacked bishop and White retreats his own bishop, the capture of the b7 pawn has not been averted, but its capture will occur at a depth too great for the programme to see at this point in the game.

But in fact 16 Rg1-g5 is White's best move. – DL]

```
16  ...                     Bh5-g6
17  Ba6-b5        0.4
```

[An aimless move. 17 h4-h5 is also not good because of 17 ... Nc5-e6. But 17 Ba6-c4 was obviously the best choice, since if Black were then to capture the b7 pawn White could play 18 h4-h5, trapping the bishop, for example 18...h7-h6 19 Rg5xg6. If Black meets 17 Ba6-c4 with 17...Kg8-h8, avoiding the pin along the g8-a2 diagonal, White wins by 18 h4-h5, for example 18... f7-f6 19 Rg5xg6 h7xg6 20 h5xg6, when it is impossible for Black to prevent White from checkmating him on h7 by Qe2-f1 followed by Qf1-h1, or by f3-f4 followed by Qe2-h5. The idea 17... Kg8-h8 18 h4-h5 h7-h6 is also no good because of 19 h5xg6 h6xg5 20 Qe2-f1 etc. It seems that Black must reply to 17 Ba6-c4 with 17... Nc5-e6 18 Bc4xe6 f7xe6, when White has an excellent position.

So it appears that Turing's programme had a clear advantage. – DL]

```
17  ...                     Nc5xb7
18  0-0-0         3.2*
```

[18 Bb5-c4, threatening 19 h4-h5, probably gives White a won game. 18 h4-h5 h7-h6 19 h5xg6 h6xg5 20 Bb5-c4 is also very difficult for Black to meet. The programme, however, is more attracted by the bonus attached to castling. – DL]

```
18  ...                     Nb7-c5
```

[Now it is too late for 19 h4-h5, which can be met by ... Nc5-e6, while 19 Bb5-c4 Nc5xa4 is also good for Black. – DL]

```
19  Bb5-c6                  Rf8-c8
```

[19 ... Nc5-e6 was essential, for obvious reasons. – DL]

```
20  Bc6-d5          Bb4xc3
21  Bd2xc3          Qa5xa4
22  Kc1-d2
```

[22 h4-h5 wins for White. – DL]

```
22  ...             Nc5-e6
```

[At last, but it ought to be too late. – DL]

```
23  Rg5-g4     -0.3
```

[23 Bc4xe6 was correct. Now Black's knight becomes a nuisance. – DL]

```
23  ...             Ne6-d4
```

[23 ... Ne6-f4 followed by 24... Nf4xd5 would have put an end to White's K-side play. – DL]

```
24  Qe2-d3          Nd4-b5
25  Bd5-b3          Qa4-a6
26  Bb3-c4
```

[26 Rd1-g1 gives White a winning attack. – DL]

```
26  ...             Bg6-h5
27  Rg4-g3
```

[It would be better to go back to g5. – DL]

```
28  ...             Qa6-a4
28  Bc4xb5          Qa4xb5
29  Qd3xd6
```

Fiddling while Rome burns!

[After 29 Qd3×b5 Rb8×b5 30 Rd1-g1 g7-g6 31 Kd2-e3, White could unravel his rooks and keep a big advantage because of the superiority of his own bishop over that of his opponent. – DL]

```
29  ...             Rc8-d8
```

[White had overlooked the strength of this move because it was too far ahead. When playing its 29th move the programme was unable to see to the position in which its queen was captured (which was at a depth of 4-ply.) – DL]

```
30  Resigns
```

On the advice of his trainer. — So ends Turing's description of this trial run.

We refer the reader to Turing's own entertaining discussion of the the 'Numerous criticisms of the machine's play [that] may be made.' And of aspects of the future role of computers in the playing of chess.

References

Turing, A.M., 1946. Proposed electronic calculator, National Physical Laboratory report. In: Carpenter, B.E., Doran, R.W. (Eds.), A.M. Turing's ACE Report of 1946 and other papers, MIT Press and Tomash Publishers (1986).
Turing, A.M., 1948. Intelligent machinery, report, National Physical Laboratory. In: Meltzer, B., Michie, D. (Eds.), Machine Intelligence 5, Edinburgh University Press, pp. 3–23.
Turing, S., 1959. Alan M. Turing. W. Heffer and Sons, Cambridge.

Can Digital Computers Think?

(BBC Third Programme radio broadcast (15 May 1951), transcript edited B. J. Copeland)

Intelligent Machinery: A Heretical Theory

(Lecture given to *51 Society* in Manchester (c.1951), transcript edited B. J. Copeland)

Can Automatic Calculating Machines Be Said To Think?

(Broadcast discussion, BBC Third Programme (14 and 23 January 1952), transcript edited B. J. Copeland)

B. Jack Copeland introduces the transcripts —

TURING AND THE PHYSICS OF THE MIND

1. Introduction

Turing's lecture 'Can Digital Computers Think?' was broadcast on BBC Radio on 15th May 1951 (repeated on 3rd July). It was the second in a series of lectures entitled 'Automatic Calculating Machines'. Other contributors to the series included Max Newman (like Turing from the University of Manchester), Douglas Hartree (University of Cambridge), Maurice Wilkes (Cambridge), and F. C. Williams (Manchester)[1]. In modern times, 'Can Digital Computers Think?' was virtually unknown until 1999, when I included it in a small collection of unpublished work by Turing ('A Lecture and Two Radio Broadcasts on Machine Intelligence by Alan Turing', in *Machine Intelligence 15*) and again in *The Essential Turing* in 2004. The previously published text, reproduced here, is from Turing's own typescript and incorporates corrections made in his hand.

In this broadcast Turing's overarching aim was to defend his view that 'it is not altogether unreasonable to describe digital computers as brains'. The broadcast contains a bouquet of fascinating arguments, and includes discussions of the Church–Turing thesis and of free will. There is a continuation of Turing's discussion of 'Lady Lovelace's dictum' (which he had begun in 'Computing Machinery and Intelligence' the previous year), and a priceless analogy that likens the attempt to program a computer to act like a brain to trying to write a treatise about family life on Mars – and moreover with insufficient paper. The broadcast makes manifest Turing's real attitude to

[1] Letter from Maurice Wilkes to Copeland (9 July 1997).

talk of machines thinking. In 'Computing Machinery and Intelligence', he famously said that the question 'Can machines think?' is 'too meaningless to deserve discussion', Turing (1950) but in this broadcast he made liberal use of phrases such as 'programming a machine to think' and 'the attempt to make a thinking machine'. In one passage he said:

> our main problem [is] how to programme a machine to imitate the brain, or as we might say more briefly, if less accurately, to think.

However, the feature of the broadcast that is of absolutely outstanding interest, and is the topic of this note, is Turing's brief discussion of the possibility that physical action is not always computable.[2] Roger Penrose attributed the opposite view to Turing. He said, 'It seems likely that he [Turing] viewed physical action in general – which would include the action of a human brain – to be always reducible to some kind of Turing-machine action' (Penrose, 1994, p. 21). Penrose even named this claim *Turing's thesis.* Yet Turing never endorsed this thesis. As 'Can Digital Computers Think?' makes clear, Turing was aware that the thesis might be false.

2. Physics and uncomputability: a brief history

This overview mentions only a few of the important milestones in the history of the current debate about physics and uncomputability. For more information, see Copeland (2002a, b, 2003) and Copeland and Sylvan (1999).

1. Scarpellini
   In an article published in German in 1963, Scarpellini speculated that non-recursive (i.e., non-Turing-machine-computable) processes might occur in nature. He wrote:

   > [O]ne may ask whether it is possible to construct an analogue-computer which is in a position to generate functions $f(x)$ for which the predicate $\int f(x) \cos nx dx > 0$ is not decidable [by Turing machine] while the machine itself decides by direct measurement whether $\int f(x) \cos nx dx$ is greater than zero or not.[3]

   Scarpellini's suggestion was de novo. He had no knowledge of Turing's 'Can Digital Computers Think?'. Working in isolation at the Battelle Research Institute in Geneva in 1961, he conceived the idea that natural processes describable by classical analysis might falsify the suggestion that every function effectively computable by machine is also computable by Turing machine. He recollects that the influence of Turing's work on his paper was twofold: 'Technically, in that Turing machines appear, explicitly or implicitly, at various points of my paper; and conceptually, in that his work caused me to perform for the continuum, i.e., analysis, constructions analogous to those that he introduced for discrete mathematics'.[4] In comments on his 1963 paper made in 2003 Scarpellini said:

   > '[I]t does not seem unreasonable to suggest that the brain may rely on analogue processes for certain types of computation and decision-making. Possible candidates which may give rise to such processes are the axons of nerve cells.... [I]t is conceivable that the mathematics of a collection of axons may lead to undecidable propositions.'[5]

[2] Aspects of Turing's broadcast not treated here are discussed in my 2004 (ch. 13), 2000, and introduction to 1999.

[3] Scarpellini 1963 (Scarpellini's translation is from Copeland 2003: 77).

[4] Letter from Bruno Scarpellini to Copeland (3 August 2011).

[5] Scarpellini in Copeland 2003:84-85.

2. Komar

   In 1964 Komar showed that the behaviour of a quantum system having an infinite number of degrees of freedom can be uncomputable, proving that, in general, the universal Turing machine cannot determine whether or not two arbitrarily chosen states of such a system are macroscopically distinguishable. He wrote:

   > '[T]here exists no [effective] procedure for determining whether two arbitrarily given physical states can be superposed to show interference effects characteristic of quantum systems. ... [I]t is rather curious ... that the issue of the macroscopic distinguishability of quantum states should be among the undecidable questions.'[6]

3. Kreisel

   Kreisel emphasised that it is an open question whether there are uncomputable natural processes. He discussed this theme in relation to classical mechanics, classical electrodynamics and quantum mechanics, in remarks throughout a series of papers spanning three decades. See, for example, Kreisel (1965, 1967, 1971, 1972, 1982, 1987), and especially the study published in 1970 and 1974. He said:

   > '*There is no evidence*that even present day quantum theory is a *mechanistic*, i.e. *recursive* theory in the sense that a recursively described system has recursive behaviour'.[7]

4. Pour-El and Richards

   In 1979 Pour-El and Richards published their paper 'A computable ordinary differential equation which possesses no computable solution', followed in 1981 by their equally transparently titled 'The wave equation with computable initial data such that its unique solution is not computable'. See also the book by Pour-El and Richards (1989). They explained their general approach as follows:

   > Our results are related to some remarks of Kreisel. In [1974], Kreisel concerns himself with the following question. Can one *predict theoretically* on the basis of some current physical theory—e.g. classical mechanics or quantum mechanics—the existence of a physical constant which is not a recursive real? Since physical theories are often expressed in terms of differential equations, it is natural to ask the following question: Are the solutions of
   > $\phi' = F(x,\phi)$, $\phi(0) = 0$, computable when F is?[8]

   Pour-El and Richards proved in their second paper that the behaviour of a certain system with computable initial conditions and evolving in accordance with the familiar three-dimensional wave equation is not computable. In a review of their two papers Kreisel wrote:

   > The authors suggest, albeit briefly and in somewhat different terms, that they have described an analogue computer that—even theoretically—cannot be simulated by a Turing machine. Here 'analogue computer' refers to any physical system, possibly with a discrete output, such as bits of computer hardware realizing whole 'subroutines'.

---

[6] Komar 1964: 543-544.

[7] Kreisel 1967: 270.

[8] Pour-El and Richards 1979: 63.

(Turing's idealized digital computer becomes an analogue computer once the physical systems are specified that realize—tacitly, according to physical theory—his basic operations.)[9]

5. Karp and Lipton
In a conference presentation in 1980, Karp and Lipton (1982) discussed infinite families of digital circuits, for example circuits consisting of boolean logic gates or McCulloch–Pitts neurons. Each individual circuit in an infinite family is finite. A family may be regarded as a representation of a piece of hardware that grows over time, each stage of growth consisting of the addition of a finite number of new nodes (e.g., neurons). The behaviour of any given circuit in a family can be calculated by some Turing machine or other (since each circuit is finite), but there may be no single Turing machine able to do this for all circuits in the family. In the case of some families, that is to say, the function computed by the growing hardware – successive members of the family computing values of the function for increasingly large inputs – is uncomputable.

6. Doyle
In 1982, Doyle suggested that the physical process of equilibriating – for example, a quantum system's settling into one of a discrete spectrum of states of equilibrium – is 'so easily, reproducibly and mindlessly accomplished' that it can be granted equal status alongside the operations usually termed effective. He wrote:

> My suspicion is that physics is easily rich enough so that ... the functions computable *in principle*, given Turing's operations and equilibriating, include non-recursive functions. For example, I think that chemistry may be rich enough that given a diophantine equation ... we plug values into [a] molecule as boundary conditions, and solve the equation iff the molecule finds an equilibrium.[10]

7. Rubel
Rubel emphasised that aspects of brain function are analogue in nature and suggested that the brain be modelled in terms of continuous mathematics, as against the discrete mathematics of what he called the 'binary model'. A proponent of analogue computation, he noted that '[A]nalog computers, besides their versatility, are extremely fast at what they do ... In principle, they act instantaneously and in real time.... Analogue computers are still unrivalled when a large number of closely related differential equations must be solved' (Rubel, 1985, p. 78–79). He maintained that not all analogue computers need be amenable to digital simulation, even in principle:

> One can easily envisage other kinds of black boxes of an input-output character that would lead to different kinds of analog computers. ... Whether digital simulation is possible for these 'extended' analog computers poses a rich and challenging set of research questions.[11]

8. Geroch and Hartle
Geroch and Hartle argued that theories describing uncomputable physical systems 'should be no more unsettling to physics than has the existence of well-posed problems unsolvable by any algorithm been to mathematics', suggesting that such theories 'may be forced upon us' in

[9] Kreisel 1982: 901.
[10] Doyle 1982 (pp. 519-520 in Copeland 2002).
[11] Rubel 1989: 1011.

the quantum domain (Geroch and Hartle, 1986, pp. 534, 549). They drew an analogy between the process of deriving predictions from such a physical theory and the process of calculating approximations to some uncomputable real number. One algorithm may deliver the first*n* digits of the decimal representation of the real number, another the next *m* digits and so on. Discovering each algorithm is, as they put it, akin to finding the theory in the first place – a creative act. They said:

> To predict to within, say, 10%, one manipulates the mathematics of the theory for a while, arriving eventually at the predicted number. To predict to within 1%, it will be necessary to work much harder, and perhaps to invent a novel way to carry out certain mathematical steps. To predict to 0.1% would require still more new ideas. ... The theory certainly 'makes definite predictions', in the sense that predictions are always in principle available. It is just that ever increasing degrees of sophistication would be necessary to extract those predictions. The prediction process would never become routine.[12]

This sophisticated analogy might have appealed to Turing. In a little noticed passage in 'On Computable Numbers' Turing defined a certain infinite binary sequence δ, which he showed to be uncomputable, and said: 'It is (so far as we know at present) possible that any assigned number of figures of δ can be calculated, but not by a uniform process. When sufficiently many figures of δ have been calculated, an essentially new method is necessary in order to obtain more figures' (1936, p. 253). In a wartime letter to Newman, Turing spoke at greater length about the necessity of introducing mathematical methods transcending any single uniform process, and he emphasised the connection between the necessity for 'essentially new methods' and the 'ordinal logics' of his 1939 paper (Turing, 1939, 1940). Placing the incompleteness results in a different light – these are usually stated in terms of there being true mathematical statements that are not provable – Turing said:

> The straightforward unsolvability or incompleteness results about systems of logic amount to this
>
> (α) One cannot expect to be able to solve the Entscheidungs problem for a system
> (β) One cannot expect that a system will cover all possible methods of proof.
>
> [W]e ... make proofs ... by hitting on one and then checking up to see that it is right. ... When one takes β) into account one has to admit that not one but many methods of checking up are needed. In writing about ordinal logics I had this kind of idea in mind."[13]

The picture of uncomputable physical systems drawn by Geroch and Hartle, and their rejection of an implication from a physical system's uncomputability to its unpredictability, fit very comfortably with Turing's thinking about the foundations of mathematics. As we shall see, though, Turing's pioneering discussion of uncomputability in physics concerned only the straightforwardly correct converse implication, from a physical system's unpredictability (over arbitrarily long spans of behaviour) to its uncomputability.

---

[12] Geroch and Hartle 1986: 549.

[13] Turing c. 1940: 212-213. See further Copeland and Shagrir 2012 (forthcoming).

9. Pitowsky

   In a conference presentation given in 1987, Pitowsky considered the question 'Can a physical machine compute a non-recursive function?' (1990, p. 82). Referring to the thesis that no non-recursive function is physically computable as *Wolfram's thesis*, see Wolfram (1985). Pitowsky said:

   > The question of whether Wolfram's thesis is valid is a problem in the physical sciences, and the answer is still unknown. Yet there are very strong indications that Wolfram's thesis may be invalid.[14]

   Pitowsky described notional physical devices, compatible with general relativity, that are able to compute functions that no Turing machine can compute. Pitowsky's proposals have been further developed by Hogarth, Shagrir and others (Hogarth, 1992, 1994; Shagrir and Pitowsky, 2003). See also Copeland and Shagrir (2007, 2011).

10. Penrose

    In 1989 the speculation that physics – and in particular the physics of the mind – may not always be computable hit the headlines, with the publication of Penrose's book *The Emperor's New Mind*. Penrose suggested that 'non-algorithmic action' may 'have a role within the physical world of very considerable importance' and that 'this role is intimately bound up with … "mind"' (1989, p. 557). See also Penrose (1994). In a précis of the book, Penrose wrote:

    > I have tried to stress that the mere fact that something may be scientifically describable in a precise way does not imply that it is computable. It is quite on the cards that the physical activity underlying our conscious thinking may be governed by precise but nonalgorithmic physical laws and our conscious thinking could indeed be the inward manifestation of some kind of nonalgorithmic physical activity.[15]

3. Turing on physics and uncomputability

Turing's early observation that there might be uncomputable physical processes has not been widely noticed (as witnessed by Penrose's attribution to him of a thesis equivalent to Wolfram's, under the name 'Turing's Thesis'). Yet Turing was one of the first, perhaps the very first, to raise the question whether there are uncomputable physical processes, and he must be regarded as a founding father of the enquiry whose origins are being sketched here.

In the course of his discussion in 'Can Digital Computers Think?', Turing considered the claim that if 'some particular machine can be described as a brain we have only to programme our digital computer to imitate it and it will also be a brain'. He observed that this 'can quite reasonably be challenged', pointing out that there is a problem if the machine's behaviour is not 'predictable by calculation'; and he drew attention to the view of physicist Arthur Eddington (expressed in Eddington's 1927 Gifford Lectures, 'The Nature of the Physical World') that in the case of the brain – and indeed the world more generally – 'no such prediction is even theoretically possible' (as Turing summarised Eddington) on account of 'the indeterminacy principle in quantum mechanics' (Eddington, 1929, ch. 14).

Turing's casual observation that something about the physics of the brain might make it impossible for a digital computer to calculate the brain's behaviour may largely have passed over the heads of his BBC radio audience. Yet, with hindsight, this observation prefaced the discussion of physics and uncomputability that would gradually unfold over the following decades.

[14] Pitowsky 1990: 86

[15] Penrose 1990: 653.

4. Uncomputability and freewill

Eddington's discussion of quantum indeterminacy was closely bound up with his discussion of the 'freedom of the human mind' (1929, p. 310). Turing, too, was much interested in the issue of free will, and seems to have believed that the mind is a partially random machine. See further Copeland (2000). We have the word of one of Turing's closest associates, Newman, that Turing 'had a deep-seated conviction that the real brain has a 'roulette wheel' somewhere in it.'[16]

Turing's principal aim in 'Can Digital Computers Think?' was not, though, to offer an analysis of free will, but to answer affirmatively the question posed by his title. He wished to guard his view that appropriate digital computers can be described as brains against any objection along the following lines: if a brain's future states cannot be predicted by computation, and if this feature of the brain is (not a detail of minor importance but) the seat of our free will, then digital computers, with their deterministic action, must be a completely different sort of beast.

Turing argued that his proposition 'If any machine can appropriately be described as a brain, then any digital computer can be so described' is entirely consistent with the possibility that the brain is the seat of free will:

> To behave like a brain seems to involve free will, but the behaviour of a digital computer, when it has been programmed, is completely determined. ... [I]t is certain that a machine which is to imitate a brain must appear to behave as if it had free will, and it may well be asked how this is to be achieved. One possibility is to make its behaviour depend on something like a roulette wheel or a supply of radium. ... It is, however, not really even necessary to do this. It is not difficult to design machines whose behaviour appears quite random to anyone who does not know the details of their construction.

Turing called machines of the latter sort 'apparently partially random'. An example that he gave elsewhere is a Turing machine in which 'the digits of the number π [are] used to determine the choices' (Turing, 1948, p. 416). although the secret cryptographic machines that he had worked on during the war, such as the Lorenz SZ40/42 ('Tunny'), would have formed much better examples of deterministic machines whose behaviour can appear quite random – if only he could have mentioned them. The story of Tunny is told in Copeland (2006); see Appendix 6 for a detailed description of Turing's main contributions to the attack on Tunny. (His involvement with all aspects of wartime code breaking was subject to the British government's Official Secrets Act.)

A genuinely partially random machine, on the other hand, is a discrete-state machine that contains a genuinely random element (Turing, 1948, p. 416). Except in the case where (even under idealisation) the machine has only a finite number N of configurations, a partially random discrete-state machine cannot be simulated by any Turing machine. This is because, as Church pointed out in 1939, if a sequence of integers $a_1, a_2, \ldots a_n, \ldots$ is random, then there is no function $f(n) = a_n$ that is calculable by Turing machine (Church, 1940). Randomness is an extreme form of uncomputability.

Apparently, partially random machines imitate partially random machines. As is well known, Turing advocated imitation as the basis of a test – the Turing test – that '[y]ou might call ... a test to see whether the machine thinks' (Turing et al., 1952, p. 495). See also Turing (1950). An appropriately programmed digital computer could give a convincing imitation of the behaviour produced by a human brain even if the brain is a partially random machine. The appearance that this deterministic machine gives of possessing free will is, Turing said, 'mere sham', but it is in his view nevertheless 'not altogether unreasonable' to describe a machine that successfully 'imitate[s] a brain' as itself being a brain.

---

[16] Newman in interview with Christopher Evans. ('The Pioneers of Computing: An Oral History of Computing'. London: Science Museum.)

Turing's strategy for dealing with what can be termed the *freewill objection* to human-level AI is elegant and provocative. For more on Turing and human-level AI, see Proudfoot and Copeland (2011).

References

Church, A., 1940. On the concept of a random sequence. Am. Math. Soc. Bull. 46, 130–135.

Copeland, B.J., 1999. A lecture and two radio broadcasts on machine intelligence by Alan Turing. In: Furukawa, K., Michie, D., Muggleton, S., (Eds.), Machine Intelligence 15, Oxford University Press, Oxford.

Copeland, B.J., 2000. Narrow versus wide mechanism. J. Philos. 96, 5–32.

Copeland, B.J., (Ed.), 2002. Hypercomputation. Part 1. Special issue of Minds and Machines, vol. 12(4).

Copeland, B.J., 2002a. 'Hypercomputation', in Copeland 2002.

Copeland, B.J., (Ed.), 2003. Hypercomputation. Part 2. Special issue of Minds and Machines, vol. 13(1).

Copeland, B.J., (Ed.), 2004. The Essential Turing, Oxford University Press, Oxford.

Copeland, B.J., et al., 2006. Colossus: The Secrets of Bletchley Park's Codebreaking Computers. Oxford University Press, Oxford.

Copeland, B.J., Shagrir, O., 2007. Physical computation: how general are Gandy's principles for mechanisms. Minds and Machines, 17, 217–231.

Copeland, B.J., Shagrir, O., 2011. Do accelerating Turing machines compute the uncomputable? Minds and Machines, 21, 221–239.

Copeland, B.J., Shagrir, O., forthcoming. Turing and Gödel on Computability and the Mind.

Copeland, B.J., Sylvan, R., 1999. Beyond the universal Turing machine. Australas. J. Philos. 77, 46–66.

Doyle, J., 1982. What is Church's Thesis? Laboratory for Computer Science, MIT. Published in Copeland 2002.

Eddington, A.S., 1929. The Nature of the Physical World. Cambridge University Press, Cambridge.

Geroch, R., Hartle, J.B., 1986. Computability and physical theories. Found. Phys. 16, 533–550.

Hogarth, M.L., 1992. Does general relativity allow an observer to view an eternity in a finite time? Found. Phys. Lett. 5, 173–181.

Hogarth, M.L., 1994. Non-Turing computers and non-Turing computability. Philos. Sci. Assoc. 1, 126–138.

Karp, R.M., Lipton, R.J., 1982. Turing machines that take advice. In: Engeler, E., et al., (Eds.), Logic and Algorithmic. L'Enseignement Mathématique, Genève.

Komar, A., 1964. Undecidability of macroscopically distinguishable states in quantum field theory. Phys. Rev. Second Ser. 133B, 542–544.

Kreisel, G., 1965. Mathematical Logic. In: Saaty, T.L. (Ed.), Lectures on Modern Mathematics, vol. 3. John Wiley, New York.

Kreisel, G., 1967. Mathematical logic: what has it done for the philosophy of mathematics? In: Schoenman, R. (Ed.), Bertrand Russell: Philosopher of the Century. George Allen and Unwin, London.

Kreisel, G., 1970. Hilbert's programme and the search for automatic proof procedures. In: Laudet, M, et al. (Eds.), Symposium on Automatic Demonstration. Lecture Notes in Mathematics, vol. 125. Springer, Berlin.

Kreisel, G., 1971. Some reasons for generalising recursion theory. In: Gandy, R.O., Yates, C.M.E. (Eds.), Logic Colloquium '69, North-Holland, Amsterdam.

Kreisel, G., 1972. Which number theoretic problems can be solved in recursive progressions on Π^1_1 – paths through $\mathcal{O}$? J. Symbol. Log. 37, 311–334.

Kreisel, G., 1974. A notion of mechanistic theory. Synthese 29, 11–26.

Kreisel, G., 1982. Review of Pour-El and Richards. J. Symbol. Log. 47, 900–902.

Kreisel, G., 1987. Church's thesis and the ideal of formal rigour. Notre Dame J. Formal Log. 28, 499–519.

Penrose, R., 1989. The Emperor's New Mind: Concerning Computers, Minds, and the Laws of Physics. Oxford University Press, Oxford.

Penrose, R., 1990. Précis of the emperor's new mind. Behavioral and Brain Sci. 13, 643–655, 692–705.

Penrose, R., 1994. Shadows of the Mind: A Search for the Missing Science of Consciousness. Oxford University Press, Oxford.

Pitowsky, I., 1990. The physical Church thesis and physical computational complexity. Iyyun 39, 81–99.

Pour-El, M.B., Richards, J.I., 1979. A computable ordinary differential equation which possesses no computable solution. Ann. Math. Log. 17, 61–90.

Pour-El, M.B., Richards, J.I., 1981. The wave equation with computable initial data such that its unique solution is not computable. Adv. Math. 39, 215–239.

Pour-El, M.B., Richards, J.I., 1989. Computability in Analysis and Physics. Springer, Berlin.

Proudfoot, D., Copeland, B.J., 2011. Artificial intelligence. In: Margolis, E., Samuels, R.I., Stich, S.P. (Eds.), The Oxford Handbook of Philosophy of Cognitive Science. Oxford University Press, New York.

Rubel, L.A., 1985. The brain as an analog computer. J. Theor. Neurobiol. 4, 73–81.

Rubel, L.A., 1989. Digital simulation of analog computation and Church's thesis. J. Symbol. Log. 54, 1011–1017.

Scarpellini, B., 1963. Zwei unentscheitbare Probleme der Analysis. Zeitschrift für mathematische Logik und Grundlagen der Mathematik 9, 265–289. Published in translation as 'Two Undecidable Problems of Analysis' (with a new commentary by Scarpellini) in Copeland 2003.

Shagrir, O., Pitowsky, I., 2003. Physical Hypercomputation and the Church-Turing Thesis, in Copeland 2003.

Turing, A.M., 1936. On computable numbers, with an application to the Entscheidungsproblem. Proc. Lond. Math. Soc. Ser. 2, 42, 230–265.

Turing, A.M., 1939. Systems of logic based on ordinals. Proc. Lond. Math. Soc. Ser. 2, 45, 161–228.

Turing, A.M., 1940. Letter from The Crown, Shenley Brook End', in Copeland 2004.

Turing, A.M., 1948. 'Intelligent Machinery'. National Physical Laboratory report. First published in a form reproducing the layout and wording of Turing's document in Copeland 2004.

Turing, A.M., 1950. Computing machinery and intelligence. Mind 59, 433–460.

Turing, A.M., Braithwaite, R.B., Jefferson, G., Newman, M.H.A., 1952. 'Can Automatic Calculating Machines Be Said To Think?', in Copeland 2004.

Wolfram, S., 1985. Undecidability and intractability in theoretical physics. Phys. Rev. Lett. 54, 735–738.

Can Digital Computers Think?[1]

A. M. TURING

Digital computers have often been described as mechanical brains. Most scientists probably regard this description as a mere newspaper stunt, but some do not. One mathematician has expressed the opposite point of view to me rather forcefully in the words 'It is commonly said that these machines are not brains, but you and I know that they are'. In this talk I shall try to explain the ideas behind the various possible points of view, though not altogether impartially. I shall give most attention to the view which I hold myself, that it is not altogether unreasonable to describe digital computers as brains. A different point of view has already been put by Professor Hartree.

First we may consider the naive point of view of the man in the street. He hears amazing accounts of what these machines can do: most of them apparently involve intellectual feats of which he would be quite incapable. He can only explain it by supposing that the machine is a sort of brain, though he may prefer simply to disbelieve what he has heard.

The majority of scientists are contemptuous of this almost superstitious attitude. They know something of the principles on which the machines are constructed and of the way in which they are used. Their outlook was well summed up by Lady Lovelace over a hundred years ago, speaking of Babbage's Analytical Engine. She said, as Hartree has already quoted, 'The Analytical Engine has no pretensions whatever to *originate* anything. It can do whatever *we know how to order it* to perform.' This very well describes the way in which digital computers are actually used at the present time, and in which they will probably mainly be used for many years to come. For any one calculation the whole procedure that the machine is to go through is planned out in advance by a mathematician. The less doubt there is about what is going to happen the better the mathematician is pleased. It is like planning a military operation. Under these circumstances it is fair to say that the machine doesn't originate anything.

There is however a third point of view, which I hold myself. I agree with Lady Lovelace's dictum as far as it goes, but I believe that its validity depends on considering how digital computers *are* used rather than how they *could be* used. In fact I believe that they could be used in such a manner that they could appropriately be described as brains. I should also say that 'If any machine can appropriately be described as a brain, then any digital computer can be so described'.

This last statement needs some explanation. It may appear rather startling, but with some reservations it appears to be an inescapable fact. It can be shown to follow from a characteristic property of digital computers, which I will call their *universality*. A digital computer is a *universal* machine in the sense that it can be made to replace any machine of a certain very wide class. It will not replace a bulldozer or a steam-engine or a telescope, but it will replace any rival design of calculating machine, that is to say any machine into which one can feed data and which will later print out results. In order to arrange for our computer to imitate a given machine it is only necessary to programme the computer to calculate what the machine in question would do under given circumstances, and in particular what answers it would print out. The computer can then be made to print out the same answers.

---

[1]First published in Copeland, B.J., 1999. A Lecture and Two Radio Broadcasts on Machine Intelligence by Alan Turing. In: Furukawa, K., Michie, D., Muggleton, S. (Eds.), Machine Intelligence 15. Oxford University Press, Oxford and New York, pp. 445–476. Reprinted in *The Essential Turing*, 2004.

If now some particular machine can be described as a brain we have only to programme our digital computer to imitate it and it will also be a brain. If it is accepted that real brains, as found in animals, and in particular in men, are a sort of machine it will follow that our digital computer suitably programmed, will behave like a brain.

This argument involves several assumptions which can quite reasonably be challenged. I have already explained that the machine to be imitated must be more like a calculator than a bulldozer. This is merely a reflection of the fact that we are speaking of mechanical analogues of brains, rather than of feet or jaws. It was also necessary that this machine should be of the sort whose behaviour is in principle predictable by calculation. We certainly do not know how any such calculation should be done, and it was even argued by Sir Arthur Eddington that on account of the indeterminacy principle in quantum mechanics no such prediction is even theoretically possible.

Another assumption was that the storage capacity of the computer used should be sufficient to carry out the prediction of the behaviour of the machine to be imitated. It should also have sufficient speed. Our present computers probably have not got the necessary storage capacity, though they may well have the speed. This means in effect that if we wish to imitate anything so complicated as the human brain we need a very much larger machine than any of the computers at present available. We probably need something at least a hundred times as large as the Manchester Computer. Alternatively of course a machine of equal size or smaller would do if sufficient progress were made in the technique of storing information.

It should be noticed that there is no need for there to be any increase in the complexity of the computers used. If we try to imitate ever more complicated machines or brains we must use larger and larger computers to do it. We do not need to use successively more complicated ones. This may appear paradoxical, but the explanation is not difficult. The imitation of a machine by a computer requires not only that we should have made the computer, but that we should have programmed it appropriately. The more complicated the machine to be imitated the more complicated must the programme be.

This may perhaps be made clearer by an analogy. Suppose two men both wanted to write their autobiographies, and that one had had an eventful life, but very little had happened to the other. There would be two difficulties troubling the man with the more eventful life more seriously than the other. He would have to spend more on paper and he would have to take more trouble over thinking what to say. The supply of paper would not be likely to be a serious difficulty, unless for instance he were on a desert island, and in any case it could only be a technical or a financial problem. The other difficulty would be more fundamental and would become more serious still if he were not writing his life but a work on something he knew nothing about, let us say about family life on Mars. Our problem of programming a computer to behave like a brain is something like trying to write this treatise on a desert island. We cannot get the storage capacity we need: in other words we cannot get enough paper to write the treatise on, and in any case we don't know what we should write down if we had it. This is a poor state of affairs, but, to continue the analogy, it is something to know how to write, and to appreciate the fact that most knowledge can be embodied in books.

In view of this it seems that the wisest ground on which to criticise the description of digital computers as 'mechanical brains' or 'electronic brains' is that, although they might be programmed to behave like brains, we do not at present know how this should be done. With this outlook I am in full agreement. It leaves open the question as to whether we will or will not eventually succeed in finding such a programme. I, personally, am inclined to believe that such a programme will be found. I think it is probable for instance that at the end of the century it will be possible to programme a machine to answer questions in such a way that it will be extremely difficult to guess whether the answers are being given by a man or by the machine. I am imagining something like a viva-voce examination, but with the questions and answers all typewritten in order that we need not consider such irrelevant matters as the faithfulness with which the human voice can be imitated. This only represents my opinion; there is plenty of room for others.

There are still some difficulties. To behave like a brain seems to involve free will, but the behaviour of a digital computer, when it has been programmed, is completely determined. These two facts must somehow be reconciled, but to do so seems to involve us in an age-old controversy, that of 'free will and determinism'. There are two ways out. It may be that the feeling of free will which we all have is an illusion. Or it may be that we really have got free will, but yet there is no way of telling from our behaviour that this is so. In the latter case, however well a machine imitates a man's behaviour it is to be regarded as a mere sham. I do not know how we can ever decide between these alternatives but whichever is the correct one it is certain that a machine which is to imitate a brain must appear to behave as if it had free will, and it may well be asked how this is to be achieved. One possibility is to make its behaviour depend on something like a roulette wheel or a supply of radium. The behaviour of these may perhaps be predictable, but if so, we do not know how to do the prediction.

It is, however, not really even necessary to do this. It is not difficult to design machines whose behaviour appears quite random to anyone who does not know the details of their construction. Naturally enough the inclusion of this random element, whichever technique is used, does not solve our main problem, how to programme a machine to imitate the brain, or as we might say more briefly, if less accurately, to think. But it gives us some indication of what the process will be like. We must not always expect to know what the computer is going to do. We should be pleased when the machine surprises us, in rather the same way as one is pleased when a pupil does something which he had not been explicitly taught to do.

Let us now reconsider Lady Lovelace's dictum. 'The machine can do whatever *we know how to order it* to perform'. The sense of the rest of the passage is such that one is tempted to say that the machine can *only* do what we know how to order it to perform. But I think this would not be true. Certainly the machine can only do what we *do* order it to perform, anything else would be a mechanical fault. But there is no need to suppose that, when we give it its orders we know what we are doing, what the consequences of these orders are going to be. One does not need to be able to understand how these orders lead to the machine's subsequent behaviour, any more than one needs to understand the mechanism of germination when one puts a seed in the ground. The plant comes up whether one understands or not. If we give the machine a programme which results in its doing something interesting which we had not anticipated I should be inclined to say that the machine *had* originated something, rather than to claim that its behaviour was implicit in the programme, and therefore that the originality lies entirely with us.

I will not attempt to say much about how this process of 'programming a machine to think' is to be done. The fact is that we know very little about it, and very little research has yet been done. There are plentiful ideas, but we do not yet know which of them are of importance. As in the detective stories, at the beginning of the investigation any trifle may be of importance to the investigator. When the problem has been solved, only the essential facts need to be told to the jury. But at present we have nothing worth putting before a jury. I will only say this, that I believe the process should bear a close relation to that of teaching.

I have tried to explain what are the main rational arguments for and against the theory that machines could be made to think, but something should also be said about the irrational arguments. Many people are extremely opposed to the idea of a machine that thinks, but I do not believe that it is for any of the reasons that I have given, or any other rational reason, but simply because they do not like the idea. One can see many features which make it unpleasant. If a machine can think, it might think more intelligently than we do, and then where should we be? Even if we could keep the machines in a subservient position, for instance by turning off the power at strategic moments, we should, as a species, feel greatly humbled. A similar danger and humiliation threatens us from the possibility that we might be superseded by the pig or the rat. This is a theoretical possibility which is hardly controversial, but we have lived with pigs and rats for so long without their intelligence much increasing, that we no longer trouble ourselves about this possibility. We feel that if it is to happen at all it will not be for several million years to come. But this new danger is much closer. If it comes at all it will almost certainly be within the next millennium. It is remote but not astronomically remote, and is certainly something which can give us anxiety.

It is customary, in a talk or article on this subject, to offer a grain of comfort, in the form of a statement that some particularly human characteristic could never be imitated by a machine. It might for instance be said that no machine could write good English, or that it could not be influenced by sex-appeal or smoke a pipe. I cannot offer any such comfort, for I believe that no such bounds can be set. But I certainly hope and believe that no great efforts will be put into making machines with the most distinctively human, but non-intellectual characteristics such as the shape of the human body; it appears to me to be quite futile to make such attempts and their results would have something like the unpleasant quality of artificial flowers. Attempts to produce a thinking machine seem to me to be in a different category. The whole thinking process is still rather mysterious to us, but I believe that the attempt to make a thinking machine will help us greatly in finding out how we think ourselves.

INTELLIGENT MACHINERY: A HERETICAL THEORY[1]

A. M. TURING

'You cannot make a machine to think for you'. This is a commonplace that is usually accepted without question. It will be the purpose of this paper to question it.

Most machinery developed for commercial purposes is intended to carry out some very specific job, and to carry it out with certainty and considerable speed. Very often it does the same series of operations over and over again without any variety. This fact about the actual machinery available is a powerful argument to many in favour of the slogan quoted above. To a mathematical logician this argument is not available, for it has been shown that there are machines theoretically possible which will do something very close to thinking. They will, for instance, test the validity of a formal proof in the system of Principia Mathematica, or even tell of a formula of that system whether it is provable or disprovable. In the case that the formula is neither provable nor disprovable such a machine certainly does not behave in a very satisfactory manner, for it continues to work indefinitely without producing any result at all, but this cannot be regarded as very different from the reaction of the mathematicians, who have for instance worked for hundreds of years on the question as to whether Fermat's last theorem is true or not. For the case of machines of this kind a more subtle argument is necessary. By Gödel's famous theorem, or some similar argument, one can show that however the machine is constructed there are bound to be cases where the machine fails to give an answer, but a mathematician would be able to. On the other hand, the machine has certain advantages over the mathematician. Whatever it does can be relied upon, assuming no mechanical 'breakdown', whereas the mathematician makes a certain proportion of mistakes. I believe that this danger of the mathematician making mistakes is an unavoidable corollary of his power of sometimes hitting upon an entirely new method. This seems to be confirmed by the well known fact that the most reliable people will not usually hit upon really new methods.

My contention is that machines can be constructed which will simulate the behaviour of the human mind very closely. They will make mistakes at times, and at times they may make new and very interesting statements, and on the whole the output of them will be worth attention to the same sort of extent as the output of a human mind. The content of this statement lies in the greater frequency expected for the true statements, and it cannot, I think, be given an exact statement. It would not, for instance, be sufficient to say simply that the machine will make any true statement sooner or later, for an example of such a machine would be one which makes all possible statements sooner or later. We know how to construct these, and as they would (probably) produce true and false statements about equally frequently, their verdicts would be quite worthless. It would be the actual reaction of the machine to circumstances that would prove my contention, if indeed it can be proved at all.

Let us go rather more carefully into the nature of this 'proof'. It is clearly possible to produce a machine which would give a very good account of itself for any range of tests, if the machine were made sufficiently elaborate. However, this again would hardly be considered an adequate proof. Such a machine would give itself away by making the same sort of mistake over and over again,

---

[1]First published in B. J. Copeland *A Lecture and Two Radio Broadcasts on Machine Intelligence by Alan Turing* in Furukawa, K., Michie, D., Muggleton, S. (eds) *Machine Intelligence 15*, Oxford and New York: Oxford University Press, 1999, pp. 445-476. Reprinted in *The Essential Turing*, 2004.

and being quite unable to correct itself, or to be corrected by argument from outside. If the machine were able in some way to 'learn by experience' it would be much more impressive. If this were the case there seems to be no real reason why one should not start from a comparatively simple machine, and, by subjecting it to a suitable range of 'experience' transform it into one which was more elaborate, and was able to deal with a far greater range of contingencies. This process could probably be hastened by a suitable selection of the experiences to which it was subjected. This might be called 'education'. But here we have to be careful. It would be quite easy to arrange the experiences in such a way that they automatically caused the structure of the machine to build up into a previously intended form, and this would obviously be a gross form of cheating, almost on a par with having a man inside the machine. Here again the criterion as to what would be considered reasonable in the way of 'education' cannot be put into mathematical terms, but I suggest that the following would be adequate in practice. Let us suppose that it is intended that the machine shall understand English, and that owing to its having no hands or feet, and not needing to eat, nor desiring to smoke, it will occupy its time mostly in playing games such as Chess and GO, and possibly Bridge. The machine is provided with a typewriter keyboard on which any remarks to it are typed, and it also types out any remarks that it wishes to make. I suggest that the education of the machine should be entrusted to some highly competent schoolmaster who is interested in the project but who is forbidden any detailed knowledge of the inner workings of the machine. The mechanic who has constructed the machine, however, is permitted to keep the machine in running order, and if he suspects that the machine has been operating incorrectly may put it back to one of its previous positions and ask the schoolmaster to repeat his lessons from that point on, but he may not take any part in the teaching. Since this procedure would only serve to test the bona fides of the mechanic, I need hardly say that it would not be adopted at the experimental stages. As I see it, this education process would in practice be an essential to the production of a reasonably intelligent machine within a reasonably short space of time. The human analogy alone suggests this.

I may now give some indication of the way in which such a machine might be expected to function. The machine would incorporate a memory. This does not need very much explanation. It would simply be a list of all the statements that had been made to it or by it, and all the moves it had made and the cards it had played in its games. This would be listed in chronological order. Besides this straightforward memory there would be a number of 'indexes of experiences'. To explain this idea I will suggest the form which one such index might possibly take. It might be an alphabetical index of the words that had been used giving the 'times' at which they had been used, so that they could be looked up in the memory. Another such index might contain patterns of men on parts of a GO board that had occurred. At comparatively late stages of education the memory might be extended to include important parts of the configuration of the machine at each moment, or in other words it would begin to remember what its thoughts had been. This would give rise to fruitful new forms of indexing. New forms of index might be introduced on account of special features observed in the indexes already used. The indexes would be used in this sort of way. Whenever a choice has to be made as to what to do next, features of the present situation are looked up in the indexes available, and the previous choice in the similar situations, and the outcome, good or bad, is discovered. The new choice is made accordingly. This raises a number of problems. If some of the indications are favourable and some are unfavourable what is one to do? The answer to this will probably differ from machine to machine and will also vary with its degree of education. At first probably some quite crude rule will suffice, e.g. to do whichever has the greatest number of votes in its favour. At a very late stage of education the whole question of procedure in such cases will probably have been investigated by the machine itself, by means of some kind of index, and this may result in some highly sophisticated, and, one hopes, highly satisfactory, form of rule. It seems probable however that the comparatively crude forms of rule will themselves be reasonably satisfactory, so that progress can on the whole be made in spite of the crudeness of the choice [of] rules.[2] This seems to be verified by the fact that engineering problems are sometimes solved by the crudest rule of thumb procedure which deals only with the most superficial aspects of the problem, e.g. whether a function increases or decreases with one of its variables. Another problem raised by this picture of the way behaviour is determined is the idea of 'favourable outcome'. Without some such idea, corresponding to the 'pleasure principle' of the psychologists, it is very difficult to see how to proceed. Certainly it would be most natural to introduce some such thing into the machine. I suggest that there should be two keys which can be manipulated by the schoolmaster, and which represent the ideas of pleasure and pain. At later stages in education the machine would recognise certain other conditions as desirable owing to their having been constantly associated in the past with pleasure, and likewise certain others as undesirable. Certain expressions of anger on the part of the schoolmaster might, for instance, be recognised as so ominous that they could never be overlooked, so that the schoolmaster would find that it became unnecessary to 'apply the cane' any more.

To make further suggestions along these lines would perhaps be unfruitful at this stage, as they are likely to consist of nothing more than an analysis of actual methods of education applied to human children. There is, however, one feature that I would like to suggest should be incorporated in the machines, and that is a 'random element'. Each machine should be supplied with a tape bearing a random series of figures, e.g. 0 and 1 in equal quantities, and this series of figures should be used in the choices made by the machine. This would result in the behaviour of the machine not being by any means completely determined by the experiences to which it was subjected, and would have some valuable uses when one was experimenting with it. By faking the choices made one would be able to control the development of the machine to some extent. One might, for instance, insist on the choice made being a particular one at, say, 10 particular places, and this would mean that about one machine in 1024 or more would develop to as high a degree as the one which had been faked. This cannot very well be given an accurate statement because of the subjective nature of the idea of 'degree of development' to say nothing of the fact that the machine that had been faked might have been also fortunate in its unfaked choices.

Let us now assume, for the sake of argument, that these machines are a genuine possibility, and look at the consequences of constructing them. To do so would of course meet with great opposition, unless we have advanced greatly in religious toleration from the days of Galileo. There would be great opposition from the intellectuals who were afraid of being put out of a job. It is probable though that the intellectuals would be mistaken about this. There would be plenty to do, [trying to understand what the machines were trying to say, e.g.][3] in trying to keep one's intelligence up to the standard set by the machines, for it seems probable that once the machine thinking method had started, it would not take long to outstrip our feeble powers. There would be no question of the machines dying, and they would be able to converse with each other to sharpen their wits. At some stage therefore we should have to expect the machines to take control, in the way that is mentioned in Samuel Butler's 'Erewhon'.

[2] Editor's note. Words enclosed in square brackets do not appear in the typescript.

[3] Editor's note. The words 'trying to understand what the machines were trying to say, e.g.' are handwritten and are marked in the margin 'Inserted from Turing's Typescript'.

CAN AUTOMATIC CALCULATING MACHINES BE SAID TO THINK?[1]

By ALAN TURING, RICHARD BRAITHWAITE, GEOFFREY JEFFERSON, MAX NEWMAN

Braithwaite: We're here today to discuss whether calculating machines can be said to think in any proper sense of the word. Thinking is ordinarily regarded as so much a speciality of man, and perhaps of other higher animals, that the question may seem too absurd to be discussed. But, of course, it all depends on what is to be included in thinking. The word is used to cover a multitude of different activities. What would you, Jefferson, as a physiologist, say were the most important elements involved in thinking?

Jefferson: I don't think that we need waste too much time on a definition of thinking since it will be hard to get beyond phrases in common usage, such as having ideas in the mind, cogitating, meditating, deliberating, solving problems or imagining. Philologists say that the word 'Man' is derived from a Sanskrit word that means 'to think,' probably in the sense of judging between one idea and another. I agree that we could no longer use the word 'thinking' in a sense that restricted it to man. No one would deny that many animals think, though in a very limited way. They lack insight. For example, a dog learns that it is wrong to get on cushions or chairs with muddy paws, but he only learns it as a venture that doesn't pay. He has no conception of the real reason, that he damages fabrics by doing that.
The average person would perhaps be content to define thinking in very general terms such as revolving ideas in the mind, of having notions in one's head, of having one's mind occupied by a problem, and so on. But it is only right to add that our minds are occupied much of the time with trivialities. One might say in the end that thinking was the general result of having a sufficiently complex nervous system. Very simple ones do not provide the creature with any problems that are not answered by simple reflex mechanisms. Thinking then becomes all the things that go on in one's brain, things that often end in an action but don't necessarily do so. I should say that it was the sum total of what the brain of man or animal does. Turing, what do you think about it? Have you a mechanical definition?

Turing: I don't want to give a definition of thinking, but if I had to I should probably be unable to say anything more about it than that it was a sort of buzzing that went on inside my head. But I don't really see that we need to agree on a definition at all. The important thing is to try to draw a line between the properties of a brain, or of a man, that we want to discuss, and those that we don't. To take an extreme case, we are not interested in the fact that the brain has the consistency of cold porridge. We don't want to say 'This machine's quite hard, so it isn't a brain, and so it can't think.' I would like to suggest a particular kind of *test* that one might apply to a machine. You might call it a test to see whether the machine thinks, but it would be better to avoid begging the question, and say that the machines that pass are (let's

[1]First published in B. J. Copeland *A Lecture and Two Radio Broadcasts on Machine Intelligence by Alan Turing* in Furukawa, K., Michie, D., Muggleton, S. (eds) *Machine Intelligence 15*, Oxford and New York: Oxford University Press, 1999, pp. 445-476. Reprinted in *The Essential Turing*, 2004.

say) 'Grade A' machines. The idea of the test is that the machine has to try and pretend to be a man, by answering questions put to it, and it will only pass if the pretence is reasonably convincing. A considerable proportion of a jury, who should not be expert about machines, must be taken in by the pretence. They aren't allowed to see the machine itself - that would make it too easy. So the machine is kept in a far away room and the jury are allowed to ask it questions, which are transmitted through to it: it sends back a typewritten answer.

Braithwaite: Would the questions have to be sums, or could I ask it what it had had for breakfast?

Turing: Oh yes, anything. And the questions don't really have to be questions, any more than questions in a law court are really questions. You know the sort of thing. 'I put it to you that you are only pretending to be a man' would be quite in order. Likewise the machine would be permitted all sorts of tricks so as to appear more man-like, such as waiting a bit before giving the answer, or making spelling mistakes, but it can't make smudges on the paper, any more than one can send smudges by telegraph. We had better suppose that each jury has to judge quite a number of times, and that sometimes they really are dealing with a man and not a machine. That will prevent them saying 'It must be a machine' every time without proper consideration.

Well, that's my test. Of course I am not saying at present either that machines really could pass the test, or that they couldn't. My suggestion is just that this is the question we should discuss. It's not the same as 'Do machines think,' but it seems near enough for our present purpose, and raises much the same difficulties.

Newman: I should like to be there when your match between a man and a machine takes place, and perhaps to try my hand at making up some of the questions. But that will be a long time from now, if the machine is to stand any chance with no questions barred?

Turing: Oh yes, at least 100 years, I should say.

Jefferson: Newman, how well would existing machines stand up to this test? What kind of things can they do now?

Newman: Of course, their strongest line is mathematical computing, which they were designed to do, but they would also do well at some questions that don't look numerical, but can easily be made so, like solving a chess problem or looking you up a train in the time-table.

Braithwaite: Could they do that?

Newman: Yes. Both these jobs can be done by trying all the possibilities, one after another. The whole of the information in an ordinary time-table would have to be written in as part of the programme, and the simplest possible routine would be one that found the trains from London to Manchester by testing every train in the time-table to see if it calls at both places, and printing out those that do. Of course, this is a dull, plodding method, and you could improve on it by using a more complicated routine, but if I have understood Turing's test properly, you are not allowed to go behind the scenes and criticise the method, but must abide by the scoring on correct answers, found reasonably quickly.

Jefferson: Yes, but all the same a man who has to look up trains frequently gets better at it, as he learns his way about the time-table. Suppose I give a machine the same problem again, can it learn to do better without going through the whole rigmarole of trying everything over every time? I'd like to have your answer to that because it's such an important point. Can machines learn to do better with practice?

Newman: Yes, it could. Perhaps the chess problem provides a better illustration of this. First I should mention that *all* the information required in any job - the numbers, times of trains, positions of pieces, or whatever it is, and also the instructions saying what is to be done with them - all this material is stored in the same way. (In the Manchester machine it is stored as a pattern on something resembling a television screen.) As the work goes on the pattern is changed. Usually it is the part of the pattern that contains the data that changes, while the instructions stay fixed. But it is just as simple to arrange that the instructions themselves shall be changed now and then. Well, now a programme could be composed that would cause the machine to do this: a 2-move chess problem is recorded into the machine in some suitable coding, and whenever the machine is started, a white move is chosen at random (there is a device for making random choices in our machine). All the consequences of this move are now analysed, and if it does *not* lead to forced mate in two moves, the machine prints, say, 'P-Q3, wrong move,' and stops. But the analysis shows that when the right move is chosen the machine not only prints, say, 'B-Q5, solution,' but it changes the instruction calling for a random choice to one that says 'Try B-Q5.' The result is that whenever the machine is started again it will immediately print out the right solution - and this without the man who made up the routine knowing beforehand what it was. Such a routine could certainly be made now, and I think this can fairly be called learning.

Jefferson: Yes, I suppose it is. Human beings learn by repeating the same exercises until they have perfected them. Of course it goes further, and at the same time we learn generally to shift the knowledge gained about one thing to another set of problems, seeing relevances and relationships. Learning means remembering. How long can a machine store information for?

Newman: Oh, at least as long as a man's lifetime, if it is refreshed occasionally.

Jefferson: Another difference would be that in the learning process there is much more frequent intervention by teachers, parental or otherwise, guiding the arts of learning. You mathematicians put the programme once into the machine and leave it to it. You wouldn't get any distance at all with human beings if that is what you did. In fact, the only time you do that in the learning period is at examinations.

Turing: It's quite true that when a child is being taught, his parents and teachers are repeatedly intervening to stop him doing this or encourage him to do that. But this will not be any the less so when one is trying to teach a machine. I have made some experiments in teaching a machine to do some simple operation, and a very great deal of such intervention was needed before I could get any results at all. In other words the machine learnt so slowly that it needed a great deal of teaching.

Jefferson: But who was learning, you or the machine?

Turing: Well, I suppose we both were. One will have to find out how to make machines that will learn more quickly if there is to be any real success. One hopes too that there will be a sort of snowball effect. The more things the machine has learnt the easier it ought to be for it to learn others. In learning to do any particular thing it will probably also be learning to learn more efficiently. I am inclined to believe that when one has taught it to do certain things one will find that some other things which one had planned to teach it are happening without any special teaching being required. This certainly happens with an intelligent human mind, and if it doesn't happen when one is teaching a machine there is something lacking in the machine. What do you think about learning possibilities, Braithwaite?

Braithwaite: No-one has mentioned what seems to me the great difficulty about learning, since we've only discussed learning to solve a particular problem. But the most important part of human learning is learning from experience - not learning from one particular kind of experience, but being able to learn from experience in general. A machine can easily be constructed with a feed-back device so that the programming of the machine is controlled by the relation of its output to some feature in its external environment - so that the working of the machine in relation to the environment is self-corrective. But this requires that it should be some particular feature of the environment to which the machine has to adjust itself. The peculiarity of men and animals is that they have the power of adjusting themselves to almost all the features. The feature to which adjustment is made on a particular occasion is the one the man is attending to and he attends to what he is *interested in*. His interests are determined, by and large, by his appetites, desires, drives, instincts - all the things that together make up his 'springs of action.' If we want to construct a machine which will vary its attention to things in its environment so that it will sometimes adjust itself to one and sometimes to another, it would seem to be necessary to equip the machine with something corresponding to a set of appetites. If the machine is built to be treated only as a domestic pet, and is spoon-fed with particular problems, it will not be able to learn in the varying way in which human beings learn. This arises from the necessity of adapting behaviour suitably to environment if human appetites are to be satisfied.

Jefferson: Turing, you spoke with great confidence about what you are going to be able to do. You make it sound as if it would be fairly easy to modify construction so that the machine reacted more like a man. But I recollect that from the time of Descartes and Borelli on people have said that it would be only a matter of a few years, perhaps 3 or 4 or maybe 50, and a replica of man would have been artificially created. We shall be wrong, I am sure, if we give the impression that these things would be easy to do.

Newman: I agree that we are getting rather far away from computing machines as they exist at present. These machines have rather restricted appetites, and they can't blush when they're embarrassed, but it's quite hard enough, and I think a very interesting problem, to discover how near these actually existing machines can get to thinking. Even if we stick to the reasoning side of thinking, it is a long way from solving chess problems to the invention of new mathematical concepts or making a generalisation that takes in ideas that were current before, but had never been brought together as instances of a single general notion.

Braithwaite: For example?

Newman: The different kinds of number. There are the integers, 0, 1, -2, and so on; there are the real numbers used in comparing lengths, for example the circumference of a circle and its diameter; and the complex numbers involving $\sqrt{-1}$; and so on. It is not at all obvious that these are instances of one thing, 'number.' The Greek mathematicians used entirely different words for the integers and the real numbers, and had no single idea to cover both. It is really only recently that the general notion of kinds of number has been abstracted from these instances and accurately defined. To make this sort of generalisation you need to have the power of recognising similarities, seeing analogies between things that had not been put together before. It is not just a matter of testing things for a specified property and classifying them accordingly. The concept itself has to be framed, something has to be created, say the idea of a number-field. Can we even guess at the way a machine could make such an invention from a programme composed by a man who had not the concept in his own mind?

Turing: It seems to me, Newman, that what you said about 'trying out possibilities' as a method applies to quite an extent, even when a machine is required to do something as advanced as finding a useful new concept. I wouldn't like to have to define the meaning of the word 'concept,' nor to give rules for rating their usefulness, but whatever they are they've got outward and visible forms, which are words and combinations of words. A machine could make up such combinations of words more or less at random, and then give them marks for various merits.

Newman: Wouldn't that take a prohibitively long time?

Turing: It would certainly be shockingly slow, but it could start on easy things, such as lumping together rain, hail, snow and sleet, under the word 'precipitation.' Perhaps it might do more difficult things later on if it was learning all the time how to improve its methods.

Braithwaite: I don't think there's much difficulty about seeing analogies that can be formally analysed and explicitly stated. It is then only a question of designing the machine so that it can recognise similarities of mathematical structure. The difficulty arises if the analogy is a vague one about which little more can be said than that one has a feeling that there is some sort of similarity between two cases but one hasn't any idea as to the respect in which the two cases are similar. A machine can't recognise similarities when there is nothing in its programme to say what are the similarities it is expected to recognise.

Turing: I think you could make a machine spot an analogy, in fact it's quite a good instance of how a machine could be made to do some of those things that one usually regards as essentially a human monopoly. Suppose that someone was trying to explain the double negative to me, for instance, that when something isn't not green it must be green, and he couldn't quite get it across. He might say 'Well, it's like crossing the road. You cross it, and then you cross it again, and you're back where you started.' This remark might just clinch it. This is one of the things one would like to work with machines, and I think it would be likely to happen with them. I imagine that the way analogy works in our brains is something like this. When two or more sets of ideas have the same pattern of logical connections, the brain may very likely economise parts by using some of them twice over, to remember the logical connections both in the one case and in the other. One must suppose that some part of my brain was used twice over in this way, once for the idea of double negation and once for crossing the road, there and back. I am really supposed to know about both these things but can't get what it is the man is driving at, so long as he is talking about all those dreary nots and not-nots. Somehow it doesn't get through to the right part of the brain. But as soon as he says his piece about crossing the road it gets through to the right part, but by a different route. If there is some such purely mechanical explanation of how this argument by analogy goes on in the brain, one could make a digital computer do the same.

Jefferson: Well, there isn't a mechanical explanation in terms of cells and connecting fibres in the brain.

Braithwaite: But could a machine really do this? How would it do it?

Turing: I've certainly left a great deal to the imagination. If I had given a longer explanation I might have made it seem more certain that what I was describing was feasible, but you would probably feel rather uneasy about it all, and you'd probably exclaim impatiently, 'Well, yes, I see that a machine could do all that, but I wouldn't call it thinking.' As soon as one can see the cause and effect working themselves out in the brain, one regards it as not being thinking, but a sort of unimaginative donkey-work. From this point of view one might be tempted to define thinking as consisting of 'those mental processes that we don't understand.' If this is right then to make a thinking machine is to make one which does interesting things without our really understanding quite how it is done.

Jefferson: If you mean that we don't know the wiring in men, as it were, that is quite true.

Turing: No, that isn't at all what I mean. We know the wiring of our machine, but it already happens there in a limited sort of way. Sometimes a computing machine does do something rather weird that we hadn't expected. In principle one could have predicted it, but in practice it's usually too much trouble. Obviously if one were to predict everything a computer was going to do one might just as well do without it.

Newman: It is quite true that people are disappointed when they discover what the big computing machines actually do, which is just to add and multiply, and use the results to decide what further additions and multiplications to do. '*That's* not thinking', is the natural comment, but this is rather begging the question. If you go into one of the ancient churches in Ravenna you see some most beautiful pictures round the walls, but if you peer at them through binoculars you might say, 'Why, they aren't really pictures at all, but just a lot of little coloured stones with cement in between.' The machine's processes are mosaics of very simple standard parts, but the designs can be of great complexity, and it is not obvious where the limit is to the patterns of thought they could imitate.

Braithwaite: But how many stones are there in your mosaic? Jefferson, is there a sufficient multiplicity of the cells in the brain for them to behave like a computing machine?

Jefferson: Yes, there are thousands, tens of thousands more cells in the brain than there are in a computing machine, because the present machine contains - how many did you say?

Turing: Half a million digits. I think we can assume that is the equivalent of half a million nerve cells.

Braithwaite: If the brain works like a computing machine then the present computing machine cannot do all the things the brain does. Agreed; but if a computing machine were made that could do all the things the brain does, wouldn't it require more digits than there is room for in the brain?

Jefferson: Well, I don't know. Suppose that it is right to equate digits in a machine with nerve cells in a brain. There are various estimates, somewhere between ten thousand million and fifteen thousand million cells are supposed to be there. Nobody knows for certain, you see. It is a colossal number. You would need 20,000 or more of your machines to equate digits with nerve cells. But it is not, surely, just a question of size. There would be too much logic in your huge machine. It wouldn't be really like a human output of thought. To make it more like, a lot of the machine parts would have to be designed quite differently to give greater flexibility and more diverse possibilities of use. It's a very tall order indeed.

Turing: It really is the size that matters in this case. It is the amount of information that can be stored up. If you think of something very complicated that you want one of these machines to do, you may find the particular machine you have got won't do, but if any machine can do it at all, then it can be done by your first computer, simply increased in its storage capacity.

Jefferson: If we are really to get near to anything that can be truly called 'thinking' the effects of external stimuli cannot be missed out; the intervention of all sorts of extraneous factors, like the worries of having to make one's living, or pay one's taxes, or get food that one likes. These are not in any sense minor factors, they are very important indeed, and worries concerned with them may greatly interfere with good thinking, especially with creative thinking. You see a machine has no environment, and man is in constant relation to his environment, which, as it were punches him whilst he punches back. There is a vast background of memories in a man's brain that each new idea or experience has to fit in with. I wonder if you could tell me how far a calculating machine meets that situation. Most people agree that man's first reaction to a new idea (such as the one we are discussing today) is one of rejection, often immediate and horrified denial of it. I don't see how a machine could as it were say 'Now Professor Newman or Mr. Turing, I don't like this programme at all that you've just put into me, in fact I'm not going to have anything to do with it.'

Newman: One difficulty about answering that is one that Turing has already mentioned. If someone says, 'Could a machine do this, e.g. could it say "I don't like the programme you have just put into me" ', and a programme for doing that very thing is duly produced, it is apt to have an artificial and ad hoc air, and appear to be more of a trick than a serious answer to the question. It is like those passages in the Bible, which worried me as a small boy, that say that such and such was done 'that the prophecy might be fulfilled which says' so and so. This always seemed to me a most unfair way of making sure that the prophecy came true. If I answer your question, Jefferson, by making a routine which simply caused the machine to say just the words 'Newman and Turing, I don't like your programme,' you would certainly feel this was a rather childish trick, and not the answer to what you really wanted to know. But yet it's hard to pin down what you want.

Jefferson: I want the machine to reject the problem because it offends it in some way. That leads me to enquire what the ingredients are of ideas that we reject because we instinctively don't care for them. I don't know why I like some pictures and some music and am bored by other sorts. But I'm not going to carry that line on because we are all different, our dislikes are based on our personal histories and probably too on small differences of construction in all of us, I mean by heredity. Your machines have no genes, no pedigrees. Mendelian inheritance means nothing to wireless valves. But I don't want to score debating points! We ought to make it clear that not even Turing thinks that all that he has to do is to put a skin on the machine and that it is alive! We've been trying for a more limited objective whether the sort of thing that machines do can be considered as thinking. But is not your machine more certain than any human being of getting its problem right at once, and infallibly?

Newman: Oh!

Turing: Computing machines aren't really infallible at all. Making up checks on their accuracy is quite an important part of the art of using them. Besides making mistakes they sometimes haven't done quite the calculation one had expected, and one gets something that might be called a 'misunderstanding.'

Jefferson: At any rate, they are not influenced by the emotions. You have only to upset a person enough and he becomes confused, he can't think of the answers and may make a fool of himself. It is high emotional content of mental processes in the human being that makes him quite different from a machine. It seems to me to come from the great complexity of his nervous system with its 10^{10} cells and also from his endocrine system which imports all sorts of emotions and instincts, such as those to do with sex. Man is essentially a chemical machine, he is much affected by hunger and fatigue, by being 'out of sorts' as we say, also by innate judgements, and by sexual urges. This chemical side is tremendously important, not the least so because the brain does exercise a remote control over the most important chemical processes that go on in our bodies. Your machines don't have to bother with that, with being tired or cold or happy or satisfied. They show no delight at having done something never done before. No, they are 'mentally' simple things. I mean that however complicated their structure is (and I know it *is* very complicated), compared with man they are very simple and perform their tasks with an absence of distracting thoughts which is quite *inhuman*.

Braithwaite: I'm not sure that I agree. I believe that it will be necessary to provide the machine with something corresponding to appetites, or other 'springs of action,' in order that it will pay enough attention to relevant features in its environment to be able to learn from experience. Many psychologists have held that the emotions in men are by-products of their appetites and that they serve a biological function in calling higher levels of mental activity into play when the lower levels are incapable of coping with an external situation. For example, one does not feel afraid when there is no danger, or a danger which can be avoided more or less automatically: fear is a symptom showing that the danger has to be met by conscious thought. Perhaps it will be impossible to build a machine capable of learning in general from experience without incorporating in it an emotional apparatus, the function of which will be to switch over to a different part of the machine when the external environment differs too much from what would satisfy the machine's appetites by more than a certain amount. I don't want to suggest that it will be necessary for the machine to be able to throw a fit of tantrums. But in humans tantrums frequently fulfil a definite function - that of escaping from responsibility; and to protect a machine against a too hostile environment it may be essential to allow it, as it were, to go to bed with a neurosis, or psychogenic illness - just as, in a simpler way, it is provided with a fuse to blow, if the electric power working it threatens its continued existence.

Turing: Well, I don't envisage teaching the machine to throw temperamental scenes. I think some such effects are likely to occur as a sort of by-product of genuine teaching, and that one will be more interested in curbing such displays than in encouraging them. Such effects would probably be distinctly different from the corresponding human ones, but recognisable as variations on them. This means that if the machine was being put through one of my imitation tests, it would have to do quite a bit of acting, but if one was comparing it with a man in a less strict sort of way the resemblance might be quite impressive.

Newman: I still feel that too much of our argument is about what hypothetical future machines will do. It is all very well to say that a machine could easily be made to do this or that, but, to take only one practical point, what about the time it would take to do it? It would only take an hour or two to make up a routine to make our Manchester machine analyse all possible variations of the game of chess right out, and find the best move that way - *if* you didn't mind its taking thousands of millions of years to run through the routine. Solving a problem on the machine doesn't mean finding a way to do it between now and eternity, but within a reasonable time. This is not just a technical detail that will be taken care of by future improvements. It's most unlikely that the engineers can ever give us a factor of more than a thousand or two times our present speeds. To assume that runs that would take thousands of millions of years on our present machines will be done in a flash on machines of the future, is to move into the realms of science fiction.

Turing: To my mind this time factor is the one question which will involve all the real technical difficulty. If one didn't know already that these things can be done by brains within a reasonable time one might think it hopeless to try with a machine. The fact that a brain *can* do it seems to suggest that the difficulties may not really be so bad as they now seem.

Braithwaite: I agree that we ought not to extend our discussion to cover whether calculating machines could be made which would do everything that a man can do. The point is, surely, whether they can do all that it is proper to call thinking. Appreciation of a picture contains elements of thinking, but it also contains elements of feeling; and we're not concerned with whether a machine can be made that will feel. Similarly with moral questions: we're only concerned with them so far as they are also intellectual ones. We haven't got to give the machine a sense of duty or anything corresponding to a will: still less need it be given temptations which it would then have to have an apparatus for resisting. All that it has got to do in order to think is to be able to solve, or to make a good attempt at solving, all the intellectual problems with which it might be confronted by the environment in which it finds itself. This environment, of course, must include Turing asking it awkward questions as well as natural events such as being rained upon, or being shaken up by an earthquake.

Newman: But I thought it was you who said that a machine wouldn't be able to learn to adjust to its environment if it hadn't been provided with a set of appetites and all that went with them?

Braithwaite: Yes, certainly. But the problems raised by a machine having appetites are not properly our concern today. It may be the case that it wouldn't be able to learn from experience without them; but we're only required to consider whether it would be able to learn at all - since I agree that being able to learn is an essential part of thinking. So oughtn't we to get back to something centred on thinking? Can a machine make up new concepts, for example?

Newman: There are really two questions that can be asked about machines and thinking, first, what do we require before we agree that the machine does *everything* that we call thinking? This is really what we have been talking about for most of the time; but there is also another interesting and important question: Where does the doubtful territory begin? What is the *nearest* thing to straight computing that the present machines perhaps can't do?

Braithwaite: And what would your own answer be?

Newman: I think perhaps to solve mathematical problems for which no method is known, in the way that men do; to find new methods. This is a much more modest aim than inventing new mathematical concepts. What happens when you try to solve a new problem in the ordinary way is that you think about it for a few seconds, or a few years, trying out all the analogies you can think of with problems that have been solved, and then you have an idea. You try it out in detail. If it is no good you must wait for another idea. This is a little like the chess-problem routine, where one move after another is tried, but with one very important difference, that if I am even a moderately good mathematician the ideas that I get are not just random ones, but are pre-selected so that there is an appreciable chance that after a few trials one of them will be successful. Henry Moore says about the studies he does for his sculpture, 'When the work is more than an exercise, inexplicable jumps occur. This is where the imagination comes in.' If a machine could really be got to imitate this sudden pounce on an idea, I believe that everyone would agree that it had begun to think, even though it didn't have appetites or worry about the income tax. And suppose that we also stuck to what we know about the physiology of human thinking, how much would that amount to, Jefferson?

Jefferson: We know a great deal about the end-product, thinking itself. Are not the contents of our libraries and museums the total up to date? Experimental psychology has taught us a lot about the way that we use memory and association of ideas, how we fill in gaps in knowledge and improvise from a few given facts. But exactly how we do it in terms of nerve cell actions we don't know. We are particularly ignorant of the very point that you mentioned just now, Newman, the actual physiology of the pounce on an idea, of the sudden inspiration. Thinking is clearly a motor activity of the brain's cells, a suggestion supported by the common experience that so many people think better with a pen in their hand than viva voce or by reverie and reflection. But you can't so far produce ideas in a man's mind by stimulating his exposed brain here or there electrically. It would have been really exciting if one could have done that - if one could have perhaps excited original thoughts by local stimulation. It can't be done. Nor does the electro-encephalograph show us how the process of thinking is carried out. It can't tell you what a man is thinking about. We can trace the course, say, of a page of print or of a stream of words into the brain, but we eventually lose them. If we could follow them to their storage places we still couldn't see how they are reassembled later as ideas. You have the great advantage of knowing how your machine was made. We only know that we have in the human nervous system a concern compact in size and in its way perfect for its job. We know a great deal about its microscopical structure and its connections. If fact, we know everything except how those myriads of cells allow us to think. But, Newman, before we say 'not only does this machine think but also here in this machine we have an exact counterpart of the wiring and circuits of human nervous systems,' I ought to ask whether machines have been built or could be built which are as it were anatomically different, and yet produce the same work.

Newman: The logical plan of all of them is rather similar, but certainly their anatomy, and I suppose you could say their physiology, varies a lot.

Jefferson: Yes, that's what I imagined - we cannot then assume that any one of these electronic machines is a replica of part of a man's brain even though the result of its actions has to be conceded as thought. The real value of the machine to you is its end results, its performance, rather than that its plan reveals to us a model of our brains and nerves. Its usefulness lies in the fact that electricity travels along wires 2 or 3 million times faster than nerve impulses pass along nerves. You can set it to do things that man would need thousands of lives to complete. But that old slow coach, man, is the one with the ideas - or so I think. It would be fun some day, Turing, to listen to a discussion, say on the Fourth Programme, between two machines on why human beings think that they think!

Examining the Work and Its Later Impact

Richard Jozsa takes us forward to —

QUANTUM COMPLEXITY AND THE FOUNDATIONS OF COMPUTING

In the radio broadcast 'Can automatic calculating machines be said to think?' Turing and his interlocutors explore the boundaries of the notion of computation itself by comparing and contrasting it with amazing (yet routinely familiar!) features of the functioning of the human brain. This key issue of what computation 'really is' or 'should be', has since the 1980s been vigorously explored in another context, that of fundamental physics, especially quantum physics. We have witnessed the emergence of a new field of scientific research known as quantum computation, having momentous implications for both the fundamental theory of computation and issues of practical computing.

Turing's original notion of computation was motivated from operational considerations, with the head and work tape of a Turing machine corresponding to a person calculating on sheets of paper. But we can entertain such operational considerations at a more fundamental level – if we recognise that any computer is a physical device and information is always represented in physical degrees of freedom then it follows that the possibilities and limitations of information processing must depend on the *laws of physics* and cannot be determined from mathematics or abstract thought alone. This fundamental connection between computation and theoretical physics, foreshadowed by earlier pioneers such as J. Wheeler, R. Feynman and R. Landauer, was perhaps first explicitly developed by David Deutsch in his seminal paper from 1985 'Quantum theory, the Church-Turing principle and the universal quantum computer' (*Proc. Roy. Soc. Lond.* A400, pp. 97-117). Deutsch writes for example that 'there is no a priori reason why the physical laws should respect the limitations of the mathematical processes we call "algorithms"' and 'Nor, conversely, is it obvious a priori that any of the familiar recursive functions is in physical reality computable ... The reason is that the laws of physics "happen to" permit the existence of physical models for [these] operations'.

The notion of computation embodied in the operation of a Turing machine may be argued to correspond to the computational possibilities of classical physics. On the other hand, quantum physics, superseding classical theory, is well known to give rise to a notoriously strange picture of the world and correspondingly it offers extraordinary novel possibilities for the representation and processing of information. Deutsch in the above paper, introduced the notion of 'quantum Turing machine' initiating the formal study of this new quantum computation.

To appreciate the significance of quantum versus classical computing it is necessary to introduce some basic notions from computational complexity theory. Instead of asking only whether a task is computable or uncomputable, we seek a finer distinction in the former case, asking how 'hard' it is to compute. Hardness is measured by consumption of resources viz. time (number of steps) and space (number of tape cells) that the computation (Turing machine) uses, as a function of input size (defined to be the length of the initial input string). Here we will consider only time complexity; we are mostly interested in one fundamental question: is the time complexity bounded by a polynomial function of the input size, or does it grow super-polynomially (exponentially)? Poly-time computations are regarded as 'feasible or *computable in practice*' whereas exponential time tasks, although computable in principle, are regarded as unfeasible or uncomputable *in practice*, as they consume resources at an unacceptable rate.

Returning to considering quantum physics from a computational perspective, we note first that all laws of quantum physics embody standardly computable prescriptions: given a description of a quantum physical system and its starting state, we can, using a classical computer, in principle calculate its state at any later time and the results of measurements of any of its physical properties. Thus we would not expect any 'quantum computation' (as a quantum process) to be able to compute any task that is Turing-uncomputable. However at the level of computational *complexity* there are known to be profound and exciting differences.

The non-classical characteristics of quantum theory may be broadly categorised into three features – quantum superposition, entanglement and measurement, and they turn out to have a singular significance for issues of computational complexity. If a quantum system S is capable of representing a generic n-bit string x, then according to the superposition principle it also has states involving the 'superposition' of all n-bit strings x with 'amplitudes' a_x. Such a generic state is also called entangled since the individual bit carriers cannot generally be assigned individual superpositions of 1-bit values, but instead they can be globally correlated ('entangled') in a peculiarly quantum fashion. The concept of entanglement is difficult to convey intuitively but a key feature is the following: a physical system that is the quantum analogue of a classical n-bit memory has states whose descriptions can be exponentially more complicated than those of the corresponding classical system, providing a rich new feature for information representation that goes beyond the possibilities of classical computing. These states can be formally interpreted as embodying the simultaneous presence of all n-bit strings, in the following sense: if we apply a physical time evolution then the final result is the same as if we had evolved each n-bit string separately and combined the final states with amplitudes a_x. For example if the evolution corresponds to the computation of some function $f(x)$ then at the expense of just a *single* run (on the superposed input), we can produce a state whose identity embodies *all* values of $f(x)$ for all 2^n inputs x. This kind of 'quantum parallel computation' is not available in the standard (Turing) notion of computation and appears to provide a new *exponentially powerful* mode of information processing i.e., one which would need an exponential amount of computing effort to simulate classically. However there is a catch, perhaps the strangest feature of quantum physics, namely the formalism of quantum measurement: if we are given a generic quantum state, then according to the laws of quantum physics it is impossible to determine the state's identity by any physical process (in contrast to classical physics where such determination is always possible while leaving the original state intact). Any extraction of classical information about the quantum state's identity is necessarily accompanied by irrevocable damage to the state (sometimes called 'collapse of the wave function in quantum measurement') and only a small amount of (generally probabilistic) information can be gained before total destruction of the given state's identity. Collecting the above, we get a very strange picture of the world: to perform quantum physical evolution, Nature updates quantum states in a remarkable way that would require an exponential overhead of classical computing effort to mimic by classical means; yet at the end, most of the updated information is kept hidden from us – we can read out as classical information, only a small part (which to some extent can be chosen by us) while the rest is irrevocably destroyed!

As an example, given a superposition state involving all 2^n values $f(x)$ of a Boolean function, we can extract very little information about the exponentially many values. But remarkably this 'small amount of information' can be *global or collective information about the set of all values*, which although small, may still require exponentially many function evaluations $f(x)$ to determine classically. In this way we can obtain a quantum computational advantage. The first explicit example of this phenomenon was the so-called Deutsch–Jozsa quantum algorithm of 1992: if f (having 1-bit values) is promised a priori to be either (a) the constant function having $f(x) = 0$ for all x or (b) a 'balanced' function, having 0 and 1 equally often amongst its 2^n values, then it may be shown that we can decide (a) versus (b) with certainty (i.e., just a single bit of global information about the set of values) from a *single* evaluation of f on an equal superposition of all x values. On the other hand, classically exponentially many evaluations are necessary to decide this task, giving the key exponential separation in time complexity. Another significant example is provided by Peter Shor's 1994 poly-time quantum algorithm for integer factorisation, perhaps to date the most celebrated result in the subject area, making the fundamentally important task of factorisation feasible on a quantum computer (whereas no feasible classical algorithm is known). Using some basic number theory the problem of integer factorisation is first converted into a problem of determining the period of an associated periodic function. This periodicity information (small compared to the total list of function values!) is then extracted by a quantum computation from a uniform superposition of the function's values, superposed across a suitably large range of inputs to capture the periodic pattern.

Perhaps the most famous outstanding issue in computational complexity theory is the so-called P versus NP problem i.e., establishing the relationship of the complexity class NP to that of (classical) poly-time computations. Thus in our present context we are compelled to ask: can quantum computation provide a poly-time algorithm for an NP-complete problem? Indeed from the foregoing discussion, we may at first sight have reason to be excitedly optimistic! Consider for example the (NP complete) satisfiability problem for Boolean functions – given a Boolean function f we ask 'does f take the value 1 or not, amongst its set of values?' i.e., we ask for just a *single bit* of of information about the full set of all values, and indeed as before, we can easily generate all of the exponentially many values in superposition. But alas the information requested, although small, is the 'wrong kind' of information – intuitively if we consider two Boolean functions requiring opposite answers viz. one uniquely satisfiable and the other unsatisfiable, then they differ in only a *single* $f(x)$ value. Hence the quantum states representing superpositions of all their values are exponentially close in the space of physical states and thus hard (i.e., require exponential effort) to distinguish by physical means. In contrast for a pattern structure such as periodicity, if the periodicity changes then a large number of values must change and the superposition states become readily distinguishable. At present (just as in classical computation theory and the class P) it is not known whether quantum computations can provide poly-time solutions to NP complete problems or not, but it generally believed that it is not possible. Indeed for the satisfiability problem, if the function (from n bits to one bit) is given as a *black box* then in 1994 it was shown by E. Bernstein, C. H. Bennett, G. Brassard and U. Vazirani that any quantum algorithm that purports to decide satisfiability (with any constant level of probability) must access the black box at least $O(\sqrt{2^n})$ times, and soon thereafter in 1996, Lov Grover gave an explicit quantum algorithm matching this lower bound. The relationship of NP to computability in the physical universe remains an infectiously fascinating subject, explored, for example in Scott Aaronson's paper 'NP-complete problems and physical reality' (*ACM SIGACT News*, March 2005).

In this commentary we have seen that the union of concepts from fundamental quantum physics and the theory of computing has a remarkable synergy. On the one hand we obtain a new (physically realistic) computational paradigm allowing the computation of some tasks in a way that consumes exponentially fewer resources than is known to be possible with standard (classical) computing. Furthermore, on the other hand concepts from computational complexity theory offer a potentially new perspective on quantum (or any post-quantum) physics with its own characteristic flavour of guiding principles. For example we might adopt as fundamental the principle that any prospective physical theory should not allow the efficient solution of an NP complete problem. This would greatly restrict the form of any prospective future theory (for example already ruling out some proposed non-linear generalisations of quantum theory) and remarkably, it appears to hold in the established formalisms of both classical and quantum physics, which were developed from entirely different considerations.

Alan Turing would surely have been delighted by the development of quantum computation. It is known (cf Andrew Hodges' celebrated biography 'Alan Turing – The Enigma') that he maintained a lifelong interest in quantum physics, but curiously he appears never to have made a connection with his fundamental work on computation. Returning finally to the subject of the radio broadcast it is interesting to mention that various authors (notably for example, Roger Penrose in his book 'The Emperor's New Mind') have controversially proposed a significance for quantum processes in the workings of the brain and even more, entertained the suggestion that non-computability may play a role there, and in fundamental physics. But even within the framework of conventionally accepted theory, the study of quantum computation is today one of the most active areas of all scientific research worldwide, and it continues to inspire imaginative developments at the forefront of fundamental science as well as offering powerful new computational capabilities.

Part IV

The Mathematics of Emergence: The Mysteries of Morphogenesis

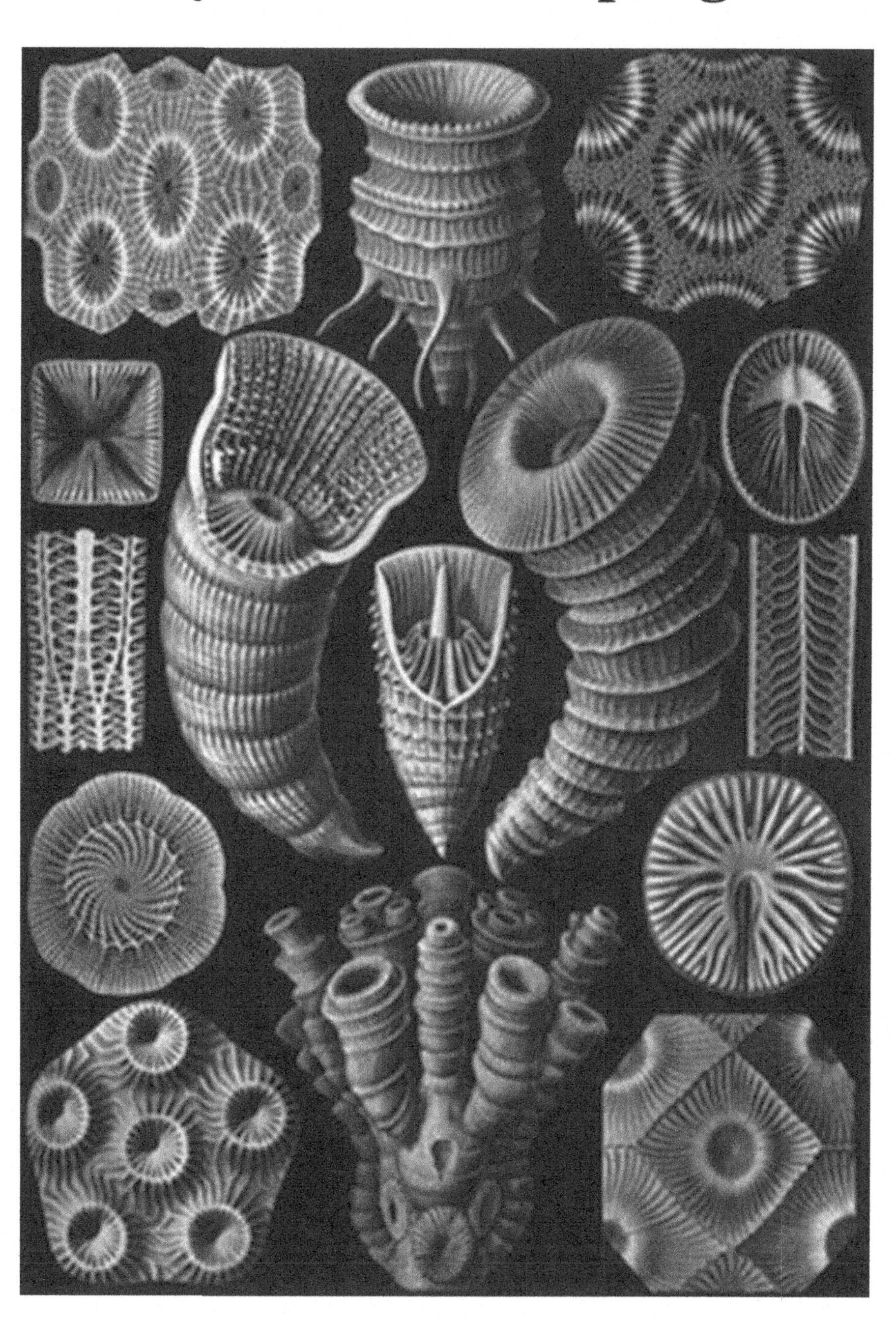

The Chemical Basis of Morphogenesis

(Phil. Trans. R. Soc. London B 237 (1952), pp. 37–72)

Peter Saunders introduces —

ALAN TURING'S WORK IN BIOLOGY[1]

Turing's work in biology illustrates just as clearly as his other work his ability to identify a fundamental problem and to approach it in a highly original way, drawing remarkably little from what others had done. He chose to work on the problem of form at a time when the majority of biologists were primarily interested in other questions. There are very few references in these papers, and most of them are for confirmation of details rather than for ideas that he was following up. In biology, as in almost everything else he did within science – or out of it – Turing was not content to accept a framework set up by others.

Even the fact that the mathematics in these papers is different from what he used in his other work is significant. For while it is not uncommon for a newcomer to make an important contribution to a subject, this is usually because he brings to it techniques and ideas that he has been using in his previous field but which are not known in the new one. Now much of Turing's career up to this point had been concerned with computers, from the hypothetical Turing machine to the real life Colossus, and this might have been expected to have led him to see the development of an organism from egg to adult as being programmed in the genes and to set out to study the structure of the programmes. This would also have been in the spirit of the times, because the combining of Darwinian natural selection and Mendelian genetics into the synthetic theory of evolution had only been completed about 10 years earlier, and it was in the very next year that Crick and Watson discovered the structure of DNA. Alternatively, Turing's experience in computing might have suggested to him something like what are now called cellular automata, models in which the fate of a cell is determined by the states of its neighbours through some simple algorithm, in a way that is very reminiscent of the Turing machine.

For Turing, however, the fundamental problem of biology had always been to account for pattern and form, and the dramatic progress that was being made at that time in genetics did not alter his view. And because he believed that the solution was to be found in physics and chemistry, it was to these subjects and the sort of mathematics that could be applied to them that he turned. In my view, he was right, but even someone who disagrees must be impressed by the way in which he went directly to what *he* saw as the most important problem and set out to attack it with the tools that he judged appropriate to the task, rather than those which were easiest to hand or which others were already using. What is more, he understood the full significance of the problem in a way that many biologists did not and still do not. We can see this in the manuscript with Wardlaw which is

[1]This brief overview of Turing's work is taken from Peter Saunders' Introduction to the Morphogenesis volume of the Collected Works, with some minor editorial updating.

included in the Collected Works, but it is clear just from the comment he made to Robin Gandy (see Hodges (1983, p. 431)) that his new ideas were 'intended to defeat the argument from design'.[2]

The development of any organism, and above all a complex one such as a human being, is a truly remarkable process. We each begin as a single cell and eventually become an adult made up of approximately 10^{15} cells of about 200 different types organised in a very complicated arrangement and able to cooperate to carry out many vital functions. This would be an impressive enough accomplishment if it were done under the supervision of an intelligent craftsman; in fact, it happens through nothing more than a series of interrelated physical and chemical processes. The genes play an important role in this, but we cannot just say that the genes create the form and let it go at that. The genes can only influence development through their effects on chemical reactions, and they themselves have to be turned on and off at appropriate times. Important though developmental genetics is, ultimately it is the physics and chemistry that we have to understand.

While the later stages of development are often complicated and hard to understand in detail, perhaps the greatest difficulty in principle is at the very beginning. Once a pattern of some sort has been established, it can serve as the basis for the next stage and so on. But how does the process start? The original cell is not, to be sure, totally symmetric; it has a polarity induced by the point of entry of the sperm, but this does not seem enough to determine the structure that is to appear. How does a pattern appear in a region that has nothing to serve as a template – or, equivalently, where does the template come from? This was what Turing saw as the fundamental problem.

Reference

Hodges, A., 1983. Alan Turing: The Enigma, Burnett Books, London.

And Philip K. Maini wonders at —

TURING'S THEORY OF MORPHOGENESIS

Alan Turing's paper, 'The chemical basis of morphogenesis' (Turing, 1952) has been hugely influential in a number of areas[3]. In this paper, Turing proposed that biological pattern formation arises in response to a chemical pre-pattern which, in turn, is set up by a process which is now known as *diffusion-driven instability*. The genius of this work was that he considered a system which was stable in the absence of diffusion and then showed that the addition of diffusion, which is naturally stablising, actually caused an instability. Thus, it was the *integration* of the parts that was as crucial to the understanding of embryological development as the parts themselves – patterns *emerged* or *self-organisd* as a result of the individual parts interacting. To see how far ahead of his time he was,

---

[2] Peter Saunders continues with further comments on this aspect of the morphogenesis work in his Introduction in the *Collected Works*. But see his contribution later in this chapter – *Defeating the Argument from Design* – for his recent thinking on the topic.

[3] 3,459 citations – ISI web of science 25/4/11.

one has to note that it is only now in the post-genomic era of systems biology that the majority of the scientific community has arrived at the conclusion he came to some 60 years ago.

Turing termed these chemicals 'morphogens'. For example, in phyllotaxis, an area in which he was interested, if the morphogen was a growth hormone, then a spatially non-uniform pre-pattern in it could cause symmetry breaking in a ring of cells leading to branching or leaf formation. More generally, if the morphogen determined cell fate, then the model could be used to describe patterning phenomena in a variety of settings (see, for example, Murray (2003)).

The Turing model was generalised in 1972 by Gierer and Meinhardt (see Meinhardt, 1982 and references therein) to the idea of activator–inhibitor systems and the general patterning principle of 'short-range activation, long-range inhibition' or 'local-activation, lateral inhibition (LALI)'. They also showed how the model could be used to explain regulation as well as normal patterning. Many subsequent models for biological pattern formation have been proposed based on very different biology, for example, (i) the theory of chemotaxis in which cells are proposed to move up gradients in chemicals (known as chemoattractants) so that the patterns are in cebu now.

Idensity (see, for example, Keller and Segel (1970)); (ii) the mechanochemical [4] theory of morphogenesis in which it is proposed that the physical interaction of cells with extracellular matrix leads to instabilities from which spatially varying patterns in cell density emerge (Murray, 2003, and references therein); (iii) neural models in which the neuroscretory system leads to growth and pigmentation (Boettiger et al., 2009). These models also have very different mathematical forms from the original parabolic system of equations proposed by Turing, some involving hyperbolic and/or elliptical terms, while others are of integro-partial differential equation type. However, they all generate pattern by the LALI principle. As a result of this, one can draw general conclusions on how patterns will depend on domain length and geometry, leading to mathematically derived notions of developmental constraints and the evolution of and morphogenetic rules for certain patterns (for example, cartilage formation in the vertebrate limb – see Murray (2003)).

The notion that something as complicated as patterning in developmental biology can be described by two equations appears fanciful but this was recognised by Turing, who stated in his paper:

> This model will be a simplification and an idealization, and consequently a falsification. It is to be hoped that the features retained for discussion are those of greatest importance in the present state of knowledge.

His model may be thought of as being applied at a certain scale, with the link to other scales appearing via the parameters (production, decay and interaction rates). So, for example, in the application to fish pigmentation patterns, while the model is assumed to operate on the tissue level, the parameters would arise from interactions occurring from the genetic level upwards (Kondo, 2002; Watanabe et al., 2006). Understanding how biological function arises through the integration of processes interacting on very different spatial and temporal scales is perhaps *the* greatest challenge facing developmental biology this century.

The Turing model for morphogenesis has provoked much controversy. For example, it does not account for factors such as chemotaxis, mechanics, etc., subsequently modelled as mentioned above. While Turing patterns have been shown to exist in chemistry (see, for example, Bánsági et al. (2011) and references therein) their existence in biology is still highly controversial. While the existence of morphogens is now widely accepted, it is still not at all clear if patterns arise by the mechanism proposed by Turing. While a vast number of organisms obey the developmental constraints arising

[4] In fact Turing mentioned that it may be possible to take mechanical effects into account within the framework he had developed.

from LALI models and a great number of experimental manipulations give results consistent with the Turing model (or extensions thereof) to actually *prove* that a pattern arises via a Turing mechanism would require one to know the precise chemical interactions of the morphogens and show that the parameters satisfy the inequalities necessary for diffusion-driven instability to occur. This is a far way off but encouraging recent progress has been made. For example, Garfinkel et al. (2004) investigated in vitro the self-organising properties of multipotential adult vascular mesenchymal cells. They identified BMP-2 and MGP as qualitatively satisfying all the conditions necessary for a Turing morphogen pair and showed that the cells aggregate into stripe, spot and labyrinthine patterns in response to various manipulations precisely as predicted by the Turing model. Sick et al. (2006) propose that WNT and DKK may be a Turing morphogen pair involved in the patterning of hair follicles in mice. They showed that experimental manipulation of the system leads to results that are qualitatively similar to those predicted by a Turing mechanism.

Adding fuel to the debate was the discovery that one of the candidates for a Turing pattern, the periodic pattern of expression of pair-rule genes in *Drosophila*, is actually not formed via diffusion-driven instability but an asymmetry is already inherited from the mother and this leads to a cascade of patterning interactions (Akam, 1989). This is very different from the Turing model which assumes that pattern arises de novo. However, again this was recognised by Turing, who said in his famous paper, 'Most of an organism, most of the time, is developing from one pattern to another rather than from homogeneity into a pattern'.

Not only are there biological issues about the Turing model, there are also mathematical concerns. It is known that the model exhibits multiple stable steady states and this means that it can be very sensitive to noise and stochasticity in, for example, initial conditions (Bard and Lauder, 1974) or slight changes in domain shape (Bunow et al., 1980). Therefore, the patterns produced are not robust. In some cases, this may not be a problem (for example, for certain pigmentation patterns), but in other cases, for example, limb development, it may be crucial. It has been shown that using different types of biologically relevant boundary conditions can improve robustness (Dillon et al., 1994) as can domain growth (Crampin et al., 1999). In the latter case, it has been shown that growth can enhance selection of certain patterning modes at the expense of others. It is observed that biological pattern often arises behind a moving front (either in growth or a critical parameter) and in the context of the Turing theory for morphogenesis, this would be explained by noting that such a process keeps the patterning domain small, and on such domains pattern selection can be easily controlled. Very recently, it has been shown that the model is highly sensitive to the inclusion of gene transcription delays in the kinetics (see Seirin Lee et al. (2011) and references therein).

Not only has the Turing model challenged biologists to prove/disprove the existence of LALI morphogens. It has also challenged several generations of mathematicians. While the equations look relativity simple, the vast array of patterning behaviours it produces (many of which were computed by Turing in his paper) is astonishing. While the linear theory is well understood (presented by Turing himself), the nonlinear theory is largely intractable. Early analysis investigated the weakly nonlinear case where amplitude equations could be derived using perturbation approaches (see, for example, the book by Britton (1986), for the general theory) and are valid in the vicinity of primary bifurcation points. Although Turing studied and computed the linear system, he does mention the nonlinear problem and, in an *ad hoc* way, derives the amplitude equation that many found later multi-timescale analysis. Group symmetry approaches have also been used to study the problem of mode interactions (see Comanici and Golubitsky (2008) and references therein). Away from the primary bifurcation points one has to resort, by and large, to numerical computation, except in certain cases of very slow activator diffusion in which spike solutions form. These solutions have complex behaviour in which they move to certain parts of the domain to stabilise and spikes may also coalesce with other spikes (see, for example, Ward et al. (2002) and references therein).

Applications of Turing's work to developmental biology are too numerous to list but include limb development, pigmentation patterning, hair and feather germ formation, tooth morphogenesis, phyllotaxis, hydra patterning and regeneration. Moreover, ideas of self-organisation now abound in biology, chemistry and ecology. The stimulus for a lot of this work stems from Turing's original ideas. Thus, his paper has significantly advanced the field and the paper is being cited almost everyday[5]. Advanced computational power now means that the model is trivial to simulate and explore (Kondo and Miura, 2010) and it is envisaged that this will result in its impact increasing instead of waning. George Box is quoted as making the very truthful statement, 'all models are wrong, but some are useful'. Although still very controversial, Turing's theory for morphogenesis provided a paradigm shift in our way of thinking, which has stimulated countless experimental programmes and resulted in novel experiments being carried out that may not otherwise have been undertaken. It has also generated new mathematical and computational problems that have advanced those fields considerably.

Acknowledgement

This work was partially supported by a Royal Society-Wolfson Research Merit award.

References

Akam, M., 1989. Making stripes inelegantly. Nature 341, 282–283.

Bánsági Jr., T., Vanag, V.K., Epstein, I.R., 2011. Tomography of reaction–diffusion microemulsions reveals three-dimensional Turing patterns. Science 331, 1309–1312.

Bard, J., Lauder, I., 1974. How well does Turing's theory of morphogenesis work? J. Theor. Biol. 45, 501–531.

Boettiger, A., Ermentrout, B., Oster, G.F., 2009. The neural origins of shell structure and pattern in aquatic mollusks. Proc. Nat. Acad. Sci. USA 106, 6837–6842.

Britton, N.F., 1986. Reaction-Diffusion Equations and Their Applications to Biology. Academic, New York.

Bunow, B., Kernevez, J.P., Joly, G., Thomas, D. 1980. Pattern formation by reaction–diffusion instabilities: Application to morphogenesis in Drosophila. J. Theor. Biol. 84, 629–649.

Comanici, A., Golubitsky, M., 2008. Patterns in growing domains via mode interactions. Dyn. Syst. 23, 167206.

Crampin, E.J., Gaffney, E.A., Maini, P.K., 1999. Reaction and diffusion on growing domains: Scenarios for robust pattern formation. Bull. Math. Biol. 61, 1093–1120.

Dillon, R., Maini, P.K., Othmer, H.G., 1994. Pattern formation in generalised Turing systems: I. Steady-state patterns in systems with mixed boundary conditions. J. Math. Biol. 32, 345–393.

Garfinkel, A., Tintut, Y., Petrasek, D., Bostrom, K., Demer, L.L., 2004. Pattern formation by vascular mesenchymal cells. Proc. Nat. Acad. Sci. USA 101, 9247–9250.

Keller, E.F., Segel, L.A., 1970. Initiation of slime mold aggregation viewed as an instability. J. Theor. Biol. 26, 399–415.

Kondo, S., 2002. The reaction–diffusion system: a mechanism for autonomous pattern formation in the animal skin. Genes. Cells. 7, 535–541.

Kondo, S., Miura, T., 2010. Reaction-diffusion model as a framework for understanding biological pattern formation. Science 329, 1616–1620.

Meinhardt, H., 1982. Models of Biological Pattern Formation. Academic Press, London.

Murray, J.D., 2003. Mathematical Biology II: Spatial Models and Biomedical Applications, Springer, Berlin, Heidelberg.

[5] In the 3 weeks between first and second drafts of this paper, the author noticed an additional 19 citations to the paper from the ISI Web of Science.

Seirin Lee, S., Gaffney, E.A., Baker, R.E., 2011. The dynamics of Turing patterns for morphogen-regulated growing domains with cellular response delays. Bull. Math. Biol. 73 (11), 2527–51.

Sick, S., Reinker, S., Timmer, J., Schlake, T., 2006. WNT and DKK determine hair follicle spacing through a reaction–diffusion mechanism. Science 314, 1447–1450.

Turing, M.A., 1952. The chemical basis of morphogenesis. Phil. Trans. R. Soc. B 237, 37–72.

Ward, M.J., McInerney, D., Houston, P., Gavaghan, D., Maini, P.K., 2002. The dynamics and pinning of a spike for a reaction–diffusion system. SIAM J. Appl. Maths. 62, 1297–1328.

Watanabe, M., Iwashita, M., Ishii, M., Kurachi, Y., Kawakami, A., Kondo, S., et al., 2006. Spot pattern of leopard Danio is caused by mutation in the zebrafish connexin 41.8 gene. EMBO Rep. 7, 893–897.

The Chemical Basis of Morphogenesis

By A. M. Turing, F.R.S. *University of Manchester*

(*Received* 9 *November* 1951–*Revised* 15 *March* 1952)

It is suggested that a system of chemical substances, called morphogens, reacting together and diffusing through a tissue, is adequate to account for the main phenomena of morphogenesis. Such a system, although it may originally be quite homogeneous, may later develop a patten or structure due to an instability of the homogeneous equilibrium, which is triggered off by random disturbances. Such reaction-diffusion systems are considered in some detail in the case of an isolated ring of cells, a mathematically convenient, though biologically unusual system. The investigation is chiefly concerned with the onset of instability. It is found that there are six essentially different forms which this may take. In the most interesting form stationary waves appear on the ring. It is suggested that this might account, for instance, for the tentacle patterns on *Hydra* and for whorled leaves. A system of reactions and diffusion on a sphere is also considered. Such a system appears to account for gastrulation. Another reaction system in two dimensions gives rise to patterns reminiscent of dappling. It is also suggested that stationary waves in two dimensions could account for the phenomena of phyllotaxis.

The purpose of this paper is to discuss a possible mechanism by which the genes of a zygote may determine the anatomical structure of the resulting organism. The theory does not make any new hypotheses; it merely suggests that certain well-known physical laws are sufficient to account for many of the facts. The full understanding of the paper requires a good knowledge of mathematics, some biology, and some elementary chemistry. Since readers cannot be expected to be experts in all of these subjects, a number of elementary facts are explained, which can be found in text-books, but whose omission would make the paper difficult reading.

1. A model of the embryo: Morphogens

In this section a mathematical model of the growing embryo will be described. This model will be a simplification and an idealization, and consequently a falsification. It is to be hoped that the features retained for discussion are those of greatest importance in the present state of knowledge.

The model takes two slightly different forms. In one of them the cell theory is recognized but the cells are idealized into geometrical points. In the other the matter of the organism is imagined as continuously distributed. The cells are not, however, completely ignored, for various physical and physico-chemical characteristics of the matter as a whole are assumed to have values appropriate to the cellular matter.

With either of the models one proceeds as with a physical theory and defines an entity called 'the state of the system'. One then describes how that state is to be determined from the state at a moment very shortly before. With either model the description of the state consists of two parts, the mechanical and the chemical. The mechanical part of the state describes the positions, masses, velocities and elastic properties of the cells, and the forces between them. In the continuous form of the theory essentially the same information is given in the form of the stress, velocity, density and elasticity of the matter. The chemical part of the state is given (in the cell form of theory) as the chemical composition of each separate cell; the diffusibility of each substance between each two adjacent cells must also be given. In the continuous form of the theory the concentrations and diffusibilities of each substance have to be given at each point. In determining the changes of state one should take into account

(i) The changes of position and velocity as given by Newton's laws of motion.
(ii) The stresses as given by the elasticities and motions, also taking into account the osmotic pressures as given from the chemical data.
(iii) The chemical reactions.
(iv) The diffusion of the chemical substances. The region in which this diffusion is possible is given from the mechanical data.

This account of the problem omits many features, e.g. electrical properties and the internal structure of the cell. But even so it is a problem of formidable mathematical complexity. One cannot at present hope to make any progress with the understanding of such systems except in very simplified cases. The interdependence of the chemical and mechanical data adds enormously to the difficulty, and attention will therefore be confined, so far as is possible, to cases where these can be separated. The mathematics of elastic solids is a well- developed subject, and has often been applied to biological systems. In this paper it is proposed to give attention rather to cases where the mechanical aspect can be ignored and the chemical aspect is the most significant. These cases promise greater interest, for the characteristic action of the genes themselves is presumably chemical. The systems actually to be considered consist therefore of masses of tissues which are not growing, but within which certain substances are reacting chemically, and through which they are diffusing. These substances will be called morphogens, the word being intended to convey the idea of a form producer. It is not intended to have any very exact meaning, but is simply the kind of substance concerned in this theory. The evocators of Waddington provide a good example of morphogens (Waddington 1940). These evocators diffusing into a tissue somehow persuade it to develop along different lines from those which would have been followed in its absence. The genes themselves may also be considered to be morphogens. But they certainly form rather a special class. They are quite indiffusible. Moreover, it is only by courtesy that genes can be regarded as separate molecules. It would be more accurate (at any rate at mitosis) to regard them as radicals of the giant molecules known as chromosomes. But presumably these radicals act almost independently, so that it is unlikely that serious errors will arise through regarding the genes as molecules. Hormones may also be regarded as quite typical morphogens. Skin pigments may be regarded as morphogens if desired. But those whose action is to be considered here do not come squarely within any of these categories.

The function of genes is presumed to be purely catalytic. They catalyze the production of other morphogens, which in turn may only be catalysts. Eventually, presumably, the chain leads to some morphogens whose duties are not purely catalytic. For instance, a substance might break down into a number of smaller molecules, thereby increasing the osmotic pressure in a cell and promoting its growth. The genes might thus be said to influence the anatomical form of the organism by determining the rates of those reactions which they catalyze. If the rates are assumed to be those determined by the genes, and if a comparison of organisms is not in question, the genes themselves may be eliminated from the discussion. Likewise any other catalysts obtained secondarily through the agency of the genes may equally be ignored, if there is no question of their concentrations varying. There may, however, be some other morphogens, of the nature of evocators, which cannot be altogether forgotten, but whose role may nevertheless be subsidiary, from the point of view of the formation of a particular organ. Suppose, for instance, that a 'leg-evocator' morphogen were being produced in a certain region of an embryo, or perhaps diffusing into it, and that an attempt was being made to explain the mechanism by which the leg was formed in the presence of the evocator. It would then be reasonable to take the distribution of the evocator in space and time as given in advance and to consider the chemical reactions set in train by it. That at any rate is the procedure adopted in the few examples considered here.

2. Mathematical background required

The greater part of this present paper requires only a very moderate knowledge of mathematics. What is chiefly required is an understanding of the solution of linear differential equations with constant coefficients. (This is also what is chiefly required for an understanding of mechanical and electrical oscillations.) The solution of such an equation takes the form of a sum $\Sigma A e^{bt}$, where the quantities A, b may be complex, i.e. of the form $\alpha + i\beta$, where α and β are ordinary (real) numbers and $i = \sqrt{-1}$. It is of great importance that the physical significance of the various possible solutions of this kind should be appreciated, for instance, that

(a) Since the solutions will normally be real one can also write them in the form $\mathcal{R}\sum Ae^{bt}$ or $\sum \mathcal{R}Ae^{bt}$ ($\mathcal{R}$ means 'real part of').

(b) That if $A = A'e^{i\phi}$ and $b = \alpha + i\beta$, where A', α, β, ϕ are real, then

$$\mathcal{R}Ae^{bt} = A'e^{\alpha t} = A'e^{\alpha t}\cos(\beta t + \phi).$$

Thus each such term represents a sinusoidal oscillation if $\alpha = 0$, a damped oscillation if $\alpha < 0$, and an oscillation of ever-increasing amplitude if $\alpha > 0$.

(c) If any one of the numbers b has a positive real part the system in question is unstable.

(d) After a sufficiently great lapse of time all the terms Ae^{bt} will be negligible in comparison with those for which b has the greatest real part, but unless this greatest real part is itself zero these dominant terms will eventually either tend to zero or to infinite values.

(e) That the indefinite growth mentioned in (*b*) and (*d*) will in any physical or biological situation eventually be arrested due to a breakdown of the assumptions under which the solution was valid. Thus, for example, the growth of a colony of bacteria will normally be taken to satisfy the equation $dy/dt = \alpha y (\alpha > 0)$, y being the number of organisms at time t, and this has the solution $y = Ae^{\alpha t}$. When, however, the factor $e^{\alpha t}$ has reached some billions the food supply can no longer be regarded as unlimited and the equation $dy/dt = \alpha y$ will no longer apply.

The following relatively elementary result will be needed, but may not be known to all readers:

$$\sum_{r=1}^{N} \exp\left[\frac{2\pi i r s}{N}\right] = 0 \quad \text{if} \quad 0 < s < N,$$

but

$$= N \quad \text{if} \quad s = 0 \quad \text{or} \quad s = N.$$

The first case can easily be proved when it is noticed that the left-hand side is a geometric progression. In the second case all the terms are equal to 1.

The relative degrees of difficulty of the various sections are believed to be as follows. Those who are unable to follow the points made in this section should only attempt §§ 3, 4, 11, 12, 14 and part of § 13. Those who can just understand this section should profit also from §§ 7, 8, 9. The remainder, §§ 5, 10, 13, will probably only be understood by those definitely trained as mathematicians.

3. Chemical reactions

It has been explained in a preceding section that the system to be considered consists of a number of chemical substances (morphogens) diffusing through a mass of tissue of given geometrical form and reacting together within it. What laws are to control the development of this situation ? They are quite simple. The diffusion follows the ordinary laws of diffusion, i.e. each morphogen moves from regions of greater to regions of less concentration, at a rate proportional to the gradient of the concentration, and also proportional to the 'diffusibility' of the substance. This is very like the conduction of heat, diffusibility taking the place of conductivity. If it were not for the walls of the cells the diffusibilities would be inversely proportional to the square roots of the molecular weights. The pores of the cell walls put a further handicap on the movement of the larger molecules in addition to that imposed by their inertia, and most of them are not able to pass through the walls at all.

The reaction rates will be assumed to obey the 'law of mass action'. This states that the rate at which a reaction takes place is proportional to the concentrations of the reacting substances. Thus, for instance, the rate at which silver chloride will be formed and precipitated from a solution of silver nitrate and sodium chloride by the reaction

$$\mathrm{Ag}^+ + \mathrm{Cl}^- \rightarrow \mathrm{AgCl}$$

will be proportional to the product of the concentrations of the silver ion Ag^+ and the chloride ion $C1^-$. It should be noticed that the equation

$$\mathrm{AgNO_3} + \mathrm{NaC1} \rightarrow \mathrm{AgC1} + \mathrm{NaNO_3}$$

is not used because it does not correspond to an actual reaction but to the final outcome of a number of reactions. The law of mass action must only be applied to the *actual* reactions. Very often certain substances appear in the individual reactions of a group, but not in the final outcome. For instance, a reaction $A \rightarrow B$ may really take the form of two steps $A+G \rightarrow C$ and $C \rightarrow B+G$. In such a case the substance G is described as a catalyst, and as catalyzing the reaction $A \rightarrow B$. (Catalysis according to this plan has been considered in detail by Michaelis & Menten (1913).) The effect of the genes is presumably achieved almost entirely by catalysis. They are certainly not permanently used up in the reactions.

Sometimes one can regard the effect of a catalyst as merely altering a reaction rate. Consider, for example, the case mentioned above, but suppose also that A can become detached from G, i.e. that the reaction $C \rightarrow A+G$ is taken into account. Also suppose that the reactions $A+G \leftrightarrows C$ both proceed much faster than $C \rightarrow B+G$. Then the concentrations of A, G, C will be related by the condition that there is equilibrium between the reactions $A+G \rightarrow C$ and $C \rightarrow A+G$, so that (denoting concentrations by square brackets) $[A][G] = k[C]$ for some constant k. The reaction $C \rightarrow B+G$ will of course proceed at a rate proportional to $[C]$, i.e. to $[A][G]$. If the amount of C is always small compared with the amount of G one can say that the presence of the catalyst and its amount merely alter the mass action constant for the reaction $A \rightarrow B$, for the whole proceeds at a rate proportional to $[A]$. This situation does not, however, hold invariably. It may well happen that nearly all of$G$ takes the combined C so long as any A is left. In this case the reaction proceeds at a rate independent of the concentration of A until A is entirely consumed. In either of these cases the rate of the complete group of reactions depends only on the concentrations of the reagents, although usually not according to the law of mass action applied crudely to the chemical equation for the whole group. The same applies in any case where all reactions of the group with one exception proceed at speeds much greater than that of the exceptional one. In these cases the rate of the reaction is a function of the concentrations of the reagents. More generally again, no such approximation is applicable. One simply has to take all the actual reactions into account.

According to the cell model then, the number and positions of the cells are given in advance, and so are the rates at which the various morphogens diffuse between the cells. Suppose that there are N cells and M morphogens. The state of the whole system is then given by MN numbers, the quantities of the M morphogens in each of N cells. These numbers change with time, partly because of the reactions, partly because of the diffusion. To determine the part of the rate of change of one of these numbers due to diffusion, at any one moment, one only needs to know the amounts of the same morphogen in the cell and its neighbours, and the diffusion coefficient for that morphogen. To find the rate of change due to chemical reaction one only needs to know the concentrations of all morphogens at that moment in the one cell concerned.

This description of the system in terms of the concentrations in the various cells is, of course, only an approximation. It would be justified if, for instance, the contents were perfectly stirred. Alternatively, it may often be justified on the understanding that the 'concentration in the cell' is the concentration at a certain representative point, although the idea of 'concentration at a point' clearly itself raises difficulties. The author believes that the approximation is a good one, whatever argument is used to justify it, and it is certainly a convenient one.

It would be possible to extend much of the theory to the case of organisms immersed in a fluid, considering the diffusion within the fluid as well as from cell to cell. Such problems are not, however, considered here.

4. The breakdown of symmetry and homogeneity

There appears superficially to be a difficulty confronting this theory of morphogenesis, or, indeed, almost any other theory of it. An embryo in its spherical blastula stage has spherical symmetry, or if there are any deviations from perfect symmetry, they cannot be regarded as of any particular importance, for the deviations vary greatly from embryo to embryo within a species, though the organisms developed from them are barely distinguishable. One may take it therefore that there is perfect spherical symmetry. But a system which has spherical symmetry, and whose state is changing because of chemical reactions and diffusion, will remain spherically symmetrical for ever. (The same would hold true if the state were changing according to the laws of electricity and magnetism, or of quantum mechanics.) It certainly cannot result in an organism such as a horse, which is not spherically symmetrical.

There is a fallacy in this argument. It was assumed that the deviations from spherical symmetry in the blastula could be ignored because it makes no particular difference what form of asymmetry there is. It is, however, important that there are *some* deviations, for the system may reach a state of instability in which these irregularities, or certain components of them, tend to grow. If this happens a new and stable equilibrium is usually reached, with the symmetry entirely gone. The variety of such new equilibria will normally not be so great as the variety of irregularities giving rise to them. In the case, for instance, of the gastrulating sphere, discussed at the end of this paper, the direction of the axis of the gastrula can vary, but nothing else.

The situation is very similar to that which arises in connexion with electrical oscillators. It is usually easy to understand how an oscillator keeps going when once it has started, but on a first acquaintance it is not obvious how the oscillation begins. The explanation is that there are random disturbances always present in the circuit. Any disturbance whose frequency is the natural frequency of the oscillator will tend to set it going. The ultimate fate of the system will be a state of oscillation at its appropriate frequency, and with an amplitude (and a wave form) which are also determined by the circuit. The phase of the oscillation alone is determined by the disturbance.

If chemical reactions and diffusion are the only forms of physical change which are taken into account the argument above can take a slightly different form. For if the system originally has no sort of geometrical symmetry but is a perfectly homogeneous and possibly irregularly shaped mass of tissue, it will continue indefinitely to be homogeneous. In practice, however, the presence of irregularities, including statistical fluctuations in the numbers of molecules undergoing the various reactions, will, if the system has an appropriate kind of instability, result in this homogeneity disappearing.

This breakdown of symmetry or homogeneity may be illustrated by the case of a pair of cells originally having the same, or very nearly the same, contents. The system is homogeneous: it is also symmetrical with respect to the operation of interchanging the cells. The contents of either cell will be supposed describable by giving the concentrations X and Y of two morphogens. The chemical reactions will be supposed such that, on balance, the first morphogen (X) is produced at the rate $5X - 6Y + 1$ and the second (Y) at the rate $6X - 7Y + 1$. When, however, the strict application of these formulae would involve the concentration of a morphogen in a cell becoming negative, it is understood that it is instead destroyed only at the rate at which it is reaching that cell by diffusion. The first morphogen will be supposed to diffuse at the rate 0.5 for unit difference of concentration between the cells, the second, for the same difference, at the rate 4.5. Now if both morphogens have unit concentration in both cells there is equilibrium. There is no resultant passage of either morphogen across the cell walls, since there is no concentration difference, and there is no resultant production (or destruction) of either morphogen in either cell since $5X - 6Y + 1$ and $6X - 7Y + 1$ both have the value zero for $X = 1, Y{=}1$. But suppose the values are $X_1 = 1.06, Y_1 = 1.02$ for the first cell and $X_2 = 0.94, Y_2 = 0.98$ for the second. Then the two morphogens will be being produced by chemical action at the rates 0.18, 0.22 respectively in the first cell and destroyed at the same rates in the second. At the same time there is a flow due to diffusion from the first cell to the second at the rate 0.06 for the first morphogen and 0.18 for the second. In sum the effect is a flow from the second cell to the first at the rates 0.12, 0.04 for the two morphogens respectively. This flow tends to accentuate the already existing differences between the two cells. More generally, if

$$X_1 = 1 + 3\xi, \quad X_2 = 1 - 3\xi, \quad Y_1 = 1 + \xi, \quad Y_2 = 1 - \xi,$$

at some moment the four concentrations continue afterwards to be expressible in this form, and ξ increases at the rate 2ξ. Thus there is an exponential drift away from the equilibrium condition. It will be appreciated that a drift away from the equilibrium occurs with almost any small displacement from the equilibrium condition, though not normally according to an exact exponential curve. A particular choice was made in the above argument in order to exhibit the drift with only very simple mathematics.

Before it can be said to follow that a two-cell system can be unstable, with inhomogeneity succeeding homogeneity, it is necessary to show that the reaction rate functions postulated really can occur. To specify actual substances, concentrations and temperatures giving rise to these functions would settle the matter finally, but would be difficult and somewhat out of the spirit of the present inquiry. Instead, it is proposed merely to mention imaginary reactions which give rise to the required functions by the law of mass action, if suitable reaction constants are assumed. It will be sufficient to describe

(i) A set of reactions producing the first morphogen at the constant rate 1, and a similar set forming the second morphogen at the same rate.
(ii) A set destroying the second morphogen (Y) at the rate $7Y$.
(iii) A set converting the first morphogen (X) into the second (Y) at the rate $6X$.
(iv) A set producing the first morphogen (X) at the rate $11X$.
(v) A set destroying the first morphogen (X) at the rate $6Y$, so long as any of it is present.

The conditions of (i) can be fulfilled by reactions of the type $A \to X$, $B \to$ Y, where A and B are substances continually present in large and invariable concentrations. The conditions of (ii) are satisfied by a reaction of the form $Y \to D$, D being an inert substance and (iii) by the reaction $X \to Y$ or $X \to Y + E$. The remaining two sets are rather more difficult. To satisfy the conditions of (iv) one may suppose that X is a catalyst for its own formation from A. The actual reactions could be the formation of an unstable compound U by the reaction $A + X \to U$, and the subsequent almost instantaneous breakdown $U \to 2X$. To destroy X at a rate proportional to Y as required in (v) one may suppose that a catalyst C is present in small but constant concentration and immediately combines with X, $X + C \to V$. The modified catalyst reacting with Y, at a rate proportional to Y, restores the catalyst but not the morphogen X, by the reactions $V + Y \to W$, $W \to C + H$, of which the latter is assumed instantaneous.

It should be emphasized that the reactions here described are by no means those which are most likely to give rise to instability in nature. The choice of the reactions to be discussed was dictated entirely by the fact that it was desirable that the argument be easy to follow. More plausible reaction systems are described in § 10.

Unstable equilibrium is not, of course, a condition which occurs very naturally. It usually requires some rather artificial interference, such as placing a marble on the top of a dome. Since systems tend to leave unstable equilibria they cannot often be in them. Such equilibria can, however, occur naturally through a stable equilibrium changing into an unstable one. For example, if a rod is hanging from a point a little above its centre of gravity it will be in stable equilibrium. If, however, a mouse climbs up the rod the equilibrium eventually becomes unstable and the rod starts to swing. A chemical analogue of this mouse-and-pendulum system would be that described above with the same diffusibilities but with the two morphogens produced at the rates

$$(3+I)X-6Y+I-1 \quad \text{and} \quad 6X-(9+I)Y-I+1.$$

This system is stable if $I<0$ but unstable if $I>0$. If I is allowed to increase, corresponding to the mouse running up the pendulum, it will eventually become positive and the equilibrium will collapse. The system which was originally discussed was the case $I=2$, and might be supposed to correspond to the mouse somehow reaching the top of the pendulum without disaster, perhaps by falling vertically on to it.

5. Left-handed and right-handed organisms

The object of this section is to discuss a certain difficulty which might be thought to show that the morphogen theory of morphogenesis cannot be right. The difficulty is mainly concerned with organisms which have not got bilateral symmetry. The argument, although carried through here without the use of mathematical formulae, may be found difficult by non-mathematicians, and these are therefore recommended to ignore it unless they are already troubled by such a difficulty.

An organism is said to have 'bilateral symmetry' if it is identical with its own reflexion in some plane. This plane of course always has to pass through some part of the organism, in particular through its centre of gravity. For the purpose of this argument it is more general to consider what may be called 'left-right symmetry'. An organism has left-right symmetry if its description in any right-handed set of rectangular Cartesian co-ordinates is identical with its description in some set of left-handed axes. An example of a body with left-right symmetry, but not bilateral symmetry, is a cylinder with the letter P printed on one end, and with the mirror image of a P on the other end, but with the two upright strokes of the two letters not parallel. The distinction may possibly be without a difference so far as the biological world is concerned, but mathematically it should not be ignored.

If the organisms of a species are sufficiently alike, and the absence of left-right symmetry sufficiently pronounced, it is possible to describe each individual as either right-handed or left-handed without there being difficulty in classifying any particular specimen. In man, for instance, one could take the X-axis in the forward direction, the Y-axis at right angles to it in the direction towards the side on which the heart is felt, and the Z-axis upwards. The specimen is classed as left-handed or right-handed according as the axes so chosen are left-handed or right-handed. A new classification has of course to be defined for each species.

The fact that there exist organisms which do not have left-right symmetry does not in itself cause any difficulty. It has already been explained how various kinds of symmetry can be lost in the development of the embryo, due to the particular disturbances (or 'noise') influencing the particular specimen not having that kind of symmetry, taken in conjunction with appropriate kinds of instability. The difficulty lies in the fact that there are species in which the proportions of left-handed and right-handed types are very unequal. It will be as well to describe first an argument which appears to show that this should not happen.

The argument is very general, and might be applied to a very wide class of theories of morphogenesis.

An entity may be described as '*P*-symmetrical' if its description in terms of one set of right-handed axes is identical with its description in terms of any other set of right-handed axes with the same origin. Thus, for instance, the totality of positions that corkscrew would take up when rotated in all possible ways about the origin has *P*-symmetry. The entity will bc said to be '*F*-symmetrical' when changes from right-handed axes handed may also be made. This would apply if the corkscrew were replaced by a bilaterally symmetrical object such as a coal scuttle, or a left-right symmetrical object. In these terms one may say that there are species such that the totality of specimens from that species, together with the rotated specimens, is *P*-symmetrical, but very far from *F*-symmetrical. On the other hand, it is reasonable to suppose that

(i) The laws of physics are *F*-symmetrical.
(ii) The initial totality of zygotes for the species is *F*-symmetrical.
(iii) The statistical distribution of disturbances is *F*-symmetrical. The individual disturbances of course will in general have neither *F*-symmetry nor *P*-symmetry.

It should be noticed that the ideas of *P*-symmetry and *F*-symmetry as defined above apply even to so elaborate an entity as 'the laws of physics'. It should also be understood that the laws are to be the laws taken into account in the theory in question rather than some ideal as yet undiscovered laws.

Now it follows from these assumptions that the statistical distribution of resulting organisms will have *F*-symmetry, or more strictly that the distribution deduced as the result of working out such a theory will have such symmetry. The distribution of observed mature organisms, however, has no such symmetry. In the first place, for instance, men are more often found standing on their feet than their heads. This may be corrected by taking gravity into account in the laws, together with an appropriate change of definition of the two kinds of symmetry. But it will be more convenient if, for the sake of argument, it is imagined that some species has been reared in the absence of gravity, and that the resulting distribution of mature organisms is found to be *P*-symmetrical but to yield more right-handed specimens than left-handed and so not to have *F*-symmetry. It remains therefore to explain this absence of *F*-symmetry.

Evidently one or other of the assumptions (i) to (iii) must be wrong, i.e. in a correct theory one of them would not apply. In the morphogen theory already described these three assumptions do all apply, and it must therefore be regarded as defective to some extent. The theory may be corrected by taking into account the fact that the morphogens do not always have an equal number of left- and right-handed molecules. According to one's point of view one may regard this as invalidating either (i), (ii) or even (iii). Simplest perhaps is to say that the totality of zygotes just is not *F*-symmetrical, and that this could be seen if one looked at the molecules. This is, however, not very satisfactory from the point of view of this paper, as it would not be consistent with describing states in terms of concentrations only. It would be preferable if it was found possible to find more accurate laws concerning reactions and diffusion. For the purpose of accounting for unequal numbers of left- and right-handed organisms it is unnecessary to do more than show that there are corrections which would not be *F*-symmetrical when there are laevo- or dextrorotatory morphogens, and which would be large enough to account for the effects observed. It is not very difficult to think of such effects. They do not have to be very large, but must, of course, be larger than the purely statistical effects, such as thermal noise or Brownian movement.

There may also be other reasons why the totality of zygotes is not F-symmetrical, e.g. an asymmetry of the chromosomes themselves. If these also produce a sufficiently large effect, so much the better.

Though these effects may be large compared with the statistical disturbances they are almost certainly small compared with the ordinary diffusion and reaction effects. This will mean that they only have an appreciable effect during a short period in which the break-down of left-right symmetry is occurring. Once their existence is admitted, whether on a theoretical or experimental basis, it is probably most convenient to give them mathematical expression by regarding them as P-symmetrically (but not F-symmetrically) distributed disturbances. However, they will not be considered further in this paper.

6. Reactions and diffusion in a ring of cells

The original reason for considering the breakdown of homogeneity was an apparent difficulty in the diffusion-reaction theory of morphogenesis. Now that the difficulty is resolved it might be supposed that there is no reason for pursuing this aspect of the problem further, and that it would be best to proceed to consider what occurs when the system is very far from homogeneous. A great deal more attention will nevertheless be given to the breakdown of homogeneity. This is largely because the assumption that the system is still nearly homogeneous brings the problem within the range of what is capable of being treated mathematically. Even so many further simplifying assumptions have to be made. Another reason for giving this phase such attention is that it is in a sense the most critical period. That is to say, that if there is any doubt as to how the organism is going to develop it is conceivable that a minute examination of it just after instability has set in might settle the matter, but an examination of it at any earlier time could never do so.

There is a great variety of geometrical arrangement of cells which might be considered, but one particular type of configuration stands out as being particularly simple in its theory, and also illustrates the general principles very well. This configuration is a ring of similar cells. One may suppose that there are N such cells. It must be admitted that there is no biological example to which the theory of the ring can be immediately applied, though it is not difficult to find ones in which the principles illustrated by the ring apply.

It will be assumed at first that there are only two morphogens, or rather only two interesting morphogens. There may be others whose concentration does not vary either in space or time, or which can be eliminated fron the discussion for one reason or another. These other morphogens may, for instance be catalysts involved in the reactions between the interesting morphogens. An example of a complete system of reactions is given in § 10. Some consideration will also be given in §§ 8,9 to the case of three morphogens. The reader should have no difficulty in extending the results to any number of morphogens, but no essentially new features appear when the number is increased beyond three.

The two morphogens will be called X and Y. These letters will also be used to denote their concentrations. This need not lead to any real confusion. The concentration of X in cell r may be written X_r, and Y_r has a similar meaning. It is convenient to regard 'cell N' and 'cell 0' as synonymous, and likewise 'cell 1' and cell '$N+1$'. One can then say that for each r satisfying $1 \leqslant r \leqslant N$ cell r exchanges material by diffusion with cells $r-1$ and $r+1$. The cell-to-cell diffusion constant for X will be called μ, and that for Y will be called ν. This means that for unit concentration difference of X, this morphogen passes at the rate μ from the cell with the higher concentration to the (neighbouring) cell with the lower concentration. It is also necessary to make assumptions about the rates of chemical reaction. The most general assumption that can be made is that for concentrations X and Y chemical reactions are tending to increase X at the rate $f(X, Y)$ and Y at the rate $g(X, Y)$. When the changes in X and Y due to diffusion are also taken into account the behaviour of the system may be described by the $2N$ differential equations

$$\left.\begin{aligned}\frac{\mathrm{d}X_r}{\mathrm{d}t} &= f(X_r, Y_r) + \mu(X_{r+1} - 2X_r + X_{r-1})\\ \frac{\mathrm{d}Y_r}{\mathrm{d}t} &= g(X_r, Y_r) + \nu(Y_{r+1} - 2Y_r + Y_{r-1})\end{aligned}\right\} \quad (r = 1, \ldots, N). \tag{6.1}$$

If $f(h, k) : g(h, k) = 0$, then an isolated cell has an equilibrium with concentrations $X = h$, $Y = k$. The ring system also has an equilibrium, stable or unstable, with each X_r equal to h and each Y_r equal to k. Assuming that the system is not very far from this equilibrium it is convenient to put $X_r = h + x_r$, $Y_r = k + y_r$. One may also write $ax + by$ for $f(h + x, y + k)$ and $cx + dy$ for $g(h + x, y + k)$. Since $f(h, k) = g(h, k) = 0$ no constant terms are required, and since x and y are supposed small the terms in higher powers of x and y will have relatively little effect and one is justified in ignoring them. The four quantities a, b, c, d may be called the 'marginal reaction rates'. Collectively they may be described as the 'marginal reaction rate matrix'. When there are M morphogens this matrix consists of M^2 numbers. A marginal reaction rate has the dimensions of the reciprocal of a time, like a radioactive decay rate, which is in fact an example of a marginal (nuclear) reaction rate.

With these assumptions the equations can be rewritten as

$$\left.\begin{aligned}\frac{\mathrm{d}x_r}{\mathrm{d}t} &= ax_r + by_r + \mu(x_{r+1} - 2x_r + x_{r-1})\\ \frac{\mathrm{d}y_r}{\mathrm{d}t} &= cx_r + dy_r + \nu(y_{+1} - 2y_r + y_{r-1})\end{aligned}\right\} \tag{6.2}$$

To solve the equations one introduces new co-ordinates $\xi_0, \ldots, \xi_{N-1}$ and $\eta_0, \ldots, \eta_{N-1}$ by putting

$$\left.\begin{aligned}x_r &= \sum_{s=0}^{N-1} \exp\left[\frac{2\pi irs}{N}\right]\xi_s,\\ y_r &= \sum_{s=0}^{N-1} \exp\left[\frac{2\pi irs}{N}\right]\eta_s.\end{aligned}\right\} \tag{6.3}$$

These relations can also be written as

$$\left.\begin{aligned}\xi_r &= \frac{1}{N}\sum_{s=1}^{N} \exp\left[-\frac{2\pi irs}{N}\right]x_s,\\ \eta_r &= \frac{1}{N}\sum_{s=1}^{N} \exp\left[-\frac{2\pi irs}{N}\right]y_s,\end{aligned}\right\} \tag{6.4}$$

as may be shown by using the equations

$$\begin{aligned}\sum_{s=1}^{N} \exp\left[\frac{2\pi irs}{N}\right] &= 0 \quad \text{if} \quad 0 < r < N,\\ &= N \quad \text{if} \quad r = 0 \quad \text{or} \quad r = N,\end{aligned} \tag{6.5}$$

(referred to in § 2). Making this substitution one obtains

$$\begin{aligned}\frac{\mathrm{d}\xi_s}{\mathrm{d}t} &= \frac{1}{N}\sum_{s=1}^{N}\exp\left[-\frac{2\pi irs}{N}\right]\left[ax_r + by_r + \mu\left(\exp\left[-\frac{2\pi is}{N}\right] - 2 + \exp\left[\frac{2\pi is}{N}\right]\right)\xi_s\right] \\ &= a\xi_s + b\eta_s + \mu\left(\exp\left[-\frac{2\pi is}{N}\right] - 2 + \exp\left[\frac{2\pi is}{N}\right]\right)\xi_s \\ &= \left(a - 4\mu\sin^2\frac{\pi s}{N}\right)\xi_\zeta + b\eta_s.\end{aligned} \tag{6.6}$$

Likewise

$$\frac{\mathrm{d}\eta_s}{\mathrm{d}t} = c\xi_s + \left(d - 4\nu\sin^2\frac{\pi s}{N}\right)\eta_s. \tag{6.7}$$

The equations have now been converted into a quite manageable form, with the variables separated. There are now two equations concerned with ξ_1 and η_1, two concerned with ξ_2 and η_2, etc. The equations themselves are also of a well-known standard form, being linear with constant coefficients. Let p_s and p'_s be the roots of the equation

$$(p - a + 4\mu\sin^2\frac{\pi s}{N})(p - d + 4\nu\sin^2\frac{\pi s}{N}) = bc \tag{6.8}$$

(with $\mathcal{R}p_s \geqslant \mathcal{R}p'_s$ for definiteness), then the solution of the equations is of the form

$$\left.\begin{aligned}\xi_s &= A_s\mathrm{e}^{p_s t} + B_s\mathrm{e}^{p'_s t}, \\ \eta_s &= C_s\mathrm{e}^{p_s t} + D_s\mathrm{e}^{p'_s t},\end{aligned}\right\} \tag{6.9}$$

where, however, the coefficients A_s, B_s, C_s, D_s are not independent but are restricted to satisfy

$$\left.\begin{aligned}A_s\left(p_s - a + 4\mu\sin^2\tfrac{\pi s}{N}\right) &= bC_s, \\ B_s\left(p'_s - a + 4\mu\sin^2\tfrac{\pi s}{N}\right) &= bD_s.\end{aligned}\right\} \tag{6.10}$$

If it should happen that $p_s = p'_s$ the equations (6.9) have to be replaced by

$$\left.\begin{aligned}\xi_s &= (A_s + B_s t)\mathrm{e}^{p_s t}, \\ \eta_s &= (C_s + D_s t)\mathrm{e}^{p_s t}.\end{aligned}\right\} \tag{6.9'}$$

and remains true. Substituting back into (6.3) and replacing the variables x_r, y_r by X_r, Y_r (the actual concentrations) the solution can be written

$$\left.\begin{aligned}X_r &= h + \sum_{s=1}^{N}(A_s\mathrm{e}^{p_s t} + B_s\mathrm{e}^{p'_s t})\exp\left[\tfrac{2\pi irs}{N}\right], \\ Y_r &= k + \sum_{s=1}^{N}(C_s\mathrm{e}^{p_s t} + D_s\mathrm{e}^{p'_s t})\exp\left[\tfrac{2\pi irs}{N}\right].\end{aligned}\right\} \tag{6.11}$$

Here A_s, B_s, C_s, D_s are still related by (6.10), but otherwise are arbitrary complex numbers; p_s and p'_s are the roots of (6.8).

The expression (6.11) gives the general solution of the equations (6.1) when one assumes that departures from homogeneity are sufficiently small that the functions $f(X, Y)$ and $g(X, Y)$ can safely be taken as linear. The form (6.11) given is not very informative. It will be considerably simplified in § 8. Another implicit assumption concerns random disturbing influences. Strictly speaking one should consider such influences to be continuously at work. This would make the mathematical treatment considerably more difficult without substantially altering the conclusions. The assumption which is implicit in the analysis, here and in § 8, is that the state of the system at $t = 0$ is not one of homogeneity, since it has been displaced from such a state by the disturbances; but after $t = 0$ further disturbances are ignored. In § 9 the theory is reconsidered without this latter assumption.

7. Continuous ring of tissue

As an alternative to a ring of separate cells one might prefer to consider a continuous ring of tissue. In this case one can describe the position of a point of the ring by the angle θ which a radius to the point makes with a fixed reference radius. Let the diffusibilities of the two substances be μ' and ν'. These are not quite the same as μ and ν of the last section, since μ and ν are in effect referred to a cell diameter as unit of length, whereas μ' and ν' are referred to a conventional unit, the same unit in which the radius ρ of the ring is measured. Then

$$\mu = \mu' \left(\frac{N}{2\pi\rho}\right)^2, \nu = \nu' \left(\frac{N}{2\pi\rho}\right)^2.$$

The equations are

$$\left.\begin{aligned} \frac{\partial X}{\partial t} &= a(X-h) + b(Y-k) + \frac{\mu'}{\rho^2}\frac{\partial^2 X}{\partial \theta^2}, \\ \frac{\partial Y}{\partial t} &= c(X-h) + d(Y-k) + \frac{\nu'}{\rho^2}\frac{\partial^2 Y}{\partial \theta^2}, \end{aligned}\right\} \qquad (7.1)$$

which will be seen to be the limiting case of (6.2). The marginal reaction rates a, b, c, d are, as before, the values at the equilibrium position of $\partial f/\partial X$, $\partial f/\partial Y$, $\partial g/\partial X$, $\partial g/\partial Y$. The general solution of the equations is

$$\left.\begin{aligned} X &= h + \sum_{s=-\infty}^{\infty} (A_s e^{p_s t} + B_s e^{p'_s t}) e^{is\theta}, \\ Y &= k + \sum_{s=-\infty}^{\infty} (C_s e^{p_s t} + D_s e^{p'_s t}) e^{is\theta}, \end{aligned}\right\} \qquad (7.2)$$

where p_s, p'_s are now roots of

$$\left(p - a + \frac{\mu' s^2}{\rho^2}\right)\left(p - d + \frac{\nu' s^2}{\rho^2}\right) = bc \qquad (7.3)$$

and

$$\left.\begin{aligned} A_s\left(p_s - a + \tfrac{\mu' s^2}{\rho^2}\right) &= bC_s, \\ B_s\left(p'_s - a + \tfrac{\mu' s}{\rho^2}\right) &= bD_s. \end{aligned}\right\} \qquad (7.4)$$

This solution may be justified by considering the limiting case of the solution (6·11). Alternatively, one may observe that the formula proposed is a solution, so that it only remains to prove that it is the most general one. This will follow if values of A_s, B_s, C_s, D_s, can be found to fit any given initial conditions. It is well known that any function of an angle (such as X) can be expanded as a 'Fourier series'

$$X(\theta) = \sum_{s=-\infty}^{\infty} G_s e^{is\theta} \; (X(\theta) \quad \text{being values of } X \text{ at } t = 0),$$

provided, for instance, that its first derivative is continuous. If also

$$Y(\theta) = \sum_{s=-\infty}^{\infty} H_s e^{is\theta} \; (Y(\theta) \quad \text{being values of } Y \text{ at } t = 0),$$

then the required initial conditions are satisfied provided $A_s + B_s = G_s$ and $C_s + D_s = H_s$. Values A_s, B_s, C_s, D_s to satisfy these conditions can be found unless $p_s = p'_s$. This is an exceptional case and its solution if required may be found as the limit of the normal case.

8. Types of asymptotic behaviour in the ring after a lapse of time

As the reader was reminded in § 2, after a lapse of time the behaviour of an expression of the form of (6.11) is eventually dominated by the terms for which the corresponding p_s has the largest real part. There may, however, be several terms for which this real part has the same value, and these terms will together dominate the situation, the other terms being ignored by comparison. There will, in fact, normally be either two or four such 'leading' terms. For if p_{s_0} is one of them then $p_{N-s_0} = p_{s_0}$, since

$$\sin^2 \frac{\pi(N - s_0)}{N} = \sin^2 \frac{\pi s_0}{N},$$

so that p_{s_0} and p_{N-s_0} are roots of the same equation (6.8). If also p_{s_0} is complex then $\mathcal{R}p_{s_0} = \mathcal{R}p_{s_0}$ and so in all

$$\mathcal{R}p_{s_0} = \mathcal{R}p'_{s_0} = \mathcal{R}p_{N-s_0} = \mathcal{R}p'_{N-s_0}.$$

One need not, however, normally anticipate that any further terms will have to be included. If p_{s_0} and p_{s_1} are to have the same real part, then, unless $s_1 = s_0$ or $s_0 + s_1 = N$ the quantities $a,\ b,\ c,\ d, \mu,\ \nu$ will be restricted to satisfy some special condition, which they would be unlikely to satisfy by chance. It is possible to find circumstances in which as many as ten terms have to be included if such special conditions *are* satisfied, but these have no particular physical or biological importance. It is assumed below that none of these chance relations hold.

It has already been seen that it is necessary to distinguish the cases where the value of p_{s_0} for one of the dominant terms is real from those where it is complex. These may be called respectively the *stationary* and the *oscillatory* cases.

Stationary case. After a sufficient lapse of time $X_r - h$ and $Y_r - k$ approach asymptotically to the forms

$$\left.\begin{aligned} X_r - h &= 2\mathcal{R}A_{s_0} \exp\left[\tfrac{2\pi i s_0 r}{N} + It\right], \\ Y_r - k &= 2\mathcal{R}C_{s_0} \exp\left[\tfrac{2\pi i s_0 r}{N} + It\right]. \end{aligned}\right\} \tag{8.1}$$

Oscillatory case. After a sufficient lapse of time $X_r - h$ and $Y_r - k$ approach the forms

$$\left.\begin{aligned} X_r - h &= 2e^{It}\mathcal{R}\left\{A_{s_0}\exp\left[\frac{2\pi i s_0 r}{N} + i\omega t\right] + A_{N-s_0}\exp\left[-\frac{2\pi i s_0 r}{N} - i\omega t\right]\right\},\\ Y_r - k &= 2e^{It}\mathcal{R}\left\{C_{s_0}\exp\left[\frac{2\pi i s_0 r}{N} + i\omega t\right] + C_{N-s_0}\exp\left[-\frac{2\pi i s_0 r}{N} - i\omega t\right]\right\}. \end{aligned}\right\} \quad (8.2)$$

The real part ofp_{s_0} has been represented by I, standing for 'instability', and in the oscillatory case its imaginary part is ω. By the use of the $\mathcal{R}$ operation (real part of), two terms have in each case been combined in one.

The meaning of these formulae may be conveniently described in terms of waves. In the stationary case there are stationary waves on the ring having s_0 lobes or crests. The coefficients A_{s_0} and C_{s_0} are in a definite ratio given by (6.10), so that the pattern for one morphogen determines that for the other. With the lapse of time the waves become more pronounced provided there is genuine instability, i.e. if I is positive. The wave-length of the waves may be obtained by dividing the number of lobes into the circumference of the ring. In the oscillatory case the interpretation is similar, but the waves are now not stationary but travelling. As well as having a wave-length they have a velocity and a frequency. The frequency is $\omega/2\pi$, and the velocity is obtained by multiplying the wave-length by the frequency. There are two wave trains moving round the ring in opposite directions.

The wave-lengths of the patterns on the ring do not depend only on the chemical data a, b, c, d, μ', ν' but on the circumference of the ring, since they must be submultiples of the latter. There is a sense, however, in which there is a 'chemical wave-length' which does not depend on the dimensions of the ring. This may be described as the limit to which the wave-lengths tend when the rings are made successively larger. Alternatively (at any rate in the case of continuous tissue), it may be described as the wave-length when the radius is chosen to give the largest possible instability I. One may picture the situation by supposing that the chemical wave-length is true wave-length which is achieved whenever possible, but that on a ring it is necessary to 'make do' with an approximation which divides exactly into the circumference.

Although all the possibilities are covered by the stationary and oscillatory altematives there are special cases of them which deserve to be treated separately. One of these occurs when $s_0 = 0$, and may be described as the 'case of extreme long wave-length', though this term may perhaps preferably be reserved to describe the chemical data when they are such that s_0 is zero whatever the dimensions of the ring. There is also the case of 'extreme short wave-length'. This means that $\sin^2(\pi s_0/N)$ is as large as possible, which is achieved by s_0 being either $\frac{1}{2}N$, or $\frac{1}{2}(N-1)$. If the remaining possibilities are regarded as forming the 'case of finite wave-length', there are six sub-cases altogether. It will be shown that each of these really can occur, although two of them require three or more morphogens for their realization.

(a) *Stationary y case with extreme long wave-length.* This occurs for instance if $\mu = \nu = \frac{1}{4}$, $b = c = 1$, $a = d$. Then $p_s = a - \sin^2\frac{\pi s}{N} + 1$. This is certainly real and is greatest when $s = 0$. In this case the contents of all the cells are the same; there is no resultant flow from cell to cell due to diffusion, so that each is behaving as if it were isolated. Each is in unstable equilibrium, and slips out of it in synchronism with the others.

(b) *Oscillatory case with extreme long wave-length.* This occurs, for instance, if $\mu = \nu = \frac{1}{4}$, $b = -c = 1$, $a = d$. Then $p_s = a - \sin^2\frac{\pi s}{N} \pm i$. This is complex and its real part is greatest when $s = 0$. As in case (a) each cell behaves as if it were isolated. The difference from case (a) is that the departure from the equilibrium is oscillatory.

(c) *Stationary waves of extreme short wave-length.* This occurs, for instance, if $\nu = 0$, $\mu = 1$, $d = I$, $a = I - 1$, $b = -c = 1$. p_s is

$$I - \frac{1}{2} - 2\sin^2\frac{\pi s}{N} + \sqrt{\left\{\left(2\sin^2\frac{\pi s}{N} + \frac{1}{2}\right)^2 - 1\right\}},$$

and is greatest when $\sin^2(\pi s/N)$ is greatest. If N is even the contents of each cell are similar to those of the next but one, but distinctly different from those of its immediate neighbours. If, however, the number of cells is odd this arrangement is impossible, and the magnitude of the difference between neighbouring cells varies round the ring, from zero at one point to a maximum at a point diametrically opposite.

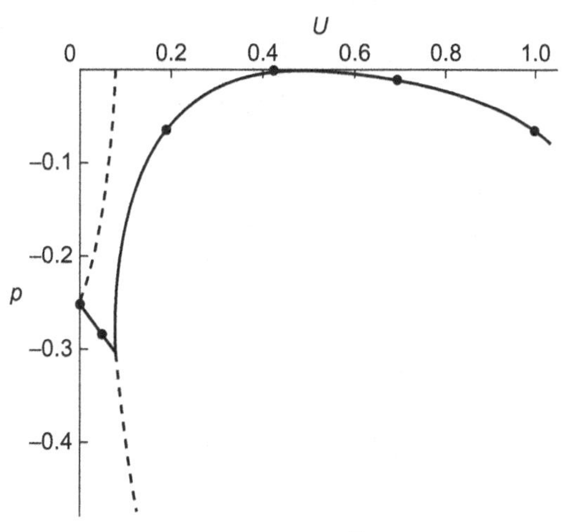

Fig. 1: Values of $\mathcal{R}p$ (instability or growth rate), and $|\mathcal{I}p|$ (radian frequency of oscillation), related to wave-length $2\pi U^{-\frac{1}{2}}$ as in the relation (8.3) with $I = 0$. This is a case of stationary waves with finite wave-length. Full line, $\mathcal{R}p$; broken line, $-|\mathcal{I}p|$ (zero for $U > 0.071$); dotted line, $\mathcal{R}p'$. The full circles on the curve for $\mathcal{R}p$ indicate the values of U, p actually achievable on the finite ring considered in § 10, with $s = 0$ on the extreme left, $s = 5$ on the right.

(d) *Stationary waves of finite wave-length.* This is the case which is of greatest interest, and has most biological application. It occurs, for instance, if $a = I - 2$, $b = 2.5$, $c = -1.25$, $d = I + 1.5$, $\mu' = 1$, $\nu' = \frac{1}{2}$, and $\frac{\mu}{\mu'}, = \frac{\nu}{\nu'}, = \left(\frac{N}{2\pi\rho}\right)^2$. As before ρ is the radius of the ring, and N the number of cells in it. If one writes U for $(\frac{N}{\pi\rho})^2 \sin^2\frac{\pi s}{N}$, then equation (6.8) can, with these special values, be written

$$(p - I)^2 + \left(\frac{1}{2} + \frac{3}{2}U\right)(p - I) + \frac{1}{2}\left(U - \frac{1}{2}\right)^2 = 0. \tag{8.3}$$

This has a solution $p = I$ if $U = \frac{1}{2}$. On the other hand, it will be shown that if U has any other (positive) value then both roots for $p - I$ have negative real parts. Their product is positive being $\frac{1}{2}(U - \frac{1}{2})^2$, so that if they are real they both have the same sign. Their sum in this case is $-\frac{1}{2} - \frac{3}{2}U$ which is negative. Their common sign is therefore negative. If, however, the roots are complex their real parts are both equal to $-\frac{1}{4} - \frac{3}{4}U$, which is negative.

If the radius ρ of the ring be chosen so that for some integer s_0, $\frac{1}{2} = U = \left(\frac{N}{\pi\rho}\right)^2 \sin^2 \frac{\pi s_0}{N}$, there will be stationary waves with s_0 lobes and a wave-length which is also equal to the chemical wave-length, for p_{s_0} will be equal to I, whereas every other p_s will have a real part smaller than I. If, however, the radius is chosen so that $\left(\frac{N}{\pi\rho}\right)^2 \sin^2 \frac{\pi s}{N} = \frac{1}{2}$ cannot hold with an integral s, then (in this example) the actual number of lobes will be one of the two integers nearest to the (non-integral) solutions of this equation, and usually *the* nearest. Examples can, however, be constructed where this simple rule does not apply.

Figure 1 shows the relation (8.3) in graphical form. The curved portions of the graphs are hyperbolae.

The two remaining possibilities can only occur with three or more morphogens. With one morphogen the only possibility is (*a*).

(e) *Oscillatory case with a finite wave-length.* This means that there are genuine travelling waves. Since the example to be given involves three morphogens it is not possible to use the formulae of § 6. Instead, one must use the corresponding three morphogen formulae. That which corresponds to (6.8) or (7.3) is most conveniently written as

$$\begin{vmatrix} a_{11} - p - \mu_1 U & a_{12} & a_{13} \\ a_{21} & a_{22} - p - \mu_2 U & a_{23} \\ a_{31} & a_{32} & a_{33} - p - \mu_3 U \end{vmatrix} = 0, \tag{8.4}$$

where again U has been written for $(\frac{N}{\pi\rho})^2 \sin^2 \frac{\pi s}{N}$. (This means essentially that $U = (\frac{2\pi}{\lambda})^2$, where λ is the wave-length.) The four marginal reactivities are superseded by nine $a_{11}, \ldots, a_{33}$, and the three diffusibilities are μ_1, μ_2, μ_3. Special values leading to travelling waves are

$$\left.\begin{array}{lll} \mu_1 = \frac{2}{3}, & \mu_2 = \frac{1}{3}, & \mu_3 = 0 \\ a_{11} = -\frac{10}{3}, & a_{12} = 3, & a_{13} = -1, \\ a_{21} = -2, & a_{22} = \frac{7}{3}, & a_{23} = 0, \\ a_{31} = 3, & a_{32} = -4, & a_{33} = 0, \end{array}\right\} \tag{8.5}$$

and with them (8.4) reduces to

$$p^3 + p^2(U+1) + p(1 + \tfrac{2}{9}(U-1)^2) + U + 1 = 0. \tag{8.6}$$

If $U = 1$ the roots are $\pm$i and -2. If U is near to I they are approximately $-1 - U$ and $\pm\text{i} + \frac{(U-1)^2}{18}(\pm\text{i} - 1)$, and all have negative real parts. If the greatest real part is not the value zero, achieved with $U = 1$, then the value zero must be reached again at some intermediate value of U. Since P is then pure imaginary the even terms of (8.6) must vanish, i.e. $(p^2+1)(U+1) = 0$. But this can only happen if $p = \pm\text{i}$, and the vanishing of the odd terms then shows that $U = 1$. Hence zero is the largest real part for any root p of (8.6). The corresponding p is $\pm$i and U is 1. This means that there are travelling waves with unit (chemical) radian frequency and unit (chemical) velocity. If I is added to a_{11}, a_{22} and a_{33}, the instability will become I in place of zero.

(f) Oscillatory case with extreme short wave-length. This means that there is metabolic oscillation with neighbouring cells nearly 180° out of phase. It can be achieved with three morphogens and the following chemical data:

$$\left.\begin{array}{llll} \mu = 1, & \mu_2 = \mu_3 = 0, & & \\ a_{11} = -1, & a_{12} = -1, & a_{13} = 0 & \\ a_{21} = 1, & a_{22} = 0, & a_{23} = -1, & \\ a_{31} = 0, & a_{32} = 1, & a_{33} = 0. & \end{array}\right\} \tag{8.7}$$

With these values (8.4) reduces to

$$p^3 + p^2(U+1) + 2p + U + 1 = 0. \tag{8.8}$$

This may be shown to have all the real parts of its roots negative if $U \geqslant 0$, for if $U = 0$ the roots are near to -0.6, $-0.2 \pm 1.3\mathrm{i}$, and if U be continuously increased the values of p will alter continuously. If they ever attain values with a positive real part they must pass through pure imaginary values (or zero). But if p is pure imaginary $p^3 + 2p$ and $(p^2+1)(U+1)$ must both vanish, which is impossible if $U \geqslant 0$. As U approaches infinity, however, one of the roots approaches i. Thus $\mathcal{R}p = 0$ can be approached as closely as desired by large values of U, but not attained.

9. Further consideration of the mathematics of the ring

In this section some of the finer points concerning the development of wave patterns are considered. These will be of interest mainly to those who wish to do further research on the subject, and can well be omitted on a first reading.

(1) *General formulae for the two morphogen case.* Taking the limiting case of a ring of large radius (or a filament), one may write $(\frac{N}{\pi\rho})^2 \sin^2 \frac{\pi s}{N} = U = (\frac{2\pi}{\lambda})^2$ in (6.11) or $\frac{s^2}{\rho^2} = U = (\frac{2\pi}{\lambda})^2$ in (7.3) and obtain

$$(p - a + \mu' U)(p - d + \nu' U) = bc, \tag{9.1}$$

which has the solution

$$p = \frac{a+d}{2} - \frac{\mu' + \nu'}{2} U \pm \sqrt{\left\{\left(\frac{\mu' - \nu'}{2} U + \frac{d-a}{2}\right)^2 + bc\right\}}. \tag{9.2}$$

One may put $I(U)$ for the real part of this, representing the instability for waves of wave-length $\lambda = 2\pi U^{-\frac{1}{2}}$. The dominant waves correspond to the maximum of $I(U)$. This maximum may either be at $U = 0$ or $U = \infty$ or at a stationary point on the part of the curve which is hyperbolic (rather than straight). When this last case occurs the values of p (or I) and U at the maximum are

$$\left.\begin{array}{l} p = I = \left(d\mu' - a\nu' - 2\surd(\mu'\nu')\,\surd(-bc)(\mu' - \nu')^{-1},\right. \\ U = (a - d + \frac{\mu'+\nu'}{\surd(\mu'\nu')}\surd(-bc))(\mu' - \nu')^{-1}. \end{array}\right\} \tag{9.3}$$

The conditions which lead to the four cases (a), (b), (c), (d) described in the last section are

(a) (Stationary waves of extreme long wave-length.) This occurs if either

$$\text{(i) } bc > 0, \quad \text{(ii) } bc < 0 \text{ and } \frac{d-a}{\surd(-bc)} > \frac{\mu' + \nu'}{\surd(\mu'\nu')}, \quad \text{(iii) } bc < 0 \text{ and } \frac{d-a}{\surd(-bc)} < -2.$$

The condition for instability in either case is that either $bc > ad$ or $a+d > 0$.

(b) (Oscillating case with extreme long wave-length, i.e. synchronized oscillations.) This occurs if

$$bc < 0 \text{ and } -2 < \frac{d-a}{\surd(-bc)} < \frac{4\surd(\mu'\nu')}{\mu'+\nu'}.$$

There is instability if in addition $a+d > 0$.

(c) (Stationary waves of extreme short wave-length.) This occurs if $bc < 0$, $\mu' > \nu' = 0$. There is instability if, in addition, $a+d > 0$.

(d) (Stationary waves of finite wave-length.) This occurs if

$$bc < 0 \text{ and } \frac{4\surd(\mu'\nu')}{\mu'+\nu'} < \frac{d-a}{\surd(-bc)} < \frac{\mu'+\nu'}{(\surd\mu'\nu')}, \tag{9.4a}$$

and there is instability if also

$$\frac{d}{\surd(-bc)}\sqrt{\frac{\mu'}{\nu'}} - \frac{a}{\surd(-bc)}\sqrt{\frac{\nu'}{\mu'}} > 2. \tag{9.4b}$$

It has been assumed that $\nu' \leqslant \mu' > 0$. The case where $\mu' \leqslant \nu' > 0$ can be obtained by interchanging the two morphogens. In the case $\mu' = \nu' = 0$ there is no co-operation between the cells whatever.

Some additional formulae will be given for the case of stationary waves of finite wave-length. The marginal reaction rates may be expressed parametrically in terms of the diffusibilities, the wave-length, the instability, and two other parameters α and χ. Of these α may be described as the ratio of $X-h$ to $Y-k$ in the waves. The expressions for the marginal reaction rates in terms of these parameters are

$$\left.\begin{aligned} a &= \mu'(\nu'-\mu')^{-1}(2\nu' U_0 + \chi) + I, \\ b &= \mu'(\nu'-\mu')^{-1}((\mu'+\nu')U_0 + \chi)\alpha, \\ c &= \nu'(\mu'-\nu')^{-1}((\mu'+\nu')U_0 + \chi)\alpha^{-1}, \\ d &= \nu'(\mu'-\nu')^{-1}(2\mu' U_0 + \chi) + I, \end{aligned}\right\} \tag{9.5}$$

and when these are substituted into (9.2) it becomes

$$p = I - \frac{1}{2}\chi - \frac{\mu'+\nu'}{2}U + \surd\left\{\left(\frac{\mu'+\nu'}{2}U + \frac{1}{2}\chi\right)^2 - \mu'\nu'(U-U_0)^2\right\}. \tag{9.6}$$

Here $2\pi U_0^{-\frac{1}{2}}$ is the chemical wave-length and $2\pi U^{-\frac{1}{2}}$ the wave-length of the Fourier component under consideration. χ must be positive for case (*d*) to apply.

If s be regarded as a continuous variable one can consider (9.2) or (9.6) as relating s to p, and $\mathrm{d}p/\mathrm{d}s$ and $\mathrm{d}^2p/\mathrm{d}s^2$ have meaning. The value of $\mathrm{d}^2p/\mathrm{d}s^2$ at the maximum is of some interest, and will be used below in this section. Its value is

$$\frac{\mathrm{d}^2p}{\mathrm{d}s^2} = -\frac{\surd(\mu'\nu')}{\rho^2}\cdot\frac{8\surd(\mu'\nu')}{\mu'+\nu'}\cos^2\frac{\pi s}{N}(1+\chi U_0^{-1}(\mu'+\nu')^{-1})^{-1}. \tag{9.7}$$

(2) In §§ 6, 7, 8 it was supposed that the disturbances were not continuously operative, and that the marginal reaction rates did not change with the passage of time. These assumptions will now be dropped, though it will be necessary to make some other, less drastic, approximations to

replace them. The (statistical) amplitude of the 'noise' disturbances will be assumed constant in time. Instead of (6.6), (6.7), one then has

$$\left.\begin{aligned}\frac{\mathrm{d}\xi}{\mathrm{d}t} &= a'\xi + b\eta + R_1(t),\\ \frac{\mathrm{d}\eta}{\mathrm{d}t} &= c\xi + d'\eta + R_2(t),\end{aligned}\right\} \tag{9.8}$$

where ξ, η have been written for ξ_s, η_s since s may now be supposed fixed. For the same reason $a - 4\mu \sin^2 \frac{\pi s}{N}$ has been replaced by a' and $d - 4\nu \sin^2 \frac{\pi s}{N}$ by d'. The noise disturbances may be supposed to constitute white noise, i.e. if $(t_1,\ t_2)$ and $(t_3,\ t_4)$ are two non-overlapping intervals then $\int_{t_1}^{t_2} R_1(t)\mathrm{d}t$ and $\int_{t_3}^{t_4} R_2(t)\mathrm{d}t$ are statistically independent and each is normally distributed with variances $\beta_1(t_2 - t_1)$ and $\beta_1(t_4 - t_3)$ respectively, β_1 being a constant describing the amplitude of the noise. Likewise for $R_2(t)$, the constant β_1 being replaced by β_2. If p and p' are the roots of $(p - a')(p - d') = bc$ and p is the greater (both being real), one can make the substitution

$$\left.\begin{aligned}\xi &= b(u + v),\\ \eta &= (p - a')u + (p' - a')v,\end{aligned}\right\} \tag{9.9}$$

which transforms (9.8) into

$$\frac{\mathrm{d}u}{\mathrm{d}t} = pu + \frac{p' - a'}{(p' - p)b}R_1(t) - \frac{R_2(t)}{p' - p} + \xi\frac{\mathrm{d}}{\mathrm{d}t}\left(\frac{p' - a'}{(p' - p)b}\right) - \eta\frac{\mathrm{d}}{\mathrm{d}t}\left(\frac{1}{p' - p}\right), \tag{9.10}$$

with a similar equation for v, of which the leading terms are $\mathrm{d}v/\mathrm{d}t = p'v$. This indicates that v will be small, or at least small in comparison with u after a lapse of time. If it is assumed that $v = 0$ holds (9.11) may be written

$$\frac{\mathrm{d}u}{\mathrm{d}t} = qu + L_1(t)R_1(t) + L_2(t)R_2(t), \tag{9.11}$$

where

$$L_1(t) = \frac{p' - a'}{(p' - p)b}, \quad L_2(t) = \frac{1}{p' - p}, \quad q = p + bL_1'(t). \tag{9.12}$$

The solution of this equation is

$$u = \int_{-\infty}^{t} (L_1(w)R_1(w) + L_2(w)R_2(w)) \exp\left[\int_{w}^{t} q(z)\mathrm{d}z\right] dw. \tag{9.13}$$

One is, however, not so much interested in such a solution in terms of the statistical disturbances as in the consequent statistical distribution of values of u, ξ and η at various times after instability has set in. In view of the properties of 'white noise' assumed above, the values of u at time t will be distributed according to the normal error law, with the variance

$$\int_{-\alpha}^{t} \left[\beta_1(L_1(w))^2 + \beta_2(L_2(w))^2\right] \exp\left[2\int_{t}^{w} q(z)\mathrm{d}z\right] dw. \tag{9.14}$$

There are two commonly occurring cases in which one can simplify this expression considerably without great loss of accuracy. If the system is in a distinctly stable state, then $q(t)$, which is near to $p(t)$, will be distinctly negative, and $\exp\left[\int_w^t q(z)\mathrm{d}z\right]$ will be small unless w is near to t. But then $L_1(w)$ and $L_2(w)$ may be replaced by $L_1(t)$ and $L_2(t)$ in the integral, and also $q(z)$ may be replaced by $q(t)$. With these approximations the variance is

$$(-2q(t))^{-1}\left[\beta_1(L_1(t))^2+\beta_2(L_2(t))^2\right]. \tag{9.15}$$

A second case where there is a convenient approximation concerns times when the system is unstable, so that $q(t)>0$. For the approximation concerned to apply $2\int_w^t q(z)\mathrm{d}z$ must have its maximum at the last moment $w(=t_0)$ when $q(t_0)=0$, and it must be the maximum by a considerable margin (e.g. at least 5) over all other local maxima. These conditions would apply for instance if $q(z)$ were always increasing and had negative values at a sufficiently early time. One also requires $q'(t_0)$ (the rate of increase of q at time t_0) to be reasonably large; it must at least be so large that over a period of time of length $(q'(t_0))^{-\frac{1}{2}}$ near to t_0 the changes in $L_1(t)$ and $L_2(t)$ are small, and $q'(t)$ itself must not appreciably alter in this period. Under these circumstances the integrand is negligible when w is considerably different from t_0, in comparison with its values at that time, and therefore one may replace $L_1(w)$ and $L_2(w)$ by $L_1(t_0)$ and $L_2(t_0)$, and $q'(w)$ by $q'(t_0)$. This gives the value

$$\sqrt{\pi}(q'(t_0))^{-\frac{1}{2}}\left[\beta_1(L_1(t_0))^2+\beta_2(L_2(t_0))^2\right]\exp\left[2\int_{t_0}^{t} q(z)\mathrm{d}z\right], \tag{9.16}$$

for the variance of u.

The physical significance of this latter approximation is that the disturbances near the time when the instability is zero are the only ones which have any appreciable ultimate effect. Those which occur earlier are damped out by the subsequent period of stability. Those which occur later have a shorter period of instability within which to develop to greater amplitude. This principle is familiar in radio, and is fundamental to the theory of the superregenerative receiver.

Naturally one does not often wish to calculate the expression (9.17), but it is valuable as justifying a common-sense point of view of the matter. The factor $\exp[\int_{t_0}^t q(z)\mathrm{d}z]$ is essentially the integrated instability and describes the extent to which one would expect disturbances of appropriate wave-length to grow between times t_0 and t. Taking the terms in β_1, β_2 into consideration separately, the factor $\sqrt{\pi}\beta_1(q'(t_0))^{-\frac{1}{2}}(L_1(t_0))^2$ indicates that the disturbances on the first morphogen should be regarded as lasting for a time

$$\sqrt{\pi}(q_1(t_0))^{-\frac{1}{2}}(bL_1(t_0))^2.$$

The dimensionless quantities $bL_1(t_0)$, $bL_2(t_0)$ will not usually be sufficiently large or small to justify their detailed calculation.

(3) The extent to which the component for which p_s is greatest may be expected to outdistance the others will now be considered in case (*d*). The greatest of the p_s will be called p_{s_0}. The two closest competitors to s_0 will be s_0-1 and s_0+1; it is required to determine how close the competition is. If the variation in the chemical data is sufficiently small it may be assumed that, although the exponents p_{s_0-1}, p_{s_0}, p_{s_0+1} may themselves vary appreciably in time, the differences $p_{s_0}-p_{s_0-1}$ and $p_{s_0}-p_{s_0+1}$ are constant. It certainly can happen that one of these differences is zero or nearly zero, and there is then 'neck and neck' competition. The weakest competition occurs when $p_{s_0-1}=p_{s_0+1}$. In this case

$$p_{s_0}-p_{s_0-1}=p_{s_0}-p_{s_0+1}=-\frac{1}{2}(p_{s_0+1}-2p_{s_0}+p_{s_0-1}).$$

But if s_0 is reasonably large $p_{s_0+1} - 2p_{s_0} + p_{s_0-1}$ can be set equal to $(\mathrm{d}^2p/\mathrm{d}s^2)_{s=s_0}$. It may be concluded that the rate at which the most quickly growing component grows cannot exceed the rate for its closest competitor by more than about $\frac{1}{2}(\mathrm{d}^2p/\mathrm{d}s^2)_{s=s_0}$. The formula (9.7), by which $\mathrm{d}^2p/\mathrm{d}s^2$ can be estimated, may be regarded as the product of two factors. The dimensionless factor never exceeds 4. The factor $\surd(\mu'\nu')/\rho^2$ may be described in very rough terms as 'the reciprocal of the time for the morphogens to diffuse a length equal to a radius'. In equally rough terms one may say that a time of this order of magnitude is required for the most quickly growing component to get a lead, amounting to a factor whose logarithm is of the order of unity, over its closest competitors, in the favourable case where $p_{s_0-1} = p_{s_0+1}$.

(4) Very little has yet been said about the effect of considering non-linear reaction rate functions when far from homogeneity. Any treatment so systematic as that given for the linear case seems to be out of the question. It is possible, however, to reach some qualitative conclusions about the effects of non-linear terms. Suppose that z_1 is the amplitude of the Fourier component which is most unstable (on a basis of the linear terms), and which may be supposed to have wave-length λ. The non-linear terms will cause components with wave-lengths $\frac{1}{2}\lambda, \frac{1}{3}\lambda, \frac{1}{4}\lambda, \ldots$ to appear as well as a space-independent component. If only quadratic terms are taken into account and if these are somewhat small, then the component of wave-length $\frac{1}{2}\lambda$ and the space-independent component will be the strongest. Suppose these have amplitudes z_2 and z_1. The state of the system is thus being described by the numbers z_0, z_1, z_2. In the absence of non-linear terms they would satisfy equations

$$\frac{\mathrm{d}z_0}{\mathrm{d}t} = p_0z_0, \quad \frac{\mathrm{d}z_1}{\mathrm{d}t} = p_1z_1, \quad \frac{\mathrm{d}z_2}{\mathrm{d}t} = p_2z_2,$$

and if there is slight instability p_1 would be a small positive number, but p_0 and p_2 distinctly negative. The effect of the non-linear terms is to replace these equations by ones of the form

$$\frac{\mathrm{d}z_0}{\mathrm{d}t} = p_0z_0 + Az_1^2 + Bz_2^2,$$

$$\frac{\mathrm{d}z_1}{\mathrm{d}t} = p_1z_1 + Cz_2z_1 + Dz_0z_1,$$

$$\frac{\mathrm{d}z_2}{\mathrm{d}t} = p_2z_2 + Ez_1^2 + Fz_0z_2.$$

As a first approximation one may put $\mathrm{d}z_0/\mathrm{d}t = \mathrm{d}z_2/\mathrm{d}t = 0$ and ignore z_1^4 and higher powers; z_0 and z_1 are then found to be proportional to z_1^2, and the equation for z_1 can be written $\mathrm{d}z_1/\mathrm{d}t = p_0z_1 - kz_1^3$. The sign of k in this differential equation is of great importance. If it is positive, then the effect of the term kz_1^3 is to arrest the exponential growth of z_1 at the value $\surd(p_1/k)$. The 'instability' is then very confined in its effect, for the waves can only reach a finite amplitude, and this amplitude tends to zero as the instability (p_1) tends to zero. If, however, k is negative the growth becomes something even faster than exponential, and, if the equation $\mathrm{d}z_1/\mathrm{d}t = p_1z_1 - kz_1^3$ held universally, it would result in the amplitude becoming infinite in a finite time. This phenomenon may be called 'catastrophic instability'. In the case of two-dimensional systems catastrophic instability is almost universal, and the corresponding equation takes the form $\mathrm{d}z_1/\mathrm{d}t = p_1z_1 + kz_1^2$. Naturally enough in the case of catastrophic instability the amplitude does not really reach infinity, but when it is sufficiently large some effect previously ignored becomes large enough to halt the growth.

(5) Case (*a*) as described in § 8 represents a most extremely featureless form of pattern development. This may be remedied quite simply by making less drastic simplifying assumptions, so that a less gross account of the pattern can be given by the theory. It was assumed in § 9 that

only the most unstable Fourier components would contribute appreciably to the pattern, though it was seen above (heading (3) of this section) that (in case (*d*)) this will only apply if the period of time involved is adequate to permit the morphogens, supposed for this purpose to be chemically inactive, to diffuse over the whole ring or organ concerned. The same may be shown to apply for case (*a*). If this assumption is dropped a much more interesting form of pattern can be accounted for. To do this it is necessary to consider not merely the components with $U = 0$ but some others with small positive values of U. One may assume the form $At - BU$ for p. Linearity in U is assumed because only small values of U are concerned, and the term At is included to represent the steady increase in instability. By measuring time from the moment of zero instability the necessity for a constant term is avoided. The formula (9.17) may be applied to estimate the statistical distribution of the amplitudes of the components. Only the factor $\exp\left[2\int_{t_0}^{t} q(z)\mathrm{d}z\right]$ will depend very much on U, and taking $q(t) = p(t) = At - BU$, t_0 must be BU/A and the factor is

$$\exp\left[A(t - BU/A)^2\right].$$

The term in U^2 can be ignored if At^2 is fairly large, for then either B^2U^2/A^2 is small or the factor e^{-BUt} is. But At^2 certainly is large if the factor e^{At^2}, applying when $U = 0$, is large. With this approximation the variance takes the form $C\mathrm{e}^{-\frac{1}{2}k^2U}$, with only the two parameters C, k to distinguish the pattern populations. By choosing appropriate units of concentration and length these pattern populations may all be reduced to a standard one, e.g. with $C = k = 1$. Random members of this population may be produced by considering any one of the type (*a*) systems to which the approximations used above apply. They are also produced, but with only a very small amplitude scale, if a homogeneous one-morphogen system undergoes random disturbances without diffusion for a period, and then diffusion without disturbance. This process is very convenient for computation, and can also be applied to two dimensions. Figure 2 shows such a pattern, obtained in a few hours by a manual computation.

To be more definite a set of numbers $u_{r,s}$ was chosen, each being ± 1, and taking the two values with equal probability. A function $f(x, y)$ is related to these numbers by the formula

$$f(x,y) = \Sigma u_{r,s} \exp[-\frac{1}{2}((x - hr)^2 + (y - hs)^2)].$$

In the actual computation a somewhat crude approximation to the function

$$\exp\left[-\frac{1}{2}(x^2 + y^2)\right]$$

was used and h was about 0.7. In the figure the set of points where $f(x, y)$ is positive is shown black. The outlines of the black patches are somewhat less irregular than they should be due to an inadequacy in the computation procedure.

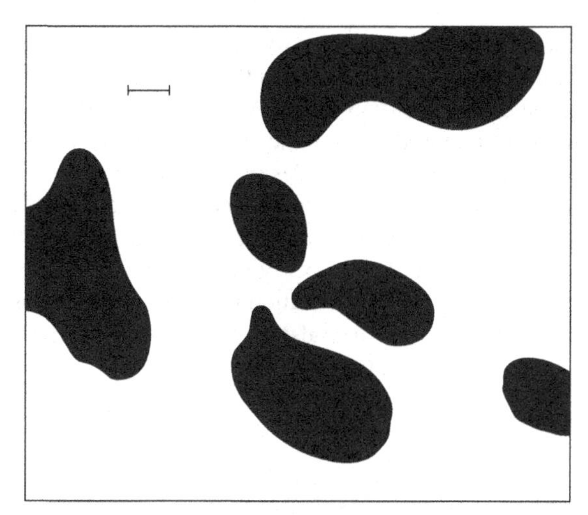

Fig. 2: An example of a 'dappled' pattern as resulting from a type (*a*) morphogen system. A marker of unit length is shown. See text, § 9, 11.

10. A numerical example

The numerous approximations and assumptions that have been made in the foregoing analysis may be rather confusing to many readers. In the present section it is proposed to consider in detail a single example of the case of most interest, (*d*). This will be made as specific as possible. It is unfortunately not possible to specify actual chemical reactions with the required properties, but it is thought that the reaction rates associated with the imagined reactions are not unreasonable.

The detail to be specified includes

(i) The number and dimensions of the cells of the ring.
(ii) The diffusibilities of the morphogens.
(iii) The reactions concemed.
(iv) The rates at which the reactions occur.
(v) Information about random disturbances.
(vi) Information about the distribution, in space and time, of those morphogens which are of the nature of evocators.

These will be taken in order.

(i) It will be assumed that there are twenty cells in the ring, and that they have a diameter of 0.1 mm each. These cells are certainly on the large rather than the small side, but by no means impossibly so. The number of cells in the ring has been chosen rather small in order that it should not be necessary to make the approximation of continuous tissue.
(ii) Two morphogens are considered. They will be called X and Y, and the same letters will be used for their concentrations. This will not lead to any real confusion. The diffusion constant for X will be assumed to be $5 \times 10^{-8}\,\mathrm{cm^2 s^{-1}}$ and that for Y to be $2.5 \times 10^{-8}\,\mathrm{cm^2 s^{-1}}$. With cells of diameter 0.01 cm this means that X flows between neighbouring cells at the rate 5×10^{-4} of the difference of X-content of the two cells per second. In other words, if there is nothing altering the concentrations but diffusion the difference of concentrations suffers an exponential decay with time constant 1000 s, or 'half-period' of 700 s. These times are doubled for Y.

If the cell membrane is regarded as the only obstacle to diffusion the permeability of the membranes to the morphogen is 5×10^{-6}cm/s or 0.018 cm/h. Values as large as 0.1 cm/h have been observed (Davson & Danielli 1943, figure 28).

(iii) The reactions are the most important part of the assumptions. Four substances A, X, Y, B are involved; these are isomeric, i.e. the molecules of the four substances are all rearrangements of the same atoms. Substances C, C', W will also be concerned. The thermodynamics of the problem will not be discussed except to say that it is contemplated that of the substances A, X, Y, B the one with the greatest free energy is A, and that with the least is B. Energy for the whole process is obtained by the degradation of A into B. The substance C is in effect a catalyst for the reaction $Y \to X$, and may also be regarded as an evocator, the system being unstable if there is a sufficient concentration of C.

The reactions postulated are

$$\begin{aligned} Y + X &\to W, \\ W + A &\to 2Y + B \quad \text{instantly}, \\ 2X &\to W, \\ A &\to X, \\ Y &\to B, \\ Y + C &\to C' \quad \text{instantly}, \\ C' &\to X + C. \end{aligned}$$

(iv) For the purpose of stating the reaction rates special units will be introduced (for the purpose of this section only). They will be based on a period of 1000 s as units of time, and 10^{-11}mole/cm^3 as concentration unit*. There will be little occasion to use any but these special units (S.U.). The concentration of A will be assumed to have the large value of 1000 S.U. and the catalyst C, together with its combined form C' the concentration $10^{-3}(1+\gamma)$ S.U., the dimensionless quantity γ being often supposed somewhat small, though values over as large a range as from -0.5 to 0.5 may be considered. The rates assumed will be

$$\begin{aligned} Y + X &\to W && \text{at the rate } \frac{25}{16}YX, \\ 2X &\to W && \text{at the rate } \frac{7}{64}X^2, \\ A &\to X && \text{at the rate } \frac{1}{16} \times 10^{-3}A, \\ C' &\to X + C && \text{at the rate } \frac{55}{32} \times 10^{+3}C', \\ Y &\to B && \text{at the rate } \frac{1}{16}Y. \end{aligned}$$

With the values assumed for A and C' the net effect of these reactions is to convert X into Y at the rate $\frac{1}{32}[50XY + 7X^2 - 55(1+\gamma)]$ at the same time producing X at the constant rate $\frac{1}{16}$, and destroying Y at the rate $Y/16$. If, however, the concentration of Y is zero and the rate of increase of

* A somewhat larger value of concentration unit (e.g. 10^{-9}mole/cm^3) is probably more suitable. The choice of unit only affects the calculations through the amplitude of the random disturbances.

Y required by these formulae is negative, the rate of conversion of Y into X is reduced sufficiently to permit Y to remain zero.

In the special units $\mu = \frac{1}{2},\ \nu = \frac{1}{4}$.

(v) Statistical theory describes in detail what irregularities arise from the molecular nature of matter. In a period in which, on the average, one should expect a reaction to occur between n pairs (or other combinations) of molecules, the actual number will differ from the mean by an amount whose mean square is also n, and is distributed according to the normal error law. Applying this to a reaction proceeding at a rate F (S.U.) and taking the volume of the cell as 10^{-8}cm^3 (assuming some elongation tangentially to the ring) it will be found that the root mean square irregularity of the quantity reacting in a period τ of time (S.U.) is $0.004\sqrt{(F\tau)}$.

Table 1 Some Stationary-Wave Patterns.

	First Specimen				Second Specimen: Incipient	'Slow Cooking': Incipient	Four-Lobed Equilibrium	
	Incipient Pattern		*Final Pattern*					
Cell Number	X	Y	X	Y	Y	Y	X	Y
0	1.130	0.929	0.741	1.463	0.834	1.057	1.747	0.000
1	1.123	0.940	0.761	1.469	0.833	0.903	1.685	0.000
2	1.154	0.885	0.954	1.255	0.766	0.813	1.445	2.500
3	1.215	0.810	1.711	0.000	0.836	0.882	0.445	2.500
4	1.249	0.753	1.707	0.000	0.930	1.088	1.685	0.000
5	1.158	0.873	0.875	1.385	0.898	1.222	1.747	0.000
6	1.074	1.003	0.700	1.622	0.770	1.173	1.685	0.000
7	1.078	1.000	0.699	1.615	0.740	0.956	0.445	2.500
8	1.148	0.896	0.885	1.382	0.846	0.775	0.445	2.500
9	1.231	0.775	1.704	0.000	0.937	0.775	1.685	0.000
10	1.204	0.820	1.708	0.000	0.986	0.969	1.747	0.000
11	1.149	0.907	0.944	1.273	1.019	1.170	1.685	0.000
12	1.156	0.886	0.766	1.451	0.899	1.203	0.445	2.500
13	1.170	0.854	0.744	1.442	0.431	1.048	0.445	2.500
14	1.131	0.904	0.756	1.478	0.485	0.868	1.685	0.000
I5	1.090	0.976	0.935	1.308	0.919	0.813	1.747	0.000
16	1.109	0.957	1.711	0.000	1.035	0.910	1.685	0.000
17	1.201	0.820	1.706	0.000	1.003	1.050	0.445	2.500
18	1.306	0.675	0.927	1.309	0.899	1.175	0.445	2.500
19	1.217	0.811	0.746	1.487	0.820	1.181	1.685	0.000

The diffusion of a morphogen from a cell to a neighbour may be treated as if the passage of a molecule from one cell to another were a monomolecular reaction; a molecule must be imagined to change its form slightly as it passes the cell wall. If the diffusion constant for a wall is μ, and quantities M_1, M_2 of the relevant morphogen lie on the two sides of it, the root-mean-square irregularity in the amount passing the wall in a period τ is

$$0.004\sqrt{\{(M_1 + M_2)\mu\tau\}}.$$

These two sources of irregularity are the most significant of those which arise from truly statistical cause, and are the only ones which are taken into account in the calculations whose results are given below. There may also be disturbances due to the presence of neighbouring anatomical structures, and other similar causes. These are of great importance, but of too great variety and complexity to be suitable for consideration here.

(vi) The only morphogen which is being treated as an evocator is C. Changes in the concentration of A might have similar effects, but the change would have to be rather great. It is preferable to assume that A is a 'fuel substance' (e.g. glucose) whose concentration does not change. The concentration of C, together with its combined form C', will be supposed the same in all cells, but it changes with the passage of time. Two different varieties of the problem will be considered, with slightly different assumptions.

The results are shown in table 1. There are eight columns, each of which gives the concentration of a morphogen in each of the twenty cells; the circumstances to which these concentrations refer differ from column to column. The first five columns all refer to the same 'variety' of the imaginary organism, but there are two specimens shown. The specimens differ merely in the chance factors which were involved. With this variety the value of γ was allowed to increase at the rate of 2^{-7} S.U. from the value $-\frac{1}{4}$ to $+\frac{1}{16}$. At this point a pattern had definitely begun to appear, and was recorded. The parameter γ was then allowed to decrease at the same rate to zero and then remained there until there was no more appreciable change. The pattern was then recorded again. The concentrations of

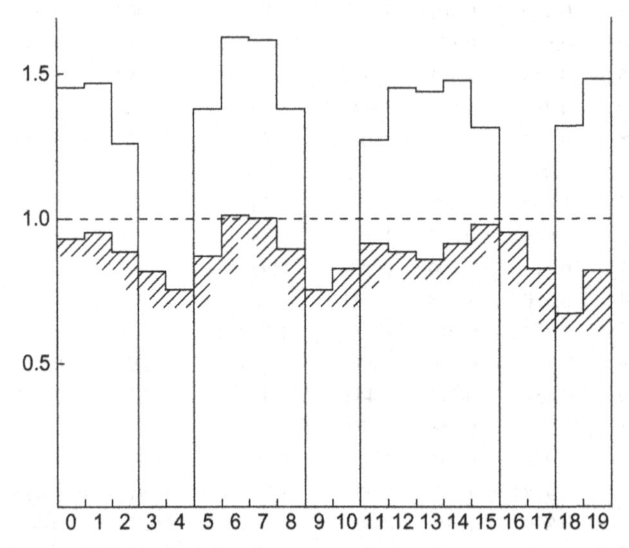

Fig. 3: Concentrations of Y in the development of the first specimen (taken from table 1). - - - - - original homogeneous equilibrium; ////// incipient pattern; —— final equilibrium.

Y in these two recordings are shown in figure 3 as well as in table 1. For the second specimen only one column of figures is given, viz. those for the Y morphogen in the incipient pattern. At this stage the X values are closely related to the Y values, as may be seen from the first specimen (or from theory). The final values can be made almost indistinguishable from those for the first specimen by renumbering the cells and have therefore not been given. These two specimens may be said to belong to the 'variety with quick cooking', because the instability is allowed to increase so quickly that the pattern appears relatively soon. The effect of this haste might be regarded as rather unsatisfactory, as the incipient pattern is very irregular. In both specimens the four-lobed component is present in considerable strength in the incipient pattern. It 'beats' with the three-lobed component producing considerable irregularity. The relative magnitudes of the three- and four-lobed components depend on chance and vary from specimen to specimen. The four-lobed component may often be the stronger, and may occasionally be so strong that the final pattern is four-lobed. How often this happens is not known, but the pattern, when it occurs, is shown in the last two columns of the table. In this case the disturbances were supposed removed for some time before recording, so as to give a perfectly regular pattern.

The remaining column refers to a second variety, one with 'slow cooking'. In this the value of γ was allowed to increase only at the rate 10^{-5}. Its initial value was -0.010, but is of no significance. The final value was 0.003. With this pattern, when shown graphically, the irregularities are definitely perceptible, but are altogether overshadowed by the three-lobed component. The possibility of the ultimate pattern being four-lobed is not to be taken seriously with this variety.

The set of reactions chosen is such that the instability becomes 'catastrophic' when the second-order terms are taken into account, i.e. the growth of the waves tends to make the whole system more unstable than ever. This effect is finally halted when (in some cells) the concentration of Y has become zero. The constant conversion of Y into X through the agency of the catalyst C can then no longer continue in these cells, and the continued growth of the amplitude of the waves is arrested. When $\gamma = 0$ there is of course an equilibrium with $X = Y = 1$ in all cells, which is very slightly stable. There are, however, also other stable equilibria with $\gamma = 0$, two of which are shown in the table. These final equilibria may, with some trouble but little difficulty, be verified to be solutions of the equations (6.1) with

$$\frac{dX}{dt} = \frac{dY}{dt} = 0,$$

and $32f(X,Y) = 57 - 50XY - 7Y^2, \quad 32g(X,Y) = 50XY + 7Y^2 - 2Y - 55.$

The morphogen concentrations recorded at the earlier times connect more directly with the theory given in §§ 6 to 9. The amplitude of the waves was then still sufficiently small for the approximation of linearity to be still appropriate, and consequently the 'catastrophic' growth had not yet set in.

The functions $f(X, Y)$ and $g(X, Y)$ of § 6 depend also on γ and are

$$f(X,Y) = \frac{1}{32}\left[-7X^2 - 50XY + 57 + 55\gamma\right],$$

$$g(X,Y) = \frac{1}{32}\left[7X^2 + 50XY - 2Y - 55 - 55\gamma\right].$$

In applying the theory it will be as well to consider principally the behaviour of the system when γ remains permanently zero. Then for equilibrium $f(X, Y) = g(X, Y) = 0$ which means that $X = Y = 1$, i.e. $h = k = 1$. One also finds the following values for various quantities mentioned in §§ 6 to 9:

$$a = -2, b = -1.5625, c = 2, d = 1.500, s = 3.333,$$

$$I = 0, \alpha = 0.625, \chi = 0.500, (d-a)(-bc)^{-\frac{1}{2}} = 1.980,$$

$$(\mu + \nu)(\mu\nu)^{-\frac{1}{2}} = 2.121, p_0 = -0.25 \pm 0.25i,$$

$$p_2 = -0.0648, p_3 = -0.0034, p_4 = -0.0118.$$

(The relation between p and U for these chemical data, and the values p_n, can be seen in figure 1, the values being so related as to make the curves apply to this example as well as that in § 8.) The value $s = 3.333$ leads one to expect a three-lobed pattern as the commonest, and this is confirmed

Table 2						
cell	0	1	2	3	4	5
X	7.5	3.5	2.5	2.5	3.5	7.5
Y	0	8	8	8	8	0

by the values p_n. The four-lobed pattern is evidently the closest competitor. The closeness of the competition may be judged from the difference $p_3 - p_4 = 0.0084$, which suggests that the three-lobed component takes about 120 S.U. or about 33 h to gain an advantage of a neper (i.e. about 2.7 : 1) over the four-lobed one. However, the fact that γ is different from 0 and is changing invalidates this calculation to some extent.

The figures in table 1 were mainly obtained with the aid of the Manchester University Computer.

Although the above example is quite adequate to illustrate the mathematical principles involved it may be thought that the chemical reaction system is somewhat artificial. The following example is perhaps less so. The same 'special units' are used. The reactions assumed are

$A \to X$	at the rate	$10^{-3}A,\ A = 10^3,$
$X + Y \to C$	at the rate	$10^3 XY,$
$C \to X + Y$	at the rate	$10^6 C,$
$C \to D$	at the rate	$62.5C,$
$B + C \to W$	at the rate	$0.125BC,\ B = 10^3,$
$W \to Y + C$	instantly,	
$Y \to E$	at the rate	$0.0625Y,$
$Y + V \to V'$	instantly,	
$V' \to E + V$	at the rate	$62.5V',\ V' = 10^{-3}\beta.$

The effect of the reactions $X + Y \leftrightarrows C$ is that $C = 10^{-3}XY$. The reaction $C \to D$ destroys C, and therefore in effect both X and Y, at the rate $\frac{1}{16}XY$. The reaction $A \to X$ forms X at the constant rate 1, and the pair $Y + V \to V' \to E + V$ destroys Y at the constant rate $\dfrac{1}{16}\beta$. The pair $B + C \to W \to Y + C$ forms Y at the rate $\frac{1}{8}XY$, and $Y \to E$ destroys it at the rate $\frac{1}{16}Y$. The total effect therefore is that X is produced at the rate $f(X, Y) = \frac{1}{16}(16 - XY)$, and Y at the rate $g(X, Y) = \frac{1}{16}(XY - Y - \beta)$. However, $g(X, Y) = 0$ if $Y \leqslant 0$. The diffusion constants will be supposed to be $\mu = \dfrac{1}{4}$, $\nu = \dfrac{1}{16}$. The homogeneity condition gives $hk = 16$, $k = 16 - \beta$. It will be seen from conditions (9.4*a*) that case (*d*) applies if and only if $\frac{4}{k} + \frac{k}{4} < 2.75$, i.e. if k lies between 1.725 and 9.257. Condition (9.4*b*) shows that there will be instability if in addition $\frac{8}{k} + \frac{k}{8} > \sqrt{3} + \frac{1}{2}$, i.e. if k does not lie between 4.98 and 12.8. It will also be found that the wave-length corresponding to $k = 4.98$ is 4.86 cell diameters.

In the case of a ring of six cells with $\beta = 12$ there is a stable equilibrium, as shown in table 2.

It should be recognized that these equilibria are only dynamic equilibria. The molecules which together make up the chemical waves are continually changing, though their concentrations in any particular cell are only undergoing small statistical fluctuations. Moreover, in order to maintain the wave pattern a continual supply of free energy is required. It is clear that this must be so since there is a continual degradation of energy through diffusion. This energy is supplied through the 'fuel substances' (A, B in the last example), which are degraded into 'waste products' (D, E).

11. Restatement and biological interpretation of the results

Certain readers may have preferred to omit the detailed mathematical treatment of §§ 6 to 10. For their benefit the assumptions and results will be briefly summarized, with some change of emphasis.

The system considered was either a ring of cells each in contact with its neighbours, or a continuous ring of tissue. The effects are extremely similar in the two cases. For the purposes of this summary it is not necessary to distinguish between them. A system with two or three morphogens only was considered, but the results apply quite generally. The system was supposed to be initially in a stable homogeneous condition, but disturbed slightly from this state by some influences unspecified, such as Brownian movement or the effects of neighbouring structures or slight irregularities of form. It was supposed also that slow changes are taking place in the reaction rates (or, possibly, the diffusibilities) of the two or three morphogens under consideration. These might, for instance, be due to changes of concentration of other morphogens acting in the role of catalyst or of fuel supply, or to a concurrent growth of the cells, or a change of temperature. Such changes are supposed ultimately to bring the system out of the stable state. The phenomena when the system is just unstable were the particular subject of the inquiry. In order to make the problem mathematically tractable it was necessary to assume that the system never deviated very far from the original homogeneous condition. This assumption was called the 'linearity assumption' because it permitted the replacement of the general reaction rate functions by linear ones. This linearity assumption is a serious one. Its justification lies in the fact that the patterns produced in the early stages when it is valid may be expected to have strong qualitative similarity to those prevailing in the later stages when it is not. Other, less important, assumptions were also made at the beginning of the mathematical theory, but the detailed effects of these were mostly considered in § 9, and were qualitatively unimportant.

The conclusions reached were as follows. After the lapse of a certain period of time from the beginning of instability, a pattern of morphogen concentrations appears which can best be described in terms of 'waves'. There are six types of possibility which may arise.

(a) The equilibrium concentrations and reaction rates may become such that there would be instability for an isolated cell with the same content as any one of the cells of the ring. If that cell drifts away from the equilibrium position, like an upright stick falling over, then, in the ring, each cell may be expected to do likewise. In neighbouring cells the drift may be expected to be in the same direction, but in distant cells, e.g. at opposite ends of a diameter there is no reason to expect this to be so.

    This is the least interesting of the cases. It is possible, however, that it might account for 'dappled' colour patterns, and an example of a pattern in two dimensions produced by this type of process is shown in figure 2 for comparison with 'dappling'. If dappled patterns are to be explained in this way they must be laid down in a latent form when the foetus is only a few inches long. Later the distances would be greater than the morphogens could travel by diffusion.

(b) This case is similar to (*a*), except that the departure from equilibrium is not a unidirectional drift, but is oscillatory. As in case (*a*) there may not be agreement between the contents of cells at great distances.

    There are probably many biological examples of this metabolic oscillation, but no really satisfactory one is known to the author.

(c) There may be a drift from equilibrium, which is in opposite directions in contiguous cells.

    No biological examples of this are known.

(d) There is a stationary wave pattern on the ring, with no time variation, apart from a slow increase in amplitude, i.e. the pattern is slowly becoming more marked. In the case of a ring of continuous tissue the pattern is sinusoidal, i.e. the concentration of one of the morphogens plotted against position on the ring is a sine curve. The peaks of the waves will be uniformly spaced round the ring. The number of such peaks can be obtained approximately by dividing the so-called 'chemical wave-length' of the system into the circumference of the ring. The chemical wave-length is given for the case of two morphogens by the formula (9.3). This formula for the number of speaks of course does not give a whole number, but the actual number of peaks will always be one of the two whole numbers nearest to it, and will usually be *the* nearest. The degree of instability is also shown in (9.3).

The mathematical conditions under which this case applies are given in equations (9.4*a*), (9.4*b*).

Biological examples of this case are discussed at some length below.

(e) For a two-morphogen system only the alternatives (*a*) to (*d*) are possible, but with three or more morphogens it is possible to have travelling waves. With a ring there would be two sets of waves, one travelling clockwise and the other anticlockwise. There is a natural chemical wave-length and wave frequency in this case as well as a wave-length; no attempt was made to develop formulae for these.

In looking for biological examples of this there is no need to consider only rings. The waves could arise in a tissue of any anatomical form. It is important to know what wave-lengths, velocities and frequencies would be consistent with the theory. These quantities are determined by the rates at which the reactions occur (more accurately by the 'marginal reaction rates', which have the dimensions of the reciprocal of a time), and the diffusibilities of the morphogens. The possible range of values of the reaction rates is so immensely wide that they do not even give an indication of orders of magnitude. The diffusibilities are more helpful. If one were to assume. that all the *dimensionless* parameters in a system of travelling waves were the same as in the example given in § 8, one could say that the product of the velocity and wave-length of the waves was 3π times the diffusibility of the most diffusible morphogen. But this assumption is certainly false, and it is by no means obvious what is the true range of possible values for the numerical constant (here 3π). The movements of the tail of a spermatozoon suggest themselves as an example of these travelling waves. That the waves are within one cell is no real difficulty. However, the speed of propagation seems to be somewhat greater than can be accounted for except with a rather large numerical constant.

(f) Metabolic oscillation with neighbouring cells in opposite phases. No biological examples of this are known to the author.

It is difficult also to find cases to which case (*d*) applies directly, but this is simply because isolated rings of tissue are very rare. On the other hand, systems that have the same kind of symmetry as a ring are extremely common, and it is to be expected that under appropriate chemical conditions, stationary waves may develop on these bodies, and that their circular symmetry will be replaced by a polygonal symmetry. Thus, for instance, a plant shoot may at one time have circular symmetry, i.e. appear essentially the same when rotated through any angle about a certain axis; this shoot may later develop a whorl of leaves, and then it will only suffer rotation through the angle which separates the leaves, or any multiple of it. This same example demonstrates the complexity of the situation when more than one dimension is involved. The leaves on the shoots may not appear in whorls, but be imbricated. This possibility is also capable of mathematical analysis, and will be considered in detail in a later paper. The cases which appear to the writer to come closest biologically to the 'isolated ring of cells' are the tentacles of (e.g.) *Hydra*, and the whorls of leaves of certain plants such as Woodruff (*Asperula odorata*).

Hydra is something like a sea-anemone but lives in fresh water and has from about five to ten tentacles. A part of a *Hydra* cut off from the rest will rearrange itself so as to form a complete new organism. At one stage of this proceeding the organism has reached the form of a tube open at the head end and closed at the other end. The external diameter is somewhat greater at the head end than over the rest of the tube. The whole still has circular symmetry. At a somewhat later stage the symmetry has gone to the extent that an appropriate stain will bring out a number of patches on the widened head end. These patches arise at the points where the tentacles are subsequently to appear (Child 1941, p. 101 and figure 30). According to morphogen theory it is natural to suppose that reactions, similar to those which were considered in connection with the ring of tissue, take place in the widened head end, leading to a similar breakdown of symmetry. The situation is more complicated than the case of the thin isolated ring, for the portion of the *Hydra* concerned is neither isolated nor very thin. It is not unreasonable to suppose that this head region is the only one in which the chemical conditions are such as to give instability. But substances produced in this region are still free to diffuse through the surrounding region of lesser activity. There is no great difficulty in extending the mathematics to cover this point in particular cases. But if the active region is too wide the system no longer approximates the behaviour of a thin ring and one can no longer expect the tentacles to form a single whorl. This also cannot be considered in detail in the present paper.

In the case of woodruff the leaves appear in whorls on the stem, the number of leaves in a whorl varying considerably, sometimes being as few as five or as many as nine. The numbers in consecutive whorls on the same stem are often equal, but by no means invariably. It is to be presumed that the whorls originate in rings of active tissue in the meristematic area, and that the rings arise at sufficiently great distance to have little influence on one another. The number of leaves in the whorl will presumably be obtainable by the rule given above, viz. by dividing the chemical wave-length into the circumference, though both these quantities will have to be given some new interpretation more appropriate to woodruff than to the ring. Another important example of a structure with polygonal symmetry is provided by young root fibres just breaking out from the parent root. Initially these are almost homogeneous in cross-section, but eventually a ring of fairly evenly spaced spots appear, and these later develop into vascular strands. In this case again the full explanation must be in terms of a two-dimensional or even a three-dimensional problem, although the analysis for the ring is still illuminating. When the cross-section is very large the strands may be in more than one ring, or more or less randomly or hexagonally arranged. The two-dimensional theory (not expounded here) also goes a long way to explain this.

Flowers might appear superficially to provide the most obvious examples of polygonal symmetry, and it is probable that there are many species for which this 'waves round a ring' theory is essentially correct. But it is certain that it does not apply for all species. If it did it would follow that, taking flowers as a whole, i.e. mixing up all species, there would be no very markedly preferred petal (or corolla, segment, stamen, etc.) numbers. For when all species are taken into account one must expect that the diameters of the rings concerned will take on nearly all values within a considerable range, and that neighbouring diameters will be almost equally common. There may also be some variation in chemical wave-length. Neighbouring values of the ratio circumferences to wave-length should therefore be more or less equally frequent, and this must mean that neighbouring petal numbers will have much the same frequency. But this is not borne out by the facts. The number five is extremely common, and the number seven rather rare. Such facts are, in the author's opinion, capable of explanation on the basis of morphogen theory, and are closely connected with the theory of phyllotaxis. They cannot be considered in detail here.

The case of a filament of tissue calls for some comment. The equilibrium patterns on such a filament will be the same as on a ring, which has been cut at a point where the concentrations of the morphogens are a maximum or a minimum. This could account for the segmentation of such filaments. It should be noticed, however, that the theory will not apply unmodified for filaments immersed in water.

12. Chemical waves on spheres: Gastrulation

The treatment of homogeneity breakdown on the surface of a sphere is not much more difficult than in the case of the ring. The theory of spherical harmonics, on which it is based, is not, however, known to many that are not mathematical specialists. Although the essential properties of spherical harmonics that are used are stated below, many readers will prefer to proceed directly to the last paragraph of this section.

The anatomical structure concerned in this problem is a hollow sphere of continuous tissue such as a blastula. It is supposed sufficiently thin that one can treat it as a 'spherical shell'. This latter assumption is merely for the purpose of mathematical simplification; the results are almost exactly similar if it is omitted. As in § 7 there are to be two morphogens, and a, b, c, d, μ', ν', h, k are also to have the same meaning as they did there. The operator ∇^2 will be used here to mean the superficial part of the Laplacian, i.e. $\nabla^2 V$ will be an abbreviation of

$$\frac{1}{\rho^2}\frac{\partial^2 V}{\partial \phi^2}+\frac{1}{\rho^2 \sin^2\theta}\frac{\partial}{\partial \theta}\left(\sin\theta\frac{\partial V}{\partial \theta}\right),$$

where θ and ϕ are spherical polar co-ordinates on the surface of the sphere and ρ is its radius. The equations corresponding to (7.1) may then be written

$$\left.\begin{aligned}\frac{\partial X}{\partial t}&=a(X-h)+b(Y-k)+\mu'\nabla^2 X,\\ \frac{\partial Y}{\partial t}&=c(X-h)+d(Y-k)+\nu'\nabla^2 Y.\end{aligned}\right\}\qquad(12.1)$$

It is well known (e.g. Jeans 1927, chapter 8) that any function on the surface of the sphere, or at least any that is likely to arise in a physical problem, can be 'expanded in spherical surface harmonics'. This means that it can be expressed in the form

$$\sum_{n=0}^{\infty}\left[\sum_{m=-n}^{n} A_n^m P_n^m(\cos\theta)e^{im\phi}\right].$$

The expression in the square bracket is described as a 'surface harmonic of degree n'. Its nearest analogue in the ring theory is a Fourier component. The essential property of a spherical harmonic of degree n is that when the operator ∇^2 is applied to it the effect is the same as multiplication by $-n(n+1)/\rho^2$. In view of this fact it is evident that a solution of (12.1) is

$$\left.\begin{aligned}X&=h+\sum_{n=0}^{\infty}\sum_{m=-n}^{n}\left(A_n^m e^{iq_n t}+B_n^m e^{iq'_n t}\right)P_n^m(\cos\theta)e^{im\phi},\\ Y&=k+\sum_{n=0}^{\infty}\sum_{m=-n}^{n}\left(C_n^m e^{iq_n t}+D_n^m e^{iq'_n t}\right)P_n^m(\cos\theta)e^{i\phi},\end{aligned}\right\}\qquad(12.2)$$

where q_n and ${q'}_n$ are the two roots of

$$\left(q-a+\frac{\mu'}{\rho^2}n(n+1)\right)\left(q-d+\frac{\nu'}{\rho^2}n(n+1)\right)=bc\qquad(12.3)$$

and

$$A_n^m\left(q_n - a + \frac{\mu'}{\rho^2}n(n+1)\right) = bC_n^m,$$
$$B_n^m\left(q'_n - a + \frac{\mu'}{\rho^2}n(n+1)\right) = cD_n^m. \tag{12.4}$$

This is the most general solution, since the coefficients A_n^m and B_n^m can be chosen to give any required values of X, Y when $t=0$, except when (12.3) has two equal roots, in which case a treatment is required which is similar to that applied in similar circumstances in § 7. The analogy with § 7 throughout will indeed be obvious, though the summation with respect to m does not appear there. The meaning of this summation is that there are a number of different patterns with the same wave-length, which can be superposed with various amplitude factors. Then supposing that, as in § 8, one particular wave-length predominates, (12.2) reduces to

$$\left.\begin{aligned} X - h &= e^{iq_{n0}t}, \sum_{m=-n_0}^{n_0} A_{n_0}^m P_{n_0}^m(\cos\theta)e^{im\phi}, \\ b(Y-k) &= \left(q_{n_0} - a + \tfrac{\mu'}{\rho^2}n(n+1)\right)(X-h). \end{aligned}\right\} \tag{12.5}$$

In other words, the concentrations of the two morphogens are proportional, and both of them are surface harmonics of the same degree n_0, viz. that which makes the greater of the roots q_{n0}, q'_{n_0} have the greatest value.

It is probable that the forms of various nearly spherical structures, such as radiolarian skeletons, are closely related to these spherical harmonic patterns. The most important application of the theory seems, however, to be to the gastrulation of a blastula. Suppose that the chemical data, including the chemical wave-length, remain constant as the radius of the blastula increases. To be quite specific suppose that

$$\mu' = 2,\ \nu' = 1,\ a = -4,\ b = -8,\ c = 4,\ d = 7.$$

With these values the system is quite stable so long as the radius is less than about 2. Near this point, however, the harmonics of degree 1 begin to develop and a pattern of form (12.5) with $n_0 = 1$ makes its appearance. Making use of the facts that

$$P_1^0(\cos\theta) = \cos\theta,\ P_1^1(\cos\theta) = P_1^{-1}(\cos\theta) = \sin\theta,$$

it is seen that $X-h$ is of the form

$$X - h = A\cos\theta + B\sin\theta\cos\phi + C\sin\theta\sin\phi, \tag{12.6}$$

which may also be interpreted as

$$X - h = A'\cos\theta', \tag{12.7}$$

where θ' is the angle which the radius θ, ϕ makes with the fixed direction having direction cosines proportional to B, C, A and $A' = \surd(A^2+B^2+C^2)$.

The outcome of the analysis therefore is quite simply this. Under certain not very restrictive conditions (which include a requirement that the sphere be relatively small but increasing in size) the pattern of the breakdown of homogeneity is axially symmetrical, not about the original axis of spherical polar co-ordinates, but about some new axis determined by the disturbing influences. The concentrations of the first morphogen are given by (12.7), where θ' is measured from this new axis; and $Y-k$ is proportional to $X-h$. Supposing that the first morphogen is, or encourages the production of, a growth hormone, one must expect the blastula to grow in an axially symmetric manner, but at a greater rate at one end of the axis than at the other This might under many circumstances lead to gastrulation, though the effects of such growth are not very easily determinable. They depend on the elastic properties of the tissue as well as on the growth rate at each point. This growth will certainly lead to a solid of revolution with a marked difference between the two poles, unless, in addition to the chemical instability, there is a mechanical instability causing the breakdown of axial symmetry. The direction of the axis of gastrulation will be quite random according to this theory. It may be that it is found experimentally that the axis is normally in some definite direction such as that of the animal pole. This is not essentially contradictory to the theory, for any small asymmetry of the zygote may be sufficient to provide the 'disturbance' which determines the axis.

13. Non-linear theory: Use of digital computers

The 'wave' theory which has been developed here depends essentially on the assumption that the reaction rates are linear functions of the concentrations, an assumption which is justifiable in the case of a system just beginning to leave a homogeneous condition. Such systems certainly have a special interest as giving the first appearance of a pattern, but they are the exception rather than the rule. Most of an organism, most of the time, is developing from one pattern into another, rather than from homogeneity into a pattern. One would like to be able to follow this more general process mathematically also. The difficulties are, however, such that one cannot hope to have any very embracing *theory* of such processes, beyond the statement of the equations. It might be possible, however, to treat a few particular cases in detail with the aid of a digital computer. This method has the advantage that it is not so necessary to make simplifying assumptions as it is when doing a more theoretical type of analysis. It might even be possible to take the mechanical aspects of the problem into account as well as the chemical, when applying this type of method. The essential disadvantage of the method is that one only gets results for particular cases. But this disadvantage is probably of comparatively little importance. Even with the ring problem: considered in this paper, for which a reasonably complete mathematical analysis was possible, the computational treatment of a particular case was most illuminating. The morphogen theory of phyllotaxis, to be described, as already mentioned, in a later paper, will be covered by this computational method. Non-linear equations will be used.

It must be admitted that the biological examples which it has been possible to give in the present paper are very limited. This can be ascribed quite simply to the fact that biological phenomena are usually very complicated. Taking this in combination with the relatively elementary mathematics used in this paper one could hardly expect to find that many observed biological phenomena would be covered. It is thought, however, that the imaginary biological systems which have been treated, and the principles which have been discussed, should be of some help in interpreting real biological forms.

References

Child, C. M. 1941 *Patterns and problems of development.* University of Chicago Press.
Davson, H. & Danielli, J. F. 1943 *The permeability of natural membranes.* Cambridge University Press.
Jeans, J. H. 1927 *The mathematical theory of elasticity and magnetism*, 5th ed. Cambridge University Press.
Michaelis, L. & Menten, M. L. 1913 Die Kinetik der Invertinwirkung. *Biochem. Z.* **49**, 333.
Thompson, Sir D'Arcy 1942 *On growth and form*, 2nd ed. Cambridge University Press.
Waddington, C. H. 1940 *Organisers and genes.* Cambridge University Press.

Examining the Work and Its Later Impact

Henri Berestycki on the visionary power of —

ALAN TURING AND REACTION–DIFFUSION EQUATIONS

A singular gem, alone it shines. The 1952 article of Turing 'The Chemical basis of morphogenesis' addresses one of the mysteries of life. A paper impossible to classify – it is not even clear on which shelf of the library one should look for it – it does not resemble anything else. The article is revolutionary, one of its kind. It is both a mathematical invention and a biological vision of how reaction and diffusion mechanisms, indeed very basic ones, can account for morphogenesis.

Turing's article (1952) is a theory aiming at explaining how form and differentiation appear in the germinal form of life when cells are undifferentiated. More generally, this paper addresses the problem of symmetry breaking. It offers indeed an explanation of how starting with a perfectly symmetric state a system can evolve into differentiated components based on simple and fundamental mechanisms that operate at very basic levels of living systems. This vision is carried through systems of reaction–diffusion equations both in spatial discrete and continuous forms. Along with several other underlying themes, the article contains a vision of the role of non-linear partial differential equations and scientific computing.

1. The article and its style

Even the background material included in this article is rather unusual and contains somewhat heteroclite material. To start with, it lists some basic properties of Fourier series. Then basics of chemical kinetics, essentially the law of mass action, are described and some examples are given. The problem of embryogenesis is introduced with some details.

An obvious effort is made in this paper to keep the mathematical level of the presentation as simple as possible so as not to discourage biologists. This goal is explicitly stated and what would be regarded as very elementary facts of differential equations with constant coefficients and Fourier series decompositions are recalled in a section entitled 'Mathematical background'. It is also suggested to non-mathematicians that they skip some specific sections of the paper.

The style of Turing, throughout the paper is that of an extreme modesty. Turing enters the world of life sciences, attacking one of the great questions of life development with extreme care. Intellectually, this is an extremely bold paper, yet it is presented with the utmost humility. Do not be disturbed because of me he seems to say to his fellow biologists. Cautionary remarks abound in the paper epitomised by the one in the second sentence: 'this model will be a simplification and an idealisation, and consequently a falsification'. In the conclusion of his article, he writes that the 'biological examples which it has been possible to give in the present paper are very limited. This can be ascribed simply to the fact that biological phenomena are usually very complicated'. Yet, in spite of the effort to spare the reader mathematical efforts, I believe that the intuition of Turing is primarily mathematical.

Morphogenesis in Turing's vision can be explained by two imaginary or archetypical chemical substances that he calls morphogens interacting by reaction and diffusion. One such morphogen is an activator and the other one an inhibitor. Those are not further specified but the plausibility of these morphogens is carefully described say in the form of ions. Turing also explains that such morphogens are not antithetic to genetics but could for instance be ascribed to genetic activity. Genes can be at work in creating the receptors that give the chemical basis for morphogens. The essence of the phenomenon is that the homogeneous distribution in a certain regime becomes unstable yielding to a non-homogeneous one thus producing patterns. This phenomenon is now universally referred to as Turing's instability mechanism.

To make his point, Turing introduces an imaginary system, an annulus of cells, hence a discrete space (but continuous in time) model. The cells are labelled $k = 1, \ldots, N$ and cell $N+1$ is the same as cell 1 while cell 0 is the same as cell N. Two morphogens, called X, Y, interact and their levels in cell r are denoted X_r, Y_r. Between contiguous cells, diffusion moves chemical substances from one cell into the neighboring one if the density is smaller in the latter. The rate of mobility between cells is given by diffusion coefficients μ, ν for each morphogen respectively. Morphogens interact through chemical reactions. The kinetic rates of these reactions, given by the law of mass action, are respectively $f(X_r, Y_r)$ and $g(X_r, Y_r)$. The resulting system of equation then simply reads

$$\begin{cases} \dfrac{\mathrm{d}X_r}{\mathrm{d}t} = f(X_r, Y_r) + \mu(X_{r+1} - 2X_r + X_{r-1}) \\ \dfrac{\mathrm{d}Y_r}{\mathrm{d}t} = g(X_r, Y_r) + \nu(Y_{r+1} - 2Y_r + Y_{r-1}), \end{cases} \tag{1.1}$$

for $r = 1, \ldots, N$. Assume that there is a homogeneous stationary solution, that is, a solution such that $X_r = h$ and $Y_r = k$ for all r, where h, k are constants. Since it must be the case that $f(h,k) = g(h,k) = 0$, it is of interest to look at the *linearised system* in the vicinity of h, k. Denoting $x_r = X_r - h$ and $y_r = Y_r - k$ viewed as perturbations from the homogeneous stationary state this system reads

$$\begin{cases} \dfrac{\mathrm{d}x_r}{\mathrm{d}t} = ax_r + by_r + \mu(x_{r+1} - 2x_r + x_{r-1}) \\ \dfrac{\mathrm{d}y_r}{\mathrm{d}t} = cx_r + dy_r + \nu(y_{r+1} - 2y_r + y_{r-1}), \end{cases} \tag{1.2}$$

where

$$a = \frac{\partial f}{\partial X}(h,k), \quad b = \frac{\partial f}{\partial Y}(h,k), \quad c = \frac{\partial g}{\partial X}(h,k), \quad d = \frac{\partial g}{\partial Y}(h,k).$$

This is an elementary system of two equations in two unknowns. Turing is able to analyse it mathematically using finite sums of cosine and sine functions. The extraordinary finding of Turing is that in a certain parameter regime, such a simple system yields morphogenesis. Starting from an homogeneous distribution of the chemicals, that is, identical densities of the chemicals in all cells, as some parameters are changed, the system may give rise to a new kind of stable stationary solutions. These correspond to differentiated cells in which the distributions of chemicals varies form one cell to another. For certain values of the parameters, the only configuration that is observed is the homogeneous one with undifferentiated cells. Then, as the parameter is changed, a critical threshold is crossed after which one observes a non-homogeneous distribution of chemicals in the cells.

A specific example of a system of the type (1.1) is mentioned in the article of Turing. It reads:

$$\begin{cases} f(X,Y) = \dfrac{1}{32}[-7X^2 - 50XY + 57] \\ g(X,Y) = \dfrac{1}{32}[7X^2 + 50XY - 2Y - 55] \end{cases} \tag{1.3}$$

In this case, the homogenous solution is $h = k = 1$. Turing mentions a computation for the case $N = 20$ and $\mu = 1/2, \nu = 1/4$. The longterm shape of the solutions of this equation starting from small perturbations (order 10^{-4}) of the homogenous state, i.e., $X_0 \equiv Y_0 \equiv 1$, is shown in Fig 1. These correspond to values of $N = 80$, $\nu = 1/4$ and different values of μ close to $\mu = 1/2$ (with the proper scaling). One curve represents X and the other one Y. This system exhibits pattern formation and bifurcation occurs at a value close to $\mu = 1/2$. When $\mu < 1/2$ with μ very close to $1/2$, the solution (X,Y) converges to $(1,1)$ (upper left diagram). As μ gets closer to $1/2$ and crosses it, this ceases to be the case, $(1,1)$ is not stable any more and one can see solutions with oscillations as is seen in the upper right diagram ($\mu = 0.497$) and the amplitude increases as shown in the lower diagrams ($\mu = 0.5$ and $\mu = 0.51$ respectively). One can see the strong parameter sensitivity close to the bifurcation point. These computations illustrate the role of the diffusion of one of the components as the bifurcation parameter. Turing's mechanism is indeed often referred to as a *diffusion–driven instability*.

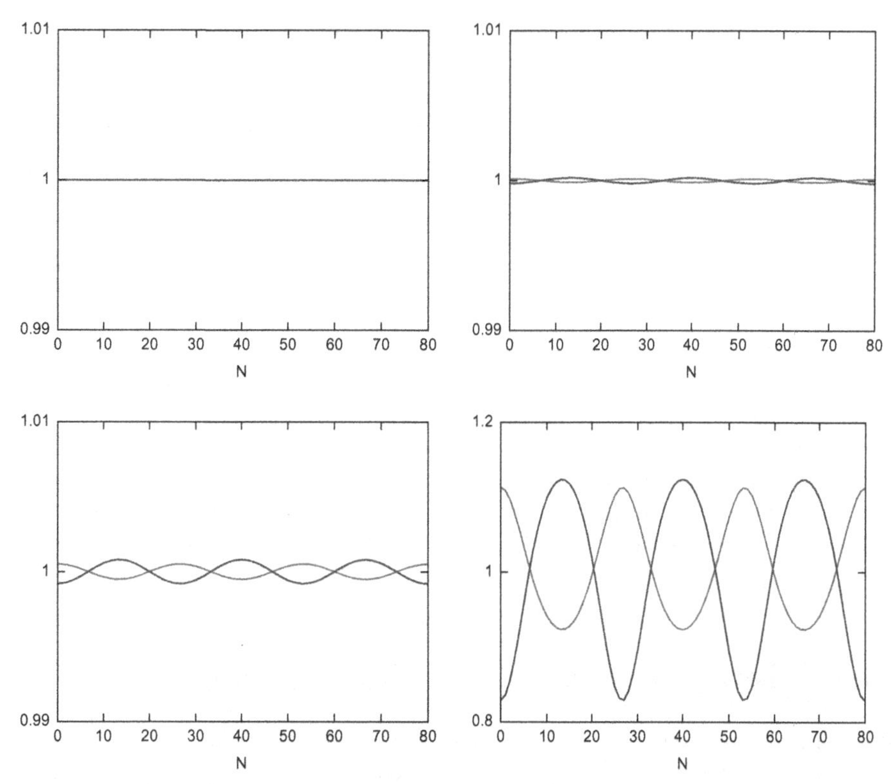

Fig. 1: Long-term ($T = 500$) solution for the system (1.1) in the case of reaction kinetics terms (1.3). In abscissa is the coordinate r of the cell in the circle. Upper left calculation: $\mu = 0.49$; upper right: $\mu = 0.497$; lower left $\mu = 0.5$; lower right: $\mu = 0.52$. (*Computations are courtesy of Lionel Roques*).

What has this to do with embryogenesis? Turing states his goal as to describe '... a mathematical model of the growing embryo'. As a mathematician, Turing tries to find the simplest form in which the mechanisms at work can be seen clearly, that is, as a mathematical truth. In the paper, Turing may appear as a dreaming mathematician (by this I mean a visionary one). He is not chiefly concerned with describing a precise situation but rather general classes of problems where morphogenesis can be seen to occur. For him the word morphogen 'is not intended to have any very exact meaning but is simply the kind of substance concerned in this theory'. And he quotes Waddington (1940) as a precursor of his line of thinking (Waddington had introduced the notion of 'evocators'). He could as well have invoked D'Arcy Thompson (1942) in the great British lineage of morphogenesis theory. He adds that genes themselves can be considered as morphogens but also, say, skin pigments. Here, as well as in all the article, Turing's writing style is quite remarkable: '... it is only by courtesy that genes may be regarded as separate molecules'.

Turing's great mathematical discovery in this paper is that a two-component system of reaction–diffusion equations even though of homogeneous form may give rise to non-homogeneous stable solutions. Going from a discrete circular assembly of cell, he moves to the continuous version. In the system of reaction–diffusion equations analogous to (1.2) above, he introduces for a circle reads in its linearised form:

$$\begin{cases} \dfrac{\partial X}{\partial t} = a(X-h) + b(Y-k) + \dfrac{\mu'}{\rho^2}\dfrac{\partial^2 X}{\partial \theta^2}, \\[2ex] \dfrac{\partial Y}{\partial t} = c(X-h) + d(Y-k) + \dfrac{\nu'}{\rho^2}\dfrac{\partial^2 Y}{\partial \theta^2} \end{cases} \tag{1.4}$$

where ρ is the radius of the circle.

This type of system is now referred to as a Turing system. More general Turing systems are of the form:

$$\begin{cases} \dfrac{\partial X}{\partial t} = f(X,Y) + \dfrac{\mu'}{\rho^2}\Delta X, \\[2ex] \dfrac{\partial Y}{\partial t} = g(X,Y) + \dfrac{\nu'}{\rho^2}\Delta Y \end{cases} \tag{1.5}$$

Actually, Turing wrote this system for chemical waves on spheres to study gastrulation. Then, the equation is set on the surface of the sphere of radius ρ and the Δ operator is to be understood as the Laplace–Beltrami operator on the sphere. That is the two-dimensional analogue of the annulus problem. Turing again studies the linearised equation, which is now somewhat more involved. By calling upon the theory of spherical harmonics rather than Fourier series, he is able to carry the same type of analysis as on the circle. Here too, bifurcation occurs. Patterns depart from spherical symmetry but will have axial symmetry.

2. A brief history of reaction–diffusion equations

The history of reaction–diffusion equations begins in the first half of the twentieth century. It is remarkable and as a matter of fact without parallel that all the founding articles that proposed this important class of non-linear partial differential equations have all been motivated by biology. It starts with the work of a great British statistician and geneticist, Ronald A. Fisher (also from Cambridge). With the aim to describe the spatial spread of a genetic trait, and dissatisfied with the Malthus law of population growth, Fisher (1937) introduced a non-linear diffusion equation with a logistic term limiting the growth of the population.

$$\partial_t u - \Delta u = f(u) \tag{2.1}$$

Here, $f(u) = u(1 - u)$.

Inspired by Fisher's work and with the genetics model in mind, Kolmogorov, Petrovsky and Piskunov (KPP) (Kolmogorov et al., 1937) took up the model to derive this non-linear parabolic (or diffusion) equation.It is classically referred to as the Fisher–Kolmogorov or Fisher–KPP equation. At about the same time and at the same place, in Moscow in 1939, Zel'dovich and Frank-Kamenetskii (1938) introduced a simplified model called the thermo-diffusive approximation to describe flame propagation. It takes the above form, but with different functions f. A general reference for reaction–diffusion equations of this kind as well as their generalisations in non-homogeneous settings can be found in Berestycki and Hamel (2012).

Only a few years passed after the Second World War until entirely new areas of applications motivated the study of further reaction–diffusion systems. In 1951, Skellam proposed to use the diffusion equation, first without the saturation or logistic term in it and then in the form (2.1), in ecology to model the dispersal of biological species in a given environment (see Skellam (1951)). With remarkable success, using this equation, he was able to account for the observations of the area colonised by muskrats, a species that was first introduced in Europe, near Prague in 1905. These Muskrat observations were akin to a *natural experiment*. Indeed, muskrats is a species native of North America. The exact day and location where it was introduced in Europe are known. Furthermore, rather accurate observations and measurements had been made yearly for the region it colonised. In particular, the surface expansion rate was available. Skellam's model explained surprisingly well the rate of growth of the area inhabited by the muskrats.

Up to then, everyone had used single equations. The next major progress came in the early 1950s with the works of Hodgkin and Huxley in Cambridge and of Turing who proposed *systems* of reaction–diffusion equations. They were achieved at about the same period, all the articles being dated 1952. As it is known, in a series of papers (in particular Hodgkin and Huxley (1952)) Hodgkin and Huxley introduced a system of reaction–diffusion equations to describe the propagation of the nerve impulse along an axone. This work opened a new field and Hodgkin and Huxley were awarded the Nobel Prize in Physiology for it.

Could this be a coincidence? Turing was probably aware of what was done in Cambridge at the time. Twice at the same time then and at the same places (first in Moscow, 1937–1938, then in Cambridge and in the United Kingdom in 1952), fundamental breakthroughs were achieved by introducing equations and then systems of reaction–diffusion equations. It is hard to know whether or to what extent these works somehow influenced each other. This is an interesting question for the historians of science.

The reason why the works of Hodgkin–Huxley and Turing were such quantum leaps when they came out rests on the nature of diffusion equations. These were generally viewed as mostly diffusing, which in some sense they are, thus leading to homogenisation of the quantities. The diffusion mechanism, by its very nature, works as an equaliser, that is, it tends towards a greater homogeneity.

Fisher and KPP first showed that they were in some sense the right model to describe transitions and 'waves of advance'. This concept that was later used even in paleoanthropology by a student of Fisher, Cavalli-Sforza with Amermann (Ammerman and Cavalli-Sforza, 1984) as well as in many other instances, such as the modelling for phase transitions in physics. As for nerve impulse, Helmholtz, for instance, thought that only wave equations could provide an adequate framework to describe such a phenomenon.

It was something of an iconoclastic view that reaction–diffusion systems could achieve this. For the same reason, the idea that equations of this kind could lead to pattern formation was hard to conceive. These phenomena were thought to belong to an entirely different realm. These discoveries in the 1930s and 1950s lead to a *Paradox of diffusion*. Indeed, contrary to what had been thought by every one until then, with Fisher and Kolmogorov (KPP), Zeldovich and Frank-Kameneskii, Skellam and Hodgkin–Huxley, one sees diffusion as generating waves with constant shapes and velocity. Then, with Turing, one sees diffusion operating so as to enhance spatial inhomogeneity.

Therefore, the uses of reaction–diffusion systems to describe propagation of nerve impulse or pattern formation were some kind of revolutions. And in fact, this could not have been achieved with a scalar (or single component) equation alone. That Hodgkin–Huxley and Turing were able to depart from the use of single equations, which was the norm at that time, was an essential step. The extraordinary intuition of Turing was that at least a two-component system of two equations was required to account for pattern formation in an otherwise homogeneous model. This is why I think that Turing's intuition was more mathematical than biological. I am convinced that Turing could see that what was needed to produce forms was a *system* of equations. How such a deep insight – really, a streak of genius – came to someone who did not seem to have practiced partial differential equations hitherto remains quite mysterious.

3. Instability and symmetry breaking

For the central preoccupation of Turing is the transition from homogeneous (undifferentiated state) to one where the distribution of densities varies along locations. In more mathematical terms, the question is that of symmetry breaking. One looks at a system which is homogeneous and in which there is a homogeneous or symmetric solution and one wishes to understand how such a system can evolve into non-symmetric state. This is the process of symmetry breaking. Turing's approach here is a way to understand such transitions and, in fact, applies to a wide range of phenomena in the life sciences.

The mechanism at the heart of this phenomenon is what is called bifurcation and exchange of stability. In essence, a non-linear system undergoes a bifurcation as some parameter varies if, say, when going through some critical value of the parameter, new types of solutions appear that are stable whereas the old ones have lost their stability. Thus, bifurcation phenomena are strongly related to the principle of exchange of stability. The most classical example of such a phenomena is the *elastica* solved by Euler in 1744. Initially at rest, an elastic rod at both ends of which a pressure is exerted, remains at rest (undeformed) when the pressure is small. However, when the pressure goes through a critical value, the rod suddenly buckles. This abrupt change of shape was first studied in a celebrated article of Euler who computed the critical pressure. It should be noted that it is not that the undeformed solution disappears but rather that it becomes unstable, with the new solution, the buckled state, acquiring as it were its stability.

The mechanism described by Turing is precisely of this nature. In his study, one of the diffusion coefficients plays the role of the bifurcation parameter. For small values of this coefficient, the homogeneous state is initially stable. As this parameter is increased beyond a critical threshold, this state loses its stability and new solutions in fact non-homogeneous states appear that are now the stable ones. This is described in the sequence of computations for various values of μ in Fig. 1.

Ideally, if a system starts from a symmetric state, it will keep this symmetry. Loss of stability means that any deviation from symmetry in the initial data, be it as small as one wishes, would lead as time evolves to solutions without symmetry. Again, in the example of the rod, if everything is perfectly symmetric, say if the rod is an ideal object, it should not be deformed even at high values of the pressure. And in fact, it stays like that even somewhat beyond criticality. This is why bending of the rod occurs in an abrupt fashion. The slightest departure from symmetry ensures that the state will leave the symmetric configuration to reach the new stable bucked state.

The same happens in the reaction–diffusion equation. Now this idea that a small departure from symmetry is needed seems to have worried Turing. Indeed, a full section of the paper is devoted to discussing the breakdown of symmetry and homogeneity, and another one to left–right symmetry. Actually, the discussions border philosophical considerations. First he argues that one can consider an embryo in its spherical blastula stage as perfectly spherically symmetric. Indeed, says Turing, 'if there are any deviations from perfect symmetry, they cannot be regarded as of any particular importance, for the deviations vary greatly from embryo to embryo within a species, though the organisms developed from them are barely distinguishable'. Then,'a system which has spherical symmetry, and whose state is changing because of chemical reactions and diffusion will remain spherically symmetric for ever. It certainly cannot result in an organism such as a horse, which is not spherically symmetric'. But then, he points out to the fallacy of the argument that deviations from symmetry could be ignored.

Surprisingly, to a certain extent, Turing does not put his discovery within the framework of bifurcation theory to which, in a modern perspective, it obviously belongs. No references are given to the Elastica of Euler, nor to any other works in mechanics where this type of ideas appeared. A more elaborate use of bifurcation and stability exchange was developed by Poincaré in 1885 for celestial mechanics. Poincaré imagined such a mechanism at work to explain how double stars are created by a transition from a spherically symmetric configuration to a state with an hourglass shape that eventually yields two stars rotating one around the other. No more general mention of bifurcation and symmetry exchange is indicated in Turing's paper.

Now, if one believes that mathematicians are general thinkers, that is, develop general models that can be used afterwards in varied contexts of applications, it is surprising that Turing shies away from explaining or mentioning this more general framework. Another problem for historians to delve in. For this, I see two possible explanations. First, Turing clearly aimed at writing a biology article, one that would carry with him the biologists. Turing surely wrote the paper as self-contained mathematically as possible. From this viewpoint, he might have seen references to very different contexts or to a general theory as leading to confusion in the readers' minds and for this reason would have preferred to stay clear of this mathematical vantage point. This concern was certainly present in Turing's intention.

However, to this explanation, I prefer to imagine that the paper of Turing came from nowhere. A specialist of mathematical logic, having worked early on in probability, and then, during the War, having accumulated a deep knowledge of statistics and signal theory, Turing was simply very far from partial differential equations and mechanics. I rather imagine Turing as having invented this system of reaction–diffusion equations out of nowhere. Not quite from scratch say, since he may have grasped some knowledge here and there at lectures about reaction–diffusion equations, but he probably was not aware of the Poincaré's paper or the use of bifurcation methods in mechanics.

Also quite remarkable then, if no prior encounter with this type of theories is to be reckoned, is the use of the linearisation of the problem. This is indeed now the established approach. First, Turing justifies his choice to concentrate on the bifurcation point itself: 'This is largely because the assumption that the system is still nearly homogeneous brings the problem within the range of what is capable of being treated mathematically'.

To concentrate on the bifurcation points means that morphogenesis really is the study of nascent shapes. This is why Turing emphasises the role of linearised equations. The linearisation theory on which Turing concentrates is justified 'in the case of a system just beginning to leave a homogeneous condition. Such systems certainly have a special interest as giving the first appearance of a pattern'. Here is a central idea, one which has been at the core of the success of Turing's approach in all the subsequent developments. By looking at the linearised equation, one can track not only the critical value at which departure from homogeneity occurs but also the shape of the new configurations emerging as a result from these transitions. Indeed, in general terms, bifurcation occurs when the linearised operator is singular (has a non-zero kernel). Then, by looking at the kernel of the linearised operator, one gets the shapes of the new solutions, at least at first order.

Even if one were to know well the general theory of bifurcation, to fathom that reaction–diffusion systems would produce Turing instabilities was far from obvious. It is remarkable that Turing could imagine this mechanism.

4. Turing's paper in perspective

What is the legacy of Turing's paper? First, for a while, the Turing mechanisms were taken as a serious explanation for morphogenesis. Since morphogens as such have not really been identified in embryogenesis and because the view of morphogenesis is now a more global one very much resting on genetic programming, Turing's mechanism is not considered a primary description of what happens in the embryo. Yet, it still captivates the imagination in its ability to describe such a complicated process by making it rest on simple chemical and diffusion mechanisms we know to operate in living systems. Surely, if we were to adopt Occam's razor, Turing's mechanism would be the widely recognised theory for morphogenesis.

For chemical reactions, however, Turing systems have been devised in a series of remarkable experiments in the early 1990s by De Kepper and collaborators (De Kepper et al., 1991; Horváth et al., 2009) as well as by Ouyang and Swinney (1991). Turing's mechanisms fit remarkably well in these observations. In particular, all the shapes predicted in the bifurcation analysis have been observed. Therefore, an experimental confirmation was thus finally given 40 years after the predictions of Turing.

In the biological fields, one can say that the richness of structures that are available from Turing-type systems is fascinating. In several works of Jim Murray and also in a series of works by Philip Maini and collaborators (see the foundation book of mathematical biology of Murray (2003)), Turing systems have been developed that could reproduce virtually all types of shapes of animal coatings. Even explaining through additional geometrical constraints specific transition regions near legs or tails. Also, Hans Meinhardt took up the challenge to use systems of this type to explain the variety of design in shells. His book (Meinhardt, 1995) is not only a collection of pictures of marine wonders but also a testimony of what can be achieved with this type of mathematical systems. All this is not a proof of this mechanism, but it is rather amazing to see the richness of the structure that one can get from reaction–diffusion equations.

Another area of success for the ideas that Turing has introduced and their generalisations are the bacterial growth colonies. The generation of patterns by bacteria is indeed fascinating (see the striking experiments reported in Ben Jacob et al. (1995, 2000); Budrene and Berg (1995)). Of particular importance in the study of bacterial colonies is the phenomenon of chemotaxis (Keller and Segel, 1970) (see Perthame (2007) for population models). In a sense, this can be also related to the paradigm of Turing. The original idea of Turing can be formulated in terms of an interaction of a short-range activator with a long-range inhibitor generating patterns. In the light of the study of A. Aotani, M. Mimura and T. Mollee Ahotani et al. (2010), it is also possible to view chemotaxis in a similar way[1]. For this, the short range activator is the density of bacteria, while chemotactic attractant can be viewed as the long range inhibitor of mobility. Indeed, the attractant also acts as though to inhibit the dispersing of bacteria owing to chemotactic aggregation.

Lastly, I would like to mention modelling visual hallucination shapes. In a remarkable series of works, Bresloff, Ermentrout, Cowan and their collaborators (Bressloff and Cowan, 2003; Bressloff et al., 2001; Ermentrout and Cowan, 1979) have developed systems of equations based on what is known of the visual cortex and the way the human brain processes visual signals. Using Turing's

[1] I owe this insight to Masayasu Mimura.

approach and the study of the linearisations, the authors were able to explain in great detail the catalogue of different shapes that patients experience in their hallucinations. Again, a remarkable achievement.

Put in modern perspective, the paper of Alan Turing is one of great visionary power. First, the use of reaction–diffusion systems to describe pattern formation was a revolutionary idea that paved the way to considerable development. Then, it has opened an approach to modelling in the life sciences that has proved extraordinarily successful. Using bifurcation analysis and linearisations to study sharp transitions and creation of shapes became a widely used method. One might view it really as the only general approach we have. Turing writes that such attention to this phase of transition is warranted because 'it is in a sense the most critical period'. In this respect, René Thom Thom (1972) has followed in the footsteps of Turing with the use of *catastrophe theory*. As we know, much of modern science has concentrated on phase transitions or critical and near critical situations.

Very modern is also the use in the same paper of discrete systems and their continuous analogues. The former are amenable to more explicit results whilst the latter lend themselves more easily to general situations (e.g., systems at the surface of a sphere). It is worthwhile noting that the ease with which Turing passes from discrete to continuous and vice versa has been a constant in his scientific life. One can say that Turing is truly the heir of both Democritus and Leibniz !

The last short section of the 1952 article is devoted to the non-linear theory. These are required since bifurcation situations 'are the exception rather than the rule'. This means that bifurcation occurs at specific *critical* values of parameters. A complete study, valid for all values of the parameters, therefore requires the study of non-linear equations that cannot be reduced to their linearised form. In a short paragraph, Turing lays out a visionary philosophy about the study of non-linear problems. Having recognised that all embracing general theories are not to be expected, Turing argues for the value of studying mathematically specific problems (which he calls examples). After the attempts at building general non-linear theories, it can be said that modern non-linear partial differential equations deal more with specific classes of equations and even specific equations than with general theories. Turing was ahead of his time in understanding this.

Then, in a true vision of what scientific computing will bring in the second half of the twentieth century and into the present one, he envisions the widespread use of digital computers for the study of non-linear problems. The richness of specific cases and the ability to solve them numerically will be 'most illuminating'. The combination in a single article of developing *ab nihilo* a model with much effort in understanding the biology, the use of rigorous mathematical analysis and the use of numerical simulations is a remarkable quality of this paper. It is in a sense a very modern attitude where boundaries between pure and applied, between discrete and continuous, between rigorous and numerical can be crossed at ease, with elegance and with points of view that reinforce one another. Such a remarkable combination, and the lack of intellectual prejudice it supposes, highlights the great freedom of thinking of Alan Turing.

In perspective, one can only agree with the last sentence of the Turing article: 'It is thought, however, that the imaginary biological systems which have been treated, and the principles which have been discussed, should be of some help in interpreting real biological forms'. A more general meaning can be attributed to what I would like to call Turing's philosophy. Indeed, this last sentence in which Turing maps out the role of mathematics in biology is well worth further meditation. The clarity of reasoning, the rigour of deductions, the possibility of isolating specific causes and effects, the precise study of elementary building blocks of living systems, are some of the benefits that mathematics brings to biological modelling. Even if no pretence is made to encompass the whole complexity of living systems, mathematics, even on 'imaginary systems' is a way to approach this complexity by looking at well-defined and specific phenomena. Keeping in mind their limits, mathematical models are to be trusted. This too is the lesson of Alan Turing.

References

Ahotani, A., Mimura, M., Mollee, T., 2010. A model aided understanding of spot pattern formation in chemotactic E. coli colonies. Japan J. Indus. Appl. Math. 27, 5–22.

Ammerman, A.J., Cavalli-Sforza L.L., 1984. The Neolithic transition and the genetics of populations in Europe, Princeton University Press, Princeton, p. 176.

Ben Jacob, E., Cohen, E., Levine, H., 2000. Cooperative self-organization of micro-organisms. Adv. Phys. 49(4), 395–554.

Ben Jacob, E., Cohen, E., Schochet, O., Aranson, I., Levine, H., Tsimring, L., et al., 1995. Complex bacterial patterns. Nature 373, 566–567.

Berestycki, H., Hamel, F., In press. Reaction-Diffusion Equations and Propagation Phenomena, Springer Verlag, New York, Berlin.

Bressloff, P.C., Cowan, J.D., 2003. Spontaneous pattern formation in primary visual cortex. In: Hogan, S.J., Champneys, A.R., Krauskopf, B., di Bernardo, M., Wilson, R.E., Osinga, H.M., Homer, M.E. (Eds.) Nonlinear Dynamics and Chaos, Where do we go from here? IoP, Bristol, (Chapter 11), pp. 269–320.

Bressloff, P.C., Cowan, J.D., Golubitsky, M., Thomas, P., Wiener, M.C., 2001. Geometric visual hallucinations, Euclidean geometry and the functional architecture of striate cortex. Phil. Trans. Royal Soc. Lond. 356, 299–330.

Budrene, E.O., Berg, H.C., 1995. Dynamics of formation of symmetrical patterns by chemotactic bacteria. Nature 376, 49–53.

De Kepper, P., Castets, V., Dulos, E., Boissonade, J., 1991. Turing-type chemical patterns in the chlorite-iodide-malonic acid reaction. Phys. D 94, 6525–6536.

Ermentrout, G.B., Cowan, J.D., 1979. A mathematical theory of visual hallucinations patterns. Biol. Cybernet. 356, 137–186.

Fisher, R.A., 1937. The advance of advantageous genes. Ann. Eugen. 7, 335–369.

Hodgkin, A., Huxley, A., 1952. A quantitative description of membrane current and its application to conduction and excitation in nerve. J. Physiol. 117(4), 500–544.

Horváth, J., Szalai, I., De Kepper, P., 2009. An experimental design method leading to chemical turing patterns. Science 324, 772–775.

Keller, E.F., Segel, L.A., 1970. Inititation of slime mold aggregation viewed as an instability. J. Theor. Biol. 26, 399–415.

Kolmogorov, A.N., Petrovsky, I.G., Piskunov, N.S., 1937. Étude de l'équation de la diffusion avec croissance de la quantité de matière et son application à un problème biologique. Bulletin Université d'Etat à Moscou (Bjul. Moskowskogo Gos. Univ.) 1, 1–26.

Meinhardt, H., 1995. The algorithmic beauty of sea shells. With contributions and images by Przemyslaw Prusinkiewicz and Deborah R. Fowler. The virtual Laboratory, Springer Verlag, Berlin.

Murray, J.D., 2003. Mathematical Biology, Springer Verlag, Berlin.

Ouyang, Q., Swinney, H.L., 1991. Transition from a uniform state to hexagonal and striped Turing patterns. Nature 352, 610–612.

Perthame, B., 2007. Transport Equations in Biology, Birkhäuser, Basel.

Skellam, J.G., 1951. Random dispersal in theoretical populations. Biometrika 38, 196–218.

Thom, R., 1972. Stabilité Structurelle et Morphogénèse, Interéditions, Paris.

Thompson, D., 1942. On Growth and Form. second ed. Cambridge University Press, Cambridge, New York.

Turing, A.M., 1952. The chemical basis of morphogenesis. Phil. Trans. Royall Soc. Lond. Ser. Biol. Sci. 237(641), 37–72.

Waddington, C.H., 1940. Organbrs and Genes, Cambridge University Press, Cambridge, New York.

Zel'dovich, Y.B., Frank-Kamenetskii, D.A., 1938. The theory of thermal flame propagation. Zhur. Fiz. Khim. 12,100 (In Russian).

Hans Meinhardt focuses on —

TRAVELLING WAVES AND OSCILLATIONS OUT OF PHASE: AN ALMOST FORGOTTEN PART OF TURING'S PAPER

1. Introduction

Turing's 1952 paper is most famous for the discovery that stable patterns can be generated by the interaction of two substances that diffuse at different rates. In a second part, he discusses interactions of three components and showed that this can lead to travelling waves and to out-of-phase oscillations – a part that became largely forgotten. Turing himself mentioned (p. 718 of this volume) that he is not aware of any example of such an out-of-phase oscillation in biology. Presumably for this reason he did not provide a set of example equations for three-component systems as he did for the reactions that form stable patterns. For both types, he did not provide an intuitive explanation of how this patterning mechanism works. This may explain why his work found initially only little attention.

By searching for mechanisms that account for the pigment pattern on tropical sea shells, we came across with a reaction type that is also based on three components and that also shows travelling waves and out-of-phase oscillations (Meinhardt, 2009; Meinhardt and Klingler, 1987). Therefore, both mechanisms are presumably mathematically equivalent, but this has still to be proven. In the present article, I will work out the requirements for reactions that display out-of-phase oscillations and travelling waves and discuss some biological applications. As elaborated in other parts of this book, a substantial part of Turing's Biology-oriented unpublished work deals with phyllotaxis. It will be shown that such three-component systems are most convenient to account for a helical arrangement of leaves or seeds, e.g., on a fire cone. Thus, considering three component systems provides an unanticipated link between these two themes of Turing's work.

2. Stable pattern and travelling waves by two-component systems

Turing derived the conditions for stable patterning by mathematical consideration only. On page 694, he gave the following set of equations as an example:

$$\frac{dX}{dt} = 5X - 6Y + 1 \tag{1a}$$

$$\frac{dY}{dt} = 6X - 7Y + 1 \quad [+\text{Diffusion}]. \tag{1b}$$

Both Eqs (1a) and (1b), look very similar. It is not immediately obvious why such an interaction allows pattern formation. In a general theory of biological pattern formation proposed 20 years later, we have shown that pattern formation is possible if and only if local self-enhancement is combined with an antagonistic reaction of longer range (Meinhardt, 1982; Gierer and Meinhardt, 1972). This condition is not inherent in Turing's paper, even not in a hidden form. It seems, however, that later he recognised the role of inhibition. In unpublished notes found after his death, the following sentence occurred: *'The amplitude of the waves is largely controlled by the concentration V of "poison"'* (see Hodges (1983, p. 494)). It is remarkable that long-range inhibition does not occur in his work since phyllotaxis was prevalent for him and the idea that the spacing of leaves is achieved by a mutual long-range inhibition of leaves was discussed since long (Schoute, 1913; Snow and Snow, 1931).

A straightforward realisation of the general mechanism we proposed is an interaction between a short-ranging autocatalytic activator a that catalyses its own production and that of a long-ranging component h that acts as an inhibitor in the self-enhancement.

$$\frac{\partial a}{\partial t} = \frac{sa^2}{h} - r_a a + D_a \frac{\partial^2 a}{\partial x^2} + \rho_a \tag{2a}$$

$$\frac{\partial h}{\partial t} = sa^2 - r_h h + D_h \frac{\partial^2 h}{\partial x^2} + \rho_h. \tag{2b}$$

Knowing the basic principle allowed us to introduce molecular realistic interactions for which non-linear interactions are indispensable and to recognise somewhat hidden molecular realisations in biological systems. Meanwhile, several biological pattern-forming systems of this type are known (for review see Meinhardt (2008)). Formation of a stable pattern is illustrated in Fig. 1a.

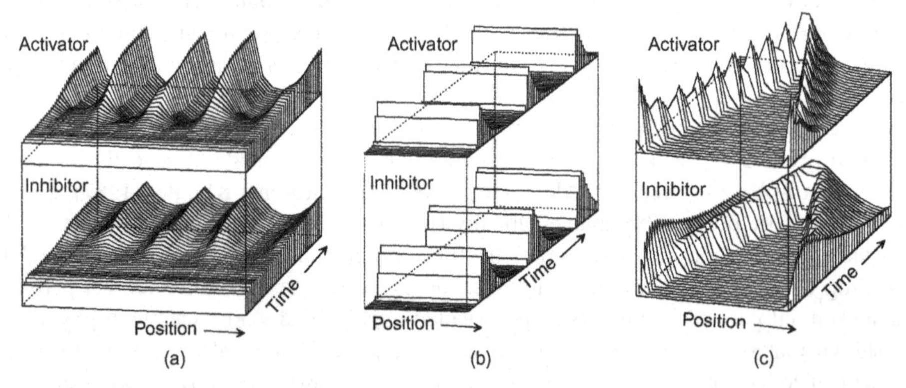

Fig. 1: Pattern formation by a two-component activator–inhibitor system. (a) Stable patterns can result if the self-enhancing activator has a short and the inhibitor has a long range. (b) Oscillations will result if the inhibitor has a longer time constant than the activator. (c) Travelling waves result if the activator has a moderate spread but the inhibitor is local. A baseline inhibitor production suppresses the spontaneous trigger of activations, otherwise synchronous oscillation would appear. Two colliding waves annihilate each other (Meinhardt, 1982, 2009).

It is easy to see that Turing's example equation satisfies our condition. X in Eq. (1a) enhances its own production and catalyses that of Y [Eq. (1b)]. In turn, Y – the diffusible component – causes a higher destruction rate of X, antagonising in this way the self-enhancement of X. The mechanism proposed by Turing has an essential drawback: its molecular basis is not reasonable. According to Eq. (1a), the number of X molecules disappearing per time unit is assumed to be independent of the number of X molecules and only proportional to the number of Y molecules. In other words, X molecules can disappear at a high rate even if no X molecule is present any longer. This leads to negative concentrations. Turing had seen this problem and proposed to ignore negative concentrations. To avoid this problem, first-order decay kinetics is needed, which requires non-linear production terms as given in Eq. (2a).

3. Oscillations and wave formation by two-component systems

The same reaction type can lead to oscillations and to travelling waves. Oscillations occur if the antagonistic reaction has a longer time constant than the self-enhancing component (Fig. 1b). Then, activation occurs in a burst-like fashion since the antagonistic reaction follows too slowly. After some time, when, e.g., sufficient inhibitor accumulated, the self-enhancement breaks down. After a refractory period, the next burst can trigger. For travelling waves, the activation has to show a moderate spread while the antagonistic reaction has to be local (Fig. 1c). Thus, for wave formation and for stable patterns, the requirements for the antagonistic reaction are just the opposite. Well-known processes of this type play a role in different areas. The wave-like spread of a forest fire or of an epidemic is an example. A characteristic feature of such waves is that they annihilate each other upon a collision (Fig. 1c). A forest fire can not spread into a region just burned down by a wave that came from the opposite direction.

Wave formation by two-component systems looks simple by a first inspection. However, more conditions have to be satisfied. There must be a spatial pattern that determines where the waves should start. For instance, for the regular contraction waves in the heart, it is crucial that the oscillations run somewhat faster in the sinus node that acts as pacemaker region. Likewise, a wave front of a fire needs a local ignition. Thus, a complete travelling wave system would require at least four components, two for the generation of the stable pattern determining where the wave should start and two components for the proper wave.

4. Wave formation by three-component systems

The pigment patterns on the shells of mollusks provide a wonderful natural picture book to study all sorts of pattern formation. They are of great diversity and frequently of great beauty. Mollusks can increase the size of their shells only by accretion of new material along the growing edge of the shell. In most species, pigment only becomes incorporated only there during growth. In these cases, the formation of the pattern is generated in a linear array of cells. The shell itself consists of dead material and preserves the pattern. Thus, the two-dimensional shell pattern is a protocol of what happens as function of time along the one-dimensional edge. Shell patterns are, therefore, space-time plots and provide a unique situation in that the complete history of a highly dynamic process is preserved.

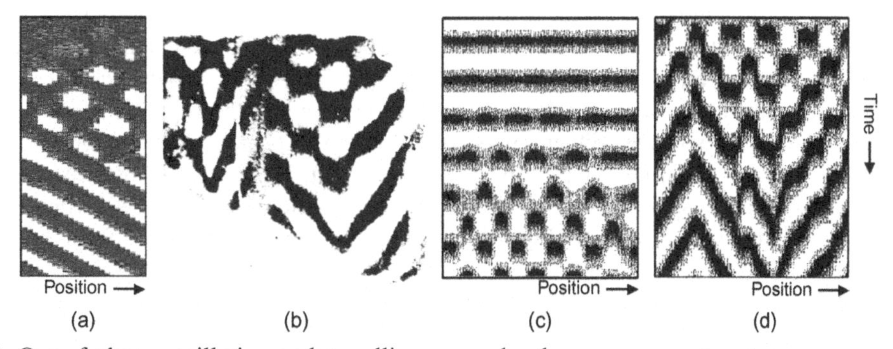

Fig. 2: Out-of-phase oscillation and travelling waves by three-component systems, presumably as anticipated by Turing. (a) The spontaneous generation of travelling waves by a self-enhancing activator and two antagonistic components [Eqs (3a) through (3c)]. (b) The chessboard–like pattern on the small shell of *Bankivia fasciata* has its origin in out-of-phase oscillations that changed into travelling waves. (c) In three-component systems, a synchronous oscillation can become instable and change to out-of-phase oscillations. (d) A transition to travelling waves can occur if the activation becomes stable enough that it does not collapse but escape into an adjacent position (Meinhardt, 2004, 2009; Meinhardt and Klingler, 1987).

In an attempt to decode the underlying mechanism, it became clear that the usual two-component systems are insufficient for an explanation. Some patterns result from out-of-phase oscillations (Fig. 2). Crossing of oblique lines are the result of travelling waves that penetrate each other (Fig. 3), a very unusual behaviour for waves in excitable media. The key was the idea that these patterns are explicable by three-component systems (Meinhardt, 2004, 2009; Meinhardt and Klingler, 1987).

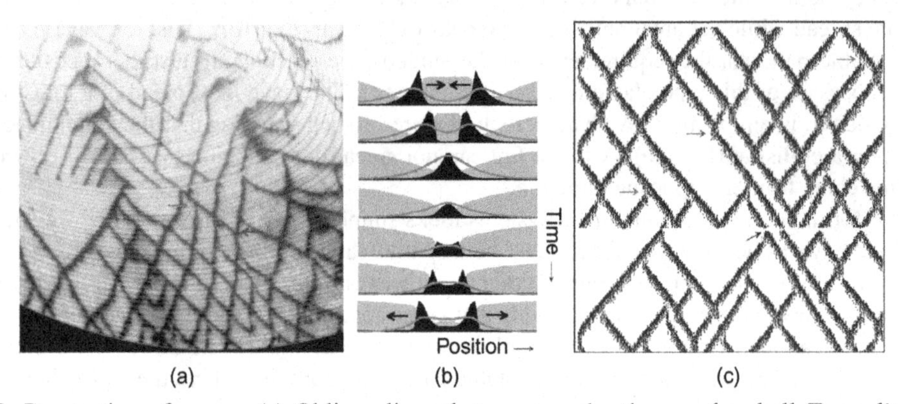

Fig. 3: Penetration of waves. (a) Oblique lines that cross each other on the shell *Tapes literatus* are time records of travelling waves at the growing edge which were penetrating each other. (b) Penetration of waves in a three-component system [Eqs (3a) through (3c)]. Assumed is an activator–inhibitor system (black area and gray line). The removal of a substrate (gray area) that is required for the autocatalysis leads to a local destabilisation. Activation shifts into a region where still sufficient substrate is available (arrows); travelling waves result. During collision, the shift is suspended. Due to the rapidly declining inhibitor, the activation survives at a low level. After the recovery of the substrate, the waves start to move again into divergent directions, completing the penetration. (c) The simulation in a larger field shows that the model accounts for the fact that sometimes only one of the two waves survive a collision, leading to an 'amputated X' and that backwards-running waves can trigger spontaneously (arrows in a and c) (*From Meinhardt (2009); Meinhardt and Klingler (1987)*).

As mentioned above, two components are required to generate a stable pattern. Imagine that a second antagonistic component – i.e., a third component – is involved that acts local and that has a long time constant. This can be an inhibitor or can result from the depletion of a necessary component, as given in Eqs (3a) through (3c). A local maximum becomes quenched some time after its generation by the slowly accumulating antagonistic reaction. Depending on the parameters, the system can react in two ways. Either a maximum shifts to an adjacent position, becomes there quenched too, and moves further, i.e., a travelling wave result. Alternatively, the maximum may vanish and reappear at a displaced position. Out-of-phase oscillation can result, even from a system that oscillated originally in synchrony (Fig. 2d). Travelling waves can be formed from an initially homogeneous situations, i.e., without a pacemaker (Fig. 2a). The equation given below provides an example that has been used for the simulations given in Figs. 2 and 3.

5. An example

In the following example reaction, the activator a is antagonised by a substrate b that becomes depleted during autocatalysis; a is produced with the same rate as b is removed. The third component c acts as an inhibitor. It is produced under the control of the activator and that slows down the autocatalysis. In this way, a higher c concentration leads also to a lower removal rate of the substrate b. With a notation as used elsewhere (Meinhardt, 2009), the reaction can be described by the following equation:

$$\frac{\partial a}{\partial t} = \frac{sb(a^2+b_a)}{(s_b+s_c c)(1+s_a a^2)} - r_a a + D_a \frac{\partial^2 a}{\partial x^2} \tag{3a}$$

$$\frac{\partial b}{\partial t} = b_b - \frac{sb(a^2+b_a)}{(s_b+s_c c)(1+s_a a^2)} - r_b b + D_b \frac{\partial^2 b}{\partial x^2} \tag{3b}$$

$$\frac{\partial c}{\partial t} = r_c(a-c) + D_c \frac{\partial^2 c}{\partial x^2}. \tag{3c}$$

Fig. 4: Flashing up of local signals at displaced positions in phyllotaxis and shell patterning. (a) Whorl-like pattern suggest that spacing around the circumference and along the axis of the shoot result from two different processes. (b, c) Signals for leaf or seedling formation appear at displaced positions. (d) Model: assumed is an autocatalytic activator (black) that is antagonised by two inhibitors. One has a short time constant and a long range (dark pixels); it keeps the maxima localised and is responsible for the spacing around the circumference. A second inhibitor (gray) acts more locally and has a long time constant; it takes care that the leaf-initiating signal disappears after a certain time interval. A new signal can only appear at a displaced region where both inhibitions are below a threshold (white); calculation on a ring, second dimension is time. (e) Oblique rows of spots on the shell of a mollusk, also resulting from the successive formation of spot-like signals at regularly displaced position. (f) Calculation on a smaller ring with space for one helix. The most recent leaf initiation sites are numbered. The angular displacement is around the golden angle of 137.5°, illustrating that the corresponding displacement of leaf initiation sites is correctly described (Meinhardt et al., 1998).

6. Phyllotaxis: initiation of leaves along spirals

The regular arrangement of leaves has fascinated people for centuries (see Kuhlemeier (2007) for a recent review and molecules involved and Smith et al. (2006) for recent simulations). As described in other parts of this book, Turing is one of them. The tip of the shoot, the apical meristem, contains undifferentiated cells that divide rapidly. Cells just leaving this zone become competent to form new leaves. Thus, the leaf-forming zone has the geometry of a ring. Similar as in shell patterning, the arrangement of leaves is a time record of the signal distribution in this leaf-forming zone. Many different models have been proposed that have the assumption in common that existing primordia have an inhibitory influence on the initiation of the next leaf. Implicit in this assumption is that the spacing around the circumference in the leaf-forming zone and the separation along the axis, i.e., the time span at which the next leaf can be formed, is based on the same signaling. However, the existence of whorl-like leaf patterns (Fig. 4a) suggests that the spacing around the circumference and spacing in time (e.g., when the next whorl will come) is based on different mechanisms. This suggests an explanation of leaf spacing by two different inhibitions, one in space and one in time, leading to models that are analogous to those proposed for shell patterning (Fig. 4).

7. Other systems and outlook

The three-component patterning, discovered for a more exotic type of pattern formation, has turned out to be very powerful to account for very different biological phenomena. In E. coli, the determination of the plane where the next cell division will take place works in this way. A high concentration of a protein (*MinD*) occurs at the poles in an alternating sequence. This out-of-phase oscillation inhibits the onset of cell division near the pole regions, resulting in the initiation of cell division precisely at the cell center (Meinhardt and de Boer, 2001). The separation of the barbs of an avian feather occurs by travelling waves that have been described with the same equations as given above (Harris et al., 2005). Chemotactic cells show a highly dynamic pattern of stretching and retracting pseudopods. Three-component systems provide on the one hand the high amplification to make the cells sensitive to minute concentration differences, and, on the other hand, allow to maintain this sensitivity by avoiding that the cells are trapped in a once-made decision (Meinhardt, 1999). Thus three-component systems seems to be used by nature for very different purposes in biology.

In conclusion, we found a mechanism that has the properties anticipated by Turing in the second part of this paper dealing with three-component systems. In the view of our model, the three substances have the following function: a self-enhancing feedback loop together with a long-ranging antagonist enables pattern formation in space. The third component, a local short-ranging antagonist, causes on a longer time scale a local destabilisation. In this mode, a stable steady state is never reached. The shell patterns provide a beautiful record of such highly dynamic interaction and were the key to disentangle the underlying complex interaction.

References

Gierer, A., Meinhardt, H., 1972. A theory of biological pattern formation. Kybernetik 12, 30–39.

Harris, P., Williamson, S., Fallon, F., Meinhardt, H., Prum, O., 2005. Molecular evidence for an activator-inhibitor mechanism in development of embryonic feather branching. PNAS 102, 11734–11739.

Hodges, A., 1983. Alan Turing: The Enigma, Simon and Schuster, New York.

Kuhlemeier, C., 2007. Phyllotaxis. Trends Plant Sci 12, 143–150.

Meinhardt, H., 1982. Models of biological pattern formation, Academic Press, London. http://www.eb.tuebingen.mpg.de/meinhardt/82-book

Meinhardt, H., 1999. Orientation of chemotactic cells and growth cones: models and mechanisms. J. Cell Sci. USA 112, 2867–2874.

Meinhardt, H., 2004. Out-of-phase oscillations and traveling waves with unusual properties: the use of three-component systems in biology. Phys. D 199, 264–277.
Meinhardt, H., 2008. Models of biological pattern formation: from elementary steps to the organization of embryonic axes. Curr. Top. Dev. Biol. 81, 1–63.
Meinhardt, H., 2009. The Algorithmic Beauty of Sea Shells, 4nd enlarged ed. (with programs on CD) Springer, Heidelberg, New York.
Meinhardt, H., de Boer, J., 2001. Pattern formation in e.coli: a model for the pole-to-pole oscillations of Min proteins and the localization of the division site. PNAS 98, 14202–14207.
Meinhardt, H., Klingler, M., 1987. A model for pattern formation on the shells of molluscs. J. Theor. Biol 126, 63–89.
Meinhardt, H., Koch, J., Bernasconi, G., 1998. Models of pattern formation applied to plant development. In: Barabe, D., Jean, R. V. (Eds.), Symmetry in Plants, World Scientific Publishing, Singapore, pp. 723–758.
Schoute, C., 1913. Beiträge zur Blattstellung. Rec. Trav. Bot. Neerl. 10, 153–325.
Smith, S., Guyomarc'h, S., Mandel, T., Reinhardt, D., Kuhlemeier, C., Prusinkiewicz, P., 2006. A plausible model of phyllotaxis. PNAS 103, 1301–1306.
Snow, M., Snow, R., 1931. Experiments on phyllotaxis: The effect of isolating a primordium. Phil. Trans. Roy. Soc. B 22, 1–43.
Turing, A., 1952. The chemical basis of morphogenesis. Phil. Trans. B. 237, 37–72.

James D. Murray on what happened —

AFTER TURING – THE BIRTH AND GROWTH OF INTERDISCIPLINARY MATHEMATICS AND BIOLOGY

1. Introduction

An accepted norm of the study of biology during much of the nineteenth century was classification and list making. Its death knell was truly sounded by D'Arcy Thompson's classic book *On Growth and Form* first published in 1917 by Thompson. There had been, of course, considerable interest in morphological patterns for a long time. The classic work of Geoffroy Saint-Hilaire (1836) is a remarkable seminal example. He was a strong supporter of Lamarck and was ridiculed in 1842 by an etching of him as an ape. Sainte-Hilaire was particularly interested in teratology and was probably the first to introduce the important concept of a developmental constraint which we come back to below. D'Arcy Thompson specifically commented on the absence of certain particular forms that implies an awareness of developmental constraints. He emphasised the parallels between the study of form in physical systems and of biological form that presages landmark paper of Turing (1952) on the chemical basis of morphogenesis, a paper with only six references, one of which is to the second edition of D'Arcy Thompson's (1917) book.

Turing's foray into biology is primarily practical but essentially mathematical: he says nothing about real biology or morphogenesis. Importantly, however, he shows how a model system of reacting and diffusing chemicals in a bounded domain can result in steady-state spatial patterning of the chemical concentrations. In this article, I shall give a very brief description of reaction–diffusion theory and an equally brief and biased history of the developments following from Turing's paper but with references for further reading. I shall describe the emergence and astonishing growth of a new field – mathematical biology. By way of illustration, I shall describe two specific biological problems I have worked on, both of which resulted in experimental research projects. Finally, I shall point out some of the limitations of Turing-type reaction–diffusion mechanisms that necessitated a new, and more experimentally verifiable, approach to biological pattern formation.

In spite of the enormous amount of research and the exploding study of genetics, the development of spatial pattern and form is still one of the central issues in embryology. In the past 20 to 30 years, it has spawned exciting, important and genuine interdisciplinary research between theoreticians and experimentalists, the common aim of which is the elucidation of the underlying mechanisms involved in embryology and medicine; most of which are essentially still unknown.

Mathematicians and theoreticians did not begin to examine the development of spatial pattern and form until almost 20 years after Turing's 1952 paper. The late 1960s and early 1970s was when the field of mathematical biology, theoretical biology, or whatever one wants to call this genuine interdisciplinary field of mathematics and biology, really got going. The first mention of Turing's paper was by Prigogine and Nicolis (1967) and the Brussels group subsequently introduced a simpler theoretical reaction–diffusion mechanism of the Turing type. Another influential activator–inhibitor reaction–diffusion system was proposed by Gierer and Meinhardt (1972). The first genuine experimentally based reaction–diffusion system that produced steady-state chemical spatial patterns in line with Turing's predictions was developed by Thomas (1975). Interestingly, when you take these model systems and look at the parameter ranges that can generate spatial pattern by far the largest ranges are those of the practical system proposed by Thomas (1975). Mimura and Murray (1978) showed mathematically how this specific reaction–diffusion system produced steady-state spatial patterns. There are numerous review articles and books: for example, Murray's books *Mathematical Biology* (Murray (1989, 2002)) in 1990, 2003 describe, among many other topics, Turing's theory of morphogenesis and its influence on modelling biological pattern and form in some detail and, more specifically, Turing's theory in a review article Murray (1990). The study by Maini (2004) specifically addresses pattern and form as do the conference proceedings by Brenner et al. (1981), Jäger and Murray (1984) and the book of articles edited by Chaplain et al. (1999) and Maini and Othmer (2001). In population ecology, the review article by Levin (1992) describes the mathematical modelling of that field from an equivalent 'reaction–diffusion' approach in which the reaction terms are population growth and interaction terms.

The basic concept, which Turing demonstrated mathematically, was that if you have two chemicals, an activator and an inhibitor, which react together and at the same time diffuse, crucially at different rates with the inhibitor having the larger diffusion coefficient, it is possible for such a coupled system of reaction–diffusion equations to produce steady-state spatial patterns in chemical concentrations of the reactants. In the early to mid 1970s, Turing's paper was rediscovered by more theoreticians with an increasing number of publications starting to appear. Closely related, but not specifically to Turing's work, is the fundamentally important experimental work on the importance of chemical gradients in embryonic development by Wolpert (1969) who introduced the concept of 'positional information'. In it cells react to a chemical gradient. His work initiated a huge amount of experimental and theoretical work, often controversial, that is still going on. For a review of his work and his views on development, see Wolpert's (2006) book on the principles of development.

To get an intuitive idea of how the patterning works, consider the following, albeit unrealistic scenario, of a field of dry grass in which there is a large number of grasshoppers that can generate a lot of moisture by sweating if they get warm. Now suppose the grass is set alight at some point and a flame front starts to propagate. We can think of the grasshopper as an inhibitor and the fire as an activator. If there were no moisture to quench the flames, the fire would simply spread over the whole field, which would result in a uniform charred area. Suppose, however, that when the grasshoppers get warm enough, they can generate enough moisture to dampen the grass so that when the flames reach such a pre-moistened area the grass will not burn. The scenario for spatial pattern of charred and uncharred grass is then the following. The fire starts to spread; it is one of the 'reactants', the activator, with a fire 'diffusion' coefficient, which quantifies how fast the fire spreads. When the grasshoppers, the inhibitor 'reactant', ahead of the flame front feel it coming, they move quickly ahead of it; that is, they have a 'diffusion' coefficient that is much larger than that of the fire front. The grasshoppers then sweat profusely and generate enough moisture to prevent the fire spreading into the moistened area. In this way, the charred area is restricted to a finite domain that depends on the 'diffusion' coefficients of the reactants – fire and grasshoppers – and the various 'reaction' parameters. If, instead of a single initial fire, there were a random scattering of them, we can see how this process would result in a final, spatially heterogeneous, steady-state distribution of charred and uncharred regions in the field and a spatial distribution of grasshoppers, since around each fire the above scenario would take place. If the grasshoppers and flame front 'diffused' at the same speed, no such spatial pattern could evolve.

2. How the leopard gets its spots

I became interested in how animal coat patterns were formed Murray (1981a, b) and used the practical reaction–diffusion mechanism of Thomas (1975) to study them. I showed that a single pre-patterning mechanism was capable of generating the geometry of mammalian coat patterns, from the mouse to the badger to the giraffe to the elephant and almost everything in between, with the end pattern governed simply by the size and shape of the embryo at the time the pattern formation process took place. In solving these reaction–diffusion systems, the domain's size and shape is important. For a given mechanism if you try and simulate solutions in a very small domain, it is not possible to obtain steady-state spatial patterns. A minimum size is needed to drive any sustainable spatial pattern. If, for example, it is a long, thin domain, you can only generate stripes. You can think about this intuitively. It is like disturbing the water surface in a long, thin tank: the only waves that persist are one dimensional along the tank. If the tank is large and the surface is disturbed, it is possible to have complex wave patterns.

Suppose the surface, which corresponds to the reaction–diffusion domain, is a rectangle. As mentioned, if the surface is very small, it cannot have any spatial pattern: a minimum size is therefore needed to exhibit spatial heterogeneity. As the size of the rectangle is increased, a series of increasingly complex spatial patterns emerge. The concept behind the model is that the simulated spatial patterns solutions of a reaction–diffusion mechanism reflect the final morphogen melanin landscape (Murray, 1981a,b, 1989, 2002) observed on animal coats. With this scenario, the cells react to a given level in morphogen concentration, thus producing melanin (or rather becoming melanocytes – cells that produce melanin). In the figures, the dark regions represent high levels of morphogen concentration. It should be emphasised that this model is a hypothetical one which has not been verified experimentally but rather circumstantial. The main purpose is to show how scale and shape play major roles in animal coat patterns as it must in other developmental processes.

An example of how the geometry constrains the possible pattern modes is when the domain is so narrow that only simple, essentially one-dimensional, modes can exist. Two-dimensional patterns require the domain to have enough breadth as well as length. Consider a tapering cylinder. If the radius at one end is large enough, two-dimensional patterns can exist on the surface. So, such a tapering cylinder can exhibit a gradation from a two-dimensional pattern to simple stripes as in Fig. 1.

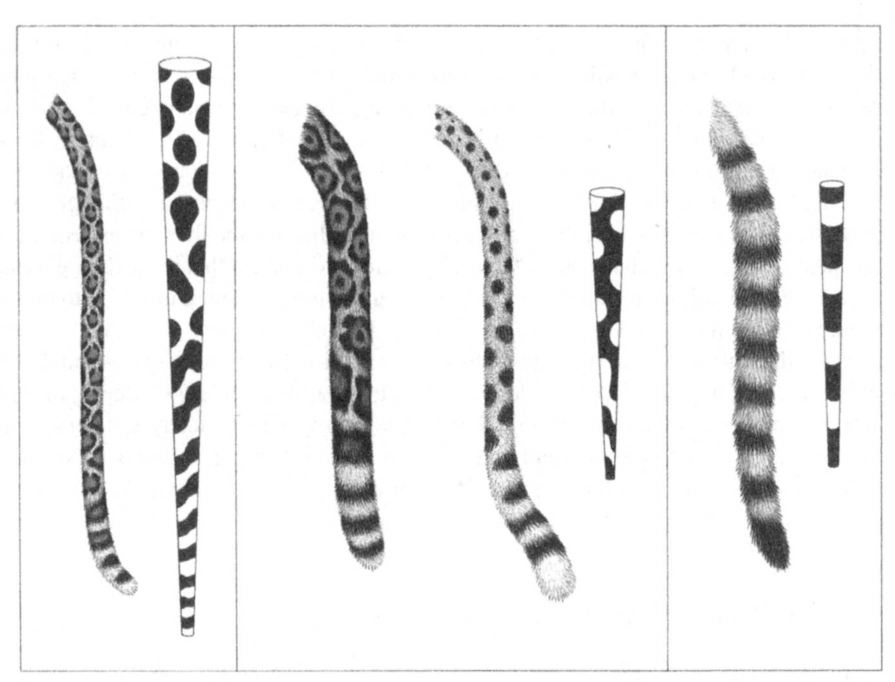

Fig. 1: Examples of a developmental constraint. Spotted animals can have striped tails but not the other way round. From left to right are typical examples of the tail of the leopard, the cheetah and the genet together with the solutions from a reaction–diffusion system which can generate steady-state spatial patterns. The geometry and scale when the pattern mechanism is activated play crucial roles in the resulting coat patterns. Dark regions represent areas of high morphogen concentration. (*Tail art work by Patricia Wynne, www.patriciawynne.com*).

Fig. 2: Geoffrey's cat ((*Leopardus geoffroyi*). An example of the developmental constraint described in Fig.1. (*Photo credit: Charles Barilleaux/Wikimedia Commons*)).

In the calculations, a set of reaction and diffusion parameters were chosen which could produce a diffusion-driven instability and kept them fixed for all the calculations. *Only* the scale and geometry of the domain were varied. The resulting patterns are colored dark and light in regions where the concentration of one of the morphogens is greater than or less than the concentration in the homogeneous steady state. Even with such limitations on the parameters, the wealth of possible patterns is remarkable.

The solutions of the reaction–diffusion system in domains shown in Fig. 1 were first computed. These domains all taper as shown. This shows that the conical domain mandates that it is not possible to have a tail with spots at its tip and stripes at its base, but only the converse: Fig. 1 shows some examples of specific animal tails. This is a genuine example of a *developmental constraint*. Cheetahs are prime examples of this, as well as other spotted animals. Geoffrey's cat (*Leopardus geoffroyi*) (see Fig. 2), named after Geoffroy Saint-Hilaire who travelled extensively in the south and east of South America, the habitat of this animal is a less well-known example of a spotted cat on which both the tail and the legs exhibit the developmental constraint. If the threshold level of morphogen is different, a different but related pattern can develop. In this way quite different, but globally similar, patterns can be formed and could be the explanation for the different types of patterns on different species of the same animal genre, such as the giraffe (see Fig. 7).

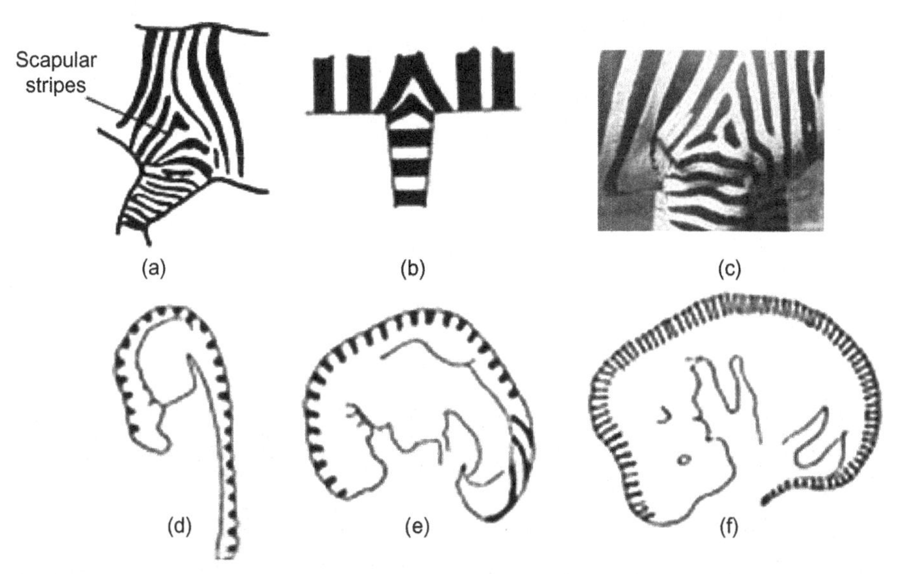

Fig. 3: (a) Typical scapular stripe pattern on a zebra (*Equus burchelli*) with (b) the pattern obtained from the same reaction–diffusion system used in Fig. 1 with the same parameters. (c) is a typical foreleg-body pattern in the zebra. (d) is a schematic dorsal striping on a 21-day embryo of *Equus burchelli*: the stripes are approximately 0.4 mm apart while (e) shows the effect on the pattern of approximately 3–4 day growth in which the stripe deformation is associated with a typical *Equus burchelli* as in Fig. 4(a). (f) This shows more typical stripe spacing associated with a 3-week old embryo of the Grevy zebra (*Equus grevyi*) which result in typical striping as in the herd photo in Fig. 4(b).

The case of the zebra exhibits another aspect. The early embryo of the zebra is, although curved, approximately linear and so stripes are formed along it as schematically shown in Fig. 3(d)–(f). The legs are also narrow and linear appendages and again the typical markings are stripes. What was encouraging for the hypothesis was the patterns obtained when the domain of simulation involved two linear stripe-sustaining domains at right angles to each other. The calculated patterns are shown in Fig. 3(b), while Fig. 3(c) is a photograph of typical zebra patterns: note the scapular stripe pattern at the junction of the body and foreleg. It is possible to quantify the number of stripes necessary at the early embryonic stage and also their spacing. Further details are given in the study by Murray (1989, 2002). Figure 4a and b are photographs of typical patterning on two species of zebra.

(a)

(b)

Fig. 4: (a) Photograph of a female plains zebra and its foal. Note the scapular stripes at the foreleg-body junction and the lack of them on the back legs. This is indicated by the initial elongation in stripe patterns in the embryo as shown in Fig. 3e. (b) Photograph of a herd of Grevy zebras in which the scapular stripes are evident in all four legs and are closer together. (*Photographs courtesy of Professor Daniel Rubenstein*)

The pattern formation mechanism for coat patterns is activated at some specific time in the embryo's development. There are minor variations in the timing and certainly small random variations in the initial morphogen concentrations. As a result, it is clear why the coat patterns on an animal are unique to that animal. It is similar in concept to the uniqueness of human fingerprints (Murray, 1989, 2002). The uniqueness of patterns is clear in the photos in Fig. 4. Such small random parameter and timing variations, however, do not explain the occasional unusual aberrations observed, such as the case of a black zebra with white stripes photographed in the Kruger National Park in 1967 or the black sheep with white stripes (Murray, 2002). Aberrations in cheetah patterns are more common, resulting in one case in the erroneous conclusion that it was a new species with a name associated with the scientist who first saw it.

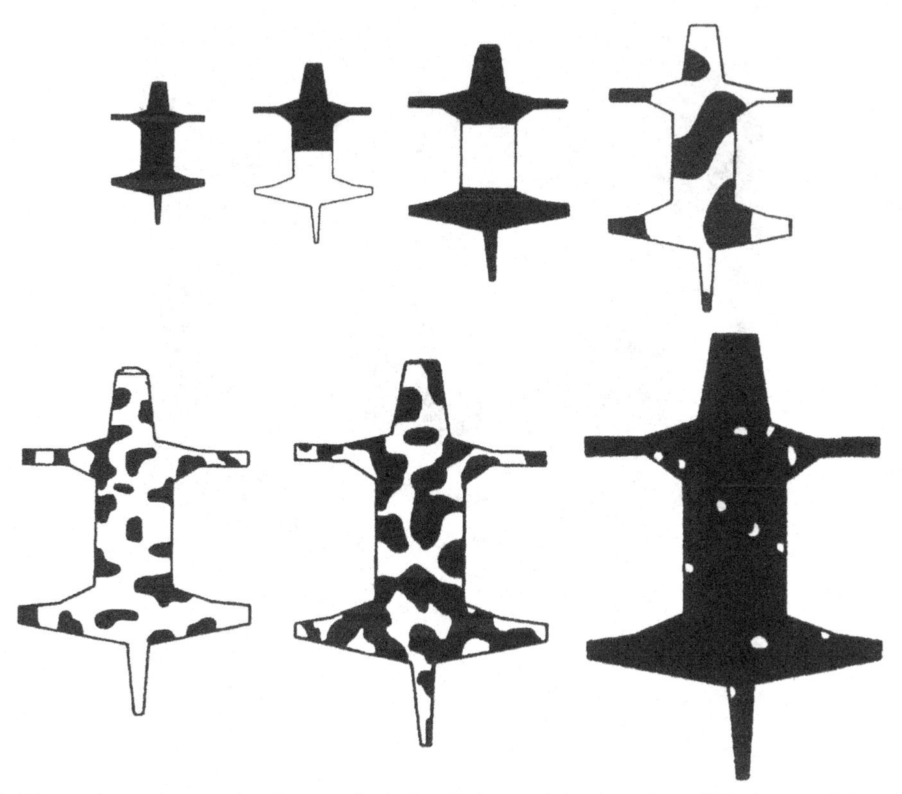

Fig. 5: These show the result of numerical simulations of the reaction–diffusion model proposed for the generation of coat markings on animals; it is the same mechanism used in Figs 1 and 3. The model parameters were also the same; only the scale parameter was varied. As before the dark regions represent high levels of morphogen concentration. The domain sizes have been reduced so as to fit in a single figure, but in the simulations, there was a scale difference of 1000 between the smallest and the largest figure. The interpretation is that if the animal skin size is too small when the patterning mechanism is activated, as in the mouse, or too large, as in the hippopotamus and elephant, then no pattern will be formed and these animals are generally uniform in colour. In between, there is a progressively more complex pattern as the size increases; the first two bifurcations are illustrated in Fig. 6 with the largest animals still showing coat pattern in Fig. 7.

Scale plays a crucial role in the type of pattern that can be generated. Incorporating scale showed that if the animal is too small when the patterning mechanism is activated (as in very small animals) or too large (as in the hippopotamus and elephant), then no distinct pattern will be formed. Figure 5 schematically illustrates these scale aspects. The idealised animal skin domains have been reduced to fit in a single figure. In the model simulations, there is a scale difference of 1000 between the smallest and the largest figure.

(a)

(b)

Fig. 6: Photographs of animals which exhibit initial bifurcating patterns: (a) Valais Blackneck goat (*Photograph by BS Thurner Hof, Wikimedia Commons*); (b) Belted Galloway cows (*Photograph courtesy of Allan Wright*).

In a recent interesting paper, Allen et al. (2011) point out that an understanding of the diversity of animal coat patterns requires an understanding of both the mechanisms that create them and their adaptive value. Among other things, they discuss the advantages of specific patterns in different environments. They use a reaction–diffusion model, but their conclusions are general and do not rely on specific reaction–diffusion models. They convincingly show how different marking relate to specific natural environments for the specific felids.

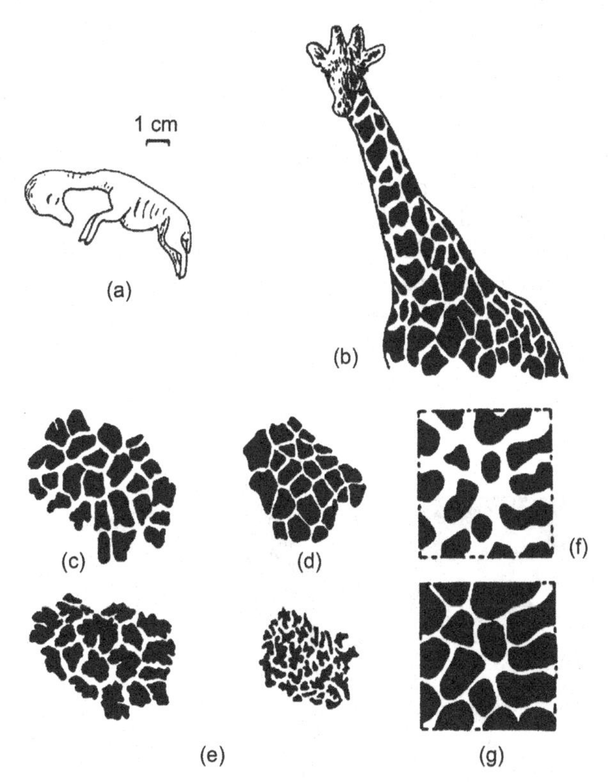

Fig. 7: (a) A typical giraffe embryo of about 35–45 days. We anticipate the pattern has been laid down by this time although the actual melanisation will not take place until very much later. (b) Typical neck markings on the reticulated giraffe (*Giraffa camelopardalis reticulata*). (c)–(e) Tracings of giraffe trunk spots of *Giraffa camelopardalis*. (c) *Giraffa rothschildi*, (d) *Giraffa reticulata* and (e) *Giraffa tippelskirchi*. (f) Spatial patterns obtained from the model reaction–diffusion mechanism. (g) Spatial patterns obtained from the same model reaction–diffusion mechanism but with a lower threshold of morphogen for cell differentiation into melancytes than that in (f) (*see Murray (1981a,b, 1989, 2002) for more details*).

3. Experimentally verified prediction of a reaction–diffusion pattern formation model and resolution of a controversy

Reaction–diffusion models have now been applied to a vast array of patterning problems in biology such as snake skin patterns, tooth formation in crocodilia, butterfly wing patterns and many many others: see Murray (1989, 2002) for numerous examples and a review of the literature. There is an interesting study by Painter et al. (1999) on the stripes on zebra fish (*Pomcanthus circulates*) that incorporates the effect of growth on the patterns obtained. They include chemotaxis in their model. This is where there is cell movement *up* a chemical concentration gradient a crucial aspect of the slime mold, *Dictyostelium*, aggregation. There are other interesting papers on the temporal growth and variation in the stripes formation on this fish, for example, by Kondo et al. (2009): see other references in their paper. They importantly relate their theory to specific experiments. Other articles such as that done by Nijhout et al. (2003) discuss butterfly wing patterns, the model results of which are also related to experiment. Travelling stripes on the skin of a mutant mouse are discussed the study by Suzuki et al. (2003) and Maini (2003). Genuinely, practical applications of modelling are now the norm in several large research groups such as the Center for Mathematical Biology in Oxford and the Department of Ecology and Evolutionary Biology in Princeton, both of which are particularly outstanding.

One of the major problems with much of the reaction–diffusion modelling research in biological pattern formation up to the mid-1990s is that it was difficult to carry out experimental verification or otherwise since, among other things, the morphogens involved were unknown and genetics was a fast growing field in which a widespread belief, among geneticists, was that genes did everything. Mechanisms, however, are needed even though they are controlled by genes. Unlike many of the patterning problems studied up to the late 1980s and early 1990s, little was done to try and model the effect of embryonic growth on the patterning process. The above mentioned works on zebra fish and mice are important examples which show that when growth and pattern formation processes are occurring simultaneously it can explain hitherto un-understood phenomena. Growth plays a crucial role.

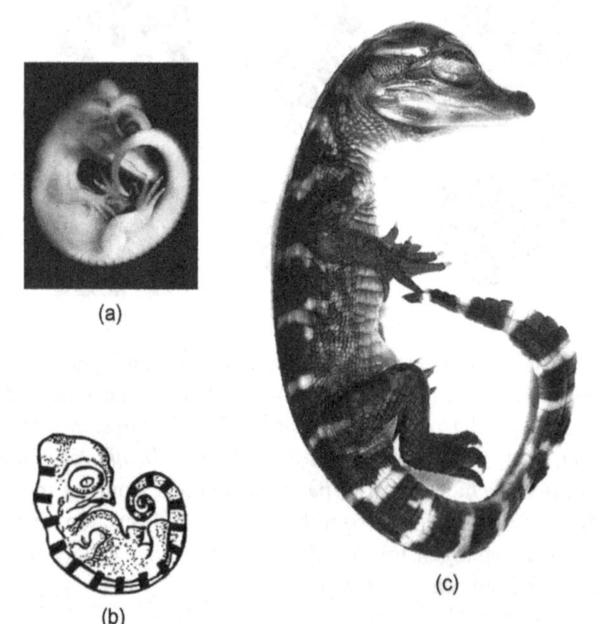

Fig. 8: Typical stripe pattern on an alligator hatchling. (a) Alligator embryo around 21 day of gestation, approximately when the stripe pattern is laid down: it is schematically shown in (b). (c) This embryo is around 51–60 days of gestation and from head to the bend in the tail is 8 mm. *(Photographs courtesy of Professor Mark W.J. Ferguson)*

A controversy arose associated with the number of stripes on alligators. It was proposed that genes controlled the number of stripes whereas developmental biologists said they were not specifically involved but that it was the developmental process and temperature that determined the number. The issue was resolved using a reaction–diffusion mechanism Deeming and Ferguson (1989c); Murray et al. (1990) suggested specific detailed experiments. Males in general have one more stripe than females and geneticists insisted this was controlled by genes although no specific gene was suggested or found. The alligator embryo is a particularly convenient embryo to study and manipulate since development takes place in the egg external to the adult. Ferguson and his coauthors (Deeming and Ferguson, 1989c; Murray et al., 1990, see also Section 8) studied this stripe patterning on alligators and their model suggested specific experiments at different stages of the embryo's development that subsequently confirmed their theoretical predictions that genes did not control the number. Figure 8(c) shows a typical embryo. There is generally 20 stripes on the female and 21 on the male.

In regard to spatial pattern, which only shows up at a later stage, we often do not know what the actual mechanism is in the developmental process. A major drawback is that we also often do not know *when* in development the pattern generating mechanism is operative; we only observe the results. Because of the relative ease of embryonic manipulation and reliability of growth data obtainable with alligator embryos, Deeming and Ferguson (1989c) and Murray et al. (1990) considered the stripe patterning on the alligator to try and determine the time of initiation of the patterning mechanism and to quantify the effect of size on the stripe pigmentation pattern. It is in part their work that we describe here. The theory (Murray et al., 1990) suggested specific experiments (Deeming and Ferguson, 1989c; Murray et al., 1990) to resolve the question in development, namely the role of genetics in pattern determination, as it applies to the stripe pattern formation on alligators. As mentioned, males generally have one more stripe than females. What they showed was that genetics does *not* play a role in the detailed patterning mechanism.

A major attribute of the *crocodilia* is that incubation temperature determines sex (Deeming and Ferguson, 1989e,c; Murray et al., 1990). Females come from eggs incubated at lower temperatures, less than approximately 32°C, and males at temperatures greater than approximately 32°C. Females at 30°C have fewer white stripes than males, incubated from eggs incubated at 33°C. During development, the pattern is first apparent at approximately 41–45 days of incubation. The gestation period is from 65 to 70 days. The optimal temperature for both females and males is in the region of 32°C.

To discover a real biological pattern formation mechanism from experiment, it is essential to know when (and of course where) during development to look for it: it is too late after we see the pattern. Murray et al. (1990) counted the number of white stripes along the dorsal (top) side of alligator hatchlings, from the nape of the neck to the tip of the tail. The number of stripes on the body (nape to rump) and on the tail (rump to tail tip) was recorded together with the colour of the tail tip. The total length of the animal, the nape–rump length and rump–tail tip length were also measured to the nearest 0.1 mm at various times during development. Hatchlings from two incubation temperatures, 30°C and 33°C (which resulted respectively in 100% female and 100% male hatchlings), were examined (these were identical animals to those examined by Deeming and Ferguson Deeming and Ferguson (1989c)).

To investigate the effects of sex on pigmentation pattern, specifically the number of stripes, hatchlings from eggs of a pulsed 'shift twice' experiment (Deeming and Ferguson, 1989c) were analysed. In these 'shift twice' experiments, eggs were incubated at 33°C, except for days 7 to 14 when they were incubated at 30°C. This incubation treatment produced 23 male and 5 female hatchlings despite the male-inducing temperature of 33°C for the rest of the incubation (Murray et al., 1990). The fact that there were some females is that sometime during this time period and with this temperature the sex was determined to be female.

In a second experiment various measurements of embryos, at 30°C and 33°C, were taken from days 10 to 50 of incubation. These included total length of the animal, nape–rump length and rump–tail tip length (Murray et al., 1990). The embryos were assigned a stage of development. Regression estimates were calculated for embryo growth at the two temperatures. Measurements were also taken for a third group of embryos incubated for 32, 36, 40, 44, 48 and 52 days. They found that temperature clearly affected the pigmentation pattern of hatchling alligators (Murray et al., 1990). There was a higher number of stripes on animals incubated at 33°C compared to those incubated at 30°C. Those animals with a white tip to their tail had, on average, one more stripe than those with a black tip at both temperatures. Generally, there are 8 stripes on the body and 12 on the tail.

The number of stripes was *not genetically sex linked*: male hatchlings from eggs incubated at 33°C (30°C between 7 and 14 days) had a mean number of stripes of 19.96 (+/− 1.15), whereas females from the same treatment had 20.00 (+/− 0.71) stripes; see Murray et al. (1990) for experimental details and data and the study by Murray (2002) for further and more detailed discussion.

4. Concluding discussion

This has been a very short and personal choice from the vast literature associated with reaction–diffusion models of biological pattern formation. With most of the research conclusions speculative, there was a decrease in the new applications of reaction–diffusion models in the 1980s since demonstrating the existence of specific morphogens was proving elusive. This resulted in a new mechanochemical theory of biological pattern and form (Murray, 1989, 2002; Murray et al., 1983; Murray and Oster, 1984; Oster et al., 1983d), which is very briefly described below. From the mid-1990s on, the practical use of reaction–diffusion models has again vastly increased as has research and developments of the Murray–Oster mechanochemical theory of pattern formation.

Models and their biological predictions encouragingly have been a stimulant for guiding critical experiments that have resulted in significant discoveries. This, of course, should be the aim of any mathematical biology modelling, namely to stimulate in any way whatsoever any endeavour that results in furthering our understanding of biology.

Although we now know a lot about pattern development, most mechanisms are still far from being fully understood. We do not know, for example, the complete mechanisms of how cartilage patterns in developing limbs are formed or the specialised structures in the skin such as feathers, scales, glands and hairs, or the myriad of patterns on butterfly wings. The list is endless. Many of the rich spectrum of patterns and structures observed in development evolve from a homogeneous mass of cells that are orchestrated by genes that initiate and control the pattern formation mechanisms: genes themselves are not involved in the actual physical process of pattern generation. The basic philosophy behind practical modelling is to try to incorporate the physico-chemical events, which from observation and experiment appear to be going on during development, within a model mechanistic framework that can then be studied mathematically and, importantly, the results related back to the biology. These morphogenetic models provide the embryologist with possible scenarios as to how, and often when, pattern is laid down, how elements in the embryo might be created and what constraints on possible patterns are imposed by different models.

These mechanochemical models – referred to as the Murray–Oster mechanochemical theory – assume that cells move in response to external physical and chemical guidance cues and form spatial patterns; see the books of Murray (1989, 2002) for a full discussion. Oster and Murray's mechanochemical approach directly brought forces and known measurable properties of biological tissue into the morphogenetic pattern formation process. The mechanisms start with known experimental facts about embryonic cells and tissue involved in development. They were used to construct model mechanisms that reflect these physical facts. Basically, they took the view that mechanical morphogenetic movements themselves create the pattern and form. The models try to quantify the co-ordinated movement and patterning of populations of cells. The models are based on the experimental observations that early embryonic dermal cells are capable of independent movement and have the ability to generate traction forces through long finger-like protrusions called filopodia. These can attach to adhesive sites on the tissue's extracellular matrix (ECM) and thus pull themselves along; at the same time, they deform the ECM. This cell traction is resisted by the viscoelasticity of the ECM. The orchestration of the various physical effects can generate spatial aggregation patterns in cell number density and the models show how the parameters affect the size and shape of the patterns and when they can form. Here pattern formation and morphogenesis occur *simultaneously* as a single process. Work on these models have resulted in understanding and suggesting numerous examples of developmental constraints that have been confirmed by experiment and have suggested new avenues of research.

Both the mechanochemical models and reaction–diffusion models have been fruitfully applied to a vast range of biological problems in morphogenesis and elsewhere, including feather primordia arrangement, wound healing, wound scarring, cartilage formation, shell and mollusk patterns and many others. It is almost certain that both mechanisms are involved in development and although they are in a sense competing theories, I do not think of them as such but rather mechanisms that complement each other. Perhaps, the most fundamental difference between the theories is that the elements involved directly in the mechanical theory are all real biological quantities, namely cells, tissue and the forces generated by the cells. *All* quantities involved are measurable. In the end, however, a crucial aspect of these mechanisms is their ability to predict the subsequent form and pattern that can be verified experimentally.

Modelling of pattern and form in biology and medicine has reached the stage where different mechanisms can generate the observed biological patterns. The question is how to distinguish between them so as to determine which may be the relevant mechanism in vivo. These different models, and explanations for how pattern arises, have suggested different experiments that have been shown to lead to a greater understanding of the biological processes involved. The final arbiter of a model's correctness and usefulness is not so much in what biological patterns it generates (although a first necessary condition for any such model is that it must be able to produce biologically observed patterns), but in how consistent it appears in the light of subsequent experiments and observations.

The explosion in biochemical techniques over the last several decades has led to a still larger increase in our biological knowledge but has partially eclipsed the study of the intermediate mechanisms that translate gene influence into chemicals, into gradients and into pattern and form. As a result, there is much still to be done in this area, both experimentally and theoretically.

We have clearly only scratched the surface of a huge, important and burgeoning interdisciplinary world. Biology, in its broadest sense, is obviously *the* science of the foreseeable future. What is clear is that the application of mathematical modelling in the biological, medical, ecological and psychological sciences is going to play an increasingly important role in future major discoveries and control strategies. There is an ever increasing number of areas where theoretical modelling is important such as social behaviour, conservation in animals when their environment is changed through human land use and so on. In the case of zebras, for example, Rubenstein (2010) shows, by unravelling how species adapt to specific environmental changes, such as land use, why, of the two types of zebra (shown in Fig. 1) the Grevy zebra is nearing extinction while the other has adapted its behaviour to survive. Behavioral ecology is another important expanding area of research. How bird flocks, schools of fish and so on reach community decisions is another exciting relatively new area. In the case of how fish reach communal decisions see, for example, the work of Leonard et al. (2010). Human social sciences is another area that will increasingly use mathematical modelling: one example is in marital interaction and divorce prediction which has already proved extremely useful in designing a new and successful scientific marital therapy (Murray, 1989). Another positive development, although still, very much in its infancy and not generally accepted, is the realisation that in medical training, medical trials and so on there is often a singular lack of true scientific process: the controversy over the use of PSA (prostate-specific antigen) tests and prostate cancer is one example (Swanson et al., 2001) and brain tumour is another (Murray, 2002). A crucially important aspect of all this type of mathematical or theoretical biological research is its genuine interdisciplinary content. There is no way mathematical modelling could solve major biological problems on its own. On the other hand, it is highly unlikely that even a reasonably complete understanding could come solely from experiment.

References

Allen, W.L., Cuthill, I.C., Scott-Samuel, N.E., Baddeley, R., 2011. Why the leopard got its spots: relating pattern development to ecology in felids. Proc. R. Soc. B 278, 1373–1380.

Brenner, S., Murray, J.D., Wolpert, L. (Eds.), Theories of Biological Pattern Formation. In: Proceedings of the Royal Society meeting of that name at the Royal Society, London, 1981. Royal Society, London.

Chaplain, M.A.J., Singh, G.D., Maclachlan, J.C., 1999. On Growth and Form Spatio-temporal Pattern Formation in Biology, John Wiley and Son, Ltd., Chichester, England.

Deeming, D.C., Ferguson, M.W.J., 1989. In the heat of the nest. New Sci. 25, 33–38.

Deeming, D.C., Ferguson, M.W.J., 1989. The mechanism of temperature dependent sex determination in crocodilians: a hypososis. Am. Zool. 29, 973–985.

Gierer, A., Meinhardt, H., 1972. A theory of biological pattern formation. Kybernetik 12, 30–39.

Jäger, W., Murray, J.D. (Eds), Modelling of Patterns in Space and Time. In: Proceedings of a workshop of that name in Heidelberg, 1983 Springer-Verlag, Heidelberg, 1984.

Kondo, S., Iwashita, M., Yamaguchi, M., 2009. How animals get their skin patterns: Fish pigment pattern as a live Turing wave. Inst. J. Dev. Biol. 53, 851–856.

Leonard, N.E., Paley, D.A., Davis, R.E., Fratantoni, D.M., Lekien, F. Zhang, F., 2010. Coordinated control of an underwater glider fleet in an adaptive ocean sampling field experiment in monterey bay, J. Field Robot. 27, 718–740.

Levin, S.A., 1992. The problem of pattern and scale in ecology, Ecology 73, 1943–1967.

Maini, P.K., 2003. How the mouse got its stripes. Proc. Nat. Acad. Sci. USA 100, 9656–9657.

Maini, P.K., 2004. Using mathematical models to help understand biological pattern formation, C. R. Biol. 327, 225–234.

Maini, P.K., Othmer, H.G. (Eds.), 2001. Mathematical Models for Biological Pattern Formation Mathematics and its Applications IMA, vol. 121. Springer, New York.

Mimura, M., Murray, J.D., 1978. Spatial structures in a model substrate-inhibition reaction diffusion system. Z. Für. Naturforsch 33c, 580–586.

Murray, J.D., 1981. A pre-pattern formation mechanism for animal coat markings. J. Theor. Biol. 88, 161–199.

Murray, J.D., 1981. On pattern formation mechanisms for lepidopteran wing patterns and mammalian coat markings. Phil. Trans. Roy. Soc. Lond. B 295, 473–496.

Murray, J.D., 1989. Mathematical Biology, Springer-Verlag, Heidelberg.

Murray, J.D., 1990. Turing's theory of morphogenesis - its influence on modelling biological pattern and form. Bull. Math. Biol. 52, 119–152.

Murray, J.D., 2003. Mathematical Biology, third ed, vol. 2. Mathematical Biology: I. An Introduction, Springer, New York, 2002. Mathematical Biology: II. Spatial Models and Biomedical Applications. Springer, New York, 2003.

Murray, J.D., Deeming, D.C., Ferguson, M.W.J., 1990. Size dependent pigmentation pattern formation in embryos of Alligator mississippiensis: time of initiation of pattern generation mechanism. Proc. Roy. Soc. Lond. B 239, 279–293.

Murray, J.D., Oster, G.F., Harris, A.K., 1983. A mechanical model for mesenchymal morphogenesis. J. Math. Biol. 17, 125–129.

Murray, J.D., Oster, G.F., 1984. Generation of biological pattern and form. IMA J. Maths. Appl. Medic. Biol. 1, 51–75.

Nijhout, H. F., Maini, P. K., Madzvamuse, Wathen, A.J., Sekimura, T. Pigmentation pattern formation in butterflies: experiments and models. , 2003. C. R. Biologies 326, 717–727.

Oster, G.F., Murray, J.D., Harris, A.K., 1983. Mechanical aspects of mesenchymal morphogenesis. J. Embryol. Exp. Morph. 78, 83–125.

Painter, K.J., Maini, P.K., Othmer, H.G., 1999. Stripe formation in juvenile Pomacanthus explained by a generalized Turing mechanism with chemotaxis. Proc. Nat. Acad. Sci. USA 96, 5549–5554.

Prigogine, I., Nicolis, G., 1967. On symmetry-breaking instabilities in dissipative systems. J. Chem. Phys. 46, 3542–3551.

Rubenstein, D.I., 2010. Ecology, social behavior, and conservation in zebras. In: Macedo, R. (Ed.), Advances in the Study Behavior: Behavioral Ecology of Tropical Animals, vol. 42. Elsevier Press, Oxford, UK, pp. 231–258.

Saint-Hilaire, G. Traité de Tératologie, vol. 1–3. Baillière, Paris, 1836.

Suzuki, N., Hirata, M., Kondo, S., 2003. Traveling stripes on the skin of a mutant mouse. Proc. Nat. Acad. Sci. USA 100, 9680–9685.

Swanson, K.R., Murray, J.D., Lin, D., True, L., Buhler, K., Vassella, R., 2001. A quantitative model for the dynamics of serum prostate specific antigen as a marker for cancerous growth: an explanation of a medical anomaly. Amer. J. Pathol. 158, 2195–2199 (Invited editorial).

Thomas, D., 1975. Artificial enzyme membranes, transport, memory, and oscillatory phenomena. In: Thomas, D., Kernevez, J.P. (Eds.), Analysis and Control of Immobilized Enzyme Systems, Springer-Verlag, Berlin-Heidelberg-New York, pp. 115–150.

Thompson, D.W. On Growth and Form, second ed. 1942. Cambridge University Press, 1917.

Turing, A.M., 1952. The chemical basis of morphogenesis. Phil. Trans. Roy. Soc. B 237, 37–72.

Wolpert, L., 1969. Positional Information and the Spatial Pattern of Cellular Differentiation. J. Theor. Biol. 25, 1–47.

Wolpert, L., 2006. Principles of Development, Oxford University Press, Oxford.

Peter T. Saunders observes Alan Turing —

DEFEATING THE ARGUMENT FROM DESIGN

1. Introduction

The obvious question to ask about 'The Chemical Basis of Morphogenesis' is why Turing took up the problem at all. Pattern formation, interesting though it may be to biologists, does not look like the sort of fundamental problem that Turing characteristically chose to devote his time and effort to. The answer is simply that he saw it not as a mere puzzle but as a way of addressing what he considered to be a crucial issue in biology. As he said to his student Robin Gandy, his aim was to 'defeat the argument from design'.

2. The argument from design

The argument from design is often put forward as scientific proof that God exists. As William Paley, one of the late eighteenth century 'natural theologians' put it, if we were to find a watch, composed as it is of a large number of parts, all fitting so precisely together and combining to keep accurate time, then we would know that somewhere there must be a watchmaker. In the same way, if we look at an organism, we cannot but conclude that there must be a Creator.

Turing was not, of course, setting out to refute Paley. That had been done almost a century before by Charles Darwin, who had shown how natural selection could (not, by the way, that it *does*) explain the evolution of organisms. Darwin had pointed out that while offspring generally resemble their parents, variations do sometimes occur. Many of these are of no importance, some are harmful. Even if the variations are completely random, however, a few are bound to be beneficial, that is, they will increase the probability of survival and reproduction. The fortunate individuals in which they occur will leave more offspring than the rest, and over many generations, the advantageous variation will spread through the population, replacing the original trait.

Note that for Darwin's argument to defeat the natural theologians, it is sufficient that the variations can be random. It is not necessary that they *must* be. Darwin himself wrote, in the *Origin of Species*:

"I have hitherto sometimes spoken as if the variations ... were due to chance. This, of course, is a wholly incorrect expression, but it serves to acknowledge plainly our ignorance of the cause of each particular variation."

Contemporary evolutionists see things differently. To be sure, few if any of them claim explicitly that the variations are totally random. On the contrary, when challenged they insist that when they speak of 'random', they mean random with respect to the needs of the organism. They do, however, carry out their research as though the variations were completely random. That leaves natural selection as the only non-random effect in evolution and so it is to natural selection alone that we are to look for explanations.

Any heritable trait is to be explained by showing how it is or might be useful to the organism and so would have been selected, that is to say by much the same sort of argument that was used by the natural theologians to justify their belief in a beneficient God. Thus despite Darwin, the argument from design persists. As Julian Huxley (1942) put it: '*Paley redivivus*, one might say, but philosophically upside down, with Natural Selection replacing the Divine Artificer as the *deus ex machina*'.

One of the most trenchant critics of this approach was the Scottish biologist D'Arcy Thompson. In his classic, *On Growth and Form,* Thompson (1917) urged biologists to seek to explain form in the same ways that physicists do:

"The waves of the sea, the little ripples on the shore, the sweeping curve of the sandy bay between the headlands, the outline of the hills, the shape of the clouds, all these are so many riddles of form, so many problems of morphology, and all of them the physicist can more or less easily read and adequately solve; solving them by reference to their antecedent phenomena, and in the material system of mechanical forces to which they belong and to which we interpret them as being due. They have also, doubtless, their *immanent* teleological significance; but it is on another plane of thought from the physicist's that we contemplate their intrinsic harmony and perfection, and 'see that they are good'.

"Cell and tissue, shell and bone, leaf and flower, are so many portions of matter, and it is in obedience to the laws of physics that their particles have been moved, moulded and conformed. They are no exception to the rule Θεὸς ἀεὶ γεωμετρεῖ. Their problems of form are in the first instance mathematical problems, their problems of growth are essentially physical problems, and the morphologist is, *ipso facto*, a student of physical science."

Turing had read *On Growth and Form* many years before. It is one of only six references in the paper, and we can hear an echo of the passage quoted above in the abstract:

'The theory does not make any new hypotheses; it merely suggests that certain well-known physical laws are sufficient to account for many of the facts'.

Turing did not expand on this remark; that was left to a non-technical paper that he was preparing with the botanist C.W. Wardlaw but which was never published. Wardlaw, who had probably done most of the actual writing, published a slightly different version (Wardlaw, 1953), but this does not seem to have been noticed by most of those who were interested in 'The Chemical Basis of Morphogenesis'. As a result, while Turing's work has been very influential in mathematical biology, it has so far had little effect on the problem he was actually addressing.

3. Defeating the argument from design

To illustrate how pattern and form can often be explained by the physical and chemical processes that produce them, Turing turned to phyllotaxis, the arrangement of leaves on the stem of a plant. This is a classic problem, and also one that he clearly expected to be able to solve with the tools available at the time. Today, when almost everyone has access to sufficient computing power to solve the reaction–diffusion equations numerically, we can find easier examples.

Think of a herd of Friesian cattle. They all have black and white coats with the colours in large irregular patches. Each individual, however, has its own unique pattern; no two are identical.

It is easy to propose an adaptationist argument to account for this. Intra-species differences in coat patterning must have evolved because of the selective advantage they give through 'kin recognition' (Hepper, 1991). The idea is that it is important for animals to be able to distinguish their close relatives from others of the same species so that they can favour them through kin selection and also avoid inbreeding.

Turing's work suggests another explanation. Murray (1982) has shown that the reaction–diffusion model can generate quite realistic mammalian coat patterns. The bifurcation parameters (two because the pattern is two dimensional) are the dimensions of the embryo at the time the pattern is laid down. Domains that are approximately the same dimensions will have patterns of the same general kind: solid colour, spots, stripes, large patches and so on. The details, however, are very sensitive to the precise shape.

Now we would expect that within a species all embryos at the relevant stage will be similar in size and shape but not precisely the same. Hence if the coat pattern is formed by a reaction–diffusion mechanism, while every individual will have the same general sort of pattern, they should all differ in the details. Which is precisely what we observe in many species, including of course Friesians. It may well be that the animals find it useful to be able to tell their neighbours apart, but that is not at all the same as saying that is why they all look slightly different from each other.

4. Conclusion

While Turing recognised that the interplay between reaction and diffusion was by no means the only significant process in morphogenesis, he thought it was likely to be one of the most important. It now appears that reaction–diffusion is not as common a mechanism in development as he thought. But there are now other models of pattern formation and these make many of the same predictions. Even the vibrations of a thin metal plate can produce patterns similar to those produced by reaction–diffusion (Xu et al., 1983).

Thus, the argument of the previous section does not depend crucially on the assumption that the coat patterns are formed through the reaction–diffusion mechanism; it holds for others as well. What we would like to know, of course, is how generic the result is, i.e., what is the class of mechanisms for which it is valid. For this we would need a classification of the kind that Thom had in mind in what he called generalised catastrophes (Thom, 1972). Even without that, however, 'The Chemical Basis of Morphogenesis' has led to research that can contribute towards Turing's aim of defeating the argument from design. It is just taking longer than he anticipated.

References

Hepper, P.G. (Ed.), 1991. Kin Recognition, Cambridge University Press, Cambridge.
Huxley, J., 1942. Evolution: The Modern Synthesis, Allen and Unwin, London.

Murray, J.D. 1982. A pre-pattern formation mechanism for animal coat markings. J. Theor. Biol. 88, 161–199.
Thom, R. 1972. Stabilité Structurelle et Morphogénèse: Essai d'une Théorie Générale des Modèles, W.A. Benjamin, Reading MA.
Thompson, D.W. 1917. On Growth and Form, Cambridge University Press, Cambridge.
Wardlaw, C.W. 1953. A commentary on Turing's diffusion-reaction theory of morphogenesis. New Phytol. 52, 40–47.
Xu, Y., Vest, C.M., Murray, J.D., 1983. Holographic information used to demonstrate a theory of pattern formation in animal coats. Appl. Optics 22, 3479–3483.

Stephen Wolfram fills out the computational view of —

THE MECHANISMS OF BIOLOGY

In the comparatively sparse history of theoretical biology, this paper stands as something of a milestone. For it provided, for essentially the first time, a potentially mechanistic mathematical model for features of biological morphogenesis. And the notion of reaction–diffusion equations that it introduced has become influential and widespread in biology and elsewhere – though the paper itself languished in almost complete obscurity into well into the 1980s.

Viewed from a modern perspective, however, there is a certain irony to the paper. Did Turing even think about Turing machines when he tried to make a model for how biological systems operate? It seems not. For the main thrust of this paper is precisely to use ideas from traditional continuum physics, and apply them to biological growth. Turing did not know about DNA and digital genomic information. But he even chose to ignore the discrete cellular structure of biological organisms. And instead treated biological systems just like pieces of bulk physical material.

As it turns out, he was partly correct. And indeed over the past 25 years his concept of morphogens has proved to be at least part of the story of biological growth and pattern formation. But what Turing missed was the incredible richness that thinking in terms of computational processes brings to studying biological phenomena. In the paper, Turing actually simulated his reaction–diffusion equations using an early electronic computer. But he viewed the mathematical equations as the true model; the computation was just an approximation, found with all sorts of effort of numerical analysis.

But are equations ultimately the appropriate raw material for modelling biological processes? From my point of view, equations represent just a tiny corner of the possible types of underlying rules by which a system may operate. And in biology, there are plenty of other types of rules that increasingly seem important.

Cellular automata are an example that I happen to have studied extensively. Each cell in a cellular automaton may represent many biological cells. But the point is that there is a rule that governs the behaviour of these cells and that can be thought of as corresponding to a simple program. When one looks at different simple programs – in the computational universe of all possibilities – there is great diversity in the behaviour one sees. But a remarkable observation is that this diversity seems not dissimilar to the kind of diversity that we observe in biological systems. And in cases such as mollusc shell patterns I have managed to show that there is indeed a detailed correspondence between the behaviour of possible cellular automaton programs and the diversity we actually observe in nature.

Fig. 1: Examples of patterns produced by the evolution of each of the simplest possible symmetrical one-dimensional cellular automaton rules, starting from a random initial condition.

The concept of Darwinian evolution might lead one to assume that whatever processes now give rise to the forms we see in biological systems, they must have been carefully shaped by natural selection.

But the surprising observation that I have made in at least several cases (Wolfram, 2002) is that instead – in the computational universe of all possible underlying rules – it seems that biology in effect just samples essentially all possibilities, distributing the results among the species of the Earth.

In the abstract, one might think that there could never be any real theory in biology, and that instead all features of current organisms must just be the result of endless historical accidents. But instead it increasingly seems that just by knowing the abstract structure of the computational universe, one can understand the different forms that occur across the biological world.

Darwinian evolution tends to imagine continuous variations from one form to another, and predicts various 'missing links' that interpolate between forms. Studying the computational universe gives much more surprising predictions about the diversity of possible forms that can occur. Since Turing's time, most of biological science has shifted from studies of overall forms and behaviour of organisms to a much more molecular scale. Reaction-diffusion processes are, if anything, potentially more directly relevant at this scale than at larger scales.

But in a sense such processes end up only being the lowest level primitives; there is more to the whole architecture of how the systems operate. There is much that we still do not know about that architecture. Genomics and digital information is one key element.

Quite possibly the various features of simple programs are another. The genome, in a sense, provides all sorts of simple programs. But we need to understand the overall characteristics of these programs, and how they lead to what we actually see in biology. Once we truly understand that, it will dramatically change medicine – and likely also much of our traditional human condition.

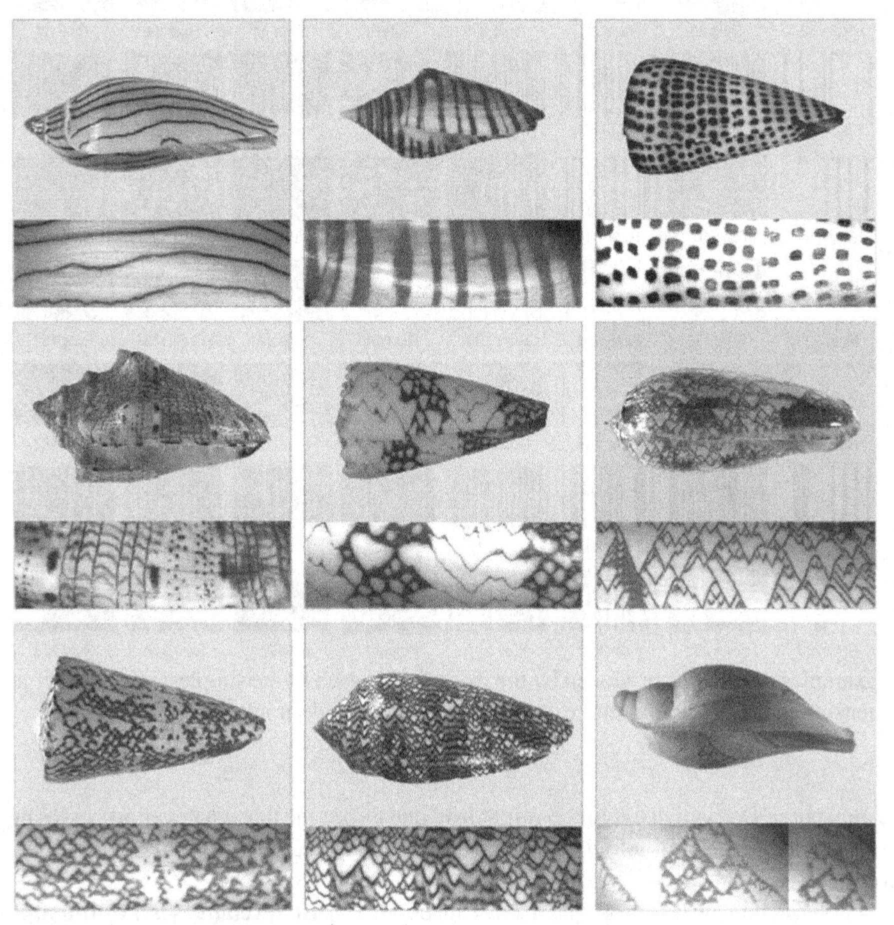

Fig. 2: Typical examples of pigmentation patterns on mollusc shells. Shell patterns show a detailed correspondence with the behaviour of cellular automata.

Morphogens and reaction–diffusion equations will be a small part of what is involved. But a much larger part will come from Turing's most important legacy: the concept of universal computation.

Turing machines are not, in detail, likely to be good models of biological processes – though ribosomes and the like on strands of RNA do seem awfully reminiscent of Turing machines. But the concept of computation is a central one for the future of biology.

Whereas it was the concept of mathematical equations that really launched theoretical physics three centuries ago, it is, I think, the concept of computation that can really launch theoretical biology today.

Reference

Wolfram, S., 2002. A New Kind of Science, Wolfram Media Inc., Champaign, Illinois.

K. Vela Velupillai connects —

FOUR TRADITIONS OF EMERGENCE: MORPHOGENESIS, ULAM-VON NEUMANN CELLULAR AUTOMATA, THE FERMI-PASTA-ULAM PROBLEM, AND BRITISH EMERGENTISM

The classic works by the trio of Mill (1890), Lewes (1891) and Lloyd Morgan (1927), together with Broad (1929) and Alexander (1920), made up what has come to be called the *'British Emergentist'* school. The concept of *emergence* came to have its current connotations as a result of these (and a few other) clearly identifiable sequence of classic works by these pioneering British philosophers. A representative sample of crucial definitions, in these classics, may provide a decent backdrop against which to proceed. In particular, the one by Lewes, the man who introduced the word *'emergent'*, from which Lloyd Morgan derived *'emergence'*:

> "[T]here are laws which, like those of chemistry and physiology, owe their existence to .. *heteropathic laws*... . The Laws of Life will never be deducible from the mere laws of the ingredients, but the prodigiously *complex* Facts of Life may all be deducible from comparatively *simple* laws of life;..."
> Mill (1890) Bk.III, Ch.VI, p.269; italics added.
>
> "Thus, although each effect is the resultant of its components, the product of its factors, we cannot always trace *the steps of the process*, so as to see in the product the mode of operation of each factor. In this latter case, I propose to call the effect *an emergent*. It arises out of the combined agencies, but in a form which does not display the agents in action."
> Lewes (1891) Problem V, Ch.III, p.368, italics added.
>
> "The concept of *emergence* was dealt with .. by J.S.Mill in his Logic .. under the discussion of *'heteropathic laws'* in causation. The word *'emergent'* as contrasted with *'resultant,'* was suggested by G.H.Lewes in his Problems of Life and Mind'. .... What makes emergents emerge? .. *What need [is there] for a directive Source of emergence. Why should it not proceed without one?"*
> **?**, pp. 2, 32; italics added.

The rise and fall of *British Emergentism* has been eloquently and almost persuasively argued in the study by Brian McLaughlin (1992), basing himself on the emerging (sic!) codification of quantum mechanics in the works of Heisenberg, Schrödinger, Dirac and Pauling, and on the philosophical critiques of the 1920s, launched primarily by Pepper (1926), Stace (1939) and Bayliss (1929). Remarkably, it is possible, with a good dose of hindsight, to pinpoint the reason for McLaughlin's premature obituary of *British Emergentism* and, at the same time, link that failure with a prescient – and typically penetrating, yet almost playfully formulated – observation by Alan Turing in his last publication before a tragically truncated life came to an end.

In what can only be called a moment of weakness, because Dirac, surely, is not capable of carelessness or flippancy, one of the great founding fathers of modern quantum mechanics, slipped badly in a pronouncement that was the fulcrum around which the premature obituary of *British Emergentism* was proclaimed:

> "The underlying physical laws necessary for the mathematical theory of a large part of physics and *the whole of chemistry are thus completely known*[1], and the difficulty is only that the exact application of these laws leads to equations much too complicated to be soluble. It therefore becomes desirable that approximate practical methods of applying quantum mechanics should be developed, which can lead to an explanation of the main features of complex atomic systems *without too much computation*."
> Dirac (1929), p. 714; italics added.

Contrast this with Turing's wonderfully laconic, yet eminently sensible precept (Turing, 1954, p. 9; italics added):

> "No mathematical method can be *useful* for *any* problem *if it involves much calculation*."

It is precisely in this sense that Scerri (1994) – and in a series of perceptive writings on the failure of reductionism, in general, and the untenability of McLaughlin's thesis (Scerri and Lee McIntyre, 1997, Scerri, 2007) – has made his case against Dirac's unfortunate claim. The *British Emergentists* were prescient in their approach to the formalisation of emergence, coupled to the dialectic between the simple and the complex, in a natural dynamic context. They *rise and rise*; there was **never any fall** of the *British Emergentisits*!

Turing's remarkable original work on *The Chemical Basis of Morphogenesis* was neither inspired by, nor influenced any later allegiance to the British Emergentist's tradition – such as the influential experimental and theoretical neurological and neurophilosophical work of Nobel Laureate, Roger Sperry. [2]

On the other hand, the structure of the experimental framework Turing chose to construct was uncannily similar to the one devised by Fermi, Pasta and Ulam (1955) although with different purposes in mind. But there was – and there remains – a deeper affinity in that the violation of the equipartition of energy principle that was observed in the Fermi–Pasta–Ulam simulation and the symmetry breaking that is intrinsic to the dynamical system behaviour of Turing's system of reaction–diffusion equations.

Turing's aim was to devise a mechanism by which a spatially homogeneous distribution of chemicals – i.e., formless or patternless structure – could give rise to form or patterns via what has

---

[1] This must rank with the celebrated but, mercifully, falsified prophetic pontifications by two other intellectual giants of the 19th century: Lord Kelvin and John Stuart Mill. The former is reputed to have *suggested*, on the eve of the works by Planck and Einstein, that *all the problems of physics had been solved*', except for just two anomalies: the Michelson-Morley experiment, on the one hand, and Black Body radiation, on the other'! As for the great and saintly John Stuart Mill, in what can only be called an unfortunate moment of weakness, he etched for posterity theese un-prophetic thoughts (Mill, 1848, [1898], Bk. III, Ch. 1, p. 266; italics added):

> "*Happily, there is nothing in the laws of Value which remains for the present writer to clear up; the theory of the subject is complete*: the only difficulty to be overcome is that of so stating it as to solve by anticipation the chief perplexities which occur in applying it: and to do this, some minuteness of exposition, and considerable demands on the patience of the reader, are inevitable."

[2] See, in particular, Sperry's outstanding Noble Prize Lecture, delivered on 8 December, 1981, on the nature of the emergence of consciousness and its relation to brain processing.

come to be called a *Turing Bifurcation*. A reaction–diffusion mechanism formalised as a (linear) dynamical system and subject to what I have referred to, in other writings, as *the linear mouse theory of self-organisation.* [3].

As young boy, Alan Turing won the *Morcom science prize*[4] for his work on the study of 'the reaction between *iodic acid* and sulphur dioxide' (cf., Hodges, 1983, p. 52; italic added). Indeed, even as a 12-year-old boy, during Christmas holidays spent at the family villa in the Rue du Casino[5], Alan Turing had 'heaved great quantities of sea-weed .. from the beach in order to extract a minute amount of iodine' (ibid, p. 18). It is, therefore, particularly satisfying to note that the kind of patterns suggested by Turing's theory of morphogenesis was first definitively established in *iodine reactions*, in the work of Castets et al. (1990) and Ouyang and Swinney (1991). How much more serendipitous could events be?

In this same vein, it is most satisfying to note the role the *Turing Bifurcation* played in the development of the *Brusselator* and the work of the 1977 Chemistry Nobel Prize winner, Ilya Prigogine (cf. Nicoils and Prigogine, 1977) on self-organisation in non-equilibrium systems. I have come to try to characterise, at least for the purpose of classifying in some systematic way, the contributions to emergence in terms of: (i) Novelty; (ii) Irreducibility; (iii) Unpredictability; (iv) Non-reductive Physicalism; (v) Downward Causation.

These are the categories that played decisive roles in the emergence literature that originated in the work of the British Emergentists. Perhaps the time is apposite for a reconsideration of the philosophical underpinnings of Turing's methodology for morphogenesis. If so, then it is the basis in the work of the British Emergentists, and their above characterising categories, that one may find the way forward. This will not be incongruent at all, given that Lloyd Morgon was, among other things, also a zoologist, a pupil of T.H. Huxley and the man who coined the word Emergence in his famous *Gifford Lectures*. Add to this the names of Sperry and Prigogine, and the trio of Fermi, Pasta and Ulam and their experimental structure, and it would be a simple completion of an honour roll when Turing's name is added to the list – and this, too, on the basis of only one of his many fundamental contributions.

I have also found it useful to utilise the following three precise notions for the classifying exercise: *Potential Surprise*, *Computation Universality*, *Mereological Confusion*. In every one of the five classifying categories and the three analytical notions used for the classifying exercise, I have been inspired by some aspect of Turing's work.

Above all, it is by now only too well known that von Neumann's contribution – in his famous joint work with Ulam – to the theory of self-reproducing automata, was almost wholly underpinned by Turing's theory of computation.

---

[3] In typically playful fashion, he summarised the mathematical mechanism he sought (Turing, 1952, pp. 43–4):

> "Unstable equilibrium is not . . . a condition which occurs very naturally. .. Since sytems tend to leave unstable equilibria they cannot often be in them. *Such equilibria can, however, occur naturally through a stable equilibrium changing into an unstable one*. For example, if a rod is hanging from a point a little above its centre of gravity it will be in stable equilibrium. If, however, a mouse climbs up the rod the equilibrium eventually becomes unstable and the rod starts to swing. . . . The system which was originally discussed . . . might be supposed to correspond to the mouse somehow reaching the top of the pendulum without disaster, perhaps by falling vertically on to it."

[4] Christopher Morcom was Alan's dear friend during the very brief period they shared at Sherborne school and the Morcom family, on the unfortunately early death of their son, had endowed, in memory of their son, 'a science prize to be awarded for work which included an element of originality', (Hodges, op.cit., p. 51).

[5] In Dinard, France.

Reference

Alexander, S., 1920. Space, Time and Deity- The Gifford Lectures at Glasgow, vol. I & vol. II. Macmillan and Co, London.

Bayliss, C.A., 1929. The philosophic functions of emergence. Philos. Rev. 38 (4), 372–384.

Broad, C.D., 1929. The Mind and its Place in Nature, The Tarner Lectures delivered at Trinity College, Cambridge, 1923. Kegan Paul, Trench. London: Trubner & Co., Ltd.

Castets, V., Dulos, E., Boissonade, J., De Kepper, P., 1990. Experimental evidence of a sustained standing Turing-type nonequilibrium chemical pattern. Phys. Rev. Lett. 64 (24), 2953–2956.

Dirac, P.A.M. 1929. Quantum mechanics of many-electron systems. In: Proceedings of the Royal Society of London, Series A, vol. 123, # 792, pp. 714–733.

Fermi, E., Pasta, J., Ulam, S., 1955. Studies of Non Linear Problems, Los Alamos Preprint, LA-1940, May.

Hodges, A., 1983. Alan Turing: The Enigma, Burnett Books, London.

Lewes, G.H. 1891. Problems of Life and Mind, Houghton, Miffin & Co, New York.

Lloyd Morgan, C., 1927. Emergent Evolution, second ed. The Gifford Lectures Williams & Norgate.

McLaughlin, B.P., 1992. The rise and fall of british emergentism. In: Beckermann, A., Flohr, H., Kim, J. (Eds.), Emergence or Reduction - Essays on the Prospects of Nonreductive Physicalism, Walter de Gruyter, Berlin, pp. 49–93.

Mill, J.S. 1848; [1898]. Principles of Political Economy with Some of Their Applications to Social Philosophy, Longmans, Green, and Co., London.

Mill, J.S. 1890. A System of Logic, eighth ed. Harper & Brothers Publishers, New York.

Nicolis, G., Prigogine, I., 1977. Self-Organization in Non-Equilibrium Systems, John Wiley & Sons, New York.

Ouyang, Q., Swinney, H.L., 1991. Transition from a uniform state to hexagonal and striped turing patterns, Nature, 352, 610–12.

Pepper, S.C., 1926. Emergence. J. Philos., 23(9) 241–245.

Scerri, E.R. 1994. Has Chemistry Been at Least Approximately Reduced to Quantum Mechanics, Proceedings of the Biennial Meeting of the Philosophy of Science Association, Volume One: Contributed Papers, pp. 160–170.

Scerri, E.R., McIntyre, L., 1997. The case for the philosophy of chemistry. Synthese 111(3), 213–232.

Scerri, E.R., 2007. Reduction and emergence in chemistry – two recent approaches. Philos. Sci. 74(5), 920–931.

Sperry, R.W., 1981. Some Effects of Disconnecting the Cerebral Hemispheres, Nobel Prize Lecture, Stockholm, 8 December, 1981.

Stace, W.T., 1939. Novelty, indeterminism and emergence. Philos. Rev. 48(3), 296–310.

Turing, A.M., 1954. Solvable and unsolvable problems. Sci. News 31, 7–23.

von Neumann, J., 1966. Theory of Self-Reproducing Automata, edited and completed by Arthur W. Burks, University of Illinois Press, Urbana.

Gregory Chaitin takes the story forward —

FROM TURING TO METABIOLOGY AND LIFE AS EVOLVING SOFTWARE

Having applied mathematical methods to the foundations of reason (mathematics) and to the question of how we think, Turing obviously could not resist attempting to increase the scope of mathematical methods in biology, a subject notably resistant to mathematical reasoning.

Newton taught us to use ordinary differential equations in physics, and Maxwell taught us to use partial differential equations. Fisher–Wright–Haldane population genetics makes use of ordinary differential equations, and Turing's work on morphogenesis puts partial differential equations to good use.

Unfortunately, like many pioneers, Turing himself was caught half-way between traditional continuous mathematics and the new world revealed by *On Computable Numbers.* Newton was much more in the Middle Ages[1] than the modern thinker he is portrayed as by Voltaire, and Darwin was less of a Darwinian extremist than many of his determined followers. In a similar manner, Turing could not, dying as he did prematurely in 1954, barely after the work of Watson and Crick in 1953, appreciate the fact that DNA is a powerful digital programming language. Indeed, DNA is presumably a universal programming language, a concept for which we are indebted to Turing.

As I have argued in my work on what I call *metabiology,* following Turing's ideas (but not his own work on biology) suggests modelling life as randomly evolving software, software that describes a random walk of increasing fitness in program space. In this manner, we can discuss biological creativity, something that has gotten lost in the accounts of Darwinian evolution emphasising competition and survival of the fittest.

Sir Ronald Fisher has been referred to as the greatest biologist of the twentieth century because his population genetics gives a mathematical basis for Darwinian evolution. But by definition there is no biological creativity in population genetics, since it deals with a fixed gene pool and merely studies the changes in gene frequencies in response to selective pressures.

How then are we to understand mathematically biological creativity such as the invention of the eye or the transition from unicellular to multicellular organisms? For that it is helpful to think of DNA as randomly mutating computer software and to study the random evolution of artificial software (computer programs) rather than natural software (DNA). In this manner, it is easy to understand the absence of intermediate forms, the fact that ontogeny recapitulates phylogeny, and I have recently even been able to prove that in such a simplified setting random mutations will drive unlimited and unending biological creativity. I achieve this by forcing my organisms to work on mathematical problems for which there are no general methods and unlimited, unending creativity is essential.

Thus, my proposed new field of metabiology, which already has some mathematical successes to its credit, deliberately mixes mathematical and biological creativity in order to enable us to prove

[1] See John Maynard Keynes' famous essay *Newton, the Man* on Newton as the last Babylonian magician.

that evolution works. In my toy model of biology, the *Against Method* moral of the unsolvability of the halting problem is used to drive evolution.

In fact, in one of the versions of metabiology, the organisms that rapidly evolve due to random mutations are better and better approximations to the halting probability Ω – lower bounds in fact. Indeed, all possible versions of the halting probability evolve in parallel. So random mutations yield mathematical creativity; intelligence emerges from randomness. I suspect that this version of metabiology may be the Platonic ideal of evolution that real, messy biological evolution can only approximate asymptotically in the limit from below.

In another version of metabiology, I can show that hierarchical structure will rapidly emerge, which is a conspicuous feature of actual biological organisms.

How would Turing view these developments? This metabiological work on life as evolving software views organisms as machines in the spirit of Turing's paper in *MIND,* but metabiology simultaneously takes advantage of the unsolvability of the halting problem from *On Computable Numbers* to show that evolution is unending. In a sense, we simultaneously use and refute Turing – and we 'refute' him by using his own methods, so that it is clear that the contradictory spirit of Turing very much lives on in this metabiological work.[2]

[2] For more on metabiology see: 'Metaphysics, metamathematics and metabiology' in H. Zenil, *Randomness Through Computation*, World Scientific, 2011, pp. 93–103; 'Life as evolving software', in H. Zenil, *A Computable Universe*, World Scientific, to appear; and G. Chaitin, *Proving Darwin: Making Biology Mathematical*, Pantheon, to appear.

The Morphogen Theory of Phyllotaxis

I. Geometrical and Descriptive Phyllotaxis
II. Chemical Theory of Morphogenesis
III. (Bernard Richards) A Solution of the Morphogenical Equations for the Case of Spherical Symmetry

(Prepared after December 1954 by N. E. Hoskin and B. Richards, using manuscripts of Turing and notes from his lectures in Manchester)

Bernard Richards recalls Alan Turing and —

RADIOLARIA: THE RESULT OF MORPHOGENESIS

1. Introduction

To some it might seem an anomaly that two topics, namely high-speed electronic computers and tiny sea creatures, at opposite ends of the scientific spectrum, can be connected by computer science.

The late Dr. Alan Turing was well known in the UK for his work on cryptography and 'Thinking Machines' but his work in the field of botany and biology is less well-known. What was this latter work? He was able to take further the work of D'Arcy Thompson 'On Growth and Form' (Thompson, 1917), and to offer explanations as to why certain visual appearances were perceived in biological creatures. In 1952 he wrote a now-famous paper entitled 'The Chemical Basis of Morphogenesis' (Turing, 1952). He was aware of some earlier work by Waddington (1940), who invented the word 'evocators', but Turing, wishing to go further, invented the word 'Morphogens'. He proposed that growth in two and three dimensions could be explained by his theories. In two dimensions he explained the black-and-white dappling on cows, and in three dimensions the shapes of some plants, e.g. the woodruff (*Asperula odorata*) and some creatures, e.g. the hydra. He proposed that the many shapes observed in minute sea creatures, the species 'Radiolaria', could be explained by postulating the diffusion of saline into a growing spherical body resulting in tentacles ('spines') growing out at equilibrium. The present author took this postulate of Turing's and set out to prove, or otherwise, the validity of this theory by solving the differential diffusion equations and examining the resultant observable shapes.

2. Radiolaria

Radiolaria are saltwater marine creatures whose unicellular body consists of two main portions supported by a membrane: an inner central capsule and an outer surface in contact with the outside world for feeding and protection. Radiolaria are found in all seas and in all climatic zones. The sphere measures about 2 mm in diameter.

During the development from the spherical newborn, 'spines' (a kind of spike) develop from the surface in pre-determined positions. These resulting shapes give rise to six sub-species of Radiolaria. The spines are usually of a length equal to the radius of the main spherical body and are to be found symmetrically placed over the sphere. For only two spines these are placed at the north and south poles, see Fig. 1. There are no three-, four-, or five spine versions. For six spines, the spines are placed 90° apart, i.e. two at the poles and four around the equator, see Fig. 2. Figure 3 shows another species with six spines. Figure 4 shows a species with 12 spines, whilst Figure 5 shows a specimen with 20 spines. The general shapes taken by these species resemble the regular mathematical solids.

Thus, Fig. 1 has two spines, one at the north pole and one at the south pole. Figure 2 shows a version with six spines. Figure 3 shows another species with six spines, whereas Fig. 4 has 12 spines. Figure 5 has 20 spines, as does Fig. 6.

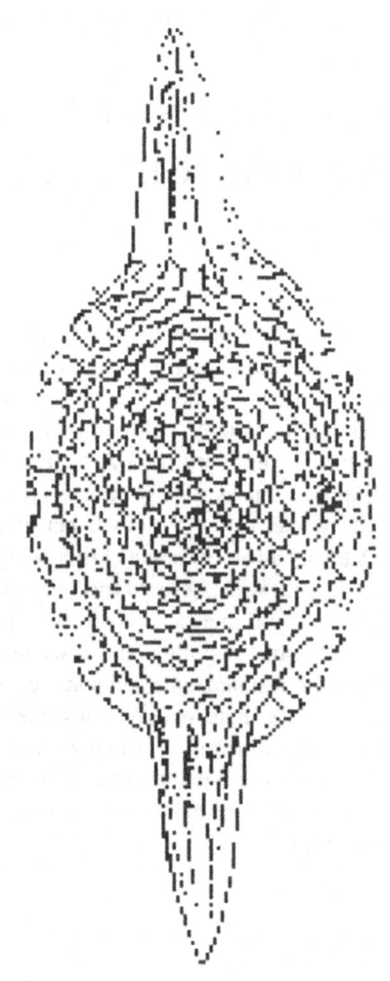

Fig. 1: Cromyatractus tetracelyphus with two spines.

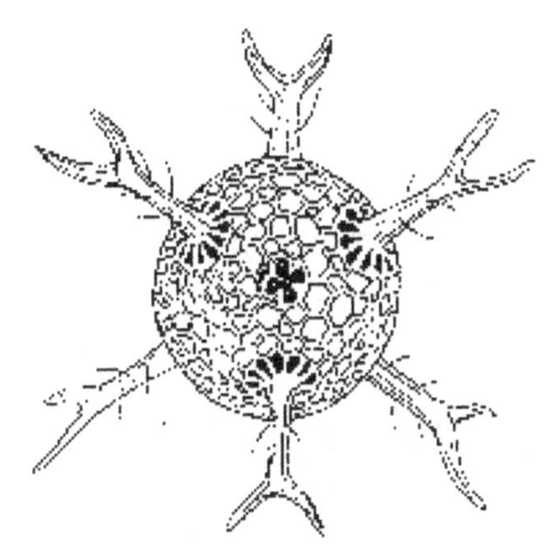

Fig. 2: Circopus sexfurcus with six spines.

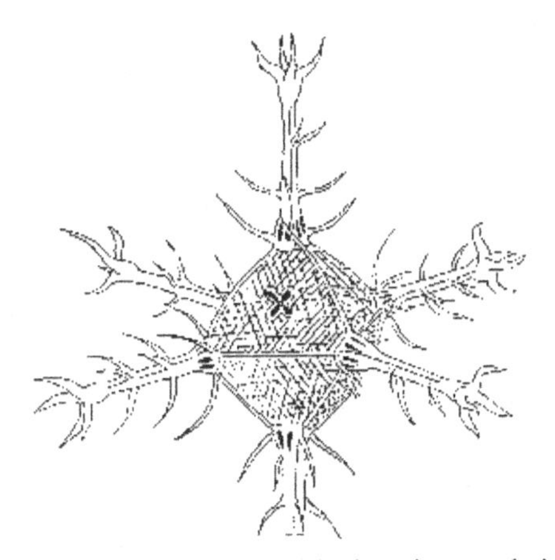

Fig. 3: Circopurus octahedrus with six spines and eight faces.

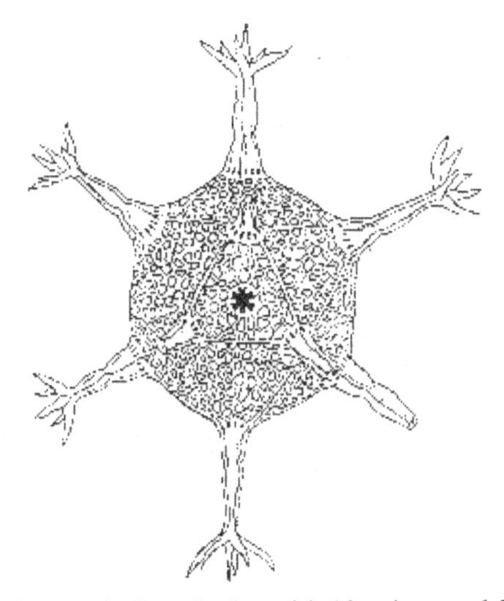

Fig. 4: Circogonia Icosahedra with 12 spines and 20 faces.

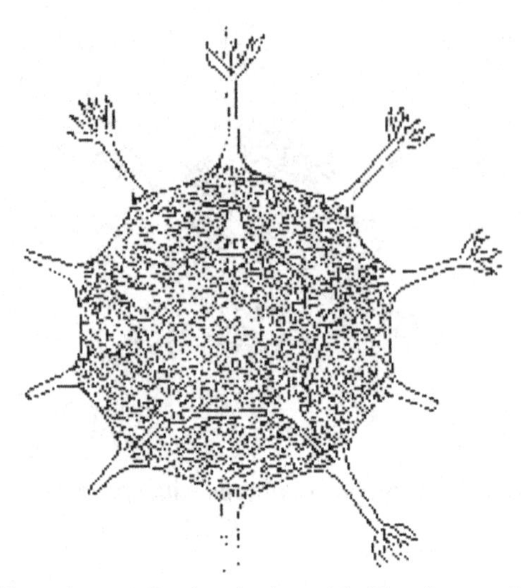

Fig. 5: Circorrhegma Dodecahedra with 20 spines and 12 faces.

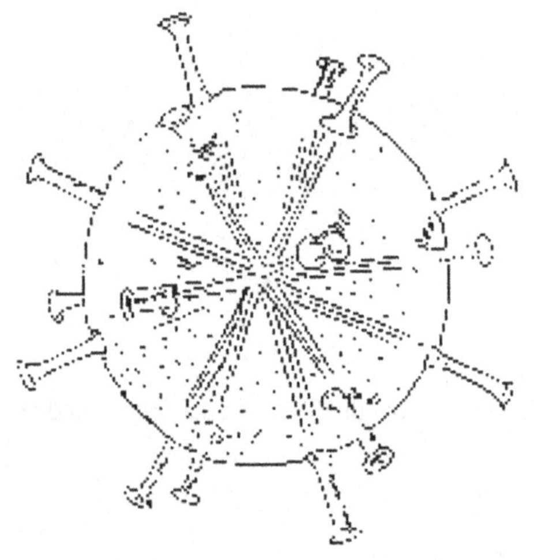

Fig. 6: Cannocapsa Stethoscopium with 20 spines.

3. The differential equation

One can introduce U, a substance which represents a 'growth dimension' e.g. the radius of the sphere, and V an alien invader-chemical which is anti-growth, a sort of poisoning factor. One then sets up the diffusion equations for the state of affairs in the single cell as regards these two substances. It is assumed that V will diffuse uniformly into the cell.

The starting point is therefore the two equations:

$$\frac{dU}{dt} = \Phi(\nabla^2)U + GU^2 - HUV \tag{3.1}$$

$$V = (\tilde{U}^2), \text{ the mean value over the sphere.} \tag{3.2}$$

It can be shown that the solution U must be a linear combination of different eigenfunctions, it is therefore appropriate to assume that the solution is of the form:

$$U(\theta,\varphi,t) = \sum S_m(t) P_n{}^m(\cos\theta) e^{im\varphi} \tag{3.3}$$

with U real and the functions $P_n{}^m(\cos\,\theta)$ being normalised Legendre Associated Functions. These functions satisfy the relation

$$\frac{1}{4\pi}\int\int P_n{}^r(\cos\theta)P_n{}^s(\cos\theta)dS = \{1 \text{ for } r = s, \text{and } 0 \text{ for } r \neq s\}$$

It is possible to evaluate V from (3.2) above by integration since

$$V = \frac{1}{4\pi}\int\int U^2 dS.$$

Now in Eq. (3.1), the function $\Phi(\nabla^2)$ can be replaced by a constant I, since for spherical surface harmonics,

$$\nabla^2 = -n(n+1)/R^2$$

So Eq. (3.1) becomes

$$\frac{dU}{dt} = IU + GU - HUV \tag{3.4}$$

To proceed one has to introduce the Legendre Functions $P_n{}^m(\mu)$, where $\mu = \cos\theta$ and the variables $L_n{}^{p,q,r} = 1/4\pi \int\int P_n{}^p(\mu)P_n{}^q(\mu)P_n{}^r(\mu)^{ei(p+q+r)\varphi}d\mu d\varphi$.

The mathematics hereafter becomes very complicated and very long. It can be found in great detail in some 30 pages in the thesis 'Morphogenesis of Radiolaria', written by the author of this contribution, and in a more concise form in the *Morphogenesis* volume of the *Collected Works* edited by P.T. Saunders (1992).

4. The solutions of the equations

The outcome is that several sets of simultaneous equations are derived depending on the parameter 'n' as used above. These in turn allow the radius U to be calculated.

The case $n = 2$

The solution is

$$\begin{aligned} U &= (7/4)(3\cos^2\theta - 1) \\ V &= 49/20. \end{aligned} \tag{4.1}$$

This is a prolate spheroid whose major axis coincides with the direction $\theta = 0$, and it resembles Fig. 1 above in a small way. The radius at the two poles is therefore 7/2. This is the only solution for the parameter n taking the value 2.

The case $n = 4$

Here there are five simultaneous equations in the S_m to solve with coefficients involving the Legendre Functions as in (3.3). There is a solution involving S_0 and S_4 being non-zero. These give rise to a solution:

$$U = 143/1728\left[6(35\mu^4 - 30\mu^2 + 3) + 30(1-\mu^2)^2\cos 4\varphi\right]. \tag{4.2}$$

This is a spheroid with spines at the two poles and four spines equidistant around the equator, the radius (length) of the polar spines being the same as those on the equator ($\mu = 0, \cos\varphi = 1$). The spine length, measured from the centre of the sphere, is 143/36.

The case $n = 6$

As before, this needs sets of seven simultaneous equations to be solved. One solution takes the form:

$$U = \left(323\sqrt{13}/1300\right)\left[P_6{}^9(\mu) + \left(323\sqrt{1001}/14300\right)P_6{}^5(\mu)\cos 5\varphi\right]. \tag{4.3}$$

It is almost impossible to conjecture what three-dimensional shape this represents. Using the computer (see below) reveals that it is a regular Icosahedron with 12 equal spines separated by the correct angular distance of 62° 24′, two of the spines being at the poles.

The case $n = 8$

This time there are nine simultaneous equations to solve but, as before, one can choose some of the S_i to be zero giving rise to many solutions of the simultaneous equations. One of these shapes will resemble the Dodecahedron analogous to Fig. 6.

5. The part played by the computer

Given a solution for U, e.g., as in Eqs (4.2) and (4.3), it is necessary to discover what three-dimensional shape this produces. This is not an easy thing to do when one not only needs to know where the spines are on the sphere but also the diameter of the sphere and the length of the spines which protrude therefrom. Here the computer played a role.

The computer involved, the Ferranti MARK I, had no visual display output facilities, but only a very primitive line-printer restricted to numerical and alphabetic characters. So it was decided to use that printer to print contour maps of the surface. On an A4 page was displayed an array with values of θ talking values from 0° to 90°, and φ taking values 0° to 350°, whilst on the second page, the values of θ from 90° to 180° were shown. The pages were covered in the teleprinter symbols, each one representing a distance from the centre (a height) on a scale of from zero to 31. Thus the whole surface of the sphere was covered. The writer was then able to draw on these sheets the contour lines, locate the spines and record their lengths. The computer outputs confirmed the statements made for $n = 2$ and 4 above, and also identified the icosahedra and dodecahedra solutions.

6. Comparisons with the marine species Radiolaria

As has been said above, the solution of the differential equations produced a class of solutions. These were evaluated in three dimensions and the corresponding shapes were set against their matches from the species Radiolaria. The matches were very good. Figure 7 shows the solution for $n = 4$, which is Eq. (4.2), superimposed upon the six-spined Circopus Sexfurcus, this having spines at the two poles and four around the equator.

Figure 8 shows the solution for $n = 8$, Eq. (4.3), the computer solution being superimposed on Circogonia Icosahedre. This latter has twelve spines equidistantly spread over its surface.

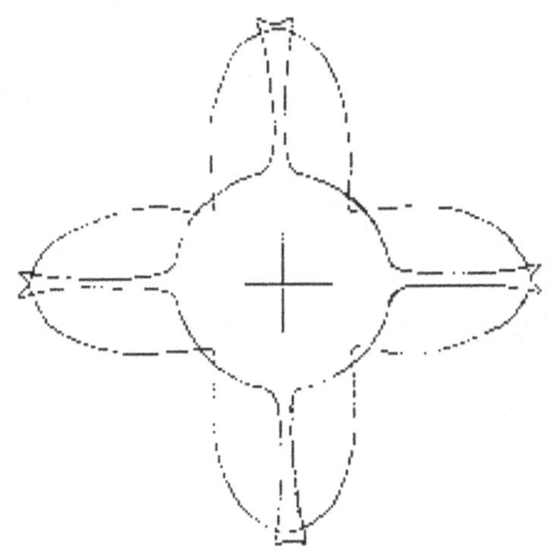

Fig. 7: The computer solution superimposed on Circopus Sexfurcus.

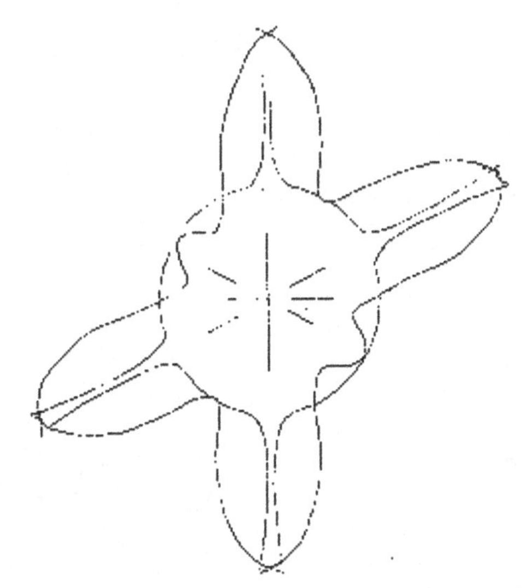

Fig. 8: The computer solution superimposed on Circogonia Icosahedra.

7. Conclusion

This work seems to give credence to Turing's ideas of Morphogenesis, and in particular that the external shapes in the species Radiolaria could be explained by diffusion. Whilst it is a triumph for Turing's theory, it is very sad that he did not see the results detailed above as he died before they were obtained. Nevertheless, they remain as a tribute to his genius, foresight and love of nature.

References

Richards, B., 1954. The Morphogenesis of Radiolaria, M.Sc. Thesis, University of Manchester.
Saunders, P.T. (ed.), 1992. Collected Works of A.M. Turing: Morphogenesis, North Holland Press, London.
Thompson, D'A. W., 1917. On Growth and Form, Cambridge University Press, Cambridge.
Turing, A.M., 1952. The chemical basis of morphogenesis. Trans. Royal Soc. 237 (641), 37–72.
Waddington, C.H., 1940. Organisers and Genes, Cambridge University Press, Cambridge.

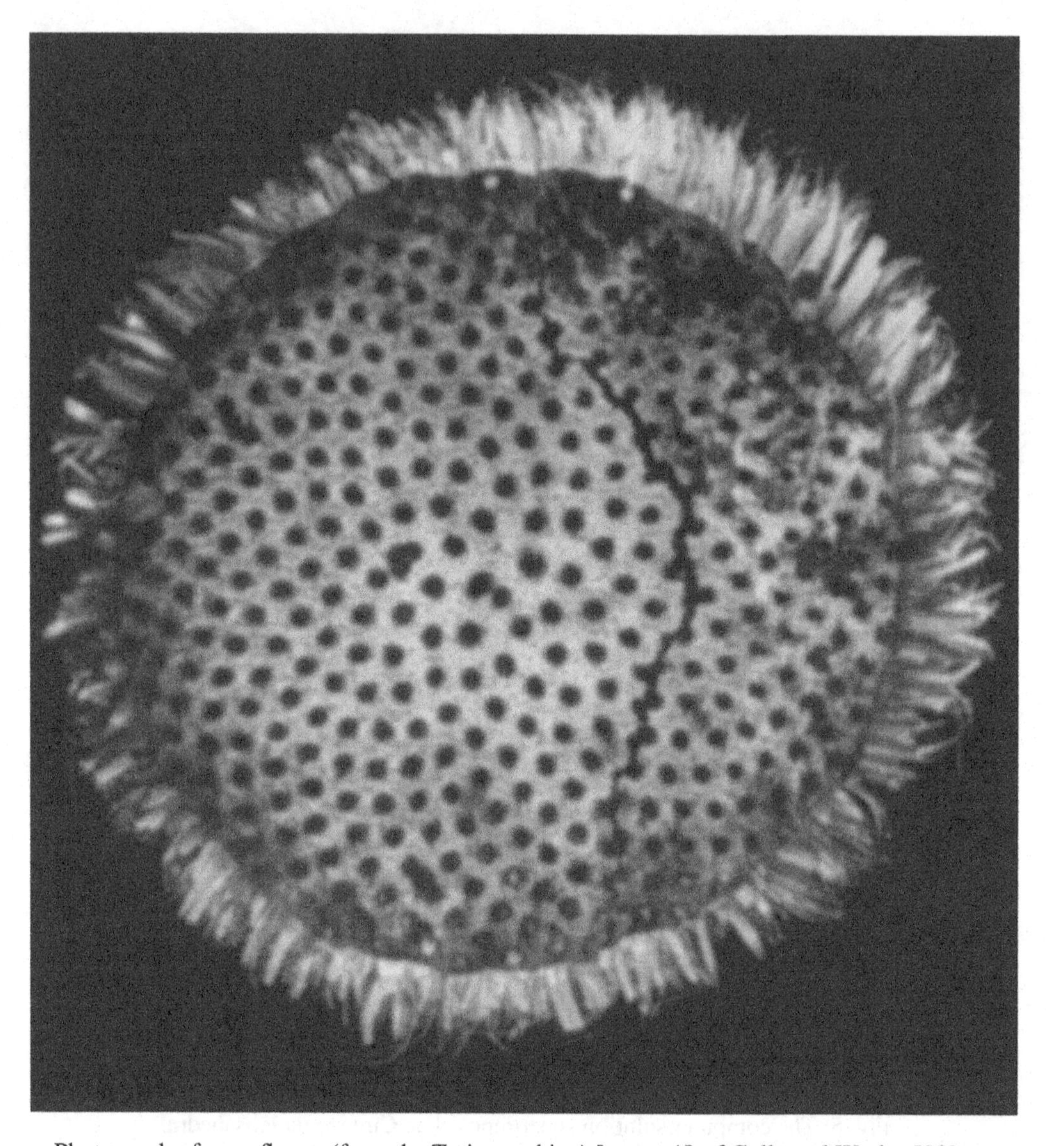

Photograph of a sunflower (from the Turing archive) [see p. 48 of Collected Works, V.3].

THE MORPHOGEN THEORY OF PHYLLOTAXIS[1]

Part I. Geometrical and Descriptive Phyllotaxis

Phyllotaxis deals with the arrangements of leaves on the stems of plants. By a liberal interpretation of the terms "leaf" and "stem" it deals also with the arrangements of florets in a head (e.g. in a sunflower) and with the leaf primordia near the growing point of a bud. All these kinds of patterns will be discussed in the present paper. In the first part, which deals with some of the more superficial problems, the leaves are usually treated as if they were geometrical points distributed on a cylinder. Such patterns on cylinders are appropriate for the description of the mature structures, but their use may be criticised on the grounds that the patterns of real importance are not those formed by the mature structures, but of the leaf primordia. I would indeed go further and say that we should not consider even the primordia but certain patterns of concentration of chemical substances ("morphogens") which are present before there is any visible growth of primordia at all. This criticism is entirely valid, and the second part of the paper takes account of it. Nevertheless, a consideration of the patterns formed by the mature structures is enormously helpful, for a number of reasons.

(1) Suitable specimens of stems with leaves, large and robust enough for convenient examination, can be found almost anywhere, whereas the primordia can only be observed with the aid of a microscope and inconvenient techniques.
(2) The leaf patterns on a mature cylindrical stem are mathematically simpler and more easily intelligible than those near the growing point.
(3) In order to describe the patterns near the growing point satisfactorily it would in any case be necessary to carry through the mathematical theory of cylindrical patterns such as those formed by the mature structures, at least as an abstract discipline.

The method of exposition will be to alternate sections of mathematical theory with sections which describe facts about plants. The purpose of the theory may be lost if it is all given at once, and before any descriptive matter. On the other hand, if an attempt is made to describe first and theorise later the necessary terminology is lacking.

[1]For the citations below, see the bibliography from the *Morphogenesis* volume of the *Collected Works of A. M. Turing*, pp.125–127, reproduced on pp. 832–833 below.

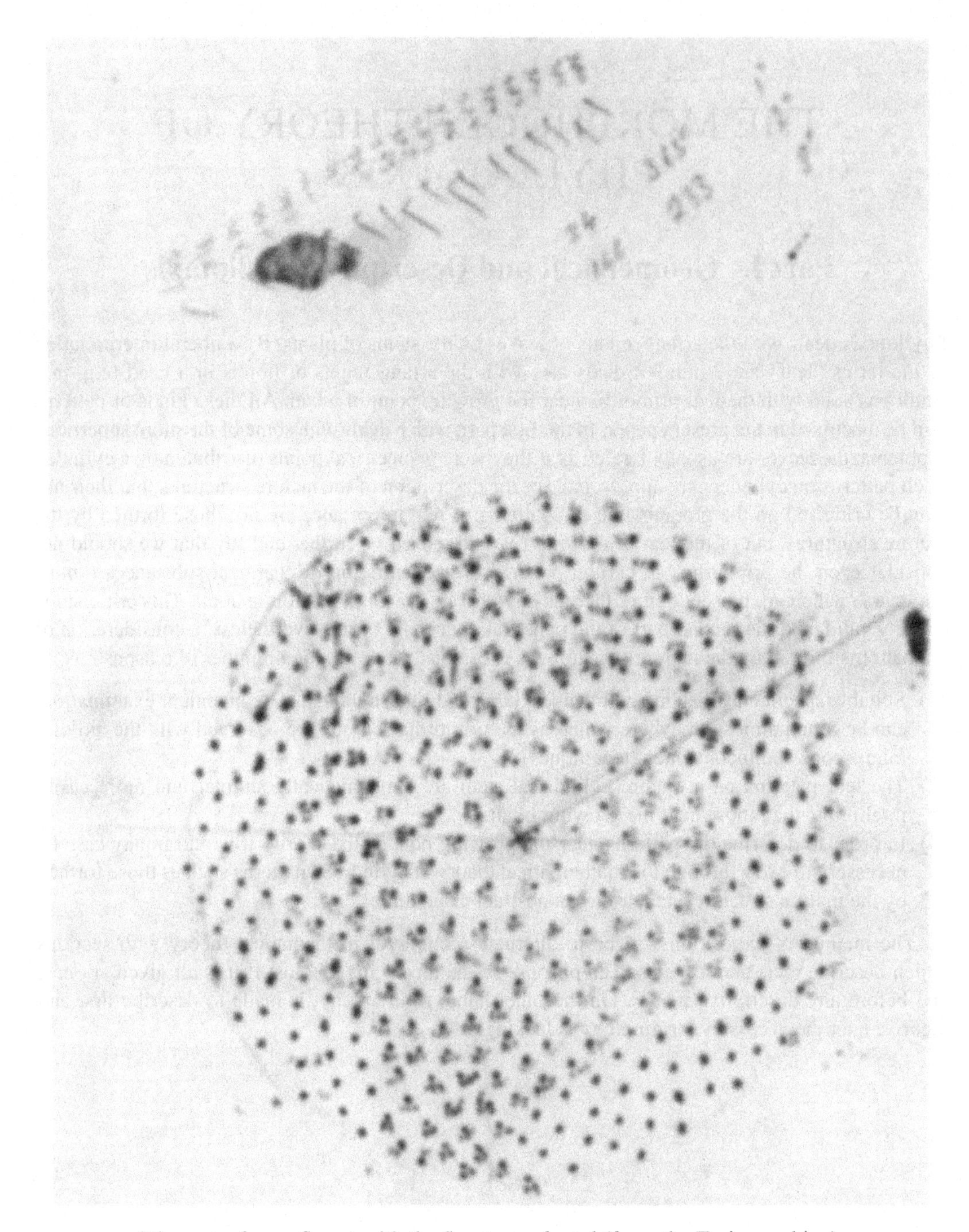

Diagram of a sunflower with the florets numbered (from the Turing archive).

1. A description of certain leaf distribution patterns

Plate 1 shows a portion of a branch of *Pinus* with a very regular arrangement of scales which at one time had supported leaves. The leaves have been removed to enable the patterns to be seen more

Plate 1: A branch of *Pinus*. This photograph does not have the precise regularity described in the text (the original cannot be found) but the parastichies can be readily seen, especially where the leaves have been removed.

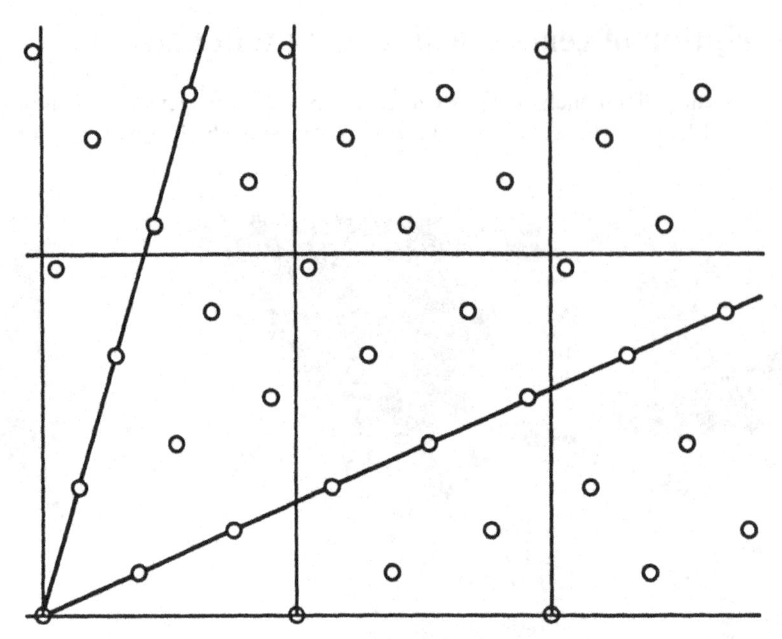

Fig. 1: An idealized plane representation of Plate 1.

clearly. The same pattern is shown in Fig. 1 diagrammatically. The surface of the cylinder has here been unrolled onto the flat paper surface, and the whole enlarged. The scales have been reduced to points and will be referred to as "leaves". The three vertical lines represent one generator of the cylinder repeated. Each point between the first pair of lines represents a leaf which is also represented by one of the points between the second pair. The pattern is remarkably regular and is seen to have the following properties:

(1) If the cylinder is rotated and at the same time shifted along its length in such a way as to make a leaf A move into the position previously occupied by a leaf B, every other leaf also moves into a position previously occupied by a leaf. This may be called the *congruence property*.
(2) All the leaves lie at equal intervals along a helix. On the specimen in Plate 1 [i.e. the lost original], the pitch of the helix is about 0.046 cm and the successive leaves differ in angular position by about 137°.

These two properties are by no means independent. All patterns with the second property have also the first; but there are many species which produce leaf patterns having the first property but not the second. Figure 2 is a diagram similarly constructed to Fig. 1, and showing the arrangement of leaves on the stem of a maiden pink. In this arrangement each leaf has a partner at the same level with it on the stem. A helix could only pass through both partners if it had zero pitch and so degenerated into a circle. However, property (1) holds for these patterns also.

It need hardly be said that for the majority of botanical species the congruence property is only very roughly satisfied. But this need not trouble us for the present. It will be sufficient if the reader will admit that the congruence property has a certain botanical importance, and is willing in consequence to give some attention to the mathematics of patterns having the property.

Fig. 2: A sketch of a maiden pink (*Dianthus deltoides*). Redrawn after ROSS-CRAIG (1951).

2. Helical coordinates for a phyllotactic system

Consider the set of congruences of the patterns formed by the leaves on a stem, i.e. the set of pairs $[\theta, z]$ such that if the stem is simultaneously rotated through an angle θ about its axis, and shifted a distance z along it, both measured algebraically, each leaf is thereby moved into the position previously occupied by another leaf. If $[\theta_1, z_1]$ is one such congruence and $[\theta_2, z_2]$ is another, then clearly $[\theta_1 + \theta_2, z_1 + z_2], [-\theta_1, -z_1]$ are also congruences, that is, the congruences form an Abelian group Γ. If n is an integer, then $[2n\pi,\ 0]$ is a congruence. Consider now those congruences which, like these, have a translation component (second coordinate) zero. The possible rotation components include 2π. Let κ be the smallest positive angle such that $[\kappa,\ 0]$ is in Γ, and let γ be any other such angle. One can write $\gamma = r\kappa + \delta$ where r is an integer and $0 \leqslant \delta \leqslant \kappa$. Then $[\delta, 0]$ is a congruence, and therefore $\delta = 0$, for otherwise the definition of κ would be contradicted. Thus every congruence which is a pure rotation is a rotation through a multiple of κ. In particular 2π must be a multiple of κ, $2\pi = J\kappa$, say. It is easily seen that J may be interpreted as the number of leaves which lie at one level on the stem. This number will be called the "jugacy", in conformity with the established practice of calling systems with $J = 2$ "bijugate" (two leaves being "yoked" together on the stem), and those with $J > 1$ "multijugate", but it is usually more convenient to use $\kappa = 2\pi/J$. If $J = 1$, i.e. $\kappa = 2\pi$, the system is described as "simple" (or in some books "alternate") but that phrase will not be used, as it suggests rather distichous. It is possible for all the congruences to have $z = 0$, but this case is too degenerate to be of much interest. Let therefore η be the smallest positive value of z occurring in any of the congruences. This quantity will be called the "plastochrone distance" on account of its relation to the "plastochrone ratio" as defined by RICHARDS (1948), p. 226. An argument similar to that above shows that all the displacements (second coordinates) are multiples of η. Now let $[\alpha,\ \eta]$ be a congruence, the angle α being chosen so as to have the smallest possible absolute value for the given η, and if this still leaves the sign in doubt, to be non-negative. Then $[n\alpha + r\kappa,\ n\eta]$ is a congruence, and indeed every congruence can be put into this form. For if $[\theta,\ n\eta]$ is a congruence, then so is $[\theta - n\alpha,\ 0]$; and since the translation component of the latter is zero, $\theta - n\alpha$ must be of the form $r\kappa$. The angle α is called the *divergence angle*.

The three parameters

(i) the jugacy J(or $\kappa = 2\pi/J$),
(ii) the plastochrone distance η,
(iii) the divergence angle α

together completely describe the phyllotactic system, i.e. the total group of congruences. These three parameters, together with the radius of the cylinder, are the helical coordinates of the phyllotactic system.

3. Parastichies and parastichy numbers

In a diagram such as Fig. 1 showing the leaves on a stem, one can distinguish numerous straight lines with leaves at uniform intervals along them. These are known as *parastichies*. The word is commonly used for those series of leaves which most readily catch the eye, but no such restriction will be imposed in the present paper. A parastichy is thus the totality of leaves obtained by repeatedly applying the same congruence to some one leaf. Thus if a leaf has coordinates (θ_0, z_0) and $[\theta, z]$ is a congruence, then the leaves with coordinates $(\theta_0 + n\theta, z_0 + nz)$ form a parastichy. If one uses a different leaf, $(\theta_1,\ z_1)$ but the same congruence, one will in general obtain a different parastichy, running parallel to the first, though it may happen that one obtains the same one again. If the congruence $[\theta, z]$ is $[n\alpha + r\kappa,\ n\eta]$, then the cylinder includes $\eta^{-1}|n|^{-1}$ leaves of the parastichy per unit length. Since there are $J\eta^{-1}$ leaves per unit length altogether, there must be $|n|\ J = 2\pi\,|n|\kappa^{-1}$ parallel parastichies generated by the congruence $[n\alpha + r\kappa,\ n\eta]$. This explains the use of the term "parastichy number", for nJ is the number of different parastichies which the congruence generates, provided that $n > 0$. If $n = 0$, the parastichies are not helices on the cylinder but circles; each contains only a finite number of leaves, and there are infinitely many of them. It is preferable however to say in these cases that the parastichy number is zero rather than that it is infinite, so that the representation of the congruence as $[n\alpha + r\kappa,\ n\eta]$ may hold for all values of n, positive, zero or negative.

It is evident from the definition that if one adds two congruences, the parastichy number for the resulting congruence is obtained by adding the parastichy numbers for the two original congruences. This simple but important property is mentioned explicitly, since it is by no means so obvious when the parastichy numbers are defined by counting.

Note: What is here called "parastichy number" is called "leaf number difference" by botanists, whose own "parastichy" is a factor of our parastichy number. A parastichy with parastichy number 3 is indicated in Fig. 1.

4. Phyllotactic systems as lattices. The principal congruences

If ϱ is the radius of the cylinder, then $[\varrho\theta, z]$ will be called the *surface coordinates* of the congruence (or point) with cylindrical polar coordinates (θ, z). The surface coordinates of the congruences $[n\alpha + m\kappa, n\eta]$ of a phyllotactic system may be described as consisting of all the vectors $m\boldsymbol{u} + n\boldsymbol{v}$ where m, n are integers and $\boldsymbol{u}, \boldsymbol{v}$ are respectively $(\varrho\kappa, 0)$ and $(\varrho\alpha, \eta)$. There are many other possible choices of $\boldsymbol{u}, \boldsymbol{v}$, e.g. $(3\varrho\alpha + 2\varrho\kappa, 3\eta)$ and $(\varrho\alpha + \varrho\kappa, \eta)$. The totality of vectors $m\boldsymbol{u} + n\boldsymbol{v}$ where m, n run over the integers and $\boldsymbol{u}, \boldsymbol{v}$ are fixed vectors is called a *lattice*. In order that a lattice should arise from a phyllotactic system on a cylinder of radius ϱ it is necessary that $(2\pi\varrho, 0)$ should be a point of the lattice. This is also sufficient, as may be seen by interpreting a vector (y, x) of the lattice as a congruence $[y/\varrho, x]$.

One may define the *first principal vector* of a lattice as being that which is of shortest non-zero length. This defines it at best with a doubtful sign, and, as will appear later, at worst there are six equally valid candidates. It will be supposed that one of these is chosen to be the first principal vector; there is no need to enquire by what criteria. One may also define similarly the second, third principal vectors, etc. Each is to be the shortest consistent with not being a multiple of one of the earlier principal vectors.

In the sequel the first three principal vectors will play an important part. They correspond more or less to the "contact parastichies" of other investigators, the correspondence being closest for the parastichies generated by the first two principal vectors; but it has been thought that confusion would best be avoided by using an entirely different terminology.

It should be observed that the first two principal vectors of a lattice generate the lattice. For if not, a lattice-parallelogram must contain other lattice points within it. But a point within a parallelogram is always closer to one of the vertices than are some pair of the vertices from one another. Hence the definition of the principal vectors would be contradicted. A small consequence is that the first two principal vectors may also be defined as those two vertices which generate the lattice, and for which (subject to this condition) the square of the scalar product $(\boldsymbol{u} \cdot \boldsymbol{v})$ has the minimum value.

Given two vectors $\boldsymbol{u}$, $\boldsymbol{v}$ which generate the lattice, the value of $(\boldsymbol{u} \cdot \boldsymbol{v})$ is increased by u^2 (or decreased by v^2) by replacing $\mathbf{u}$ by $\boldsymbol{u} \pm \boldsymbol{v}$ (or $\boldsymbol{v}$ by $\boldsymbol{v} \pm \boldsymbol{u}$). By repeatedly modifying the vectors in this way and reducing $|(\boldsymbol{u} \cdot \boldsymbol{v})|$ without changing the sign of $(\boldsymbol{u} \cdot \boldsymbol{v})$ one must eventually come to a pair for which $(\boldsymbol{u} \cdot \boldsymbol{v})$ has the same sign as it had originally and $|(\boldsymbol{u} \cdot \boldsymbol{v})| \leqslant u^2 \leqslant v^2$. By changing the sign of one of the original vectors if necessary, one may suppose this scalar product to be negative. Then

$$0 \leqslant -(\boldsymbol{u} \cdot \boldsymbol{v}) \leqslant u^2 \leqslant v^2$$

from which it follows that all three of the scalar products which can be formed from the vectors $\boldsymbol{u}, \boldsymbol{v}, -(\boldsymbol{u} + \boldsymbol{v})$ are negative, i.e. the vectors form an acute angled triangle. If one had started with the principal vectors, no reduction would have been possible at all, so that the first three vectors must form an acute angled triangle. Conversely, *three vectors forming an acute angled triangle, any two of which generate the lattice, are the principal vectors.* For if $\boldsymbol{u}, \boldsymbol{v}$ are the shortest and second shortest sides of the triangle respectively, then $|(\boldsymbol{u} \cdot \boldsymbol{v})| \leqslant \frac{1}{2}u^2$, since $\boldsymbol{v}$ is shorter than $(\boldsymbol{u} \pm \boldsymbol{v})$. Then if m, n are any two non-zero integers,

$$\begin{aligned}
(mu + nv)^2 - v^2 &= m^2u^2 + 2mn(\boldsymbol{u} \cdot \boldsymbol{v}) + (n^2 - 1)v^2 \\
&\geqslant (m^2 - |mn|)u^2 + (n^2 - 1)v^2 \\
&\geqslant (m^2 - |mn| + n^2 - 1)u^2 \quad \text{since } n^2 \geqslant 1 \text{ and } |\boldsymbol{v}| \geqslant |\boldsymbol{u}| \\
&= ((|m| - |n|)^2 + |mn| - 1)u^2 \\
&\geqslant 0 \qquad\qquad\qquad\qquad \text{since } |m| \geqslant 1,\ |n| \geqslant 1.
\end{aligned}$$

Thus only vectors for which $m = 0$ or $n = 0$, i.e. only multiples of $\boldsymbol{u}$ or of $\boldsymbol{v}$, can be shorter than $\boldsymbol{v}$. Consequently $\boldsymbol{u}$ and $\boldsymbol{v}$ are the first two principal vectors.

These results may be summed up in the theorem on principal vectors:

The principal vectors form an acute angled triangle, and are the only vectors generating the lattice which do so. The first two principal vectors are also characterised by the property that they are the pair of vectors which generate the lattice and minimise the modulus of their scalar product.

In a phyllotactic lattice one may speak of the first, second, etc. principal parastichies and parastichy numbers. One then has the following simple consequence of the fact that $\boldsymbol{u} \pm \boldsymbol{v}$ is the third principal vector:

Corollary. *The third principal parastichy number is the sum or difference of the first and second parastichy numbers.*

Although it is not intended to enter into the matter yet in any detail, it may be mentioned that for a very large proportion of those plants which show sufficient regularity for parastichies to be counted, the principal parastichy numbers are all numbers of the Fibonacci series, in which each number after the first two is the sum of its two predecessors: $0, 1, 1, 2, 3, 5, 8, 13, 21, 34, 55, 89, \ldots$.

Clearly, if (say) the first two principal parastichy numbers are consecutive members of the series, the third and fourth must be also.

5. The measurement of the phyllotaxis parameters

It was explained in §2 that a phyllotaxis scheme is described by the parameters $\kappa = 2\pi/J$, α, η, ϱ. On almost any specimen it is as well to measure the radius ϱ directly. On specimens on which the leaves are not very closely packed the jugacy $J = 2\pi/\kappa$ may be determined by counting how many leaves there are at any level on the stem. With more closely packed leaves this is not feasible, and it is best to choose two vectors which generate the lattice. The jugacy may then be determined as the highest common factor of two corresponding parastichy numbers. On specimens such as the stem shown in Plate 1, it is convenient to measure the distance and the angle between two leaves which are at a considerable distance apart.

To complete the calculation one must find the parastichy number corresponding to the congruence chosen, and the number of complete revolutions of the helix, which must be added to the angle measured. The parastichy number is obtained conveniently not by a direct count, but by counting two of the principle parastichies and combining the results by the addition rule. The divergence angle in such a case is best obtained by first making a less accurate measurement based on leaves which are not so far apart.

On more closely packed specimens it is better to choose two congruences (preferably principal congruences which generate the whole lattice) say $[m\alpha + r\kappa,\ m\eta]$ and $[n\alpha + s\kappa,\ n\eta]$, and measure the angles ψ_1, ψ_2 which the corresponding parastichies make with the generators of the cylinder. Then the area of the parallelogram generated by the first two principal vectors is

$$\Delta = mn\eta^2|\tau_2 - \tau_1|$$

where $\tau_1 = \tan\psi_1$ and $\tau_2 = \tan\psi_2$. This area is also $\kappa\varrho\eta$, and therefore, since $\eta > 0$,

$$\eta = \kappa\varrho/mn|\tau_2 - \tau_1|.$$

The angle α satisfies

$$m\alpha = m\eta\varrho^{-1}\tau_1 \bmod \kappa \tag{I.5.1}$$

$$n\alpha = n\eta\varrho^{-1}\tau_2 \bmod \kappa \tag{I.5.2}$$

and since m, n are co-prime, positive integers, k, l can be found satisfying $km - ln = 1$. Therefore,

$$\begin{aligned}\alpha &\equiv (km\tau_1 - ln\tau_2)\eta\varrho^{-1} \bmod \kappa \\ &\equiv \pm\left(\frac{2\pi k}{n}\frac{-\tau_1}{\tau_2 - \tau_1} + \frac{2\pi l}{m}\frac{\tau_2}{\tau_2 - t_1}\right).\end{aligned} \tag{I.5.3}$$

The following rule expresses this formula in a convenient form. *Choose two vectors which generate the lattice and whose parastichy helices turn in opposite directions. Calculate (or look up in Table* 1) *what would be the divergence angle if either one of these parastichies were an orthostichy, i.e. parallel to the axis of the cylinder. The correct divergence angle may be obtained as a weighted average of these two. Each is to be weighted in proportion to the modulus of the cotangent of the angle which the corresponding parastichy makes with the generators of the cylinder.*

Fraction of 2π	Deg.,	Min.,	Sec.	Degrees
1/2	180°			
1/3	120°			
2/5	144°			
3/8	135°			
5/13	138°	27′	41.5″	138.46154
8/21	137°	8′	34.3″	137.14286
13/34	137°	38′	49.4″	137.64706
21/55	137°	27′	16.4″	137.45454
34/89	137°	31′	41.1″	137.52809
55/144	137°	30′	00.0″	137.50000
89/233	137°	30′	38.6′	137.51073
Limiting value	137°	30′	27.9′	137.50778

Table 1: Divergence angles.

These angles are given with greater accuracy than can be used, though perhaps not so much greater as might be supposed. Since the angles given for the higher parastichy numbers differ by less than a minute, and since the angles ψ_1, ψ_2 can be measured to a few degrees, the divergence angle can be determined in such cases to within a few seconds. It need hardly be said that the value so obtained is not accurately repeated from leaf to leaf, and may vary by a degree or more, and it is only the averages over a considerable number of plastochrones that behave consistently. This insensitivity of the divergence angle to errors in the angles ψ_1, ψ_2 may be expressed in the equation

$$\left|\frac{d\alpha}{d\xi}\right| = \frac{\kappa}{mn}\frac{1}{(\xi - 1)^2}.$$

In the case of limiting divergence-angle phyllotaxis ($\xi = 0$) this has the value κ/mn.

6. Phyllotaxis on surfaces of revolution

The patterns of leaves so far considered have been on the surface of a cylinder, and remain essentially the same on parts of the cylinder far removed from one another. Although species can be found for which, for the stems, this is a good approximation, some broader point of view is necessary to deal with the majority of phyllotactic patterns.

In general one may suppose that the specimen is a solid of revolution on which the lateral organs, idealised into points, are distributed. A common case is a capitulum, e.g. a sunflower or a daisy. The "leaves" are then florets and the surface of revolution is a disc, or nearly so.

For our purposes the geometry of the surface of revolution can be conveniently described as follows. The position of a point on the surface is fixed by two coordinates (θ, z) as on the cylinder. The coordinate z is measured *along the surface* (and not, as might be considered most natural, parallel to the axis). The shape of the surface is determined by giving the radius ϱ for each z. On such a surface one may define a phyllotactic system in which the parameters vary continuously with the coordinate z. The jugacy J, being an integer, cannot of course be allowed to vary at all. But suppose that at each z a value is assigned for the plastochrone distance η and the divergence angle α, as well as for the radius ϱ: what would be the positions of the leaves that correspond to arbitrary values of the parameters?

A natural answer can be given to this question if the formula $(n\alpha + r\kappa,\ n\eta)$ is extended to non-integral values of n, which we rename u. The formula is replaced by the two differential equations:

$$\frac{dz(u)}{du} = \eta(z), \qquad \frac{d\theta(u)}{du} = \alpha(z), \tag{I.6.1}$$

the positions of the leaves being given by $(\theta(u) + r\kappa, z(u))$ for integral values of u and r.

With these conventions, one may obtain values of $z(n)$ and $\theta(n)$ by measurement and could, in theory, infer values of $\eta(z)$, $\alpha(z)$ by ordinary finite difference methods. In practice there will be such errors of measurement, and irregularities in the positions of the leaves, that the use of differentiation formulae involving high differences is inappropriate. The method which the author finds most convenient is to draw freehand the principal parastichies in the neighbourhood of the value of z in question, measure the angles ψ_1, ψ_2 which these curves make with a plane through the axis (i.e. in practice with the intersection of this plane with the surface), count the parastichy numbers, and apply the formula of §5.

According to the point of view of this section there is a complete phyllotactic system corresponding to each value of the parameter z, described by parameters α, η, ϱ varying continuously with z. It will be convenient to continue to speak of such systems as if they were given on a cylinder, although η, α are defined by (I.6.1); and to consider the phyllotactic system as the lattice of points $(n\alpha + r\kappa,\ n\eta)$.

Attention will be given later (§11) to phyllotactic systems varying with a parameter.

7. The bracket and the fractional notations

When describing a specimen one may not always wish to make sufficient measurements to give a complete description of the lattice at some level on the stem: an indication of the principal parastichy numbers would often be enough. For this purpose the notation of CHURCH (1904) is appropriate. He used such notations as $(8+13)$, which in this paper will be used to signify that the principal parastichy numbers are 8, 13 and 21.

Some latitude must be allowed when the third and fourth principal vectors are of nearly equal length, and the three numbers may consequently be the first, second and fourth parastichy numbers. This cannot have been Church's intention, for he believed that two of the principal parastichies are always at right angles, an assumption which is not always correct.

Another, less happy notation is the use of fractions of a revolution as measures of the divergence angle. The most satisfactory approximations are of course the continued fraction convergents, and these will normally be the ratio of two Fibonacci numbers. That such ratios were good approximations to the divergence angle was first observed by Schimper and Braun (BRAUN 1835), and was an important discovery. However the use of different fractions of this kind to distinguish phyllotactic systems must be deplored. For instance, in the case where the divergence angle has the limiting value $2\pi\omega^{-2} = 137^\circ\ 30'27.9''$, all of these ratios are good approximations to the divergence angle. What then is the significance of the choice of one rather than another? On the whole, the tendency seems to be to choose larger denominators for smaller plastochrone ratios, but no very definite rule seems to have been formulated. In cases where there is some real reason for regarding the divergence angle as a rational fraction of a revolution, the use of such fractions is admissible. Such cases arise with distichous ($\alpha = 180^\circ$, $J = 1$) and decussate ($\alpha = 90^\circ$, $J = 2$) systems, and in fact with all symmetrical systems. Another example is provided by the genus *Carex*, where the stem itself is a triangular prism, thus ensuring that the divergence angle does not, on average, wander far from 120°. There are likewise species with a pentagonal stem (e.g. Plumbago) where the angle may be supposed to be 144°.

8. Naturally occurring phyllotactic patterns

It is found that the numbers in the Fibonacci series $0, 1, 1, 2, 3, 5, 8, 13, \ldots$ are by far the commonest parastichy numbers and a phyllotactic system with these numbers is described as normal. In these cases the divergence angle α is in the region 135° to 140° and if the principal parastichy numbers are large, α is near to $137^\circ 30'28''$. However this system is not universal and other types of phyllotaxis mentioned below may be encountered.

(a) There are cases where the Fibonacci series is to be replaced by the series 1, 3, 4, 7, 11, ... (the "anomalous" series of Church). In some species this series is fairly common, but in others it appears only in a small proportion of specimens. For these cases the divergence angle is found to be in the neighbourhood of $99^\circ 30'$.

(b) There are species (e.g. *Dipsacus sylvestris*) for which the principal parastichy numbers are taken from the double Fibonacci series $0, 2, 2, 4, 6, 10, 16, 26, \ldots$, and the divergence angle is half the normal, i.e. about $68^\circ 45' 14''$.

(c) Some species have mirror symmetry, and indeed, this is true in the majority of cases where there are opposite leaves (i.e. bijugate phyllotaxis, $J = 2$). Commonest among these are the "decussate" leaf patterns, i.e. those for which the divergence angle is 90°, and the first four principal parastichy numbers are $0, 2, 2, 4$ (not necessarily respectively). There are also cases where these parastichy numbers are $0, 1, 1, 2$; and, relatively rarely, such combinations as $0, 6, 6, 12$. The latter occur in species where the parastichy numbers are very variable, e.g. one might find within the species $0, 6, 6, 12$; $1, 6, 7, 13$; $0, 7, 7, 14$; $0, 5, 5, 10$; each forming a considerable fraction of the whole.

It is the main purpose of the present paper to explain, in part at any rate, the phenomena described above. The explanations given will be at two levels. In this first part of the paper the arguments are entirely geometrical. The geometrical arguments do not exactly give a theory of the development of a phyllotactic pattern. It is merely shown that if the development satisfies certain not very artificial conditions, then when once a phyllotactic pattern has started it will develop into patterns of the kind observed. These geometrical arguments have been expounded by some previous writers, but often in a rather unsatisfactory form, and with the emphasis misplaced. The writer has consequently considered it appropriate to give a new exposition of these arguments. This first part of the paper is not however entirely old material in a new form. In particular the use of the inverse lattice and other ideas related to Fourier analysis appears to be new. This first and purely geometrical part of the paper must however be considered as merely a preliminary to the second part, which expounds a chemical explanation of the same phenomena. The chemical theory will be much more complex than the geometrical theory, and, in effect, justifies the assumptions of the latter. Although it might have been possible to expound the chemical theory totally independently of the geometrical, it was not thought advisable to do so, because of the insight which the geometrical theory gives.

9. Lattice parameters

It has been seen that the two principal vectors (a_0, b_0) and (c_0, d_0) generate the lattice. If these vectors, so far not uniquely specified, are precisely defined, their four coordinates can be used as parameters for describing the lattice, as alternatives to the helical parameters $\alpha, \eta, \varrho, J (= 2\pi/\kappa)$. Owing to the close connection with the principal parastichy numbers, the new parameters will be found more useful in theories of the origin of phyllotaxis. In order to make the definitions unique it is necessary to specify the signs that are to be given to the two vectors, and the order in which they are to be taken. It is convenient to require that the *second* coordinate of each vector should be non-negative. In phyllotactic systems this second coordinate, being an integral multiple of η, must be either at least as large as η or zero. In the latter case the convention will be that the first coordinate shall be positive, but this is rarely used in what follows. The ordering is to be such that $a_0 d_0 - b_0 c_0 > 0$. (In view of $b_0 \geqslant 0, d_0 \geqslant 0$, the condition for this is that the first vector can be made parallel to the second by turning it to the left through an angle of less than 180°.)

The conditions on the four numbers a_0, b_0, c_0, d_0 are thus that for every pair of non-zero integers m, n

$$a_0^2 + b_0^2 \leqslant (ma_0 + nc_0)^2 + (mb_0 + nd_0)^2, \tag{I.9.1}$$

$$c_0^2 + d_0^2 \leqslant (ma_0 + nc_0)^2 + (mb_0 + nd_0)^2, \tag{I.9.2}$$

$$b_0 > 0 \text{ or } b_0 = 0 \text{ and } a_0 > 0, \tag{I.9.3}$$

$$d_0 > 0 \text{ or } d_0 = 0 \text{ and } c_0 > 0, \tag{I.9.4}$$

$$a_0 d_0 - b_0 c_0 > 0. \tag{I.9.5}$$

Since the matrix

$$\begin{pmatrix} a_0 & b_0 \\ c_0 & d_0 \end{pmatrix}$$

plays a prominent part, the numbers (a_0, b_0, c_0, d_0) are called the *principal matrix coordinates* of the lattice. They are unique so long as the third principal vector is longer than the second. A further set of parameters suggested by these, having considerable intuitive appeal, are:

$$\Delta = a_0 d_0 - b_0 c_0, \tag{I.9.6}$$

$$\zeta = \left(\frac{a_0^2 + b_0^2}{c_0^2 + d_0^2} \right)^{1/2}, \tag{I.9.7}$$

$$\phi = -\sin^{-1} \left(\frac{a_0 c_0 + b_0 d_0}{[(a_0^2 + b_0^2)(c_0^2 + d_0^2)]^{1/2}} \right), \tag{I.9.8}$$

$$\psi = \tan^{-1} \left(\frac{a_0 + c_0}{b_0 + d_0} \right). \tag{I.9.9}$$

The letter Δ has already been used in its present sense of denoting the area of the parallelogram generated by the first two principal vectors, i.e., the area occupied by each leaf, and it may accordingly be called the *leaf area*; ζ is the ratio of the first two vectors; and ϕ is the angle between them reduced by 90°. The parameter ψ describes the direction of the sum of the first two principal vectors. It follows from the theorem on principal vectors that ϕ lies between $-30°$ and $30°$, and ψ between $-90°$ and $90°$. In practice, in phyllotactic lattices, $|\psi|$ does not often exceed 30°, while ζ is usually close to 1.

A lattice can be described by any pair of vectors which generate it. If $(a,\ b)$ and $(c,\ d)$ are two such vectors, the matrix

$$\begin{pmatrix} a & b \\ c & d \end{pmatrix}$$

will be called a *matrix representation* of the lattice. A necessary and sufficient condition that two matrices should describe the same lattice is that one should be obtainable from the other by left multiplication with a matrix with integral coefficients and determinant ± 1. By the second part of the theorem on principal vectors the principal representation of a lattice can be recognised by the fact that the vectors $(a_0,\ b_0)$ and (c_0, d_0) form two of the sides of an acute angled triangle. It must of course also satisfy the conditions $b_0 \geqslant 0,\ d_0 \geqslant 0,\ a_0 d_0 - b_0 c_0 \geqslant 0$. If mJ and nJ are two parastichy numbers and if the corresponding vectors generate the lattice, and the parastichies make angles $\psi_1,\ \psi_2$ with the generators of the cylinder, then the matrix

$$\frac{\varrho\kappa}{\tau_2 - \tau_1} \begin{pmatrix} -\tau_1/n & 1/n \\ -\tau_2/m & 1/m \end{pmatrix}$$

is one of the matrix representations of the lattice. Here $\kappa = 2\pi/J,\ \tau_1 = \tan\psi_1,\ \tau_2 = \tan\psi_2$, and (cf. I.5.1)

$$\Delta = \frac{(\varrho\kappa)^2}{mn|\tau_2 - \tau_1|}. \tag{I.9.10}$$

To convert any matrix coordinates (a,b,c,d) for a lattice into helical coordinates η,α,ϱ,J one proceeds as follows. The value of η is easily obtained as the highest common factor of b and d. It is not possible to find the value of J since the same lattice may be wrapped around cylinders of various radii. For the present we suppose it given. To obtain $2\pi\varrho$ one must find the vectors of the lattice which have their first coordinates zero. If $b = m\eta$ and $d = n\eta$ then these vectors are clearly multiples of $(na - mc, 0)$, i.e. $\varrho\kappa = |na - mc|$. To obtain α let $km - ln = 1$, then $\alpha = (ka - lc)/\varrho$ modulo 2π.

10. Continued fraction properties

The procedure by which any matrix description of a lattice may be made to yield the principal description was described in effect in §4. Suppose that the scalar product of two vectors is negative. Then one repeatedly adds one vector to another, and continues until the modulus of the scalar product can no longer be reduced. Suppose that k is the largest integer such that the scalar product $(ka+c,\ kb+d)\cdot(a,b)$ is negative. Then after adding the first row (a,b) of the matrix k times to the second, it will be necessary to interchange the two rows, if the first is always to be added to the second. The effect of the combined addition and interchange is expressed by the multiplication of the matrix

$$\begin{pmatrix} a & b \\ c & d \end{pmatrix} \text{ on the left by } \begin{pmatrix} k & 1 \\ 1 & 0 \end{pmatrix}.$$

The reduction process as a whole is then expressed by left multiplication by a product of a number of such matrices $C_{k_0}, C_{k_1}, \ldots, C_{k_r}$, where C_k represents

$$\begin{pmatrix} k & 1 \\ 1 & 0 \end{pmatrix}.$$

In order finally to bring the matrix to the form agreed as standard it may be necessary to left-multiply by one of the matrices

$$\begin{pmatrix} 0 & \pm1 \\ \pm1 & 0 \end{pmatrix} \text{ or } \begin{pmatrix} \pm1 & 1 \\ 0 & \pm1 \end{pmatrix}.$$

Every unimodular matrix of order 2 can be expressed as a product

$$\begin{pmatrix} \pm1 & 0 \\ 0 & \pm1 \end{pmatrix} C_{k_1} C_{k_2} \cdots C_{k_r}.$$

Products $C_{k_0} C_{k_1} \cdots C_{k_r}$ are very closely related to continued fractions. In fact it can be shown by a simple inductive argument that if

$$K_0 + \frac{1}{K_1} + \frac{1}{K_2} + \frac{1}{K_3} + \cdots \frac{1}{K_r} = \frac{p_r}{q_r}$$

is in its lowest terms, then

$$C_{k_0} C_{k_1} \cdots C_{k_r} = \begin{pmatrix} p_r & p_{r-1} \\ q_r & q_{r-1} \end{pmatrix}.$$

This shows in effect that

Every improper unimodular matrix may be expressed in the form

$$\begin{pmatrix} p_r & q_r \\ p_{r-1} & q_{r-1} \end{pmatrix}$$

where p_r/q_r, p_{r-1}/q_{r-1} are two consecutive convergents of the continued fraction of some number. If the unimodular matrix is obtained by reduction of a matrix representation of a lattice then the partial quotients are given in reverse order as the number of times one vector is to be subtracted from the other without interchange.

This result may be applied to the lattice as described by the helical coordinates α, η, ϱ, J. One representative matrix is

$$2\pi\varrho J^{-1}\begin{pmatrix}-1 & 0\\ x & \sigma\end{pmatrix}$$

where $\sigma = \eta J/2\pi\varrho$, $x = \alpha J/2\pi$. The lattice vectors are

$$2\pi\varrho J^{-1}(p \quad q)\begin{pmatrix}-1 & 0\\ X & \sigma\end{pmatrix}$$

where p, q are any integers. It will be convenient to represent this vector by the expression (p/q): this notation is intended to suggest a connection with fractions, but the brackets are always to be retained to prevent any confusion. By what has been shown above, a standard representative, apart from the order of the rows, can be written in the form

$$2\pi\varrho J^{-1}\begin{pmatrix}q_n x - p_n & \sigma q_n\\ q_{n-1}x - p_{n-1} & \sigma q_{n-1}\end{pmatrix}.$$

The first two principal vectors are then, not necessarily in order,

$$\boldsymbol{u} = (p_n/q_n), \quad \boldsymbol{v} = (p_{n-1}/q_{n-1}).$$

Denoting the third vector by (p/q), the theorem on principal vectors and acute angled triangles gives

$$(p/q) = (p_n \pm p_{n-1}/q_n \pm q_{n-1}).$$

Now the three quantities $q_n x - p_n$, $q_{n-1}x - p_{n-1}$, $qx\text{-}p$ cannot all have the same sign, for if they did the three scalar products (p_n/q_n), (p_{n-1}/q_{n-1}), (p/q) would all be positive, contrary to the results of the same theorem. If $q_n x - p_n$ and $q_{n-1}x - p_{n-1}$ have opposite signs, then p_{n-1}/q_{n-1} is a convergent of x. If, however, they have the same sign, (p/q) must be $(p_n - p_{n-1}/q_n - q_{n-1})$ and $|q_{n-1}x - p_{n-1}| > |q_n x - p_n|$. Since

$$p_n = K_n p_{n-1} + p_{n-2}, \quad q_n = K_n q_{n-1} + q_{n-2}, \text{ and } K_n \geqslant 1,$$

it follows that $q_{n-2}x - p_{n-2}$ also has the opposite sign to $q_{n-1}x - p_{n-1}$, and therefore p_{n-2}/q_{n-2} is a convergent of x. If $K_n = 1$ then (p_{n-2}/q_{n-2}) is the third vector. If, however, $K_n > 1$ then

$$\begin{aligned}|q_{n-2}x - p_{n-2}| &= (K_n - 1)|q_{n-1}x - p_{n-1}| + |qx - p|\\ &> |q_{n-1}x - p_{n-1}|.\end{aligned} \tag{I.10.1}$$

Thus in any case one of the three principal vectors, that with the smallest or second smallest parastichy number, corresponds to a convergent of x.

Rather more may be asserted in the conditions which normally apply in real phyllotaxis. Let p_{r-1}/q_{r-1}, p_r/q_r, p_{r+1}/q_{r+1} now represent three consecutive convergents of $x = \alpha J/2\pi$, and suppose that

$$\begin{aligned}-\left(x - \frac{p_{r-1}}{q_{r-1}}\right)\left(x - \frac{p_r}{q_r}\right) &> \sigma^2\\ &> -\left(x - \frac{p_r}{q_r}\right)\left(x - \frac{p_{r+1}}{q_{r+1}}\right) > 0,\end{aligned} \tag{I.10.2}$$

$$p_{r+1} = p_r + p_{r-1}, \quad q_{r+1} = q_r + q_{r-1}, \tag{I.10.3}$$

then it follows at once that the scalar product $((p_{r-1}/q_{r-1})\cdot(p_r/q_r))$ is negative and that $((p_r/q_r)\cdot(p_{r+1}/q_{r+1}))$ and $((p_{r-1}/q_{r-1})\cdot(p_{r+1}/q_{r+1}))$ are positive. Also $(p_{r+1}/q_{r+1}) = (p_r/q_r) + (p_{r-1}/q_{r-1})$. Thus these three vectors form an acute angled triangle, and therefore are the three principal vectors of the lattice. Thus

If the divergence angle as a fraction of $2\pi/J$ has one of its partial quotients a_{r_0+1} equal to unity, so that

$$p_{r_0+1} = p_{r_0} + p_{r_0-1}, \quad q_{r_0+1} = q_{r_0} + q_{r_0-1}$$

then for some values of the plastochrone ratio η/ϱ the principal vectors correspond to these three consecutive convergents of the divergence angle, though not necessarily in order. If there are a number of consecutive unit partial quotients, the corresponding ranges of values of the plastochrone ratio are consecutive intervals.

In particular, if all the partial quotients are unity from some point onwards, then for all sufficiently small plastochrone ratios the principal vectors all correspond to convergents of the divergence angle. Such a divergence angle may be called a *limiting divergence angle*. The reduced parastichy numbers from some point onwards are the denominators q_r and satisfy $q_{r+1} = q_r + q_{r-1}$. They determine the value of the limiting divergence angle, apart of course for sign and for additive multiples of 2π. For the numerators p_r must satisfy $p_{r+1} = p_r + p_{r-1}$ and $q_r p_{r-1} - q_{r-1} p_r = \pm 1$. A change of sign in the value of $q_r p_{r-1} - q_{r-1} p_r$ may be accomplished by changing the sign of each p_r and therefore of the limiting angle

$$\lim_{r\to\infty} \frac{2\pi}{J} \frac{p_r}{q_r}.$$

Suppose then that

$$q_{r_0} p_{r_0+1} - q_{r_0+1} p_{r_0} = 1$$

and that p_r' is a second solution so that

$$q_r(p_{r+1} - p_{r+1}') = q_{r-1}(p_r - p_r').$$

Then since q_r, q_{r+1} have no common factor, for some integer m one must have

$$p_{r+1}' = p_{r+1} + m q_{r+1}, \quad p_r' = p_r + m q_r$$

and p_r/q_r differs from p_r'/q_r' only by the integer m. If p_r'/q_r' tends to a limit it can only differ from the limit of p_r/q_r by an integer, and hence from the corresponding estimates of α by a multiple of κ, which may be eliminated by the minimal condition on α. For this reason, the limiting angle may be described as "the limiting divergence angle corresponding to the series q_{r_0-1}, q_{r_0}, q_{r_0+1} of reduced parastichy numbers".

The value of the limiting divergence angle may be expressed as

$$\kappa \frac{p_{r_0} + \omega p_{r_0+1}}{q_{r_0} + \omega q_{r_0+1}}.$$

The cases of chief interest are:

Normal Fibonacci phyllotaxis. The parastichy numbers belong to the series $0, 1, 1, 2, 3, 5, \ldots$ and the numerators to the same series displaced: $-1, 1, 0, 1, 1, 2, \ldots$. The limiting angle is $2\pi(1-\omega^{-1})$, i.e. 137° 30′28″.

Normal bijugate phyllotaxis. The parastichy numbers are $0, 2, 2, 4, \ldots$, i.e. the Fibonacci numbers doubled. The jugacy is two, and the reduced parastichy numbers are again the Fibonacci numbers. The limiting angle is $\pi(1-\omega^{-1})$ i.e. 68°45′14.0″.

Phyllotaxis of the so-called anomalous series. The principle parastichy numbers belong to the series $1, 3, 4, 7, 11, \ldots$. The limiting angle is $2\pi\omega/(1+3\omega)$, i.e. 99°30′6″.

Examples were also collected by the brothers BRAVAIS (1838) of the occurrence of other series.

11. Continuously changing phyllotaxis

There are at least two ways in which one may be concerned with a phyllotactic system which depends continuously on some real parameter. On a growing specimen the dimensions of the stem may be altering and the leaves moving relative to one another. In this way one is concerned with a phyllotaxis which depends on time. One may equally well be concerned with a phyllotaxis which changes continuously along the length of a stem, or towards the centre of a capitulum. Variation of phyllotaxis in time will be given greater attention here than variation in space, though it can be less conveniently demonstrated. For the purpose of the present section either kind of variation will be described by allowing α, η, ϱ and other quantities describing the phyllotaxis to depend upon a parameter t, which will be described as "the time" regardless of the fact that other interpretations would be equally appropriate.

Until comparatively recently, phyllotaxis continuously varying in space has only been considered in the rather trivial case in which although the radius ϱ is allowed to vary with position (as indeed on a disc it is obliged to do), the dimensionless quantities $\alpha, \eta/\varrho, J$ remain constant. In such a system all other dimensionless quantities, such as angles, principal parastichy numbers, etc., also remain constant, and it may be considered that one has effectively the same phyllotaxis at each radius. This hypothesis may be most familiar to the reader in the form in which it is stated that the parastichies are logarithmic spirals. It may be appropriate for certain conditions of exponential growth, but there are also many conditions for which it is entirely inappropriate. For instance in the case of the florets on a mature capitulum of *Taraxacum officinale* the area per floret, so far from varying inversely as the square of the radius, as would be required on this hypothesis, actually increases towards the centre. It is also inappropriate for the neighbourhood of the growing point, at least if there be any truth in the theory expounded in part II. It would be particularly inappropriate to restrict consideration of continuously varying phyllotaxis to this case, which does not admit of changes of principal parastichy numbers, since it is intended to explain in part these changes in terms of continuously changing phyllotaxis.

More recently RICHARDS (1948) has considered phyllotactic systems on a disc with the divergence angle always equal to the Fibonacci angle, i.e. the limiting angle for the Fibonacci series, viz. $2\pi\omega^{-2}$, and the radius varying with the leaf number in various ways. In general if $f(u)$ is some increasing function of the real variable u one may consider leaves to be placed at the points described in plane polar coordinates by $(f(u),\ 2\pi\omega^{-2}u)$.

According to the principles of §6, one has $z = \varrho$ and therefore $z(u) = f(u)$. The value of η for radius ϱ is $f'(u)$ where $\varrho = f(u)$. A particularly interesting example is provided by putting $f(u) = A(u+B)^{1/2}$. Then $\eta = A^2/2\varrho$, and the leaf area has the same value πA^2 at each radius: this is quite a good approximation for the distribution of mature florets in the head of a member of the compositae. Figure 3 shows such a phyllotaxis. It is taken from RICHARDS (1948) and in it $B = 1/2$.

It follows from the results of the last section that the principal parastichy numbers for this phyllotactic pattern are Fibonacci numbers at each radius. One must be cautious here and avoid assuming causal connections groundlessly. Certainly if the divergence angle is exactly the Fibonacci angle then the principal parastichy numbers must be Fibonacci numbers, and conversely, if the principal parastichy numbers are Fibonacci numbers the divergence angle cannot differ much from the Fibonacci angle. If one of these phenomena has to be the cause of the other, then the less objectionable assumption is that the value of the angle is the effect. But there is no need to adopt either hypothesis. In view of the corollary to the theorem on principal parastichy numbers and acute angled triangles (§4) it is not possible for the first two of the principal parastichy numbers to be Fibonacci numbers without the third being one also. It is to be expected that this fact should go a long way to explain the great preponderance of Fibonacci numbers amongst the principal parastichy numbers. Clearly, however, some other hypothesis is necessary in addition, for it is possible to produce

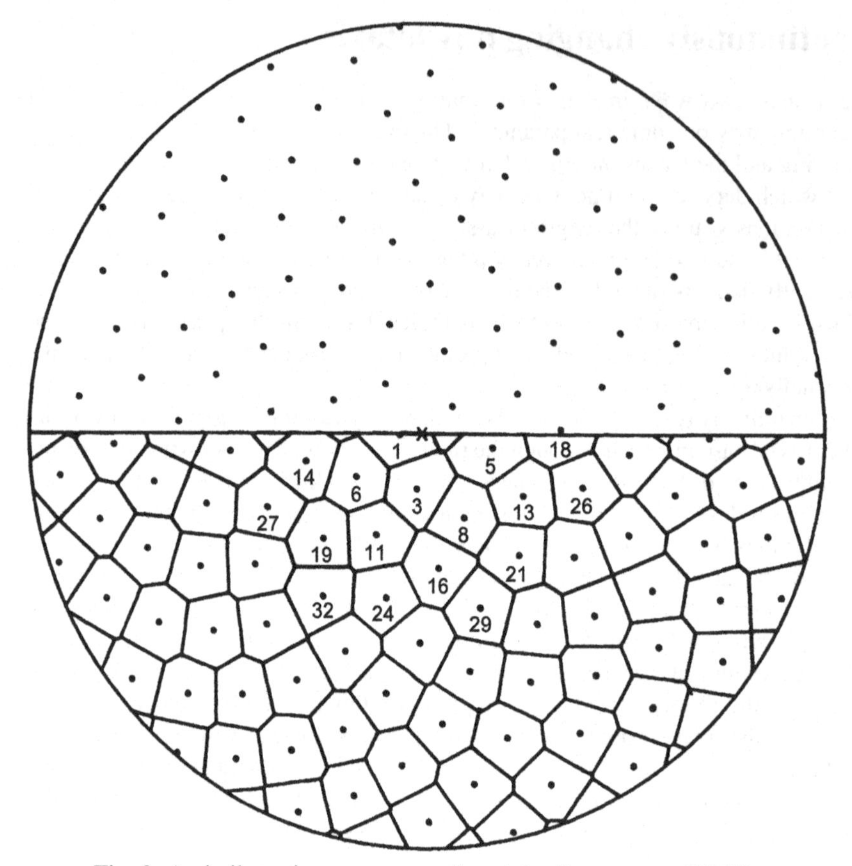

Fig. 3: A phyllotactic system on a disc. After RICHARDS (1948).

(as mathematical constructions) phyllotactic systems with any two given integers as the first two principal parastichy numbers. An appropriate hypothesis for the purpose is the following:

The third principal parastichy number does not lie numerically between the first and second.

This will be known as the *hypothesis of geometrical phyllotaxis*. It can easily be seen by examination of Fig. 3 that it applies for Fibonacci angle phyllotaxis, and it may also be shown to apply for any limiting angle phyllotaxis. It appears also to apply to all naturally occurring cases. (To verify it on a specimen, look for the acute angled triangles. The longest side should either join the uppermost and lowermost points of the triangle, or else be sufficiently nearly horizontal for its projection onto a vertical line to be less than half the projection of the whole triangle.) It is adequate as a subsidiary condition for Fibonacci phyllotaxis for

If a phyllotactic system varies in time whilst satisfying the hypothesis of geometrical phyllotaxis, then the three principal parastichy numbers of the system always belong to the same sequence p_r obeying the Fibonacci law $p_{r+1} = p_r + p_{r-1}$.

If this be not so then there must be a time when the principal parastichy numbers change from a set p_{r-1}, p_r, p_{r+1} obeying the rule to a set which do not. This must arise through the original third parastichy number being dropped out and being replaced by another number. It cannot be p_{r-1} which is dropped, for then p_r, p_{r+1} would remain, and by §4 the third parastichy number after the change must be either $p_{r+1} - p_r$ or $p_{r+1} + p_r$, i.e. either p_{r-1} or p_{r+2}, either of which would belong to the series contrary to hypothesis. Likewise it cannot be p_{r+1} which is dropped. It cannot be p_r for this would contradict the hypothesis of geometrical phyllotaxis. Thus each alternative is impossible, and the assumption that it is possible to reach principal parastichy numbers not in the series p_r is contradicted.

However, this hypothesis cannot be regarded as entirely satisfactory. However true it is, and however logically it follows that the principal parastichy numbers remain in a Fibonacci-like series, the hypothesis is itself quite arbitrary and unexplained. Its merit is that it replaces an empirical law, of a rather weird and magical appearance, by something simpler and much less mysterious. The question remains "Why should the hypothesis of geometrical phyllotaxis hold?" and this is a question which the geometrical approach is not capable of answering. There are other questions which it is also unable to answer, such as "Why should leaf patterns take the form of lattices at all?", "How does the lattice pattern develop on previously undifferentiated pieces of tissue?" and "How does a mirror symmetrical pattern develop into an unsymmetrical one?" and "Why are the principal vectors of the lattice of such importance?" All of these topics must be left until part II. There remain however a number of questions which can be treated by the methods of geometrical phyllotaxis, which have not yet been considered. Some of these are preparatory for the work of part II; others throw further light on the mathematics of continuously changing phyllotaxis and the mathematical description of phyllotactic lattices, and yet others are concerned with the packing of mature and semi-mature leaves.

12. The inverse lattice

Suppose that one does not consider the leaves as geometrical points but as described by some function on the surface of a cylinder, or in the plane obtained by unrolling the surface of the cylinder. The points of the cylinder may be expressed in the form

$$(\xi_1, \xi_2)\begin{pmatrix} a & b \\ c & d \end{pmatrix} = (x, z)$$

since the matrix

$$\begin{pmatrix} a & b \\ c & d \end{pmatrix}$$

is non-singular. When the function is described in terms of the variables (ξ_1, ξ_2) it is periodic with unit period in both variables, i.e. of the (x, z) it is periodic with unit period in both variables, i.e. of the form

$$\sum A_{m,n} e^{2\pi i(m\xi_1 + n\xi_2)}.$$

To express this in terms of the original variables one must express (ξ_1, ξ_2) in terms of (x, z). If the inverse of $\begin{pmatrix} a & b \\ c & d \end{pmatrix}$ is $\begin{pmatrix} A & C \\ B & D \end{pmatrix}$ then

$$\xi_1 = Ax + Bz, \quad \xi_2 = Cx + Dz, \tag{I.12.1}$$

and the function can be written as

$$\sum A_{m,n} e^{2\pi i((mA+nC)x+(mB+nD)z)}$$

or as

$$f(\boldsymbol{x}) = \sum \boldsymbol{A}\boldsymbol{u}\ e^{2\pi i(\boldsymbol{u},\boldsymbol{x})} \tag{I.12.2}$$

where the summation is over the lattice described by the matrix $\begin{pmatrix} A & B \\ C & D \end{pmatrix}$. This lattice may be called the inverse lattice, because it is described by the transposed inverse of the matrix describing the lattice arising from the congruences.

In the inverse lattice it is only the lattice points relatively close to the origin that are of any particular importance. Consider for example a function in the plane having the symmetry of the lattice, and of the form $\sum_{y \in L} g(\boldsymbol{x}+\boldsymbol{y})$ where L is the lattice $\begin{pmatrix} a & b \\ c & d \end{pmatrix}$. If

$$g(\boldsymbol{x}) = \int \phi(\boldsymbol{u}) e^{2\pi i(\boldsymbol{u},\boldsymbol{x})} d\boldsymbol{u}$$

then the coefficients $\boldsymbol{Au}$ are proportional to $\phi(\boldsymbol{u})$.

If mJ and nJ are two parastichy numbers, if the corresponding vectors generate the lattice, and if the parastichies make angles ψ_1, ψ_2 with the generators of the cylinder, then the matrix

$$\frac{J}{2\pi\varrho} \begin{pmatrix} n & n\tau_2 \\ -m & -m\tau_1 \end{pmatrix}$$

is one of the matrix descriptions of the inverse lattice. Here

$$\tau_1 = \tan\psi_1, \qquad \tau_2 = \tan\psi_2. \tag{I.12.3}$$

It will be seen that the coefficients A, C are bound to be multiples of $J/2\pi\varrho$. In other words, the first coordinates of the points of the inverse lattice are all multiples of $J/2\pi\varrho$. This simply represents the fact that the leaf pattern is unaltered by rotating the cylinder through the angle $2\pi/J$.

When drawing diagrams to describe a phyllotaxis through its inverse lattice, it is helpful to draw a number of vertical lines, $u = m/2\pi\varrho$ or possibly only the lines $u = mJ/2\pi\varrho$. The points of the inverse lattice are bound to lie on these lines, and may be imagined as beads sliding up and down on them. Owing to the fact that only relatively few points of the inverse lattice can be of importance it is appropriate to use a relatively large scale for such diagrams, i.e. a larger area of paper may be used per lattice point in the case of an inverse lattice than would be appropriate for the primary lattice. It may be noticed that, apart from scale, the two lattices are obtainable from one another by rotating through 90°; the scale factor is Δ or Δ^{-1}.

In part II ideas involved in the inverse lattice will be found of immense importance. To a large extent, however, it will no longer be possible to work in terms of lattices. To assume that one has a lattice is an approximation which is no longer appropriate when discussing the origin of the phyllotactic patterns. It is nevertheless still appropriate to describe functions on the surface of the cylinder by a Fourier analysis of some kind. The appropriate kind of Fourier analysis for functions defined on a cylinder is of the form of a Fourier series in one variable and a Fourier integral in the other:

$$f(x,y) = \sum_{m=-\infty}^{\infty} \int_{-\infty}^{\infty} F_m(v) e^{(imx/P)+ivy} dy. \tag{I.12.4}$$

The function $F_m(v)$ may be regarded as being defined on the lines $u = m/2\pi\varrho$ of the diagram mentioned above. In general $F_m(v)$ will be complex, and cannot be very easily represented on the diagram. It may happen however (and in practice it always happens), that $f(x,y)$ has a centre of symmetry, $f(x,y) = f(x_0 - x, y_0 - y)$. In this case, if (x_0, y_0) is made the origin, the function $F_m(v)$ is real. If in addition, as often happens, the function is nowhere negative, the function may be conveniently shown diagrammatically by representing the vertical lines of varying thickness. The case of a lattice arises when this widening of the lines is restricted to isolated points.

The diagrams for the inverse lattice also include a circle whose radius is $K_0 = 2\pi/\lambda_0$ where λ_0 is the so-called "optimum wavelength". K_0 may be called the "optimum radian wave number". There is a tendency for the lattice points with the greatest coefficients in (I.12.2) to lie not very far from this circle.

The expression (I.12.2) is a very familiar one when applied to three-dimensional lattices. It then gives the relation between the electron density in a crystal and the X-ray reflection pattern. The similarity of phyllotactic patterns to crystal lattices was first observed by the brothers Bravais.

13. Flow matrices

If a lattice is changing, the manner of its change may be described by some matrix description $\boldsymbol{A}(t_0)$ of the lattice at some time t_0, together with the product $(\boldsymbol{A}(t))^{-1}\boldsymbol{A}'(t)$ at other times, the dash here representing differentiation. This product will be called the "flow matrix". It is independent of the matrix description chosen, for if $\boldsymbol{B}(t)$ is another matrix description there is an improper unimodular matrix $\boldsymbol{L}$ such that $\boldsymbol{B}(t) = \boldsymbol{LA}(t)$, and if $\boldsymbol{A}(t)$ and $\boldsymbol{B}(t)$ are continuous, $\boldsymbol{L}$ must be constant. But then

$$(\boldsymbol{B}(t))^{-1}\boldsymbol{B}'(t) = (\boldsymbol{LA}(t))^{-1}\boldsymbol{LA}'(t) = (\boldsymbol{A}(t))^{-1}\boldsymbol{A}'(t).$$

If one uses helical coordinates the matrix description may be written as

$$\begin{pmatrix} 2\pi\varrho J^{-1} & 0 \\ \alpha\varrho & \eta \end{pmatrix}$$

and the flow matrix is

$$F = \begin{pmatrix} F_{11} & F_{12} \\ F_{21} & F_{22} \end{pmatrix} = \begin{bmatrix} \dfrac{\mathrm{d}\log\varrho}{\mathrm{d}t} & 0 \\ \dfrac{\varrho}{\eta}\dfrac{\mathrm{d}\alpha}{\mathrm{d}t} & \dfrac{\mathrm{d}\log\eta}{\mathrm{d}t} \end{bmatrix}. \qquad \text{(I.13.1)}$$

A convenient way of picturing flow matrices is to imagine the change in the lattice as being due to the leaves being carried over the surface of the lattice by a fluid whose velocity is a linear function of position. The flow matrix then gives the relation between the velocity and the position. This point of view is particularly suitable when one is concerned with leaves which are sufficiently mature to be no longer moving with respect to the surrounding tissue, but only have movement due to the growth of that tissue. The coefficient F_{11} then represents the exponential rate of growth in girth of the stem, and the coefficient F_{22} the exponential rate of increase of the stem in length. The sum of these, the trace of the flow matrix, is the exponential rate of increase of the leaf area. The coefficient F_{21} represents whatever tendency there is for the stem to twist. It should be small, or in other words the divergence angle should not be appreciably affected by such growth. If this coefficient F_{21} is zero, the flow may be described as being "without twist". A flow without twist and with $F_{11} = F_{22}$, i.e. a scalar flow matrix, may be described as a "compression". One with $F_{11} + F_{22} = 0$ may be described as "area preserving". One may also consider flows with $F_{12} \neq 0$ but these of course cannot apply to phyllotactic lattices but only to more general lattices. A flow with $F_{11} = F_{22} = 0$, $F_{12} + F_{21} = 0$ represents a rotation.

If a lattice has a representation which changes continuously with time from $\boldsymbol{A}(t_1)$ to $\boldsymbol{A}(t_2)$, then the ratio $(\boldsymbol{A}(t_1))^{-1}\boldsymbol{A}(t_2)$ may be called the "finite flow matrix" for the period t_1 to t_2. Finite flows are also independent of the representation.

If a flow matrix is independent of time, the corresponding finite flow may be expressed as the exponential of the product of the flow and the time for which it acts. Particular cases of exponentials of matrices that may be relevant in this connection are*

$$\exp\begin{pmatrix}0 & \theta\\ -\theta & 0\end{pmatrix} = \begin{pmatrix}\cos\theta & \sin\theta\\ -\sin\theta & \cos\theta\end{pmatrix}, \tag{I.13.2}$$

$$\exp\begin{pmatrix}\kappa_1 & 0\\ 0 & \kappa_2\end{pmatrix} = \begin{pmatrix}e^{\kappa_1} & 0\\ 0 & e^{\kappa_2}\end{pmatrix}, \tag{I.13.3}$$

$$\exp\begin{pmatrix}0 & 0\\ \chi & 0\end{pmatrix} = \begin{pmatrix}1 & 0\\ \chi & 1\end{pmatrix}. \tag{I.13.4}$$

They give the effects of constant rotations, twistless flows and pure twists.

If a finite flow matrix $\boldsymbol{G}$ arises from the continuous change of a *phyllotactic* lattice then $G_{12} = 0$ since the point $(2\pi\varrho, 0)$ must transform into another of the same form. Also G_{11} and G_{22} must be positive, since they are continuous, initially both unity, and can never vanish without the leaf area vanishing. On the other hand, every matrix $\boldsymbol{G}$ with $G_{12} = 0,\ G_{11} > 0,\ G_{22} > 0$ is a possible finite flow matrix, as may be seen by writing it in the form

$$\begin{pmatrix}G_{11} & 0\\ 0 & G_{22}\end{pmatrix}\begin{pmatrix}1 & 0\\ G_{21}/G_{11} & 1\end{pmatrix}.$$

It can now easily be seen that if a continuous change of a phyllotactic system is described by a finite flow matrix equal to the matrix $\boldsymbol{G}$ then the corresponding curve in the space of lattices is deformable to zero; for the set of points satisfying $G_{12} = 0,\ G_{11} > 0,\ G_{22} > 0$ is simply connected.

14. The touching circles phyllotaxis

The lattice patterns which arise from packing circles as tightly as possible on the surface of a cylinder have been considered as models for phyllotaxis (v. ITERSON 1907). The cylinder is supposed to increase in diameter and the lattice continuously adjusts itself, without major alteration, to allow as many circles as possible per unit area. It is not difficult to see that in tightest packing every circle is touching two others. If a formal proof is desired, one may argue as follows. If no vectors of the lattice are as short as the diameter l of the circles, then η may be decreased until one of them has this length. Thereafter one may continue to decrease η until a second vector has length l, but during this second phase of the process the divergence angle must be continually modified to ensure that the first vector remains of length l. Such a lattice can be described by the matrix

$$l\begin{pmatrix}\sin\psi_1 & \cos\psi_1\\ \sin\psi_2 & \cos\psi_2\end{pmatrix}$$

---

* If $\boldsymbol{A}$ is a matrix, $\exp(\boldsymbol{A}) = \boldsymbol{I} + \boldsymbol{A} + \boldsymbol{A}^2/2 + \cdots + \boldsymbol{A}^n/n! + \cdots$. The formula (I.13.2) follows from

$$\begin{pmatrix}0 & \theta\\ -\theta & 0\end{pmatrix}^n = \begin{cases}\begin{pmatrix}0 & \theta^n\\ -\theta^n & 0\end{pmatrix}(-1)^{(n-1)/2} & \text{if } n \text{ is odd,}\\[2ex] \begin{pmatrix}\theta^n & 0\\ 0 & \theta^n\end{pmatrix}(-1)^{n/2} & \text{if } n \text{ is even.}\end{cases}$$

where

$$|\psi_1| < \pi/2, \quad |\psi_2| < \pi/2, \quad \pi/3 \leq \psi_1 - \psi_2 < 2\pi/3$$

and the flow matrix will be found to be

$$\frac{\mathrm{d}\log\varrho}{\mathrm{d}t}\begin{pmatrix} 1 & 0 \\ -(\tan\psi_1 + \tan\psi_2) & \tan\psi_1\tan\psi_2 \end{pmatrix}.$$

The angles ψ_1, ψ_2 are obliged to satisfy

$$\dot{\psi}_1 \tan\psi_1 = \dot{\psi}_2 \tan\psi_2$$

in order that $F_{12} = 0$.

At certain values of the radius the third principal vector is also of length l. The lattice may then be called "equilateral". For such a lattice $\psi_1 - \psi_2$ must be either $\pi/3$ or $2\pi/3$. It may be supposed to be $2\pi/3$; otherwise one may interchange the first two vectors and change the sign of one of them. If one then puts $\psi_3 = \psi_2 + 2\pi/3$, all the angles $\psi_1 - \psi_2, \psi_2 - \psi_3, \psi_3 - \psi_1$ will be $2\pi/3$. If the radius of the cylinder is allowed to alter still further, the directions of the three sides of the principal triangle will still be given by the angles ψ_1, ψ_2, ψ_3 or angles very close to them, but their roles will have been interchanged. The first two vectors will be in directions ψ_1' and ψ_2' and the third in direction ψ_3' and

$$\psi_1' - \psi_2' = \psi_2' - \psi_3' = \psi_3' - \psi_1' = 2\pi/3$$

and ψ_1'; ψ_2', ψ_3' will be a permutation of ψ_1, ψ_2, ψ_3. The permutation has to be cyclic to ensure $\psi_1' - \psi_2' = \psi_2' - \psi_3' = \psi_3' - \psi_1' = 2\pi/3$. The question as to which of the three angles ψ_1, ψ_2, ψ_3 is to be ψ_3' must be decided by the condition that the subsequent change of the lattice shall, for a short time at any rate, satisfy $\pi/3 \leqslant \psi_1' - \psi_2' < 2\pi/3$ and, subject to this, the area $\Delta = l^2 \sin(\psi_1' - \psi_2')$ shall be as small as possible. This means to say that $\mathrm{d}\Delta/\mathrm{d}t$ must change sign but, subject to this, have the smallest available modulus. Now in the period before the change

$$\begin{aligned}\frac{1}{\Delta}\frac{\mathrm{d}\Delta}{\mathrm{d}t} &= (1 + \tan\psi_1\tan\psi_2)\frac{\mathrm{d}\log\varrho}{\mathrm{d}t} \\ &= \frac{\cos(\psi_1 - \psi_2)}{\cos\psi_1\cos\psi_2\cos\psi_3}\cos\psi_3\frac{\mathrm{d}\log\varrho}{\mathrm{d}t} \\ &= \frac{-\mathrm{d}\log\varrho/\mathrm{d}t}{2\cos\psi_1\cos\psi_2\cos\psi_3}\cos\psi_3\end{aligned}$$

whereas afterwards

$$\frac{1}{\Delta}\frac{\mathrm{d}\Delta}{\mathrm{d}t} = \frac{-\mathrm{d}\log\varrho/\mathrm{d}t}{2\cos\psi_1'\cos\psi_2'\cos\psi_3'}\cos\psi_3'.$$

Then $\cos\psi_3'$ has the opposite sign to $\cos\psi_3$ and $|\cos\psi_3'|$ must equal the smaller of $|\cos\psi_1|$ and $|\cos\psi_2|$. Since $\cos\psi_1 + \cos\psi_2 + \cos\psi_3 = 0$ the change of sign is certainly possible. In the case that $\cos\psi_1$ and $\cos\psi_2$ have the same sign,

$$|\cos\psi_3'| = \mathrm{Min}\,|\cos\psi_r|$$

whereas if they have the opposite sign

$$|\cos\psi_3'| = \mathrm{Max}\,|\cos\psi_r|.$$

In no case is $|\cos\psi_3'|$ intermediate between $|\cos\psi_1'|$ and $|\cos\psi_2'|$. But the three principal parastichy numbers (after the change) are proportional to $|\cos\psi_1'|$, $|\cos\psi_2'|$, $|\cos\psi_3'|$. The third principal parastichy number is therefore either the smallest or the greatest.

In the case that $d\log\varrho/dt > 0$, $\cos\psi_1'\cos\psi_2' < 0$ (since $d\Delta/dt > 0$) and the third parastichy number is the smallest. Then the hypothesis of geometrical phyllotaxis is satisfied immediately after the lattice is equiangular. It will continue to be so until there is a change in the principal parastichy numbers. The first two can only change at equiangular lattices but the third can also change when the lattice is square. Now if q_1, q_2 are the first two parastichy numbers the third is either q_1+q_2 or $|q_1-q_2|$. If the change is upward, i.e. if $d\log\varrho/dt > 0$, and the third parastichy number is the smallest, the new value of the third parastichy number must be q_1+q_2 and the hypothesis is still satisfied. If the parastichy number is decreasing, the new value of the third parastichy number is $|q_1-q_2|$ and the hypothesis will be satisfied if and only if q_1/q_2 lies between 1/2 and 2. Thus the hypothesis of geometrical phyllotaxis is satisfied in the case of "touching circles phyllotaxis" from the point when the lattice first becomes equiangular and can only cease to do so when $d\varrho/dt < 0$, and the third parastichy number cannot be decreased to become the difference of the first two without contradicting the hypothesis, therefore

In a continuously varying touching circles phyllotaxis the hypothesis of geometrical phyllotaxis is satisfied from the time when an equilateral triangle first appears onwards.

The difficulty in the proof given above lay largely in deciding which side of the equilateral triangle increases when the diameter of the cylinder increases. The following not very rigorous argument may be found helpful. Consider three of the circles forming an equilateral triangle of the lattice (Fig.4). The circles are being pressed downwards to ensure the closest packing. The downward pressure of the upper circle will tend to wedge the lower circles apart, whilst at the same time holding it in contact with the other two. Thus if ϱ is increasing it is the most nearly horizontal of the sides which increases. When a touching circles lattice with decreasing ϱ reaches a state when $\tan\psi_1\tan\psi_2 = \infty$ it is no longer possible for the lattice to continue. This happens if ψ_1 or ψ_2 is 90°.

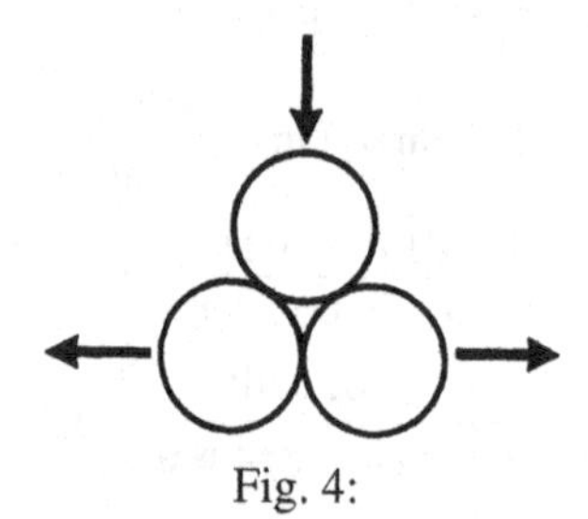

Fig. 4:

It is of some interest to know what the angles ψ_1, ψ_2, ψ_3 will be when the lattice is equilateral. Writing $l\cos\psi_i = \eta m_j$ one obtains

$$\tan\psi_j = (m_{i+1} - m_{i-1})/m_j\sqrt{3}$$

(the suffixes being reckoned mod 3). When the parastichy numbers are 1, 1, 2 the directions are 0°mod 60°. When they are 1, 2, 3 the directions are $10^\circ 54'$mod 60°. When the parastichy numbers are 2, 3, 3 the directions are $6^\circ 35'$mod 60°. For the limit of large parastichy numbers of the Fibonacci series the directions are $\tan^{-1}(\omega^{-3}/\sqrt{3})$, i.e. $7^\circ 46'$mod 60°. It will be seen that these lattices have vectors lying very nearly along the generators of the cylinder. This will be of importance in §18, where evidence will be brought to show that the touching circles theory is unlikely to be valid. The arguments apply also equally well to discredit any theory requiring that for some values of the radius of the cylinder the lattice be equiangular.

The divergence angles for the equiangular lattices are as follows:

1,2,2	180°
1,2,3	128°34′
2,3,5	142°6′
F_{n-1}, F_n, F_{n+1}	$2\pi\left(\dfrac{F_{n-1}F_{n+1}+F_{n-1}^2+F_{n-2}^2}{F_{n+1}^2+F_n^2+F_{n-1}^2}\right)$.

Closely related to touching circles phyllotaxis is a phyllotaxis whose inverse lattice is a touching circles lattice. The importance of such a lattice is that its points may be regarded as the maxima of two sets of waves of fixed wavelength superimposed on one another.

15. The lattice described by its twist and other coordinates

Any phyllotactic lattice may be described by its helical coordinates in the form of the matrix

$$\begin{pmatrix} 2\pi\varrho J^{-1} & 0 \\ \alpha\varrho & \eta \end{pmatrix}.$$

Alternatively, it may be expressed as a product of the matrix

$$\begin{pmatrix} 2\pi\varrho J^{-1} & 0 \\ \alpha_0\varrho & \eta \end{pmatrix}$$

describing a lattice with a limiting divergence angle α_0, and a finite flow matrix

$$\begin{pmatrix} 1 & 0 \\ \chi & 1 \end{pmatrix}$$

describing a twist. The matrix describing the lattice may be further broken down into factors as follows

$$\Delta^{1/2}\begin{pmatrix} 1 & 0 \\ \alpha_0 J/2\pi & 1 \end{pmatrix}\begin{pmatrix} \kappa & 0 \\ 0 & \kappa^{-1} \end{pmatrix}\begin{pmatrix} 1 & 0 \\ \chi & 1 \end{pmatrix} \tag{I.15.1}$$

where $\kappa = (2\pi\varrho/\eta J)^{1/2} = 2\pi\varrho/J\Delta^{1/2}$. One may of course multiply it on the left by any unimodular matrix, for instance

$$\begin{pmatrix} p_n & q_n \\ p_{n+1} & q_{n+1} \end{pmatrix}$$

where p_n/q_n and p_{n+1}/q_{n+1} are two successive convergents of $\alpha_0 J/2\pi$.

Since $\alpha_0 J/2\pi$ is a limiting divergence angle, the partial quotients of $\alpha_0 J/2\pi$ are all 1 beyond some point and it will be supposed that p_n/q_n and p_{n+1}/q_{n+1} are obtained by taking sufficiently many partial quotients to ensure that some of these 1's are included. Then

$$q_n\frac{\alpha J}{2\pi} - p_n = (-1)^n\omega^{-n}A \quad \text{where} \quad A = \frac{\omega J}{q_0+\omega q_1}$$

and the lattice is described by the matrix

$$\Delta^{1/2}\begin{pmatrix} A\kappa(-\omega^{-1})^n & \kappa^{-1}q_n \\ A\kappa(-\omega^{-1})^{n+1} & \kappa^{-1}q_{n+1}\end{pmatrix}\begin{pmatrix}1 & 0\\ \chi & 1\end{pmatrix}. \tag{I.15.2}$$

In the case that $\chi = 0$, for some n, Jq_n, Jq_{n+1}, Jq_{n-1} are the principal parastichy numbers, and (I.15.2) is the principal matrix representation. In this way the lattice is described by the leaf area Δ, the parameter κ which is directly related to the plastochrone ratio, and the twist χ. The parastichy series is supposed known so that the value of $A = \omega J/(q_0 + q_1\omega)$ is determined. In theory one can refer the phyllotaxis to any parastichy series, but in practice if one refers it to the "wrong" series monstrously large twists are required.

For large parastichy numbers, q_n is approximately $(q_0 + \omega q_1)\omega^{n-1}/\sqrt{5}$ i.e. $J\omega^n/A\sqrt{5}$, and the matrix approximates the form

$$\frac{\Delta^{1/2}}{5^{1/4}}\begin{pmatrix}\omega^{2\theta+1/2} & \omega^{-2\theta-1/2}\\ -\omega^{2\theta-1/2} & \omega^{-2\theta+1/2}\end{pmatrix}\begin{pmatrix}1 & 0\\ \chi & 1\end{pmatrix} \tag{I.15.3}$$

where n has been supposed even and $\theta = -n/2 - 1/4 + \frac{1}{2}\log_\omega(A\kappa)$. Lattices of the form (I.15.3) with $\chi = 0$ may be called "ideal phyllotactic lattices". Apart from the leaf area, which is only a scale factor, the only parameter is θ, and the lattice is a periodic function of θ with unit period. If 1/2 is added to θ the lattice is transformed into the mirror image in a generator of the cylinder for

$$\begin{pmatrix}\omega^{2\theta+3/2} & \omega^{-2\theta-3/2}\\ -\omega^{2\theta+1/2} & \omega^{-2\theta-1/2}\end{pmatrix} = \begin{pmatrix}-1 & 1\\ 1 & 0\end{pmatrix}\begin{pmatrix}\omega^{2\theta+1/2} & \omega^{-2\theta-1/2}\\ -\omega^{2\theta-1/2} & \omega^{-2\theta+1/2}\end{pmatrix}.$$

A very natural way of describing a general lattice is to express it by means of a matrix which is the product of one which describes an ideal lattice and another matrix of unit determinant and with two parameters, thus for instance

$$\frac{\Delta^{1/2}}{5^{1/4}}\begin{pmatrix}\omega^{2\theta+1/2} & \omega^{-2\theta-1/2}\\ -\omega^{2\theta-1/2} & \omega^{-2\theta+1/2}\end{pmatrix}\begin{pmatrix}1 & \mu\\ 0 & 1\end{pmatrix}\begin{pmatrix}1 & 0\\ \chi & 1\end{pmatrix}. \tag{I.15.4}$$

The order of the matrices $\left(\begin{smallmatrix}1 & 0\\ \chi & 1\end{smallmatrix}\right)$ and $\left(\begin{smallmatrix}1 & \mu\\ 0 & 1\end{smallmatrix}\right)$ is conveniently immaterial. As a finite flow, the first of these represents a pure twist and the second a pure shear. The shear coordinate is closely related to the plastochrone ratio by the relation

$$\mu = \pm(-1)^n \begin{vmatrix} q_{n-1} & q_n\\ q_n & q_{n+1}\end{vmatrix}\frac{\Delta^2}{2\pi\varrho}$$

which holds for any n, with sign independent of n.

A convenient way of describing the effect of a finite twisting flow

$$\begin{pmatrix}1 & 0\\ \chi & 1\end{pmatrix}$$

is to say that it adds χ to the tangent of all the angles which lattice vectors make with the generators of the cylinder. This idea at once provides a method of determining the twist of a phyllotactic system. Suppose that the parastichies with parastichy numbers q_n, q_{n+1} make angles ψ_n, ψ_{n+1} with the generators, and that the limiting angle concerned is that which corresponds to the sequence with q_n, q_{n+1} as consecutive terms. Then if the relation $\tan\psi_n = -\omega\tan\psi_{n+1}$ were satisfied, the divergence angle would have the limiting angle. Since this relation can be ensured by subtracting χ from both $\tan\psi_n$ and $\tan\psi_{n+1}$, the equation

$$\tan\psi_n - \chi = -\omega(\tan\psi_{n+1} - \chi)$$

must hold, i.e.

$$\chi = (\tan\psi_n + \omega\tan\psi_{n+1})/(1+\omega).$$

In the case of an equilateral lattice of high parastichy number, one may take $\tan\psi_n = (3\omega^{-1})/2\sqrt{3}$, $\tan\psi_{n+1} = \omega^{-3}/\sqrt{3}$, $\chi = 5/(2\omega^3\sqrt{3}) = 0.341$.

In the case of an ideal lattice with $\theta = 0$, the matrix is orthogonal and so represents a square lattice, which is then also a touching circles lattice. Thus the touching circles lattice is ideal when both are square. The sides of the square make an angle $\tan^{-1}\omega^{-1}$, i.e. $31°43'$ with the coordinate axes.

16. The optimum packing problem

If one submits a lattice, with matrix description

$$\begin{pmatrix} a & b \\ c & d \end{pmatrix}$$

to an area preserving twistless flow

$$\begin{pmatrix} \kappa & 0 \\ 0 & \kappa^{-1} \end{pmatrix}$$

then the points of the lattice all move along rectangular hyperbolae, and have a non-zero minimum distance from the origin. The point

$$(m,n)\begin{pmatrix} a & b \\ c & d \end{pmatrix}$$

will have minimum distance $[2(ma+nc)(mb+nd)]^{1/2}$. The minimum of this distance taken over all the points of the lattice is of a certain interest and importance. It is the shortest distance to which any leaves approach one another during the flow. If this distance is small in comparison with $\Delta^{1/2}$ then the leaves will, at some stage of the flow, become awkwardly close. It is interesting in this connection that *the ideal lattices are optimum in the sense that with them the minimum distance has the maximum value for the given value of* Δ. To prove this, let

$$\begin{pmatrix} a & b \\ c & d \end{pmatrix}$$

be a matrix representation of the lattice in which $ac < 0$, $bd > 0$. Such a representation can be made from two of the principal vectors. Then the minimum of $|(ma+nc)(mb+nd)|$ for pairs of integers m, n of the same sign (say for positive m, n) occurs when m/n is one of the convergents of the continued fraction of $-c/a$. If not, let $p_r/q_r, p_{r+1}/q_{r+1}$ be two consecutive convergents such that $q_r \leq n \leq q_{r+1}$. Since the matrix

$$\begin{pmatrix} p_r & q_r \\ p_{r+1} & q_{r+1} \end{pmatrix}$$

has determinant ± 1, one can find integers k, l such that $m = kp_r + lp_{r+1}, n = kq_r + lq_{r+1}$. Since m/n is not a convergent, neither k nor l can be zero. They must be of opposite sign since $n < q_{r+1}$. Now if $m \geq p_r$ then

$$|mb + nd| > |p_r b + q_r d|$$

and

$$|ma + nc| = |k(p_r a + q_r c) + l(p_{r+1} a + q_{r+1} c)| > |p_r a + q_r c|.$$

But then m, n could not then give the minimum for p_r, q_r would do better. Then $m < p_r$ and $n > q_r$. But in this case by reducing q_r by 1 one would certainly reduce $|mb + nd|$ and one would reduce $|ma + nc|$ for

$$ma/c + n > ma/c + (n - 1) \geq (-a/c)(p_r + q_r a/c + 1) \geq 0.$$

This shows that m/n must be a convergent of $-c/a$. Likewise if m/n is negative it must be a convergent of $-d/b$. In the case of the ideal lattice, $-a/c = \omega$, $b/d = \omega^{-1}$ and m and n are consequently Fibonacci numbers. Then

$$|F_r a + F_{r+1} c| = |c||\omega F_r - F_{r+1}| = |c|\omega^{-r},$$

$$|F_r b + F_{r+1} d| = |b||\omega F_r + F_{r+1}| = b\omega^{r+1} = |d|\omega^r,$$

$$|F_r a + F_{r+1} c||F_r b + F_{r+1} d| = |cd| = \Delta/\sqrt{5}.$$

Thus the shortest distance is

$$5^{-1/4}(2\Delta)^{1/2} = 0.8945\sqrt{\Delta}.$$

When a lattice of such a family has worst packing then $\pm(ma + nc) = mb + nd$ and $(ma + nc,\ mb + nd)$ is a principal vector, i.e. one may take $a = b$ and $m = 1$, $n = 0$. The worst packing ideal lattice is

$$\left(\frac{\Delta}{\sqrt{5}}\right)\begin{pmatrix} 1 & 1 \\ -\omega & \omega^{-1} \end{pmatrix}.$$

It must still be shown that no other lattice gives so large a minimum. Write $A_r = p_r a + q_r c$, $B_r = p_r b + q_r d$, then the determinants $A_r B_{r-1} - A_{r-1} B_r$ all have the value $\pm\Delta$. If the partial quotients of $-a/c$ are K_r, then

$$A_{r+1} = K_{r+1} A_r + A_{r-1}, \quad B_{r+1} = K_{r+1} B_r + B_{r-1}.$$

Now suppose $K_{r+1} \geq 3$, then for $B_r/B_{r+1} > 0$, $A_r/A_{r+1} < 0$

$$\frac{|\Delta|}{|A_r B_r|} = \left|\frac{B_{r+1}}{B_r} - \frac{A_{r+1}}{A_r}\right| \geq 3.$$

Since

$$\frac{|\Delta|}{|A_r B_r|} < \sqrt{5} < 3$$

it follows that there can be no partial quotients as large as 3 in the partial fraction for $-c/a$. Then $B_{r+1}/B_r \leq 3$ for each r, and so if $K_r = 2$ then

$$\frac{|\Delta|}{|A_r B_r|} \geq \left|\frac{B_r}{B_{r+1}}\right| = \left|K_r + \frac{B_{r-1}}{B_r}\right| > 2\tfrac{1}{3} > \sqrt{5}.$$

There can therefore not be any partial quotients as large as 2, so they are all 1, i.e. $-c/a$ is ω or ω^{-1}. Likewise b/d is ω or ω^{-1}. [There is a handwritten marginal note here that "more argument is needed for $r = 0$". The above proof is indeed not valid for K_0, but the value of K_0 does not affect the divergence angle.]

Though this optimum property is of considerable mathematical interest, its biological importance is perhaps rather secondary. Above all it would be quite unjustified to suppose that the appearance in nature of nearly ideal lattices is due to a search for the best lattice. On the contrary it seems probable that the effect of such a search would be to defeat its own ends. It would be likely that the evolutionary process would lead to some not too bad lattice which was a local optimum and remain there. It will be realised that if the ratio a/c is allowed to change continuously it has to pass through rational values, and these are the very worst from the point of view of packing. More specifically, if one wishes to alter one of the partial quotients of a number by altering the number continuously, then it is necessary to allow the next partial quotient to take unlimited large values. Of course not all the partial quotients concerned in a phyllotactic lattice can be of importance, but if any one of them is of sufficient importance that it must be kept down to a moderate size then this fact prevents any of the previous partial quotients from being altered. However, although it is unreasonable to suppose that there is any such evolutionary search for the best lattice from the point of view of packing in spite of twistless area-preserving flow, the fact that the naturally occurring lattices have, or very nearly have, the optimum property, still has its advantages. It means in effect that if there are mutations which modify the twistless growth, disadvantageous packing effects will arise.

17. Comparison of methods of describing lattices

A considerable number of different sets of parameters have now been introduced for the description of phyllotactic lattices. Their various purposes, merits and defects will now be compared.

(1) The most fundamental way of describing lattices is by a matrix

$$\begin{pmatrix} a & b \\ c & d \end{pmatrix}.$$

The other methods described may all be related to it. Its main advantage is its generality, and its main disadvantage its lack of uniqueness.

(2) Closely related to the matrix describing the lattice is that which describes the inverse lattice

$$\begin{pmatrix} A & B \\ C & D \end{pmatrix}.$$

This will be found particularly useful in part II.

(3) Amongst methods of making the matrix description unique is the use of helical coordinates, i.e. the use of the matrix description

$$\begin{pmatrix} 2\pi\varrho/J & 0 \\ \alpha\varrho & \eta \end{pmatrix}.$$

If the plastochrone distance η is required to be positive and the modulus of the divergence angle α, satisfies $-\pi < \alpha \leq \pi$, this representation is unique. This form of description is more suitable for the description of the lattice as a group of congruences of the cylinder, but is not very helpful where theories of the origin of the phyllotactic pattern are concerned. The divergence angle and the plastochrone displacement are not easily measured or even appreciated on specimens with high parastichy numbers. As compared to the method next to be described the helical coordinates at least have the advantage of changing continuously with continuously changing lattices.

(4) Another method of making the matrix description unique is to make the vectors represented by the two rows of the matrix be the first two principal vectors. This is the principal matrix representation. This representation is, one might say, the most natural matrix representation, i.e. that which one would be most likely to choose if asked to give a matrix representation of a lattice. Its main disadvantage is that it undergoes discontinuous changes when the lattice changes continuously.

(5) There are methods of describing a lattice by means of parameters which vary continuously with change of lattice, and in such a way that lattices whose principal matrix representations are near to one another are represented by neighbouring sets of parameters. Such descriptions can for instance be based on the theory of elliptic functions. The disadvantage of these methods is that these sets of parameters are most unmanageable from the point of view of their algebraic properties. They are not further discussed elsewhere in this paper.

(6) When one wishes to measure the parameters of a lattice, suitable quantities are the radius ϱ and the two angles ψ_1, ψ_2 which the two parastichies make with the generators. In addition to these measured quantities one needs to know the relevant parastichy numbers.

(7) When one is concerned with phyllotactic lattices belonging to a known series of parastichy numbers, rather different parameters are appropriate. These are the parameters Δ, θ, μ, χ (expression I.15.4). These parameters vary continuously in a continuously varying lattice. In theory it is possible for very similar lattices to be described by very different parameters, but this does not cause any genuine misunderstanding.

18. Variation principle theories. Equilateral lattices

One rather attractive type of theory to account for the change of phyllotaxis with changing radius is to suppose that there is some function of the lattice that the plant attempts to minimise. It is only able to achieve local minima, and is restricted to phyllotactic lattices that can be fitted to the cylinder available. This "potential" function should of course be defined for lattices other than phyllotactic lattices, and should be unchanged on rotating the lattice. The touching circles phyllotaxis can be defined by such a potential function, viz. the ratio of the length of the shortest lattice vector to l, so long as this ratio exceeds 1, and by $\Delta/2l$ otherwise. Likewise the fixed wavelength lattices can be defined by a similar potential. Other potential functions may be defined in the form

$$\phi(\Lambda) = \sum_{\boldsymbol{u} \in \Lambda} f(|\boldsymbol{u}|).$$

The function $f(r)$ should preferably tend to zero quickly as r tends to infinity, and have a negative minimum at some positive value, l. When the minimum is very sharp, one approaches the touching circles lattices again.

Nearly all such theories require that at certain values of the plastochrone ratio the lattice must be hexagonal, for if the equilateral lattice is optimum with a certain vector length in the infinite plane it will also be optimum on any circle on which it can be fitted. Now suppose that the vector length is chosen so that a lattice (in the unrestricted plane) is equilateral and gives the minimum potential for equilateral lattices. Then although this lattice may not be an optimum or even a local optimum it is at least locally stationary in the space of lattices. For suppose the lattice begins to change with flow matrix $\boldsymbol{F}$. Then the rate of change of potential will be linear in the coefficients of $\boldsymbol{F}$, e.g trace($\boldsymbol{FH}$). Now if the whole lattice is rotated by a matrix $\boldsymbol{U}$ then $\boldsymbol{F}$ becomes $\boldsymbol{UFU}^{-1}$. Since for rotations of 60° the rate of change of potential due to the flow $\boldsymbol{F}$ will be unaltered by the rotation, it follows that, if $\boldsymbol{U}$ is such a rotation, trace $(\boldsymbol{UFU}^{-1}\boldsymbol{H}) =$ trace $(\boldsymbol{FH})$ for any $\boldsymbol{F}$, i.e. $\boldsymbol{U}^{-1}\boldsymbol{HU} = \boldsymbol{H}$. Thus $\boldsymbol{H}$ commutes with the matrices

$$\begin{pmatrix} \cos r\pi/3 & \sin r\pi/3 \\ -\sin r\pi/3 & \cos r\pi/3 \end{pmatrix}$$

and is therefore of the form

$$\begin{pmatrix} A & B \\ -B & A \end{pmatrix}.$$

But since the potential is stationary for pure compressions, $A = 0$. It is also stationary for pure rotations, and therefore $B = 0$. Hence the equilateral lattice is a stationary point of the potential. It may not be a minimum, and if it is a minimum it may still not be one which really ever comes into play. But these possibilities on the whole seem rather unlikely. One must expect that theories depending on a variation principle will involve equilateral lattices for appropriate radii.

It seems however that no such theory can be right, for in the experimental material there seems to be no trace of any equilateral lattices. As has been mentioned, equilateral lattices of high parastichy number have a twist χ of about 0.34. In actual mature specimens one seldom finds values of χ even as large as 0.1. This might possibly be explained by the lattice being subjected to a squashing flow

$$\begin{pmatrix} \kappa & 0 \\ 0 & \kappa^{-1} \end{pmatrix}$$

in the growth after the lattice has been formed, which results in the twist getting magnified by the factor κ^2. It is difficult however to estimate the values of κ which might apply. But the same applies with growing points.

An equilateral lattice, if it occurred, would be rather unsatisfactory from the point of view of packing. A not very great degree of squashing applied to an equilateral lattice gives a lattice with very poor packing indeed.

Part II. Chemical Theory of Morphogenesis

1. Morphogen equations for an assembly of cells. The linear case

In TURING (1952) the theory of a reaction and diffusion system was developed for the case where the geometrical form of the organism was a ring of cells, and where the reaction rates might be considered as linear functions of the concentrations. The equations that will be found in this part are applicable to arbitrary geometrical forms and reaction rate functions. In this investigation, as in the previous one, the geometrical form is assumed to remain unchanged throughout. This assumption cannot of course always be satisfied—indeed variations of chemical concentrations would be of little importance if they did not ultimately affect growth—but the rates of growth are likely to be slow enough for the equilibria of chemical concentration that are reached not to be appreciably affected by the growth.

The description of the organism may be divided into a geometrical and a chemical part, concerned respectively with the diffusion and the reactions of the morphogens which are to be found in it. The word "morphogen", which was introduced in TURING (1952), was there, in effect, defined to mean essentially "chemical substance relevant to morphogenesis". In the present paper it will be given a slightly more restrictive meaning, viz. "chemical substance, the variation of whose concentration is described by a variable in the mathematical theory". The state of the organism at any time t may be described by MN numbers $\Gamma_{mn}(m = 1,2,\ldots,M; n = 1,2,\ldots,N)$, where Γ_{mn} is the concentration of the mth morphogen in the nth cell. This description supposes that there is no need to distinguish one point of a cell from another, an assumption which is probably true, as there is usually considerable protoplasmic flow in the interior of cells, which will result in good mixing of the contents. It is not necessary to assume the cells of equal volume, and one may suppose the rth cell to have volume v_r. The rate of flow from one cell to another will of course be proportional to the difference of concentrations of the flowing substance, i.e. the rate of flow of the mth morphogen from cell r to cell s will be proportional to $\Gamma_{mr} - \Gamma_{ms}$. It must also be proportional to a quantity g_{rs} dependent on the geometry of the wall of separation between cells r and s, but independent of the substance flowing, and there will be a further factor μ_m, the diffusion constant for the morphogen in question, diffusing through the material of which all the cell walls are assumed to be made.

Ignoring the chemical reactions, the equations of the system are

$$v_r \frac{\mathrm{d}\Gamma_{mr}}{\mathrm{d}t} = \mu_m \sum_{s \neq r} g_{rs}(\Gamma_{rs} - \Gamma_{ms}). \tag{II.1.1}$$

If g_{rr} is defined to be

$$-\sum_{s \neq r} g_{rs}$$

then (II.1.1) may be written in the more convenient form

$$v_r \frac{\mathrm{d}\Gamma_{mr}}{\mathrm{d}t} = -\mu_m \sum_{s} g_{rs}\Gamma_{ms} \tag{II.1.2}$$

or, by putting $\Gamma_{mr}^{(1)} = v_r^{1/2}\Gamma_{mr}$, in the form

$$\frac{\mathrm{d}\Gamma_{mr}^{(1)}}{\mathrm{d}t} = -\mu_m \sum_{s} \frac{g_{rs}}{\sqrt{(v_r v_s)}}\Gamma_{ms}^{(1)}. \tag{II.1.3}$$

Since the matrix $g_{rs}/\sqrt{(v_r v_s)}$ is symmetrical, it may be brought to diagonal form by an orthogonal transformation:

$$\frac{g_{rs}}{\sqrt{(v_r v_s)}} = \sum_k \alpha_k l_{rk} l_{sk}, \tag{II.1.4a}$$

$$\sum_k l_{rk} l_{sk} = \delta_{rs} \tag{II.1.4b}$$

and if one then puts

$$\Gamma_{mj}^{(2)} = \sum_r \Gamma_{mr}^{(1)} l_{rj} \tag{II.1.5}$$

and consequently

$$\Gamma_{mr}^{(1)} = \sum_j \Gamma_{mj}^{(2)} l_{rj} \tag{II.1.6}$$

the equations become simply

$$\frac{\mathrm{d}\Gamma_{mj}^{(2)}}{\mathrm{d}t} = -\mu_m \alpha_j \Gamma_{mj}^{(2)}. \tag{II.1.7}$$

The characteristic values α_k are real, since g_{rs} is real and symmetric. None of them is negative, as can be seen on physical grounds. If one of them were negative, then there would be solutions of (II. 1.7) in which concentration differences increase with time. There can be only one α_k which is zero, provided the organism is connected, for there is then, for each m, only one linearly independent solution of the equations which is constant in time.

As regards the chemical reactions, the one essential point is that they proceed at rates which depend only on the concentrations of the various morphogens in the same cell. In TURING (1952) the main interest centred around the case in which the reaction rates are linear functions of the concentrations, an assumption which is reasonably valid so long as only small variations of concentration are concerned. The theory of this linear case was carried through with the rather special geometrical assumption that the organism consisted of a ring of cells. This restriction was, however, an altogether unnecessary one. It was made merely in order to make the problem under consideration a quite definite one, and so make the argument more generally intelligible. As will be seen very shortly, the conclusions which were obtained in that case can be directly taken over to any arrangement of cells.

Suppose that when the concentrations of the M morphogens are $\Gamma_1, \Gamma_2, \ldots, \Gamma_M$ the rate of production of the mth morphogen is $f_m(\Gamma_1, \ldots, \Gamma_M)$ per unit volume. In this case the equations describing the effect of diffusion and reaction together are

$$v_r \frac{\mathrm{d}\Gamma_{mr}}{\mathrm{d}t} = -\mu_m \sum_s g_{rs} \Gamma_{ms} + v_r f_m(\Gamma_{1r}, \ldots, \Gamma_{Mr}). \tag{II.1.8}$$

The equations may be transformed by the substitutions (II.1.3), (II.1.6) to give a result analogous to (II.1.7).

If $\Gamma_{mr}^{(3)}$ are variables which are similar to $\Gamma_{mr}^{(2)}$, except that they refer to differences from the equilibrium, and if these differences are sufficiently small for it to be admissable to treat the reaction rates as linear functions of them, then the transformed equations become

$$\frac{\mathrm{d}\Gamma_{mj}^{(3)}}{\mathrm{d}t} = -\mu_m \alpha_j \Gamma_{mj}^{(3)} + \sum_k a_{mk} \Gamma_{kj}^{(3)} \tag{II.1.9}$$

where a_{mk} is the value of $\partial f_m/\partial \Gamma_k$ for the equilibrium concentrations. It will be seen that these equations separate into N independent sets of M equations each. In each set of equations the geometry of the system comes into the problem only through the characteristic values α_j of the diffusion matrix $g_{rs}/\sqrt{(v_r v_s)}$. With the rings of cells considered in TURING (1952) every possible non-negative value of α (there written U) could arise, and was allowed for, and no greater variety of values of α can arise with any other geometrical arrangement.

The solution of (II.1.9) can be written in the form

$$\Gamma_{mj}^{(3)}(t) = \sum_r \Gamma_{rj}^{(4)}(t) S_{mrj} \tag{II.1.10a}$$

where

$$\Gamma_{rj}^{(4)}(t) = \Gamma_{rj}^{(4)}(0) e^{p_r(\alpha_j)t} \tag{II.1.10b}$$

and

$$(p_r(\alpha_j) + \mu_m \alpha_j) S_{mrj} = \sum_s a_{ms} S_{srj}. \tag{II.1.11}$$

On eliminating the S_{mrj} from equations (II.1.11) one obtains an algebraic relation between p and α. For each value of α there will be M values of p satisfying the relation, and M corresponding solutions (II.1.10). TURING (1952) was very largely concerned with this relation between α and p, and the various forms it can take with different chemical conditions. It would not be appropriate to repeat that analysis here, but some of the main points of relevance to the present problem may be mentioned. Evidently the terms in (II.1.10b) of greatest importance are those for which the real part of p is greatest, for these are the ones which grow fastest. Ultimately, in fact, provided that the linearity assumption remains true, one may ignore all components of the solution except those with the largest Re(p). In TURING (1952) the possibilities were classified according to the values of p and of α when Re(p) has its maximum value. Since α may be zero, finite or infinite, and p may be real or complex, there are six alternatives, each of which was shown to occur with appropriate imaginary chemical reactions. The case of chief interest, and the only one to be considered here, was where p is real at the maximum, and α is finite and different from zero. This is described as the "case of stationary waves".

When the organism contains only a finite number of cells, or where the cells are infinitesimal but the whole organism of finite volume, the characteristic values α_k will be discrete. There can then only be a finite number for which Re(p) has its greatest value. It will however be by no means unusual for this number to exceed unity. In fact, if the system has some geometrical symmetry, multiple roots of the characteristic equation of (II.1.11) will not be at all exceptional, for when a symmetry operation is applied to the S_{mrj} they will be converted into characteristic vectors corresponding to the same characteristic value. It may happen that the vector $\Gamma_m^{(3)}$ itself has the symmetry in question, but if it has not (and there is no reason why it should) then the symmetry operation transforms it into *another* solution, and the characteristic root must be multiple. If there is a k-fold root corresponding to the greatest Re(p), then the limiting solution has k parameters. This situation is greatly modified when the quadratic terms are taken into account.

For many purposes the description of the organism in terms of cells is inconvenient. It may be mathematically more satisfactory to take the limiting case in which the volume of each cell is allowed to shrink to zero and the number of cells is allowed to increase correspondingly. There is no need to assume that the resulting continuous tissue is either homogeneous or isotropic, nor indeed is there any need to carry the theory through in any detail. In any particular case there will be a certain linear operator represented by "∇^2", such that the diffusion of a substance with concentration c and diffusion constant μ obeys the law

$$\frac{dc}{dt} = \mu \nabla^2 c. \tag{II.1.12}$$

When reactions are taken into account, the equations become

$$\frac{d\Gamma_m}{dt} = \mu_m \nabla^2 \Gamma_m + f_m(\Gamma_1, \ldots, \Gamma_m). \tag{II.1.13}$$

As in the finite case, one may expand any function of position in the organism in terms of characteristic functions of the operator ∇^2 analogously to (II.1.5), and the remainder of the theory of this section may be applied to this expansion.

2. Assumptions concerning the chemistry of phyllotaxis

The behaviour of the solutions (II.1.13) can take very different forms according to the functions f_m and the diffusibilities μ_m involved. This variety is even further increased if the concentrations of the morphogens affect the growth of the organism. It was suggested in TURING (1952) that this might be the main means by which the chemical information contained in the genes was converted into a geometrical form. If this be so, then any particular type of anatomical structure can only arise from a relatively small fraction of the possible chemical reaction systems: if it arose from all then no other anatomical forms would be possible. In particular, not all reaction systems can be appropriate for the description of phyllotactic phenomena. It is the purpose of this second part of the paper to describe conditions under which the chemical reactions are appropriate for that purpose, to find a partial differential equation which describes the progress of those reactions, and to investigate the behaviour of that partial differential equation. It is unfortunately not possible to say *a priori* how great is the variety of behaviour which could be described by the other reaction systems. It seems to be capable of giving rise to an enormous variety of possible solutions, but whether this variety is, or is not, sufficient to describe all the variety of living forms cannot easily be settled.

The principal assumptions to be made are:

(a) The reaction system is such that there is a homogeneous equilibrium, and small deviations from this equilibrium satisfy the conditions for stationary waves (see TURING (1952) p. 52).
(b) The deviations from the homogeneous state are not very great. They are sufficiently large for the linear approximation to be inapplicable, but nevertheless sufficiently small for the effects of the quadratic and higher terms to be regarded as little more than perturbations.
(c) The only wavelengths which are significant are those which are either very long or fairly near to the optimum.

When it is not intended to assume that the reaction rate functions are linear one may

(i) make some quite definite assumptions about the chemistry of the system and obtain definite reaction rate functions, or,
(ii) admit that the reaction rate functions may be any functions of the concentrations, or,
(iii) assume that the reaction rates are polynomials in the concentrations, or,
(iv) assume that the reaction rates are quadratic functions of the concentrations.

The assumption that the reaction rates are polynomials can be directly justified by the law of mass action. However it gives one very little advantage to know this. Moreover, if one is trying to deal with a particular case one will often prefer to use functions which are not polynomials, since by doing so one can often, if some of the reactions proceed quickly, reduce the number of morphogens, i.e. the number of variables which need to be considered. The assumption that the reaction rates are quadratic functions of the concentrations is a more useful one, although superficially it appears to be rather arbitrary and of limited validity. It can, however, be justified quite as well as the assumption of polynomial reaction rates. The justifying argument may be expressed either in a purely mathematical form or in physical terms, although it is really essentially the same argument in either case. The physical argument is that all chemical reactions in dilute solution are either monomolecular or bimolecular. Where three molecules react there will be two of them which meet and form some, probably more or less unstable, combination, before reacting with the third substance. When these unstable substances are all included as morphogens, there will only be monomolecular and bimolecular reactions, and consequently only linear and quadratic terms in the reaction rates. It may be necessary, if this point of view is to be maintained, to treat excited states of molecules as being molecules of a different compound, but there is clearly no objection to this.

The mathematical argument also proceeds by introducing new variables, and these variables might, if one wished, be imagined as representing the concentrations of imaginary intermediate products according to some possible theory of how the reactions are to be broken down into monomolecular and bimolecular reactions.

The additional variables which are necessary if one is to have quadratic rate functions are clearly no disadvantage in the sort of theoretical discussion where one is in any case committed to an indefinite but finite number of concentration variables.

Although the equations are no longer linear it is convenient to use the same substitutions as in the linear theory, viz. first

$$\Gamma_{mj} = h_m + \sum_r \Gamma_{mj}^{(3)} v_j^{-1/2} l_{rj} \tag{II.2.1}$$

where $f_m(h_1, \ldots, h_M) = 0$ for all m, i.e. $h_1, \ldots, h_M$ represents a homogeneous equilibrium. Naturally, if there are variables for every substance which occurs (in solution) in the system, then $h_1 = h_2 = \cdots = h_M = 0$ will be such an equilibrium, but it is preferable for a number of reasons not to suppose that $h_1, \ldots, h_M$ have these values. If some of the substances whose concentrations are effectively constant have not been assigned variables, the values zero may not be an equilibrium. Moreover, the zero equilibrium is by no means a representative one, and not all the phenomena described in TURING (1952) will occur with this trivial form of equilibrium.

The M different roots of the equation resulting from eliminating the $\Gamma_{mr}^{(3)}(0)$ from (II.1.10) may be called $p_l(\alpha)$, $l = 1, 2, \ldots, M$, and there will be numbers $W_{ml}(\alpha)$ such that

$$(p_l(\alpha) + \mu_m \alpha) W_{ml}(\alpha) = \sum_s a_{ms} W_{ms}(\alpha). \tag{II.2.2}$$

If p_l is a k-fold repeated latent root, there will be k corresponding $W_{ml}(\alpha)$. The matrix $W_{ms}(\alpha)$ is accordingly always non-singular and has an inverse $W_{st}^{-1}(\alpha)$. Consequently the matrix $\Gamma_{mr}^{(3)}(t)$ may be expressed in the form

$$\Gamma_{mr}^{(3)}(t) = \sum_l W_{ml}(\alpha_r) X_{lr}(t) \tag{II.2.3}$$

and so, by (II.2.1)

$$\Gamma_{mj}(t) - h_m = \sum_{r,l} W_{ml}(\alpha_r) X_{lr}(t) v_j^{-1/2} l_{rj}. \tag{II.2.4}$$

The assumption (c) above can now be brought into play. Only terms for which α is close to zero or to k_0^2, where $2\pi/k_0$ is the optimum wavelength, must be considered. It is assumed also, that in view of the analogy of the linear case, and assumption (ii) that the non-linear terms are only of secondary importance, one may ignore all terms on the right-hand side of (II.2.4) which do not arise from the largest $\mathrm{Re}\{p_l(\alpha_r)\}$ for the α_r in question. For α_r near to zero this will be supposed to be $p_l(0)(\alpha_r)$, and when α_r is near to k_0^2 it will be supposed to be $l^{(1)}$.

With these assumptions, (II.2.4) becomes

$$\Gamma_{mj}(t) - h_m = \sum_r (\alpha_r \text{ near } 0) W_{ml^{(0)}}(\alpha_r) X_{l^{(0)}r}(t) v_j^{-1/2} l_{rj} + \sum_r (\alpha_r \text{ near } k_0^2) W_{ml^{(1)}}(\alpha_r) X_{l^{(1)}r}(t) v_j^{-1/2} l_{rj}. \quad \text{(II.2.5)}$$

A further, and rather drastic, approximation is the assumption that over the two short ranges of values of α concerned, the functions $W_{ml^{(0)}}(\alpha)$ and $W_{ml^{(1)}}(\alpha)$ may be treated as constants, and so

$$\begin{aligned}\Gamma_{mj}(t) - h_m &= W_{ml^{(0)}}(0) \sum_r X_{l^{(0)}r}(t) v_j^{-1/2} l_{rj} + W_{ml^{(1)}}(k_0^2) \sum_r X_{l^{(1)}r}(t) v_j^{-1/2} l_{rj} \\ &= W_{ml^{(0)}}(0) V_j + W_{ml^{(1)}}(k_0^2) U_j. \end{aligned} \quad \text{(II.2.6)}$$

Now assuming, in agreement with (iv), that the reaction rates are quadratic functions of the concentrations, the differential equations controlling the behaviour of $\Gamma_{mj}(t)$ can be written in the form

$$\frac{\mathrm{d}\Gamma_{mj}}{\mathrm{d}t} = \sum a_{ms}(\Gamma_{ms} - h_m) + \sum K_{mrs}(\Gamma_{rj} - h_r)(\Gamma_{Sj} - h_s) - \frac{\mu_m}{v_j} \sum g_{jj} \Gamma_{mj} \quad \text{(II.2.7)}$$

the constant terms being absent by the definition of the h_r. If these equations are expressed in terms of the variables $X_{lr}(t)$ by means of the relation (II.2.4), one obtains

$$\frac{\mathrm{d}X_{lr}}{\mathrm{d}t} = \sum_{m,j} W_{lm}^{-1}(\alpha_r) l_{rj} v_j^{-1/2} \frac{\mathrm{d}\Gamma_{mj}}{\mathrm{d}t}. \quad \text{(II.2.8)}$$

The linear terms will of course be the same as in the purely linear theory, and the quadratic terms may be evaluated by the approximation (II.2.6). There is evidently no need to give the equations except where $l = l^{(0)}$ or $l = l^{(1)}$. The equation for $X_{l^{(0)}r}$ is

$$\frac{\mathrm{d}X_{l^{(0)}r}}{\mathrm{d}t} = p_{l^{(0)}}(\alpha_r) X_{l^{(0)}r} + \sum_{m,j,s} W_{l^{(0)}m}^{-1} l_{rj} v_j^{-1/2} K_{mrs} [W_{rl^{(0)}}(0) V_j + W_{rl^{(1)}}(k_0^2) U_j] \times [W_{sl^{(0)}}(0) V_j + W_{sl^{(1)}}(k_0^2) U_j] \quad \text{(II.2.9)}$$

and may be written

$$\frac{\mathrm{d}X_{l^{(0)}r}}{\mathrm{d}t} = p_{l^{(0)}}(\alpha_r) X_{l^{(0)}r} + \sum_{m,j} W_{l^{(0)}m}^{-1} l_{rj} v_j^{-1/2} (F_m^{(1)} V_j^2 + 2F_m^{(2)} V_j U_j + F_m^{(3)} U_j^2). \quad \text{(II.2.10)}$$

There is of course a similar equation for $X_{l(1)_r}$. One can now change the variables to U_j, V_j, and one obtains

$$\frac{\mathrm{d}V_j}{\mathrm{d}t} = [p_{l(0)}(-\nabla^2)V]_j + \sum W^{-1}_{l(0)_m}(\alpha_r) l_{rk} v_k^{-1/2} v_j^{1/2} l_{rj} (F_m^{(1)} V_j^2 + 2F_m^{(2)} V_j U_j + F_m^{(3)} U_j^2). \tag{II.2.11}$$

Here $p(-\nabla^2)$ represents a certain linear operator. As implied by the notation used, the characteristic vectors of these operators are the same as those of the operator ∇^2, but where $-\nabla^2$ has the characteristic value $\alpha, p(-\nabla^2)$ is to have the characteristic value $p(\alpha)$. If now the matrix $W^{-1}_{lm}(\alpha)$ is treated as independent of α over the range involved, as has already been supposed for its inverse, the equation reduces to the much simplified form

$$\frac{\mathrm{d}V_j}{\mathrm{d}t} = [p_{l(0)}(-\nabla^2)V]_j + F^{(4)} V_j^2 + 2F^{(5)} V_j U_j + F^{(6)} U_j^2. \tag{II.2.12a}$$

Likewise the equation for U_j takes the similar form

$$\frac{\mathrm{d}U_j}{\mathrm{d}t} = [p_{l(1)}(-\nabla^2)U]_j + G^{(4)} V_j^2 + 2G^{(5)} V_j U_j + G^{(6)} U_j^2. \tag{II.2.12b}$$

The assumptions which led to these equations have been somewhat drastic, but it must be remembered that it is by no means claimed that these approximations are appropriate for all problems of morphogenesis, but merely that, in the cases where these approximations hold good one obtains equations suitable for the description of the phyllotactic phenomena. When considering the validity of the approximations, therefore, one should ask whether there are any cases in which they hold good, rather than whether they are universally valid.

There will be still further assumptions to be made, but the equations (II.2.12) may be regarded as a convenient bridge linking the more or less chemical theory with the rest of the problem.

A number of additional assumptions will be made with even less appearance of generality. First, the coefficients $F^{(4)}, F^{(5)}, G^{(4)}$ will be assumed to be zero. One may, if one wishes, explain the ignoring of the term $G^{(4)} V_j^2$ on the grounds that, in a Euclidean space at any rate, it does not contain components near to the optimum wavelength; and the ignoring of $2F^{(5)} U_j V_j$ on the grounds that it does not contain terms of long wavelength. The ignoring of $F^{(4)} V_j^2$ might be justified by the view that V_j is small. But whatever justification be offered, these assumptions are made. It is assumed further that there is effective equilibrium in the equation (II.2.12a), i.e. that the right-hand side of this equation vanishes, so that

$$V = [p_{l(0)}(-\nabla^2)]^{-1} U^2 = \psi(-\nabla^2) U^2 \tag{II.2.13}$$

putting $F^{(6)} = -1$ by suitable scaling of the variables.

The equation for U may be written

$$\frac{\mathrm{d}U_j}{\mathrm{d}t} = [\phi(-\nabla^2)U]_j G U_j^2 - H U_j V_j \tag{II.2.14}$$

The two functions ϕ, ψ are bound in theory to be algebraic, but this is of course no real restriction. Any analytic function can be approximated as closely as one pleases by an algebraic function. The essential point about the function $\phi(\alpha)$ is that it has a maximum near $\alpha = k_0^2$. Since it has been supposed in any case that it is only the components of U with wavelengths near to $2\pi/k_0$ which are significant, it can only be the values of $\phi(\alpha)$ for values of α near to k_0^2 which matter, i.e. only the values near to the maximum. An appropriate approximation for $\phi(\alpha)$ therefore seems to be $I(\alpha - \alpha_0)^2$. As regards the function $\psi(\alpha)$, the most natural assumption seems to be that $p_{l(0)}(\alpha)$ increases linearly with increasing α and so that $\psi(\alpha)$ is of the form $A/(B+\alpha)$. However, since computations are greatly simplified if $\psi(\alpha)$ can be put equal to zero for the majority of values of α, other forms for this function will sometimes be used.

3. Equations for small organisms

If the dimensions of an organism are not too large in comparison with the optimum wavelength, the characteristic values α_r will usually be well spaced out, except where they are multiple. In this case some of the approximations of the last section will be rather more convincingly justifiable, for there may be only one value of α which needs to be considered in each of the two important ranges, viz. that near the optimum wavelength and that near zero. The value near the optimum wavelength will probably be multiple, but, as has already been mentioned, for a connected organism the root zero is simple: $V_j(t)$ is therefore independent of position. According to the point of view in which V represents the concentration of a diffusing poison, the organism is sufficiently small that the poison may be assumed to be uniformly distributed over it. The function $U_j(t)$, on the other hand, must be a linear combination of diffusion eigenfunctions all with the same eigenvalue, or, in other words, waves with the same wavelength. If $\mathcal{F}$ is a linear operator which removes all components which have not the appropriate wavelength, then the equations may be written

$$\frac{\mathrm{d}U_j}{\mathrm{d}t} = (P - HV)U_j + G[\mathcal{F}U^2]_j, \qquad \text{(II.3.1)}$$

$$V = \overline{U^2}. \qquad \text{(II.3.2)}$$

Here P is the value of $p_{I(1)}(\alpha)$ for the particular α (near the optimum) which is concerned. The equilibrium solutions of these equations take the form of multiples of solutions of the equations $U = \mathcal{F}(U^2)$.

This problem may be illustrated by the case where the organism forms a spherical shell. It is not of course possible to build up a spherical shell out of a large number of similarly shaped areas. But if it be built up of a large number of cells which are not quite the same, the effect of the irregularities will become small as the number of cells increases. In any case it will be assumed that the "∇^2" for the shell is the ordinary three-dimensional Laplacian in spherical polar coordinates, with the radial term omitted, and so has the spherical harmonics as characteristic functions. The operator $\mathcal{F}$ will then be one which removes from a function on a sphere all spherical harmonic components except those of a particular degree. To solve the equation $U = \mathcal{F}(U^2)$ is to find a spherical harmonic of that degree which, when it is squared and again has the appropriate orders removed, is unchanged. For each degree there is only a finite number of essentially different solutions of this problem, i.e. solutions which are inequivalent under rotations of the sphere. They have been investigated by B. Richards, and the results are described in Part III.

4. The equations applied to a plane

In the case that the cells form an isotropic homogeneous plane (apart from local variations over regions containing not very many cells), the diffusion characteristic functions are plane waves of the form $\mathrm{e}^{\mathrm{i}(Xx+Yy)}$. Over any finite area of the plane the functions may be approximated as closely as one pleases by Fourier sums

$$\sum C_n \mathrm{e}^{\mathrm{i}(X_n x + Y_n y)}. \qquad \text{(II.4.1)}$$

When a function has two independent periods it may be accurately represented by such a series. As was seen in Part I, the exponents (X_n, Y_n) then form a lattice. Such a series can also represent any function accurately within a bounded region, by choosing sufficiently large periods. Functions in the plane can also be approximated by Fourier integrals, and sometimes accurately expressed by them. There is no need, however, to go into these problems of analysis here. There are no organisms consisting of infinite planes of tissue, and it is only of value to consider such imaginary organisms for the light that they throw on other systems more nearly represented in nature. This light will not be appreciably dimmed by making the assumption that the functions concerned have two independent periods, possibly rather large, and the expression (II.4.1) will therefore be supposed exact.

Before going on to the non-linear theory of the plane it will be worth while to ask what sorts of patterns one would get if the linear theory applied throughout. The value of his question lies in the fact that it has a fairly definite answer, and is valuable as an illustration of certain points. Whether the patterns which arise from it can fairly be claimed to occur in nature is another matter.

The equations for the linear case can be obtained by omitting the quadratic terms in (II.2.12b) and so become

$$\frac{\mathrm{d}U}{\mathrm{d}t} = p_{I(1)}(-\nabla^2)U. \tag{II.4.2}$$

This equation may also be inferred fairly directly from (II.2.9) and (II.2.10) by ignoring all but the largest roots p. The equation (II.4.2) can only be regarded as valid over a limited range of time. Apart from the question (which is being intentionally ignored) of the effects of quadratic terms as t increases, there is the effect of Brownian movement and similar "noise" effects when t has large negative values and U is consequently very small. If this Brownian movement is taken into account, the character of the problem changes. One is no longer trying to find the totality of solutions, or the time development of a solution, but rather to find statistical information about the "ensemble" of solutions.

5. "Noise" effects

The equation (II.1.8) can be regarded as accurate only so long as one assumes that Avogadro's number is infinite. Otherwise it will be necessary to admit that the concentration of a morphogen in a cell can only have discrete values, corresponding to the various numbers of molecules that may occur in it. It will also not be possible to predict the actual new concentration at any future time, but only to give probabilities. In some applications of the theory it may be important to consider seriously the possibility that there may be only one or two molecules present, or even none. This would apply, for instance, in the case that the same theory is applied to the spread of epidemics, the "molecules" now being infected and uninfected men, rats, corpses, fleas, etc. It may even apply to the morphogenetic problem for some of the morphogens, e.g. the genes themselves. However, since these statistical effects are in any case of rather secondary importance in this problem it is appropriate to make some simplifying assumptions. It will be supposed in fact that over intervals of time short enough for the concentrations not to undergo appreciable proportionate changes, the number of molecules undergoing any one of the reactions in any one of the cells, or passing through any of the membranes, is large enough that one is justified in using a normal distribution instead of a Poisson distribution for it. Let Γ_{mr} be expressed in gram molecules per unit volume and N be Avogadro's number, then ... [the manuscript breaks off here]

6. Effects of random disturbances

If there is one value $p_k(\alpha_r)$ which exceeds all the others, then for almost all initial values the term $\exp\{ip_k(\alpha_r)t\}$ will eventually be far greater than any of the other terms which contribute to Γ_{ms}. In this case then the ultimate condition of the system is almost independent of the initial conditions. Even if there are non-linear terms which eventually have to be taken into account, this will quite possibly not apply until this dominant term has outdistanced the others. However it is not at all uncommon for there to be several different characteristic vectors for which the corresponding diffusion characteristic values α_r are all equal. In this case the "argument by outdistancing" merely tells us that ultimately the only characteristic vectors which need be included in the sum are those for which the $p_k(\alpha_r)$ has its maximum value. It does not, however, say anything about the proportions in which those characteristic vectors are to be taken. According to the theory expounded in §II.1, this could be settled by giving all the concentrations of all the morphogens at some early time. These would then determine the concentrations at all later times. The theory of §II.1 is, however, in error about this. For there are some small effects which are ignored. This does not matter under most circumstances, but sometimes it will, as for instance when the system is near to an unstable equilibrium. The main effects concerned are probably

(1) The statistical nature of the chemical reactions.
(2) The statistical nature of the diffusion.
(3) Variations of the reaction rates from cell to cell, due for instance to the presence of different concentrations of indiffusible catalysts (not reckoned as morphogens).
(4) The irregularity of the cell pattern in the cases where this cell pattern has been idealised in some way, e.g. where it is regarded as forming a spherical surface.

It is pointless to attempt to give any complete list of the effects ignored because of the very approximate nature of the whole investigation, but the above list may be used to illustrate the nature of these effects. So long as one is interested in the manner of departure from homogeneous equilibrium, one can fairly say that even if non-linear effects are eventually to be considered, nevertheless during the period when these minor disturbances are of significance, the non-linear terms can be ignored. It can happen that a non-homogeneous equilibrium, for whose maintenance the non-linear terms are of importance, may become unstable, and then again these small effects will determine the course that the system will actually take. In these cases, however, there are usually very few alternative courses which can be taken. There is also usually a symmetry which shows that the alternative possible courses are all equally probable: in fact no detailed theory is necessary. In the present section therefore it will be supposed that the linear theory applies. This really means that it is being assumed that the linear theory applies whilst the disturbances operate, but the progress of the differential equation will not be followed up to the time when the non-linear terms become effective. It will also be assumed in this investigation that the multiplicity of the roots of the diffusion characteristic equation is entirely due to symmetry. The situation is very similar to that which arises in the theory of spectra, where the multiplicity of an energy level is almost always due to symmetry.

It seems probable that effects (3) and (4) would normally be of somewhat greater magnitude than the atomistic effects (1) and (2). The atomistic effects have, however, not been ignored: this is mainly because the theory of them is much more satisfactory.

One may, if one wishes, regard a morphogenetic system as described by the MN values of the concentrations of the various morphogens in the various cells, without attempting to classify these variables. If $y_1, \ldots, y_{MN}$ are these variables, the equations may be written in the form

$$\frac{\mathrm{d}y_r}{\mathrm{d}t} = F_r(y_1, \ldots, y_{MN}, t) + N_r(t), \tag{II.6.1}$$

where the functions $F_r(y_1, \ldots, y_{MN}, t)$ describe both the diffusion and the reactions, and $N_r(t)$ describe the statistical effects.* The choice of the values of these functions is to be imagined as independently chosen at different times, although the various values at a particular time may be statistically or even functionally dependent.

* The use of these $N_r(t)$ is strictly speaking quite unjustifiable; they could not themselves be given finite values at each time, though their integrals can, and also the integrals below in which they appear.

The value of the integral of $N_r(t)$ over any period of time being a sum of a large number of statistically independent components, arising from any way in which the interval of integration may be broken up into subintervals, has a normal distribution. Its variance is expressible in the form

$$\int_a^b \sigma_r(t)\mathrm{d}t$$

where (a,b) is the time interval concerned. Likewise, if $g(t)$ is a function statistically independent of $N_r(t)$, then

$$\int_a^b N_r(t)g(t)\mathrm{d}t$$

is also normally distributed and its variance is

$$\int_a^b \sigma_r(t)[g(t)]^2\mathrm{d}t.$$

Now suppose that $y_1(t),\ldots,y_{MN}(t)$ would be an equilibrium at time t if only the conditions were not changing, i.e. that

$$F_r(y_1^{(0)}(t),\ldots,y_{MN}^{(0)}(t),t) = 0 \qquad \text{(II.6.2)}$$

and let the partial derivative $\partial F_r/\partial y_s$ have the value $b_{rs}(t)$ for the arguments $y_1^{(0)}(t),\ldots,y_{MN}^{(0)}(t),t$. Then putting $y_r^{(1)}(t) = y_r(t) - y_r^{(0)}(t)$ one has

$$\frac{\mathrm{d}y_r^{(1)}}{\mathrm{d}t} = \sum_s b_{rs}(t)y_s^{(1)}(t) + N_r(t) - \dot{y}_r^{(0)}(t) \qquad \text{(II.6.3)}$$

with a dot denoting differentiation with respect to t. Similar equations hold if the variables are subjected to a linear transformation. The solution of the equation may be expressed in the form

$$y_r^{(1)}(t) = \sum_s \int_{-\infty}^t (N_s(u) - \dot{y}_s^{(0)}(u))K_{rs}(t,u)\mathrm{d}u \qquad \text{(II.6.4)}$$

where

$$\frac{\partial K_{rs}}{\partial t} = \sum_q b_{rq}(t)K_{qs}(t) \quad \text{if } t > u \qquad \text{(II.6.5)}$$

and

$$K_{rs}(u,u) = \delta_{rs}. \qquad \text{(II.6.6)}$$

One may regard $b_{rs}(t)$ as differing rather little from some "ideal" coefficients $b_{rs}^{(0)}(t)$ which have the appropriate symmetry. Similarly, one may have $K_{rs}^{(0)}$ related to $b_{rs}^{(0)}$ as in (II.6.5). In the case where the b_{rs} are independent of time, the $K_{rs}(t)$ are sums of time exponentials with the $p_k(\alpha_r)$ as growth rates. These $p_k(\alpha_r)$ are of course the characteristic values of the matrix b_{rq}. The small difference between b_{rs} and $b_{rs}^{(0)}$ will make only a small difference in the characteristic values, but if the time lapse is large the effect on the corresponding exponential growth factor may be considerable. With the actual $b_{rs}(t)$ which may be encountered, which are not constant in time, the same considerations of magnitude will apply.

It will not be feasible to give a very complete account of the behaviour of the solutions, but some general conditions may be considered. The case in which there are only atomistic disturbances, i.e. of kinds (1) and (2), can be dealt with fairly fully. In this case the term

$$-\int_{-\infty}^{t} \dot{y}_s^{(0)} K_{rs}(t,u)\mathrm{d}u$$

in (II.6.4) represents an effect independent of position. That this is so is not at all obvious from the equation, which does not in any case distinguish space and morphogen effects. It may be seen from the general principle that in the absence of any of the effects (1) to (4) a solution homogeneous at one time will remain so thereafter. Concentrating on the space dependent term

$$\int_{-\infty}^{t} N_s(u)K_{rs}(t,u)\mathrm{d}u = y_r^{(2)}(t)$$

one sees at once that each of its components is normally distributed, and that every linear combination of components is also normally distributed. The vector $y_1^{(2)},\ldots,y_{MN}^{(2)}$ consequently has a frequency function of the form

$$(\Delta(t))^{-1/2}(2\pi)^{-MN/2}\exp\left[-\frac{1}{2}\sum_{i,j}\theta_{ij}(t)y_i^{(2)}y_j^{(2)}\right] \qquad \text{(II.6.7)}$$

where the quadratic form $\sum\theta_{ij}(t)y_i^{(2)}y_j^{(2)}$ is positive definite and $\Delta(t)$ is the determinant of its coefficients.

It is proposed to consider this matrix θ_{ij} in respect of its asymptotic behaviour in time, and the symmetry of the system. For any given t, the function $K_{rs}(t,u)$ will normally be small compared with its maximum except over a comparatively narrow range of values of u, viz. near to the point where the largest real part of a characteristic value of the matrix $b_{rs}(u)$ is zero (cf. TURING (1952) equation (9.15)). For large values of t, therefore, one may replace the lower limit of integration in (II.6.4) by some fixed t_0, and then $y_r^{(2)}(t)$ will satisfy the differential equation

$$\frac{\mathrm{d}y_r^{(2)}}{\mathrm{d}t} = \sum_s b_{rs}(t)y_s^{(2)}$$

and consequently the matrix θ_{ij} will satisfy

$$\frac{\mathrm{d}\theta_{ij}}{\mathrm{d}t} = -\sum_p(\theta_{ip}b_{jp} - \theta_{jp}b_{ip}). \qquad \text{(II.6.8)}$$

If one makes a linear transformation to variables $y_r^{(3)}(t)$ such that the corresponding matrix $b_{rs}^{(3)}(t)$ is diagonal, these variables will be, or could be, the same as the $\Gamma_{rs}^{(3)}$, and the characteristic values $b_{ii}^{(3)}$ are the $p_k(\alpha_r)$. Writing I for the largest of the real parts of these characteristic values and $y_r^{(4)}(t)$ for $y_r^{(3)}(t)\mathrm{e}^{-It}$, then $y_r^{(4)}(t)$ tends to a limit $y_r^{(5)}$ as t tends to infinity. This limit is zero unless $\mathrm{Re}(b_{rr}^{(3)} - I) = 0$, and when this happens the corresponding imaginary part $\mathrm{Im}(b_{rr}^{(3)})$ is zero, since it is assumed that the stationary wave case applies. The differential equation (II.6.8) becomes, when applied to the coefficients $\theta_{ij}^{(4)}$ which describe the distribution of $y_r^{(4)}$,

$$\frac{\mathrm{d}\theta_{ii}^{(4)}}{\mathrm{d}t} = -2(b_{ii}^{(3)} - I)\theta_{ii}$$

($\theta_{ij} = 0$ if $i \neq j$), and so θ_{ii} tends to infinity unless $b_{ii}^{(3)} = I$. This merely expresses that it is infinitely improbable that $y_r^{(5)}$ will not be zero if $b_{rr}^{(5)} = I$. The $\theta_{rr}^{(4)}$ with $b_{rr}^{(3)} = I$ give the frequency distribution of the $y_r^{(5)}$. They are normally distributed and independent.

The above argument gives an existence theorem about the distribution of the $y_r^{(5)}$, i.e. of those linear combinations of the coordinates which tend to infinity fastest. The actual distribution is best obtained by symmetry requirements. Although the full details of these arguments must be related to the particular symmetry group involved, a little can be said which is generally applicable.

It has already been mentioned that the degeneracy of the characteristic equation associated with diffusion was to be supposed entirely due to symmetry. This assumption needs some clarification before it can be used. Let $\boldsymbol{y}$ be any vector (i.e. an assignment of morphogen concentrations to the cells), then this vector can be transformed into various others, e.g. $\boldsymbol{S}_p\boldsymbol{y}$ by the symmetry operations, i.e. the permutations τ of the cells which satisfy

$$g_{p_r p_s} = g_{rs}.$$

If $\boldsymbol{y}$ is a characteristic vector of the operator ∇^2 with characteristic value α, then $\boldsymbol{S}_p\boldsymbol{y}$ is also a characteristic vector with the same characteristic value. The various vectors

$$\boldsymbol{S}_{p_1}\boldsymbol{y}, \boldsymbol{S}_{p_2}\boldsymbol{y}, \ldots$$

span a space of vectors with characteristic value α. In the metaphor of the assumption on p. 102, the equality of these characteristic values is “due to the symmetry”. That assumption states that the characteristic values are equal only if equality is due to symmetry, i.e. if $\boldsymbol{y}, \boldsymbol{y}'$ are both characteristic vectors with characteristic value α then $\boldsymbol{y}'$ can be expressed in the form $\sum \boldsymbol{\beta}_p \boldsymbol{S}_p \boldsymbol{y}$. Another way of expressing the condition is to say that when the group of symmetry is represented on the vectors with characteristic value α, this representation is irreducible. What is actually required below is a little more, viz. that the degeneracy of the largest roots of the matrix $b_{rs}(t)$ (for large t) should be entirely due to symmetry. These roots are the $p_k(\alpha_r)$. The degeneracies of the diffusion matrix mean that various sets of the α_r are equal. It is required that if $I = \mathrm{Max}[p_k(\alpha_r)]$ then $p_k(\alpha_r) = I$ and $p_{k'}(\alpha_{r'}) = I$ imply that $\alpha_r = \alpha_{r'}$ and $k = k'$. The cases in which this is most likely to be incorrect are when either there is some kind of symmetry in the chemical reaction system, as for instance when the dextro and laevo forms of the morphogens are distinguished, or when $\alpha_r \neq \alpha_{r'}$, $k = k'$, but $p_k(\alpha_r) = p_k(\alpha_{r'})$ due to fortuitous values of the geometrical dimensions. With these assumptions, the representation of the symmetry group on the vectors for which b_{rs} has characteristic value I is irreducible. Now let the values of r for which $b_{rr}^{(3)} = I$ be $1, 2, \ldots, J$, and let the effect of the symmetry operation on $y_1^{(5)}, \ldots, y_5^{(5)}, 0, 0, \ldots, 0$ be to convert it to the vector whose rth coordinate ($r \leqslant J$) is

$$\sum_{s=1}^{J} u_{rs}(p) y_s^{(5)}.$$

Also let the frequency distribution of the values of $y_1^{(5)}, \ldots, y_J^{(5)}$ be

$$(2\pi)^{-J/2}(\det\theta^{(5)})^{-J/2}\exp\left[-\frac{1}{2}\sum_{i,j}\theta_{ij}^{(5)} y_i^{(5)} y_j^{(5)}\right] \mathrm{d}y_1^{(5)}, \ldots, \mathrm{d}y_J^{(5)}.$$

The frequency distribution for the transformed coordinates will have

$$\sum_{k,l} u_{ik}(\tau^{-1}) u_{jl}(\tau^{-1}) \theta_{kl}$$

in place of $\theta_{ij}^{(5)}$, but by the symmetry condition the frequency distribution is unaltered, i.e. $\sum u_{ik}(\tau^{-1})u_{jl}(\tau^{-1})\theta_{kl} = \theta_{ij}^{(5)}$.

Now for a finite group of symmetry, the basis of the vectors may be chosen in such a way that the matrices of the representation are unitary. If this is done, the condition may be expressed by saying that the matrix θ_{ij} commutes with each representative matrix $u_{ij}(\tau^{-1})$. But by a well known theorem on representations, this means that the matrix θ_{ij} is a multiple of the unit matrix.

When the coordinates are chosen in the space of vectors of greatest growth rate in such a way that the representatives of the symmetry group are unitary, then the frequency distribution is normal in each coordinate and the coordinates are independent of one another and have equal variances.

The consequences of the non-atomistic effects (3) and (4) are not capable of so satisfactory a treatment. If one were justified in supposing that the final effects were linear functions of independent causes the outcome of effects (3) and (4) would be of the same kind as the outcome of effects (1) and (2). It is likely however that this is not usually a good approximation. Although it would be fair to regard the matrix $(b_{rs}) = \boldsymbol{B}$ as influenced by a number of linearly independent causes, this matrix takes effect mainly through an exponential $e^{\boldsymbol{B}\tau}$ with a fairly large value of τ. Unless $\varepsilon\tau$ is small, $e^{(\boldsymbol{B}+\varepsilon\boldsymbol{B}')\tau}$ cannot be satisfactorily approximated by a linear function of $\boldsymbol{B}'$.

It is not proposed to enter into these effects here in great detail, although it is probable that they are of greater importance than those of atomistic origin. It is only intended to deal with an extreme case, one which is very different from the case of the atomistic effects, viz. that in which the time period is so long that in the exponential $e^{(\boldsymbol{B}+\varepsilon\boldsymbol{B}')\tau}$ the differences of the characteristic values of the matrix, when multiplied by the time interval, are large. In this case one comes back to the situation in §II.1, where only the largest characteristic value need be considered. The actual characteristic value is of no particular interest, as it will merely determine the time before a pattern of a given amplitude appears. Greater interest attaches to the characteristic vector, as it determines the pattern.

Part III. A Solution of the Morphogenetical Equations for the Case of Spherical Symmetry

1. Reduction of the differential equation

In the previous part the equations for small organisms were deduced, and an attempt was made to justify the assumptions on a chemical basis. One is interested in finding the concentration function $U(\theta,\phi,t)$ satisfying the equations:

$$\frac{dU}{dt} = \Phi(\nabla^2)U + GU^2 - HUV, \tag{III.1.1}$$

$$V = \overline{U_2}. \tag{III.1.2}$$

It has been shown that the solution U must be a linear combination of diffusion eigenfunctions; it is therefore appropriate to assume that the solution of (III.1.1) is of the form

$$U(\theta,\phi,t) = \sum_{m=-n}^{m=n} S_m(t)\overline{P_n^m}(\cos\theta)e^{im\phi} \tag{III.1.3}$$

with U real and the functions $\overline{P_n^m}(\cos\theta)$ being the normalised Legendre Associated Functions (see Appendix). These functions satisfy the relation

$$\frac{1}{4\pi}\int\int \overline{P_n^r}(\cos\theta)\overline{P_n^s}(\cos\theta)dS = \begin{cases} 1 & \text{for } r = s, \\ 0 & \text{for } r \neq s. \end{cases} \tag{III.1.4}$$

Since U is real, it is equal to its complex conjugate U^*, and since

$$\overline{P_n^m}(\cos\theta) = \overline{P_n^{-m}}(\cos\theta)$$

it follows that

$$S_{-m}(t) = S_m^*(t). \tag{III.1.5}$$

It is then possible to evaluate V from (III.1.2) by integration, since

$$\begin{aligned} V &= \frac{1}{4\pi}\int\int U^2 dS \\ &= \frac{1}{4\pi}\int\int \sum_{k=-n}^{k=n}\sum_{m=-n}^{m=n} S_k S_m \overline{P_n^k}(\cos\theta)\overline{P_n^m}(\cos\theta)e^{i(k+m)\phi} d\cos\theta d\phi \\ &= \sum_{k=-n}^{k=n}\sum_{m=-n}^{m=n} S_k S_m \frac{1}{4\pi}\int\int \overline{P_n^k}(\cos\theta)\overline{P_n^m}(\cos\theta)e^{i(k+m)\phi} d\cos\theta d\phi. \end{aligned}$$

Now using the relations (III.1.4) and (III.1.5), it follows that

$$V = \sum_{m=-n}^{m=n} |S_m|^2. \tag{III.1.6}$$

In equation (III.1.1), the function $\Phi(\nabla^2)$ is replaced by a constant I, since for the case of spherical surface harmonics,

$$\nabla^2 = -n(n+1)/R^2.$$

Equation (III.1.1) thus becomes:

$$\frac{dU}{dt} = IU + GU^2 - HUV. \tag{III.1.7}$$

Substituting the series solution into this equation leads to:

$$\frac{dS_m}{dt} = (I - HV)S_m + G\sum_{i=-n}^{i=n}\sum_{j=-n}^{j=n} S_iS_jL_n^{i,j,-m} \tag{III.1.8}$$

the coefficients $L_n^{p,q,r}$ being defined by:

$$L_n^{p,q,r} = \frac{1}{4\pi}\int\int \overline{P_n^p}(\mu)\overline{P_n^q}(\mu)\overline{P_n^r}(\mu)e^{i(p+q+r)\phi}\,d\mu\,d\phi, \tag{III.1.9}$$

$-\mu$ being written for $\cos\theta$.

As time goes on, $S_m(t)$ will reach its equilibrium value as determined by equation (III.1.8). Therefore, assuming that equilibrium has been reached, this equation becomes

$$(I - HV)S_m + G\sum_{i=-n}^{i=n}\sum_{j=-n}^{j=n} S_iS_jL_n^{i,j,-m} = 0 \tag{III.1.10}$$

which can be written in the form

$$-\left(\frac{I-HV}{G}\right)S_m = \sum_{i=-n}^{i=n}\sum_{j=-n}^{j=n} S_iS_jL_n^{i,j,-m}. \tag{III.1.11}$$

For the purposes of obtaining a solution it is more convenient to solve the equation

$$T_m = \sum_{i=-n}^{i=n}\sum_{j=-n}^{j=n} T_iT_jL_n^{i,j,-m} \tag{III.1.12}$$

which does not contain any of the arbitrary constants. If T_m is a solution of this equation, then $S_m = \lambda T_m$ will satisfy the equation

$$\lambda S_m = \sum_{i=-n}^{i=n}\sum_{j=-n}^{j=n} S_iS_jL_n^{i,j,-m} \tag{III.1.13}$$

and thus λT_m will satisfy (III.1.11) provided one chooses λ so that

$$\lambda = -(I - HV)/G. \tag{III.1.14}$$

Here the constant V is given by

$$V = \sum |S_m|^2 = \lambda^2 \sum |T_m|^2 \tag{III.1.15}$$

so that the final equation for λ is

$$\lambda = -\frac{I - \lambda^2 H \sum |T_m|^2}{G}. \tag{III.1.16}$$

Since λ is the same for each of the T_m, there will only be a constant factor between the distribution given by the T's and that given by the S's.

One thus arrives at the set of equations

$$S_m = \sum_{i=-n}^{i=n} \sum_{j=-n}^{j=n} S_i S_j L_n^{i,j,-m} \tag{III.1.17}$$

which are to be solved for the unknown S_m. The solutions are in general complex, but it is possible to choose solutions which are purely real, since it can be shown that the complex solutions differ only by a rotation.

It is possible to enumerate the $(n+1)$ equations as given by (III.1.17) for each particular value of n, in the forms given below. The relations

$$\begin{aligned} S_{-m} &= S_m^*, \\ L_n^{r,s,-m} &= E_n^{|r|,|s|,|m|} \quad \text{where } r+s-m=0, \\ E_n^{p,q,r} &= 0 \quad \text{if any of } p,q,r \text{ is } > n \end{aligned} \tag{III.1.18}$$

have been used.

For $n \leqslant 8$ the equations are

$$S_0 = S_0^2 E_n^{0,0,0} + 2\sum_1^n |S_k|^2 E_n^{0,0,0},$$

$$S_1 = 2\sum_0^{n-1} S_k^* S_{k+1} E_n^{1,k,k+1},$$

$$S_2 = 2\sum_0^{n-2} S_k^* S_{k+2} E_n^{2,k,k+2} + S_1^2 E_n^{1,1,2},$$

$$S_3 = 2\sum_0^{n-3} S_k^* S_{k+3} E_n^{3,k,k+3} + 2S_1 S_2 E_n^{1,2,3},$$

$$S_4 = 2\sum_0^{n-4} S_k^* S_{k+4} E_n^{4,k,k+4} + 2S_1 S_3 E_n^{1,3,4} + S_2^2 E_n^{2,2,4},$$

$$S_5 = 2\sum_0^{n-5} S_k^* S_{k+5} E_n^{5,k,k+5} + 2S_1 S_4 E_n^{1,4,5} + 2S_2 S_3 E_n^{2,3,5},$$

$$S_6 = 2\sum_{0}^{n-6} S_k^* S_{k+6} E_n^{6,k,k+6} + 2S_1 S_5 E_n^{1,5,6}$$
$$+ 2S_2 S_4 E_n^{2,4,6} + 2S_3^2 E_n^{3,3,6},$$
$$S_7 = 2\sum_{0}^{n-7} S_k^* S_{k+7} E_n^{7,k,k+7} + 2S_1 S_6 E_n^{1,6,7}$$
$$+ 2S_2 S_5 E_n^{2,5,7} + 2S_3 S_4 E_n^{3,4,7},$$
$$S_8 = 2\sum_{0}^{n-8} S_k^* S_{k+8} E_n^{8,k,k+8} + 2S_1 S_7 E_n^{1,7,8}$$
$$+ 2S_2 S_6 E_n^{2,6,8} + 2S_3 S_5 E_n^{3,5,8} + S_4^2 E_n^{4,4,8}$$

where for $n < 8$ only the first $n + 1$ equations hold.

These equations have algebraic solutions, though they are not necessarily real. The solutions $S = (S_0, S_1, \ldots, S_n)$ for given n are not unique, and indeed it is sometimes possible to specify n of the $n + 1$ components of S and to vary the remaining component, keeping its modulus fixed. Solutions have been sought with $S_1 = 0$, since in any solution with $S_1 \neq 0$ it is possible to rotate the coordinate axes so as to eliminate S_1' in the resulting transformation $S \to S'$. In doing this, the same physical solution is preserved.

It will be seen that in the solutions obtained, some values are negative. It is to be remembered that the solutions represent deviations from the sphere. The dimensions are somewhat arbitrary, but a correct balance between the oscillations of the function U and the radius of the initial sphere can be obtained by reference to suitable biological species.

2. Solutions of the simultaneous equations

The solutions of the sets of equations conform, in general, to a set pattern. If there is a solution in which S_r is non-zero, then S_{2r}, S_{3r}, etc. will also occur in that particular solution. Thus there will be a solution with S_0, S_3, S_6 etc. all non-zero; or again, S_0, S_2, S_4, S_6 etc. non-zero. There will be additional solutions if certain coincidences in numerical values hold among the integrals $L_n^{i,J,k}$.

The equations are solved for the values of $n = 2, 4, 6$, together with the restriction mentioned previously, namely that $S_1 \equiv 0$.

Case $n = 2$. The equations are:

$$S_0 = S_0^2 E_2^{0,0,0} + 2S_1^* S_1 E_2^{1,1,0} + 2S_2^* S_2 E_2^{2,2,0},$$
$$S_1 = 2S_1 S_0 E_2^{1,1,0} + 2S_1^* S_0 E_2^{1,1,2}, \tag{III.2.1}$$
$$S_2 = 2S_2 S_0 E_2^{2,2,0} + S_1^2 E_2^{1,1,2}$$

while the coefficients E_n have numerical values

$$E_2^{0,0,0} = 2\sqrt{5}/7, \quad E_2^{1,1,0} = \sqrt{5}/7,$$
$$E_2^{2,2,0} = -2\sqrt{5}/7 \quad E_2^{1,1,2} = \sqrt{30}/7.$$

The simplest solution of this set (III.2.1) is that for which S_0 is the only non-zero variable. This solution is

$$S_0 = (E_2^{0,0,0})^{-1} = 7\sqrt{5}/10$$

and hence

$$U = S_0\overline{P_2^0}(\cos\theta) = (7/4)(3\cos^2\theta - 1),$$
$$V = S_0^2 = 49/20.$$

This gives rise to a prolate spheroid whose major axis coincides with the direction $\theta = 0$. All other solutions of this set of equations do not introduce any physical solutions but only rotate the spheroid through varying angles about different axes. For example, consider the solution with $S_0 \neq 0$, $S_2 \neq 0$. We can choose S_2 to be positive, for the case S_2 negative is obtained by rotation through $\pi/2$ about the axis $\phi = 0$, since

$$-S_2\overline{P_2^2}(\mu)\cos 2\phi = S_2\overline{P_2^2}\cos 2(\phi \pm \pi/2).$$

The solution is then

$$U = (7/8)[(1+3\cos 2\phi) - 3\mu^2(1+\cos 2\phi)],$$
$$V = 49/20.$$

This is seen to be the same prolate spheroid as the original, but merely rotated through $\pi/2$ about the axis $\phi = \pi/2$.

Case $n = 4$. (i) Here again the simplest solution is a fundamental one physically; the solution being that for which S_0 is the only non-zero coefficient. It is

$$U_r = (E_4^{0,0,0})^{-1}\overline{P_4^0}(\cos\theta) = (1001/1296)(35\cos^4\theta - 30\cos^2\theta + 3),$$
$$V = (1001/486)^2.$$

This is a solid of revolution about the polar axis and resembles a discoid elongated in the polar directions.

In identifying the physical solutions, the value of the integral V, as calculated from (II.3.1), serves as an identity number, since for two physical solutions to be the same they must have the same wavelength predominating and must have the same value for V.

(ii) If one assumes two non-zero coefficients, namely S_0 and S_4, the solution is

$$U = S_0\overline{P_4^0}(\mu) + 2S_4\overline{P_4^4}(\mu)\cos 4\phi$$
$$= (143/108)[(3/8)(35\mu^4 - 30\mu^2 + 3) + (30/16)(1-\mu^2)^2\cos 4\phi],$$
$$V = 20449/6804.$$

This solid is a spheroid with a spine at each pole and four around the equator; in fact it has six-point symmetry. It is thus totally different from the previous solution.

(iii) The solution with S_0 and S_3 as the only non-zero coefficients yields the same solid, the solution being

$$U = S_0\overline{P_4^0}(\mu) + 2S_3\overline{P_4^3}(\mu)\cos 3\phi$$
$$= -(143/324)[(3/4)(35\mu^4 - 30\mu^2 + 3) - 15\sqrt{2}(1-\mu^2)^{3/2}\mu\cos 3\phi],$$
$$V = 20449/6804.$$

In this case, however, the pole has been rotated through $\cos^{-1}(1/\sqrt{3}) = 54^\circ\ 44'$.

(iv) In seeking a further solution one makes the assumption that $S_0 \neq 0$, $S_2 \neq 0$. If one further assumes that these two are real, one can deduce that S_4 is also real and non-zero. Since the solution of a quadratic in S_0 is required, one expects two solutions; in fact these two solutions are those previously obtained, namely, the discoid and the six-spined spheroid, the former having the line $\phi = \pi/2$ for its axis of rotation, while the latter has been rotated about the axis $\phi = \pi/2$. The solutions are

$$S_0 = -1001/3024, \qquad S_2 = 1001\sqrt{10}/3024,$$
$$S_4 = 1001\sqrt{70}/144112, \qquad v = 20449/6804,$$

and

$$S_0 = 1001/1296, \qquad S_2 = 1001\sqrt{90}/11664,$$
$$S_4 = 1001\sqrt{70}/7776, \quad V = (1001/486)^2.$$

(v) The only other solution in the pattem results from the assumption that S_2 is real and S_3 is non-zero. This most general solution is obtained due to a certain numerical relationship among the $E's$. The value of the integral is

$$V = (1001/486)^2$$

and thus it is the discoid, having, in fact, been rotated about the axis $\phi = \pi/2$ through an angle of 50°.

For this case, $n = 4$, there are thus only the two physical solutions, the discoid and the six-spined spheroid.

Case $n = 6$. (i) For the case $n = 6$, the solid of revolution is again the easiest solution to obtain and corresponds to the function $\overline{P_6^0}(\cos\theta)$ in that it has two ridges equidistant from the equator. Thus the solution is

$$U = (3553\sqrt{13}/5200)\overline{P_6^0}(\cos\theta),$$
$$V = 13(3553/5200)^2.$$

(ii) A more interesting solution is that for which S_0 and S_5 are non-zero, the solution being

$$U = (323\sqrt{13}/1300)[\overline{P_6^0}(\mu) + (323\sqrt{1001}/14300)\overline{P_6^5}(\mu)\cos 5\phi],$$
$$V = 104329/57200.$$

This corresponds to a regular icosahedron, the twelve equal spines being separated by the correct angular distance of $63^\circ\ 24'$. this particular solution having spines at the poles.

(iii) Assume S_3, S_5 to be zero and S_2, S_4, S_6 to be real. Now because $E_6^{2,2,0} = E_6^{6,6,0}$ it is possible to find a simpler solution in which $S_4 = 0$ because the equations are compatible. The solution of the reduced equations then leads to

$$U = -\frac{323\sqrt{13}}{520}\left[\overline{P_6^0}(\mu) + \sqrt{\frac{21}{29}}\overline{P_6^2}(\mu)\cos 2\phi + \sqrt{\frac{378}{319}}\overline{P_6^6}(\mu)\cos 6\phi\right],$$
$$V = \frac{5\,529\,437}{228\,800}.$$

This gives rise to a somewhat irregular solid with ten spines.

(iv) The solution in which S_0 and S_6 are assumed to be non-zero gives a similar solution with the same value of V.

(v) The icosahedron mentioned above appears again in the solution with S_0, S_3, S_6 all non-zero.

The solution with S_0 and S_4 non-zero has twelve equal spines situated about the polar axis as the axis of symmetry, and is really solution (i) slightly modified by the $\cos 4\phi$ term.

3. Comparison with physical species

The biological group which best illustrates the spherical harmonic pattern is that of the Radiolaria. These marine organisms are unicellular, and are surrounded by a skeleton for support and protection, this latter being generally composed of silica. These small Radiolarian cells are about a millimetre in diameter, and are found in all the seas of the world, in all climactic zones, and at all depths, but are not found in fresh water. Their most interesting property from the present point of view is that of possessing radial spines which radiate from the outer shell of the skeleton. From a morphological aspect the number, the arrangement and disposition of the spines is usually the determining factor regarding the general form of the skeleton. Physiologically they discharge distinct functions as organs of protection and support.

The life of a single cell is essentially individual and its growth is influenced by the surroundings. Thus it can be conceived that the numerous forms that abound are due to various concentrations of diffusing materials both organic and inorganic. The salinity of the water or the silica content may be likened to the poison morphogen, as these are known to influence the growth.

For the purposes of biological classification, the Radiolaria are divided into two subclasses, the "Porulosa" and the "Osculosa", and are further subdivided into four legions. The subclass Porulosa includes the two legions SPUMELLARIA and ACANTHARIA, which have the following characteristics:

(1) The central capsule is a sphere and retains this form throughout the majority of the species.

(2) The equilibrium of the floating unicellular body is either pantostatic (indifferent) or polystatic (plural-stable), since a vertical axis is either absent, or if present has its two poles similarly constituted.

(3) The ground forms of the skeleton are therefore almost always spherotypic or isopolar-monaxon, very rarely zygotypic.

The subclass Osculosa comprises the two legions NASSELARIA and PHAEODARIA, which agree in similar and constant characteristics.

(1) The central capsule is a sphere and retains this ground form in most of the species.

(2) The equilibrium of the floating body is monostatic and unistable, since the two poles of the main axis are always more or less different from each other.

(3) The ground forms of the skeleton are, therefore, for the most part grammotypic (centraxon) or zygotypic, rarely spherotypic.

The four principal groups of Radiolaria, which have been given the name "legions", are natural units; when, however, the attempt is made to bring them all into a phylogenetic relationship it undoubtedly appears that the SPUMELLARIA are the primitive stem. The other three have developed, probably independently, from the most ancient stem form of the SPUMELLARIA, the spherical "Actissa".

As our main interest will lie in the development of the individual ground forms, it will be appropriate to give a survey of the various types. We can classify the great variety of the geometrical ground forms into four principal groups: the "Centrostigma" or Spherotypic, the "Centraxonia" or Grammotypic, the "Centroplana" or Zygotypic, and the "Acentrica" or Atypic. The natural centre of the body, about which all its parts are regularly arranged, is in the first group a point (stigma), in the second a straight line (principle axis), in the third a plane (sagittal plane) and in the fourth a centre is, of course, wanting.

The Spherical or Homaxon ground form is the only absolutely regular ground form, since only in it are all axes which pass through the centre equal. It is very often realised among the Radiolaria, especially in the SPUMELLARIA and in the ACANTHARIA, where it furnishes the common original ground form, but it is often to be seen in the shells of many PHAEODARIA. On the other hand, it is never found among the NASSELLARIA.

The endospherical polyhedron or polyaxon ground form naturally follows the spherical or homaxon. Under it are included all polyhedra whose angles fall on the surface of a sphere; this ground form is especially common among the SPUMELLARIA but it is also found among the ACANTHARIA. Strictly speaking, all those lattice shells which have been incorrectly called "spherical" belong to this category, for none of them are true spheres in the geometrical sense, but rather endospherical polyhedra, whose angles are indicated by the nodal points of the lattice shell or the radial spines which spring from them. These polyhedra may be divided into three groups, regular, subregular and irregular. Of the regular polyhedra, properly so-called, only five can exist, namely the regular tetrahedron, cube, dodecahedron, octahedron and icosahedron. All these are actually manifested among the Radiolaria, although the subregular endospherical polyhedra are much more common.

The ground form whose geometrical type is the regular icosahedron (bounded by twenty equilateral triangles) occurs among the PHAEODARIA (e.g. "Circongonia") and also in certain Aulosphaerida. This ground form may also be assumed to occur in those Sphaeroidea whose spherical lattice shells bear twelve equal and equidistant radial spines; the basal points of these spines indicate the twelve angles of the regular icosahedron.

The regular octahedron (bounded by eight equilateral triangles) commonly appears among the SPUMELLARIA. In these Sphaeroidea the typical ground form is usually indicated by six equal radial spines which lie on three perpendicular axes. Occasionally the spherical form of the lattice shell passes over into that of the regular octahedron. The same form recurs in "Circoporus" among the PHAEODARIA.

The regular cubic ground form and the regular tetrahedral ground form also occur. The former may be regarded as occurring in those species whose spherical lattice shell bears eight equal and equidistant radial spines. The isopolar-monaxon ground form is characterised by the possession of a vertical main axis with equal poles, whilst no transverse axes are differentiated. All horizontal planes which cut the axis at right angles are circles. The most important ground forms of this group are the "phacoids" (the lens or oblate spheroid) and the ellipsoid (or prolate spheroid) e.g. *Phacodiscus rotula* and *Cromyactractus tetracelyphus*.

Appendix

The functions $\overline{P_n^m}(\cos\theta)$ are the normalised Legendre Associated Functions defined by

$$\overline{P_n^m}(\mu) = A_n^m P_n^m(\mu),$$

where μ is written for $\cos\theta$ and $P_n^m(\mu)$ represents the usual Legendre Associated Functions, with the condition that $P_n^m(\mu) = P_n^{-m}(\mu)$ and A_n^m is chosen so that

$$\frac{1}{4\pi}\int\int_S \overline{P}_n^{m_1}(\mu)\overline{P}_n^{m_2}(\mu)\mathrm{e}^{\mathrm{i}(m_1+m_2)\phi}\mathrm{d}\cos\theta\ \mathrm{d}\phi = 1.$$

One therefore finds that

$$A_n^m = \sqrt{\frac{(2n+1)(n-m)!}{(n+m)!}}$$

(see HOBSON (1931) p. 162).

It is convenient to introduce the functions $L_n^{p,q,r}$ and $E_n^{p,q,r}$, defined by:

$$L_n^{p,q,r} = \frac{1}{4\pi}\int_0^{2\pi}\int_{-1}^{1} \overline{P_n^p}(\mu)\overline{P_n^q}(\mu)\overline{P_n^r}(\mu)\mathrm{e}^{\mathrm{i}(p+q+r)\phi}\,\mathrm{d}\mu\,\mathrm{d}\phi$$

and

$$E_n^{p,q,r} = \frac{1}{2}\int_{-1}^{1} \overline{P_n^p}(\mu)\overline{P_n^q}(\mu)\overline{P_n^r}(\mu)\mathrm{d}\mu.$$

It follows that

$$L_n^{p,q,r} = \begin{cases} E_n^{p,q,r} = E_n^{|p|,|q|,|r|} & \text{for } p+q+r=0,\\ 0 & \text{for } p+q+r\neq 0.\end{cases}$$

Since the definition of $L_n^{p,q,r}$ is independent of choice of axis, it is possible to rotate the coordinate axis and keep the integral constant. This enables one to obtain a form of recurrence relation for the L's:

$${}^{m_1}K_n^{m_1-1}L_n^{m_1-1,m_2,m_3} + {}^{m_2}K_n^{m_2-1}L_n^{m_1,m_2-1,m_3} + {}^{m_3}K_n^{m_3-1}L_n^{m_1,m_2,m_3-1} = 0$$

subject to the condition that $m_1+m_2+m_3=1$; the constants ${}^pK_n^q$ being given by

$${}^{p+1}K_n^p = -\frac{1}{2}\sqrt{(n-p)(n+p+1)},$$

$${}^pK_n^{p+1} = \frac{1}{2}\sqrt{(n-p)(n+p+1)},$$

$${}^pK_n^q \equiv 0 \quad \text{unless } |p-q|=1,$$

$${}^pK_n^q = {}^{|p|}K_n^{|q|}.$$

The set of equations which arises from the recurrence relation is not unique, and it is therefore possible to derive several checking equations. For the case $n=2$ the set for solution contains three equations since, in all cases, it is necessary to determine $E_n^{0,0,0}$ beforehand. This is best done by direct calculation from the integral or by use of the formula:

$$E_n^{0,0,0} = \frac{1}{3n+1}\left(\frac{1\cdot 3\cdot\ldots\cdot(n-1)}{2\cdot 4\cdot\ldots\cdot n}\right)^3 \frac{2\cdot 4\cdot\ldots\cdot(3n)}{1\cdot 3\cdot\ldots\cdot(3n-1)}$$

$$= \frac{1}{3n+1}\left(\frac{n!}{(n/2)!}\right)^3 \frac{(3n/2)!}{(3n)!}$$

(see HOBSON (1931) p. 86).

For $n=4$ there are nine equations, and for $n=6, 17$. It is possible to check the results by evaluating $E_n^{n,n,0}$ using the polynomial and the Beta function, and also by means of the relation

$$\sum_{-n}^{n} E_n^{i,i,0} = 0.$$

Examining the Work and Its Later Impact

Peter Saunders comments* on the background to —

TURING'S MORPHOGEN THEORY OF PHYLLOTAXIS

It is not hard to imagine why Turing chose the arrangement of leaves on plants as the first application of his theory. Phyllotaxis is a classical problem which remains unsolved to this day, despite the efforts of many workers. Yet at the same time it is hard to believe that it does not have a straightforward solution, if only one were clever enough to find it. The phenomenon to be explained is the occurrence of a small number of regular patterns on a simply shaped and accessible surface. The pattern on a mature specimen is essentially that which is laid down in the first place, which is not so in many other developmental processes. And for a mathematician there is the additional twist that the Fibonacci sequence is involved.

Turing's attempt on the problem consists of two parts. The first is a detailed geometrical analysis of the patterns, and the second is the beginning of an application of the (1952) theory to explain them. While the latter, though incomplete, is quite straightforward and self-contained, the former requires some further explanation. Many readers may not know much about phyllotaxis, and most of those that do will probably be accustomed to accounts written by botanists, who usually approach the problem slightly differently. In particular, because Turing was setting out to investigate as deeply as possible the patterns he was hoping to explain, he chose to represent the leaves as the points of a lattice. This amounts to considering the mature stem as a cylinder, unrolling the surface onto the plane, and then repeating the pattern infinitely many times. There are obvious mathematical advantages in this, and Turing is not the only author to have done it, but it does mean that there are some differences between his approach and the usual botanists' picture, which is based on a cross section. Above all, the connection with the Fibonacci sequence is far less obvious. To assist the reader, therefore, I provide below an outline introduction to phyllotaxis with definitions of the terms that are used, referring where necessary to both representations. I also include a brief account of continued fractions and an explanation of how the Fibonacci numbers enter into the problem and how they are connected with the Fibonacci angle.

Phyllotaxis:

There are several common forms of phyllotaxis. In some plants, such as grasses and peas, each leaf is at an angle of 180° from the one before it on the stem. This is called distichous phyllotaxis. In another form, known as decussate phyllotaxis and found in, for example, trees like the ash and horse chestnut, the leaves occur in opposing pairs, with each pair in a plane at right angles to the one before. In most flowering plants, however, and in conifers and various other families, the leaves are arranged around the stem in such a way that it is possible to draw a single spiral, called the

* Comments reproduced from the *Morphogenesis* volume of the *Collected Works of A. M. Turing*, pp. XVII–XXIV.

fundamental (also *genetic, ontogenetic* or *generative*) spiral which passes through the centres of all of them in the order in which they appeared. Since the time interval between successive appearances of primordia (the *plastochrone*) is approximately constant, so too are the distances and the angles between them.

The angle between successive primordia, or leaf centres in the mature plant, is called the *divergence angle.* To specify the arrangement completely a second coordinate is required and if we are studying the cross section a convenient choice is the *plastochrone ratio* (RICHARDS 1948), the ratio of the transverse distances from the centre of successive primordia. In a uniform system it is a measure of the radial expansion of the apex during one plastochrone. Because he was concerned with the side view, Turing (Section 2) used instead what he called the *plastochrone distance*, which is measured along the surface of the stem; in the case of a cylindrical stem it is along one of the generators.

On account of the regularity, many spirals other than the fundamental one can be drawn through primordia; any such spiral is called a *parastichy.* Two examples are shown in Figs. 1 and 2. These are illustrations of the kind commonly found in biological works and show a transection of the apical bud. Drawn in this way, one's eye is immediately caught not by the fundamental spiral but by the two parastichies, one spiralling clockwise and one anticlockwise, that pass through primordia that are actually in contact and are therefore called *contact parastichies.* In the mature plant they are the parastichies which pass through a given leaf and one of the two or three adjacent leaves above or below it, and they are then sometimes referred to as the *conspicuous opposed parastichy pair,* because the leaves are not actually in contact as the primordia were. If the primordia are numbered in order of formation, or in the case of a mature plant if the leaves are numbered in order along the stem, it is easily seen that the difference in number between successive primordia on a parastichy will be constant. What is surprising is that in the vast majority of cases, these numbers, called the numbers or orders of the parastichies, are members of the Fibonacci sequence, 1, 1, 2, 3, 5, 8, 13, 21, ... The numbers of the two contact parastichies are successive terms in the sequence.

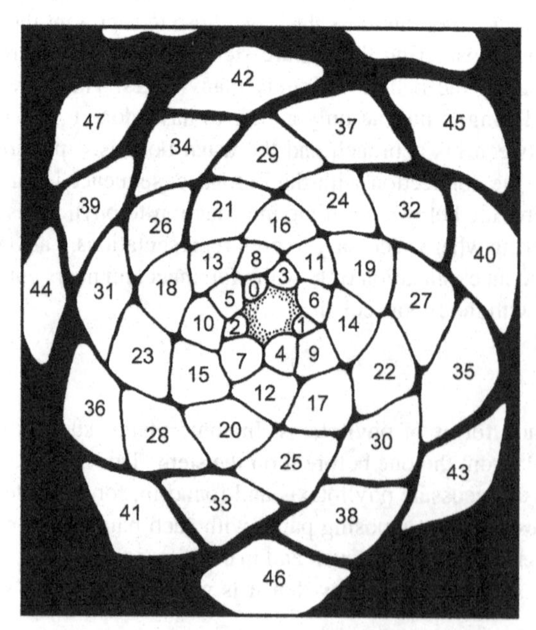

Fig. 1: Transection of the apical bud of a young seedling of *Prinus pinea*. The leaves are numbered in the order of formation. The contact parastichy numbers are 5 and 8. Redrawn after Church (1920).

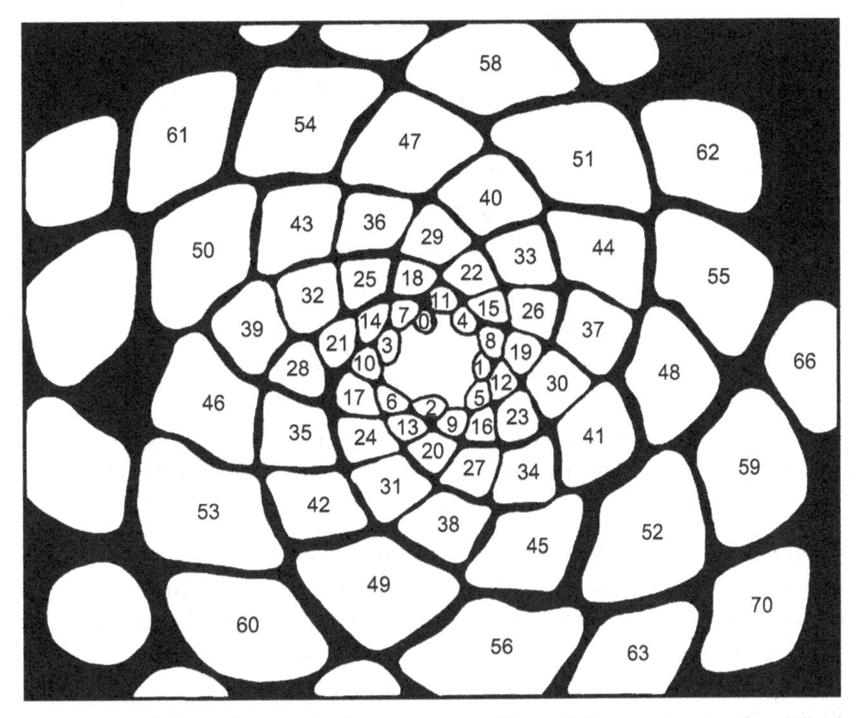

Fig. 2: Transection of the apical bud of a young seedling of *Araucaria excelsa*. The leaves are numbered in order of formation. The contact parastichy numbers are 7 and 11. Redrawn after Church (1920).

The significance of the Fibonacci numbers in phyllotaxis has been recognized for a long time; according to ADLER (1974), Kepler was the first to comment on it. Kepler also suggested that the appearance of the sequence in biology might be connected with its property that each of the terms is the sum of the two which precede it. And indeed in spiral phyllotaxis, even if the parastichy numbers are not from the Fibonacci sequence they are often from another sequence formed by a similar rule, such as 1, 3, 4, 7, ... , 1, 4, 5, 9, ... or 2, 5, 7, 12, ... etc. Even if there is no simple explanation, that so many plants should have this curious property does strongly suggest a common underlying process which is regular enough that we can hope to elucidate it, which is doubtless why so many workers have been attracted to the problem.

In the side view, the contact parastichies are less obvious, but it is easier to see whether or not there is more than one leaf at each level. Usually there is not, but when there is then there is also more than one fundamental spiral. Turing denoted the number of leaves at each level by J, and called it the *jugacy*, because the cases $J = 2$, $J = 3$, $J > 3$ are commonly referred to as bijugate, trijugate and multijugate, respectively. In Fig. 3 we have supposed that $J = 1$. There is consequently only one fundamental spiral, and we take this to be a helix with the leaves at equal intervals along it so that the leaves form a cylindrical lattice. Figure 3(b) shows the equivalent plane lattice.

A parastichy will not in general pass through a leaf at every level. If it passes through a leaf at every nth level only, then it is called a parastichy of order n. Turing called n the *parastichy number*. There are altogether nJ parastichies of order n in a phyllotaxis of jugacy J. They are all parallel, and every leaf lies on exactly one of them, so they partition the set of leaves. A collection of mJ m-order parastichies and nJ n-order parastichies with m and n is chosen, so that one set is clockwise and the one in anticlockwise is called an *opposed parastichy pair* of order (m, n). There need not be a

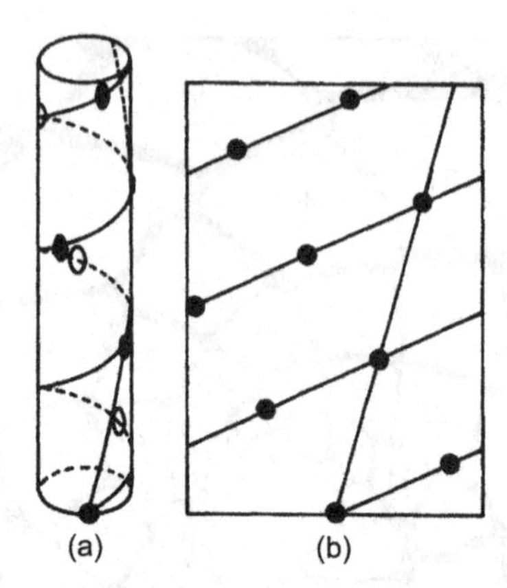

Fig. 3: (a) Side view of an idealised stem and (b) the equivalent plane lattice. The contact parastichy numbers are 1 and 2. Both the generative spiral and a parastichy of order 3 are shown.

leaf at every intersection of the two spirals of such a pair, but if there is, the pair is called *visible.* In the lattice representation the contact parastichies are defined as the parastichy pair defined by a leaf together with its nearest neighbours to the right and to the left. This is of course equivalent to the definition given above.

The Fibonacci sequence:

Almost everyone who writes about phyllotaxis points out the striking property that the contact parastichies typically have numbers that are members of the Fibonacci sequence and that the divergence angle (i.e., the angle between successive primordia) is then close to the Fibonacci angle, approximately 137.51°. They also generally mention that other divergence angles occur, though less frequently, that these are approximately 99.50° and 77.96°, and that the contact parastichies then have numbers from the series 1, 3, 4, 7, 11,... (the anomalous or first accessory series) or 1, 4, 5, 9, 14, ..., respectively. The latter two series satisfy the same recurrence relation as the Fibonacci series, viz. $u_n = u_{n-1} + u_{n-2}$.

These statements are usually made without any explanation. At the time when Turing was writing, it may have been safe to suppose that most readers, at least those who were mathematicians, would be familiar with the connection between the Fibonacci sequence and the particular angle. Because this is less true today, and even continued fractions (which are used in Section 10) are no longer a standard part of mathematics syllabi, we give a brief account here.

A continued fraction is a fraction of the form

$$a_0 + \cfrac{1}{a_1 + \cfrac{1}{a_2 + \cfrac{1}{a_3 + \cfrac{1}{a_4 + \dots}}}}$$

where $a_0, a_1, a_2, \dots, a_n, \dots$ are real numbers all of which, with the possible exception of a_0, are positive. Because this is an awkward expression to write or set in type it is usual to employ a conventional notation, such as the one Turing used in Section 10 or the even simpler $[a_0; a_1, a_2, \dots]$. The numbers a_n are called the partial quotients of the fraction. If there are only a finite number of non-zero partial quotients, the continued fraction is said to be finite; if all the partial quotients are integers, it is said to be simple. The finite continued fraction obtained from an infinite one by cutting off the expansion after the nth partial quotient, a_n, is called the nth convergent of the continued fraction.

It is not difficult to show (see, e.g. BURTON (1976), p. 306) that the nth convergent of the simple continued fraction $[a_0; a_1, a_2, \ldots]$ is given by

$$C_n = p_n/q_n$$

where

$$\begin{aligned} p_0 &= a_0, & q_0 &= 1, \\ p_1 &= a_1 a_0 + 1, & q_1 &= a_1, \\ p_k &= a_k p_{k-1} + p_{k-2}, & q_k &= a_k q_{k-1} + q_{k-2},\ k > 1 \end{aligned}$$

It can also be shown that p_k and q_k are relatively prime, so the formula gives the convergents in their lowest terms. If $a_0 = 0$ and $a_k = 1$ for all $k > 0$, the recurrence relations for p_k and q_k generate Fibonacci series, and the successive convergents are

$$C_n = u_{n+1}/u_n$$

where u_n denotes the nth Fibonacci number.

Every rational number can be written as a finite simple continued fraction in two closely related ways. Every irrational number has a unique representation as an infinite simple continued fraction. Much of the importance of continued fractions arises from the fact that if an irrational number x is expressed as a continuous fraction, the convergents p_n/q_n are the best approximations to x in the sense that each of them gives the closest approximation to x among all rational numbers with denominators q_n or less. This property makes continued fractions useful in numerical analysis, and it is also the reason that they are connected with phyllotaxis.

The contact parastichies are the parastichies formed by adjacent leaf bases. Now which bases are adjacent to a given base O depends not only on the angular separation between them but also on the pitch of the helix (see Fig. 4). On the other hand, it is clear that a leaf base is a candidate for being adjacent only if it is closer to the generator (i.e. the vertical line) through O than is any previous base on the generative spiral, or at least any previous base on the same side of the generator.

Now the leaves are generally equally spaced along the generative spiral. Let their angular separation, i.e. the divergence angle, be θ, and let $\alpha = \theta/2\pi$. In finding leaves that are close to the original vertical line we are looking for integers a, b such that $a/b \doteqdot 2\alpha\pi$. Here b is the number of the leaf in sequence along the generative spiral and a is the number of rotations the spiral has made around the stem. The closest rational approximations to α are the convergents of its expansion as a continued fraction. Given any α, therefore, we can immediately calculate the convergents, and because the convergents are automatically in their lowest terms this gives us p_k and q_k separately. The sequence p_k gives the sequence of leaves successively closest to the vertical line, i.e. the set of possible contact parastichy numbers for the given divergence angle.

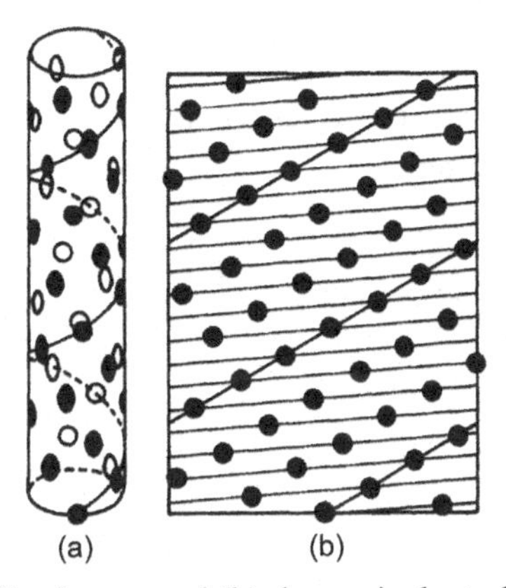

Fig. 4: (a) Side view of an idealised stem and (b) the equivalent plane lattice. As Fig. 3 except for the pitch of the helix. The contact parastichy numbers are 3 and 5.

Conversely, if we are given the complete set of possible contact parastichy numbers we can work out the divergence angle by solving the recurrence relation for the a_k. In particular, if $p_k = u_k$, the kth Fibonacci number, then the continued fraction is $[0, 1, 1, 1, \ldots]$, which can be shown to be equal to $(1 + \sqrt{5})/2$, or about 1.618, the so-called 'golden mean'. This implies a divergence angle of about 582.5°, or, equivalently, 137.5°. For this divergence angle, therefore, the contact parastichy numbers will always be Fibonacci numbers, though which ones they will be will depend on the pitch of the helix. The angle 137.5° is called the Fibonacci angle.

The next most common spirals have parastichy numbers from the subsidiary series 1, 3, 4, 7, 11, ... This corresponds to the continued fraction [0, 3, 1, 1, 1, ...] which implies a divergence angle of approximately 99.5°. Also observed is 1, 4, 5, 9, 14, ... , with the continued fraction [0, 4, 1, 1, 1, ...] and angle 77.96°.

For more on continued fractions and the related sequences, see BURTON (1976) or almost any other elementary book on number theory. The 'simple inductive argument' referred to in the text (p. 66) and proofs of the results mentioned above can be found in Burton's book.

BIBLIOGRAPHY

ADLER, I.
1974 A model of contact pressure in phyllotaxis
J. Theor. Biol. **45,** 1–79.

ARBER, A.
1950 *The Natural Philosophy of Plant Form*
(Cambridge University Press, Cambridge).

BRAUN, A.
1835 *Flora*, 145.

BRAVAIS, L. and A. BRAVAIS
1837 *Ann. Sci. Nat. Botanique* (2) **7,** 42 and 8, 193.

BURTON, D.M.
1976 *Elementary Number Theory*
(Allyn and Bacon, London).

CASTETS, V., E. DULOS, J. BOISSONADE and P. KEPPER
1990 Experimental evidence of a sustained standing Turing type non-equilibrium chemical pattern
Phys. Rev. Lett. **64,** 2953–2956.

CHURCH, A.H.
1904 *On the Relation of Phyllotaxis to Mechanical Laws*
(Williams & Norgate, London).

CHURCH, A.H.
1920 *Oxf. Bot. Mem.*, No. 6.

DAWKINS, R.
1986 *The Blind Watchmaker*
(Longmans, London).

DE BEER, G.
1951 *Embryos and Ancestors*
(Oxford University Press, Oxford).

GOEBEL, K.
1922 Gesetzmässigkeiten in Blattaufbau
Bot. Abh. (Jena) **1.**

HARLAND, S. C.
1936 The genetical conception of the species
Biol. Rev. **11,** 83.

HOBSON, E.W.
1931 *The Theory of Spherical and Ellipsoidal Harmonics*
(Cambridge University Press, Cambridge).

HODGES, A.
1983 *Alan Turing: The Enigma of Intelligence*
(Hutchinson, London).

MURRAY, J.D.
1981 A pre-pattern mechanism for animal coat markings
J. Theor. Biol. **88,** 161–199.

PALEY, W.
1802 *Natural Theology*
(F. & J. Rivington, London).

RICHARDS, F.J.
1948 The geometry of phyllotaxis and its origin
Symp. Soc. Exp. Biol. **2,** 217–245.

ROSS-CRAIG, S.
1951 *Drawings of British Plants, Part V*
(Bell and Sons, London).

THODAY, D.
1939 The interpretation of plant structure
Nature **144,** 571.

THOM, R.
1972 *Stabilité Structurelle et Morphogénèse*
(Benjamin, Reading).

THOMPSON, D' A.W.
1917 *On Growth and Form*
(Cambridge University Press, Cambridge).

VAN ITERSON, G.
1907 *Mathematische und Mikroskopische-Anatomische Studien der Blattstellungen*
(Fischer, Jena).

WARDLAW, C.W.
1953 A commentary on Turing's diffusion-reaction theory of morpho-genisis
New Phytol. 52, 40-47. Republished in *Essays on Form in Plants*
(Manchester University Press, Manchester, 1968).

Jonathan Swinton explores further —

TURING, MORPHOGENESIS, AND FIBONACCI PHYLLOTAXIS: LIFE IN PICTURES

After completing *The Chemical Basis of Morphogenesis*, and at some point in last three years of his life, Turing produced the pen-and-ink pictures reproduced as figures throughout this commentary and published here for the first time. When his editors came to publish the Morphogenesis volume of the Collected Works in 1992 they struggled with the fragmentary nature of *The Morphogen Theory of Phyllotaxis* and *Outline of the Development of the Daisy* and these images, without captions or a clear place, were not included. But they tell a fascinating story of Turing's quest to explain, among much else, the appearance of Fibonacci numbers in the natural world. They have considerable historical interest as very early graphics of computational biology, but the work they represent remained in advance of published theory for decades. It is one goal of this commentary to represent these figures and to explain their origin in the context of Turing's phyllotactic theory. The other goal is to provide a partial guide to those aspects of the surviving fragmentary works which can be understood from this perspective.

1. Introduction

> ... *I think I can account for the appearance of Fibonacci numbers in connection with fir-cones*[1] ...

In 1951 Turing had submitted what is now seen as a seminal paper on the generation of form in biological systems (Turing, 1952). Turing was to publish no more in this field in life though at his death three years later he did leave behind a mass of working notes and two related but uncompleted manuscripts, *The Morphogen Theory of Phyllotaxis* and the much sketchier *Outline of the Development of the Daisy*, both published posthumously in the Collected Works (Turing, 1992), together with a mass of working notes including a set of figures.[2] The figures cannot be slotted directly into those manuscripts but they are illustrations of their central concerns and were probably intended for subsequent drafts. I have published elsewhere an account of my own attempts to understand that work (Swinton, 2003), and here I repeat and extend that material, taking the opportunity to resurrect some of Turing's rather elegant approach to mathematical phyllotaxis.[3]

Although so much of the discussion of Turing's heritage in mathematical biology has focussed on the reaction–diffusion mechanism (Allaerts, 2003; Murray, 1993), which Turing attempted to put

[1] A. M. Turing, letter to M. Woodger, February 1951, quoted in Hodges (1992).

[2] Although in print here for the first time they been available online for a decade at http://www.turingarchive.org/ and visible to view for visitors to King's College, Cambridge. The archives in Cambridge, and the other archive of Turing material at the UK National Archive for the History of Computing (NAHC) in Manchester include more material than is described here.

[3] This section draws heavily on Swinton (2003) but see also Allaerts (2003); Hodges (1992); Nanjundiah (2003).

on a firm footing in Part II of his *The Morphogen Theory of Phyllotaxis*, it is helpful in understanding the figures, and the unpublished manuscripts, to distinguish that mechanism from a second broad mathematical idea that Turing was also thinking and writing about in the morphogenesis work, and which makes up much of Part I. Reaction–diffusion was a mechanism for a generation of form—something that could create spots or strips where only uniformity had been before as Meinhardt and Richards elaborate on elsewhere in this volume (Meinhardt, 2013; Richards, 2013). Reaction–diffusion systems have become part of the mathematical mainstream[4] and though an understanding of them is necessary to fully understand the archive material I refer the reader to Murray (1993) as an introduction. The second, in some ways mathematically simpler, idea was the question of what happens when a mechanism of, say spot-generation, is allowed to repeat itself over and over again, constrained by the existence of previous spots, and with slowly changing parameters. It is this second idea that lies at the heart of the explanation of Fibonacci phyllotaxis that Turing sought, partly found, and which has subsequently been discovered.

2. Modelling Fibonacci phyllotaxis

Fig. 1: Florets on a sunflower. The most visually obvious spirals have been identified, and then starting from about 2 o'clock, every 10th such spiral has been removed showing that there are 55 of them. The spirals in the opposite direction, which are almost radial lines and harder to count, number 89: this is a (55,89) parastichy pair.

[4] Although their utility as models of developmental biology has remained contested (Fox Keller, 2002).

The Fibonacci series is the series of numbers

$$0, 1, 1, 2, 3, 5, 8, 13, 21, 34, 55, 89, 144, \ldots,$$

where each number after the second is the sum of the previous two. Phyllotaxis means the arrangement of forms in plants, and Fig. 1 shows a sunflower in which the two come together as Fibonacci phyllotaxis; the appearance of numbers from this sequence in plant structures. Much has been written in and beyond the mathematical literature on the problem of explaining this appearance (Adler and Barabé, 1997; Jean, 1994; Jean and Barabé, 1998 and Swinton, 2003) and this contribution will not attempt to review it all, but to understand the figures we do need two ingredients of an explanation of this problem. The first is a caricature of how structures appear on growing plant stems, and the second is a mathematical model of that process.

2.1. *A caricature of plant growth*

A caricature description of stem growth and floret formation is to view it as taking place on a stem with a growing tip, just below which is a ring called the apical meristem. In this region, a decision is from time to time made to generate the cells which will subsequently differentiate into bud or leaf formations. After having writ, the growing tip moves on and upward, leaving the bud behind. As well as the decision of a time to commit to bud formation, which corresponds via the speed of growth to the spatial distance between buds, a decision is made as to the angular position of the bud around the quasi-cylindrical, apical meristem. If we also imagine that the stem is thickening with age, then looking down at this process from above, we can make sense of the first of Turing's figures: (Fig. 2). This figure starts with a Fibonacci structure in that it uses the Fibonacci angle and so it is inevitable that Fibonacci numbers will appear (Jean and Barabé, 1998). Turing's task was to understand why these structures appeared without the imposition of a particular angle.

2.2. *Lattice dynamics*

Assume further that bud positioning within the apical meristem is only a function of effects acting within the meristem, and that the positioning process reaches equilibrium on a time scale shorter than that on which the apical meristem itself enlarges. By approximating the ring of the apical meristem to be exactly cylindrical, and supposing that it has enough vertical extent that edge effects can be neglected, we arrive at a mathematical model for pattern formation on a tall vertical cylinder. This cylinder may be slowly enlarging with time, and as it does so the evolved pattern rapidly come to a new equilibrium.[5] The first central question of such a model is what symmetries are to be expected of pattern formation processes on cylinders, and the second is how those symmetries will *evolve* over the growth process.

3. Lattices on cylinders

Turing had his reaction–diffusion process quite unambiguously in mind as the pattern formation process, and was certainly using it for his numerical work. But much progress can be made in understanding the kinds of patterns it must generate on cylinders by recognising that it is enough to require a mechanism that produces spots separated in space by a characteristic distance. Indeed, the reaction–diffusion mechanism is in some ways a more complicated approach than those of others which have largely focused on packings based on rigid circles (Atela et al., 2002; Mitchison, 1977; van Iterson, 1907), or equivalently electrically charged particles (Douady, 1996; Kunz, 1995).

[5] For more discussion of this approximation see Turing (1992).

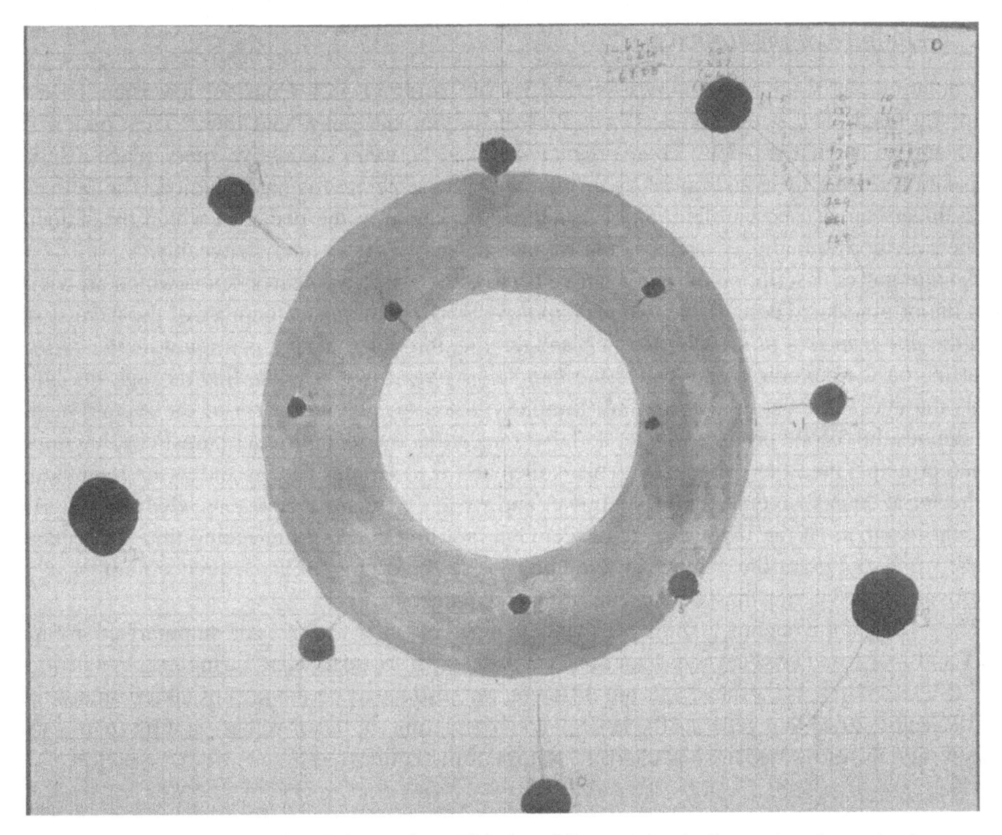

Fig. 2: An idealised view of bud formation. This is a Fibonacci spiral; starting from the innermost spot at about 3 o'clock, labelled (0), each successor spot is rotated a further angle $2\pi/\omega^2 \approx 137^\circ$; ω is the limiting ratio of successive terms of the Fibonacci series and solves $\omega^2 + \omega = 1$. This is calculated in the first pencil column at the top left. The difference in radial distance between each spot is a factor of 1.1—calculated in the second column at the top left. It can be seen that the nearest spots to spot 0 are those numbered 3 and 5, so this displays a Fibonacci phyllotaxis in which the principal parastichy numbers, (3,5) are Fibonacci ones. This is by construction, for when the rotation angle is $2\pi/\omega^2$ the parastichies are forced to be Fibonacci numbers (Jean, 1994; Turing, 1992). However even with this constraint the particular members of the Fibonacci sequence which appear are not fixed but depend on the *rise*, the difference in radial distance, here the factor 1.1. Moreover, the third nearest spot to spot 0 has number 8: the third parastichy number is the sum of the first two. From the Modern Archives Centre, King's College, Cambridge archive at AMT/K3/3. The archive is online at `http://www.turingarchive.org/`. ©PN Furbank.

But in any case Turing needed a theory of lattices on cylinders, and characteristically regenerated his own rather than building on earlier attempts (Church, 1904; van Iterson, 1907). The implications of that theory have largely been rediscovered and enlarged subsequently (Jean, 1994), but it has some elegant corners, particularly in its treatment of patterns with lattice symmetry and their Fourier transforms, which are essential to understand the figures and which have not been subsequently emphasised. The theory was written up in Part I of *The Morphogen Theory of Phyllotaxis*, and so published in 1992. However, the manuscript was never a complete one, and Turing's style never the most transparent, which might explain why it has never even been cited in the mathematical phyllotaxis literature. To redress that, and to explain what we need to understand the figures, I give a summary of the theory here.

3.1. *Geometrical phyllotaxis*

This notation and discussion follows Section I.4 of Turing (1992). Consider a vertical cylinder of infinite height and circumference $2\pi\rho$. Pick an origin, and put a spot there. Then pick a *rise* $\eta > 0$, and an angular rotation (the *divergence angle* α) between successive spots, place a spot at the position rotated by α and raised by η, then repeat to generate the entire lattice. The lattice on the cylinder can also be unrolled into a $2\pi\rho$-periodic lattice in the plane, in which the sequence numbers are correspondingly mapped, and we take these two views interchangeably.

An alternative description of such a lattice is to pick out the two nearest spots which are visible from the origin spot. These define two *principal vectors*. The sequence number of the nearest spot is m, the *first principal parastichy number*, and the line through that spot is a spiral on the cylinder (or a line on the periodic plane) called the *first order parastichy*: it is the line through the lattice points the eye will most easily pick out. Similarly, the sequence number n of the second nearest spot defines the second principal vector and, by translation, the second-order parastichy, having the second principal parastichy number n. This is then an (m,n) lattice. The ms and ns are then models for the spiral counts we make in the sunflower and we are trying to see reasons why they are often adjacent members of the Fibonacci sequence. One first hint comes from seeing that the sequence number of a spot gives the number of rise units above the origin, so the sequence number of the vector sum of two points must be the sum of their sequence numbers.

The theory of this section is similar to that developed by other workers and summarised in Chapters 1 and 2 of Jean (1994), where Jean uses a 'visible pair' to mean what Turing terms 'parastichy pair' (Swinton, in press). The details are different, but also based on the idea of taking an arbitrary pair of vectors $(\mathbf{u},\mathbf{v})$ that generate the lattice and 'contracting' it, that is replacing it by $(\mathbf{u},\mathbf{u}-\mathbf{v})$ in order to find the most fundamental characterisation of the lattice.

3.2. *Finding the principal parastichy vectors*

Since we will later want to know how m and n change as the lattice parameters, particularly ρ, change, it is useful to be able to compute m and n from the lattice parameters, and Turing (1992) provides a mechanism to do so, which also sheds further light on why Fibonacci structures may emerge. The converse of this problem, computing the lattice parameters from (m,n), is Adler and Jean's Fundamental Theorem of Phyllotaxis (Adler, 1998; Jean, 1994).

Firstly, Turing observes that the first and second principal vectors must generate the lattice. This 'small consequence' follows from a section of the manuscript[6] that was not included by Hoskin: because the area of the parallelogram formed by any two vectors $(\mathbf{u},\mathbf{v})$ in the lattice is $\Delta(\mathbf{u},\mathbf{v})$, with $\Delta^2(\mathbf{u},\mathbf{v}) = \mathbf{u}^2\mathbf{v}^2 - (\mathbf{u}.\mathbf{v})^2$ and for vectors which generate the lattice this can be interpreted as the area per point, and so must be the same for any choice of generating vectors. Since the principal vectors minimise $\mathbf{u}^2\mathbf{v}^2$ among generating vectors, they must also minimise $(\mathbf{u}.\mathbf{v})^2 = \mathbf{u}^2\mathbf{v}^2 - \Delta^2$.

Turing then gives an algorithm to construct these principal vectors from any pair of generating vectors. Now the third principal parastichy vector has to be one of $\mathbf{u}\pm\mathbf{v}$. Turing's reasoning can be verified by considering the tiling of the plane by the $\mathbf{u},\mathbf{v}$ parallelogram. If we fix the sign of $\mathbf{u}$ by making its vertical component positive (and equal to $m\eta$ where m is the first parastichy number), there is a choice of sign for $\mathbf{v}$ and Turing makes the choice in which $\mathbf{u}.\mathbf{v} < 0$. With this sign choice the third parastichy vector must be $\mathbf{u}-\mathbf{v}$ because $|\mathbf{u}-\mathbf{v}| < |\mathbf{u}+\mathbf{v}|$. But while $\mathbf{v}$ may have a positive or negative vertical component, the second parastichy numbers n is by definition positive and so the second parastichy vector has vertical component equal to $\pm n\eta$. Thus the parastichy number of the third parastichy vector can be $|m\pm n|$: the sum or difference of the first two.

[6] Kings/AMT-C25-14.

3.3. *Change of parastichy numbers as a lattice changes: the hypothesis of geometrical phyllotaxis*

So far the discussion has been about a single, fixed lattice. However as modelled in Section 2.2 the phyllotactic pattern is laid down in an arena that is smoothly changing, affecting the circumference, and in response to a spot-generation pattern that is also smoothly changing, affecting the rise. Turing develops a system of flow matrices to describe these processes in general, which he uses in his subsequent discussion of optimal packings. But the key idea is captured by observing that these changes in geometry can be thought of simply as smooth deformations of the original lattice.

The observation that the third parastichy number must be the sum or difference of the first two is then particularly important when we consider how the first two parastichy numbers must change as the lattice is smoothly changed as they can only change when the third nearest point becomes closer than the second nearest point.

Now suppose our lattice at some point has first three parastichy numbers, members of the Fibonacci series (p_{r-1}, p_r, p_{r+1}). Under smooth change, the order may change but eventually one of the three may drop out. If p_{r-1} drops out it must be replaced by a new third parastichy number x satisfying $x = p_r \pm p_{r+1} = p_{r-1}$ or p_{r+2}; if p_{r+1} drops out it must similarly be replaced by p_{r-2} or p_{r+1}. So if we could forbid lattice transitions that allowed p_r to be dropped, we could ensure that lattices which began as Fibonacci remained Fibonacci. This *Hypothesis of Geometrical Phyllotaxis* is that the third parastichy number has to be larger or smaller than the first two. If Turing could have demonstrated this he would essentially have solved his Fibonacci Phyllotaxis problem.

4. Exploring the Hypothesis of Geometrical Phyllotaxis

The Hypothesis of Geometrical Phyllotaxis (H. of G. P.) is not necessarily true in general: as Turing (1992), p.72 (p.791 this volume) states, 'the hypothesis is itself quite arbitrary and unexplained' and why it should hold is a question that 'the geometrical approach is not capable of answering'. He writes that he will discuss this in Part II, but in fact only two relevant fragments survive. One is in the discussion of what Turing calls 'touching-circles' phyllotaxis, following van Iterson (1907) and in some ways anticipating Mitchison (1977), in which the spot-formation mechanism at any lattice geometry is a packing of rigid disks. Turing found this an unsatisfactory mechanism for a model, perhaps because it provides no mechanism to ensure optimal packing, but he notes (see pp.77–80; pp.794–797 this volume) that in such a model the H. of G. P. holds. The second, more relevant pages are in the brief but important *Outline of the Development of the Daisy*, which gives, among other things, a description of numerical simulations that had already been carried out, and which should produce exactly the kind of smoothly changing lattice to which the H. of G. P. should apply. In the absence of an analytical proof, it was likely Turing's plan to show that lattices emerging as solutions of reaction–diffusion equations (i.e. with nodes more-or-less a fixed distance apart) would satisfy the H. of G. P., or at least to rely on its truth to have confidence that the emerging solutions would have Fibonacci structure. To do that, he needed a lattice representation, not of spots, but of patterns with lattice symmetry.

The standard account of the Turing instability sets up equations for two chemicals describing their mutual reactions and diffusions. For some sets of parameters, the solution to the equation is a homogeneous one, but as the parameters vary that state loses stability and solutions appear which are periodic in space. In one dimension x, this is analysed by plotting the dispersion relation $\lambda(k)$ between the growth rate λ in time t and the wavenumber k of solutions $e^{\lambda(k)t+ikx}$ of the linearised equation. For certain classes of equations, this dispersion relation can have a maximum at say k_0, and the value of this maximum can be changed from negative to positive by changes in parameters. Then solutions with wavelengths in the range near $2\pi/k_0$ can be expected. On a plane, the same reaction–diffusion system has no preferred direction, so all possible wavevectors $\mathbf{u}$ with modulus near k_0 are those which are promoted by the dynamics. On a cylinder with a particular lattice symmetry there are more constraints. Turing proceeds to an elegant representation of these constraints by observing the relationship between them and the lattice symmetry of the pattern.

4.1. *The lattice matrix and inverse lattice*

First, although, we need another way of representing lattices. In Section 9 of *MTP*, Turing introduces a matrix representation. If this is an (m,n) lattice with rise in η and the first and second principal vectors make angles $\tan^{-1}\tau_1$ and $\tan^{-1}\tau_1$ with the vertical, then the principal parastichy vectors are $\mathbf{u}_1 = (a,b)$ and $\mathbf{u}_2 = (c,d)$ with

$$L = \begin{pmatrix} a & b \\ c & d \end{pmatrix} = \begin{pmatrix} m\eta\tau_1 & m\eta \\ n\eta\tau_2 & n\eta \end{pmatrix} \tag{4.1.1}$$

and the area of the parallelogram generated by $\mathbf{u}_1$ and $\mathbf{u}_2$ is

$$\Delta = |ad - bc| \tag{4.1.2}$$

$$= mn\eta^2 |\tau_2 - \tau_1| \tag{4.1.3}$$

$$= 2\pi\eta \tag{4.1.4}$$

In Section 12 of *MTP*, Turing introduces coordinates (ξ_1, ξ_2) which are normalised to the parallelogram of the principal parastichy vectors, in which, e.g., the first principal parastichy vector is of the form (1,0), and all points of the lattice can be generated by allowing ξ_1, ξ_2 to run over the integers. We can transform to these coordinates with $(x,z) = (\xi_1, \xi_2)L$ and transform back by inverting and then transpose the lattice matrix L to obtain

$$M = \begin{pmatrix} A & B \\ C & D \end{pmatrix} = \begin{pmatrix} a & b \\ c & d \end{pmatrix}^{-1T} \tag{4.1.5}$$

$$= \frac{1}{2\pi\rho} \begin{pmatrix} n & -n\tau_2 \\ m & m\tau_1 \end{pmatrix}. \tag{4.1.6}$$

Just as the matrix L generates the original lattice, the lattice generated by M is the *inverse* lattice.

4.2. *Fourier representations of functions with lattice symmetries*

Turing introduced a wholly new dimension into the analysis by considering not merely the lattice points themselves, but functions which have the symmetry of the lattice. This is a natural consequence of thinking of the spots as high-points of a surface generated by a reaction–diffusion system, or from the technique of solving such reaction–diffusion systems in Fourier space.

If $f(x)$, arising perhaps from such a reaction–diffusion system, has the symmetry of the lattice, then it is defined, in the lattice coordinates above, by specifying $f(\xi_1, \xi_2)$ for $0 \le \xi_1,\ \xi_2 \le 1$ within the parallelogram of the first two principal vectors, and as a periodic function can be represented as a Fourier sum, given in Section 12 as

$$f(\xi_1, \xi_2) = \sum_{m,n} A_{m,n} e^{2\pi i(m\xi_1 + n\xi_2)} \tag{4.1.7}$$

and then converted into cylindrical coordinates as

$$f(\mathbf{x}) = \sum_{m,n} A_{m,n} e^{2\pi i[(mA+nC)x+(mB+nD)z]} \tag{4.1.8}$$

$$= \sum_{\mathbf{u}\in L(M)} A(\mathbf{u}) e^{2\pi i \mathbf{u}.\mathbf{x}}, \tag{4.1.9}$$

where the sum is over the lattice generated by the matrix M.

So f is entirely determined by (i) the lattice generated by M and (ii) the values of A at those lattice points. But we see from Eq. (4.1.6) that A and C are integer multiples of $1/2\pi\rho$, constraining the inverse lattice points to lie on those vertical lines. This constraint corresponds to the cylindrical periodicity of the original lattice. The additional constraint that f had lattice symmetry corresponds to the coefficients A only being non-zero at various 'beads', which can be imagined as sliding up and down on those vertical lines.

But now we combine with the central insight from reaction–diffusion that only wavenumbers near k_0 will be seen in an evolving pattern, and we find that only the coefficients of beads near to the circle of radius k_0 will be promoted by the dynamics, or conversely that the pattern on the cylinder at the nonlinear steady state will be defined by the values of the coefficients at the beads. This allows the entire evolving pattern to be represented by only a small numbers of parameters which is crucial for both analysis and numerical simulation.

5. The figures

We can now revisit the figures with a little more understanding. It's now plausible that they are representations of, at the bottom, a pattern with a lattice symmetry on a cylinder, and above the inverse lattice as defined in Section 4.1. The cylinder itself is vertically orientated, and we are seeing two repeats of the pattern around the cylinder. The pattern is not symmetric in the vertical axis z.

The circle, representing the maximally unstable wavenumber, has a constant radius of 12 grid squares across the three diagrams, while the 'wires' narrow from 5 in Figs 3 to 4 in Fig. 4 to 10/3 in Fig. 5. The cylindrical patterns have horizontal symmetries at intervals of about 30, 24 and 28 grid squares, but the cylindrical segment in Fig. 3 has height 24 in contrast to those of Figs 4 and 5. If that is indeed drawn at a scale 50% larger than the other two the symmetry widths are in the proportion 20:24:28 consistent with the narrowing of the wires. It is fairly easy to see that the two lattices are indeed inverses of each other by visualling the parallelogram of the first two principal vectors and seeing that they are rotated by 90°. Figure 5 is a more complicated figure; the spots on the cylinder are a superposition of two different lattices, with the corresponding inverse lattices as indicated; moreover the lattice with the larger red spots has a horizontal periodicity half that of the cylinder. The lattices in both Fig. 3 and Fig. 4 have parastichy numbers (1,2).

Although there is a reference to figures of this form on page 74 of Turing (1992), I do not think these figures were intended to be placed at that point in the manuscript. We have a clear explanation of the spots in all of the figures, and indeed a suggestion that they show the evolution of the symmetry of the pattern over the slow expansion of cylinder width. But what is not clear is how the pattern within each symmetry unit is defined, or in other words what the Fourier coefficients at the beads on the wires are. That they are the solution of a reaction–diffusion equation seems very likely; that each one has been derived from its predecessor by using it (or at least its symmetry cell) as an initial condition is also probable. Moreover, we have a number of computer outputs from the 1951–54 period, which clearly solve reaction–diffusion systems such as Fig. 8 (Swinton, 2003). So it is a reasonable inference that these were computer-generated solutions of partial differential equations with rather carefully chosen initial symmetry and geometry conditions. Indeed much of Turing's Manchester Mark I code to generate such patterns has survived in the Manchester Archive. So much for an explanation of what these pictures are. But why were they generated?

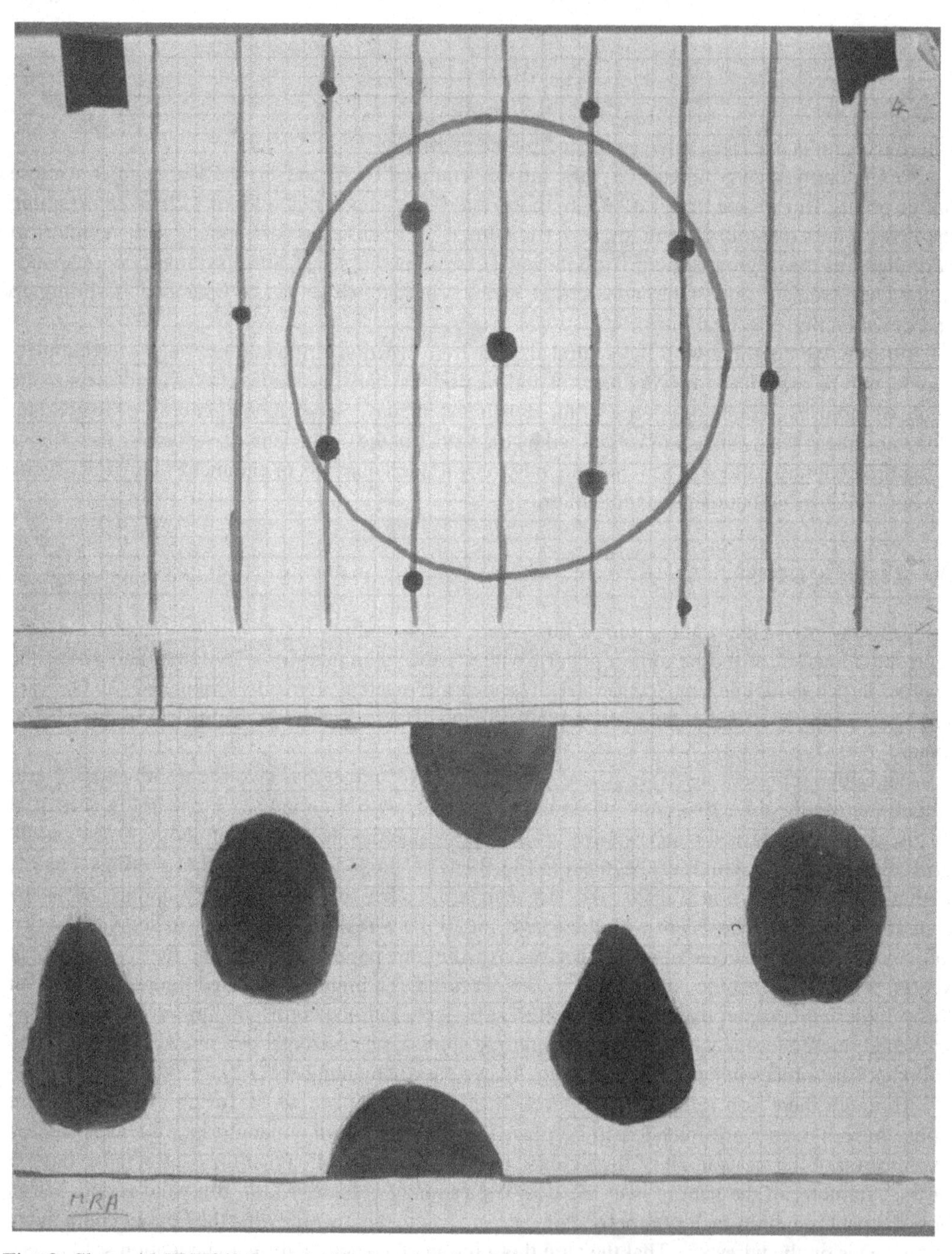

Fig. 3: Sheet AMT/K3/4 from the Turing Archive at King's College Cambridge. Double foolscap (320×230 mm) with each grid line at 1/4-inch intervals; the upper half is formed from a second single foolscap sheet fixed on top at the large dark tape marks. For more details, see Section 5 ©PN Furbank.

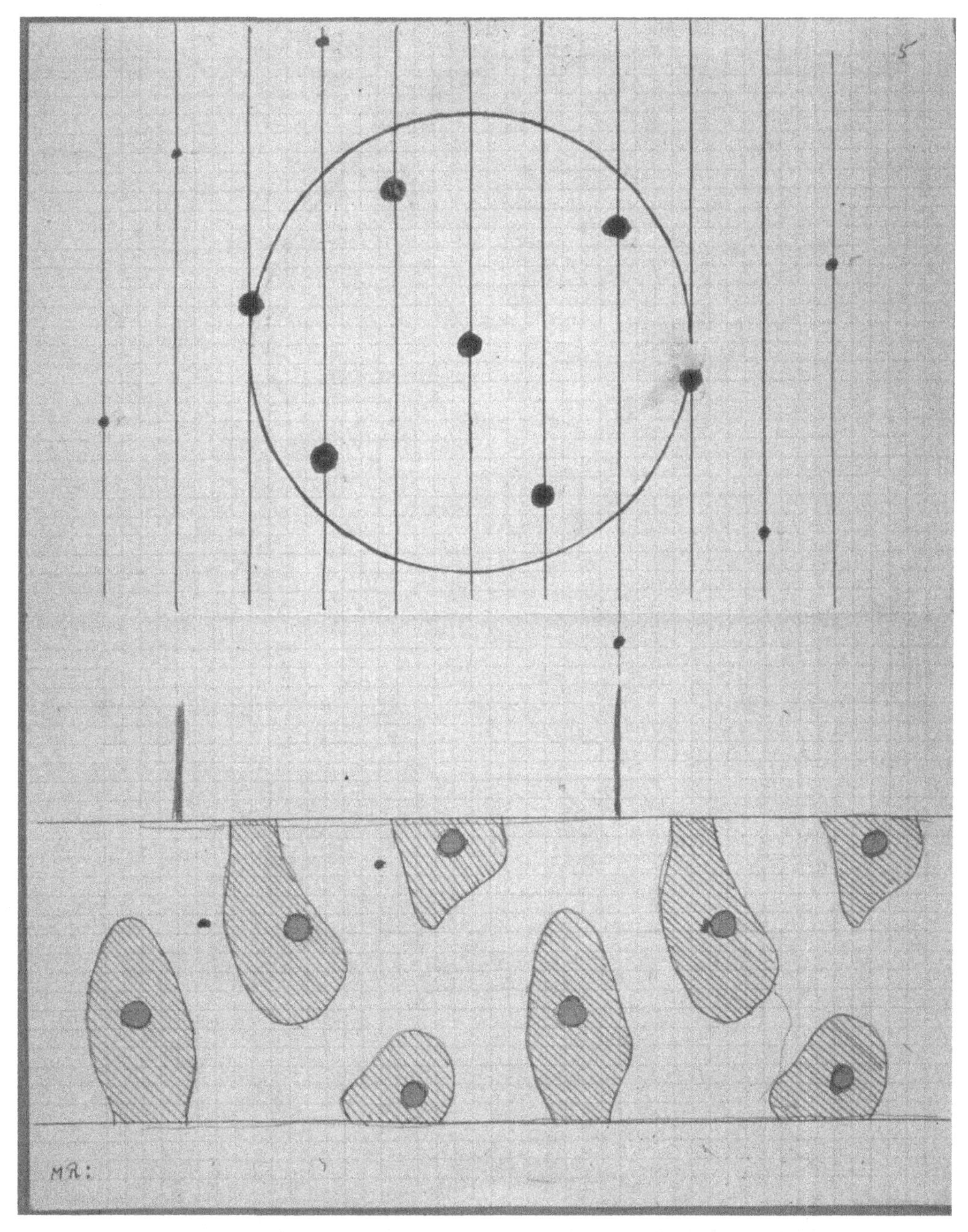

Fig. 4: Sheet AMT/K3/5 from the Turing Archive at King's College Cambridge. Double foolscap. For more details see Section 5 ©PN Furbank.

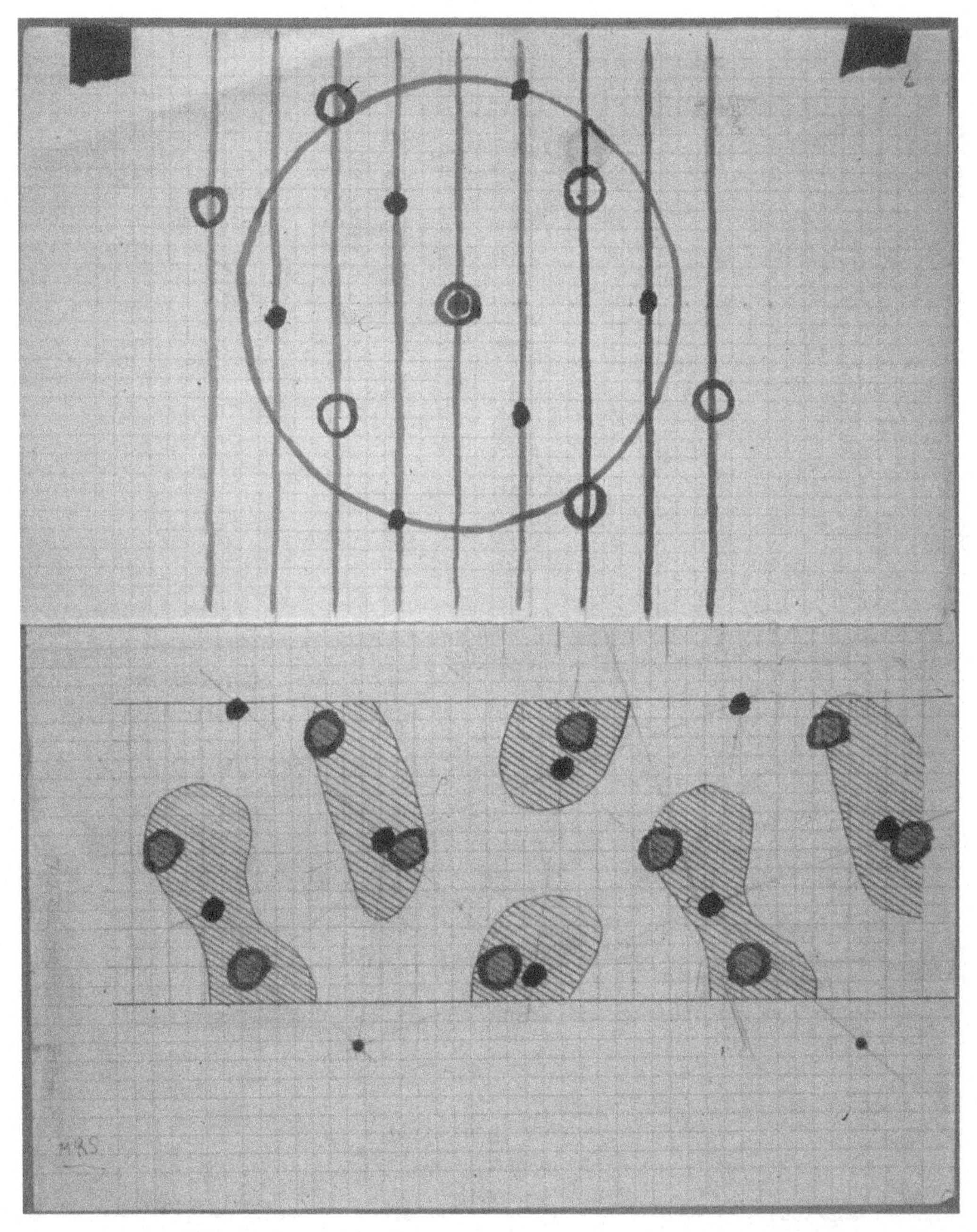

Fig. 5: Sheet AMT/K3/6 from the Turing Archive at King's College Cambridge. Like Fig. 3, the upper half is formed from a second single foolscap sheet taped over. For more details see section 5 ©PN Furbank.

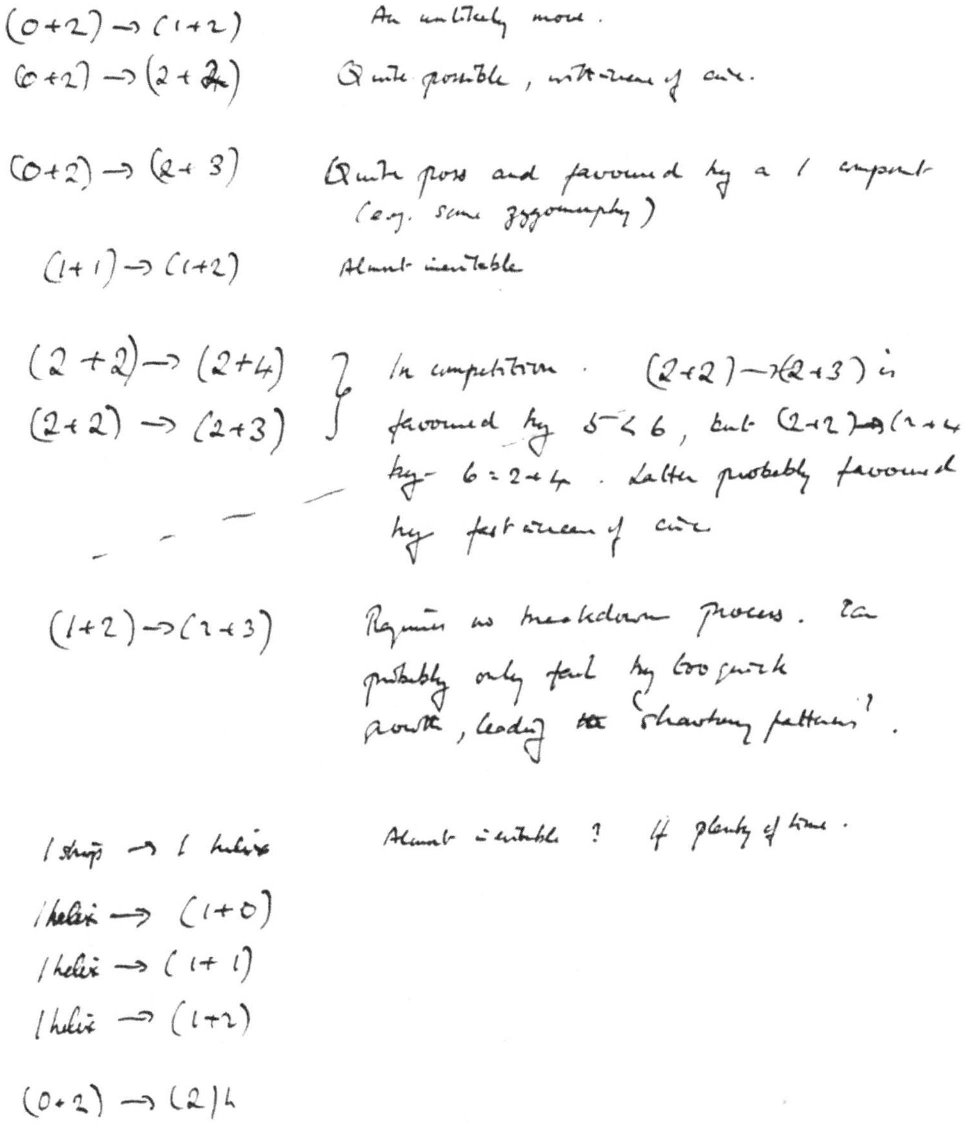

(0+2) → (1+2) An unlikely move.

(0+2) → (2+2) Quite possible, with increase of curv.

(0+2) → (2+3) Quite poss and favoured by a 1 component (e.g. some zygomorphy)

(1+1) → (1+2) Almost inevitable

(2+2) → (2+4)
(2+2) → (2+3) } In competition. (2+2)→(2+3) is favoured by 5<6, but (2+2)→(2+4) by 6=2+4. Latter probably favoured by fast increase of curv.

(1+2) → (2+3) Requires no breakdown process. Can probably only fail by too quick growth, leading to 'shattering pattern'.

1 stripe → 1 helix Almost inevitable? If plenty of time.

1 helix → (1+0)

1 helix → (1+1)

1 helix → (1+2)

(0+2) → (2)4

Fig. 6: Description of likely phyllotactic transitions. From the Turing papers NAHC/TUR/C2,3 in the UK National Archive for the History of Computing, University of Manchester, http://www.chstm.manchester.ac.uk/research/nahc/. ©PN Furbank.

What Turing needed was an understanding of which bifurcations were allowed and which are not: to find a basis for the hypothesis of geometrical phyllotaxis. Evidence for this comes mainly from material in the Manchester Archive such as Figs 6 and 7. Like the majority of the archive this note is undated, but I have speculated that (Swinton, 2003) the majority of the Manchester material was probably created close to the end of Turing's work on this project and it is unlikely that much

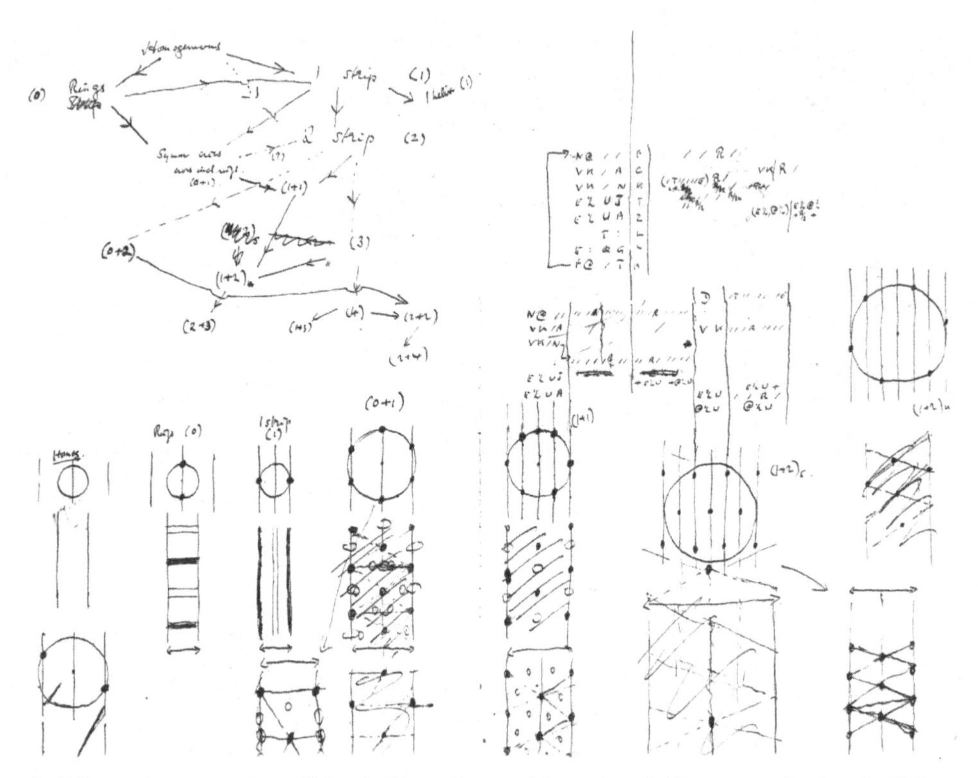

Fig. 7: Bifurcation tree of possible phyllotactic transitions (top left), inverse lattice (middle row) and actual phyllotactic patterns (bottom row). The significance of the code and data (top right) is unknown. From the Manchester Turing papers NAHC/TUR/C2,3. ©PN Furbank.

of the 'planned' work took place, for reasons we are unlikely to fully understand although the sheer difficulty of programming this sophisticated mathematical modelling on the Manchester computer must have played a significant part. Based on the fragmentary archive records there is no evidence he fully completed a numerical confirmation of his intuitive speculations about the likely transitions, much less was able to deploy them in support of a demonstration of the hypothesis of geometrical phyllotaxis. That was to be the work of others decades later.

6. Seeing spots and making sense of life

I first saw these figures as a mathematics student in the early 1980s, in an attic room in King's College at a time when the narrow circles where Alan Turing's reputation had persisted were already beginning to enlarge into progressively larger and less marginalised groups. Propelled by a brilliant biography (Hodges, 1992), the awareness of Turing as an iconic, field-defining figure, passed from logicians and philosophers, through gay rights activists, applied mathematicians, cryptographers, artificial lifers, mainstream computer scientists, and into popular scientific culture where he is now firmly established as a boffin of fiction, stage and screen. One culture where he might have, but has not, had a major impact is in developmental biology. He spent years on this at the end of his life; he introduced technical approaches that are now a reflex for modellers; but the program of 'chemical embryology' implied by that late work of his is nearly invisible by contrast with the wildly successful reductionist program of late twentieth century molecular developmental biology. Carroll (2005) represents a majority opinion in that field about Turing when

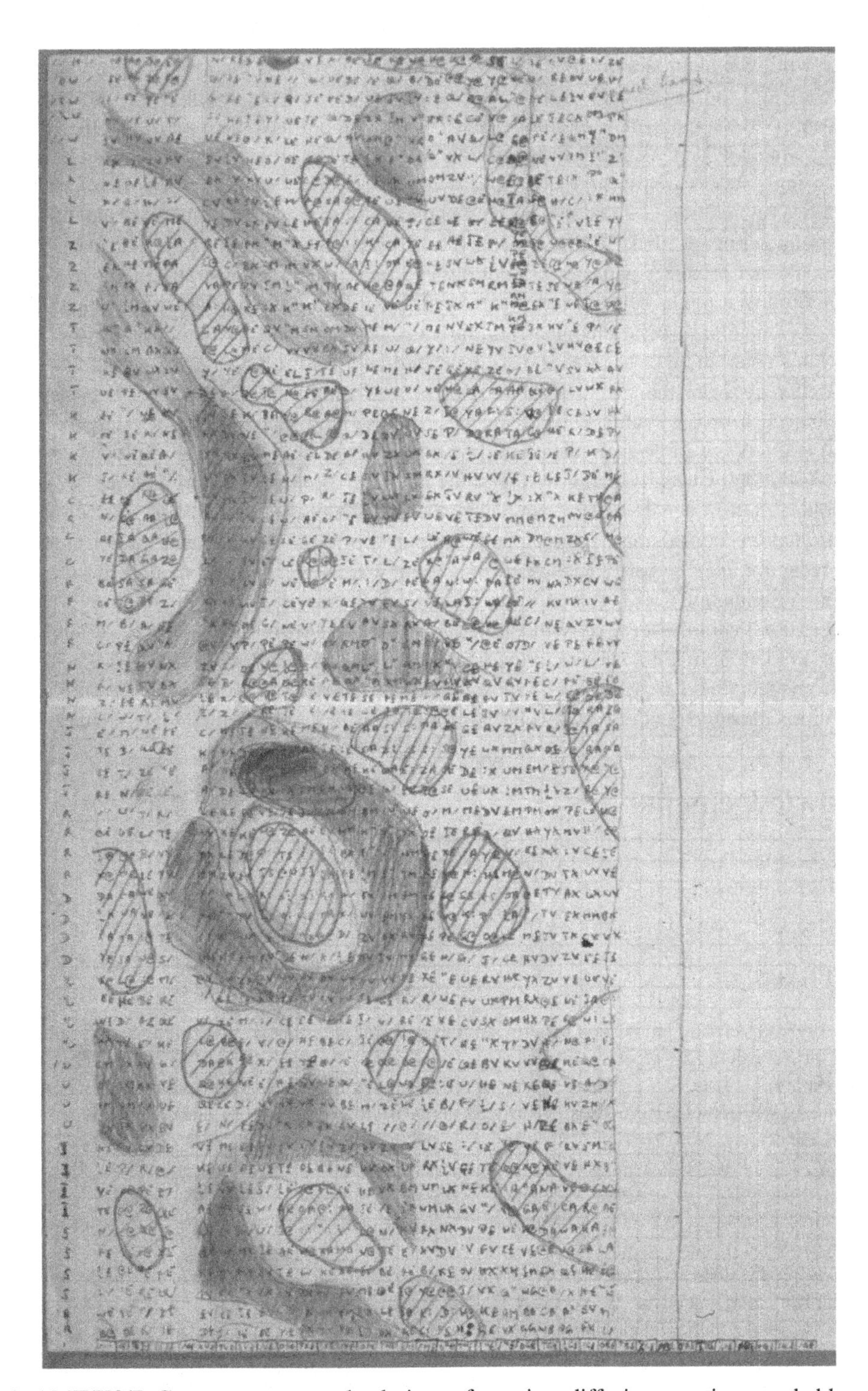

Fig. 8: AMT/K3/7. Computer-generated solutions of reaction–diffusion equations; probably generated by the Kjell family of routines (Swinton, 2003).

he writes 'while the math and the models are beautiful, none of this theory has been borne out by the discoveries of the last twenty years'. Fox Keller (2002) reflects at length on the reasons for this. She observes the mathematical culture that Turing, and perhaps the majority of the readers of this volume, inhabit. It is a culture to which the student in the attic was just then being inducted into by Turing's successors in the Fellowship, with observations such as 'the rest is just botany' for any finite enumeration task and in which an explanation is powerful to the very extent to which it is not grounded in specifics. In this world view, showing that a morphogen with particular diffusivity exists and produces patterns is much less valuable than demonstrating that, in general, a fictional morphogen might be expected to demonstrate pattern under certain conditions.[7] By contrast, Fox Keller (2002), p.99, characterises the view of many biologists then and now as 'what possible value could there be to an explanation posited on an entirely imaginary system?'

Fox Keller goes on, though, to wonder if (p.112) 'Turing may simply have been too soon'. There is indeed a newer synthesis between these fields, now that essentially mathematical and computational tools of large-scale data handling and analysis are now being deployed in a newly quantitative biology currently going by the name '*Systems Biology*' (Alberghina and Westerhoff, 2007; Noble, 2006), which makes extensive use of models. Despite some not unwarranted scepticism, there is considerable interest from biologists in these models, and the difference from Turing's reception may be partly accounted for by the way in which those models are represented as 'computer simulations' as opposed to mathematical abstractions. It is interesting to speculate on the reception that a finished, fully illustrated *Morphogen Theory of Phyllotaxis*, complete with computer simulation, might have had in the late 1950s. Would the turning away of developmental biology from theory have lasted so long with such a clear demonstration of the information locked in the algorithmic development of plants: what Stewart (1999, 2011) has called 'Life's Other Secret'?

Acknowledgements

The sunflower in Fig. 1 was grown by Justine Parker. Permission to reproduce Turing's previously unpublished work was granted by P.N. Furbank.

References

Adler, I., 1998. Generating phyllotaxis patterns on a cylindrical point lattice. In Symmetry in Plants, World Scientific Publishing, Singapore.

Adler, I., Barabé, D., Jean, R.V., 1997. A history of the study of phyllotaxis. Ann. Botany 80, 231–244.

Alberghina, L., Westerhoff, H.V., (Eds.), 2007. Systems Biology: Definitions and Perspectives, Springer, Berlin, Heidelberg, New York.

Allaerts, W., 2003. Fifty years after Alan M. Turing. An extraordinary theory of morphogenesis. Belg. J. Zool. 133(1), 3–14.

Atela, P., Golé, C., Hotton, S., 2002. A dynamical system for plant pattern formation: A rigorous analysis. J. Nonlinear Sci. 12(6), 641–676.

Carroll, S.B., 2005. Endless Forms Most Beautiful: The New Science of Evo Devo and the Making of the Animal Kingdom, first ed. W. W. Norton & Company, New York.

Church, A.H., 1904. On the Relation of Phyllotaxis to Mechanical Laws, Williams and Norgate, London.

Douady, S., Couder, Y., 1996. Phyllotaxis as a dynamical self organizing process (part I, II, III). J. Theor. Biol. 178.

Fox Keller, E.F., 2002. Making Sense of Life, Harvard University Press, Cambridge MA, London.

[7] Specifically, 'giving bicycles to the missionaries' (Swinton, 2003).

Hodges, A., 1992. Alan Turing: The Enigma, Vintage, London.
Jean, R.V., 1994. Phyllotaxis: A Systemic Study in Plant Morphogenesis (Cambridge Studies in Mathematics), Cambridge University Press, Cambridge.
Jean, R.V., Barabé, D., 1998. Symmetry in Plants, World Scientific, Singapore.
Kunz, M., 1995. Some analytical results about two physical models of phyllotaxis. Communications Math. Phys. 169(2), 261–295.
Meinhardt. Travelling waves and oscillations out of phase: An almost forgotten part of Turing's paper. This volume.
Mitchison, G.J., 1977. Phyllotaxis and the Fibonacci series. Science, 196, 270–275.
Murray, J., 1993. Mathematical Biology, Springer-Verlag.
Nanjundiah, V., 2003. Alan Turing and 'The Chemical Basis of Morphogenesis'. In: Sekimura, T. (Eds.), Morphogenesis and Pattern Formation in Biological Systems, Springer-Verlag, Tokyo.
Noble, D., 2006.. The Music of Life, Oxford University Press, Oxford.
Richards, R., Radiolaria: The result of morphogenesis. This volume.
Stewart, I., 1999. Life's Other Secret, Wiley, New York.
Stewart, I., 2011. Mathematics of Life: Unlocking the Secrets of Existence, Profile Books, London.
Swinton, J., 2003. Watching the daisies grow: Turing and Fibonacci phyllotaxis In: Teuscher, C.A. (Ed.), Alan Turing: Life and Legacy of a Great Thinker, Springer, Berlin, Heidelberg.
Swinton, J., in press. The Fundamental Theorem of Phyllotaxis revisited. Preprint.
Turing, A.M., 1952. The chemical basis of morphogenesis. Philos. Trans. Royal Soc. Lond. B 237, 37–72.
Turing, A.M., 1992. The Morphogen Theory of Phyllotaxis, vol. 3. North-Holland/Elsevier, Amsterdam.
van Iterson, G., 1907. Mathematische und Microscopisch-Anatomische Studien ber Blattstellungen, nebst Betraschung ber den Schalebau der Milionen, Gustav-Fischer Verlag, Jena.

Aaron Sloman travels forward to —

VIRTUAL MACHINERY AND EVOLUTION OF MIND (PART 3)*

META-MORPHOGENESIS: EVOLUTION OF INFORMATION-PROCESSING MACHINERY

1. Introduction: Types of emergence

In 1936 Turing made major contributions to our understanding of certain types of emergence, by showing how a Turing machine can be set up so as to generate large numbers of very simple processes that cumulatively produce qualitatively new large scale results, e.g. TM operations producing results related to results of human mathematical reasoning. Later work by Turing and others led to electronic computing machinery enabling a small set of very simple operations to produce very many kinds of novel, useful, complex, and qualitatively varied results—now impacting on many

* This is Part 3 of a sequence of related contributions to *Alan Turing – His Work and Impact*. The previous chapters will be referred to as 'Part 1' and 'Part 2'. The ideas presented here are developed further in: http://www.cs.bham.ac.uk/research/projects/cogaff/misc/meta-morphogenesis.html.

aspects of everyday life. Universal TMs showed that both the construction of *mechanisms* and the construction of *things on which mechanisms operate* can in some cases be handled in a uniform way, by having mechanisms that can construct and manipulate mechanisms (e.g. self-modifying computer programs). A similar theme was implicit in his 1950 paper. As far as I know, Turing's last work on micro–macro emergence was the 1952 paper on morphogenesis, explaining how micro-interactions in physicochemical structures could account for global transformations from a fertilised egg to an animal or plant, within a single organism.

All those ways in which complex configurations of simple structures and processes can have qualitatively new features are examples of micro–macro relationships that can be labelled as 'emergent' (Cohen and Stewart, 1994).

It is now clear that physical and chemical mechanisms involved in biological reproduction can, like computational machinery, include specifications not only for (partially) controlled construction of new physical mechanisms (where some of the control comes from the environment) but also for the production of new construction specifications, and new mechanisms for using such specifications, as well as development and learning mechanisms for growing and modifying already functioning machinery, and mechanisms for detecting damage and producing repairs. The combined products of all these mechanisms, ecosystems and socio-economic-political systems, together constitute the most complex known example of emergence.

Much research on evolution and development has focused on production of new physical forms and new physical behaviours. We also need to understand micro–macro relationships involving creation and use of *new forms of information-processing*, without which much of the complexity could not have arisen. There is much knowledge and expertise about information processing in computer science, software engineering and more generally computer systems engineering, but relatively little understanding of biological 'meta-morphogenesis' (MM), the information processing mechanisms involved in producing biological novelty, including new forms of information processing.[2]

2. Layered computational emergence

Computing systems developers create new micro–macro relationships, using a set of micro components: types of hardware or software structure, a small collection of possible processes, and ways of ways of combining processes and structures using syntactic composition methods. The resulting new macro components (e.g. electronic circuits or computer programs) have complex and varied structures, and can support yet more new types of complex and varied processes, some of which provide 'platforms' for constructing further layers of complexity. As argued in Part 1 (this volume), the functions, states and processes in the new layers often cannot be defined in the language of physics and chemistry, or digital circuits. In that sense, although the new layers may be fully implemented in the old ones, they are not reduced to them. For example, the concepts 'win' and 'lose', required for describing a running chess program, are not definable in the language of physics. So the chess machine is implemented in, but not reducible to physical machinery.

Achieving such micro–macro bridges requires understanding the deep and unobvious generative potential of the initial fragments and their possible relationships. The full potential was unobvious in the early days of computing, but new programming languages, new development environments, new operating systems, new re-usable packages and, above all, new problems, have continually revealed new, more complex, achievable targets. The complexity we now take for granted was achievable only through *layered* development of tools and techniques, often depending on use of earlier layers. Similar constraints apply to biological evolutionary and developmental trajectories. Many biological mechanisms, structures and functions that developed recently could not have occurred in earlier times, despite availability of all the *physical* materials, because many small intermediate changes were required in order to produce the *infrastructure* for newer more complex mechanisms.

[2] For an answer to 'What is information?', see Sloman (2011).

New layers of computing machinery were in part a response to *external* pressures from application domains, with which new computing systems had to interact, e.g. using sensors (e.g. cameras, pressure sensors etc.), effectors (e.g. grippers, wheels, paint sprayers etc.), or network connections. Similar, still unidentified, environmental pressures led to new emergent mechanisms and processes in biological evolution. Other pressures can come from *internal* requirements to improve speed, reliability, energy efficiency, easy of monitoring, ease of debugging and ease of extension.

The physical universe produces objects of varying complexity, from subatomic particles through molecules, planets, galaxies and beyond. Large lumps of solid or liquid matter can result from materials being brought together concurrently. But many intermediate-sized structures of great complexity, including organic molecules and organisms of many kinds, require special construction mechanisms, or intermediate scale components, that are not always directly available even when the physical materials are available. Such complex systems need to be assembled over time using precisely controlled selections from among physically and chemically possible alternatives. For example, there was no way the matter on this planet several billion years ago could have immediately reorganised itself into an oak tree or an orangutan.

A Tornado could not assemble a 747 airliner from a junkyard full of the required parts. Assembling an airliner requires not only prior assembly of smaller parts, but also machinery for producing the intermediate structures, and maintaining them in relationships required for subsequent operations. Biological evolution also requires intermediate stages including intermediate mechanisms of reproduction and development. Intermediate stages in evolution require increasingly complex forms of information processing, so biological information processing mechanisms, like computer systems engineering, must have involved many intermediate forms of information processing. Compare how later stages of a mathematical proof depend on earlier stages, preventing simultaneous discovery of all parts of the proof.

Successive information-processing mechanisms must have had successively more complex physical components, forms of representation, ontologies, algorithms, architectures and functions, especially information processing functions relating to the environment. We need to understand the mechanisms of meta-morphogenesis.

Some thinkers assume that there must be a single master designer controlling such processes of assembly of complex living structures from inanimate matter. But development of software engineering sophistication over the last six decades did not require some super-engineer controlling the whole process. There was only a large collection of successively discovered or created bootstrapping processes in a multitude of forms of competition and co-operation partly driven by a plethora of new more complex goals that became visible as horizons receded. Humans mostly stumbled across more and more complex ways in which previous achievements could be extended. Natural selection had much in common with this, except that there were no designers detecting new targets – until species emerged with sufficient intelligence to engage in mate selection and other selective breeding activities.

3. Meta-morphogenesis and biological complexity

A feature of growth of complexity is that as new mechanisms are developed some of them transform, and simplify, opportunities for subsequent developments, as illustrated for individual cognitive development in Chappell and Sloman (2007). Related points were made in Cohen and Stewart (1994). New mechanisms, new forms of representation and new architectures can sometimes be combined to provide new 'platforms' bringing entire new spaces within (relatively) easy reach. Examples in the history of computing include new operating systems, new programming languages, new interfacing protocols, new networking technologies, new constraints and requirements from users, including requirements for reliability, modifiability, security, ease of learning, ease of use etc. We do not know all the new pressures that influenced developments of biological information-processing mechanisms, both in evolution and in individual development, though we can guess some of them.

Evolution of biological information processing was much slower, and did not need goal direction, only random 'implicit' search (implicit because there were no explicit goals, only opportunities that allowed some changes to be relatively advantageous). Identifying those opportunities and the evolutionary changes they influenced is hard. An example is the difference between organisms in an amorphous chemical soup and organisms whose environment has distinct enduring parts with different properties (e.g. providing different, persistent, nutrients and dangers, in different locations). Only the second type could benefit from mechanisms for acquiring and storing information about those enduring structures. Such information would necessarily have to be built up piecemeal over time. If the organism had visual mechanisms it could rapidly take in information about complex structures at different distances. If it only had tactile/haptic sensors the information would have to be acquired in much smaller doses with more movements required.

Some computing developments, such as a new notation, or a new ontology (e.g. for types of communication, or types of event handler, or types of data-structure), or creation of a new type of operating system, can provide a 'platform' supporting a very wide range of further developments. There were probably also many different kinds of platform-producing transitions in biological evolution, e.g., development of new means of locomotion, new sensors, new manipulators and new forms of learning. Some of these were changes in physical form or structure or forms of motion, or types of connectivity, while others were changes concerned with information processing. Smith and Szathmáry (1995) discussed changes in forms of *communication*, but there must have been many more transitions in information processing capabilities and mechanisms, some discussed by Sloman (2008).

When a new multifunction platform is developed, searches starting from the new platform can (relatively) quickly reach results that would previously have involved intractable search spaces. After learning a powerful language like Prolog, a programmer can often quickly produce programs that would have been very difficult to express using earlier languages. New high level languages add new opportunities for rapid advances. Likewise, as Dawkins and others have pointed out, some biological developments, including new forms of information processing, could, in principle, dramatically shorten timespans required for subsequent developments, even though there is no goal-directed design going on. Even random search (though not a tornado?) can benefit from a billion-fold reduction in size of a search space.

4. Less blind evolutionary transitions

Some animals can formulate explicit goals and preferences and select actions accordingly. The evolution of that capability can provide a basis for selecting actions that influence reproductive processes, e.g. selecting mates, or favouring some offspring over others, e.g. bigger, stronger or more creative offspring. When animals acquire such cognitive capabilities, such choices can be used, explicitly or as a side-effect of other choices, to influence selective breeding, in ways that may be as effective as explicit selective breeding of another species, e.g. domestic cattle or hunting dogs. Which types of selective breeding a species is capable of will depend on which features they are capable of recognising. If all they can distinguish among prospective mates or their offspring is size or patterns of motion, that could speed up evolution of physical strength and prowess. If they can distinguish differences in information processing capabilities that could lead to kinds of selective breeding of kinds of intelligence. (N.B. I am not endorsing eugenics.)

These are examples of ways in which production of a new platform can transform something impossible into something possible, overcoming limitations of pre-existing mechanisms of composition. That includes bringing yet more new platforms within reach, as has happened repeatedly in computer systems engineering when new tools allowed the construction of even more powerful tools, e.g. using each new generation of processor design to help with production of subsequent designs.

A major research task is to identify evolutionary and developmental transitions that facilitate new subsequent evolutionary and developmental transitions. Innate learning capabilities produced at a late stage in evolution may include important pre-compiled partial information about the environment that facilitates specific kinds of learning about that sort of environment. (Compare Chomsky's claims about human language learning, and Karmiloff-Smith (1992).) Special-purpose evolved learning systems may, on this planet, outstrip all *totally general*, domain-neutral, mechanisms of learning or evolution sought investigated by some researchers. In his 1950 paper Turing suggested that 'blank slate' learning would be possible, which I find surprising. In contrast, McCarthy (2008) argues convincingly that evolution produced new, specialised, learning capabilities, required for human learning in a human lifetime, in certain sorts of changing 3-D environments.

5. From morphogenesis to meta-morphogenesis

In the same general spirit as Turing's paper on morphogenesis, I have tried to sketch a rudimentary theory of 'meta-morphogenesis' showing how kinds of development that are possible in a complex system can change dramatically after new 'platforms' (for evolution or development) are produced by pre-existing mechanisms. Biological evolution is constantly confronted with environmental changes that reduce or remove, or in some cases enhance, the usefulness of previously developed systems, while blocking some opportunities for change and opening up new opportunities. In that sense the environment (our planet) is something like a very capricious teacher guiding a pupil. Initially the 'teacher' could change only physical aspects of the environment, through climate changes, earthquakes, volcanic eruptions, asteroid collisions, solar changes, and a host of local changes in chemical soups and terrain features. Later, the teacher itself was transformed by products of biological evolution, including global changes in the composition of the atmosphere, seas, lakes, and the land–water distribution influenced by evolution of microbes that transformed the matter with which they interacted.

As more complex organisms evolved, they formed increasingly significant parts of the environment for other organisms, of the same or different types, providing passive or active food (e.g. prey trying to escape being caught), new materials for use in various forms of construction (e.g. building shelters, protective clothing, or tools) active predators, mates, and competitors for food, territory, or even mates. As a species evolved new physical forms and new information-processing mechanisms, those new developments could make possible new developments that were previously out of reach, e.g. modification of a control mechanism might allow legs that had originally evolved for locomotion to be used for digging, fighting or manipulation. As new control subsystems evolved, they could have produced new opportunities for system architectures containing those subsystems to develop, allowing old competences to be combined in novel ways. So developments in the 'learner' can be seen as developments in the 'teacher', the environment. Two concepts used in educational theory, Vygotsky's concept of *Zone of proximal development* (ZPD) and Bruner's notion of 'scaffolding' can therefore be generalised to evolution. Evolutionary and other changes can modify the ZPD of an existing species and provide scaffolding that encourages or supports new evolutionary developments. Further details would contribute to a theory of meta-morphogenesis.

6. Evolved information processing: beyond Gibson

Almost all organisms are control systems, using stored energy (sometimes externally supplemented, e.g. when birds use up-draughts) to produce internal and external changes that serve their needs. The control details depend on information acquired through sensors of various kinds, at various times. So organisms are 'informed control systems'. Information available, and also the control possibilities, vary enormously: from the simplest micro-organisms, mostly responding passively in chemical soups, to animals with articulated bodies and multiple sensors, who were capable of performing many different sorts of action, and requiring increasingly complex information processing to notice opportunities, to select goals, to select ways of achieving goals, to carry out those selected actions, to deal with unexpected details of the environment detected during execution, and to learn both from experiences of performing successful and unsuccessful actions and from observation of other things occurring in the environment. A full account of these transitions requires several generalisations of James Gibson's notion of 'affordance' (Sloman, 2009).

We need to extend not only Turing's work but also the work of Maynard Smith and Szathmáry, on transitions in evolution, to include detailed investigation of transitions in types of *information processing*. Transitions in forms of communication are often noted, for instance the development in humans of communication using syntactic structures, but there are far more biological processes involving information than communication (internal or external). The need for them will be obvious to experienced designers of intelligent, autonomous robots. The information processing requirements include interpreting sensory information, controlling sensors, learning, forming plans, dealing with conflicts, evaluating options and many more (Sloman, 2006). Many of the requirements are not obvious; so researchers often notice only a tiny subset and therefore underestimate the problems to be solved—as has happened repeatedly in the history of AI. An extreme example is assuming that the function of animal vision is to provide geometric information about the surfaces in view (Marr, 1982), ignoring the functions concerned with detecting affordances, interpreting communications and continuous control of actions (Gibson, 1979).

A particularly pernicious type of myopia occurs in research in robotics, biology, psychology, neuroscience and philosophy that focuses entirely on the continuous or discrete *online* interactions between an organism (or robot) and the immediate environment, ignoring requirements for planning, explaining and reasoning about things going on in other locations, and past and possible future events discussed in Sloman (2006,2009). Overcoming this myopia can be very difficult, but progress is possible if instead of focusing attention on single organisms or particular designs, we examine *spaces of possibilities*: possible sets of requirements for organisms and robots, and possible sets of design features capable of meeting those requirements. For example, noticing an organism or individual failing to do something may draw attention to the problem of explaining how others succeed—a requirement that may previously have gone unnoticed. A special case of this is the work of Jean Piaget on the many partial or missing competences of young children, which help to draw attention to the hidden complexities in the competences of (normal) adults. Likewise events following brain damage can expose unobvious aspects of normal cognition.

Simply observing or dissecting organisms will not reveal their information-processing: we also need to engage in detailed analysis of differences between environments and morphologies, showing how, as environments change, a succession of increasingly complex demands and opportunities can make possible cumulative changes not only in physical structure, size, strength, and behaviours, but also in the kinds of information available, the kinds of information processing mechanisms, and the uses of such information.

We also need to identify different requirements for belief-like and desire-like states that inform behaviours as discussed (incompletely) in Sloman et al. (2005). Changes in the environment can affect the goals that are essential or useful for an organism to pursue. In some cases goals remain the same, but the information processing and behaviours required to achieve them change: for example if drought or competition makes a certain kind of fruit more scarce, requiring the animals to travel further, climb higher up trees, or physically engage with competitors after the same food. In other cases, changes in the environment may produce new constraints or new opportunities, making it useful to acquire new types of goal. For example, a new kind of food may become available, and if food is scarce the species that acquire desires to find and consume the new food will benefit. However, the physical actions required to obtain and consume that food (e.g. breaking open a shell) may benefit from new forms of control, thereby allowing yet another genetic change to be useful.

Even when environment and sensorimotor morphology remain the same, changes in the *mode of processing* of the information available may provide benefits, e.g.,

- acquiring new ways of learning correlations between sensorimotor signals,
- acquiring new actions that provide or refine information about the environment, e.g. approaching objects, viewing them from new locations, rotating them, acting on them by prodding, pushing, squeezing, twisting, pulling apart, etc. (Gibson, 1979),
- developing a new ontology extending old semantic contents (e.g. developing an exosomatic ontology of 3-D structures and processes that exist independently of being sensed, or developing an ontology that allows information about the past or the future or states of affairs out of sight to be represented),
- developing new explanatory theories about the materials, structures, processes, and causal interactions in the environment,
- developing ways of exploring future possible actions to find good plans before initiating behaviours (Craik, 1943; Sloman, 2006),
- developing new meta-semantic competences that allow the information processing of other organisms to be taken into account (e.g. prey, predators, conspecifics, offspring, mates).

7. Monitoring and controlling virtual machinery

Some changes produce new opportunities for informed control of monitoring and other processes, including operations on the intermediate virtual machine structures in perceptual sub-systems. Parts 1 and 2 of this commentary point out that such biological developments involving virtual machinery can explain philosophically puzzling features of animal (including human) minds, such as the existence of 'qualia'. This can enhance our understanding of requirements for future machines rivalling biological intelligence. We need to explore the space of *possible* minds, and the different requirements different sorts of minds need to satisfy—a very difficult task, since many of the requirements are unobvious. In particular, I hope it is now clear that not all the requirements for embodied organisms (and future robots) are concerned with real-time, continuous, online interactions with the immediate environment, except for very simple organisms with very simple sensory–motor capabilities.

Turing was interested in evolution and epigenesis and made pioneering suggestions regarding morphogenesis–differentiation of cells to form diverse body parts during development. As far as I know he did not do any work on how a genome can produce *behavioural competences* of the complete organism, including behaviours with complex conditional structures so that what is done depends on internal and external sensory information, nor internal behaviours that extend or modify previously developed information processing architectures, as discussed in Karmiloff-Smith (1992).

Even if we can understand in the abstract that evolution produces behavioural competences by selecting brain mechanisms that provide those competences, explaining how it actually works raises many deep problems, especially where the competences are not themselves behavioural.

The human-produced mechanisms for constructing more and more complex computing systems from a relatively small set of relatively simple types of components are all examples of 'emergence' of qualitatively new large-scale structures and processes from combinations of much simpler building blocks.[3] Perhaps a deeper study of the evolution of tools, techniques, concepts and theories for designing complex systems in the last half century will stimulate new conjectures about the evolution of natural information processing systems, including those that build themselves only partly on the basis of an inherited specification. I suspect that people who predict imminent singularities underestimate the extent of our ignorance about what evolution has achieved, and some of the difficulties of replicating it using known mechanisms. Most biological meta-morphogenesis remains undetected.

References

Chappell, J.M., Sloman, A., 2007. Natural and artificial meta-configured altricial information-processing systems. Int. J. Unconventional Comput. 3(3), 211–239.

Cohen, J., Stewart, I., 1994. The Collapse of Chaos, Penguin Books, New York.

Craik, K., 1943. The Nature of Explanation, Cambridge University Press, London, New York.

Gibson, J.J., 1979. The Ecological Approach to Visual Perception, Houghton Mifflin, Boston, MA.

Karmiloff-Smith, A., 1992. Beyond Modularity: A Developmental Perspective on Cognitive Science, MIT Press, Cambridge MA

Marr, D., 1982. Vision, W.H.Freeman, San Francisco.

McCarthy, J., 2008. The well-designed child. Artif. Intell. 172(18), 2003–2014.

Sloman, A., 2006. Requirements for a Fully Deliberative Architecture (Or component of an architecture). Research Note COSY-DP-0604, School of Computer Science, University of Birmingham, Birmingham, UK.

Sloman, A., 2008. Evolution of minds and languages. What evolved first and develops first in children: Languages for communicating, or languages for thinking (Generalised Languages: GLs)?

Sloman, A. 2009. Some requirements for human-like robots: Why the recent over-emphasis on embodiment has held up progress. In: Sendhoff, B., Koerner, E., Sporns, O., Ritter, H., Doya, K. (Eds.), Creating Brain-like Intelligence, Springer-Verlag, Berlin, pp. 248–277.

Sloman, A., 2011. What's information, for an organism or intelligent machine? How can a machine or organism mean? In: Dodig-Crnkovic, G., Burgin, M. (Eds.), Information and Computation, World Scientific, New Jersey.

Sloman, A., Chrislet, R., and Scheutz, M., 2005. The architectural basis of affective states and processes. In M. Arbib and J.-M. Fellous (Eds.), Who Needs Emotions?: The Brain Meets the Robot, Oxford University Press, New York, pp. 203–244.

Smith, J.M., Szathmáry, E., (1995). The Major Transitions in Evolution, Oxford University Press, Oxford, England.

[3] Part 1 introduced a distinction between implementation and reduction, where a Running Virtual Machine (RVM) can be fully implemented in physical machinery (PM) even though the concepts required to describe the processes in the RVM cannot be defined in terms of concepts of physics. In that case the RVM is implemented in but not reduced to physical machinery. Part 2 showed how this might account for some of the special properties of mental phenomena such as qualia.

Drawing of Alan Turing by his mother, at his preparatory school, Hazelhurst, Sussex, 1923.
Courtesy of Sherborne School.

Outline of the Development of the Daisy

(Prepared from Turing's notes by P.T. Saunders for the *Collected Works* and updated by J. Swinton.)

Jonathan Swinton's updating of the texts —

AN EDITORIAL NOTE

Timed as it is for the centenary of the birth of Alan Turing in 2012, the editors have called on many who have detailed knowledge of Turing's life and work, including Turing's ex-student Bernard Richards. In this case, the familiarity of Jonathan Swinton with the original papers kept at King's College, Cambridge, has been invaluable. His work in this area has continued over a number of years, and both Bernard Richards and Jonathan Swinton retain their connection with Manchester, where this seminal work originated. The timely centenary return to Turing's writings contained in this book forms but a small stepping-stone to a fuller appreciation of these visionary but sadly fragmented notes relating to morphogenesis.

Peter Saunders commented in his Introduction to the morphogenesis volume of the *Collected Works*:

> Finally there is the incomplete *Outline of the Development of the Daisy*. As I explained in the preface, I am not including everything in the Archive, but I felt that this material does stand on its own. It gives us an idea of how Turing meant to proceed and it also reminds us that Turing was interested not just in mathematics but also in real flowers, an interest that goes back to his childhood if we may judge by the sketch that serves as the frontispiece.[1]

Peter Saunders observes in this Preface mentioned above:

> I found reading the archive material a fascinating experience. For while at first glance Turing's work on biology appears quite different from his other writings, it actually exhibits the features typical of all his work: his ability to identify a crucial problem in a field, his comparative lack of interest in what others were doing, his selection of an appropriate mathematical approach, and the great skill and evident ease with which he handled a wide range of mathematical techniques.

Particularly relevant here, Peter goes on in relation to the various unpublished fragments represented by *Outline of Development of the Daisy*:

> [They] were never edited into a form ready for publication and so I have had to undertake this task myself. I have made some obvious minor corrections and filled in a few gaps where it was clear what was missing, but there are no significant alterations. My aim has been to produce as nearly as possible the papers that would have appeared had Turing lived. To avoid cluttering the text with indications of trivial deviations from the manuscript, I have not marked the corrections. Readers whose primary interest is historical are therefore warned that not only does the archive contain more material than is in this volume, but not everything that is here is word for word as it appears in the manuscripts.

Here now is a brief outline of the editorial changes to the later morphogenesis material from the *Collected Works*, kindly provided by Jonathan Swinton:

---

[1] Editors note: Reproduced opposite.

The three parts of *Morphogen Theory of Phyllotaxis*, ('*MTPI*', '*MTPII*', '*MTPIII*') and *Outline of the Development of the Daisy* ('*ODD*'), are derived from Turing's typescripts in the King's College archive. *MTPI* was first prepared for publication from AMT/C25 by Nick Hoskin, probably during the 1950s, with a number of minor editorial changes: for example changing $2\pi/J$ to κ throughout. Hoskin's typescript was preserved as AMT/C8, and was adopted almost unchanged by Saunders in his preparation of the 1992 *Collected Works*. Similarly, Sections 1–3 of *MTPII* were prepared by Hoskin from AMT/C26 into AMT/C9. However, Sections 4–7 of the Saunders text were taken direct from AMT/C26. The work behind *MTPIII* is described by Richards in this volume.

ODD is drawn from AMT/C24, primarily sheets 1–15. These sheets were numbered by 'ROG' (Robin Gandy), but this ordering was not adopted in full by Saunders who omitted a number of sheets and figures. There are a number of clues elsewhere in the archive of how to fill in the gaps in both papers, but given the availability of the original material online there seems little value in attempting to establish a canonical text of papers which were in any case drafts. Accordingly, we have made a small number of silent minor changes from the Saunders volume where there has been clear typographical error.

The exception to this is in *ODD* where we have reordered the second half based largely on Gandy's ordering which is in our view more coherent and included Turing's original and informative, if rough, diagrams. Thus the version of *ODD* here includes previously unpublished material.

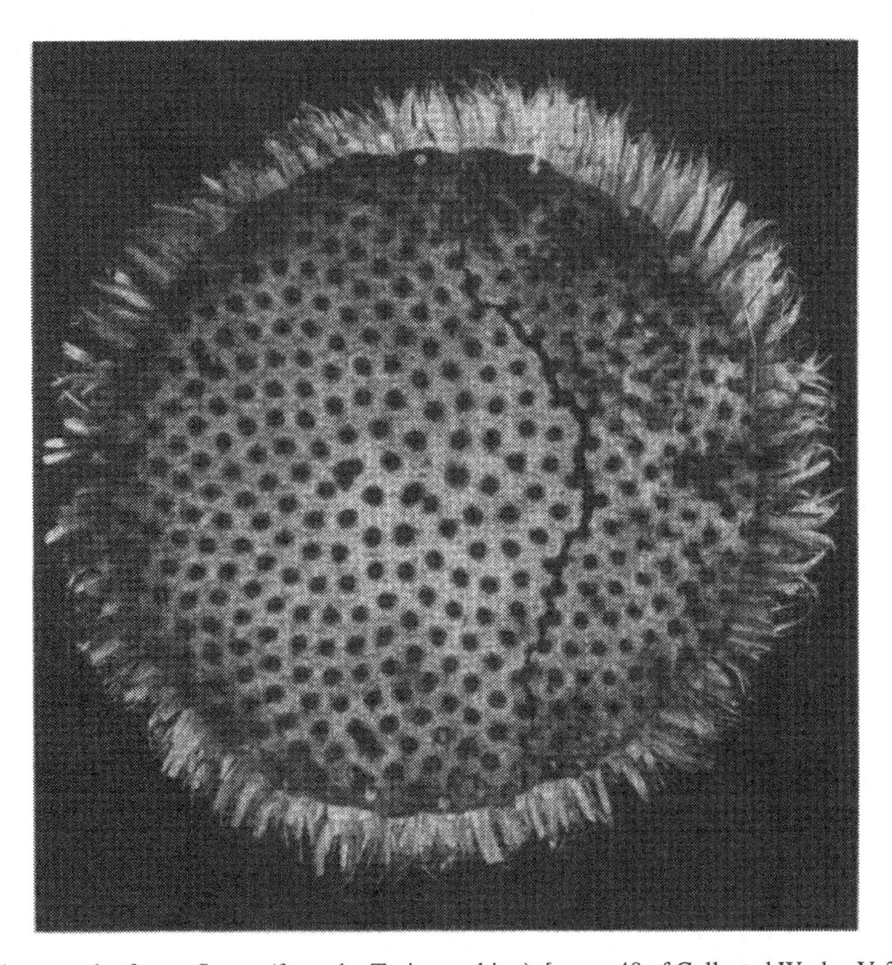

Photograph of a sunflower (from the Turing archive). [see p. 48 of Collected Works, V. 3]

OUTLINE OF THE DEVELOPMENT OF THE DAISY

The theory developed in this paper is limited by a number of assumptions which are by no means always satisfied. Two are of special importance:

(i) That the pattern passes through a long developmental period without forming any visible structures, and indeed without the chemical patterns modifying in any way the geometry of the system. When the visible structures are finally formed, this is done without essential alteration of the chemical pattern.
(ii) That the pattern is always developed within a ring so narrow that it may reasonably be treated as a portion of a cylinder.

The first of these assumptions is one which it would be very difficult to avoid. It would be exceedingly difficult to know what to assume about the anatomical changes. For the majority of plants this assumption is probably false. In the development of the capitulum of a daisy it seems to be more or less correct, however. The capitulum is appreciably separated from the rest of the plant by a length of petiole before the development of the capitulum starts. Thus a new start is made in the development of the capitulum. It is not appreciably influenced by the proximal structures. That this is the case is confirmed by the following facts:

(a) The directions of the generating spirals of the rosette and of the capitulum are statistically independent. Thus of 15 capitula and corresponding rosettes examined by the author, four cases had both rosette and capitulum left handed. In five cases the rosette was left handed but the capitulum right handed, and in four the rosette right handed and the capitulum left handed; in one case both were right handed. Thus in nine out of the fifteen cases, the rosette and the capitulum were in different directions.
(b) Beneath the 13 bracts enclosing the capitulum, there are no other distinguishable structures.

It is suggested that the development of the daisy proceeds essentially as follows. First, the petiole grows up from the rosette without any differentiation either of a visible anatomical form or of an invisible chemical form. Subsequently, the distal end of the petiole undergoes two kinds of change. Its diameter increases, and at the same time a chemical pattern is determined by purely chemical considerations, and there is therefore little reason to expect the wavelength to change much. As the diameter increases further, therefore, the pattern will have to change in order that it may continue to fit on the petiole with its new diameter. A very rough description of the concentration patterns during this process may be described as follows:

The concentration U of one of the morphogen concentrations $x = (\varrho\theta, z)$ is to be given by the formula

$$U = \sum_{\eta} e^{i(\eta,x)} G(\eta^2) W(x) \tag{1}$$

where the summation is to be over the lattice $\begin{pmatrix} A & B \\ C & D \end{pmatrix}$ reciprocal to $\begin{pmatrix} a & b \\ c & d \end{pmatrix}$.

The function $G(\eta^2)$ is to have a maximum near the square of the shortest vector of the lattice $\begin{pmatrix} A & B \\ C & D \end{pmatrix}$. A suitable form for $G(\eta^2)$ and the suitable range for the shortest vectors of the reciprocal lattice $\begin{pmatrix} a & b \\ c & d \end{pmatrix}$ are given in Fig. 1.

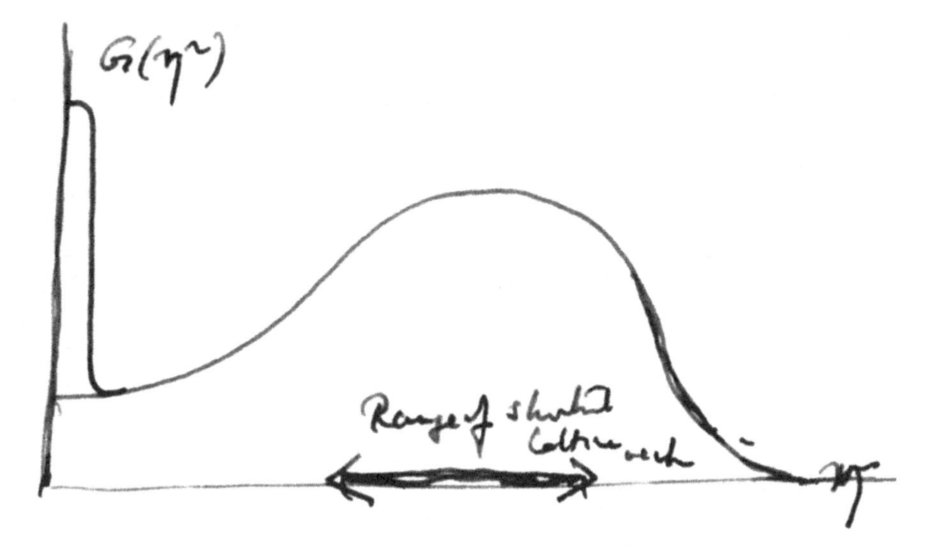

Fig. 1: [Editors note: From AMT/C24/6. Graph of η^2 against $G(\eta^2)$. The annotation reads 'range of shortest lattice vector'.]

The function $W(x)$ should depend only on z and typically may be of the form $\exp(-z^2/2\sigma^2)$. The ratio of the standard deviation σ to the shortest vectors of the lattice $\begin{pmatrix} a & b \\ c & d \end{pmatrix}$ is probably between 2 and 5. The inclusion of this factor $W(x)$ of course results in the pattern not having the symmetry of the lattice $\begin{pmatrix} a & b \\ c & d \end{pmatrix}$, or of any other lattice. But it is nevertheless possible to use the lattice $\begin{pmatrix} a & b \\ c & d \end{pmatrix}$ applying to the formula (I1) to describe the pattern instead of the symmetry lattice. It remains only then to describe what in the lattice is to be used for each value of the diameter of the petiole. A suitable form for the lattice is the limiting divergence angle lattice described in Part I. Clearly this description cannot hold at all times. It breaks down for the period during which the pattern is beginning to form. There may also be a period during which there is a pattern with reflexion symmetry (e.g., a decussate pattern), and the formula above will be invalid for this period also. The sections which follow are concerned with considering the chemical conditions under which this sort of description of the pattern very broadly holds.

At a certain point in the development of the daisy, the anatomical changes begin. From this point, as has been mentioned, it becomes hopelessly impracticable to follow the process mathematically; nevertheless it will be as well to describe how the process does proceed (at least in the author's opinion). In the regions of high concentration of one of the morphogens, growth is accelerated, and subsequently florets appear. Also, the chemical pattern begins to spread inwards towards the apex, and the florets follow it. The wave length of course remains essentially unaltered during this inward movement, and therefore, as the apex is approached the parastichy numbers fall, producing the usual disc pattern, possibly with some slight irregularity at the very centre. There may still be some growth of the capitulum itself, but the pattern can no longer adjust itself to keep the wavelength constant. Either the chemical pattern has lost all its importance and gives way to the relatively unchangeable anatomical pattern or else secretions from the new structures ensure that the wavelength of the chemical pattern increases with that of the anatomical pattern.

A special point arises in connection with the daisy, the formation of the ring of 13 bracts. This number is very constant. The author does not recall finding any specimen with a different number of bracts, excepting a very few deformed or damaged specimens. It is suggested that this ring of bracts is formed as follows. Within the band of lattice pattern there appears at some stage a ring of reduced activity, so that the band becomes divided into two separate bands. The more distal of these bands continues its development and eventually forms the floret pattern. The proximal band, however, is rather narrow and weak (it is pointless to enquire why). This process is described in.... The number of maxima in the ring under these circumstances will be one of the three principal parastichy numbers, usually the largest of the three. In view of the fact that the daisy develops according to the normal Fibonacci pattern, this number must be expected to be a Fibonacci number, as it is.

In order to justify this account it is necessary to describe a chemical system for which the pattern develops accordingly. No actual system will actually be described, nor even imaginary chemical reactions as described in Turing (1951). However a partial differential equation will be obtained which is thought to give a good approximation to mark the behaviour of certain kinds of chemical system. The differential equation has a number of parameters and it is necessary to find values for these parameters which will make the differential equation behave appropriately. The choice of parameters is largely made on theoretical grounds, described in this paper, but in order to be sure that the differential equation does really describe a development such as that mentioned above, it is necessary to follow its behaviour by computation.

1. Considerations governing the choice of parameters

The assumptions to be made concerning the development of the pattern are

(i) That the pattern is described by functions U, V of position on the cylinder and of time, satisfying the partial differential equations

$$\frac{\partial U}{\partial t} = \phi(\nabla^2)U + I(x,t)U + GU^2 - HUV,$$
$$V = \psi(\nabla^2)U^2.$$

(ii) The operator $\phi(\nabla^2)$ is supposed to take the form

$$\phi(\nabla^2) = I_2\left(1 + \frac{\nabla^2}{k_0^2}\right)^2.$$

(iii) The operator $\psi(\nabla^2)$ is supposed to take the form

$$\psi(\nabla^2) = \frac{1}{1 - \nabla^2/R^2}$$

though in the computations other forms may be used, taking the value zero outside a finite region.

(iv) A quasi-steady state is assumed to hold, i.e. the time derivative $\partial U/\partial t$ is supposed to be zero, or so near zero as is consistent with slow changes in the radius of the cylinder. This assumption of course implies that certain details as to the effect of the growth on the equation need not be considered.

(v) The function $I(x,t)$ is supposed given in advance. At each time it may be supposed to take the form $I_0 - I_2 z^2/l^2$. The quantity I_0 is initially supposed to be negative and to increase to an asymptotic value, reaching very near to it when the optimum wavelength is about one third of a circumference. The quantity l can remain very nearly constant or increase slightly with increasing radius. Clearly in view of (iv), it is only the variation of I_0 and l with radius which is significant, not the variation with time.

If we concentrate our attention on the period of time in which the optimum wavelength is less than a third of a circumference, I_0 and l may be taken as constants, i.e., on a par with G, H, I_2, k_0, R. We have to consider what are appropriate values for these seven quantities. Of the seven quantities there are really only four that are dimensionless. In other words, if we are quite uninterested in the units of time, length and concentration, new units may be introduced which will result in three of these parameters taking the value unity. Actually it is not advisable to do this reduction in every context. A certain amount of interest attaches to the relation of the time and space scales of the phenomena and the diffusion constants for the morphogens in the tissue. The enormous variety of possible reaction constants, and the fact that exceedingly weak concentrations of morphogens could be effective to influence growth, mean that our ignorance of the other two dimensionful quantities is too great for there to be any value in considering them in detail.

If three of the parameters are to be taken as unity, appropriate ones seem to be k_0, fixing the unit of length as the optimum radian wavelength, I_2, fixing the unit of time, and G, fixing the unit of concentration.

The parameters required are thus reduced to four, viz R, H, I_0, l. When actual computations are being carried out, the number of quantities to be specified is again increased to seven by the inclusion of the radius ρ, and two other quantities I_1 and h concerned with the method of calculation. Of these, only the role of h need be mentioned here. In the actual calculations the function $I_0 - z^2/l^2$ is replaced by $I_0 - (h^2/\pi^2\rho^2)\sin^2 \pi zh$, and the pattern is periodic in z with period h. But this is of course only a mathematical device. The calculations are applied to the Fourier coefficients of U and the number of these that has to be considered is proportional to h. One therefore has to make h as small as possible without the pattern $I_0 - (h^2/\pi^2\rho^2)\sin^2 \pi zh$ differing too much from $I_0 - z^2/l^2$ and, what is more important without the bands of pattern becoming so close as to influence one another appreciably.

The main consideration governing the choice of the quantity R is that an excessively small value has the effect that large areas of more or less uniform pattern tend to be unstable and to break up into a number of separate patches. This phenomenon may be explained as follows. The amplitude of the waves is largely controlled by the concentration V of 'poison'. If the quantity R is small, it means that the poison diffuses very fast. This reduces its power of control, for if the U values are large in a patch and large quantities of poison are produced, the effect of the poison will mainly be to diffuse out of the patch and prevent the increase of U in the neighbourhood.

Another way of expressing the effect is that the poison, acting through the HUV term, prevents the growth of waves whose wave vectors are near to that of a strong wave train. The quantity R expresses essentially the range of action in the wave-vector space. If it is too small, there will be liberty for 'side bands' to develop round the strong components. These side bands will represent the modulation of the patchiness. If R is allowed to become too large, it can happen that this 'side band surpression' effect even prevents the formation of a hexagonal lattice; neighbouring points around the hexagon of wave-vectors surpress one another. This however happens only with certain values of the other parameters.

In the actual calculations (initially, at any rate) the function chosen for $\psi(\nabla^2)$ was

$$\psi(r^2) = \begin{cases} \left(1 - \left(\dfrac{r}{r_{\max}}\right)^2\right)^2, & r \leq r_{\max}, \\ 0, & r \geq r_{\max} \end{cases} \tag{2}$$

with $r_{\max}/k_0$ usually about $1/\sqrt{2}$. (This function calculated in 'Subgroup smooth'.)

The choice of the parameters H, I_0, l is assisted by obtaining an approximate form of solution valid for patterns covering a large area, i.e. in effect with l very large. One may then, as a very crude approximation, suppose that when $I(x,t)$ varies from place to place one may find near each point more or less the solution which would apply over the whole plane if the value of I appropriate for that point were applied to the whole plane. A nomogram for this purpose is given elsewhere.[1] Another approach to the problem is provided by considering the effect of the terms $\phi(\nabla^2)U$ and $I(x,t)U$ taken in conjunction in the absence of the terms $GU^2 - HUV$. The terms $I(x,t)U$ may then be regarded as modifying the effect of the $\phi(\nabla^2)U$ term, so that $\phi(\nabla^2)$ has to be replaced by another function of the wave vector, no longer dependent on the length alone. Having expressed the effect of the $I(x,t)$ term in this way, it may be assumed, as another (alternative) crude approximation, that the effect of this term is the same even in the presence of the terms $GU^2 - HUV$. Clearly this approximation will not be too unreasonable if the really important term is $\phi(\nabla^2)U$.

2. Early stages in pattern formation

The most probable course of pattern formation in its early stages is something as follows. The value of I_0 remains sufficiently small to preclude the formation of any pattern until ρk_0 has a value somewhere between 2 and 3. At this stage, when I_0 reaches the appropriate value the homogeneous distribution (or at least θ-independent) breaks up and gives rise to a pattern which is symmetrical under rotation through 120°, i.e., which has three maxima and a reciprocal lattice pattern as shown in Fig. 2.

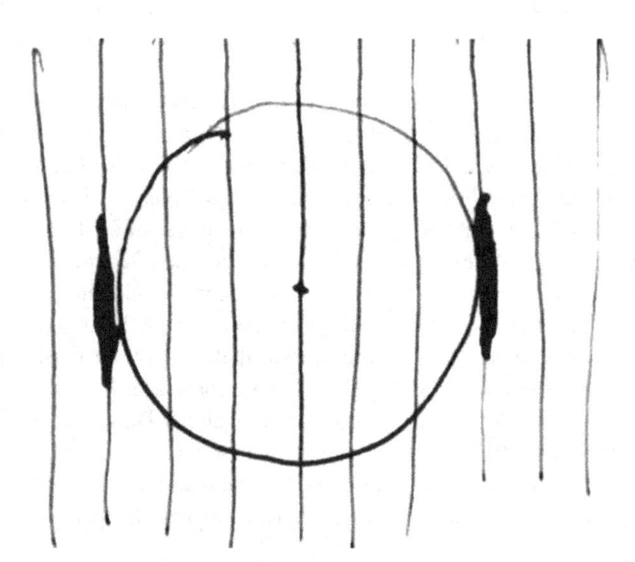

Fig. 2: [Editors note: From AMT/C24/15].

[1] Editors note: There is a nomogram surviving in the archives at AMT/C24/71 but it is not obviously the one referred to here, which appears to have been lost.

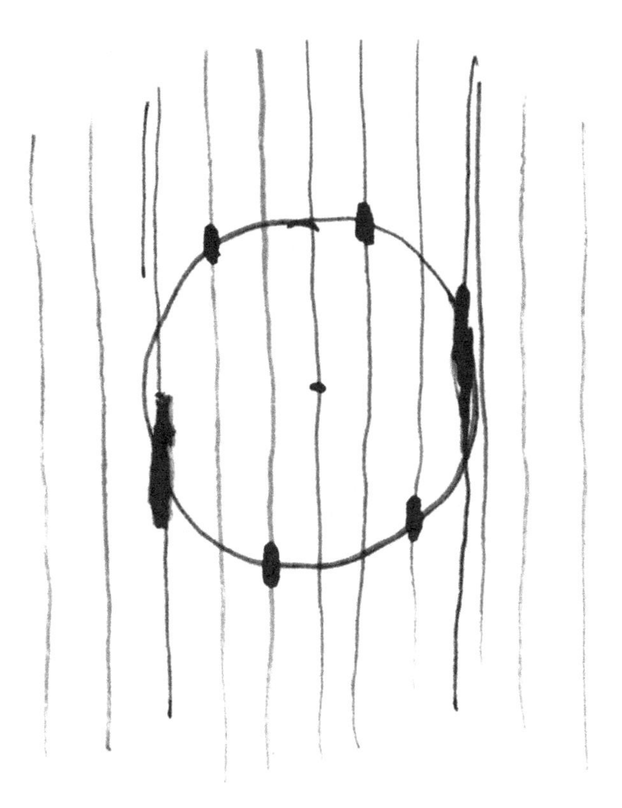

Fig. 3: [Editors note: From AMT/C24/15].

If I_0 increases further, this pattern itself becomes unstable and develops into a more or less hexagonal pattern without reflection symmetry, as shown in Fig. 3.

Afterword

Einar Fredriksson Recalls[1] the —

HISTORY OF THE PUBLICATION OF THE COLLECTED WORKS OF ALAN M. TURING

The history of the project to publish the *Collected Works of Alan M. Turing* is a long drawn-out one and dates back to the founding of a book series *Studies in Logic*, which took place at the 10th Congress of Philosophy held in Amsterdam in 1948. The aims of this series were to publish outstanding work in pure and applied logic, an area to which Turing had made fundamental contributions. This Congress was important as being the first such post-war international event; its secretary was E. W. Beth, who later became Professor in Logic at the University of Amsterdam. Also present were Turing's US colleagues S. C. Kleene, who mentioned Turing's work in his lecture, and H. B. Curry, although neither Turing himself nor his close UK logic colleagues, such as Robin Gandy, were present. After Turing's death in 1954 Beth took the initiative to organise a publication in honour of Turing, and this initiative resulted in a visit by M. D. Frank, series publisher of North-Holland Publishing Co., to Turing's mother, Mrs E. Sara Turing in Spring 1958.

This led to a publishing agreement for Turing's Collected Works, dated 29 July 1958 and signed by P. N. (Nick) Furbank as Executor of AMT's will, dated 13 February 1954 and by the other four beneficiaries of the will. The Works were to be edited by M. H. A. (Max) Newman, Turing's superior at Manchester University, where Turing held a readership at the time of his death. Gandy was, by the terms of the will, recipient of all mathematical books and papers. There is a correspondence from 1959 to 1963 between Newman and the publisher, but Newman in 1963 retired from his professorship and gave up the editorship. The editorship was passed on to Gandy in the same year.

In 1969 I began a new career as a publishing editor at North-Holland and soon received the dossier for Turing's *Collected Works* from M. D. Frank with the request to see the project through to publication. Sara Turing had been in touch annually with Frank since their first meeting. Beth had died in 1964 and the series editors in 1969 now included Andzrej Mostowski of Warsaw and Abraham Robinson of Yale University. Mostowski was spending a sabbatical in Oxford as a Fellow of All Souls, while Gandy had moved from Manchester and was also in Oxford, having accepted a readership there in the same year. Gandy was asked by me to take up editing of the Works and meetings to arrange practical steps were organised in Oxford, starting in February 1970. Later that year a complete outline of works by Turing, mostly published, was prepared and circulated among the editors of *Studies in Logic*. The Works were divided into four parts: (i) Pure Mathematics, (ii) Mathematical Logic, (iii) Mechanical Intelligence and (iv) Morphogenesis.

Gandy's editing was designed to include the collection of errata to all articles through experts in the different areas. Some articles were published with many typesetting errors and other mistakes of various kinds. On top of this, Gandy was to write a short introduction and comment on each paper. He cautioned that he would need time to go through all this material and to consult with specialists where necessary. He was further led into new areas of research of his own whilst working on the Turing papers, and there were often other interruptions due to his many other conflicting commitments. By 1971 it became clear that progress would be very slow and that more drastic action was necessary.

Sara Turing kept calling to check whether the publisher was doing his utmost to get the Works published in her lifetime (she was 90 in 1971, with a very clear mind). The publisher complained of the slow editing process and Sara proposed to call Gandy's mother to help getting matters moving

[1] The help of Peter Brown in verifying some of the historical details is gratefully acknowledged.

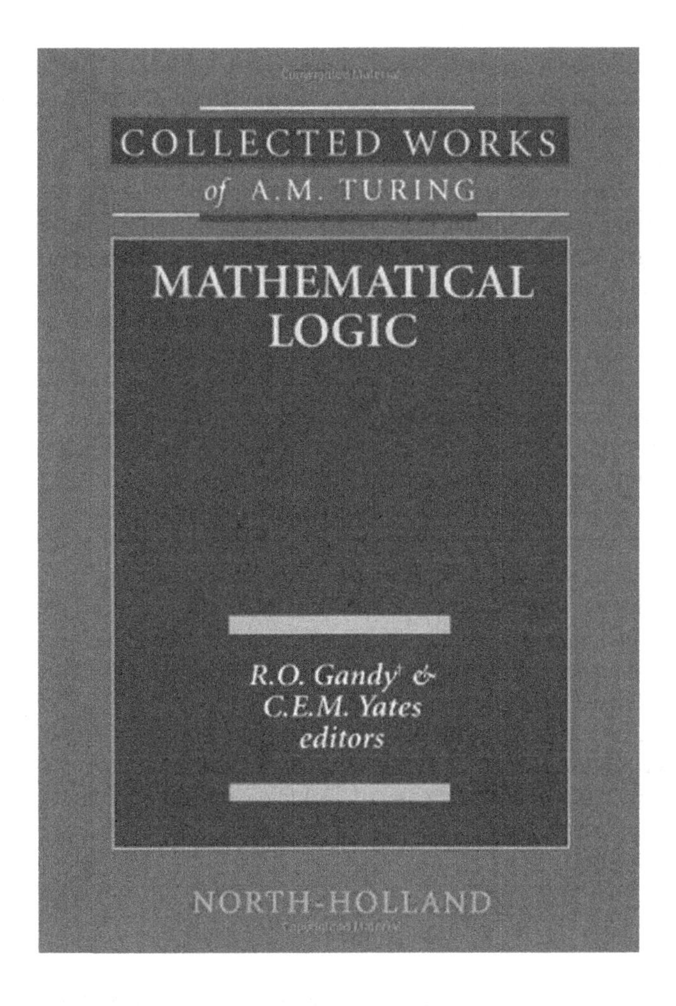

more rapidly. Repeated visits to the editor were not followed by measurable success and finally Gandy was asked what was needed to keep him concentrated on the project. He then suggested that one of his Oxford students should be paid a grant to assist him. Sara liked this idea and offered to provide a fund of £1000 to give assistance to Gandy. The publisher previously had repeatedly offered Gandy financial assistance for his time on the project, but Gandy had always declined. However, work with the student did not give the desired result and progress with editing the Works remained slow, despite further visits, prodding by colleagues and other pressure. In 1974 Sara called the publisher to say that Donald Michie (Professor at Edinburgh University and an authority on Artificial Intelligence) and Mrs Michie (Dr Anne McLaren, a leading geneticist) had visited her and proposed that the £1000 fund should form the basis for an 'A.M. Turing Trust' under Michie's chairmanship. Sara asked the publisher whether we had any objection to this transfer of funds, and of course we did not. The publishing agreement with the beneficiaries of Turing's will was entirely separate from the creation of a Trust and there was no possibility that the student would play an effective role in Gandy's editorial work. Sara passed the matter of the fund transfer to her solicitor and this was her last contact with the publisher. In March 1976 we were informed by the solicitor of the death of Sara Turing.

Interest in Turing's work increased with growth of the Turing archive set up at Kings College, Cambridge and resulted in continuous demands for the publication of the *Collected Works*. But during the period 1977–83 Gandy was frequently being interviewed by Andrew Hodges to provide information for the biography that Hodges was writing, and these calls on his time, coupled with interruptions due to repeated attempts by Michie to become involved in the publication of the Works, gave Gandy good reason to postpone active work on the editorship, so that it is fair to say that during this period it had come virtually to a standstill. The publishers sought additional editorial help, visiting the National Physical Laboratory at Teddington (Donald W. Davies and Mike Woodger) and other scholars who had worked with Turing, but to no effect.

Fortunately a turning point resolving these difficulties was at hand. Unexpectedly, Nick Furbank, Executor of Turing's will, got in touch with us as the contracted publishers of the Works, stressing his role as Executor. Furbank had not had any effective contact with Gandy since the 1950s, and we then related to him the entire frustrating editorial history in some detail. Gandy was not very happy when we visited him in 1986, accompanied by Furbank, with the purpose of rejuvenating the editorial process. He himself had not attempted to contact Furbank in all those years.

Thanks to Furbank's intervention an editorial meeting was organised at King's College, London in January 1987 with experts invited according to the original organisational lines defined by Gandy. Those attending were J. L. Britton (Part I, Mathematics), Gandy (Part II, Logic), D. C. Ince (Part III, Machine Intelligence) and P. T. Saunders (Part IV, Morphogenesis). Dr Saunders was host to the meeting, which was chaired by Furbank, and I attended as a publisher. The contents of each volume of a four-volume series were defined and deadlines set for submission of the manuscripts. Three volumes (i), (iii) and (iv) appeared in 1992. In 1991 Gandys former colleague at Manchester, C. E. M. Yates, was invited to be co-editor of part (ii) with Gandy and the work was interrupted by Gandy's death in November 1995. A new part, *Enigmas, Mysteries and Loose Ends* was added to this last published volume, which appeared in 2001.

The Publisher, Elsevier (North-Holland), completed the publication of the *Collected Works*, as originally announced in the preface of the Sara Turing biography of 1959, after 43 years. In the Turing Centenary, 2012, this new edited volume, including a broad commentary by experts, is published by the same company.

Mike Yates Writing in *The Independent*, Friday 24 November 1995 —

OBITUARY: ROBIN GANDY

Robin Gandy was one of the grand old men in the international community of mathematical logicians. 'Old man' is not an expression, however, that sits happily with Gandy, who until his last few months seemed forever youthful, a friend as well as a mentor to his endless stream of PhD students. A colourful and complex character who, when I first met him, would arrive at Manchester University in motor-cycle leathers, and later dominate a crowd in the nearest pub with his foghorn voice, plumes of smoke and witty anecdotes, Robin Gandy had immense intellectual and personal qualities and utter dedication to his subject. He was born in Peppard, Oxfordshire, where his father, Thomas Gandy, was in general practice. His mother, Ida Gandy, earned a reputation for a sequence of books based on her early life in Wiltshire. Educated at Abbotsholme, a progressive public school, he went on to join that special elite at King's College, Cambridge. In 1940, his graduation year, he met Alan Turing, famed now for breaking the German Enigma code, and in 1944 they started working together at Hanslope Park, in Buckinghamshire, by which time Gandy had become an expert on military radio and radar.

His friendship with Turing continued. In 1946 he returned to King's to take Part III of the Mathematical Tripos with distinction, then began studying for a PhD under Turing's supervision; his successful thesis on the logical foundation of physics, entitled 'On Axiomatic Systems in Mathematics and Theories in Physics' and presented in 1953, can now be seen as a bridge between his early expertise and later career. When Turing died in 1954 he left his mathematical books and papers to Gandy, who between 1950 and 1961 held lectureships in Applied Mathematics at first Leicester, then Leeds. During this period his commitment to logic evolved and he developed a Mathematics–Philosophy course at Leeds, with Martin Löb.

In 1961, Gandy moved to Manchester, where the seemingly retiring but extremely astute Max Newman had with (now Sir) James Lighthill built up what was then the best mathematics department in the country. Newman had brought Turing to Manchester, and he selected Gandy to develop logic and start up a Mathematics–Philosophy course.

Now officially a logician, Gandy appointed new staff and invited many visitors from abroad. He was promoted to a chair in 1967, and organised the European summer meeting of the Association for Symbolic Logic in Manchester in 1969, supported as was usual then by Nato funds. Turing had gently chided Gandy in 1940 for his left-wing beliefs; now, ironically, be came to be attacked as right-wing for his support of Nato funding. These were halcyon days for mathematical logic, with unexpected connections being made between the principal areas of research. Gandy's own research into functionals of higher types had made him prominent, quite aside from his high motivating qualities.

In 1969 he gave up his chair in Manchester for a readership in Mathematical Logic at Oxford, where he was to be based for the rest of his life. He was adopted by the young Wolfson College and soon had rooms in the college's fine new building in North Oxford. He occasionally complained about the 'tedious beat of heavy metal' from some other room but generally found college life very congenial.

He was responsible for the Mathematics–Philosophy course, and with John Sheperdson from Bristol brought the British Logic Colloquium into being. Dana Scott was appointed to a new chair of Mathematical Logic in 1972, Michael Dummett succeeded Sir Alfred Ayer to the Wykeham chair of logic in 1979 and Ronald Jensen came to All Souls in 1981. Mathematical logic came into its own in Oxford and Gandy's list of PhD students grew from three to around 30.

He retired in 1986 amongst fireworks and full moon at the University of Wales's retreat at Gregynog in Powys, feted at a conference in his honour by an international gathering and most of his PhD students. He continued to publish with great vigour, and was a familiar figure at international conferences until shortly before his death.

He had seemed more fragile recently, but in earlier years he loved walking the Snowdonian hills, especially his beloved Cnicht, or, based at his cottage on the Portmeirion estate, combing the forests for fungi: one favourite memory is of him perched on top of a wall in his jodhpurs, pipe in hand and turning his craggy face to a red-faced farmer to say, 'There is nothing to worry about. I'm used to climbing your walls'.

He made a number of appearances on radio and television, especially to reminisce about Alan Turing. When asked about Turing's motives if he really did commit suicide, Gandy would become quite heated: 'Some things are too deep and private and should not be pried into'. Himself, he was much loved and his generosity, tolerance, hospitality, kindliness, good-humour, irreverence, erudition and mouth-watering home-made ice-cream will be sorely missed. He would often chide himself as a 'silly old owl', but then the owl is by repute the wisest of birds.

Robin Oliver Gandy, mathematical logician: born Peppard, Oxfordshire 22 September 1919; Lecturer in Applied Mathematics, Leicester University 1950–56; Lecturer in Applied Mathematics, Leeds University 1956–61; Senior Lecturer in Mathematical Logic, Manchester University 1961–64, Reader 1964–67, Professor 1967–69; Reader in Mathematical Logic, Oxford University 1969–86; died Oxford 20 November 1995.

Bernard Richards shares with us —

RECOLLECTIONS OF LIFE IN THE LABORATORY WITH ALAN TURING

This commentary describes the man, Alan Turing, through the eyes of a co-worker and it reveals a man who was warm, friendly and kind, and who had an exciting and vibrant personality. It explains too why it is right to consider him as one of the 'Wonders of the Modern World', and to liken him to the Great Pyramid at Giza, which is one of the Wonders of the Ancient World.

1. Turing as a man

As a man, Turing was somewhat shy. At times his voice would suddenly rise two octaves. But he was always very polite and very patient, at least with me, his student. He was a good athlete, so much so that occasionally he was known to have run to the Computing Laboratory from his home in Wilmslow, some 10 miles. I did not know it, but according to his mother (see Section 4) he was well recognised by his family as being a very generous man, showering gifts on his many nieces.

But it would be wrong to give the impression that Turing was insignificant – someone of no importance. I want to liken him to the Great Pyramid of the Pharaoh Khufu at Giza. That pyramid, the largest in the world, is 480ft high and each of the four sides has a baseline 756ft long. It is one of the Seven Wonders of the World. Do not think it is wrong of me to compare Turing to this Pyramid. He was a genius: he was a 'Wonder of the world', and he had four sides, or facets, to his work. He was essentially (i) a computer designer and user, (ii) a mathematician, (iii) a pseudo-biologist, and (iv) a philosopher who thought about 'Thinking Machines'.

2. Turing as a computer designer and user: the first facet of a genius

During the Second World War, at Bletchley Park, Turing was engaged in breaking the enemy military codes. To this end he designed a computer known as the 'Bombe' to take forward the work already done by the Poles, one Lieutenant Marian Rejewski in 1932. Rejewski was born in Bydgoszcz in northern Poland and obtained a degree in Mathematics from Poznan University. He then secretly joined the Polish General Staff's Cipher Bureau. He was able to crack the codes on the German Enigma I cipher machine. His discoveries were told to the British in July 1939. The Bombe was a complicated machine electrically, but simple in concept. Soon Turing realised that he could do much better and his ideas led to the building of the 'Colossus' computer. This was fed a 5-hole tape containing text to be decoded and that machine was very successful and was said by some to have 'won the war for Britain'. But so too did Rejewski win the war for Britain by revealing his work and bring a Cipher Machine to Britain. It is fitting that both men are honoured in the same way. Rejewski was honoured by a statue of him sitting on a bench in his home city of Bydgoszcz, and Turing is honoured by a statue of him sitting on a bench in Sackville Gardens in his home city of Manchester.

To use the Ferranti Mark I computer, installed in February 1951 in the Computing Machine Laboratory of Manchester University, one had to have the brains and skills to write the program and the eyes and hands to use the Operator's Console, and to be able to use these latter to prepare one's program on 5-hole paper tape, the input media for the computer.

As Turing was the originator of some of the concepts of the Mark I and of some of its circuits, and was responsible for its Instruction Set, it made his own life easier when it came to solving his own mathematical problems on the machine. The Ferranti Mark I, sometimes called the Manchester Mark I, came into being as the commercial production version of the Prototype, the experimental machine built in the Electrical Engineering Department of the University of Manchester, the experimental machine which many years later became known as the 'Baby'. Whilst there were many changes as regards the hardware of the new machine, and the engineering implementation, there are two interesting changes to the machine's 'Order Code' – the Instruction Set – introduced by Alan Turing showing his foresight. The first was the instruction to generate a Random Number. Numbers in the machine were stored as a 40-bit integer, and so the Random Number Generator would produce a random pattern in those 40 digits. This instruction was useful in many spheres of mathematics, statistics and in game playing. The other instruction invented by Turing was the 'Sideways Adder'. This instruction counted the number of 1s in the 40 binary digits – the 40 bits.

When it came to using the Operator's Console, Turing was very adept. He could manipulate the switches to enable him to 'call-down his program from the drum' – the backing store – and then change them to enable him to look at both the sequence of instructions in his program and at his data, an exercise known as 'peeping'.

He was also very adept at preparing the five-hole paper tape which was the standard input for the Mark I machine. He converted all his data into binary numbers before typing them into the paper tape. One day he saw me typing my Radiolaria data into the teleprinter – in *decimals* ! He looked skeptical and somewhat shocked, and he asked me what I was doing. So I explained. The reason for his question was because he was using the original Input Scheme – Scheme A – which he devised and for which he wrote 'The Programming Manual: Scheme A' in 1951. This Scheme required all data to be input in *binary form.* Later, his colleague, Tony Brooker, wrote the 'Programming Manual for Scheme B' that contained a decimal input routine. I was using the latter, but Turing was still using Scheme A which he knew well and loved.

Fig. 1: Alan Turing and the console of the Manchester Mark I computer, c. 1951.

3. Turing's researches in mathematics: the second facet of a genius

Turing had been using the Mark I prototype – the BABY – to search for Mersenne Primes. These are prime numbers of the form $M = 2^p - 1$. These numbers take their name from work done by a French monk, Marin Mersenne, in the 17th century. Thus, if $p = 2$, we find that $M = 3$ which is a prime number, and if $p = 4$, $M = 15$ which is not prime. But if $p = 5$, $M = 31$ which is prime. The search was on to find the largest value of p which would still give a prime number. Mersenne conjectured the values of p for which M would still be prime, as far as $p = 257$. In fact, $p = 127$ is the largest number in his list to give a prime number, viz. for $p = 127$, M is a number with 39 decimal digits. In 2009, a value of $p = 43{,}112{,}609$ gave a value for M with 12,978,189 digits! But in 1950, the largest p to give a prime number was still $p = 127$, and Alan Turing wanted to find the next value.

He also considered the potential of the machine for illustrating the Riemann Hypothesis, since as early as 1938, whilst at Cambridge, he has dreamt of building a mechanical device to test the Riemann Hypothesis. Then, in his Handbook, he mentions the value of Tchebysheff Polynomials for mapping functions. This was yet another illustration of his mathematical ability and interest.

4. Turing's work on Morphogenesis: the third facet of a genius

I spoke with Turing's mother, Sara Turing, just after Alan died and she told me that he had always been interested in the shape, form, and growth of plants and flowers. He used to pick the flowers, especially the daisies, and examine them minutely. This must have been the start of his interest in the growth and form of botanical, and later, biological, species.

Turing was aware of the work of D'Arcy Thompson who wrote on 'Growth and Form' (Thompson, 1917). Turing's own work on growth resulted in his paper 'The Chemical Basis of Morphogenesis' (Turing, 1952). He had used his ideas of Morphogenesis to explain the black-and-white dappling on cows.

Then I came along and he took me on to work alongside him in his work on Morphogenesis. He showed me his Diffusion Equation: the basis of his Morphogenesis in biological species. I set about solving that equation. It involved many, many, pages of degree-level mathematics. One such equation, used to calculate $L_n^{p,q,r}$ involved the Legendre Functions:

$$L_n^{p,q,r} = \frac{1}{4\pi} \iint P_n^p(\mu) P_n^q(\mu) P_n^r(\mu)^{ei(p+q+r)\varphi} d\mu d\varphi$$

Following through with the equations gave rise to four solutions with n taking the values 2, 4, 6, 8. The computer was then called upon to plot, using the teleprinter characters, the surface contours – contours on the surface of a sphere – which then gave the shape of the resulting 'computer species'. What emerged were spherical bodies having 2, 6, 12 and 20 tentacles – spikes – extending the length of one radius above the surface of the sphere.

The theoretical results – the solutions of the Diffusion Equation – matched very closely the marine biological species Radiolaria. The finer details are given in the chapter 'Radiolaria: the Results of Morphogenesis', and in the thesis 'Morphogenesis of Radiolaria' (Richards, 1954).

It seems that Turing knew of the existence of Radiolaria in that he directed me to a research publication which detailed the findings of the Research Ship 'HMS Challenger' which sailed the Pacific Ocean in the 19th century.

One day I told Turing that I had some results and that I would go to find the Challenger Report. We made an arrangement to meet again in 3 days time. Alas, he died 2 days later before he could see the excellent correspondence between my theoretical creatures and the real biological species Radiolaria. There is no doubt that these two, taken together, vindicated his Theory of Morphogenesis. It is very sad that he did not live to see this happen. However, I did show the results to his mother when we met just after Turing's death. She was, understandably, very sad, but felt so proud on seeing those results.

5. Turing's work on 'Thinking Machines': the fourth facet of a genius

It is now well-known that, during the Second World War, Turing worked at Bletchley Park in the Government Code and Cipher School decoding enemy messages. But before that, in 1935, whilst at Cambridge University, Turing conceived the idea of the modern computer. He had the idea that the computer could store a program whose reaction to the external world would be to improve that program based on the responses it was getting. That concept of Turing's became known as the 'Universal Turing Machine'. Then the war came, but he was able and keen to return to this topic again more formally after the war.

Whilst at Bletchley Park he was committed to breaking the enemy codes. But when he was not using the Colossus, in his rest periods Turing would talk with colleagues about chess and the possibility of building a computer to play chess. He wanted to write programs which would give the computer the idea of 'Intelligence'. And then the title of 'Artificial Intelligence' was given to those concepts.

It was after the war, and when at Manchester in 1950 that he wrote a famous paper entitled 'Computing Machinery and Intelligence'. This topic dealt with his previous ideas on Artificial Intelligence and considered the question 'Can machines think?' His target was to build a system such that a human could not discover whether, in a conversation, he or she was talking to a human or the computer. Then, in the expectation that it would happen, people adopted the label 'Turing Machine' to this concept.

In Manchester, Turing turned his mind again to the thoughts he had at Bletchley on chess. He began to play with a colleague, one Dr Dietrich Prinz, an employee of the computer company Ferranti Limited, the builder of the Mark I computer. Rapidly the conclusion was reached that the present computer did not have the speed or capacity to play a complete game of chess. The pair agreed that perhaps the computer could be programmed to play 'End-Games' of the type 'Mate-in two'. Then Dr Prinz, or DP as he was known, went away and in 1951 programmed the Mark I to play 'End-Games'. Turing felt that his ideas had been vindicated.

6. The departure of a Genius

In June 1954 I had in my diary a date for a meeting with Alan Turing. I believed that he too had that same date in his diary. Alas he was not able to attend for that meeting. Everyone in the Laboratory, not least myself, felt a great loss. It was particularly poignant in my case as I had lost my Supervisor. But more sadness was to come. Turing's mother, Sara Ethel Turing, came to see me and we shared his loss together, each in our own way. She told me about his early life and his interest in plants, and his athleticism. She told me about his time at Sherborne School, and some years later I was able to see the staircase where he exercised his then already scientific mind.

We parted recognising that we both had suffered a great loss.

References

Richards, B. 1954. The Morphogenesis of Radiolaria, M.Sc. Thesis.
Thompson, D'A, W. 1917. On Growth and Form, Cambridge University Press.
Turing, A.M. 1952. The chemical basis of morphogenesis, Trans. Royal Soc. 237 (641), 37–72.

Bibliography

A very comprehensive and detailed bibliography of works of or relating to Alan Turing is to be found as part of the BibNet Project and TeX User Group bibliography archives, which have been in existence for approximately two decades now. Created and maintained by Professor Nelson H. F. Beebe of the University of Utah, one can find:

A BIBLIOGRAPHY OF PUBLICATIONS OF ALAN MATHISON TURING

in BibTex form at:

http://ftp.math.utah.edu/pub//bibnet/authors/t/turing-alan-mathison.html

The advantage of this valuable and very impressive online resource is that it is updated weekly, and amendments or additions can be notified to Professor Beebe (contact details in the bibliography).

Further information on the BibNet project can be found at:

http://ftp.math.utah.edu/pub/bibnet/faq.html

Index

Page numbers followed by "*f*" indicate figures, "*t*" indicate tables, and "*n*" indicate footnotes.

D

F

M

O

P

T

X

Y

Z

Printed in the United States
By Bookmasters